Probability Theory and Stochastic Modelling

Volume 82

The **Probability Theory and Stochastic Modelling** series is a merger and continuation of Springer's two well established series Stochastic Modelling and Applied Probability and Probability and Its Applications series. It publishes research monographs that make a significant contribution to probability theory or an applications domain in which advanced probability methods are fundamental. Books in this series are expected to follow rigorous mathematical standards, while also displaying the expository quality necessary to make them useful and accessible to advanced students as well as researchers. The series covers all aspects of modern probability theory including

- Gaussian processes
- Markov processes
- Random fields, point processes and random sets
- Random matrices
- Statistical mechanics and random media
- Stochastic analysis

as well as applications that include (but are not restricted to):

- Branching processes and other models of population growth
- Communications and processing networks
- Computational methods in probability and stochastic processes, including simulation
- Genetics and other stochastic models in biology and the life sciences
- Information theory, signal processing, and image synthesis
- Mathematical economics and finance
- Statistical methods (e.g. empirical processes, MCMC)
- Statistics for stochastic processes
- Stochastic control
- Stochastic models in operations research and stochastic optimization
- Stochastic models in the physical sciences

More information about this series at http://www.springer.com/series/13205

Giorgio Fabbri · Fausto Gozzi
Andrzej Święch

Stochastic Optimal Control in Infinite Dimension

Dynamic Programming and HJB Equations

With a Contribution by
Marco Fuhrman and Gianmario Tessitore

 Springer

Giorgio Fabbri
Aix-Marseille School of Economics
CNRS, Aix-Marseille University, EHESS,
    Centrale Marseille
Marseille
France

Andrzej Święch
School of Mathematics
Georgia Institute of Technology
Atlanta, GA
USA

Fausto Gozzi
Dipartimento di Economia e Finanza
Università LUISS – Guido Carli
Rome
Italy

ISSN 2199-3130   ISSN 2199-3149 (electronic)
Probability Theory and Stochastic Modelling
ISBN 978-3-319-53066-6   ISBN 978-3-319-53067-3 (eBook)
DOI 10.1007/978-3-319-53067-3

Library of Congress Control Number: 2017934613

Mathematics Subject Classification (2010): 49Lxx, 93E20, 49L20, 35R15, 35Q93, 49L25, 65H15, 37L55

Printed on acid-free paper

This Springer imprint is published by Springer Nature
The registered company is Springer International Publishing AG
The registered company address is: Gewerbestrasse 11, 6330 Cham, Switzerland

To my parents and to Sara
– G.F.
To my parents and to my family
– F.G.
To my parents Franciszka Święch and Jerzy Święch
– A.Ś.

Preface

The main objective of this book is to give an overview of the theory of Hamilton–Jacobi–Bellman (HJB) partial differential equations (PDEs) in infinite-dimensional Hilbert spaces and its applications to stochastic optimal control of infinite-dimensional processes and related fields. Both areas have developed very rapidly in the last few decades. While there exist several excellent monographs on this subject in finite-dimensional spaces (see e.g., [263, 264, 385, 453, 468, 490, 576]), much less has been written in infinite-dimensional spaces. A good account of the infinite-dimensional case in the deterministic context can be found in [404] (see also [562] on optimal control of deterministic PDEs). Other books that touch on the subject are [29, 179, 468]. We attempt to fill this gap in the literature. Infinite-dimensional diffusion processes appear naturally and are used to model phenomena in physics, biology, chemistry, economics, mathematical finance, engineering, and many other areas (see e.g., [124, 177, 180, 372, 569]). This book investigates the PDE approach to their stochastic optimal control; however, infinite-dimensional PDEs can also be used to study other properties of such processes as large deviations, invariant measures, stochastic viability, stochastic differential games for infinite-dimensional diffusions, etc. (see [86, 177, 179, 249, 251, 261, 465, 467, 542, 544]).

To illustrate the main theme of the book, let us begin with a model distributed parameter stochastic optimal control problem. We want to control a process (called the state) given by an abstract stochastic differential equation in a real, separable Hilbert space H

$$\begin{cases} dX(s) = (AX(s) + b(s, X(s), a(s)))ds + \sigma(s, X(s), a(s))dW(s), & s > t \geq 0 \\ X(t) = x \in H, \end{cases}$$

where A is the generator of a C_0 semigroup in H, b, σ are some functions, and W is a so-called Q-Wiener process[1] in H. The functions $a(\cdot)$, called controls, are

---

[1] Q is a suitable self-adjoint positive operator in H, the covariance operator for W.

stochastic processes with values in some metric space Λ, which satisfy certain measurability properties. The above abstract stochastic differential equation is very general and includes various semilinear stochastic PDEs, as well as other equations which can be rewritten as stochastic functional evolution equations, for instance, stochastic differential delay equations. In a most typical optimal control problem we want to find a control $a(\cdot)$, called optimal, which minimizes a cost functional

$$J(t,x;a(\cdot)) = \mathbb{E}\left[\int_t^T l(s,X(s),a(s))ds + g(X(T))\right]$$

(for some $T > t$) among all admissible controls for some functions $l : [0,T] \times H \times \Lambda \to \mathbb{R}$, $g : H \to \mathbb{R}$.

The dynamic programming approach to the above problem is based on studying the properties of the so-called value function

$$V(t,x) = \inf_{a(\cdot)} J(t,x;a(\cdot))$$

and characterizing it as a solution of a fully nonlinear PDE, the associated HJB equation. Since the state $X(s)$ evolves in the infinite-dimensional space H, this PDE is defined in $[0,T] \times H$. The link between the value function V and the HJB equation is established by the Bellman principle of optimality known as the dynamic programming principle (DPP),

$$V(t,x) = \inf_{a(\cdot)} \mathbb{E}\left[\int_t^\eta l(s,X(s),a(s))ds + V(\eta,X(\eta))\right], \quad \text{for all } \eta \in [t,T].$$

Heuristically, the DPP can be used to define a two-parameter nonlinear evolution system and the associated HJB equation

$$\begin{cases} V_t + \langle Ax, DV \rangle + \inf_{a \in \Lambda}\left\{\frac{1}{2}\mathrm{Tr}\left[(\sigma(t,x,a)Q^{\frac{1}{2}})(\sigma(t,x,a)Q^{\frac{1}{2}})^* D^2 V\right]\right. \\ \qquad\qquad\qquad\qquad\qquad\qquad \left. + \langle b(t,x,a), DV \rangle + l(t,x,a)\right\} = 0, \\ V(T,x) = g(x) \end{cases}$$

$$(1)$$

is its generating equation. Such a PDE is called infinite-dimensional or a PDE in infinitely many variables. We also call it unbounded since it has a term with an unbounded operator A which is well defined only on the domain of A. Other terms may also be undefined for some values of DV and $D^2 V$, the Fréchet derivatives of V, which we may identify with elements of H and with bounded, self-adjoint operators in H respectively. In particular, the term $\mathrm{Tr}[(\sigma Q^{\frac{1}{2}})(\sigma Q^{\frac{1}{2}})^* D^2 V]$ is well defined only if $(\sigma Q^{\frac{1}{2}})(\sigma Q^{\frac{1}{2}})^* D^2 V$ is of trace class.

The main idea is to use the HJB equation to study the properties of the value function, find conditions for optimality, obtain formulas for synthesis of optimal feedback controls, etc. This approach turned out to be very successful for finite-dimensional problems because of its clarity and simplicity and thanks to the developments of the theory of fully nonlinear elliptic and parabolic PDEs, in particular the introduction of the notion of a viscosity solution and advances in regularity theory. However, even there many open questions remain, especially if the HJB equations are degenerate. We hope the dynamic programming approach will be equally valuable for infinite-dimensional problems even though a complete theory is not available yet.

Equation (1) is an example of a fully nonlinear second-order PDE of (degenerate) parabolic type. In this book, we will deal with more general and different versions of such equations and their degenerate elliptic counterparts. If Λ is a singleton, (1) is just a terminal value problem for a linear Kolmogorov equation. If Λ is not a singleton but the diffusion coefficient σ is independent of the control parameter a, (1) is semilinear. The theory of linear equations (and some special semilinear equations) has been studied by many authors and can be found in the books [29, 106, 179, 583]. The emphasis of this book is on semilinear and fully nonlinear equations.

There are several notions of solution applicable to PDEs in Hilbert spaces which are discussed in this book: classical solutions, strong solutions, mild solutions in the space of continuous functions, solutions in $L^2(\mu)$, and viscosity solutions. Classical solutions are the most regular ones. This notion of solution requires $C^{1,2}$ regularity in the Fréchet sense and imposes additional conditions so that all terms in the equation make sense pointwise for $(t,x) \in [0,T] \times H$. When classical solutions exist, we can apply the classical dynamic programming approach to obtain verification theorems and the synthesis of optimal feedback controls. Unfortunately, in almost all interesting cases it is not possible to find such solutions; however, they are very useful as a theoretical tool in the theory. The notions of strong solutions, mild solutions in the space of continuous functions, and solutions in $L^2(\mu)$ are introduced and studied only for semilinear equations and define solutions which have at least first derivative (in some suitable sense). Verification theorems and synthesis of optimal feedback controls can still be developed within their framework. The notion of viscosity solutions is the most general and applies to fully nonlinear equations; however, at the current stage there are no results on verification theorems and synthesis of optimal feedback controls.

Infinite-dimensional problems present unique challenges, and among them are the lack of local compactness and no equivalent of Lebesgue measure. This means that standard finite-dimensional elliptic and parabolic techniques which are based on measure theory cannot be carried over to the infinite-dimensional case. Moreover, the equations are mostly degenerate and contain unbounded terms which are singular. So the methods to find regular solutions to PDEs in infinite dimension like ours tend to be global and are based on semigroup theory, smoothing properties of transition semigroups (like the Ornstein–Uhlenbeck semigroups), fixed point techniques, and stochastic analysis. These methods are mostly restricted to equations of semilinear type. On the other hand, the notion of a viscosity solution is perfectly suited for fully nonlinear equations. It is local, and it does not require any regularity of solutions except continuity. As in finite dimension, it is based on a maximum principle through the idea of "differentiation by parts," i.e., replacing the nonexisting derivatives of viscosity subsolutions (respectively, supersolutions) by the derivatives of smooth test functions at points where their graphs touch the graphs of subsolutions (respectively, supersolutions) from above (respectively, below). However, as the readers will see, this idea has to be carried out very carefully in infinite dimension.

This book includes chapters on the most important topics in HJB equations and the DPP approach to infinite-dimensional stochastic optimal control.

Chapter 1 contains the basic material on infinite-dimensional stochastic calculus which is needed in subsequent chapters. It is, however, not intended to be an introduction to stochastic calculus, which the reader is expected to have some familiarity with. Chapter 1 is included to make the book more self-contained. Most of the results presented there are well known; hence, we only provide references where the reader can find proofs and more information about concepts, examples, etc. We provide proofs only in cases where we could not find good references in the literature.

In Chap. 2, we introduce a general stochastic optimal control problem and prove a key result in the theory, namely the dynamic programming principle. We formulate it in an abstract and general form so that it can be used in many cases without the need to prove it again. Solutions of stochastic PDEs must be interpreted in various ways (strong, mild, variational, etc.), and our formulation of the DPP tries to capture this phenomenon. Our proof of the DPP is based on standard ideas; however, we have tried to avoid heavy probabilistic methods regarding weak uniqueness of solutions of stochastic differential equations. Our proof is thus more analytical.

We also introduce many examples of stochastic optimal control problems which can be studied in the framework of the approach presented in the book. They should give the readers an idea of the range and applicability of the material.

Chapter 3 is devoted to the theory of viscosity solutions. The reader should keep in mind the following principle when it comes to unbounded PDEs in infinite dimension: There is no single definition of viscosity solutions that applies to all equations. This is due to the fact that there are many different PDEs which contain different unbounded operators and terms which are continuous in various norms. Also the solutions have to be continuous with respect to weaker topologies. However, the main idea of the notion of viscosity solutions is always the same as we described before. What changes is the choice of test functions, spaces, topologies, and the interpretation of various terms in the equation. In this book, we focus on the notion of a so-called *B*-continuous viscosity solution which was introduced by Crandall and Lions in [141, 142] for first-order equations and later adapted to second-order equations in [539]. The key result in the theory is the comparison principle, which is very technical. Its main component is the so-called maximum principle for semicontinuous functions. The proof of such a result in finite dimension was first obtained in [370] and was later simplified and generalized in [137–139, 360]. It is heavily based on measure theory and is not applicable to infinite dimension. Thus, the theory uses a finite-dimensional reduction technique introduced by Lions in [413]. It restricts the class of equations which can be considered; in particular, they have to be highly degenerated in the second-order terms. We present three techniques to obtain the existence of viscosity solutions. The first and most important for this book is the DPP and the stochastic optimal control interpretation, showing directly that the value function is a viscosity solution. This technique applies to HJB equations. The other techniques are finite-dimensional approximations and Perron's method. Both can be applied to more general equations, for instance, Isaacs equations associated to two-player, zero-sum stochastic differential games; however, they have limitations of their own. Moreover, we discuss other topics in the theory of viscosity solutions such as consistency and singular perturbations. Several special equations are also studied in this book because of their importance and because they are good examples to show how the definition of viscosity solutions and some techniques can be adjusted to particular cases. They are the HJB equations for the optimal control of the Duncan–Mortensen–Zakai equation, stochastic Navier–Stokes equations, and stochastic boundary control. In particular, the last one also contains ideas on how to handle HJB equations which may be nondegenerate, for instance, if Q is not of trace class. Finally, we present applications to the infinite-dimensional Black–Scholes–Barenblatt equations of mathematical finance.

Chapter 4 is devoted to the theory of mild and strong solutions in spaces of continuous functions through fixed point techniques based on the smoothing properties of transition semigroups such as Ornstein–Uhlenbeck semigroups. This theory applies only to semilinear equations, i.e., when the coefficient σ does not depend on the control parameter a, and historically it was the first approach introduced in the literature. The theory was initiated by Barbu and Da Prato [29] and later improved and developed in various papers, see e.g., [89, 90, 105, 107, 302, 307, 308, 311].

Chapter 4 is divided into four main parts. In the first part (Sects. 4.2 and 4.3), we present the basic tools needed for the analysis: the theory of generalized gradients and the smoothing of transition semigroups. In the second part (Sects. 4.4–4.7), we develop the theory for a general type of semilinear HJB equation (parabolic and elliptic) without connection with optimal control problems. The main idea behind this approach is the following. Consider the HJB equation (1) in the semilinear case when the coefficient σ is time-independent:

$$\begin{cases} V_t + \mathcal{A}V + \inf_{a \in \Lambda}\{\langle b(t,x,a), DV \rangle + l(t,x,a)\} = 0, \\ V(T,x) = g(x), \end{cases} \tag{2}$$

where $\mathcal{A}$ is the linear operator

$$\mathcal{A}\varphi = \langle Ax, D\varphi \rangle + \frac{1}{2}\mathrm{Tr}\left[(\sigma(x)Q^{\frac{1}{2}})(\sigma(x)Q^{\frac{1}{2}})^* D^2\varphi\right].$$

If such an operator generates a semigroup e^{tA} then, by the variation of constants formula, one can rewrite Eq. (2) in the integral form as

$$V(t,x) = e^{(T-t)A}g(x) + \int_t^T \left(e^{(T-s)A}F(s,\cdot)\right)(x)\, s,$$

where $F(s,x) := \inf_{a \in A}\{\langle b(s,x,a), DV \rangle + l(s,x,a)\}$. The solution of this integral equation is called a mild solution and is obtained by fixed point techniques. To define it, the solution must at least have a first-order spatial Gâteaux derivative, possibly only in some directions needed to give sense to the nonlinear term, the so-called G-derivative. Thus, one needs suitable smoothing properties of the semigroup e^{tA} (which is the Ornstein–Uhlenbeck semigroup in the simplest case). Since this semigroup is not strongly continuous, except in very special cases, one needs to use the theory of π-semigroups introduced in [493] or that of weakly continuous (or $\mathcal{K}$-continuous) semigroups [101, 108, 301]. Sects. 4.4 and 4.5 consider a general type of operator $\mathcal{A}$, possibly depending on t, while Sects. 4.6 and 4.7 focus on the case when $\mathcal{A}$ is of Ornstein–Uhlenbeck type, where stronger results can be proved.

In the third part (Sect. 4.8), we develop a connection with stochastic optimal control problems. The fact that mild solutions have a first-order spatial derivative allows us to give a meaning to formulae for optimal feedbacks. However, the proofs of the verification theorems and optimal feedback formulae cannot be done straightforwardly as one needs to apply Itô's formula in infinite dimension, which requires smooth functions. For this reason (following [307]), we introduce the notion of a strong solution of the HJB equation (2) as a suitable limit of classical solutions and prove that any mild solution is also a strong solution.

The fourth and last part of the chapter (Sects. 4.9 and 4.10) deals with some special equations. In Sect. 4.9, we show how the techniques developed in the previous sections can be adapted to HJB equations and analysis of optimal control problems for the stochastic Burgers equation, stochastic Navier–Stokes equations and stochastic reaction diffusion equations. In Sect. 4.10, we discuss some equations for which explicit representations of the solutions can be found. Such cases are always of interest in applications.

Chapter 5 is devoted to a relatively new and promising theory of mild and strong solutions in spaces of L^2 functions with respect to a suitable measure μ (see [3, 4, 125, 299]). The contents of this chapter are similar to the previous one as the main ideas behind the definition of mild and strong solutions of HJB equations are the same. The difference is in the fact that the reference space is not the space of continuous functions but the space of square-integrable functions with respect to the measure μ. The results are similar: existence and uniqueness of solutions of HJB equations through fixed point arguments, verification theorem through approximations, and existence of optimal feedbacks. The advantage of this approach is that the results require weaker assumptions on the data, thus enlarging the range of possible applications, including the control of delay equations; however, at a cost of

weaker statements, for example, the first-order spatial derivative is now defined in a
Sobolev weak sense and is not in general a Gâteaux or Fréchet derivative. The main
tools used here are the theory of invariant measures for infinite-dimensional
stochastic differential equations and the properties of transition semigroups in the
space of integrable functions with respect to such measures.

Chapter 6 is devoted to a different and in many respects complementary technique of Backward Stochastic Differential Equations (BSDEs). The chapter was
written independently and autonomously by M. Fuhrman and G. Tessitore, who are
well-recognized experts in the field. We are grateful for their invaluable contribution. BSDEs are Itô type equations in which the initial condition is replaced by a
final condition and a new unknown process appears corresponding to a suitable
martingale term. In the nonlinear, finite-dimensional case BSDEs were introduced
in [476] while their direct connection with optimal stochastic control was first
investigated in [212] and [483]. Since then, the general theory of BSDEs has
developed considerably, see [78, 80, 210, 378, 421, 475]. Besides stochastic control, applications were given to many fields, for instance, to optimal stopping,
stochastic differential games, nonlinear partial differential equations and many
topics related to mathematical finance. Infinite-dimensional BSDEs have also been
considered, see for instance, [130, 285, 331, 351, 477]. The interest for us is that
BSDEs provide an alternative way to represent the value function of an optimal
control problem and consequently to study the corresponding HJB equation and to
solve the control problem. It turns out that the most suitable notion of solution for
the HJB equation is, in this context, that of a mild solution on spaces of continuous
functions but, unlike in Chap. 4, the BSDE method seems particularly adapted to
treating degenerate cases in which the transition semigroup has no smoothing
properties. The price to pay is that normally we need more regular coefficients and a
structural condition (imposing, roughly speaking, that the control acts within the
image of the noise). If these requirements are satisfied, the BSDE techniques are
revealed to be very flexible. In particular, in Chap. 6 we will show how they allow
us to treat both parabolic and elliptic HJB equations (see [77, 286, 352, 436, 478]).
The parabolic case is treated for nonconstant diffusion and Lipschitz nonlinearity,
while the elliptic case is considered for a constant diffusion operator with locally
Lipschitz (with respect to the gradient) nonlinearity and a mild dissipativity
assumption (with respect to the solution). We also report (without proofs) the
results of [286] concerning elliptic HJB equations with nonconstant diffusion, a
globally Lipschitz Hamiltonian and strong dissipativity. A detailed discussion of the
literature on BSDEs in infinite dimension is contained in the bibliographic notes
of Chap. 6.

It is impossible to cover all aspects of the theory of HJB equations in infinite
dimension and its connections to stochastic optimal control. In particular, the theory
of integro-PDEs is an emerging area which is not presented in the book. We do not
discuss first-order equations and extensions to Banach spaces. Equations in the
space of probability measures is another emerging topic. We have chosen a
selection of topics which give a broad overview of the field and enough information
so that the readers can start exploring the subject on their own. There are already

enough important applications to justify the interest in the subject. The readers should not be restricted to the boundaries drawn by the book. We hope that this book will spur interest and research in the field among theoretical and applied mathematicians, and that it will be useful to all kinds of scientists and researchers working in areas related to stochastic control.

Suggestions for reading. The readers who are familiar with probability and stochastic analysis in infinite dimension can skip Chap. 1 and go directly to Chap. 2. Chapter 2 is needed for the understanding of the other chapters; however, some material in Sect. 2.3 related to technical details of the proof of the dynamic programming principle can be omitted during the first reading. Chaps. 3–6 are to a large extent independent of each other, and hence the reader can pass from Chap. 2 directly to any of them.

Marseille, France   Giorgio Fabbri
Rome, Italy   Fausto Gozzi
Atlanta, GA, USA   Andrzej Święch

Acknowledgements

The writing of this book was a daunting task which took several years to complete. We greatly benefited from comments, remarks, and advice from many people who read parts of the manuscript or provided useful suggestions regarding the book. Their input improved the content of this book and the presentation of the material and reduced the number of mistakes and errors. The list, in alphabetical order, includes Elena Bandini, Daniel Bauer, Enrico Biffis, Sandra Cerrai, Andrea Cosso, Giuseppe Da Prato, Cristina Di Girolami, Salvatore Federico, Ben Goldys, Carlo Marinelli, Federica Masiero, Chenchen Mou, Mauro Rosestolato, Nizar Touzi, and Jerzy Zabczyk. We thank all of them for their valuable help.

G. Fabbri wishes to express his gratitude to his wife Sara for her constant understanding and encouragement.

F. Gozzi is grateful to his family (Enrica, Matteo, and Marta) who supported him in this long work and to all his friends who encouraged him to accomplish this book. He also expresses special thanks to G. Da Prato, for introducing him to the theory of HJB equations in infinite dimension, for constant encouragement in this work, and also for reading part of the manuscript.

A. Święch would also like to express his gratitude to M.G. Crandall who introduced him to viscosity solutions and PDEs in infinite-dimensional spaces and who greatly influenced his mathematical career.

M. Fuhrman and G. Tessitore would like to thank G. Da Prato and J. Zabczyk for introducing them to stochastic analysis and for their constant help and support.

Finally, we are grateful to Boris Rozovski, the former editor of the series, for his support and Marina Reizakis and the Springer production team for their patience with us and for their very professional handling of the project.

Contents

About the Authors

Giorgio Fabbri is a CNRS Researcher at the Aix-Marseille School of Economics, Marseille, France. He works on optimal control of deterministic and stochastic systems, notably in infinite dimensions, with applications to economics. He has also published various papers in several economic areas, in particular in growth theory and development economics.

Fausto Gozzi is a Full Professor of Mathematics for Economics and Finance at Luiss University, Roma, Italy. His main research field is the optimal control of finite and infinite-dimensional systems and its economic and financial applications. He is the author of many papers in various areas, from Mathematics, to Economics and Finance.

Andrzej Święch is a Full Professor at the School of Mathematics, Georgia Institute of Technology, Atlanta, USA. His main research interests are in nonlinear PDEs and integro-PDEs, PDEs in infinite-dimensional spaces, viscosity solutions, stochastic and deterministic optimal control, stochastic PDEs, differential games, mean-field games, and calculus of variations.

About the Contributors

Marco Fuhrman is a Full Professor of Probability and Mathematical Statistics at the University of Milano, Italy. His main research topics are stochastic differential equations in infinite dimensions and backward stochastic differential equations for optimal control of stochastic processes.

Gianmario Tessitore is a Full Professor of Probability and Mathematical Statistics at Milano-Bicocca University. He is the author of several scientific papers on control of stochastic differential equations in finite and infinite dimensions. He is, in particular, interested in the applications of backward stochastic differential equations in stochastic control.

Chapter 1
Preliminaries on Stochastic Calculus in Infinite Dimension

1.1 Basic Probability

We recall some basic notions of measure theory and give a short introduction to random variables and the theory of the Bochner integral.

1.1.1 Probability Spaces, σ-Fields

Definition 1.1 (*π-system, σ-field*) Consider a set Ω and denote by $\mathcal{P}(\Omega)$ the power set of Ω.

(i) A non-empty class of subsets of Ω, $\mathscr{F} \subset \mathcal{P}(\Omega)$, is called a *π-system* if it is closed under finite intersections.
(ii) A class of subsets of Ω, $\mathscr{F} \subset \mathcal{P}(\Omega)$, is called a *σ-field in* Ω if $\Omega \in \mathscr{F}$ and $\mathscr{F}$ is closed under complements and countable unions.
(iii) A class of subsets of Ω, $\mathscr{F} \subset \mathcal{P}(\Omega)$, is called a *λ-system* if:

- $\Omega \in \mathscr{F}$;
- if $A, B \in \mathscr{F}$, $A \subset B$, then $B \setminus A \in \mathscr{F}$;
- if $A_i \in \mathscr{F}$, $i = 1, 2, ..., A_i \uparrow A$, then $A \in \mathscr{F}$.

If $\mathscr{G}$ and $\mathscr{F}$ are two σ-fields in Ω and $\mathscr{G} \subset \mathscr{F}$, we say that $\mathscr{G}$ is a sub-σ-field of $\mathscr{F}$. Given a class $\mathscr{C} \subset \mathcal{P}(\Omega)$, the smallest σ-field containing $\mathscr{C}$ is called the *σ-field generated by* $\mathscr{C}$. It is denoted by $\sigma(\mathscr{C})$. A σ-field $\mathscr{F}$ in Ω is said to be *countably generated* if there exists a countable class of subsets $\mathscr{C} \subset \mathcal{P}(\Omega)$ such that $\sigma(\mathscr{C}) = \mathscr{F}$.

If $\mathscr{C} \subset \mathcal{P}(\Omega)$ and $A \subset \Omega$ we define $\mathscr{C} \cap A := \{B \cap A : B \in \mathscr{C}\}$. We denote by $\sigma_A(\mathscr{C} \cap A)$ the σ-field of subsets of A generated by $\mathscr{C} \cap A$. It is easy to see that $\sigma_A(\mathscr{C} \cap A) = \sigma(\mathscr{C}) \cap A$ (see, for instance, [18], p. 5).

© Springer International Publishing AG 2017
G. Fabbri et al., *Stochastic Optimal Control in Infinite Dimension*,
Probability Theory and Stochastic Modelling 82,
DOI 10.1007/978-3-319-53067-3_1

For $A \subset \Omega$ we denote its *complement* by $A^c := \Omega \setminus A$, and for $A, B \subset \Omega$ we denote their *symmetric difference* by $A \triangle B := (A \setminus B) \cup (B \setminus A)$. We will write $\mathbb{R}^+ = [0, +\infty), \overline{\mathbb{R}}^+ = [0, +\infty) \cup \{+\infty\}, \overline{\mathbb{R}} = \mathbb{R} \cup \{\pm\infty\}$.

Theorem 1.2 *Let $\mathscr{G}$ be a π-system and $\mathscr{F}$ be a λ-system in some set Ω, such that $\mathscr{G} \subset \mathscr{F}$. Then $\sigma(\mathscr{G}) \subset \mathscr{F}$.*

Proof See [370], Theorem 1.1, p. 2. $\square$

Corollary 1.3 *Let $\mathscr{G}$ be a π-system and $\mathscr{F}$ be the smallest family of subsets of Ω such that:*

- $\mathscr{G} \subset \mathscr{F}$;
- *if $A \in \mathscr{F}$ then $A^c \in \mathscr{F}$;*
- *if $A_i \in \mathscr{F}$, $A_i \cap A_j = \emptyset$ for $i, j = 1, 2, ..., i \neq j$, then $\cup_{i=1}^{\infty} A_i \in \mathscr{F}$.*

   Then $\sigma(\mathscr{G}) = \mathscr{F}$.

Proof Since $\sigma(\mathscr{G})$ satisfies the three conditions for $\mathscr{F}$, we obviously have $\mathscr{F} \subset \sigma(\mathscr{G})$. For the opposite inclusion it remains to observe that $\mathscr{F}$ is a λ-system. (For a self-contained proof, see also [180], Proposition 1.4, p. 17.) $\square$

Definition 1.4 (*Measurable space*) If Ω is a set and $\mathscr{F}$ is a σ-field in Ω, the pair $(\Omega, \mathscr{F})$ is called a *measurable space*.

Definition 1.5 (*Probability measure, probability space*) Consider a measurable space $(\Omega, \mathscr{F})$. A function $\mu : \mathscr{F} \to [0, +\infty) \cup \{+\infty\}$ is called a *measure* on $(\Omega, \mathscr{F})$ if $\mu(\emptyset) = 0$, and whenever $A_i \in \mathscr{F}$, $A_i \cap A_j = \emptyset$ for $i, j = 1, 2, ..., i \neq j$, then

$$\mu\left(\bigcup_{i=1}^{\infty} A_i\right) = \sum_{i=1}^{\infty} \mu(A_i).$$

The triplet $(\Omega, \mathscr{F}, \mu)$ is called a *measure space*. If $\mu(\Omega) < +\infty$ we say that μ is a *bounded measure*. If $\Omega = \bigcup_{n=1}^{\infty} A_n$, where $A_n \in \mathscr{F}$, $\mu(A_n) < +\infty, n = 1, 2, ...,$ we say that μ is a *σ-finite measure*. If $\mu(\Omega) = 1$ we say that μ is a *probability measure*. We will use the symbol $\mathbb{P}$ to denote probability measures. The triplet $(\Omega, \mathscr{F}, \mathbb{P})$ is called a *probability space*.

   Thus a probability measure is a σ-additive function $\mathbb{P} : \mathscr{F} \to [0, 1]$ such that $\mathbb{P}(\Omega) = 1$.

   Given a measure space $(\Omega, \mathscr{F}, \mu)$, we define $\mathscr{N} := \{F \subset \Omega : \exists G \in \mathscr{F}, F \subset G, \mu(G) = 0\}$. The elements of $\mathscr{N}$ are called *μ-null sets*. If $\mathscr{N} \subset \mathscr{F}$, the measure space $(\Omega, \mathscr{F}, \mu)$ is said to be *complete*. The σ-field $\overline{\mathscr{F}} := \sigma(\mathscr{F}, \mathscr{N})$ is called the *completion* of $\mathscr{F}$ (with respect to μ). It is easy to see that $\sigma(\mathscr{F}, \mathscr{N}) = \{A \cup B : A \in \mathscr{F}, B \in \mathscr{N}\}$. If $\mathscr{G} \subset \mathscr{F}$ is another σ-field then $\sigma(\mathscr{G}, \mathscr{N})$ is called the *augmentation* of $\mathscr{G}$ by the null sets of $\mathscr{F}$. The augmentation of $\mathscr{G}$ may be different from its completion, as the latter is just the augmentation of $\mathscr{G}$ by the subsets of the sets of measure zero in $\mathscr{G}$. We also have $\sigma(\mathscr{G}, \mathscr{N}) = \{A \subset \Omega : A \triangle B \in \mathscr{N} \text{ for some } B \in \mathscr{G}\}$.

Let μ, ν be two measures on a measurable space $(\Omega, \mathscr{F})$. We say that μ is *absolutely continuous* with respect to ν (we write $\mu << \nu$) if for every $A \in \mathscr{F}$ such that $\nu(A) = 0$ we have $\mu(A) = 0$. If $\mu << \nu$ and $\nu << \mu$, we say that the measures μ and ν are *equivalent* (we write $\mu \sim \nu$). If there exists a set $A \in \mathscr{F}$ such that for every $B \in \mathscr{F}$ we have $\mu(B) = \mu(A \cap B)$, we say that μ is *concentrated on the set* A. If μ and ν are concentrated on disjoint sets we say that μ and ν are (mutually) *singular* and we write $\mu \perp \nu$.

Lemma 1.6 *Let μ_1, μ_2 be two bounded measures on a measurable space $(\Omega, \mathscr{F})$, and let $\mathscr{G}$ be a π-system in Ω such that $\Omega \in \mathscr{G}$ and $\sigma(\mathscr{G}) = \mathscr{F}$. Then $\mu_1 = \mu_2$ if and only if $\mu_1(A) = \mu_2(A)$ for every $A \in \mathscr{G}$.*

Proof See [370], Lemma 1.17, p. 9. □

Let $\Omega_t, t \in T$ be a family of sets. We will denote the Cartesian product of the family Ω_t by $\times_{t \in T} \Omega_t$. If T is finite $(T = \{1, ..., n\})$ or countable $(T = \mathbb{N})$, we will also write $\Omega_1 \times ... \times \Omega_n$, respectively $\Omega_1 \times \Omega_2 \times ....$ If each Ω_t is a topological space, we endow $\times_{t \in T} \Omega_t$ with the product topology. If each Ω_t has a σ-field $\mathscr{F}_t$, we define the *product σ-field* $\otimes_{t \in T} \mathscr{F}_t$ in $\times_{t \in T} \Omega_t$ as the σ-field generated by the one-dimensional cylinder sets $A_t \times (\times_{s \neq t} \Omega_s)$. If $T = \{1, ..., n\}$ (respectively, $T = \mathbb{N}$) we will just write $\otimes_{t \in T} \mathscr{F}_t = \mathscr{F}_1 \otimes ... \otimes \mathscr{F}_n$ (respectively, $\otimes_{t \in T} \mathscr{F}_t = \mathscr{F}_1 \otimes \mathscr{F}_2 \otimes ...$).

If S is a topological space, the σ-field generated by the open sets of S is called the *Borel σ-field*. It will be denoted by $\mathcal{B}(S)$. If S is a metric space, unless stated otherwise, its default σ-field will always be $\mathcal{B}(S)$. It is not difficult to see that if $S_1, S_2, ...$ are separable metric spaces, then

$$\mathcal{B}(S_1 \times S_2 \times ...) = \mathcal{B}(S_1) \otimes \mathcal{B}(S_2) \otimes ....$$

If (S, ρ) is a metric space, $A \subset S$, and we consider (A, ρ) as a metric space, then $\mathcal{B}(A) = A \cap \mathcal{B}(S)$. A complete separable metric space is called a *Polish space*. Also $\mathcal{B}(\overline{\mathbb{R}}^+) = \sigma(\mathcal{B}(\mathbb{R}^+), \{+\infty\}), \mathcal{B}(\overline{\mathbb{R}}) = \sigma(\mathcal{B}(\mathbb{R}), \{-\infty\}, \{+\infty\})$.

A measurable space $(\Omega, \mathscr{F})$ is called *countably determined* (or $\mathscr{F}$ is called countably determined) if there is a countable set $\mathscr{F}_0 \subset \mathscr{F}$ such that any two probability measures on $(\Omega, \mathscr{F})$ that agree on $\mathscr{F}_0$ must be the same. It follows from Lemma 1.6 that if $\mathscr{F}$ is countably generated then $\mathscr{F}$ is countably determined. If S is a Polish space then $\mathcal{B}(S)$ is countably generated.

If $(\Omega_i, \mathscr{F}_i, \mu_i), i = 1, ..., n$, are measure spaces, their product measure on $(\Omega_1 \times ... \times \Omega_n, \mathscr{F}_1 \otimes ... \otimes \mathscr{F}_n)$ is denoted by $\mu_1 \otimes ... \otimes \mu_n$.

If S is a metric space, a bounded measure μ on $(S, \mathcal{B}(S))$ is called *regular* if

$$\mu(A) = \sup\{\mu(C) : C \subset A, C \text{ closed}\} = \inf\{\mu(U) : A \subset U, U \text{ open}\} \quad \forall A \in \mathcal{B}(S).$$

Every bounded measure on $(S, \mathcal{B}(S))$ is regular (see [478], Chap. II, Theorem 1.2). A bounded measure μ on $(S, \mathcal{B}(S))$ is called *tight* if for every $\varepsilon > 0$ there exists a compact set $K_\varepsilon \subset S$ such that $\mu(S \setminus K_\varepsilon) < \varepsilon$. If S is a Polish space then every bounded measure on $(S, \mathcal{B}(S))$ is tight (see [478], Chap. II, Theorem 3.2).

We refer to [58, 61, 267, 370, 478] for more on the general theory of measure and probability.

1.1.2 Random Variables

Definition 1.7 (*Random variable*) A *measurable* map X between two measurable spaces $(\Omega, \mathscr{F})$ and $(\tilde{\Omega}, \mathscr{G})$ is a called a *random variable*. This means that X is a random variable if $X^{-1}(A) \in \mathscr{F}$ for every $A \in \mathscr{G}$. We write it shortly as $X^{-1}(\mathscr{G}) \subset \mathscr{F}$. Sometimes we will just say that X is $\mathscr{F}/\mathscr{G}$-measurable.

If $\tilde{\Omega} = \mathbb{R}$ (resp. $\mathbb{R}^+$) and $\mathscr{G}$ is the Borel σ-field $\mathcal{B}(\mathbb{R})$ (resp. $\mathcal{B}(\mathbb{R}^+)$) then X is said to be a *real random variable* (resp. *positive random variable*).

If $\Omega, \tilde{\Omega}$ are topological spaces and $\mathscr{F}, \mathscr{G}$ are the Borel σ-fields then X is said to be *Borel measurable*.

If $(\Omega, \mathscr{F}, \mu)$ is a measure space and $X, X_1 : \Omega \to \tilde{\Omega}$, we say that X_1 is a *version* of X if $X = X_1$ μ-a.e.

Given a random variable $X : (\Omega, \mathscr{F}) \to (\tilde{\Omega}, \mathscr{G})$ we denote by $\sigma(X)$ the smallest sub-σ-field of $\mathscr{F}$ that makes X measurable, i.e. $\sigma(X) := X^{-1}(\mathscr{G})$. It is called the *$\sigma$-field generated by X*. Given a set of indices I and a family of random variables $X_i : (\Omega, \mathscr{F}) \to (\tilde{\Omega}, \mathscr{G}), i \in I$, the σ-field $\sigma(X_i : i \in I)$ generated by $\{X_i\}_{i \in I}$ is the smallest sub-σ-field of $\mathscr{F}$ that makes all the functions $X_i : (\Omega, \sigma(X_i : i \in I)) \to (\tilde{\Omega}, \mathscr{G})$ measurable, i.e. $\sigma(X_i : i \in I) = \sigma\left(X_i^{-1}(\mathscr{G}) : i \in I\right)$.

Lemma 1.8 *Let $(\Omega, \mathscr{F})$ be a measurable space. Then:*

(i) If $(\tilde{\Omega}, \mathscr{G})$ is a measurable space, $X : \Omega \to \tilde{\Omega}$, and $\mathscr{C} \subset \mathscr{G}$ is such that $\sigma(\mathscr{C}) = \mathscr{G}$, then X is $\mathscr{F}/\mathscr{G}$-measurable if and only if $X^{-1}(\mathscr{C}) \subset \mathscr{F}$. Moreover, $\sigma(X) = \sigma(X^{-1}(\mathscr{C}))$.

(ii) If $X_n : \Omega \to \overline{\mathbb{R}}, n = 1, 2, ...,$ are random variables, then $\sup_n X_n, \inf_n X_n, \limsup_n X_n, \liminf_n X_n$ are random variables.

(iii) Let $X_n : \Omega \to S, n = 1, 2, ...,$ be random variables, where S is a metric space. Then:

- *if S is complete then $\{\omega : X_n(\omega) \text{ converges}\} \in \mathscr{F}$;*
- *if $X_n \to X$ on Ω, then X is a random variable.*

(iv) Let $(\Omega_i, \mathscr{F}_i), i = 1, 2,$ be measurable spaces, and $X : \Omega_1 \times \Omega_2 \to \Omega$ be $(\mathscr{F}_1 \otimes \mathscr{F}_2)/\mathscr{F}$-measurable. Then, for every $\omega_1 \in \Omega_1, X_{\omega_1}(\cdot) = X(\omega_1, \cdot)$ is $\mathscr{F}_2/\mathscr{F}$-measurable, and, for every $\omega_2 \in \Omega_2, X_{\omega_2}(\cdot) = X(\cdot, \omega_2)$ is $\mathscr{F}_1/\mathscr{F}$-measurable. □

Proof See, for instance, [370], Lemmas 1.4, 1.9, 1.10, and [520], Theorem 7.5, p. 138. □

Theorem 1.9 *Let $(\Omega, \mathscr{F})$ and $(\tilde{\Omega}, \mathscr{G})$ be two measurable spaces and (S, d) a Polish space. Let $X : (\Omega, \mathscr{F}) \to (\tilde{\Omega}, \mathscr{G})$ and $\phi : (\Omega, \mathscr{F}) \to (S, \mathcal{B}(S))$ be two random variables. Then ϕ is measurable as a map from $(\Omega, \sigma(X))$ to $(S, \mathcal{B}(S))$ if and only if there exists a measurable map $\eta : (\tilde{\Omega}, \mathscr{G}) \to (S, \mathcal{B}(S))$ such that $\phi = \eta \circ X$.*

Proof See [370], Lemma 1.13, p. 7, or [575] Theorem 1.7, p. 5.   □

We refer to [58, 267, 370, 520] for more on measurability and for the general theory of integration.

Definition 1.10 *(Borel isomorphism)* Let $(\Omega, \mathscr{F})$ and $(\tilde{\Omega}, \mathscr{G})$ be two measurable spaces. A bijection f from Ω onto $\tilde{\Omega}$ is called a *Borel isomorphism* if f is $\mathscr{F}/\mathscr{G}$-measurable and f^{-1} is $\mathscr{G}/\mathscr{F}$-measurable. We then say that $(\Omega, \mathscr{F})$ and $(\tilde{\Omega}, \mathscr{G})$ are Borel isomorphic.

Definition 1.11 *(Standard measurable space)* A measurable space $(\Omega, \mathscr{F})$ is called *standard* if it is Borel isomorphic to one of the following spaces:

(i) $(\{1, .., n\}, \mathcal{B}(\{1, .., n\}))$,
(ii) $(\mathbb{N}, \mathcal{B}(\mathbb{N}))$,
(iii) $(\{0, 1\}^{\mathbb{N}}, \mathcal{B}(\{0, 1\}^{\mathbb{N}}))$,

where we have the discrete topologies in $\{1, .., n\}$ and $\mathbb{N}$, and the product topology in $\{0, 1\}^{\mathbb{N}}$.

The following theorem collects results that can be found in [478] (Chap. I, Theorems 2.8 and 2.12).

Theorem 1.12 *If S is a Polish space, then $(S, \mathcal{B}(S))$ is standard. If a Borel subset of S is uncountable, then it is Borel isomorphic to $\{0, 1\}^{\mathbb{N}}$. Two Borel subsets of S are Borel isomorphic if and only if they have the same cardinality. If $(\Omega, \mathscr{F})$ is standard and $A \in \mathscr{F}$, then $(A, \mathscr{F} \cap A)$ is standard.*

In particular, we have the following result.

Theorem 1.13 *If $(\Omega, \mathscr{F})$ is standard, then it is Borel isomorphic to a closed subset of $[0, 1]$ (with its induced Borel sigma field).*

Definition 1.14 *(Simple random variable)* Let $(\Omega, \mathscr{F})$ be a measurable space, and (S, d) be a metric space (endowed with the Borel σ-field induced by the distance). A random variable $X \colon (\Omega, \mathscr{F}) \to (S, \mathcal{B}(S))$ is called simple (or a simple function) if it has a finite number of values.

Lemma 1.15 *Let $f \colon (\Omega, \mathscr{F}) \to S$ be a measurable function between a measurable space $(\Omega, \mathscr{F})$ and a separable metric space (S, d) (endowed with the Borel σ-field induced by the distance). Then there exists a sequence $f_n \colon \Omega \to S$ of simple, $\mathscr{F}/\mathcal{B}(S)$-measurable functions, such that $d(f(\omega), f_n(\omega))$ is monotonically decreasing to 0 for every $\omega \in \Omega$.*

Proof See [180], Lemma 1.3, p. 16.   □

Lemma 1.16 *Let S be a Polish space with metric d. Let $(\Omega, \mathscr{F}, \mathbb{P})$ be a complete probability space and let $\mathscr{G}_1, \mathscr{G}_2 \subset \mathscr{F}$ be two σ-fields with the following property: for every $A \in \mathscr{G}_2$ there exists a $B \in \mathscr{G}_1$ such that $\mathbb{P}(A \triangle B) = 0$. Let $f \colon (\Omega, \mathscr{G}_2) \to (S, \mathcal{B}(S))$ be a measurable function. Then there exists a function $g \colon (\Omega, \mathscr{G}_1) \to (S, \mathcal{B}(S))$ such that $f = g$, $\mathbb{P}$-a.e., and simple functions $g_n \colon (\Omega, \mathscr{G}_1) \to (S, \mathcal{B}(S))$ such that $d(f(\omega), g_n(\omega))$ monotonically decreases to 0, $\mathbb{P}$-a.e.*

Proof The proof follows the lines of the proof of Lemma 1.25, p. 13, in [370].

Step 1: Let us assume first that $f = x\mathbf{1}_A$ ($\mathbf{1}_A$ denotes the characteristic function of the set A) for some $A \in \mathscr{G}_2$ and $x \in S$. By hypothesis, we can find $B \in \mathscr{G}_1$ s.t. $\mathbb{P}(A \triangle B) = 0$ and then the claim is proved if we choose $g_n \equiv g = x\mathbf{1}_B$. The same argument holds for a simple function f.

Step 2: For the case of a general f, thanks to Lemma 1.15 we can find a sequence of simple, $\mathscr{G}_2$-measurable functions f_n such that $d(f(\omega), f_n(\omega))$ monotonically decreases to 0. By Step 1, we can find simple, $\mathscr{G}_1$-measurable functions g_n such that $f_n = g_n$, $\mathbb{P}$-a.e. Thus the claim follows by taking $g(\omega) := \lim g_n(\omega)$ if the limit exists and $g(\omega) = s$ (for some $s \in S$) otherwise. □

Lemma 1.17 *Let $(\Omega, \mathscr{F})$ be a measurable space, and $V \subset E$ be two real separable Banach spaces such that the embedding of V into E is continuous. Then:*

(i) $\mathcal{B}(E) \cap V \subset \mathcal{B}(V)$ and $\mathcal{B}(V) \subset \mathcal{B}(E)$.

(ii) *If $X : \Omega \to V$ is $\mathscr{F}/\mathcal{B}(V)$-measurable, then it is $\mathscr{F}/\mathcal{B}(E)$-measurable.*

(iii) *If $X : \Omega \to E$ is $\mathscr{F}/\mathcal{B}(E)$-measurable, then $X \cdot \mathbf{1}_{\{X \in V\}}$ is $\mathscr{F}/\mathcal{B}(V)$-measurable.*

(iv) *$X : \Omega \to E$ is $\mathscr{F}/\mathcal{B}(E)$-measurable if and only if for every $f \in E^*$, $f \circ X$ is $\mathscr{F}/\mathcal{B}(\mathbb{R})$-measurable.*

Proof The embedding of V into E is continuous, so $\mathcal{B}(E) \cap V \subset \mathcal{B}(V)$. Since the embedding is also one-to-one, it follows from [478], Theorem 3.9, p. 21, that $\mathcal{B}(V) \subset \mathcal{B}(E)$, which completes the proof of (i). Parts (ii) and (iii) are direct consequences of (i). $f(\Omega)$ is separable because E is separable, so Part (iv) is a particular case of the Pettis theorem, see [488] Theorem 1.1. □

Lemma 1.18 *Let $(\Omega, \mathscr{F})$ be a measurable space and (S_1, ρ_1), (S_2, ρ_2) be two metric spaces with S_1 separable. Let $f : \Omega \times S_1 \to S_2$ be such that*

(i) *for each $x \in S_1$, the function $f(\cdot, x) : \Omega \to S_2$ is $\mathscr{F}/\mathcal{B}(S_2)$-measurable;*

(ii) *for each $\omega \in \Omega$ the function $f(\omega, \cdot) : S_1 \to S_2$ is continuous.*

Then $f : \Omega \times S_1 \to S_2$ is $\mathscr{F} \otimes \mathcal{B}(S_1)/\mathcal{B}(S_2)$-measurable.

Proof See Lemma 4.51, p. 153 of [8]. □

Notation 1.19 If E is a Banach space we denote by $|\cdot|_E$ its norm. Given two Banach spaces E and F, we denote by $\mathcal{L}(E, F)$ the Banach space of all continuous linear operators from E to F. If $E = F$ we will usually write $\mathcal{L}(E)$ instead of $\mathcal{L}(E, F)$. If H is a Hilbert space we denote by $\langle \cdot, \cdot \rangle$ its inner product. We will always identify H with its dual via Riesz representation theorem. If V, H are two real separable Hilbert spaces, we denote by $\mathcal{L}_2(V, H)$ the space of Hilbert–Schmidt operators from V to H (see Appendix B.3). The space $\mathcal{L}_2(V, H)$ is a real separable Hilbert space with the inner product $\langle \cdot, \cdot \rangle_2$, see Proposition B.25. ∎

Lemma 1.20 *Let $(\Omega, \mathscr{F})$ be a measurable space and V, H be real separable Hilbert spaces. Suppose that $F : \Omega \to \mathcal{L}_2(V, H)$ is a map such that for every $v \in V$, $F(\cdot)v$ is $\mathscr{F}/\mathcal{B}(H)$-measurable. Then F is $\mathscr{F}/\mathcal{B}(\mathcal{L}_2(V, H))$-measurable.*

Proof Since $\mathcal{L}_2(V, H)$ is separable, by Lemma 1.17-(iv) it is enough to show that for every $T \in \mathcal{L}_2(V, H)$

$$\omega \rightarrow \langle F(\omega), T \rangle_2 = \sum_{k=1}^{+\infty} \langle F(\omega)e_k, T e_k \rangle$$

is $\mathscr{F}/\mathcal{B}(\mathbb{R})$-measurable, where $\{e_k\}$ is any orthonormal basis of V. But this is clear since for every ω

$$\langle F(\omega), T \rangle_2 = \lim_{n \to +\infty} F_n^T(\omega),$$

where

$$F_n^T(\omega) = \sum_{k=1}^{n} \langle F(\omega)e_k, T e_k \rangle$$

and $F_n^T(\omega)$ is $\mathscr{F}/\mathcal{B}(\mathbb{R})$-measurable because it is a finite sum of functions that are $\mathscr{F}/\mathcal{B}(\mathbb{R})$-measurable. $\square$

Let I be an interval in $\mathbb{R}$, E, F be two real Banach spaces, and let E be separable. If $f : I \times E \to F$ is Borel measurable then for every $t \in I$ the function $f(t, \cdot) : E \to F$ is Borel measurable (by Lemma 1.8-(iv)).

Assume now that, for all $t \in I$ and for some $m \geq 0$, $f(t, \cdot) \in B_m(E, F)$ (the space of Borel measurable functions with polynomial growth m, see Appendix A.2 for the precise definition). It is not true in general that the function

$$I \to B_m(E, F), \qquad t \to f(t, \cdot)$$

is Borel measurable. As a counterexample[1] one can take the function

$$[0, 1] \times L^2(\mathbb{R}) \to L^2(\mathbb{R}), \qquad (t, x) \to S_t x,$$

where $(S_t)_{t \geq 0}$ is the semigroup of left translations. Indeed, the map

$$[0, 1] \to \mathcal{L}(L^2(\mathbb{R})), \qquad t \to S_t$$

is not measurable (see e.g. [180], Sect. 1.2). Since $\mathcal{L}(L^2(\mathbb{R})) \subset B_1(L^2(\mathbb{R}), L^2(\mathbb{R}))$ and the norm in $\mathcal{L}(L^2(\mathbb{R}))$ is equivalent to the one induced by $B_1(L^2(\mathbb{R}), L^2(\mathbb{R}))$, the claim follows in a straightforward way.

On the other hand, we have the following useful result.

Lemma 1.21 *Let I and Λ be two Polish spaces. Let μ be a measure defined on the Borel σ-field $\mathcal{B}(I)$ and denote by $\overline{\mathcal{B}(I)}$ the completion of $\mathcal{B}(I)$ with respect to μ. Let $f : I \times \Lambda \to \mathbb{R}$ be Borel measurable and such that for every $t \in I$, $f(t, \cdot)$ is bounded from below (respectively, above). Then the function*

---

[1] This example has been suggested to us by Mauro Rosestolato.

$$\underline{f} : I \to \mathbb{R}, \qquad t \to \inf_{a \in \Lambda} f(t, a) \tag{1.1}$$

(respectively, $\overline{f} : I \to \mathbb{R}, t \to \sup_{a \in \Lambda} f(t, a)$) is $\overline{\mathcal{B}(I)}/\mathcal{B}(\mathbb{R})$-measurable.[2]

In particular, if I is an interval in $\mathbb{R}$, E, F are two real Banach spaces with E separable, if $\rho : I \times E \to F$ is Borel measurable and, for all $t \in I$ and for some $m \geq 0$, $\rho(t, \cdot) \in B_m(E, F)$, then the function

$$\rho_1 : I \to \mathbb{R}, \qquad t \to \| f(t, \cdot) \|_{B_m(E, F)} \tag{1.2}$$

is Lebesgue measurable.

Proof The first part is Example 7.4.2 in Volume 2 of [61] (recall that Polish spaces are Souslin spaces, see [61], Definition 6.6.1, and so $I \times \Lambda$ is a Souslin space).

For the second claim, observe that since f is Borel measurable, the function

$$f : I \times E \to \mathbb{R}, \qquad f(t, x) := \frac{|\rho(t, x)|_F}{1 + |x|_E^m}$$

is also Borel measurable (since it is the product of a continuous function with the composition of a continuous function and a Borel measurable function). The result thus follows from part one.   □

Definition 1.22 (*Independence*) Consider a probability space $(\Omega, \mathcal{F}, \mathbb{P})$. Let I be a set of indices, and $\mathscr{C}_i \subset \mathscr{F}$ for all $i \in I$. We say that the families $\mathscr{C}_i, i \in I$, are independent if, for every finite subset J of I and every choice of $A_i \in \mathscr{C}_i, (i \in J)$, we have

$$\mathbb{P}\left(\bigcap_{i \in J} A_i \right) = \prod_{i \in J} \mathbb{P}(A_i).$$

If $\mathscr{C}_i \subset \mathscr{F}$ is, for all $i \in I$, a π-system (resp. σ-field), the definition above gives in particular the notion of *independent π-systems* (resp. σ-fields). Random variables are said to be independent if they generate independent σ-fields. A random variable X is independent of some σ-field $\mathscr{G}$ if $\sigma(X)$ and $\mathscr{G}$ are independent σ-fields.

Lemma 1.23 *Consider a probability space* $(\Omega, \mathcal{F}, \mathbb{P})$. *Let* $\mathscr{C}_i \subset \mathscr{F}$ *be a π-system for every $i \in I$. If $\mathscr{C}_i, i \in I$, are independent, then* $\sigma(\mathscr{C}_i), i \in I$, *are independent.*

Proof See [370] Lemma 2.6, p. 27.   □

---

[2]Note that $\underline{f}$ is not always Borel measurable, see [61] Volume 2, Exercise 6.10.42(ii), p. 59.

1.1.3 The Bochner Integral

Throughout this section $(\Omega, \mathscr{F}, \mu)$ is a measure space where μ is σ-finite, and E is a separable Banach space with norm $|\cdot|_E$. We endow E with the Borel σ-field $\mathcal{B}(E)$.

Lemma 1.24 *Let* $X: (\Omega, \mathscr{F}) \to E$ *be a random variable. Then the real-valued function* $|X|_E$ *is measurable.*

Proof See [180] Lemma 1.2, p. 16. $\square$

Let $p \geq 1$. We denote by $L^p(\Omega, \mathscr{F}, \mu; E)$ the quotient space of the set

$$\tilde{L}^p(\Omega, \mathscr{F}, \mu; E) := \left\{ X: (\Omega, \mathscr{F}) \to (E, \mathcal{B}(E)) \text{ measurable} : \int_\Omega |X(\omega)|_E^p \, d\mu(\omega) < +\infty \right\}$$

with respect to the equivalence relation of equality μ-a.e. $L^p(\Omega, \mathscr{F}, \mu; E)$ is a Banach space when endowed with the norm

$$|X|_{L^p(\Omega, \mathscr{F}, \mu; E)} = \left(\int_\Omega |X(\omega)|_E^p \, d\mu(\omega) \right)^{1/p}$$

(see e.g. [191] Theorem 7.17 p. 104). We will often write $L^p(\Omega, \mu; E)$ or $L^p(\Omega; E)$ for $L^p(\Omega, \mathscr{F}, \mu; E)$ and denote the norm by $|X|_{L^p}$ when the context is clear. If H is a separable Hilbert space, then $L^2(\Omega, \mathscr{F}, \mu; H)$ is a Hilbert space as well, equipped with the scalar product $\langle X, Y \rangle_{L^2(\Omega, \mathscr{F}, \mu; H)} = \int_\Omega \langle X(\omega), Y(\omega) \rangle_H \, d\mu(\omega)$.

The space $L^\infty(\Omega, \mathscr{F}, \mu; E)$ is the quotient space of the space of bounded $\mathscr{F}/\mathcal{B}(E)$-measurable functions with respect to the relation of being equal a.e. It is a Banach space equipped with the norm

$$|X|_{L^\infty(\Omega, \mathscr{F}, \mu; E)} = \operatorname*{ess\,sup}_{\Omega} |X(\omega)|_E .$$

In the special case when $\Omega = I$ is an interval with endpoints a and b with $a < b$ (which may be $\pm\infty$), $\mathscr{F}$ is the Borel σ-field of I, and μ is the Lebesgue measure on I, we will simply write $L^p(I; E)$ or $L^p(a, b; E)$ for $L^p(I, \mathscr{F}, \mu; E)$. Finally, we denote by $L^p_{\text{loc}}(I; E)$ the set of measurable functions $f: I \to E$ such that $\int_K |f(s)|_E^p ds$ is finite for every compact subset K of I.

Lemma 1.25 *If* $\mathscr{F}$ *is countably generated apart from null sets then* $L^p(\Omega, \mathscr{F}, \mu; E)$ *is a separable Banach space.*

Proof See [194], p. 92. $\square$

Definition 1.26 (*Bochner integral*) Let $X: (\Omega, \mathscr{F}, \mu) \to E$ be a simple random variable $X = \sum_{i=1}^N x_i \mathbf{1}_{A_i}$, where $x_i \in E$, $A_i \in \mathscr{F}$, $\mu(A_i) < +\infty$. The *Bochner integral* of X is defined as

$$\int_\Omega X(\omega)d\mu(\omega) := \sum_{i=1}^{N} x_i\mu(A_i).$$

Let X be in $L^1(\Omega, \mathscr{F}, \mu; E)$. The *Bochner integral* of X is defined as

$$\int_\Omega X(\omega)d\mu(\omega) := \lim_{n\to+\infty} \int_\Omega X_n(\omega)d\mu(\omega),$$

where $X_n \colon (\Omega, \mathscr{F}, \mu) \to E$ are simple random variables such that

$$\lim_{n\to+\infty} \int_\Omega |X(\omega) - X_n(\omega)|_E d\mu(\omega) = 0. \tag{1.3}$$

Remark 1.27 It follows easily from Lemma 1.15 that, for $X \in L^1(\Omega, \mathscr{F}, \mu; E)$, there always exists a sequence of simple random variables $X_n \colon (\Omega, \mathscr{F}, \mu) \to E$ as in Definition 1.26, satisfying (1.3). ∎

Proposition 1.28 *Let $X \in L^1(\Omega, \mathscr{F}, \mu; E)$. Then the Bochner integral of X is well defined and does not depend on the choice of the sequence. Moreover,*

$$\left| \int_\Omega X(\omega)d\mu(\omega) \right|_E \leq \int_\Omega |X(\omega)|_E d\mu(\omega). \tag{1.4}$$

Proof See [180] Sect. 1.1 (in particular inequality (1.6), p. 19, and the part below Lemma 1.5). The proof there is done for a probability measure μ, but the general case is identical. □

Proposition 1.29 *Assume that $(\Omega, \mathscr{F}, \mu)$ is a complete measure space, E and F are separable Banach spaces and $A \colon D(A) \subset E \to F$ is a closed operator (see Definition B.3). If $X \in L^1(\Omega, \mathscr{F}, \mu; E)$ and $X \in D(A)$ a.s., then AX is an F-valued random variable, and X is a $D(A)$-valued random variable, where $D(A)$ is endowed with the graph norm of A (see Definition B.3). If, moreover, $\int_\Omega |AX(\omega)|_F d\mu(\omega) < +\infty$, then*

$$A \int_\Omega X(\omega)d\mu(\omega) = \int_\Omega AX(\omega)d\mu(\omega).$$

Proof The facts that X is a $D(A)$-valued random variable and AX is an F-valued random variable follow from Lemma 1.17-(ii). For the last part, see the proof of Proposition 1.6, Chap. 1 of [180]. □

Corollary 1.30 *Assume that E and F are separable Banach spaces and $T \colon E \to F$ is a continuous linear operator. If $X \in L^1(\Omega, \mathscr{F}, \mu; E)$, then*

$$T \int_\Omega X(\omega)d\mu(\omega) = \int_\Omega TX(\omega)d\mu(\omega).$$

Proof This is a particular case of Proposition 1.29. $\square$

Remark 1.31 In this subsection we assumed that the space E is separable. This was done for simplicity and since we will only need this case in the vast majority of the book. However, the Bochner integral of a random variable $X : (\Omega, \mathscr{F}, \mu) \to E$ can also be defined when E is non-separable, see Sect. II.2 of [190]. If E is non-separable the definition of measurability is different. The random variable X is called measurable if there exists a sequence of simple random variables $X_n : (\Omega, \mathscr{F}, \mu) \to E$ such that $\lim_{n\to+\infty} |X(\omega) - X_n(\omega)|_E = 0$ μ-a.e. When E is separable this definition of measurability is equivalent to ours. Most of the results on the Bochner integral still hold in the non-separable case. In particular, Proposition 1.29 (hence also Corollary 1.30) still holds in the following form, which we will use later in Chap. 4 (see, for example, the proof of Corollary 4.14 and of Theorem 4.80).

Let $(\Omega, \mathscr{F}, \mu)$ be a complete measure space, E and F be Banach spaces and $A : D(A) \subset E \to F$ be a closed operator. If $X \in L^1(\Omega, \mathscr{F}, \mu; E)$ and $AX \in L^1(\Omega, \mathscr{F}, \mu; F)$, then

$$A \int_\Omega X(\omega)d\mu(\omega) = \int_\Omega AX(\omega)d\mu(\omega).$$

This is Theorem 6, p. 47 of [190].   ∎

Theorem 1.32 Let $(\Omega_1, \mathscr{F}_1)$ and $(\Omega_2, \mathscr{F}_2)$ be two measurable spaces and μ_1 (respectively μ_2) be a σ-finite measure on $(\Omega_1, \mathscr{F}_1)$ (respectively on $(\Omega_2, \mathscr{F}_2)$). Then there exists a unique measure $\mu_1 \otimes \mu_2$ on $\mathscr{F}_1 \otimes \mathscr{F}_2$ such that, for every $A \in \mathscr{F}_1$ and $B \in \mathscr{F}_2$ with finite measure,

$$(\mu_1 \otimes \mu_2)(A \times B) = \mu_1(A)\mu_2(B).$$

The measure $\mu_1 \otimes \mu_2$ is σ-finite.

Proof See Theorem 8.2, p. 160 in Chap. VI, Sect. 8 of [397]. $\square$

Theorem 1.33 (Fubini's Theorem) Let $(\Omega_1, \mathscr{F}_1)$ and $(\Omega_2, \mathscr{F}_2)$ be two measurable spaces and μ_1 (respectively μ_2) be a σ-finite measure on $(\Omega_1, \mathscr{F}_1)$ (respectively on $(\Omega_2, \mathscr{F}_2)$). Let E be a separable Banach space with norm $|\cdot|_E$.

(i) Let X be in $L^1(\Omega_1 \times \Omega_2, \mathscr{F}_1 \otimes \mathscr{F}_2, \mu_1 \otimes \mu_2; E)$. Then, for μ_1-almost every $\omega_1 \in \Omega_1$, the function $X(\omega_1, \cdot)$ is in $L^1(\Omega_2, \mathscr{F}_2, \mu_2; E)$, and the function given by

$$\omega_1 \to \int_{\Omega_2} X(\omega_1, \omega_2)d\mu_2(\omega_2)$$

for μ_1-almost all ω_1 (and defined arbitrarily for other ω_1) is in $L^1(\Omega_1, \mathscr{F}_1, \mu_1; E)$. Moreover, we have

$$\int_{\Omega_1 \times \Omega_2} X(\omega_1, \omega_2)d(\mu_1 \otimes \mu_2)(\omega_1, \omega_2) = \int_{\Omega_1}\int_{\Omega_2} X(\omega_1, \omega_2)d\mu_1(\omega_1)d\mu_2(\omega_2).$$

(ii) *Let $X\colon \Omega_1 \times \Omega_2 \to E$ be an $\mathscr{F}_1 \otimes \mathscr{F}_2$-measurable map. Assume that, for μ_1-almost every $\omega_1 \in \Omega_1$, the function $X(\omega_1, \cdot)$ is in $L^1(\Omega_2, \mathscr{F}_2, \mu_2; E)$ and that the map given by*

$$\omega_1 \to \int_{\Omega_2} |X(\omega_1, \omega_2)| d\mu_2(\omega_2)$$

for μ_1-almost all ω_1 (and defined arbitrarily for other ω_1) is in $L^1(\Omega_1, \mathbb{R})$. Then X is in $L^1(\Omega_1 \times \Omega_2, \mathscr{F}_1 \otimes \mathscr{F}_2, \mu_1 \otimes \mu_2; E)$ and part (i) of the theorem applies.

Proof See Theorems 8.4, p. 162, and 8.7, p. 165 in Chap. VI, Sect. 8 of [397]. □

Theorem 1.34 *Let E be a separable Banach space and μ be a bounded measure on $(E, \mathcal{B}(E))$. Then the set of uniformly continuous and bounded functions $UC_b(E)$ is dense in $L^p(E, \mathcal{B}(E), \mu)$ for $1 \le p < +\infty$.*

Proof By Lemma 1.15 and the monotone convergence theorem it is enough to prove that every characteristic function $\mathbf{1}_A$ for some $A \in \mathcal{B}(E)$ can be approximated by functions in $UC_b(E)$. Since μ is regular, for every $\varepsilon > 0$ we can find a closed set $C, C \subset A$, and an open set $U, A \subset U$, such that $\mu(U \setminus C) < \varepsilon^p$. Moreover, considering sets $U_n = \{x \in U : \text{dist}(x : A) > 1/n\}$ if necessary, we can assume that $\text{dist}(C, U) > 0$. Then the function

$$f_\varepsilon(x) := \frac{\text{dist}(x, U)}{\text{dist}(x, A) + \text{dist}(x, U)}$$

belongs to $UC_b(E)$ and $|\mathbf{1}_A - f_\varepsilon|_{L^p} < \varepsilon$. □

1.1.4 Expectation, Covariance and Correlation

Let $(\Omega, \mathscr{F}, \mathbb{P})$ be a probability space and E be a separable Banach space with norm $|\cdot|_E$.

Definition 1.35 (*Expectation*) Given X in $L^1(\Omega, \mathscr{F}, \mathbb{P}; E)$, we denote by $\mathbb{E}[X]$ the (Bochner) integral $\int_\Omega X(\omega) d\mathbb{P}(\omega)$. $\mathbb{E}[X]$ is said to be the *expectation* (or the *mean*) of X.

To define the covariance operator, we recall first that if $x \in E, y \in F$, where E, F are Hilbert spaces, the operator $x \otimes y : F \to E$ is defined by

$$(x \otimes y)h = x\langle y, h\rangle_F.$$

Definition 1.36 (*Covariance operator, correlation*) Given a real, separable Hilbert space H and $X \in L^2(\Omega, \mathscr{F}, \mathbb{P}; H)$, the *covariance operator* of X is defined by

$$Cov(X) := \mathbb{E}\Big[(X - \mathbb{E}[X]) \otimes (X - \mathbb{E}[X])\Big].$$

For $X, Y \in L^2(\Omega, \mathscr{F}, \mathbb{P}; H)$, the *correlation* of X and Y is the operator defined by

$$Cor(X, Y) := \mathbb{E}\Big[(X - \mathbb{E}[X]) \otimes (Y - \mathbb{E}[Y])\Big].$$

Remark 1.37 For $X \in L^2(\Omega, \mathscr{F}, \mathbb{P}; H)$, the operator $Cov(X)$ is positive, symmetric and nuclear (see [180], p. 26). ∎

1.1.5 Conditional Expectation and Conditional Probability

Theorem 1.38 *Consider a separable Banach space E, a probability space $(\Omega, \mathscr{F}, \mathbb{P})$ and a sub-σ-field $\mathscr{G} \subset \mathscr{F}$. There exists a unique contractive linear operator $\mathbb{E}[\cdot|\mathscr{G}]: L^1(\Omega, \mathscr{F}, \mathbb{P}; E) \to L^1(\Omega, \mathscr{G}, \mathbb{P}; E)$ such that*

$$\int_A \mathbb{E}[\xi|\mathscr{G}](\omega)d\mathbb{P}(\omega) = \int_A \xi(\omega)d\mathbb{P}(\omega) \quad \text{for all } A \in \mathscr{G} \text{ and } \xi \in L^1(\Omega, \mathscr{F}, \mathbb{P}; E).$$

If $E = H$ is a Hilbert space the restriction of $\mathbb{E}[\cdot|\mathscr{G}]$ to $L^2(\Omega, \mathscr{F}, \mathbb{P}; H)$ is the orthogonal projection $L^2(\Omega, \mathscr{F}, \mathbb{P}; H) \to L^2(\Omega, \mathscr{G}, \mathbb{P}; H)$.

Proof See [180] Proposition 1.10, p. 26, and [458] Proposition V-2-5, pp. 102–103. □

Definition 1.39 *(Conditional expectation)* Given $X \in L^1(\Omega, \mathscr{F}, \mathbb{P}; E)$, the random variable $\mathbb{E}[X|\mathscr{G}] \in L^1(\Omega, \mathscr{G}, \mathbb{P}; E)$, defined by Theorem 1.38, is called the *conditional expectation* of X given $\mathscr{G}$.

Definition 1.40 Let $(\Omega, \mathscr{F}, \mathbb{P})$ be a probability space and let E be a separable Banach space. A family $\mathcal{H}$ of integrable random variables $X \in L^1(\Omega, \mathscr{F}, \mathbb{P}; E)$ is called *uniformly integrable* if

$$\lim_{R \to \infty} \sup_{X \in \mathcal{H}} \int_{|X|_E \geq R} |X(\omega)|_E d\mathbb{P}(\omega) = 0.$$

The following proposition collects various properties of conditional expectation (see e.g. [487] Proposition 3.15, p. 25, see also [572] Sect. 9.7, p. 88, for similar properties for real-valued random variables).

Proposition 1.41 *Let $(\Omega, \mathscr{F}, \mathbb{P})$ be a probability space and let E be a separable Banach space. The conditional expectation has the following properties:*

(i) If $X \in L^1(\Omega, \mathscr{F}, \mathbb{P}; E)$ is $\mathscr{G}$-measurable, then $\mathbb{E}[X|\mathscr{G}] = X$ $\mathbb{P}$-a.s.

(ii) Given $X \in L^1(\Omega, \mathscr{F}, \mathbb{P}; E)$ and two σ-fields $\mathscr{G}_1$ and $\mathscr{G}_2$ such that $\mathscr{G}_1 \subset \mathscr{G}_2 \subset \mathscr{F}$,

$$\mathbb{E}\Big[\mathbb{E}[X|\mathscr{G}_1]\big|\mathscr{G}_2\Big] = \mathbb{E}\Big[\mathbb{E}[X|\mathscr{G}_2]\big|\mathscr{G}_1\Big] = \mathbb{E}[X|\mathscr{G}_1] \quad \mathbb{P}\text{-}a.s.$$

(iii) Let $X \in L^1(\Omega, \mathscr{F}, \mathbb{P}; E)$. If X is independent of $\mathscr{G}$, then $\mathbb{E}[X|\mathscr{G}] = \mathbb{E}[X]$ $\mathbb{P}$-a.s. Moreover, X is independent of $\mathscr{G}$ if and only if, for any bounded, Borel measurable $f: E \to \mathbb{R}$, $\mathbb{E}[f(X)|\mathscr{G}] = \mathbb{E}f(X)$ $\mathbb{P}$-a.s.

(iv) If X is $\mathscr{G}$-measurable and ζ is a real-valued integrable random variable such that $\zeta X \in L^1(\Omega, \mathscr{F}, \mathbb{P}; E)$, then

$$\mathbb{E}\Big[\zeta X|\mathscr{G}\Big] = X\mathbb{E}\Big[\zeta|\mathscr{G}\Big] \quad \mathbb{P}\text{-}a.s.$$

(v) If $X \in L^1(\Omega, \mathscr{F}, \mathbb{P}; E)$ and ζ is an integrable, real-valued, $\mathscr{G}$-measurable random variable such that $\zeta X \in L^1(\Omega, \mathscr{F}, \mathbb{P}; E)$, then

$$\mathbb{E}\Big[\zeta X|\mathscr{G}\Big] = \zeta\mathbb{E}\Big[X|\mathscr{G}\Big] \quad \mathbb{P}\text{-}a.s.$$

(vi) If $X \in L^1(\Omega, \mathscr{F}, \mathbb{P}; E)$ and $f: \mathbb{R} \to \mathbb{R}$ is a convex function such that $\mathbb{E}[|f(|X|_E)|] < +\infty$, then

$$f\left(\Big|\mathbb{E}\big[X|\mathscr{G}\big]\Big|_E\right) \le \mathbb{E}\Big[f(|X|_E)\big|\mathscr{G}\Big] \quad \mathbb{P}\text{-}a.s.$$

(vii) If $X, X_n \in L^1(\Omega, \mathscr{F}, \mathbb{P}; E)$ for every $n \in \mathbb{N}$, the family $(X_n)_{n\in\mathbb{N}}$ is uniformly integrable and $X_n \xrightarrow{n\to\infty} X$, $\mathbb{P}$-a.s., then

$$\mathbb{E}\Big[X_n|\mathscr{G}\Big] \xrightarrow{n\to\infty} \mathbb{E}\Big[X|\mathscr{G}\Big] \quad \mathbb{P}\text{-}a.s.$$

(viii) Let $X \in L^1(\Omega, \mathscr{F}, \mathbb{P}; E)$. Assume that $\mathscr{G}_n$ for $n \in \mathbb{N}$ is an increasing family of σ-fields such that $\mathscr{G} = \sigma(\mathscr{G}_n : n \in \mathbb{N})$ is a sub-σ-field of $\mathscr{F}$. Then

$$\mathbb{E}\Big[X|\mathscr{G}_n\Big] \xrightarrow{n\to\infty} \mathbb{E}\Big[X|\mathscr{G}\Big] \quad \mathbb{P}\text{-}a.s.$$

(ix) Let Z be a separable Banach space and let $T \in \mathcal{L}(E, Z)$. Then

$$\mathbb{E}[TX|\mathscr{G}] = T\mathbb{E}[X|\mathscr{G}] \quad \mathbb{P}\text{-}a.s.$$

Proposition 1.42 Let $(\Omega, \mathscr{F}, \mathbb{P})$ be a probability space. Then:

(i) If $X, Y \in L^1(\Omega, \mathscr{F}, \mathbb{P}; \mathbb{R})$ and $X \ge Y$, then

$$\mathbb{E}[X|\mathscr{G}] \ge \mathbb{E}[Y|\mathscr{G}].$$

(ii) *(Conditional Fatou Lemma) If* $X_n \in L^1(\Omega, \mathscr{F}, \mathbb{P}; \mathbb{R})$ *and* $X_n \geq 0$, *then*

$$\mathbb{E}[\liminf_{n \to \infty} X_n | \mathscr{G}] \leq \liminf_{n \to \infty} \mathbb{E}[X_n | \mathscr{G}] \quad \mathbb{P}\text{-}a.s.$$

Proof See [572], Sect. 9.7, p. 88. □

Proposition 1.43 *Let* $(E_1, \mathscr{E}_1)$ *and* $(E_2, \mathscr{E}_2)$ *be two measurable spaces and* $\psi : E_1 \times E_2 \to \mathbb{R}$ *be a bounded measurable function. Let* X_1, X_2 *be two random variables in a probability space* $(\Omega, \mathscr{F}, \mathbb{P})$ *with values in* $(E_1, \mathscr{E}_1)$ *and* $(E_2, \mathscr{E}_2)$ *respectively, and let* $\mathscr{G} \subset \mathscr{F}$ *be a* σ-*field. If* X_1 *is* $\mathscr{G}$-*measurable and* X_2 *is independent of* $\mathscr{G}$, *then*

$$\mathbb{E}[\psi(X_1, X_2) | \mathscr{G}] = \widehat{\psi}(X_1), \quad \mathbb{P}\text{-}a.s., \tag{1.5}$$

where

$$\widehat{\psi}(x_1) = \mathbb{E}[\psi(x_1, X_2)], \quad x_1 \in E_1. \tag{1.6}$$

Proof See Proposition 1.12, p. 28 of [180]. □

Let $(\Omega, \mathscr{F}, \mathbb{P})$ be a probability space, and $\mathscr{G}$ be a sub-σ-field of $\mathscr{F}$. The conditional probability of $A \in \mathscr{F}$ given $\mathscr{G}$ is defined by

$$\mathbb{P}(A | \mathscr{G})(\omega) := \mathbb{E}[\mathbf{1}_A | \mathscr{G}](\omega).$$

Definition 1.44 Let $(\Omega, \mathscr{F}, \mathbb{P})$ be a probability space, and $\mathscr{G}$ be a sub-σ-field of $\mathscr{F}$. A function $p \colon \Omega \times \mathscr{F} \to [0, 1]$ is called a *regular conditional probability* given $\mathscr{G}$ if it satisfies the following conditions:

(i) for each $\omega \in \Omega$, $p(\omega, \cdot)$ is a probability measure on $(\Omega, \mathscr{F})$;
(ii) for each $B \in \mathscr{F}$, the function $p(\cdot, B)$ is $\mathscr{G}$-measurable;
(iii) for every $A \in \mathscr{F}$, $\mathbb{P}(A | \mathscr{G})(\omega) = p(\omega, A)$, $\mathbb{P}$-a.s.

It thus follows that, if $X \in L^1(\Omega, \mathscr{F}, \mathbb{P}; E)$, where E is a separable Banach space, then

$$\mathbb{E}[X | \mathscr{G}](\omega) = \int_{\Omega} X(\omega') p(\omega, d\omega') \quad \mathbb{P} \ a.s.$$

Theorem 1.45 *Let* $(\Omega, \mathscr{F}, \mathbb{P})$ *be a probability space, where* $(\Omega, \mathscr{F})$ *is a standard measurable space. Then, for every sub-*σ-*field* $\mathscr{G} \subset \mathscr{F}$, *there exists a regular conditional probability* $p(\cdot, \cdot)$ *given* $\mathscr{G}$. *Moreover, if* $p'(\cdot, \cdot)$ *is another regular conditional probability given* $\mathscr{G}$, *then there exists a set* $N \in \mathscr{G}, \mathbb{P}(N) = 0$, *such that, if* $\omega \notin N$ *then* $p(\omega, A) = p'(\omega, A)$ *for all* $A \in \mathscr{F}$.
Moreover, if $\mathscr{H}$ *is a countably determined sub-*σ-*field of* $\mathscr{G}$, *then there exists a* $\mathbb{P}$-*null set* $N \in \mathscr{G}$ *such that, if* $\omega \notin N$ *then* $p(\omega, A) = \mathbf{1}_A(\omega)$ *for every* $A \in \mathscr{H}$. *In particular, if* $(\Omega_1, \mathscr{F}_1)$ *is a measurable space,* $\mathscr{F}_1$ *is countably determined,* $\{x\} \in \mathscr{F}_1$ *for all* $x \in \Omega_1$ *and* $\xi \colon (\Omega, \mathscr{F}) \to (\Omega_1, \mathscr{F}_1)$ *is a* $\mathscr{G}/\mathscr{F}_1$-*random variable, then* $p\left(\omega, \{\omega' : \xi(\omega) = \xi(\omega')\}\right) = 1$ *for* $\mathbb{P}$-a.e. ω.

Proof See Theorem 8.1, p. 147 in [478], or Theorems 3.1, 3.2, and the corollary following them in [356] (see also [575] Proposition 1.9, p. 11).   □

Notation 1.46 If the regular conditional probability exists, we will often write $\mathbb{P}(\cdot|\mathscr{G})(\omega)$ or $\mathbb{P}_\omega$ for $p(\omega, \cdot)$.   ∎

Definition 1.47 (*Law of a random variable*) Given a probability space $(\Omega, \mathscr{F}, \mathbb{P})$, a measurable space $(\Omega_1, \mathscr{F}_1)$, and a random variable $X \colon (\Omega, \mathscr{F}) \to (\Omega_1, \mathscr{F}_1)$, the probability measure on $(\Omega_1, \mathscr{F}_1)$ defined by

$$\mathcal{L}_\mathbb{P}(X)(A) := \mathbb{P}(\{\omega \in \Omega \ : \ X(\omega) \in A\})$$

is called the *law* (or *distribution*)[3] of X. We denote the law of X by $\mathcal{L}_\mathbb{P}(X)$.

Proposition 1.48 (Change of variables) *Given a probability space $(\Omega, \mathscr{F}, \mathbb{P})$, a measurable space $(\Omega_1, \mathscr{F}_1)$, a random variable $X \colon (\Omega, \mathscr{F}) \to (\Omega_1, \mathscr{F}_1)$, and a bounded Borel function $\varphi : \Omega_1 \to \mathbb{R}$ we have*

$$\int_\Omega \varphi(X(\omega)) d\mathbb{P}(\omega) = \int_{\Omega_1} \varphi(\omega') d\mathcal{L}_\mathbb{P}(X)(\omega').$$

Definition 1.49 (*Convergence of random variables*) Consider a probability space $(\Omega, \mathscr{F}, \mathbb{P})$ and a Polish space (S, d) endowed with the Borel σ-field. Let $X_n \colon \Omega \to S$ and $X \colon \Omega \to S$ be random variables. We say that:

(i) X_n converges to X $\mathbb{P}$-*a.s.* (and we write $X_n \to X$ $\mathbb{P}$-a.s.) if $\lim_{n\to\infty} d$ $(X_n(\omega), X(\omega)) = 0$ $\mathbb{P}$-a.s.

(ii) X_n converges to X *in probability* if, for every $\varepsilon > 0$, $\lim_{n\to+\infty} \mathbb{P}$ $\{\omega \in \Omega \ : \ d(X_n(\omega), X(\omega)) > \varepsilon\} = 0$.

(iii) X_n converges to X *in law* if, for every bounded and continuous $f \colon S \to \mathbb{R}$, $\int_S f(u) d\mathcal{L}_\mathbb{P}(X)(u) = \lim_{n\to\infty} \int_S f(u) d\mathcal{L}_\mathbb{P}(X_n)(u)$ (i.e.   if $\mathbb{E}[f(X)] = \lim_{n\to\infty} \mathbb{E}[f(X_n)]$).

Lemma 1.50 *Consider a probability space $(\Omega, \mathscr{F}, \mathbb{P})$ and a Polish space (S, d) endowed with the Borel σ-field. Let $X_n \colon \Omega \to S$ and $X \colon \Omega \to S$ be random variables.*

(i) *If X_n converges to X $\mathbb{P}$-a.s. then X_n converges to X in probability.*

(ii) *If X_n converges to X in probability then X_n converges to X in law.*

(iii) *If X_n converges to X in probability then it contains a subsequence X_{n_k} such that X_{n_k} converges to X $\mathbb{P}$-a.s.*

(iv) *(Egoroff's theorem) If X_n converges to X $\mathbb{P}$-a.s. then for every $\varepsilon > 0$, there exists an $\tilde{\Omega} \in \mathscr{F}$ such that $\mathbb{P}(\Omega \setminus \tilde{\Omega}) < \varepsilon$, and X_n converges uniformly to X on $\tilde{\Omega}$.*

---

[3]In measure theory it is more often called the *push-forward of* $\mathbb{P}$ and denoted by $X_\#\mathbb{P}$.

(v) Let $X, X_n \in L^p(\Omega, \mathscr{F}, \mathbb{P}; E), n \in \mathbb{N}, p \geq 1$, and E be a separable Banach space. If X_n converges to X in $L^p(\Omega, \mathscr{F}, \mathbb{P}; E)$, then X_n converges to X in probability.

Proof For (i), (ii) and (iii) see, for instance, [370] Lemmas 4.2, p. 63 and 4.7, p. 66. Part (iv) can be found, for instance, in [73] Theorem 2, p. 170, Sect. 4.5.4. Property (v) is straightforward. $\square$

Lemma 1.51 *Let $p > 1$ and $X, X_n \in L^p(\Omega, \mathscr{F}, \mathbb{P}; E), n \in \mathbb{N}$, for some separable Banach space E. Suppose that, for some $M > 0$, $\mathbb{E}\left[|X_n|_E^p\right] \leq M$ for all $n \in \mathbb{N}$. If $X_n \to X$ in probability, then $\mathbb{E}\left[|X - X_n|_E\right] \to 0$.*

Proof Since the sequence (X_n) is bounded in $L^p(\Omega, \mathscr{F}, \mathbb{P}; E)$, it is uniformly integrable (see e.g. [572], p. 127, Sect. 13.3). The claim follows, for example, from Theorem 13.7, p. 131 of [572]. $\square$

1.1.6 Gaussian Measures on Hilbert Spaces and the Fourier Transform

In this section we recall the notions of Gaussian measure and the Fourier transform for Hilbert space-valued random variables. For an extensive treatment of the subject we refer to [180], Chap. 2, [153], Chap. 1 or [154], Chap. 1.

For a real separable Hilbert space H we denote by $\mathcal{L}_1(H)$ the Banach space of the trace class operators on H, by $\mathcal{L}^+(H)$ the subspace (of $\mathcal{L}(H)$) of all bounded, linear, self-adjoint, positive operators, and we set $\mathcal{L}_1^+(H) := \mathcal{L}_1(H) \cap \mathcal{L}^+(H)$ (see Appendix B.3). We will denote by $M_1(H)$ the set of probability measures on $(H, \mathcal{B}(H))$.

Proposition 1.52 *Consider a real, separable Hilbert space H with the Borel σ-field $\mathcal{B}(H)$ and a probability measure $\mathbb{P}$ on $(H, \mathcal{B}(H))$. If $\int_H |y| \, d\mathbb{P}(y) < +\infty$, then we can define*

$$m := \int_H y \, d\mathbb{P}(y) \in H.$$

If $\int_H |y|^2 d\mathbb{P}(y) < +\infty$, then there exists a unique $Q \in \mathcal{L}_1^+(H)$ such that

$$\langle Qx, y \rangle := \int_H \langle x, h - m \rangle \, \langle y, h - m \rangle \, d\mathbb{P}(h).$$

Proof See [153], p. 7. $\square$

Definition 1.53 (*Mean and covariance of a measure on H*) We call m and Q, defined by Proposition 1.52, respectively the *mean* and the *covariance* of $\mathbb{P}$. In other words, the mean (respectively covariance) of $\mathbb{P}$ is the mean (respectively covariance) of the identity random variable $I : (H, \mathcal{B}(H), \mathbb{P}) \to (H, \mathcal{B}(H))$.

Definition 1.54 (*Fourier transform of a measure*) Let H be a Hilbert space and $\mathcal{B}(H)$ be its Borel σ-field. Given a probability measure $\mathbb{P}$ on $(H, \mathcal{B}(H))$ we define, for $x \in H$,

$$\hat{\mathbb{P}}(x) := \int_H e^{i\langle y, x \rangle} d\mathbb{P}(y).$$

We call $\hat{\mathbb{P}} \colon H \to \mathbb{C}$ the *Fourier transform* of $\mathbb{P}$.

Proposition 1.55 *Let H be a real, separable Hilbert space, $\mathcal{B}(H)$ be its Borel σ-field, and $\mathbb{P}_1$ and $\mathbb{P}_2$ be two probability measures on $(H, \mathcal{B}(H))$. If $\hat{\mathbb{P}}_1(x) = \hat{\mathbb{P}}_2(x)$ for all $x \in H$, then $\mathbb{P}_1 = \mathbb{P}_2$.*

Proof See [153] Proposition 1.7, p. 6, or [180], Proposition 2.5, p. 35. □

Theorem 1.56 *Let $X_1, ..., X_n$ be random variables in a real, separable Hilbert space H. The random variables are independent if and only if for every $y_1, ..., y_n \in H$*

$$\mathbb{E}\left[e^{\left[i \sum_{i=1}^n \langle X_i, y_i \rangle\right]}\right] = \prod_{i=1}^n \mathbb{E}\left[e^{\left[i \langle X_i, y_i \rangle\right]}\right]. \tag{1.7}$$

Proof Obviously if $X_1, ..., X_n$ are independent then (1.7) holds. Also, Theorem 1.56 is well known if $H = \mathbb{R}^k$. Let now $k \in \mathbb{N}$ and $y_i^j \in H, i = 1, ..., n, j = 1, ..., k$, and consider random variables $X_i^k = (\langle X_i, y_i^1 \rangle, ..., \langle X_i, y_i^k \rangle), i = 1, ..., n$ in $\mathbb{R}^k$. Therefore, if (1.7) holds then $X_i^k, i = 1, ..., n$, are independent for every $k \in \mathbb{N}$ and $y_i^j \in H$, $j = 1, ..., k$. Since cylindrical sets of the form $\{x : (\langle x, y_i^1 \rangle, ..., \langle x, y_i^k \rangle) \in A \in \mathcal{B}(\mathbb{R}^k)\}$ generate $\mathcal{B}(H)$ and are a π-system, the collection of sets $\{\omega : (\langle X_i, y_i^1 \rangle, ..., \langle X_i, y_i^k \rangle) \in A \in \mathcal{B}(\mathbb{R}^k)\}$ over all $k \in \mathbb{N}$ and $y_i^j \in H, i = 1, ..., n, j = 1, ..., k, A \in \mathcal{B}(\mathbb{R}^k)$ is a π-system generating $\sigma(X_i)$. Thus, by Lemma 1.23, the sigma algebras $\sigma(X_1), ..., \sigma(X_n)$ are independent. □

Theorem 1.57 *Let H be a real, separable Hilbert space, $\mathcal{B}(H)$ be its Borel σ-field, $a \in H$, and $Q \in \mathcal{L}_1^+(H)$. Then there exists a unique probability measure $\mathbb{P}$ on $(H, \mathcal{B}(H))$ such that*

$$\hat{\mathbb{P}}(x) = e^{i\langle a, x \rangle - \frac{1}{2}\langle Qx, x \rangle}.$$

The measure $\mathbb{P}$ has mean a and covariance Q.

Proof See [153] Theorem 1.12, p. 12. □

Definition 1.58 (*Gaussian measure on H*) Let H be a real, separable Hilbert space, $\mathcal{B}(H)$ be its Borel σ-field, $a \in H$, and $Q \in \mathcal{L}_1^+(H)$. The unique probability measure $\mathbb{P}$ identified by Theorem 1.57 is called the *Gaussian measure* with mean a and covariance Q, and is denoted by $\mathcal{N}(a, Q)$. When $a = 0$ we will denote it by $\mathcal{N}_Q$ and call it a centered Gaussian measure.

We now provide two useful results about Gaussian measures.

Proposition 1.59 *Let $Q \in \mathcal{L}_1^+(H)$. Then for all $y, z \in H$*

$$\int_H \langle x, y \rangle \langle x, z \rangle \mathcal{N}_Q(dx) = \langle Qy, z \rangle. \tag{1.8}$$

Define, for $y \in Q^{1/2}(H)$, $\mathcal{Q}_y \in L^2(H, \mathcal{N}_Q)$ as

$$\mathcal{Q}_y(x) := \langle Q^{-1/2}y, x \rangle, \tag{1.9}$$

where $Q^{-1/2}$ is the pseudoinverse of $Q^{1/2}$ (see Definition B.1). The map (called the "white noise function", see e.g. [154] Sect. 2.5)

$$y \in Q^{1/2}(H) \to \mathcal{Q}_y \in L^2(H, \mathcal{N}_Q)$$

can be extended to $H_0 = \overline{Q^{1/2}(H)} = (\ker Q)^\perp$ and it satisfies

$$\int_H \mathcal{Q}_y(x)\mathcal{Q}_z(x)\mathcal{N}_Q(dx) = \langle y, z \rangle, \qquad y, z \in H_0.$$

Moreover, for all $m > 0$ we have

$$\int_H |x|^{2m} \mathcal{N}_Q(dx) \le K(m)[\mathrm{Tr}(Q)]^m \tag{1.10}$$

for some $K(m) > 0$, independent of Q.

Proof Formula (1.8) follows from Proposition 1.2.4 in [179].

The second statement is proved, when $\ker Q = \{0\}$, in [154] Sect. 2.5.2 (see also Sect. 1.2.4 of [179]). Since here we do not assume $\ker Q = \{0\}$, we provide a proof. First we observe that $\ker Q = \ker Q^{1/2}$ and that $Q^{1/2}(H)$ is dense in $(\ker Q)^\perp$ since $Q^{1/2}$ is self-adjoint. Moreover, by Definition B.1, the pseudoinverse of $Q^{1/2}$ is the operator $Q^{-1/2} : Q^{1/2}(H) \to (\ker Q)^\perp$, hence the map $y \to \mathcal{Q}_y = \langle Q^{-1/2}y, x \rangle$ is well defined for all $y \in Q^{1/2}(H)$. Furthermore, thanks to formula (1.8), we have, for $y_1, y_2 \in Q^{1/2}(H)$

$$\int_H \langle Q^{-1/2}y_1, x \rangle \langle Q^{-1/2}y_2, x \rangle \mathcal{N}_Q(dx) = \langle Q(Q^{-1/2}y_1), Q^{-1/2}y_2 \rangle = \langle y_1, y_2 \rangle,$$

where we used that $Q^{1/2}Q^{-1/2}y = y$ for all $y \in Q^{1/2}(H)$. Hence, for $y_1, y_2 \in Q^{1/2}(H)$,

$$\int_H \mathcal{Q}_{y_1}(x)\mathcal{Q}_{y_2}(x)\mathcal{N}_Q(dx) = \langle y_1, y_2 \rangle. \tag{1.11}$$

In view of the above the map $y \to \mathcal{Q}_y = \langle Q^{-1/2}y, x \rangle$ is an isometry and can be extended to $\overline{Q^{1/2}(H)} = (\ker Q)^\perp$ (endowed with the inner product inherited from H) and (1.11) extends to all $y_1, y_2 \in (\ker Q)^\perp$.

We remark that as pointed out in [154] Sect. 2.5.2, for a generic $y \in (\ker Q)^{\perp}$ the image $\mathcal{Q}_y$ is an element of $L^2(H, \mathcal{N}_Q)$, hence an equivalence class of random variables defined $\mathcal{N}_Q$-a.e.; in particular, writing $\mathcal{Q}_y(x) = \langle y, Q^{-1/2}x \rangle$, $\mathcal{N}_Q$-a.e., would be misleading since, as proved in [154] Proposition 2.22, $\mathcal{N}_Q(Q^{1/2}(H)) = 0$.

Concerning the third claim, by Proposition 2.19, p. 50, of [180], it holds for $m \in \mathbb{N}$. If $k - 1 < m < k$ for $k = 1, 2, \ldots$, we use

$$\int_H |x|^{2m} \mathcal{N}_Q(dx) \le \left[\int_H |x|^{2k} \mathcal{N}_Q(dx) \right]^{m/k}.$$

$\square$

Theorem 1.60 (Cameron–Martin formula) *Let H be a real, separable Hilbert space. Let $a_1, a_2 \in H$ and $Q \in \mathcal{L}_1^+(H)$. Then:*

(1) *The Gaussian measures $\mathcal{N}(a_1, Q)$ and $\mathcal{N}(a_2, Q)$ are either singular or equivalent.*

(2) *They are equivalent if and only if $a_1 - a_2 \in Q^{1/2}(H)$ and in this case*

$$\frac{d\mathcal{N}(a_1, Q)}{d\mathcal{N}(a_2, Q)}(x) = \exp\left(\langle Q^{-1/2}(a_1 - a_2), Q^{-1/2}(x - a_2) \rangle - \frac{1}{2} \left| Q^{-1/2}(a_1 - a_2) \right|^2 \right)$$

for $\mathcal{N}(a_2, Q)$-a.e. $x \in H$.

Proof See Theorem 2.23, p. 53 of [180]. $\square$

We now recall some results concerning compactness of a family of measures in $M_1(H)$ (see e.g. Sect. 2.1 in [180] or [219, 478] for more on this).

Definition 1.61

(i) A sequence $(\mathbb{P}_n)$ in $M_1(H)$ is said to be weakly convergent to some $\mathbb{P} \in M_1(H)$ if, for every $\phi \in C_b(H)$,

$$\lim_{n \to +\infty} \int_H \phi(x)\mathbb{P}_n(dx) = \int_H \phi(x)\mathbb{P}(dx).$$

(ii) A family $\Lambda \subset M_1(H)$ is said to be compact (respectively, relatively compact) if an arbitrary sequence $\mathbb{P}_n$ of elements of Λ contains a subsequence $\mathbb{P}_{n_k}$ weakly convergent to a measure $\mathbb{P} \in \Lambda$ (respectively, to a measure $\mathbb{P} \in M_1(H)$).

(iii) A family $\Lambda \subset M_1(H)$ is said to be tight if for any $\varepsilon > 0$ there exists a compact set K_ε such that, for every $\mathbb{P} \in \Lambda$,

$$\mathbb{P}(K_\varepsilon) > 1 - \varepsilon.$$

The following theorem (which also holds when H is a Polish space) is due to Prokhorov.

Theorem 1.62 *Let H be a real separable Hilbert space. A family $\Lambda \subset M_1(H)$ is relatively compact if and only if it is tight.*

Proof See [180], the proof of Theorem 2.3. □

The next theorem gives a useful sufficient condition for compactness.

Theorem 1.63 *Let H be a real separable Hilbert space and let $\{e_i\}_{i\in\mathbb{N}}$ be an orthonormal basis in H. A family $\Lambda\subset M_1(H)$ is relatively compact if*

$$\lim_{N\to+\infty}\sup_{\mathbb{P}\in\Lambda}\int_H\sum_{i=N}^{+\infty}\langle x,e_i\rangle^2\mathbb{P}(dx)=0.$$

Proof See [478], the proof of Theorem VI.2.2. □

Concerning Gaussian measures, we have the following result (see Proposition 1.1.5 of [493]).

Proposition 1.64 *Let $\mathcal{N}_{Q_n}$ ($n\in\mathbb{N}$) and $\mathcal{N}_Q$ be centered Gaussian measures on H. If $\lim_{n\to+\infty}\|Q_n-Q\|_{\mathcal{L}_1(H)}=0$, then the measures $\mathcal{N}_{Q_n}$ converge weakly to $\mathcal{N}_Q$.*

Proof Observe that if $\{e_i\}_i$ is an orthonormal basis in H, it follows from (1.8) that for any $N\in\mathbb{N}$,

$$\int_H\sum_{i=N}^{+\infty}\langle x,e_i\rangle^2\mathcal{N}_{Q_n}(dx)=\sum_{i=N}^{+\infty}\langle Q_ne_i,e_i\rangle.$$

Since $\lim_{n\to+\infty}\|Q_n-Q\|_{\mathcal{L}_1(H)}=0$, the above formula implies in particular that Theorem 1.63 applies and thus the sequence $(\mathcal{N}_{Q_n})$ is relatively compact.

Moreover, from Theorem 1.57 and Definition 1.58 it is immediate that, as $n\to+\infty$,

$$\widehat{\mathcal{N}_{Q_n}}(x)=e^{-\frac{1}{2}\langle Q_nx,x\rangle}\longrightarrow e^{-\frac{1}{2}\langle Qx,x\rangle}=\widehat{\mathcal{N}_Q}(x),\qquad\forall x\in H.$$

Take now a subsequence $\mathcal{N}_{Q_{n_k}}$ weakly convergent to a probability measure $\mathbb{P}_0$. By Definition 1.54 we must have

$$\widehat{\mathcal{N}_{Q_{n_k}}}(x)\to\widehat{\mathbb{P}_0}(x),\qquad\forall x\in H.$$

This implies that $\widehat{\mathbb{P}_0}=\widehat{\mathcal{N}_Q}$ and hence, by Proposition 1.55, that $\mathbb{P}_0=\mathcal{N}_Q$. Since this is true for any convergent subsequence, the claim now follows by a standard contradiction argument. □

We conclude with a useful result about uniformity of weak convergence. The result is also true if H is a Polish space, see [478], Theorem II.6.8.

Theorem 1.65 *Let $\mathbb{P}_n$ be a sequence in $M_1(H)$ and $\mathbb{P}\in M_1(H)$. Then $\mathbb{P}_n$ is weakly convergent to $\mathbb{P}$ if and only if*

$$\lim_{n\to+\infty}\sup_{\phi\in\mathcal{C}_0}\left|\int_H\phi(x)\mathbb{P}_n(dx)-\int_H\phi(x)\mathbb{P}(dx)\right|=0$$

for every family $\mathcal{C}_0 \subset C_b(H)$ which is equicontinuous at all points $x \in H$ and uniformly bounded, i.e., for some constant $M > 0$, $|f(x)| \leq M$ for all $x \in H$ and $f \in \mathcal{C}_0$.

Proof See [478], the proof of Theorem II.6.8. □

1.2 Stochastic Processes and Brownian Motion

1.2.1 Stochastic Processes

Definition 1.66 (*Filtration, usual conditions*) Let $t \geq 0$. A filtration $\{\mathcal{F}_s^t\}_{s \geq t}$ in a complete probability space $(\Omega, \mathcal{F}, \mathbb{P})$ is a family of σ-fields such that $\mathcal{F}_s^t \subset \mathcal{F}_r^t \subset \mathcal{F}$ whenever $t \leq s \leq r$.

(i) We say that $\{\mathcal{F}_s^t\}_{s \geq t}$ is *right-continuous* if, for all $s \geq t$, $\mathcal{F}_{s+}^t := \bigcap_{r > s} \mathcal{F}_r^t = \mathcal{F}_s^t$.

(ii) We say that $\{\mathcal{F}_s^t\}_{s \geq t}$ is *left-continuous* if, for all $s > t$, $\mathcal{F}_{s-}^t := \sigma\left(\bigcup_{r < s} \mathcal{F}_r^t\right) = \mathcal{F}_s^t$. We say that $\{\mathcal{F}_s^t\}_{s \geq t}$ is *continuous* if it is both left and right-continuous.

(iii) We say that $\{\mathcal{F}_s^t\}_{s \geq t}$ satisfies the *usual conditions* if it is right-continuous and complete, i.e. if $\mathcal{F}_s^t$ contains all $\mathbb{P}$-null sets of $\mathcal{F}$ for every $s \geq t$.

We will often write $\mathcal{F}_s^t$ instead of $\{\mathcal{F}_s^t\}_{s \geq t}$. We also set $\mathcal{F}_{+\infty}^t := \sigma\left(\bigcup_{r < +\infty} \mathcal{F}_r^t\right)$. Since we will mostly deal with filtrations satisfying the usual conditions we will assume from now on that this property holds unless explicitly stated otherwise. For this reason we include the usual conditions in the definition of a filtered probability space.

Definition 1.67 (*Filtered probability space*) Let $\mathcal{F}_s^t$ be a filtration satisfying the usual conditions on a complete probability space $(\Omega, \mathcal{F}, \mathbb{P})$. The 4-tuple $(\Omega, \mathcal{F}, \mathcal{F}_s^t, \mathbb{P})$ is called a *filtered probability space*.

Notation 1.68 We use the following convention in this section. When we write $s \in [t, T]$ we mean that $s \in [t, T]$ if $T \in \mathbb{R}$, and $s \in [t, +\infty)$ if $T = +\infty$. So $[t, T]$ is understood to be $[t, +\infty)$ if $T = +\infty$. ∎

Definition 1.69 (*Stochastic process*) Let $T \in (0, +\infty]$, $t \in [0, T)$ and $(\Omega, \mathcal{F})$ and $(\Omega_1, \mathcal{F}_1)$ be two measurable spaces. A family of random variables $X(\cdot) = \{X(s)\}_{s \in [t,T]}$, $X(s): \Omega \to \Omega_1$, is called a *stochastic process* in $[t, T]$. If $(\Omega_1, \mathcal{F}_1) = (\mathbb{R}, \mathcal{B}(\mathbb{R}))$ then $X(\cdot)$ is called a *real stochastic process*.

Definition 1.70 Let $\left(\Omega, \mathcal{F}, \{\mathcal{F}_s^t\}_{s \geq t}, \mathbb{P}\right)$ be a filtered probability space and $(\Omega_1, \mathcal{F}_1)$ be a measurable space. A stochastic process $\{X(s)\}_{s \in [t,T]}: [t, T] \times \Omega \to \Omega_1$ is said to be:

(i) *Measurable*, if the map $(s, \omega) \to X(s)(\omega)$ is $\mathcal{B}([t, T]) \otimes \mathscr{F}/\mathscr{F}_1$-measurable.

(ii) *Adapted*, if, for each $s \in [t, T]$, $X(s) \colon \Omega \to \Omega_1$ is an $\mathscr{F}_s^t/\mathscr{F}_1$-measurable random variable.

(iii) *Progressively measurable*, if for all $s \in (t, T]$, the restriction of $X(\cdot)$ to $[t, s] \times \Omega$ is $\mathcal{B}([t, s]) \otimes \mathscr{F}_s^t/\mathscr{F}_1$-measurable.

(iv) *Predictable*, if the map $(s, \omega) \to X(s)(\omega)$ is $\mathcal{P}_{[t,T]}/\mathscr{F}_1$-measurable, where $\mathcal{P}_{[t,T]}$ is the σ-field (the predictable σ-field) in $[t, T] \times \Omega$ generated by all sets of the form $(s, r] \times A, t \leq s < r \leq T, A \in \mathscr{F}_s^t$ and $\{t\} \times A, A \in \mathscr{F}_t^t$.

(v) If E is a separable Banach space (endowed with its Borel σ-field), the process $\{X(s)\}_{s\in[t,T]} \colon [t, T] \times \Omega \to E$ is called *stochastically continuous* at $s \in [t, T]$ if for every $\varepsilon, \delta > 0$ there exists $\rho > 0$ such that

$$\mathbb{P}\left(|X(r) - X(s)| \geq \varepsilon\right) \leq \delta, \qquad \text{for all } r \in (s - \rho, s + \rho) \cap [t, T].$$

(vi) If (S, d) is a metric space (endowed with its Borel σ-field), the process $\{X(s)\}_{s\in[t,T]} \colon [t, T] \times \Omega \to S$ is called *continuous* (respectively, right-continuous, left-continuous), if for $\mathbb{P}$-a.e. $\omega \in \Omega$, the function $s \to X(s)(\omega)$ is continuous (respectively, right-continuous, left-continuous).

(vii) If E is a separable Banach space (endowed with its Borel σ-field), the process $\{X(s)\}_{s\in[t,T]} \colon [t, T] \times \Omega \to E$ is called *integrable* (respectively square-integrable) if $\mathbb{E}[|X(s)|] < +\infty$ (respectively $\mathbb{E}[|X(s)|^2] < +\infty$) for all $s \in [t, T]$. The process is called *uniformly integrable* if it is integrable and the family $\{X(s)\}_{s\in[t,T]}$ is uniformly integrable (see Definition 1.40).

(viii) If E is a separable Banach space (endowed with the Borel σ-field induced by the norm), the process $\{X(s)\}_{s\in[t,T]} \colon [t, T] \times \Omega \to E$ is said to be *mean square continuous* if $\mathbb{E}[|X(s)|^2] < +\infty$ for all $s \in [t, T]$ and $\lim_{r \to s} \mathbb{E}[|X(r) - X(s)|^2] = 0$ for all $s \in [t, T]$.

It is easy to see that if a process is mean square continuous then it is stochastically continuous.

The concepts of adapted, progressively measurable, and predictable processes can be defined for any filtration $\mathscr{G}_s^t$. To emphasize the filtration used, we will refer to the processes as $\mathscr{G}_s^t$-adapted, $\mathscr{G}_s^t$-progressively measurable, and $\mathscr{G}_s^t$-predictable.

Progressive measurability can also be defined using the concept of progressively measurable sets, see e.g. [447], p. 4, or [219], p. 71. We say that a set $A \subset [t, T] \times \Omega$ is $\mathscr{F}_s^t$-progressively measurable if the function $\mathbf{1}_A$ is a progressively measurable process. Equivalently this means that $A \cap ([t, s] \times \Omega) \in \mathcal{B}([t, s]) \otimes \mathscr{F}_s^t$ for every $s \in [t, T]$. It can be proved that the $\mathscr{F}_s^t$-progressively measurable sets form a σ-field and that a process $X(\cdot)$ is progressively measurable if and only if it is measurable with respect to the σ-field of $\mathscr{F}_s^t$-progressively measurable sets.

Definition 1.71 (*Stochastic equivalence, modification*) Let $(\Omega, \mathscr{F}, \mathbb{P})$ be a probability space, and $(\Omega_1, \mathscr{F}_1)$ be a measurable space. Processes $X(\cdot), Y(\cdot) \colon [t, T] \times \Omega \to \Omega_1$ are called *stochastically equivalent* if for all $s \in [t, T]$, $\mathbb{P}(X(s) = Y(s)) = 1$. In this case, $Y(\cdot)$ is said to be a *modification* or *version* of $X(\cdot)$. The processes $X(\cdot)$ and

$Y(\cdot)$ are called *indistinguishable* if $\mathbb{P}(X(s) = Y(s) : \forall s \in [t, T]) = 1$. We will also say that $Y(\cdot)$ is an *indistinguishable version* of $X(\cdot)$.

Lemma 1.72 *Let* $\left(\Omega, \mathscr{F}, \{\mathscr{F}^t_s\}_{s \geq t}, \mathbb{P}\right)$ *be a filtered probability space and let* $\{X(s)\}_{s \geq t}$ *be a process with values in a Polish space* (S, d), *endowed with the Borel* σ-*field induced by the distance.*

(i) *If* $X(\cdot)$ *is* $\mathcal{B}([t, T]) \otimes \mathscr{F}/\mathcal{B}(S)$-*measurable and* $\mathscr{F}^t_s$-*adapted, then* $X(\cdot)$ *has an* $\mathscr{F}^t_s$-*progressively measurable modification.*

(ii) *If* $X(\cdot)$ *is* $\mathscr{F}^t_s$-*adapted and* $X(\cdot)$ *is left- (or right-) continuous for every* ω, *then* $X(\cdot)$ *itself is* $\mathscr{F}^t_s$-*progressively measurable.*

Proof Part (i): Since S is Borel isomorphic to a Borel subset A of $\mathbb{R}$, without loss of generality we can consider $X(\cdot)$ to be an $\mathbb{R}$-valued process with values in A. By [449], Theorem T46, p. 68, $X(\cdot)$ has an $\mathbb{R}$-valued, $\mathscr{F}^t_s$-progressively measurable modification $\tilde{X}(\cdot)$. Let $a \in A$. We define a process $Y(\cdot)$ by $Y(s) := \tilde{X}(s)\mathbf{1}_{\tilde{X}(s) \in A} + a\mathbf{1}_{\tilde{X}(s) \in (\mathbb{R} \setminus A)}$. The process $Y(\cdot)$ is $\mathscr{F}^t_s$-progressively measurable. Moreover, if $\tilde{X}(s) = X(s)$, then $Y(s) = X(s)$, so $Y(\cdot)$ is a modification of $X(\cdot)$. *Part (ii):* See [449], Theorem T47, p. 70, or [372], Proposition 1.13, p. 5. □

Lemma 1.73 *Let* $(\Omega, \mathscr{F}, \mathbb{P})$ *be a complete probability space and let* $\{X(s)\}_{s \geq t}$ *be a stochastic process with values in a separable Banach space* E *endowed with the Borel* σ-*field. If* $X(\cdot)$ *is stochastically continuous then it has a measurable modification.*

Proof See [180], Proposition 3.2. □

Lemma 1.74 *Let* $\left(\Omega, \mathscr{F}, \{\mathscr{F}^t_s\}_{s \geq t}, \mathbb{P}\right)$ *be a filtered probability space and let* $\{X(s)\}_{s \geq t}$ *be an adapted process with values in a separable Banach space* E *endowed with the Borel* σ-*field. If* $X(\cdot)$ *is stochastically continuous then it has an* $\mathscr{F}^t_s$-*progressively measurable modification.*

Proof See [180], Proposition 3.6. It is also a corollary of Lemmas 1.72-(i) and 1.73. □

1.2.2 Martingales

Notation 1.75 Unless specified otherwise, any Banach space E and any metric space (S, d) will be understood to be endowed with the Borel σ-field induced respectively by the norm and by the distance. ∎

Definition 1.76 (*Martingale*) Let $\left(\Omega, \mathscr{F}, \mathscr{F}^t_s, \mathbb{P}\right)$ be a filtered probability space, and let $M(\cdot)$ be an $\mathscr{F}^t_s$-adapted and integrable process with values in a separable Banach space E. Then $M(\cdot)$ is said to be a *martingale* if, for all $r, s \in [t, T], s \leq r$,

$$\mathbb{E}\left[M(r)|\mathscr{F}^t_s\right] = M(s) \qquad \mathbb{P} - a.s.$$

If $E = \mathbb{R}$, we say that $M(s)$ is a *submartingale* (respectively, *supermartingale*) if

$$\mathbb{E}\left[M(r)|\mathscr{F}_s^t\right] \geq M(s), \quad (\text{respectively,} \ \mathbb{E}\left[M(r)|\mathscr{F}_s^t\right] \leq M(s)) \ \mathbb{P} - a.s.$$

Theorem 1.77 (Doob's maximal inequalities) *Let* $T > 0$, $(\Omega, \mathscr{F}, \mathscr{F}_s^t, \mathbb{P})$ *be a filtered probability space, and H be a separable Hilbert space. Let $M(\cdot)$ be a right-continuous H-valued martingale such that $M(s) \in L^p(\Omega, \mathscr{F}, \mathbb{P}; H)$ for all $s \in [t, T]$. Then:*

(i) If $p \geq 1$, $\mathbb{P}\left(\sup_{s \in [t,T]} |M(s)| > \lambda\right) \leq \frac{1}{\lambda^p}\mathbb{E}\left[|M(T)|^p\right]$, for all $\lambda > 0$.

(ii) If $p > 1$, $\mathbb{E}\left[\sup_{s \in [t,T]} |M(s)|^p\right] \leq \left(\frac{p}{p-1}\right)^p \mathbb{E}\left[|M(T)|^p\right]$.

Proof We observe that, if $M(\cdot)$ is a right-continuous H-valued martingale such that $M(s) \in L^p(\Omega, \mathscr{F}, \mathbb{P}; H)$, $p \geq 1$, for all $s \in [t, T]$, then by Proposition 1.41-(vi), $|M(\cdot)|^p$ is a right-continuous $\mathbb{R}$-valued submartingale with $|M(s)| \in L^p$ $(\Omega, \mathscr{F}, \mathbb{P}; \mathbb{R})$ for all $s \in [t, T]$. The claims now easily follow from [372] Theorem 3.8 (i) and (iii), pp. 13–14. $\qquad\square$

In particular, we see that a right-continuous E-valued martingale $M(\cdot)$ is square-integrable if and only if $\mathbb{E}|M(T)|^2 < +\infty$.

Notation 1.78 (*Square-integrable martingales*) Let $T \in (0, +\infty)$, $t \in [0, T)$, let $(\Omega, \mathscr{F}, \mathscr{F}_s^t, \mathbb{P})$ be a filtered probability space, and E be a separable Banach space. The class of all continuous square-integrable martingales $M: [t, T] \times \Omega \to E$ is denoted by $\mathcal{M}_{t,T}^2(E)$. $\qquad\blacksquare$

If H is a separable Hilbert space then $\mathcal{M}_{t,T}^2(H)$ endowed with the scalar product

$$\langle M, N \rangle_{\mathcal{M}_{t,T}^2} := \mathbb{E}\left[\langle M(T), N(T) \rangle\right].$$

is a Hilbert space (see [294], p. 22).

Theorem 1.79 (Angle bracket process, Quadratic variation process) *Let* $T > 0, t \in [0, T)$, H *be a separable Hilbert space, and $(\Omega, \mathscr{F}, \mathscr{F}_s^t, \mathbb{P})$ be a filtered probability space. For every $M \in \mathcal{M}_{t,T}^2(H)$ there exists a unique (real) increasing, adapted, continuous process starting from 0 at t, called the* angle bracket process, *and denoted by $\langle M \rangle_t$, such that $|M_s|^2 - \langle M \rangle_s$ is a continuous martingale. Moreover, there exists a unique $\mathcal{L}_1^+(H)$-valued continuous adapted process starting from 0 at t, called the* quadratic variation *of M, and denoted by $\langle\langle M \rangle\rangle_s$, such that, for all $x, y \in H$, the process*

$$\langle M_s, x \rangle \langle M_s, y \rangle - \left\langle \langle\langle M \rangle\rangle_s(x), y \right\rangle, \quad s \in [t, T]$$

is a continuous martingale. Moreover, $\langle M \rangle_s = \mathrm{Tr}(\langle\langle M \rangle\rangle_s)$.

Proof See [294], Definition 2.9 and Lemma 2.1, p. 22. $\qquad\square$

Theorem 1.80 (Burkholder–Davis–Gundy inequality) *Let $T > 0, t \in [0, T)$, H be a separable Hilbert space, and $(\Omega, \mathscr{F}, \mathscr{F}_s^t, \mathbb{P})$ be a filtered probability space. For every $p > 0$ there exists a $c_p > 0$ such that, for every $M \in \mathcal{M}_{t,T}^2(H)$ with $M(0) = 0$,*

$$c_p^{-1} \mathbb{E} \left[\langle M \rangle_T^{p/2} \right] \leq \mathbb{E} \left[\sup_{s \in [t,T]} |M(s)|^p \right] \leq c_p \mathbb{E} \left[\langle M \rangle_T^{p/2} \right].$$

Proof See [487], Theorem 3.49, p. 37. $\qquad\qquad\qquad\qquad\qquad\qquad\qquad\qquad\qquad\quad$ $\square$

1.2.3 Stopping Times

Definition 1.81 (*Stopping time*) Consider a probability space $(\Omega, \mathscr{F}, \mathbb{P})$ and a filtration $\{\mathscr{F}_s^t\}_{s \geq t}$ on Ω. A random variable $\tau : (\Omega, \mathscr{F}) \to [t, +\infty]$ is said to be an $\mathscr{F}_s^t$-*stopping time* if, for all $s \geq t$,

$$\{\tau \leq s\} := \{\omega \in \Omega \; : \; \tau(\omega) \leq s\} \in \mathscr{F}_s^t.$$

Given a stopping time τ we denote by $\mathscr{F}_\tau$ the sub-σ-field of $\mathscr{F}$ defined by

$$\mathscr{F}_\tau := \left\{ A \in \mathscr{F} \; : \; A \cap \{\tau \leq s\} \in \mathscr{F}_s^t \text{ for all } s \geq t \right\}.$$

Proposition 1.82 *Let $(\Omega, \mathscr{F}, \mathscr{F}_s^t, \mathbb{P})$ be a filtered probability space.*

(i) *If τ and σ are $\mathscr{F}_s^t$-stopping times, so are $\tau \wedge \sigma$, $\tau \vee \sigma$ and $\tau + \sigma$.*

(ii) *If σ_n (for $n = 1, 2...$) are $\mathscr{F}_s^t$-stopping times, then*

$$\sup_n \sigma_n, \quad \inf_n \sigma_n, \quad \limsup_n \sigma_n, \quad \liminf_n \sigma_n$$

are $\mathscr{F}_s^t$-stopping times.

(iii) *For any $\mathscr{F}_s^t$-stopping time τ there exists a decreasing sequence of discrete-valued $\mathscr{F}_s^t$-stopping times τ_n, such that $\lim_{n \to \infty} \tau_n = \tau$.*

(iv) *Let (S, d) be a metric space (endowed with the Borel σ-field induced by the distance), and $X : [t, +\infty) \times \Omega \to S$ be a continuous and $\mathscr{F}_s^t$-adapted process. Let $A \subset S$ be an open or a closed set. Then the* hitting time

$$\tau_A := \inf\{s \geq t \; : \; X(s) \in A\}$$

is a stopping time. (It is understood that $\inf\{\emptyset\} = +\infty$.)

Proof (i) and (ii) see [372], Lemmas 2.9 and 2.11, p. 7. (iii) see [370], Lemma 7.4, p. 122. (iv) see [575], Example 3.3, p. 24, or [452], Proposition 1.3.2, p. 12 (there $S = \mathbb{R}^n$, but the proofs are the same). $\qquad\qquad\qquad\qquad\qquad\qquad\qquad\quad$ $\square$

Proposition 1.83 *Let* $\left(\Omega, \mathscr{F}, \left\{\mathscr{F}_s^t\right\}_{s \geq t}, \mathbb{P}\right)$ *be a filtered probability space,* $(\Omega_1, \mathscr{F}_1)$ *be a measurable space,* $X: [t, +\infty) \times \Omega \rightarrow \Omega_1$ *be an* $\mathscr{F}_s^t$-*progressively measurable process, and* τ *be an* $\mathscr{F}_s^t$-*stopping time. Then the random variable* $X(\tau)$, *(where* $X(\tau)(\omega) := X(\tau(\omega), \omega)$), *is* $\mathscr{F}_\tau$-*measurable and the process defined, for any* $s \in [t, +\infty)$, *by* $X(s \wedge \tau)$ *is* $\mathscr{F}_s^t$-*progressively measurable.*

Proof See [452], Proposition 1.3.5, p. 13, or [575], Proposition 3.5, p. 25. $\square$

Theorem 1.84 (Doob's optional sampling theorem) *Let* $\left(\Omega, \mathscr{F}, \left\{\mathscr{F}_s^t\right\}_{s \geq t}, \mathbb{P}\right)$ *be a filtered probability space,* $X: [t, +\infty) \times \Omega \rightarrow \mathbb{R}$ *be a right-continuous* $\mathscr{F}_s^t$-*submartingale, and* τ, σ *be two* $\mathscr{F}_s^t$-*stopping times with* τ *bounded. Then* X_τ *is integrable and*

$$\mathbb{E}[X_\tau | \mathscr{F}_\sigma^t] \geq X_{\tau \wedge \sigma}, \quad \mathbb{P} \text{ a.s.}$$

If X^+ *(the positive part of the process) is uniformly integrable then the statement extends to unbounded* τ.

Proof See [370], Theorem 7.29, p. 135. $\square$

Definition 1.85 (*Local martingale*) *Let* $\left(\Omega, \mathscr{F}, \left\{\mathscr{F}_s^t\right\}_{s \geq t}, \mathbb{P}\right)$ *be a filtered probability space. An* $\left\{\mathscr{F}_s^t\right\}_{s \geq t}$-*adapted process* $\{X(s)\}_{s \geq t}$ *with values in a separable Banach space* E *is said to be a* local martingale *if there exists an increasing sequence of stopping times* $(\tau_n)_{n \in \mathbb{N}}$ *with* $\mathbb{P}(\tau_n \uparrow +\infty) = 1$, *such that the process* $\{X(s \wedge \tau_n)\}_{s \geq t}$ *is a martingale for every* $n \in \mathbb{N}$.

1.2.4 Q-Wiener Processes

Definition 1.86 (*Real Brownian motion*) *Given* $t \in \mathbb{R}$, *a real stochastic process* $\beta: [t, +\infty) \times \Omega \rightarrow \mathbb{R}$ *on a complete probability space* $(\Omega, \mathscr{F}, \mathbb{P})$ *is a* standard *(one-dimensional) real Brownian motion on* $[t, +\infty)$ *starting at* 0, *if*

(1) β is continuous and $\beta(t) = 0$;
(2) for all $t \leq t_1 < t_2 < ... < t_n$ the random variables $\beta(t_1)$, $\beta(t_2) - \beta(t_1)$, ..., $\beta(t_n) - \beta(t_{n-1})$ are independent;
(3) for all $t \leq t_1 \leq t_2$, $\beta(t_2) - \beta(t_1)$ has a Gaussian distribution with mean 0 and covariance $t_2 - t_1$.

Consider a real, separable Hilbert space Ξ and $Q \in \mathcal{L}^+(\Xi)$. Define $\Xi_0 := Q^{1/2}(\Xi)$ and let $Q^{-1/2}$ be the pseudo-inverse of $Q^{1/2}$ (see Definition B.1). Ξ_0 is a separable Hilbert space when endowed with the inner product $\langle x, y \rangle_{\Xi_0} := \left\langle Q^{-1/2}x, Q^{-1/2}y \right\rangle_\Xi$. Let Ξ_1 be an arbitrary real, separable Hilbert space such that $\Xi \subset \Xi_1$ with continuous embedding and $\Xi_0 \subset \Xi_1$ with Hilbert–Schmidt embedding $J: \Xi_0 \hookrightarrow \Xi_1$ (see Appendix B.3 on Hilbert–Schmidt operators). The operator

$Q_1 := JJ^*$ belongs to $\mathcal{L}_1^+(\Xi_1)$ and Ξ_0 is identical with the space $Q_1^{\frac{1}{2}}(\Xi_1)$ (see [180] Proposition 4.7, p. 85).

Theorem 1.87 *Consider the setting described above. Let $\{g_k\}_{k\in\mathbb{N}}$ be an orthonormal basis of Ξ_0 and $(\beta_k)_{k\in\mathbb{N}}$ be a sequence of mutually independent, standard one-dimensional Brownian motions $\beta_k \colon [t, +\infty) \times \Omega \to \mathbb{R}$ on $[t, +\infty)$ starting at 0. Then for every $s \in [t, +\infty)$ the series*

$$W_Q(s) := \sum_{k=1}^{\infty} g_k \beta_k(s) \tag{1.12}$$

is convergent in $L^2(\Omega, \mathscr{F}, \mathbb{P}; \Xi_1)$.

Proof See [180] Propositions 4.3, p. 82, and 4.7, p. 85. $\qquad\qquad\square$

Definition 1.88 (*Q-Wiener process*) The process W_Q defined by (1.12) is called a *Q-Wiener process* on $[t, +\infty)$ starting at 0.

Remark 1.89 We will use the notation W_Q to denote a Q-Wiener process. If Q is trace-class, $\Xi_1 = \Xi$ is a canonical choice and it will be understood that W_Q is a Ξ-valued process. If Q is not trace-class, writing W_Q and calling it a Q-Wiener process is a slight abuse of notation as it would be more precise to write W_{Q_1} and call it a Q_1-Wiener process with values in Ξ_1. However, even though the construction we have described is not canonical if $\mathrm{Tr}(Q) = +\infty$, and the choice of Ξ_1 is not unique, the class of the integrable processes is independent of the choice of Ξ_1 (see [180] Sect. 4.1 and in particular Proposition 4.7). Moreover (see [180] Sect. 4.1.2), for arbitrary $a \in \Xi$ the stochastic process

$$< a, W(s) > := \sum_{k=1}^{\infty} \langle a, g_k \rangle \beta_k(s), \quad s \geq t,$$

is a real-valued Wiener process and

$$\mathbb{E} < a, W(s_1) > < b, W(s_2) > = ((s_1 - t) \wedge (s_2 - t))\langle Qa, b \rangle, \quad a, b \in \Xi.$$

For these reasons, even when $\mathrm{Tr}(Q) = +\infty$, we will still use the notation W_Q. When Q is the identity on Ξ we will call it a *cylindrical Wiener process in Ξ*. $\qquad\blacksquare$

Proposition 1.90 *Let Ξ be a real, separable Hilbert space, $Q \in \mathcal{L}^+(\Xi)$ and let Ξ_0, Ξ_1 and J be as described above. Let $(\Omega, \mathscr{F}, \mathbb{P})$ be a complete probability space and $B \colon [t, +\infty) \times \Omega \to \Xi_1$ be a stochastic process. Denote by $\mathscr{F}_s^{t,0}$ the filtration generated by B, i.e.*

$$\mathscr{F}_s^{t,0} = \sigma(B(r) \colon t \leq r \leq s),$$

and $\mathscr{F}_s^t := \sigma(\mathscr{F}_s^{t,0}, \mathcal{N})$, where $\mathcal{N}$ is the class of the $\mathbb{P}$-null sets. Then B is a Q-Wiener process on $[t, +\infty)$ starting at 0 if and only if:

(1) $B(t) = 0$.
(2) *B has continuous trajectories.*
(3) *For all $t \leq t_1 \leq t_2$ the random variable $B(t_2) - B(t_1)$ is independent of $\mathscr{F}_{t_1}^t$.*
(4) $\mathcal{L}_{\mathbb{P}}(B(t_2) - B(t_1)) = \mathcal{N}(0, (t_2 - t_1)Q_1)$, *where $Q_1 = JJ^*$.*

Proof The "only if" part follows from [180], Proposition 4.7, p. 85 (observe that in [180] a Wiener process is in fact defined using the four properties (1)–(4)). The "if" part is proved in [180] Proposition 4.3-(ii), p. 81 (if $\mathrm{Tr}(Q) = +\infty$ we apply the proposition in the space Ξ_1). $\qquad\square$

The existence of a process satisfying conditions (1)–(4) above can also be proved using the Kolmogorov extension theorem (see [180], Proposition 4.4).

Remark 1.91 If $W_Q(s) = \sum_{k=1}^{\infty} g_k \beta_k(s)$ for some orthonormal basis $\{g_k\}_{k \in \mathbb{N}}$ of Ξ_0, it is easy to see that regardless of the choice of Ξ_1, $\mathscr{F}_s^{t,0} = \sigma(\beta_k(r) : t \leq r \leq s$, $k \in \mathbb{N})$. Thus the filtration generated by W_Q does not depend on the choice of Ξ_1. ∎

Definition 1.92 *(Translated $\mathscr{G}_s^t$-Q-Wiener process)* Let $0 \leq t < T \leq +\infty$. Let Ξ be a real, separable Hilbert space, $Q \in \mathcal{L}^+(\Xi)$ and let Ξ_0, Ξ_1 and J be as described above. Let $(\Omega, \mathscr{F}, \mathscr{G}_s^t, \mathbb{P})$ be a filtered probability space. We say that a stochastic process $B \colon [t, T] \times \Omega \to \Xi_1$ is a *translated $\mathscr{G}_s^t$-Q-Wiener process* on $[t, T]$ if:

(1) *B has continuous trajectories.*
(2) *B is adapted to $\mathscr{G}_s^t$.*
(3) *For all $t \leq t_1 < t_2 \leq T$, $B(t_2) - B(t_1)$ is independent of $\mathscr{G}_{t_1}^t$.*
(4) $\mathcal{L}_{\mathbb{P}}(B(t_2) - B(t_1)) = \mathcal{N}(0, (t_2 - t_1)Q_1)$, *where $Q_1 = JJ^*$.*

If we also have $B(t) = 0$ then we call B a *$\mathscr{G}_s^t$-Q-Wiener process* on $[t, T]$.

We remark that if B is a translated $\mathscr{G}_s^t$-Q-Wiener process, then it is also a translated $\mathscr{F}_s^t$-Q-Wiener process, where $\mathscr{F}_s^t$ is the augmented filtration generated by B. Moreover, if W_Q is a Q-Wiener process as in Definition 1.88 then it is also a $\mathscr{F}_s^t$-Q-Wiener process, where $\mathscr{F}_s^t$ is the augmented filtration generated by B.

Lemma 1.93 *Let $0 \leq t < T \leq +\infty$. Let Ξ be a real, separable Hilbert space, $Q \in \mathcal{L}^+(\Xi)$ and let Ξ_0 and Ξ_1 be as described above. Let $(\Omega, \mathscr{F}, \mathbb{P})$ be a complete probability space. Let $B \colon [t, T] \times \Omega \to \Xi_1$ be a continuous stochastic process such that $B(t) = 0$. Then B is a Q-Wiener process on $[t, T]$ if and only if, for all $a \in \Xi_1$, $t \leq t_1 \leq t_2 \leq T$, we have*

$$\mathbb{E}\left[e^{i\langle a, B(t_2) - B(t_1) \rangle_{\Xi_1}} \,\middle|\, \mathscr{F}_{t_1}^t \right] = e^{-\frac{\langle Q_1 a, a \rangle_{\Xi_1}}{2}(t_2 - t_1)}. \tag{1.13}$$

Proof (The proof uses the same arguments as in the finite-dimensional case, see Proposition 1.2.7 of [452].)

The "only if" part: if B is a Q-Wiener process then, by Proposition 1.90-(4), Theorem 1.57 and Definition 1.58,

$$\mathbb{E}\left[e^{i\langle a,B(t_2)-B(t_1)\rangle_{\Xi_1}}\right] = e^{-\frac{\langle Q_1 a,a\rangle_{\Xi_1}}{2}(t_2-t_1)}.$$

Moreover, since $B(t_2) - B(t_1)$ is independent of $\mathscr{F}_{t_1}^t$,

$$\mathbb{E}\left[e^{i\langle a,B(t_2)-B(t_1)\rangle_{\Xi_1}}\right] = \mathbb{E}\left[e^{i\langle a,B(t_2)-B(t_1)\rangle_{\Xi_1}}|\mathscr{F}_{t_1}^t\right].$$

The "if" part: We have to prove the four conditions in Proposition 1.90: (1) and (2) are already in the assumptions of the lemma. Condition (4) follows easily from (1.13), Theorem 1.57 and Definition 1.58. To prove condition (3), i.e. that $Y := B(t_2) - B(t_1)$ is independent of $\mathscr{F}_{t_1}^t$, observe that, for all $Z: \Omega \to \Xi_1$ which are $\mathscr{F}_{t_1}^t$-measurable, one has, for all $a, b \in \Xi_1$,

$$\mathbb{E}\left[e^{i\langle a,Y\rangle_{\Xi_1}}e^{i\langle b,Z\rangle_{\Xi_1}}\right] = \mathbb{E}\left[\mathbb{E}\left[e^{i\langle a,Y\rangle_{\Xi_1}}|\mathscr{F}_{t_1}^t\right]e^{i\langle b,Z\rangle_{\Xi_1}}\right]$$
$$= e^{-\frac{\langle Q_1 a,a\rangle_{\Xi_1}}{2}(t_2-t_1)}\mathbb{E}\left[e^{i\langle b,Z\rangle_{\Xi_1}}\right] = \mathbb{E}\left[e^{i\langle a,Y\rangle_{\Xi_1}}\right]\mathbb{E}\left[e^{i\langle b,Z\rangle_{\Xi_1}}\right].$$

Since the above holds for all $Z: \Omega \to \Xi_1$ which are $\mathscr{F}_{t_1}^t$-measurable, and for all $a, b \in \Xi_1$, we conclude that Y is independent of $\mathscr{F}_{t_1}^t$ by Theorem 1.56. $\square$

Lemma 1.94 Let $\mathscr{F}_s^{t,0}$ and $\mathscr{F}_s^t$ be the filtrations defined in Proposition 1.90 for a Q-Wiener process W_Q. Then $\mathscr{F}_s^t$ is right-continuous. Moreover, for all $T > t$, $\mathscr{F}_T^{t,0}$, and consequently $\mathscr{F}_T^t$, are countably generated up to sets of measure zero. If the trajectories of W_Q are everywhere continuous then

$$\mathscr{F}_T^{t,0} = \mathscr{F}_{T-}^{t,0} = \sigma\left(W_Q(s_i) : i = 1, 2, ...\right), \tag{1.14}$$

where (s_i), $i = 1, 2, ...$ is any dense sequence in $[t, T)$, and hence the filtration $\mathscr{F}_s^{t,0}$ is countably generated and left-continuous.

Proof The proof follows arguments from [513] and [372] (Sect. 2.7-A). Consider $\tau > s$ and $\varepsilon > 0$. Since $W_Q(\tau + \varepsilon) - W_Q(s + \varepsilon)$ is independent of $\mathscr{F}_{s+}^{t,0}$, for every $A \in \mathscr{F}_{s+}^{t,0}$ and $f \in C_b(\Xi_1)$

$$\mathbb{E}\left(1_A f(W_Q(\tau + \varepsilon) - W_Q(s + \varepsilon))\right) = \mathbb{P}(A)\mathbb{E}f(W_Q(\tau + \varepsilon) - W_Q(s + \varepsilon)).$$

Letting $\varepsilon \to 0$ we thus have by the dominated convergence theorem that

$$\mathbb{E}\left(1_A f(W_Q(\tau) - W_Q(s))\right) = \mathbb{P}(A)\mathbb{E}f(W_Q(\tau) - W_Q(s)). \tag{1.15}$$

Now if $B = \overline{B} \subset \Xi_1$ then there exist functions $f_n \in C_b(\Xi_1), 0 \le f_n \le 1$, such that $f_n(x) \to 1_B(x)$ as $n \to +\infty$ for every $x \in \Xi_1$. Therefore (1.15) implies that

$$\mathbb{P}(A \cap \{W_Q(\tau) - W_Q(s) \in B\}) = \mathbb{P}(A)\mathbb{P}(\{W_Q(\tau) - W_Q(s) \in B\})$$

and since the sets $\{\{W_Q(\tau) - W_Q(s) \in B\} : B = \overline{B} \subset \Xi_1\}$ are a π-system generating $\sigma(W_Q(\tau) - W_Q(s))$, it follows from Lemma 1.23 that $\mathscr{F}_{s+}^{t,0}$ and $\sigma(W_Q(\tau) - W_Q(s))$ are independent.

Now let $s = \tau_0 < \tau_1 < ... < \tau_k \leq T$. We have $\sigma(W_Q(\tau_i) - W_Q(s) : i = 1, ..., k) = \sigma(W_Q(\tau_i) - W_Q(\tau_{i-1}) : i = 1, ..., k)$. Let now $A \in \mathscr{F}_{s+}^{t,0}$ and $B_i \in \sigma(W_Q(\tau_i) - W_Q(\tau_{i-1}))$, $i = 1, ..., k$. Since B_i is independent of $A \cap B_1 \cap ... \cap B_{i-1} \in \mathscr{F}_{\tau_{i-1}}^{t,0}$, $i = 1, ..., k$ and $B_1, ..., B_k$ are independent

$$\mathbb{P}(A \cap B_1 \cap ... \cap B_k) = \mathbb{P}(A \cap B_1 \cap ... \cap B_{k-1})\mathbb{P}(B_k) = ...$$
$$= \mathbb{P}(A \cap B_1) \prod_{i=2}^{k} \mathbb{P}(B_i) = \mathbb{P}(A) \prod_{i=1}^{k} \mathbb{P}(B_i) = \mathbb{P}(A)\mathbb{P}(B_1 \cap ... \cap B_k).$$

Therefore $\bigcup \sigma(W_Q(\tau_i) - W_Q(s) : i = 1, ..., k)$ (where the union is taken over all partitions $s = \tau_0 < \tau_1 < ... < \tau_k \leq T$) is a π-system independent of $\mathscr{F}_{s+}^{t,0}$ and thus $\mathscr{G}_s = \sigma(W_Q(\tau) - W_Q(s) : s \leq \tau \leq T)$ is independent of $\mathscr{F}_{s+}^{t,0}$.

Since $\mathscr{F}_T^{t,0} = \sigma(\mathscr{F}_s^{t,0}, \mathscr{G}_s)$, the family $\{A_s \cap B_s : A_s \in \mathscr{F}_s^{t,0}, B_s \in \mathscr{G}_s\}$ is a π-system generating $\mathscr{F}_T^{t,0}$. Let now $A \in \mathscr{F}_{s+}^{t,0}$ and let ξ be a version of $\mathbf{1}_A - \mathbb{E}(\mathbf{1}_A|\mathscr{F}_s^{t,0})$. Since ξ is $\mathscr{F}_{s+}^{t,0}$-measurable, it is independent of $\mathscr{G}_s$, so if $A_s \in \mathscr{F}_s^{t,0}$, $B_s \in \mathscr{G}_s$ then

$$\mathbb{E}\left(\xi \mathbf{1}_{A_s \cap B_s}\right) = \mathbb{E}\left(\xi \mathbf{1}_{A_s} \mathbf{1}_{B_s}\right) = \mathbb{P}(B_s)\mathbb{E}\left(\xi \mathbf{1}_{A_s}\right)$$
$$= \mathbb{P}(B_s) \int_{A_s} \xi d\mathbb{P} = \mathbb{P}(B_s) \left[\int_{A_s} \mathbf{1}_A d\mathbb{P} - \int_{A_s} \mathbb{E}(\mathbf{1}_A|\mathscr{F}_s^{t,0})d\mathbb{P}\right] = 0$$

by the definition of conditional expectation. This implies that $\int_D \xi d\mathbb{P} = 0$ for every $D \in \mathscr{F}_T^t$ and thus $\xi = 0$, $\mathbb{P}$-a.e. Therefore $\mathbf{1}_A = \mathbb{E}(\mathbf{1}_A|\mathscr{F}_s^{t,0})$, $\mathbb{P}$-a.e., i.e. if $\tilde{A} = \mathbb{E}(\mathbf{1}_A|\mathscr{F}_s^{t,0})^{-1}(1)$ then $\tilde{A} \in \mathscr{F}_s^{t,0}$ and $\mathbb{P}(A \triangle \tilde{A}) = 0$. This shows that $\mathscr{F}_{s+}^{t,0} \subset \mathscr{F}_s^t$.

Now let $A \in \mathscr{F}_{s+}^t$, which means that for every $n \geq 1$, $A \in \mathscr{F}_{s+1/n}^t$ and there exists a $B_n \in \mathscr{F}_{s+1/n}^{t,0}$ such that $A \triangle B_n \in \mathcal{N}$. Set

$$B = \bigcap_{m=1}^{+\infty} \bigcup_{n=m}^{+\infty} B_n.$$

Then $B \in \mathscr{F}_{s+}^{t,0} \subset \mathscr{F}_s^t$ and

$$B \setminus A \subset \left(\bigcup_{n=1}^{+\infty} B_n\right) \setminus A = \bigcup_{n=1}^{+\infty}(B_n \setminus A) \in \mathcal{N}.$$

Moreover,

$$A \setminus B = A \cap \left(\bigcap_{m=1}^{+\infty} \bigcup_{n=m}^{+\infty} B_n \right)^c = A \cap \left(\bigcup_{m=1}^{+\infty} \bigcap_{n=m}^{+\infty} B_n^c \right)$$

$$= \bigcup_{m=1}^{+\infty} \bigcap_{n=m}^{+\infty} (A \cap B_n^c) \subset \bigcup_{m=1}^{+\infty} (A \cap B_m^c) = \bigcup_{m=1}^{+\infty} (A \setminus B_m) \in \mathcal{N}.$$

Thus $A \triangle B \in \mathcal{N}$, which implies that $A \in \mathscr{F}_s^t$, which completes the proof of the right continuity.

To show that $\mathscr{F}_T^{t,0}$ is countably generated up to sets of measure zero we take a dense sequence (s_i), $i = 1, 2, \ldots$, in $[t, T)$. Since $\mathcal{B}(\Xi_1)$ is countably generated (for instance by open balls with rational radii centered at points of a countable dense set), each $\sigma(W_Q(s_i))$ is countably generated and so $\sigma(W_Q(s_i) : i \geq 1)$ is countably generated. It remains to show that for every $s \in (t, T]$, $\sigma(W_Q(s)) \subset \sigma(\mathcal{N}, W_Q(s_i) : s_i < s)$. Let $\Omega_0 \subset \Omega$, $\mathbb{P}(\Omega_0) = 1$ be such that W_Q has continuous trajectories on $[t, T]$ for $\omega \in \Omega_0$. Let A be an open subset of Ξ_1 and set $A_n = \{x \in A : \text{dist}(x, A^c) > 1/n\}$, $n = 1, 2, \ldots$. Then A_n is open, $\overline{A}_n \subset A_{n+1}$, and $\bigcup_{n=1}^{+\infty} A_n = A$. Let s_{i_k} be a sequence of s_i such that $s_{i_k} < s$ and $s_{i_k} \to s$ as $k \to +\infty$. Then, using the continuity of the trajectories of W_Q, it is easy to see that

$$\Omega_0 \cap W_Q(s)^{-1}(A) = \Omega_0 \cap \bigcup_{n=1}^{+\infty} \bigcap_{k=n}^{+\infty} W_Q(s_{i_k})^{-1}(A_n) \in \sigma(\mathcal{N}, W_Q(s_i) : s_i < s).$$

Therefore $W_Q(s)^{-1}(A) \in \sigma(\mathcal{N}, W_Q(s_i) : s_i < s)$ and since the sets $\{W_Q(s)^{-1}(A) : A \text{ is an open subset of } \Xi_1\}$ generate $\sigma(W_Q(s))$, the result follows. If $\Omega_0 = \Omega$ then we have above

$$W_Q(s)^{-1}(A) = \bigcup_{n=1}^{+\infty} \bigcap_{k=n}^{+\infty} W_Q(s_{i_k})^{-1}(A_n) \in \sigma(W_Q(s_i) : s_i < s).$$

The argument that $\sigma(W_Q(t)) \subset \sigma(W_Q(s_i) : i = 1, 2, \ldots)$ is similar (or we can just assume that $s_1 = t$). This yields (1.14). $\qquad \square$

In fact the above argument shows that if S is a Polish space, $T > t$, and $X : [t, T] \times \Omega \to S$ is a stochastic process with everywhere continuous trajectories, then the filtration generated by X, $\mathscr{F}_s^X := \sigma(X(\tau) : t \leq \tau \leq s)$ is countably generated and left-continuous.

1.2.5   Simple and Elementary Processes

Definition 1.95 ($\mathscr{F}_s^t$-*simple process*) Let E be a Banach space (endowed with the Borel σ-field) and let $(\Omega, \mathscr{F}, \{\mathscr{F}_s^t\}_{s \in [t,T]}, \mathbb{P})$ be a filtered probability space. A process $X : [t, T] \times (\Omega, \mathscr{F}, \mathbb{P}) \to E$ is called $\mathscr{F}_s^t$-*simple* if:

(i) Case $T = +\infty$: there exists a sequence of real numbers $(t_n)_{n \in \mathbb{N}}$ with $t = t_0 < t_1 < ... < t_n < ...$ and $\lim_{n \to \infty} t_n = +\infty$, a constant $C < +\infty$, and a sequence of random variables $\xi_n : \Omega \to E$ with $\sup_{n \geq 0} |\xi_n(\omega)|_E \leq C$ for every $\omega \in \Omega$, such that ξ_n is $\mathscr{F}_{t_n}^t$-measurable for every $n \geq 0$, and

$$X(s)(\omega) = \begin{cases} \xi_0(\omega) & \text{if } s = t \\ \xi_i(\omega) & \text{if } s \in (t_i, t_{i+1}]. \end{cases}$$

(ii) Case $T < +\infty$: there exist $t = t_0 < t_1 < ... < t_N = T$, a constant $C < +\infty$, and random variables $\xi_n : \Omega \to E$ for $n = 0, ..., N - 1$ with $\sup_{0 \leq n \leq N-1} |\xi_n(\omega)|_E \leq C$ for every $\omega \in \Omega$, such that ξ_n is $\mathscr{F}_{t_n}^t$-measurable, and

$$X(s)(\omega) = \begin{cases} \xi_0(\omega) & \text{if } s = t \\ \xi_i(\omega) & \text{if } s \in (t_i, t_{i+1}]. \end{cases}$$

Definition 1.96 ($\mathscr{F}_s^t$-*elementary process*) Let $T \in (0, +\infty)$, $t \in [0, T)$. Let (S, d) be a complete metric space (endowed with the Borel σ-field), and $(\Omega, \mathscr{F}, \{\mathscr{F}_s^t\}_{s \in [t,T]}, \mathbb{P})$ be a filtered probability space. We say that a process $X : [t, T] \times (\Omega, \mathscr{F}, \mathbb{P}) \to S$ is $\mathscr{F}_s^t$-*elementary* if there exist S-valued random variables $\xi_0, \xi_1, .., \xi_{N-1}$, and a sequence $t = t_0 < t_1 < .. < t_N = T$, such that

(1) ξ_i has a finite numbers of values for every $i \in \{0, ..N - 1\}$.
(2) ξ_i is $\mathscr{F}_{t_i}^t$-measurable for every $i \in \{0, ..N - 1\}$.
(3) $X(s)(\omega) = \xi_i(\omega)$ for $s \in (t_i, t_{i+1}]$ for $i \in \{0, ..N - 1\}$, and $X(t) = \xi_0$.

Finally, we say that a process $X : [t, +\infty) \times (\Omega, \mathscr{F}, \mathbb{P}) \to S$ is $\mathscr{F}_s^t$-*elementary* if there exists $T_1 > t$ such that the restriction of X to $[t, T_1]$ is $\mathscr{F}_s^t$-elementary and $X(s) = 0$ for $s > T_1$.

It is immediate from the definitions that simple and elementary processes are progressively measurable and predictable.

Remark 1.97 In Definitions 1.14, 1.95 and 1.96 we introduced the concepts of a *simple* random variable, $\mathscr{F}_s^t$-*simple* process, and $\mathscr{F}_s^t$-*elementary* process. The reader should be aware that in the literature the use of these terms varies and the same word is often used by different authors to mean different things. ∎

Lemma 1.98 *Let E be a separable Banach space endowed with the Borel σ-field, $(\Omega, \mathscr{F}, \mathscr{F}_s^t, \mathbb{P})$ be a filtered probability space and $X : [t, T] \times \Omega \to E$ be a bounded, measurable, $\mathscr{F}_s^t$-adapted process, where $T \in [t, +\infty) \cup \{+\infty\}$. There exists a sequence $X^m(\cdot)$ of $\mathscr{F}_s^t$-elementary E-valued processes on $[t, T]$ such that, for every $1 \leq p < +\infty$ and $R > t$,*

$$\lim_{m \to +\infty} \mathbb{E} \int_t^{R \wedge T} |X^m(s) - X(s)|_E^p \, ds = 0. \tag{1.16}$$

The same claim holds if, instead of the Banach space, we consider E to be an interval $[a, b] \subset \mathbb{R}$ or a countable closed subset of $[a, b]$. In these cases the norm $|\cdot|_E$ in (1.16) is replaced by $|\cdot|_{\mathbb{R}}$.

Proof It is enough to prove the result for a single $p \geq 1$. To obtain a sequence of $\mathscr{F}_s^t$-simple processes $X^m(\cdot)$ with the required properties, the proof follows exactly the proof of Lemma 3.2.4, p. 132, in [372] with obvious technical modifications as we now have to deal with Bochner integrals in E. We then use Lemma 1.16 to approximate the random variables ξ_i defining $X^m(\cdot)$ by simple random variables to obtain $\mathscr{F}_s^t$-elementary approximating processes.

If E is a countable closed subset of $[a, b]$, we first produce $[a, b]$-valued $\mathscr{F}_s^t$-elementary approximating processes $X^m(\cdot)$. We then construct an E-valued $\mathscr{F}_s^t$-elementary process $Y^m(\cdot)$ from $X^m(\cdot)$ as follows. Let $X^m(s) = \xi_i$ for $s \in (t_i, t_{i+1}]$ for $i \in \{0, ..N-1\}$, and $X(t) = \xi_0$. Let $\tilde{\xi}_i$ be defined in the following way. If $\xi_i(\omega) \in E$, we set $\tilde{\xi}_i(\omega) = \xi_i(\omega)$. If $\xi_i(\omega) \notin E$, we set $\tilde{\xi}_i(\omega) = \arg\min_{x \in E} |\xi(\omega) - x|$ if $\arg\min_{x \in E} |\xi(\omega) - x|$ is a singleton. If $\arg\min_{x \in E} |\xi(\omega) - x|$ has two points $x_1 < x_2$, we set $\tilde{\xi}_i(\omega) = x_1$. Obviously $\tilde{\xi}_i$ is a simple, $\mathscr{F}_{t_i}^t$-measurable process. We now define $Y^m(s) = \tilde{\xi}_i$ for $s \in (t_i, t_{i+1}]$ for $i \in \{0, ..N-1\}$, and $X(t) = \tilde{\xi}_0$. Then, since $X(\cdot)$ has values in E, it is easy to see that $|Y^m(s) - X(s)| \leq 2|X^m(s) - X(s)|$ for any $s \in [t, +\infty)$ and $\omega \in \Omega$. Therefore the result follows. $\qquad \square$

Lemma 1.99 *Let $\mathscr{F}_s^{t,0}$ and $\mathscr{F}_s^t$ be as in Proposition 1.90, $T \in [t, +\infty) \cup \{+\infty\}$, and let $a(\cdot) : [t, T] \times \Omega \to S$ be an $\mathscr{F}_s^t$-progressively measurable process, where (S, d) is a Polish space endowed with the Borel σ-field. Then there exists an $\mathscr{F}_s^{t,0}$-progressively measurable and $\mathscr{F}_s^{t,0}$-predictable process $a_1(\cdot) : [t, T] \times \Omega \to S$, such that $a(\cdot) = a_1(\cdot)$, $dt \otimes \mathbb{P}$-a.e. on $[t, T] \times \Omega$.*

Proof In light of Theorems 1.12 and 1.13 we can assume that $S = [0, 1]$ or S is a countable closed subset of $[0, 1]$. Using Lemma 1.98, we can find approximating $\mathscr{F}_s^t$-elementary processes $a^n(\cdot)$ on $[t, T]$ of the form

$$a^n(t)(\omega) = \begin{cases} \xi_0^n(\omega) & \text{if } s = t \\ \xi_i^n(\omega) & \text{if } s \in (t_i, t_{i+1}] \end{cases}$$

such that

$$\sup_{R \geq t} \lim_{n \to \infty} \mathbb{E} \int_t^{R \wedge T} |a(s) - a^n(s)|_{\mathbb{R}}^2 \, ds = 0.$$

Using Lemma 1.16, we can change every ξ_i^n on a null-set to obtain a sequence of $\mathscr{F}_s^{t,0}$-elementary processes $a_1^n(\cdot)$ that still satisfy

$$\sup_{R \geq t} \lim_{n \to \infty} \mathbb{E} \int_t^{R \wedge T} |a(s) - a_1^n(s)|_{\mathbb{R}}^2 \, ds = 0.$$

Obviously the processes $a_1^n(\cdot)$ are $\mathscr{F}_s^{t,0}$-progressively measurable. We can now extract a subsequence (still denoted by $a_1^n(\cdot)$) such that $a_1^n(\cdot) \to a(\cdot)$ $dt \otimes \mathbb{P}$-

a.e. on $[t, T] \times \Omega$, and define $a_1(\cdot) := \liminf_{n \to +\infty} a_1^n(\cdot)$. The process $a_1(\cdot)$ is $\mathscr{F}_s^{t,0}$-progressively measurable, $\mathscr{F}_s^{t,0}$-predictable, and $a(\cdot) = a_1(\cdot)$, $dt \otimes \mathbb{P}$-a.e. on $[t, T] \times \Omega$. $\square$

1.3 The Stochastic Integral

Let $T \in (0, +\infty)$, and $t \in [0, T)$. Throughout the whole section Ξ and H will be two real, separable Hilbert spaces, Q will be an operator in $\mathcal{L}^+(\Xi)$, $\left(\Omega, \mathscr{F}, \left\{\mathscr{F}_s^t\right\}_{s \in [t,T]}, \mathbb{P}\right)$ will be a filtered probability space, and W_Q will be a translated $\mathscr{F}_s^t$-Q-Wiener process on Ω on $[0, T]$. The following concept will be used in Chap. 2.

Definition 1.100 A 5-tuple $\mu := \left(\Omega, \mathscr{F}, \left\{\mathscr{F}_s^t\right\}_{s \in [t,T]}, \mathbb{P}, W_Q\right)$ described above is called a *generalized reference probability space*.

A process $X(\cdot)$ will always be assumed to be defined on Ω, and the expressions "adapted" and "progressively measurable" will always refer to the filtration $\mathscr{F}_s^t$.

1.3.1 Definition of the Stochastic Integral

In this section we will assume that $\mathrm{Tr}(Q) < +\infty$. If $\mathrm{Tr}(Q) = +\infty$, the construction of the stochastic integral is the same, we just have to consider W_Q as a Ξ_1-valued Wiener process with nuclear covariance Q_1 (see Sect. 1.2.4). This way W_Q is not uniquely determined but $Q_1^{1/2}(\Xi_1) = \Xi_0 = Q^{1/2}(\Xi)$, $|x|_{\Xi_0} = |Q_1^{-1/2}x|_{\Xi_1}$ for all possible extensions Ξ_1 and the class of integrands and the value of the integrals are independent of the choice of the space Ξ_1 (see [180], Proposition 4.7 and Sect. 4.1.2).

We recall that we denote by $\mathcal{L}_2(\Xi_0, H)$ the space of Hilbert–Schmidt operators from Ξ_0 to H (see Appendix B.3). It is equipped with its Borel σ-field $\mathcal{B}(\mathcal{L}_2(\Xi_0, H))$. $\mathcal{L}_2(\Xi_0, H)$ is a real, separable Hilbert space (see Proposition B.25), and $\mathcal{L}(\Xi, H)$ is dense in $\mathcal{L}_2(\Xi_0, H)$ (see e.g. [294], pp. 24–25).

Definition 1.101 (*The space* $\mathcal{N}_Q^p(t, T; H)$) Given $p \geq 1$, we denote by $\mathcal{N}_Q^p(t, T; H)$ the space of all $\mathcal{L}_2(\Xi_0, H)$-valued, progressively measurable processes $X(\cdot)$, such that

$$|X(\cdot)|_{\mathcal{N}_Q^p(t,T;H)} := \left(\mathbb{E}\int_t^T \|X(s)\|_{\mathcal{L}_2(\Xi_0,H)}^p ds\right)^{1/p} < \infty.$$

$\mathcal{N}_Q^p(t, T; H)$ is a Banach space if it is endowed with the norm $|\cdot|_{\mathcal{N}_Q^p(t,T;H)}$.

We remark that, as always, two processes in $\mathcal{N}_Q^p(t, T; H)$ are identified if they are equal $\mathbb{P} \otimes dt$-a.e.

Remark 1.102 In several classical references (see e.g. [180] or [491]), the theory of stochastic integration is developed for predictable processes instead of progressively measurable ones like in our case. However, it follows for instance from Lemma 1.99, that for every $\mathcal{L}_2(\Xi_0, H)$-valued progressively measurable process X there exists a predictable process X_1 which is $\mathbb{P} \otimes dt$-a.e. equal to X. Thus, since we are working with stochastic integrals with respect to Wiener processes (which are continuous), the two concepts coincide. ■

For an $\mathcal{L}(\Xi, H)$-valued, $\mathscr{F}_s^t$-simple process Φ on $[t, T]$, $\Phi(s) = \Phi_0 \mathbf{1}_{\{t\}}(s) + \sum_{i=0}^{i=N-1} \mathbf{1}_{(t_i, t_{i+1}]}(s) \Phi_i$, the stochastic integral with respect to W_Q is defined by

$$\int_t^T \Phi(s) dW_Q(s) := \sum_{i=0}^{N-1} \Phi_i (W_Q(t_{i+1}) - W_Q(t_i)) \in L^2(\Omega; H).$$

Note that if we take Φ to be $\mathcal{L}_2(\Xi_0, H)$-valued, we cannot guarantee that the expression above is well defined, since $\mathcal{L}_2(\Xi_0, H)$ contains genuinely unbounded operators in Ξ (see e.g. [294], p. 25, Exercise 2.7).

We now extend the stochastic integral to all processes in $\mathcal{N}_Q^2(t, T; H)$ by the following theorem.

Theorem 1.103 (Itô isometry) *For every $\mathcal{L}(\Xi, H)$-valued, $\mathscr{F}_s^t$-simple process Φ we have*

$$\mathbb{E} \left| \int_t^T \Phi(s) dW_Q(s) \right|_H^2 = \mathbb{E} \int_t^T \|\Phi(s)\|_{\mathcal{L}_2(\Xi_0, H)}^2 ds.$$

Thus the stochastic integral is an isometry between the set of $\mathcal{L}(\Xi, H)$-valued, $\mathscr{F}_s^t$-simple processes in $\mathcal{N}_Q^2(t, T; H)$ and its image in $L^2(\Omega; H)$. Moreover, since $\mathcal{L}(\Xi, H)$-valued, $\mathscr{F}_s^t$-simple (and in fact elementary) processes are dense in $\mathcal{N}_Q^2(t, T; H)$, it can be uniquely extended to all processes in $\mathcal{N}_Q^2(t, T; H)$. We denote this unique extension by

$$\int_t^T \Phi(s) dW_Q(s)$$

and call it the stochastic integral of Φ with respect to W_Q.

Proof See [294], Propositions 2.1, 2.2, and Definition 2.10. See also [180], Proposition 4.22 in the context of predictable processes. □

Proposition 1.104 *For $\Phi \in \mathcal{N}_Q^2(t, T; H)$, consider the process*

$$\begin{cases} I(\Phi) : [t, T] \times \Omega \to H \\ I(\Phi)(r) := \int_t^r \Phi(s) dW_Q(s) := \int_t^T \Phi(s) \mathbf{1}_{[t,r]} dW_Q(s). \end{cases}$$

$I(\Phi)$ *is a continuous square-integrable martingale and $I : \mathcal{N}_Q^2(t, T; H) \to \mathcal{M}_{t,T}^2(H)$ is an isometry. Moreover,*

$$\langle\langle I(\Phi)\rangle\rangle_s = \int_t^s \left(\Phi(s)Q^{\frac{1}{2}}\right)\left(\Phi(s)Q^{\frac{1}{2}}\right)^* ds,$$

$$\langle I(\Phi)\rangle_s = \int_t^s \|\Phi(s)\|_{\mathcal{L}_2(\Xi_0, H)}^2 ds.$$

Proof See [294] Theorem 2.3, p. 34. □

The definition of stochastic integral can be further extended to all $\mathcal{L}_2(\Xi_0, H)$-valued progressively measurable processes $\Phi(\cdot)$ such that

$$\mathbb{P}\left(\int_t^T \|\Phi(s)\|_{\mathcal{L}_2(\Xi_0, H)}^2 ds < +\infty\right) = 1. \tag{1.17}$$

Lemma 1.105 *Let $\{\Phi(s)\}_{s\in[t,T]}$ be an $\mathcal{L}_2(\Xi_0, H)$-valued progressively measurable process satisfying (1.17). Then there exists a sequence Φ_n of $\mathcal{L}(\Xi, H)$-valued $\mathscr{F}_s^t$-simple processes such that*

$$\lim_{n\to\infty} \int_t^T \|\Phi(s) - \Phi_n(s)\|_{\mathcal{L}_2(\Xi_0, H)}^2 ds = 0 \quad \mathbb{P} - a.s. \tag{1.18}$$

Moreover, there exists an H-valued random variable, denoted by $\mathcal{I}$, such that

$$\lim_{n\to\infty} \int_t^T \Phi_n(s) dW_Q(s) = \mathcal{I} \quad \text{in probability.}$$

$\mathcal{I}$ does not depend on the choice of approximating sequence, more precisely, given Φ_n^1 and Φ_n^2 satisfying (1.18), if $\mathcal{I}_1 := \lim_{n\to\infty} \int_t^T \Phi_n^1(s) dW_Q(s)$ and $\mathcal{I}_2 := \lim_{n\to\infty} \int_t^T \Phi_n^2(s) dW_Q(s)$, then $\mathcal{I}_1 = \mathcal{I}_2 \mathbb{P} - a.s.$

Proof See [294], Lemmas 2.3, p. 39, and 2.6, p. 41. □

The process $\mathcal{I}$ defined by Lemma 1.105 is called the stochastic integral of Φ with respect to W_Q, and is denoted by $\int_t^T \Phi(s) dW_Q(s)$. We also set $\int_t^r \Phi(s) dW_Q(s) := \int_t^T \Phi(s)\mathbf{1}_{[t,r]} dW_Q(s)$.

Proposition 1.106 *Let $\{\Phi(s)\}_{s\in[t,T]}$ be an $\mathcal{L}_2(\Xi_0, H)$-valued progressively measurable process satisfying (1.17). Then the process*

$$\begin{cases} I(\Phi): [t, T] \times \Omega \to H \\ I(\Phi)(r) := \int_t^r \Phi(s) dW_Q(s) \end{cases}$$

is a continuous local martingale.

Proof See [294], pp. 42–44. □

Finally, we may extend the definition of stochastic integral to all processes (not necessarily progressively measurable) that are $dt \otimes \mathbb{P}$-equivalent to progressively measurable processes satisfying (1.17) in the sense of the following definition (see also [372], p. 130).

Definition 1.107 We say that two processes Φ_1 and Φ_2 are $dt \otimes \mathbb{P}$-equivalent if $\Phi_1 = \Phi_2$, $dt \otimes \mathbb{P}$-a.e. If Φ belongs to the equivalence class of a progressively measurable process Φ_1 satisfying (1.17),[4] we set

$$\int_t^T \Phi(s) dW_Q(s) := \int_t^T \Phi_1(s) dW_Q(s).$$

This definition is obviously independent of the choice of a representative process Φ_1. Thus a representative process defines the stochastic integral for the whole equivalence class.

Example 1.108 Every $\mathcal{L}_2(\Xi_0, H)$-valued, $\mathscr{F}_s^t$-adapted, and $\overline{\mathcal{B}([t, T]) \otimes \mathscr{F}}$-measurable process Φ satisfying (1.17) is stochastically integrable, where $\overline{\mathcal{B}([t, T]) \otimes \mathscr{F}}$ is the completion of $\mathcal{B}([t, T]) \otimes \mathscr{F}$ with respect to $dt \otimes \mathbb{P}$. To see this we need to find a progressively measurable process Φ_1 which is equivalent to Φ. First, let Φ_2 be a $\mathcal{B}([t, T]) \otimes \mathscr{F}$-measurable process equivalent to Φ (which exists by Lemma 1.16). Then, for a.e. $s \in [t, T]$, we have $\Phi_2(s, \cdot) = \Phi(s, \cdot)$ $\mathbb{P}$-a.s. and, since every $\mathscr{F}_s^t$ is complete, also $\Phi_2(s, \cdot)$ is $\mathscr{F}_s^t$-measurable for a.e. s. Thus there exists an $A \in \mathcal{B}([t, T])$ of full measure such that $\Phi_2(s, \cdot)$ is $\mathscr{F}_s^t$-measurable for $s \in A$. We then define $\Phi_3 = \Phi_2 \mathbf{1}_A$. Φ_3 is $\mathcal{B}([t, T]) \otimes \mathscr{F}$-measurable and $\mathscr{F}_s^t$-adapted, thanks to Lemma 1.72 it has a progressively measurable modification Φ_1 which is clearly equivalent to Φ. ∎

Theorem 1.109 *Let* $(E, \mathscr{G}, \mu)$ *be a measure space with bounded measure. Let* $\Phi : [t, T] \times \Omega \times E \to \mathcal{L}_2(\Xi_0, H)$ *be* $(\mathcal{B}([t, T]) \otimes \mathscr{F}_T^t \otimes \mathscr{G})/\mathcal{B}(\mathcal{L}_2(\Xi_0, H))$-*measurable. Suppose that, for any* $x \in E$, $\{\Phi(s, \cdot, x)\}_{s \in [t,T]}$ *is progressively measurable and*

$$\int_E |\Phi(\cdot, \cdot, x)|_{\mathcal{N}_Q^2(t, T; H)} d\mu(x) < +\infty.$$

Then:

(i) $\int_t^T \Phi(s, \cdot, \cdot) dW_Q(s)$ *has an* $\mathscr{F}_T^t \otimes \mathscr{G}/\mathcal{B}(H)$-*measurable version.*

(ii) $\int_E \Phi(\cdot, \cdot, x) d\mu(x)$ *is progressively measurable.*

(iii) *The following equality holds* $\mathbb{P}$-*a.s.:*

$$\int_E \int_t^T \Phi(s, \cdot, x) dW_Q(s) d\mu(x) = \int_t^T \int_E \Phi(s, \cdot, x) d\mu(x) dW_Q(s).$$

---

[4]Note that if a process X is progressively measurable and satisfies (1.17) and Y is $dt \otimes \mathbb{P}$-equivalent to X, then Y must also satisfy (1.17) since for $\mathbb{P}$-a.s. ω, $X(\cdot, \omega) = Y(\cdot, \omega)$, a.e. on [t,T].

Proof See Theorem 2.8, Sect. 2.2.6, p. 57 of [294] and Theorem 4.33, Sect. 4.5, p. 110 of [180].   □

1.3.2   Basic Properties and Estimates

Lemma 1.110 *Let* $T > 0$ *and* $t \in [0, T)$. *Assume that* Φ *is in* $\mathcal{N}_Q^2(t, T; H)$ *and that* τ *is an* $\mathscr{F}_s^t$-*stopping time such that* $\mathbb{P}(\tau \leq T) = 1$. *Then* $\mathbb{P}$-*a.s.*

$$\int_t^T \mathbf{1}_{[t,\tau]}(r)\Phi(r)dW_Q(r) = \int_t^\tau \Phi(r)dW_Q(r).$$

Proof See [294], Lemma 2.7, p. 43 (also [180], Lemma 4.24, p. 99).   □

As a consequence of Theorem 1.80 and Proposition 1.104 we obtain the following theorem (see also e.g. [177], Theorem 5.2.4, p. 58).

Theorem 1.111 (Burkholder–Davis–Gundy inequality for stochastic integrals) *Let* $T > 0$ *and* $t \in [0, T)$. *For every* $p \geq 2$, *there exists a constant* c_p *such that, for every* Φ *in* $\mathcal{N}_Q^p(t, T; H)$,

$$\mathbb{E}\left[\sup_{s \in [t,T]} \left|\int_t^s \Phi(r)dW_Q(r)\right|^p\right] \leq c_p \mathbb{E}\left[\int_t^T \|\Phi(r)\|_{\mathcal{L}_2(\Xi_0, H)}^2 dr\right]^{p/2}$$

$$\leq c_p(T - t)^{\frac{p}{2}-1}\mathbb{E}\left[\int_t^T \|\Phi(r)\|_{\mathcal{L}_2(\Xi_0, H)}^p dr\right].$$

Proposition 1.112 *Let* $T > 0$ *and* $t \in [0, T)$. *Let* A *be the generator of a* C_0-*semigroup* $\{e^{rA}, \ r \geq 0\}$ *on* H *such that* $\|e^{rA}\| \leq Me^{\alpha r}$ *for every* $r \geq 0$ *for some* $\alpha \in \mathbb{R}$, $M > 0$. *Let* $p > 2$ *and* $\Phi \in \mathcal{N}_Q^p(t, T; H)$. *Let* A_n *be the Yosida approximation of* A. *Then the stochastic convolution process*

$$\Psi(s) := \int_t^s e^{(s-r)A}\Phi(r)dW_Q(r), \qquad s \in [t, T], \tag{1.19}$$

has a continuous modification,

$$\mathbb{E}\left[\sup_{s \in [t,T]} \left|\int_t^s e^{(s-r)A}\Phi(r)dW_Q(r)\right|^p\right] \leq C\mathbb{E}\left[\int_t^T \|\Phi(r)\|_{\mathcal{L}_2(\Xi_0, H)}^p dr\right], \tag{1.20}$$

where the constants c *and* C *depend only on* $T - t$, p, M, α, *and*

$$\lim_{n\to\infty} \mathbb{E}\left[\sup_{s \in [t,T]} \left|\int_t^s \left(e^{(s-r)A_n} - e^{(s-r)A}\right)\Phi(r)dW_Q(r)\right|^p\right] = 0. \tag{1.21}$$

If, moreover, A generates a C_0-pseudo-contraction semigroup (i.e. $M = 1$ above, see Appendix B.4) then the claims are also true for $p = 2$.

Proof See [294], Lemma 3.3, p. 87. The claims for p=2 can be proved by repeating the arguments of the proof of Proposition 3.3 of [543], which uses the Unitary Dilation Theorem. □

Proposition 1.113 *Let A be the generator of a C_0-semigroup on H, $T > 0$, and $t \in [0, T]$. Assume that $\Phi : [t, T] \times \Omega \to \mathcal{L}_2(\Xi_0, H)$ is a progressively measurable process such that $\Phi(s) \in \mathcal{L}_2(\Xi_0, D(A))$ $\mathbb{P}$-a.s., for a.e. $s \in [t, T]$. Assume that*

$$\mathbb{P}\left(\int_t^T \|\Phi(s)\|^2_{\mathcal{L}_2(\Xi_0, D(A))} ds < +\infty\right) = 1.$$

Then

$$\mathbb{P}\left(\int_t^T \Phi(s)dW_Q(s) \in D(A)\right) = 1 \qquad (1.22)$$

and

$$A\int_t^T \Phi(s)dW_Q(s) = \int_t^T A\Phi(s)dW_Q(s), \qquad \mathbb{P}-a.s. \qquad (1.23)$$

Proof We can assume without loss of generality that $Q \in \mathcal{L}_1^+(\Xi)$. The proof follows the proof of Proposition 3.1 (p. 76) of [294], however we present it here to clarify a measurability issue. Indeed, we first need to show that Φ is an $\mathcal{L}_2(\Xi_0, D(A))$-valued, progressively measurable process. To do this we take $\Psi_n = J_n\Phi$, where $J_n = n(nI - A)^{-1}$ (see Definition B.40). Since $J_n \in \mathcal{L}(H, D(A))$, Ψ_n is an $\mathcal{L}_2(\Xi_0, D(A))$-valued, progressively measurable process. Moreover, it is easy to see that if, for some $s \in [t, T]$ and $\omega \in \Omega$, $\Phi(s)(\omega) \in \mathcal{L}_2(\Xi_0, D(A))$, then $\Psi_n(s)(\omega) \to \Phi(s)(\omega)$ in $\mathcal{L}_2(\Xi_0, D(A))$. Therefore, defining $V := \{(s, \omega) : \Psi_n(s)(\omega)$ converges in $\mathcal{L}_2(\Xi_0, D(A))\}$, it follows from Lemma 1.8-(iii) that Φ is equivalent to a progressively measurable process $\lim_{n\to+\infty} 1_V \Psi_n$. The proof is now done in two steps.

Step 1: The claim is true for $\mathscr{F}_s^t$-simple $\mathcal{L}(\Xi, D(A))$-valued processes.

Step 2: If Φ is a $\mathcal{L}_2(\Xi_0, D(A))$-valued progressively measurable process satisfying the hypotheses of this proposition, we take a sequence of $\mathscr{F}_s^t$-simple $\mathcal{L}(\Xi, D(A))$-valued processes Φ_n approximating Φ in the sense of (1.18) so that

$$\lim_{n\to+\infty} \int_t^T \|\Phi(s) - \Phi_n(s)\|^2_{\mathcal{L}_2(\Xi_0, D(A))} ds = 0 \qquad \mathbb{P}-a.s.$$

In particular we have

$$\int_t^T \Phi_n(s)dW_Q(s) \xrightarrow{n\to\infty} \int_t^T \Phi(s)dW_Q(s),$$

$$A \int_t^T \Phi_n(s)dW_Q(s) = \int_t^T A\Phi_n(s)dW_Q(s) \xrightarrow{n\to\infty} \int_t^T A\Phi(s)dW_Q(s)$$

in probability, so the claim follows since A is a closed operator. □

In the rest of this section we explain how the factorization method is used to prove continuity of trajectories of stochastic convolution processes.

Lemma 1.114 (Factorization Lemma) *Let $T > 0$, $t \in [0, T)$, and $0 < \alpha < 1$. Let A be the generator of a C_0-semigroup $\{e^{rA}, r \geq 0\}$ on H. Consider a linear, densely defined, closed operator $A_1 : D(A_1) \subset H \to H$ such that, for any $r > 0$, $e^{rA}H \subset D(A_1)$, $A_1 e^{rA}$ is bounded and $A_1 e^{rA} = e^{rA}A_1$ on $D(A_1)$. Let $\Phi : [t, T] \times \Omega \to \mathcal{L}_2(\Xi_0, H)$ be progressively measurable and such that for every $s \in [t, T]$*

$$\mathbb{E} \int_t^s \left\| A_1 e^{(s-r)A}\Phi(r) \right\|_{\mathcal{L}_2(\Xi_0, H)}^2 \, dr < +\infty.$$

Assume that, for all $s \in [t, T]$,

$$\int_t^s (s-r)^{\alpha-1} \left(\int_t^r (r-h)^{-2\alpha} \mathbb{E}\left[\left\| A_1 e^{(r-h)A}\Phi(h) \right\|_{\mathcal{L}_2(\Xi_0, H)}^2 \right] dh \right)^{1/2} dr < +\infty. \tag{1.24}$$

Then

$$\int_t^s A_1 e^{(s-r)A}\Phi(r)dW_Q(r) = \frac{\sin(\alpha\pi)}{\pi} \int_t^s (s-r)^{\alpha-1} e^{(s-r)A} Y_\alpha^\Phi(r)dr \qquad \mathbb{P}-a.s.$$

for all $s \in [t, T]$, where $Y_\alpha^\Phi(\cdot)$ is a $\mathcal{B}([t, T]) \otimes \mathscr{F}_T^t/\mathcal{B}(H)$-measurable process which is $dt \otimes \mathbb{P}$-equivalent to

$$\int_t^r (r-h)^{-\alpha} A_1 e^{(r-h)A}\Phi(h)dW_Q(h).$$

Proof The statement is similar to [177], Theorem 5.2.5, p. 58, Sect. 5.2.1. We give the proof for completeness.

We use the identity

$$\int_\sigma^t (t-s)^{\alpha-1}(s-\sigma)^{-\alpha}ds = \frac{\pi}{\sin(\pi\alpha)}, \qquad \text{for all } \sigma \leq s \leq t, \ 0 < \alpha < 1$$

(which can be proved by a simple direct computation). Define

$$X(r, h) = \mathbf{1}_{[t,r]}(h)(r-h)^{-\alpha} A_1 e^{(r-h)A}\Phi(h).$$

Since (1.24) implies

$$\int_t^T \left(\mathbb{E} \int_t^T \|X(r,h)\|^2_{\mathcal{L}_2(\Xi_0, H)} \, dh \right)^{1/2} dr < +\infty,$$

by the stochastic Fubini Theorem 1.109 (see also Theorem 4.33, p. 110 of [180] or Theorem 2.8, p. 57 of [294]) there exists a $\mathcal{B}([t,T]) \otimes \mathscr{F}_T^t / \mathcal{B}(H)$-measurable process $Y_\alpha^\Phi : [t,T] \times \Omega \to H$ such that

$$\int_t^T X(r,h) dW_Q(h) = \int_t^r (r-h)^{-\alpha} A_1 e^{(r-h)A} \Phi(h) dW_Q(h) = Y_\alpha^\Phi(r), \quad dt \otimes \mathbb{P}\text{-a.e.}$$

Then for every $s \in [t,T]$ the process $Z_\alpha^{\Phi,s}(\cdot)$, defined for any $r \in [t,s]$ by $Z_\alpha^{\Phi,s}(r) = (s-r)^{\alpha-1} e^{(s-r)A} Y_\alpha^\Phi(r)$, is jointly measurable and $dt \otimes \mathbb{P}$-equivalent to

$$(s-r)^{\alpha-1} e^{(s-r)A} \int_t^r (r-h)^{-\alpha} A_1 e^{(r-h)A} \Phi(h) dW_Q(h)$$

on $[t,s] \times \Omega$. Thus fixing any $s \in [t,T]$ and applying the stochastic Fubini Theorem on $[t,s] \times [t,s] \times \Omega$ (whose assumptions are satisfied by (1.24)) and noticing that we can use the process $Z_\alpha^{\Phi,s}(\cdot)$ in place of a process provided by the stochastic Fubini Theorem (since it will give $\mathbb{P}$-a.e. the same integrals) we obtain for $\mathbb{P}$-a.e. ω

$$\frac{\pi}{\sin(\pi\alpha)} \int_t^s A_1 e^{(s-h)A} \Phi(h) dW_Q(h)$$

$$= \int_t^s \int_t^s \mathbf{1}_{[h,s]}(r)(s-r)^{\alpha-1} e^{(s-r)A} (r-h)^{-\alpha} A_1 e^{(r-h)A} \Phi(h) dr dW_Q(h)$$

$$= \int_t^s (s-r)^{\alpha-1} e^{(s-r)A} Y_\alpha^\Phi(r) dr.$$

$\square$

Lemma 1.115 *Let A be the generator of a C_0-semigroup $\{e^{rA}, \ r \geq 0\}$ on H, $T > 0$, $t \in [0,T)$ and $f \in L^p(t,T;H)$, $p \geq 1$. Then:*

(i) *If either $1/p < \alpha \leq 1$, or $p = \alpha = 1$, then the function*

$$G_\alpha f(s) := \int_t^s (s-r)^{\alpha-1} e^{(s-r)A} f(r) dr$$

is in $C([t,T],H)$.

(ii) *If the semigroup e^{tA} is analytic, $\lambda \in \mathbb{R}$ is such that $(\lambda I - A)^{-1} \in \mathcal{L}(H)$, $\beta > 0$ and $\alpha > \beta + 1/p$, then the function*

$$G_{\alpha,\beta} f(s) := \int_t^s (s-r)^{\alpha-1} (\lambda I - A)^\beta e^{(s-r)A} f(r) dr$$

is in $C([t,T],H)$.

Proof Part (i): Let $1/p < \alpha \le 1$. Let $t \le s_1 \le s_2 \le T$ and put $h = s_2 - s_1$. We have

$$\left| \int_t^{s_2} (s_2 - r)^{\alpha-1} e^{(s_2-r)A} f(r) dr - \int_t^{s_1} (s_1 - r)^{\alpha-1} e^{(s_1-r)A} f(r) dr \right|$$

$$\le I_1 + I_2 := \int_t^{t+h} \left| (s_2 - r)^{\alpha-1} e^{(s_2-r)A} f(r) \right| dr$$

$$+ \left| \int_{t+h}^{s_2} (s_2 - r)^{\alpha-1} e^{(s_2-r)A} f(r) dr - \int_t^{s_1} (s_1 - r)^{\alpha-1} e^{(s_1-r)A} f(r) dr \right|.$$

Set $q := \frac{p}{p-1}$ and let $R > 0$ be such that $\left\| e^{sA} \right\| \le R$ for all $s \in [0, T]$. Then

$$I_1 \le R \left(\int_0^h (h - r)^{q(\alpha-1)} dr \right)^{1/q} \left(\int_t^T |f(r)|^p dr \right)^{1/p} \to 0 \text{ as } h \to 0$$

since $0 \ge q(\alpha - 1) > -1$. As regards I_2, after a change of variables we have

$$I_2 \le \int_t^{s_1} (s_1 - r)^{\alpha-1} e^{(s_1-r)A} |f(r + h) - f(r)| dr$$

$$\le R \left(\int_t^T (T - r)^{q(\alpha-1)} dr \right)^{1/q} \left(\int_t^{T-h} |f(r + h) - f(r)|^p dr \right)^{1/p} \to 0 \text{ as } h \to 0.$$

The proof in the case $p = \alpha = 1$ is straightforward.

Part (ii) follows from Proposition A.1.1 in Appendix A, p. 307 of [177].   □

Proposition 1.116 *Let $T > 0$ and $t \in [0, T)$. Let A, A_1, Φ satisfy the assumptions of Lemma 1.114 except (1.24). Assume that there exist $0 < \alpha < 1$, $C > 0$ and $p > \frac{1}{\alpha}$, $p \ge 2$ such that*

$$\int_t^T \mathbb{E} \left(\int_t^r \| (r - h)^{-\alpha} A_1 e^{(r-h)A} \Phi(h) \|^2_{\mathcal{L}_2(\Xi_0, H)} dh \right)^{p/2} dr < C. \tag{1.25}$$

Then

$$\Psi(s) := \int_t^s A_1 e^{(s-r)A} \Phi(r) dW_Q(r), \qquad s \in [t, T],$$

has a continuous modification.

Proof We follow the scheme of the proof of Theorem 5.2.6 in [177] (p. 59, Sect. 5.2.1). We give some details because our claim is slightly more general. Observe that using Hölder's and Jensen's inequalities we obtain

$$\int_t^s (s-r)^{\alpha-1} \left(\int_t^r (r-h)^{-2\alpha} \mathbb{E}\left[\left\| A_1 e^{(r-h)A} \Phi(h) \right\|^2_{\mathcal{L}_2(\Xi_0, H)} \right] dh \right)^{1/2} dr$$

$$\leq \left(\int_t^s (s-r)^{\frac{(\alpha-1)p}{p-1}} \right)^{\frac{p-1}{p}} \left(\int_t^s \mathbb{E}\left(\int_t^r (r-h)^{-2\alpha} \left\| A_1 e^{(r-h)A} \Phi(h) \right\|^2_{\mathcal{L}_2(\Xi_0, H)} dh \right)^{p/2} \right)^{\frac{1}{p}}$$

$$< +\infty,$$

where we used (1.25) and that $\frac{(1-\alpha)p}{p-1} < 1$, which follows from $p > 1/\alpha$. Therefore the hypotheses of Lemma 1.114 are satisfied and thus we have

$$\int_t^s A_1 e^{(s-r)A} \Phi(r) dW_Q(r) = \frac{\sin(\alpha\pi)}{\pi} \int_t^s (s-r)^{\alpha-1} e^{(s-r)A} Y_\alpha^\Phi(r) dr \qquad \mathbb{P} - a.s.$$

for all $s \in [t, T]$, where $Y_\alpha^\Phi(\cdot)$ is defined in Lemma 1.114. The claim will follow from Lemma 1.115-(i) applied to a.e. trajectory. Thus we need to know that the process $Y_\alpha^\Phi(\cdot)$ has p-integrable trajectories a.s. This is guaranteed if

$$\mathbb{E} \int_t^T \left| Y_\alpha^\Phi(s) \right|^p ds < +\infty.$$

However, from Theorem 1.111, we have

$$\int_t^T \mathbb{E}\left(\left[\left| Y_\alpha^\Phi(s) \right|^p \right] \right) ds \leq c_p \int_t^T \mathbb{E}\left(\int_t^s \| (s-r)^{-\alpha} A_1 e^{(s-r)A} \Phi(r) \|^2_{\mathcal{L}_2(\Xi_0, H)} dr \right)^{p/2} ds,$$

$$(1.26)$$

which is bounded thanks to (1.25). $\qquad\qquad\qquad\qquad\qquad\qquad\qquad\qquad\qquad\qquad\qquad\qquad\quad\square$

The factorization method can also be used to show the continuity of deterministic convolution integrals. The following lemma deals with a case which arises in Sects. 1.5.2 and 1.5.3.

Lemma 1.117 *Let $T > 0$, $t \in [0, T)$, and $0 < \alpha < 1$. Let A be the generator of a C_0-semigroup $\{e^{rA}, r \geq 0\}$ on H. Let ϕ be a function defined on $[t, T]$ such that, for every $s \in (0, T - t]$, $e^{sA}\phi : [t, T] \to H$ is well defined, measurable and*

$$|e^{sA}\phi(r)| \leq s^{-\beta} g(r) \text{ for } r \in [t, T], \qquad (1.27)$$

where $0 \leq \beta < 1, g \in L^q(t, T; H), q > \frac{1}{1-\beta}$. Then the function

$$\psi(s) = \int_t^s e^{(s-r)A} \phi(r) dr$$

belongs to $C([t, T], H)$.

Proof Let $0 < \alpha$ be such that $\alpha + \beta < 1$ and $q > \frac{1}{1-(\alpha+\beta)}$. We have, by the Fubini Theorem 1.33,

$$\int_t^s e^{(s-r)A}\phi(r)dr = \frac{\sin(\pi\alpha)}{\pi}\int_t^s (s-r)^{\alpha-1}e^{(s-r)A}Y(r)dr,$$

where

$$Y(r) = \int_t^r (r-h)^{-\alpha}e^{(r-h)A}\phi(h)dh.$$

It remains to notice that, using (1.27) and Hölder's inequality, we have for $t \le r \le T$

$$|Y(r)| \le \int_t^r (r-h)^{-(\alpha+\beta)}g(h)dh \le C_T|g|_{L^q(t,T;H)}.$$

Thus the result follows from Lemma 1.115-(i).   □

1.4 Stochastic Differential Equations

In this section we consider $T > 0$ and take H, Ξ, Q, and a generalized reference probability space $\mu = (\Omega, \mathscr{F}, \{\mathscr{F}_s\}_{s\in[0,T]}, \mathbb{P}, W_Q)$ as in Sect. 1.3 (with $t = 0$). A is the infinitesimal generator of a C_0-semigroup on H, and Λ is a Polish space. We will look at stochastic differential equations (SDEs) on the interval $[0, T]$, however all results would be the same if, instead of $[0, T]$, we took an interval $[t, T]$, for $0 \le t < T$.

1.4.1 Mild and Strong Solutions

Let $b\colon [0, T] \times H \times \Omega \to H$ and $\sigma\colon [0, T] \times H \times \Omega \to \mathcal{L}_2(\Xi_0, H)$. We consider the following general stochastic differential equation (SDE)

$$\begin{cases} dX(s) = (AX(s) + b(s, X(s)))ds + \sigma(s, X(s))dW_Q(s) & s \in (0, T] \\ X(0) = \xi, \end{cases} \tag{1.28}$$

where ξ is an H-valued $\mathscr{F}_0$-measurable random variable. To simplify the notation we dropped the ω variable in (1.28) and we use this convention throughout the section.

Definition 1.118 (*Strong solution of* (1.28)) An H-valued progressively measurable process $X(\cdot)$ is called a *strong solution* of (1.28) if:

(i) For $dt \otimes \mathbb{P}$-a.e. $(s, \omega) \in [0, T] \times \Omega$, $X(s)(\omega) \in D(A)$.

(ii) $\mathbb{P}\left(\int_0^T (|X(s)| + |AX(s)| + |b(s, X(s))|)\, ds < +\infty \right) = 1$ and

$$\mathbb{P}\left(\int_0^T \|\sigma(s, X(s))\|^2_{\mathcal{L}_2(\Xi_0, H)}ds < +\infty \right) = 1.$$

(iii) For every $t \in [0, T]$

$$X(t) = \xi + \int_0^t AX(s) + b(s, X(s))ds + \int_0^t \sigma(s, X(s))dW_Q(s) \quad \mathbb{P}\text{-a.e.}$$

Definition 1.119 (*Mild solution of* (1.28)) An H-valued progressively measurable process $X(\cdot)$ is called a *mild solution* of (1.28) if:

(i) For every $t \in [0, T]$

$$\mathbb{P}\left(\int_0^t \left(|X(s)| + |e^{(t-s)A}b(s, X(s))| \right) ds < +\infty \right) = 1$$

and

$$\mathbb{P}\left(\int_0^t \|e^{(t-s)A}\sigma(s, X(s))\|_{\mathcal{L}_2(\Xi_0, H)}^2 ds < +\infty \right) = 1.$$

(ii) For every $t \in [0, T]$

$$X(t) = e^{tA}\xi + \int_0^t e^{(t-s)A}b(s, X(s))ds + \int_0^t e^{(t-s)A}\sigma(s, X(s))dW_Q(s) \quad \mathbb{P}\text{-a.e.}$$

In order for the above definitions to be meaningful, all the processes involved must be well defined and have proper measurability properties so that the integrals that appear in the definitions make sense. We do not want to analyze here the required measurability properties in the most generality. Instead, we discuss one case which will frequently appear in applications to optimal control in Remark 1.123 below. Moreover, note that if A_n is the Yosida approximation of A, since by Lemma 1.17-(i) $D(A) \in \mathcal{B}(H)$, it follows that the processes $\mathbf{1}_{X(\cdot) \in D(A)} A_n X(\cdot)$ are progressively measurable and they converge as $n \to +\infty$ to $\mathbf{1}_{X(\cdot) \in D(A)} AX(\cdot)$ for every (s, ω). Thus the process $AX(\cdot)$ (understood as $\mathbf{1}_{X(\cdot) \in D(A)} AX(\cdot)$) is progressively measurable.

Remark 1.120 In the definition of a mild solution we assumed that $b \colon [0, T] \times H \times \Omega \to H$ and $\sigma \colon [0, T] \times H \times \Omega \to \mathcal{L}_2(\Xi_0, H)$. However, Definition 1.119 may still make sense even if b and σ do not have values in H and $\mathcal{L}_2(\Xi_0, H)$, provided that the terms $e^{(t-s)A}b(s, X(s))$ and $e^{(t-s)A}\sigma(s, X(s))$ have values in these spaces when they are interpreted properly (see, for instance, Sect. 1.5.1 and also Remark 1.123). Therefore in the future when we are dealing with such cases, we will not repeat the definition of a mild solution, instead we will just explain how to interpret the above terms.   ∎

Definition 1.121 (*Weak mild solution of* (1.28)) Assume that in (1.28) we have $b \colon [0, T] \times H \to H$ and $\sigma \colon [0, T] \times H \to \mathcal{L}_2(\Xi_0, H)$. A *weak mild solution* of (1.28) is defined to be any 6-tuple $(\Omega, \mathscr{F}, \mathscr{F}_s, W_Q, \mathbb{P}, X(\cdot))$, where $(\Omega, \mathscr{F}, \mathscr{F}_s, \mathbb{P})$ is a filtered probability space, W_Q is a translated $\mathscr{F}_s$-Q-Wiener process on Ω, and $X(\cdot)$ is a mild solution for (1.28) in the generalized reference probability space $(\Omega, \mathscr{F}, \mathscr{F}_s, W_Q, \mathbb{P})$.

Notation 1.122 In the existing literature, different authors often give different names to the same notion of solution, and the same name does not always correspond to the same definition. For instance, the *weak mild solution* introduced above is often called a weak solution and in [180] Chap. 8 it is called a *martingale solution.* ∎

Remark 1.123 Let Λ be a Polish space. Suppose that $\sigma : [0, T] \times H \times \Lambda \to \mathcal{L}(\Xi_0, H)$ is such that for every $u \in \Xi_0$, the map $(t, x, a) \to \sigma(t, x, a)u$ is $\mathcal{B}([0, T]) \otimes \mathcal{B}(H) \otimes \mathcal{B}(\Lambda)/\mathcal{B}(H)$-measurable, and $e^{sA}\sigma(t, x, a) \in \mathcal{L}_2(\Xi_0, H)$ for every (t, x, a) and $s > 0$. It then follows from Lemma 1.20 that, after possibly redefining it at $s = 0$, the map $(s, t, x, a) \to e^{sA}\sigma(t, x, a)$ is $\mathcal{B}([0, T]) \otimes \mathcal{B}([0, T]) \otimes \mathcal{B}(H) \otimes \mathcal{B}(\Lambda)/\mathcal{B}(\mathcal{L}_2(\Xi_0, H))$-measurable. Now, if $X(\cdot) : [0, T] \times \Omega \to H, a(\cdot) : [0, T] \times \Omega \to \Lambda$ are $\mathscr{F}_s$-progressively measurable, then for every $t \in [0, T]$,

$$(s, \omega) \to e^{(t-s)A}\sigma(s, X(s), a(s))$$

is an $\mathcal{L}_2(\Xi_0, H)$-valued $\mathscr{F}_s$-progressively measurable process on $[0, t] \times \Omega$. If this process is in $\mathcal{N}_Q^2(0, t; H)$ for every t then the process

$$Z(t) = \int_0^t e^{(t-s)A}\sigma(s, X(s), a(s))dW_Q(s), \qquad t \in [0, T]$$

is an H-valued $\mathscr{F}_t$-adapted process. One way to argue that $Z(\cdot)$ has a progressively measurable modification is the following.

Suppose that there is a constant $K \geq 0$ such that

$$\mathbb{E}|Z(t)| \leq K \quad \text{for all } t \in [0, T]$$

and that for all $0 \leq t \leq h \leq T$

$$\mathbb{E}\int_t^h \left\| e^{(h-s)A}\sigma(s, X(s), a(s)) \right\|_{\mathcal{L}_2(\Xi_0, H)}^2 ds \leq \rho(h - t)$$

for some modulus ρ. We have for $0 \leq t \leq h \leq T$

$$Z(h) - Z(t) = \left(e^{(h-t)A} - I\right)Z(t) + \int_t^h e^{(h-s)A}\sigma(s, X(s), a(s))dW_Q(s).$$

Let $\{e_n\}$ be an orthonormal basis of H. Then

$$\langle Z(h) - Z(t), e_n \rangle = \left\langle Z(t), e^{(h-t)A}e_n - e_n \right\rangle + \left\langle \int_t^h e^{(h-s)A}\sigma(s, X(s), a(s))dW_Q(s), e_n \right\rangle$$

and hence

$$\mathbb{E}\,|\langle Z(h) - Z(t), e_n \rangle| \leq K|e^{(h-t)A^*}e_n - e_n| + \sqrt{\rho(h - t)} \leq \rho_n(h - t)$$

for some modulus ρ_n. Therefore it is easy to see that the process $\langle Z(t), e_n \rangle$ is stochastically continuous and thus, by Lemma 1.74, it has a progressively measurable modification which we denote by $Z_n(\cdot)$. The process $\tilde{Z}(\cdot)$ defined, for $t \in [0, T]$, by

$$\tilde{Z}(t) = \begin{cases} \sum_{n=1}^{+\infty} Z_n(t) e_n & \text{if the limit exists,} \\ 0 & \text{otherwise} \end{cases}$$

is a progressively measurable modification of $Z(\cdot)$. ∎

1.4.2 Existence and Uniqueness of Solutions

Definition 1.124 (*The space* $M_\mu^p(t, T; E)$) In this definition $T \in (0, +\infty) \cup \{+\infty\}$. Let $p \geq 1$ and $0 \leq t < T$. Given a Banach space E, we denote by $M_\mu^p(t, T; E)$ the space of all E-valued progressively measurable processes $X(\cdot)$ such that

$$|X(\cdot)|_{M_\mu^p(t,T;E)} := \left(\mathbb{E} \left(\int_t^T |X(s)|^p ds \right) \right)^{1/p} < +\infty. \tag{1.29}$$

$M_\mu^p(t, T; E)$ is a Banach space endowed with the norm $|\cdot|_{M_\mu^p(t,T;E)}$.

Note that in the notation $M_\mu^p(t, T; E)$ we emphasize the dependence on the generalized reference probability space μ. Processes in $M_\mu^p(t, T; E)$ are identified if they are equal $\mathbb{P} \otimes dt$-a.e.

Let $a: [0, T] \times \Omega \to \Lambda$ be an $\mathscr{F}_s$-progressively measurable process (a control process), where Λ is, as before, a Polish space. We consider the controlled SDE

$$\begin{cases} dX(s) = (AX(s) + b(s, X(s), a(s))) \, ds + \sigma(s, X(s), a(s)) dW_Q(s) \\ X(0) = \xi. \end{cases} \tag{1.30}$$

This equation falls into the category of equations (1.28) with $b(s, x, \omega) := b(s, x, a(s, \omega))$ and $\sigma(s, x, \omega) := \sigma(s, x, a(s, \omega))$. Thus strong, mild and weak mild solutions of (1.30) are defined using the definitions for Eq. (1.28).

Hypothesis 1.125 The operator A is the generator of a strongly continuous semigroup e^{sA} on H. The function $b: [0, T] \times H \times \Lambda \to H$ is $\mathcal{B}([0, T]) \otimes \mathcal{B}(H) \otimes \mathcal{B}(\Lambda)/\mathcal{B}(H)$-measurable, $\sigma: [0, T] \times H \times \Lambda \to \mathcal{L}_2(\Xi_0, H)$ is $\mathcal{B}([0, T]) \otimes \mathcal{B}(H) \otimes \mathcal{B}(\Lambda)/\mathcal{B}(\mathcal{L}_2(\Xi_0, H))$-measurable, and there exists a constant $C > 0$ such that

$$|b(s, x, a) - b(s, y, a)| \leq C|x - y| \qquad \forall x, y \in H, s \in [0, T], a \in \Lambda, \tag{1.31}$$

$$\|\sigma(s, x, a) - \sigma(s, y, a)\|_{\mathcal{L}_2(\Xi_0, H)} \leq C|x - y| \qquad \forall x, y \in H, s \in [0, T], a \in \Lambda, \tag{1.32}$$

$$|b(s, x, a)| \leq C(1 + |x|) \qquad \forall x \in H, s \in [0, T], a \in \Lambda, \tag{1.33}$$

$$\|\sigma(s, x, a)\|_{\mathcal{L}_2(\Xi_0, H)} \leq C(1 + |x|) \qquad \forall x \in H, s \in [0, T], a \in \Lambda. \tag{1.34}$$

Definition 1.126 (*The space* $\mathcal{H}_p^\mu(t, T; E)$) Let $p \geq 1$ and $0 \leq t < T$. Given a Banach space E, we denote by $\mathcal{H}_p^\mu(t, T; E)$ the set of all progressively measurable processes $X \colon [t, T] \times \Omega \to E$ such that

$$|X(\cdot)|_{\mathcal{H}_p^\mu(t,T;E)} := \left(\sup_{s \in [t,T]} \mathbb{E}|X(s)|^p \right)^{1/p} < +\infty. \tag{1.35}$$

It is a Banach space with the norm $|\cdot|_{\mathcal{H}_p^\mu(t,T;E)}$.

Processes in $\mathcal{H}_p^\mu(t, T; E)$ are identified if they are equal $\mathbb{P} \otimes dt$-a.e. Therefore the *sup* in the definition of $\mathcal{H}_p^\mu(t, T; E)$ must be understood as *esssup*. However, we will keep the notation *sup* here and in all subsequent uses of this space. If the generalized reference probability space μ is clear we will just write $M^p(t, T; E)$ and $\mathcal{H}_p(t, T; E)$ for simplicity.

Mild solutions in $\mathcal{H}_p^\mu(0, T; E)$ (or $M_\mu^p(0, T; E)$) of various versions of (1.30) will be obtained as fixed points in these spaces of some maps. We point out that this will not imply that every representative of the equivalence class is a mild solution. Since a mild solution $X(\cdot)$ satisfies the integral equality in Definition 1.119-(ii) for every $t \in [0, T]$, $X(t)$ is prescribed by the right-hand side of this equality, which does not depend on the choice of a representative of the equivalence class. Thus there is a unique (up to a modification) representative of the equivalence class which is a mild solution. We will then always be able to evaluate $\mathbb{E}|X(t)|^p$ for the mild solution $X(\cdot)$ for every $t \in [0, T]$ (and in fact compute the $\mathcal{H}_p^\mu(0, T; E)$ norm of this representative by taking the *sup* over all $t \in [0, T]$ instead of the *esssup*).

Theorem 1.127 *Let* $\xi \in L^p(\Omega, \mathscr{F}_0, \mathbb{P})$ *for some* $p \geq 2$, *and let* A, b *and* σ *satisfy Hypothesis 1.125. Let* $a(\cdot) \colon [0, T] \to \Lambda$ *be an* $\mathscr{F}_s$*-progressively measurable process. Then the SDE (1.30) has a unique, up to a modification, mild solution* $X(\cdot) \in \mathcal{H}_p(0, T; H)$. *The solution is in fact unique among all processes such that* $\mathbb{P}\left(\int_0^T |X(s)|^2 ds < +\infty \right) = 1$, *in particular among the processes in* $M_\mu^2(0, T; H)$. $X(\cdot)$ *has a continuous modification. Given two continuous versions* $X_1(\cdot)$, $X_2(\cdot)$ *of the solution, there exists a* $\tilde{\Omega} \subset \Omega$ *with* $\mathbb{P}(\tilde{\Omega}) = 1$ *s.t.* $X_1(s) = X_2(s)$ *for all* $s \in [0, T]$ *and* $\omega \in \tilde{\Omega}$, *i.e. they are indistinguishable.*

Proof The proof can be found, for instance, in [180], Theorem 7.2, p. 188 or [294], Theorems 3.3, p. 97, and 3.5, p. 105. For the last claim, we can take

$$\tilde{\Omega} := \bigcap_{s \in \mathbb{Q} \cap [0,T]} \{\omega \in \Omega \: : \: X_1(s)(\omega) = X_2(s)(\omega)\}.$$

Since $X_1(\cdot)$ is a modification of $X_2(\cdot)$, we have $\mathbb{P}(\tilde{\Omega}) = 1$, and since $X_1(\cdot)$ and $X_2(\cdot)$ are continuous, it follows that $X_1(s)(\omega) = X_2(s)(\omega)$ for all $s \in [0, T]$, $\omega \in \tilde{\Omega}$. $\square$

We will denote the solution of (1.30) by $X(\cdot; \xi, a(\cdot))$ if we want to emphasize the dependence on the initial datum and the control.

Corollary 1.128 *Let $\xi \in L^p(\Omega, \mathscr{F}_0, \mathbb{P})$ for some $p \geq 2$, let A, b and σ satisfy Hypothesis 1.125. If $a_1(\cdot), a_2(\cdot) \colon [0, T] \times \Omega \to \Lambda$ are two progressively measurable processes such that $a_1(\cdot) = a_2(\cdot)$, $dt \otimes \mathbb{P}$-a.e. on $[0, T] \times \Omega$, then, $\mathbb{P} - a.e.$,*

$$X(s; \xi, a_1(\cdot)) = X(s; \xi, a_2(\cdot)) \quad \text{for all } s \in [0, T].$$

Proof Define $X_i(\cdot) := X(\cdot; \xi, a_i(\cdot))$. Using Theorem 1.103, Jensen's inequality, and $\sup_{s \in [0,T]} \|e^{sA}\| \leq C$ for some $C \geq 0$, it follows that, for suitable positive C_1 and C_2:

$$\mathbb{E}\left[|X_1(s) - X_2(s)|^2\right] \leq C_1 \left(\int_0^s \mathbb{E}|b(r, X_1(r), a_1(r)) - b(r, X_2(r), a_2(r))|^2 dr \right.$$

$$\left. + \int_0^s \mathbb{E}\|\sigma(r, X_1(r), a_1(r)) - \sigma(r, X_2(r), a_2(r))\|^2_{\mathcal{L}_2(\Xi_0, H)} dr \right)$$

$$\leq C_2 \int_0^s \mathbb{E}|X_1(r) - X_2(r)|^2 dr, \qquad s \in [0, T],$$

and the claim follows by using Gronwall's lemma and the continuity of the trajectories. $\square$

Remark 1.129 Above we assumed that the σ always takes values in $\mathcal{L}_2(\Xi_0, H)$. Existence and uniqueness results for SDEs with more general σ can be found, for instance, in [294] Theorem 3.15, p. 143, or in [180] Theorem 7.5, p. 197. To treat some specific examples we will also prove more general results in Sect. 1.5. $\blacksquare$

1.4.3 Properties of Solutions

Theorem 1.130 *Let $\xi \in L^p(\Omega, \mathscr{F}_0, \mathbb{P})$ for some $p \geq 2$, $a \colon [0, T] \times \Omega \to \Lambda$ be $\mathscr{F}_s$-progressively measurable, and let A, b and σ satisfy Hypothesis 1.125.*

(i) Let $X(\cdot) = X(\cdot; \xi, a(\cdot))$ be the unique mild solution of (1.30) (provided by Theorem 1.127). Then, for any $s \in [0, T]$,

$$\sup_{s\in[0,T]} \mathbb{E}\left[|X(s)|^p\right] \le C_p(T)(1+\mathbb{E}|\xi|^p) \quad \text{if } p \ge 2, \tag{1.36}$$

$$\mathbb{E}\left[\sup_{s\in[0,T]} |X(s)|^p\right] \le C_p(T)(1+\mathbb{E}|\xi|^p) \quad \text{if } p > 2, \tag{1.37}$$

and

$$\mathbb{E}\left[\sup_{r\in[0,s]} |X(r)-\xi|^p\right] \le \omega_\xi(s) \quad \text{if } p > 2, \tag{1.38}$$

where $C_p(T)$ is a constant depending on p, T, C (from Hypothesis 1.125) and M, α (where $\|e^{rA}\| \le Me^{r\alpha}$ for $r \ge 0$), and ω_ξ is a modulus depending on the same constants and on ξ (in particular they are independent of the process $a(\cdot)$ and of the generalized reference probability space).

(ii) *If $\xi, \eta \in L^p(\Omega, \mathscr{F}_0, \mathbb{P})$ for $p > 2$, and $X(\cdot) = X(\cdot; \xi, a(\cdot))$, $Y(\cdot) = Y(\cdot; \eta, a(\cdot))$ are the solutions of (1.30), then, for all $s \in [0, T]$,*

$$\mathbb{E}\left[\sup_{s\in[0,T]} |X(s)-Y(s)|^2\right] \le C_T \left(\mathbb{E}\left[|\xi-\eta|^p\right]\right)^{\frac{2}{p}}, \tag{1.39}$$

where C_T depends only on p, T, C, M, α.

Proof Part (i): For (1.36) and (1.37) we refer, for instance, to [180] Theorem 9.1, p. 235, or [294], Lemma 3.6, p. 102, and Corollary 3.3, p. 104. Regarding (1.38), we have that there is a constant c_1 depending only on p and $\sup_{t\in[0,T]} \|e^{tA}\|$, such that

$$\mathbb{E}\left[\sup_{r\in[0,s]} |X(r)-\xi|^p\right] \le c\left(\mathbb{E}\left[\sup_{r\in[0,s]} |e^{rA}\xi-\xi|^p\right]\right.$$
$$+ \mathbb{E}\left[\sup_{r\in[0,s]} \left(\int_0^r |b(u, X(u), a(u))| du\right)^p\right]$$
$$\left.+ \mathbb{E}\left[\sup_{r\in[0,s]} \left|\int_0^r e^{(r-u)A}\sigma(u, X(u), a(u))dW_Q(u)\right|^p\right]\right).$$

Using Hypothesis 1.125, (1.37), Hölder's inequality, and Proposition 1.112, we see that

$$\mathbb{E}\left[\sup_{r\in[0,s]} |X(r)-\xi|^p\right] \le c_2\left(\mathbb{E}\left[\sup_{r\in[0,s]} |e^{rA}\xi-\xi|^p\right] + \int_0^s (1+\mathbb{E}|\xi|^p)\, dr\right).$$

Since $\sup_{r\in[0,s]} |e^{rA}\xi-\xi|^p \xrightarrow{s\to 0^+} 0$ a.e., and $\sup_{r\in[0,s]} |e^{rA}\xi-\xi|^p \le C_1|\xi|^p$, the result follows by the Lebesgue dominated convergence theorem.

Part (ii): See [180] Theorem 9.1, p. 235. □

Theorem 1.131 *Let $\xi \in L^p(\Omega, \mathscr{F}_0, \mathbb{P})$ for some $p > 2$, and let A, b and σ satisfy Hypothesis 1.125. Let $a: [0, T] \times \Omega \to \Lambda$ be a progressively measurable process. Let $X(\cdot)$ be the unique mild solution of (1.30). Consider the approximating equations*

$$\begin{cases} dX^n(s) = (A_n X^n(s) + b(s, X^n(s), a(s))) \, ds + \sigma(s, X^n(s), a(s)) dW_Q(s) \\ X^n(0) = \xi, \end{cases}$$

$$(1.40)$$

where A_n is the Yosida approximation of A. Let $X_n(\cdot)$ be the solution of (1.40). Then

$$\lim_{n \to \infty} \mathbb{E} \left[\sup_{s \in [0,T]} |X^n(s) - X(s)|^p \right] = 0. \qquad (1.41)$$

Proof See [180] Proposition 7.4, p. 196, or [294], Proposition 3.2, p. 101. □

The next proposition is a simpler version of Theorem 1.131 which will be useful in the proofs of the results of Sect. 1.7.

Proposition 1.132 *Let $\xi \in L^p(\Omega, \mathscr{F}_0, \mathbb{P})$, $f \in M_\mu^p(0, T; H)$, and $\Phi \in \mathcal{N}_Q^p(0, T; H)$ for some $p \geq 2$. Let $X(\cdot)$ be the mild solution of*

$$\begin{cases} dX(s) = (AX(s) + f(s)) \, ds + \Phi(s) dW_Q(s) \\ X(0) = \xi \end{cases} \qquad (1.42)$$

and $X^n(\cdot)$ be the solution of

$$\begin{cases} dX^n(s) = (A_n X^n(s) + f(s)) \, ds + \Phi(s) dW_Q(s) \\ X^n(0) = \xi, \end{cases} \qquad (1.43)$$

where A generates a C_0-semigroup and A_n is the Yosida approximation of A. Then, if $p > 2$,

$$\lim_{n \to \infty} \mathbb{E} \left[\sup_{s \in [0,T]} |X^n(s) - X(s)|^p \right] = 0. \qquad (1.44)$$

Moreover, for $p \geq 2$, there exists an $M > 0$, independent of n, such that

$$\sup_{s \in [0,T]} \mathbb{E} \left[|X^n(s)|^p \right] \leq M, \qquad \sup_{s \in [0,T]} \mathbb{E} \left[|X(s)|^p \right] \leq M. \qquad (1.45)$$

Proof Observe first that the mild solution of (1.42) is well defined thanks to the assumptions on ξ, f and Φ, and

$$X(s) = e^{sA} \xi + \int_0^s e^{(s-r)A} f(r) dr + \int_0^s e^{(s-r)A} \Phi(r) dW_Q(r), \qquad s \in [0, T].$$

The same is true for the mild solution of (1.43) (which is also a strong solution).
  To prove (1.44), we write, for $s \in [0, T]$,

$$X^n(s) - X(s) = \left(e^{sA_n} - e^{sA}\right)\xi + \int_0^s \left(e^{(s-r)A_n} - e^{(s-r)A}\right) f(r)dr$$

$$+ \int_0^s \left(e^{(s-r)A_n} - e^{(s-r)A}\right) \Phi(r)dW_Q(r) =: I_1^n(s) + I_2^n(s) + I_3^n(s).$$

It is enough to show that $\lim_{n\to\infty} \mathbb{E}\left[\sup_{s\in[0,T]} |I_i^n(s)|^p\right] = 0$ for $i \in \{1, 2, 3\}$. For $i = 3$ this follows from (1.21). To prove it for $i = 2$, we observe that (B.15) implies that if

$$\psi_n(r) := \sup_{s\in[r,T]} \left|\left(e^{(s-r)A_n} - e^{(s-r)A}\right) f(r)\right|,$$

then $\psi_n(r) \xrightarrow{n\to\infty} 0$ a.e. on Ω. Moreover, thanks to (B.14), there exists a C_1 such that, for all $t \in [0, T]$ and all n, $\left\|e^{tA_n}\right\| \le C_1$, so $\psi_n(r) \le 2C_1|f(r)|$ for all n. Since $\int_t^T |f(r)|dr < +\infty$ for almost every $\omega \in \Omega$, by the Lebesgue dominated convergence theorem we have

$$\sup_{s\in[0,T]} \left|\int_0^s \left|\left(e^{(s-r)A_n} - e^{(s-r)A}\right) f(r)\right| dr\right|^p$$

$$\le \sup_{s\in[0,T]} \left|\int_0^s \psi_n(r)dr\right|^p \le \left|\int_0^T \psi_n(r)dr\right|^p \xrightarrow{n\to\infty} 0$$

for a.e. $\omega \in \Omega$. Now observe that

$$\sup_{s\in[0,T]} \left|\int_0^s \left|\left(e^{(s-r)A_n} - e^{(s-r)A}\right) f(r)\right| dr\right|^p$$

$$\le \sup_{s\in[0,T]} \int_0^s (2C_1)^p |f(r)|^p dr \le \int_0^T (2C_1)^p |f(r)|^p dr,$$

and the last expression is integrable (on Ω), since $f \in M_\mu^p(0, T; H)$. Therefore we can apply the Lebesgue dominated convergence theorem, obtaining $\lim_{n\to\infty} \mathbb{E}\left[\sup_{s\in[0,T]} |I_2^n(s)|^p\right] = 0$. The claim for $i = 1$ follows again from (B.15) and the Lebesgue dominated convergence theorem.

Estimates (1.45) are easy consequences of (B.14) and the assumptions on ξ, f, Φ. $\qquad\square$

1.4.4 Uniqueness in Law

Definition 1.133 (*Finite-dimensional distributions*) Let $T > 0$ and $t \in [0, T)$. Consider a measurable space $(\Omega, \mathscr{F})$, two probability spaces $(\Omega_i, \mathscr{F}_i, \mathbb{P}_i)$ for $i = 1, 2$, and two processes $\{X_i(s)\}_{s\in[t,T]} : (\Omega_i, \mathscr{F}_i, \mathbb{P}_i) \to (\Omega, \mathscr{F})$. We say that $X_1(\cdot)$ and $X_2(\cdot)$ have the same *finite-dimensional distributions* on $D \subset [t, T]$ if for any

$t \leq t_1 < t_2 < ... < t_n \leq T, t_i \in D$ and $A \in \underbrace{\mathscr{F} \otimes \mathscr{F} \otimes ... \otimes \mathscr{F}}_{n \text{ times}}$, we have

$$\mathbb{P}_1 \{\omega_1 : (X_1(t_1), ..X_1(t_n))(\omega_1) \in A\} = \mathbb{P}_2 \{\omega_2 : (X_2(t_1), ..X_2(t_n))(\omega_2) \in A\}.$$

In this case we write $\mathcal{L}_{\mathbb{P}_1}(X_1(\cdot)) = \mathcal{L}_{\mathbb{P}_2}(X_2(\cdot))$ on D. Often we will just write $\mathcal{L}_{\mathbb{P}_1}(X_1(\cdot)) = \mathcal{L}_{\mathbb{P}_2}(X_2(\cdot))$, which should be understood as meaning that the finite-dimensional distributions are the same on some set of full measure.

Theorem 1.134 *Let H be a separable Hilbert space. Let $(\Omega_i, \mathscr{F}_i, \mathbb{P}_i)$ for $i = 1, 2$ be two complete probability spaces, and $(\tilde{\Omega}, \tilde{\mathscr{F}})$ be a measurable space. Let $\xi_i : \Omega_i \to \tilde{\Omega}, i = 1, 2$ be two random variables, and $f_i : [t, T] \times \Omega_i \to H, i = 1, 2$, be two processes satisfying*

$$\mathbb{P}_1 \left(\int_t^T |f_1(s)| ds < +\infty \right) = \mathbb{P}_2 \left(\int_t^T |f_2(s)| ds < +\infty \right) = 1$$

and, for some subset $D \subset [t, T]$ of full measure,

$$\mathcal{L}_{\mathbb{P}_1} (f_1(\cdot), \xi_1) = \mathcal{L}_{\mathbb{P}_2} (f_2(\cdot), \xi_2) \quad \text{on } D.$$

Then

$$\mathcal{L}_{\mathbb{P}_1} \left(\int_t^{\cdot} f_1(s) ds, \xi_1 \right) = \mathcal{L}_{\mathbb{P}_2} \left(\int_t^{\cdot} f_2(s) ds, \xi_2 \right) \quad \text{on } [t, T]. \qquad (1.46)$$

Proof See [471] Theorem 8.3, where the theorem was proved for a more general case of Banach space-valued processes. □

Theorem 1.135 *Let $\left(\Omega_1, \mathscr{F}_1, \mathscr{F}_s^{1,t}, \mathbb{P}_1, W_{Q,1}\right)$ and $\left(\Omega_2, \mathscr{F}_2, \mathscr{F}_s^{2,t}, \mathbb{P}_2, W_{Q,2}\right)$ be two generalized reference probability spaces. Let $\Phi_i : [t, T] \times \Omega_i \to \mathcal{L}_2(\Xi_0, H)$, $i = 1, 2$, be two $\mathscr{F}_s^{i,t}$-progressively measurable processes satisfying*

$$\mathbb{P}_1 \left(\int_t^T \|\Phi_1(s)\|^2_{\mathcal{L}_2(\Xi_0, H)} ds < +\infty \right) = \mathbb{P}_2 \left(\int_t^T \|\Phi_2(s)\|^2_{\mathcal{L}_2(\Xi_0, H)} ds < +\infty \right) = 1.$$

Let $(\tilde{\Omega}, \tilde{\mathscr{F}})$ be a measurable space and $\xi_i : \Omega_i \to \tilde{\Omega}, i = 1, 2$, be two random variables. Assume that, for some subset $D \subset [t, T]$ of full measure,

$$\mathcal{L}_{\mathbb{P}_1} \left(\Phi_1(\cdot), W_{Q,1}(\cdot), \xi_1 \right) = \mathcal{L}_{\mathbb{P}_2} \left(\Phi_2(\cdot), W_{Q,2}(\cdot), \xi_2 \right) \quad \text{on } D.$$

Then

$$\mathcal{L}_{\mathbb{P}_1} \left(\int_t^{\cdot} \Phi_1(s) dW_{Q,1}(s), \xi_1 \right) = \mathcal{L}_{\mathbb{P}_2} \left(\int_t^{\cdot} \Phi_2(s) dW_{Q,2}(s), \xi_2 \right) \quad \text{on } [t, T]. \quad (1.47)$$

Proof See [471] Theorem 8.6. □

Consider now an operator A and mappings b, σ satisfying Hypothesis 1.125, and $x \in H$. Let $\left(\Omega_1, \mathscr{F}_1, \mathscr{F}_s^{1,t}, \mathbb{P}_1, W_{Q,1}\right)$ and $\left(\Omega_2, \mathscr{F}_2, \mathscr{F}_s^{2,t}, \mathbb{P}_2, W_{Q,2}\right)$ be as in Theorem 1.135. For $i = 1, 2$ consider an $\mathscr{F}_s^{i,t}$-progressively measurable process $a_i : [t, T] \times \Omega_i \to \Lambda$.

Let $p > 2$ and let $\zeta_i \in L^p(\Omega_i, \mathscr{F}_t^{i,t}, \mathbb{P}_i)$, $i = 1, 2$. Denote by $\mathcal{H}_{p,i}$ the Banach space of all $\mathscr{F}_s^{i,t}$-progressively measurable processes $Z_i : [t, T] \times \Omega_i \to H$ such that

$$\left(\sup_{s \in [t,T]} \mathbb{E}_i |Z_i(s)|^p \right)^{1/p} < +\infty.$$

Let $\mathcal{K}_i : \mathcal{H}_{p,i} \to \mathcal{H}_{p,i}$ be the continuous map (see [180], p. 189) defined as

$$\mathcal{K}_i(Z_i(\cdot))(s) := e^{(s-t)A}\zeta_i + \int_t^s e^{(s-r)A} b(r, Z_i(r), a_i(r)) dr$$
$$+ \int_t^s e^{(s-r)A} \sigma(r, Z_i(r), a_i(r)) dW_{Q,i}(r).$$
(1.48)

Lemma 1.136 *Consider the setting described above, and let $\theta_i : [t, T] \times \Omega_i \to H, i = 1, 2$, be stochastic processes. If*

$$\mathcal{L}_{\mathbb{P}_1}(Z_1(\cdot), a_1(\cdot), W_{Q,1}(\cdot), \theta_1(\cdot), \zeta_1) = \mathcal{L}_{\mathbb{P}_2}(Z_2(\cdot), a_2(\cdot), W_{Q,2}(\cdot), \theta_2(\cdot), \zeta_2)$$

on some subset $D \subset [t, T]$ of full measure, then

$$\mathcal{L}_{\mathbb{P}_1}(\mathcal{K}_1(Z_1(\cdot))(\cdot), a_1(\cdot), W_{Q,1}(\cdot), \theta_1(\cdot), \zeta_1)$$
$$= \mathcal{L}_{\mathbb{P}_2}(\mathcal{K}_2(Z_2(\cdot))(\cdot), a_2(\cdot), W_{Q,2}(\cdot), \theta_2(\cdot), \zeta_2) \quad \text{on } D.$$

Proof Observe that, since we only have to check the finite-dimensional distributions, the claims of Theorems 1.134 and 1.135 hold even if ξ_1 and ξ_2 are stochastic processes, with (1.46) and (1.47) then being true on some set of full measure. Let us choose a partition $(t_1, .., t_n)$, with $t \le t_1 < t_2 < ... < t_n \le T, t_k \in D, k = 1, ..., n$. We need to show that

$$\mathcal{L}_{\mathbb{P}_1}(\mathcal{K}_1(Z_1(\cdot))(t_k), a_1(t_k), W_{Q,1}(t_k), \theta_1(t_k), \zeta_1 : k = 1, ..., n)$$
$$= \mathcal{L}_{\mathbb{P}_2}(\mathcal{K}_2(Z_2(\cdot))(t_k), a_2(t_k), W_{Q,1}(t_k), \theta_2(t_k), \zeta_2 : k = 1, ..., n).$$
(1.49)

Define $f^i(r) := \mathbf{1}_{[t,t_1]}(r) e^{(t_1-r)A} b(r, Z_i(r), a_i(r))$ and $\Phi^i(r) := \mathbf{1}_{[t,t_1]}(r) e^{(t_1-r)A} \sigma(r, Z_i(r), a_i(r)), i = 1, 2$. We have

$$\mathcal{L}_{\mathbb{P}_1}(f^1(\cdot), \Phi_1(\cdot), Z_1(\cdot), a_1(\cdot), W_{Q,1}(\cdot), \theta_1(\cdot), \zeta_1)$$
$$= \mathcal{L}_{\mathbb{P}_2}(f^2(\cdot), \Phi_2(\cdot), Z_2(\cdot), a_2(\cdot), W_{Q,2}(\cdot), \theta_2(\cdot), \zeta_2) \text{ on } D,$$

and thus, by Theorem 1.134 applied with

$$\xi_1(\cdot) = (f^1(\cdot), \Phi^1(\cdot), Z_1(\cdot), a_1(\cdot), W_{Q,1}(\cdot), \theta_1(\cdot), \zeta_1),$$

$$\xi_2(\cdot) = (f^2(\cdot), \Phi^2(\cdot), Z_2(\cdot), a_2(\cdot), W_{Q,2}(\cdot), \theta_2(\cdot), \zeta_2),$$

$$\mathcal{L}_{\mathbb{P}_1}\left(\int_t^{t_1} f^1(s)ds, f^1(\cdot), \Phi^1(\cdot), Z_1(\cdot), a_1(\cdot), W_{Q,1}(\cdot), \theta_1(\cdot), \zeta_1\right)$$
$$= \mathcal{L}_{\mathbb{P}_2}\left(\int_t^{t_1} f^2(s)ds, f^2(\cdot), \Phi^2(\cdot), Z_2(\cdot), a_2(\cdot), W_{Q,2}(\cdot), \theta_2(\cdot), \zeta_2\right) \text{ on } D.$$

Now, applying Theorem 1.135 with

$$\xi_1(\cdot) = \left(\int_t^{t_1} f^1(s)ds, f^1(\cdot), \Phi^1(\cdot), Z_1(\cdot), a_1(\cdot), W_{Q,1}(\cdot), \theta_1(\cdot), \zeta_1\right),$$

$$\xi_2(\cdot) = \left(\int_t^{t_1} f^2(s)ds, f^2(\cdot), \Phi^2(\cdot), Z_2(\cdot), a_2(\cdot), W_{Q,2}(\cdot), \theta_2(\cdot), \zeta_2\right),$$

we obtain

$$\mathcal{L}_{\mathbb{P}_1}\left(\int_t^{t_1} f^1(s)ds, \int_t^{t_1} \Phi^1(s)dW_{Q,1}(s), f^1(\cdot), \Phi^1(\cdot), Z_1(\cdot), a_1(\cdot), W_{Q,1}(\cdot), \theta_1(\cdot), \zeta_1\right)$$
$$= \mathcal{L}_{\mathbb{P}_2}\left(\int_t^{t_1} f^2(s)ds, \int_t^{t_1} \Phi^2(s)dW_{Q,2}(s), f^2(\cdot), \Phi^2(\cdot), Z_2(\cdot), a_2(\cdot), W_{Q,2}(\cdot), \theta_2(\cdot), \zeta_2\right)$$

on D (we recall that the stochastic convolution terms in (1.48) and the stochastic integrals above have continuous trajectories a.e.). In particular, this implies that

$$\mathcal{L}_{\mathbb{P}_1}(\mathcal{K}_1(Z_1(\cdot))(t_1), f^1(\cdot), \Phi^1(\cdot), Z_1(\cdot), a_1(\cdot), W_{Q,1}(\cdot), \theta_1(\cdot), \zeta_1)$$
$$= \mathcal{L}_{\mathbb{P}_2}(\mathcal{K}_2(Z_2(\cdot))(t_1), f^2(\cdot), \Phi^2(\cdot), Z_2(\cdot), a_2(\cdot), W_{Q,2}(\cdot), \theta_2(\cdot), \zeta_2) \text{ on } D.$$

We now repeat the above procedure for $t_2, ..., t_n$ which will yield (1.49) as its consequence.   □

Proposition 1.137 *Let the operator A and the mappings b, σ satisfy Hypothesis 1.125. Let $\left(\Omega_1, \mathcal{F}_1, \mathcal{F}_s^{1,t}, \mathbb{P}_1, W_{Q,1}\right)$ and $\left(\Omega_2, \mathcal{F}_2, \mathcal{F}_s^{2,t}, \mathbb{P}_2, W_{Q,2}\right)$ be two generalized reference probability spaces. Let $a_i : [t, T] \times \Omega_i \to \Lambda, i = 1, 2$ be an $\mathcal{F}_s^{i,t}$- progressively measurable process, and let $\zeta_i \in L^p(\Omega_i, \mathcal{F}_t^{i,t}, \mathbb{P}_i), i = 1, 2, p > 2$. Let $\mathcal{L}_{\mathbb{P}_1}(a_1(\cdot), W_{Q,1}(\cdot), \zeta_1) = \mathcal{L}_{\mathbb{P}_2}(a_2(\cdot), W_{Q,1}(\cdot), \zeta_2)$ on some subset $D \subset [0, T]$ of full measure. Denote by $X_i(\cdot), i = 1, 2$, the unique mild solution of*

$$\begin{cases} dX_i(s) = (AX_i(s) + b(s, X_i(s), a_i(s))) \, ds + \sigma(s, X_i(s), a_i(s)) dW_{Q,i}(s) \\ X_i(t) = \zeta_i \end{cases}$$

(1.50)

on $[t, T]$. *Then* $\mathcal{L}_{\mathbb{P}_1}(X_1(\cdot), a_1(\cdot)) = \mathcal{L}_{\mathbb{P}_2}(X_2(\cdot), a_2(\cdot))$ *on* D.

Proof It is known (see [180], proof of Theorem 7.2, pp. 188–193) that the map $\mathcal{K}_i$ is a contraction in $\mathcal{H}_{p,i}$ if $[t, T]$ is small enough. Thus if we divide $[t, T]$ into such small intervals $[t, T_1], \dots[T_k, T]$, $X_i(\cdot)$ on $[t, T_1]$ is obtained as the limit in $\mathcal{H}_{p,i}$ (restricted to $[t, T_1]$) of the iterates $(\mathcal{K}_i^n(x))(\cdot)$. Therefore, using Lemma 1.136 and passing to the limit as $n \to +\infty$ we obtain

$$\mathcal{L}_{\mathbb{P}_1}(\mathbf{1}_{[t,T_1]}(\cdot)X_1(\cdot), a_1(\cdot), W_{Q,1}(\cdot)) = \mathcal{L}_{\mathbb{P}_2}(\mathbf{1}_{[t,T_1]}(\cdot)X_2(\cdot), a_2(\cdot), W_{Q,1}(\cdot)) \text{ on } D.$$

Without loss of generality we may assume that $T_1 \in D$. The solutions on $[T_1, T_2]$ are obtained as the limits in $\mathcal{H}_{p,i}$ (restricted to $[T_1, T_2]$) of the iterates $(\mathcal{K}_i^n(X_i(T_1)))(\cdot)$, where now

$$\mathcal{K}_i(Z_i(\cdot))(s) := e^{(s-T_1)A} X_i(T_1) + \int_{T_1}^s e^{(s-r)A} b(r, Z_i(r), a_i(r)) dr$$

$$+ \int_{T_1}^s e^{(s-r)A} \sigma(r, Z_i(r), a_i(r)) dW_{Q,i}(r).$$

Thus, again using Lemma 1.136 and passing to the limit as $n \to +\infty$, it follows that

$$\mathcal{L}_{\mathbb{P}_1}(\mathbf{1}_{[t,T_2]}(\cdot)X_1(\cdot), a_1(\cdot), W_{Q,1}(\cdot)) = \mathcal{L}_{\mathbb{P}_2}(\mathbf{1}_{[t,T_2]}(\cdot)X_2(\cdot), a_2(\cdot), W_{Q,1}(\cdot)) \text{ on } D.$$

We repeat the procedure to obtain the required claim. $\qquad\square$

1.5 Further Existence and Uniqueness Results in Special Cases

Throughout this section $T > 0$ is a fixed constant, H, Ξ, Q, and the generalized reference probability space $\mu = (\Omega, \mathscr{F}, \{\mathscr{F}_s\}_{s \in [0,T]}, \mathbb{P}, W_Q)$ are as in Sect. 1.3 (with $t = 0$), A is the infinitesimal generator of a C_0-semigroup on H, and Λ is a Polish space. As in previous sections we will only consider equations on the interval $[0, T]$, however all results would be the same if instead of $[0, T]$ we took an interval $[t, T]$, for $0 \le t < T$.

1.5.1 SDEs Coming from Boundary Control Problems

In this section we study SDEs that include equations coming from optimal control
problems with boundary control and noise. To see how they arise the reader can look
at the examples in Sects. 2.6.2 and 2.6.3, and Appendix C. We consider the following
SDE in H:

$$\begin{cases} dX(s) = \big(AX(s) + b(s, X(s), a(s)) + (\lambda I - A)^{\beta} Ga_b(s)\big)\, ds \\ \qquad\qquad\qquad + \sigma(s, X(s), a(s)) dW_Q(s), \qquad s \in (0, T] \quad (1.51) \\ X(0) = \xi. \end{cases}$$

Hypothesis 1.138

(i) A generates an analytic semigroup e^{tA} for $t \geq 0$ and λ is a real constant such
   that $(\lambda I - A)^{-1} \in \mathcal{L}(H)$.
(ii) $a : [0, T] \times \Omega \to \Lambda$ is progressively measurable, $b(\cdot, \cdot, \cdot)$ satisfies (1.31) and
   (1.33).
(iii) Λ_b is a Hilbert space and $a_b(\cdot) : [0, T] \times \Omega \to \Lambda_b$ is progressively measurable.
(iv) $G \in \mathcal{L}(\Lambda_b, H)$.
(v) $\beta \in [0, 1)$.
(vi) γ is a constant belonging to the interval $\left[0, \frac{1}{2}\right)$, σ is a mapping such that $(\lambda I - A)^{-\gamma}\sigma : [0, T] \times H \times \Lambda_b \to \mathcal{L}_2(\Xi_0, H)$ is continuous. There exists a constant
   $C > 0$ such that

$$\|(\lambda I - A)^{-\gamma}\sigma(s, x, a)\|_{\mathcal{L}_2(\Xi_0, H)} \leq C(1 + |x|)$$

   for all $s \in [0, T]$, $x \in H$, $a \in \Lambda$ and

$$\|(\lambda I - A)^{-\gamma}[\sigma(s, x_1, a) - \sigma(s, x_2, a)]\|_{\mathcal{L}_2(\Xi_0, H)} \leq C|x_1 - x_2|$$

   for all $s \in [0, T]$, $x_1, x_2 \in H$, $a \in \Lambda$.

Remark 1.139 Part (i) of Hypothesis 1.138 implies, thanks to (B.18), that for every
$\theta \geq 0$ there exists an $M_\theta > 0$ such that

$$|(\lambda I - A)^{\theta} e^{tA} x| \leq \frac{M_\theta}{t^{\theta}} |x|, \quad \text{for every } t \in (0, T], \ x \in H. \qquad (1.52)$$

∎

Following Remark 1.120, the definition of a mild solution of (1.51) is given by
Definition 1.119 in which the term

$$\int_0^s e^{(s-r)A}(\lambda I - A)^{\beta} Ga_b(r) dr$$

is interpreted as

$$\int_0^s (\lambda I - A)^\beta e^{(s-r)A} G a_b(r) dr,$$

and the term

$$\int_0^s e^{(s-r)A} \sigma(r, X(r), a(r)) dW_Q(r)$$

as

$$\int_0^s (\lambda I - A)^\gamma e^{(s-r)A} (\lambda I - A)^{-\gamma} \sigma(r, X(r), a(r)) dW_Q(r).$$

This is natural since $(\lambda I - A)^\beta e^{(s-r)A}$ is an extension of $e^{(s-r)A}(\lambda I - A)^\beta$ and $(\lambda I - A)^\gamma e^{(s-r)A} (\lambda I - A)^{-\gamma} = e^{(s-r)A}$.

Remark 1.140 SDEs of type (1.51) appear most frequently in optimal control problems of parabolic equations on a domain $\mathcal{O} \subset \mathbb{R}^n$ with boundary control/noise, see Sect. 2.6.2. More precisely, the cases $\beta \in \left(\frac{3}{4}, 1\right)$ and $\beta \in \left(\frac{1}{4}, \frac{1}{2}\right)$ are related respectively to the Dirichlet and Neumann boundary control problems when one takes $\Lambda_b = L^2(\partial\mathcal{O})$ (or some subset of it) and $H = L^2(\mathcal{O})$. $\gamma \in \left(\frac{1}{4}, \frac{1}{2}\right)$ arises when one treats problems with boundary noise of Neumann type where again $\Lambda_b = L^2(\partial\mathcal{O})$ and $H = L^2(\mathcal{O})$. $\gamma, \beta \in \left(\frac{1}{2} - \varepsilon, \frac{1}{2}\right)$ arise in some specific Dirichlet boundary control/noise problems when one considers $\Lambda_b = L^2(\partial\mathcal{O})$ and a suitable weighted L^2 space as H. ∎

Theorem 1.141 *Assume that Hypothesis 1.138 holds, $p \geq 2$, and let $\alpha := \frac{1}{2} - \gamma$. Suppose that*

$$p > \frac{1}{\alpha} \tag{1.53}$$

and $a_b(\cdot) \in M_\mu^q(0, T; \Lambda_b)$ for some $q \geq p, q > \frac{1}{1-\beta}$. Then, for every initial condition $\xi \in L^2(\Omega, \mathscr{F}_0, \mathbb{P})$, there exists a unique mild solution $X(\cdot) = X(\cdot; 0, \xi, a(\cdot), a_b(\cdot))$ of (1.51) in $\mathcal{H}_2(0, T; H)$ with continuous trajectories $\mathbb{P}$-a.s. If there exists a constant $C > 0$ such that

$$\|(\lambda I - A)^{-\gamma} \sigma(s, x, a)\|_{\mathcal{L}_2(\Xi_0, H)} \leq C \tag{1.54}$$

for all $s \in [0, T]$, $x \in H$, $a \in \Lambda$, then the solution has continuous trajectories $\mathbb{P}$-a.s. without the restriction $p > \frac{1}{\alpha}$. If $\xi \in L^p(\Omega, \mathscr{F}_0, \mathbb{P})$ then $X(\cdot) \in \mathcal{H}_p(0, T; H)$ and there exists a constant $C_{T,p}$ independent of ξ such that

$$\sup_{s \in [0,T]} \mathbb{E}|X(s)|^p \leq C_{T,p}(1 + \mathbb{E}|\xi|^p). \tag{1.55}$$

Proof Assume first that $\xi \in L^p(\Omega, \mathscr{F}_0, \mathbb{P})$ where $p \geq 2$ without the restriction (1.53). Similarly to the proof of Theorem 1.127, we will show that for some $T_0 \in (0, T]$ the map

$$\begin{cases} \mathcal{K}\colon \mathcal{H}_p(0, T_0) \to \mathcal{H}_p(0, T_0), \\ \mathcal{K}(Y)(s) = e^{sA}\xi + \displaystyle\int_0^s e^{(s-r)A}b(r, Y(r), a(r))dr + \int_0^s (\lambda I - A)^\beta e^{(s-r)A}Ga_b(r)dr \\ \qquad\qquad + \displaystyle\int_0^s (\lambda I - A)^\gamma e^{(s-r)A}(\lambda I - A)^{-\gamma}\sigma(r, Y(r), a(r))dW_Q(r) \end{cases}$$

$$(1.56)$$

is well defined and is a contraction. The only difference between our case here and that considered in Theorem 1.127 is the last two terms in (1.56).

First we prove that $\mathcal{K}$ maps $\mathcal{H}_p(0, T_0)$ into $\mathcal{H}_p(0, T_0)$. We only show how to deal with the non-standard terms. For the third term in (1.56) we can argue as follows. If M_β is the constant from (1.52) for $\theta = \beta$, using (1.52), Hölder and Jensen's inequalities, and $q \geq p, q > \frac{1}{1-\beta}$, we obtain

$$\sup_{s\in[0,T_0]} \mathbb{E}\left| \int_0^s (\lambda I - A)^\beta e^{(s-r)A}Ga_b(r)dr \right|^p$$

$$\leq \sup_{s\in[0,T_0]} M_\beta^p \|G\|^p \mathbb{E}\left(\int_0^s \frac{1}{(s-r)^\beta}|a_b(r)|dr \right)^p$$

$$\leq M_\beta^p \|G\|^p \left(\int_0^{T_0} \frac{1}{(T_0-r)^{\frac{\beta q}{q-1}}}dr \right)^{\frac{p(q-1)}{q}} \mathbb{E}\left[\int_0^{T_0} |a_b(r)|^q dr \right]^{\frac{p}{q}}$$

$$\leq C_1 \left(\mathbb{E}\left[\int_0^{T_0} |a_b(r)|^q dr \right] \right)^{\frac{p}{q}} < +\infty.$$

$$(1.57)$$

As regards the stochastic integral term, using Theorem 1.111, (1.52), and Hypothesis 1.138-(vi), we estimate

$$\sup_{s\in[0,T_0]} \mathbb{E}\left| \int_0^s (\lambda I - A)^\gamma e^{(s-r)A}(\lambda I - A)^{-\gamma}\sigma(r, Y(r), a(r))dW_Q(r) \right|^p$$

$$\leq \sup_{s\in[0,T_0]} C_1\mathbb{E}\left| \int_0^s \frac{1}{(s-r)^{2\gamma}}\|(\lambda I - A)^{-\gamma}\sigma(r, Y(r), a(r))\|^2_{\mathcal{L}_2(\Xi_0, H)}dr \right|^{\frac{p}{2}}$$

$$\leq \sup_{s\in[0,T_0]} C_2 \left(\int_0^{T_0} \frac{1}{(T_0-r)^{2\gamma}}dr \right)^{\frac{p}{2}-1} \int_0^s \frac{1}{(s-r)^{2\gamma}}\mathbb{E}[(1 + |Y(r)|)^p]dr$$

$$\leq C_3\left(1 + |Y|^p_{\mathcal{H}_p(0,T_0)} \right) \quad (1.58)$$

for some constant C_3. Progressive measurability of all the terms appearing in the definition of $\mathcal{K}(Y)(\cdot)$ can be proved by using estimates similar to (1.57) and (1.58) and arguing as in Remark 1.123.

Regarding the proof that, for T_0 small enough, $\mathcal{K}$ is a contraction, the only non-standard term to check is the stochastic convolution term, since the third term in

(1.56) does not depend on X. Arguing as before we have that for $X, Y \in \mathcal{H}_p(0, T_0)$, thanks to Theorem 1.111, (1.52), Hypothesis 1.138-(vi), and Jensen's inequality,

$$
\sup_{s \in [0, T_0]} \mathbb{E} \left| \int_0^s (\lambda I - A)^\gamma e^{(s-r)A} (\lambda I - A)^{-\gamma} \left[\sigma(r, X(r), a(r)) - \sigma(r, Y(r), a(r)) \right] dW_Q(r) \right|^p
$$

$$
\leq \sup_{s \in [0, T_0]} C_1 \mathbb{E} \left(\int_0^s \frac{1}{(s-r)^{2\gamma}} \left\| (\lambda I - A)^{-\gamma} \left[\sigma(r, X(r), a(r)) - \sigma(r, Y(r), a(r)) \right] \right\|_{\mathcal{L}_2(\Xi_0, H)}^2 dr \right)^{\frac{p}{2}}
$$

$$
\leq \sup_{s \in [0, T_0]} C_2 \mathbb{E} \left(\int_0^s \frac{1}{(s-r)^{2\gamma}} |X(r) - Y(r)|^2 dr \right)^{\frac{p}{2}}
$$

$$
\leq \sup_{s \in [0, T_0]} C_2 \left(\int_0^{T_0} \frac{1}{(T_0-r)^{2\gamma}} dr \right)^{\frac{p}{2}-1} \int_0^s \frac{1}{(s-r)^{2\gamma}} \mathbb{E}[|X(r) - Y(r)|^p] dr
$$

$$
\leq \omega(T_0) |X - Y|_{\mathcal{H}_p(0, T_0)}^p , \tag{1.59}
$$

where $\omega(r) \xrightarrow{r \to 0^+} 0$. So for T_0 small enough (which is independent of the initial condition) we can apply the Banach fixed point theorem in $\mathcal{H}_p(0, T_0)$ as in the proof of Theorem 1.127 (see also the proof of [180], Theorem 7.2, p. 188). The process can now be reapplied on intervals $[T_0, 2T_0], \ldots, [kT_0, T]$, where $k = [T/T_0]$, to obtain the existence of a unique mild solution in $\mathcal{H}_p(0, T)$ in the sense of the integral equality being satisfied for a.e. $s \in [0, T]$.

Estimate (1.55) follows from similar arguments using the growth assumptions on b, σ in Hypothesis 1.138 and Gronwall's lemma in the form given in Proposition D.30.

We will now prove the continuity of the trajectories if condition (1.53) is satisfied. We will only prove the continuity of the stochastic convolution term in (1.56) since the continuity of the other terms is easier to show. In particular, the continuity of the trajectories of the third term in (1.56) follows from Lemma 1.115-(ii).

Let now $p > \frac{1}{\alpha}$. Hence there is an $0 < \alpha' < \alpha$ such that $p > \frac{1}{\alpha'}$. Then, for $r \in [t, T]$, using (1.52), (1.55), Hypothesis 1.138-(vi), and Jensen's inequality

$$
\mathbb{E} \left(\int_0^r (r-h)^{-2\alpha'} \left\| (\lambda I - A)^\gamma e^{(r-h)A} (\lambda I - A)^{-\gamma} \sigma(h, X(h), a(h)) \right\|_{\mathcal{L}_2(\Xi_0, H)}^2 dh \right)^{\frac{p}{2}}
$$

$$
\leq \mathbb{E} \left(\int_0^r (r-h)^{-2\alpha'} \left\| (\lambda I - A)^\gamma e^{(r-h)A} \right\|_{\mathcal{L}(H)}^2 \left\| (\lambda I - A)^{-\gamma} \sigma(h, X(h), a(h)) \right\|_{\mathcal{L}_2(\Xi_0, H)}^2 ds \right)^{\frac{p}{2}}
$$

$$
\leq C_1 \mathbb{E} \left(\int_0^r (r-h)^{-2\alpha'} (r-h)^{-2\gamma} (1 + |X(h)|)^2 dh \right)^{\frac{p}{2}}
$$

$$
\leq C_1 \left(\int_0^T (T-h)^{-2\alpha'} (T-h)^{-2\gamma} dh \right)^{\frac{p}{2}} \sup_{h \in [0, T]} \mathbb{E}[(1 + |X(h)|)^p] =: C_2 < +\infty. \tag{1.60}
$$

Observe that C_2 does not depend on $r \in [0, T]$. This proves (1.25) and thus the claim follows from Proposition 1.116. When (1.54) holds, estimate (1.60) is easier and can be done for any exponent $p' > 1/\alpha$ in place of p, and thus (1.25) is always satisfied.

Finally, we need to discuss the continuity of the trajectories if $\xi \in L^2(\Omega, \mathscr{F}_0, \mathbb{P})$. We argue as in the proof of Theorem 7.2 of [180]. For $n \geq 1$ we define the random variables

$$\xi_n = \begin{cases} \xi & \text{if } |\xi| \leq n \\ 0 & \text{if } |\xi| > n. \end{cases}$$

The solutions $X(\cdot; 0, \xi, a(\cdot), a_b(\cdot))$ and $X(\cdot; 0, \xi_n, a(\cdot), a_b(\cdot))$ on $[0, T_0]$ are obtained as fixed points in $\mathcal{H}_2(0, T_0)$ and $\mathcal{H}_p(0, T_0)$, with p large enough, of the same contraction map (1.56) with the second map having the term $e^{sA}\xi_n$ in place of $e^{sA}\xi$. Therefore both solutions can be obtained as limits of successive iterations starting, say, from processes $e^{sA}\xi$ and $e^{sA}\xi_n$, respectively. It is then easy to see that we have $X(\cdot; 0, \xi, a(\cdot), a_b(\cdot)) = X(\cdot; 0, \xi_n, a(\cdot), a_b(\cdot))$, $\mathbb{P}$-a.s. on $\{\omega : |\xi(\omega)| \leq n\}$. However, the solutions $X(\cdot; 0, \xi_n, a(\cdot), a_b(\cdot))$ have continuous trajectories. Thus $X(\cdot; 0, \xi, a(\cdot), a_b(\cdot))$ has continuous trajectories $\mathbb{P}$-a.s. on $[0, T_0]$ and we can then continue the argument on intervals $[T_0, 2T_0], \ldots$. $\qquad\square$

Proposition 1.142 *Let the assumptions of Theorem 1.141 be satisfied. Denote the unique mild solution of (1.51) in $\mathcal{H}_p(0, T; H)$ by $X(\cdot) = X(\cdot; 0, \xi, a(\cdot), a_b(\cdot))$.*

(i) *If $\xi^1 = \xi^2$ $\mathbb{P}$-a.s., $a^1(\cdot) = a^2(\cdot)$ $dt \otimes \mathbb{P}$-a.s. $a_b^1(\cdot) = a_b^2(\cdot)$ $dt \otimes \mathbb{P}$-a.s., then $\mathbb{P}$-a.s., $X(\cdot; 0, \xi^1, a^1(\cdot), a_b^1(\cdot)) = X(\cdot; 0, \xi^2, a^2(\cdot), a_b^2(\cdot))$ on $[0, T]$.*

(ii) *Let $(\Omega_1, \mathscr{F}_1, \mathscr{F}_s^1, \mathbb{P}_1, W_{Q,1})$ and $(\Omega_2, \mathscr{F}_2, \mathscr{F}_s^2, \mathbb{P}_2, W_{Q,2})$ be two generalized reference probability spaces. Let $\zeta_i \in L^p(\Omega_i, \mathscr{F}_0^i, \mathbb{P}_i)$, $i = 1, 2$. Let (a^i, a_b^i): $[0, T] \times \Omega_i \to \Lambda \times \Lambda_b$, $i = 1, 2$ be $\mathscr{F}_s^i$-progressively measurable processes satisfying the assumptions of Theorem 1.141. Suppose that $\mathcal{L}_{\mathbb{P}_1}(a^1(\cdot), a_b^1(\cdot), W_{Q,1}(\cdot), \zeta_1) = \mathcal{L}_{\mathbb{P}_2}(a^2(\cdot), a_b^2(\cdot), W_{Q,1}(\cdot), \zeta_2)$ on some subset $D \subset [t, T]$ of full measure. Then $\mathcal{L}_{\mathbb{P}_1}(X(\cdot; 0, \zeta_1, a^1(\cdot), a_b^1(\cdot)), a^1(\cdot), a_b^1(\cdot)) = \mathcal{L}_{\mathbb{P}_2}(X(\cdot; 0, \zeta_2, a^2(\cdot), a_b^2(\cdot)), a^2(\cdot), a_b^2(\cdot))$ on D.*

(iii) *The solution of (1.51) is unique in $M_\mu^p(0, T; H)$ as well.*

Proof (i) If $X_i(\cdot) := X(\cdot; 0, \xi^i, a^i(\cdot), a_b^i(\cdot))$, arguing as in (1.59) and using Hölder's inequality, we obtain, for $s \in [0, T]$,

$$\mathbb{E}|X_1(s) - X_2(s)|^p \leq C_T \int_0^s \mathbb{E}|X_1(r) - X_2(r)|^p dr,$$

and the claim follows by using Gronwall's lemma (Proposition D.29), and the continuity of the trajectories.

(ii) The argument is the same as the one used to prove Lemma 1.136 and Proposition 1.137, since in the current case the solution is also found by iterating the map $\mathcal{K}$.

(iii) The uniqueness in $M_\mu^p(0, T_0; H)$ follows from the estimate in Part (i) above and Proposition D.29. $\qquad\square$

1.5.2 Semilinear SDEs with Additive Noise

In this section we give more precise results for some semilinear SDEs with additive noise, i.e. for Eq. (1.28) when the coefficient σ is constant and we have possible unboundedness in the drift.

Hypothesis 1.143

(i) The linear operator A is the generator of a strongly continuous semigroup $\{e^{tA}, \, , t \geq 0\}$ in H and, for suitable $M \geq 1$ and $\omega \in \mathbb{R}$,

$$|e^{tA}x| \leq Me^{\omega t}|x|, \quad \forall t \geq 0, \ x \in H. \tag{1.61}$$

(ii) $Q \in \mathcal{L}^+(\Xi), \sigma \in \mathcal{L}(\Xi, H)$ and $e^{sA}\sigma Q\sigma^* e^{sA^*} \in \mathcal{L}_1(H)$ for all $s > 0$. Moreover, for all $t \geq 0$,

$$\int_0^t \mathrm{Tr}\left[e^{sA}\sigma Q\sigma^* e^{sA^*}\right] ds < +\infty,$$

so the symmetric positive operator

$$Q_t : H \to H, \qquad Q_t := \int_0^t e^{sA}\sigma Q\sigma^* e^{sA^*} ds, \tag{1.62}$$

is of trace class for every $t \geq 0$, i.e.

$$\mathrm{Tr}\left[Q_s\right] < +\infty. \tag{1.63}$$

Let W_Q be a Q-Wiener process in Ξ and consider the stochastic convolution process defined, for $s \geq 0$, as follows:

$$W^A(s) = \int_0^s e^{(s-r)A}\sigma dW_Q(r). \tag{1.64}$$

Proposition 1.144 *Suppose that Hypothesis 1.143 is satisfied. Then the process $W^A(\cdot)$ defined in (1.64) is a Gaussian process with mean 0 and covariance operator Q_s, is mean square continuous and $W^A(\cdot) \in \mathcal{H}_p^\mu(0, T; H)$ for every $p \geq 2$. Moreover, if there exists a $\gamma > 0$ such that*

$$\int_0^T s^{-\gamma} \mathrm{Tr}\left[e^{sA}\sigma Q\sigma^* e^{sA^*}\right] ds < \infty, \tag{1.65}$$

then $W^A(\cdot)$ has continuous trajectories[5] and, for $p > 0$,

---

[5] Without assuming (1.65) such continuity of trajectories may fail to hold, see e.g. [357].

$$\mathbb{E}\left[\sup_{0\le s\le T}|W^A(s)|^p\right] < +\infty.$$

Proof See [180] Chap. 5, Theorems 5.2 and 5.11. The fact that $W^A(\cdot) \in \mathcal{H}_p^\mu(0, T; H)$ for every $p \ge 2$ follows from Theorem 1.111. The last estimate can be found, for example, as a particular case of Proposition 3.2 in [284]. $\qquad\square$

A completely analogous result holds for the stochastic convolution starting at a point $t \ge 0$, i.e.

$$W^A(t, s) := \int_t^s e^{(s-r)A}\sigma dW_Q(r), \qquad s \ge t. \tag{1.66}$$

Let $T > 0$. We consider the SDE

$$\begin{cases} dX(s) = (AX(s) + b(s, X(s)))\, ds + \sigma dW_Q(s), & s > 0 \\ X(0) = \xi. \end{cases} \tag{1.67}$$

Hypothesis 1.145 $p \ge 1$ and $b(s, x) = b_0(s, x, a_1(s)) + a_2(s)$, where:

(i) The process $a_1(\cdot) : [0, T] \times \Omega \to \Lambda$ (where Λ is a given Polish space) is $\mathscr{F}_s$-progressively measurable. The map $b_0 : [0, T] \times H \times \Lambda \to H$ is Borel measurable and there exists a non-negative function $f \in L^1(0, T; \mathbb{R})$ such that

$$|b_0(s, x, a_1)| \le f(s)(1 + |x|) \qquad \forall s \in [0, T],\ x \in H\ \text{and}\ a_1 \in \Lambda.$$

$$|b_0(s, x_1, a_1) - b_0(s, x_2, a_1)| \le f(s)|x_1 - x_2|$$
$$\forall s \in [0, T],\ x_1, x_2 \in H\ \text{and}\ a_1 \in \Lambda.$$

(ii) The process $a_2(\cdot)$ is such that for all $t > 0$, the process $(s, \omega) \to e^{tA}a_2(s, \omega)$, when interpreted properly, is $\mathscr{F}_s$-progressively measurable on $[0, T] \times \Omega$ with values in H, and

$$|e^{tA}a_2(s, \omega)| \le t^{-\beta}g(s, \omega) \qquad \forall(t, s, \omega) \in [0, T] \times [0, T] \times \Omega, \tag{1.68}$$

for some $\beta \in [0, 1)$ and $g \in M_\mu^q(0, T; \mathbb{R})$, where $q \ge p$ and $q > \frac{1}{1-\beta}$.

Hypothesis 1.145 covers some cases which are not standard and for which a separate proof of existence and uniqueness of mild solutions of (1.67) is required.

Remark 1.146 Hypothesis 1.145-(ii) is satisfied, for example, when A is the generator of an analytic C_0-semigroup and the process $a_2(\cdot)$ is of the form $a_2(s) = (\lambda I - A)^\beta a_3(s)$, where $\lambda \in \mathbb{R}$ is such that $(\lambda I - A)$ is invertible, $\beta \in (0, 1)$, $a_3(\cdot) \in M_\mu^q(0, T; H)$, $q \ge p, q > \frac{1}{1-\beta}$. In such cases the definition of a mild solution of (1.67) is given by Definition 1.119 in which the formal term

$$\int_0^s e^{(s-r)A}a_2(r)dr = \int_0^s e^{(s-r)A}(\lambda I - A)^\beta a_3(r)dr$$

appearing in the definition of a mild solution is interpreted as

$$\int_0^s (\lambda I - A)^\beta e^{(s-r)A} a_3(r) dr.$$

This is natural since $(\lambda I - A)^\beta e^{(s-r)A}$ is an extension of $e^{(s-r)A}(\lambda I - A)^\beta$.

Another more general case where Hypothesis 1.145-(ii) is satisfied is when $a_2(\cdot)$: $[0, T] \times \Omega \to V^*$ is progressively measurable, where V^* denotes the topological dual of $V = D(A^*)$. In such a case the semigroup e^{tA} may be extended, by a standard construction (see e.g. [232]), to the space V^*. Denoting this extension still by e^{tA}, the process $e^{tA} a_2(\cdot) : [0, T] \times \Omega \to V^*$ is well defined. If we further assume that $e^{tA} a_2(\cdot)$ takes values in H and satisfies (1.68) for some $\beta \in (0, 1)$, then Hypothesis 1.145-(ii) is satisfied. A similar and even slightly more general case has been studied in [232] in a deterministic context. ∎

Proposition 1.147 *Let $\xi \in L^p(\Omega, \mathscr{F}_0, \mathbb{P})$ and Hypotheses 1.143 and 1.145 be satisfied. Then Eq. (1.67) has a unique mild solution $X(\cdot; 0, \xi) \in \mathcal{H}_p^\mu(0, T; H)$. The solution satisfies, for some $C_p(T) > 0$ independent of ξ,*

$$\sup_{s \in [0,T]} \mathbb{E}\left[|X(s; 0, \xi)|^p\right] \le C_p(T)(1 + \mathbb{E}[|\xi|^p]). \tag{1.69}$$

Moreover, if $\xi_1, \xi_2 \in L^p(\Omega, \mathscr{F}_0, \mathbb{P})$, we have, $\mathbb{P}$-a.s.,

$$|X(s; 0, \xi_1) - X(s; 0, \xi_2)| \le M e^{\omega T} |\xi_1 - \xi_2| e^{M e^{\omega T} \int_0^s f(r) dr}, \quad s \in [0, T]. \tag{1.70}$$

Finally, if (1.65) also holds for some $\gamma > 0$, then the solution $X(\cdot; 0, \xi)$ has $\mathbb{P}$-a.s. continuous trajectories, and if $\xi = x \in H$ is deterministic we then have

$$\mathbb{E}(\sup_{s \in [0,T]} |X(s)|^p) \le C_p(T)(1 + |x|^p) \tag{1.71}$$

for a suitable constant $C_p(T) > 0$ independent of x. In particular, if g in Hypothesis 1.145-(ii) is in $M_\mu^q(0, T; \mathbb{R})$ for every $q \ge 1$, then estimate (1.69) holds for every $p > 0$ and the same is true for (1.71) if $\xi = x \in H$.

Proof The proof of existence and uniqueness uses the same techniques employed in the Lipschitz case (Theorem 1.127) but contains a small additional difficulty due the presence of the term $a_2(\cdot)$ and possible singularities in s of the Lipschitz norm of $b_0(s, \cdot)$. We will write $\mathcal{H}_p(0, T)$ for $\mathcal{H}_p^\mu(0, T; H)$. For $Y \in \mathcal{H}_p(0, T)$ we set

$$\mathcal{K}(Y)(s) = e^{(s-t)A} \xi + \int_0^s e^{(s-r)A} b_0(r, Y(r), a_1(r)) dr + \int_0^s e^{(s-r)A} a_2(r) dr + W^A(s).$$
$$\tag{1.72}$$

W^A belongs to $\mathcal{H}_p(0, T)$ thanks to Proposition 1.144. Hypotheses 1.145-(i) and 1.145-(ii) ensure, respectively, that the second and third term in the definition of the map $\mathcal{K}$ belong to $\mathcal{H}_p(0, T)$ as well (one can use the same arguments as these to obtain

(1.57) when $\beta \in (0, 1)$ and Hölder's inequality if $\beta = 0$). So $\mathcal{K}$ maps $\mathcal{H}_p(0, T)$ into itself. For $Y_1, Y_2 \in \mathcal{H}_p(0, T)$, $s \in [0, T]$,

$$|\mathcal{K}(Y_1)(s) - \mathcal{K}(Y_2)(s)| \le M e^{\omega T} \int_0^s f(r)|Y_1(r) - Y_2(r)| dr,$$

which yields, for $T_0 \in (0, T]$,

$$\begin{aligned}
|\mathcal{K}(Y_1) - \mathcal{K}(Y_2)|_{\mathcal{H}_p(0,T_0)}^p &\le M e^{\omega T} \sup_{s \in [0,T_0]} \mathbb{E} \left[\int_0^s f(r)|Y_1(r) - Y_2(r)| dr \right]^p \\
&\le M e^{\omega T} \left[\int_0^{T_0} f(r) dr \right]^p \sup_{s \in [0,T_0]} \mathbb{E}|Y_1(s) - Y_2(s)|^p \\
&= M e^{\omega T} \left[\int_0^{T_0} f(r) dr \right]^p |Y_1 - Y_2|_{\mathcal{H}_p(0,T_0)}^p.
\end{aligned}$$

(1.73)

Therefore, if T_0 is sufficiently small, we can apply the contraction mapping principle to find the unique mild solution of (1.67) in $\mathcal{H}_p(0, T_0)$. The existence and uniqueness of a solution on the whole interval $[0, T]$ follows, as usual, by repeating the procedure a finite number of times, since the estimate (1.73) does not depend on the initial data, and the number of steps does not blow up since f is integrable. Estimate (1.69) follows from (1.72) applied to the solution X if we perform estimates similar to those above and use Gronwall's Lemma.

To show (1.70) we observe that if $Z(s) = X(s; 0, \xi_1) - X(s; 0, \xi_2)$, then for $s \in [0, T]$

$$Z(s) = e^{sA}(\xi_1 - \xi_2) + \int_0^s e^{(s-r)A}[b_0(r, X(r; 0, \xi_1), a_1(r)) - b_0(r, X(r; 0, \xi_2), a_1(r))] dr.$$

By Hypothesis 1.145 we thus have

$$|Z(s)| \le M e^{\omega T}|\xi_1 - \xi_2| + M e^{\omega T} \int_0^s f(r)|Z(r)| dr, \quad s \in [0, T]$$

so that, by Gronwall's inequality (see Proposition D.29),

$$|Z(s)| \le M e^{\omega T}|\xi_1 - \xi_2| e^{M e^{\omega T} \int_0^s f(r) dr},$$

which gives the claim. The continuity of trajectories follows from Proposition 1.144, Hypothesis 1.145 and Lemma 1.115 for the second and fourth terms in (1.72), and from Lemma 1.117 for the $\int_0^s e^{(s-r)A} a_2(r) dr$ term.

The last estimate (1.71) follows by standard arguments (see the proof of (1.37) in Theorem 1.130) if we use Proposition 1.144. This implies that if $g \in M_\mu^q(0, T; \mathbb{R})$ for any $q > 0$, (1.71) holds for any $p \ge 2$. For $p \in (0, 2)$, defining $Z_r(s) :=$

$\sup_{s \in [0,T]} |X(s)|^r$, we have

$$\mathbb{E}(Z_p(s)) \leq [\mathbb{E}(Z_p(s)^{2/p})]^{p/2} \leq (C(1 + |x|^2))^{p/2} \leq C_1(1 + |x|^p).$$

$\square$

Proposition 1.148 *Assume that Hypotheses 1.143, 1.145, together with (1.65), are satisfied, and let $a_2(\cdot)$ be as in Remark 1.146. Then:*

(i) *Let $\xi_1, \xi_2 \in L^2(\Omega, \mathscr{F}_0, \mathbb{P})$, $\xi_1 = \xi_2$ $\mathbb{P}$-a.s. Let $(a_1^1(\cdot), a_3^1(\cdot)), (a_1^2(\cdot), a_3^2(\cdot))$ be two processes satisfying Hypothesis 1.145, together with Remark 1.146, such that $(a_1^1(\cdot), a_3^1(\cdot)) = (a_1^2(\cdot), a_3^2(\cdot))$, $dt \otimes \mathbb{P}$-a.s. Then, denoting by $X^i(\cdot; 0, \xi_i)$ the solution of (1.67) for $b(s, x) = (\lambda - A)^\beta a_3^i(s) + b_0(s, x, a_1^i(s))$, we have $X^1(\cdot; 0, \xi_1) = X^2(\cdot; 0, \xi_2)$, $\mathbb{P}$-a.s. on $[0, T]$.*

(ii) *Let $(\Omega_1, \mathscr{F}_1, \mathscr{F}_s^1, \mathbb{P}_1, W_{Q,1})$ and $(\Omega_2, \mathscr{F}_2, \mathscr{F}_s^2, \mathbb{P}_2, W_{Q,2})$ be two generalized reference probability spaces. Let $\xi_i \in L^2(\Omega_i, \mathscr{F}_0^i, \mathbb{P}_i), i = 1, 2$. Let $a_1^i(\cdot), a_3^i(\cdot), i = 1, 2$, be processes on $[0, T] \times \Omega_i$ satisfying Hypothesis 1.145, together with Remark 1.146. Suppose that $\mathcal{L}_{\mathbb{P}_1}(a_1^1(\cdot), a_3^1(\cdot), W_{Q,1}(\cdot), \xi_1) = \mathcal{L}_{\mathbb{P}_2}(a_1^2(\cdot), a_3^2(\cdot), W_{Q,2}(\cdot), \xi_2)$. Then $\mathcal{L}_{\mathbb{P}_1}(X^1(\cdot; 0, \xi_1), a_1^1(\cdot), a_3^1(\cdot)) = \mathcal{L}_{\mathbb{P}_2}(X(\cdot; 0, \xi_2), a_1^2(\cdot), a_3^2(\cdot))$.*

(iii) *If $f \in L^2(0, T; \mathbb{R})$ then the solution of (1.67) ensured by Proposition 1.147 is unique in $M_\mu^2(0, T; H)$ as well.*

Proof Parts (i) and (ii) are proved similarly as Proposition 1.142 (i)–(ii). Part (iii) follows from (1.70), which is also true in this case. We also point out that if $p = 2$, $f \in L^2(0, T; \mathbb{R})$ then $\mathcal{K}$ maps $M_\mu^2(0, T; H)$ into itself and is a contraction in $M_\mu^2(0, T_0; H)$ for small T_0. $\square$

1.5.3 Semilinear SDEs with Multiplicative Noise

This section contains a result for a class of semilinear SDEs with multiplicative noise. Let $T > 0$, and let H, Ξ, Λ and a generalized reference probability space $(\Omega, \mathscr{F}, \{\mathscr{F}_s\}_{s \in [0,T]}, \mathbb{P}, W)$ be as in Sect. 1.3, where $W(t), t \in [0, T]$, is a cylindrical Wiener process (so here $\Xi_0 = \Xi$). We consider the following SDE in H for $s \in [0, T]$:

$$\begin{cases} dX(s) = AX(s)\,ds + b(s, X(s), a(s))\,ds + \sigma(s, X(s), a(s))\,dW(s), \\ X(0) = \xi. \end{cases} \quad (1.74)$$

Hypothesis 1.149

(i) The operator A generates a strongly continuous semigroup e^{tA} for $t \geq 0$ in H.

(ii) $a(\cdot)$ is a Λ-valued progressively measurable process.

(iii) b is a function such that, for all $s \in (0, T]$, $e^{sA}b : [0, T] \times H \times \Lambda \to H$ is measurable and there exist $L \geq 0$ and $\gamma_1 \in [0, 1)$ such that, with $f_1(s) = Ls^{-\gamma_1}$,

$$|e^{sA}b(t, x, a)| \leq f_1(s)(1 + |x|), \tag{1.75}$$

$$|e^{sA}(b(t, x, a) - b(t, y, a))| \leq f_1(s)|x - y|, \tag{1.76}$$

for any $s \in (0, T]$, $t \in [0, T]$, $x, y \in H$, $a \in \Lambda$.

(iv) The function $\sigma : [0, T] \times H \times \Lambda \to \mathcal{L}(\Xi, H)$ is such that, for every $v \in \Xi$, the map $\sigma(\cdot, \cdot, \cdot)v : [0, T] \times H \times \Lambda \to H$ is measurable and, for every $s > 0$, $t \in [0, T]$, $a \in \Lambda$ and $x \in H$, $e^{sA}\sigma(t, x, a)$ belongs to $\mathcal{L}_2(\Xi, H)$. Moreover, there exists a $\gamma_2 \in [0, 1/2)$ such that, with $f_2(s) = Ls^{-\gamma_2}$,

$$|e^{sA}\sigma(t, x, a)|_{\mathcal{L}_2(\Xi, H)} \leq f_2(s)(1 + |x|), \tag{1.77}$$

$$|e^{sA}\sigma(t, x, a) - e^{sA}\sigma(t, y, a)|_{\mathcal{L}_2(\Xi, H)} \leq f_2(s)|x - y|, \tag{1.78}$$

for every $s \in (0, T]$, $t \in [0, T]$, $x, y \in H$, $a \in \Lambda$.

Remark 1.150 Hypothesis 1.149-(iii) covers some cases where the term b is unbounded, which arise, for example, from a stochastic heat equation with a non-zero boundary condition which may also depend on the state variable x (see the last part of Example 4.222).

Moreover, Hypothesis 1.149-(iv) applies to cases, such as reaction-diffusion equations (see e.g. [177], Chap. 11 or, in our Chap. 2, Sect. 2.6.1 and, in particular, Eqs. (2.79) and (2.83), where the operator σ is a nonlinear Nemytskii type operator. Indeed, in such cases it is known that, when the underlying space is $L^2(\mathcal{O})$ ($\mathcal{O} \subset \mathbb{R}^n$, open), the operator $\sigma(t, \cdot) : H \to \mathcal{L}(H)$ is never Lipschitz continuous while $e^{sA}\sigma(t, \cdot) : H \to \mathcal{L}_2(H)$ is so (see e.g. [177], proof of Theorem 11.2.4 and Sect. 11.2.1, or [283], Remark 2.2). ∎

Remark 1.151 If in Hypothesis 1.125 we set $W_Q = Q^{1/2}\tilde{W}$ for a suitable cylindrical Wiener process $\tilde{W}$ in $\tilde{\Xi} = R(Q^{-1/2})$ and we substitute σ with $\tilde{\sigma} = \sigma Q^{1/2}$, it is easy to see that Hypothesis 1.149 is more general. However, we need to replace Ξ by $\tilde{\Xi}$. A cylindrical Wiener process W in Ξ may not be adapted to the original filtration. Similarly, Hypothesis 1.149 is more general than Hypotheses 1.143 and 1.145, together with (1.65), if we take f bounded and $a_2(\cdot) \equiv 0$ there. ∎

The solution of Eq. (1.74) is defined in the mild sense of Definition 1.119, where the convolution term

$$\int_0^s e^{(s-r)A}\sigma(r, X(r), a(r)) \, dW(r), \qquad s \in [0, T],$$

makes sense thanks to (1.77) and Remark 1.123. Moreover, since $s \to e^{sA}b(t, x, a)$ is continuous on $(0, T]$ for every $t \in [0, T], x \in H, a \in \Lambda$, we have from Lemma 1.18 that $e^{\cdot A}b$ is $\mathcal{B}([0, T]) \otimes \mathcal{B}([0, T]) \otimes \mathcal{B}(H) \otimes \mathcal{B}(\Lambda)/\mathcal{B}(H)$-measurable.

Theorem 1.152 *Let Hypothesis 1.149 hold and let $a(\cdot)$ be a Λ-valued, progressively measurable process. Let $p \in [2, \infty)$. Then, for every initial condition $\xi \in L^p(\Omega, \mathscr{F}_0, \mathbb{P})$, the SDE (1.74) has a unique mild solution $X(\cdot)$ in $\mathcal{H}_p(0, T; H)$. The solution satisfies*

$$\sup_{s \in [0,T]} \mathbb{E}\left[|X(s)|^p\right] \leq C_0(1 + \mathbb{E}[|\xi|^p]) \tag{1.79}$$

for some constant $C_0 > 0$ independent of ξ and $a(\cdot)$. The mild solution $X(\cdot)$ has continuous trajectories and, when $\xi \equiv x \in H$, we have

$$\mathbb{E}\left[\sup_{s \in [0,T]} |X(s)|^p\right] \leq C(1 + |x|^p), \quad \text{for all } p > 0, \tag{1.80}$$

for some constant C depending only on p, γ_1, γ_2, T, L and $M_T := \sup_{s \in [0,T]} |e^{sA}|$.

Finally, when b and σ do not depend on a, mild solutions of (1.74) defined on different generalized reference probability spaces have the same laws.

Proof Let $p \geq 2$. The existence of a unique solution is proved using the Banach contraction mapping theorem in $\mathcal{H}_p(0, T_0)$ for some $T_0 \in (0, T)$ small enough. We define $\mathcal{K} \colon \mathcal{H}_p(0, T) \to \mathcal{H}_p(0, T)$ by

$$\mathcal{K}(Y)(s) := e^{sA}\xi + \int_0^s e^{(s-r)A}b(r, Y(r), a(r))dr + \int_0^s e^{(s-r)A}\sigma(r, Y(r), a(r))dW(r). \tag{1.81}$$

We observe first that this expression belongs to $\mathcal{H}_p(0, T)$. Thanks to (1.75), (1.77) and Theorem 1.111, we have

$$\mathbb{E}\left|\int_0^s e^{(s-r)A}b(r, Y(r), a(r))dr + \int_0^s e^{(s-r)A}\sigma(r, Y(r), a(r))dW(r)\right|^p$$

$$\leq C_p\left(\mathbb{E}\left|\int_0^s [f_1(s-r)(1+|Y(r)|)]dr\right|^p\right.$$

$$\left. + \mathbb{E}\left|\int_0^s e^{(s-r)A}\sigma(r, Y(r), a(r))dW(r)\right|^p\right)$$

$$\leq C_p\left[\int_0^T f_1(r)dr\right]^p \sup_{r \in [0,T]} \mathbb{E}(1+|Y(r)|)^p$$

$$+ C_p\left[\int_0^T f_2^2(r)dr\right]^{\frac{p}{2}} \sup_{r \in [0,T]} \mathbb{E}(1+|Y(r)|)^p, \tag{1.82}$$

where the constant C_p depends only on p. Therefore, for any $Y \in \mathcal{H}_p(0, T)$, $\mathcal{K}(Y) \in \mathcal{H}_p(0, T)$. The estimates showing that $\mathcal{K}$ is a contraction on $\mathcal{H}_p(0, T_0)$ for $T_0 \in (0, T]$ small enough are essentially the same. Using (1.76) and (1.78) instead of (1.75) and (1.77) we obtain, for all $Y_1, Y_2 \in \mathcal{H}_p(0, T_0)$,

$$|\mathcal{K}(Y_1) - \mathcal{K}(Y_2)|^p_{\mathcal{H}_p(0,T_0)} \leq C_p \left(\left[\int_0^{T_0} f_1(r)dr \right]^p \right.$$

$$\left. + \left[\int_0^{T_0} f_2^2(r)dr \right]^{\frac{p}{2}} \right) \sup_{r \in [0,T_0]} \mathbb{E}(|Y_1(r) - Y_2(r)|^p),$$

and thus $\mathcal{K}$ is a contraction in $\mathcal{H}_p(0, T_0)$ if $T_0 \in (0, T]$ is small enough. The existence and uniqueness of solution in $\mathcal{H}_p(0, T)$ follows, as usual, by repeating the procedure a finite number of times, since the estimate does not depend on the initial data, and the number of steps does not blow up since f_1 and f_2^2 are integrable. Estimate (1.79) follows in a standard way by applying estimates like those in (1.82) to the fixed point of the map $\mathcal{K}$ and using Gronwall's lemma (see also the proof of Theorem 7.5 in [180]).

The continuity of the trajectories and (1.80) are proved using the factorization method similarly to the way it is done in the proof of Proposition 6.9 for $p > 2$. We extend (1.80) to $0 < p \leq 2$ in the same way as in the proof of Proposition 1.147. Uniqueness in law is proved similarly as in Proposition 1.137. □

Proposition 1.153 *Assume that Hypothesis 1.149 holds. Let* (t_1, x_1), $(t_2, x_2) \in [0, T] \times H$ *with* $t_1 \leq t_2$. *Denote by* $X(\cdot; t_1, x_1, a(\cdot))$, $X(\cdot; t_2, x_2, a(\cdot))$ *the corresponding mild solutions of (1.74) with the same progressively measurable process* $a(\cdot)$ *and initial conditions* $X(t_i) = x_i \in H$, $i = 1, 2$. *Then, for all* $s \in [t_2, T]$ *we have, setting* $\gamma_3 := [2(1 - \gamma_1)] \wedge [1 - 2\gamma_2]$,

$$\mathbb{E}[|X(s; t_1, x_1, a(\cdot)) - X(s; t_2, x_2, a(\cdot))|^2] \leq$$

$$\leq C_2 \left[|x_1 - x_2|^2 + (1 + |x_1|^2)|t_2 - t_1|^{\gamma_3} + |e^{(t_2-t_1)A}x_1 - x_1|^2 \right] \tag{1.83}$$

for some constant C_2 *depending only on* γ_1, γ_2, T, L *and* $M := \sup_{s \in [0,T]} |e^{sA}|$.
Moreover, the term $|e^{(t_2-t_1)A}x_1 - x_1|^2$ *can be replaced by* $|e^{(t_2-t_1)A}x_2 - x_2|^2$.

Proof To simplify the notation we define $X_i(s) := X(s; t_i, x_i, a(\cdot))$, $b(r, X_i(r)) := b(r, X_i(r), a(r))$, $\sigma(r, X_i(r)) := \sigma(r, X_i(r), a(r))$, $i = 1, 2$. By the definition of a mild solution we have, for $s \in [t_i, T]$,

$$X_i(s) = e^{(s-t_i)A}x_i + \int_{t_i}^s e^{(s-r)A}b(r, X_i(r))dr + \int_{t_i}^s e^{(s-r)A}\sigma(r, X_i(r))dW(r),$$

hence

$$|X_1(s) - X_2(s)| \leq |e^{(s-t_1)A}x_1 - e^{(s-t_2)A}x_2|$$

$$+ \left| \int_{t_1}^{t_2} e^{(s-r)A}b(r, X_1(r))dr \right| + \left| \int_{t_2}^s e^{(s-r)A}(b(r, X_1(r)) - b(r, X_2(r)))dr \right|$$

$$+ \left| \int_{t_1}^{t_2} e^{(s-r)A}\sigma(r, X_1(r))dW(r) \right| + \left| \int_{t_2}^s e^{(s-r)A}(\sigma(r, X_1(r)) - \sigma(r, X_2(r)))dW(r) \right|.$$

Therefore

$$\mathbb{E}|X_1(s) - X_2(s)|^2 \leq 5|e^{(s-t_1)A}x_1 - e^{(s-t_2)A}x_2|^2$$

$$+ 5\mathbb{E}\left|\int_{t_1}^{t_2} e^{(s-r)A}b(r, X_1(r))dr\right|^2 + 5\mathbb{E}\left|\int_{t_2}^{s} e^{(s-r)A}\left(b(r, X_1(r)) - b(r, X_2(r))\right)dr\right|^2$$

$$+ 5\mathbb{E}\left|\int_{t_1}^{t_2} e^{(s-r)A}\sigma(r, X_1(r))dW(r)\right|^2$$

$$+ 5\mathbb{E}\left|\int_{t_2}^{s} e^{(s-r)A}\left(\sigma(r, X_1(r)) - \sigma(r, X_2(r))\right)dW(r)\right|^2. \qquad (1.84)$$

To estimate the second and the third terms we use Jensen's inequality applied to the inner integral. Using Hypothesis 1.149-(ii) and (1.80) we then obtain

$$\mathbb{E}\left|\int_{t_1}^{t_2} e^{(s-r)A}b(r, X_1(r))dr\right|^2 \leq L^2\mathbb{E}\left|\int_{t_1}^{t_2} (s-r)^{-\gamma_1}(1 + |X_1(r)|)dr\right|^2$$

$$\leq L^2\left(\int_{t_1}^{t_2} (s-r)^{-\gamma_1}dr\right)\int_{t_1}^{t_2} (s-r)^{-\gamma_1}\mathbb{E}(1 + |X_1(r)|)^2 dr$$

$$\leq 2L^2[1 + C(1 + |x_1|^2)]\left(\int_{t_1}^{t_2} (s-r)^{-\gamma_1}dr\right)^2$$

$$\leq 2L^2[1 + C(1 + |x_1|^2)]\frac{1}{1-\gamma_1}(t_1 - t_2)^{2(1-\gamma_1)}.$$

In the same way we estimate the third term obtaining, by Hypothesis 1.149-(ii),

$$\mathbb{E}\left|\int_{t_2}^{s} e^{(s-r)A}\left(b(r, X_1(r)) - b(r, X_2(r))\right)dr\right|^2$$

$$\leq L^2\left(\int_{t_2}^{s} (s-r)^{-\gamma_1}dr\right)\int_{t_2}^{s} (s-r)^{-\gamma_1}\mathbb{E}|X_1(r) - X_2(r)|^2 dr$$

$$\leq \frac{L^2(s-t_2)^{1-\gamma_1}}{1-\gamma_1}\int_{t_2}^{s} (s-r)^{-\gamma_1}\mathbb{E}|X_1(r) - X_2(r)|^2 dr.$$

The fourth and the fifth term of (1.84) are estimated using the isometry formula. We have

$$\mathbb{E}\left|\int_{t_1}^{t_2} e^{(s-r)A}\sigma(r, X_1(r))dW(r)\right|^2 = \int_{t_1}^{t_2} \mathbb{E}|e^{(s-r)A}\sigma(r, X_1(r))|^2_{\mathcal{L}_2(\Xi, H)}dr$$

$$\leq L^2\int_{t_1}^{t_2} (s-r)^{-2\gamma_2}\mathbb{E}(1 + |X_1(r)|)^2 dr \leq 2L^2[1 + C(1 + |x_1|^2)]\int_{t_1}^{t_2} (s-r)^{-2\gamma_2}dr$$

$$\leq 2L^2[1 + C(1 + |x_1|^2)] \frac{1}{1 - 2\gamma_2}(t_1 - t_2)^{1-2\gamma_2}$$

and

$$\mathbb{E}\left|\int_{t_2}^s e^{(s-r)A}\left(\sigma(r, X_1(r)) - \sigma(r, X_2(r))\right) dW(r)\right|^2$$

$$= \int_{t_2}^s \mathbb{E}|e^{(s-r)A}\left(\sigma(r, X_1(r)) - \sigma(r, X_2(r))\right)|^2_{\mathcal{L}_2(\Xi,H)} dr$$

$$\leq L^2 \int_{t_2}^s (s - r)^{-2\gamma_2}\mathbb{E}|X_1(r) - X_2(r)|^2 dr.$$

Using all these estimates in (1.84) we obtain, for a suitable constant $C_1 > 0$, for $\gamma_3 := [2(1 - \gamma_1)] \wedge [1 - 2\gamma_2]$ and $\gamma_4 := \gamma_1 \vee [2\gamma_2]$,

$$\mathbb{E}|X_1(s) - X_2(s)|^2 \leq 5|e^{(s-t_1)A}x_1 - e^{(s-t_2)A}x_2|^2 + C_1(1 + |x_1|^2)|t_2 - t_1|^{\gamma_3} +$$

$$+ C_1 \int_{t_2}^s (s - r)^{-\gamma_4}\mathbb{E}|X_1(r) - X_2(r)|^2 dr.$$

Observing that

$$|e^{(s-t_1)A}x_1 - e^{(s-t_2)A}x_2| \leq M|x_1 - x_2| + |e^{(s-t_2)A}(e^{(t_2-t_1)A}x_1 - x_1)|,$$

we can thus apply Gronwall's lemma in the form of Proposition D.30. It gives us

$$\mathbb{E}|X_1(s) - X_2(s)|^2 \leq C_2\left[|x_1 - x_2|^2 + (1 + |x_1|^2)|t_2 - t_1|^{\gamma_3} + |e^{(t_2-t_1)A}x_1 - x_1|^2\right]$$

for some $C_2 > 0$ with the required properties.   □

Lemma 1.154 *Assume that Hypothesis 1.149 holds. Fix a Λ-valued progressively measurable process $a(\cdot)$. Let X be the unique mild solution of (1.74) described in Theorem 1.152 with initial condition $X(0) = x \in H$. Define, for $s \in [0, T]$, $\psi(s) = b(s, X(s), a(s))$, $\Phi(s) = \sigma(s, X(s), a(s))$. Let $\{e_i\}_{i\in\mathbb{N}}$ be an orthonormal basis of Ξ and, for any $k \in \mathbb{N}$, let $P^k \colon \Xi \to \Xi$ be the orthogonal projection onto $\mathrm{span}\{e_1, ..., e_k\}$. Let X^k be the unique mild solution of*

$$\begin{cases} dX^k(s) = (AX^k(s) + e^{\frac{1}{k}A}\psi(s))ds + e^{\frac{1}{k}A}\Phi(s)P^k dW(s), \\ X^k(0) = x. \end{cases} \tag{1.85}$$

Then, for any $p > 0$, there exists an $M_p > 0$ such that

$$\sup_{k\in\mathbb{N}} \mathbb{E}\left[\sup_{s\in[0,T]} |X^k(s)|^p\right] \leq M_p. \tag{1.86}$$

Moreover, for every $s \in [0, T]$,

$$\lim_{k \to \infty} \mathbb{E}\left[|X^k(s) - X(s)|^2\right] = 0 \tag{1.87}$$

and, for every $\varphi \in C_m(H)$ $(m \geq 0)$,

$$\lim_{k \to \infty} \mathbb{E}\left[\varphi(X^k(s))\right] = \mathbb{E}\left[\varphi(X(s))\right], \qquad s \in [0, T]. \tag{1.88}$$

Proof It is easy to see, by using (1.80), that (1.86) is satisfied.

We now prove (1.87). We have, for $s \in [0, T]$,

$$\mathbb{E}\left|X(s) - X^k(s)\right|^2 \leq 2\mathbb{E}\left|\int_0^s e^{(s-r)A}\left(\psi(r) - e^{\frac{1}{k}A}\psi(r)\right) dr\right|^2$$

$$+ 4\mathbb{E}\left|\int_0^s e^{(s-r)A}\Phi(r)(I - P^k)dW(r)\right|^2$$

$$+ 4\mathbb{E}\left|\int_0^s (I - e^{\frac{1}{k}A})e^{(s-r)A}\Phi(r)P^k dW(r)\right|^2 = I_1 + I_2 + I_3.$$

We have for any k,

$$\left|e^{(s-r)A}\left(\psi(r) - e^{\frac{1}{k}A}\psi(r)\right)\right| \leq 2L(s-r)^{-\gamma_1}(1 + |X(r)|)$$

which is integrable on $[0, s]$ for a.e. ω. Moreover,

$$\left|e^{(s-r)A}\left(\psi(r) - e^{\frac{1}{k}A}\psi(r)\right)\right| \to 0 \quad \text{as } k \to +\infty$$

$dr \otimes \mathbb{P}$-a.s. Therefore it follows from the dominated convergence theorem that

$$\int_0^s e^{(s-r)A}\left(\psi(r) - e^{\frac{1}{k}A}\psi(r)\right) dr \to 0 \quad \text{as } k \to +\infty$$

$\mathbb{P}$-a.s. Now by Hölder's inequality

$$\left|\int_0^s e^{(s-r)A}\left(\psi(r) - e^{\frac{1}{k}A}\psi(r)\right) dr\right|^2$$

$$\leq 4L^2 \left(\int_0^s (s-r)^{-\gamma_1} dr\right)\left(\int_0^s (s-r)^{-\gamma_1}(1 + |X(r)|)^2 dr\right)$$

which is integrable on Ω. Thus, using the dominated convergence theorem again we conclude that $\lim_{k \to \infty} I_1 = 0$.

Recall that $\Xi_0 = \Xi$. To estimate I_2, we set $Q^k := I - P^k$. We have

$$I_2 = 4\mathbb{E}\left|\int_0^s e^{(s-r)A}\Phi(r)(I - P^k)dW(r)\right|^2$$

$$= 4\int_0^s \mathbb{E}\left\|e^{(s-r)A}\Phi(r)Q^k\right\|_{\mathcal{L}_2(\Xi,H)}^2 dr$$

$$= 4\int_0^s \mathbb{E}\sum_{i\in\mathbb{N}}\left\langle e^{(s-r)A}\Phi(r)Q^k e_i, e^{(s-r)A}\Phi(r)Q^k e_i\right\rangle dr =: \eta(k).$$

Observe that

$$\sum_{i\in\mathbb{N}}\left\langle e^{(s-r)A}\Phi(r)Q^k e_i, e^{(s-r)A}\Phi(r)Q^k e_i\right\rangle$$

$$= \sum_{i=k+1}^{+\infty}\left\langle e^{(s-r)A}\Phi(r)e_i, e^{(s-r)A}\Phi(r)e_i\right\rangle$$

$$\leq \sum_{i\in\mathbb{N}}\left\langle e^{(s-r)A}\Phi(r)e_i, e^{(s-r)A}\Phi(r)e_i\right\rangle = \left\|e^{(s-r)A}\Phi(r)\right\|_{\mathcal{L}_2(\Xi,H)}^2.$$

Since the series above has nonnegative terms, we obtain

$$\lim_{k\to\infty}\left\|e^{(s-r)A}\Phi(r)Q^k\right\|_{\mathcal{L}_2(\Xi,H)}^2 = 0 \quad dr\otimes\mathbb{P}\text{-a.s.}$$

Therefore, thanks to (1.80), Hypothesis 1.149 and the dominated convergence theorem, we obtain

$$\lim_{k\to\infty} I_2 \leq \lim_{k\to\infty} \eta(k) = 0.$$

The term I_3 is estimated similarly.

Thanks to (1.87), for any subsequence of $X^k(s)$ we can extract a sub-subsequence converging to $X(s)$ almost everywhere and then, thanks to (1.86), (1.80) and the dominated convergence theorem, we obtain (1.88) along the sub-subsequence. This implies (1.88) for the whole sequence $X^k(s)$.   □

Remark 1.155 Observe that if b and σ satisfy Hypothesis 1.149, the functions $e^{\frac{1}{k}A}b(s, x, a)$ and $e^{\frac{1}{k}A}\sigma(s, x, a)P_k$ satisfy Hypothesis 1.125.   ■

The last lemma concerns the additive noise case of Sect. 1.5.2, however we included it here since its proof is similar to the proof of Lemma 1.154.

Let W_Q be from Sect. 1.5.2. We know (see (1.12)) that $W_Q(s) = \sum_{n=1}^{+\infty} g_n \beta_n(s)$, $s \geq 0$, where $\{g_n\}$ is an orthonormal basis of Ξ_0. Define $e_n = Q^{-1/2}g_n$, $n \in \mathbb{N}$. Then $\{e_n\}$ is an orthonormal basis of Ξ. Let $\tilde{P}^k$ be the orthogonal projection in Ξ_0 onto span$\{g_1, ..., g_k\}$ and P^k be the orthogonal projection in Ξ onto span$\{e_1, ..., e_k\}, k \in \mathbb{N}$. It is easy to see that $\tilde{P}^k Q^{1/2} = Q^{1/2}P^k$ as operators on Ξ.

Lemma 1.156 *Let Hypotheses 1.143 and 1.145 be satisfied and let $q \geq 2$. Let X be the unique mild solution of (1.67) described in Proposition 1.147 with initial condition $X(0) = x \in H$. Define for $k, m \in \mathbb{N}$, $B_k = \{(s, \omega) : |b_0(s, X(s), a_1(s))| \leq k\}$,*

$D_m = \{(s, \omega) : |g(s, \omega)| \leq m\}$. *There exists a sequence m_k such that the sequence X^k of the solutions of the SDE*

$$\begin{cases} dX^k(s) = \left(AX^k(s) + \psi_k(s)\right) ds + \sigma \tilde{P}^k dW_Q(s), & s > 0, \\ X^k(0) = x, \end{cases} \tag{1.89}$$

where $\psi_k(s) = b_0(s, X(s), a_1(s))\mathbf{1}_{B_k}(s, \omega) + e^{\frac{1}{k}A}a_2(s)\mathbf{1}_{D_{m_k}}(s, \omega)$, satisfies the following.

(i) For any $p \in [2, q]$ there exists an $M_p > 0$ such that

$$\sup_{k} \sup_{s \in [0,T]} \mathbb{E}\left[|X^k(s)|^p\right], \ \sup_{s \in [0,T]} \mathbb{E}\left[|X(s)|^p\right] \leq M_p. \tag{1.90}$$

(ii) For every $s \in [0, T]$

$$\lim_{k \to \infty} \mathbb{E}\left[|X^k(s) - X(s)|^2\right] = 0.$$

Proof Part (i). The moment estimates are uniform in k (regardless of the choice of m_k) thanks to the following facts:

(a) Define $W^{A,k}(s) := \int_0^s e^{(s-r)A}\sigma\tilde{P}^k dW_Q(r)$, $s \in [0, T]$. Given an orthonormal basis $\{w_n\}$ of H, for any $k \in \mathbb{N}$ and $s \in [0, T]$, we have

$$\begin{aligned} 0 \leq &\operatorname{Tr}\left(\left(e^{sA}\sigma\tilde{P}^k Q^{1/2}\right)\left(e^{sA}\sigma\tilde{P}^k Q^{1/2}\right)^*\right) \\ = &\operatorname{Tr}\left(\left(e^{sA}\sigma Q^{1/2}P^k\right)\left(e^{sA}\sigma Q^{1/2}P^k\right)^*\right) \\ = &\sum_{n \in \mathbb{N}} |P_k Q^{1/2}\sigma^* e^{sA^*} w_n|^2 \leq \sum_{n \in \mathbb{N}} |Q^{1/2}\sigma^* e^{sA^*} w_n|^2 = \sum_{n \in \mathbb{N}} \operatorname{Tr}\left(e^{sA}\sigma Q\sigma^* e^{sA^*}\right). \end{aligned} \tag{1.91}$$

Thus, by Theorem 1.111, it follows that for any $k \in \mathbb{N}$ and $p \geq 1$,

$$\sup_{k} \sup_{s \in [0,T]} \mathbb{E}\left[|W^{A,k}(s)|^p\right] < +\infty.$$

Using (1.91) we also have, by the Lebesgue dominated convergence theorem,

$$\int_0^T \|e^{sA}\sigma\tilde{P}^k - e^{sA}\sigma\|_{\mathcal{L}_2(\Xi_0, H)}^2 ds = \int_0^T \sum_{n \in \mathbb{N}} |(P_k - I)Q^{1/2}\sigma^* e^{sA^*} w_n|^2 ds \to 0. \tag{1.92}$$

(b) By the definition

$$|e^{tA}\psi_k(s)| \leq f(s)(1 + |X(s)|) + t^{-\beta}g(s, \omega) \quad \text{for } t, s \in [0, T], \omega \in \Omega.$$

Part (ii). The scheme of the proof is similar to that of (1.87). We choose m_k such that

$$\mathbb{E}\left|\int_0^T k^\beta g(r, \omega)|1 - \mathbf{1}_{D_{m_k}}(r, \omega)|dr\right|^2 \leq \frac{1}{k}. \tag{1.93}$$

We have for every $s \in [0, T]$,

$$\mathbb{E}\left|X(s) - X^k(s)\right|^2 \leq 4\mathbb{E}\left|\int_0^s e^{(s-r)A}b_0(r, X(r), a_1(r))(1 - \mathbf{1}_{B_k}(r, \omega))dr\right|^2$$

$$+ 4\mathbb{E}\left|\int_0^s e^{(s-r)A}(a_2(r) - e^{\frac{1}{k}A}a_2(r))dr\right|^2$$

$$+ 4\mathbb{E}\left|\int_0^s e^{(\frac{1}{k}+s-r)A}a_2(r)(1 - \mathbf{1}_{D_{m_k}}(r, \omega))dr\right|^2$$

$$+ 4\mathbb{E}\left|W^{A,k}(s) - W^A(s)\right|^2 = J_1 + J_2 + J_3 + J_4.$$

The term J_1 converges to 0 as $k \to +\infty$ by Hypothesis 1.145, Hölder's inequality, (1.69) for $p = 2$ and the dominated convergence theorem. The term J_2 converges to 0 by the same arguments as for the term I_1 in the proof of Lemma 1.154. The term J_3 converges to 0 by (1.93) and finally $J_4 \to 0$ by (1.92). $\qquad\square$

1.6 Transition Semigroups

Let $T \in (0, +\infty]$ and recall that, as before, when $T = +\infty$ the notation $[0, T]$ and $[t, T]$ means $[0, +\infty)$ and $[t, +\infty)$. Let H, Ξ, Q, and the generalized reference probability space $\mu = (\Omega, \mathscr{F}, \{\mathscr{F}_s\}_{s\in[0,T]}, \mathbb{P}, W_Q)$ be the same as in Sect. 1.3. Consider for $t \in [0, T]$ the following SDE with non-random coefficients

$$\begin{cases} dX(s) = (AX(s) + b(s, X(s)))\, ds + \sigma(s, X(s))dW_Q(s) \\ X(t) = x \in H, \end{cases} \tag{1.94}$$

where $b: [0, T] \times H \to H$ and $\sigma: [0, T] \times H \to \mathcal{L}_2(\Xi_0, H)$. If Hypothesis 1.125, where we drop the dependence on a in all conditions, (respectively, Hypotheses 1.143 and 1.145 with $a_2(\cdot) \equiv 0$ and with no dependence on a_1, respectively, Hypothesis 1.149 with no dependence on a) is satisfied, then Theorem 1.127 (respectively, Proposition 1.147, respectively, Theorem 1.152) ensures that (1.94) has a unique mild solution $X(\cdot; t, x)$. Moreover, we also have uniqueness in law of the solutions.

We will be using the spaces $B_b(H)$ of bounded Borel measurable functions on H and $B_m(H)$, $m > 0$, of Borel measurable functions on H with at most polynomial growth of order m, defined in Appendix A.2.

For any $\phi \in B_b(H)$ and $t \geq 0$, $s \in [t, T]$, we define

$$\begin{cases} P_{t,s}[\phi]: H \to \mathbb{R} \\ P_{t,s}[\phi]: x \to \mathbb{E}[\phi(X(s;t,x))]. \end{cases} \tag{1.95}$$

It is not obvious that $P_{t,s}[\phi] \in B_b(H)$ and it has to be checked in each case. The general argument is the following and we illustrate it in the case when Hypothesis 1.149 is satisfied. First, using (1.83) it is easy to see that $P_{t,s}[\phi] \in C_b(H)$ if $\phi \in UC_b(H)$. Then, using the functions constructed in the proof of Theorem 1.34 and the dominated convergence theorem, we get that $P_{t,s}[\phi] \in B_b(H)$ for every $\phi = 1_A, A = \overline{A} \subset H$. This, together with Corollary 1.3 and the dominated convergence theorem, allows us to extend $P_{t,s}[\phi] \in B_b(H)$ to every $\phi = 1_A, A \in B(H)$. We can then use Lemma 1.15 to conclude that $P_{t,s}[\phi] \in B_b(H)$ for every $\phi \in B_b(H)$. Similar arguments can be applied in the cases when Hypotheses 1.143 and 1.145 hold or if Hypothesis 1.125 is satisfied. Moreover, thanks to estimates (1.36), (1.69) and (1.80), $P_{t,s}[\phi]$ is then also well defined for any $\phi \in B_m(H), m > 0$.

Theorem 1.157 (Markov property) *Let $T \in (0, +\infty]$. Let Hypothesis 1.149 be satisfied with b and σ independent of a. Then for every $\phi \in B_m(H)$ ($m \geq 0$) and $0 \leq t \leq s \leq r \leq T$ (with the last inequality strict when $T = +\infty$),*

$$\mathbb{E}\phi(X(r;t,x)|\mathscr{F}_s) = P_{s,r}[\phi](X(s;t,x)) \quad \mathbb{P} - almost \ surely,$$

and

$$P_{t,r}[\phi](x) = P_{t,s}\left[P_{s,r}[\phi]\right](x) \quad for \ all \ x \in H. \tag{1.96}$$

The same result is true if Hypotheses 1.143 and 1.145 hold without dependence on a_1 and with $a_2(\cdot) = 0$ or if Hypothesis 1.125 holds without the dependence on a in all conditions.

Proof See [180], Theorem 9.14, p. 248, and Corollary 9.15, p. 249. The hypotheses are a little different from these in [180], however the same arguments can be easily adapted using the proof of Proposition 1.153. The proof in [180] is given for $\phi \in B_b(H)$ but the argument is exactly the same when $\phi \in B_m(H)$ ($m > 0$) simply recalling that the operator $P_{t,s}$ is well defined on such functions thanks to estimate (1.80). $\qquad \square$

It follows from the uniqueness in law of the solutions of (1.94) that the operators $P_{t,s}$ do not depend on the choice of a generalized reference probability space μ. As a consequence of the uniqueness in law we also have the following corollary.

Corollary 1.158 *Let Hypothesis 1.149 be satisfied with b and σ independent of a and of the time variable s. Equation (1.94) then reduces to*

$$\begin{cases} dX(s) = (AX(s) + b(X(s))) \, ds + \sigma(X(s)) dW_Q(s), \\ X(t) = x \in H. \end{cases} \tag{1.97}$$

Denote by $X(\cdot; t, x)$ the unique mild solution of this equation (defined on $[t, +\infty)$). In this case, for any $\phi \in B_m(H)$ ($m \geq 0$) and $0 \leq t \leq s$, we have

$$P_{t,s}[\phi](x) = P_{0,s-t}[\phi]. \tag{1.98}$$

Hence, defining $P_s[\phi]$ as follows,

$$\begin{cases} P_s[\phi]: H \to \mathbb{R} \\ P_s[\phi]: x \to \mathbb{E}\phi(X(s; 0, x)), \end{cases} \tag{1.99}$$

we have

$$P_{s+r}[\phi](x) = P_s\left[P_r[\phi]\right](x) \quad \text{for all } x \in H, s, r \geq 0. \tag{1.100}$$

The same result is true if Hypotheses 1.143 and 1.145 hold without dependence on a_1 and with $a_2(\cdot) = 0$ or if Hypothesis 1.125 holds without the dependence on a in all conditions.

Proof We only need to prove (1.98), which is an immediate consequence of the uniqueness in law of the mild solutions of (1.97). Indeed, by the uniqueness in law, for all $s \geq t \geq 0$ and $x \in H$, the random variables $X(s; t, x)$ and $X(s - t; 0, x)$ have the same distributions, hence

$$P_{t,s}[\phi](x) = \mathbb{E}[\phi(X(s; t, x))] = \mathbb{E}[\phi(X(s - t; 0, x))] = P_{0,s-t}[\phi](x). \quad \square$$

Definition 1.159 (*Transition semigroup, (strong) Feller property*) If (1.96) (respectively, (1.100)) is satisfied we call $P_{t,s}$ (respectively, P_t) the two-parameter *transition semigroup* (respectively, one-parameter *transition semigroup*) associated to Eq. (1.94).

We say that $P_{t,s}$ (respectively, P_t) possesses the *Feller property* if

$$P_{t,s}(C_b(H)) \subset C_b(H) \text{ (respectively, } P_t(C_b(H)) \subset C_b(H))$$

and that $P_{t,s}$ (respectively, P_t) possesses the *strong Feller property* if

$$P_{t,s}(B_b(H)) \subset C_b(H) \text{ (respectively, } P_t(B_b(H)) \subset C_b(H))$$

for all $0 \leq t < s \leq T$ (respectively, $t \in (0, T]$).

Lemma 1.160 *Assume that (1.94) has unique mild solutions $X(\cdot; t, x)$ which satisfy, for every $m \geq 0$, the estimate*

$$\mathbb{E}[|X(s; t, x)|^m] \leq C(m)(1 + |x|^m), \quad t \geq 0, \ s \in [t, T], \ x \in H, \tag{1.101}$$

for some constant $C(m)$. If the Feller property holds for the associated two-parameter transition semigroup $P_{t,s}$ ($t \geq 0$, $s \in [t, T]$), then we also have

$$P_{t,s}(C_m(H)) \subset C_m(H) \qquad \forall m \geq 0$$

while, if the strong Feller property holds, we also have

$$P_{t,s}(B_m(H)) \subset C_m(H) \qquad \forall m \geq 0.$$

Proof Let $\phi \in B_m(H)$ and define, for $k \in \mathbb{N}$,

$$\phi_k(x) = \phi(x)\mathbf{1}_{|x|\leq k} + \phi\left(k\frac{x}{|x|}\right)\mathbf{1}_{|x|>k}.$$

It is clear that $\phi_k \in B_b(H)$, it coincides with ϕ on $\{|x| \leq k\}$ and if ϕ is continuous so is ϕ_k. Moreover, when $k \to +\infty$, ϕ_k converges to ϕ uniformly on bounded sets. Assume now that the strong Feller property holds (the argument for the Feller property is exactly the same). In this case $P_{t,s}[\phi_k]$ is continuous, hence, to get the claim, it is enough to show that $P_{t,s}[\phi_k]$ converges to $P_{t,s}[\phi_k]$ uniformly on bounded sets. Indeed,

$$P_{t,s}[\phi_k - \phi](x) = \mathbb{E}\left[(\phi_k - \phi)(X(s;t,x))\right]$$

$$= \mathbb{E}\left[\left(\phi\left(k\frac{X(s;t,x)}{|X(s;t,x)|}\right) - \phi(X(s;t,x))\right)\mathbf{1}_{|X(s;t,x)|>k}\right]$$

$$\leq 2\mathbb{E}\left[\|\phi\|_{B_m}(1 + |X(s;t,x)|^m)\mathbf{1}_{|X(s;t,x)|\geq k}\right].$$

Hence, for any $p > 1$ we have by (1.101)

$$P_{t,s}[\phi_k - \phi](x) \leq 2\|\phi\|_{B_m}\left[\mathbb{E}(1 + |X(s;t,x)|^m)^p\right]^{1/p}\left[\mathbb{E}\mathbf{1}_{|X(s;t,x)|\geq k}\right]^{1-1/p}$$

$$\leq C(1 + |x|^m)\left[\frac{\mathbb{E}|X(s;t,x)|}{k}\right]^{1-1/p} \leq C(1 + |x|^m)\left[\frac{1 + |x|}{k}\right]^{1-1/p}$$

which converges to 0 uniformly on bounded sets. □

Remark 1.161 Estimate (1.101) is satisfied in two important cases:

- when Hypothesis 1.149 is satisfied with b and σ independent of a;
- when Hypotheses 1.143 and 1.145 hold without dependence on a_1 and with $a_2(\cdot) = 0$.

This follows from the growth estimates of Theorem 1.152 and Proposition 1.147.∎

Theorem 1.162 *Assume that Hypothesis 1.149 is satisfied. Then for every $\phi \in C_m(H)$ ($m \geq 0$), the function $P_{t,s}[\phi]\colon H \to \mathbb{R}$ belongs to $C_m(H)$. The same holds if we assume that Hypotheses 1.143 and 1.145 hold without dependence on a_1 and with $a_2(\cdot) = 0$.*

Proof The result is a consequence of the continuous dependence and growth estimates of Theorem 1.152 and Propositions 1.153 and 1.147. □

1.7 Itô's and Dynkin's Formulae

In this section we assume that $T > 0$, H, Ξ, Q, and the generalized reference probability space $\mu = (\Omega, \mathscr{F}, \{\mathscr{F}_s\}_{s \in [0,T]}, \mathbb{P}, W_Q)$ are the same as in Sect. 1.3. The operator A is the generator of a C_0-semigroup on H, and Λ is a Polish space. The various Itô's and Dynkin's formulae presented in this section are used in proving existence of viscosity solutions (Chap. 3) and verification theorems (Chaps. 4 and 5).

Given a function $F : [0, T] \times H \to \mathbb{R}$, we denote by F_t the derivative of $F(t, x)$ with respect to t and by DF and D^2F the first and second-order Fréchet derivatives with respect to x.

Theorem 1.163 (Itô's Formula) *Assume that Φ is a process in $\mathcal{N}_Q^2(0, T; H)$, f is an H-valued progressively measurable ($\mathbb{P}$-a.s.) Bochner integrable process on $[0, T]$, and define, for $s \in [0, T]$,*

$$X(s) := X(0) + \int_0^s f(r)dr + \int_0^s \Phi(r)dW_Q(r),$$

where $X(0)$ is an $\mathscr{F}_0$-measurable H-valued random variable. Consider $F : [0, T] \times H \to \mathbb{R}$ and assume that F and its derivatives F_t, DF, D^2F are continuous and bounded on bounded subsets of $[0, T] \times H$. Let τ be an $\mathscr{F}_s$-stopping time. Then, for $\mathbb{P}$-a.e. ω,

$$F(s \wedge \tau, X(s \wedge \tau)) = F(0, X(0)) + \int_0^{s \wedge \tau} F_t(r, X(r))dr$$

$$+ \int_0^{s \wedge \tau} \langle DF(r, X(r)), f(r) \rangle \, dr + \int_0^{s \wedge \tau} \langle DF(r, X(r)), \Phi(r)dW_Q(r) \rangle$$

$$+ \frac{1}{2} \int_0^{s \wedge \tau} \mathrm{Tr}\left[\left(\Phi(r)Q^{1/2}\right)\left(\Phi(r)Q^{1/2}\right)^* D^2F(r, X(r)) \right]dr \quad on \ [0, T].$$

$$(1.102)$$

Proof See [294], Theorems 2.9 and 2.10. See also, under the assumption of uniform continuity on bounded sets of F and its derivatives, [180] Theorem 4.32, p. 106. □

Proposition 1.164 *Let $F : [0, T] \times H \to \mathbb{R}$ and $x \in H$. Assume that F and its derivatives F_t, DF, D^2F are continuous and bounded on bounded subsets of $[0, T] \times H$. Suppose that $DF : [0, T] \times H \to D(A^*)$ and that A^*DF is continuous and bounded on bounded subsets of $[0, T] \times H$. Let $f \in M_\mu^p(0, T; H)$, $\Phi \in \mathcal{N}_Q^p(0, T; H)$ for some $p > 2$. Let $X(\cdot)$ be the unique mild solution of (1.42) such that $X(0) = x$ and τ be an $\mathscr{F}_s$-stopping time. Then, for $\mathbb{P}$-a.e. ω,*

$$F(s \wedge \tau, X(s \wedge \tau)) = F(0, x) + \int_0^{s \wedge \tau} F_t(r, X(r)) dr$$

$$+ \int_0^{s \wedge \tau} \langle A^* DF(r, X(r)), X(r) \rangle dr + \int_0^{s \wedge \tau} \langle DF(r, X(r)), f(r) \rangle dr$$

$$+ \frac{1}{2} \int_0^{s \wedge \tau} \mathrm{Tr} \left[\left(\Phi(r) Q^{1/2} \right) \left(\Phi(r) Q^{1/2} \right)^* D^2 F(r, X(r)) \right] dr$$

$$+ \int_0^{s \wedge \tau} \langle DF(r, X(r)), \Phi(r) dW_Q(r) \rangle \qquad on \ [0, T].$$

$$(1.103)$$

Proof Since both sides of (1.103) are continuous processes, it is enough to prove the formula for a single s. We approximate $X(\cdot)$ by the sequence $X^n(\cdot)$ introduced in Proposition 1.132. By definition $X^n(\cdot)$ solves the integral equation

$$X^n(s) = \int_0^s \left(A_n X^n(r) + f(r) \right) dr + \int_0^s \Phi(r) dW_Q(r).$$

For any $R > 0$ such that $|x| < R$ define the stopping times

$$\hat{\tau}^R := \inf \{ s \in [0, T] \ : \ |X(s)| > R \}, \quad \hat{\tau}_n^R := \inf \{ s \in [0, T] \ : \ |X_n(s)| > R + 1 \}$$

and denote by τ^R and τ_n^R, respectively,

$$\tau^R := \min(\tau, \hat{\tau}^R), \quad \tau_n^R := \min(\tau, \hat{\tau}^R, \hat{\tau}_n^R).$$

Observe that, thanks to (1.44), up to extracting a subsequence of X_n (still denoted by X_n), $\sup_{s \in [0,T]} |X^n(s) - X(s)|^p$ converges to 0 on some set $\tilde{\Omega}$ with $\mathbb{P}(\tilde{\Omega}) = 1$. It is then easy to see that on $\tilde{\Omega}$ we have

$$\lim_{n \to \infty} \tau_n^R = \tau^R.$$

We deduce that, for $\omega \in \tilde{\Omega}$,

$$\lim_{n \to \infty} \mathbf{1}_{[0, s \wedge \tau_n^R]} = \mathbf{1}_{[0, s \wedge \tau^R]}, \quad \text{pointwise on } [0, T]. \qquad (1.104)$$

We can apply Itô's formula (1.102) to the approximating problem (A_n^* is the adjoint of A_n) obtaining, once we rewrite it using Lemma 1.110,

$$F(s \wedge \tau_n^R, X^n(s \wedge \tau_n^R)) = F(0, x) + \int_0^s \mathbf{1}_{[0, s \wedge \tau_n^R]}(r) F_t(r, X^n(r)) dr$$

$$+ \int_0^s \mathbf{1}_{[0, s \wedge \tau_n^R]}(r) \langle A_n^* DF(r, X^n(r)), X^n(r) \rangle dr$$

$$+ \int_0^s \mathbf{1}_{[0,s \wedge \tau_n^R]}(r) \langle DF(r, X^n(r)), \, f(r) \rangle dr$$

$$+ \frac{1}{2} \int_0^s \mathbf{1}_{[0,s \wedge \tau_n^R]}(r) \text{Tr} \left[\left(\Phi(r) Q^{1/2} \right) \left(\Phi(r) Q^{1/2} \right)^* D^2 F(r, X^n(r)) \right] dr$$

$$+ \int_0^s \mathbf{1}_{[0,s \wedge \tau_n^R]}(r) \langle DF(r, X^n(r)), \, \Phi(r) dW_Q(r) \rangle. \quad (1.105)$$

By the local boundedness of F and its derivatives, it follows that for $\mathbb{P}$-a.e. ω all the integrands of the deterministic integrals in (1.105) are dominated for $n \in \mathbb{N}$ by integrable functions. Regarding the term containing $A_n^* DF(r, X^n(r))$, recall from (B.11) that $A_n = J_n A$ are uniformly bounded as linear operators from $D(A)$ (endowed with the graph norm) to H. Moreover, thanks to (1.104), (1.44) and the continuity of F and its derivatives, we know that these integrands converge to the corresponding ones in (1.103) (with τ_R instead of τ) on $[0, s]$, $\mathbb{P}$-a.s. We can thus conclude, by using the Lebesgue dominated convergence theorem, that the deterministic integrals in (1.105) converge to their counterparts in (1.103).

To justify the convergence of the stochastic integral we observe that, with

$$I_n := \int_0^s \mathbf{1}_{[0,s \wedge \tau_n^R]}(r) \langle DF(r, X^n(r)), \, \Phi(r) dW_Q(r) \rangle,$$

$$I := \int_0^s \mathbf{1}_{[0,s \wedge \tau^R]}(r) \langle DF(r, X(r)), \, \Phi(r) dW_Q(r) \rangle,$$

we have

$$\mathbb{E} |I_n - I|^2$$
$$\leq \int_0^s \mathbb{E} \| \Phi(r) \|^2_{\mathcal{L}_2(\Xi_0, H)} \left| \mathbf{1}_{[0,s \wedge \tau_n^R]}(r) DF(r, X^n(r)) - \mathbf{1}_{[0,s \wedge \tau^R]}(r) DF(r, X(r)) \right|^2 dr \to 0$$

as $n \to +\infty$ by the dominated convergence theorem. Therefore, up to a subsequence, we have $\lim_{n \to +\infty} I_n = I$, $\mathbb{P}$-a.s.

It now remains to let $R \to +\infty$ to obtain the claim.   □

Proposition 1.165 *Let b and σ satisfy Hypothesis 1.125 and let $a: [t, T] \to \Lambda$ be a progressively measurable process. Let $X(\cdot)$ be the unique mild solution of (1.30) such that $X(0) = x \in H$. Consider $F: [0, T] \times H \to \mathbb{R}$. Assume that F and its derivatives $F_t, DF, D^2 F$ are continuous on $[0, T] \times H$. Suppose that $DF: [0, T] \times H \to D(A^*)$ and that $A^* DF$ is continuous on $[0, T] \times H$. Moreover, suppose that there exist $C \geq 0, N \geq 0$ such that*

$$|F(s, x)| + |DF(s, x)| + |F_t(s, x)| + \|D^2 F(s, x)\|$$
$$+ |A^* DF(s, x)| \leq C(1 + |x|)^N \quad (1.106)$$

for all $x \in H$, $s \in [0, T]$. Let τ be an $\mathscr{F}_s$-stopping time. Then:

(i) For $\mathbb{P}$-a.e. ω,

$$F(s \wedge \tau, X(s \wedge \tau)) = F(0, x) + \int_0^{s \wedge \tau} F_t(r, X(r))dr$$

$$+ \int_0^{s \wedge \tau} \langle A^* DF(r, X(r)), X(r) \rangle dr + \int_0^{s \wedge \tau} \langle DF(r, X(r)), b(r, X(r), a(r)) \rangle dr$$

$$+ \frac{1}{2} \int_0^{s \wedge \tau} \text{Tr} \left[\left(\sigma(r, X(r), a(r)) Q^{1/2} \right) \left(\sigma(r, X(r), a(r)) Q^{1/2} \right)^* D^2 F(r, X(r)) \right] dr$$

$$+ \int_0^{s \wedge \tau} \langle DF(r, X(r)), \sigma(r, X(r), a(r)) dW_Q(r) \rangle \quad \text{on } [0, T]. \quad (1.107)$$

(ii) Let η be a real process solving

$$\begin{cases} d\eta(s) = \tilde{b}(s)ds \\ \eta(0) = \eta_0 \in \mathbb{R}, \end{cases}$$

where $\tilde{b} : [0, T] \to \mathbb{R}$ is bounded and progressively measurable. Then, for $\mathbb{P}$-a.e. ω,

$$F(s \wedge \tau, X(s \wedge \tau))\eta(s \wedge \tau) = F(0, x)\eta_0 + \int_0^{s \wedge \tau} (F_t(r, X(r))\eta(r) + F(r, X(r))\tilde{b}(r))dr$$

$$+ \int_0^{s \wedge \tau} \langle A^* DF(r, X(r)), X(r) \rangle \eta(r)dr + \int_0^{s \wedge \tau} \langle DF(r, X(r)), b(r, X(r), a(r)) \rangle \eta(r)dr$$

$$+ \frac{1}{2} \int_0^{s \wedge \tau} \text{Tr} \left[\left(\sigma(r, X(r), a(r)) Q^{\frac{1}{2}} \right) \left(\sigma(r, X(r), a(r)) Q^{\frac{1}{2}} \right)^* D^2 F(r, X(r)) \right] \eta(r)dr$$

$$+ \int_0^{s \wedge \tau} \langle DF(r, X(r))\eta(r), \sigma(r, X(r), a(r)) dW_Q(r) \rangle \quad \text{on } [0, T]. \quad (1.108)$$

In particular, for $s \in [0, T]$,

$$\mathbb{E}\left[F(s \wedge \tau, X(s \wedge \tau))\eta(s \wedge \tau) \right] = F(0, x)\eta_0 + \mathbb{E} \int_0^{s \wedge \tau} (F_t(r, X(r))\eta(r) + F(r, X(r))\tilde{b}(r))dr$$

$$+ \mathbb{E} \int_0^{s \wedge \tau} \langle A^* DF(r, X(r)), X(r) \rangle \eta(r)dr + \mathbb{E} \int_0^{s \wedge \tau} \langle DF(r, X(r)), b(r, X(r), a(r)) \rangle \eta(r)dr$$

$$+ \frac{1}{2} \mathbb{E} \int_0^{s \wedge \tau} \text{Tr} \left[\left(\sigma(r, X(r), a(r)) Q^{\frac{1}{2}} \right) \left(\sigma(r, X(r), a(r)) Q^{\frac{1}{2}} \right)^* D^2 F(r, X(r)) \right] \eta(r)dr.$$

$$(1.109)$$

Proof Part *(i)* follows directly from Proposition 1.164 applied with $f(s) := b(s, a(s), X(s))$ and $\Phi(s) := \sigma(s, a(s), X(s))$, $s \in [0, T]$, by noticing that, thanks to (1.33), (1.34) and (1.37), we have $f \in M_\mu^p(0, T; H)$ and $\Phi \in \mathcal{N}_Q^p(0, T; H)$ for every $p \geq 1$.

Part *(ii)* is a corollary of *(i)*. We introduce the Hilbert space $\hat{H} := H \times \mathbb{R}$ (with the usual inner product), and set

$$\hat{A} = \begin{pmatrix} A \\ 0 \end{pmatrix}, \quad \hat{b} = \begin{pmatrix} b \\ \tilde{b} \end{pmatrix}, \quad \hat{\sigma} = \begin{pmatrix} \sigma & 0 \\ 0 & 0 \end{pmatrix}.$$

Then the process

$$\hat{X}(s) = \begin{pmatrix} X(s) \\ \eta(s) \end{pmatrix}$$

is the mild solution of the SDE

$$\begin{cases} d\hat{X}(s) = \left(\hat{A}\hat{X}(s) + \hat{b}(s, \hat{X}(s), a(s))\right) ds + \hat{\sigma}(s, \hat{X}(s), a(s)) dW_Q(s) \\ \hat{X}(0) = \begin{pmatrix} x \\ \eta_0 \end{pmatrix}. \end{cases}$$

Therefore, (1.108) follows from (1.107) applied to the function $\hat{F}(s, \hat{x}) = F(s, x)\eta_0$, where $\hat{x} = (x, \eta_0)$. Taking expectation in (1.108) we obtain (1.109).   □

Proposition 1.166 *Let Hypothesis 1.125 be satisfied and A be maximal dissipative. Let $a: [t, T] \to \Lambda$ be a progressively measurable process. Let $X(\cdot)$ be the unique mild solution of (1.30) such that $X(0) = x \in H$. Let $F \in C^{1,2}([0, T] \times H)$ be of the form $F(t, x) = \varphi(t, |x|)$ for some $\varphi(t, r) \in C^{1,2}([0, T] \times \mathbb{R})$, where $\varphi(t, \cdot)$ is even and non-decreasing on $[0, +\infty)$. Moreover, suppose that there exist $C \geq 0, N \geq 0$ such that*

$$|F(s, x)| + |DF(s, x)| + |F_t(s, x)| + \|D^2 F(s, x)\| \leq C(1 + |x|)^N \qquad (1.110)$$

for all $x \in H$, $s \in [0, T]$. Let τ be an $\mathscr{F}_s$-stopping time. Then:

(i) For $\mathbb{P}$-a.e. ω,

$$F(s \wedge \tau, X(s \wedge \tau)) \leq F(0, x) + \int_0^{s \wedge \tau} \left[F_t(r, X(r)) + \langle b(r, X(r), a(r)), DF(r, X(r)) \rangle \right.$$

$$\left. + \frac{1}{2} \mathrm{Tr} \left[\left(\sigma(r, X(r), a(r))Q^{\frac{1}{2}}\right) \left(\sigma(r, X(r), a(r))Q^{\frac{1}{2}}\right)^* D^2 F(r, X(r)) \right] \right] dr$$

$$+ \int_0^{s \wedge \tau} \langle DF(r, X(r)), b(r, X(r), a(r)) dW_Q(r) \rangle \qquad \text{on } [0, T]. \qquad (1.111)$$

(ii) If η is as in part (ii) of Proposition 1.165 and η is positive then, for $\mathbb{P}$-a.e. ω,

$$F(s \wedge \tau, X(s \wedge \tau))\eta(s \wedge \tau) \leq F(0, x)\eta_0 + \int_0^{s \wedge \tau} (F_t(r, X(r))\eta(r) + F(r, X(r))\tilde{b}(r)) dr$$

$$+ \int_0^{s \wedge \tau} \langle DF(r, X(r)), b(r, X(r), a(r)) \rangle \, \eta(r) dr$$

$$+ \frac{1}{2} \int_0^{s \wedge \tau} \mathrm{Tr} \left[\left(\sigma(r, X(r), a(r))Q^{\frac{1}{2}}\right) \left(\sigma(r, X(r), a(r))Q^{\frac{1}{2}}\right)^* D^2 F(r, X(r)) \right] \eta(r) dr$$

$$+ \int_0^{s \wedge \tau} \langle DF(r, X(r))\eta(r), \sigma(r, X(r), a(r)) dW_Q(r) \rangle \qquad \text{on } [0, T]. \qquad (1.112)$$

In particular, for $s \in [0, T]$,

$$\mathbb{E}\left[F(s \wedge \tau, X(s \wedge \tau))\eta(s \wedge \tau)\right] \leq F(0, x)\eta_0$$

$$+ \mathbb{E} \int_0^{s \wedge \tau} (F_t(r, X(r))\eta(r) + F(r, X(r))\tilde{b}(r))dr$$

$$+ \mathbb{E} \int_0^{s \wedge \tau} \langle DF(r, X(r)), b(r, X(r), a(r)) \rangle \, \eta(r)dr$$

$$+ \frac{1}{2}\mathbb{E} \int_0^{s \wedge \tau} \operatorname{Tr}\left[\left(\sigma(r, X(r), a(r))Q^{\frac{1}{2}}\right) \left(\sigma(r, X(r), a(r))Q^{\frac{1}{2}}\right)^* D^2 F(r, X(r)) \right] \eta(r)dr.$$
$$(1.113)$$

Proof (*i*) We set, for $s \in [0, T]$, $f(s) := b(s, a(s), X(s))$ and $\Phi(s) := \sigma(s, a(s), X(s))$ and consider the approximation $X^n(\cdot)$ of $X(\cdot)$ as in Proposition 1.132. Observe that, thanks to (1.33), (1.34) and (1.37) we have $f \in M_\mu^p(0, T; H)$ and $\Phi \in \mathcal{N}_Q^p(0, T; H)$ for every $p \geq 1$ so the assumptions of Proposition 1.132 are satisfied.

We observe that $DF(s, x) = \frac{\partial \varphi}{\partial r}(s, |x|)\frac{x}{|x|}$ and, since $\varphi(s, \cdot)$ is non-decreasing on $[0, +\infty)$, $\frac{\partial \varphi}{\partial r}(s, r) \geq 0$. Therefore, since A, and thus A_n, is dissipative,

$$\langle A_n X^n(s), DF(r, X^n(s)) \rangle = \frac{\partial \varphi}{\partial r}(s, |X^n(s)|)\frac{1}{|X^n(s)|}\langle A_n X^n(s), X^n(s) \rangle \leq 0$$
$$(1.114)$$

for every $s \geq 0$.

Hence, defining for any $R > |x|$ the stopping times τ_n^R as in Proposition 1.164, applying Itô's formula for $X^n(\cdot)$ and using (1.114), we obtain

$$F(s \wedge \tau_n^R, X^n(s \wedge \tau_n^R)) = F(0, x) + \int_0^{s \wedge \tau_n^R} \left[F_t(r, X^n(r)) + \langle A_n X^n(r), DF(r, X^n(r)) \rangle \right.$$

$$+ \langle f(r), DF(r, X^n(r)) \rangle + \frac{1}{2}\operatorname{Tr}\left[\left(\Phi(r)Q^{\frac{1}{2}}\right)\left(\Phi(r)Q^{\frac{1}{2}}\right)^* D^2 F(r, X^n(r)) \right] \right]dr$$

$$+ \int_0^{s \wedge \tau_n^R} \langle DF(r, X^n(r)), b(r, X^n(r), a(r))dW_Q(r) \rangle$$

$$\leq F(0, x) + \int_0^{s \wedge \tau_n^R} \left[F_t(r, X^n(r)) + \langle f(r), DF(r, X^n(r)) \rangle \right.$$

$$+ \frac{1}{2}\operatorname{Tr}\left[\left(\Phi(r)Q^{\frac{1}{2}}\right)\left(\Phi(r)Q^{\frac{1}{2}}\right)^* D^2 F(r, X^n(r)) \right] \right]dr$$

$$+ \int_0^{s \wedge \tau_n^R} \langle DF(r, X^n(r)), b(r, X^n(r), a(r))dW_Q(r) \rangle. \qquad (1.115)$$

It remains to pass to the limit as $n \to +\infty$ and $R \to +\infty$ in (1.115). This is done following the same arguments as in the proof of Proposition 1.164.

(*ii*) The proof combines the proof of (*i*) with the arguments used in the proof of Proposition 1.165-(*ii*). □

Remark 1.167 Propositions 1.165 and 1.166 are used to work with viscosity solution test functions in Chap. 3. In particular, parts (ii) of them are useful when discount factors are present (see e.g. Lemma 3.65). ∎

The next two non-standard versions of Dynkin's formula will be used to prove verification theorems in Chaps. 4 and 5.

Proposition 1.168 *Let $Q = I$. Assume that Hypothesis 1.149 is satisfied. Assume that there exists $\lambda \in \mathbb{R}$, $\lambda \in \varrho(A)$ such that $(\lambda I - A)^{-1}b \colon [0, T] \times H \times \Lambda \to H$ is measurable. Suppose moreover that there exists a $C > 0$ such that, for all $(t, x, a) \in [0, T] \times H \times \Lambda$,*

$$\begin{cases} |(\lambda I - A)^{-1}b(t, x, a)| \leq C(1 + |x|) \\ \|\sigma(t, x, a)\|_{\mathcal{L}(\Xi, H)} \leq C(1 + |x|). \end{cases} \tag{1.116}$$

Fix a Λ-valued progressively measurable process $a(\cdot)$. Let X be the unique mild solution of (1.74) described in Theorem 1.152 such that $X(0) = x \in H$. Let $F \colon [0, T] \times H \to \mathbb{R}$ be such that F and its derivatives F_t, DF, D^2F are continuous in $[0, T] \times H$. Suppose that $DF \colon [0, T] \times H \to D(A^*)$, that A^*DF is continuous in $[0, T] \times H$, that $D^2F \colon [0, T] \times H \to \mathcal{L}_1(H)$ is continuous, and that there exist $C > 0$ and $N \geq 1$ such that

$$|F(s, x)| + |DF(s, x)| + |F_t(s, x)| + \|D^2F(s, x)\|_{\mathcal{L}_1(H)}$$
$$+ |A^*DF(s, x)| \leq C(1 + |x|)^N. \tag{1.117}$$

Then, for any $s \in [0, T]$,

$$\mathbb{E}\,[F(s, X(s))] = F(0, x) + \mathbb{E} \int_0^s F_t(r, X(r))dr + \mathbb{E} \int_0^s \langle A^*DF(r, X(r)), X(r)\rangle dr$$
$$+ \mathbb{E} \int_0^s \Big\langle (\lambda I - A^*)DF(r, X(r)), (\lambda I - A)^{-1}b(r, X(r), a(r))\Big\rangle dr$$
$$+ \frac{1}{2}\mathbb{E} \int_0^s \mathrm{Tr}\Big[\sigma(r, X(r), a(r))\sigma(r, X(r), a(r))^* D^2F(r, X(r))\Big]dr. \tag{1.118}$$

Proof We approximate the process $X(\cdot)$ by the processes $X^k(\cdot)$ from Lemma 1.154.

Observe that, thanks to Hypothesis 1.149 and to (1.80), the processes $r \to e^{\frac{1}{k}A}b(r, X(r), a(r))$ and $r \to e^{\frac{1}{k}A}\sigma(r, X(r), a(r))$ belong respectively to $M_\mu^p(0, T; H)$ and $\mathcal{N}_I^p(0, T; H)$ for all $p \geq 1$. Thus we can apply Proposition 1.164 obtaining, for $s \in [0, T]$,

$$\mathbb{E}\left[F(s, X^k(s))\right] = F(0, x) + \int_0^s \mathbb{E}\, F_t(r, X^k(r))dr$$
$$+ \int_0^s \mathbb{E}\langle A^*DF(r, X^k(r)), X^k(r)\rangle dr + \int_0^s \mathbb{E}\Big\langle DF(r, X^k(r)), e^{\frac{1}{k}A}b(r)\Big\rangle dr$$
$$+ \frac{1}{2}\int_0^s \mathbb{E}\,\mathrm{Tr}\left[\left(e^{\frac{1}{k}A}\sigma(r)P^k\right)\left(e^{\frac{1}{k}A}\sigma(r)P^k\right)^* D^2F(r, X^k(r))\right]dr,$$
$$\tag{1.119}$$

where we use the notation $b(r) := b(r, X(r), a(r))$, $\sigma(r) := \sigma(r, X(r), a(r))$. The claim will follow if we can pass to the limit as $k \to +\infty$ in each term of this expression. We will only show how to prove the convergence of the last two terms since the arguments for the other terms are similar and simpler.

Using (1.80), (1.86), (1.87) and the dominated convergence theorem it is easy to see that

$$\lim_{k \to \infty} |X(\cdot) - X^k(\cdot)|_{M^2_\mu(0,T;H)} = 0.$$

Therefore we can find a subsequence, still denoted by $X^k(\cdot)$, that converges to $X(\cdot)$ $dt \otimes \mathbb{P}$-a.e.

Using the assumptions it is obvious that

$$\left\langle DF(r, X^k(r)), e^{\frac{1}{k}A}b(r) \right\rangle = \left\langle (\lambda I - A^*)DF(r, X^k(r)), e^{\frac{1}{k}A}(\lambda I - A)^{-1}b(r) \right\rangle$$
$$\to \left\langle (\lambda I - A^*)DF(r, X(r)), (\lambda I - A)^{-1}b(r) \right\rangle \quad dt \otimes \mathbb{P} - a.e.$$

as $k \to +\infty$. Moreover, thanks to (1.80), (1.86), (1.116) and (1.117),

$$\int_0^s \mathbb{E} \left| \left\langle (\lambda I - A^*)DF(r, X^k(r)), e^{\frac{1}{k}A}(\lambda I - A)^{-1}b(r) \right\rangle \right|^2 dr$$
$$\leq C_1 \int_0^s \mathbb{E} \left[\left(1 + |X^k(r)|^{2N}\right) \left(1 + |X(r)|^2\right) \right] dr \leq C_2$$

for some C_1 and C_2 independent of k. Similarly we obtain

$$\int_0^s \mathbb{E} \left| \left\langle (\lambda I - A^*)DF(r, X(r)), (\lambda I - A)^{-1}b(r) \right\rangle \right|^2 dr \leq C_3$$

for some C_3. Therefore it follows from Lemma 1.51 that

$$\lim_{k \to +\infty} \int_0^s \mathbb{E} \left\langle DF(r, X^k(r)), e^{\frac{1}{k}A}b(r) \right\rangle dr$$
$$= \int_0^s \mathbb{E} \left\langle (\lambda I - A^*)DF(r, X(r)), (\lambda I - A)^{-1}b(r) \right\rangle dr.$$

Regarding the last term in (1.119),

$$\text{Tr} \left[e^{\frac{1}{k}A}\sigma(r)P^k(e^{\frac{1}{k}A}\sigma(r)P^k)^* D^2 F(r, X^k(r)) \right] - \text{Tr} \left[\sigma(r)\sigma(r)^* D^2 F(r, X(r)) \right]$$
$$= I_1 + I_2 := \text{Tr} \left[e^{\frac{1}{k}A}\sigma(r)P^k(e^{\frac{1}{k}A}\sigma(r)P^k)^* \left(D^2 F(r, X^k(r)) - D^2 F(r, X(r)) \right) \right]$$
$$+ \text{Tr} \left[\left(e^{\frac{1}{k}A}\sigma(r)P^k(e^{\frac{1}{k}A}\sigma(r)P^k)^* - \sigma(r)\sigma(r)^* \right) D^2 F(r, X(r)) \right].$$

By Proposition B.26, (1.116) and the assumptions for D^2F we have

$$|I_1| \leq C_4(1 + |X(r)|)^2 \|D^2F(r, X^k(r)) - D^2F(r, X(r))\|_{\mathcal{L}_1(H)} \to 0 \quad \text{as} \quad k \to +\infty$$

$dt \otimes \mathbb{P}$-a.e. Let $\{e_1, e_2, ...\}$ be an orthonormal basis of eigenvectors of $D^2F(r, X(r))$ and $\lambda_1, \lambda_2, ...$ be the corresponding eigenvalues. Then

$$\text{Tr}\left[e^{\frac{1}{k}A}\sigma(r)P^k(e^{\frac{1}{k}A}\sigma(r)P^k)^*D^2F(r, X(r)) \right]$$

$$= \sum_{n=1}^{\infty} \lambda_n \left| P^k\sigma(r)^*e^{\frac{1}{k}A^*}e_n \right|_\Xi^2 \to \sum_{n=1}^{\infty} \lambda_n \left| \sigma(r)^*e_n \right|_\Xi^2$$

$$= \text{Tr}\left[\sigma(r)\sigma(r)^*D^2F(r, X(r)) \right] \quad \text{as} \quad k \to +\infty$$

$dt \otimes \mathbb{P}$-a.e. Therefore $\lim_{k \to +\infty}(I_1 + I_2) = 0$ $dt \otimes \mathbb{P}$-a.e. Since, by (1.80), (1.86), (1.116) and (1.117), we also have

$$\int_0^s \mathbb{E}|I_1 + I_2|^2 dr \leq C_5$$

for some constant C_5 independent of k, the convergence of the last term in (1.119) now follows from Lemma 1.51. $\square$

Proposition 1.169 *Let Hypotheses 1.143 and 1.145 be satisfied and let $q \geq 2$. Consider $\lambda \in \mathbb{R}$ such that $(\lambda I - A)$ is invertible and $(\lambda I - A)^{-1} \in \mathcal{L}(H)$. Assume that $(\lambda I - A)^{-1}a_2(\cdot) \in M_\mu^1(0, T; H)$. Let X be the unique mild solution of (1.67) described in Proposition 1.147 such that $X(0) = x \in H$. Let $F: [0, T] \times H \to \mathbb{R}$ be such that F and its derivatives F_t, DF, D^2F are continuous in $[0, T] \times H$. Suppose that $DF: [0, T] \times H \to D(A^*)$, A^*DF is continuous in $[0, T] \times H$, $D^2F: [0, T] \times H \to \mathcal{L}_1(H)$ is continuous and there exists a $C > 0$ such that (1.117) holds with $N = 0$. Then, for any $s \in [0, T]$,*

$$\mathbb{E}[F(s, X(s))] = F(0, x) + \mathbb{E}\int_0^s F_t(r, X(r))dr$$

$$+ \mathbb{E}\int_0^s \langle A^*DF(r, X(r)), X(r)\rangle dr + \mathbb{E}\int_0^s \langle DF(r, X(r)), b_0(r, X(r), a_1(r))\rangle dr$$

$$+ \mathbb{E}\int_0^s \langle (\lambda I - A^*)DF(r, X(r)), (\lambda I - A)^{-1}a_2(r)\rangle dr$$

$$+ \frac{1}{2}\mathbb{E}\int_0^s \text{Tr}\left[\sigma Q\sigma^*D^2F(r, X(r)) \right] dr.$$

Proof We approximate X using the processes X^k defined in Lemma 1.156. It is immediate to see that $\psi_k \in M_\mu^p(0, T; H)$ and $\sigma \tilde{P}^k \in \mathcal{N}_Q^p(0, T; H)$ for all $p \geq 1$ so we can apply Proposition 1.164 obtaining for every $s \in [0, T]$,

$$\mathbb{E}\Big[F(s, X^k(s))\Big] = F(0, x) + \mathbb{E}\int_0^s F_t(r, X^k(r))dr \ + \mathbb{E}\int_0^s \Big\langle A^* DF(r, X^k(r)), X^k(r)\Big\rangle dr$$

$$+ \mathbb{E}\int_0^s \mathbf{1}_{B_k}(r, \omega)\Big\langle DF(r, X^k(r)), b_0(r, X(r), a_1(r))\Big\rangle dr$$

$$+ \mathbb{E}\int_0^s \mathbf{1}_{D_{m_k}}(r, \omega)\Big\langle(\lambda I - A^*)DF(r, X^k(r)), e^{\frac{1}{k}A}(\lambda I - A)^{-1}a_2(r)\Big\rangle dr$$

$$+ \frac{1}{2}\mathbb{E}\int_0^s \mathrm{Tr}\Big[(\sigma Q^{1/2} P^k)(\sigma Q^{1/2} P^k)^* D^2 F(r, X^k(r))\Big]dr, \qquad (1.120)$$

where B_k, D_{m_k} and P^k are introduced in Lemma 1.156 and in the paragraph before it.

We need to check the convergence of each term of this expression. Using parts (i) and (ii) of Lemma 1.156 we have

$$\lim_{k\to\infty} |X(\cdot) - X^k(\cdot)|_{M_\mu^1(0,T;H)} = 0.$$

Therefore we can find a subsequence of X^k, still denoted by X^k, that converges $dt \otimes \mathbb{P}$-a.e. to X. The proof proceeds using the same arguments (and even simpler) as those in the proof of Proposition 1.169. We only look at the two middle terms of the right-hand side of (1.120) that are a little different. We observe that

$$\left|\mathbb{E}\int_0^s \Big\langle(\lambda I - A^*)DF(r, X^k(r)), \Big(1 - \mathbf{1}_{D_{m_k}}(r, \omega)e^{\frac{1}{k}A}\Big)(\lambda I - A)^{-1}a_2(r)\Big\rangle dr\right|$$

converges to zero thanks to the dominated convergence (recall that, by assumption, (1.117) holds with $N = 0$). Regarding the fourth term observe that

$$\mathbf{1}_{B_k}(r, \omega)\Big\langle DF(r, X^k(r)), b_0(r, X(r), a_1(r))\Big\rangle$$

converges to

$$\langle DF(r, X(r)), b_0(r, X(r), a_1(r))\rangle$$

$dt \otimes \mathbb{P}$-a.e. as $k \to +\infty$. Moreover, since DF is bounded, Hypothesis 1.145-(i) implies

$$\Big|\mathbf{1}_{B_k}(r, \omega)\Big\langle DF(r, X^k(r)), b_0(r, X(r), a_1(r))\Big\rangle\Big| \le Cf(r)(1 + |X(r)|)$$

for all $k \in \mathbb{N}$. Thus the result follows by the dominated convergence theorem. □

1.8 Bibliographical Notes

Section 1.1 contains elements of basic probability and measure theory. Classical references include, for example, [18, 58, 61, 267, 370, 478, 520]. We refer in particular to [58, 61, 267, 370] for the general theory of measure and probability (Sect. 1.1.1) and to [58, 61, 267, 520] for results on measurability (Sects. 1.1.2 and 1.1.3). For the Bochner integral and the integration of Banach-valued functions (Sect. 1.1.3), the reader can consult [190, 191, 194, 397]; some results, useful from the stochastic calculus perspective are contained in [180]. For Sects. 1.1.4 and 1.1.5 conditional expectation for Banach-valued random variables the reader can refer to [180, 356, 370, 478, 572]. Gaussian measure in Hilbert spaces (Sect. 1.1.6) and Fourier transform are nicely introduced in [153, 180] and a more extended study of the subject is contained in [391].

Generalities about stochastic processes, martingales and stopping times in Sect. 1.2 can be found in many different books, e.g. [356, 372, 384, 447–449, 503, 508, 572], while for Hilbert-valued martingales (Sect. 1.2.2) the reader may consult [180, 294, 487]. For standard Wiener and Q-Wiener processes and related results we refer to [124, 180, 294, 372, 447, 448, 452]. The material of Sect. 1.2.4 is based on [180]. Definition 1.92, which not contained in the standard literature, is introduced here because it is useful to study stochastic control problems. The presentation of Lemma 1.94 is based on [372, 513]. The material of Sect. 1.2.5 is loosely based on [180, 294, 372].

The material of Sect. 1.3 is based on [177, 180, 294] (see also [124, 491]). These books present the theory in Hilbert spaces while [447, 448] (see also [192]) present the Banach space case.

The presentation of Sect. 1.4 on solutions of stochastic differential equations in Hilbert spaces is also based on [180, 294]. In particular, [180] is a standard reference in the theory. Other references on strong and mild solutions are, for example, in [124, 177, 413] while a good introduction to variational solutions is in [124, 387, 413, 491, 519]. The reader is also referred to [180] for more on weak mild solutions. Section 1.4.4, containing some results about uniqueness in law, uses the approach of [471]. For a different approach to weak uniqueness based on the theorem of Yamada–Watanabe, we refer the reader to [491], Appendix E.

Section 1.5 contains existence and uniqueness results for stochastic differential equations with special unbounded terms and cylindrical additive noise. They are more or less common knowledge, however we presented proofs since no complete references seem to be available in the literature.

Classical results on transition semigroups (Sect. 1.6) can be found in [180]. The statements here are a little modified and extended so that they may be used in our applications to optimal control, mainly in Chap. 4.

Section 1.7 contains various versions of Itô's and Dynkin's formulae (Propositions 1.164–1.166) in connection with mild solutions for functions that have properties of test functions used in the definition of a viscosity solution (Definition 3.32). Such results have been known and used in the viscosity solution literature, however complete proofs are available only in [374]. The statements here are slightly more general than those in [374] and we presented proofs for the reader's convenience. The last two results of Sect. 1.7 (Propositions 1.168 and 1.169) are used to prove the verification theorems of Sects. 4.8 and 5.5. They have been used in the literature (e.g. in [306]) but without complete proofs, hence we provide them for completeness. We finally recall that Itô's formula related to variational solutions of linear stochastic parabolic equations is proved in [467].

Chapter 2
Optimal Control Problems and Examples

In this chapter we discuss the connection between the study of infinite-dimensional stochastic optimal control problems and that of second-order Hamilton–Jacobi–Bellman (HJB) equations in Hilbert spaces. This so-called "dynamic programming approach" to optimal control problems is based on two main results:

- The *dynamic programming principle* (DPP), which is a functional equation for the value function of the control problem, and whose differential form is the HJB equation. This is the core result in the dynamic programming approach.
- The *verification theorem*, which gives a sufficient (and sometimes necessary) condition for optimality. Verification theorems rely on the HJB equation and open the way to the so-called *optimal synthesis*, i.e. the expression of the optimal control strategy as a function of the current state trajectory (the *feedback form*).

To carry out this dynamic programming approach one needs suitable existence, uniqueness, and regularity results for the solutions of the HJB equation. With this in mind we organize the chapter as follows.

In Sect. 2.1 we describe a general stochastic infinite-dimensional optimal control problem (in both strong and weak formulations).

Sections 2.2 and 2.3 contain the dynamic programming principle (with a complete proof) and the equivalence between weak and strong formulations when the underlying "information structure" of the problem is given by a *reference probability space*, see Definition 2.7. These formulations are a little less general than the one of Sect. 2.1 which uses the notion of the so-called *generalized reference probability space*, see Definition 1.100.

The problem and the statement of the dynamic programming principle are formulated in an abstract form so that they can be used in many cases when the solutions of the state equation (which is an infinite-dimensional SDE) are interpreted in various ways (strong, mild, variational, etc.). Since this increases the level of technicality, we recommend that the readers assume on first reading that the state equation in the

© Springer International Publishing AG 2017
G. Fabbri et al., *Stochastic Optimal Control in Infinite Dimension*,
Probability Theory and Stochastic Modelling 82,
DOI 10.1007/978-3-319-53067-3_2

control problem is the one described in Sect. 2.2.3 with solutions defined in the mild sense, as this is the most common case in this book and the theory then applies more straightforwardly. Section 2.4 is devoted to the infinite horizon problem.

In Sect. 2.5 we present classical verification theorems and the optimal synthesis when the value function is regular, in both the finite and the infinite horizon cases.

Finally, in Sect. 2.6 we discuss various examples of stochastic infinite-dimensional optimal control problems, which arise in applications, and which can be studied in the framework of the theory presented in this book.

The material on the dynamic programming principle and the examples are presented for optimal control problems defined on the whole space. We do not discuss in this book problems in bounded domains with more general cost functionals, including cost of exiting through the boundary, problems with optimal stopping, state constraint problems, singular control problems, risk sensitive control problems, ergodic control problems, and stochastic differential games. Some references to results for such problems are scattered throughout other sections.

2.1 Stochastic Optimal Control Problems: General Formulation

2.1.1 Strong Formulation

We start with a description of a general stochastic optimal control problem in an infinite-dimensional Hilbert space. We will be using the convention of Notation 1.68.

We make the following assumptions:

Hypothesis 2.1

(i) The *state space* H and the *noise space* Ξ are real separable Hilbert spaces.
(ii) The *control space* Λ is a Polish space.
(iii) The horizon of the problem is $T \in (0, +\infty) \cup \{+\infty\}$, and the initial time is $t \in [0, T)$.
(iv) $\mu := \left(\Omega^\mu, \mathscr{F}^\mu, \{\mathscr{F}^t_{\mu,s}\}_{s \in [t,T]}, \mathbb{P}^\mu, W^\mu_Q \right)$ is a generalized reference probability space from Definition 1.100 with $W_Q(t) = 0$, $\mathbb{P}$-a.s.

We introduce the set of *admissible controls*

$$\mathcal{U}^\mu_t := \left\{ a(\cdot) \colon [t, T] \times \Omega \to \Lambda \; : \; a(\cdot) \text{ is } \mathscr{F}^t_{\mu,s}\text{-progressively measurable} \right\}. \quad (2.1)$$

The notation $\mathcal{U}^\mu_t$ emphasizes the dependence on the generalized reference probability space. Sometimes additional conditions (e.g. state constraints) are imposed on the admissible controls.

In a general infinite-dimensional stochastic optimal control problem, we consider, for every $a^\mu(\cdot) \in \mathcal{U}_t^\mu$, a system driven by an abstract stochastic differential equation in H

$$\begin{cases} dX(s) = \beta(s, X(s), a^\mu(s))ds + \sigma(s, X(s), a^\mu(s))dW_Q^\mu(s), & s \in [t, T], \\ X(t) = x \in H, \end{cases} \tag{2.2}$$

where β, σ are appropriate functions for which the above equation is well posed (in a sense to be made precise, see Remark 2.2) for every admissible control. Such an equation is called the *state equation* and we denote by $X(\cdot; t, x, a^\mu(\cdot)) : [t, T] \to H$ (or simply by $X(\cdot)$ when its meaning is clear) its unique solution. This is the *state trajectory* of the system. The pair $(a^\mu(\cdot), X(\cdot; t, x, a^\mu(\cdot)))$ will be called an *admissible couple* (or *admissible pair*).

The goal is to *minimize*, over all $a^\mu(\cdot) \in \mathcal{U}_t^\mu$, the *cost functional*

$$J^\mu(t, x; a^\mu(\cdot)) = \mathbb{E}^\mu \Bigg[\int_t^T e^{-\int_t^s c(X(\tau; t, x, a^\mu(\cdot)))d\tau} l(s, X(s; t, x, a^\mu(\cdot)), a^\mu(s))ds$$
$$+ e^{-\int_t^T c(X(\tau; t, x, a^\mu(\cdot)))d\tau} g(X(T; t, x, a^\mu(\cdot))) \Bigg], \tag{2.3}$$

where $l : [t, T] \times H \times \Lambda \to \mathbb{R}, c, g : H \to \mathbb{R}$ are Borel measurable functions, and c is bounded from below. The function l is the so-called running cost, g is the terminal cost, and c is a function responsible for discounting. When $T = +\infty$ the standing convention will be that $g = 0$, i.e. the cost functional only depends on the running cost and discounting. When T is finite the problem is called a *finite horizon problem*, and when $T = +\infty$ it is called an *infinite horizon problem*. The expectation $\mathbb{E}^\mu$ is computed with respect to the probability measure $\mathbb{P}^\mu$, so it depends on the generalized reference probability space. When the generalized reference probability space is clear we will often drop the superscript μ in our notation. We will refer to the above problem as the *strong formulation* of the stochastic optimal control problem (2.2) and (2.3) on $[t, T]$. Here 'strong' means that the generalized reference probability space is fixed.

The discounting function c may also depend on the control variable. The results of this chapter can be easily extended to cover such a case. However, we chose not to include this dependence in order not to overcomplicate the presentation, which is already very technical.

Remark 2.2 In the infinite-dimensional case the state equation (2.2) can have different forms, which may call for various definitions of solutions (strong solutions, mild solutions, variational solutions, etc.) and various approaches to solve them. For this reason, in our general formulation we do not specify the concept of solution of (2.2) and we do not specify the required assumptions. Later, in Sect. 2.2, we will formulate and prove the dynamic programming principle (DPP) in a general form so that it can be applied in these different situations. Thus we will make a series of rather abstract assumptions (see Hypotheses 2.11 and 2.12) about (2.2) that are satisfied in various cases for different concepts of solutions and which are sufficient to prove the DPP.

However, the reader should keep in mind that our primary guiding examples are the control problems of the type (2.2) and (2.3) where the state equation is a stochastic evolution equation with solutions interpreted in the mild sense. In such cases we have $\beta = A + b$, where A is the generator of a C_0-semigroup on H, and b, σ are functions satisfying suitable Lipschitz conditions. This case requires a less general formulation to prove the DPP and will be discussed separately in Sect. 2.2.3. The cases which do not use mild solutions include optimal control problems for the Duncan–Mortensen–Zakai, Burgers, Navier–Stokes, and reaction diffusion equations. ∎

The value function for problem (2.2) and (2.3) in the strong formulation with initial time t is defined as

$$V_t^{\mu}(x) = \inf_{a^{\mu}(\cdot) \in \mathcal{U}_t^{\mu}} J^{\mu}(t, x; a^{\mu}(\cdot)). \tag{2.4}$$

Notice, however, that in this strong formulation the generalized reference probability space changes when we change t and so does the control set $\mathcal{U}_t^{\mu}$.

Definition 2.3 (*Optimal control/couple*) If, for given initial data (t, x), $a^*(\cdot) \in \mathcal{U}_t^{\mu}$ minimizes (2.3), i.e. if $J^{\mu}(t, x; a^*(\cdot)) = V_t^{\mu}(x)$, we say that $a^*(\cdot)$ is a μ-*optimal control* at (t, x). The associated state trajectory $X^*(\cdot) := X(\cdot; t, x, a^*(\cdot))$ (i.e. the solution of (2.2) with control $a^*(\cdot)$) is an *optimal state* at (t, x). The pair $(a^*(\cdot), X^*(\cdot))$ is called a μ-*optimal couple* (or μ-*optimal pair*) at (t, x).

To perform the dynamic programming approach in the strong formulation we need to consider a family of problems (2.2) and (2.3) parameterized by the initial time t which are defined on a common generalized reference probability space, and introduce a value function defined on $[0, T] \times H$. To do this we take a generalized reference probability space $\mu = \left(\Omega^{\mu}, \mathscr{F}^{\mu}, \{\mathscr{F}^0_{\mu,s}\}_{s \in [0,T]}, \mathbb{P}^{\mu}, W_Q^{\mu} \right)$ on $[0, T]$ with $W_Q(0) = 0$ (i.e. μ satisfies Hypothesis 2.1 with initial time $t = 0$). We then define the value function

$$V^{\mu}(t, x) = \inf_{a^{\mu}(\cdot) \in \mathcal{U}_0^{\mu}} J^{\mu}(t, x; a^{\mu}(\cdot)), \tag{2.5}$$

where $J^{\mu}(t, x; a^{\mu}(\cdot))$ is defined by (2.3) with $X(\cdot; t, x, a^{\mu}(\cdot)))$ solving (2.2). We notice that for μ as above, the generalized reference probability spaces $\mu_t := \left(\Omega^{\mu}, \mathscr{F}^{\mu}, \{\mathscr{F}^0_{\mu,s}\}_{s \in [t,T]}, \mathbb{P}^{\mu}, W_Q^{\mu}(\cdot) - W_Q^{\mu}(t) \right)$ satisfy Hypothesis 2.1 with initial time t. Thus it is reasonable to expect that $V^{\mu}(t, x)$ should be equal to $V_t^{\mu_t}(x)$ for $(t, x) \in [0, T] \times H$. This is indeed the case for control problems considered in this book and it is a simple consequence of the properties of the stochastic integral (see e.g. (2.14)). Thus the requirement that $W_Q(t) = 0$ in Hypothesis 2.1-(iv) can be dropped for all practical purposes.

2.1.2   Weak Formulation

In the strong formulation of the optimal control problem (2.2) and (2.3), the generalized reference probability space $\mu := \left(\Omega, \mathscr{F}, \mathscr{F}_s^t, \mathbb{P}, W_Q\right)$ is fixed. However, it is often more convenient or necessary to include the generalized reference probability space as part of the control. In particular this approach is used to prove the dynamic programming principle and to construct optimal feedback controls (see Sects. 2.2 and 2.5). This leads us to the weak formulation of the stochastic optimal control problem.

To be clear from the beginning we must say that the weak formulation we introduce in this subsection is not the only possible one. Indeed, we will use two other types of weak formulations, the weak formulation for the DPP (see Sect. 2.2) which will be used to state and prove the dynamic programming principle, and the extended weak formulation (see Remark 2.6) which will be used in Sect. 6.5 to prove existence of optimal feedback controls in some cases. Roughly speaking the difference between the three is the following:

- the weak formulation for the DPP contains fewer controls than the one used in this subsection as they must be progressively measurable with respect to the augmented filtration generated by the underlying Wiener process W_Q and so they do not input any additional uncertainty into the system.
- the weak formulation of this section allows controls to be progressively measurable with respect to a possibly wider filtration than the one generated by W_Q so the controls can introduce more uncertainty into the system.
- the extended weak formulation used in Sect. 6.5 is like the weak formulation described, however we do not require that solutions of the state equation are unique. This formulation is very useful in the construction of optimal feedback controls, where solutions of the so-called closed loop equation are understood in the weak probabilistic sense where the filtered probability space becomes part of the solution.

We now introduce the details of the weak formulation of this section. The state equation and the cost functionals are the same as in Sect. 2.1.1, however, for each fixed $t \in [0, T]$, any generalized reference probability space μ is allowed and so the class of admissible controls is enlarged. We define

$$\overline{\mathcal{U}}_t := \bigcup_\mu \mathcal{U}_t^\mu, \tag{2.6}$$

where the union is taken over all generalized reference probability spaces μ satisfying Hypothesis 2.1-(iv). We say that $a(\cdot)$ is an admissible control if $a(\cdot) \in \overline{\mathcal{U}}_t$, i.e. if there exists a generalized reference probability space $\mu = \left(\Omega^\mu, \mathscr{F}^\mu, \mathscr{F}_s^{\mu,t}, \mathbb{P}^\mu, W_Q^\mu\right)$ satisfying Hypothesis 2.1-(iv) such that $a(\cdot): [t, T] \times \Omega^\mu \to \Lambda$ is $\mathscr{F}_s^{\mu,t}$-progressively measurable. We will often write $a^\mu(\cdot)$ to indicate the dependence of $a(\cdot)$ on the generalized reference probability space. This way, choosing an admissible control also means choosing a generalized reference space μ so, with a slight abuse of notation, we will often write $a(\cdot) = \left(\Omega, \mathscr{F}, \mathscr{F}_s^t, \mathbb{P}, W_Q, a(\cdot)\right)$.

Given a control $a^\mu(\cdot) \in \overline{\mathcal{U}}_t$ and the related trajectory $X(\cdot; t, x, a^\mu(\cdot))$,[1] we call the couple $(a^\mu(\cdot), X(\cdot; t, x, a^\mu(\cdot)))$ an *admissible couple* (or *admissible pair*) in the weak sense. If $a^*(\cdot) \in \overline{\mathcal{U}}_t$ minimizes (2.7), we say that $a^*(\cdot)$ is an *optimal control* at (t, x) for the weak formulation and the pair $(a^*(\cdot), X(\cdot; t, x, a^*(\cdot)))$ is called an *optimal couple* (or *optimal pair*) at (t, x) for the weak formulation. We will sometimes just say optimal control, optimal couple or optimal pair when the context is clear.

Remark 2.4 To avoid misunderstandings we clarify that the use of the term "weak" in the "weak formulation" of our stochastic control problem refers only to the fact that the generalized reference probability spaces vary with the controls and not to the concept of solution of the state equation. Indeed, in our framework, once the control $a^\mu(\cdot)$ is fixed (and with it also the generalized reference space), the solution is taken in the same generalized reference space (i.e. in the so-called strong probabilistic sense). Solutions of the state equation in the weak probabilistic sense will be used only to treat the closed loop equations in some cases, see Remark 2.6. ∎

In the weak formulation the goal is to minimize the same cost functional (2.3), however now over all controls $a^\mu(\cdot) \in \overline{\mathcal{U}}_t$. Consequently, the value function for the weak formulation is now defined by

$$\overline{V}(t, x) = \inf_{a^\mu(\cdot) \in \overline{\mathcal{U}}_t} J^\mu(t, x; a^\mu(\cdot)), \quad (t, x) \in [0, T) \times H, \qquad (2.7)$$

and we set $\overline{V}(T, x) := g(x)$ for $x \in H$ if $T < +\infty$. From the above definition we clearly have

$$\overline{V}(t, x) = \inf_\mu \inf_{a(\cdot) \in \mathcal{U}_t^\mu} J^\mu(t, x; a(\cdot)) = \inf_\mu V_t^\mu(x).$$

Remark 2.5 For the optimal control problem we could also have required the controls in $\mathcal{U}_t^\mu$ to be measurable and adapted instead of progressively measurable, since, by Lemma 1.72, every adapted $a(\cdot)$ has a progressively measurable modification $\tilde{a}(\cdot)$. We chose to deal with progressively measurable controls to avoid unnecessary technical issues. In light of Lemma 1.99, we could have chosen to work with predictable controls as well.

Moreover, in the definition of $\mathcal{U}_t^\mu$ we did not specify possible further restrictions on the control and on the state (state constraints, integrability conditions on the controls, etc.), which commonly arise in examples, see Sect. 2.6. Such kinds of restrictions usually lead to more complicated problems, however in principle they can be treated in the same framework. ∎

---

[1] To be sure that such a trajectory exists and is unique we need to assume that the state equation (2.2) is well posed for every admissible control $a^\mu(\cdot)$, so in particular for every generalized reference space.

Remark 2.6 To study problems where neither existence nor uniqueness of solutions of the state equation is guaranteed for arbitrary control process $a(\cdot)$ (in particular to study the existence of optimal feedback controls) it is useful to extend the formulation of an optimal control problem. In such cases the *extended* formulation of the control problem can be given as follows. Given a generalized reference probability space μ, we call $(a(\cdot), X(\cdot))$ an *admissible control pair* if $a(\cdot)$ is an $\mathscr{F}_s^t$-progressively measurable process with values in Λ and $X(\cdot)$ is a (not necessarily unique) solution of (2.2) corresponding to $a(\cdot)$. To every admissible control pair we associate the cost (2.3). The optimal control problem in the *extended* strong formulation consists in minimizing the functional $J^\mu(t, x; a(\cdot), X(\cdot))$ over all admissible control pairs $(a(\cdot), X(\cdot))$, and in characterizing the value function (where we use the same notation for simplicity)

$$V_t^\mu(x) = \inf_{(a(\cdot),X(\cdot))} J^\mu(t, x; a(\cdot), X(\cdot)).$$

The optimal control problem in the *extended* weak formulation consists in further minimizing with respect to all generalized reference probability spaces, i.e. in characterizing the value function (where again we use the same notation for simplicity)

$$\overline{V}(t, x) = \inf_\mu V_t^\mu(x).$$

Such a formulation is often much more suitable for construction of optimal feedback controls, see Corollary 2.38. In this book it is employed in Chapter 6, Sects. 6.5 and 6.10, but it may also be used to extend results of Chap. 4 (in particular Propositions 4.199 and 4.218) and of Chap. 5 (in particular Corollary 5.60) to more general cases when the function R used there is not Lipschitz continuous. ∎

2.2 The Dynamic Programming Principle: Setup and Assumptions

In this section we introduce the dynamic programming principle (DPP). This is one of the fundamental results of stochastic optimal control, whose formulation and proof are very technical, here even more so since we are dealing with the infinite-dimensional case. We first present the stochastic setup and the main assumptions, and then follow with the statement of the DPP and the proof.

2.2.1 The Setup

Definition 2.7 A *reference probability space* is a generalized reference probability space $\nu := \left(\Omega, \mathscr{F}, \mathscr{F}_s^t, \mathbb{P}, W_Q\right)$ (see Definition 1.100), where $W_Q(t) = 0, \mathbb{P}$-a.s., and

$\mathscr{F}_s^t = \sigma(\mathscr{F}_s^{t,0}, \mathcal{N})$, where $\mathscr{F}_s^{t,0} = \sigma(W_Q(\tau) : t \le \tau \le s)$ is the filtration generated by W_Q, and $\mathcal{N}$ is the collection of the $\mathbb{P}$-null sets in $\mathscr{F}$.

Definition 2.8 We will say that a reference probability space ν is *standard* if there exists a σ-field $\mathscr{F}'$ such that $\mathscr{F}_T^{t,0} \subset \mathscr{F}' \subset \mathscr{F}$, $\mathscr{F}$ is the completion of $\mathscr{F}'$, and $(\Omega, \mathscr{F}')$ is a standard measurable space (see Sect. 1.1).

We will consider control problem (2.2) and (2.3) in the weak formulation in which *generalized reference probability spaces* are replaced by *reference probability spaces*. This means that we are restricting the set of admissible controls. The set of all admissible controls is now defined by

$$\mathcal{U}_t := \bigcup_\nu \mathcal{U}_t^\nu, \tag{2.8}$$

where the union is taken over all reference probability spaces ν. Obviously $\mathcal{U}_t \subset \overline{\mathcal{U}}_t$, where $\overline{\mathcal{U}}_t$ is defined by (2.6). Thus $a(\cdot)$ is an admissible control now if there exists a reference probability space $\nu = (\Omega^\nu, \mathscr{F}^\nu, \mathscr{F}_s^{\nu,t}, \mathbb{P}^\nu, W_Q^\nu)$ such that $a(\cdot) \colon [t, T] \times \Omega^\nu \to \Lambda$ is $\mathscr{F}_s^{\nu,t}$-progressively measurable. As before we will often write $a^\nu(\cdot)$ to indicate the dependence of $a(\cdot)$ on the reference probability space.

The reason why we choose this setup is twofold. On the one hand the dynamic programming requires comparing problems with different initial times, which is accomplished well by using regular conditional probabilities and changing the underlying probability space. On the other hand, the use of reference probability spaces allows us to represent control processes as functions of the Wiener processes which allows us to pass easily from one reference probability space to another, hence the restrictions to reference probability spaces. Other approaches are possible, see the bibliographical notes at the end of this chapter.

With this definition the value function is now defined by

$$V(t, x) = \inf_{a^\nu(\cdot) \in \mathcal{U}_t} J^\nu(t, x; a^\nu(\cdot)) \tag{2.9}$$

(with the same convention that $V(T, x) := g(x)$ if $T < +\infty$) and, clearly, if $\overline{V}$ is the value function defined in (2.7),

$$\overline{V}(t, x) \le V(t, x) \le V_t^\nu(x), \qquad \text{for every reference probability space } \nu.$$

We will later see (Theorem 2.22) that the last inequality is indeed an equality under our assumptions. When solutions of the HJB equations are regular enough to allow construction of optimal feedbacks we will also see in Chaps. 4, 5 and 6 that both inequalities become equalities (see e.g. Theorems 4.201, 4.204 and 4.220). We do not study this issue here but the reader may check [467] for an argument that for control problems considered in [467] the first inequality is an equality. It is possible that the approach from [467] can be applied to the control problems in this book.

2.2.2 The General Assumptions

We make the following general assumption.

Hypothesis 2.9

(i) The *state space* H and the *noise space* Ξ are real separable Hilbert spaces.

(ii) The *control space* Λ is a Polish space.

(iii) The horizon of the problem is $T \in (0, +\infty) \cup \{+\infty\}$ and the initial time is $t \in [0, T)$.

(iv) $Q \in \mathcal{L}_1^+(\Xi)$.

Remark 2.10 We assume here that $Q \in \mathcal{L}_1^+(\Xi)$ (i.e. $\mathrm{Tr}\,(Q) < +\infty$), which implies that the processes W_Q in the reference probability spaces are Ξ-valued Q-Wiener processes. We do this for technical reasons, because in our proof of the DPP it is important that the Q-Wiener processes always have values in some (fixed) space Ξ. However, in many examples discussed in this chapter we will encounter Q-Wiener processes for which $\mathrm{Tr}\,(Q) = +\infty$. Recalling the definition of a Q-Wiener process (see Definition 1.88), we then have to choose and fix a space Ξ_1 such that W_Q is a Ξ_1-valued Q_1-Wiener process. (Since the space Ξ_1 is often not important, abusing notation, we will still call such process a W_Q-Wiener process, see Remark 1.89.) This puts us in the framework developed in this chapter and this is how the reader should understand such control problems, as it will not be repeated in the future when we discuss the examples in this chapter, unless it is essential. However, as mentioned in Remark 1.91, if $W_Q(s) = \sum_{k=1}^{\infty} g_k \beta_k(s)$ for some orthonormal basis $\{g_k\}_{k \in \mathbb{N}}$ of Ξ_0 (see Definition 1.88), then regardless of the choice of Ξ_1, $\mathscr{F}_s^{t,0} = \sigma\,(\beta_k(r) : t \le r \le s, k \in \mathbb{N})$. Thus the filtration does not depend on the choice of Ξ_1, and then also the class of integrable processes is independent of Ξ_1. Therefore the control problems discussed in the examples are independent of the choice of Ξ_1 and the theory can be applied to optimal control problems for which $Q \in \mathcal{L}^+(\Xi)$. ■

The following comment is important. The Q-Wiener processes in the reference probability spaces in general have trajectories which are only $\mathbb{P}$-a.e. continuous. However, we can always modify them on a set of measure zero so that the trajectories are continuous everywhere. Moreover, it is obvious that such a modified Q-Wiener process generates the same filtration $\mathscr{F}_s^t$ as the original one, so the set of admissible controls does not change. Furthermore, the solutions of the stochastic differential equations for control problems considered in this book are indistinguishable after this modification of the Q-Wiener processes, so the cost functional will be the same (see assumption $(A1)$ of Hypothesis 2.12). Therefore, unless specified otherwise, without loss of generality, **we will always assume that the Q-Wiener processes in the reference probability spaces have everywhere continuous paths**, however we will point it out explicitly if it is important to avoid any misunderstandings.

The assumptions about existence and uniqueness of solutions of the state equation are the following.

Hypothesis 2.11 Assume that, for every $0 \leq t < T$, reference probability space $\nu := (\Omega, \mathscr{F}, \mathscr{F}_s^t, \mathbb{P}, W_Q), a(\cdot) \in \mathcal{U}_t^\nu$, and an H-valued $\mathscr{F}_t^t$-measurable random variable ζ (i.e., $\zeta = x$, $\mathbb{P}$-a.s. for some $x \in H$), we have a unique, up to a modification, solution (in a certain sense) $X(\cdot) = X(\cdot; t, \zeta, a(\cdot))$ on $[t, T]$ of the abstract stochastic differential equation

$$\begin{cases} dX(s) = \beta(s, X(s), a(s))ds + \sigma(s, X(s), a(s))dW_Q(s), \\ X(t_1) = \zeta. \end{cases} \tag{2.10}$$

The solution $X(\cdot; t, \zeta, a(\cdot))$ is $\mathscr{F}_s^t$-progressively measurable, has continuous trajectories in H and $X(t; t, \zeta, a(\cdot)) = \zeta$, $\mathbb{P}$-a.s.

The above hypothesis particularly implies that any modification of a solution is still a solution of the same equation as long as it has continuous trajectories. To emphasize the dependence of the solution on the reference probability space we will sometimes use the notation $X^\nu(\cdot; t, \zeta, a(\cdot))$.

Hypothesis 2.12 collects assumptions about the properties of the family of solutions of the state equation. To simplify the notation we will write W instead of W_Q in Hypothesis 2.12 and in other places when the notation becomes cumbersome and when the meaning of it is clear.

Hypothesis 2.12 Assume that Hypothesis 2.11 holds. For every $0 \leq t \leq \eta < T, x \in H$, reference probability space $\nu = (\Omega, \mathscr{F}, \mathscr{F}_s^t, \mathbb{P}, W), a(\cdot) \in \mathcal{U}_t^\mu$, and an H-valued $\mathscr{F}_t^t$-measurable random variable ζ such that $\zeta = x$, $\mathbb{P}$-a.s., we have the following properties:

(A0) $X(\cdot; t, \zeta, a(\cdot)) = X(\cdot; t, x, a(\cdot))$ on $[t, T]$, $\mathbb{P}$-a.s.

(A1)   If $\nu_1 = (\Omega_1, \mathscr{F}_1, \mathscr{F}_{1,s}^t, \mathbb{P}_1, W_1)$, $\nu_2 = (\Omega_2, \mathscr{F}_2, \mathscr{F}_{2,s}^t, \mathbb{P}_2, W_2)$ are two reference probability spaces, $a_1(\cdot) \in \mathcal{U}_t^{\nu_1}, a_2(\cdot) \in \mathcal{U}_t^{\nu_2}$, and $\mathcal{L}_{\mathbb{P}_1}(a_1(\cdot), W_1(\cdot)) = \mathcal{L}_{\mathbb{P}_2}(a_2(\cdot), W_2(\cdot))$ (see Definition 1.133), then

$$\mathcal{L}_{\mathbb{P}_1}(X(\cdot; t, x, a_1(\cdot)), a_1(\cdot)) = \mathcal{L}_{\mathbb{P}_2}(X(\cdot; t, x, a_2(\cdot)), a_2(\cdot)).$$

(A2)   If $a_1(\cdot), a_2(\cdot) \in \mathcal{U}_t^\nu$ are such that $a_1(\cdot) = a_2(\cdot)$, $dt \otimes \mathbb{P}$-a.e. on $[t, \eta] \times \Omega$, then

$$X(\cdot; t, x, a_1(\cdot)) = X(\cdot; t, x, a_2(\cdot)) \text{ on } [t, \eta], \ \mathbb{P}\text{-a.s.}$$

(A3)   Let $\nu = (\Omega, \mathscr{F}, \mathscr{F}_s^t, \mathbb{P}, W)$ be a standard reference probability space (Definition 2.8) with W having everywhere continuous trajectories. Let $\nu_{\omega_0} = (\Omega, \mathscr{F}_{\omega_0}, \mathscr{F}_{\omega_0,s}^\eta, \mathbb{P}_{\omega_0}, W_\eta)$, where $\mathbb{P}_{\omega_0} = \mathbb{P}(\cdot|\mathscr{F}_\eta^{t,0})(\omega_0)$ is the regular conditional probability, $\mathscr{F}_{\omega_0}$ is the augmentation of $\mathscr{F}'$ by the $\mathbb{P}_{\omega_0}$ null sets, and $\mathscr{F}_{\omega_0,s}^\eta$ is the augmented filtration generated by W_η.[2] Let $a(\cdot) \in \mathcal{U}_t^\nu$ and $a_{|[\eta,T]}(\cdot) \in \mathcal{U}_\eta^{\nu_{\omega_0}}$ for $\mathbb{P}$-a.e. ω_0. Then the process $X^\nu(\cdot; t, x, a(\cdot))$ has an

---

[2]We remark that ν_{ω_0} in (A3) is a reference probability space for $\mathbb{P}$-a.e. ω_0 by Lemma 2.25. $W_\eta(s) := W(s) - W(\eta)$.

indistinguishable version such that, for $\mathbb{P}$-a.e. ω_0, $X^{\nu_{\omega_0}}(\cdot; \eta, X^{\nu}(\eta), a(\cdot)) = X^{\nu}(\cdot; t, x, a(\cdot))$ on $[\eta, T]$, $\mathbb{P}_{\omega_0}$-a.s.

Remark 2.13 It is possible to relax and slightly simplify Hypothesis 2.12 by combining conditions $(A0)$–$(A1)$ into one condition

$(A1')$ If $\nu_1 = \left(\Omega_1, \mathscr{F}_1, \mathscr{F}_{1,s}^t, \mathbb{P}_1, W_1\right)$, $\nu_2 = \left(\Omega_2, \mathscr{F}_2, \mathscr{F}_{2,s}^t, \mathbb{P}_2, W_2\right)$ are two reference probability spaces, $x \in H$, ζ is an H-valued $\mathscr{F}_{1,t}^t$-measurable random variable such that $\zeta = x$, $\mathbb{P}_1$-a.s., $a_1(\cdot) \in \mathcal{U}_t^{\nu_1}$, $a_2(\cdot) \in \mathcal{U}_t^{\nu_2}$, and $\mathcal{L}_{\mathbb{P}_1}(a_1(\cdot), W_1(\cdot)) = \mathcal{L}_{\mathbb{P}_2}(a_2(\cdot), W_2(\cdot))$, then

$$\mathcal{L}_{\mathbb{P}_1}(X(\cdot; t, \zeta, a_1(\cdot)), a_1(\cdot)) = \mathcal{L}_{\mathbb{P}_2}(X(\cdot; t, x, a_2(\cdot)), a_2(\cdot)).$$

With this change the proof of the dynamic programming principle is virtually unchanged. However, since condition $(A0)$ is standard and is satisfied by control problems considered in this book, we opted to keep it in the formulation of Hypothesis 2.12 hoping that the proof of the dynamic programming principle will be slightly easier to follow. ∎

Remark 2.14 We point out that since the trajectories of the solutions are continuous, $(A1)$ implies in particular that

$$\mathcal{L}_{\mathbb{P}_1}(X(\cdot; t, x, a_1(\cdot))) = \mathcal{L}_{\mathbb{P}_2}(X(\cdot; t, x, a_2(\cdot))) \quad \text{on } [t, T].$$

∎

Let us briefly explain the nature of the abstract assumptions of Hypothesis 2.12. Condition $(A0)$ guarantees pathwise uniqueness for our solutions with almost deterministic initial conditions, while condition $(A1)$ is a statement about uniqueness in law which guarantees that the joint law of $(X(\cdot; t, x, a(\cdot)), a(\cdot))$ only depends on the joint law of $(a(\cdot), W(\cdot))$. Condition $(A2)$ is a requirement that if the controls are "almost the same" then the solutions do not change. Finally, the most complicated condition $(A3)$ is a technical assumption which is needed since we do not define precisely what we mean by a solution. It guarantees, roughly speaking, that if we have a solution X in one reference probability space, then, for $\mathbb{P}$-a.e. ω_0, X is still a solution in reference probability spaces equipped with measures $\mathbb{P}_{\omega_0}$ provided certain conditions are satisfied. We remark that for $\mathbb{P}$-a.e. ω_0, $X^{\nu}(\eta)$ is $\mathbb{P}_{\omega_0}$-a.e. constant and is equal to $X^{\nu}(\eta)(\omega_0)$. We remark that condition $(A3)$ in particular implies that the version of $X^{\nu}(\cdot; t, x, a(\cdot))$ is $\mathscr{F}_{\omega_0,s}^{\eta}$-progressively measurable on $[\eta, T]$ for $\mathbb{P}$-a.e. ω_0, and has continuous trajectories $\mathbb{P}_{\omega_0}$-a.e. for $\mathbb{P}$-a.e. ω_0. These two properties can be proved for every solution X^{ν} satisfying our Hypothesis 2.11. We required that ν is a standard reference probability space to guarantee the existence of the regular conditional probability $\mathbb{P}_{\omega_0}$. The requirement that $a_{|[\eta,T]}(\cdot) \in \mathcal{U}_\eta^{\nu_{\omega_0}}$ can always be assumed since we will see in Lemma 2.26 that for every $a(\cdot) \in \mathcal{U}_t^{\nu}$ there is an $a_1(\cdot) \in \mathcal{U}_t^{\nu}$ such that $a(\cdot) = a_1(\cdot)$, $\mathbb{P} \otimes dt$-a.e. and $a_{1|[\eta,T]}(\cdot) \in \mathcal{U}_\eta^{\nu_{\omega_0}}$ for $\mathbb{P}$-a.e. ω_0.

Remark 2.15 This is a very important remark regarding optimal control problems with additional conditions on the set of admissible controls. In the proof of the dynamic programming principle we will use the following property of admissible controls.

(A4)   If ν is a standard reference probability space as in $(A3)$ and $a(\cdot) \in \mathcal{U}_t^\nu$, then there exists an $a_1(\cdot) \in \mathcal{U}_t^\nu$ such that $a(\cdot) = a_1(\cdot)\,\mathbb{P} \otimes dt$-a.e. and $a_{\vert[\eta,T]}(\cdot) \in \mathcal{U}_\eta^{\nu_{\omega_0}}$ for $\mathbb{P}$-a.e. ω_0, where ν_{ω_0} is as in $(A3)$.

This property is always true for our abstract optimal control problem and it is shown in Lemma 2.26, whose proof is only based on considerations of measurability. However, if the set of admissible controls is characterized by additional conditions, for instance some integrability conditions, property $(A4)$ must be established in each particular case, see for instance Remark 2.27. Therefore the reader should be very careful when adapting the abstract proof of the dynamic programming principle to such cases.   ∎

We close with some important remarks about standard reference probability spaces and regular conditional probabilities. Let $\nu = \left(\Omega, \mathscr{F}, \mathscr{F}_s^t, \mathbb{P}, W_Q\right)$ be a standard reference probability space. Regular conditional probabilities will be denoted by $\mathbb{P}_\omega$, and to indicate that $\mathbb{P}_{\omega_0}$ is the regular conditional probability given a sigma field $\mathscr{F}_s^{t,0}$ we will write $\mathbb{P}_{\omega_0} = \mathbb{P}(\cdot|\mathscr{F}_s^{t,0})(\omega_0)$ even though this is a slight abuse of notation. The expectation with respect to $\mathbb{P}_{\omega_0}$ will be denoted by $\mathbb{E}_{\omega_0}$.

For every $\Omega_1 \in \mathscr{F}$ such that $\mathbb{P}(\Omega_1) = 1$ there exists $\Omega_2 \subset \Omega_1$, $\Omega_2 \in \mathscr{F}'$ (from Definition 2.8) such that $\mathbb{P}(\Omega_2) = 1$. Therefore

$$1 = \mathbb{P}(\Omega_2) = \mathbb{E}\left[\mathbb{E}\left[\mathbf{1}_{\Omega_2}|\mathscr{F}_s^{t,0}\right](\omega_0)\right] = \mathbb{E}\left[\mathbb{P}_{\omega_0}(\Omega_2)\right].$$

Thus we obtain that $\Omega_1 \in \mathscr{F}_{\omega_0}$ for $\mathbb{P}$-a.e. ω_0 and

$$1 = \mathbb{P}_{\omega_0}(\Omega_2) = \mathbb{P}_{\omega_0}(\Omega_1).$$

Now suppose that $Y \in L^1(\Omega, \mathscr{F}, \mathbb{P})$. Let $Y' \in L^1(\Omega, \mathscr{F}', \mathbb{P})$ be such that $Y = Y'$, $\mathbb{P}$-a.s. Hence also $Y = Y'\,\mathbb{P}_{\omega_0}$-a.s. for $\mathbb{P}$-a.s. ω_0, which implies that Y is $\mathscr{F}_{\omega_0}$-measurable for $\mathbb{P}$-a.s. ω_0. Thus for $\mathbb{P}$-a.s. ω_0

$$\mathbb{E}\left[Y|\mathscr{F}_s^t\right](\omega_0) = \mathbb{E}\left[Y'|\mathscr{F}_s^{t,0}\right](\omega_0) = \int Y'(\omega)d\mathbb{P}_{\omega_0}(\omega) = \int Y(\omega)d\mathbb{P}_{\omega_0}(\omega) = \mathbb{E}_{\omega_0}[Y].$$

Therefore $\mathbb{E}_{\omega_0}[Y]$ (as a function of ω_0) is in $L^1(\Omega, \mathscr{F}, \mathbb{P})$ and

$$\mathbb{E}[Y] = \mathbb{E}\left[\mathbb{E}\left[Y|\mathscr{F}_s^t\right]\right] = \mathbb{E}\left[\mathbb{E}_{\omega_0}[Y]\right].$$

This fact will be used frequently in the following chapters without repeating the technical details.

2.2.3 The Assumptions in the Case of Control Problems for Mild Solutions

In this subsection we briefly illustrate the abstract setup for the case which is the most frequent among the problems treated in the book, namely optimal control problems driven by general stochastic evolution equations (with Lipschitz coefficients) with solutions interpreted in the mild sense, and explain how Hypotheses 2.11 and 2.12 are satisfied. In this case the state equation (2.10) is of type (1.30), i.e.

$$\begin{cases} dX(s) = (AX(s) + b(s, X(s), a(s)))\, ds + \sigma(s, X(s), a(s))dW_Q(s), \\ X(t) = \zeta, \end{cases} \tag{2.11}$$

where A, b, and σ satisfy Hypothesis 1.125 and its solution is understood in the mild sense of Definition 1.119, i.e. we have

$$X(s) = e^{(s-t)A}\zeta + \int_t^s e^{(s-r)A}b(r, X(r), a(r))dr + \int_t^s e^{(s-r)A}\sigma(r, X(r), a(r))dW_Q(r) \tag{2.12}$$

on $[t, T]$, $\mathbb{P}$-a.e.

Proposition 2.16 *Consider Eq. (2.11) under Hypotheses 1.125 and 2.9. Then Hypotheses 2.11 and 2.12 are satisfied for its mild solutions.*

Proof Hypothesis 2.11 follows from Theorem 1.127. Regarding Hypothesis 2.12, condition $(A0)$ follows from the fact that $e^{sA}\zeta = e^{sA}x$ $\mathbb{P}$-a.e. for all $s \in [t, T]$ and from (1.39). Condition $(A1)$ follows from Proposition 1.137 and $(A2)$ follows from Corollary 1.128.

To show $(A3)$, we will first show that $X^\nu(\cdot) := X^\nu(\cdot; t, x, a(\cdot))$ has a modification X_1^ν which is everywhere continuous and, for $\mathbb{P}$-a.e. ω_0, is $\mathscr{F}_{\omega_0,s}^\eta$-progressively measurable on $[\eta, T]$. In general X^ν is only $\mathscr{F}_s^t$-progressively measurable. Let Ω_0 be such that $\mathbb{P}(\Omega_0) = 1$, and for $\omega \in \Omega_0$, $X^\nu(\cdot, \omega)$ is continuous on $[t, T]$. Let $\{s_k\}, k \geq 1, s_1 = t$, be a countable dense set in $[t, T]$, and let $A_k \in \mathcal{F}_{s_k}^{t,0}, k \geq 1$ be sets such that $\mathbb{P}(A_k) = 1$, and $X^\nu(s_k) = \xi_k$ on A_k for some $\mathcal{F}_{s_k}^{t,0}$-measurable random variable ξ_k. Set $\Omega_1 = \Omega_0 \cap \bigcap_{k+1}^\infty A_k$. Then $\mathbb{P}(\Omega_1) = 1$ and thus for $\mathbb{P}$-a.e. ω_0, $\mathbb{P}_{\omega_0}(\Omega_1) = 1$, which implies that $\Omega_1 \in \mathscr{F}_{\omega_0,s}^\eta$ for $\mathbb{P}$-a.e. ω_0. We now define $X_1^\nu(s) = X^\nu(s)$ for $s \in [t, T]$, $\omega \in \Omega_1$ and $X_1^\nu(s) = 0$ for $s \in [t, T]$, $\omega \in \Omega \setminus \Omega_1$. The process X^ν has continuous trajectories. Since for $\omega \in \Omega_1$, $X_1^\nu(s) = \lim_{s_k \to s, s_k \leq s} \xi_k$, X_1^ν is $\sigma(\mathscr{F}_s^{t,0}, \Omega_1)$-adapted. However, thanks to Lemma 2.26, $\mathscr{F}_s^{t,0} \subset \mathscr{F}_{\omega_0,s}^\eta$ for $\mathbb{P}$-a.s. ω_0, and so it follows that X_1^ν is $\mathscr{F}_{\omega_0,s}^\eta$-adapted, which, since it has continuous trajectories, implies by Lemma 1.72 that it is $\mathscr{F}_{\omega_0,s}^\eta$-progressively measurable for $\mathbb{P}$-a.s. ω_0. From now on we will write $X^\nu(\cdot)$ for $X_1^\nu(\cdot)$.

We observe that $X^\nu(\cdot) \in M^p_\nu(t, T; H)$, $p > 2$, so in particular

$$\mathbb{E}\left[\mathbb{E}\left[\int_\eta^T |X^\nu(s)|^p ds | \mathscr{F}^{t,0}_\eta\right]\right] = \mathbb{E}\left[\int_\eta^T |X^\nu(s)|^p ds\right] < +\infty, \qquad (2.13)$$

so for $\mathbb{P}$-a.e. ω_0, $X^\nu(\cdot) \in M^p_{\nu_{\omega_0}}(\eta, T; H)$. Thus by uniqueness of mild solutions given by Theorem 1.127 we will be done, provided we know that, for $\mathbb{P}$-a.e. ω_0, $X^\nu(\cdot)$ is a mild solution in the interval $[\eta, T]$ in the reference probability space ν_{ω_0}.

We have the flow property

$$X^\nu(s) = e^{(s-\eta)A}\left[\zeta + \int_t^\eta e^{(\eta-r)A} b(r, X^\nu(r), a(r)) dr\right.$$
$$\left. + \int_t^\eta e^{(\eta-r)A}\sigma(r, X^\nu(r), a(r)) dW(r)\right] + \int_\eta^s e^{(s-r)A} b(r, X^\nu(r), a(r)) dr$$
$$+ \int_\eta^s e^{(s-r)A}\sigma(r, X^\nu(r), a(r)) dW(r) = e^{(s-\eta)A} X^\nu(\eta)$$
$$+ \int_\eta^s e^{(s-r)A} b(r, X^\nu(r), a(r)) dr + \int_\eta^s e^{(s-r)A}\sigma(r, X^\nu(r), a(r)) dW(r)$$
$$= X^\nu(s; \eta, X(\eta; t, \zeta, a(\cdot)), a(\cdot)), \qquad s \in [\eta, T].$$

Since $\mathbb{P}$-a.s.

$$\int_\eta^s e^{(s-r)A}\sigma(r, X^\nu(r), a(r)) dW(r) = \int_\eta^s e^{(s-r)A}\sigma(r, X^\nu(r), a(r)) dW_\eta(r)$$

$$(2.14)$$

on $[\eta, T]$, and since for every set Ω_1 such that $\mathbb{P}(\Omega_1) = 1$ we have $\mathbb{P}_{\omega_0}(\Omega_1) = 1$ for $\mathbb{P}$-a.e. ω_0, the equality

$$X^\nu(s) = e^{(s-\eta)A} X^\nu(\eta) + \int_\eta^s e^{(s-r)A} b(r, X^\nu(r), a(r)) dr$$
$$+ \int_\eta^s e^{(s-r)A}\sigma(r, X^\nu(r), a(r)) dW_\eta(r)$$

is satisfied $\mathbb{P}_{\omega_0}$-a.s. for $\mathbb{P}$-a.e. ω_0. This equality is exactly the same for the mild solution in the interval $[\eta, T]$ in the reference probability space ν_{ω_0} except for the fact that there the stochastic integral is taken in the reference probability space ν_{ω_0} instead of ν. So, to conclude, it is enough to show that for $\mathbb{P}$-a.e. ω_0

$$I_\nu(s) := \int_\eta^s e^{(s-r)A}\sigma(r, X^\nu(r), a(r)) dW_\eta(r)$$

is $\mathbb{P}_{\omega_0}$-a.e. equal on $[\eta, T]$ to the stochastic integral in the reference probability space ν_{ω_0}, which we denote by $I_{\nu_{\omega_0}}(s)$.[3] By continuity of the paths of the stochastic convolution (see Proposition 1.112) it is enough to show it for a single s. Define $\Phi(r) = e^{(s-r)A}\sigma(r, X^\nu(r), a(r))$, $r \in [\eta, s]$. By Lemma 1.98 and the proof of Lemma 1.99 there exist a sequence of elementary and $\mathscr{F}_r^{t,0}$-progressively measurable processes Φ_n with values in $\mathcal{L}(\Xi, H)$, such that $\|\Phi - \Phi_n\|_{\mathcal{N}_{Q,\nu}^2(\eta,s;H)} \to 0$, where we indicated the dependence on the reference probability space in the notation for the norm. By Lemma 2.26, the Φ_n are also $\mathscr{F}_{\omega_0,r}^\eta$-progressively measurable. Since $\mathbb{E}^\nu |\int_\eta^s [\Phi_n(r) - \Phi(r)]dW_\eta(r)|^2 \to 0$, passing to a subsequence if necessary, we can assume that

$$\int_\eta^s \Phi_n(r)dW_\eta(r) \to I_\nu(s), \quad \mathbb{P} - a.e., \tag{2.15}$$

say on a set Ω_2, where $\mathbb{P}(\Omega_2) = 1$ and we can assume that $\mathbb{P}_{\omega_0}(\Omega_2) = 1$ for $\mathbb{P}$-a.e. ω_0.

On the other hand, again by using conditional expectation as in (2.13), we know that, up to a subsequence, for $\mathbb{P}$-a.e. ω_0 we have $\|\Phi - \Phi_n\|_{\mathcal{N}_{Q,\nu_{\omega_0}}^2(\eta,s;H)} \to 0$. So, for $\mathbb{P}$-a.e. ω_0, we have that there exists a subsequence of Φ_n such that

$$\int_\eta^s \Phi_n(r)dW_\eta(r) \to I_{\nu_{\omega_0}}(s), \quad \mathbb{P}_{\omega_0}\text{-a.e.} \tag{2.16}$$

Since $\mathbb{P}_{\omega_0}(\Omega_2) = 1$ for $\mathbb{P}$-a.e. ω_0, (2.15) and (2.16) imply that, for $\mathbb{P}$-a.e. ω_0, $I_\nu(s) = I_{\nu_{\omega_0}}(s)$, $\mathbb{P}_{\omega_0}$-a.e.

A different approach to proving $(A3)$ can be found in [545]. □

Remark 2.17 Two further examples of systems satisfying Hypotheses 2.11 and 2.12 are given by the boundary control system described in Sect. 1.5.1 (Theorem 1.141) and by the semilinear system with non-nuclear covariance described in Sect. 1.5.2 (Proposition 1.147). We briefly sketch how one can show that Hypotheses 2.11 and 2.12 hold in these two cases. However, we point out that these cases do not fully conform to our general abstract control problem as additional integrability conditions on the controls must be assumed to guarantee the existence and uniqueness of a unique mild solution. Thus the formulation of the control problem must be slightly adjusted in an obvious way.

Concerning the case of Sect. 1.5.1, suppose that the assumptions of Theorem 1.141 including (1.54) are satisfied. Then Hypothesis 2.11 follows from Theorem 1.141. Regarding Hypothesis 2.12, $(A0)$ and $(A2)$ follow from part (i) of Proposition 1.142, $(A1)$ follows from part (ii) of Proposition 1.142 while for $(A3)$ one can argue as in Proposition 2.16, using $(A4)$ which holds by Remark 2.27.

---

[3]Observe that we can compute the integral $I_{\nu_{\omega_0}}(s)$ since the control and the trajectory $X^\nu(\cdot)$ are $\mathscr{F}_{\omega_0,s}^\eta$-progressively measurable on $[\eta, T]$ for $\mathbb{P}$-a.e. ω_0.

As regards the case of Sect. 1.5.2, suppose that the assumptions of Propositions 1.147 and 1.148 are satisfied and $a_2(\cdot)$ is as in Remark 1.146. Hypothesis 2.11 follows from Propositions 1.147. For Hypothesis 2.12, $(A0)$ and $(A2)$ follow from part (i) of Proposition 1.148, $(A1)$ follows from part (ii) of Proposition 1.148, while for $(A3)$ one can again argue as in Proposition 2.16 using $(A4)$.   ∎

2.3   The Dynamic Programming Principle: Statement and Proof

This section is devoted to the formulation and the proof of the dynamic programming principle. Throughout the whole section we always assume that Hypothesis 2.9 is satisfied. We begin with a technical subsection.

2.3.1   Pullback to the Canonical Reference Probability Space

Fix $t \in [0, T]$. The canonical reference probability space is the 5-tuple $\nu_W :=$ $(\mathbf{W}, \mathcal{F}_*, \mathbb{P}_*, \mathcal{B}_s^t, \mathcal{W})$, where $\mathbf{W} := \{\omega \in C([t, T], \Xi) : \omega(t) = 0\}$ equipped with the usual sup-norm, $\mathbb{P}_*$ is the Wiener measure on $(\mathbf{W}, \mathcal{B}(\mathbf{W}))$ (where $\mathcal{B}(\mathbf{W})$ is the Borel σ-field), i.e. the unique probability measure on $\mathbf{W}$ that makes the mapping

$$\begin{cases} \mathcal{W} : [t, T] \times \mathbf{W} \to \Xi \\ \mathcal{W}(s, \omega) = \omega(s) \end{cases} \tag{2.17}$$

a Q-Wiener process in Ξ (see [391]), $\mathcal{F}_*$ is the completion of $\mathcal{B}(\mathbf{W})$, and for $s \in [t, T]$, $\mathcal{B}_s^{t,0} = \sigma(\mathcal{W}(\tau) : t \leq \tau \leq s)$, $\mathcal{B}_s^t = \sigma\left(\mathcal{B}_s^{t,0}, \mathcal{N}^*\right)$, where $\mathcal{N}^*$ are the $\mathbb{P}_*$-null sets. $\mathbf{W}$ is a Polish space.

It is easy to see that $\mathcal{B}(\mathbf{W})$ is generated by the one-dimensional cylinder sets $C = \{\omega : \omega(s) \in A\}$, where $s \in [t, T]$, A is open in Ξ, and that $\mathcal{B}_T^{t,0} = \mathcal{B}(\mathbf{W})$ (Lemma 2.18). Theorem 1.12 thus guarantees that ν_W is a standard reference probability space.

The canonical reference probability space on $[t, +\infty)$ is defined in the same way except that now $\mathbf{W} := \{\omega \in C([t, +\infty), \Xi) : \omega(t) = 0\}$ is equipped with the metric

$$\rho(w_1, w_2) = \sum_{n=1}^{\infty} 2^{-n} (\|w_1 - w_2\|_{C([t,t+n], \Xi)} \wedge 1),$$

which makes it a Polish space.

Lemma 2.18 *Let for $s \in [t, T]$ the map $\varphi_s : \mathbf{W} \to \mathbf{W}$ be defined by $\varphi_s(\omega)(\tau) = \omega(\tau \wedge s)$. Then*

$$\mathcal{B}_s^{t,0} = \varphi_s^{-1}(\mathcal{B}(\mathbf{W})).$$

In particular, $\mathcal{B}_T^{t,0} = \mathcal{B}(\mathbf{W})$.

Proof Observe that for a one-dimensional cylinder $C = \{\omega : \omega(r) \in A\}$, where $r \in [t, T]$ and A is open in Ξ, we have

$$\varphi_s^{-1}(C) = \{\omega \in \mathbf{W} : \varphi_s(\omega)(r) \in C\} = \{\omega \in \mathbf{W} : \omega(r \wedge s) \in C\} \in \mathcal{B}_s^{t,0}.$$

Since the cylinder sets C generate $\mathcal{B}(\mathbf{W})$, we thus obtain $\varphi_s^{-1}(\mathcal{B}(\mathbf{W})) \subset \mathcal{B}_s^{t,0}$.

For the opposite inclusion, since $\mathcal{B}_s^{t,0}$ is generated by sets of the form $B = \mathcal{W}^{-1}(r, \cdot)(V)$, where $r \in [t, s]$ and V is open in Ξ, we have

$$B = \{\omega \in \mathbf{W} : \omega(r) \in V\} = \{\omega \in \mathbf{W} : \omega(r \wedge s) \in V\} = \varphi_s^{-1}(\{\omega \in \mathbf{W} : \omega(r) \in V\}).$$

Thus $\mathcal{B}_s^{t,0} \subset \varphi_s^{-1}(\mathcal{B}(\mathbf{W}))$. □

Lemma 2.19 *Let $(\Omega, \mathscr{F}, \mathscr{F}_s^t, \mathbb{P}, W)$ be a reference probability space (i.e. it satisfies Definition 2.7), and let the paths of the Q-Wiener process $W(\cdot, \omega)$ be continuous for every $\omega \in \Omega$. Then, for $s \in [t, T]$,*

$$\mathscr{F}_s^{t,0} = W(\cdot \wedge s)^{-1}(\mathcal{B}(\mathbf{W})).$$

Proof The proof is similar to that of Lemma 2.18. □

We denote by $\mathcal{P}_{[t,T]}^{\mathbf{W}}$ the sigma field of $\mathcal{B}_s^{t,0}$-predictable sets, i.e. the sigma field generated by the sets of the form $(s, r] \times A, t \le s < r \le T, A \in \mathcal{B}_s^{t,0}$ and $\{t\} \times A, A \in \mathcal{B}_t^{t,0}$. For a reference probability space $(\Omega, \mathscr{F}, \mathscr{F}_s^t, \mathbb{P}, W_Q)$ we denote by $\mathcal{P}_{[t,T]}^{\Omega}$ the sigma field of $\mathscr{F}_s^{t,0}$-predictable sets.

We will use the following simple representation lemma from [545].

Lemma 2.20 *Let $a(\cdot) = (\Omega, \mathscr{F}, \mathscr{F}_s^t, \mathbb{P}, W, a(\cdot)) \in \mathcal{U}_t$ (defined by (2.8)) be $\mathscr{F}_s^{t,0}$-predictable, and let the paths of the Q-Wiener process $W(\cdot, \omega)$ be continuous for every $\omega \in \Omega$. Then there exists a $\mathcal{P}_{[t,T]}^{\mathbf{W}}/\mathcal{B}(\Lambda)$-measurable function $f : [t, T] \times \mathbf{W} \to \Lambda$ such that*

$$a(s, \omega) = f(s, W(\cdot, \omega)), \quad for \; \omega \in \Omega, \; s \in [t, T]. \tag{2.18}$$

Proof Define the process

$$\begin{cases} \beta: [t, T] \times \Omega \to [t, T] \times \mathbf{W} \\ \beta(\tau, \omega) = (\tau, W(\cdot, \omega)). \end{cases}$$

The sets of the form $A_1 = (s, r] \times \{\omega \in \Omega : W(\eta, \omega) \in B\}, t \le \eta \le s < r \le T,$ $B \in \mathcal{B}(\Xi)$, and $A_2 = \{t\} \times \{\omega \in \Omega : W(t, \omega) \in B\}, B \in \mathcal{B}(\Xi)$, generate $\mathcal{P}^\Omega_{[t,T]}$. But $(\tau, \omega) \in A_1$ if and only if $\tau \in (s, r]$ and $W(\cdot, \omega) \in \tilde{B}_1 = \{\xi \in \mathbf{W} : \xi(\eta) \in B\} \in \mathcal{B}_s^{t,0}$, and $(t, \omega) \in A_2$ if and only if $W(\cdot, \omega) \in \tilde{B}_2 = \{\xi \in \mathbf{W} : \xi(t) \in B\} \in \mathcal{B}_t^{t,0}$. Therefore, $A_1 = \beta^{-1}((s, r] \times \tilde{B}_1), A_2 = \beta^{-1}(\{t\} \times \tilde{B}_2)$. Since the sets of the form $(s, r] \times \{\xi \in \mathbf{W} : \xi(\eta) \in B\}, t \le \eta \le s < r \le T, B \in \mathcal{B}(\Xi)$, and $\{t\} \times \{\xi \in \mathbf{W} : \xi(t) \in B\}, B \in \mathcal{B}(\Xi)$, generate $\mathcal{P}^\mathbf{W}_{[t,T]}$, we have $\mathcal{P}^\Omega_{[t,T]} = \beta^{-1}(\mathcal{P}^\mathbf{W}_{[t,T]})$. Therefore, by Theorem 1.9, there exists a $\mathcal{P}^\mathbf{W}_{[t,T]}/\mathcal{B}(\Lambda)$-measurable function $f : [t, T] \times \mathbf{W} \to \Lambda$ such that (2.18) is satisfied.   □

Corollary 2.21 *Let* $a(\cdot) = (\Omega, \mathscr{F}, \mathscr{F}_s^t, \mathbb{P}, W, a(\cdot)) \in \mathcal{U}_t$ *be* $\mathscr{F}_s^{t,0}$-*predictable. Let* $(\Omega_1, \mathscr{F}_1, \mathscr{F}_{1,s}^t, \mathbb{P}_1, W_1)$ *be another reference probability space. Suppose that* W *and* W_1 *have everywhere continuous trajectories. Let* $f : [t, T] \times \mathbf{W} \to \Lambda$ *be the function from Lemma 2.20 satisfying (2.18). Then the process*

$$\tilde{a}(s, \omega_1) = f(s, W_1(\cdot, \omega_1)), \qquad s \in [t, T], \tag{2.19}$$

is $\mathscr{F}_{1,s}^{t,0}$-*predictable and hence* $\mathscr{F}_{1,s}^{t,0}$-*progressively measurable on* $[t, T] \times \Omega_1$, *and*

$$\mathcal{L}_\mathbb{P}(a(\cdot), W(\cdot)) = \mathcal{L}_{\mathbb{P}_1}(\tilde{a}(\cdot), W_1(\cdot)). \tag{2.20}$$

2.3.2 Independence of Reference Probability Spaces

To prove the dynamic programming principle we have formulated our optimal control problem in a special weak form in which we only use reference probability spaces. Here we show that the control problem does not depend on the choice of the reference probability space ν and thus the strong and weak formulations are equivalent.

We will formulate the result only for the case $T < +\infty$, however the reader can easily modify the assumptions so that the result also holds for $T = +\infty$.

Theorem 2.22 (Independence of the reference probability space) *Let* $T \in (0, +\infty)$. *Let Hypotheses 2.9, 2.11 and 2.12 be satisfied. Let the functions* $l : [0, T] \times H \times \Lambda \to \mathbb{R}, g : H \to \mathbb{R}, c : H \to \mathbb{R}$ *be Borel measurable,* c *be bounded from below, and let, for every* $0 \le t < T, x \in H$, *every reference probability space* $\nu = (\Omega, \mathscr{F}, \mathscr{F}_s^t, \mathbb{P}, W)$ *and* $a(\cdot) \in \mathcal{U}_t^\nu$,

$$l(\cdot, X(\cdot; t, x, a(\cdot)), a(\cdot)) \in M_\nu^1(t, T; \mathbb{R}), \quad g(X(T; t, x, a(\cdot))) \in L^1(\Omega, \mathscr{F}, \mathbb{P}).$$

Then, for every $0 \le t < T, x \in H$, *every two reference probability spaces* $\nu_1 = (\Omega_1, \mathscr{F}_1, \mathscr{F}_{1,s}^t, \mathbb{P}_1, W_1), \nu_2 = (\Omega_2, \mathscr{F}_2, \mathscr{F}_{2,s}^t, \mathbb{P}_2, W_2)$, *and* $a(\cdot) \in \mathcal{U}_t^{\nu_1}$, *there exists an* $a_2(\cdot) \in \mathcal{U}_t^{\nu_2}$ *such that*

$$\mathcal{L}_{\mathbb{P}_1}(X^{\nu_1}(\cdot; t, x, a(\cdot)), a(\cdot)) = \mathcal{L}_{\mathbb{P}_2}(X^{\nu_2}(\cdot; t, x, a_2(\cdot)), a_2(\cdot)).$$

In particular, for every reference probability space ν,

$$V_t^\nu(x) = V(t, x).$$

Proof Let $a(\cdot) \in \mathcal{U}_t^{\nu_1}$. Let $a_1(\cdot)$ be the $\mathscr{F}_s^{t,0}$-predictable process from Lemma 1.99 such that $a_1(\cdot) = a(\cdot)$, $\mathbb{P}_1 \otimes dt$-a.e. Let $\tilde{a}_1(\cdot) \in \mathcal{U}_t^{\nu_2}$ be the process from Corollary 2.21. (Without loss of generality we can assume that W_1, W_2 have every-where continuous trajectories.) Since $\mathcal{L}_{\mathbb{P}_1}(a_1(\cdot), W_1(\cdot)) = \mathcal{L}_{\mathbb{P}_2}(\tilde{a}_1(\cdot), W_2(\cdot))$, it fol-lows from $(A1)$, $(A2)$ and Theorem 1.134 that, with $X^{\nu_1}(\cdot) = X^{\nu_1}(\cdot; t, x, a_1(\cdot))$, $X^{\nu_2}(\cdot) = X^{\nu_2}(\cdot; t, x, \tilde{a}_1(\cdot))$,

$$f_1(s) = e^{-\int_t^s c(X^{\nu_1}(\tau))d\tau}, \quad f_2(s) = e^{-\int_t^s c(X^{\nu_2}(\tau))d\tau}, \quad s \in [t, T],$$

we have

$$\mathcal{L}_{\mathbb{P}_1}(f_1(\cdot), X^{\nu_1}(\cdot), a(\cdot)) = \mathcal{L}_{\mathbb{P}_2}(f_2(\cdot), X^{\nu_2}(\cdot), \tilde{a}_1(\cdot)). \tag{2.21}$$

This proves the first claim.

Using (2.21) we thus obtain

$$J^{\nu_1}(t, x; a(\cdot)) = J^{\nu_1}(t, x; a_1(\cdot)) = J^{\nu_2}(t, x; \tilde{a}_1(\cdot)),$$

which implies

$$\inf_{a(\cdot) \in \mathcal{U}_t^{\nu_1}} J^{\nu_1}(t, x; a(\cdot)) \geq \inf_{a(\cdot) \in \mathcal{U}_t^{\nu_2}} J^{\nu_2}(t, x; a(\cdot)).$$

The opposite inequality is obtained in the same way and thus it follows that

$$\inf_{a(\cdot) \in \mathcal{U}_t^{\nu_1}} J^{\nu_1}(t, x; a(\cdot)) = \inf_{a(\cdot) \in \mathcal{U}_t^{\nu_2}} J^{\nu_2}(t, x; a(\cdot)). \tag{2.22}$$

This completes the proof. $\square$

2.3.3 The Proof of the Abstract Principle of Optimality

We now state and prove the dynamic programming principle (DPP) in an abstract formulation. We will do this only for the finite horizon problem. However, the same proof applies to the infinite horizon case if Hypothesis 2.23 is slightly changed. A special infinite horizon case is discussed in more detail in Sect. 2.4.

Hypothesis 2.23 Let $T \in (0, +\infty)$. The functions $l : [0, T] \times H \times \Lambda \to \mathbb{R}$, $g : H \to \mathbb{R}$, $c : H \to \mathbb{R}$ are Borel measurable, and c is bounded from below. For every $0 \leq t \leq \eta < T, x \in H$, every reference probability space $\nu = (\Omega, \mathscr{F}, \mathscr{F}_s^t, \mathbb{P}, W)$, and $a(\cdot) \in \mathcal{U}_t^\nu$

$$l(\cdot, X(\cdot; t, x, a(\cdot)), a(\cdot)) \in M_\nu^1(t, T; \mathbb{R}), \quad g(X(T; t, x, a(\cdot))) \in L^1(\Omega, \mathscr{F}, \mathbb{P}),$$

$$V(\eta, X(\eta; t, x, a(\cdot))) \in L^1(\Omega, \mathscr{F}, \mathbb{P}).$$

Moreover, the functional $J(t, y; a(\cdot))$ is uniformly continuous in the variable y on bounded sets of H, uniformly for $a(\cdot) \in \mathcal{U}_t$.

Hypothesis 2.23 in particular ensures that the value function V is finite.

Theorem 2.24 (Dynamic programming principle) *Assume that Hypotheses 2.9, 2.11, 2.12, and 2.23 are satisfied. Let $0 \le t < \eta < T, x \in H$. Then*

$$V(t, x) = \inf_{a(\cdot) \in \mathcal{U}_t} \mathbb{E}\left[\int_t^\eta e^{-\int_t^s c(X(\tau))d\tau} l(s, X(s), a(s))\, ds + e^{-\int_t^\eta c(X(\tau))d\tau} V(\eta, X(\eta))\right].$$
(2.23)

The proof is very technical so we will proceed slowly. We begin with two simple lemmas.

Lemma 2.25 *Let $0 \le t \le \eta < T$. Let $\left(\Omega, \mathscr{F}, \mathscr{F}_s^t, \mathbb{P}, W\right)$ be a standard reference probability space, and let W have everywhere continuous trajectories. Define, for $s \in [\eta, T]$, $W_\eta(s) := W(s) - W(\eta)$. Then for $\mathbb{P}$-a.e. $\omega_0 \in \Omega$, W_η is a Q-Wiener process on $\left(\Omega, \mathscr{F}_{\omega_0}, \mathscr{F}_{\omega_0,s}^\eta, \mathbb{P}_{\omega_0}\right)$, where $\mathbb{P}_{\omega_0} = \mathbb{P}(\cdot | \mathscr{F}_\eta^{t,0})(\omega_0)$ is the regular conditional probability, $\mathscr{F}_{\omega_0}$ is the augmentation of $\mathscr{F}^t$ (see Definition 2.8) by the $\mathbb{P}_{\omega_0}$ null sets, and $\mathscr{F}_{\omega_0,s}^\eta$ is the augmented filtration generated by W_η.*

Proof We notice that for $\eta \le s \le T$

$$\mathscr{F}_{\omega_0,s}^{\eta,0} = \sigma\left(W_\eta(\tau) : \eta \le \tau \le s\right) \subset \mathscr{F}_s^{t,0}$$

(observe that $\mathscr{F}_{\omega_0,t_1}^{\eta,0}$ is independent of ω_0) and, by Lemma 2.26-(i), for $\mathbb{P}$-a.e. ω_0, $\mathscr{F}_s^{t,0} \subset \mathscr{F}_{\omega_0,s}^\eta$ for every $\eta \le s \le T$. Thus for $\mathbb{P}$-a.e. ω_0, $\mathscr{F}_{\omega_0,s}^\eta$ is the augmentation of $\mathscr{F}_s^{t,0}$ by the $\mathbb{P}_{\omega_0}$ null sets for every $\eta \le s \le T$.

We fix $\eta \le t_1 < t_2, y \in \Xi$. We want to apply Lemma 1.93 (with $\Xi_1 = \Xi$ and $Q_1 = Q$) so we need to compute for $\mathbb{P}$-a.e. ω_0,

$$h := \mathbb{E}_{\omega_0}\left[e^{i\langle y, W_\eta(t_2) - W_\eta(t_1)\rangle_\Xi} | \mathscr{F}_{\omega_0,t_1}^\eta\right] = \mathbb{E}_{\omega_0}\left[e^{i\langle y, W_\eta(t_2) - W_\eta(t_1)\rangle_\Xi} | \mathscr{F}_{t_1}^{t,0}\right] \quad \mathbb{P}_{\omega_0}\text{-a.s.}$$

Thus we can assume that h is $\mathscr{F}_{t_1}^{t,0}$-measurable. We have

$$\int_A h(\omega) d\mathbb{P}_{\omega_0}(\omega) = \int_A e^{i\langle y, W_\eta(t_2) - W_\eta(t_1)\rangle_\Xi}(\omega) d\mathbb{P}_{\omega_0}(\omega) \quad \forall A \in \mathscr{F}_{t_1}^{t,0},$$

which (by the definition of $\mathbb{P}_{\omega_0}$) means that for $\mathbb{P}$-a.e. ω_0

$$\int_A h(\omega)d\mathbb{P}_{\omega_0}(\omega) = \mathbb{E}\left[e^{i\langle y, W_\eta(t_2)-W_\eta(t_1)\rangle_\Xi}\mathbf{1}_A|\mathscr{F}_\eta^{t,0}\right](\omega_0)$$

$$= \mathbb{E}\left[\mathbf{1}_A\mathbb{E}\left[e^{i\langle y, W(t_2)-W(t_1)\rangle_\Xi}|\mathscr{F}_{t_1}^t\right]|\mathscr{F}_\eta^{t,0}\right](\omega_0)$$

$$= \mathbb{E}\left[e^{-\frac{\langle Qy,y\rangle_\Xi}{2}(t_2-t_1)}\mathbf{1}_A|\mathscr{F}_\eta^{t,0}\right](\omega_0) = e^{-\frac{\langle Qy,y\rangle_\Xi}{2}(t_2-t_1)}\mathbb{P}_{\omega_0}(A).$$

Therefore, since $\mathscr{F}_{t_1}^{t,0}$ is countably generated, it follows that $h = e^{-\frac{\langle Qy,y\rangle_\Xi}{2}(t_2-t_1)}$ for $\mathbb{P}$-a.s. ω_0. Thus by the separability of Ξ, for $\mathbb{P}$-a.s. ω_0 we have $h = e^{-\frac{\langle Qy,y\rangle_\Xi}{2}(t_2-t_1)}$ for all $y \in \Xi$. Consider now all pairs (t_1^k, t_2^k), $k = 1, 2, ...$, where $t_1^k = \eta$ or t_1^k is rational, t_2^k is rational and $\eta \le t_1^k < t_2^k \le T$. We can conclude from the above that there is a set Ω_0 such that $\mathbb{P}(\Omega_0) = 1$ and such that for every $\omega_0 \in \Omega_0$, $y \in \Xi$ and $k = 1, 2, ...$

$$\mathbb{E}_{\omega_0}\left[e^{i\langle y, W_\eta(t_2^k)-W_\eta(t_1^k)\rangle_\Xi}|\mathscr{F}_{t_1^k}^{t,0}\right] = e^{-\frac{\langle Qy,y\rangle_\Xi}{2}(t_2^k-t_1^k)}. \tag{2.24}$$

It remains to prove that if $\omega_0 \in \Omega_0$, $y \in \Xi$ and $\eta \le t_1 < t_2 \le T$, then

$$\mathbb{E}_{\omega_0}\left[e^{i\langle y, W_\eta(t_2)-W_\eta(t_1)\rangle_\Xi}|\mathscr{F}_{t_1}^{t,0}\right] = e^{-\frac{\langle Qy,y\rangle_\Xi}{2}(t_2-t_1)}. \tag{2.25}$$

So let $\omega_0 \in \Omega_0$, $y \in \Xi$ and $\eta \le t_1 < t_2 \le T$. We will assume that $t_1 \ne \eta$ and t_1, t_2 are not rational since in such cases the argument is similar and easier. Then for some subsequence of our sequence of pairs, which we will still denote by (t_1^k, t_2^k), we have $t_1^k \to t_1$, $t_2^k \to t_2$ and $t_1^k < t_1$, $t_2^k < t_2$. We claim that $\mathbb{P}_{\omega_0}$-a.s.

$$\lim_{k\to+\infty} \mathbb{E}_{\omega_0}\left[e^{i\langle y, W_\eta(t_2^k)-W_\eta(t_1^k)\rangle_\Xi}|\mathscr{F}_{t_1^k}^{t,0}\right] = \mathbb{E}_{\omega_0}\left[e^{i\langle y, W_\eta(t_2)-W_\eta(t_1)\rangle_\Xi}|\mathscr{F}_{t_1}^{t,0}\right] \tag{2.26}$$

which, together with (2.24), will establish (2.25). First we notice that, since the filtration $\mathscr{F}_s^{t,0}$ is left-continuous, by Proposition 1.41-(viii)

$$\lim_{k\to+\infty} \mathbb{E}_{\omega_0}\left[e^{i\langle y, W_\eta(t_2)-W_\eta(t_1)\rangle_\Xi}|\mathscr{F}_{t_1^k}^{t,0}\right] = \mathbb{E}_{\omega_0}\left[e^{i\langle y, W_\eta(t_2)-W_\eta(t_1)\rangle_\Xi}|\mathscr{F}_{t_1}^{t,0}\right] \quad \mathbb{P}_{\omega_0}\text{-a.s.}$$

Then we observe that by Proposition 1.41-(vi)

$$\mathbb{E}_{\omega_0}\left|\mathbb{E}_{\omega_0}\left[e^{i\langle y, W_\eta(t_2^k)-W_\eta(t_1^k)\rangle_\Xi}|\mathscr{F}_{t_1^k}^{t,0}\right] - \mathbb{E}_{\omega_0}\left[e^{i\langle y, W_\eta(t_2)-W_\eta(t_1)\rangle_\Xi}|\mathscr{F}_{t_1^k}^{t,0}\right]\right|$$

$$\le \sqrt{2}\mathbb{E}_{\omega_0}\left|e^{i\langle y, W_\eta(t_2^k)-W_\eta(t_1^k)\rangle_\Xi} - e^{i\langle y, W_\eta(t_2)-W_\eta(t_1)\rangle_\Xi}\right| \to 0 \text{ as } k \to +\infty.$$

Thus for some subsequence, $\mathbb{P}_{\omega_0}$-a.s.

$$\lim_{k\to+\infty}\left|\mathbb{E}_{\omega_0}\left[e^{i\langle y, W_\eta(t_2^k)-W_\eta(t_1^k)\rangle_\Xi}|\mathscr{F}_{t_1^k}^{t,0}\right] - \mathbb{E}_{\omega_0}\left[e^{i\langle y, W_\eta(t_2)-W_\eta(t_1)\rangle_\Xi}|\mathscr{F}_{t_1^k}^{t,0}\right]\right| = 0.$$

These two convergences prove (2.26).   □

The reader can consult [545] for a different argument to prove Lemma 2.25.

Lemma 2.26 *Let* $0 \leq t \leq \eta < T$. *Let* $\nu = \left(\Omega, \mathscr{F}, \mathscr{F}_s^t, \mathbb{P}, W\right)$ *be a standard reference probability space and let* W *have everywhere continuous trajectories. Let* $a(\cdot) \in \mathcal{U}_t^\nu$, *and let* $a_1(\cdot)$ *be from Lemma 1.99. Then we have the following.*

(i) For $\mathbb{P}$-*a.e.* $\omega_0 \in \Omega$, $\mathscr{F}_s^{t,0} \subset \mathscr{F}_{\omega_0,s}^\eta$ *for every* $\eta \leq s \leq T$.
(ii) For $\mathbb{P}$-*a.e.* $\omega_0 \in \Omega$, $a^{\omega_0}(\cdot) := \left(\Omega, \mathscr{F}_{\omega_0}, \mathscr{F}_{\omega_0,s}^\eta, \mathbb{P}_{\omega_0}, W_\eta, a_1|_{[\eta,T]}(\cdot)\right) \in \mathcal{U}_\eta$.

Proof To prove Part *(i)*, we take a countable generating family $\{A_k\}$ of $\mathcal{B}(\Xi)$ and a countable dense subset $\{s_m\}$ in $[t, T]$. We will show that for a.e. $\omega_0 \in \Omega$, $W(s_m)^{-1}(A_k) \in \mathscr{F}_{\omega_0,s}^\eta$ for all $k \geq 1$, $s_m \leq s$. If $s_m \leq \eta$, since $\mathscr{F}_\eta^{t,0}$ is countably generated, we obtain by Theorem 1.45 that $W(s_m)(\omega) = W(s_m)(\omega_0)$, $\mathbb{P}_{\omega_0}$-a.e., for $\mathbb{P}$-a.e. $\omega_0 \in \Omega$. Thus, up to a set of $\mathbb{P}_{\omega_0}$ measure 0, $W(s_m)^{-1}(A_k)$ is either empty or is equal to Ω and so it is in $\mathscr{F}_{\omega_0,s}^\eta$. If $s_m > \eta$ then again up to a set of $\mathbb{P}_{\omega_0}$ measure 0, $W(s_m)^{-1}(A_k) = W_\eta(s_m)^{-1}(A_k - W_\eta(\omega_0))$ and so it is in $\mathscr{F}_{\omega_0,s}^\eta$ for $\mathbb{P}$-a.e. $\omega_0 \in \Omega$. This implies that $\sigma\left(W(s_m) : s_m \leq s\right) \subset \mathscr{F}_{\omega_0,s}^\eta$ for $\mathbb{P}$-a.e. $\omega_0 \in \Omega$. It remains to observe that $\mathscr{F}_s^{t,0} = \sigma\left(W(s_m) : s_m \leq s\right)$.

Part *(ii)*: In view of Lemma 2.25 it is enough to show that for $\mathbb{P}$-a.e. $\omega_0 \in \Omega$, $a_1(\cdot)$ is $\mathscr{F}_{\omega_0,s}^\eta$-progressively measurable on $[\eta, T]$. This fact follows from Part *(i)*.   □

Remark 2.27 If $a(\cdot)$ in Lemma 2.26 is such that $a(\cdot) \in M_\nu^p(t, T; E)$, $p \geq 1$, for some Banach space E, it is easy to see that we also have $a_1|_{[\eta,T]}(\cdot) \in M_{\nu_{\omega_0}}^p(\eta, T; E)$, for $\mathbb{P}$-a.e. ω_0, where $\nu_{\omega_0} = \left(\Omega, \mathscr{F}_{\omega_0}, \mathscr{F}_{\omega_0,s}^\eta, \mathbb{P}_{\omega_0}, W_\eta\right)$ is as in Lemma 2.26.   ∎

Proof of Theorem 2.24 We first perform the following reduction. Define (similarly to (2.8))

$$\tilde{\mathcal{U}}_t = \left\{ \bigcup_\nu \mathcal{U}_t^\nu : \nu \text{ is a standard reference probability space} \right\}.$$

The set $\tilde{\mathcal{U}}_t$ is non-empty since, for instance, the canonical reference probability space ν_W is in it. It is clear from Theorem 2.22 (and the same argument used to justify (2.21) there) that (2.23) will follow if we can prove it with $\mathcal{U}_t$ replaced by $\tilde{\mathcal{U}}_t$. Therefore it remains to show that

$$V(t, x) = \inf_{a(\cdot) \in \tilde{\mathcal{U}}_t} \mathbb{E}\left[\int_t^\eta e^{-\int_t^s c(X(\tau))d\tau} l(s, X(s), a(s)) \, ds \right.$$

$$\left. + e^{-\int_t^\eta c(X(\tau))d\tau} V(\eta, X(\eta)) \right]. \quad (2.27)$$

(In fact, using Theorem 2.22 it would be enough to replace $\mathcal{U}_t$ by $\mathcal{U}_t^{\nu_W}$ and do everything on canonical reference probability spaces, however we will prove the theorem in the more general setup since the arguments and technicalities are similar.) We recall that we assume that all Q-Wiener processes in the reference probability spaces have everywhere continuous trajectories.

Part 1. (inequality $\geq$ in (2.27)): Let $a(\cdot) \in \tilde{\mathcal{U}}_t$. Defining $X(\cdot) = X(\cdot; t, x, a(\cdot))$ we have

$$J(t, x; a(\cdot)) = \mathbb{E}\left[\int_t^\eta e^{-\int_t^s c(X(\tau))d\tau} l(s, X(s), a(s))\, ds\right]$$

$$+\mathbb{E}\left[\int_\eta^T e^{-\int_t^s c(X(\tau))d\tau} l(s, X(s), a(s))\, ds + e^{-\int_t^T c(X(\tau))d\tau} g(X(T))\right]. \quad (2.28)$$

Let $a_1(\cdot)$ be from Lemma 1.99, and for $\mathbb{P}$-a.e. $\omega_0 \in \Omega$, $a^{\omega_0}(\cdot)$ be the control from Lemma 2.26. Let $\mathbb{P}_{\omega_0} = \mathbb{P}\left(\cdot|\mathscr{F}_\eta^{t,0}\right)(\omega_0)$ and $\mathbb{E}_{\omega_0}$ be the expectation with respect to $\mathbb{P}_{\omega_0}$.

Let $X_1(s) = X(s; t, x, a_1(\cdot))$, $s \in [t, T]$. By (A2), X_1 and X are indistinguishable. Thus we obtain by (A3) that (up to an indistinguishable modification), $X_1(s) = X(s; \eta, X(\eta), a^{\omega_0}(\cdot))$ in $\left(\Omega, \mathscr{F}_{\omega_0}, \mathscr{F}_{\omega_0, s}^\eta, \mathbb{P}_{\omega_0}, W_\eta\right)$ for $\mathbb{P}$-a.e. ω_0.

Therefore, using this, (A0) and the fact that for $\mathbb{P}$-a.e. ω_0, $\mathbb{P}_{\omega_0}(\{\omega : X_1(\eta, \omega) = X_1(\eta, \omega_0)\}) = 1$, we have

$$\mathbb{E}\left[\int_\eta^T e^{-\int_t^s c(X(\tau))d\tau} l(s, X(s), a(s))\, ds + e^{-\int_t^T c(X(\tau))d\tau} g(X(T))\right]$$

$$= \mathbb{E}\left[\int_\eta^T e^{-\int_t^s c(X_1(\tau))d\tau} l(s, X_1(s), a_1(s))\, ds + e^{-\int_t^T c(X_1(\tau))d\tau} g(X_1(T))\right]$$

$$= \mathbb{E}\left[e^{-\int_t^\eta c(X(\tau))d\tau} \mathbb{E}\left[\int_\eta^T e^{-\int_\eta^s c(X_1(\tau))d\tau} l(s, X_1(s), a_1(s))\, ds\right.\right.$$

$$\left.\left. + e^{-\int_\eta^T c(X_1(\tau))d\tau} g(X_1(T)) |\mathscr{F}_\eta^t\right]\right]$$

$$= \mathbb{E}\left[e^{-\int_t^\eta c(X(\tau))d\tau} J(\eta, X_1(\eta, \omega_0); a^{\omega_0}(\cdot))\right] \geq \mathbb{E}\left[e^{-\int_t^\eta c(X(\tau))d\tau} V(\eta, X_1(\eta, \omega_0))\right]$$

$$= \mathbb{E}\left[e^{-\int_t^\eta c(X(\tau))d\tau} V(\eta, X(\eta))\right], \quad (2.29)$$

where we used the remarks at the end of Sect. 2.2.2. Thus, using (2.28), we obtain

$$J(t, x; a(\cdot)) \geq \mathbb{E}\left[\int_t^\eta e^{-\int_t^s c(X(\tau))d\tau} l(s, X(s), a(s))\, ds + e^{-\int_t^\eta c(X(\tau))d\tau} V(\eta, X(\eta))\right]$$

and the claim follows by taking the infimum over all $a(\cdot) \in \tilde{\mathcal{U}}_t$.

Part 2. (inequality $\leq$ in (2.27)): Let $t \leq \eta \leq T$. We fix $a(\cdot) = \left(\Omega, \mathscr{F}, \mathscr{F}_s^t, \mathbb{P}, W, a(\cdot)\right) \in \tilde{\mathcal{U}}_t$.

We can choose $\delta_1 > 0$ so that, for $|x - \bar{x}| < \delta_1$ and $|x|, |\bar{x}| \leq 1$, we have for each $\tilde{a}(\cdot) \in \mathcal{U}_\eta$

$$|J\left(\eta, x; \tilde{a}(\cdot)\right) - J\left(\eta, \bar{x}; \tilde{a}(\cdot)\right)| + |V\left(\eta, x\right) - V\left(\eta, \bar{x}\right)| < \varepsilon.$$

Since H is separable we can choose a partition $\left\{D_j^1\right\}_{j \in \mathbb{N}}$ of $B_H(0, 1)$ into countable disjoint Borel subsets with $diam(D_j^1) < \delta_1$. Similarly we can choose a (possibly smaller) $\delta_2 > 0$ such that, for $|x - \bar{x}| < \delta_2$ and $|x|, |\bar{x}| \leq 2$, we have for each $\tilde{a}(\cdot) \in \mathcal{U}_\eta$

$$|J\left(\eta, x; \tilde{a}(\cdot)\right) - J\left(\eta, \bar{x}; \tilde{a}(\cdot)\right)| + |V\left(\eta, x\right) - V\left(\eta, \bar{x}\right)| < \varepsilon,$$

and we can choose a partition $\left\{D_j^2\right\}_{j \in \mathbb{N}}$ of $B_H(0, 2) \backslash B_H(0, 1)$ into countable disjoint Borel subsets with $diam(D_j^2) < \delta_2$.

Iterating the argument we can find a partition $\left\{D_j\right\}_{j \in \mathbb{N}}$ of H into countable disjoint Borel subsets with the following property: for all D_j and all $x, \bar{x} \in D_j$ we have, for each $\tilde{a}(\cdot) \in \mathcal{U}_\eta$,

$$|J\left(\eta, x; \tilde{a}(\cdot)\right) - J\left(\eta, \bar{x}; \tilde{a}(\cdot)\right)| + |V\left(\eta, x\right) - V\left(\eta, \bar{x}\right)| < \varepsilon.$$

For each $j \in \mathbb{N}$ we choose $x_j \in D_j$ and $a_j(\cdot) = \left(\Omega_j, \mathscr{F}_j, \mathscr{F}_{j,s}^\eta, \mathbb{P}_j, W_j, a_j(\cdot)\right) \in \mathcal{U}_\eta^{\nu_j}$ such that

$$J\left(\eta, x_j; a_j(\cdot)\right) < V\left(\eta, x_j\right) + \varepsilon. \tag{2.30}$$

We define a new control $a^\eta(\cdot) \in \tilde{\mathcal{U}}_t$ on the probability space $\left(\Omega, \mathscr{F}, \mathscr{F}_s^t, \mathbb{P}, W\right)$ as follows. Let $a_{j,1}(\cdot)$ be the $\mathscr{F}_{j,s}^{\eta,0}$-predictable processes from Lemma 1.99 such that $a_{j,1}(\cdot) = a_j(\cdot)$, $\mathbb{P}_j \otimes dt$-a.e. and let $f_j : [\eta, T] \times C\left([\eta, T], \Xi\right) \to \Lambda$ be the functions from Lemma 2.20 such that

$$f_j\left(s, W_j(\cdot, \omega)\right) = a_{j,1}\left(s, \omega\right), \qquad \text{for } \omega \in \Omega_j, \ s \in [\eta, T].$$

We now set $\tilde{a}_j\left(s, \omega\right) = f_j\left(s, W_\eta\left(\cdot, \omega\right)\right)$. By Corollary 2.21 and Lemma 2.26 the process $\tilde{a}_j(\cdot)$ is $\mathscr{F}_s^{t,0}$-progressively measurable and, for $\mathbb{P}$-a.e. ω_0, is $\mathscr{F}_{\omega_0,s}^\eta$-progressively measurable in the reference probability spaces $\nu_{\omega_0} := \left(\Omega, \mathscr{F}_{\omega_0}, \mathscr{F}_{\omega_0,s}^\eta, \mathbb{P}_{\omega_0}, W_\eta\right)$ defined in Lemma 2.25. Moreover, $\mathcal{L}_{\mathbb{P}_{\omega_0}}(\tilde{a}_j(\cdot), W_\eta(\cdot)) = \mathcal{L}_{\mathbb{P}_j}(a_{j,1}(\cdot), W_j(\cdot))$. We define

$$a^\eta\left(s, \omega\right) = a\left(s, \omega\right) \mathbf{1}_{\{t \leq s < \eta\}} + \mathbf{1}_{\{s \geq \eta\}} \sum_{j \in \mathbb{N}} \tilde{a}_j\left(s, \omega\right) \mathbf{1}_{\{X(\eta; t, x, a(\cdot)) \in D_j\}}. \tag{2.31}$$

Obviously $\left(\Omega, \mathscr{F}, \mathscr{F}_s^t, \mathbb{P}, W, a^\eta(\cdot)\right) \in \tilde{\mathcal{U}}_t$.

Let $X(s) = X(s; t, x, a^\eta(\cdot))$. Notice that $X(s; t, x, a^\eta(\cdot)) = X(s; t, x, a(\cdot))$ on $[t, \eta]$, $\mathbb{P}$-a.e.

Define $O_j := \{\omega : X(\eta; t, x, a(\cdot)) \in D_j\}$. Since for $\mathbb{P}$-a.e. ω_0, $\mathbb{P}_{\omega_0}(\{\omega : X(\eta, \omega) = X(\eta, \omega_0)\}) = 1$, if $\omega_0 \in O_j$, then $\mathbb{P}_{\omega_0}(\Omega \backslash O_j) = 0$, which implies that in this case $\tilde{a}_j(\cdot) = a^\eta(\cdot)$ on $[\eta, T]$, $\mathbb{P}_{\omega_0}$-a.s., and thus, for $\mathbb{P}$-a.e. ω_0, $a^\eta|_{[\eta,T]} \in \mathcal{U}_\eta^{\nu_{\omega_0}}$, and

$$\mathcal{L}_{\mathbb{P}_{\omega_0}}(a^\eta(\cdot), W_\eta(\cdot)) = \mathcal{L}_{\mathbb{P}_j}(a_{j,1}(\cdot), W_j(\cdot)), \quad j \in \mathbb{N}. \tag{2.32}$$

Moreover, we can assume, by $(A3)$, that for $\mathbb{P}$-a.e. ω_0, $X(\cdot) = X^{\nu_{\omega_0}}(\cdot; \eta, X(\eta), a^\eta(\cdot))$ on $[\eta, T]$ $\mathbb{P}_{\omega_0}$-a.s.

By the definition of V,

$$V(t, x) \leq \mathbb{E}\left[\int_t^T e^{-\int_t^s c(X(\tau))d\tau} l(s, X(s), a^\eta(s)) ds + e^{-\int_t^T c(X(\tau))d\tau} g(X(T))\right]$$

$$= \mathbb{E}\left[\int_t^\eta e^{-\int_t^s c(X(\tau))d\tau} l(s, X(s), a(s)) ds\right]$$

$$+ \mathbb{E}\left[\int_\eta^T e^{-\int_t^s c(X(\tau))d\tau} l(s, X(s), a^\eta(s)) ds + e^{-\int_t^T c(X(\tau))d\tau} g(X(T))\right]. \tag{2.33}$$

We have

$$\mathbb{E}\left[\int_\eta^T e^{-\int_t^s c(X(\tau))d\tau} l(s, X(s), a^\eta(s)) ds + e^{-\int_t^T c(X(\tau))d\tau} g(X(T))\right]$$

$$= \mathbb{E}\left[e^{-\int_t^\eta c(X(\tau))d\tau} \mathbb{E}\left[\int_\eta^T e^{-\int_\eta^s c(X(\tau))d\tau} l(s, X(s), a^\eta(s)) ds\right.\right.$$

$$\left.\left. + e^{-\int_\eta^T c(X(\tau))d\tau} g(X(T)) | \mathscr{F}_\eta^t\right]\right]$$

$$= \sum_{j \in \mathbb{N}} \int_{O_j} e^{-\int_t^\eta c(X(\tau))d\tau} \mathbb{E}_{\omega_0}\left[\int_\eta^T e^{-\int_\eta^s c(X(\tau))d\tau} l(s, X(s), a^\eta(s)) ds\right.$$

$$\left. + e^{-\int_\eta^T c(X(\tau))d\tau} g(X(T))\right] d\mathbb{P}(\omega_0).$$

By (2.32) and $(A0)$–$(A1)$ we obtain

$$\mathcal{L}_{\mathbb{P}_{\omega_0}}(X(\cdot), a^\eta(\cdot)) = \mathcal{L}_{\mathbb{P}_j}(X^{\nu_j}(\cdot), a_{j,1}(\cdot)), \quad j \in \mathbb{N},$$

where $X^{\nu_j}(s) = X^{\nu_j}(s; \eta, X(\eta; t, x, a(\cdot))(\omega_0), a_{j,1}(\cdot))$. Thus, it follows from Theorem 1.134 that, with

$$f(s) = e^{-\int_\eta^s c(X(\tau))d\tau}, \quad f_j(s) = e^{-\int_\eta^s c(X^{\nu_j}(\tau))d\tau},$$

$$\mathcal{L}_{\mathbb{P}_{\omega_0}}(f(\cdot), X(\cdot), a^\eta(\cdot)) = \mathcal{L}_{\mathbb{P}_j}(f_j(\cdot), X^{\nu_j}(\cdot), a_{j,1}(\cdot)), \quad j \in \mathbb{N}.$$

Therefore,

$$
\mathbb{E}\left[\int_\eta^T e^{-\int_t^s c(X(\tau))d\tau} l\left(s, X\left(s\right), a^\eta\left(s\right)\right) ds + e^{-\int_t^T c(X(\tau))d\tau} g\left(X\left(T\right)\right)\right]
$$

$$
= \sum_{j\in\mathbb{N}} \int_{O_j} e^{-\int_t^\eta c(X(\tau))d\tau} J_{\mathbb{P}_{\omega_0}}\left(\eta, X(\eta; t, x, a(\cdot))(\omega_0); a^\eta(\cdot)\right) d\mathbb{P}(\omega_0)
$$

$$
= \sum_{j\in\mathbb{N}} \int_{O_j} e^{-\int_t^\eta c(X(\tau))d\tau} J_{\mathbb{P}_j}\left(\eta, X(\eta; t, x, a(\cdot))(\omega_0); a_{j,1}(\cdot)\right) d\mathbb{P}(\omega_0).
$$

Moreover, using (2.30), we get for a.e. $\omega_0 \in O_j$

$$
J_{\mathbb{P}_j}\left(\eta, X\left(\eta; t, x, a(\cdot)\right)(\omega_0); a_j(\cdot)\right) \leq J_{\mathbb{P}_j}\left(\eta, x_j; a_j(\cdot)\right) + \varepsilon
$$
$$
\leq V\left(\eta, x_j\right) + 2\varepsilon \leq V\left(\eta, X\left(\eta; t, x, a(\cdot)\right)(\omega_0)\right) + 3\varepsilon,
$$

so we finally obtain

$$
\mathbb{E}\left[\int_\eta^T e^{-\int_t^s c(X(\tau))d\tau} l\left(s, X\left(s\right), a^\eta\left(s\right)\right) ds + e^{-\int_t^T c(X(\tau))d\tau} g\left(X\left(T\right)\right)\right]
$$
$$
\leq \mathbb{E}\left[e^{-\int_t^\eta c(X(\tau))d\tau} V\left(\eta, X\left(\eta; t, x, a(\cdot)\right)\right)\right] + C\varepsilon.
$$

Therefore, by (2.33) and the arbitrariness of $a(\cdot)$,

$$
V\left(t, x\right) \leq \inf_{a(\cdot)\in\mathcal{U}_t} \mathbb{E}\left[\int_t^\eta e^{-\int_t^s c(X(\tau))d\tau} l\left(s, X\left(s\right), a\left(s\right)\right) ds \right.
$$
$$
\left. + e^{-\int_t^\eta c(X(\tau))d\tau} V\left(\eta, X\left(\eta\right)\right)\right] + C\varepsilon
$$

and the claim follows by letting $\varepsilon \to 0$. $\square$

If we know more information about the value function and the control problem, in particular that the value function is continuous in both variables, the dynamic programming principle can be strengthened to include stopping times. We do not do it here in the abstract case. We explain how to obtain such a formulation of the dynamic programming principle for a control problem for mild solutions in Sect. 3.6, Theorem 3.70.

2.4 Infinite Horizon Problems

In this section we consider a special infinite horizon problem described by an evolution equation

$$\begin{cases} dX(s) = \beta(X(s), a(s))ds + \sigma(X(s), a(s))dW_Q(s) \\ X(t) = x, \end{cases} \tag{2.34}$$

with a cost functional of the form

$$J(t, x; a(\cdot)) = \mathbb{E}\left[\int_t^{+\infty} e^{-\int_t^s c(X(\tau;t,x,a(\cdot)))d\tau} l(X(s; t, x, a(\cdot)), a(s))ds\right], \tag{2.35}$$

where $c \geq \lambda > 0$. We are really only interested in the case $t = 0$, but we will keep the dependence on t for a while.

This is a very important class of problems which are semi-"autonomous" in the sense that the coefficients β, σ and the cost l do not depend explicitly on time. In this case the value function does not depend on time and the DPP takes on a simpler form.

We define the value function for $t \geq 0$ as

$$V(t, x) = \inf_{a(\cdot) \in \mathcal{U}_t} \mathbb{E}\left[\int_t^{+\infty} e^{-\int_t^s c(X(\tau;t,x,a(\cdot)))d\tau} l(X(s; t, x, a(\cdot)), a(s))ds\right] \tag{2.36}$$

and set

$$J(x; a(\cdot)) := J(0, x; a(\cdot)), \quad V(x) := V(0, x). \tag{2.37}$$

We assume now that Hypotheses 2.9, 2.11 and 2.12 are satisfied with $T = +\infty$ (i.e. reference probability spaces and solutions are defined on $[t, +\infty)$). We also replace Hypothesis 2.23 by the following one.

Hypothesis 2.28 The functions $l : H \times \Lambda \to \mathbb{R}$, $c : H \to \mathbb{R}$ are Borel measurable and there exists a $\lambda > 0$ such that $c(x) \geq \lambda$ for every $x \in H$. Moreover, for every $0 \leq \eta < +\infty$, $x \in H$, reference probability space ν, $a(\cdot) \in \mathcal{U}_0^\nu$

$$e^{-\int_0^\cdot c(X(\cdot;0,x,a(\cdot)))d\tau} l(X(\cdot; 0, x, a(\cdot)), a(\cdot)) \in M_\nu^1(0, +\infty; \mathbb{R}),$$

$$V(X(\eta; 0, x, a(\cdot))) \in L^1(\Omega, \mathscr{F}, \mathbb{P}).$$

Finally, $J(\cdot; a(\cdot))$ is uniformly continuous on bounded sets of H, uniformly for $a(\cdot) \in \mathcal{U}_0$.

Since we are dealing with an abstract state equation we have to add another hypothesis which reflects the "autonomous" nature of the system. First we observe that

$$\left(\Omega, \mathscr{F}, \{\mathscr{F}_s^t\}_{s \geq t}, \mathbb{P}, W_Q(\cdot), a(\cdot)\right) \in \mathcal{U}_t$$

$$\iff \left(\Omega, \mathscr{F}, \{\mathscr{F}_{s+t}^t\}_{s \geq 0}, \mathbb{P}, W_Q(t + \cdot), a(t + \cdot)\right) \in \mathcal{U}_0.$$

Hypothesis 2.29 Assume that the family of solutions $X(\cdot; t, x, a(\cdot))$ of (2.34) satisfies the following property. For every $\left(\Omega, \mathscr{F}, \mathscr{F}_s^t, \mathbb{P}, W_Q(\cdot), a(\cdot)\right) \in \mathcal{U}_t$

(A5) $\mathcal{L}_{\mathbb{P}}(X(t + \cdot; t, x, a(\cdot)), a(t + \cdot)) = \mathcal{L}_{\mathbb{P}}(X(\cdot; 0, x, a(t + \cdot)), a(t + \cdot))$ on $[0, +\infty)$, where $X(\cdot; 0, x, a(t + \cdot))$ is the solution of (2.34) with $W_Q(\cdot)$ replaced by $W_Q(t + \cdot)$.

Remark 2.30 An example of a state equation satisfying Hypothesis 2.29 is given by the mild solution of an SDE

$$\begin{cases} dX(s) = AX(s)ds + b(X(s), a(s))ds + \sigma(X(s), a(s))dW_Q(s) \\ X(t) = x, \end{cases} \quad (2.38)$$

where A, b and σ satisfy the assumptions described in Hypothesis 1.125.   ∎

Using Hypothesis 2.29, by a change of variable and $(A5)$, we observe that

$$\begin{aligned} V(t, x) &= \inf_{a(\cdot)\in\mathcal{U}_t} \mathbb{E}\left[\int_t^{+\infty} e^{-\int_t^s c(X(\tau;t,x,a(\cdot)))d\tau} l(X(s; t, x, a(\cdot)), a(s))ds\right] \\ &= \inf_{a(\cdot)\in\mathcal{U}_t} \mathbb{E}\left[\int_0^{+\infty} e^{-\int_0^s c(X(t+\tau;t,x,a(\cdot)))d\tau} l(X(t + s; t, x, a(\cdot)), a(t + s))ds\right] \\ &= \inf_{a(\cdot)\in\mathcal{U}_t} \mathbb{E}\left[\int_0^{+\infty} e^{-\int_0^s c(X(\tau;0,x,a(t+\cdot)))d\tau} l(X(s; 0, x, a(t + \cdot)), a(t + s))ds\right] \\ &= \inf_{a(\cdot)\in\mathcal{U}_0} \mathbb{E}\left[\int_0^{+\infty} e^{-\int_0^s c(X(\tau;0,x,a(\cdot)))d\tau} l(X(s; 0, x, a(\cdot))a(s))ds\right] = V(x). \end{aligned}$$

We thus have the following theorem, whose proof is obtained by a simple modification of the proofs of Theorems 2.22 and 2.24.

Theorem 2.31 (DPP, infinite horizon case) *Assume that Hypotheses 2.9, 2.11 and 2.12 for $T = +\infty$ hold, and that Hypotheses 2.28 and 2.29 are satisfied. Then the value function V satisfies the dynamic programming principle: For every $\eta > 0$, $x \in H$,*

$$V(x) = \inf_{a(\cdot)\in\mathcal{U}_0} \mathbb{E}\left[\int_0^{\eta} e^{-\int_0^s c(X(\tau))d\tau} l(X(s), a(s)) ds + e^{-\int_0^{\eta} c(X(\tau))d\tau} V(X(\eta))\right]. \quad (2.39)$$

Moreover,

$$V(x) = V^{\nu}(x) := \inf_{a(\cdot)\in\mathcal{U}_0^{\nu}} \mathbb{E}\left[\int_0^{+\infty} e^{-\int_0^s c(X(\tau))d\tau} l(X(s), a(s)) ds\right]$$

for every reference probability space ν.

2.5 The HJB Equation and Optimal Synthesis in the Smooth Case

Once we know that the dynamic programming principle (DPP) holds, we want to use it to solve the control problem, i.e. to find optimal pairs and, possibly, to study their properties. In the dynamic programming approach the path to do this consists of:

- formulating a differential form of the DPP (the HJB equation);
- finding a solution v of the HJB equation (which we do not know ex ante to be the value function);
- using such a solution v to prove a verification theorem, i.e. sufficient, and possibly necessary, conditions for optimality, which expresses optimal controls as functions of the current state (feedback controls);
- performing the optimal synthesis, i.e. using the optimality conditions of the previous step to find optimal feedback controls: this will also imply that v is indeed the value function.

Such a program can be performed if we know in advance that the HJB equation has a smooth solution or if we know that the value function is sufficiently regular, both of which may not be true even in finite dimension. However, it is still useful to present how the program works in the smooth case to understand the machinery of the dynamic programming approach. We do it for our model problem, when the state equation admits a solution in the mild sense, assuming that the value function is smooth. We prove the following three results (in both finite and infinite horizon cases):

- The value function solves the HJB equation.
- The verification theorem (necessary and sufficient conditions for optimality).
- The existence of optimal pairs in feedback form.

One of the main goals of the theory presented in this book is to obtain some of these results under more realistic assumptions.

It is important to note that, if one finds a sufficiently smooth solution of the HJB equation, then the verification theorem and the existence of optimal feedbacks can be done without using the DPP, and this is done in Chaps. 4–6.

We will present everything for a control problem in the weak formulation of Sect. 2.1.2, i.e. when the set of admissible controls is equal to $\overline{\mathcal{U}}_t$, as this setup is more convenient when discussing optimal feedback controls. However, the same results are also true for control problems in the weak formulation of Sect. 2.2 used to prove the DPP, with $\overline{\mathcal{U}}_t$ replaced by $\mathcal{U}_t$, or in the strong formulation of Sect. 2.1.1.

2.5.1   The Finite Horizon Problem: Parabolic HJB Equation

Let Hypothesis 2.1 hold. Consider an optimal control problem of minimizing the cost functional (2.3) for the system governed by (2.11), where for simplicity we do not have discounting in (2.3), i.e. we set $c = 0$. We rewrite it here for the reader's convenience. The state equation is

$$\begin{cases} dX(s) = (AX(s) + b(s, X(s), a^{\mu}(s))) \, ds + \sigma(s, X(s), a^{\mu}(s)) dW_Q(s) \\ X(t) = x, \end{cases} \tag{2.40}$$

where $a^{\mu}(\cdot) \in \mathcal{U}_t^{\mu}$ for some generalized reference probability space μ satisfying Hypothesis 2.1, and the cost functional

$$J(t, x; a^{\mu}(\cdot)) = \mathbb{E}^{\mu} \left[\int_t^T l(s, X(s; t, x; a^{\mu}(\cdot)), a^{\mu}(s)) ds + g(X(T; t, x, a^{\mu}(\cdot))) \right]. \tag{2.41}$$

We consider the control problem in the weak formulation of Sect. 2.1.2, and assume that Hypothesis 1.125 is satisfied. The HJB equation associated with this problem is

$$\begin{cases} v_t + \langle Dv, Ax \rangle + \inf_{a \in \Lambda} \left\{ \frac{1}{2} \mathrm{Tr} \left[\left(\sigma(t, x, a) Q^{\frac{1}{2}} \right) \left(\sigma(t, x, a) Q^{\frac{1}{2}} \right)^* D^2 v \right] \right. \\ \qquad \left. + \langle Dv, b(t, x, a) \rangle + l(t, x, a) \right\} = 0, \\ v(T, x) = g(x). \end{cases} \tag{2.42}$$

In the above equation Dv, $D^2 v$ are the Fréchet derivatives of v with respect to x, which are identified respectively with elements of H and $S(H)$, the set of bounded, self-adjoint operators in the Hilbert space H. For $(t, x, p, S, a) \in [0, T] \times H \times H \times S(H) \times \Lambda$, the term

$$F_{CV}(t, x, p, S, a) := \frac{1}{2} \mathrm{Tr} \left[\left(\sigma(t, x, a) Q^{\frac{1}{2}} \right) \left(\sigma(t, x, a) Q^{\frac{1}{2}} \right)^* S \right] + \langle p, b(t, x, a) \rangle + l(t, x, a) \tag{2.43}$$

will be called the *current value Hamiltonian* of the system and its infimum over $a \in \Lambda$

$$F(t, x, p, S) := \inf_{a \in \Lambda} \left\{ \frac{1}{2} \mathrm{Tr} \left[\left(\sigma(t, x, a) Q^{\frac{1}{2}} \right) \left(\sigma(t, x, a) Q^{\frac{1}{2}} \right)^* S \right] \right.$$
$$\left. + \langle p, b(t, x, a) \rangle + l(t, x, a) \right\} \tag{2.44}$$

will be called the *Hamiltonian*.[4] Using this notation, the HJB equation (2.42) can be rewritten as

$$\begin{cases} v_t + \langle Dv, Ax \rangle + F(t, x, Dv, D^2v) = 0, \\ v(T, x) = g(x). \end{cases} \qquad (2.45)$$

The HJB equation (2.42) can be viewed as a differential form of the DPP.

Definition 2.32 (*Classical solution, parabolic case*) A function $v: (0, T] \times H \to \mathbb{R}$ is a *classical solution* of (2.42) if $v \in C^{1,2}((0, T) \times H) \cap C((0, T] \times H)$, $Dv: (0, T) \times H \to D(A^*)$, $A^*Dv \in C((0, T) \times H, H)$ and v satisfies

$$\begin{cases} v_t + \langle A^*Dv, x \rangle + \inf_{a \in \Lambda} \left\{ \frac{1}{2} \mathrm{Tr} \left[\left(\sigma(t, x, a) Q^{\frac{1}{2}} \right) \left(\sigma(t, x, a) Q^{\frac{1}{2}} \right)^* D^2v \right] \right. \\ \qquad \left. + \langle Dv, b(t, x, a) \rangle + l(t, x, a) \right\} = 0, \qquad (t, x) \in (0, T) \times H, \\ v(T, x) = g(x), \qquad x \in H, \end{cases}$$

pointwise.

We will use the following assumption.

Hypothesis 2.33

(i) The functions $\sigma(t, x, a)$, $b(t, x, a)$ and $l(t, x, a)$ are uniformly continuous in t on $[0, T]$, uniformly for $(x, a) \in B(0, R) \times \Lambda$ for every $R > 0$.

(ii) There exist $C, N > 0$ such that

$$|l(t, x, a)| \le C(1 + |x|)^N \qquad (2.46)$$

for all $(t, x, a) \in [0, T] \times H \times \Lambda$.

(iii) The function $v: [0, T] \times H \to \mathbb{R}$ is uniformly continuous on bounded subsets of $[0, T] \times H$, and its derivatives Dv, D^2v, v_t are uniformly continuous on bounded subsets of $(0, T) \times H$. Moreover, $Dv: (0, T) \times H \to D(A^*)$ and A^*Dv is uniformly continuous on bounded subsets of $(0, T) \times H$. Finally, there exist $C, N > 0$ such that

$$|v(t, x)| + |Dv(t, x)| + |v_t(t, x)| + \|D^2v(t, x)\| + |A^*Dv(t, x)| \le C(1 + |x|)^N \qquad (2.47)$$

for all $(t, x) \in (0, T) \times H$.

Theorem 2.34 *Let Hypotheses 1.125, 2.1 and 2.33 be satisfied, $v(T, x) = g(x)$ for every $x \in H$, and let the function v satisfy the DPP, i.e. for every $0 < t < \eta < T, x \in H$,*

---

[4]Sometimes it is called the *minimum value Hamiltonian*.

$$v(t, x) = \inf_{a(\cdot) \in \overline{\mathcal{U}}_t} \mathbb{E}\left[\int_t^\eta l(s, X(s), a(s))ds + v(\eta, X(\eta))\right]. \qquad (2.48)$$

Then v is a classical solution of (2.42).

Proof To prove that Eq. (2.42) is satisfied we show separately the two inequalities. We will not present all the details here as the proof follows the lines of the proof of Theorem 3.66, where it is shown that the value function is a viscosity solution by applying Dynkin's formula to a suitable family of test functions.

Part 1. (Supersolution inequality). We fix $(t, x) \in (0, T) \times H$. By (2.48), for every $\varepsilon \in (0, T - t)$ we can choose a control $a^{\mu_\varepsilon}(\cdot) \in \mathcal{U}_t^{\mu_\varepsilon}$ such that

$$v(t, x) + \varepsilon^2 \geq \mathbb{E}^{\mu_\varepsilon}\left[\int_t^{t+\varepsilon} l(s, X^{\mu_\varepsilon}(s), a^{\mu_\varepsilon}(s))ds + v(t + \varepsilon, X^{\mu_\varepsilon}(t + \varepsilon))\right],$$

where $X^{\mu_\varepsilon}(\cdot)$ is the trajectory starting at (t, x) driven by $a^{\mu_\varepsilon}(\cdot)$. Dividing the above by ε we have

$$\varepsilon \geq \mathbb{E}^{\mu_\varepsilon} \frac{v(t + \varepsilon, X^{\mu_\varepsilon}(t + \varepsilon)) - v^{\mu_\varepsilon}(t, x)}{\varepsilon} + \frac{1}{\varepsilon}\mathbb{E}^{\mu_\varepsilon}\int_t^{t+\varepsilon} l(s, X^{\mu_\varepsilon}(s), a^{\mu_\varepsilon}(s))ds$$

and, using Dynkin's formula from Proposition 1.165,

$$\varepsilon \geq \frac{1}{\varepsilon}\mathbb{E}^{\mu_\varepsilon}\left[\int_t^{t+\varepsilon}\left[v_t(s, X^{\mu_\varepsilon}(s)) + \langle A^*Dv(s, X^{\mu_\varepsilon}(s)), X^{\mu_\varepsilon}(s)\rangle \right.\right.$$
$$+ \langle Dv(s, X^{\mu_\varepsilon}(s)), b(s, X^{\mu_\varepsilon}(s), a^{\mu_\varepsilon}(s))\rangle$$
$$+ \frac{1}{2}\text{Tr}\left[\left(\sigma(s, X^{\mu_\varepsilon}(s), a^{\mu_\varepsilon}(s))Q^{1/2}\right)\left(\sigma(s, X^{\mu_\varepsilon}(s), a^{\mu_\varepsilon}(s))Q^{1/2}\right)^* D^2v(s, X^{\mu_\varepsilon}(s))\right]$$
$$\left.\left.+ l(s, X^{\mu_\varepsilon}(s), a^{\mu_\varepsilon}(s))\right]ds\right] = \frac{1}{\varepsilon}\mathbb{E}^{\mu_\varepsilon}\int_t^{t+\varepsilon}\Psi(s, X^{\mu_\varepsilon}(s), a^{\mu_\varepsilon}(s))ds,$$

$$(2.49)$$

where

$$\Psi(s, y, a) := v_t(s, y) + \langle A^*Dv(s, y), y\rangle + \langle Dv(s, y), b(s, y, a)\rangle$$
$$+ \frac{1}{2}\text{Tr}\left[\left(\sigma(s, y, a)Q^{1/2}\right)\left(\sigma(s, y, a)Q^{1/2}\right)^* D^2v(s, y)\right] + l(s, y, a).$$

By our assumptions we have, for some $h > 0$ and modulus ρ, depending on t, x,

$$|\Psi(s, y, a) - \Psi(t, x, a)| \leq \rho(|s - t| + |y - x|) \quad \text{for all } (s, y) \in [t, t + h] \times B_1(x), a \in \Lambda,$$
$$(2.50)$$

and, for some C and $M \geq 0$,

$$|\Psi(s, X^{\mu_\varepsilon}(s), a^{\mu_\varepsilon}(s))| \leq C(1 + |X^{\mu_\varepsilon}(s)|^M). \tag{2.51}$$

Moreover, it follows from (1.38) that there is an $r_\varepsilon > 0$, independent of a^{μ_ε}, such that $r_\varepsilon \to 0$ as $\varepsilon \to 0$, and with

$$\Omega_1^\varepsilon = \{\omega \in \Omega^{\mu_\varepsilon} : \sup_{s \in [t, t+\varepsilon]} |X^{\mu_\varepsilon}(s) - x| \leq r_\varepsilon\},$$

we have

$$\mathbb{P}^{\mu_\varepsilon}(\Omega_1^\varepsilon) \geq \gamma(\varepsilon) \to 1 \quad \text{as } \varepsilon \to 0. \tag{2.52}$$

Thus, using (1.37), (2.49)–(2.52) we obtain (see the proof of Theorem 3.66 for more details) that there exists a modulus $\rho_1(\varepsilon)$, depending on t and x, such that

$$
\begin{aligned}
\rho_1(\varepsilon) &\geq \frac{1}{\varepsilon} \mathbb{E}^{\mu_\varepsilon} \int_t^{t+\varepsilon} \Psi(t, x, a^{\mu_\varepsilon}(s)) ds \\
&\geq v_t(t, x) + \langle A^* Dv(t, x), x \rangle + \frac{1}{\varepsilon} \mathbb{E}^{\mu_\varepsilon} \Bigg[\int_t^{t+\varepsilon} \inf_{a \in \Lambda} \Big\{ \frac{1}{2} \mathrm{Tr}\Big[\big(\sigma(t, x, a) Q^{1/2}\big) \\
&\quad \times \big(\sigma(t, x, a) Q^{1/2}\big)^* D^2 v(t, x)\Big] + \langle Dv(t, x), b(t, x, a) \rangle + l(t, x, a) \Big\} ds \Bigg] \\
&= v_t(t, x) + \langle A^* Dv(t, x), x \rangle + \inf_{a \in \Lambda} \Big\{ \frac{1}{2} \mathrm{Tr}\Big[\big(\sigma(t, x, a) Q^{\frac{1}{2}}\big) \big(\sigma(t, x, a) Q^{\frac{1}{2}}\big)^* D^2 v(t, x)\Big] \\
&\quad + \langle Dv(t, x), b(t, x, a) \rangle + l(t, x, a) \Big\}.
\end{aligned}
$$

The inequality follows letting $\varepsilon \to 0$.

Part 2. (Subsolution inequality). Choose $a \in \Lambda$ and consider the constant control $\bar{a}(\cdot) \equiv a \in \Lambda$ for some generalized reference probability space μ. Denote by $X(s)$ the trajectory starting from (t, x) driven by the control $\bar{a}(\cdot)$. From (2.48) we have for $\varepsilon \in (0, T - t)$

$$v(t, x) \leq \mathbb{E}^\mu \left[\int_t^{t+\varepsilon} l(s, X(s), a) ds + v(t + \varepsilon, X(t + \varepsilon)) \right].$$

Again using Dynkin's formula from Proposition 1.165, we thus obtain

$$
\begin{aligned}
0 &\leq \frac{\mathbb{E}^\mu [v(t + \varepsilon, X(t + \varepsilon)) - v(t, x)]}{\varepsilon} + \frac{1}{\varepsilon} \mathbb{E}^\mu \int_t^{t+\varepsilon} l(s, X(s), a) ds \\
&= \frac{1}{\varepsilon} \mathbb{E}^\mu \left[\int_t^{t+h} \Big[v_t(s, X(s)) + \langle A^* Dv(s, X(s)), X(s) \rangle \right. \\
&\quad + \langle Dv(s, X(s)), b(s, X(s), a) \rangle
\end{aligned}
$$

$$+ \frac{1}{2} \mathrm{Tr} \left[\left(\sigma(s, X(s), a) Q^{1/2} \right) \left(\sigma(s, X(s), a) Q^{1/2} \right)^* D^2 v(s, X(s)) \right]$$

$$+ l(s, X(s), a) \bigg] ds \bigg].$$

(2.53)

We can now pass to the limit as $\varepsilon \to 0$ above, as in Part 1, to obtain that for every $a \in \Lambda$

$$0 \le v_t(t, x) + \langle A^* Dv(t, x), x \rangle + \frac{1}{2} \mathrm{Tr} \left[\left(\sigma(t, x, a) Q^{\frac{1}{2}} \right) \left(\sigma(t, x, a) Q^{\frac{1}{2}} \right)^* D^2 v(t, x) \right]$$

$$+ \langle Dv(t, x), b(t, x, a) \rangle + l(t, x, a), \ t \in (0, T), \ x \in H.$$

The inequality follows by taking the infimum over $a \in \Lambda$ above. $\qquad \square$

Remark 2.35 It is clear from the proof that Theorem 2.34 still holds if $\overline{\mathcal{U}}_t$ in (2.48) is replaced by $\mathcal{U}_t$ or if (2.48) is stated in the strong formulation, i.e. if $\overline{\mathcal{U}}_t$ in (2.48) is replaced by $\mathcal{U}_0^\mu$ for some fixed generalized reference probability space μ on $[0, T]$. $\qquad \blacksquare$

We now show how to use the HJB equation to characterize optimal controls. First we prove the so-called verification theorem. It could also be stated in the strong formulation.

Theorem 2.36 (Smooth Verification Theorem, Sufficient Condition) *Let $v : [0, T] \times H \to \mathbb{R}$ be a classical solution of (2.42) as defined in Definition 2.32. Let Hypotheses 1.125, 2.1 and 2.33-(ii)(iii) be satisfied. Then:*

(i) We have

$$v(t, x) \le \overline{V}(t, x) \quad \textit{for all } (t, x) \in [0, T] \times H. \tag{2.54}$$

(ii) Let $(a^*(\cdot), X^*(\cdot))$ be an admissible pair at (t, x) such that

$$a^*(s) \in \arg \min_{a \in \Lambda} F_{CV}(s, X^*(s), Dv(s, X^*(s)), D^2 v(s, X^*(s)), a),$$

(2.55)

for almost every $s \in [t, T]$ and $\mathbb{P}$-almost surely. Then the pair $(a^*(\cdot), X^*(\cdot))$ is optimal at (t, x), and $v(t, x) = \overline{V}(t, x)$.

Proof We first prove the following identity.[5] For every $a(\cdot) \in \overline{\mathcal{U}}_t$:

$$v(t, x) = J(t, x; a(\cdot))$$

$$- \mathbb{E} \int_t^T \Big[F_{CV} \left(r, X(r), Dv(r, X(r)), D^2 v(r, X(r)), a(r) \right)$$

$$- F \left(r, X(r), Dv(r, X(r)), D^2 v(r, X(r)) \right) \Big] dr.$$

(2.56)

---

[5]This is often called the *fundamental identity* for the optimal control problem.

Indeed, consider $a(\cdot) \in \overline{\mathcal{U}}_t$ and the corresponding trajectory $X(\cdot)$ starting at x at time t. We apply Proposition 1.165, to the process $v(s, X(s))$, $s \in [t, T]$, obtaining

$$\mathbb{E}v(T, X(T)) = v(t, x) + \mathbb{E} \int_t^T v_t(r, X(r))dr$$

$$+ \mathbb{E} \int_t^T \langle A^* Dv(r, X(r)), X(r) \rangle dr + \mathbb{E} \int_t^T \langle Dv(r, X(r)), b(r, X(r), a(r)) \rangle dr$$

$$+ \frac{1}{2} \mathbb{E} \int_t^T \mathrm{Tr} \Big[\big(\sigma(r, X(r), a(r)) Q^{1/2} \big) \big(\sigma(r, X(r), a(r)) Q^{1/2} \big)^* D^2 v(r, X(r)) \Big] dr.$$

$$(2.57)$$

We now use that $v(T, \cdot) = g$, rearrange the terms, and we add and subtract $\mathbb{E} \int_t^T l(r, X(r), a(r))dr$ obtaining, by the definition of the current value Hamiltonian F_{CV} in (2.43),

$$v(t, x) = \mathbb{E}g(X(T)) + \mathbb{E} \int_t^T l(r, X(r), a(r))dr$$

$$- \mathbb{E} \int_t^T \Big[v_t(r, X(r)) + \langle A^* Dv(r, X(r)), X(r) \rangle \Big] dr$$

$$- \mathbb{E} \int_t^T F_{CV}(r, X(r), Dv(r, X(r)), Dv(r, X(r)), a(r))dr.$$

$$(2.58)$$

Equality (2.56) is now a consequence of the definition of the functional J and the fact that v is a classical solution of the HJB equation (2.42).

Therefore (i) follows by observing that, by definition, $F_{CV} - F \geq 0$ everywhere, and by taking the infimum over $a(\cdot) \in \overline{\mathcal{U}}_t$ in the right-hand side of (2.56).

Regarding (ii), let $(a^*(\cdot), X^*(\cdot))$ be an admissible pair at (t, x) satisfying (2.55) for almost every $s \in [t, T]$ and $\mathbb{P}$-almost surely. We then have

$$\mathbb{E} \int_t^T \Big[F_{CV}(r, X^*(r), Dv(r, X^*(r)), D^2 v(r, X^*(r)), a^*(r))$$

$$- F\left(r, X^*(r), Dv(r, X^*(r)), D^2 v(r, X^*(r))\right) \Big] dr = 0.$$

Thus, by (2.56), we get

$$v(t, x) = J(t, x; a^*(\cdot)), \tag{2.59}$$

which, together with (i), implies that $(a^*(\cdot), X^*(\cdot))$ is optimal at (t, x) and $v(t, x) = \overline{V}(t, x)$. $\qquad\square$

Note that part (i) of the above theorem remains true if v is any classical subsolution of the HJB equation (2.42)[6] with the required regularity.

If we know from the beginning that the solution v in Theorem 2.36 is the value function $\overline{V}$ then (2.55) also becomes a necessary condition for optimality.

Corollary 2.37 (Smooth Verification Theorem, Necessary Condition) *Let the assumptions of Theorem 2.36 hold for $v = \overline{V}$. Let $(a^*(\cdot), X^*(\cdot))$ be an optimal pair at (t, x). Then we must have*

$$a^*(s) \in \arg\min_{a \in \Lambda} F_{CV}(s, X^*(s), D\overline{V}(s, X^*(s)), D^2\overline{V}(s, X^*(s)), a), \qquad (2.60)$$

for almost every $s \in [t, T]$ and $\mathbb{P}$-almost surely.

Proof Now the function $v = \overline{V}$ satisfies (2.56). Since $(a^*(\cdot), X^*(\cdot))$ is an optimal pair at (t, x), we have $\overline{V}(t, x) = J(t, x; a^*(\cdot))$. Therefore, (2.56) for $\overline{V}$ implies that the integrand of the last term of (2.56) is zero $dt \otimes \mathbb{P}$-a.e. and the claim follows. $\square$

Assume now that we have a classical solution v of the HJB equation (2.42). Define the multivalued function

$$\begin{cases} \Phi: (0, T) \times H \to \mathcal{P}(\Lambda), \\ \Phi: (t, x) \to \arg\min_{a \in \Lambda} F_{CV}(t, x, Dv(t, x), D^2v(t, x), a). \end{cases} \qquad (2.61)$$

The *Closed Loop Equation* (CLE) associated with our problem and v is then formally defined as

$$\begin{cases} dX(s) \in AX(s)dt + b(s, X(s), \Phi(s, X(s)))ds + \sigma(s, X(s), \Phi(s, X(s)))dW_Q(s) \\ X(t) = x. \end{cases} \qquad (2.62)$$

If we can find a solution (in a suitable sense) $X_\Phi(\cdot)$ of such a stochastic differential inclusion, we expect that, if $a_\Phi(\cdot)$ is a suitable measurable selection of $\Phi(\cdot, X(\cdot))$, then the pair $(a_\Phi(\cdot), X_\Phi(\cdot))$ is optimal at (t, x). This is indeed the statement of the next corollary.

Corollary 2.38 (Optimal Feedback Controls) *Let the assumptions of Theorem 2.36 hold and let $t \in [0, T]$. Assume, moreover, that the feedback map Φ defined in (2.61) admits a measurable selection $\phi_t : (t, T) \times H \to \Lambda$ such that the Closed Loop Equation*

$$\begin{cases} dX(s) = AX(s)ds + b(s, X(s), \phi_t(s, X(s)))ds + \sigma(s, X(s), \phi_t(s, X(s)))dW_Q(s) \\ X(t) = x \end{cases}$$

$$(2.63)$$

---

[6]This is in the sense that $v(T, x) \leq g(x)$ and it satisfies (2.42) with the inequality $\geq$.

has a weak mild solution (see Definition 1.121) $X_{\phi_t}(\cdot)$ in some generalized reference probability space μ satisfying Hypothesis 2.1-(iv). Then the pair $(a_{\phi_t}(\cdot), X_{\phi_t}(\cdot))$, where the control $a_{\phi_t}(\cdot)$ is defined, for $s \in [t, T]$, by the feedback law $a_{\phi_t}(s) = \phi(s, X_{\phi_t}(s))$, is admissible and it is optimal at (t, x).

Proof By construction the pair $(a_{\phi_t}(\cdot), X_{\phi_t}(\cdot))$ satisfies (2.55). Then, by Theorem 2.36-(ii) we obtain that such pair is optimal. Observe that since the assumptions of Theorem 2.36 are satisfied, $X_{\phi_t}(\cdot)$ is the unique mild solution (in the strong probabilistic sense) of the state equation associated to the control $a_{\phi_t}(\cdot)$ in the generalized reference probability space μ. □

In the above corollary we assumed that the closed loop equation has a weak mild solution to obtain the existence of an optimal feedback control for the weak formulation. If we consider the control problem (2.40) and (2.41) in the strong formulation with the value function defined by (2.5), all the results above remain true except for Corollary 2.38. Indeed, the weak mild solution $X_{\phi_t}(\cdot)$, and hence the control $a_{\phi_t}(\cdot)$, may be defined in a different reference probability space than the starting one. To get existence of an optimal feedback control in the strong formulation one needs to have existence of solutions of the closed loop equation (2.63) in the strong probabilistic sense (i.e. in the mild or strong sense of Definitions 1.118 and 1.119).

To conclude let us reiterate the three step process to carry out the so-called synthesis of optimal control for the problem (2.40) and (2.41) once we have a classical solution v of our HJB equation.

1. Introduce the feedback function Φ, depending on v, as in (2.61). It provides a candidate-optimal control in terms of the state.
2. Look for a solution $X^*(\cdot)$ of the closed loop equation (2.63) for some measurable selection ϕ_t of the map Φ.
3. Define, for $s \in [t, T]$, $a^*(s) := \phi_t(s, X^*(s))$. Then the pair $(a^*(\cdot), X^*(\cdot))$ is optimal at (t, x) thanks to Theorem 2.36.

Of course, given an optimal control problem like (2.40) and (2.41), it may not be possible to perform the above steps as they are. However, even if the HJB equation does not have a classical solution, we may still be able to synthesize optimal controls. This will be explained in later chapters for some special cases. The general synthesis of optimal controls is still a largely open problem.

As was explained in Remark 2.6, the extended weak formulation may be more suitable for Corollary 2.38 (and also Corollary 2.44 in the infinite horizon case). This is done in Chap. 6, Sects. 6.5 and 6.10 (see also Chap. 4, Propositions 4.199 and 4.218, and Chap. 5, Sect. 5.5.5).

2.5.2 The Infinite Horizon Problem: Elliptic HJB Equation

Consider the optimal control problem of minimizing the infinite horizon functional (2.35) for the system governed by the state equation (2.38) with $t = 0$ and the set of admissible controls equal to $\overline{\mathcal{U}}_0$. For simplicity we will assume that $c(\cdot) \equiv \lambda > 0$. This is a typical infinite horizon problem with constant discounting.

The current value Hamiltonian is now defined by

$$F_{CV}(x, p, S, a) := \frac{1}{2}\mathrm{Tr}\left[\left(\sigma(x, a)Q^{\frac{1}{2}}\right)\left(\sigma(x, a)Q^{\frac{1}{2}}\right)^* S\right] + \langle p, b(x, a)\rangle + l(x, a),$$

$$(2.64)$$

the Hamiltonian is given by

$$F(x, p, S) := \inf_{a \in \Lambda}\left\{\frac{1}{2}\mathrm{Tr}\left[\left(\sigma(x, a)Q^{\frac{1}{2}}\right)\left(\sigma(x, a)Q^{\frac{1}{2}}\right)^* S\right] + \langle p, b(x, a)\rangle + l(x, a)\right\},$$

$$(2.65)$$

and the HJB equation associated to our infinite horizon optimal control problem is

$$\lambda v - \langle Dv, Ax\rangle - F(x, Dv, D^2v) = 0 \qquad (2.66)$$

for the unknown function $v : H \to \mathbb{R}$.

We present below the infinite horizon versions of the results of the previous subsection.

Definition 2.39 (*Classical solution, elliptic case*) A function $v : H \to \mathbb{R}$ is a *classical solution* of (2.66) if $v \in C^2(H)$, $A^*Dv \in C(H, H)$ and v satisfies

$$\lambda v - \langle A^*Dv, x\rangle - F(x, Dv, D^2v) = 0$$

pointwise.

Similarly to the previous section we will need the following assumption.

Hypothesis 2.40

(i) There exist $C, N > 0$ such that

$$|l(x, a)| \leq C(1 + |x|)^N \qquad (2.67)$$

for all $(x, a) \in H \times \Lambda$.

(ii) The function $v : H \to \mathbb{R}$ and its derivatives Dv, D^2v, v_t are uniformly continuous on bounded subsets of H. Moreover, $Dv : H \to D(A^*)$ and A^*Dv is uniformly continuous on bounded subsets of H, and

$$|v(x)| + |Dv(x)| + \|D^2v(x)\| + |A^*Dv(x)| \leq C(1 + |x|)^N \qquad (2.68)$$

for all $x \in H$.

Theorem 2.41 *Let Hypotheses 1.125, 2.1 for $T = +\infty$, and Hypothesis 2.40 be satisfied. Assume that the function v satisfies the DPP, i.e. for every $0 < \eta < +\infty$, $x \in H$,*

$$v(x) = \inf_{a(\cdot) \in \overline{\mathcal{U}}_0} \mathbb{E}\left[\int_0^\eta l(s, X(s), a(s))ds + v(X(\eta))\right]. \qquad (2.69)$$

Then, the function v is a classical solution of (2.66).

Proof The proof follows the lines of the proof of Theorem 2.34.   □

We now pass to the verification theorem, the necessary conditions and the closed loop equation. In these results we may encounter integrability problems. To avoid technical complications, here we consider the case where the discount factor λ is sufficiently big.

Theorem 2.42 (Smooth Verification, Sufficient Condition, Infinite Horizon) *Let* $v: H \to \mathbb{R}$ *be a classical solution of (2.66) as defined in Definition 2.39. Let Hypotheses 1.125, 2.1 for $T = +\infty$, and Hypothesis 2.40 be satisfied, and let $\lambda >$ $\overline{\lambda} = (N + 2)(C + \frac{1}{2}(N + 1)C^2)$, where C is the constant from (1.33) and (1.34) (see Proposition 3.24 for $m = N + 2$). Then:*

(i) For all $x \in H$

$$v(x) \le \overline{V}(x). \qquad (2.70)$$

(ii) Let $(a^*(\cdot), X^*(\cdot))$ be an admissible pair at x such that

$$a^*(s) \in \arg\min_{a \in \Lambda} F_{CV}(X^*(s), Dv(s, X^*(s)), D^2v(s, X^*(s)), a) \qquad (2.71)$$

for almost every $s \in [0, +\infty)$ and $\mathbb{P}$-almost surely. Then the pair $(a^*(\cdot), X^*(\cdot))$ is optimal at x, and $v(x) = \overline{V}(x)$.

Proof The proof is similar to that of Theorem 2.36 except for the fact that we now have to take the limit as $T \to +\infty$, in (2.56). Indeed, arguing as in the proof of Theorem 2.36, we obtain that for every $a(\cdot) \in \overline{\mathcal{U}}_0$, and every $T > 0$,

$$v(x) = e^{-\lambda T}\mathbb{E}v(X(T)) + \int_0^T e^{-\lambda r}l(X(r), a(r))dr$$

$$-\mathbb{E}\int_0^T e^{-\lambda r}\Big[F_{CV}\big(X(r), Dv(r, X(r)), D^2v(r, X(r)), a(r)\big)$$

$$- F\big(X(r), Dv(r, X(r)), D^2v(r, X(r))\big)\Big]dr. \qquad (2.72)$$

The condition $\lambda > \overline{\lambda}$ guarantees, due to estimate (3.32), that we can pass to the limit as $T \to +\infty$ above, obtaining the *fundamental identity*:

$$v(x) = \int_0^{+\infty} e^{-\lambda r} l(X(r), a(r)) dr$$

$$- \mathbb{E} \int_0^{+\infty} e^{-\lambda r} \Big[F_{CV}\left(X(r), Dv(r, X(r)), D^2 v(r, X(r)), a(r)\right)$$

$$- F\left(X(r), Dv(r, X(r)), D^2 v(r, X(r))\right) \Big] dr.$$

$$(2.73)$$

The claims now follow as in the proof of Theorem 2.36.   □

Corollary 2.43 (Smooth Verification, Necessary Cond., Infinite Horizon) *Let the assumptions of Theorem 2.42 hold for* $v = \overline{V}$ *and let* $(a^*(\cdot), X^*(\cdot))$ *be an optimal pair at* x. *Then we must have*

$$a^*(s) \in \arg\min_{a \in \Lambda} F_{CV}(X^*(s), D\overline{V}(s, X^*(s)), D^2\overline{V}(s, X^*(s)), a) \qquad (2.74)$$

for almost every $s \in [0, +\infty)$ *and* $\mathbb{P}$-*almost surely.*

Proof The same as Corollary 2.37 using (2.73).   □

As in the finite horizon case we assume that we have a classical solution v of the HJB equation (2.66). We define the multivalued function

$$\begin{cases} \Phi \colon H \to \mathcal{P}(\Lambda) \\ \Phi \colon x \to \arg\min_{a \in \Lambda} F_{CV}(x, Dv(t, x), D^2 v(t, x), a). \end{cases} \qquad (2.75)$$

The *Closed Loop Equation* (CLE) associated with our problem and v is then formally defined as

$$\begin{cases} dX(s) \in AX(s)dt + b(X(s), \Phi(X(s)))ds + \sigma(X(s), \Phi(X(s)))dW_Q(s), \\ X(0) = x. \end{cases}$$

$$(2.76)$$

Again, if a solution $X_\Phi(\cdot)$ of this stochastic differential inclusion can be found, and we can find $a_\Phi(\cdot)$, a suitable measurable selection of $\Phi(\cdot, X(\cdot))$, we would expect the pair $(a_\Phi(\cdot), X_\Phi(\cdot))$ to be optimal at x.

Corollary 2.44 (Optimal Feedback Controls, Infinite Horizon) *Let the assumptions of Theorem 2.42 hold. Assume, moreover, that the feedback map* Φ *defined in* (2.75) *admits a measurable selection* $\phi : H \to \Lambda$ *such that the Closed Loop Equation*

$$\begin{cases} dX(s) = AX(s)dt + b(X(s), \phi(X(s)))ds + \sigma(X(s), \phi(X(s)))dW(s) \\ X(0) = x \end{cases}$$

$$(2.77)$$

has a weak mild solution (see Definition 1.121) $X_\phi(\cdot)$ in some generalized reference probability space satisfying Hypothesis 2.1-(iv). Then the pair $(a_\phi(\cdot), X_\phi(\cdot))$, where the control $a_\phi(\cdot)$ is defined by the feedback law $a_\phi(s) = \phi(s, X_\phi(s))$, is admissible and it is optimal at (t, x).

Proof The proof is the same as that of Corollary 2.38.   □

The optimal synthesis is performed in the same way as in the finite horizon case.

Remark 2.45 In Sect. 2.4 and in this subsection we have considered infinite horizon problems satisfying Hypothesis 2.29, which substantially means that the data b, σ, l and c must be time-independent. We made this restriction partly for simplicity of exposition and partly because such cases are very common in applied models. However, with little effort, it is possible to apply the dynamic programming approach to infinite horizon problems with "non-autonomous" data (so without Hypothesis 2.29). In such cases the value function would be a function of (t, x), the DPP would have the form (2.23), and the HJB equation would be a parabolic equation on $(0, +\infty) \times H$ like (2.45) but with a zeroth order term coming from the discount factor, and without a terminal condition:

$$v_t - \lambda v + \langle Dv, Ax \rangle + F(t, x, Dv, D^2v) = 0. \tag{2.78}$$

Such problems are more difficult but are still interesting, for example, in some financial applications (see [145, 237, 290]).   ■

2.6 Some Motivating Examples

In this section we describe several examples that motivate the study of stochastic optimal control problems in infinite dimension. Our goal here is to show how various applied problems, arising in different areas of science and engineering, are naturally modeled within the framework of infinite-dimensional stochastic analysis. The first five examples are concerned with the control of various kinds of stochastic PDEs, while the last deals with the control of stochastic delay equations. Despite similarities, these examples are very different from each other and it is difficult to find a general theory that includes all of them. This is an unpleasant feature of infinite-dimensional optimal control which will force us to apply different approaches and adaptations of the main general theory.

We have chosen our examples, among many others, since they are representative of interesting applied models and since they motivate the four different approaches described in this book (viscosity solutions, strong solutions, L^2_m solutions, solutions via BSDE). In all examples the criteria to maximize/minimize are given as the expectation of a Bolza-type functional (in finite or infinite horizon cases). We leave aside other types of criteria (mean-variance, risk sensitive, ergodic, etc.) for which we will refer the reader to the existing literature. For each example we provide a short motivation, the main mathematical framework (*state equation, objective functional and constraints*), and show how to translate the problem into the abstract framework of infinite-dimensional stochastic optimal control introduced before. Finally, we discuss the issue of the dynamic programming principle, and we present the associated HJB equations with references to further material in the book and in the literature. For purposes of the DPP we always take the optimal control problems in the weak formulation of Sect. 2.2, however the control problems can be studied with other formulations.

We remark that the existing theory is far from providing a satisfactory treatment of all problems: many challenging questions remain open and call for further research. An important issue in this respect is that of constraints: to obtain more realistic control problems it is often necessary to impose suitable constraints on the state and on the control variables. Such constraints strongly depend on the particular problems. Since the addition of state constraints makes the dynamic programming approach much harder and not much is known at the present stage, we will mention how state constraints arise in specific problems, however we will not deal with state constraint problems.

Finally, we mention a few things about the notation.

- To be consistent with the general setting introduced before, the initial time is always $t \geq 0$.
- In all examples we start first with the finite-dimensional notation. Thus the state equation is first written as a PDE (or a functional equation) in finite dimension with solutions defined informally, then in a subsequent section the state equation is rewritten as an evolution equation in an infinite-dimensional space. To distinguish the two cases we write $y(\cdot)$ to denote the finite-dimensional state and $\alpha(\cdot)$ for the finite-dimensional control, while the infinite-dimensional state and control will be denoted, as before, by $X(\cdot)$ and $a(\cdot)$, respectively.
- Following the usual convention, even if all variables depend on the "scenario" ω, we will always drop such dependence unless needed in the context.

WARNING: The HJB equations that appear in the examples in this section are formal and are all written using the convention adapted from the natural way the equation was written in (2.42)–(2.44) in Sect. 2.5.1. This form is preferable from the PDE point of view as all the terms (when they are defined) use only the reference Hilbert space H of the independent variable x. We want to focus on the examples and leave the details of how the equations are interpreted and solved to later chapters. In some cases the Hamiltonians appearing here are always well defined and will not need any interpretation. In some cases some terms may not make sense the way they are written here and they will need special interpretation which would take too long to explain here. This is especially true of equations discussed in Sects. 2.6.2 and 2.6.3. The formal versions are enough to point out the main difficulties posed by the equations, however the reader should be careful as the equations may not be what they appear.

2.6.1 Stochastic Controlled Heat Equation: Distributed Control

Our first example concerns the problem of controlling a nonlinear stochastic heat equation in a given space region $\mathcal{O} \subset \mathbb{R}^N$. This is a very popular example and here, under "reasonable" assumptions, the theory applies quite well giving rise to results concerning the HJB equation and the synthesis of optimal controls.

Optimal control problems of this type arise in various applied contexts. We recall some of them.

- Optimal control of the heat distribution of a given body (the region $\mathcal{O}$). The deterministic case is described, for instance, in the monograph [403] (pp. 3–5). The presence of the stochastic additive term in the equation can be justified by the presence of (small) random perturbations in the system. (It may be of interest to see what happens when such term goes to zero, see e.g. [89].)
- Optimal control of stochastic reaction diffusion equations, where the white noise term describes the internal fluctuation of the system due to its many-particle nature (see e.g. p. 8 of [180], and [16] for the model without control and [105, 107] for the model with control).
- Optimal control of the motion of an elastic string in a random viscous environment (see e.g. p. 4 of [180] and [287] for the model without control).
- Optimal control of the stochastic cable equation (arising also in neurosciences, see p. 9 of [180] and [568] for the stochastic model, and e.g. [74] for the optimal control problem in the deterministic case).
- Optimal advertising problems (see [429]) or spatial growth problems (see [68, 223]) arising in economics.

2.6.1.1 Setting of the Problem

We are given an open, connected and bounded set $\mathcal{O} \subset \mathbb{R}^N$ with C^1 boundary $\partial\mathcal{O} \subset \mathbb{R}^N$ and a reference probability space $(\Omega, \mathscr{F}, (\mathscr{F}_s)_{s \in [t,T]}, \mathbb{P}, W_Q)$. We consider a controlled dynamical system driven by the following stochastic PDE in the time interval $[t, T]$, for $0 \leq t \leq T < +\infty$

$$
\begin{cases}
dy(s, \xi) = \big(\Delta_\xi y(s, \xi) + f(y(s, \xi)) + a(s, \xi)\big)\, ds + dW_Q(s)(\xi), & s \in (t, T], \xi \in \mathcal{O} \\[2mm]
y(s, \xi) = 0 \quad (s, \xi) \in (t, T] \times \partial\mathcal{O} \\[2mm]
y(t, \xi) = x(\xi) \in L^2(\mathcal{O}), \xi \in \mathcal{O},
\end{cases}
$$

$$(2.79)$$

where:

- the function $y : [t, T] \times \mathcal{O} \times \Omega \to \mathbb{R}, (s, \xi, \omega) \to y(s, \xi, \omega)$ is a stochastic process that describes, for example, the evolution of the temperature distribution and is the *state trajectory* of the system;
- the function $\alpha : [t, T] \times \mathcal{O} \times \Omega \to \mathbb{R}, (s, \xi, \omega) \to \alpha(s, \xi, \omega)$ is a stochastic process giving, for example, the dynamics of the external source of heat acting at every interior point of $\mathcal{O}$ and is the *control strategy* of the system.

We will omit the variable ω writing simply $y(s, \xi)$ and $\alpha(s, \xi)$. Moreover:

- Δ_ξ is the Laplace operator. We consider the Dirichlet boundary condition, however the problem can be studied similarly with the Neumann boundary condition. Conditions of mixed type are also possible, see on this, for example, Chaps. 3 and 5 of [416];
- $f \in C(\mathbb{R})$ is a nonlinear function of the state (which may represent a "reaction" term);
- W_Q is a Q-Wiener process with $Q \in \mathcal{L}^+(L^2(\mathcal{O}))$ and $(\mathscr{F}_s)_{s \in [t,T]}$ is the augmented filtration generated by W_Q[7] (see Remark 2.10 if $\mathrm{Tr}(Q) = +\infty$);
- $x(\cdot) \in L^2(\mathcal{O})$ is the initial state (e.g. temperature distribution) in the region $\mathcal{O}$.

The solution[8] of (2.79) will be denoted by $y^{\alpha(\cdot),t,x}(\cdot)$ to underline the dependence of the state $y(\cdot)$ on the control $\alpha(\cdot)$ and on the initial data t, x. Having in mind the control of the temperature distribution, a reasonable objective of the controller here is that of getting such distribution $y^{\alpha(\cdot),t,x}(\cdot)$ to be close to a required distribution $\bar{y}$ (for each time $s \in [t, T]$ or only at the final time T) while spending the least amount of energy doing this. In such case a reasonable cost functional may be of the form (for suitable constants $c_0, c_1, c_2 \in \mathbb{R}$)

$$I_1(t, x; \alpha(\cdot)) = \mathbb{E}\left\{ \int_t^T \int_{\mathcal{O}} \left[c_0 |y^{\alpha(\cdot),t,x}(s, \xi) - \bar{y}(s, \xi)|^2 + c_1 |\alpha(s, \xi)|^2 \right] d\xi ds \right.$$

$$\left. + \int_{\mathcal{O}} c_2 |y^{\alpha(\cdot),t,x}(T, \xi) - \bar{y}(T, \xi)|^2 d\xi \right\}, \quad (2.80)$$

and the objective would be to minimize the functional I_1 above over all control strategies $\alpha(\cdot)$, progressively measurable with respect to the filtration generated by W_Q, and satisfying suitable constraints and integrability conditions (e.g. such that the state equation and above integrals make sense). More generally, one could consider a cost functional

---

[7]Indeed, stochastic PDEs with more general types of noise can also be treated, see e.g. [487], but this is beyond the scope of this book.

[8]For the concept of solution and the assumptions on the data f, Q and on the control strategy $\alpha(\cdot)$ that guarantee the existence and uniqueness of it, see the next section.

$$I_2(t, x; \alpha(\cdot)) = \mathbb{E}\left\{ \int_t^T \int_{\mathcal{O}} \beta(y^{\alpha(\cdot),t,x}(s, \xi), \alpha(s, \xi)) d\xi ds + \int_{\mathcal{O}} \gamma(y^{\alpha(\cdot),t,x}(T, \xi)) d\xi \right\}$$

$$(2.81)$$

where $\beta \in C(\mathbb{R}^2)$ and $\gamma \in C(\mathbb{R})$ are given functions depending on the objective of the controller.

Finally, the constraints: If the state is the absolute temperature, it is natural to require the positivity of $y^{\alpha(\cdot),t,x}(\cdot)$; moreover, it is reasonable to assume bounds on the control strategies depending on the physical device used to control the system (e.g., for any $s \in [t, T]$ and $\xi \in \mathcal{O}$, $\alpha(s, \xi) \in [m, M]$ for given $m < M$). The constraints depend on a particular problem.

2.6.1.2   The Infinite-Dimensional Setting and the HJB Equation

Take $H = \Xi = L^2(\mathcal{O})$ and let Λ be a closed, bounded subset of $L^2(\mathcal{O})$. For instance, if $\alpha(s, \xi) \in [m, M]$ for every $(s, \xi) \in [t, T] \times \mathcal{O}$ then we would take $\Lambda = \{ f \in L^2(\mathcal{O}) : f(\xi) \in [m, M], \forall \xi \in \mathcal{O} \}$. Consider the Laplace operator with Dirichlet boundary condition defined as (see e.g. [548], Sect. 5.2 p. 180):

$$\begin{cases} D(A) = H^2(\mathcal{O}) \cap H_0^1(\mathcal{O}), \\ Ax = \Delta x, \text{ for } x \in D(A) \end{cases} \qquad (2.82)$$

that generates an analytic semigroup of compact operators e^{tA}, $t \geq 0$. Moreover, define the Nemytskii operator $b : H \to H$ as

$$b(x)(\xi) = f(x(\xi)). \qquad (2.83)$$

Defining, for any $s \in [t, T]$, $X(s) := y(s, \cdot) \in L^2(\mathcal{O})$ and $a(s) := \alpha(s, \cdot) \in \Lambda$, the state equation (2.79) can be rewritten as an SDE in H as follows

$$\begin{cases} dX(s) = (AX(s) + b(X(s)) + a(s)) ds + dW_Q(s) \\ X(t) = x \in H. \end{cases} \qquad (2.84)$$

We know from Proposition 1.147 that, when f is Lipschitz[9] (and so is b) and $Q_r := \int_0^r e^{\tau A} Q e^{\tau A^*} d\tau$ is trace class for all $r > 0$, the above equation admits a unique mild solution denoted by $X(\cdot; t, x, a(\cdot))$ (or simply $X(\cdot)$ when no confusion is possible). If (1.65) also holds, then such solution has continuous trajectories.[10] If $\text{Tr}(Q) < +\infty$ then, thanks to Proposition 2.16, Hypotheses 2.11 and 2.12 are satisfied. If $\text{Tr}(Q) = +\infty$ the claim is still true as outlined in Remark 2.17.

---

[9]In the case studied in [105, 107], b is not Lipschitz, see Sect. 4.9.2 for more details.

[10]Such assumptions are true, for example, when $N = 1$ and Q is the identity, or when $N = 2$ and $Q = (-A)^{-\delta}$ for some $\delta > 0$.

Defining $l\colon H \times \Lambda \to \mathbb{R}$

$$l(x, a) = \int_{\mathcal{O}} \beta(x(\xi), a(\xi))d\xi,$$

and $g\colon H \to \mathbb{R}$ as

$$g(x) = \int_{\mathcal{O}} \gamma(x(\xi))d\xi,$$

the functional I_2 of (2.81) can be rewritten in the Hilbert space setting as

$$J_2(t, x; a(\cdot)) = \mathbb{E}\left\{\int_t^T l(X(s), a(s))ds + g(X(T))\right\}. \qquad (2.85)$$

Suppose that β and γ satisfy the right conditions so that Hypothesis 2.23 holds. This is the case, for instance, in Sect. 3.6, Propositions 3.61 and 3.62. Then, all the assumptions of Theorem 2.24 are satisfied and hence the dynamic programming principle holds.

The Hamilton–Jacobi–Bellman equation associated with problem (2.84) and (2.85) is the following:

$$\begin{cases} v_t + \dfrac{1}{2}\,\mathrm{Tr}\,[QD^2v] + \langle Ax + b(x), Dv\rangle + \inf_{a\in\Lambda}\{\langle a, Dv\rangle + l(x, a)\} = 0, \\ v(T, x) = g(x). \end{cases} \qquad (2.86)$$

This problem falls into the classes studied for instance in [374, 537, 538][11] by the viscosity solution approach, in [29, 89, 90, 105, 189, 306, 307, 432, 434, 438][12] by the mild/strong solutions approach, in [125, 298] by the L^2 approach, and in [284, 436][13] by the BSDE approach. The theory of such HJB equations is described in Chaps. 3–6. The theory of viscosity solutions presented in Chap. 3 applies when $\mathrm{Tr}(Q) < +\infty$ and we refer in particular to Sect. 3.6. Concerning Chap. 4 we refer to Sects. 4.8.3.1 (Examples 4.222 and 4.227), 4.8.3.2, and 4.9.2 for specific examples and to Sect. 4.10.1 for a case where an explicit Feynman–Kac formula for the solution of the HJB equation (2.86) is found in the case of a quadratic Hamiltonian. The setting of Chap. 5 can be applied to the present example, in particular in Sect. 5.6.3 a control problem with a state equation like (2.79) for $\mathcal{O} = \mathbb{R}^N$ is discussed.

---

[11] These papers treat the fully nonlinear case.

[12] Reference [105] deals with non-Lipschitz continuous b, [432, 434, 438] also treat the case of multiplicative noise, and [434] treats a Banach space case.

[13] These papers also treat the case of multiplicative noise and [436] considers it in a Banach space.

2.6.1.3   The Infinite Horizon Case

For the infinite horizon case we again rewrite the state equation (2.79) as (2.84), starting at time 0 at the point $x \in H$. The cost functional

$$I_3(x; a(\cdot)) = \mathbb{E}\left\{ \int_0^{+\infty} e^{-\rho s} \int_{\mathcal{O}} \beta(y^{\alpha(\cdot),0,x}(s, \xi), \alpha(s, \xi))d\xi ds \right\} \qquad (2.87)$$

is then expressed as

$$J_3(x; a(\cdot)) = \mathbb{E}\left\{ \int_0^{+\infty} e^{-\rho s} l(X(s; 0, x, a(\cdot)), a(s))ds \right\}. \qquad (2.88)$$

Hypotheses 2.11 and 2.12 are satisfied as in the finite horizon part. Hypothesis 2.29 holds thanks to Remark 2.30. Hypothesis 2.28 holds if β satisfies proper conditions. This is discussed in Sect. 3.6, Propositions 3.73 and 3.74. The Hamilton–Jacobi–Bellman equation associated with the problem is now

$$\rho v - \frac{1}{2} \text{Tr}\, [QD^2 v] - \langle Ax + b(x), Dv \rangle - \inf_{a \in \Lambda}\{\langle a, Dv \rangle + l(x, a)\} = 0. \quad (2.89)$$

As regards the literature, we refer to [374, 537, 538] for the viscosity solution approach in the fully nonlinear case and with multiplicative noise which is discussed in Chap. 3, to [107, 241, 317, 433][14] for the mild/strong solution approach which is presented in Chap. 4 (see, in particular, Sects. 4.8.3.1, 4.8.3.2 and 4.9.2.2), and to [285] for the BSDE approach in the case of multiplicative noise, which is presented in Chap. 6, see Sect. 6.10. In [301], an ergodic control problem is studied, using the results for the infinite horizon problem.

2.6.2   Stochastic Controlled Heat Equation: Boundary Control

The second example is also concerned with the control of a nonlinear stochastic heat equation in a given space region $\mathcal{O}$ but perhaps in a more realistic case, when the control can be exercised only at the boundary of $\mathcal{O}$ or in a subset of $\mathcal{O}$. We present only the case of the control at the boundary, remarking that the case of the control on a subdomain of $\mathcal{O}$ (that may even reduce to a point) gives rise to very similar mathematical difficulties that are treated, for example, in [277, 435]. We consider two cases that are the most standard and commonly used: the first when the control at the boundary enters through the Dirichlet boundary condition, and the second when one controls the flow, i.e. the Neumann boundary condition.

---

[14]The paper [433] also treats the case of a multiplicative noise.

2.6.2.1 Setting of the Problem: Dirichlet Case

As in the previous example, assume $\mathcal{O}$ to be an open, connected, bounded subset of $\mathbb{R}^N$ with smooth boundary $\partial\mathcal{O}$. We consider the controlled dynamical system driven by the following stochastic PDE on the time interval $[t, T]$, for $0 \le t \le T < +\infty$,

$$\begin{cases} dy(s, \xi) = \big(\Delta_\xi y(s, \xi) + f\left(y(s, \xi)\right)\big) ds + dW_Q(s)(\xi) & \text{in } (t, T] \times \mathcal{O} \\ y(t, \xi) = x(\xi) & \text{on } \mathcal{O} \\ y(s, \xi) = \alpha(s, \xi) & \text{on } (t, T] \times \partial\mathcal{O}, \end{cases}$$
(2.90)

where Δ_ξ, f, W_Q, x, $y(\cdot)$ are as in Eq. (2.79). The difference with respect to Eq. (2.79) is that here the control is no longer in the drift term of the state equation but it influences the system through its values at the boundary (the so-called Dirichlet boundary condition). So here the *control strategy* of the system is the function $\alpha : [t, T] \times \partial\mathcal{O} \times \Omega \to \mathbb{R}$, which may be interpreted as the dynamics of an external source of heat acting at every boundary point of $\mathcal{O}$.

Following the notation of Sect. 2.6.1.1, we denote the unique solution (whenever it exists, see the next subsection for a more precise setting) of (2.90) by $y^{\alpha(\cdot),t,x}(\cdot)$ to underline the dependence of the state y, on the control $\alpha(\cdot)$ and on the initial data t, x.

Similarly to the distributed control case, a reasonable objective of the controller is that of minimizing a functional

$$I(t, x; \alpha(\cdot)) = \mathbb{E}\left\{ \int_t^T \left(\int_{\mathcal{O}} \beta_1(y^{\alpha(\cdot),t,x}(s, \xi))d\xi + \int_{\partial\mathcal{O}} \beta_2(\alpha(s, \xi))d\xi \right) ds + \int_{\mathcal{O}} \gamma(y^{\alpha(\cdot),t,x}(T, \xi))d\xi \right\}$$
(2.91)

where $\beta_1, \beta_2, \gamma \in C(\mathbb{R})$ are given functions depending on the objective of the controller. Observe that the difference with respect to the cost functional I_2 in (2.81) is that here we take the integral on $\partial\mathcal{O}$ when the control $\alpha(\cdot)$ is involved. The goal of the controller here would be to minimize the functional I above, over all control strategies $\alpha(\cdot)$ which are progressively measurable with respect to the augmented filtration generated by W_Q, and such that the above integrals make sense. The constraints can be the same as in the distributed control case in Sect. 2.6.1.1.

2.6.2.2 Setting of the Problem: Neumann Case

In this case the boundary condition in (2.90) is replaced by

$$\frac{\partial y(s, \xi)}{\partial n} = \alpha(s, \xi) \text{ on } (t, T] \times \partial\mathcal{O},$$
(2.92)

where n is the outward unit normal vector to $\partial\mathcal{O}$. This means that one controls the heat flow across the boundary. The goal again is to minimize a cost functional of type (2.91) over all admissible controls $\alpha(\cdot)$.

2.6.2.3 The Infinite-Dimensional Setting and the HJB Equation

To rewrite the state equation (2.90) (with either Dirichlet or Neumann boundary condition) in an infinite-dimensional setting we take $H = L^2(\mathcal{O})$ and Λ to be a closed subset of $L^2(\partial\mathcal{O})$ depending on the control constraints.

We first consider the Dirichlet case. Let A be the operator defined in (2.82) and $b :$ $H \to H$ be the Nemytskii operator defined in (2.83). Let D be the Dirichlet operator defined in (C.2). We define, as before, for $s \in [t, T]$, $X(s) := y(s, \cdot) \in L^2(\mathcal{O})$, and $a(s) := \alpha(s, \cdot) \in \Lambda$. We assume in addition that Λ is bounded in $L^2(\partial\mathcal{O})$. Then, as explained in Appendix C, Sect. 1.2 (Notation C.15), the state equation (2.90) can be formally rewritten as

$$\begin{cases} dX(s) = (AX(s) + b(X(s)) - ADa(s))\,ds + dW_Q(s), & t < s \le T \\ X(t) = x, & x \in H. \end{cases} \tag{2.93}$$

Now, thanks to (C.2), if we write the term $-AD$ as $(-A)^\beta B$ for $\beta \in (3/4, 1)$, where $B := (-A)^{1-\beta}D$, the operator B is bounded from $L^2(\partial\mathcal{O})$ to H. Thus, passing to the integral form (see again Notation C.15), we can write (2.93) as follows

$$X(s) = e^{(s-t)A}x + \int_t^s e^{(s-r)A}b(X(r))dr + \int_t^s (-A)^\beta e^{(s-r)A}Ba(r)dr$$
$$+ \int_t^s e^{(s-r)A}dW_Q(s). \tag{2.94}$$

The Neumann boundary control case is handled similarly. Here we take $\Lambda = L^2(\partial\mathcal{O})$ and $\mathcal{U}_t^\nu = M_\nu^2(t, T; L^2(\partial\mathcal{O}))$. Let A be the Laplace operator with Neumann boundary condition (see e.g. [548], Sect. 5.2, p. 180):

$$\begin{cases} D(A) = \{x \in H^2(\mathcal{O}) : \frac{\partial x}{\partial n} = 0 \text{ on } \partial\mathcal{O}\}, \\ Ax = \Delta x, \text{ for } x \in D(A). \end{cases} \tag{2.95}$$

It generates an analytic semigroup of compact operators $e^{tA}, t \ge 0$, in H. We consider, for fixed $\lambda > 0$, the Neumann operator N_λ defined in (C.7). Similarly to the Dirichlet boundary control case, as explained in Appendix C, Sect. 1.3 (Notation C.18), the state equation can be formally expressed as an evolution equation as follows:

$$\begin{cases} dX(s) = \Big(AX(s) + b(X(s)) + (\lambda I - A)N_\lambda a(s)\Big)ds + dW_Q(s), \quad t < s \le T \\[2mm] X(t) = x, \quad x \in H. \end{cases}$$

(2.96)

Now, thanks to (C.10), if we write the term $(\lambda I - A)N_\lambda$ as $(\lambda I - A)^\beta B_\lambda$, for $\beta \in (1/4, 1/2)$ and $B_\lambda := (\lambda I - A)^{1-\beta}N_\lambda$, the operator B_λ is bounded from $L^2(\partial\mathcal{O})$ to H. Then, passing to the integral form (see again Notation C.18), we can rewrite (2.96) as

$$X(s) = e^{(s-t)A}x + \int_t^s e^{(s-r)A}b(X(r))dr$$
$$+ \int_t^s (\lambda I - A)^\beta e^{(s-r)A}B_\lambda a(r)dr + \int_t^s e^{(s-r)A}dW_Q(s). \quad (2.97)$$

If f (and thus b) is Lipschitz[15] and (1.65) holds (if $\mathrm{Tr}(Q) = +\infty$) both integral equations (2.94) and (2.97) have unique mild solutions (see Theorem 1.141 and Proposition 1.147) with continuous trajectories, which we denote by $X(\cdot, ; t, x, a(\cdot))$ (or simply $X(\cdot)$ if its meaning is clear).

Thus it follows from the discussion in Remark 2.17 that Hypotheses 2.11 and 2.12, and condition $(A4)$ in the Neumann case, needed for the dynamic programming principle, hold for both problems.

We now define $l_1\colon H \to \mathbb{R}$ by

$$l_1(x) = \int_\mathcal{O} \beta_1(x(\xi))d\xi,$$

$l_2\colon \Lambda \to \mathbb{R}$ by

$$l_2(a) = \int_{\partial\mathcal{O}} \beta_2(a(\xi))d\xi,$$

and $g\colon H \to \mathbb{R}$ by

$$g(x) = \int_\mathcal{O} \gamma(x(\xi))d\xi.$$

The functional I in (2.91) can be rewritten in the Hilbert space setting as

$$J(t, x; a(\cdot)) = \mathbb{E}\left\{\int_t^T [l_1(X(s)) + l_2(a(s))]\, ds + g(X(T))\right\}. \quad (2.98)$$

Thus, if β_1, β_2, γ satisfy proper continuity and growth conditions that guarantee Hypothesis 2.23, then the hypotheses of Theorem 2.24 are satisfied, and thus the dynamic programming principle stated there holds.

---

[15]More general assumptions on f could be used, as in [105, 107] in the distributed control case, for example.

The associated HJB equation in both cases can be written as

$$
\begin{cases}
v_t + \dfrac{1}{2}\, \mathrm{Tr}\,[QD^2v] + \langle Ax + b(x), Dv\rangle \\
\qquad + \inf_{a\in\Lambda}\left\{\langle(\lambda I - A)^\beta B_\lambda a, Dv\rangle + l_2(a)\right\} + l_1(x) = 0, \qquad (2.99)\\[2mm]
v(T, x) = g(x),
\end{cases}
$$

where in the Dirichlet case we take $\lambda = 0$ and $\beta \in (3/4, 1)$, while in the Neumann case we take $\lambda > 0$ and $\beta \in (1/4, 1/2)$.

Observe that the term $\langle(\lambda I - A)^\beta B_\lambda a, Dv\rangle$ caused by the presence of the boundary control term in the state equation does not make sense in general. However, if this term is interpreted as $\langle B_\lambda a, (\lambda I - A)^\beta Dv\rangle$ then (writing $l(x, a) = l_1(x) + l_2(a)$) the Hamiltonian F defined by

$$
F(x, p) = \inf_{a\in\Lambda}\left\{\langle B_\lambda a, (\lambda I - A)^\beta p\rangle + l(x, a)\right\} \qquad (2.100)
$$

is well defined on $H \times D\left((\lambda I - A)^\beta\right)$. Such unboundedness of F is difficult to treat and typically requires better regularity properties of the solution, e.g. that the $Dv(t, x)$ belongs to the narrower space $D\left((\lambda I - A)^\beta\right)$. Since $D\left((\lambda I - A)^\beta\right)$ is larger in the Neumann case, the regularity needed for the value function to solve the HJB equation in the Neumann case is weaker than in the Dirichlet case. Thus the Neumann case can be studied under weaker assumptions and/or with better results. In the framework of viscosity solutions, the unboundedness of F may require additional conditions on test functions. Overall this problem is much more difficult to study than that of the previous section.

The theory of viscosity solutions has been developed for such equations in [318].[16] It is presented in Sect. 3.12, where existence and uniqueness of viscosity solutions for stationary HJB equations is proved in great generality, covering Dirichlet boundary conditions and very general drift and diffusion coefficients allowing for possibly fully nonlinear Hamiltonians. However no results about feedback controls exist with this approach. A Cauchy problem was also studied in [560] using the techniques of [318]. A related finite horizon problem has been studied partly with a viscosity solution approach in [577] in a case with boundary control and boundary noise: uniqueness of solutions is not proved.

Regarding the mild/strong solution approach, only the Neumann case has been investigated in [189, 241, 310] when the term b is zero or regular (see also Chap. 4 and, in particular, Example 4.222). The existence and uniqueness of a regular solution of the HJB equation, and the existence of feedback controls were obtained there.[17] The L^2 approach presented in Chap. 5 and the BSDE approach of Chap. 6 have not

---

[16]See also, for the deterministic case, the papers [93, 96, 97, 221, 222].

[17]See also, for the deterministic case, [229, 230, 234].

yet been applied to such equations, however there are results when the boundary
control comes together with the boundary noise, see Sect. 2.6.3.

Remark 2.46 A problem related to the one presented in this subsection is where the
state equation is the same as (2.79) (possibly substituting the Dirichlet boundary
condition by the Neumann boundary condition) but where the boundary condition
depends on the state. For example, we can have[18]

$$y(s, \xi) = b_1(y(s, \cdot))(\xi) \quad \text{on } (t, T] \times \partial\mathcal{O}, \tag{2.101}$$

where $b_1 : L^2(\mathcal{O}) \to L^2(\partial\mathcal{O})$ is a given Lipschitz continuous function taking account
of the influence of the internal state on its boundary values, a term which may arise
in many applications (see e.g. [13, 353] in population dynamics problems). With
this modification, the state equation (2.79) can still be solved using Theorem 1.152.
Indeed, its infinite-dimensional rewriting is exactly as in (4.339) with the additional
term $-ADb_1(X(s))$ (or $(\lambda I - A)Nb_1(X(s))$) in the case of the Neumann boundary
condition) in the drift. This term satisfies the requirements of Hypothesis 4.149-(iii).

Similar considerations apply if in the problems presented in this subsection
((2.90) or (2.92)) one substitutes the boundary condition there with the following
one:

$$y(s, \xi) \quad or \quad \frac{\partial y(s, \xi)}{\partial n} = b_1(y(s, \cdot))(\xi) + \alpha(s, \xi) \quad \text{on } (t, T] \times \partial\mathcal{O},$$

where b_1 is as above. See also Remark 4.226.   ∎

2.6.3 Stochastic Controlled Heat Equation: Boundary Control and Boundary Noise

The third example still concerns the control of the stochastic heat equation in a given
space region $\mathcal{O}$. In this case we assume that both the noise and the control act only at
the boundary of that region. The problem is very hard since the presence of the noise
at the boundary introduces a strong unboundedness in the model. For this reason,
up to now, only one-dimensional cases have been studied, and so we present here
a one-dimensional example with a Neumann boundary condition taken from [181].
We will also mention what happens in a more difficult Dirichlet boundary condition
case (see [225, 437]).

---

[18]In the Neumann boundary condition case we simply replace the left-hand side of (2.101) by
$\frac{\partial y(s, \xi)}{\partial n}$.

2.6.3.1 Setting of the Problem

We consider an optimal control problem for a state equation of parabolic type on a bounded interval, which, for convenience, we take to be $[0, \pi]$. We consider a Neumann boundary condition in which the derivative of the unknown function is equal to the sum of the control and of a white noise in time, namely:

$$\begin{cases} \dfrac{\partial y}{\partial s}(s, \xi) = \Delta_\xi y(s, \xi) + f\,(y(s, \xi)) & \text{in } (t, T] \times (0, \pi), \\[2ex] y(t, \xi) = x(\xi) & \text{on } (0, \pi), \\[2ex] \dfrac{\partial y(s, 0)}{\partial n} = a_1(s) + \dot{W}_1(s), \quad \dfrac{\partial y(s, \pi)}{\partial n} = a_2(s) + \dot{W}_2(s) & \text{on } (t, T]. \end{cases}$$

$$(2.102)$$

In the above equation, $\{W_i(t)\}_{t \geq 0}$, $i = 1, 2$, are independent standard real Wiener processes; the unknown $(s, \xi, \omega) \to y(s, \xi, \omega)$, representing the state of the system, is a real-valued process; the control is modeled by the real-valued processes $(s, \xi) \to a_i(s, \omega)$, $i = 1, 2$ acting, respectively, at $\xi = 0$ and $\xi = \pi$ which are progressively measurable with respect to the augmented filtration generated by $W = (W_1, W_2)$; and x is in $L^2(0, \pi)$. The function f belongs to $C_b(\mathbb{R})$ and is globally Lipschitz continuous. As in the previous subsection, the solution of (2.102) is denoted by $y^{a(\cdot), t, x}(\cdot)$ to underline the dependence of the state on the control $a(\cdot) = (a_1(\cdot), a_2(\cdot))$ and on the initial data t, x.

The functional to minimize is

$$I(t, x; a_1(\cdot), a_2(\cdot)) = \mathbb{E}\left[\int_t^T \left(\int_0^\pi \beta_1(\xi, y^{a(\cdot), t, x}(s, \xi)) d\xi \right. \right.$$
$$\left. \left. + \beta_2(a_1(s), a_2(s)) \right) ds + \int_0^\pi \gamma(\xi, y^{a(\cdot), t, x}(T, \xi)) d\xi \right].$$

$$(2.103)$$

2.6.3.2 The Infinite-Dimensional Setting

To rewrite the problem in an infinite-dimensional setting we take $H = L^2(0, \pi)$, $\Lambda = \Xi = \mathbb{R}^2$, Q is the identity operator on Ξ, $W_Q = W$, and $a(\cdot) = (a_1(\cdot), a_2(\cdot))$. As in the previous example, using the results of Sect. 1.4 and Appendix C, Sect. 1.3, we get, formally, the following infinite-dimensional state equation for the variable $X(s) = y(s, \cdot)$, $s \in [t, T]$:

$$\begin{cases} dX(s) = (AX(s) + b(X(s)) + (\lambda I - A)N_\lambda a(s)) \, ds + (\lambda I - A)N_\lambda dW_Q(s), & s \in (t, T] \\[2ex] X(t) = x, \quad x \in H, \end{cases}$$

$$(2.104)$$

where b and N_λ are as in Sect. 2.6.2.3. This equation is interpreted in the mild form as

$$X(s) = e^{(s-t)A}x + \int_t^s e^{(s-r)A}[b(X(s)) + (\lambda I - A)N_\lambda a(s)]dr$$
$$+ \int_t^s (\lambda I - A)^\beta e^{(s-r)A} B_\lambda dW_Q(r), \qquad (2.105)$$

where $\beta \in (1/4, 1/2)$ and $B_\lambda := (\lambda I - A)^{1-\beta}N_\lambda$ is a bounded operator. We take $\mathcal{U}_t$ to be the set of processes $a(\cdot)$ belonging to $M_\nu^2(t, T; \mathbb{R}^2)$ for a given reference probability space.

Theorem 1.141 guarantees the existence and uniqueness of a mild solution $X(\cdot) := X(\cdot; t, x, a(\cdot))$ of (2.105) with continuous trajectories. The validity of Hypotheses 2.11, 2.12 and (A4) needed for the dynamic programming principle is discussed in Remark 2.17.

We now define $l: H \to \mathbb{R}$ by

$$l_1(x) = \int_0^\pi \beta_1(\xi, x(\xi))d\xi,$$

$l_2: \Lambda \to \mathbb{R}$ by

$$l_2(a) = \beta_2(a_1, a_2),$$

and $g: H \to \mathbb{R}$ by

$$g(x) = \int_0^\pi \gamma(\xi, x(\xi))d\xi.$$

The functional I in (2.91) can thus be rewritten as

$$J(t, x; a(\cdot)) = \mathbb{E}\left\{\int_t^T [l_1(X(s)) + l_2(a(s))]\,ds + g(X(T))\right\}. \qquad (2.106)$$

Again, β_1, β_2, γ must satisfy the right continuity and growth assumptions to guarantee Hypothesis 2.23, so that we can claim that the dynamic programming principle is satisfied.

2.6.3.3 The HJB Equation

The HJB equation associated with problem (2.105) and (2.106) is

$$\begin{cases} v_t + \dfrac{1}{2} \operatorname{Tr}\left[(\lambda I - A)N_\lambda \left[(\lambda I - A)N_\lambda\right]^* D^2 v\right] + \langle Ax, Dv \rangle + F(Dv) + l_2(x) = 0, \\ v(T, x) = g(x), \ x \in H, \end{cases}$$

where the Hamiltonian F is given by

$$F(p) = \inf_{a \in \mathbb{R}^2} \{ \langle (\lambda I - A) N_\lambda a, p \rangle + l_2(a) \}.$$

Similarly to the boundary control case, here the Hamiltonian makes sense when one rewrites the term $\langle (\lambda I - A) N_\lambda a, p \rangle$ as $\langle B_\lambda a, (\lambda I - A)^\beta p \rangle$, and then F is unbounded with respect to the variable p as it is only defined if $p \in D((\lambda I - A)^\beta)$. However, an extra difficulty arises due to the second-order term $\mathrm{Tr}[(\lambda I - A)N_\lambda [(\lambda I - A)N_\lambda]^* D^2 v]$, which is written here in a formal way and needs to be given special interpretation. We notice that the same "operator" $(\lambda I - A)N_\lambda$ acts on the control and on the Wiener process in (2.104), and thus the control acts on the solution in the same way as the noise. This allows us to use the BSDE approach to mild solutions (see [181] and, later, [574, 591, 592]). The L^2 approach is in principle applicable to this problem but, up to now, it has not been developed. Concerning the viscosity solution approach, we mention the paper [577], where the authors show that the value function is a viscosity solution of the HJB equation but without proving uniqueness. At the present stage, the mild/strong solution approach presented in Chap. 4 does not seem to be applicable here.

Remark 2.47 A one-dimensional control problem in the half-line $[0, +\infty)$ with boundary control and noise in the Dirichlet case (i.e. with the boundary condition of the type $y(s, 0) = a(s) + \dot{W}(s)$ for $s \in (t, T]$) has been studied in [225, 437].

However, the choice of the infinite-dimensional setting in this case presents a problem, since, choosing as the state space $H = L^2(0, +\infty)$, the continuity of the trajectories in $L^2(0, +\infty)$ is not ensured (see for example [175] Proposition 3.1, p. 176). We can have the continuity only in some spaces of distributions extending $L^2(0, +\infty)$, or in L^2 with a suitable weight (see [225] Proposition 2.2, Lemma 2.2 and Theorem 2.7).

In [225] the linear quadratic case is studied while in [437] a more general case is studied by the BSDE approach. The problem has not yet been studied using other methods.

We remark that, similarly to the case of boundary control, the HJB equation for the Dirichlet boundary noise case is more difficult than that for the Neumann boundary noise (as the unbounded operators arising in the first and second-order terms contain "higher powers of A"). ∎

2.6.4 Optimal Control of the Stochastic Burgers Equation

Our fourth example concerns optimal control of the stochastic Burgers equation. The deterministic Burgers equation was introduced by J.M. Burgers (see e.g. [87, 88]) as a model in fluid mechanics and has subsequently been used in various areas of applied mathematics such as acoustics, dispersive water waves, gas dynamics, traffic flow, heat conduction, etc. As explained in [176] (p. 255) the deterministic Burgers equation is not a good model for turbulence since it does not display any chaotic phenomena; even when a force is added to the right hand side, all solutions converge to a unique stationary solution as time goes to infinity. The situation is different when the force is random. Indeed, several authors have suggested using the stochastic Burgers equation as a simple model for turbulence, as [111, 117, 368]. In [373] it is used to model the growth of a one-dimensional interface. Among other papers on the subject we mention [162, 163, 570].

Here we present a simple optimal control problem for the one-dimensional stochastic Burgers equation motivated by a model of the control of turbulence formulated in [117] and studied in [155, 156].

2.6.4.1 Setting of the Problem

The state equation is the following stochastic controlled viscous Burgers equation

$$
\begin{cases}
dy(s, \xi) = \left[\dfrac{\partial^2 y(s, \xi)}{\partial \xi^2} + \dfrac{1}{2} \dfrac{\partial}{\partial \xi} y^2(s, \xi) + \sqrt{Q} \alpha(s, \cdot)(\xi) \right] ds + dW_Q(s)(\xi), \\
\qquad\qquad\qquad\qquad\qquad\qquad\qquad\qquad\qquad s \in (t, T], \xi \in (0, 1), \\[2mm]
y(t, \xi) = x(\xi), \quad \xi \in [0, 1], \\[2mm]
y(s, 0) = y(s, 1) = 0, \quad s \in [t, T].
\end{cases}
\tag{2.107}
$$

Here:

- the function $y : [t, T] \times [0, 1] \times \Omega \to \mathbb{R}$, $(s, \xi, \omega) \to y(s, \xi, \omega)$ describes the evolution of the velocity field of the fluid;
- the control $\alpha : [t, T] \times (0, 1) \times \Omega \to \mathbb{R}$, $(s, \xi) \to \alpha(s, \xi, \omega)$ gives the dynamics of the external force acting at every point of $(0, 1)$;
- W_Q is a Q-Wiener process with $Q \in \mathcal{L}_1^+(L^2(0, 1))$ and $(\mathscr{F}_s^t)_{s \in [t, T]}$ is the augmented filtration generated by W_Q;
- $x \in L^2(0, 1)$ gives the distribution of the initial velocity field.

As before, the solution of (2.107) is denoted by $y^{\alpha(\cdot), t, x}(\cdot)$. A possible objective of the controller (used in [117, 155, 156]) is to minimize a functional

$$
I(t, x; \alpha(\cdot)) = \mathbb{E}\left\{ \int_t^T \int_0^1 \left[\left| \frac{\partial y^{\alpha(\cdot), t, x}(s, \xi)}{\partial \xi} \right|^2 + \frac{1}{2} |\alpha(s, \xi)|^2 \right] d\xi ds \right.
$$
$$
\left. + \int_0^1 \frac{1}{2} |y^{\alpha(\cdot), t, x}(T, \xi) - \bar{y}(\xi)|^2 d\xi \right\}, \tag{2.108}
$$

where $\bar{y}$ is a given "desired" velocity profile. The main idea behind this form of the cost functional is that we are trying to get the final velocity field to be close to $\bar{y}$ while minimizing the "vorticity" of the flow (measured here by the integral of the space derivative) and the energy spent controlling the system.

We then minimize the functional I over all control strategies $\alpha(\cdot)$ which are progressively measurable with respect to $\mathscr{F}_s^t$, and such that $\mathbb{E} \int_t^T \int_0^1 |\alpha(s, \xi)|^2 d\xi ds < +\infty$. Sometimes we may require some additional bounds on the control strategies (e.g. $\alpha(s, \xi) \in [m, M]$ for any $s \in [t, T]$ and $\xi \in (0, 1)$ for some $m < M$).

Note that the operator $\sqrt{Q}$ acting on the control is the square root of the covariance operator of the Wiener process. This can be interpreted as "the noise acting on the control".

2.6.4.2 The Infinite-Dimensional Setting and the HJB Equation

We take $H = \Lambda = L^2(0, 1)$. The state equation (2.107) and the functional (2.108) can be rewritten as an abstract evolution equation in H using the operator

$$\begin{cases} D(A) = H^2(0, 1) \cap H_0^1(0, 1), \\ Ax = \frac{\partial^2}{\partial \xi^2} x, \text{ for } x \in D(A), \end{cases} \tag{2.109}$$

and the nonlinear operator

$$\begin{cases} D(B) = H^1(0, 1) \\ B(x)(\xi) = x(\xi) \frac{\partial}{\partial \xi} x(\xi), \text{ for } x \in D(B). \end{cases} \tag{2.110}$$

Indeed, once we set, for any $s \in [t, T]$, $X(s) = y(s, \cdot) \in L^2(0, 1)$, $a(s) = \alpha(s, \cdot) \in L^2(0, 1)$, the state equation (2.107) becomes

$$\begin{cases} dX(s) = \big(AX(s) + B(X(s)) + \sqrt{Q}a(s)\big) ds + dW_Q(s) \\ X(t) = x, \end{cases} \tag{2.111}$$

and (2.108) is equivalent to

$$J(t, x; a(\cdot)) = \mathbb{E} \left\{ \int_t^T \left[|(-A)^{1/2} X(s)|_H^2 + \frac{1}{2} |a(s)|_H^2 \right] ds + \frac{1}{2} |X(T) - \bar{y}|_H^2 \right\}. \tag{2.112}$$

In contrast to the previous examples, the standard mild solution approach does not work for Eq. (2.111). Therefore, the existence and uniqueness results require a different framework, see [163, 177] Chap. 14 (the result is also stated in Sect. 4.9). An unpleasant consequence of this is the fact that it is not obvious that Hypotheses 2.11 and 2.12 needed for the dynamic programming principle are satisfied. We will not deal explicitly with this problem in this book, but we will see in Chap. 3 how to show the DPP for a much more difficult problem, namely the optimal control of the 2-D stochastic Navier–Stokes equations (see the next section).

The HJB equation related to our control problem is (see [156] Eq. (2.4))

$$
\begin{cases}
v_t(t, x) + \dfrac{1}{2}\mathrm{Tr}\left[QD^2v(t, x)\right] + \langle Dv(t, x),\, Ax + B(x)\rangle \\
\qquad\qquad - \dfrac{1}{2}\left|\sqrt{Q}Dv(t, x)\right|^2 + \left|(-A)^{1/2}x\right|^2 = 0, \qquad (2.113) \\
v(T, x) = \dfrac{1}{2}|x - \bar{y}|^2.
\end{cases}
$$

This equation is difficult to investigate due to the presence of the nonlinear unbounded term $\langle Dv(t, x), B(x)\rangle$, coming from the state equation, and the term $\left|(-A)^{-1/2}x\right|^2$, coming from the objective functional. It was studied in [156] by a Hopf-type change of variable and in [155, 157] using a variant of the mild/strong solution approach. In these papers the authors use finite-dimensional approximations of the state equation and are able to obtain existence and uniqueness of regular solutions to the HJB equation and to find optimal control in feedback form (see Sect. 4.9.1.1).

It is interesting to note that this technique has been extended to the case of control of stochastic Navier–Stokes equations in dimensions 2 and 3 (see next section). Equation (2.113) can also be investigated using the viscosity solution approach even though there are no explicit results. However, we refer the readers to Chap. 3 and [322].

2.6.5   Optimal Control of the Stochastic Navier–Stokes Equations

The stochastic Navier–Stokes equations are used to model turbulent flows. We refer the reader to the books [390, 567], the survey article [254], Chap. 15 of [176], and the paper [52] for more on this.

The optimal control of the Navier–Stokes equations, in the deterministic and stochastic cases, is a very challenging problem, both from the theoretical and applied points of view. For a survey on this subject we refer to the book [288] for the deterministic case and the paper [534] for the stochastic case. The dynamic programming approach to the optimal control of stochastic Navier–Stokes equations has been investigated in the papers [158, 322, 424].

We consider a model problem for a control of turbulent flow governed by the stochastic two-dimensional Navier–Stokes equations for incompressible fluids. We mainly follow the paper [158] with some changes borrowed, for example, from [534] and [322]. It has been partly generalized to the three-dimensional case in [424].

2.6.5.1   Setting of the Problem

We are given an open domain $\mathcal{O} \subset \mathbb{R}^2$ with locally Lipschitz boundary: it includes, for example, the case where $\mathcal{O}$ is a rectangle (which is quite common in the literature, see e.g. [160, 322, 555]). Given $\xi = (\xi_1, \ldots, \xi_n), \eta = (\eta_1, \ldots, \eta_n) \in \mathbb{R}^n$ we use the notation $\xi \cdot \eta := \sum_{i=1}^{n} \xi_i \eta_i$.

We take any reference probability space $(\Omega, \mathscr{F}, \mathscr{F}_s^t, \mathbb{P}, W_Q)$ satisfying the usual conditions, where W_Q is an $L^2(\mathcal{O}; \mathbb{R}^2)$)-valued Q-Wiener process with $Q \in \mathcal{L}_1^+$ $(L^2(\mathcal{O}; \mathbb{R}^2))$.

The control variable is an external force $\alpha(s, \xi)$ acting at every point ξ of $\mathcal{O}$ and at every time $s \in [t, T]$; for models with control on subdomains see e.g. [347] and [534] (p. 3). The controls are stochastic processes progressively measurable with respect to the filtration $\mathscr{F}_s^t$ for a given reference probability space, and such that $|\alpha(s, \cdot)|_{L^2(\mathcal{O}; \mathbb{R}^2))} \le R$ for some fixed $R > 0$, for all $(s, \omega) \in [t, T] \times \Omega$. The unknowns are the velocity vector field $(s, \xi) \to y(s, \xi) = (y_1(s, \xi), y_2(s, \xi))$ and the pressure $(s, \xi) \to p(s, \xi)$ (we omit the dependence on $\omega \in \Omega$ in the notation). They satisfy the system

$$
\begin{cases}
dy(s) + [(y(s, \xi) \cdot \nabla)y(s, \xi) + \nabla p(s, \xi)]\, ds \\
\qquad\qquad = [\nu \Delta y(s, \xi) + \alpha(s, \xi)]\, ds + dW_Q(s)(\xi) \ \text{ in } (t, T] \times \mathcal{O} \\[4pt]
\operatorname{div}(y(s, \xi)) = 0 \ \text{ in } [t, T] \times \mathcal{O} \\[4pt]
y(s, \xi) = 0 \ \text{ on } [t, T] \times \partial\mathcal{O} \\[4pt]
y(0, \xi) = x(\xi) \ \text{ on } \mathcal{O}.
\end{cases}
$$

$$(2.114)$$

Here ∇ denotes $(\partial_{\xi_1}, \partial_{\xi_2})$ and $y \cdot \nabla$ denotes $y_1 \partial_{\xi_1} + y_2 \partial_{\xi_2}$. The positive constant ν represents the kinematic viscosity. We remark that distributed control can be approximately realized for electrically conducting fluids (like salt water, liquid metals, etc.) by a suitable Lorentz force distribution. The boundary control, which is not present in this example, is typically implemented by blowing and suction at the boundary.

Suppose, as in the previous section, that we want to achieve the desired profile $\bar{y}$ of the flow while minimizing the turbulence of the flow and the amount of energy used to control it. This is common in engineering applications. We recall that we can measure how turbulent a flow is by evaluating the time averaged enstrophy, which is defined by

$$
\int_{\mathcal{O}} |\operatorname{curl} y(s, \xi)|^2 \, d\xi, \qquad s \in [t, T],
$$

where the rotational operator curl in dimension 2 is defined as

$$
\operatorname{curl}(y_1, y_2) = \frac{\partial y_1}{\partial \xi_2} - \frac{\partial y_2}{\partial \xi_1}.
$$

$$(2.115)$$

As in previous subsections, the solution of (2.114) is denoted by $y^{\alpha(\cdot),t,x}(\cdot)$ to underline the dependence of the state on the control $\alpha(\cdot)$ and on the initial data t, x.

Thus we consider the problem of minimizing the following functional over all control strategies $\alpha(\cdot)$:

$$I(t, x; \alpha(\cdot)) = \mathbb{E}\left[\int_t^T \int_{\mathcal{O}} \left[|\text{curl } y^{\alpha(\cdot),t,x}(s, \xi)|^2 + \frac{1}{2}|\alpha(s, \xi)|^2 \right] d\xi ds \right.$$

$$\left. + \int_{\mathcal{O}} |y^{\alpha(\cdot),t,x}(T, \xi) - \bar{y}(\xi)|^2 d\xi \right]. \quad (2.116)$$

In areas like combustion the goal may be to maximize mixing (and hence turbulence) of the flow. As remarked in [534] (p. 3) in some flow control problems and in data assimilation problems in meteorology one may also minimize the functional

$$I_1(t, x; \alpha(\cdot)) = \mathbb{E}\left[\int_t^T \int_{\mathcal{O}} \left[|\text{curl } (y^{\alpha(\cdot),t,x}(s, \xi) - \bar{y}_d(s, \xi))|^2 \right. \right.$$

$$\left. \left. + \frac{1}{2}|\alpha(s, \xi)|^2 \right] d\xi ds \right]$$

$$(2.117)$$

for a given velocity field $\bar{y}_d(s, \xi)$. (See also [288], p. 167, formula (1.15), for a similar type of functional, in the deterministic case.)

2.6.5.2 The Infinite-Dimensional Setting and the HJB Equation

Define

$$\mathcal{V} := \left\{ f \in C_0^\infty(\mathcal{O}, \mathbb{R}^2) : \text{div}(f) = 0 \right\}, \quad (2.118)$$

$$H := \text{the closure of } \mathcal{V} \text{ in } L^2(\mathcal{O}; \mathbb{R}^2), \quad (2.119)$$

and

$$V := \text{the closure of } \mathcal{V} \text{ in } H^1(\mathcal{O}; \mathbb{R}^2). \quad (2.120)$$

Recall that we have an orthogonal decomposition

$$L^2(\mathcal{O}; \mathbb{R}^2) = H \times H^\perp,$$

where $H^\perp = \{f = \nabla p : \text{for some } p \in H^1(\mathcal{O})\}$.

We define the unbounded operator in H

$$\begin{cases} D(A) := \left(H^2(\mathcal{O})\right)^2 \cap V \subset H \\ A := P\Delta, \end{cases}$$

where P is the orthogonal projection in $L^2(\mathcal{O}; \mathbb{R}^2)$ onto H. The operator A is self-adjoint and strictly negative (see [556]), generates a C_0-semigroup on H, and moreover, $V = D((-A)^{1/2})$. We also define the bilinear operator

$$\begin{cases} B: V \times V \to H \\ B(x, y) = P(x \cdot \nabla)y \end{cases}$$

and set $B(x) := B(x, x)$.

Applying the projection P to Eq. (2.114) and setting, for $s \in [t, T]$, $X(s) = y(s, \cdot) \in H$, $a(s) = P\alpha(s, \cdot) \in H$, we obtain

$$\begin{cases} dX(s) = (\nu AX(s) - B(X(s)) + a(s))\, ds + P\, dW_Q(s) \\ X(t) = x. \end{cases} \tag{2.121}$$

Since, for $s \in [t, T]$, $|a(s)|_H = |a(s)|_{L^2(\mathcal{O};\mathbb{R}^2)} \le |\alpha(s, \cdot)|_{L^2(\mathcal{O};\mathbb{R}^2)}$, we can obviously restrict the set of controls to those with values in H. Moreover, for $x \in V$, $|\operatorname{curl} x|_{L^2(\mathcal{O})} = |\nabla x|_{L^2(\mathcal{O};\mathbb{R}^4)} = |(-A)^{1/2}x|_H$. Thus the minimization of the functional I in (2.116) is equivalent to the minimization of

$$J(t, x; a(\cdot)) = \mathbb{E}\left[\int_t^T \left[|(-A)^{1/2}X(s)|_H^2 + \frac{1}{2}|a(s)|_H^2\right] ds + |X(T) - \bar{y}|_H^2\right] \tag{2.122}$$

over all $a(\cdot) \in \mathcal{U}_t$, which is defined as in Sect. 2.2.1 with $\Lambda := B_H(0, R)$.

There are various ways to define solutions of (2.121) and for existence and uniqueness results we refer, for instance, to [124, 177, 444, 567] and to Chap. 3 and Sect. 4.9 here. Unfortunately, we cannot apply the definition of a mild solution, and thus showing that Hypotheses 2.11 and 2.12 are satisfied requires a different argument. We explain it in Chap. 3.

The Hamiltonian for the control problem is

$$F(p) := \begin{cases} -\frac{1}{2}|p|^2 & \text{if } |p| \le R \\ -|p|R + \frac{1}{2}R^2 & \text{if } |p| > R, \end{cases}$$

and the Hamilton–Jacobi–Bellman equation for the system becomes

$$
\begin{cases}
v_t + \dfrac{1}{2}\mathrm{Tr}\left[PQP^*D^2v\right] + \langle Dv, \nu Ax - B(x)\rangle \\
\qquad\qquad + F(Dv) + \left|(-A)^{1/2}x\right|^2 = 0 \\
v(T, x) = |x - \bar{y}|^2.
\end{cases}
\tag{2.123}
$$

We note that one can also associate a different control problem with (2.123) by considering PW_Q to be a $\tilde{Q}$-Wiener process in H with $\tilde{Q} = PQP^*$ and taking controls to be the progressively measurable processes with respect to the augmented filtration generated by PW_Q.

Similarly to the case of the control of the stochastic Burgers equation discussed in the previous section, the difficulty of the HJB equation (2.123) comes from the presence of the unbounded terms $\langle Dv, B(x)\rangle$ and $\left|(-A)^{1/2}x\right|^2$. The unboundedness caused by the operator B is much worse now. However, the approach used for the Burgers case by Da Prato and Debussche still allows us to obtain satisfactory results on existence and uniqueness of regular (mild) solutions [158] under some conditions on $\tilde{Q}$. These results are described in Sect. 4.9.1.2. Such results have been partly extended in [424] to the case of three-dimensional stochastic Navier–Stokes equations.

The viscosity solution approach can be applied to the two-dimensional case with more general cost functionals and noise and it yields existence and uniqueness of viscosity solutions (see [322] and Chap. 3). The viscosity solution approach for the three-dimensional case is still open. Some results in the deterministic case are in [526].

2.6.6 Optimal Control of the Duncan–Mortensen–Zakai Equation

This example concerns a class of finite-dimensional stochastic optimal control problems with partial observation and correlated noises. We present the problem and we briefly show its connection with the so-called "separated" problem (see e.g. [46, 213, 214, 261, 473]) which is a fully observable infinite-dimensional stochastic optimal control problem. The setting of the partially observed control system we describe here is the same as in [323] and is borrowed from [473, 597, 598] (see also [342–344, 411]). The Duncan–Mortensen–Zakai (DMZ) equation, separated problem and optimal control of the DMZ equation are also discussed in detail in [467]. The presentation in [467] relies on [519] which also discusses filtering problems.

2.6.6.1 An Optimal Control Problem with Partial Observation

Consider, in the interval $[t, T]$, a random state process $y(\cdot)$ in $\mathbb{R}^d$ and a random observation process $y_1(\cdot)$ in $\mathbb{R}^m$. The state-observation equation is

$$\begin{cases} dy(s) = b^1(y(s), a(s))ds + \sigma^1(y(s), a(s))dW^1(s) + \sigma^2(y(s), a(s))dW^2(s), \\ y(t) = \eta, \\ dy_1(s) = h(y(s))ds + dW^2(s), \\ y_1(t) = 0, \end{cases}$$

(2.124)

where W^1 and W^2 are two independent Brownian motions in $\mathbb{R}^d$ and $\mathbb{R}^m$ respectively on some stochastic basis $(\Omega, \mathscr{F}, \{\mathscr{F}_s\}_{s\in[t,T]}, \mathbb{P})$ which is a complete probability space with the filtration satisfying the usual conditions. The initial condition η is assumed to be $\mathscr{F}_t$-measurable and square-integrable. The control set $\Lambda \subset \mathbb{R}^n$, and admissible controls are the processes $a(\cdot) : [t, T] \times \Omega \to \Lambda$ that are progressively measurable with respect to the filtration $\{\mathscr{F}_s^{y_1}\}_{s\in[t,T]}$, which is the augmented filtration of the filtration $\{\mathscr{F}_s^{y_1,0}\}_{s\in[t,T]}$ generated by the observation process $y_1(\cdot)$. We assume the following.

Hypothesis 2.48 The set Λ is a closed subset of $\mathbb{R}^n$. The functions

$$b^1 : \mathbb{R}^d \times \Lambda \to \mathbb{R}^d, \qquad h : \mathbb{R}^d \to \mathbb{R}^m$$

are uniformly continuous and the $C^2(\mathbb{R}^d)$ norms of $b^1(\cdot, a)$ and h are bounded, uniformly for $a \in \Lambda$. Moreover, the functions

$$\sigma^1 : \mathbb{R}^d \times \Lambda \to \mathcal{L}(\mathbb{R}^d, \mathbb{R}^d), \qquad \sigma^2 : \mathbb{R}^d \times \Lambda \to \mathcal{L}(\mathbb{R}^m, \mathbb{R}^d)$$

are uniformly continuous and the $C^3(\mathbb{R}^d)$ norms of $\sigma^1(\cdot, a)$ and $\sigma^2(\cdot, a)$ are bounded, uniformly for $a \in \Lambda$, and

$$\sigma^1(\xi, a) [\sigma^1(\xi, a)]^T \geq \lambda I$$

for some $\lambda > 0$ and all $\xi \in \mathbb{R}^d, a \in \Lambda$.

This assumption in particular guarantees the existence of a unique strong solution of the state equation (2.124), see e.g. Theorem 1.127. We denote its solution at time s by $(y(s; t, \eta, a(\cdot)), y_1(s; t, a(\cdot)))$ or simply by $(y(s), y_1(s))$.

We now consider the problem of minimizing the cost functional

$$I(t, \eta; a(\cdot)) = \mathbb{E}\left\{\int_t^T l_1(y(s), a(s))ds + g_1(y(T))\right\} \qquad (2.125)$$

over all admissible controls, where the cost functions

$$l_1 : \mathbb{R}^d \times \Lambda \to \mathbb{R}; \qquad g_1 : \mathbb{R}^d \to \mathbb{R}$$

are suitable continuous functions, say with at most polynomial growth at infinity in the variable y, uniformly with respect to the variable a. A control strategy $a^*(\cdot)$ minimizing the cost I in (2.125) is called an *optimal control in the strict sense*.

2.6.6.2   The Separated Problem

The optimal control of partially observed diffusions is a very difficult problem with many open questions (e.g. the existence of optimal controls in the strict sense, see e.g. [261], p. 261). One way of dealing with it is through the so-called "separated" problem where one looks at the associated problem of controlling the unnormalized conditional probability density $Y(\cdot) : [t, T] \to L^1\left(\mathbb{R}^d\right)$ of the state process $y(\cdot)$ given the observation $y_1(\cdot)$. This idea, that arises from well known results in nonlinear filtering (see e.g. [200, 453, 585]), was first introduced in [261] to prove existence of optimal controls in a suitable weak sense. Here, following mainly [473], we briefly and informally explain the separated problem and how it arises.

To introduce the new state Y and to compute the equation for it we consider for each $s \in [t, T]$ the conditional law Π_s of the random variable $y(s)$ given the path of y_1 up to time s, i.e., in our setting, given the σ-field $\mathscr{F}_s^{y_1}$, and look at its density with respect to the Lebesgue measure in $\mathbb{R}^d$. This density, up to a normalizing factor, will be the new state $Y(s)$ at time s. The conditional law Π_s is a measure-valued process such that for every $f \in C_b(\mathbb{R}^d)$, the conditional expectation

$$\mathbb{E}\left[f(y(s))|\mathscr{F}_s^{y_1}\right] = \int_{\mathbb{R}^d} f(y)d\Pi_s(y) \quad \mathbb{P}\text{-a.s.}$$

Using the notation of [473], the above expression will be denoted by $\Pi_s(f)$. The process Π_s exists if there exists a regular conditional probability given $\mathscr{F}_s^{y_1,0}$. To compute it, it is more convenient (as explained, for example, in [473] at the end of Sect. 1.5) to change the probability measure. We define the new probability measure $\overline{\mathbb{P}}$ by

$$d\overline{\mathbb{P}} = \kappa^{-1}(T)d\mathbb{P},$$

where

$$\kappa(s) = \exp\left[\int_t^s \langle h(y(r)), dy_1(r)\rangle_{\mathbb{R}^m} - \frac{1}{2}\int_t^s |h(y(r))|^2_{\mathbb{R}^m} dr\right].$$

Since $\kappa(s)$ is a martingale we have

$$d\overline{\mathbb{P}} = \kappa^{-1}(s)d\mathbb{P}, \quad \text{on } \mathscr{F}_s. \tag{2.126}$$

It follows from the Girsanov Theorem (see e.g. [372], Sect. 3.5) that the processes W^1 and y_1 become two independent Brownian motions, respectively in $\mathbb{R}^d$ and $\mathbb{R}^m$, in the new probability space $\left(\Omega, \mathscr{F}, (\mathscr{F}_s)_{s\in[t,T]}, \overline{\mathbb{P}}\right)$. In this space the equation for the process $y(\cdot)$ becomes

$$\begin{cases} dy(s) = \left[b^1(y(s), a(s)) - \sigma^2(y(s), a(s))h(y(s))\right]ds \\ \qquad\quad + \sigma^1(y(s), a(s))dW^1(s) + \sigma^2(y(s), a(s))dy_1(s), \tag{2.127} \\ y(t) = \eta. \end{cases}$$

It follows from Bayes' formula (see [519], p. 225 or [467], Proposition 1.3, p. 18), see Lemma 3.1 in [473], that for $s \in [t, T]$ and $f \in C_b(\mathbb{R}^d)$,

$$\Pi_s(f) = \mathbb{E}\left[f(y(s))|\mathscr{F}_s^{y_1}\right] = \frac{\overline{\mathbb{E}}\left[f(y(s))\kappa(s)|\mathscr{F}_s^{y_1}\right]}{\overline{\mathbb{E}}\left[\kappa(s)|\mathscr{F}_s^{y_1}\right]}.$$

So, if we are able to compute, for every $s \in [t, T]$ and for every $f \in C_b(\mathbb{R}^d)$, the quantity $\overline{\mathbb{E}}\left[f(y(s))\kappa(s)|\mathscr{F}_s^{y_1}\right]$, then we can also find $\Pi_s(f)$ for every such f.

By Itô's formula we have, for $f \in C_b^2(\mathbb{R}^d)$, $s \in [t, T]$,

$$f(y(s))\kappa(s) = f(\eta) + \int_t^s \kappa(r)L_{a(r)}f(y(r))dr$$

$$+ \int_t^s \kappa(r)\langle\nabla f(y(r)), \sigma^1(y(r), a(r))dW^1(r)\rangle_{\mathbb{R}^d}$$

$$+ \int_t^s \kappa(r)\langle\nabla f(y(r)), \sigma^2(y(r), a(r))dy_1(r)\rangle_{\mathbb{R}^d} + \int_t^s \kappa(r)f(y(r))\langle h(y(r)), dy_1(r)\rangle_{\mathbb{R}^m},$$

where for every $a \in \Lambda$, $L_a : C_b^2(\mathbb{R}^d) \to C_b(\mathbb{R}^d)$ is given by

$$(L_a f)(\xi) = \langle b^1(\xi, a), \nabla f(\xi)\rangle_{\mathbb{R}^d}$$

$$+ \frac{1}{2}\text{Tr}\left[\left(\sigma^1(\xi, a)[\sigma^1(\xi, a)]^T + \sigma^2(\xi, a)[\sigma^2(\xi, a)]^T\right)D^2 f(\xi)\right].$$

Now, computing the conditional expectation (see [473], Sect. 1.4) we have, for $s \in [t, T]$,

$$\overline{\mathbb{E}}\left[f(y(s))\kappa(s)|\mathscr{F}_s^{y_1}\right] = \overline{\mathbb{E}}[f(\eta)] + \int_t^s \overline{\mathbb{E}}\left[\kappa(r)L_{a(r)}f(y(r))|\mathscr{F}_r^{y_1}\right]dr$$

$$+\int_t^s \langle \overline{\mathbb{E}}\left[\kappa(r)[\sigma^2(y(r),a(r))]^T\nabla f(y(r)) + \kappa(r)f(y(r))h(y(r))|\mathscr{F}_r^{y_1}\right], dy_1(r)\rangle_{\mathbb{R}^m}.$$

If $\overline{\Pi}_s$ is a measure-valued process such that, for every $f \in C_b(\mathbb{R}^d)$, $\overline{\Pi}_s(f) = \overline{\mathbb{E}}\left[\kappa(s)f(y(s))|\mathscr{F}_s^{y_1}\right]$ (which exists if Π_s exists), then the equation above implies that $\overline{\Pi}_s$ must satisfy the equation

$$\overline{\Pi}_s(f) = \overline{\Pi}_t(f) + \int_t^s \overline{\Pi}_r(L_{a(r)}f)dr + \int_t^s \langle\overline{\Pi}_r(B_{a(r)}f), dy_1(r)\rangle_{\mathbb{R}^m}, \qquad s \in [t,T],$$

where for every $a \in \Lambda$, $B_a : C_b^1(\mathbb{R}^d) \to C_b(\mathbb{R}^d, \mathbb{R}^m)$ is given by

$$(B_a f)(\xi) = [\sigma^2(\xi,a)]^T\nabla f(\xi) + f(\xi)h(\xi).$$

(To justify that $\overline{\mathbb{E}}\left[\kappa(r)F(y(r),a(r))|\mathscr{F}_r^{y_1}\right] = \overline{\Pi}_t(F(\cdot,a(r)), \overline{\mathbb{P}}$-a.s. for a bounded and continuous function $F(\xi,a)$, one can use approximation by step functions, Proposition 1.41-(vii) and the Lebesgue dominated convergence theorem, since the equality is true for functions $F(\xi,a) = \mathbf{1}_A(a)\mathbf{1}_B(\xi)$, where A, B are Borel subsets of Λ, $\mathbb{R}^d$ respectively, as $a(r)$ is $\mathscr{F}_r^{y_1}$-measurable.)

If $\overline{\Pi}_s$ has a density $Y(s) \in L^1(\mathbb{R}^d)$ with respect to the Lebesgue measure, then the process $Y(\cdot)$ should satisfy, at least in a weak sense, the so-called Duncan–Mortensen–Zakai (DMZ) equation (introduced in [200, 453, 585])

$$dY(s) = L_{a(s)}^*Y(s)ds + \langle B_{a(s)}^*Y(s), dy_1(s)\rangle, \qquad Y(t) = x, \qquad (2.128)$$

where x is the density of the law of the initial datum η of Eq. (2.124). The process $Y(\cdot)$ is called the unnormalized conditional density of the state with respect of the observation process. If one can prove that Eq. (2.128) has a solution, it is the density with respect to the Lebesgue measure of $\overline{\Pi}_s$, see [473], Sect. 1.4 for more on this.

Now it is possible to rewrite the functional I in (2.125) in terms of the new probability space and infinite-dimensional state $Y(\cdot)$. Indeed, assuming that the process $Y(\cdot)$ takes values in $L^2(\mathbb{R}^d)$, using (2.126) we have

$$I(t,\eta;a(\cdot)) = \int_t^T \overline{\mathbb{E}}\left[\kappa(s)l_1(y(s),a(s))\right]ds + \overline{\mathbb{E}}[\kappa((T)g_1(y(T))]$$

$$= \overline{\mathbb{E}}\left[\int_t^T \overline{\mathbb{E}}\left[\kappa(s)l_1(y(s),a(s))|\mathscr{F}_s^{y_1}\right]ds\right] + \overline{\mathbb{E}}\left[\overline{\mathbb{E}}\left[\kappa(T)g_1(y(T))|\mathscr{F}_s^{y_1}\right]\right]$$

$$= \overline{\mathbb{E}}\left\{\int_t^T \langle l_1(\cdot,a(s)), Y(s)\rangle_{L^2}\, ds + \langle g_1(\cdot), Y(T)\rangle_{L^2}\right\} =: J(t,x;a(\cdot)).$$

$$(2.129)$$

Computing the adjoint operators, L_a^*, B_a^*, we can rewrite (2.128) in an explicit and more familiar form

$$dY(s) = A_{a(s)}Y(s)ds + \sum_{k=1}^{m} S_{a(s)}^k Y(s)dy_{1,k}(s), \qquad Y(t) = x, \qquad (2.130)$$

where for every $a \in \Lambda$, A_a and S_a^k $(k = 1, \ldots, m)$ are the following differential operators

$$(A_a x)(\xi) = \sum_{i,j=1}^{d} \partial_i [a_{i,j}(\xi, a)\partial_j x(\xi)] + \sum_{i=1}^{d} \partial_i [b_i(\xi, a)x(\xi)], \qquad (2.131)$$

and

$$\left(S_a^k x\right)(\xi) = \sum_{i=1}^{d} d_{ik}(\xi, a)\partial_i x(\xi) + e_k(\xi, a)x(\xi); \qquad k = 1, \ldots, m, \qquad (2.132)$$

where

$$a(\xi, a) = \sigma^1(\xi, a)\left[\sigma^1(\xi, a)\right]^T + \sigma^2(\xi, a)\left[\sigma^2(\xi, a)\right]^T,$$

$$b_i(\xi, a) = -b_i^1(\xi, a) + \partial_j a_{i,j}(\xi, a); \qquad i = 1, \ldots, d,$$

$$d(\xi, a) = -\sigma^2(\xi, a),$$

$$e_k(\xi, a) = h_k(\xi) - \partial_i \sigma_{ik}^2(\xi, a); \qquad k = 1, \ldots, m.$$

The *separated problem* is thus the problem of minimizing the functional J over all admissible controls $a(\cdot)$, with state equation (2.130). It is an infinite-dimensional optimal control problem which can be studied within the framework of this book in the state space $H = L^2(\mathbb{R}^d)$ or other spaces. In Sect. 3.11 we will investigate it in suitable weighted spaces. It is worth noting that even though the original control problem was nonlinear, the DMZ equation (2.130) is linear and the cost functional J is also linear in the state variable x.

The mild solution approach cannot be applied to (2.130). Existence and uniqueness of solutions in a variational sense was proved in [387] (see also [472]). We discuss variational solutions in Sect. 3.11 where we show that under suitable assumptions Eq. (2.130) is well posed in $L^2(\mathbb{R}^d)$ and in weighted versions of it. We also explain how to show that Hypotheses 2.11 and 2.12 hold.

The HJB equation for the infinite-dimensional problem has the form

$$\begin{cases} v_t + \inf_{a \in \Lambda} \left\{ \frac{1}{2} \sum_{k=1}^{m} \left\langle D^2 v S_a^k x, S_a^k x \right\rangle + \langle A_a x, Dv \rangle + f(x, a) \right\} = 0, \\ v(T, x) = g(x). \end{cases} \qquad (2.133)$$

This is a fully nonlinear equation with unbounded first- and second-order terms, and up to now only a viscosity solution approach has given some results on existence and uniqueness of solutions [22, 323, 344, 411]. In [411] the equation was studied in a standard L^2 space when the operators S_a^k were bounded multiplication operators. In [344] it was shown that the value function is a viscosity solution in a very weak sense when the HJB equation was considered in the space of measures (see also [258, 346]). The results of [323] are presented in Sect. 3.11, where (2.133) will be studied in a weighted L^2 space. The optimal control problem for the DMZ equation and the HJB equation (2.133) are also discussed in [467].

2.6.7 Super-Hedging of Forward Rates

We now present a stochastic optimal control problem arising in finance, in pricing derivatives. When the market is incomplete there is no unique way to price a derivative product. It is then useful, in some cases, to find the range of all possible prices, i.e. the maximum and the minimum of possible prices, called the super-hedging and the sub-hedging price. Such prices are defined as value functions of suitable optimal control problems and the finite-dimensional theory of this problem has been widely studied: see e.g. [21, 208] for the one-dimensional case, [326, 327, 418, 514], and [115, 529–532] in the multidimensional case; see also [564] for a first idea of the method in the infinite-dimensional case.

When the underlying asset is a forward rate the natural model for it is the so-called Musiela model introduced in [455] that describes the dynamics of forward rates in terms of the evolution of an infinite-dimensional diffusion process. Consequently, the super-hedging problem in such case is naturally formulated as a stochastic optimal control problem in infinite dimension. We present now such a problem, taken from the paper [375].

The Musiela model of interest rates [455] is a reparametrization of the Heath–Jarrow–Morton (HJM) model. In this model the forward rate process $\{r(t, \sigma)\}_{\sigma, t \geq 0}$ evolves according to a stochastic differential equation

$$dr(t, \sigma) = \left(\frac{\partial}{\partial \sigma} r(t, \sigma) + \sum_{i=1}^{d} \tau_i(t, \sigma) \int_0^{\sigma} \tau_i(t, \mu) d\mu \right) dt + \sum_{i=1}^{d} \tau_i(t, \sigma) dw(t)^i,$$

where $W = (w^1, ..., w^d)$ is a standard d-dimensional Brownian motion, and τ_i are certain functions. Using the notation $A = \frac{d}{d\sigma}$ and, for $t, \sigma \geq 0$, $\tau(t)(\sigma) = (\tau_1(t, \sigma), ..., \tau_d(t, \sigma))$,

$$b(\tau(t))(\sigma) = \sum_{i=1}^{d} \tau_i(t, \sigma) \int_0^{\sigma} \tau_i(t, \mu) d\mu,$$

the above equation can be written as an abstract infinite-dimensional stochastic differential equation

$$dr(t) = (Ar(t) + b(\tau(t)))dt + \tau(t) \cdot dW(t), \quad r(0) \in H, \tag{2.134}$$

where H is some separable Hilbert space of functions on $\mathbb{R}^+$ (for instance $H = L^2(\mathbb{R}^+)$, $H = H^1(\mathbb{R}^+)$ or their weighted versions), and $\cdot$ is the inner product in $\mathbb{R}^d$ (see [303, 455, 564, 565]). We call Eq. (2.134) the Heath–Jarrow–Morton–Musiela (HJMM) equation. Given the right choice of the space H and proper assumptions on τ, the equation has a unique mild solution, see Sect. 3.10. Using the process r, the price at time t of a zero-coupon bond (see [456]) with maturity T is

$$B_T(t) = e^{-\int_0^{T-t} r(t,\sigma)d\sigma}, \quad 0 \le t \le T.$$

This model can be used to price swaptions, caps, and other interest rates and currency derivatives.

Consider first a case of European options. Given a contingent claim with the payoff function $g : H \to \mathbb{R}$ and an initial curve at time t, $x(\sigma)$, $\sigma > 0$, the rational price of the option maturing at time T is

$$V(t, x) = \mathbb{E}\left(e^{-\int_t^T r(s,0)ds} g(r(T)) : r(t)(\sigma) = x(\sigma)\right). \tag{2.135}$$

For instance for a European swaption on a swap with cash-flows C_i, $i = 1, ..., n$, at times $T < T_1 < ... < T_n$,

$$g(z) = \left(K - \sum_{i=1}^n C_i e^{\int_0^{T_i - T} z(\sigma)d\sigma}\right)^+ \tag{2.136}$$

for some $K > 0$. The function V given by the (Feynman–Kac) formula (2.135) should satisfy the partial differential equation

$$\begin{cases} \dfrac{\partial u}{\partial t} + \dfrac{1}{2} \sum_{i=1}^d \langle D^2 u\, \tau_i(t), \tau_i(t)\rangle + \langle b(\tau(t)) + Ax, Du\rangle - x(0)u = 0 \\[2mm] u(T, x) = g(x), \end{cases} \tag{2.137}$$

where $\langle \cdot, \cdot \rangle$ is the inner product in H. The above equation is called an infinite-dimensional Black–Scholes equation. It was analyzed in [302] in the space $H = L^2((0, +\infty))$ where the existence of smooth solutions was proved for smooth g and τ independent of time but possibly depending on the state variable. The existence of solutions was also shown for some non-smooth g when τ was a constant by an argument that allowed a parallel between (2.137) and a finite-dimensional Black–Scholes equation (see also [295, 564]).

The problem of pricing of American options in the framework of the Musiela model can be rephrased as an optimal stopping problem for the above infinite-dimensional diffusion process and is connected to an obstacle problem

$$
\max \left\{ \frac{\partial u}{\partial t} + \frac{1}{2} \sum_{i=1}^{d} \langle D^2 u\, \tau_i(t), \tau_i(t) \rangle + \langle b(\tau(t)) + Ax, Du \rangle - x(0)u, u - \varphi \right\} = 0
$$

$$(2.138)$$

for some function φ. This equation was studied in [583] from the point of view of Bellman's inclusions. Similar obstacle problems in infinite dimension were also investigated in [38, 116, 293].

One of the drawbacks of the Musiela model is that it does not guarantee the positivity of rates and in some cases it is almost certain that they are not positive (see [564]). To avoid such possibilities the term $x(0)$ was replaced by $x^+(0)$ in [583]. We do the same here and throughout the section we always take the positive part of the rates.

Let us explain now the super-hedging problem. Suppose that the dynamics of the forward rates are given by Eq. (2.134), however we are not able to determine precisely the process $\tau(\cdot)$ that describes the volatility of the market. We only know that it takes values in some set $\Lambda \subset H^d$. We consider an agent who wants to price and hedge a European contingent claim with payoff $g(r(T))$ that depends on the value of the forward rate curve at the maturity time T (note that in cases of interest in finance the payoff function g is not even C^1).

To find the super-hedging price, given an initial condition $r(t) \in H$, we try to maximize the payoff

$$
\mathbb{E}\left(e^{- \int_t^T r^+(s,0)ds} g(r(T)) \right)
$$

$$(2.139)$$

with respect to all progressively measurable stochastic processes $\tau(\cdot)$ taking values in Λ. The processes $\tau(\cdot)$ become controls and the maximization of (2.139) gives the value function V which should provide the super-hedging price at time t as $C(t) = V(t, r(t))$. Following the standard finite-dimensional theory for such problems (see e.g. [208, 418, 514]), a super-hedging strategy in such context (i.e. an investment strategy that replicates the super-hedging price) "should" then be given by the process $\pi(\cdot) := DV(\cdot, r(\cdot))$, so in terms of the space-like derivative (i.e. with respect to r) of the value function V. In the infinite-dimensional case similar results have not been proved yet as the method of proof requires strong regularity properties of the value function V which are not known. However, the problem provides a strong motivation for studying the following optimal control problem: maximize (2.139) over all processes $\tau(\cdot) \in \mathcal{U}_t$, where the state equation is given by (2.134). For this optimal control problem Hypotheses 2.11 and 2.12 are satisfied thanks to Proposition 2.16. Moreover, if g is locally uniformly continuous and has at most polynomial growth, it can be proved that Hypothesis 2.23 also holds and so the dynamic programming principle holds. This is explained in Sect. 3.10. The associated HJB equation is

$$\begin{cases} \dfrac{\partial u}{\partial t} + \langle Ax, Du \rangle + F(x, u, Du, D^2 u) = 0 & \text{in } (0, T) \times H \\[3mm] u(T, x) = g(x) & \text{in } H, \end{cases} \qquad (2.140)$$

where for $x \in H, s \in \mathbb{R}, p \in H$ and $X \in S(H)$,

$$F(x, s, p, X) = \sup_{\tau \in \Lambda} \left\{ \frac{1}{2} \sum_{i=1}^{d} \langle X \tau_i, \tau_i \rangle + \langle b(\tau), p \rangle - x^+(0)s \right\}.$$

Equation (2.140) is called an infinite-dimensional Black–Scholes–Barenblatt (BSB) equation associated to the contingent claim g. In cases of interest in finance the payoff function g is not even C^1 and a notion of a generalized solution is needed. It was studied in [375] using viscosity solutions. In this context (2.140) has a unique viscosity solution that coincides with the value function provided by the maximization of (2.139). This is discussed in Sect. 3.10. The results are shown in the space $H = H^1(\mathbb{R}^+)$ which makes the term $x^+(0)$ continuous. One can also investigate the problem in weighted versions of $H^1(\mathbb{R}^+)$.

The problem of pricing derivatives in the HJMM model when the Gaussian noise is replace by a Lévy noise, and the analysis of the associated non-local BSB equation is studied in [545]. Also a Kolmogorov equation related to the problem of hedging of a derivative of a risky asset is investigated [517].

2.6.8 Optimal Control of Stochastic Delay Equations

In this last example we consider finite-dimensional stochastic controlled systems with delay in the state and/or in the control variables. Such control systems arise in many applications (for example in optimal advertising theory, see [313, 314, 428], optimal portfolio management of pension funds, see e.g. [235]) and can be rephrased, using a well known procedure (see e.g. [46] for the deterministic case and [118, 313] for the stochastic case), as infinite-dimensional controlled systems without delay. We present two cases: the first is a system with pointwise delay only in the state variable (taken from [235, 298], see also a special case in [313, 314]), while the second one displays delays (pointwise or distributed) both in the state and in the control variable (taken from [313, 314]). We separate the two cases, since they give rise to different settings with different mathematical difficulties.

2.6.8.1 Delay in the State Variable Only

Let us consider a simple controlled one-dimensional linear stochastic differential equation with a delay $r > 0$ in the state variable:

$$\begin{cases} dy(s) = (\beta_0 y(s) + \beta_1 y(s-r) + \alpha(s))\,ds + \sigma\,dW_0(s), \\ y(t) = x_0, \\ y(t+\theta) = x_1(\theta), \ \theta \in [-r,0), \end{cases} \quad (2.141)$$

where $\sigma > 0$, $\beta_0, \beta_1 \in \mathbb{R}$ are given constants, W_0 is a one-dimensional standard Brownian motion defined on a complete probability space $(\Omega, \mathscr{F}, \mathbb{P})$, and $\mathscr{F}_s^t$ is the augmented filtration generated by W_0. The control $\alpha(\cdot)$ is an $\mathscr{F}_s^t$-progressively measurable process with values in an interval $[0, R]$ for some $R > 0$. We assume that $x_1 \in L^2(-r, 0)$. This type of equation is used, for example, in optimal portfolio management of pension funds (see e.g. [235], where the state variable is the wealth of the fund and the control variable is the investment strategy), and in optimal advertising (see e.g. [313, 314, 428], where the state variable is the "goodwill" of a given product and the control is the investment in advertising). Such equations also seem to be relevant for some models arising in studying economic growth in a stochastic environment (see [24, 25, 69, 226] in the deterministic case).

Given three real functions $\varphi_0, h_0, g_0 : \mathbb{R} \to \mathbb{R}$, we consider the problem of minimizing a functional

$$I(t, y_0, y_1; \alpha(\cdot)) = \mathbb{E}\left\{ \int_t^T [\varphi_0(y(s)) + h_0(\alpha(s))]\,ds + g_0(y(T)) \right\} \quad (2.142)$$

over all control strategies $\alpha(\cdot) \in \mathcal{U}_t$. For example, in the optimal advertising problem the function h_0 represents the cost of advertising while $-\varphi_0$ and $-g_0$ represent the profit coming from the so-called "goodwill" associated to a given product. In such an applied problem it is reasonable to assume the positivity of the state variable (the "goodwill" $y(\cdot) \geq 0$) and of the control variable (the investment $\alpha(\cdot) \geq 0$), and a constraint on the control space (for example, $\sup_{s \in [t,T]} \alpha(s) \leq R$ for some $R > 0$ as we have done for other examples).

Existence, uniqueness and properties of solutions of delay equations like (2.141) can be studied either directly (see e.g. [363] Sect. 5, [524], or the survey [364]) or by introducing an equivalent infinite-dimensional formulation. If we follow the former direction, the dynamic programming approach can be used only for special problems where the HJB equation reduces to a finite-dimensional differential equation (see [398], one can find similar ideas in [215, 245]), while rephrasing the state equation and therefore the whole optimization problem in infinite dimension allows us to study a larger class of problems. There are different ways to rewrite stochastic delay differential equations in the form (2.141) as evolution equations in Hilbert or Banach spaces. Here we present the approach of [118] which allows us to rewrite equation (2.141) in the Hilbert space $\mathbb{R} \times L^2(-r, 0)$. Regarding other choices of state spaces we refer, for example, to [450, 451], where the state space is $C([-r, 0])$, or to the recent paper [257] (see, in particular, Theorem 2.2), where more general spaces are used.

The setting of [118] is the following. Denote by H the space $\mathbb{R} \times L^2(-r, 0)$ and consider the linear operator A_1 on H defined by:

$$
\begin{cases}
D(A_1) = \left\{ \begin{pmatrix} x_0 \\ x_1(\cdot) \end{pmatrix} \in \mathbb{R} \times W^{1,2}(-r, 0; \mathbb{R}), \; x_0 = x_1(0) \right\} \\[2ex]
A_1 \begin{pmatrix} x_0 \\ x_1(\cdot) \end{pmatrix} = \begin{pmatrix} \beta_0 x_0 + \beta_1 x_1(-r) \\ x_1'(\cdot) \end{pmatrix}.
\end{cases}
$$

The operator A_1 generates a strongly continuous semigroup $S_1(t)$ on H and, for $z = (z_0, z_1(\cdot)) \in H$, $S_1(t)z$ can be written in terms of the solution of the linear deterministic delay equation

$$
\begin{cases}
\dot{y}(t) = \beta_0 y(t) + \beta_1 y(t - r), \\
y(0) = z_0, \; y(\theta) = z_1(\theta), \; \theta \in [-r, 0),
\end{cases}
\tag{2.143}
$$

as follows:

$$
S_1(t)z = \begin{pmatrix} y(t) \\ y(t + \cdot) \end{pmatrix} \in H, \quad t \geq 0
$$

(see [118, 177] and also [42]). Set now $\Xi = \mathbb{R}$, $\Lambda = [0, R]$ for a suitable $R > 0$, and define $Q : \Xi \to \Xi$ and $B : \Lambda \to H, G : \Xi \to H$ by

$$
Q w_0 = w_0, \qquad B_1 w_0 = \begin{pmatrix} w_0 \\ 0 \end{pmatrix}, \qquad G w_0 = \begin{pmatrix} \sigma w_0 \\ 0 \end{pmatrix}.
$$

Then, setting, for $s \in [t, T]$, $X(s) = (y(s), y(s + \cdot))$, $a(s) = \alpha(s)$, and $W_Q = W_0$, the controlled stochastic delay Eq. (2.141) can be rewritten (see again [118, 177]) as the following linear evolution equation in H:

$$
\begin{cases}
dX(s) = [A_1 X(s) + B_1 a(s)] \, dt + G \, dW_Q(s), \\
X(t) = \begin{pmatrix} x_0 \\ x_1 \end{pmatrix} := \begin{pmatrix} y_0 \\ y_1 \end{pmatrix} \in H.
\end{cases}
\tag{2.144}
$$

Thanks to Theorem 1.127 the state equation (2.144) admits a unique mild solution (denoted by $X(\cdot; t, x, a(\cdot))$ or simply by $X(\cdot)$), and thus, thanks to Proposition 2.16, Hypotheses 2.11 and 2.12 are satisfied.

Moreover, the functional (2.142) can be rewritten as follows. Set

$$
\begin{cases}
\varphi(x_0, x_1) := \varphi_0(x_0) \\
g(x_0, x_1) := g_0(x_0)
\end{cases}
$$

so that, for a given initial datum $x \in H$, the functional I becomes

$$J(t, x; a(\cdot)) := \mathbb{E}\left\{\int_t^T [\varphi(X(s)) + h_0(a(s))]\,ds + g(X(T))\right\}. \quad (2.145)$$

If φ_0, h_0 and g_0 satisfy proper continuity and growth conditions that ensure Hypothesis 2.23, then the dynamic programming principle holds (see Sect. 3.6 on this).

The associated Hamilton–Jacobi–Bellman equation is

$$\begin{cases} v_t + \dfrac{1}{2}\mathrm{Tr}(GG^*D^2v) + \langle Dv, A_1x\rangle + F(Dv) + \varphi(x) = 0, \\ v(T, x) = g(x), \end{cases} \quad (2.146)$$

where

$$F(p) := \inf_{0\le a\le R}\{h_0(a) + \langle p, B_1a\rangle\} = \inf_{0\le a\le R}\{h_0(a) + p_0a\}. \quad (2.147)$$

Since the second component of B_1 is always zero, the Hamiltonian F only depends on the one-dimensional component p_0 (i.e. on the first derivative D_0v of v with respect to the "present" component). Similarly, in the second-order term, the fact that the second component of G is zero implies that this term only depends on the second derivative D_{00}^2v of v with respect to the "present" component. Thus we can just write $\frac{1}{2}\mathrm{Tr}(GG^*D^2v) = \frac{1}{2}\sigma^2 D_{00}^2v$.

This kind of equation was studied in [298] by the L^2 approach, where the existence of weakly differentiable solutions in Sobolev spaces with respect to a suitable invariant measure μ was proved (see Chap. 5 and, in particular, Sect. 5.6). Also the BSDE approach, which produces Gâteaux differentiable solutions, can be applied here, since the so-called *structure condition* $R(B_1) \subset R(G)$ holds. It was developed, also for more general equations, first in [281, 436], then in [595, 596], and finally in [591, 592], including boundary control/noise. For more on this, see Sect. 6.5. Some results for the viscosity solution approach have been obtained in [235, 236, 517], see also the bibliographical notes in Sect. 3.14.

We also mention that, in the deterministic case, using the fact that the Hamiltonian F only depends on the derivative with respect to the "present", in [238] a regularity result was proved for the viscosity solution of a first-order HJB equation of type (2.146) for a case with nonlinear state equation and with state constraints.

2.6.8.2 Delay in the State and Control

We now consider a stochastic optimal control problem whose state equation has delay in both the state and the control. Such equations are used, for example, to model the evolution of the goodwill stock in advertising models (see e.g. [313, 314]). Suppose we have a controlled stochastic delay differential equation

$$
\begin{cases}
dy(s) = \left[\beta_0 y(s) + \int_{-r}^{0} \beta_1(\theta) y(s+\theta)\, d\theta + \gamma_0 \alpha(s) + \int_{-r}^{0} \gamma_1(\theta)\alpha(s+\theta)\, d\theta \right] ds \\[2mm]
\hspace{6cm} + \sigma\, dW_0(s), \quad t \le s \le T, \\[2mm]
y(t) = y_0, \\[2mm]
y(t+\theta) = y_1(\theta), \quad \alpha(t+\theta) = \delta(\theta), \quad \theta \in [-r, 0),
\end{cases}
$$

$$(2.148)$$

where $\sigma > 0$, $\beta_0, \gamma_0 \in \mathbb{R}$ are given real numbers, $\beta_1, \gamma_1 \in L^2(-r, 0)$, and $W_0, \alpha(\cdot)$ are as in the previous subsection. Since β_1, γ_1 are functions, we rule out the case of pointwise delay. In fact, the pointwise delay case can also be studied, however it gives rise to an unbounded control operator B_2 in the state equation (2.149) below. For the moment we do not consider this case. We will say more about it in the comments after the HJB equation.

The initial data (y_0, y_1, δ) are taken in $\mathbb{R} \times L^2(-r, 0) \times L^2(-r, 0)$. We again try to minimize the functional I defined by (2.142), over all controls $\alpha(\cdot) \in \mathcal{U}_t$.

The problem can be rewritten in an infinite-dimensional setting using a technique which is slightly different from that of the previous subsection; the results we use are proved in [313] and they generalize those proved in the deterministic setting in [566].

We take, as before, $H := \mathbb{R} \times L^2(-r, 0)$, $\Xi = R$, $W_Q = W_0$ and $\Lambda = [0, R]$ for a suitable $R > 0$. We define the operator $A_2 : D(A_2) \subset H \to H$ as follows:

$$
\begin{cases}
D(A_2) := \left\{ x \in H \ : \ x_1 \in W^{1,2}(-r, 0), \ x_1(-r) = 0 \right\} \\[2mm]
A_2 : (x_0, x_1) \to \left(\beta_0 x_0 + x_1(0), \ \beta_1 x_0 - \dfrac{dx_1}{d\theta} \right).
\end{cases}
$$

Moreover, we define the bounded linear control operator B_2 by

$$
\begin{cases}
B_2 : \mathbb{R} \to H \\[2mm]
B_2 : a \to a(\gamma_0, \gamma_1),
\end{cases}
$$

and the operator $G : \mathbb{R} \to H$ by $G : w_0 \to (\sigma w_0, 0)$, as in the case of delay only in the state. The control variable will remain the same in the new system, so $a(s) := \alpha(s)$, $s \in [t, T]$. The state variable is called the *structural state* and is defined using the following proposition, proved in [313].

Proposition 2.49 *Let $X(\cdot)$ be the mild solution of the abstract evolution equation*

$$\begin{cases} dX(s) = (AX(s) + B_2 a(s))\, dt + G\, dW_Q(s) \\ X(t) = x \in H, \end{cases} \tag{2.149}$$

with arbitrary initial datum $x \in H$ and control $a(\cdot) \in M^2_\mu(t, T; \mathbb{R})$. Then, for $s \geq t$, one has, $\mathbb{P}$-a.s.,

$$X(s) = M(X_0(s), X_0(s + \cdot), a(s + \cdot)),$$

where

$$\begin{cases} M: H \times L^2(-r, 0) \to H \\ M: (x_0, x_1(\cdot), v(\cdot)) \mapsto (x_0, m(\cdot)), \end{cases}$$

$(X_0(s)$ is the first component of $X(s))$ and

$$m(\theta) := \int_{-r}^{\theta} \beta_1(\zeta) x_1(\zeta - \theta)\, d\zeta + \int_{-r}^{\theta} \gamma_1(\zeta) v(\zeta - \theta)\, d\zeta, \qquad \theta \in [-r, 0).$$

Moreover, let $\{y(s)\}_{s \geq t}$ be a continuous solution of the stochastic delay differential equation (2.148), and $X(\cdot)$ be the mild solution of the abstract evolution equation (2.149) with initial condition

$$x = M(y_0, y_1, \delta(\cdot)).$$

Then, for $s \geq t$, one has, $\mathbb{P}$-a.s.,

$$X(s) = M(y(s), y(s + \cdot), a(s + \cdot)),$$

hence $y(s) = X_0(s)$, $\mathbb{P}$-a.s., for all $s \geq 0$.

Using this equivalence result, we can now give a reformulation of our problem in the Hilbert space H. The state equation is (2.149) with initial condition $x := M(y_0, y_1, \delta(\cdot))$ and we denote its mild solution (which exists and is unique thanks to Theorem 1.127) by $X(s) := X(s; t, x, a(\cdot))$. The objective functional to minimize is the same J given by (2.145), where g and φ have the same meaning. Therefore, in this setup, Hypotheses 2.11 and 2.12 are satisfied thanks to Proposition 2.16. Thus, again, if φ_0, h_0 and g_0 satisfy proper continuity and growth conditions, we can ensure that the dynamic programming principle holds.

The Hamilton–Jacobi–Bellman equation in the infinite-dimensional setting is

$$\begin{cases} v_t + \frac{1}{2}\mathrm{Tr}(GG^* D^2 v) + \langle Dv, A_2 x \rangle + \inf_{0 \leq a \leq R} \{h_0(a) + \langle Dv, B_2 a \rangle\} = 0, \\ v(T, x) = g(x). \end{cases}$$

$$\tag{2.150}$$

This kind of HJB equation is more difficult than (2.146) since the so-called structure condition $(R(B_2) \subset R(G))$ is no longer true, and thus it is impossible to use the BSDE approach of [281, 436], and the approach of strong solutions in Sobolev spaces is used in [298]. However, in a special case with no delay in the state, a clever variant of the mild/strong solution approach can be applied, see [316].

Concerning a viscosity solution approach we are not aware of any results in the stochastic case. For the deterministic case, please see [244] where regularity results for viscosity solutions are also proved (see also [238] for such results).

Finally, we remark that (as can be seen, for example, in [313]) in the case of pointwise delay (i.e. when β_1 is the Dirac delta at $-r$), the operator B_2 above is unbounded. This unboundedness is similar to the one arising in boundary control problems, and up to now HJB equations of this kind have been investigated only in a special case in [316].

2.7 Bibliographical Notes

The stochastic optimal control problem introduced in Sect. 2.1 is an abstract infinite-dimensional version of problems studied in the literature. We refer to [51, 66, 206, 262, 263, 384, 408, 409, 452, 460, 467, 489, 575] for the finite-dimensional theory.

For deterministic optimal control problems and their connection with HJB equations the reader may consult [40, 53, 54, 95, 127, 128, 407, 584] and the books [29, 403] for the infinite-dimensional case. Some aspects of the theory of stochastic optimal control in infinite dimension and second-order HJB equations can be found in the books [179, 467].

We present the optimal control problem in its weak (Sect. 2.1.2) and strong (Sect. 2.1.1) formulations. The two distinct forms had already appeared in the sixties, in the early days of the studies of finite-dimensional stochastic optimal control problems (see e.g. [259, 394]); we follow the terminology of [575]. We recall that for us the "weak" in "the weak formulation" refers only to the fact that the generalized reference probability spaces vary with the controls and not to the concept of solution that in this context is always strong in the probabilistic sense (see Remark 2.4). In Sect. 2.1.2 we also mention the "extended weak" formulation which is only used in Sect. 6.5 and which, in contrast to the weak formulation we use, does not require uniqueness of solutions of the state equation. The weak formulation is also different from that used in [265] where the word "weak" is meant in the sense of the convex duality.

In Sect. 2.2 and, more precisely, in Sect. 2.2.1 we introduce a third formulation that we use to prove the DPP. We can call this third setup the *weak DPP* formulation. In this framework, as in the weak formulation, we allow the probability spaces and Q-Wiener processes W_Q to vary but we only consider the (augmented) filtration generated by the Q-Wiener processes. Thus the difference is that we pass from generalized reference probability spaces (Definition 1.100) to reference probability spaces (Definition 2.7). Other formulations of stochastic optimal control problems have been proposed in the literature with various notions of control processes. Markov (feedback) controls, i.e. controls of the form $a(t, X(t))$, where $X(t)$ is the state of the system at time t, have been considered, for example, in [66, 262, 263, 384,

395]. So-called *natural strategies*, i.e. controls that can be expressed at time t as functions of the state trajectory up to time t, are considered in [384] where it is also shown that, under suitable hypotheses, the value function of the problem for natural strategies equals the value function of the problem in the strong formulation (Theorem 7, p. 132 of [384]). Relaxed controls have been considered (see e.g. [66, 210, 339, 392]) mostly to prove the existence of optimal controls. Our formulations of control problems follow most closely those of [575]. In Theorem 2.22 we show that the weak DPP formulation and the strong formulation, if a reference probability space is used, are equivalent in the sense that the problems in the two forms have the same value function. For similar results in the finite-dimensional case, see [212, 263, 384]. Nisio in [467] proves that the weak formulations using the reference and the generalized reference probability spaces give the same value functions assuming that the control set is a convex subset of $\mathbb{R}^q$.

The DPP proved in Sect. 2.3 (Theorem 2.24) is very abstract and general. We follow to a large extent the strategy from [575]. The proof uses the continuity of the value function in the spatial variable, however this assumption can be relaxed with very little change in the proof (see e.g. [291]). Theorem 3.70, Sect. 3.6 (next chapter) contains a version of the DPP for mild solutions in the formulation with stopping times when the value function is continuous. The DPP is often considered a standard result, however we have included complete proofs since even in finite dimension it is very technical and many of the proofs available in the literature miss a lot of details.

Several other approaches to the proof of the DPP are available in the literature. Krylov [384] uses approximation of controls by step controls. A PDE-based proof is provided in Fleming and Soner [263], where the DPP is first proved for a uniformly parabolic case where the HJB equation has a smooth solution, and then the value function is approximated by smooth value functions solving uniformly parabolic HJB equations. The proof in [66] uses Markov controls. Nisio [460] uses approximations with switching controls at binary times and a reduction to the canonical reference probability space, while the proof in [459] uses so-called non-anticipative controls and approximations by controls with continuous trajectories. A proof based on discrete time dynamic programming principle and approximation by switching controls is presented in [467]. In [467] the DPP is proved for the weak formulation of control problem from Sect. 2.1.2. The proof in [452] is based on a reduction to the canonical reference probability space. In [489] a sketch of the proof is given for a measurable value function, which however omits delicate measurability issues. Soner and Touzi [530] use deep measurable selection theorems to show the DPP for stochastic target problems without continuity assumptions on the value function. We also mention recent papers [212, 601] which prove the DPP under general assumptions, and [126] which proves, in finite dimension and in a canonical sample space setting, a pseudo-Markov property which is a basic tool to prove the DPP. Other proofs (see e.g. [210, 339, 392]) use relaxed controls and compactness of the set of admissible controls. In [322] the authors adapt to the infinite-dimensional case the arguments used in [264], where the DPP was shown for a two player, zero sum stochastic differential game in finite dimension, to prove the DPP for a control problem for stochastic Navier–Stokes equations in the canonical reference probability space.

A different approach to the DPP has been introduced, for the finite-dimensional case, in [72]. In that paper, the authors introduce the notion of *weak dynamic programming*: roughly speaking, instead of proving a result similar to (2.23) they prove the following, weaker, fact: for any pair of continuous test functions ϕ and ψ (satisfying some growth conditions to guarantee integrability) such that $\phi \leq V \leq \psi$,

$$\inf_{a(\cdot) \in \mathcal{U}_t} \mathbb{E}\left[\int_t^\eta e^{-\int_t^s c(X(\tau))d\tau} l(s, X(s), a(s))\, ds + e^{-\int_t^\eta c(X(\tau))d\tau} \phi(\eta, X(\eta)) \right]$$

$$\leq V(t, x)$$

$$\leq \inf_{a(\cdot) \in \mathcal{U}_t} \mathbb{E}\left[\int_t^\eta e^{-\int_t^s c(X(\tau))d\tau} l(s, X(s), a(s))\, ds + e^{-\int_t^\eta c(X(\tau))d\tau} \psi(\eta, X(\eta)) \right],$$

see [72] or [71] for the precise statements. In this way the difficulties due to the possible lack of continuity of the value function V are avoided because the condition deals with test functions that are continuous. This formulation is of course tailored to the study of viscosity solutions of HJB equations which are defined in terms of regular test functions (see Chap. 3). Weak dynamic programming approach introduced in [72] has been generalized to the case of expectation constraints and state constraints in [71], where an abstract dynamic programming result was stated. The weak dynamic programming was also used for a class of finite-dimensional impulsive problems in [70].

The verification theorem and construction of optimal feedback controls for a smooth value function presented in Sect. 2.5 follow similar standard results for the finite-dimensional case, which can be found, for instance, in Chap. 4 of [262], Chap. 3 of [263] or in Chap. 5 of [575]. In infinite-dimensional Hilbert spaces, for the case of a quadratic Hamiltonian, the reader is referred to Chap. 13 of [179]. When the value function does not satisfy the strong regularity conditions of Sect. 2.5 ($C^{1,2}$ regularity and the derivative in the domain of A^*), only a few specific results are available:

- There are no results in infinite dimension for viscosity solutions that are only continuous. A finite-dimensional verification theorem can be found in [324, 325, 575]. In the deterministic case, some verification results for a Hilbert space case can be found in [92, 227, 403].
- For mild solutions in spaces of continuous functions there are several contributions [105, 107, 155, 156, 158, 306, 307, 310, 313, 314, 317, 432, 433], some of which will be discussed in Chap. 4.
- For mild solutions in the space of L^2 functions we refer to [3, 4, 298, 301], see also Chap. 5.
- For optimal synthesis obtained via backward stochastic differential equations the reader is referred to [276, 281, 283–285] and to Chap. 6.

A different approach to optimal control problems that is not developed in this book is the use of a maximum principle; it is closely related to the study of backward stochastic differential equations (BSDEs). A general result for the finite-dimensional case is given in [480], see also [575]. A generalization to problems with noises with jumps is addressed in [550, 551], where the authors first characterize the adjoint process of the second variation as the solution of a BSDE in the Hilbert space of Hilbert–Schmidt operators.

In infinite dimension the problem was initially studied in [45, 349] for the case of diffusion independent of the controls, and in [598] for a problem with linear state equation and cost functional. Recently, thanks to developments in the study of backward stochastic differential equations in infinite dimension, new results on the maximum principle for stochastic infinite-dimensional problems appeared. In [196, 279] the second variation is characterized as a certain stochastic bilinear form defined on $L^4(\Omega; H)$, while in [414] a general case when the coefficients are Fréchet-differentiable (twice for non-convex control domain) is treated and the second variation is characterized as a solution of a BSDE "in the sense of transposition". The approach of [196, 279] is used in [280] where regularity conditions on the coefficients are weakened to study a large class of optimal control problems driven by stochastic PDEs of parabolic type on a bounded open set of $\mathbb{R}^n$. Other results for specific classes of equations include [331], for a one-dimensional heat equation with noise and control on the boundary, and [470], for a class of problems with delay state equation (both with distributed and discrete delay). The papers [196, 470] also include, respectively, an unbounded diffusion term and Lévy noise. In general, the maximum principle approach only gives necessary conditions for optimality, however under suitable convexity assumptions sufficiency can also be proved. Such results for finite-dimensional systems can be found in [23, 575, 599]. A sufficiency result for a class of infinite-dimensional systems is proved in [445], while a sufficient condition for certain delay systems with diffusion independent of the controls is characterized in [352].

Chapter 3
Viscosity Solutions

This chapter is devoted to the theory of viscosity solutions of Hamilton–Jacobi–Bellman equations in Hilbert spaces. At its core is the notion of the so-called B-continuous viscosity solution which was introduced for first-order equations by M.G. Crandall and P.L. Lions in [141, 142] and later extended to second-order equations in [538]. The theory applies to fully nonlinear equations with various unbounded terms. This is its main advantage over the notions of mild and strong solutions discussed in Chap. 4, mild solutions in L^2 spaces discussed in Chap. 5 and the BSDE techniques of Chap. 6. After the introduction of the core theory we discuss several special cases which require various adjustments in the definition of viscosity solution. The material of the chapter is arranged in the following way:

- In Sect. 3.1 we introduce the notion of B-continuity, the spaces $H_{-\alpha}$ defined by a strictly positive self-adjoint operator B, and we present several estimates involving $|\cdot|_{-1}$ norms for solutions of deterministic and stochastic evolution equations. We also discuss a smooth perturbed optimization principle in Hilbert spaces.
- In Sect. 3.2 we present a maximum principle for B-upper semicontinuous functions in Hilbert spaces. This is a key technical result needed in the proofs of uniqueness of viscosity solutions.
- In Sect. 3.3 we introduce the definition of a viscosity solution and in Sect. 3.4 we discuss basic convergence properties of viscosity solutions.
- Section 3.5 is devoted to uniqueness of viscosity solutions. We prove several comparison theorems for degenerate parabolic and elliptic equations.
- In Sects. 3.6 and 3.7 we present results on existence of viscosity solutions. In Sect. 3.6 we study properties of value functions of stochastic optimal control problems and prove that they are viscosity solutions of the associated HJB equations. In Sect. 3.7 we discuss how to obtain existence of viscosity solutions for more general equations, for instance of Isaacs type, by the method of finite-dimensional approximations.

© Springer International Publishing AG 2017
G. Fabbri et al., *Stochastic Optimal Control in Infinite Dimension*,
Probability Theory and Stochastic Modelling 82,
DOI 10.1007/978-3-319-53067-3_3

- In Sect. 3.9 another method to prove existence of viscosity solutions, Perron's method, is presented. In this section we also explain how, in certain cases, the method of half-relaxed limits of Barles–Perthame can be adapted to viscosity solutions in Hilbert spaces. A classical limiting problem of singular perturbations is discussed in Sect. 3.8.
- In Sect. 3.10 we explain how the theory of viscosity solutions is applied to the infinite-dimensional Black–Scholes–Barenblatt equation originating in the theory of bond markets.
- Sections 3.11–3.13 discuss three special cases, the HJB equation related to the optimal control of the Duncan–Mortensen–Zakai equation, the HJB equation for a boundary optimal control problem and the HJB equation for optimal control of stochastic Navier–Stokes equations. These cases require modifications of the definition of a viscosity solution. We explain how the basic theory of Sects. 3.5 and 3.6 can be extended and adapted to equations containing special unbounded terms.

Throughout this chapter H is a real, separable Hilbert space with inner product $\langle \cdot, \cdot \rangle$ and norm $| \cdot |$. We recall that we identify H with its dual. We denote by $S(H)$ the set of bounded, self-adjoint operators on H.

3.1 Preliminary Results

3.1.1 B-Continuity and Weak and Strong B-Conditions

Definition 3.1 Given a strictly positive $B \in S(H)$ and $\alpha > 0$, we define the space $H_{-\alpha}$ as the completion of H with respect to the norm

$$|x|^2_{-\alpha} := \langle B^\alpha x, x \rangle .$$

The strict positivity of B ensures that the operator $B^{\alpha/2}$ extends to an isometry of $H_{-\alpha}$ onto H that we denote again by $B^{\alpha/2}$. $H_{-\alpha}$ is a Hilbert space when endowed with the inner product induced by $B^{\alpha/2}$:

$$\langle x, y \rangle_{-\alpha} := \left\langle B^{\alpha/2} x, B^{\alpha/2} y \right\rangle .$$

Definition 3.2 If B, α are as in Definition 3.1, we denote by H_α the space $H_\alpha := B^{\alpha/2}(H)$ endowed with the Hilbert space structure characterized by the following inner product:

$$\langle x, y \rangle_\alpha := \left\langle B^{-\alpha/2} x, B^{-\alpha/2} y \right\rangle .$$

Thanks to the strict positivity of B, $B^{-\alpha/2} \colon H_\alpha \to H$ is an isometry onto H; H_α can be identified with the dual of $H_{-\alpha}$.

Of course, even if not explicitly emphasized by the notation, the spaces H_α depend on the choice of B.

We will often use a notion of continuity, called B-continuity, which is stronger than the usual continuity and weaker than weak sequential continuity.

Definition 3.3 (*B-upper/lower semicontinuity*) Let $B \in S(H)$ be a strictly positive operator on H. Given $I \subset \mathbb{R}$ and $U \subset H$, we say that a function $u: I \times U \to \mathbb{R} \cup \{\pm\infty\}$ is B-upper semicontinuous (respectively, B-lower semicontinuous) if, for any sequences $(t_n)_{n\in\mathbb{N}}$ in I and $(x_n)_{n\in\mathbb{N}}$ in U such that $t_n \to t \in I$, $x_n \rightharpoonup x \in U$ and $Bx_n \to Bx$ as $n \to \infty$, we have

$$\limsup_{n\to\infty} u(t_n, x_n) \le u(t, x) \quad (\text{respectively, } \liminf_{n\to\infty} u(t_n, x_n) \ge u(t, x)).$$

Definition 3.4 (*B-continuity*) Given B, I and U as in Definition 3.3, we say that a function $u: I \times U \to \mathbb{R}$ is B-continuous if it is both B-upper semicontinuous and B-lower semicontinuous.

Remark 3.5 It is easy to see that one gets the same definition of B-upper/lower-semicontinuity if the condition $x_n \rightharpoonup x \in U$ in Definition 3.3 is replaced by the requirement that $(x_n)_{n\in\mathbb{N}}$ is bounded and $x \in U$. ∎

Lemma 3.6 *Let B be as in Definition 3.3. Then:*

(i) *If B is compact then u is B-upper semicontinuous (respectively, B-lower semicontinuous, B-continuous) if and only if u is weakly sequentially upper semicontinuous (respectively, weakly sequentially lower semicontinuous, weakly sequentially continuous).*

(ii) *Let $\alpha > 0$. Then u is B-upper semicontinuous (respectively, B-lower semicontinuous, B-continuous) if and only if u is B^α-upper semicontinuous (respectively, B^α-lower semicontinuous, B^α-continuous).*

(iii) *Let U be weakly sequentially closed, and $\alpha > 0$. Then u is B-continuous on $I \times U$ if and only if u is continuous in the $|\cdot| \times |\cdot|_{-\alpha}$ norm on bounded subsets of $I \times U$. If B is compact and $I = [a, b]$, then u is B-continuous on $[a, b] \times U$ if and only if u is uniformly continuous in the $|\cdot| \times |\cdot|_{-\alpha}$ norm on $[a, b] \times (U \cap B_R)$ for every $R > 0$. Finally, if u is weakly sequentially continuous on $[a, b] \times U$ then u is uniformly continuous in the $|\cdot| \times |\cdot|_{-\alpha}$ norm on $[a, b] \times (U \cap B_R)$ for every $R > 0$.*

(iv) *Let B_1, $B_2 \in S(H)$ be two strictly positive operators on H such that $B_1(H) = B_2(H)$ and let $(x_n)_{n\in\mathbb{N}}$ be a sequence in H. Then $B_1 x_n \to B_1 x$ if and only if $B_2 x_n \to B_2 x$. In particular, the notions of B_1-continuity and B_2-continuity are equivalent.*

Proof Part (i) is obvious.

(ii) We show that, for any $\alpha \ge \beta > 0$, u is B^α-continuous (resp. B^α-lower semicontinuous, B^α-upper semicontinuous) if and only if it is B^β-continuous (resp. B^β-lower semicontinuous, B^β-upper semicontinuous). To show this fact it is enough to prove that for a given weakly convergent sequence $x_n \rightharpoonup x \in H$ we have that $|B^\alpha(x_n - x)| \to 0$ if and only if $|B^\beta(x_n - x)| \to 0$. Since $\alpha \geq \beta$ the "if" part is obvious. For the "only if" part, assume that $|B^\alpha(x_n - x)| \to 0$ and observe that, since x_n is weakly convergent and hence bounded,

$$|B^{\alpha/2}(x_n - x)|^2 = \langle x_n, B^\alpha(x_n - x)\rangle - \langle x, B^\alpha(x_n - x)\rangle \to 0.$$

So, if $\alpha/2 \leq \beta$, this fact and the "if" part allow to conclude the proof, otherwise one can conclude iterating the argument.

(iii) The first statement follows from (ii). The only nontrivial statement of the second claim is the "only if" part of it. So let B be compact. From (ii) we can assume $\alpha = 2$. Assume by contradiction that, for some $\varepsilon > 0$, there exist two sequences (t_n, x_n) and (s_n, y_n) in $[a, b] \times (U \cap B_R)$ s.t.

$$|t_n - s_n| + |x_n - y_n|_{-2} \to 0 \quad \text{and} \quad |u(t_n, x_n) - u(s_n, y_n)| > \varepsilon. \tag{3.1}$$

Since $[a, b] \times (U \cap B_R)$ is weakly sequentially compact we can assume that $x_n \rightharpoonup x$ and $y_n \rightharpoonup y$ for some $x, y \in U \cap B_R$ and that $t_n, s_n \to s$ for some $s \in [a, b]$. So we have $B(x_n - y_n) \to 0$ and $(x_n - y_n) \rightharpoonup (x - y)$ and thus (since the graph of a continuous operator is weakly closed), $B(x - y) = 0$ which implies that $x = y$. Since B is compact we also have $Bx_n \to Bx$ and $By_n \to By = Bx$. So, since u is B-continuous, $u(t_n, x_n) \to u(s, x)$ and $u(s_n, y_n) \to u(s, x)$ and this contradicts (3.1). If the third claim is not true then again there must exist sequences (t_n, x_n) and (s_n, y_n) s.t. (3.1) holds. But then again, up to a subsequence, $t_n, s_n \to s$ for some $s \in [a, b]$ and $x_n, y_n \rightharpoonup x$ for some $x \in U \cap B_R$ and this, together with the weak sequential continuity of u, contradicts (3.1).

(iv) Let $B_1 x_n \to B_1 x$ as $n \to \infty$. It follows easily from the closed graph theorem that $B_1^{-1} B_2$ is bounded. Thus $B_2 B_1^{-1} = (B_1^{-1} B_2)^*$ on $B_1(H)$ and $(B_1^{-1} B_2)^*$ is a bounded operator. Therefore

$$B_2 x_n = B_2 B_1^{-1} B_1 x_n = (B_1^{-1} B_2)^* B_1 x_n \to (B_1^{-1} B_2)^* B_1 x = B_2 B_1^{-1} B_1 x = B_2 x.$$

The other implication is proved similarly. $\qquad\square$

Definition 3.7 (*B-closed set*) We will say that a set $U \subset H$ is B-closed if whenever $x_n \in U, x_n \rightharpoonup x, Bx_n \to Bx$ then $x \in U$.

Remark 3.8 Every weakly sequentially closed subset of H is B-closed, in particular every convex closed subset of H is B-closed. $\qquad\blacksquare$

The following weak and strong B-conditions were introduced in [141, 142].

Definition 3.9 (*Weak B-condition*) Let A be a linear, densely defined, closed operator in H. We say that an operator $B \in \mathcal{L}(H)$ satisfies the *weak B-condition* for A if B is strictly positive, self-adjoint, $A^*B \in \mathcal{L}(H)$, and

$$- A^*B + c_0 B \geq 0 \quad \text{for some } c_0 \geq 0. \tag{3.2}$$

Definition 3.10 (*Strong B-condition*) Let A be a linear, densely defined, closed operator in H. We say that an operator $B \in \mathcal{L}(H)$ satisfies the *strong B-condition* for A if B is strictly positive, self-adjoint, $A^*B \in \mathcal{L}(H)$, and

$$- A^*B + c_0 B \geq I \quad \text{for some } c_0 \geq 0. \tag{3.3}$$

It is well known that if A is a densely defined closed operator in H then the operator $B = (I + AA^*)^{-1/2}$ is bounded, strictly positive, self-adjoint and $A^*B \in \mathcal{L}(H)$. The strong and weak B-conditions require a little more. We will apply them when the operator A is maximal dissipative. The following result has been shown in [506].

Theorem 3.11 *If A is a linear, densely defined maximal dissipative operator in H then the weak B condition is satisfied with $B = ((\mu I - A)(\mu I - A)^*)^{-1/2}$ and $c_0 = \mu$, where $\mu \geq 0$ is any constant such that $\mu I - A^* \geq \delta I$ for some $\delta > 0$.*

Proof Let $C = (\mu I - A)(\mu I - A)^*$. By our assumptions, C^{-1} exists and $C^{-1} \in \mathcal{L}(H)$. It is also easy to see that $C = C^* > 0$. We set $B = C^{-1/2}$. Then $B = B^* > 0$, and we have, for $x \in H$,

$$|Bx|^2 = \langle ((\mu I - A)^{-1})^* (\mu I - A)^{-1} x, x \rangle = |(\mu I - A)^{-1} x|^2.$$

Therefore, by Proposition B.2-(i) (see also [180], Proposition B.1, p. 429 or [584], Theorem 2.2, p. 208), it follows that $R(B) = R(((\mu I - A)^{-1})^*) = R((\mu I - A^*)^{-1}) = D(A^*)$.

Let $S = (\mu I - A)^*B$. Then $S \in \mathcal{L}(H)$ and it is unitary. In fact SB^{-1} is the polar decomposition of $\mu I - A^*$. It remains to show that $S \geq 0$.

To this end we complexify the space and the operators. Let $H_c = \{\tilde{x} = x + iy : x, y \in H\}$ with standard operations $(x + iy) + (z + iw) = (x + z) + i(y + w)$, $(a + ib)(x + iy) = (ax - by) + i(bx + ay)$ and the inner product $\langle (x + iy), (z + iw) \rangle_c = \langle x, z \rangle + \langle y, w \rangle + i\langle y, z \rangle - i\langle x, w \rangle$. An operator T in H is complexified by setting $T_c(x + iy) = Tx + iTy$ and then $(T_c)^* = (T^*)_c$. It is easy to see that we still have $\mu I_c - A_c^* \geq 0$ in the sense that $\text{Re}\langle (\mu I_c - A_c^*)\tilde{x}, \tilde{x} \rangle_c \geq 0$, $B_c = B_c^* > 0$ and S_c is unitary (and thus normal). It is enough to show that $S_c \geq 0$.

Suppose that S_c is not nonnegative. Since S_c is normal it then follows from the spectral representation theorem that there is a nontrivial closed subspace K of H_c which is invariant for S_c and S_c is strongly dissipative on K, i.e. $\text{Re}\langle S_c\tilde{x}, \tilde{x} \rangle_c \leq -\nu|\tilde{x}|_c^2$ for some $\nu > 0$. Let P_K be the orthogonal projection onto K. Then the operator $P_K B_c : K \to K$ is self-adjoint and strictly positive. We choose $\lambda > 0$ in the spectrum of $P_K B_c$. Then there exists a $\tilde{y} \neq 0$ in K such that

$$|P_K B_c \tilde{y} - \lambda \tilde{y}|_c \leq \frac{\lambda \nu}{2\|S_c\|} |\tilde{y}|_c.$$

We set $\tilde{x} = B_c \tilde{y}$. Then

$$\langle \mu I_c - A_c^* \tilde{x}, \tilde{x} \rangle_c = \langle S_c \tilde{y}, \tilde{x} \rangle_c = \langle S_c \tilde{y}, P_K \tilde{x} \rangle_c = \langle S_c \tilde{y}, P_K B_c \tilde{y} \rangle_c$$
$$= \lambda \langle S_c \tilde{y}, \tilde{y} \rangle_c + \langle S_c \tilde{y}, P_K B_c \tilde{y} - \lambda \tilde{y} \rangle_c.$$

Taking the real part of the above relation and using that $\mathrm{Re}\langle (\mu I_c - A_c^*) \tilde{x}, \tilde{x} \rangle_c \geq 0$ we obtain

$$0 \leq -\lambda \nu |\tilde{y}|_c^2 + \|S_c\| |\tilde{y}|_c \frac{\lambda \nu}{2\|S_c\|} |\tilde{y}|_c = -\frac{\lambda \nu}{2} |\tilde{y}|_c^2,$$

which is a contradiction. Therefore $S_c \geq 0$ and thus $S \geq 0$. In fact, one can show (see [506]) that $S > 0$. $\square$

The following are two concrete examples of operators satisfying the weak B-condition:

Example 3.12 If the operator A is maximal dissipative and skew-adjoint, i.e. $A^* = -A$, the above implies that we can take $B = (\mu I - A^2)^{-1/2}$ for every $\mu > 0$. However, in such case a compact B cannot satisfy the strong B-condition, since if it did, then for every eigenvalue λ of B with an eigenvector e we would have $|e|^2 \leq \langle (-A^* B + c_0 B)e, e \rangle = \lambda \langle (A + c_0 I)e, e \rangle \leq \lambda c_0 |e|^2$, which is impossible since the eigenvalues accumulate at zero. ∎

Example 3.13 (*Operators coming from hyperbolic equations*) Let A be a maximal dissipative, self-adjoint operator in a Hilbert space H with a bounded inverse. It is then well known (see e.g. A.5.4 in [180]) that the operator

$$\mathcal{D}(\mathcal{A}) = \begin{pmatrix} D(A) \\ \times \\ D((-A)^{1/2}) \end{pmatrix}, \qquad \mathcal{A} = \begin{pmatrix} 0 & I \\ A & 0 \end{pmatrix},$$

is maximal dissipative in the Hilbert space $\mathcal{H} = \begin{pmatrix} D((-A)^{1/2}) \\ \times \\ H \end{pmatrix}$, equipped with the following "energy" type inner product

$$\left\langle \begin{pmatrix} u \\ v \end{pmatrix}, \begin{pmatrix} \bar{u} \\ \bar{v} \end{pmatrix} \right\rangle_{\mathcal{H}} = \langle (-A)^{1/2} u, (-A)^{1/2} \bar{u} \rangle_H + \langle v, \bar{v} \rangle_H, \qquad \begin{pmatrix} u \\ v \end{pmatrix}, \begin{pmatrix} \bar{u} \\ \bar{v} \end{pmatrix} \in \mathcal{H}.$$

Moreover, $\mathcal{A}^* = -\mathcal{A}$.

It is easy to check that the operator

$$\mathcal{B} = \begin{pmatrix} (-A)^{-1/2} & 0 \\ 0 & (-A)^{-1/2} \end{pmatrix}$$

is bounded, positive, self-adjoint on $\mathcal{H}$, and such that $\mathcal{A}^*\mathcal{B}$ is bounded and the weak
B-condition holds with constant $c_0 = 0$. In fact

$$\left\langle -\mathcal{A}^*\mathcal{B}\begin{pmatrix} u \\ v \end{pmatrix}, \begin{pmatrix} u \\ v \end{pmatrix} \right\rangle_{\mathcal{H}} = 0.$$

Moreover, we have

$$\left| \begin{pmatrix} u \\ v \end{pmatrix} \right|_{-1} = \left(|(-A)^{1/4}u|^2 + |(-A)^{-1/4}v|^2 \right)^{1/2}. \qquad \blacksquare$$

Let us now examine the strong B-condition in some examples.

Example 3.14 If A is maximal dissipative and self-adjoint in H, it satisfies the strong
B-condition with $B = (I - A)^{-1}$ and $c_0 = 1$. $\qquad \blacksquare$

Example 3.15 Suppose now that A_0 is a densely defined, closed operator in H which
satisfies the strong B-condition for some operator B_0 and constant c_0. Let A_1 be
another densely defined, closed operator in H such that $A_1^* B_0$ is bounded and

$$- A_1^* B_0 + c_1 B_0 \geq -\nu I \qquad (3.4)$$

for some $\nu \in (0, 1)$ and some constant c_1. It is then clear that $A = A_0 + A_1$
satisfies the strong B-condition with $B = (1/(1 - \nu))B_0$ and the new constant
$c := c_0 + c_1$. Obviously (3.4) holds if $\|A_1^* B_0\| < 1$. Also rather standard arguments
show that (3.4) is satisfied for every $\nu \in (0, 1)$ and some constant c_1 if $A_1^* B_0$ is
compact. To see this, let $\{e_1, e_2, ...\}$ be an orthonormal basis of H. For $N \geq 1$ we let
$H_N = \text{span}\{e_1, ..., e_N\}$, P_N be the orthogonal projection onto H_N, and $Q_N := I - P_N$.
For $x \in H$ we will write $x_N := P_N x$, $x_N^{\perp} := Q_N x$. Since $A_1^* B_0$ is compact, there
is an $N_1 \geq 1$ such that $\|A_1^* B_0 - P_{N_1} A_1^* B_0 P_{N_1}\| \leq \nu/2$. Therefore it is enough
to prove that there is a c_1 such that $-P_{N_1} A_1^* B_0 P_{N_1} + c_1 B \geq -\nu/2I$ which, since
$\langle P_{N_1} A_1^* B_0 P_{N_1} x, x \rangle \leq C|x_{N_1}|^2$, will be true if

$$C|x_{N_1}|^2 \leq c_1 \langle B_0 x, x \rangle + \nu|x|^2/2,$$

i.e. if

$$C \leq c_1 \langle B_0 x, x \rangle + \nu|x|^2/2 \quad \text{for any } x \text{ such that } |x_{N_1}| = 1.$$

The above is certainly satisfied if $|x_{N_1}^{\perp}| \geq (2C/\nu)^{1/2}$. Moreover, it is easy to see that

$$\inf_{\{x:|x_{N_1}|=1, |x_{N_1}^{\perp}|\leq(2C/\nu)^{1/2}\}} \langle B_0 x, x \rangle = \delta_{N_1} > 0.$$

Thus it is enough to take $c_1 = C/\delta_{N_1}$. $\qquad \blacksquare$

Example 3.16 (Operators coming from elliptic equations) Let $\mathcal{O}$ be a bounded (regular enough) domain in $\mathbb{R}^n$. Let

$$\begin{cases} A := \sum_{i,j}^n \partial_i(a_{ij}\partial_j) + \sum_i^n b_i\partial_i + c \\ D(A) := H_0^1(\mathcal{O}) \cap H^2(\mathcal{O}), \end{cases}$$

where $a_{ij} = a_{ji}, b_i, c \in L^\infty(\mathcal{O})$ for $i, j \in \{1, .., n\}$, and there exists a $\theta > 0$ such that

$$\sum_{i,j}^n a_{ij}\xi_i\xi_j \geq \theta|\xi|^2 \quad \forall \xi \in \mathbb{R}^n$$

a.e. in $\mathcal{O}$. We observe that if A_0 is the operator A with $c = b_i = 0, i = 1, ..., n$, then A_0 is maximal dissipative and self-adjoint in $H = L^2(\mathcal{O})$ and the strong B-condition holds for A_0 with $B_0 = (I - A_0)^{-1}$ and $c_0 = 1$. Moreover, B_0 is compact as an operator from $L^2(\mathcal{O})$ to $H_0^1(\mathcal{O})$ and thus, if $A_1 = A - A_0$, it follows that $A_1^* B_0$ is compact. Thus the strong B-condition is satisfied for A with $B = \lambda B_0$ for some constant λ.

If, in addition, $a_{ij} \in W^{1,\infty}(\mathcal{O}), b_i = 0, i, j = 1, ..., n$ one can also take $B_0 = \lambda(\hat{A})^{-1}$ above, where

$$\begin{cases} \hat{A}f := -\Delta f \\ D(\hat{A}) := H_0^1(\mathcal{O}) \cap H^2(\mathcal{O}) \end{cases}$$

for λ big enough. This follows from an application of the Sobolevskii inequality (see, for instance, Theorem 1.1 in [406], see also [396, 528]).   ∎

Lemma 3.17 *Let $B \in S(H)$ be a strictly positive operator on H and A be a linear, densely defined, maximal dissipative operator. Then:*

(i) *If $D(A^*) = D(B^{-1})$, then the operator $S = -A^*B + c_0B$ is invertible for any $c_0 > 0$, and $S^{-1} \in \mathcal{L}(H)$.*

(ii) *If B satisfies the strong B-condition for A, then $D(A^*) = D(B^{-1})$.*

Proof (i) The statement is obvious since $B^{-1}(-A^* + c_0I)^{-1}$ is bounded and it is the inverse of S.

(ii) Let S be defined as in part (i) but with c_0 being the constant from the strong B-condition for A. Since S is bounded and $S \geq I$, S^{-1} exists and it is bounded. Moreover, we have $B = (-A^* + c_0I)^{-1}S$ which, by the invertibility of S, implies that $D(B^{-1}) = R(B) = R(-A^* + c_0I)^{-1} = D(A^*)$.   □

We refer the reader to [506] for an abstract condition involving interpolation spaces which ensures that the strong B-condition is satisfied and to [141, 142] for other comments about B-continuity and strong and weak B-conditions.

3.1.2 Estimates for Solutions of Stochastic Differential Equations

Let $T > 0$. Let A be a linear, densely defined, maximal dissipative operator in H, and $Q \in \mathcal{L}^+(\Xi)$. Let $(\Omega, \mathcal{F}, \{\mathcal{F}_s\}_{s \in [0,T]}, \mathbb{P}, W_Q)$ be a generalized reference probability space. Let Λ be a Polish space. Let $b: [0, T] \times H \times \Lambda \to H$ be $\mathcal{B}([0, T]) \otimes \mathcal{B}(H) \otimes \mathcal{B}(\Lambda)/\mathcal{B}(H)$-measurable, and $\sigma: [0, T] \times H \times \Lambda \to \mathcal{L}_2(\Xi_0, H)$ be $\mathcal{B}([0, T]) \otimes \mathcal{B}(H) \otimes \mathcal{B}(\Lambda)/\mathcal{B}(\mathcal{L}_2(\Xi_0, H))$-measurable. Let $a(\cdot): [0, T] \times \Omega \to \Lambda$ be $\mathcal{F}_s$-progressively measurable. For $x \in H$ we consider the following SDE

$$\begin{cases} dX(s) = (AX(s) + b(s, X(s), a(s))) \, dt + \sigma(s, X(s), a(s)) dW_Q(s) \\ X(0) = x \end{cases} \tag{3.5}$$

and its approximation

$$\begin{cases} dX^n(s) = (A_n X^n(s) + b(s, X^n(s), a(s))) \, dt + \sigma(s, X^n(s), a(s)) dW_Q(s) \\ X^n(0) = x, \end{cases}$$
$$\tag{3.6}$$

where A_n is the Yosida approximation of A defined in (B.10). The approximating Eq. (3.6) was already introduced in Chap. 1. Here we discuss some more specific results that will be needed in later chapters.

Let $C \geq 0$ and $\gamma \in [0, 1]$. We will make use of the following assumptions.

$$|b(s, x, a) - b(s, y, a)| \leq C|x - y| \qquad \forall x, y \in H, s \in [0, T], a \in \Lambda, \tag{3.7}$$

$$\|\sigma(s, x, a) - \sigma(s, y, a)\|_{\mathcal{L}_2(\Xi_0, H)} \leq C|x - y| \quad \forall x, y \in H, s \in [0, T], a \in \Lambda, \tag{3.8}$$

$$|b(s, x, a)| \leq C(1 + |x|) \qquad \forall x \in H, s \in [0, T], a \in \Lambda, \tag{3.9}$$

$$\|\sigma(s, x, a)\|_{\mathcal{L}_2(\Xi_0, H)} \leq C(1 + |x|^\gamma) \qquad \forall x \in H, s \in [0, T], a \in \Lambda. \tag{3.10}$$

Recall that, thanks to Theorem 1.127, assumptions (3.7)–(3.10) ensure the existence of unique mild solutions $X(\cdot)$ and $X^n(\cdot)$ of (3.5) and (3.6).

Proposition 3.18 *Let $T > 0$ and $\gamma \in [0, 1]$. Assume that (3.7)–(3.10) hold. Let $X(\cdot)$ be the mild solution of (3.5). Then there exist constants $c_1 > 0, c_2 > 0$ (depending only on T, C, γ) such that*

$$\mathbb{E}\left(\sup_{0 \leq s \leq T} e^{c_1(1+|X(s)|^2)^{(1-\gamma)}}\right) \leq c_2 e^{(1+|x|^2)^{(1-\gamma)}} \qquad \text{if } \gamma \in [0, 1) \tag{3.11}$$

and

$$\mathbb{E}\left(\sup_{0\leq s\leq T} e^{c_1(\log(2+|X(s)|^2))^2}\right) \leq c_2 e^{(\log(2+|x|^2))^2} \qquad \text{if } \gamma = 1. \qquad (3.12)$$

Proof We first consider the case $0 \leq \gamma < 1$. Let X^n be the mild solution of the approximating Eq. (3.6) and τ_k be the minimum of T and the first exit time of X^n from the set $\{|z| \leq k\}$. Let $\beta > 0$ and $\alpha > 0$ be numbers to be specified later. Since A_n is bounded, X^n solves the integral equation

$$X^n(s) = x + \int_0^s A_n X^n(r) + b(r, X^n(r), a(r))dr$$

$$+ \int_0^s \sigma(r, X^n(r), a(r))dW_Q(r), \qquad s \in [0, T].$$

$$(3.13)$$

Thus we can apply Itô's formula (see Theorem 1.163) to the function

$$\begin{cases} \Phi: [0, T] \times H \to \mathbb{R} \\ \Phi(s, x) = e^{\beta e^{-\alpha s}(1+|x|^2)^{1-\gamma}} \end{cases}$$

and obtain, for $s \in [0, T]$,

$$e^{\beta e^{-\alpha(s\wedge\tau_k)}(1+|X^n(s\wedge\tau_k)|^2)^{1-\gamma}}$$

$$= e^{\beta(1+|x|^2)^{1-\gamma}} - \int_0^{s\wedge\tau_k} \alpha\beta e^{-\alpha r}(1+|X^n(r)|^2)^{1-\gamma} e^{\beta e^{-\alpha r}(1+|X^n(r)|^2)^{1-\gamma}} dr$$

$$+ \int_0^{s\wedge\tau_k} 2(1-\gamma)\beta e^{-\alpha r} e^{\beta e^{-\alpha r}(1+|X^n(r)|^2)^{1-\gamma}}(1+|X^n(r)|^2)^{-\gamma}$$

$$\times \langle A_n X^n(r) + b(r, X^n(r), a(r)), X^n(r)\rangle dr$$

$$+ \int_0^{s\wedge\tau_k} 2(1-\gamma)\beta e^{-\alpha r} e^{\beta e^{-\alpha r}(1+|X^n(r)|^2)^{1-\gamma}}(1+|X^n(r)|^2)^{-\gamma}$$

$$\times \langle X^n(r), \sigma(r, X^n(r), a(r))dW_Q(r)\rangle$$

$$+ \frac{1}{2}\int_0^{s\wedge\tau_k} e^{\beta e^{-\alpha r}(1+|X^n(r)|^2)^{1-\gamma}} \text{Tr}\left(\left(\sigma(r, X^n(r), a(r))Q^{\frac{1}{2}}\right)\left(\sigma(r, X^n(r), a(r))Q^{\frac{1}{2}}\right)^*\right.$$

$$\times 2\left[2\beta^2 e^{-2\alpha r}(1+|X^n(r)|^2)^{-2\gamma}(1-\gamma)^2 X^n(r) \otimes X^n(r)\right.$$

$$- 2\beta\gamma e^{-\alpha r}(1+|X^n(r)|^2)^{-\gamma-1}(1-\gamma)X^n(r) \otimes X^n(r)$$

$$\left.\left. + \beta e^{-\alpha r}(1+|X^n(r)|^2)^{-\gamma}(1-\gamma)I\right]\right)dr$$

$$\leq e^{\beta(1+|x|^2)^{1-\gamma}} + \int_0^{s\wedge T_k} (1+|X^n(r)|^2)^{1-\gamma} e^{\beta e^{-\alpha r}(1+|X^n(r)|^2)^{1-\gamma}}$$

$$\times (-\alpha + C(\beta))\beta e^{-\alpha r} dr$$

$$+ 2\int_0^s \mathbf{1}_{[0,T_k]}(1-\gamma)\beta e^{-\alpha r} e^{\beta e^{-\alpha r}(1+|X^n(r)|^2)^{1-\gamma}}(1+|X^n(r)|^2)^{-\gamma}$$

$$\times \langle X^n(r), \sigma(r, X^n(r), a(r))dW_Q(r)\rangle \tag{3.14}$$

for some absolute constant $C(\beta)$, nondecreasing in β and also depending on C, γ, where we used Lemma 1.110 in the last line of (3.14).

Therefore, choosing $\alpha = C(\beta) + 1$ in (3.14), we obtain

$$e^{\beta e^{-\alpha(s\wedge T_k)}(1+|X^n(s\wedge T_k)|^2)^{1-\gamma}} + \int_0^{s\wedge T_k} \beta e^{-\alpha r}(1+|X^n(r)|^2)^{1-\gamma} e^{\beta e^{-\alpha r}(1+|X^n(r)|^2)^{1-\gamma}} dr$$

$$\leq e^{\beta(1+|x|^2)^{1-\gamma}} + 2\int_0^s \mathbf{1}_{[0,T_k]}(1-\gamma)\beta e^{-\alpha r} e^{\beta e^{-\alpha r}(1+|X^n(r)|^2)^{1-\gamma}}(1+|X^n(r)|^2)^{-\gamma}$$

$$\times \langle X^n(r), \sigma(r, X^n(r), a(r))dW_Q(r)\rangle, \qquad s \in [0, T]. \tag{3.15}$$

Therefore, taking expectation in (3.15) yields

$$\mathbb{E}e^{\beta e^{-\alpha(s\wedge T_k)}(1+|X^n(s\wedge T_k)|^2)^{1-\gamma}}$$

$$+ \mathbb{E}\int_0^{s\wedge T_k} \beta e^{-\alpha r}(1+|X^n(r)|^2)^{1-\gamma} e^{\beta e^{-\alpha r}(1+|X^n(r)|^2)^{1-\gamma}} dr$$

$$\leq e^{\beta(1+|x|^2)^{1-\gamma}}, \qquad s \in [0, T]. \tag{3.16}$$

Now we choose $\alpha = C(2) + 1$ in (3.14) so that (3.15) and (3.16) are satisfied for $\beta = 2$. Moreover, we can observe that, since $C(\beta)$ is an increasing function of β and since the term $\beta e^{-\alpha r} e^{\beta e^{-\alpha r}(1+|X^n(r)|^2)^{1-\gamma}}(1+|X^n(r)|^2)^{1-\gamma}$ is always positive, (3.15) and (3.16) are also satisfied when we choose $\beta = 1$ and $\alpha = C(2) + 1$. Using (3.15) with this last choice of α and β and observing that the integral in the left-hand side of (3.15) is positive, we get, for $s \in [0, T]$,

$$\sup_{0\leq u\leq s} e^{e^{-\alpha(u\wedge T_k)}(1+|X^n(u\wedge T_k)|^2)^{1-\gamma}} \leq e^{(1+|x|^2)^{1-\gamma}}$$

$$+ \sup_{0\leq u\leq s}\left| \int_0^u 2e^{-\alpha r} e^{e^{-\alpha r}(1+|X^n(r)|^2)^{1-\gamma}}\mathbf{1}_{[0,T_k]} \right.$$

$$\left. \times (1-\gamma)(1+|X^n(r)|^2)^{-\gamma}\langle X^n(r), \sigma(r, X^n(r), a(r))dW_Q(r)\rangle \right|,$$

and therefore, using the Burkholder–Davis–Gundy inequality (see Theorem 1.111), we have

$$\mathbb{E} \sup_{0 \le u \le s} e^{e^{-\alpha(u \wedge \tau_k)}(1+|X^n(u \wedge \tau_k)|^2)^{1-\gamma}} \le e^{(1+|x|^2)^{1-\gamma}}$$

$$+ \left(\mathbb{E} \sup_{0 \le u \le s} \left| \int_0^u 2 e^{-\alpha r} e^{e^{-\alpha r}(1+|X^n(r)|^2)^{1-\gamma}} \mathbf{1}_{[0,\tau_k]} \right. \right.$$

$$\left. \left. \times (1-\gamma)(1+|X^n(r)|^2)^{-\gamma} \langle X^n(r), \sigma(r, X^n(r), a(r)) dW_Q(r) \rangle \right| \right)$$

$$\le e^{(1+|x|^2)^{1-\gamma}}$$

$$+ \left(\mathbb{E} \int_0^s C_1 e^{-2\alpha r} e^{2 e^{-\alpha r}(1+|X^n(r)|^2)^{1-\gamma}} (1+|X^n(r)|^2)^{1-\gamma} \mathbf{1}_{[0,\tau_k]} dr \right)^{\frac{1}{2}}, \quad s \in [0, T].$$

$$(3.17)$$

Using (3.16) with $\beta = 2$ we see that the last two lines of (3.17) are less than or equal to

$$C_2 e^{(1+|x|^2)^{1-\gamma}}.$$

Above, the constants C_i, $i = 1, 2$, only depend on C, γ, T. Thus we have obtained

$$\mathbb{E} \sup_{0 \le s \le T} e^{e^{-\alpha T}(1+|X^n(s \wedge \tau_k)|^2)^{1-\gamma}} \le C_2 e^{(1+|x|^2)^{1-\gamma}}. \qquad (3.18)$$

Since $\lim_{k \to +\infty} \tau_k = T$ a.s., letting $k \to +\infty$ in (3.18) and using Fatou's lemma, we obtain

$$\mathbb{E} \sup_{0 \le s \le T} e^{e^{-\alpha T}(1+|X^n(s)|^2)^{1-\gamma}} \le C_2 e^{(1+|x|^2)^{1-\gamma}}.$$

It now remains to use (see Theorem 1.131) that

$$\lim_{n \to \infty} \mathbb{E} \left(\sup_{0 \le s \le T} |X^n(s) - X(s)|^2 \right) = 0.$$

This implies the existence of a subsequence X^{n_k} satisfying $\lim_{n \to \infty} \sup_{0 \le s \le T} |X^{n_k}(s) - X(s)|^2 = 0$ almost surely and then it ensures that, a.s.,

$$\lim_{n_k \to \infty} \sup_{0 \le s \le T} e^{e^{-\alpha T}(1+|X^{n_k}(s)|^2)^{1-\gamma}} = \sup_{0 \le s \le T} e^{e^{-\alpha T}(1+|X(s)|^2)^{1-\gamma}}.$$

We can then apply Fatou's lemma again to obtain the claim.

For $\gamma = 1$ we can repeat the same arguments applied to the function

$$e^{\beta e^{-\alpha s}(\log(2+|x|^2))^2}. \qquad \qquad \square$$

Lemma 3.19 *Let A be a linear, densely defined maximal dissipative operator in H and B an operator satisfying the weak B-condition for A for some constant $c_0 > 0$. Then:*

(i) For any $R > 0$ there exists a constant $C(R)$ such that, for $x \in H$, $|x| \le R$ and $t \ge 0$,

$$|e^{tA}x - x|_{-1} \le C(R)\sqrt{t}. \tag{3.19}$$

(ii) If B satisfies the strong B-condition for A with constant c_0 then, for $x \in H$ and $t \ge 0$,

$$|e^{tA}x|_{-1}^2 + 2t|e^{tA}x|^2 \le e^{2c_0t}|x|_{-1}^2. \tag{3.20}$$

Proof (i) Let $Z(t) = e^{tA}x$. If $x \in D(A)$, using that A is maximal dissipative, Theorem B.45, and (3.2) we have

$$|Z(t) - x|_{-1}^2 = \int_0^t \langle 2B(Z(s) - x), AZ(s)\rangle ds$$

$$\le 2\int_0^t \langle A^*BZ(s), Z(s)\rangle ds + 2\|A^*B\||x|^2 t$$

$$\le 2c_0 \int_0^t |Z(s)|_{-1}^2 ds + 2\|A^*B\||x|^2 t \le (2c_0\|B\||x|^2 + 2\|A^*B\||x|^2)t.$$

The estimate now follows by density of $D(A)$.

(ii) Again it is enough to show the estimate for $x \in D(A)$. We then have by (3.3)

$$\frac{d}{ds}|Z(s)|_{-1}^2 = 2\langle A^*BZ(s), Z(s)\rangle \le 2c_0|Z(s)|_{-1}^2 - 2|Z(s)|^2,$$

and thus

$$\frac{d}{ds}\left(e^{-2c_0s}|Z(s)|_{-1}^2\right) = -2c_0e^{-2c_0s}|Z(s)|_{-1}^2$$

$$+ e^{-2c_0s}\frac{d}{ds}|Z(s)|_{-1}^2 \le -2e^{-2c_0s}|Z(s)|^2.$$

Integrating we obtain

$$e^{-2c_0t}|Z(t)|_{-1}^2 + 2\int_0^t e^{-2c_0s}|Z(s)|^2 ds \le |x|_{-1}^2.$$

The inequality now follows upon noticing that $e^{-2c_0t}|Z(t)|^2 \le e^{-2c_0s}|Z(s)|^2$ for $0 \le s \le t$, since e^{sA} is a semigroup of contractions. $\square$

Lemma 3.20 *Let A be a linear, densely defined maximal dissipative operator in H and B an operator satisfying the weak B-condition for A for some $c_0 \geq 0$. Let (3.7), (3.9) and (3.10) with $\gamma = 1$ hold, and let*

$$\langle b(s, x, a) - b(s, y, a), B(x - y) \rangle \leq C|x - y|^2_{-1} \tag{3.21}$$

$$\|\sigma(s, x, a) - \sigma(s, y, a)\|_{\mathcal{L}_2(\Xi_0, H)} \leq C|x - y|_{-1}, \tag{3.22}$$

for all $x, y \in H$, $s \in [0, T]$ and $a \in \Lambda$. If $X(\cdot)$ and $Y(\cdot)$ are the mild solutions of (3.5) with initial conditions $X(0) = x$ and $Y(0) = y$ respectively, driven by the same progressively measurable process $a(\cdot) : [0, T] \times \Omega \to \Lambda$, then

$$\sup_{s \in [0,T]} \mathbb{E}\left[|X(s) - Y(s)|^2_{-1}\right] \leq C(T)|x - y|^2_{-1}, \tag{3.23}$$

where $C(T)$ is a constant depending only on $T, C, c_0, \|B\|$.

Proof We define the function

$$\begin{cases} F : H \to \mathbb{R} \\ F(z) = |z|^2_{-1} = \langle Bz, z \rangle . \end{cases}$$

We notice that $DF(z) = 2Bz$ and $D^2 F(z) = 2B$. We will apply Itô's formula to F along the trajectories of the process $Z(\cdot) := X(\cdot) - Y(\cdot)$, which is a mild solution of

$$\begin{cases} dZ(s) = (AZ(s) + f(s))\, ds + \Phi(s) dW_Q(s), \\ Z(0) = x - y, \end{cases}$$

where, for any $s \in [0, T]$, $f(s) := b(s, X(s), a(s)) - b(s, Y(s), a(s))$ and $\Phi(s) := \sigma(s, X(s), a(s)) - \sigma(s, Y(s), a(s))$. Thanks to (3.7), (3.9), (3.10) and (3.22), the assumptions of Theorem 1.130 are satisfied and thus we have (1.37) and the hypotheses of Proposition 1.164 are satisfied. Therefore we have, for all $s \in [0, T]$,

$$\mathbb{E}\left[|Z(s)|^2_{-1}\right] = |x - y|^2_{-1} + \int_0^s \mathbb{E}\left[\langle 2A^* BZ(r), Z(r) \rangle + \langle 2BZ(r), f(r) \rangle\right] ds$$

$$+ \int_0^s \mathbb{E}\left[\text{Tr}\left(\left(\Phi(r)Q^{\frac{1}{2}}\right)\left(\Phi(r)Q^{\frac{1}{2}}\right)^* B\right)\right] dr, \tag{3.24}$$

and using (3.2), (3.21) and (3.22), we find

$$\mathbb{E}\left[|Z(s)|^2_{-1}\right] \leq |x - y|^2_{-1} + \int_0^s \mathbb{E}\left[2c_0|Z(r)|^2_{-1} + 2C|Z(r)|^2_{-1}\right] dr$$

$$+ \int_0^s \mathbb{E}\left[\|B\|C^2|Z(r)|^2_{-1}\right] dr.$$

Applying Gronwall's lemma we obtain (3.23). □

Remark 3.21 Condition (3.21) is obviously satisfied if

$$|b(s, x, a) - b(s, y, a)|_{-1} \leq C|x - y|_{-1}$$

for all $x, y \in H$, $s \in [0, T]$ and $a \in \Lambda$.

Condition (3.22) is satisfied if $\sigma(s, x, a) = \sigma_0(s, Kx, a)$ for some $\sigma_0 : [0, T] \times H \times \Lambda \to \mathcal{L}_2(\Xi_0, H)$ which satisfies (3.8) and $K \in \mathcal{L}(H)$ such that $|Kx| \leq L|x|_{-1}$ for some $L \geq 0$ and all $x \in H$. This requirement is also necessary since it is easy to see that the function $\sigma_0(t, x, a) := \sigma(t, B^{-1/2}x, a)$ satisfies (3.8) on $[0, T] \times R(B^{1/2}) \times \Lambda$ and thus it can be uniquely extended to a function σ_0 on $[0, T] \times H \times \Lambda$ which satisfies (3.8). Then $\sigma(s, x, a) = \sigma_0(s, B^{1/2}x, a)$. We also remark that, by Proposition B.2-(i), $|Kx| \leq L|x|_{-1}$ for all $x \in H$ is equivalent to $R(K^*) \subset R(B^{1/2})$. ∎

Lemma 3.22 *Let A be a linear, densely defined maximal dissipative operator in H. Let B be a bounded, strictly positive, self-adjoint operator on H such that $A^* B$ is bounded. Let (3.7)–(3.10) with $\gamma = 1$ hold. If $X(\cdot)$ is the mild solution of (3.5) with initial condition $X(0) = x$ driven by a progressively measurable process $a(\cdot) : [0, T] \times \Omega \to \Lambda$, then*

$$\mathbb{E}\left[|X(s) - x|^2_{-1}\right] \leq C(|x|, T)s, \quad \text{for all } s \in [0, T], \tag{3.25}$$

where $C(|x|, T)$ is a constant depending only on $T, |x|, C, \|B\|, \|A^* B\|$.

Proof We define the function

$$\begin{cases} F : H \to \mathbb{R} \\ F(z) = |z - x|_{-1} = \langle B(z - x), z - x \rangle. \end{cases}$$

We have $DF(z) = 2B(z - x)$ and $D^2 F(z) = 2B$, and applying Proposition 1.165 yields

$$\mathbb{E}\left[|X(s) - x|^2_{-1}\right]$$
$$= 2 \int_0^s \mathbb{E}\left[\langle X(r), A^* B(X(r) - x)\rangle + \langle b(r, X(r), a(r)), B(X(r) - x)\rangle\right] dr$$
$$+ \int_0^s \mathbb{E}\left[\mathrm{Tr}\left(\left(\sigma(r, X(r), a(r))Q^{\frac{1}{2}}\right)\left(\sigma(r, X(r), a(r))Q^{\frac{1}{2}}\right)^* B\right)\right] dr, \quad s \in [0, T]. \tag{3.26}$$

Using (3.9), (3.10), the boundedness of $A^* B$ and (1.37), we easily deduce using the Cauchy–Schwarz inequality, that the absolute values of the integrands in the right-hand side of (3.26) remain bounded by some constant $C(T, |x|)$ depending only on $T, |x|, C, \|B\|, \|A^* B\|$. This concludes the proof of (3.25). □

Lemma 3.23 *Let A be a linear, densely defined maximal dissipative operator in H and B an operator satisfying the strong B-condition for A for some $c_0 \geq 0$. Let (3.7), (3.9), (3.10) and (3.22) hold. If $X(\cdot)$ and $Y(\cdot)$ are the mild solutions of (3.5) with initial conditions $X(0) = x$ and $Y(0) = y$ respectively, driven by the same progressively measurable process $a(\cdot) : [0, T] \times \Omega \to \Lambda$, then*

$$\sup_{s \in [0,T]} \left(\mathbb{E}\left[|X(s) - Y(s)|^2_{-1} \right] + \mathbb{E} \int_0^s |X(r) - Y(r)|^2 dr \right) \leq C(T)|x - y|^2_{-1}$$
(3.27)

and, for any $s \in (0, T)$,

$$\mathbb{E}\left[|X(s) - Y(s)|^2 \right] \leq \frac{C(T)}{s}|x - y|^2_{-1},$$
(3.28)

where $C(T)$ is a constant depending only on $T, C, c_0, \|B\|$.

Proof Following the proof of Lemma 3.20 if we define, for any $s \in [0, T]$, $Z(s) = X(s) - Y(s)$, $f(s) = b(s, X(s), a(s)) - b(s, Y(s), a(s))$ and $\Phi(s) = \sigma(s, X(s), a(s)) - \sigma(s, Y(s), a(s))$, we have (as in (3.24)):

$$\mathbb{E}\left[|Z(s)|^2_{-1} \right] = |x - y|^2_{-1} + \int_0^s \mathbb{E}\left[\langle 2A^*BZ(r), Z(r) \rangle + \langle 2BZ(r), f(r) \rangle \right] ds$$
$$+ \int_0^s \mathbb{E}\left[\operatorname{Tr}\left(\left(\Phi(r)Q^{\frac{1}{2}} \right) \left(\Phi(r)Q^{\frac{1}{2}} \right)^* B \right) \right] dr, \quad s \in [0, T].$$
(3.29)

We observe that (3.3) implies

$$\langle 2A^*BZ(r), Z(r) \rangle + \langle 2Z(r), Z(r) \rangle \leq \langle 2c_0 BZ(r), Z(r) \rangle ,$$

which, together with (3.7), gives

$$\langle 2A^*BZ(r), Z(r) \rangle + \langle 2BZ(r), f(r) \rangle$$
$$\leq 2c_0|Z(r)|^2_{-1} - 2|Z(r)|^2 + 2C\|B\|^{1/2}|Z(r)|_{-1}|Z(r)| \leq c_1|Z(r)|^2_{-1} - |Z(r)|^2,$$
(3.30)

where c_1 depends on c_0, $\|B\|$ and C. Using (3.22) we have

$$\operatorname{Tr}\left(\left(\Phi(r)Q^{\frac{1}{2}} \right) \left(\Phi(r)Q^{\frac{1}{2}} \right)^* B \right) \leq \|B\|C^2|Z(r)|^2_{-1} = c_2|Z(r)|^2_{-1}.$$
(3.31)

It thus follows from (3.29)–(3.31) that, for $s \in [0, T]$,

$$\mathbb{E}\left[|Z(s)|^2_{-1} \right] + \int_0^s \mathbb{E}\left[|Z(r)|^2 \right] dr \leq |x - y|^2_{-1} + (c_1 + c_2) \int_0^s \mathbb{E}\left[|Z(r)|^2_{-1} \right] dr.$$

Such an inequality holds, of course, also dropping the positive term $\int_0^s \mathbb{E}\left[|Z(r)|^2\right] dr$ and then (3.27) follows easily from Gronwall's lemma. Regarding (3.28), using the definition of mild solution, (3.20), (3.27), and elementary computations, we have, for $s \in [0, T]$,

$$\mathbb{E}\left[|X(s) - Y(s)|^2\right] \le C_1(|e^{sA}(x - y)|^2 + |x - y|^2_{-1}) \le \frac{C_2}{s}|x - y|^2_{-1},$$

where C_1 and C_2 only depend on $T, C, c_0, \|B\|$. $\square$

Proposition 3.24 *Let $m > 0$. Let (3.7) and (3.8), (3.9) and (3.10) with $\gamma = 1$ hold for all $s \in [0, +\infty)$. Let $X(\cdot)$ be the mild solution of (3.5), driven by a progressively measurable process $a(\cdot) : [0, +\infty) \times \Omega \to \Lambda$ and let C be the constant appearing in (3.9) and (3.10). Let*

$$\bar{\lambda} = Cm + \frac{1}{2}C^2m(m - 1) \quad \text{if } m \ge 2$$

and

$$\bar{\lambda} = Cm + \frac{1}{2}C^2m \quad \text{if } 0 < m < 2.$$

Then for every $\lambda > \bar{\lambda}$ there exists a constant C_λ such that

$$\mathbb{E}\left[\left(C_\lambda + |X(s)|^2\right)^{\frac{m}{2}}\right] \le (C_\lambda + |x|^2)^{\frac{m}{2}} e^{\lambda s} \quad \text{for all } s \ge 0. \tag{3.32}$$

Proof Let $\lambda > \bar{\lambda}$ and let $\varepsilon = \varepsilon(\lambda) > 0$ be such that $\bar{\lambda}(1 + \varepsilon) = \lambda$. We set $C_\lambda > 1$ to be a number such that $2r \le C_\lambda - 1 + \varepsilon r^2$ for all $r \ge 0$. It is then easy to see that

$$Cmr(1 + r) + \frac{1}{2}C^2m(m - 1)(1 + r)^2 \le \lambda(C_\lambda + r^2) \quad \text{for all } r \ge 0.$$

Define $F(z) = (C_\lambda + |z|^2)^{\frac{m}{2}}$. Then $DF(z) = m(C_\lambda + |z|^2)^{\frac{m-2}{2}}z$ and $D^2F(z) = m(m - 2)(C_\lambda + |z|^2)^{\frac{m-4}{2}}z \otimes z + m(C_\lambda + |z|^2)^{\frac{m-2}{2}}I$.

Assume first that $m \ge 2$. Using Proposition 1.166 and (3.9), (3.10) we then have

$$\mathbb{E}\left[\left(C_\lambda + |X(s)|^2\right)^{\frac{m}{2}}\right] \le (C_\lambda + |x|^2)^{\frac{m}{2}}$$
$$+ \int_0^s \mathbb{E}\left[m(C_\lambda + |X(r)|^2)^{\frac{m-2}{2}}\langle X(r), b(r, X(r), a(r))\rangle\right.$$
$$+ \frac{1}{2}\text{Tr}\left(\left(\sigma(r, X(r), a(r))Q^{\frac{1}{2}}\right)\left(\sigma(r, X(r), a(r))Q^{\frac{1}{2}}\right)^*\right.$$
$$\times \left(m(m - 2)(C_\lambda + |X(r)|^2)^{\frac{m-4}{2}}X(r) \otimes X(r) + m(C_\lambda + |X(r)|^2)^{\frac{m-2}{2}}I\right)\right)\bigg]dr$$

$$\leq (C_\lambda + |x|^2)^{\frac{m}{2}} + \lambda \int_0^s \mathbb{E}\left[(C_\lambda + |X(r)|^2)^{\frac{m}{2}}\right] dr \qquad (3.33)$$

and we conclude applying Gronwall's lemma.

For $0 < m < 2$ the first term in the fourth line of (3.33) can be dropped and we argue as before since

$$Cmr(1+r) + \frac{1}{2}C^2 m(1+r)^2 \leq \lambda(C_\lambda + r^2) \quad \text{for all } r \geq 0. \qquad \square$$

3.1.3   Perturbed Optimization

The following is a classical result of Ekeland and Lebourg [204], see also [535] and [403], Lemma 4.2, p. 245, for a more general formulation.

Theorem 3.25 (Ekeland–Lebourg Theorem) *Let D be a bounded closed subset of a real Hilbert space K and $f: D \to \mathbb{R} \cup \{-\infty\}$ be upper semicontinuous and such that $\mathrm{dom}\,(f) := \{x \in D : f(x) \in \mathbb{R}\} \neq \emptyset$. Suppose that f is bounded from above. Then, for any $\delta > 0$, there exist $y \in K, \hat{x} \in D$ such that $|y|_K < \delta$ and the function*

$$x \to f(x) + \langle y, x \rangle_K$$

has a strict maximum over D at $\hat{x}$.

Corollary 3.26 *Let H be a real, separable Hilbert space with inner product $\langle \cdot, \cdot \rangle$, let B be a strictly positive operator in $S(H)$ and $D \subset H$ be a bounded, B-closed subset of H. Let $f: D \to \mathbb{R} \cup \{-\infty\}$ be a B-upper semicontinuous function, bounded from above. Then, for any $\delta > 0$, there exist $p \in H, \hat{x} \in D$ such that $|p| < \delta$, and the function*

$$x \to f(x) + \langle Bp, x \rangle$$

attains a maximum over D at $\hat{x}$, which is strict in the topology of H_{-2}.

Proof We want to apply Theorem 3.25 to D endowed with the topology induced by H_{-2}.

D is obviously bounded in H_{-2} and it is easy to see that D is closed in H_{-2}. To prove this, let $(x_n)_{n \in \mathbb{N}}$ be a sequence in D such that $x_n \xrightarrow[H_{-2}]{n \to \infty} x \in H_{-2}$, i.e. $Bx_n \to z$ for some $z \in H$. Since D is bounded in H, there is a subsequence, still denoted by x_n, such that $x_n \xrightarrow{H} \tilde{x} \in H$ for some $\tilde{x} \in D$. But the graph of B is weakly sequentially closed, so we obtain $B\tilde{x} = z$, which implies that $|x_n - \tilde{x}|_{-2} \to 0$, and thus $x = \tilde{x}$. Since D is B-closed, we thus have $x \in D$.

In particular, we showed that if (x_n) is a sequence in D such that $x_n \xrightarrow[H_{-2}]{n \to \infty} x$, then $x \in D$ and $x_n \rightharpoonup x$. Since f is B-upper semicontinuous, this shows that f is upper semicontinuous on D considered as a subset of H_{-2}.

We can now apply Theorem 3.25 to obtain that for all $\delta > 0$ there exists a $y \in H_{-2}$ with $|y|_{H_{-2}} < \delta$ such that $x \to f(x) + \langle y, x \rangle_{H_{-2}}$ attains a strict maximum (in the topology of H_{-2}) on D at some point $\hat{x}$. Define $p := By \in H$. Since $B: H_{-2} \to H$ is an isometry, we have that $|p| = |y|_{-2} < \delta$. Therefore for $x \in H$, $\langle y, x \rangle_{H_{-2}} = \langle B^2 y, x \rangle_H = \langle Bp, x \rangle_H$, which completes the proof. $\square$

3.2 A Maximum Principle

From now on, throughout the rest of this chapter, unless stated otherwise, A is a linear, densely defined, maximal dissipative operator in H.

In this section B is any strictly positive operator in $S(H)$. Let $\{e_1, e_2, ...\}$ be an orthonormal basis in H_{-1} (see Definition 3.1) made of elements of H. For $N > 2$ we let $H_N = \text{span}\{e_1, ..., e_N\}$. Let $P_N : H_{-1} \to H_{-1}$ be the orthogonal projection onto H_N. It is clear that P_N is also a bounded operator on H and therefore so is $Q_N := I - P_N$, i.e. $P_N, Q_N \in \mathcal{L}(H)$. It is also easy to see that $BP_N = P_N^* BP_N = P_N^* B$, $BQ_N = Q_N^* BQ_N = Q_N^* B$, where P_N^*, Q_N^* are adjoints of P_N, Q_N as operators in $\mathcal{L}(H)$. For $x \in H$ we will write $x_N := P_N x$, $x_N^\perp := Q_N x$.

We remark that if B is compact then $\|B^\gamma Q_N\| \to 0$ as $N \to +\infty$ for every $\gamma > 0$. Also in this case a natural choice for the basis $\{e_1, e_2, ...\}$ is to take $e_i = B^{-\frac{1}{2}} f_i$, where $\{f_1, f_2, ...\}$ is an orthonormal basis of H composed of eigenvectors of B. This choice of basis has the property that it is orthogonal in H_{-1} and H.

For a function $w \in C^2(H_{-1})$ we will write $D_{H_{-1}} w$, $D_{H_{-1}}^2 w$ to denote the Fréchet derivatives of w when w is considered as a function in $C^2(H_{-1})$ whereas Dw, $D^2 w$ mean the Fréchet derivatives of w when w is considered as a function in $C^2(H)$. We remark that the spaces H_1, H_2 in Theorem 3.27 are the spaces introduced in Sect. 3.1.1, not one of the spaces H_N defined above. This is why we put the restriction $N > 2$.

Theorem 3.27 (Maximum Principle) *Let $B \in S(H)$ be strictly positive and let $N > 2$, $\kappa > 0$. Let $u, v : H \to \mathbb{R} \cup \{-\infty\}$ be B-upper semicontinuous functions bounded from above and such that*

$$\limsup_{|x| \to +\infty} \frac{u(x)}{|x|} < 0 \quad and \quad \limsup_{|x| \to +\infty} \frac{v(x)}{|x|} < 0. \tag{3.34}$$

Let $\Phi \in C^2(H_N \times H_N)$ be such that

$$u(x_N + x_N^\perp) + v(y_N + y_N^\perp) - \Phi(x_N, y_N)$$

has a strict global maximum over $H \times H$ at a point $(\bar{x}, \bar{y})$. Then there exist functions $\varphi_k, \psi_k \in C^2(H)$ for $k = 1, 2, ...$ such that $\varphi_k, B^{-1} D\varphi_k, D^2\varphi_k, \psi_k, B^{-1} D\psi_k, D^2\psi_k$ are bounded and uniformly continuous, and such that

$$u(x) - \varphi_k(x)$$

has a global maximum at some point x_k,

$$v(y) - \psi_k(y)$$

has a global maximum at some point y_k, *and*

$$\left(x_k, u(x_k), D\varphi_k(x_k), D^2\varphi_k(x_k)\right) \xrightarrow{k \to +\infty} \left(\bar{x}, u(\bar{x}), D_x\Phi(\bar{x}_N, \bar{y}_N), X_N\right)$$
$$\text{in } H \times \mathbb{R} \times H_2 \times \mathcal{L}(H_{-1}, H_1), \quad (3.35)$$

$$\left(y_k, v(y_k), D\psi_k(y_k), D^2\psi_k(y_k)\right) \xrightarrow{k \to +\infty} \left(\bar{y}, v(\bar{y}), D_y\Phi(\bar{x}_N, \bar{y}_N), Y_N\right)$$
$$\text{in } H \times \mathbb{R} \times H_2 \times \mathcal{L}(H_{-1}, H_1), \quad (3.36)$$

where $X_N, Y_N \in S(H)$, $X_N = P_N^* X_N P_N$, $Y_N = P_N^* Y_N P_N$,

$$-\left(\frac{1}{\kappa} + \|C\|_{\mathcal{L}(H_{-1} \times H_{-1})}\right)\begin{pmatrix} BP_N & 0 \\ 0 & BP_N \end{pmatrix}$$
$$\leq \begin{pmatrix} X_N & 0 \\ 0 & Y_N \end{pmatrix} \leq \begin{pmatrix} B & 0 \\ 0 & B \end{pmatrix}(C + \kappa C^2) \quad (3.37)$$

and $C = D^2_{H_{-1} \times H_{-1}}\Phi(\bar{x}_N, \bar{y}_N)$.

We remark that in fact $\varphi_k, \psi_k \in C^2(H_{-1})$.

Proof Define

$$\tilde{u}(x_N) := \sup_{x_N^\perp \in Q_N H} u(x_N + x_N^\perp),$$

$$\tilde{v}(y_N) := \sup_{y_N^\perp \in Q_N H} v(y_N + y_N^\perp),$$

the partial sup-convolutions of u and v respectively, and let $\tilde{u}^*$ and $\tilde{v}^*$ be their upper semicontinuous envelopes (see Definition D.10). We remark that $\tilde{u}, \tilde{v}$ do not need to be upper semicontinuous (see [140]). Since $u + v - \Phi$ has a strict global maximum at $(\bar{x}, \bar{y})$ it easily follows that

$$\tilde{u}^*(x_N) + \tilde{v}^*(y_N) - \Phi(x_N, y_N) \quad (3.38)$$

has a strict global maximum over $H_N \times H_N$ at $(\bar{x}_N, \bar{y}_N)$. Moreover, we have $\tilde{u}^*(\bar{x}_N) = u(\bar{x})$, $\tilde{v}^*(\bar{y}_N) = v(\bar{y})$.

We can now apply the finite-dimensional maximum principle (see Theorem E.10, which is a particular case of Theorem 3.2 in [139]) when we consider H_N as a subspace of H_{-1}. (We recall that in H_N the topology of H_{-1} is equivalent to the topology of H.) Denote H_N with this topology by $\tilde{H}_N$. We also note that $\Phi \in C^2(\tilde{H}_N \times \tilde{H}_N)$ and thus we can consider it as a function in $C^2(H_{-1} \times H_{-1})$ by setting $\Phi(x, y) := \Phi(P_N x, P_N y)$.

Therefore there exist bounded functions $\varphi_k, \psi_k \in C^2(\tilde{H}_N)$ with bounded and uniformly continuous derivatives (which we can consider as functions in $C^2(H_{-1})$ by setting $\varphi_k(x) := \varphi_k(P_N x)$ and $\psi_k(y) := \psi_k(P_N y)$) such that $\tilde{u}^*(x_N) - \varphi_k(x_N)$ has a strict global maximum at some point x_N^k, $\tilde{v}^*(y_N) - \psi_k(y_N)$ has a strict global minimum at some point y_N^k, and such that

$$
\left(x_N^k, \tilde{u}^*(x_N^k), D_{H_{-1}}\varphi_k(x_N^k), D^2_{H_{-1}}\varphi_k(x_N^k) \right)
$$
$$
\xrightarrow{k \to \infty} \left(\bar{x}_N, u(\bar{x}), D_{H_{-1},x}\Phi(\bar{x}_N, \bar{y}_N), \tilde{X}_N \right), \quad (3.39)
$$

$$
\left(y_N^k, \tilde{v}^*(y_N^k), D_{H_{-1}}\psi_k(y_N^k), D^2_{H_{-1}}\psi_k(y_N^k) \right)
$$
$$
\xrightarrow{k \to \infty} \left(\bar{y}_N, v(\bar{y}), D_{H_{-1},y}\Phi(\bar{x}_N, \bar{y}_N), \tilde{Y}_N \right), \quad (3.40)
$$

and

$$
-\left(\frac{1}{\kappa} + \|C\|_{\mathcal{L}(H_{-1} \times H_{-1})} \right) \begin{pmatrix} P_N & 0 \\ 0 & P_N \end{pmatrix}
$$
$$
\leq \begin{pmatrix} \tilde{X}_N & 0 \\ 0 & \tilde{Y}_N \end{pmatrix} \leq C + \kappa C^2 \quad \text{in } H_{-1} \times H_{-1} \quad (3.41)
$$

for some $\tilde{X}_N, \tilde{Y}_N \in S(H_{-1})$ that satisfy $\tilde{X}_N = P_N \tilde{X}_N P_N$, $\tilde{Y}_N = P_N \tilde{Y}_N P_N$ as operators in $\mathcal{L}(H_{-1})$ and, since

$$
D_{H_{-1}}\varphi_k(x_N^k) = P_N D_{H_{-1}}\varphi_k(x_N^k), \quad D^2_{H_{-1}}\varphi_k(x_N^k) = P_N D^2_{H_{-1}}\varphi_k(x_N^k) P_N,
$$
$$
D_{H_{-1}}\psi_k(x_N^k) = P_N D_{H_{-1}}\psi_k(y_N^k), \quad D^2_{H_{-1}}\psi_k(y_N^k) = P_N D^2_{H_{-1}}\psi_k(y_N^k) P_N,
$$

and in H_N the topology of H_{-1} is equivalent to the topology of H, the convergences (3.39), (3.40) hold in $H \times \mathbb{R} \times H \times \mathcal{L}(H_{-1})$.

It is easy to see that

$$
D\varphi_k(x) = B D_{H_{-1}}\varphi_k(x), \quad D^2\varphi_k(x) = B D^2_{H_{-1}}\varphi_k(x),
$$
$$
D\psi_k(y) = B D_{H_{-1}}\psi_k(y), \quad D^2\psi_k(y) = B D^2_{H_{-1}}\psi_k(y).
$$

(Note that if $X \in S(H_{-1})$ then $BX \in S(H)$ since for $x, y \in H$ we have $\langle BXx, y \rangle = \langle Xx, y \rangle_{-1} = \langle x, Xy \rangle_{-1} = \langle x, BXy \rangle$.) Therefore, setting $X_N = B\tilde{X}_N$, $Y_N = B\tilde{Y}_N$, we obtain from (3.39) and (3.40) that

$$\left(x_N^k, \tilde{u}^*(x_N^k), D\varphi_k(x_N^k), D^2\varphi_k(x_N^k)\right) \xrightarrow{k\to\infty} \left(\bar{x}_N, u(\bar{x}), D_x\Phi(\bar{x}_N, \bar{y}_N), X_N\right)$$
$$\text{in } H \times \mathbb{R} \times H_2 \times \mathcal{L}(H_{-1}, H_1), \quad (3.42)$$

$$\left(y_N^k, \tilde{v}^*(y_N^k), D\psi_k(y_N^k), D^2\psi_k(y_N^k)\right) \xrightarrow{k\to\infty} \left(\bar{y}_N, v(\bar{y}), D_y\Phi(\bar{x}_N, \bar{y}_N), Y_N\right)$$
$$\text{in } H \times \mathbb{R} \times H_2 \times \mathcal{L}(H_{-1}, H_1), \quad (3.43)$$

$X_N = P_N^* X_N P_N, Y_N = P_N^* Y_N P_N$, and (3.37) is satisfied.

Now, using Corollary 3.26 and (3.34), for every k and j big enough we can find $p_j^k, q_j^k \in H$ such that $|p_j^k| + |q_j^k| \le 1/j$, and

$$u(x) - \varphi_k(x) - \langle Bp_j^k, x \rangle \quad \text{has a global maximum at some point } x_j^k, \quad (3.44)$$

and

$$v(y) - \psi_k(y) - \langle Bq_j^k, y \rangle \quad \text{has a global maximum at some point } y_j^k, \quad (3.45)$$

where, because of (3.34), $|x_j^k| + |y_j^k| \le R_k$ for some $R_k > 0$. Then if $|x|, |y| \le R$ for $R \ge R_k$

$$\tilde{u}^*((x_j^k)_N) + \tilde{v}^*((y_j^k)_N) - \varphi_k(x_j^k) - \psi_k(y_j^k)$$
$$\ge u(x_j^k) + v(y_j^k) - \varphi_k(x_j^k) - \psi_k(y_j^k)$$
$$\ge u(x) + v(y) - \varphi_k(x) - \psi_k(y) - \langle Bp_j^k, x - x_j^k \rangle - \langle Bq_j^k, y - y_j^k \rangle \quad (3.46)$$
$$\ge u(x) + v(y) - \varphi_k(x) - \psi_k(y) - \frac{4R\|B\|}{j}.$$

Since by (3.34) if j is big enough, $u(x) - \varphi_k(x) - \langle Bp_j^k, x - x_j^k \rangle \to -\infty$ as $|x| \to +\infty$, and $v(y) - \psi_k(y) - \langle Bq_j^k, y - y_j^k \rangle \to -\infty$ as $|y| \to +\infty$, choosing R big enough and taking suprema over $x_N^\perp, y_N^\perp$ in (3.46) and then envelopes at x_N^k, y_N^k we obtain for sufficiently big j that

$$\tilde{u}^*((x_j^k)_N) + \tilde{v}^*((y_j^k)_N) - \varphi_k(x_j^k) - \psi_k(y_j^k)$$
$$\ge u(x_j^k) + v(y_j^k) - \varphi_k(x_j^k) - \psi_k(y_j^k)$$
$$\ge \tilde{u}^*(x_N^k) + \tilde{v}^*(y_N^k) - \varphi_k(x^k) - \psi_k(y^k) - \frac{4R\|B\|}{j}. \quad (3.47)$$

Since $\tilde{u}^*(x_N) + \tilde{v}^*(y_N) - \varphi_k(x_N) - \psi_k(y_N)$ has a strict global maximum at (x_N^k, y_N^k), we deduce from (3.47) that

$$(x_j^k)_N \to x_N^k,\ (y_j^k)_N \to y_N^k, \quad \tilde{u}^*((x_j^k)_N) \to \tilde{u}^*(x_N^k), \quad \tilde{v}^*((y_j^k)_N) \to \tilde{v}^*(y_N^k)$$
(3.48)

as $j \to +\infty$ and then also

$$u(x_j^k) \to \tilde{u}^*(x_N^k), \quad v(y_j^k) \to \tilde{v}^*(y_N^k) \quad \text{as } j \to +\infty. \tag{3.49}$$

Using these and (3.42) and (3.43) we can therefore select a subsequence j_k such that

$$\left((x_{j_k}^k)_N, u(x_{j_k}^k), D\varphi_k(x_{j_k}^k), D^2\varphi_k(x_{j_k}^k)\right) \xrightarrow{k\to\infty} \left(\bar{x}_N, u(\bar{x}), D_x\Phi(\bar{x}_N, \bar{y}_N), X_N\right)$$
$$\text{in } H \times \mathbb{R} \times H_2 \times \mathcal{L}(H_{-1}, H_1),$$

$$\left((y_{j_k}^k)_N, v(y_{j_k}^k), D\psi_k(y_{j_k}^k), D^2\psi_k(y_{j_k}^k)\right) \xrightarrow{k\to\infty} \left(\bar{y}_N, v(\bar{y}), D_y\Phi(\bar{x}_N, \bar{y}_N), Y_N\right)$$
$$\text{in } H \times \mathbb{R} \times H_2 \times \mathcal{L}(H_{-1}, H_1).$$

It remains to show that $x_{j_k}^k \to \bar{x}$ and $y_{j_k}^k \to \bar{y}$. However, this is now obvious since

$$u(x_{j_k}^k) + v(x_{j_k}^k) - \Phi((x_{j_k}^k)_N, (y_{j_k}^k)_N) \to u(\bar{x}) + v(\bar{y}) - \Phi(\bar{x}_N, \bar{y}_N)$$

and by assumption this function has a strict global maximum at $(\bar{x}, \bar{y})$. Therefore the lemma holds with $\varphi_k(x) := \varphi_k(x) + \langle Bp_{j_k}^k, x\rangle$, $\psi_k(y) := \psi_k(y) + \langle Bq_{j_k}^k, y\rangle$ and $x_k := x_{j_k}^k$, $y_k := y_{j_k}^k$. □

Theorem 3.27 applied to $\Phi(x_N, y_N) = \frac{1}{2\varepsilon}|x_N - y_N|_{-1}^2$ for $\varepsilon > 0$ yields the following result which will be used in the proofs of comparison theorems.

Corollary 3.28 *Let $B \in S(H)$ be strictly positive and let $N \geq 2, \varepsilon > 0$. Let $u, -v : H \to \mathbb{R} \cup \{-\infty\}$ be B-upper semicontinuous functions bounded from above and satisfying (3.34). Let*

$$u(x_N + x_N^\perp) - v(y_N + y_N^\perp) - \frac{|x_N - y_N|_{-1}^2}{2\varepsilon}$$

have a strict global maximum over $H \times H$ at a point $(\bar{x}, \bar{y})$. Then there exist functions $\varphi_k, \psi_k \in C^2(H)$ for $k = 1, 2, ...$ such that $\varphi_k, B^{-1}D\varphi_k, D^2\varphi_k, \psi_k, B^{-1}D\psi_k, D^2\psi_k$ are bounded and uniformly continuous, and such that

$$u(x) - \varphi_k(x)$$

has a global maximum at some point x_k,

$$v(y) - \psi_k(y)$$

has a global minimum at some point y_k, and

$$\left(x_k, u(x_k), D\varphi_k(x_k), D^2\varphi_k(x_k)\right) \xrightarrow{k\to\infty} \left(\bar{x}, u(\bar{x}), \frac{B(\bar{x}_N - \bar{y}_N)}{\varepsilon}, X_N\right)$$

$$\text{in } H \times \mathbb{R} \times H_2 \times \mathcal{L}(H_{-1}, H_1), \quad (3.50)$$

$$\left(y_k, v(y_k), D\psi_k(y_k), D^2\psi_k(y_k)\right) \xrightarrow{k\to\infty} \left(\bar{y}, v(\bar{y}), \frac{B(\bar{x}_N - \bar{y}_N)}{\varepsilon}, Y_N\right)$$

$$\text{in } H \times \mathbb{R} \times H_2 \times \mathcal{L}(H_{-1}, H_1), \quad (3.51)$$

where $X_N = P_N^* X_N P_N$, $Y_N = P_N^* Y_N P_N$,

$$-\frac{3}{\varepsilon}\begin{pmatrix} BP_N & 0 \\ 0 & BP_N \end{pmatrix} \le \begin{pmatrix} X_N & 0 \\ 0 & -Y_N \end{pmatrix} \le \frac{3}{\varepsilon}\begin{pmatrix} BP_N & -BP_N \\ -BP_N & BP_N \end{pmatrix}. \quad (3.52)$$

Proof Observe that if $\Phi(x_N, y_N) = \frac{1}{2\varepsilon}|x_N - y_N|_{-1}^2$ then

$$C = D^2_{H_{-1}\times H_{-1}}\Phi(x_N, y_N) = \frac{1}{\varepsilon}\begin{pmatrix} P_N & -P_N \\ -P_N & P_N \end{pmatrix}$$

and thus $\kappa C^2 = \frac{2\kappa}{\varepsilon}C$ and $\|C\|_{\mathcal{L}(H_{-1}\times H_{-1})} = \frac{2}{\varepsilon}$. Then (3.52) follows from (3.37) choosing $\kappa = \varepsilon$. □

We remark that the convergence in $\mathcal{L}(H_{-1}, H_1)$ in particular implies convergence in $\mathcal{L}(H)$.

The time-dependent analogue of Corollary 3.28 is the following.

Corollary 3.29 *Let $B \in S(H)$ be strictly positive and let $N \ge 2, \varepsilon, \beta > 0$. Let $u, -v : (0, T) \times H \to \mathbb{R} \cup \{-\infty\}$ be B-upper semicontinuous functions bounded from above and satisfying*

$$\limsup_{|x|\to+\infty} \sup_{t\in(0,T)} \frac{u(t, x)}{|x|} < 0 \quad \text{and} \quad \limsup_{|x|\to+\infty} \sup_{t\in(0,T)} \frac{-v(t, x)}{|x|} < 0. \quad (3.53)$$

Let

$$u(t, x_N + x_N^\perp) - v(s, y_N + y_N^\perp) - \frac{|x_N - y_N|_{-1}^2}{2\varepsilon} - \frac{(t - s)^2}{2\beta}$$

have a strict global maximum over $(0, T) \times H \times (0, T) \times H$ at a point $(\bar{t}, \bar{x}, \bar{s}, \bar{y})$. Then there exist functions $\varphi_k, \psi_k \in C^2((0, T) \times H)$ for $k = 1, 2, \ldots$ such that $\varphi_k, (\varphi_k)_t, B^{-1}D\varphi_k, D^2\varphi_k, \psi_k, (\psi_k)_t, B^{-1}D\psi_k, D^2\psi_k$ are bounded and uniformly continuous, and such that

$$u(t, x) - \varphi_k(t, x)$$

has a global maximum at some point (t_k, x_k),

$$v(s, y) - \psi_k(s, y)$$

has a global minimum at some point (s_k, y_k), *and*

$$\left(t_k, x_k, u(t_k, x_k), (\varphi_k)_t(t_k, x_k), D\varphi_k(t_k, x_k), D^2\varphi_k(t_k, x_k)\right)$$
$$\xrightarrow[\mathbb{R} \times H \times \mathbb{R} \times \mathbb{R} \times H_2 \times \mathcal{L}(H_{-1}, H_1)]{k \to \infty} \left(\bar{t}, \bar{x}, u(\bar{t}, \bar{x}), \frac{\bar{t} - \bar{s}}{\beta}, \frac{B(\bar{x}_N - \bar{y}_N)}{\varepsilon}, X_N\right) \quad (3.54)$$

$$\left(s_k, y_k, v(s_k, y_k), (\psi_k)_t(s_k, y_k), D\psi_k(s_k, y_k), D^2\psi_k(s_k, y_k)\right)$$
$$\xrightarrow[\mathbb{R} \times H \times \mathbb{R} \times \mathbb{R} \times H_2 \times \mathcal{L}(H_{-1}, H_1)]{k \to \infty} \left(\bar{s}, \bar{y}, v(\bar{s}, \bar{y}), \frac{\bar{t} - \bar{s}}{\beta}, \frac{B(\bar{x}_N - \bar{y}_N)}{\varepsilon}, Y_N\right) \quad (3.55)$$

where $X_N = P_N^* X_N P_N, Y_N = P_N^* Y_N P_N$ *and they satisfy* (3.52).

Proof We can obviously extend u, v to $\mathbb{R} \times H$ preserving all the properties of the functions and the strict global maximum at $(\bar{t}, \bar{x}, \bar{s}, \bar{y})$. We consider the space $\tilde{H} = \mathbb{R} \times H$ and the operator $\tilde{B} := I_{\mathbb{R}} \times B$. Writing (t, x) for elements of this extended space we now consider the function $\Phi(t, x_N, s, y_N) = \frac{1}{2\varepsilon}|x_N - y_N|_{-1}^2 + \frac{1}{2\beta}(t - s)^2$. We now rescale time by setting

$$\tilde{u}(t, x) = u\left(\left(\frac{\beta}{\varepsilon}\right)^{\frac{1}{2}} t, x\right), \quad \tilde{v}(s, y) = u\left(\left(\frac{\beta}{\varepsilon}\right)^{\frac{1}{2}} s, y\right),$$

$$\tilde{\Phi}(t, x_N, s, y_N) = \Phi\left(\left(\frac{\beta}{\varepsilon}\right)^{\frac{1}{2}} t, x_N, \left(\frac{\beta}{\varepsilon}\right)^{\frac{1}{2}} s, y_N\right) = \frac{|x_N - y_N|_{-1}^2}{2\varepsilon} + \frac{(t - s)^2}{2\varepsilon}.$$

Then

$$\tilde{u}(t, x) - \tilde{v}(s, y) - \tilde{\Phi}(t, x_N, s, y_N)$$

has a strict global maximum over $\tilde{H} \times \tilde{H}$ at the point $((\frac{\varepsilon}{\beta})^{\frac{1}{2}}\bar{t}, \bar{x}, (\frac{\varepsilon}{\beta})^{\frac{1}{2}}\bar{s}, \bar{y})$. We can now apply Corollary 3.28 to produce the required functions φ_k, ψ_k. We now have operators $\tilde{X}_N, \tilde{Y}_N$ satisfying a version of (3.52) on $\tilde{H} \times \tilde{H}$. However, it is easy to see that its restriction to $\{0\} \times H \times \{0\} \times H$ gives (3.52). The claim follows after rescaling time back to the original variables which will only change the time derivatives of φ_k, ψ_k. □

Remark 3.30 A different type of time-dependent maximum principle can be obtained which relies on the finite-dimensional parabolic maximum principle presented in Theorem E.11. Such a result can be found in [140], Theorem 3.2, however it is stated there in a version which uses second-order parabolic jets and is applicable to equations with bounded terms (see Sect. 3.3.1). Theorem 3.2 in [140] also imposes an

additional condition on the functions, which is satisfied when they are viscosity sub-
and supersolutions of bounded time-dependent second-order equations in a Hilbert
space. The maximum principle stated in Corollary 3.29 does not impose extra conditions and thus it is more universal and can be applied more easily, which is why
we prefer it here. However, the other maximum principle has certain advantages. For
instance, one can use it to prove comparison principles for bounded time-dependent
second-order equations in a Hilbert space without the assumption that the viscosity
subsolutions and supersolutions attain the initial/terminal values locally uniformly.
Such results can be found in [378]. This type of maximum principle was also implicitly used in [537, 538].   ∎

3.3   Viscosity Solutions

Throughout this section U is an open subset of H and the operator B satisfies the
following assumption (see Sect. 3.1.1).

Hypothesis 3.31 $B \in S(H)$ is a strictly positive operator such that A^*B is bounded.

Contrary to the finite-dimensional case, in infinite dimension there is no one
universal definition of viscosity solution. The basic idea of using a pointwise maximum principle and replacing nonexistent derivatives of a solution by derivatives
of test functions is still the same. However, because of the presence of unbounded
terms and operators, the choice of test functions and the interpretation of unbounded
terms often must be adjusted for different types of equations. In this section we
present a generic definition of viscosity solution for a general class of stationary equations and time-dependent Cauchy problems. The solutions defined below are called
B-continuous viscosity solutions.

We consider the following boundary and terminal boundary value problems:

$$\begin{cases} -\langle Ax, Du \rangle + F(x, u, Du, D^2u) = 0 & \text{in } U \\ u(x) = f(x) & \text{on } \partial U \end{cases} \tag{3.56}$$

and

$$\begin{cases} u_t - \langle Ax, Du \rangle + F(t, x, u, Du, D^2u) = 0 & \text{in } (0, T) \times U \\ u(0, x) = g(x) & \text{for } x \in U, \\ u(t, x) = f_1(t, x) & \text{for } (t, x) \in (0, T) \times \partial U, \end{cases} \tag{3.57}$$

where $F : (0, T) \times U \times \mathbb{R} \times H \times S(H) \to \mathbb{R}$, $g : U \to \mathbb{R}$, $f : \partial U \to \mathbb{R}$, and
$f_1 : (0, T) \times \partial U \to \mathbb{R}$ are continuous.

Definition 3.32 A function ψ is a test function if $\psi = \varphi + h(t, |x|)$, where:

(i) $\varphi \in C^{1,2}((0, T) \times U)$ is locally bounded, and is such that φ is B-lower semicontinuous, and φ_t, $A^* D\varphi$, $D\varphi$, $D^2\varphi$ are uniformly continuous on $(0, T) \times U$.

(ii) $h \in C^{1,2}((0, T) \times \mathbb{R})$ and is such that for every $t \in (0, T)$, $h(t, \cdot)$ is even and $h(t, \cdot)$ is non-decreasing on $[0, +\infty)$.

For stationary equations φ and h are independent of t.

We remark that even though $|x|$ is not differentiable at 0, the function $h(t, |x|) \in C^{1,2}((0, T) \times H)$. The requirement that φ_t, $A^* D\varphi$, $D\varphi$, $D^2\varphi$ are uniformly continuous (and hence grow at most linearly at infinity) is a little arbitrary. It can be replaced by a requirement that they are locally uniformly continuous and have some prescribed growth at infinity, for instance at most polynomial. The growth restriction can also be removed, however it is useful in applications to stochastic optimal control since it is not clear if one can modify a test function φ outside a fixed set so that the modification has a required growth at infinity, while preserving the property that $A^* D\varphi$ is continuous. Thus the radial part h of test functions, which can be modified at will, plays the role of a cut-off function which takes care of the growth at infinity. We can thus require φ to be as nice as we want as long as our choice gives us enough test functions which are needed to build a good theory. The requirement that $A^* D\varphi$ is uniformly continuous can also be replaced by a requirement that $B^{-1} D\varphi$ is continuous. This, however, would make the definition more dependent on the choice of B. The reader can experiment with various modifications of the above definition and we will later see how the choice of test functions must be adjusted to particular cases.

Definition 3.33 A locally bounded B-upper semicontinuous (see Definition 3.3) function u on $\overline{U}$ is a viscosity subsolution of (3.56) if $u \leq f$ on ∂U and whenever $u - \psi$ has a local maximum at a point x for a test function $\psi = \varphi + h(|x|)$ then

$$- \langle x, A^* D\varphi(x) \rangle + F(x, u(x), D\psi(x), D^2\psi(x)) \leq 0. \qquad (3.58)$$

A locally bounded B-lower semicontinuous function u on $\overline{U}$ is a viscosity supersolution of (3.56) if $u \geq f$ on ∂U and whenever $u + \psi$ has a local minimum at a point x for a test function $\psi = \varphi + h(|x|)$ then

$$\langle x, A^* D\varphi(x) \rangle + F(x, u(x), -D\psi(x), -D^2\psi(x)) \geq 0. \qquad (3.59)$$

A viscosity solution of (3.56) is a function which is both a viscosity subsolution and a viscosity supersolution of (3.56).

Definition 3.34 A locally bounded B-upper semicontinuous function u on $[0, T) \times \overline{U}$ is a viscosity subsolution of (3.57) if $u(0, y) \leq g(y)$ for $y \in U$, $u \leq f_1$ on $(0, T) \times \partial U$ and whenever $u - \psi$ has a local maximum at a point $(t, x) \in (0, T) \times U$ for a test function $\psi(s, y) = \varphi(s, y) + h(s, |y|)$ then

$$\psi_t(t, x) - \langle x, A^* D\varphi(t, x) \rangle + F(t, x, u(t, x), D\psi(t, x), D^2\psi(t, x)) \le 0. \quad (3.60)$$

A locally bounded B-lower semicontinuous function u on $[0, T) \times \overline{U}$ is a viscosity supersolution of (3.57) if $u(0, y) \ge g(y)$ for $y \in U$, $u \ge f_1$ on $(0, T) \times \partial U$ and whenever $u + \psi$ has a local minimum at a point $(t, x) \in (0, T) \times U$ for a test function $\psi(s, y) = \varphi(s, y) + h(s, |y|)$ then

$$- \psi_t(t, x) + \langle x, A^* D\varphi(t, x) \rangle + F(t, x, u(t, x), -D\psi(t, x), -D^2\psi(t, x)) \ge 0. \quad (3.61)$$

A viscosity solution of (3.57) is a function which is both a viscosity subsolution and a viscosity supersolution of (3.57).

The main idea behind this definition of solution is the following. Test functions are split into two categories. Good test functions φ provide enough functions to apply the doubling argument in the proof of comparison and produce maxima and minima using perturbed optimization by functions in this class. Radial functions h are needed as cut-off functions to be able to produce local/global maxima and minima and to confine the region of their possible locations. As always in the theory of viscosity solutions, non-existing derivatives of u are replaced by existing derivatives of test functions. The term $\langle Ax, D\varphi(t, x) \rangle$ is interpreted as $\langle x, A^* D\varphi(t, x) \rangle$. We cannot do the same with the term $\langle Ax, Dh(t, |x|) \rangle$ since $Dh(t, |x|) = h_r(t, |x|) \frac{x}{|x|}$ (where h_r is the partial derivative of h with respect to the second variable) and we cannot hope in general that $x \in D(A^*)$ (nor that $x \in D(A)$). Therefore this term is dropped. This can be done effectively since the term $\frac{h_r(t, |x|)}{|x|} \langle Ax, x \rangle$ (or $\frac{h_r(t, |x|)}{|x|} \langle A^*x, x \rangle$) would be non-positive if it were well defined. Thus the definition is consistent with what the definition of viscosity solution should be under ideal conditions.

In applications to control problems it is more natural to work with terminal value problems instead of initial value problems. A terminal value problem can be converted into an initial value problem by a change of variable $\tilde{t} := T - t$. Thus a terminal boundary value problem corresponding to (3.57) is

$$\begin{cases} u_t + \langle Ax, Du \rangle - F(t, x, u, Du, D^2u) = 0 & \text{in } (0, T) \times U \\ u(T, x) = g(x) & \text{for } x \in U, \\ u(t, x) = f(t, x) & \text{for } (t, x) \in (0, T) \times \partial U, \end{cases} \quad (3.62)$$

where $f(t, x) = f_1(T - t, x)$. Since we will be working with terminal value problems we state below the definition of a viscosity solution adapted to this case, which is a consequence of Definition 3.34. (We keep the minus sign in front of the Hamiltonian F since we will formulate the conditions for F that will apply to both stationary and time-dependent terminal value problems.)

Definition 3.35 A locally bounded B-upper semicontinuous function u on $(0, T] \times \overline{U}$ is a viscosity subsolution of (3.62) if $u(T, y) \le g(y)$ for $y \in U$, $u \le f$ on $(0, T) \times \partial U$ and whenever $u - \psi$ has a local maximum at a point $(t, x) \in (0, T) \times U$ for a test function $\psi(s, y) = \varphi(s, y) + h(s, |y|)$ then

$$\psi_t(t, x) + \langle x, A^*D\varphi(t, x) \rangle - F(t, x, u(t, x), D\psi(t, x), D^2\psi(t, x)) \geq 0. \quad (3.63)$$

A locally bounded B-lower semicontinuous function u on $(0, T] \times \overline{U}$ is a viscosity supersolution of (3.62) if $u(T, y) \geq g(y)$ for $y \in U$, $u \geq f$ on $(0, T) \times \partial U$ and whenever $u + \psi$ has a local minimum at a point $(t, x) \in (0, T) \times U$ for a test function $\psi(s, y) = \varphi(s, y) + h(s, |y|)$ then

$$- \psi_t(t, x) - \langle x, A^*D\varphi(t, x) \rangle - F(t, x, u(t, x), -D\psi(t, x), -D^2\psi(t, x)) \leq 0.$$
$$(3.64)$$

A viscosity solution of (3.62) is a function which is both a viscosity subsolution and a viscosity supersolution of (3.62).

Remark 3.36 It is easy to see that if u is a viscosity subsolution (respectively, super-solution) of (3.62) on $(0, T] \times \overline{U}$ then it is a viscosity subsolution (respectively, supersolution) of (3.62) on $(T_1, T] \times \overline{U}$ for every $0 < T_1 < T$. ∎

Lemma 3.37 *Without loss of generality the maxima and minima in Definitions 3.33–3.35 can be assumed to be global and strict.*

Proof We will only show this in the case of a subsolution in Definition 3.33 as the other cases are similar. Let

$$u(x) - \varphi(x) - h(|x|) \geq u(y) - \varphi(y) - h(|y|) \quad \text{for } y \in B_R(x) \subset U$$

for some $R > 0$.

We will show that there exist test functions $\tilde{\varphi}$ and $\tilde{h}(|\cdot|)$ such that $D\tilde{\varphi}(x) = D\varphi(x)$, $D^2\tilde{\varphi}(x) = D^2\varphi(x)$, $D\tilde{h}(|x|) = Dh(|x|)$, $D^2\tilde{h}(|x|) = D^2h(|x|)$, and $u - \tilde{\varphi} - \tilde{h}(|\cdot|)$ has a strict global maximum at x. Let $\eta \in C^2([0, \infty))$ be an increasing function such that

$$r + \sup_{|y| \leq r, \, y \in U} \{|u(y)| + |\varphi(y)|\} \leq \eta(r).$$

Let $g_1 \in C^2((0, \infty))$ be a function such that

$$g_1(r) = \begin{cases} 0 & \text{if } r \leq |x| \\ (r - |x|)^4 & \text{if } |x| < r < |x| + 1 \\ \text{increasing} & \text{if } |x| + 1 \leq r \leq |x| + 2 \\ \eta(r) & \text{if } r > |x| + 2. \end{cases}$$

Let $\varphi_1 \in C^2([0, \infty))$ be defined by

$$\varphi_1(r) = \begin{cases} r^4 & r \leq 1, \\ \text{increasing} & 1 < r < 2, \\ 2 & r \geq 2. \end{cases}$$

Now for $n \geq 1$ consider the function

$$\Phi_n(y) = u(y) - \varphi(y) - n\varphi_1(|x - y|_{-1}) - h(|y|) - g_1(|y|).$$

Obviously we have

$$\Phi_n(x) = u(x) - \varphi(x) - h(|x|).$$

Suppose there is a subsequence $n_k \to +\infty$ and $y_{n_k} \in U$, $y_{n_k} \neq x$ such that $\Phi_{n_k}(y_{n_k}) \geq \Phi_{n_k}(x)$. Then we must have $|x - y_{n_k}|_{-1} \to 0$ as $k \to \infty$ and $|y_{n_k}| \leq C_1$ for some $C_1 > 0$, i.e. $y_{n_k} \to x$ and $By_{n_k} \to Bx$. Since $u, -\varphi$ are B-upper semi-continuous, and $h + g_1$ is increasing on $[|x|, +\infty)$, this implies that $|y_n| \to |x|$, and therefore $y_{n_k} \to x$ in H. But then $y_{n_k} \in B_R(x)$ for big k and so we get

$$\Phi_{n_k}(y_{n_k}) < u(y_{n_k}) - \varphi(y_{n_k}) - h(|y_{n_k}|) \leq u(x) - \varphi(x) - h(|x|),$$

which is a contradiction. Therefore there must exist $\bar{n}$ such that $\Phi_{\bar{n}}(y) < \Phi_{\bar{n}}(x)$ for all $y \in U, y \neq x$. It then easily follows that $\Phi_{\bar{n}+1}$ has a strict global maximum at x. Therefore the conclusion follows with $\tilde{\varphi}(y) = \varphi(y) + (\bar{n} + 1)\varphi_1(|x - y|_{-1})$ and $\tilde{h}(|y|) = h(|y|) + g_1(|y|)$. $\square$

It follows from the proof of Lemma 3.37 that if we know a priori that u has certain growth at ∞ (at least quadratic) and $U = H$, we can then obtain the same growth for $\tilde{h}$. (We notice that if $U = H$ then φ has at most quadratic growth at infinity.) For instance, if u has a polynomial growth at ∞ we can have $\tilde{h}$ which is a polynomial of some special form for big $|x|$. This can be important in applications to stochastic optimal control where we may want to impose additional conditions on test functions to be able to apply stochastic calculus. In these applications it may also be useful to assume that $h'(r)/r$ is globally positive for the radial test functions h. To avoid technical difficulties it may then be more convenient to choose h belonging to one particular class of functions, say certain polynomials with growth depending on the growth of sub- and supersolutions we are dealing with. However, when using such narrow classes of radial test functions one may be forced to require that the maxima and minima in the definition of viscosity solution be global as the definitions using global and local maxima and minima may no longer be equivalent.

Remark 3.38 We assumed in Definitions 3.33–3.35 that A was a linear, densely defined, maximal dissipative operator in H, i.e. that it generated a C_0-semigroup of contractions e^{tA}. The definitions can be used to cover the case when $A - \omega I$ is maximal dissipative for some $\omega > 0$, i.e. if

$$\|e^{tA}\| \leq e^{\omega t} \quad \text{for all } t \geq 0. \tag{3.65}$$

One way to do this is to replace A by $\tilde{A} = A - \omega I$ and F by $\tilde{F}(t, x, r, p, X) = F(t, x, r, p, X) - \omega\langle x, p\rangle$. Another way is by making a change of variables $\tilde{u}(t, x) = u(t, e^{\omega t}x)$ in the equation which reduces Eq. (3.62) to an equation with A replaced by $A - \omega I$ and F replaced by $\tilde{F}(t, x, r, p, X) = F(t, e^{\omega t}x, r, e^{-\omega t}p, e^{-2\omega t}X)$. ∎

Lemma 3.39 *Let $F : [0, T) \times U \times \mathbb{R} \times H \times S(H) \to \mathbb{R}$ be continuous. Definition 3.35 is equivalent to the definition in which we only require that the maxima and minima be one-sided (i.e. that $u \mp \psi$ has a local maximum/minimum at a point (t, x) restricted to $[t, T) \times U$), if we also require that the subsolutions and supersolutions are continuous. In particular, the equation is also satisfied at $t = 0$ if we in addition require in the case when a one-sided maximum/minimum is attained at $(0, x)$ that the test functions $\varphi \in C^{1,2}([0, T) \times U)$ and $h \in C^{1,2}([0, T) \times \mathbb{R})$.*

Proof Suppose that u is a viscosity subsolution of (3.62) in the sense of Definition 3.35 and let $u(s, y) - \varphi(s, y) - h(s, |y|) = u(s, y) - \psi(s, y)$ have a one-sided local maximum at (t, x) over $[t, t+\varepsilon) \times B_\varepsilon(x)$. Arguing as in the proof of Lemma 3.37, we can assume that the one-sided local maximum is strict. Then for big n there exist, by Corollary 3.26 (which we can apply because any closed convex subset of H is B-closed, see Remark 3.8), $a_n \in \mathbb{R}$, $p_n \in H$, $|a_n| + |p_n| \leq 1/n$ such that the function

$$u(s, y) - \psi(s, y) - \frac{1}{n(s - t)} - a_n s - \langle Bp_n, y \rangle$$

has a local (two-sided) maximum at $(s_n, y_n) \in (t, t + \varepsilon) \times B_\varepsilon(x)$. Since the initial local maximum was strict we have that $(s_n, y_n) \to (t, x)$. Without loss of generality we can assume that $\frac{1}{n(s-t)}$ is a test function by modifying it around $s = t$ and then extending it to $(0, T)$. Therefore we obtain, using Definition 3.35, that

$$\psi_t(s_n, y_n) + a_n - \frac{1}{n(s_n - t)^2} + \langle y_n, A^*(D\varphi(s_n, y_n) + Bp_n) \rangle$$
$$- F(s_n, y_n, u(s_n, y_n), D\psi(s_n, y_n) + Bp_n, D^2\psi(s_n, y_n)) \geq 0,$$

which gives us

$$\psi_t(t, x) + \langle x, A^* D\varphi(t, x) \rangle - F(t, x, u(t, x), D\psi(t, x), D^2\psi(t, x)) \geq 0$$

after letting $n \to +\infty$. □

3.3.1 Bounded Equations

If $A = 0$ there is no need to use the notion of B-continuity. Viscosity solutions can then be defined in the same way as for finite-dimensional problems. We present the definition for the time-independent problem

$$\begin{cases} F(x, u, Du, D^2u) = 0 & \text{in } U \\ u(x) = f(x) & \text{on } \partial U. \end{cases} \tag{3.66}$$

The definition for time-dependent problems is similar. We call such equations "bounded" since they do not contain any unbounded terms.

Definition 3.40 A locally bounded upper semicontinuous function u on $\overline{U}$ is a viscosity subsolution of (3.66) if $u \leq f$ on ∂U and whenever $u - \varphi$ has a local maximum at a point x for a test function $\varphi \in C^2(U)$ then

$$F(x, u(x), D\varphi(x), D^2\varphi(x)) \leq 0.$$

A locally bounded lower semicontinuous function u on $\overline{U}$ is a viscosity supersolution of (3.66) if $u \geq f$ on ∂U and whenever $u - \varphi$ has a local minimum at a point x for a test function $\varphi \in C^2(U)$ then

$$F(x, u(x), D\varphi(x), D^2\varphi(x)) \geq 0.$$

A viscosity solution of (3.66) is a function which is both a viscosity subsolution and a viscosity supersolution of (3.66).

Equation (3.66) and its parabolic version were studied, together with the associated control problems, in [410, 412]. In [410] a stronger definition of viscosity solution was introduced, allowing for more general test functions which are not necessarily twice Fréchet differentiable. One can also replace Definition 3.40 with a definition using second-order jets (see [139, 412]). Regularity results for bounded equations and their obstacle problems have been obtained in [410, 542]. Existence and uniqueness results for such equations can also be found in [378].

3.4   Consistency of Viscosity Solutions

The consistency property of viscosity solutions, i.e. the ability to pass to limits in the equations under minimal assumptions on the solutions, is one of the greatest strengths of the notion of viscosity solution.

Let B be an operator satisfying Hypothesis 3.31. Let $A_n, n = 1, 2, \ldots$, be linear, densely defined, maximal dissipative operators in H such that $D(A^*) \subset D(A_n^*)$. We consider equations

$$u_t + \langle A_n x, Du \rangle - F_n(t, x, u, Du, D^2u) = 0 \quad \text{in } (0, T) \times U, \qquad (3.67)$$

where U is an open subset of H. We assume that viscosity sub- and supersolutions of (3.67) are B-upper (respectively, lower) semicontinuous with the same fixed B. We note that if φ is a test function in Definition 3.32-(i), then

$$A_n^* D\varphi = A_n^*(I - A^*)^{-1}(I - A^*)D\varphi,$$

and thus, since $A_n^*(I - A^*)^{-1} \in \mathcal{L}(H)$, φ is a test function of type (i) for Eq. (3.67).

We have the following result.

Theorem 3.41 (Consistency of viscosity solutions) *Let the above assumptions about $A_n, n = 1, 2, ...,$ be satisfied. Let $u_n, n = 1, 2, ...,$ be viscosity subsolutions (respectively, supersolutions) of (3.67) (with some terminal and boundary conditions which are not essential here). Suppose that $F_n : (0, T) \times U \times \mathbb{R} \times H \times S(H) \to \mathbb{R}, n = 1, 2, ...$ are continuous, and*

$$if\ x, x_n \in D(A^*),\ x_n \to x,\ and\ A^* x_n \to A^* x,\ then\ A_n^* x_n \to A^* x. \quad (3.68)$$

Let u_n converge locally uniformly to a function u on $(0, T) \times U$. Then u is a viscosity subsolution of

$$u_t + \langle Ax, Du \rangle - F_-(t, x, u, Du, D^2 u) = 0\ \ in\ (0, T) \times U$$

(respectively, supersolution of

$$u_t + \langle Ax, Du \rangle - F^+(t, x, u, Du, D^2 u) = 0\ \ in\ (0, T) \times U),$$

where

$$F_-(t, x, r, p, X) = \lim_{i \to +\infty} \inf \Big\{ F_n(\tau, y, s, q, Y) : n \geq i,$$
$$|t - \tau| + |x - y| + |r - s| + |p - q| + \|X - Y\| \leq \frac{1}{i} \Big\}, \quad (3.69)$$

$$F^+(t, x, r, p, X) = \lim_{i \to +\infty} \sup \Big\{ F_n(\tau, y, s, q, Y) : n \geq i,$$
$$|t - \tau| + |x - y| + |r - s| + |p - q| + \|X - Y\| \leq \frac{1}{i} \Big\}. \quad (3.70)$$

Notation 3.42 We denote the right-hand side of (3.69) and (3.70) respectively by

$$\liminf_{n \to +\infty} {}_* F_n(t, x, r, p, X),$$

and

$$\limsup_{n \to +\infty} {}^* F_n(t, x, r, p, X).$$

Obviously

$$\limsup_{n \to +\infty} {}^*(-F_n(t, x, r, p, X)) = -\liminf_{n \to +\infty} {}_* F_n(t, x, r, p, X). \quad \blacksquare$$

Proof We will only prove the theorem for the subsolution case. The function u is obviously locally bounded and B-upper semicontinuous. Suppose that $u(s, y) - \psi(s, y) = u(s, y) - \varphi(s, y) - h(s, |y|)$ has a local maximum at a point (t, x). By Lemma 3.37 the maximum can be assumed to be strict. Let $D = \{(s, y) : |t - s| \le \delta, |x - y| \le \delta\}$ for some $\delta > 0$. Applying Corollary 3.26 on D we obtain, for every n, $a_n \in \mathbb{R}$, $p_n \in H$ such that $|a_n| + |p_n| \le \frac{1}{n}$ and such that

$$u_n(s, y) - (\psi(s, y) + a_n s + \langle Bp_n, y \rangle)$$

has maximum over D at some point (t_n, x_n). Since the original maximum at (t, x) was strict and the u_n converge uniformly on D to u, it is easy to see that we must have $(t_n, x_n) \to (t, x)$ as $n \to +\infty$. Since we have

$$\psi_t(t_n, x_n) + a_n + \langle x_n, A_n^*(D\varphi(t_n, x_n) + Bp_n) \rangle$$
$$- F_n(t_n, x_n, u_n(t_n, x_n), D\psi(t_n, x_n) + Bp_n, D^2\psi(t_n, x_n)) \ge 0,$$

the claim follows passing to the $\limsup^*_{n \to +\infty}$ in the above inequality. □

The result for time-independent equations is similar. In finite-dimensional spaces one can pass to weaker limits with viscosity solutions. In particular, the method of half-relaxed limits of Barles–Perthame (see [40, 139, 263]) allows us to conclude that for a family of subsolutions (respectively, supersolutions) u_n, the function $u^+ = \limsup^*_{n \to +\infty} u_n$ is a subsolution and $u_- = \liminf_{n \to +\infty} {}_* u_n$ is a supersolution. Unfortunately this is no longer true in infinite dimension due to the lack of local compactness. The following simple example from [540] illustrates this phenomenon. Half-relaxed limits in a special case are discussed in Sect. 3.9.

Example 3.43 Let H be the real l^2 space. Let $F(p) = 1 - |p|$, and $u_n(x) = x_n$, where $x = (x_1, ..., x_n, ...)$. Then the functions u_n are classical (and thus viscosity) solutions of

$$F(Du_n) = 0.$$

However $u^+ \equiv 0$ and therefore $F(Du^+) = 1$, i.e. u^+ is not a subsolution of $F(Du^+) = 0$. To see that $u^+ \equiv 0$ we observe that if $n \ge i$ and $|x - y| \le 1/i$ then

$$|u_n(y)| = |y_n| \le |x_n| + \frac{1}{i} \to 0 \quad \text{as } n, i \to \infty. \qquad ■$$

3.5 Comparison Theorems

In this section we present comparison results for viscosity solutions. They are proved under either the weak or the strong B-condition for A (see Definitions 3.9 and 3.10).

We use the notation from Sect. 3.2. In particular, we recall that $\{e_1, e_2, ...\}$ is an orthonormal basis in H_{-1} made of elements of H, and for $N > 2$, $H_N = \text{span}\{e_1, ..., e_N\}$, P_N is the orthogonal projection in H_{-1} onto H_N, and $Q_N := I - P_N$.

We will make the following assumptions about the function $F : (0, T) \times U \times \mathbb{R} \times H \times S(H) \to \mathbb{R}$.

Hypothesis 3.44 F is uniformly continuous on bounded subsets of $(0, T) \times U \times \mathbb{R} \times H \times S(H)$.

Hypothesis 3.45 There exists a $\nu \geq 0$ such that for every $(t, x, p, X) \in (0, T) \times U \times H \times S(H)$

$$F(t, x, r, p, X) - F(t, x, s, p, X) \geq \nu(r - s) \quad \text{when } r \geq s.$$

Hypothesis 3.46 For every $(t, x, r, p) \in (0, T) \times U \times \mathbb{R} \times H$

$$F(t, x, r, p, X) \geq F(t, x, r, p, Y) \quad \text{when } X \leq Y.$$

Hypothesis 3.47 For all $t \in (0, T), r \in \mathbb{R}, x \in U, p \in H, R > 0$,

$$\sup \left\{ |F(t, x, p, X + \lambda B Q_N) - F(t, x, p, X)| : \right.$$

$$\left. \|X\|, |\lambda| \leq R, X = P_N^* X P_N \right\} \xrightarrow{N \to +\infty} 0.$$

Hypothesis 3.48 For every $R > 0$ there exists a modulus ω_R such that, for all $(t, x, y, r) \in (0, T) \times U \times U \times \mathbb{R}$ such that $|r|, |x|, |y| \leq R$, for any $\varepsilon > 0$, for all $X, Y \in S(H)$ such that $X = P_N^* X P_N, Y = P_N^* Y P_N$ for some N and satisfying (3.52), we have

$$F\left(t, x, r, \frac{B(x - y)}{\varepsilon}, X\right) - F\left(t, y, r, \frac{B(x - y)}{\varepsilon}, Y\right)$$

$$\geq -\omega_R \left(|x - y|_{-1} \left(1 + \frac{|x - y|_{-1}}{\varepsilon}\right)\right).$$

Hypothesis 3.49 There exist $\gamma \in [0, 1]$ and a constant $M_F \geq 0$ such that

$$|F(t, x, r, p + q, X + Y) - F(t, x, r, p, X)|$$
$$\leq M_F \left((1 + |x|)|q| + (1 + |x|^\gamma)^2 \|Y\|\right)$$

for all $(t, x, r) \in (0, T) \times U \times \mathbb{R}$, $p, q \in H$, $X, Y \in S(H)$.

Hypothesis 3.45 guarantees that F is nondecreasing in the zeroth order variable. If $\nu > 0$ we say that F is proper. Hypothesis 3.46 ensures that F is monotone in the second-order variable. When it is satisfied we say that F (and therefore the equation) is degenerate elliptic/parabolic.

3.5.1 Degenerate Parabolic Equations

In Theorem 3.50, the boundary and terminal value functions f and g are not explicitly mentioned since they are not relevant. However, the subsolution function u and the supersolution function v are defined on $(0, T] \times \overline{U}$ and conditions (3.71) and (3.72) describe their joint behavior along the boundary ∂U and the terminal value T.

Theorem 3.50 (Comparison under weak B-condition) *Let $U \subset H$ be open and $\overline{U}$ be B-closed (see Definition 3.7). Let (3.2) hold and let F satisfy Hypotheses 3.44, 3.46–3.49 and 3.45 with $\nu = 0$. Let u be a viscosity subsolution of (3.62), and v be a viscosity supersolution of (3.62). Suppose that for every $R > 0$ there exists a modulus $\tilde{\omega}_R$ such that*

$$(u(t, x) - v(s, y))^+ + (u(t, y) - v(s, x))^+ \leq \tilde{\omega}_R(|t - s| + |x - y|_{-1}) \quad (3.71)$$

for $t, s \in (0, T)$, $x \in \partial U$, $y \in \overline{U}$, $|x|, |y| \leq R$, and that

$$\lim_{R \to +\infty} \lim_{r \to 0} \lim_{\eta \to 0} \sup \Big\{ u(t, x) - v(s, y) : |x - y|_{-1} < r$$
$$x, y \in \overline{U} \cap B_R, T - \eta \leq t, s \leq T \Big\} \leq 0. \quad (3.72)$$

Moreover, suppose that there exist constants $C, a > 0$ such that

$$u, -v \leq Ce^{a|x|^{2-2\gamma}} \quad (t, x) \in (0, T) \times H, \quad \text{if } \gamma \in [0, 1), \quad (3.73)$$

and

$$u, -v \leq Ce^{a(\log(1+|x|))^2} \quad (t, x) \in (0, T) \times H, \quad \text{if } \gamma = 1. \quad (3.74)$$

Then for every $\kappa > 0$

$$\lim_{R \to +\infty} \lim_{r \to 0} \lim_{\eta \to 0} \sup \Big\{ u(t, x) - v(s, y) : |x - y|_{-1} < r, |t - s| < \eta$$
$$x, y \in \overline{U} \cap B_R, \kappa < t, s \leq T \Big\} \leq 0. \quad (3.75)$$

In particular, $u \le v$.

Remark 3.51 It is easy to see that (3.75) implies that for every $\kappa > 0$ and $R > 0$ there exists a modulus $\tilde{\omega}_{\kappa,R}$ such that

$$u(t, x) - v(s, y) \le \tilde{\omega}_{\kappa,R}(|x - y|_{-1} + |t - s|) \quad \text{for } x, y \in \overline{U} \cap B_R, \kappa < t, s \le T.$$
(3.76)

∎

Proof of Theorem 3.50. We will prove the theorem for the case $\gamma = 1$.

Let $0 < \tau < 1$ be such that $a < 1/\sqrt{\tau}$. Additional conditions on τ will be given later. Set $T_1 = T - \tau$. The proof will be done in several steps. We will first show (3.75) on $[T_1, T]$, i.e. when we have $T_1 + \kappa \le t, s \le T$ in (3.75), and then reapply the procedure to intervals $[T - 3\tau/2, T - \tau/2], [T - 4\tau/2, T - 2\tau/2],....$

We argue by contradiction and assume that (3.75) is not true. Then there is a $\kappa > 0$ such that

$$m = \lim_{R \to +\infty} \lim_{r \to 0} \lim_{\eta \to 0} \sup \Big\{ u(t, x) - v(s, y) : |x - y|_{-1} < r, |t - s| < \eta$$
$$x, y \in \overline{U} \cap B_R, \, T_1 + \kappa \le t, s \le T \Big\} > 0.$$

We note that m can be $+\infty$. Define

$$m_\delta := \lim_{r \to 0} \lim_{\eta \to 0} \sup \Big\{ u(t, x) - v(s, y) - \delta e^{\frac{(\log(2+|x|^2))^2}{\sqrt{t - T_1}}} - \delta e^{\frac{(\log(2+|y|^2))^2}{\sqrt{s - T_1}}} :$$
$$|x - y|_{-1} < r, \, |t - s| < \eta, \, x, y \in \overline{U}, \, T_1 < t, s \le T \Big\},$$

$$m_{\delta,\varepsilon} := \lim_{\eta \to 0} \sup \Big\{ u(t, x) - v(s, y) - \delta e^{\frac{(\log(2+|x|^2))^2}{\sqrt{t - T_1}}} - \delta e^{\frac{(\log(2+|y|^2))^2}{\sqrt{s - T_1}}}$$
$$- \frac{|x - y|_{-1}^2}{2\varepsilon} : |t - s| < \eta, \, x, y \in \overline{U}, \, T_1 < t, s \le T \Big\},$$

$$m_{\delta,\varepsilon,\beta} := \sup \Big\{ u(t, x) - v(s, y) - \delta e^{\frac{(\log(2+|x|^2))^2}{\sqrt{t - T_1}}} - \delta e^{\frac{(\log(2+|y|^2))^2}{\sqrt{s - T_1}}}$$
$$- \frac{|x - y|_{-1}^2}{2\varepsilon} - \frac{(t - s)^2}{2\beta} : x, y \in \overline{U}, \, T_1 < t, s \le T \Big\}.$$

It is very easy to see that

$$m \leq \lim_{\delta \to 0} m_\delta, \tag{3.77}$$

$$m_\delta = \lim_{\varepsilon \to 0} m_{\delta,\varepsilon}, \tag{3.78}$$

$$m_{\delta,\varepsilon} = \lim_{\beta \to 0} m_{\delta,\varepsilon,\beta}. \tag{3.79}$$

Setting $u(t, x) = -\infty$ if $x \notin \overline{U}$ and $v(t, x) = +\infty$ if $x \notin \overline{U}$ we can consider u and v to be defined on $(T_1, T] \times H$. Since $\overline{U}$ is B-closed such extended u is B-upper semicontinuous on $(T_1, T] \times H$ and v is B-lower semicontinuous on $(T_1, T] \times H$.
   Define

$$\Psi(t, s, x, y) = u(t, x) - v(s, y) - \delta e^{\frac{(\log(2+|x|^2))^2}{\sqrt{t-T_1}}} - \delta e^{\frac{(\log(2+|y|^2))^2}{\sqrt{s-T_1}}} - \frac{|x-y|^2_{-1}}{2\varepsilon} - \frac{(t-s)^2}{2\beta}.$$

We notice that by (3.74) we obtain, for instance,

$$\Psi(t, s, x, y) \leq -(|x|^2 + |y|^2) \quad \text{if } |x| + |y| \geq K_\delta \tag{3.80}$$

for some $K_\delta > 0$. Therefore, using Corollary 3.26, for every $n \geq 1$ we can find $a_n, b_n \in \mathbb{R}$, and $p_n, q_n \in H$ such that $|a_n| + |b_n| + |p_n| + |q_n| \leq \frac{1}{n}$ and such that

$$\Psi(t, s, x, y) + a_n t + b_n s + \langle Bp_n, x \rangle + \langle Bq_n, y \rangle$$

achieves a strict global maximum at some point $(\bar{t}, \bar{s}, \bar{x}, \bar{y}) \in [T_1, T] \times [T_1, T] \times H \times H$. (The maximum is initially strict in the $|\cdot|_{-2}$ norm but since the radial functions are strictly increasing the maximum is in fact strict in the $|\cdot|$ norm.) Moreover, for a fixed δ

$$|\bar{x}|, |\bar{y}|, |u(\bar{t}, \bar{x})|, |v(\bar{s}, \bar{y})| \leq R_\delta \tag{3.81}$$

for some R_δ independently of ε, β, n. Obviously $(\bar{t}, \bar{s}, \bar{x}, \bar{y}) \in (T_1, T] \times (T_1, T] \times \overline{U} \times \overline{U}$. It follows from (3.80) that

$$m_{\delta,\varepsilon,\beta} \leq \Psi(\bar{t}, \bar{s}, \bar{x}, \bar{y}) + \frac{C_\delta}{n} \tag{3.82}$$

for some constant $C_\delta > 0$. Therefore, it follows that

$$m_{\delta,\varepsilon,\beta} + \frac{|\bar{t} - \bar{s}|^2}{4\beta} \leq \Psi(\bar{t}, \bar{s}, \bar{x}, \bar{y}) + \frac{|\bar{t} - \bar{s}|^2}{4\beta} + \frac{C_\delta}{n} \leq m_{\delta,\varepsilon,2\beta} + \frac{C_\delta}{n} \tag{3.83}$$

and

$$m_{\delta,\varepsilon,\beta} + \frac{|\bar{x} - \bar{y}|^2_{-1}}{4\varepsilon} + \frac{|\bar{t} - \bar{s}|^2}{4\beta} \leq m_{\delta,2\varepsilon,2\beta} + \frac{C_\delta}{n}. \tag{3.84}$$

Inequalities (3.83) and (3.79) imply

$$\lim_{\beta \to 0} \limsup_{n \to \infty} \frac{|\bar{t} - \bar{s}|^2}{\beta} = 0 \quad \text{for every } \delta, \varepsilon > 0, \tag{3.85}$$

and then (3.85), (3.84) and (3.78) imply

$$\lim_{\varepsilon \to 0} \limsup_{\beta \to 0} \limsup_{n \to \infty} \frac{|\bar{x} - \bar{y}|^2_{-1}}{\varepsilon} = 0 \quad \text{for every } \delta > 0. \tag{3.86}$$

In particular, it follows from (3.77)–(3.79), (3.82), (3.85) and (3.86) that there exists a $\delta_0 > 0$ such that for all $\delta < \delta_0$

$$\liminf_{\varepsilon \to 0} \liminf_{\beta \to 0} \liminf_{n \to \infty} (u(\bar{t}, \bar{x}) - v(\bar{s}, \bar{y})) \geq \bar{m} = \min\left(\frac{m}{2}, 1\right). \tag{3.87}$$

Conditions (3.71), (3.72), together with (3.85) and (3.86), imply that if $\delta, \varepsilon, \beta$ are small enough and n is sufficiently big we must have $(\bar{t}, \bar{s}, \bar{x}, \bar{y}) \in (T_1, T) \times (T_1, T) \times U \times U$.

We now have for $N > 2$

$$|x - y|^2_{-1} = |P_N(x - y)|^2_{-1} + |Q_N(x - y)|^2_{-1}$$

and

$$|Q_N(x - y)|^2_{-1} \leq 2\langle BQ_N(\bar{x} - \bar{y}), x - y\rangle + 2|Q_N(x - \bar{x})|^2_{-1}$$
$$+ 2|Q_N(y - \bar{y})|^2_{-1} - |Q_N(\bar{x} - \bar{y})|^2_{-1}$$

with equality at $\bar{x}, \bar{y}$. Therefore, defining

$$u_1(t, x) = u(t, x) - \delta e^{\frac{(\log(2+|x|^2))^2}{\sqrt{t-T_1}}} - \frac{\langle BQ_N(\bar{x} - \bar{y}), x\rangle}{\varepsilon} - \frac{|Q_N(x - \bar{x})|^2_{-1}}{\varepsilon}$$
$$+ \frac{|Q_N(\bar{x} - \bar{y})|^2_{-1}}{2\varepsilon} + a_n t + \langle B p_n, x\rangle$$

and

$$v_1(s, y) = v(s, y) + \delta e^{\frac{(\log(2+|y|^2))^2}{\sqrt{s-T_1}}} - \frac{\langle BQ_N(\bar{x} - \bar{y}), y\rangle}{\varepsilon} + \frac{|Q_N(y - \bar{y})|^2_{-1}}{\varepsilon}$$
$$- b_n s - \langle B q_n, y\rangle,$$

we see that

$$u_1(t, x) - v_1(s, y) - \frac{1}{2\varepsilon}|P_N(x - y)|^2_{-1} - \frac{1}{2\beta}|t - s|^2$$

has a strict global maximum at $(\bar{t}, \bar{s}, \bar{x}, \bar{y})$ over $[T_1, T] \times [T_1, T] \times H \times H$. We can therefore apply Corollary 3.29 to obtain test functions φ_k, ψ_k and points

(t_k, x_k), (s_k, y_k) such that $u_1(t, x) - \varphi_k(t, x)$ has a maximum at (t_k, x_k), $v_1(s, y) - \psi_k(s, y)$ has a minimum at (s_k, y_k), and such that (3.54), (3.55) are satisfied for u_1, v_1, respectively. In particular, (t_k, x_k), $(s_k, y_k) \in (T_1, T) \times U$ for big k.

Define

$$\varphi(t, x) = \varphi_k(t, x) + \frac{\langle BQ_N(\bar{x} - \bar{y}), x\rangle}{\varepsilon} + \frac{|Q_N(x - \bar{x})|^2_{-1}}{\varepsilon}$$
$$- \frac{|Q_N(\bar{x} - \bar{y})|^2_{-1}}{2\varepsilon} - a_n t - \langle Bp_n, x\rangle,$$

and

$$h(t, |x|) = \delta e^{\frac{(\log(2+|x|^2))^2}{\sqrt{t-T_1}}}.$$

Since u is a viscosity subsolution of (3.62) on $(T_1, T] \times \overline{U}$, using the definition of a viscosity subsolution we have

$$\varphi_t(t_k, x_k) + h_t(t_k, |x_k|) + \langle x_k, A^*D\varphi(t_k, x_k)\rangle$$
$$- F(t_k, x_k, u(t_k, x_k), D\varphi(t_k, x_k) + Dh(t_k, |x_k|), D^2\varphi(t_k, x_k) + D^2h(t_k, |x_k|)) \geq 0. \tag{3.88}$$

Letting $k \to +\infty$ in (3.88) and using (3.54) yields

$$\frac{\bar{t} - \bar{s}}{\beta} - a_n + h_t(\bar{t}, |\bar{x}|) + \left\langle \bar{x}, A^*\left(\frac{B(\bar{x} - \bar{y})}{\varepsilon} - Bp_n\right)\right\rangle$$
$$- F\left(\bar{t}, \bar{x}, u(\bar{t}, \bar{x}), \frac{B(\bar{x} - \bar{y})}{\varepsilon} - Bp_n + Dh(\bar{t}, |\bar{x}|), X_N + \frac{2BQ_N}{\varepsilon} + D^2h(\bar{t}, |\bar{x}|)\right) \geq 0. \tag{3.89}$$

We now compute

$$h_t(t, |x|) = -\frac{\delta(\log(2 + |x|^2))^2}{2(t - T_1)^{\frac{3}{2}}} e^{\frac{(\log(2+|x|^2))^2}{\sqrt{t-T_1}}},$$

$$Dh(t, |x|) = e^{\frac{(\log(2+|x|^2))^2}{\sqrt{t-T_1}}} \frac{4\delta \log(2 + |x|^2)}{\sqrt{t - T_1}} \frac{x}{2 + |x|^2},$$

and

$$D^2h(t, |x|) = \frac{4\delta}{\sqrt{t - T_1}} e^{\frac{(\log(2+|x|^2))^2}{\sqrt{t-T_1}}} \left[\left(\frac{4(\log(2 + |x|^2))^2}{\sqrt{t - T_1}(2 + |x|^2)^2}\right. \right.$$
$$\left. + \frac{2}{(2 + |x|^2)^2} - \frac{2\log(2 + |x|^2)}{(2 + |x|^2)^2}\right) x \otimes x + \left. \frac{\log(2 + |x|^2)}{2 + |x|^2} I\right].$$

We have

$$\frac{|Dh(t,|x|)|}{1+|x|} + \|D^2h(t,|x|)\| \le \frac{C_1\delta(\log(2+|x|^2))^2}{(t-T_1)(1+|x|)^2} e^{\frac{(\log(2+|x|^2))^2}{\sqrt{t-T_1}}}$$

for some absolute constant C_1. Therefore we obtain from Hypothesis 3.49

$$\left| F\left(\bar{t}, \bar{x}, u(\bar{t},\bar{x}), \frac{B(\bar{x}-\bar{y})}{\varepsilon} - Bp_n + Dh(\bar{t},|\bar{x}|), X_N + \frac{2BQ_N}{\varepsilon} + D^2h(\bar{t},|\bar{x}|)\right) \right.$$
$$\left. - F\left(\bar{t}, \bar{x}, u(\bar{t},\bar{x}), \frac{B(\bar{x}-\bar{y})}{\varepsilon} - Bp_n, X_N + \frac{2BQ_N}{\varepsilon}\right) \right|$$
$$\le M_F((1+|\bar{x}|)|Dh(\bar{t},|\bar{x}|)| + (1+|\bar{x}|)^2\|D^2h(\bar{t},|\bar{x}|)\|)$$
$$\le \frac{M_F C_1\delta(\log(2+|\bar{x}|^2))^2}{\bar{t}-T_1} e^{\frac{(\log(2+|\bar{x}|^2))^2}{\sqrt{t-T_1}}} \le -\frac{1}{2}h_t(\bar{t},|\bar{x}|)$$

if

$$\tau \le \frac{1}{(4M_F C_1)^2}. \tag{3.90}$$

Hence if (3.90) is satisfied, using that $\frac{1}{2}h_t(\bar{t},|\bar{x}|) \le -C_\tau\delta$ for some $C_\tau > 0$, it follows from (3.89) that

$$-C_\tau\delta + \frac{\bar{t}-\bar{s}}{\beta} - a_n + \left\langle \bar{x}, A^*\left(\frac{B(\bar{x}-\bar{y})}{\varepsilon} - Bp_n\right)\right\rangle$$
$$- F\left(\bar{t},\bar{x},u(\bar{t},\bar{x}), \frac{B(\bar{x}-\bar{y})}{\varepsilon} - Bp_n, X_N + \frac{2BQ_N}{\varepsilon}\right) \ge 0. \tag{3.91}$$

Arguing similarly we obtain from the fact that $v_1(s,y) - \psi_k(s,y)$ has a minimum at (s_k, y_k) that

$$\frac{\bar{t}-\bar{s}}{\beta} + b_n + \left\langle \bar{y}, A^*\left(\frac{B(\bar{x}-\bar{y})}{\varepsilon} + Bq_n\right)\right\rangle$$
$$- F\left(\bar{s},\bar{y},v(\bar{s},\bar{y}), \frac{B(\bar{x}-\bar{y})}{\varepsilon} + Bq_n, Y_N - \frac{2BQ_N}{\varepsilon}\right) \le 0. \tag{3.92}$$

Therefore subtracting (3.91) from (3.92) and using (3.81), Hypothesis 3.47 yields

$$C_\tau \delta - \left\langle \bar{x} - \bar{y}, \frac{A^* B(\bar{x} - \bar{y})}{\varepsilon} \right\rangle$$

$$+ F\left(\bar{t}, \bar{x}, u(\bar{t}, \bar{x}), \frac{B(\bar{x} - \bar{y})}{\varepsilon}, X_N\right) - F\left(\bar{s}, \bar{y}, v(\bar{s}, \bar{y}), \frac{B(\bar{x} - \bar{y})}{\varepsilon}, Y_N\right)$$

$$\leq \omega_1(\delta, \varepsilon, \beta; n, N),$$

$$(3.93)$$

where $\lim_{n \to +\infty} \lim_{N \to +\infty} \omega_1(\delta, \varepsilon, \beta; n, N) = 0$ for fixed $\delta, \varepsilon, \beta$. Now Hypothesis 3.45, (3.85), (3.87) and (3.93) imply

$$C_\tau \delta - \left\langle \bar{x} - \bar{y}, \frac{A^* B(\bar{x} - \bar{y})}{\varepsilon} \right\rangle$$

$$+ F\left(\bar{t}, \bar{x}, u(\bar{t}, \bar{x}), \frac{B(\bar{x} - \bar{y})}{\varepsilon}, X_N\right) - F\left(\bar{t}, \bar{y}, u(\bar{t}, \bar{x}), \frac{B(\bar{x} - \bar{y})}{\varepsilon}, Y_N\right)$$

$$\leq \omega_2(\delta; \varepsilon, \beta, n, N),$$

$$(3.94)$$

where $\limsup_{\varepsilon \to 0} \limsup_{\beta \to 0} \limsup_{n \to +\infty} \limsup_{N \to +\infty} \omega_2(\delta; \varepsilon, \beta, n, N) = 0$ for sufficiently small δ. We recall that X_N, Y_N satisfy (3.52). We can now use (3.2), Hypothesis 3.48, (3.81) and then invoke (3.86) to get

$$C_\tau \delta \leq c_0 \frac{|\bar{x} - \bar{y}|_{-1}^2}{\varepsilon} + \omega_{R_\delta}\left(|\bar{x} - \bar{y}|_{-1}\left(1 + \frac{|\bar{x} - \bar{y}|_{-1}}{\varepsilon}\right)\right)$$

$$+ \omega_2(\delta; \varepsilon, \beta, n, N) \leq \omega_3(\delta; \varepsilon, \beta, n, N),$$

where $\limsup_{\varepsilon \to 0} \limsup_{\beta \to 0} \limsup_{n \to +\infty} \limsup_{N \to +\infty} \omega_3(\delta; \varepsilon, \beta, n, N) = 0$ for sufficiently small δ. This yields a contradiction for small δ.

Thus we obtain that $m \leq 0$ and this allows us to reapply the procedure to intervals $[T - 3\tau/2, T - \tau/2]$, $[T - 4\tau/2, T - 2\tau/2]$, ..., $[0, T - k\tau/2]$, where k is such that $T - k\tau/2 > 0 \geq T - (k + 2)\tau/2$.

For $\gamma \in [0, 1)$ the proof is the same but we have to replace the functions

$$\delta e^{\frac{(\log(2+|x|^2))^2}{\sqrt{t-T_1}}} \quad \text{and} \quad \delta e^{\frac{(\log(2+|y|^2))^2}{\sqrt{s-T_1}}}$$

by

$$\delta e^{\frac{(1+|x|^2)^{1-\gamma}}{\sqrt{t-T_1}}} \quad \text{and} \quad \delta e^{\frac{(1+|y|^2)^{1-\gamma}}{\sqrt{s-T_1}}},$$

respectively.   □

The assumptions of the comparison theorems can be weakened if we replace the weak B-condition by the strong B-condition, i.e. if we replace (3.2) by (3.3). In this case we will use the following assumption instead of Hypothesis 3.48.

Hypothesis 3.52 For every $R > 0$ there exists a modulus ω_R such that, for all $(t, x, y, r) \in (0, T) \times U \times U \times \mathbb{R}$ such that $|r|, |x|, |y| \leq R$, for any $\varepsilon > 0$, for all $X, Y \in S(H)$ such that $X = P_N^* X P_N$, $Y = P_N^* Y P_N$ for some N and satisfying (3.52), we have

$$F\left(t, x, r, \frac{B(x-y)}{\varepsilon}, X\right) - F\left(t, y, r, \frac{B(x-y)}{\varepsilon}, Y\right)$$

$$\geq -\omega_R\left(|x-y|\left(1 + \frac{|x-y|_{-1}}{\varepsilon}\right)\right).$$

Hypothesis 3.53 For every $R > 0$ there exists a modulus ω_R such that

$$|g(x) - g(y)| \leq \omega_R(|x-y|) \quad \text{if } x, y \in H, |x|, |y| \leq R.$$

Theorem 3.54 (Comparison under strong B-condition) *Let $U = H$. Let (3.3) hold and let F satisfy Hypotheses 3.44, 3.46, 3.47, 3.52, 3.49 and 3.45 with $\nu = 0$. Let g satisfy Hypothesis 3.53. Let u be a viscosity subsolution of (3.62) in $(0, T] \times H$, and v be a viscosity supersolution of (3.62) in $(0, T] \times H$. Suppose that*

$$\lim_{t \to T}\left[(u(t, x) - g(e^{(T-t)A}x))^+ + (v(t, x) - g(e^{(T-t)A}x))^-\right] = 0 \qquad (3.95)$$

uniformly on bounded subsets of H and that either of (3.73) or (3.74) is satisfied. Then for every $0 < \mu < T$

$$m_\mu = \lim_{R \to +\infty} \lim_{r \to 0} \lim_{\eta \to 0} \sup\Big\{u(t, x) - v(s, y) : |x-y|_{-1} < r, \, |t-s| < \eta$$

$$x, y \in B_R, \, \mu < t, s \leq T - \mu\Big\} \leq 0. \qquad (3.96)$$

In particular, $u \leq v$.

Proof Let us again assume that $\gamma = 1$ and that (3.73) is satisfied. As in the proof of Theorem 3.50 we take $0 < \tau \leq \min(1, 1/\sqrt{a}, 1/(4M_F C_1)^2)$, where C_1 is the constant appearing in that proof, and we set $T_1 = T - \tau$. Define for $0 < \mu < \tau$

$$m_\mu = \lim_{R \to +\infty} \lim_{r \to 0} \lim_{\eta \to 0} \sup\Big\{u(t, x) - v(s, y) : |x-y|_{-1} < r, \, |t-s| < \eta$$

$$x, y \in B_R, \, T_1 + \mu \leq t, s \leq T - \mu\Big\}.$$

If there is a μ_0 such that $m_{\mu_0} > \tilde{m} > 0$ then $m_\mu > \tilde{m} > 0$ for all $0 < \mu < \mu_0$. Defining

$$\Psi(t,s,x,y) = u(t,x) - v(s,y) - \delta e^{\frac{(\log(2+|x|^2))^2}{\sqrt{t-T_1}}} - \delta e^{\frac{(\log(2+|y|^2))^2}{\sqrt{s-T_1}}} - \frac{|x-y|^2_{-1}}{2\varepsilon} - \frac{(t-s)^2}{2\beta}$$

we again have that for every $n \geq 1$ there exist $a_n, b_n \in \mathbb{R}$, $p_n, q_n \in H$ such that $|a_n| + |b_n| + |p_n| + |q_n| \leq \frac{1}{n}$ and that

$$\Psi(t,s,x,y) + a_n t + b_n s + \langle Bp_n, x \rangle + \langle Bq_n, y \rangle$$

achieves a strict global maximum over $[T_1, T-\mu] \times [T_1, T-\mu] \times H \times H$ at some point $(\bar{t}_\mu, \bar{s}_\mu, \bar{x}_\mu, \bar{y}_\mu) \in (T_1, T-\mu] \times (T_1, T-\mu] \times H \times H$, and that (3.81), (3.85), (3.86) hold (note that the constant R_δ in (3.81) is independent of μ). Moreover, since for $\mu < \mu_0$, $\Psi(\bar{t}_\mu, \bar{s}_\mu, \bar{x}_\mu, \bar{y}_\mu) \geq \Psi(\bar{t}_{\mu_0}, \bar{s}_{\mu_0}, \bar{x}_{\mu_0}, \bar{y}_{\mu_0}) - \frac{C_\delta}{n}$ for some $C_\delta > 0$ independent of μ, it is easy to see that for every $0 < \mu < \mu_0$

$$\liminf_{\varepsilon \to 0} \liminf_{\beta \to 0} \liminf_{n \to +\infty} (u(\bar{t}_\mu, \bar{x}_\mu) - v(\bar{s}_\mu, \bar{y}_\mu))$$

$$\geq \liminf_{\varepsilon \to 0} \liminf_{\beta \to 0} \liminf_{n \to +\infty} \Psi(\bar{t}_\mu, \bar{s}_\mu, \bar{x}_\mu, \bar{y}_\mu)$$

$$\geq \liminf_{\varepsilon \to 0} \liminf_{\beta \to 0} \liminf_{n \to +\infty} \Psi(\bar{t}_{\mu_0}, \bar{s}_{\mu_0}, \bar{x}_{\mu_0}, \bar{y}_{\mu_0})$$

$$= \liminf_{\varepsilon \to 0} \liminf_{\beta \to 0} \liminf_{n \to +\infty} \left(u(\bar{t}_{\mu_0}, \bar{x}_{\mu_0}) - v(\bar{s}_{\mu_0}, \bar{y}_{\mu_0}) \right.$$

$$\left. - \delta e^{\frac{(\log(2+|\bar{x}_{\mu_0}|^2))^2}{\sqrt{\bar{t}_{\mu_0}-T_1}}} - \delta e^{\frac{(\log(2+|\bar{y}_{\mu_0}|^2))^2}{\sqrt{\bar{s}_{\mu_0}-T_1}}} \right) \geq \tilde{m} \qquad (3.97)$$

if $\delta < \delta_0$ for some $\delta_0 > 0$ (depending only on μ_0).

If for all $\delta, \varepsilon, \beta, n$ we have $(\bar{t}_\mu, \bar{s}_\mu, \bar{x}_\mu, \bar{y}_\mu) \in (T_1, T-\mu) \times (T_1, T-\mu) \times H \times H$ then as in the proof of Theorem 3.50 and using the notation there we arrive at (3.94), i.e. that

$$C_T \delta - \left\langle \bar{x}_\mu - \bar{y}_\mu, \frac{A^* B(\bar{x}_\mu - \bar{y}_\mu)}{\varepsilon} \right\rangle$$

$$+ F\left(\bar{t}_\mu, \bar{x}_\mu, u(\bar{t}_\mu, \bar{x}_\mu), \frac{B(\bar{x}_\mu - \bar{y}_\mu)}{\varepsilon}, X_N \right) - F\left(\bar{t}_\mu, \bar{y}_\mu, u(\bar{t}_\mu, \bar{x}_\mu), \frac{B(\bar{x}_\mu - \bar{y}_\mu)}{\varepsilon}, Y_N \right)$$

$$\leq \omega_2(\delta; \varepsilon, \beta, n, N),$$

where $\limsup_{\varepsilon \to 0} \limsup_{\beta \to 0} \limsup_{n \to +\infty} \limsup_{N \to +\infty} \omega_2(\delta; \varepsilon, \beta, n, N) = 0$ for sufficiently small δ. We then use (3.2), Hypothesis 3.52 and (3.81) to get

$$C_T \delta \leq c_0 \frac{|\bar{x}_\mu - \bar{y}_\mu|^2_{-1}}{\varepsilon} - \frac{|\bar{x}_\mu - \bar{y}_\mu|^2}{\varepsilon} + \omega_{R_\delta}\left(|\bar{x}_\mu - \bar{y}_\mu| \left(1 + \frac{|\bar{x}_\mu - \bar{y}_\mu|_{-1}}{\varepsilon} \right) \right)$$

$$+ \omega_2(\delta; \varepsilon, \beta, n, N).$$

Let K_δ be a constant such that $\omega_{R_\delta}(r) \leq C_T \delta/4 + K_\delta r$. Then

$$\omega_{R_\delta}\left(|\bar{x}_\mu - \bar{y}_\mu|\left(1 + \frac{|\bar{x}_\mu - \bar{y}_\mu|_{-1}}{\varepsilon}\right)\right) \leq C_\tau \delta/4 + K_\delta |\bar{x}_\mu - \bar{y}_\mu|\left(1 + \frac{|\bar{x}_\mu - \bar{y}_\mu|_{-1}}{\varepsilon}\right)$$

$$\leq C_\tau \delta/2 + \frac{|\bar{x}_\mu - \bar{y}_\mu|^2}{\varepsilon} + \tilde{K}_\delta \frac{|\bar{x}_\mu - \bar{y}_\mu|^2_{-1}}{\varepsilon}$$

for some $\tilde{K}_\delta > 0$ and small enough ε. Therefore we obtain that

$$\frac{C_\tau \delta}{2} \leq (c_0 + \tilde{K}_\delta)\frac{|\bar{x}_\mu - \bar{y}_\mu|^2_{-1}}{\varepsilon} + \omega_2(\delta; \varepsilon, \beta, n, N)$$

and this yields a contradiction in light of (3.86).

Therefore for small δ, ε, β and large n we must have $\bar{t}_\mu = T - \mu$ or $\bar{s}_\mu = T - \mu$. Without loss of generality suppose that $\bar{s}_\mu = T - \mu$. Recalling that $|\bar{x}_\mu|, |\bar{y}_\mu| \leq R_\delta$ for some $R_\delta > 0$ and using (3.95) we have

$$u(\bar{t}_\mu, \bar{x}_\mu) - v(\bar{s}_\mu, \bar{y}_\mu) = (u(\bar{t}_\mu, \bar{x}_\mu) - g(e^{(T-\bar{t}_\mu)A}\bar{x}_\mu))_+$$
$$+ (g(e^{(T-\bar{t}_\mu)A}\bar{x}_\mu) - g(e^{(T-\bar{s}_\mu)A}\bar{y}_\mu)) + (g(e^{(T-\bar{s}_\mu)A}\bar{y}_\mu) - v(\bar{s}_\mu, \bar{y}_\mu))_+$$
$$\leq \tilde{\omega}_{R_\delta}(\mu + |\bar{t}_\mu - \bar{s}_\mu|) + |g(e^{(T-\bar{t}_\mu)A}\bar{x}_\mu) - g(e^{(T-\bar{s}_\mu)A}\bar{y}_\mu)|,$$

where $\tilde{\omega}_{R_\delta}$ is a modulus for every δ, depending on (3.95). Then by (3.19), (3.20) and Hypothesis 3.53

$$u(\bar{t}_\mu, \bar{x}_\mu) - v(\bar{s}_\mu, \bar{y}_\mu) \leq \tilde{\omega}_{R_\delta}(\mu + |\bar{t}_\mu - \bar{s}_\mu|) + \omega_{R_\delta}(|e^{(T-\bar{t}_\mu)A}\bar{x}_\mu - e^{(T-\bar{s}_\mu)A}\bar{y}_\mu|)$$

$$\leq \tilde{\omega}_{R_\delta}(\mu + |\bar{t}_\mu - \bar{s}_\mu|) + \omega_{R_\delta}\left(\frac{e^{c_0\mu}}{(2\mu)^{\frac{1}{2}}}|e^{(\bar{s}_\mu-\bar{t}_\mu)A}\bar{x}_\mu - \bar{y}_\mu|_{-1}\right)$$

$$\leq \tilde{\omega}_{R_\delta}(\mu + |\bar{t}_\mu - \bar{s}_\mu|) + \omega_{R_\delta}\left(\frac{e^{c_0\mu}}{(2\mu)^{\frac{1}{2}}}(C(R_\delta)|\bar{s}_\mu - \bar{t}_\mu|^{\frac{1}{2}} + |\bar{x}_\mu - \bar{y}_\mu|_{-1})\right).$$

Therefore it follows from this, (3.85) and (3.86) that for $\delta < \delta_0$, and $\mu < \mu_0$ such that $\tilde{\omega}_{R_\delta}(\mu) \leq \tilde{m}/2$

$$\limsup_{\varepsilon \to 0} \limsup_{\beta \to 0} \limsup_{n \to +\infty}(u(\bar{t}_\mu, \bar{x}_\mu) - v(\bar{s}_\mu, \bar{y}_\mu)) \leq \frac{\tilde{m}}{2}.$$

This is impossible in light of (3.97).

Thus we obtain that $m_\mu \leq 0$ for all $\mu < \tau$ and this allows us to reapply the procedure to intervals $[T - 3\tau/2, T - \tau/2]$, $[T - 4\tau/2, T - 2\tau/2]$,... directly as in the proof of Theorem 3.50 since we now have

$$\lim_{R\to+\infty} \lim_{r\to 0} \lim_{\eta\to 0} \sup \Big\{ u(t, x) - v(s, y) : \ |x - y|_{-1} < r$$

$$x, y \in B_R, \ T - \tau/2 - \eta \le t, s \le T - \tau/2 \Big\} \le 0. \quad (3.98)$$

For $\gamma \in [0, 1)$ the proof again uses the same modifications as indicated in the proof of Theorem 3.50. $\qquad\square$

3.5.2 Degenerate Elliptic Equations

In this subsection we consider the degenerate elliptic case. We first introduce a slightly different version of Hypothesis 3.49.

Hypothesis 3.55 For $\gamma \in [0, 1]$ there exist $M_F, N_F \ge 0$ such that

$$|F (x, r, p + q, X + Y) - F (x, r, p, X)|$$
$$\le M_F(1 + |x|^\gamma)|q| + N_F(1 + |x|^\gamma)^2\|Y\|$$

for all $(x, r) \in U \times \mathbb{R}, \ p, q \in H, \ X, Y \in S(H)$.

Theorem 3.56 (Comparison under weak B-condition) *Let $U \subset H$ be open and $\overline{U}$ be B-closed. Let (3.2) hold and let F satisfy Hypotheses 3.44, 3.46–3.48, 3.55 and 3.45 with $\nu > 0$. Let u and v be, respectively, a viscosity subsolution and a viscosity supersolution of (3.56). Suppose that for every $R > 0$ there exists a modulus $\tilde{\omega}_R$ such that*

$$(u(x) - v(y))^+ + (u(y) - v(x))^+ \le \tilde{\omega}_R(|x - y|_{-1}) \quad (3.99)$$

for $x \in \partial U, \ y \in \overline{U}, \ |x|, |y| \le R$. Moreover, suppose that there exist constants $C, a > 0$ such that one of the following conditions is satisfied

1. $\gamma \in (0, 1), \ \exists \bar{k} \ge 0 \ s.t. \ u, -v \le C(1 + |x|^{\bar{k}}) \ \forall x \in H,$ (3.100)
2. $\gamma = 0, \ 2M_F a + 4N_F(a + a^2) < \nu, \ and \ u, -v \le Ce^{a|x|} \ \forall x \in H,$ (3.101)
3. $\gamma = 1, \ \exists \bar{k} \ge 0 \ s.t. \ M_F\bar{k} + N_F\bar{k}(\bar{k} - 1) < \nu \ if \ \bar{k} \ge 2,$
     $\bar{k}(M_F + N_F) < \nu \ if \ \bar{k} < 2, \ and \ u, -v \le C(1 + |x|^{\bar{k}}) \ \forall x \in H.$ (3.102)

Then

$$m = \lim_{R\to+\infty} \lim_{r\to 0} \sup \Big\{ u(x) - v(y) : \ |x - y|_{-1} < r, \ x, y \in \overline{U} \cap B_R \Big\} \le 0. \quad (3.103)$$

In particular, $u \le v$.

Proof We will first prove the theorem in the case $\gamma \in (0, 1)$. We argue by contradiction and assume that $m > 0$. Let $k > \bar{k}, k \ge 2$. Denote

$$m_\delta := \lim_{r \to 0} \sup \left\{ u(x) - v(y) - \delta|x|^k - \delta|y|^k \ : \ |x - y|_{-1} < r, x, y \in \overline{U} \right\},$$

$$m_{\delta,\varepsilon} := \sup \left\{ u(x) - v(y) - \delta|x|^k - \delta|y|^k - \frac{|x - y|^2_{-1}}{2\varepsilon} \ : \ x, y \in \overline{U} \right\}.$$

As in the proof of Theorem 3.50, it is easy to see that

$$m = \lim_{\delta \to 0} m_\delta, \tag{3.104}$$

$$m_\delta = \lim_{\varepsilon \to 0} m_{\delta,\varepsilon}. \tag{3.105}$$

Again, setting $u(x) = -\infty$ if $x \notin \overline{U}$ and $v(x) = +\infty$ if $x \notin \overline{U}$ we can consider u and v to be defined on H. Since $\overline{U}$ is B-closed such extended u is B-upper semicontinuous on H and v is B-lower semicontinuous on H.

Define

$$\Psi(x, y) = u(x) - v(y) - \delta|x|^k - \delta|y|^k - \frac{|x - y|^2_{-1}}{2\varepsilon}.$$

By (3.100) we can apply Corollary 3.26 to produce for every $n \geq 1$ elements $p_n, q_n \in H$ such that $|p_n| + |q_n| \leq \frac{1}{n}$ and such that

$$\Psi(x, y) + \langle Bp_n, x \rangle + \langle Bq_n, y \rangle$$

achieves a strict global maximum over $H \times H$ at some point $(\bar{x}, \bar{y}) \in \overline{U} \times \overline{U}$. Moreover, we have

$$m_{\delta,\varepsilon} \leq \Psi(\bar{x}, \bar{y}) + \frac{C_\delta}{n}$$

for some constant $C_\delta > 0$. Therefore it follows that

$$m_{\delta,\varepsilon} + \frac{|\bar{x} - \bar{y}|^2_{-1}}{4\varepsilon} \leq m_{\delta,2\varepsilon} + \frac{C_\delta}{n}. \tag{3.106}$$

Inequalities (3.106) and (3.105) imply

$$\lim_{\varepsilon \to 0} \limsup_{n \to \infty} \frac{|\bar{x} - \bar{y}|^2_{-1}}{\varepsilon} = 0 \quad \text{for every } \delta > 0. \tag{3.107}$$

Condition (3.99), together with (3.107), now imply that if δ, ε are small enough and n is sufficiently big we must have $(\bar{x}, \bar{y}) \in U \times U$.

Similarly to the proof of Theorem 3.56 we now have for $N > 2$ that if we define

$$u_1(x) = u(x) - \delta|x|^k - \frac{\langle BQ_N(\bar{x} - \bar{y}), x \rangle}{\varepsilon} - \frac{|Q_N(x - \bar{x})|^2_{-1}}{\varepsilon}$$

$$+ \frac{|Q_N(\bar{x} - \bar{y})|^2_{-1}}{2\varepsilon} + \langle Bp_n, x \rangle$$

and

$$v_1(y) = v(y) + \delta|y|^k - \frac{\langle BQ_N(\bar{x} - \bar{y}), y \rangle}{\varepsilon} + \frac{|Q_N(y - \bar{y})|^2_{-1}}{\varepsilon} - \langle Bq_n, y \rangle,$$

then

$$u_1(x) - v_1(y) - \frac{1}{2\varepsilon}|P_N(x - y)|^2_{-1}$$

has a strict global maximum at $(\bar{x}, \bar{y})$ over $H \times H$. We can therefore apply Corollary 3.28 to obtain test functions φ_k, ψ_k and points x_k, y_k such that $u_1(x) - \varphi_k(x)$ has a maximum at x_k, $v_1(y) - \psi_k(y)$ has a minimum at y_k, and such that (3.50), (3.51) are satisfied for u_1, v_1, respectively. In particular, $x_k, y_k \in U$ for big k.

Therefore, since u is a viscosity subsolution of (3.56) in U, using the definition of a viscosity subsolution, letting $k \to +\infty$ and using (3.50) we obtain

$$-\left\langle \bar{x}, A^* \left(\frac{B(\bar{x} - \bar{y})}{\varepsilon} - Bp_n \right) \right\rangle + F(\bar{x}, u(\bar{x}), p_{n,\delta,\varepsilon}, X_{n,\delta,\varepsilon}) \le 0, \qquad (3.108)$$

where

$$p_{n,\delta,\varepsilon} = \frac{B(\bar{x} - \bar{y})}{\varepsilon} - Bp_n + \delta k|\bar{x}|^{k-2}\bar{x},$$

and

$$X_{n,\delta,\varepsilon} = X_N + \frac{2BQ_N}{\varepsilon} + \delta k|\bar{x}|^{k-2}((k - 2)\frac{\bar{x} \otimes \bar{x}}{|\bar{x}|^2} + I)).$$

Hence we obtain from Hypothesis 3.55 that

$$-\left\langle \bar{x}, A^* \left(\frac{B(\bar{x} - \bar{y})}{\varepsilon} - Bp_n \right) \right\rangle + F\left(\bar{x}, u(\bar{x}), \frac{B(\bar{x} - \bar{y})}{\varepsilon} - Bp_n, X_N + \frac{2BQ_N}{\varepsilon} \right)$$

$$- \delta M_F \left((1 + |\bar{x}|^\gamma)k|\bar{x}|^{k-1} + (1 + |\bar{x}|^\gamma)^2 k(k - 1)|\bar{x}|^{k-2} \right) \le 0. \qquad (3.109)$$

Arguing similarly we obtain that

$$-\left\langle \bar{y}, A^* \left(\frac{B(\bar{x} - \bar{y})}{\varepsilon} + Bq_n \right) \right\rangle + F\left(\bar{y}, v(\bar{y}), \frac{B(\bar{x} - \bar{y})}{\varepsilon} + Bq_n, X_N - \frac{2BQ_N}{\varepsilon} \right)$$

$$+ \delta M_F((1 + |\bar{y}|^\gamma)k|\bar{y}|^{k-1} + (1 + |\bar{y}|^\gamma)^2 k(k - 1)|\bar{y}|^{k-2}) \ge 0. \quad (3.110)$$

Therefore subtracting (3.110) from (3.109) and using Hypothesis 3.47 yields

$$- \left\langle \bar{x} - \bar{y}, \frac{A^* B(\bar{x} - \bar{y})}{\varepsilon} \right\rangle$$

$$+ F\left(\bar{x}, u(\bar{x}), \frac{B(\bar{x} - \bar{y})}{\varepsilon}, X_N \right) - F\left(\bar{y}, v(\bar{y}), \frac{B(\bar{x} - \bar{y})}{\varepsilon}, Y_N \right)$$

$$- C\delta \left(1 + |\bar{x}|^{k-1+\gamma} + |\bar{y}|^{k-1+\gamma} \right) \leq \omega_1(\delta, \varepsilon; n, N), \quad (3.111)$$

for some $C = C(M_F, k, \gamma)$, where $\lim_{n \to +\infty} \lim_{N \to +\infty} \omega_1(\delta, \varepsilon; n, N) = 0$ for fixed δ, ε.

It now follows from (3.104)–(3.106) that

$$\liminf_{\delta \to 0} \liminf_{\varepsilon \to 0} \liminf_{n \to +\infty} (u(\bar{x}) - v(\bar{y}) - \delta|\bar{x}|^k - \delta|\bar{y}|^k) > \bar{m} = \min\left(\frac{m}{2}, 1 \right) > 0. \tag{3.112}$$

Thus, Hypothesis 3.45 and (3.111) imply

$$- \left\langle \bar{x} - \bar{y}, \frac{A^* B(\bar{x} - \bar{y})}{\varepsilon} \right\rangle + \nu(u(\bar{x}) - v(\bar{y}))$$

$$+ F\left(\bar{x}, v(\bar{y}), \frac{B(\bar{x} - \bar{y})}{\varepsilon}, X_N \right) - F\left(\bar{y}, v(\bar{y}), \frac{B(\bar{x} - \bar{y})}{\varepsilon}, Y_N \right)$$

$$- C\delta \left(1 + |\bar{x}|^{k-1+\gamma} + |\bar{y}|^{k-1+\gamma} \right) \leq \omega_1(\delta, \varepsilon; n, N). \quad (3.113)$$

We recall that X_N, Y_N satisfy (3.52). We can now use (3.2), Hypothesis 3.48, and the fact that $|\bar{x}|, |\bar{y}|, |u(\bar{x})|, |v(\bar{y})| \leq R_\delta$ for some R_δ independent of ε, n to get

$$\nu(u(\bar{x}) - v(\bar{y})) - C\delta(1 + |\bar{x}|^{k-1+\gamma} + |\bar{y}|^{k-1+\gamma})$$

$$\leq c_0 \frac{|\bar{x} - \bar{y}|_{-1}^2}{\varepsilon} + \omega_{R_\delta} \left(|\bar{x} - \bar{y}|_{-1} \left(1 + \frac{|\bar{x} - \bar{y}|_{-1}}{\varepsilon} \right) \right) + \omega_1(\delta, \varepsilon; n, N) \quad (3.114)$$

$$\leq \omega_2(\delta; \varepsilon, n, N),$$

where $\limsup_{\varepsilon \to 0} \limsup_{n \to +\infty} \limsup_{N \to +\infty} \omega_2(\delta; \varepsilon, n, N) = 0$ for sufficiently small δ. Therefore we have from (3.112) and (3.114) that

$$\nu \bar{m} \leq -\nu\delta(|\bar{x}|^k + |\bar{y}|^k) + C\delta(1 + |\bar{x}|^{k-1+\gamma} + |\bar{y}|^{k-1+\gamma}) + \omega_3(\delta, \varepsilon, n, N), \quad (3.115)$$

where $\limsup_{\delta \to 0} \limsup_{\varepsilon \to 0} \limsup_{n \to +\infty} \limsup_{N \to +\infty} \omega_3(\delta, \varepsilon, n, N) = 0$. Since

$$\max_{r \geq 0} \left(-\nu\delta r^k + C\delta r^{k-1+\gamma} \right) \leq C_1\delta,$$

taking $\limsup_{\delta \to 0} \limsup_{\varepsilon \to 0} \limsup_{n \to +\infty} \limsup_{N \to +\infty}$ in (3.115), we conclude that

$$\nu\bar{m} \leq 0,$$

which is a contradiction unless $m \leq 0$.

For $\gamma = 0$ the proof is almost the same. We replace the functions

$$\delta|x|^k \quad \text{and} \quad \delta|x|^k$$

by

$$\delta e^{b\sqrt{1+|x|^2}} \quad \text{and} \quad \delta e^{b\sqrt{1+|y|^2}},$$

respectively, where $b > a$ is such that $\nu > 2M_F b + 4N_F(b + b^2)$. We then obtain, in place of (3.114),

$$\nu(u(\bar{x}) - v(\bar{y})) - \delta(2M_F b + 4N_F(b + b^2))\left(e^{b\sqrt{1+|\bar{x}|^2}} + e^{b\sqrt{1+|\bar{y}|^2}}\right) \leq \omega_2(\delta; \varepsilon, n, N)$$

which, using $\nu > 2M_F b + 4N_F(b + b^2)$ and the fact that now

$$\liminf_{\delta \to 0} \liminf_{\varepsilon \to 0} \liminf_{n \to +\infty}\left(u(\bar{x}) - v(\bar{y}) - \delta e^{b\sqrt{1+|\bar{x}|^2}} - \delta e^{b\sqrt{1+|\bar{y}|^2}}\right) > \bar{m} > 0,$$

produces again

$$\nu\bar{m} \leq \omega_4(\delta, \varepsilon, n, N), \tag{3.116}$$

where $\limsup_{\delta \to 0} \limsup_{\varepsilon \to 0} \limsup_{n \to +\infty} \limsup_{N \to +\infty} \omega_4(\delta, \varepsilon, n, N) = 0$.

For $\gamma = 1$ the proof is also very similar. Let $\bar{k} \geq 2$. We take $k_1 > k > \bar{k}$ such that $\nu > M_F k_1 + N_F k_1(k_1 - 1)$ and replace the functions $\delta|x|^k$ and $\delta|x|^k$ in the definition of m_δ respectively by $h(x) = \delta(1 + |x|^2)^{\frac{k}{2}}$ and $h(y) = \delta(1 + |y|^2)^{\frac{k}{2}}$. It is easy to check that

$$|Dh(x)| \leq \delta k(1 + |x|^2)^{\frac{k}{2}-1}|x|, \quad |D^2h(x)| \leq \delta k(k - 1)(1 + |x|^2)^{\frac{k}{2}-1}$$

and so there exists an $r > 0$ such that

$$(1 + |x|)|Dh(x)| \leq \delta k_1(1 + |x|^2)^{\frac{k}{2}} \quad \text{if } |x| \geq r,$$

$$(1 + |x|)^2\|D^2h(x)\| \leq \delta k_1(k_1 - 1)(1 + |x|^2)^{\frac{k}{2}} \quad \text{if } |x| \geq r.$$

If we now repeat the arguments of the proof and use the above estimates we obtain, in place of (3.114),

$$\nu(u(\bar{x}) - v(\bar{y})) - \delta(M_F k_1 + N_F k_1(k_1 - 1))\left((1 + |\bar{x}|^2)^{\frac{k}{2}} + (1 + |\bar{y}|^2)^{\frac{k}{2}}\right)$$

$$\leq \omega_2(\delta; \varepsilon, n, N) + \omega_3(\delta)$$

for some modulus ω_3 which depends on r. The result now follows upon noticing that

$$\liminf_{\delta \to 0} \liminf_{\varepsilon \to 0} \liminf_{n \to +\infty} (u(\bar{x}) - v(\bar{y}) - \delta(1 + |\bar{x}|^2)^{\frac{k}{2}} + \delta(1 + |\bar{y}|^2)^{\frac{k}{2}}) > \bar{m} > 0$$

and using $\nu > M_F k_1 + N_F k_1(k_1 - 1)$.

If $\bar{k} < 2$ we proceed in the same way as for $\bar{k} \geq 2$. We take $k_1 > k > \bar{k}$ such that $\nu > M_F k_1 + N_F k_1$ and as before take $h(x) = \delta(1 + |x|^2)^{\frac{k}{2}}$ and $h(y) = \delta(1 + |y|^2)^{\frac{k}{2}}$. However now

$$D^2 h(x) = \delta k(k-2)(1 + |x|^2)^{\frac{k}{2}-2} x \otimes x + \delta k(1 + |x|^2)^{\frac{k}{2}-1} I \leq \delta k(1 + |x|^2)^{\frac{k}{2}-1} I.$$

Thus when we plug the derivatives of h into the equation in the proof of comparison, using Hypothesis 3.46 we can replace $D^2 h(\bar{x})$ by $\delta k(1 + |\bar{x}|^2)^{\frac{k}{2}-1} I$ and also do similarly for $D^2 h(\bar{y})$. The rest of the arguments are the same. $\qquad\square$

Remark 3.57 The conditions in (3.100)–(3.102) may not be optimal for some equations due to the rather general assumption Hypothesis 3.55 and the way it is written. However they are optimal in some cases. Consider a simple first-order equation $u - xu' = 0$ in $\mathbb{R}$ which has two obvious classical solutions $u_1 \equiv 0$ and $u_2(x) = x$, and the second-order equation $2u - x^2 u'' = 0$ which has solutions $u_1 \equiv 0$ and $u_2(x) = x^2$. For $u - xu' = 0$, (3.102) produces $\bar{k} < 1$, and for $2u - x^2 u'' = 0$ we obtain $\bar{k} < 2$. Equation $u - \mu u' = 0$, $\mu > 0$, has two classical solutions $u_1 \equiv 0$ and $u_2(x) = e^{x/\mu}$. Notice that here $M_F = \mu/2$ and (3.101) gives $a < 1/\mu$. $\qquad\blacksquare$

Theorem 3.58 (Comparison under strong B-condition) *The conclusions of Theorem 3.56 hold if (3.2) is replaced by (3.3) and Hypothesis 3.48 is replaced by Hypothesis 3.52.*

Proof The proof is exactly the same as the proof of Theorem 3.56 with one modification. Using the notation of this proof, instead of (3.114) (for $\gamma \in (0, 1)$), by (3.3) and Hypothesis 3.52 we now have

$$\nu(u(\bar{x}) - v(\bar{y})) - C\delta(1 + |\bar{x}|^{k-1+\gamma} + |\bar{y}|^{k-1+\gamma}) \leq c_0 \frac{|\bar{x} - \bar{y}|^2_{-1}}{\varepsilon}$$

$$- \frac{|\bar{x} - \bar{y}|^2}{\varepsilon} + \omega_{R_\delta} \left(|\bar{x} - \bar{y}| \left(1 + \frac{|\bar{x} - \bar{y}|_{-1}}{\varepsilon} \right) \right) + \omega_1(\delta, \varepsilon; n, N).$$

$$(3.117)$$

If $\omega_{R_\delta}(r) \leq \nu \bar{m}/4 + K_\delta r$ for some $K_\delta > 0$ we obtain

$$\omega_{R_\delta} \left(|\bar{x} - \bar{y}| \left(1 + \frac{|\bar{x} - \bar{y}|_{-1}}{\varepsilon} \right) \right) \leq \frac{\nu \bar{m}}{4} + K_\delta |\bar{x} - \bar{y}| \left(1 + \frac{|\bar{x} - \bar{y}|_{-1}}{\varepsilon} \right)$$

$$\leq \frac{\nu \bar{m}}{2} + \frac{|\bar{x} - \bar{y}|^2}{\varepsilon} + \tilde{K}_\delta \frac{|\bar{x} - \bar{y}|^2_{-1}}{\varepsilon}$$

for some $\tilde{K}_\delta > 0$ and small enough ε. Therefore putting this in (3.117) and applying (3.107) yields

$$\nu(u(\bar{x}) - v(\bar{y})) - C\delta(1 + |\bar{x}|^{k-1+\gamma} + |\bar{y}|^{k-1+\gamma}) \le \frac{\nu\bar{m}}{2} + \omega_2(\delta; \varepsilon, n, N),$$

$$(3.118)$$

where $\lim\sup_{\varepsilon\to 0} \lim\sup_{n\to+\infty} \lim\sup_{N\to+\infty} \omega_2(\delta; \varepsilon, n, N) = 0$ for sufficiently small δ. This allows us to continue and conclude the proof using the same arguments as those used in the proof of Theorem 3.56. The other cases are similar. □

Remark 3.59 We remark that all results of this section extend to equations of the form

$$u_t + \inf_{\alpha\in\mathcal{A}} \sup_{\beta\in\mathcal{B}} \left\{\langle A_{\alpha,\beta}x, Du\rangle - F_{\alpha,\beta}(t, x, u, Du, D^2u)\right\} = 0$$

and

$$\inf_{\alpha\in\mathcal{A}} \sup_{\beta\in\mathcal{B}} \left\{-\langle A_{\alpha,\beta}x, Du\rangle + F_{\alpha,\beta}(t, x, u, Du, D^2u)\right\} = 0,$$

where $\mathcal{A}$, $\mathcal{B}$ are arbitrary sets, provided that all assumptions are satisfied by $A_{\alpha,\beta}$, $F_{\alpha,\beta}$, uniformly in α and β. Since the definition of a viscosity solution depended on the operators A and B, we have to assume that there exist a linear, densely defined, maximal dissipative operator A in H such that $D(A^*) \subset D(A^*_{\alpha,\beta})$ for all α, β, and a bounded, strictly positive, self-adjoint operator B on H such that A^*B is bounded. To ensure the uniformity of test functions, they are now defined by Definition 3.32 for A, and the notion of B-continuity is defined using our fixed B which works for all $A_{\alpha,\beta}$. All operators $A_{\alpha,\beta}$ must then satisfy either the weak or strong B-condition with this B and a constant c_0 independent of α and β. ■

3.6   Existence of Solutions: Value Function

In this section we investigate the existence of viscosity solutions for Hamilton–Jacobi–Bellman equations associated with stochastic optimal control problems. In such cases the Hamiltonians F in Eqs. (3.56) and (3.62) are convex/concave in u, Du, D^2u. We show that, under suitable hypotheses, the unique viscosity solution of (3.62) (respectively, (3.56)) is the value function of the associated finite horizon (respectively, infinite horizon) optimal control problem. A key ingredient in the proof will be the use of the dynamic programming principle (Theorem 2.24). We recall briefly the weak formulation of a stochastic optimal control problem that has been introduced in Chap. 2.

We fix a final time $0 < T \le +\infty$, a Polish space Λ (the control space), a real, separable Hilbert space Ξ (the space of the noise) and $Q \in \mathcal{L}_1^+(\Xi)$.

Following Definition 2.7, for $t \in [0, T)$, we say that the 5-tuple $\nu := \left(\Omega^\nu, \mathscr{F}^\nu, \mathscr{F}^{\nu,t}_s, \mathbb{P}^\nu, W^\nu_Q\right)$ is a *reference probability space* if:

(i) $(\Omega^\nu, \mathscr{F}^\nu, \mathbb{P}^\nu)$ is a complete probability space.

(ii) $W_Q^\nu = \{W_Q^\nu(s)\}_{s \in [t,T]}$ is a Ξ-valued Q-Wiener process on $(\Omega^\nu, \mathscr{F}^\nu, \mathbb{P}^\nu)$ (with $W_Q^\nu(t) = 0$, $\mathbb{P}^\nu$-a.s.).

(iii) The filtration $\mathscr{F}_s^{\nu,t} = \sigma\left(\mathscr{F}_s^{\nu,t,0}, \mathcal{N}\right)$, where $\mathscr{F}_s^{\nu,t,0} = \sigma\left(W_Q^\nu(\tau) : t \le \tau \le s\right)$ and $\mathcal{N}$ are the $\mathbb{P}^\nu$-null sets in $\mathscr{F}^\nu$.

We say that a process $a(\cdot)$ is an admissible control on $[t, T]$ (respectively on $[t, +\infty)$ if $T = +\infty$) if there exists a reference probability space $\nu = \left(\Omega^\nu, \mathscr{F}^\nu, \mathscr{F}_s^{\nu,t}, \mathbb{P}^\nu, W_Q^\nu\right)$ such that $a(\cdot) : [t, T] \times \Omega^\nu \to \Lambda$ (respectively $a(\cdot) : [t, +\infty) \times \Omega^\nu \to \Lambda$) is $\mathscr{F}_s^{\nu,t}$-progressively measurable. To indicate the dependence of $a(\cdot)$ on the reference probability space we will write $a^\nu(\cdot)$ and, with a slight abuse of notation, we will often write $a^\nu(\cdot)$ to denote the whole 6-tuple $\left(\Omega^\nu, \mathscr{F}^\nu, \mathscr{F}_s^{\nu,t}, \mathbb{P}^\nu, W_Q^\nu, a^\nu(\cdot)\right)$. We denote the set of all admissible controls $a^\nu(\cdot)$ by $\mathcal{U}_t$.

The finite horizon problem: Let $T < +\infty$. For any $a^\nu(\cdot) \in \mathcal{U}_t$ we consider the system evolving according to the following state equation

$$\begin{cases} dX(s) = (AX(s) + b(s, X(s), a^\nu(s)))\, ds + \sigma(s, X(s), a^\nu(s)) dW_Q^\nu(s) \\ X(t) = x, \end{cases}$$

(3.119)

where A is a linear, densely defined, maximal dissipative operator in H generating a C_0-semigroup of contractions e^{tA}. The functions b and σ satisfy conditions that will be specified below. Our hypotheses will guarantee that (3.119) admits, for any $a^\nu(\cdot) \in \mathcal{U}_t$, a unique mild solution (see Definition 1.119) denoted by $X(\cdot; t, x, a^\nu(\cdot))$. We consider the problem of minimizing a cost functional

$$J(t, x; a^\nu(\cdot)) = \mathbb{E}^\nu \left[\int_t^T e^{-\int_t^s c(X(\tau; t, x, a^\nu(\cdot)))d\tau} l(s, X(s; t, x, a^\nu(\cdot)), a^\nu(s)) ds \right.$$
$$\left. + e^{-\int_t^T c(X(\tau; t, x, a^\nu(\cdot)))d\tau} g(X(T; t, x, a^\nu(\cdot))) \right]$$

(3.120)

over all $a^\nu(\cdot) \in \mathcal{U}_t$. The value function of this minimization problem is defined as follows:

$$V(t, x) := \inf_{a^\nu(\cdot) \in \mathcal{U}_t} J(t, x; a^\nu(\cdot)),$$

(3.121)

while the associated Hamilton–Jacobi–Bellman equation is given by

$$\begin{cases} v_t + \langle Ax, Dv \rangle - F(t, x, v, Dv, D^2v) = 0 \\ v(T, x) = g(x), \end{cases}$$

(3.122)

where

$$F(t, x, r, p, X) := \sup_{a \in \Lambda} \left\{ -\frac{1}{2} \text{Tr} \left(\left(\sigma(t, x, a) Q^{\frac{1}{2}} \right) \left(\sigma(t, x, a) Q^{\frac{1}{2}} \right)^* X \right) \right.$$
$$\left. - \langle p, b(t, x, a) \rangle + c(x) r - l(t, x, a) \right\}.$$
(3.123)

Remark 3.60 We point out that if $\sigma(t, x, a) \in \mathcal{L}(\Xi, H)$, then the term

$$\left(\sigma(t, x, a) Q^{\frac{1}{2}} \right) \left(\sigma(t, x, a) Q^{\frac{1}{2}} \right)^*$$

can be written in a more common and convenient form

$$\sigma(t, x, a) Q \sigma(t, x, a)^*. \qquad \blacksquare$$

The infinite horizon problem: We only study the case of constant discounting, i.e. we assume that $c = \lambda > 0$. However, under suitable assumptions, the results we prove could be adapted to a more general case of non-constant c. For any $a^\nu(\cdot) \in \mathcal{U}_0$ we consider a system described by a stochastic differential equation

$$\begin{cases} dX(s) = (AX(s) + b(X(s), a^\nu(s))) \, ds + \sigma(X(s), a^\nu(s)) dW^\nu_Q(s) \\ X(0) = x. \end{cases}$$
(3.124)

The mild solution of (3.124) will be denoted by $X(\cdot; 0, x, a^\nu(\cdot))$. The infinite horizon problem consists in minimizing a cost functional

$$J(x; a^\nu(\cdot)) = \mathbb{E}^\nu \left[\int_0^{+\infty} e^{-\lambda t} l(X(s; 0, x, a^\nu(\cdot)), a^\nu(s)) ds \right]$$
(3.125)

over all controls $a^\nu(\cdot) \in \mathcal{U}_0$. The value function is given by

$$V(x) := \inf_{a^\nu(\cdot) \in \mathcal{U}_0} J(x; a^\nu(\cdot)),$$
(3.126)

and the corresponding Hamilton–Jacobi–Bellman equation is

$$\lambda v(x) - \langle Ax, Dv \rangle + F(x, v, Dv, D^2 v) = 0,$$
(3.127)

where

$$F(x, r, p, X) := \sup_{a \in \Lambda} \left\{ -\frac{1}{2} \text{Tr} \left(\left(\sigma(x, a) Q^{\frac{1}{2}} \right) \left(\sigma(x, a) Q^{\frac{1}{2}} \right)^* X \right) - \langle p, b(x, a) \rangle - l(x, a) \right\}.$$
(3.128)

3.6.1 Finite Horizon Problem

In this subsection we prove that, under suitable hypotheses, the value function (3.121) of the finite horizon problem is the unique viscosity solution of (3.122). We obtain the results under two sets of hypotheses: in the first the generator A satisfies the weak B-condition for some operator B (see Definition 3.9) and in the second it satisfies the strong B-condition (Definition 3.10) which allows us to put milder assumptions on the coefficients of the state equation and of the cost functional. Note that the words *strong* and *weak* used to describe the B-conditions have nothing to do with the strong and weak formulation of the optimal control problem (see Sects. 2.1.1 and 2.1.2).

To avoid cumbersome notation we will drop the index "ν" whenever it does not cause any confusion.

Proposition 3.61 (Regularity of V under weak B-condition) *Let B satisfy the weak B-condition for A (Definition 3.9) and b: $[0, T] \times H \times \Lambda \to H, \sigma$: $[0, T] \times H \times \Lambda \to \mathcal{L}_2(\Xi_0, H)$, l: $[0, T] \times H \times \Lambda \to \mathbb{R}$ be continuous. Assume that b and σ satisfy (3.7), (3.9), (3.10) with $\gamma = 1$, (3.21) and (3.22), and let c be bounded from below. Suppose that there exist local moduli $\omega_l(\cdot, \cdot)$ and $\omega(\cdot, \cdot)$ such that*

$$|l(t, x, a) - l(s, y, a)| \leq \omega_l(|x - y|_{-1} + |s - t|, R)$$
$$\text{for all } x, y \in B(0, R), a \in \Lambda, s, t \in [0, T]$$
$$(3.129)$$

and

$$|g(x) - g(y)|, |c(x) - c(y)| \leq \omega(|x - y|_{-1}, R) \quad \text{for all } x, y \in B(0, R). \quad (3.130)$$

Moreover, assume that there exist two nonnegative constants C, m such that

$$|c(x)|, |g(x)|, |l(t, x, a)| \leq C(1 + |x|^m) \quad (3.131)$$

for all $x \in H, a \in \Lambda$ and $t \in [0, T]$. Then:

(i) There exists a local modulus $\sigma_1(\cdot, \cdot)$ such that

$$|J(t, x; a(\cdot)) - J(t, y; a(\cdot))| \leq \sigma_1(|x - y|_{-1}, R) \quad (3.132)$$

for all $x, y \in B(0, R), t \in [0, T]$ and $a(\cdot) \in \mathcal{U}_t$.

(ii) There exists a nonnegative constant $\tilde{C}$ and a local modulus $\sigma_2(\cdot, \cdot)$ such that

$$|J(t, x; a(\cdot))|, |V(t, x)| \leq \tilde{C}(1 + |x|^m), \quad (3.133)$$

for all $(t, x) \in [0, T] \times H$ and $a(\cdot) \in \mathcal{U}_t$, and

$$|V(t, x) - V(s, y)| \leq \sigma_2(|t - s| + |x - y|_{-1}, R) \quad (3.134)$$

for all $x, y \in B(0, R)$, $t, s \in [0, T]$.

Proof Part (i): Let L be a constant such that $c(x) \geq L$ for all $x \in H$. We will assume that $L < 0$. Choose $x, y \in B(0, R)$, $a(\cdot) \in \mathcal{U}_t$ and denote $X(\cdot; t, x, a(\cdot))$ and $X(\cdot; t, y, a(\cdot))$ respectively by $X(\cdot)$ and $Y(\cdot)$. We have

$$|J(t, y; a(\cdot)) - J(t, x; a(\cdot))| \leq I_1 + I_2$$

$$:= \left(\int_t^T \mathbb{E} \left| e^{-\int_t^r c(X(\tau))d\tau} l(r, X(r), a(r)) - e^{-\int_t^r c(Y(\tau))d\tau} l(r, Y(r), a(r)) \right| dr \right)$$

$$+ \left(\mathbb{E} |e^{-\int_t^T c(X(\tau))d\tau} g(X(T)) - e^{-\int_t^T c(Y(\tau))d\tau} g(Y(T))| \right). \quad (3.135)$$

We first consider I_1.

$$I_1 \leq I_{11} + I_{12}$$

$$:= \int_t^T \mathbb{E} \left[e^{-\int_t^r c(X(\tau))d\tau} |l(r, X(r), a(r)) - l(r, Y(r), a(r))| \right] dr$$

$$+ \int_t^T \mathbb{E} \left[|l(r, Y(r), a(r))| \left| e^{-\int_t^r c(X(\tau))d\tau} - e^{-\int_t^r c(Y(\tau))d\tau} \right| \right] dr.$$

In the following we will denote by M any absolute constant independent of R and of the control. Given $\varepsilon > 0$ we can find, thanks to (D.1), a positive constant K_ε (non-increasing in ε) such that, for any $s > 0$, $\omega_l(s, \frac{1}{\varepsilon}) \leq \varepsilon + K_\varepsilon s$. Using (3.129) and (3.131), we obtain

$$I_{11} \leq e^{-TL} \int_t^T \mathbb{E} |l(r, X(r), a(r)) - l(r, Y(r), a(r))| dr$$

$$\leq e^{-TL} \int_t^T \int_{\{|X(r)| < \frac{1}{\varepsilon} \text{ and } |Y(r)| < \frac{1}{\varepsilon}\}} \omega_l \left(|X(r) - Y(r)|_{-1}, \frac{1}{\varepsilon} \right) d\mathbb{P} dr$$

$$+ e^{-TL} \int_t^T \int_{\{|X(r)| \geq \frac{1}{\varepsilon} \text{ or } |Y(r)| \geq \frac{1}{\varepsilon}\}} M(2 + |X(r)|^m + |Y(r)|^m) d\mathbb{P} dr.$$

$$\leq M \int_t^T (\varepsilon + K_\varepsilon \mathbb{E} |X(r) - Y(r)|_{-1}) dr$$

$$+ M \int_t^T \left(\mathbb{E}(1 + |X(r)|^{2m} + |Y(r)|^{2m}) \right)^{\frac{1}{2}} \left(\mathbb{P} \left(|X(r)| \geq \frac{1}{\varepsilon} \right) + \mathbb{P} \left(|Y(r)| \geq \frac{1}{\varepsilon} \right) \right)^{\frac{1}{2}} dr.$$

It follows from (1.37) that we have

$$\mathbb{E}(\sup_{t \leq r \leq T} |X(r)|^{2m} + \sup_{t \leq r \leq T} |Y(r)|^{2m}) \leq C_R, \quad (3.136)$$

where C_R is a constant independent of the control but depending on R. In particular, this implies by Chebychev's inequality that

$$\left(\mathbb{P}\left(\sup_{t \le r \le T} |X(r)| \ge \frac{1}{\varepsilon}\right) + \mathbb{P}\left(\sup_{t \le r \le T} |Y(r)| \ge \frac{1}{\varepsilon}\right)\right)^{\frac{1}{2}} \le \gamma(\varepsilon, R), \qquad (3.137)$$

for some local modulus γ. Thus, using (3.136), (3.137) and (3.23), we obtain

$$I_{11} \le M(\varepsilon + K_\varepsilon |x - y|_{-1}) + \gamma_1(\varepsilon, R) \qquad (3.138)$$

for some local modulus γ_1. Taking the infimum of the right-hand side of (3.138) over $\varepsilon > 0$ produces a local modulus $\varrho(\cdot, R)$ such that

$$I_{11} \le \varrho(|x - y|_{-1}, R). \qquad (3.139)$$

To estimate I_{12} observe that

$$\int_t^T \mathbb{E}\left[|l(r, Y(r), a(r))| \left| e^{-\int_t^s c(X(\tau))d\tau} - e^{-\int_t^s c(Y(\tau))d\tau}\right|\right] dr$$

$$\le \left(\int_t^T \mathbb{E}\left[|l(r, Y(r), a(r))|^2\right] dr\right)^{\frac{1}{2}}$$

$$\times \left(\int_t^T \mathbb{E}\left[\left|e^{-\int_t^s c(X(\tau))d\tau} - e^{-\int_t^s c(Y(\tau))d\tau}\right|^2\right] dr\right)^{\frac{1}{2}}.$$

We observe that for $a, b \in \mathbb{R}, a, b \ge TL$ one has $\left|e^{-a} - e^{-b}\right| \le e^{-TL}|a - b|$. Therefore, using (3.136), it follows similarly as before that, for some $C_R > 0$ depending on R but independent of the choice of the control,

$$I_{12} \le C_R\left[\mathbb{E}\left(\int_t^T \omega\left(|X(r) - Y(r)|_{-1}, \frac{1}{\varepsilon}\right) dr\right)^2\right.$$

$$\left. + \mathbb{P}\left(\sup_{t \le r \le T} |X(r)| \ge \frac{1}{\varepsilon}\right) + \mathbb{P}\left(\sup_{t \le r \le T} |Y(r)| \ge \frac{1}{\varepsilon}\right)\right]^{\frac{1}{2}}.$$

We now use again (3.137), (3.23), and argue as for I_{11} to find that there exists a local modulus $\varrho(\cdot, R)$ such that

$$I_{12} \le \varrho(|x - y|_{-1}, R). \qquad (3.140)$$

The term I_2 in (3.135) can be estimated similarly. Thus we obtain claim (i).

Part (ii): Estimate (3.133) follows directly from (3.131) and (1.37). Moreover, by (3.132), we have

$$|V(t, x) - V(t, y)| \le \sigma_1(|x - y|_{-1}, R) \qquad \forall x, y \in B(0, R), t \in [0, T]. \quad (3.141)$$

It remains to prove that

$$|V(t, x) - V(s, x)| \leq \tilde{\sigma}(|t - s|, R) \qquad \forall x \in B(0, R), t, s \in [0, T] \qquad (3.142)$$

for some local modulus $\tilde{\sigma}$.

We notice that it follows from our assumptions, (3.132), (3.133), and Proposition 2.16, that the assumptions of Theorem 2.24 are satisfied and thus the dynamic programming principle (2.23) holds. Let now (3.142), let $0 \leq t < s \leq T$ and $x \in B(0, R)$. Let $X(\cdot)$ be the solution of (3.119).

Using (2.23) we have, for some constant C_R depending on R,

$$
\begin{aligned}
|V(s, x) - V(t, x)| &\leq \sup_{a(\cdot) \in \mathcal{U}_t} e^{-LT} \mathbb{E} \int_t^s |l(r, X(r), a(r))| dr \\
&\quad + \sup_{a(\cdot) \in \mathcal{U}_t} \mathbb{E} \left| V(s, x) - V(s, X(s)) e^{-\int_t^s c(X(r)) dr} \right| \\
&\leq C_R |t - s| + e^{-LT} \sup_{a(\cdot) \in \mathcal{U}_t} \mathbb{E} |V(s, x) - V(s, X(s))| \\
&\quad + \sup_{a(\cdot) \in \mathcal{U}_t} \mathbb{E} \left| V(s, x) \left(e^{-\int_t^s c(X(r)) dr} - 1 \right) \right| =: C_R |t - s| + D_1 + D_2,
\end{aligned}
$$
$$(3.143)$$

where we have used (1.37), (3.131).

Since V satisfies (3.133) and (3.141), arguing as in the estimates for I_{11} and using (3.25) and (3.137), we obtain for every $\varepsilon > 0$, $a(\cdot) \in \mathcal{U}_t$

$$
\begin{aligned}
\mathbb{E} |V(s, x) - V(s, X(s))| &\leq \mathbb{E} \sigma_1 (|x - X(s)|_{-1}, \frac{1}{\varepsilon}) + C_R \left(\mathbb{P}(\sup_{t \leq r \leq T} |X(r)| \geq \frac{1}{\varepsilon}) \right)^{\frac{1}{2}} \\
&\leq \varepsilon + C_{R,\varepsilon} |t - s|^{\frac{1}{2}} + \gamma(\varepsilon, R),
\end{aligned}
$$
$$(3.144)$$

thus the same estimate holds for D_1.

To estimate D_2, let $C_\varepsilon \geq 0$ be a constant such that $c(y) \leq C_\varepsilon$ when $|y| \leq \frac{1}{\varepsilon}$. Then, for every $\varepsilon > 0$, $a(\cdot) \in \mathcal{U}_t$,

$$
\begin{aligned}
\mathbb{E} \left(|V(s, x)| \left| e^{-\int_t^s c(X(r)) dr} - 1 \right| \right) & \\
&\hspace{-6em} \leq C_R \max \left(e^{-|t-s|L} - 1, \left(1 - e^{-C_\varepsilon |t-s|} \right) + \mathbb{P}(\sup_{t \leq r \leq T} |X(r)| \geq \frac{1}{\varepsilon}) \right) \\
&\hspace{-6em} \leq C_R \max \left(e^{-|t-s|L} - 1, \left(1 - e^{-C_\varepsilon |t-s|} \right) + \gamma(\varepsilon, R) \right), \quad (3.145)
\end{aligned}
$$

and D_2 satisfies the same estimate. Plugging (3.144) and (3.145) into (3.143) and taking the infimum over $\varepsilon > 0$ provides (3.142).   □

Proposition 3.62 (Regularity of V under strong B-condition) *Let B satisfy the strong B-condition for A (Definition 3.10). Let $b: [0, T] \times H \times \Lambda \to H$, $\sigma: [0, T] \times H \times \Lambda \to \mathcal{L}_2(\Xi_0, H)$, $l: [0, T] \times H \times \Lambda \to \mathbb{R}$ be continuous, let b and σ satisfy*

(3.7), (3.9), (3.10) with $\gamma = 1$ and (3.22), and let c be bounded from below. Suppose that there exist local moduli $\omega_l(\cdot, \cdot)$ and $\omega(\cdot, \cdot)$ such that

$$|l(t, x, a) - l(s, y, a)| \leq \omega_l(|x - y| + |s - t|, R), \qquad (3.146)$$

for all $x, y \in B(0, R)$, $a \in \Lambda$, $s, t \in [0, T]$ and

$$|g(x) - g(y)|, |c(x) - c(y)| \leq \omega(|x - y|, R), \qquad (3.147)$$

for all $x, y \in B(0, R)$, and that (3.131) holds.
Then:

(i) The functions J and V satisfy (3.133) and there exists a local modulus $\sigma(\cdot, \cdot)$ such that
$$|J(t, x; a(\cdot)) - J(t, y; a(\cdot))| \leq \sigma(|x - y|, R) \qquad (3.148)$$

    for all $x, y \in B(0, R)$, $t \in [0, T]$, $a(\cdot) \in \mathcal{U}_t$.
(ii) For any $\tau \in (0, T)$, there exists a local modulus $\sigma_\tau(\cdot, \cdot)$ such that

$$|V(t, x) - V(t, y)| \leq \sigma_\tau(|x - y|_{-1}, R) \qquad (3.149)$$

    for all $x, y \in B(0, R)$, $t \in [0, \tau]$.
(iii) There exists a local modulus $\rho(\cdot, \cdot)$ such that

$$|V(t, x) - V(s, e^{(s-t)A}x)| \leq \rho(s - t, R) \qquad (3.150)$$

    for all $x \in B(0, R)$, $0 \leq t \leq s \leq T$.

Proof Obviously J and V satisfy (3.133) as in Proposition 3.61. Also (3.148) is proved exactly as (3.132) in Proposition 3.61. The only difference is that, since l, g, c are now continuous in the usual norm of H instead of the $|\cdot|_{-1}$ norm, we have to replace (3.23) by (1.39).

To show (3.149) we begin as in (3.135). The term I_1 is estimated in exactly the same way as in the proof of Proposition 3.61 using (3.27) instead of (3.23). For the term I_2 we have

$$I_2 \leq I_{21} + I_{22} := \mathbb{E}\left[e^{-\int_t^T c(X(r))dr} \, |g(X(T)) - g(Y(T))|\right]$$
$$+ \mathbb{E}\left[|g(Y(T))| \left|e^{-\int_t^T c(X(r))dr} - e^{-\int_t^T c(Y(r))dr}\right|\right].$$

The term I_{22} is again standard if we use (3.27). If g satisfied (3.130) we could also proceed as before with the term I_{21} to obtain (3.152) (see Remark 3.63). Since g only satisfies (3.147) we have to proceed slightly differently. We have, by (3.131), (3.147), (3.136), (3.137), (3.28)

$$I_{21} \le e^{-TL}\mathbb{E}\,|g(X(T)) - g(Y(T))| \le e^{-TL}\mathbb{E}\,\omega\left(|X(T) - Y(T)|, \frac{1}{\varepsilon}\right)$$

$$+ e^{-TL}\int_{\{|X(T)|\ge\frac{1}{\varepsilon}\ \text{or}\ |Y(T)|\ge\frac{1}{\varepsilon}\}} C(2 + |X(T)|^m + |Y(T)|^m)d\mathbb{P}.$$

$$\le e^{-TL}(\varepsilon + K_\varepsilon\mathbb{E}|X(T) - Y(T)|) + \gamma(\varepsilon, R)$$

$$\le e^{-TL}\varepsilon + e^{-TL}\left(\frac{C(T)}{T - \tau}\right)^{\frac{1}{2}}|x - y|_{-1} + \gamma(\varepsilon, R),$$

where $C(T)$ is the constant from (3.28) and γ is a local modulus. It remains to take the infimum over all $\varepsilon > 0$.

The proof of (3.150) is also very similar to the proof of (3.142). We can now claim that the dynamic programming principle is satisfied and thus, as in (3.151), if $x \in B(0, R)$ and $0 \le t \le s \le T$, we have

$$|V(t, x) - V(s, e^{(s-t)A}x)| \le \sup_{a(\cdot)\in\mathcal{U}_t} e^{-LT}\mathbb{E}\int_t^s |l(r, X(r), a(r))|dr$$

$$+ e^{-LT}\sup_{a(\cdot)\in\mathcal{U}_t}\mathbb{E}\left|V(s, e^{(s-t)A}x) - V(s, X(s))\right|$$

$$+ \sup_{a(\cdot)\in\mathcal{U}_t}\mathbb{E}\left|V(s, e^{(s-t)A}x)\left(e^{-\int_t^s c(X(r))dr} - 1\right)\right|, \quad (3.151)$$

where $X(r)$ is the solution of (3.119). The first and the third term above are estimated as in (3.143) and (3.145). For the middle term we first notice that there exists some constant C_R depending on R but independent of the control such that

$$\mathbb{E}|X(s) - e^{(s-t)A}x|^2 \le C_R(s - t).$$

Therefore, using (3.148),

$$\mathbb{E}\left|V(s, e^{(s-t)A}x) - V(s, X(s))\right| \le \mathbb{E}\,\sigma\left(|X(s) - e^{(s-t)A}x|, \frac{1}{\varepsilon}\right)$$

$$+ C_R\left(\mathbb{P}(\sup_{t\le r\le T}|X(r)| \ge \frac{1}{\varepsilon})\right)^{\frac{1}{2}} \le \varepsilon + C_{R,\varepsilon}|t - s|^{\frac{1}{2}} + \gamma(\varepsilon, R),$$

which implies the claim as all the constants and the local modulus γ are independent of t and the controls. $\qquad\square$

Remark 3.63 It follows easily from the above proof that if g satisfies (3.130) instead of (3.147), then

$$|V(t, x) - V(s, y)| \le \omega(|t - s| + |x - y|_{-1}, R) \qquad \forall x, y \in B(0, R), t, s \in [0, T]$$
$$(3.152)$$

for some local modulus $\omega(\cdot, \cdot)$. $\qquad\blacksquare$

Remark 3.64 It is clear from the proof that (3.148), and the same estimate for V, still holds if (3.22) is replaced by (3.8) and the strong B-condition for A is replaced by a standard requirement that A generates a C_0-semigroup. ∎

In the next lemma we provide Itô's-like formulae for test functions $\psi = \varphi + h(t, |x|)$ introduced in Definition 3.32. As we remarked after this definition, even though $|x|$ is not differentiable at 0, the function $h_0(t, x) := h(t, |x|) \in C^{1,2}((0, T) \times H)$, so with a slight abuse of notation, in the following we will write $h(t, x)$ instead of $h(t, |x|)$, $Dh(t, x)$ instead of $Dh_0(t, x) = \frac{x}{|x|}\frac{d}{dr}h(t, r)|_{r=|x|}$ (which is 0 when $x = 0$), and $D^2h(t, x)$ instead of $D^2h_0(t, x)$.

Lemma 3.65 *Let b and σ be continuous, satisfy (3.7)–(3.10), and let $c : H \to \mathbb{R}$ be continuous, bounded from below and satisfy (3.131). Consider a test function (in the sense of Definition 3.32) $\psi = \varphi + h$. Suppose that h satisfies (1.110) and consider the solution $X(\cdot)$ of (3.119) for a given control $a(\cdot) \in \mathcal{U}_t$. Then, for any $s \in [t, T]$,*

$$\mathbb{E}\left[e^{-\int_t^s c(X(\tau))d\tau}\varphi(s, X(s))\right] = \varphi(t, x)$$

$$\mathbb{E}\left[\int_t^s e^{-\int_t^r c(X(\tau))d\tau}\left(\varphi_t(r, X(r)) + \langle X(r), A^*D\varphi(r, X(r))\rangle\right.\right.$$

$$+ \langle b(r, X(r), a(r)), D\varphi(r, X(r))\rangle - c(X(r))\varphi(r, X(r))$$

$$\left.\left.+\frac{1}{2}\text{Tr}\left[\left(\sigma(r, X(r), a(r))Q^{\frac{1}{2}}\right)\left(\sigma(r, X(r), a(r))Q^{\frac{1}{2}}\right)^* D^2\varphi(r, X(r))\right]\right)dr\right]$$

$$\tag{3.153}$$

and

$$\mathbb{E}\left[e^{-\int_t^s c(X(\tau))d\tau}h(s, X(s))\right] \leq h(t, x) + \mathbb{E}\left[\int_t^s e^{-\int_t^r c(X(\tau))d\tau}\left(h_t(r, X(r))\right.\right.$$

$$+ \langle b(r, X(r), a(r)), Dh(r, X(r))\rangle - c(X(r))h(r, X(r))$$

$$\left.\left.+\frac{1}{2}\text{Tr}\left[\left(\sigma(r, X(r), a(r))Q^{\frac{1}{2}}\right)\left(\sigma(r, X(r), a(r))Q^{\frac{1}{2}}\right)^* D^2h(r, X(r))\right]\right)dr\right].$$

$$\tag{3.154}$$

Proof We define $c_n(y) := \min(c(y), n)$ and observe that $\eta_n(r) := e^{-\int_t^r c_n(X(\tau))d\tau}$ is the unique solution of $d\eta_n(r) = b_n(r)dr$ (and $\eta_n(t) = 1$), where $b_n(r) = -\eta_n(r)c_n(X(\tau))$ is bounded. We can thus use Propositions 1.165 and 1.166 and send $n \to +\infty$ to obtain the claim. □

Theorem 3.66 (Existence under weak B-condition) *Let the assumptions of Proposition 3.61 be satisfied, and let in addition $b(\cdot, x, a)$ and $\sigma(\cdot, x, a)$ be uniformly continuous on $[0, T]$, uniformly in $(x, a) \in B(0, R) \times \Lambda$ for every $R > 0$. Suppose also that, for every (t, x),*

$$\lim_{N \to +\infty} \sup_{a \in \Lambda} \mathrm{Tr} \left[\left(\sigma(t, x, a) Q^{\frac{1}{2}} \right) \left(\sigma(t, x, a) Q^{\frac{1}{2}} \right)^* B Q_N \right] = 0. \tag{3.155}$$

Then the value function $V(t, x)$, defined in (3.121), is the unique viscosity solution of (3.122) among functions in the set

$$S := \left\{ u \colon [0, T] \times H \to \mathbb{R} \ : \ |u(t, x)| \le C(1 + |x|^k) \text{ for some } k \ge 0, \right.$$

$$\left. \lim_{t \to T} |u(t, x) - g(x)| = 0 \text{ uniformly on bounded subsets of } H \right\}.$$

Proof Without loss of generality we can assume that c is positive since if $c \ge L$ for $L < 0$ then V is a viscosity solution of (3.122) if and only if $\tilde{V} = e^{L(T-t)} V$ is a viscosity solution of (3.122) with c replaced by $\tilde{c} = c - L$ and l replaced by $e^{L(T-t)} l$.

Existence: Proposition 3.61 ensures that V is B-continuous and that it belongs to S. We first prove that V is a viscosity supersolution of (3.122). Let $V + \psi$ have a local minimum at $(t, x) \in (0, T) \times H$ for a test function $\psi = \varphi + h$ (in the sense of Definition 3.32). We can assume that h and its derivatives Dh, $D^2 h$, h_t have polynomial growth (see on this the discussion following Lemma 3.37), that the minimum is global (see Lemma 3.37), and that $V(t, x) + \psi(t, x) = 0$, so for all (s, y) we have $V(s, y) \ge -\psi(s, y)$.

By Proposition 2.16, Theorem 2.24, and Proposition 3.61, the dynamic programming principle (2.23) is satisfied. Thus for $\varepsilon > 0$ there exists a control $a^{v_\varepsilon}(\cdot) \in \mathcal{U}_t$ such that, with $X^{v_\varepsilon}(\cdot) := X(\cdot; t, x, a^{v_\varepsilon}(\cdot))$,

$$V(t, x) + \varepsilon^2 \ge \mathbb{E}^{v_\varepsilon} \left[\int_t^{t+\varepsilon} e^{-\int_t^r c(X^{v_\varepsilon}(\tau)) d\tau} l(r, X^{v_\varepsilon}(r), a^{v_\varepsilon}(r)) dr \right.$$

$$\left. + e^{-\int_t^{t+\varepsilon} c(X^{v_\varepsilon}(\tau)) d\tau} V(t + \varepsilon, X^{v_\varepsilon}(t + \varepsilon)) \right].$$

This implies that

$$\varepsilon^2 - \varphi(t, x) - h(t, x) \ge \mathbb{E}^{v_\varepsilon} \left[\int_t^{t+\varepsilon} e^{-\int_t^r c(X^{v_\varepsilon}(\tau)) d\tau} l(r, X^{v_\varepsilon}(r), a^{v_\varepsilon}(r)) dr \right.$$

$$\left. - e^{-\int_t^{t+\varepsilon} c(X^{v_\varepsilon}(\tau)) d\tau} \varphi(t + \varepsilon, X^{v_\varepsilon}(t + \varepsilon)) - e^{-\int_t^{t+\varepsilon} c(X^{v_\varepsilon}(\tau)) d\tau} h(t + \varepsilon, X^{v_\varepsilon}(t + \varepsilon)) \right].$$

$$\tag{3.156}$$

Using (3.153), (3.154) and (3.156) we find

$$0 \le \varepsilon + \frac{1}{\varepsilon} \mathbb{E}^{v_\varepsilon} \left[\int_t^{t+\varepsilon} e^{-\int_t^r c(X^{v_\varepsilon}(\tau)) d\tau} \left(-l(r, X^{v_\varepsilon}(r), a^{v_\varepsilon}(r)) \right. \right.$$

$$\left. + \psi_t(r, X^{v_\varepsilon}(r)) + \langle b(r, X^{v_\varepsilon}(r), a^{v_\varepsilon}(r)), D\psi(r, X^{v_\varepsilon}(r)) \rangle \right.$$

$$+ \frac{1}{2} \mathrm{Tr} \left[\left(\sigma(r, X^{\nu_\varepsilon}(r), a^{\nu_\varepsilon}(r)) Q^{\frac{1}{2}} \right) \left(\sigma(r, X^{\nu_\varepsilon}(r), a^{\nu_\varepsilon}(r)) Q^{\frac{1}{2}} \right)^* D^2 \psi(r, X^{\nu_\varepsilon}(r)) \right]$$

$$- c(X^{\nu_\varepsilon}(r)) \psi(r, X^{\nu_\varepsilon}(r)) + \left\langle X^{\nu_\varepsilon}(r), A^* D\varphi(r, X^{\nu_\varepsilon}(r)) \right\rangle \bigg) dr \bigg].$$

(3.157)

Now we observe that, thanks to (1.38), we can find a constant $r_\varepsilon > 0$, depending on $\varepsilon > 0$ but independent of the control $a^{\nu_\varepsilon}(\cdot)$, such that $r_\varepsilon \xrightarrow{\varepsilon \to 0^+} 0$, and the set

$$\Omega_1^\varepsilon = \left\{ \omega \in \Omega^{\nu_\varepsilon} : \sup_{r \in [t, t+\varepsilon]} |X^{\nu_\varepsilon}(r) - x| \leq r_\varepsilon \right\},$$

satisfies

$$\mathbb{P}^{\nu_\varepsilon}(\Omega_1^\varepsilon) \to 1 \quad \text{as } \varepsilon \to 0.$$

(3.158)

We set $\Omega_2^\varepsilon = \Omega^{\nu_\varepsilon} \setminus \Omega_1^\varepsilon$. If we denote by $\Psi_\varepsilon(r)$ the integrand in (3.157), the assumptions and properties of test functions imply

$$|\Psi_\varepsilon(r)| \leq C(1 + |X^{\nu_\varepsilon}(r)|^N)$$

(3.159)

for some $N \geq 0$ and C independent of ε. Thus, by (1.37), (3.158), (3.159), and the continuity of the functions in the integrand, we obtain

$$0 \leq \varepsilon + \frac{1}{\varepsilon} \mathbb{E}^{\nu_\varepsilon} \bigg[\int_t^{t+\varepsilon} \bigg(-l(t, x, a^{\nu_\varepsilon}(r)) + \psi_t(t, x) + \langle b(t, x, a^{\nu_\varepsilon}(r)), D\psi(t, x) \rangle$$

$$+ \frac{1}{2} \mathrm{Tr} \left[\left(\sigma(t, x, a^{\nu_\varepsilon}(r)) Q^{\frac{1}{2}} \right) \left(\sigma(t, x, a^{\nu_\varepsilon}(r)) Q^{\frac{1}{2}} \right)^* D^2 \psi(t, x) \right]$$

$$- c(x) \psi(t, x) + \langle x, A^* D\varphi(t, x) \rangle \bigg) \mathbf{1}_{\Omega_1^\varepsilon} dr \bigg]$$

$$+ C \frac{1}{\varepsilon} \int_t^{t+\varepsilon} (\mathbb{P}(\Omega_2^\varepsilon))^{\frac{1}{2}} (\mathbb{E}[1 + |X^{\nu_\varepsilon}(r)|^N]^2)^{\frac{1}{2}} dr + \gamma_1(\varepsilon)$$

$$\leq \frac{1}{\varepsilon} \mathbb{E}^{\nu_\varepsilon} \bigg[\int_t^{t+\varepsilon} \bigg(-l(t, x, a^{\nu_\varepsilon}(r)) + \psi_t(t, x) + \langle b(t, x, a^{\nu_\varepsilon}(r)), D\psi(t, x) \rangle$$

$$+ \frac{1}{2} \mathrm{Tr} \left[\left(\sigma(t, x, a^{\nu_\varepsilon}(r)) Q^{\frac{1}{2}} \right) \left(\sigma(t, x, a^{\nu_\varepsilon}(r)) Q^{\frac{1}{2}} \right)^* D^2 \psi(t, x) \right]$$

$$+ c(x) V(t, x) + \langle x, A^* D\varphi(t, x) \rangle \bigg) dr \bigg] + \gamma_2(\varepsilon)$$

$$\leq \frac{1}{\varepsilon} \int_t^{t+\varepsilon} \mathbb{E}^{\nu_\varepsilon} \big[\psi_t(t, x) + \langle x, A^* D\varphi(t, x) \rangle$$

$$+ F(t, x, V(t, x), -D\psi(t, x), -D^2 \psi(t, x)) \big] dr + \gamma_2(\varepsilon)$$

$$= \psi_t(t, x) + \langle x, A^* D\varphi(t, x) \rangle + F(t, x, V(t, x), -D\psi(t, x), -D^2 \psi(t, x)) + \gamma_2(\varepsilon),$$

where γ_1, γ_2 above are such that $\lim_{\varepsilon \to 0} \gamma_i(\varepsilon) = 0, i = 1, 2$, and are independent of the control $a^{\nu_\varepsilon}(r)$ and of the reference probability space ν_ε. The claim follows after we let $\varepsilon \to 0$.

To show the subsolution property, let $V - \psi$ have a global maximum at (t, x), where h and its derivatives Dh, D^2h, h_t have polynomial growth, and $V(t, x) = \psi(t, x)$. We choose $a \in \Lambda$ and take a constant control $a(\cdot) \equiv a$ defined on some reference probability space, and we define $X(\cdot) := X(\cdot; t, x, a(\cdot))$. Using the dynamic programming principle (2.23) we have

$$
\begin{aligned}
\psi(t, x) &= V(t, x) \\
&\leq \mathbb{E}\left[\int_t^{t+\varepsilon} e^{-\int_t^r c(X(\tau))d\tau} l(r, X(r), a)dr + e^{-\int_t^{t+\varepsilon} c(X(\tau))d\tau} V(t+\varepsilon, X(t+\varepsilon)) \right] \\
&\leq \mathbb{E}\left[\int_t^{t+\varepsilon} e^{-\int_t^r c(X(\tau))d\tau} l(r, X(r), a)dr + e^{-\int_t^{t+\varepsilon} c(X(\tau))d\tau} \psi(t+\varepsilon, X(t+\varepsilon)) \right]
\end{aligned}
$$

and then as before we get

$$
\begin{aligned}
\frac{1}{\varepsilon} \mathbb{E}\Bigg[\int_t^{t+\varepsilon} e^{-\int_t^r c(X(\tau))d\tau} \Bigg(&l(r, X(r), a) + \psi_t(r, X(r)) + \langle X(r), A^* D\varphi(r, X(r))\rangle \\
&- c(X(r))\psi(r, X(r)) + \langle b(r, X(r), a), D\varphi(r, X(r))\rangle \\
&+ \frac{1}{2}\operatorname{Tr}\left[\left(\sigma(r, X(r), a)Q^{\frac{1}{2}}\right)\left(\sigma(r, X(r), a)Q^{\frac{1}{2}}\right)^* D^2\psi(r, X(r)) \right] \Bigg) dr \Bigg] \geq 0.
\end{aligned}
$$

The same argument as in the proof of the supersolution part now yields

$$
\begin{aligned}
l(t, x, a) + \psi_t(t, x) &+ \langle x, A^* D\varphi(t, x)\rangle - c(x)V(t, x) \\
&+ \langle b(t, x, a), D\varphi(t, x)\rangle + \frac{1}{2}\operatorname{Tr}\left[\left(\sigma(t, x, a)Q^{\frac{1}{2}}\right)\left(\sigma(t, x, a)Q^{\frac{1}{2}}\right)^* D^2\psi(t, x) \right] \geq 0
\end{aligned}
$$
$$\tag{3.160}$$

and the claim follows after we take the $\inf_{a \in \Lambda}$ in (3.160).

Uniqueness: To prove the uniqueness of the solution we need to show that the hypotheses of Theorem 3.50 are satisfied with the set $U = H$.

Hypothesis 3.44 follows from the local uniform continuity of $b(\cdot, \cdot, a), \sigma(\cdot, \cdot, a)$, $l(\cdot, \cdot, a), c(\cdot)$, uniform in $a \in \Lambda$, (3.9), (3.10), and (3.131). Hypothesis 3.45 follows from the positivity of c. For Hypothesis 3.46 we can argue as follows: since $\left(\sigma(t, x, a)Q^{\frac{1}{2}}\right)\left(\sigma(t, x, a)Q^{\frac{1}{2}}\right)^*$ is a positive, self-adjoint, trace class operator, it is obvious that, for $X, Y \in S(H)$ with $X \leq Y$,

$$
-\operatorname{Tr}\left(\left(\sigma(t, x, a)Q^{\frac{1}{2}}\right)\left(\sigma(t, x, a)Q^{\frac{1}{2}}\right)^* X \right) \geq -\operatorname{Tr}\left(\left(\sigma(t, x, a)Q^{\frac{1}{2}}\right)\left(\sigma(t, x, a)Q^{\frac{1}{2}}\right)^* Y \right),
$$

and then taking the supremum over $a \in \Lambda$ we see that Hypothesis 3.46 is satisfied. Hypothesis 3.47 follows from (3.155) (see the further comments about it after the end of the proof).

To show that Hypothesis 3.48 holds observe that, using (3.21), (3.130) and (3.129), we have, for $|r|, |x|, |y| \le R$,

$$F\left(t, x, r, \frac{B(x-y)}{\varepsilon}, X\right) - F\left(t, y, r, \frac{B(x-y)}{\varepsilon}, Y\right)$$

$$\ge -\sup_{a\in\Lambda}\left(l(t,x,a) - l(t,y,a) + \left\langle \frac{B(x-y)}{\varepsilon}, b(t,x,a)\right\rangle - \left\langle \frac{B(x-y)}{\varepsilon}, b(t,y,a)\right\rangle\right.$$

$$-r(c(x)-c(y)) + \frac{1}{2}\mathrm{Tr}\left(\left(\sigma(t,x,a)Q^{\frac{1}{2}}\right)\left(\sigma(t,x,a)Q^{\frac{1}{2}}\right)^* X\right)$$

$$\left.-\frac{1}{2}\mathrm{Tr}\left(\left(\sigma(t,y,a)Q^{\frac{1}{2}}\right)\left(\sigma(t,y,a)Q^{\frac{1}{2}}\right)^* Y\right)\right)$$

$$\ge -\sup_{a\in\Lambda}|l(t,x,a) - l(t,y,a)| - R\sup_{a\in\Lambda}|c(x)-c(y)|$$

$$-\sup_{a\in\Lambda}\left\langle \frac{B(x-y)}{\varepsilon}, b(t,x,a) - b(t,y,a)\right\rangle$$

$$-\sup_{a\in\Lambda}\left(\frac{1}{2}\mathrm{Tr}\left(\left(\sigma(t,x,a)Q^{\frac{1}{2}}\right)\left(\sigma(t,x,a)Q^{\frac{1}{2}}\right)^* X\right)\right.$$

$$\left.-\frac{1}{2}\mathrm{Tr}\left(\left(\sigma(t,y,a)Q^{\frac{1}{2}}\right)\left(\sigma(t,y,a)Q^{\frac{1}{2}}\right)^* Y\right)\right)$$

$$\ge -\omega_l(|x-y|_{-1}, R) - C\frac{|x-y|_{-1}^2}{\varepsilon} - R\omega_c(|x-y|_{-1}, R)$$

$$-\sup_{a\in\Lambda}\left(\frac{1}{2}\mathrm{Tr}\left(\left(\sigma(t,x,a)Q^{\frac{1}{2}}\right)\left(\sigma(t,x,a)Q^{\frac{1}{2}}\right)^* X\right)\right.$$

$$\left.-\frac{1}{2}\mathrm{Tr}\left(\left(\sigma(t,y,a)Q^{\frac{1}{2}}\right)\left(\sigma(t,y,a)Q^{\frac{1}{2}}\right)^* Y\right)\right).$$

To estimate the last term we use that X and Y satisfy (3.52). In particular we have

$$\begin{pmatrix} X & 0 \\ 0 & -Y \end{pmatrix} \le \frac{3}{\varepsilon}\begin{pmatrix} B & -B \\ -B & B \end{pmatrix}.$$

Multiplying both sides of this inequality by the operator

$$Z = \begin{pmatrix} \left(\sigma(t,x,a)Q^{\frac{1}{2}}\right)\left(\sigma(t,x,a)Q^{\frac{1}{2}}\right)^* & \left(\sigma(t,x,a)Q^{\frac{1}{2}}\right)\left(\sigma(t,y,a)Q^{\frac{1}{2}}\right)^* \\ \left(\sigma(t,y,a)Q^{\frac{1}{2}}\right)\left(\sigma(t,x,a)Q^{\frac{1}{2}}\right)^* & \left(\sigma(t,y,a)Q^{\frac{1}{2}}\right)\left(\sigma(t,y,a)Q^{\frac{1}{2}}\right)^* \end{pmatrix}$$

and taking the trace preserves the inequality. This can be seen by evaluating the trace on the basis of eigenvectors of Z as it is a compact, self-adjoint, and positive operator. Therefore, thanks to (3.22),

$$\mathrm{Tr}\left(\left(\sigma(t,x,a)Q^{\frac{1}{2}}\right)\left(\sigma(t,x,a)Q^{\frac{1}{2}}\right)^* X\right) - \mathrm{Tr}\left(\left(\sigma(t,y,a)Q^{\frac{1}{2}}\right)\left(\sigma(t,y,a)Q^{\frac{1}{2}}\right)^* Y\right)$$

$$\leq \frac{3}{\varepsilon}\mathrm{Tr}\left[\left((\sigma(t,x,a)-\sigma(t,y,a))Q^{\frac{1}{2}}\right)\left((\sigma(t,x,a)-\sigma(t,y,a))Q^{\frac{1}{2}}\right)^* B\right]$$

$$\leq C\frac{|x-y|^2_{-1}}{\varepsilon},$$

for all $a \in \Lambda$ for some C. We thus conclude that

$$F\left(t,x,r,\frac{B(x-y)}{\varepsilon},X\right) - F\left(t,y,r,\frac{B(x-y)}{\varepsilon},Y\right)$$

$$\geq -\omega_l(|x-y|_{-1},R) - R\omega_c(|x-y|_{-1},R) - C\frac{|x-y|^2_{-1}}{\varepsilon}$$

for some constant C, and so Hypothesis 3.48 is satisfied. Hypothesis 3.49 with $\gamma = 2$ follows from (3.9) and (3.10). This concludes the proof of the uniqueness. $\square$

Let us analyze condition (3.155). Let $\{u_1, u_2, \ldots\}$ be any orthonormal basis of Ξ. Let $\{e_1, e_2, \ldots\}$ be an orthonormal basis in H_{-1} made of elements of H as in Sect. 3.2. Then $\{f_1, f_2, \ldots\}$, where $f_i = B^{\frac{1}{2}}e_i$ is an orthonormal basis of H.

We have

$$\mathrm{Tr}\left[\left(\sigma(t,x,a)Q^{\frac{1}{2}}\right)\left(\sigma(t,x,a)Q^{\frac{1}{2}}\right)^* BQ_N\right]$$

$$= \mathrm{Tr}\left[(\sigma(t,x,a)Q^{\frac{1}{2}})^* BQ_N(\sigma(t,x,a)Q^{\frac{1}{2}})\right]$$

$$= \mathrm{Tr}\left[(\sigma(t,x,a)Q^{\frac{1}{2}})^* Q_N^* BQ_N(\sigma(t,x,a)Q^{\frac{1}{2}})\right]$$

$$= \sum_{i=1}^{\infty}\left\langle BQ_N\sigma(t,x,a)Q^{\frac{1}{2}}u_i, Q_N\sigma(t,x,a)Q^{\frac{1}{2}}u_i\right\rangle = \sum_{i=1}^{\infty}|Q_N\sigma(t,x,a)Q^{\frac{1}{2}}u_i|^2_{-1}$$

$$= \sum_{i=1}^{\infty}|B^{\frac{1}{2}}Q_N\sigma(t,x,a)Q^{\frac{1}{2}}u_i|^2 = \sum_{i=1}^{\infty}|(\sigma(t,x,a)Q^{\frac{1}{2}})^* Q_N^* B^{\frac{1}{2}}f_i|^2_{\Xi}$$

$$= \sum_{i=1}^{\infty}|(\sigma(t,x,a)Q^{\frac{1}{2}})^* Q_N^* Be_i|^2_{\Xi} = \sum_{i=N+1}^{\infty}|(\sigma(t,x,a)Q^{\frac{1}{2}})^* Be_i|^2_{\Xi}$$

$$= \sum_{i=N+1}^{\infty}|(\sigma(t,x,a)Q^{\frac{1}{2}})^* B^{\frac{1}{2}}f_i|^2_{\Xi} = \sum_{i=N+1}^{\infty}|(B^{\frac{1}{2}}\sigma(t,x,a)Q^{\frac{1}{2}})^* f_i|^2_{\Xi}$$

(see also [374], p. 33). Therefore, we have

$$f_N(a) := \mathrm{Tr}\left[\left(\sigma(t, x, a)Q^{\frac{1}{2}}\right)\left(\sigma(t, x, a)Q^{\frac{1}{2}}\right)^* BQ_N\right] = \sum_{i=1}^{\infty} |Q_N\sigma(t, x, a)Q^{\frac{1}{2}}u_i|^2_{-1}$$

$$= \sum_{i=N+1}^{\infty} |(B^{\frac{1}{2}}\sigma(t, x, a)Q^{\frac{1}{2}})^* f_i|^2_{\Xi}. \qquad (3.161)$$

The functions $f_N : \Lambda \to \mathbb{R}$, $N \geq 1$, are continuous, nonnegative, and since by (3.10),

$$\sum_{i=1}^{\infty} |(B^{\frac{1}{2}}\sigma(t, x, a)Q^{\frac{1}{2}})^* f_i|^2_{\Xi} = \mathrm{Tr}\left[\left(\sigma(t, x, a)Q^{\frac{1}{2}}\right)\left(\sigma(t, x, a)Q^{\frac{1}{2}}\right)^* B\right] \leq C_1,$$

it follows from (3.161) that for every $a \in \Lambda$, $f_N(a) \downarrow 0$ as $N \to +\infty$. Thus, if Λ is compact, we must have $f_N(a) \to 0$ uniformly on Λ as $N \to +\infty$, which means that (3.155) is satisfied.

Another case when (3.155) is satisfied is when B is compact. This is an obvious consequence of the fact that in this case $\|BQ_N\| \to 0$ as $N \to +\infty$.

One can use (3.161) to obtain other criteria for (3.155) to hold. For instance, it will be satisfied if

$$\sum_{i=1}^{\infty} a_i < +\infty,$$

where

$$a_i := \sup_{a \in \Lambda} |\sigma(t, x, a)Q^{\frac{1}{2}}u_i|^2_{-1},$$

and if for every i

$$\lim_{N \to +\infty} \sup_{a \in \Lambda} |Q_N\sigma(t, x, a)Q^{\frac{1}{2}}u_i|_{-1} = 0.$$

Theorem 3.67 (Existence under strong B-condition) *Let the assumptions of Proposition 3.62 and (3.155) be satisfied, and let in addition $b(\cdot, x, a)$, $\sigma(\cdot, x, a)$ be uniformly continuous on $[0, T]$, uniformly for $(x, a) \in B(0, R) \times \Lambda$ for every $R > 0$. Then the value function $V(t, x)$, defined in (3.121), is the unique viscosity solution of (3.122) among functions in the set*

$$S := \Big\{u : [0, T] \times H \to \mathbb{R} : |u(t, x)| \leq C(1 + |x|^k) \text{ for some } k \geq 0,$$

$$\lim_{t \to T} |u(t, x) - g(e^{(T-t)A}x)| = 0 \text{ uniformly on bounded subsets of } H\Big\}.$$

Proof The proof follows the lines of the proof for the weak case. To prove uniqueness we now use Theorem 3.54 instead of Theorem 3.50 so we need to verify

Hypothesis 3.52 instead of Hypothesis 3.48. This can be done arguing as before using
(3.7) instead of (3.21).   □

Remark 3.68 B-continuity is built into the definition of a viscosity solution, however
it is clear from the proof of existence that B-continuity of the value function is not
needed to show that it satisfies the sub- and supersolution conditions required by
the definition. Thus, if we disregard the requirement of B-continuity, we can still
prove that the value function is a "viscosity solution" under much weaker sets of
assumptions than those of Theorems 3.66 and 3.67.   ■

Example 3.69 (Controlled stochastic wave equation) Consider a control problem for
the stochastic wave equation

$$
\begin{cases}
\frac{\partial^2 y}{\partial s^2}(s, \xi) = \Delta y(s, \xi) + f(\xi, y(s, \xi), a(s)) & \\
\qquad + h(\xi, y(s, \xi), a(s)) \frac{\partial}{\partial s} \tilde{W}_{\tilde{Q}}(s, \xi), & s > t, \ \xi \in \mathcal{O}, \\
y(s, \xi) = 0, & s > t, \ \xi \in \partial\mathcal{O}, \\
y(t, \xi) = y_0(\xi), & \xi \in \mathcal{O}, \\
\frac{\partial y}{\partial t}(t, \xi) = z_0(\xi), & \xi \in \mathcal{O},
\end{cases}
\tag{3.162}
$$

where $\mathcal{O}$ is a bounded regular domain in $\mathbb{R}^d$, $y_0 \in H_0^1(\mathcal{O})$, $z_0 \in L^2(\mathcal{O})$, $\tilde{Q}$ is an
operator in $\mathcal{L}_1^+(L^2(\mathcal{O}))$ and $\tilde{W}_{\tilde{Q}}$ is a $\tilde{Q}$-Wiener process, Λ is a Polish space and
$a(\cdot) \in \mathcal{U}_t$. In addition, $f, h : \mathcal{O} \times \mathbb{R} \times \Lambda \to \mathbb{R}$. Suppose we want to minimize the
cost functional

$$
I(t, y_0, z_0; a(\cdot)) = \mathbb{E}\left[\int_t^T \int_{\mathcal{O}} \beta(s, y(s, \xi), a(s)) d\xi ds + \int_{\mathcal{O}} \gamma(y(T, \xi)) d\xi \right]
$$

over all $a(\cdot) \in \mathcal{U}_t$, where $\beta : [0, T] \times \mathbb{R} \times \Lambda \to \mathbb{R}, \gamma : \mathbb{R} \to \mathbb{R}$.

Let Δ_ξ be the Laplace operator with the domain $D(\Delta_\xi) = H^2(\mathcal{O}) \cap H_0^1(\mathcal{O})$. Then
(see Sect. C.1 and in particular (C.11)) $D((-\Delta_\xi)^{\frac{1}{2}}) = H_0^1(\mathcal{O})$. We set[1]

$$
H = \begin{pmatrix} H_0^1(\mathcal{O}) \\ \times \\ L^2(\mathcal{O}) \end{pmatrix}
$$

equipped with the inner product

$$
\left\langle \begin{pmatrix} y \\ z \end{pmatrix}, \begin{pmatrix} \bar{y} \\ \bar{z} \end{pmatrix} \right\rangle_H = \langle (-\Delta_\xi)^{1/2} y, (-\Delta_\xi)^{1/2} \bar{y} \rangle_{L^2(\mathcal{O})} + \langle z, \bar{z} \rangle_{L^2(\mathcal{O})}, \qquad \begin{pmatrix} y \\ z \end{pmatrix}, \begin{pmatrix} \bar{y} \\ \bar{z} \end{pmatrix} \in H.
$$

---

[1]We remark that if $\mathrm{Tr}(\tilde{Q}) = +\infty$ then the right choice of the state space for the stochastic wave
equation is $L^2(\mathcal{O}) \times D((-\Delta_\xi)^{-\frac{1}{2}})$, at least for additive noise, see [180] Example 5.8, p. 127.

The operator

$$D(A) = \begin{pmatrix} H^2(\mathcal{O}) \cap H_0^1(\mathcal{O}) \\ \times \\ H_0^1(\mathcal{O}) \end{pmatrix}, \qquad A = \begin{pmatrix} 0 & I \\ \Delta_\xi & 0 \end{pmatrix},$$

is maximal dissipative in H and $A^* = -A$. Equation (3.162) can then be rewritten as the following evolution equation

$$dX(s) = (AX(s) + b(X(s), a(s)))\, dt + \sigma(X(s), a(s))dW_Q(s), \quad X(t) = x := \begin{pmatrix} y_0 \\ z_0 \end{pmatrix},$$
(3.163)

in H, where

$$b\left(\begin{pmatrix} y \\ z \end{pmatrix}, a\right) = \begin{pmatrix} 0 \\ f(\cdot, y(\cdot), a) \end{pmatrix}, \quad \sigma\left(\begin{pmatrix} y \\ z \end{pmatrix}, a\right)\begin{pmatrix} \bar{y} \\ \bar{z} \end{pmatrix} = \begin{pmatrix} 0 \\ h(\cdot, y(\cdot), a)\bar{z} \end{pmatrix},$$
(3.164)

$$W_Q = \begin{pmatrix} 0 \\ \tilde{W}_{\tilde{Q}} \end{pmatrix}, \quad Q\begin{pmatrix} y \\ z \end{pmatrix} = \begin{pmatrix} 0 \\ \tilde{Q}z \end{pmatrix}.$$

We consider the operator

$$B = \begin{pmatrix} (-\Delta_\xi)^{-1/2} & 0 \\ 0 & (-\Delta_\xi)^{-1/2} \end{pmatrix}.$$

It is bounded, positive, self-adjoint on H, A^*B is bounded and

$$\left\langle A^*B\begin{pmatrix} y \\ z \end{pmatrix}, \begin{pmatrix} y \\ z \end{pmatrix} \right\rangle_H = 0.$$

Moreover, we have

$$\left| \begin{pmatrix} y \\ z \end{pmatrix} \right|_{-1} = \left(|(-\Delta_\xi)^{1/4}y|^2 + |(-\Delta_\xi)^{-1/4}z|^2 \right)^{1/2}.$$

Assume that f, h are continuous in all variables, $f(\xi, \cdot, a), h(\xi, \cdot, a)$ are Lipschitz continuous with Lipschitz constant L independent of ξ, a, and $f(\cdot, 0, \cdot), h(\cdot, 0, \cdot)$ are bounded. Then

$$\left| b\left(\begin{pmatrix} y \\ z \end{pmatrix}, a\right) - b\left(\begin{pmatrix} \tilde{y} \\ \tilde{z} \end{pmatrix}, a\right) \right|_H = |f(\cdot, y(\cdot), a) - f(\cdot, \tilde{y}(\cdot), a)|_{L^2(\mathcal{O})}$$

$$\leq L\left(\int_{\mathcal{O}} |y(\xi) - \tilde{y}(\xi)|^2 d\xi\right)^{\frac{1}{2}} = L|y - \tilde{y}|_{L^2(\mathcal{O})}$$

$$\leq C|(-\Delta_\xi)^{1/4}(y - \tilde{y})|_{L^2(\mathcal{O})} \leq C\left| \begin{pmatrix} y \\ z \end{pmatrix} - \begin{pmatrix} \tilde{y} \\ \tilde{z} \end{pmatrix} \right|_{-1}$$

for some constant C which follows from embeddings $D((-\Delta_\xi)^{1/4}) \hookrightarrow H^{1/2}(\mathcal{O}) \hookrightarrow L^2(\mathcal{O})$ (see Sect. C.1). Thus the function b satisfies (3.9) and (3.21). Unfortunately we need more in order for σ to satisfy (3.22) (or even (3.8)). We present one sufficient condition. Obviously other conditions are possible. Let $\Xi_0 = Q^{1/2}H$. Suppose that $d > 1$ and $(-\Delta_\xi)^{(d-1)/4}\tilde{Q}^{1/2} \in \mathcal{L}_2(L^2(\mathcal{O}))$. Then we have by Proposition B.26

$$
\left\| \sigma\left(\begin{pmatrix} y \\ z \end{pmatrix}, a \right) - \sigma\left(\begin{pmatrix} \tilde{y} \\ \tilde{z} \end{pmatrix}, a \right) \right\|_{\mathcal{L}_2(\Xi_0, H)}
$$

$$
= \left\| (h(\cdot, y(\cdot), a) - h(\cdot, \tilde{y}(\cdot), a))\tilde{Q}^{1/2} \right\|_{\mathcal{L}_2(L^2(\mathcal{O}))}
$$

$$
\leq \left\| (h(\cdot, y(\cdot), a) - h(\cdot, \tilde{y}(\cdot), a))(-\Delta_\xi)^{(1-d)/4} \right\|_{\mathcal{L}(L^2(\mathcal{O}))} \left\| (-\Delta_\xi)^{(d-1)/4}\tilde{Q}^{1/2} \right\|_{\mathcal{L}_2(L^2(\mathcal{O}))}
$$

$$
\leq L \sup_{|z|_{L^2(\mathcal{O})} \leq 1} \left(\int_{\mathcal{O}} |y(\xi) - \tilde{y}(\xi)|^2 |(-\Delta_\xi)^{(1-d)/4} z(\xi)|^2 d\xi \right)^{\frac{1}{2}} \left\| (-\Delta_\xi)^{(d-1)/4}\tilde{Q}^{1/2} \right\|_{\mathcal{L}_2(L^2(\mathcal{O}))}
$$

$$
\leq L|y - \tilde{y}|_{L^{2d/(d-1)}(\mathcal{O})} \sup_{|z|_{L^2(\mathcal{O})} \leq 1} |(-\Delta_\xi)^{(1-d)/4}z|_{L^{2d}(\mathcal{O})} \left\| (-\Delta_\xi)^{(d-1)/4}\tilde{Q}^{1/2} \right\|_{\mathcal{L}_2(L^2(\mathcal{O}))}
$$

$$
\leq C|(-\Delta_\xi)^{1/4}(y - \tilde{y})|_{L^2(\mathcal{O})} \left\| (-\Delta_\xi)^{(d-1)/4}\tilde{Q}^{1/2} \right\|_{\mathcal{L}_2(L^2(\mathcal{O}))}
$$

$$
\leq C \left| \begin{pmatrix} y \\ z \end{pmatrix} - \begin{pmatrix} \tilde{y} \\ \tilde{z} \end{pmatrix} \right|_{-1} \left\| (-\Delta_\xi)^{(d-1)/4}\tilde{Q}^{1/2} \right\|_{\mathcal{L}_2(L^2(\mathcal{O}))},
$$

where we have used Sobolev embeddings $H^{1/2}(\mathcal{O}) \hookrightarrow L^{2d/(d-1)}(\mathcal{O})$ and $|(-\Delta_\xi)^{(1-d)/4}z|_{L^{2d}(\mathcal{O})} \leq C_1|z|_{L^2(\mathcal{O})}$ if $d > 1$ (see Sect. C.1). Thus σ satisfies (3.22). The same calculation shows that for $d = 1$ it is enough that $(-\Delta_\xi)^\alpha \tilde{Q}^{1/2} \in \mathcal{L}_2(L^2(\mathcal{O}))$ for some $\alpha > 0$. It is now easy to check that (3.10) is true with $\gamma = 1$.

We now define

$$
l\left(s, \begin{pmatrix} y \\ z \end{pmatrix}, a \right) = \int_{\mathcal{O}} \beta(s, y(\xi), a)d\xi, \quad g\left(\begin{pmatrix} y \\ z \end{pmatrix} \right) = \int_{\mathcal{O}} \gamma(y(\xi))d\xi,
$$

and rewrite the cost functional I as

$$
J(t, x; a(\cdot)) = \mathbb{E}\left[\int_t^T l(s, X(s), a(s))ds + g(X(T)) \right].
$$

It is easy to see by calculations similar to these for b that if β is continuous and $\beta(\cdot, \cdot, a), \gamma$ are uniformly continuous with a modulus of continuity independent of a then

$$
|l(s, x_1, a) - l(t, x_2, a)| + |g(x_1) - g(x_2)| \leq \omega(|x_1 - x_2|_{-1} + |s - t|),
$$

$$
s, t \in [0, T], x_1, x_2 \in H, a \in \Lambda
$$

for some modulus ω. If $\beta(0, 0, \cdot)$ is bounded then l satisfies (3.131) with $m = 1$. ∎

3.6.2 Improved Version of the Dynamic Programming Principle

Once we know that the value function is continuous in both variables, a stronger version of the dynamic programming principle involving stopping times can be proved. We will only do this for the finite horizon problem discussed in the previous section, however the same result would be true for other optimal control problems, including those with infinite horizon.

We define the set $\mathcal{V}_t$ in the following way. For every $a(\cdot) \in \mathcal{U}_t^\nu$ for some reference probability space $\nu = (\Omega, \mathcal{F}, \mathcal{F}_s^t, \mathbb{P}, W_Q)$, we choose an $\mathcal{F}_s^t$-stopping time $t \leq \tau_{a(\cdot)} \leq T$. The set $\mathcal{V}_t$ is the set of all such pairs $(a(\cdot), \tau_{a(\cdot)})$. We also define $\tilde{\mathcal{V}}_t$ to be the set of those pairs in $\mathcal{V}_t$ for which the underlying reference probability space is standard, i.e. $\tilde{\mathcal{V}}_t = (a(\cdot), \tau_{a(\cdot)})$, where $a(\cdot) \in \tilde{\mathcal{U}}_t^\nu$. To simplify the notation we will just write $(a(\cdot), \tau)$ instead of $(a(\cdot), \tau_{a(\cdot)})$.

Theorem 3.70 (Dynamic programming principle) *Let $b \colon [0, T] \times H \times \Lambda \to H$, $\sigma \colon [0, T] \times H \times \Lambda \to \mathcal{L}_2(\Xi_0, H)$, $l \colon [0, T] \times H \times \Lambda \to \mathbb{R}$ be continuous, let b and σ satisfy (3.7)–(3.10) with $\gamma = 1$. Let c be bounded from below, and let l, g, c satisfy (3.131), (3.146), (3.147). Then, for all $(t, x) \in [0, T] \times H$,*

$$V(t, x) = \inf_{(a(\cdot), \tau) \in \mathcal{V}_t} \mathbb{E}\left[\int_t^\tau e^{-\int_t^s c(X(r))dr} l(s, X(s), a(s)) ds + e^{-\int_t^\tau c(X(r))dr} V(\tau, X(\tau))\right]. \tag{3.165}$$

Proof Without loss of generality we always assume that the Q-Wiener processes in the reference probability spaces have everywhere continuous paths. We recall that, by Proposition 3.62 and Remark 3.64, J satisfies (3.148) and V satisfies

$$|V(t, x) - V(t, y)| \leq \omega(|x - y|, R) \qquad \forall\, x, y \in B(0, R), t \in [0, T]$$

for some local modulus ω. Moreover, using (1.38) and arguing as in the proof of Proposition 3.61 it is easy to see that $V(\cdot, x)$ is continuous for every $x \in H$ and consequently V is continuous on $[0, T] \times H$.

Let $a(\cdot) \in \tilde{\mathcal{U}}_t^\nu$ for some standard reference probability space $\nu = (\Omega, \mathcal{F}, \mathcal{F}_s^t, \mathbb{P}, W_Q)$ and let τ_n be an $\mathcal{F}_s^t$-stopping time which has a finite number of values, i.e.

$$\tau_n = \sum_{i=1}^k \mathbf{1}_{A_i} t_i$$

for some pairwise disjoint sets $A_1, ..., A_k$ such that $\bigcup_{i=1}^n A_i = \Omega$ and $A_i \in \mathcal{F}_{t_i}^t, i = 1, ..., k$. It then follows from the proof of the first part of Theorem 2.24 (see (2.29)) that,

$$
J(t, x; a(\cdot)) = \sum_{i=1}^{k} \mathbb{E}\left[\mathbf{1}_{A_i} \int_t^{t_i} e^{-\int_t^s c(X(r))dr} l(s, X(s), a(s))\, ds\right.
$$

$$
\left. + \mathbf{1}_{A_i} \mathbb{E}\left[\int_{t_i}^T e^{-\int_t^s c(X(r))dr} l(s, X(s), a(s))\, ds + e^{-\int_t^T c(X(r))dr} g(X(T)) \,|\, \mathscr{F}_{t_i}^t\right]\right]
$$

$$
\geq \mathbb{E}\left[\sum_{i=1}^{k} \mathbf{1}_{A_i} \left[\int_t^{t_i} e^{-\int_t^s c(X(r))dr} l(s, X(s), a(s))\, ds + e^{-\int_t^{t_i} c(X(r))dr} V(t_i, X(t_i))\right]\right]
$$

$$
= \mathbb{E}\left[\int_t^{\tau_n} e^{-\int_t^s c(X(r))dr} l(s, X(s), a(s))\, ds + e^{-\int_t^{\tau_n} c(X(r))dr} V(\tau_n, X(\tau_n))\right].
$$

$$\tag{3.166}$$

By Proposition 1.82, every stopping time can be approximated by stopping times τ_n with a finite number of values. Therefore, thanks to (1.37), (3.131), (3.133), (3.146) and (3.147) we can apply the dominated convergence theorem to obtain in the limit that (3.166) is satisfied for every $(a(\cdot), \tau) \in \tilde{\mathcal{V}}_t$. Since $\tilde{\mathcal{V}}_t \subset \mathcal{V}_t$, it thus follows that

$$
V(t, x) = \inf_{a(\cdot) \in \mathcal{U}_t} J(t, x; a(\cdot)) = \inf_{a(\cdot) \in \tilde{\mathcal{U}}_t} J(t, x; a(\cdot))
$$

$$
\geq \inf_{(a(\cdot), \tau) \in \mathcal{V}_t} \mathbb{E}\left[\int_t^\tau e^{-\int_t^s c(X(r))dr} l(s, X(s), a(s))\, ds + e^{-\int_t^\tau c(X(r))dr} V(\tau, X(\tau))\right],
$$

where we used Theorem 2.22 to obtain the second equality.

To show the reverse inequality, let $t \leq \eta \leq T$, $a(\cdot) \in \tilde{\mathcal{U}}_t^\nu$ for some standard reference probability space $\nu = (\Omega, \mathscr{F}, \mathscr{F}_s^t, \mathbb{P}, W_Q)$, and let $X(s) = X(s; t, x, a(\cdot))$, $s \in [t, T]$. Let, for $\omega_0 \in \Omega$, $\nu_{\omega_0} = (\Omega, \mathscr{F}_{\omega_0}, \mathscr{F}_{\omega_0, s}^\eta, \mathbb{P}_{\omega_0}, W_{Q,\eta})$, where $W_{Q,\eta}(s) = W_Q(s) - W_Q(\eta)$, and $a^{\omega_0}(\cdot) = a_1(\cdot)$ (on $[\eta, T]$) be from Lemma 2.26-(ii).[2] We have $X(\cdot) = X(\cdot; \eta, X(\eta), a_1(\cdot))$, on $[\eta, T]$, $\mathbb{P}$-a.e., and (see the argument in the proof of (A3) in Proposition 2.16) that it is indistinguishable from a process, still denoted by $X(\cdot)$, such that, for $\mathbb{P}$-a.e. ω_0,

$$
X(\cdot) = X^{\nu_{\omega_0}}(\cdot; \eta, X(\eta)(\omega_0), a^{\omega_0}(\cdot)), \quad \text{on } [\eta, T], \mathbb{P}_{\omega_0} - a.e.
$$

Therefore, as a consequence of Theorem 2.24, applied to the reference probability space ν_{ω_0}, we have that, for $\mathbb{P}$-a.e. ω_0,

$$
V(\eta, X(\eta)(\omega_0)) \leq \mathbb{E}_{\omega_0}\left[\int_\eta^s e^{-\int_\eta^r c(X(\theta))d\theta} l(r, X(r), a^{\omega_0}(\cdot)(r))\, dr\right.
$$

$$
\left. + e^{-\int_\eta^s c(X(\theta))d\theta} V(s, X(s))\right]. \tag{3.167}
$$

---

[2]In Lemma 2.26-(ii) $a^{\omega_0}(\cdot)$ denotes the 6-tuple $(\Omega, \mathscr{F}_{\omega_0}, \mathscr{F}_{\omega_0, s}^\eta, \mathbb{P}_{\omega_0}, W_\eta, a_1|_{[\eta, T]}(\cdot))$ while here we use $a^{\omega_0}(\cdot)$ only to indicate the process.

Set, for $s \in [t, T]$,

$$M(s) = \int_t^s e^{-\int_t^r c(X(\theta))d\theta} l(r, X(r), a(r)) dr + e^{-\int_t^s c(X(\theta))d\theta} V(s, X(s)).$$

Then, by (3.167) and the definition of $M(s)$ (see also the remarks at the end of Sect. 2.2.2), for $\mathbb{P}$-a.e. ω_0

$$M(\eta)(\omega_0) \leq \left(\int_t^\eta e^{-\int_t^r c(X(\theta))d\theta} l(r, X(r), a(r)) dr \right)(\omega_0)$$
$$+ \left(e^{-\int_t^\eta c(X(\theta))d\theta} \right)(\omega_0) \mathbb{E}_{\omega_0} \left[\int_\eta^s e^{-\int_\eta^r c(X(\theta))d\theta} l(r, X(r), a^{\omega_0}(r)) dr \right.$$
$$\left. + e^{-\int_\eta^s c(X(\theta))d\theta} V(s, X(s)) \right]$$
$$= \left(\int_t^\eta e^{-\int_t^r c(X(\theta))d\theta} l(r, X(r), a(r)) dr \right)(\omega_0)$$
$$+ \left(e^{-\int_t^\eta c(X(\theta))d\theta} \right)(\omega_0) \mathbb{E} \left[\int_\eta^s e^{-\int_\eta^r c(X(\theta))d\theta} l(r, X(r), a(r)) dr \right.$$
$$\left. + e^{-\int_\eta^s c(X(\theta))d\theta} V(s, X(s)) \Big| \mathscr{F}_\eta^t \right](\omega_0) = \mathbb{E}\left[M(s) | \mathscr{F}_\eta^t\right](\omega_0) \quad (3.168)$$

for every $s \in [\eta, T]$. Therefore, M is a submartingale, and thus, by the Optional Sampling Theorem (Theorem 1.84), if τ is an $\mathscr{F}_s^t$-stopping time,

$$V(t, x) = M(t) \leq \mathbb{E}\left[M(\tau) | \mathscr{F}_t^t\right]$$
$$= \mathbb{E} \left[\int_t^\tau e^{-\int_t^s c(X(r))dr} l(s, X(s), a(s)) ds + e^{-\int_t^\tau c(X(r))dr} V(\tau, X(\tau)) \Big| \mathscr{F}_t^t \right].$$

Taking the expectation above (or noticing that $\mathscr{F}_t^t$ is trivial), we thus obtain

$$V(t, x) \leq \mathbb{E} \left[\int_t^\tau e^{-\int_t^s c(X(r))dr} l(s, X(s), a(s)) ds + e^{-\int_t^\tau c(X(r))dr} V(\tau, X(\tau)) \right]$$
$$(3.169)$$

for every $a(\cdot) \in \tilde{\mathcal{U}}_t^\nu$ for any standard reference probability space $\nu = (\Omega, \mathscr{F}, \mathscr{F}_s^t, \mathbb{P}, W_Q)$, and every $\mathscr{F}_s^t$-stopping time τ.

It remains to justify that (3.169) is true for every $(a(\cdot), \tau) \in \mathcal{V}_t$. We sketch the argument. Let $a(\cdot) \in \mathcal{U}^{\nu_1}$, where $\nu_1 = (\Omega_1, \mathscr{F}_1, \mathscr{F}_{1,s}^t, \mathbb{P}_1, W_Q^1)$, and let τ_n be $\mathscr{F}_{1,s}^t$-stopping times with a finite number of values approximating τ. Let $\nu_2 = (\Omega_2, \mathscr{F}_2, \mathscr{F}_{2,s}^t, \mathbb{P}_2, W_Q^2)$ be a standard reference probability space. We proceed as in the proof of Theorem 2.22. Let $a_1(\cdot)$, $\tilde{a}_1(\cdot)$ be as in the proof of Theorem 2.22. We define $X^{\nu_1}(\cdot) = X^{\nu_1}(\cdot; t, x, a(\cdot)) = X^{\nu_1}(\cdot; t, x, a_1(\cdot))$, $X^{\nu_2}(\cdot) = X^{\nu_2}(\cdot; t, x, \tilde{a}_1(\cdot))$. Since τ_n has a finite number of values, we can assume that

$$\tau_n = \sum_{i=1}^{k} \mathbf{1}_{A_i} t_i$$

for some pairwise disjoint sets $A_1, ..., A_k$ such that $\bigcup_{i=1}^{n} A_i = \Omega$ and $A_i \in \mathscr{F}_{1,t_i}^{t,0}, i = 1, ..., k$. Let $B_1, ..., B_k \in \mathcal{B}(\mathbf{W})$ be such that $W_Q^1(\cdot \wedge t_i)^{-1}(B_i) = A_i, i = 1, ..., k$ (see Lemma 2.19). We set

$$\tilde{\tau}_n = \sum_{i=1}^{k} \mathbf{1}_{W_Q^2(\cdot \wedge t_i)^{-1}(B_i)} t_i.$$

Then $\tilde{\tau}_n$ is an $\mathscr{F}_{2,s}^t$-stopping time with a finite number of values, and it follows that

$$\mathcal{L}_{\mathbb{P}_1}(\tau_n \wedge \cdot, X^{\nu_1}(\cdot), a(\cdot)) = \mathcal{L}_{\mathbb{P}_2}(\tilde{\tau}_n \wedge \cdot, X^{\nu_2}(\cdot), \tilde{a}_1(\cdot)).$$

One can then conclude that

$$\mathbb{E}^{\nu_1} \left[\int_t^{\tau_n} e^{-\int_t^s c(X^{\nu_1}(r))dr} l\left(s, X^{\nu_1}(s), a(s)\right) ds + e^{-\int_t^{\tau_n} c(X^{\nu_1}(r))dr} V\left(\tau_n, X^{\nu_1}(\tau_n)\right) \right]$$
$$= \mathbb{E}^{\nu_2} \left[\int_t^{\tilde{\tau}_n} e^{-\int_t^s c(X^{\nu_2}(r))dr} l\left(s, X^{\nu_2}(s), \tilde{a}_1(s)\right) ds + e^{-\int_t^{\tilde{\tau}_n} c(X^{\nu_2}(r))dr} V\left(\tilde{\tau}_n, X^{\nu_2}(\tilde{\tau}_n)\right) \right].$$
$$(3.170)$$

Combining (3.169) with (3.170), we thus obtain that

$$V(t, x) \le \mathbb{E}^{\nu_1} \left[\int_t^{\tau_n} e^{-\int_t^s c(X^{\nu_1}(r))dr} l\left(s, X^{\nu_1}(s), a(s)\right) ds + e^{-\int_t^{\tau_n} c(X^{\nu_1}(r))dr} V\left(\tau_n, X^{\nu_1}(\tau_n)\right) \right].$$

It remains to let $n \to +\infty$ above to conclude that (3.169) holds for every pair $(a(\cdot), \tau) \in \mathcal{V}_t$. $\square$

Remark 3.71 The proof of existence in Theorem 3.66 works in almost exactly the same way (and is in fact easier) if we use Theorem 3.70 and replace $t + \varepsilon$ by $\min(t + \varepsilon, \tau)$, where τ is the exit time of $X(\cdot)$ from some ball $B_{r_0}(x)$ for some $r_0 > 0$ (or from $B_{r_\varepsilon}(x)$ for some $r_\varepsilon > 0$). In this way one can always work with local maxima and minima in the definition of viscosity solution and avoid the requirements about global uniform continuity (and hence growth at infinity) of test functions and their derivatives. We do not pursue this here and leave the details of such a version of viscosity solution to the interested readers.   ∎

3.6.3   The Infinite Horizon Problem

In this subsection we characterize the value function of the infinite horizon optimal control problem (3.126) as the unique solution of the associated HJB equation
(3.127). We consider the following set of assumptions for $b : H \times \Lambda \to H, \sigma :
H \times \Lambda \to \mathcal{L}_2(\Xi_0, H), l : H \times \Lambda \to \mathbb{R}$.
    There exist constants $C, m \geq 0$, and a local modulus ω_l such that:

$$|b(x,a) - b(y,a)| \leq C|x - y| \qquad\qquad \forall x, y \in H, a \in \Lambda, \quad (3.171)$$
$$\|\sigma(x,a) - \sigma(y,a)\|_{\mathcal{L}_2(\Xi_0,H)} \leq C|x - y| \qquad \forall x, y \in H, a \in \Lambda, \quad (3.172)$$
$$|b(x,a)| \leq C(1 + |x|) \qquad\qquad \forall x, y \in H, a \in \Lambda, \quad (3.173)$$
$$\|\sigma(x,a)\|_{\mathcal{L}_2(\Xi_0,H)} \leq C(1 + |x|) \qquad \forall x, y \in H, a \in \Lambda, \quad (3.174)$$
$$\langle b(x,a) - b(y,a), B(x - y)\rangle \leq C|x - y|^2_{-1} \qquad \forall x, y \in H, a \in \Lambda, \quad (3.175)$$
$$\|\sigma(x,a) - \sigma(y,a)\|_{\mathcal{L}_2(\Xi_0,H)} \leq C|x - y|_{-1} \qquad \forall x, y \in H, a \in \Lambda, \quad (3.176)$$
$$|l(x,a) - l(y,a)| \leq \omega_l(|x - y|, R) \qquad \forall x, y \in B(0,R), a \in \Lambda, \quad (3.177)$$
$$|l(x,a) - l(y,a)| \leq \omega_l(|x - y|_{-1}, R) \qquad \forall x, y \in B(0,R), a \in \Lambda, \quad (3.178)$$
$$|l(x,a)| \leq C(1 + |x|^m) \qquad\qquad \forall x \in H, a \in \Lambda. \quad (3.179)$$

    Proposition 3.24 suggests that in order for the value function to be well defined
we need λ to be sufficiently big. We thus impose the following hypothesis.

Hypothesis 3.72 If $m > 0$, the discount constant λ in the functional (3.125) satisfies
$\lambda > \bar{\lambda}$, where $\bar{\lambda}$ is the constant from Proposition 3.24, where C is the constant from
(3.173) and (3.174) and m is the constant appearing in (3.179). If $m = 0$, we have
$\lambda > 0$.

Proposition 3.73 (Regularity of V under weak B-condition) *Suppose that (3.2)
holds, that b and σ are continuous, and that b, σ and l satisfy (3.171), (3.173)–
(3.176), (3.178) and (3.179). Assume that Hypotheses 2.28 and 3.72 hold. Then there
exists a local modulus ω such that:*

  (i) *The cost functional (3.125) satisfies*

$$|J(x,a(\cdot)) - J(y,a(\cdot))| \leq \omega(|x - y|_{-1}, R), \quad (3.180)$$

   for all $x, y \in B(0, R), a(\cdot) \in \mathcal{U}_0$.
 (ii) *There exists a constant $\tilde{C}$ such that*

$$|J(x; a(\cdot))|, |V(x)| \leq \tilde{C}(1 + |x|^m) \quad (3.181)$$

   for all $x \in H, a(\cdot) \in \mathcal{U}_0$.
(iii) *The value function V defined in (3.126) satisfies*

$$|V(x) - V(y)| \leq \omega(|x - y|_{-1}, R) \quad \forall x, y \in B(0, R). \quad (3.182)$$

Proof Part (i): Let $R > 0$, $x, y \in B(0, R)$, and $a(\cdot) \in \mathcal{U}_0$. Set $X(\cdot) :=$ $X(\cdot; 0, x, a(\cdot))$, $Y(\cdot) := X(\cdot; 0, y, a(\cdot))$.

Choose $\varepsilon > 0$. Thanks to (3.32) and Hypothesis 3.72, there exists T_ε, also depending on C, m, λ, R but independent of $a(\cdot)$, such that

$$\mathbb{E} \int_{T_\varepsilon}^\infty e^{-\lambda r} |l(X(r), a(r)) - l(Y(r), a(r))| dr \le \varepsilon. \qquad (3.183)$$

We now proceed as in the proof of Proposition 3.61.

$$\int_0^{T_\varepsilon} e^{-\lambda r} \int_{\{|X(r)| < \frac{1}{\varepsilon} \text{ and } |Y(r)| < \frac{1}{\varepsilon}\}} |l(X(r), a(r)) - l(Y(r), a(r))| d\mathbb{P} dr$$

$$+ \int_0^{T_\varepsilon} e^{-\lambda r} \int_{\{|X(r)| \ge \frac{1}{\varepsilon} \text{ or } |Y(r)| \ge \frac{1}{\varepsilon}\}} |l(X(r), a(r)) - l(Y(r), a(r))| d\mathbb{P} dr$$

$$\le \int_0^{T_\varepsilon} e^{-\lambda r} \int_{\{|X(r)| < \frac{1}{\varepsilon} \text{ and } |Y(r)| < \frac{1}{\varepsilon}\}} \omega_l(|X(r) - Y(r)|_{-1}, \frac{1}{\varepsilon}) d\mathbb{P} dr$$

$$+ \int_0^{T_\varepsilon} e^{-\lambda r} \int_{\{|X(r)| \ge \frac{1}{\varepsilon} \text{ or } |Y(r)| \ge \frac{1}{\varepsilon}\}} C(2 + |X(r)|^m + |Y(r)|^m) d\mathbb{P} dr$$

$$= J_1 + J_2. \qquad (3.184)$$

Thanks to (1.37), arguing as in the proof of Proposition 3.61, we have

$$J_2 \le \gamma_1(\varepsilon, R) \qquad (3.185)$$

for some local modulus γ_1, independent of $a(\cdot)$.

Let K_ε be such $\omega_l(s, \frac{1}{\varepsilon}) \le \varepsilon + K_\varepsilon s$. Using (3.23) we obtain

$$J_1 \le \frac{\varepsilon}{\lambda} + K_\varepsilon \int_0^{T_\varepsilon} e^{-\lambda r} \mathbb{E}|X(r) - Y(r)|_{-1} dr \le \frac{\varepsilon}{\lambda} + C_\varepsilon |x - y|_{-1} \qquad (3.186)$$

for some C_ε independent of $a(\cdot)$.

Therefore, (3.183)–(3.186) yield

$$|J(x, a(\cdot)) - J(y, a(\cdot))| \le \varepsilon + \frac{\varepsilon}{\lambda} + C_\varepsilon |x - y|_{-1} + \gamma_1(\varepsilon, R), \qquad (3.187)$$

and (3.180) follows by taking the infimum above over $\varepsilon > 0$.

Estimate (3.181) follows from (3.32) and Hypothesis 3.72 and (3.182) is an obvious consequence of (3.180). □

Proposition 3.74 (Regularity of V under strong B-condition) *Let (3.3) hold, let b and σ be continuous, satisfy (3.171), (3.173), (3.174) and (3.176), and let l be continuous and satisfy (3.177), (3.179). Assume that Hypotheses 2.28 and 3.72 hold. Then*

3.6 Existence of Solutions: Value Function

there exist a local modulus ω and a constant $\tilde{C}$ such that (i)–(iii) of Proposition 3.73 are satisfied.

Proof The proof is exactly the same as the proof of Proposition 3.73. We just have to replace the term $|X(r) - Y(r)|_{-1}$ by $|X(r) - Y(r)|$ in (3.184) and (3.186), and then use (3.27) instead of (3.23).   □

Theorem 3.75 (Existence under weak B-condition) *Let the assumptions of Proposition 3.73 be satisfied, and let, for every x,*

$$\lim_{N \to +\infty} \sup_{a \in \Lambda} \mathrm{Tr}\left[\left(\sigma(x,a)Q^{\frac{1}{2}}\right)\left(\sigma(x,a)Q^{\frac{1}{2}}\right)^* BQ_N\right] = 0. \qquad (3.188)$$

Then the value function V defined in (3.126) is the unique viscosity solution of (3.127) among functions in the set

$$S := \{u \colon H \to \mathbb{R} \ : |u(x)| \le C_1(1 + |x|^k)$$
$$\text{for some } C_1 \ge 0 \text{ and } k \ge 0 \text{ satisfying } (3.190)\}. \qquad (3.189)$$

$$\begin{cases} k < \frac{\lambda}{C + \frac{1}{2}C^2} & \text{if } \frac{\lambda}{C + \frac{1}{2}C^2} \le 2, \\ Ck + \frac{1}{2}C^2 k(k-1) < \lambda & \text{if } \frac{\lambda}{C + \frac{1}{2}C^2} > 2, \\ k \text{ can be any positive number if } C = 0. \end{cases} \qquad (3.190)$$

(C is the constant from (3.173)–(3.174).)

Proof The proof follows the lines of the proof of Theorem 3.66. Proposition 3.73 and Hypothesis 3.72 guarantee that V is B-continuous and that it belongs to S. By Proposition 2.16, Theorem 2.31, and Proposition 3.73 (which ensures that Hypothesis 2.28 holds), the dynamic programming principle (2.39) is satisfied.

To show that V is a viscosity supersolution of (3.127), suppose that there exist a test function $\psi = \varphi + h$ and a point $x \in H$ such that $V + \psi$ has a local minimum at x. Without loss of generality we can assume that h, Dh, D^2h have at most polynomial growth at infinity, and that the minimum is global. We can also require that $V(x) + \psi(x) = 0$, so for all y we have $V(y) \ge -\psi(y)$.

For $\varepsilon > 0$, by the dynamic programming principle, we can find $a^{\nu_\varepsilon}(\cdot) \in \mathcal{U}_0$ such that, defining $X(\cdot) := X(\cdot; 0, x, a^{\nu_\varepsilon}(\cdot))$,

$$V(x) + \varepsilon^2 \ge \mathbb{E}^{\nu_\varepsilon}\left[\int_0^\varepsilon e^{-\lambda s}l(X(s), a^{\nu_\varepsilon}(s))ds + e^{-\lambda \varepsilon}V(X(\varepsilon))\right].$$

Therefore we have

$$\varepsilon^2 - \varphi(x) - h(x) \ge \mathbb{E}^{\nu_\varepsilon}\left[\int_0^\varepsilon e^{-\lambda r}l(X(r), a^{\nu_\varepsilon}(r))dr - e^{-\lambda \varepsilon}(\varphi(X(\varepsilon)) + h(X(\varepsilon)))\right],$$

which, upon using (3.153), (3.154), yields

$$
\begin{aligned}
\varepsilon + \frac{1}{\varepsilon} \mathbb{E}^{\nu_\varepsilon} \Bigg[\int_0^\varepsilon e^{-\lambda r} \Bigg(& - l(X(r), a^{\nu_\varepsilon}(r)) + \langle b(X(r), a^{\nu_\varepsilon}(r)), D\psi(X(r)) \rangle \\
& + \frac{1}{2} \mathrm{Tr} \Big[\big(\sigma(X(r), a^{\nu_\varepsilon}(r)) Q^{\frac{1}{2}} \big) \big(\sigma(X(r), a^{\nu_\varepsilon}(r)) Q^{\frac{1}{2}} \big)^* D^2 \psi(X(r)) \Big] \\
& - \lambda \psi(X(r)) + \langle X(r), A^* D\varphi(X(r)) \rangle \Bigg) dr \Bigg] \geq 0.
\end{aligned}
$$

Using exactly the same arguments as in the proof of Theorem 3.66, it follows that there exists a modulus $\tilde{\rho}$, independent of the control $a^\nu(\cdot)$, such that:

$$
\begin{aligned}
\frac{1}{\varepsilon} \mathbb{E}^{\nu_\varepsilon} \Bigg[\int_0^\varepsilon \lambda V(x) - l(x, a^{\nu_\varepsilon}(r)) & + \langle b(x, a^{\nu_\varepsilon}(r)), D\psi(x) \rangle \\
+ \frac{1}{2} \mathrm{Tr} \Big[\big(\sigma(x, a^{\nu_\varepsilon}(r)) Q^{\frac{1}{2}} \big) \big(\sigma(x, a^{\nu_\varepsilon}(r)) Q^{\frac{1}{2}} \big)^* D^2 \psi(x) \Big] & + \langle x, A^* D\varphi(x) \rangle dr \Bigg] \geq -\tilde{\rho}(\varepsilon).
\end{aligned}
$$

Therefore, taking the supremum over $a \in \Lambda$ inside the integral and then letting $\varepsilon \to 0$ we obtain

$$
\begin{aligned}
\lambda V(x) + \langle x, A^* D\varphi(x) \rangle + \sup_{a \in \Lambda} \Big\{ & - l(x, a) + \langle b(x, a), D\psi(x) \rangle \\
& + \frac{1}{2} \mathrm{Tr} \Big[\big(\sigma(x, a) Q^{\frac{1}{2}} \big) \big(\sigma(x, a) Q^{\frac{1}{2}} \big)^* D^2 \psi(x) \Big] \Big\} \geq 0.
\end{aligned}
$$

This shows that V is a viscosity supersolution of (3.127).

To show that V is a viscosity subsolution we take a constant control, apply the DPP, and again argue as in the proof of Theorem 3.66. We leave this to the reader.

To prove that V is the unique viscosity solution among functions in S we need to show that the hypotheses of Theorem 3.56 are satisfied. This has already been done in the proof of Theorem 3.66, apart from Hypotheses 3.45 and 3.55 for $\gamma = 1$, and condition (3.102). Hypothesis 3.45 is obviously true with $\nu = \lambda$. As regards Hypothesis 3.55 for $\gamma = 1$, by (3.173) and (3.174), we obtain for all $(x, r) \in H \times \mathbb{R}$, $p, q \in H$, $X, Y \in S(H)$,

$$
\begin{aligned}
|F(x, r, p + q, X + Y) - F(x, r, p, X)| \\
\leq C(1 + |x|)|q| + \frac{1}{2} C^2 (1 + |x|)^2 \|Y\|,
\end{aligned}
$$

i.e. Hypothesis 3.55 holds with $M_F = C$, $N_F = \frac{1}{2} C^2$. Condition (3.102) thus follows from the definition of the set S. □

Theorem 3.76 (Existence under strong B-condition) *Let the assumptions of Proposition 3.74 and (3.188) be satisfied. Then the value function V defined in (3.126) is the unique viscosity solution of (3.127) among functions in S defined in Theorem 3.75.*

Proof The only difference with respect to the proof for the weak case is that to show uniqueness we now use Theorem 3.58 instead of Theorem 3.56. The fact that Hypothesis 3.52 is satisfied was observed in the proof of Theorem 3.67. □

When conditions (3.173) and (3.174) are replaced by $\|\sigma(x, a)\|_{\mathcal{L}_2(\Xi_0, H)} \leq C(1 + |x|^\gamma)$ and $|b(x, a)| \leq C(1 + |x|^\gamma)$ for some $\gamma \in [0, 1)$, conditions which must be imposed on a set of functions to guarantee that the value function is the unique viscosity solution among them can be easily deduced from (3.100) and (3.101).

3.7 Existence of Solutions: Finite-Dimensional Approximations

We have shown in Sect. 3.6 that value functions of stochastic optimal control problems are viscosity solutions of their associated HJB equations. This gives a direct method of establishing existence of viscosity solutions for a large class of equations where we have an explicit representation formula for a solution. However, many interesting equations cannot be linked to a stochastic optimal control problem. The best examples are Isaacs equations which are associated to zero-sum, two-player, stochastic differential games. For Isaacs equations, one way of showing existence of viscosity solutions is by proving directly that the associated (upper or lower) value of the game is the solution. Such results can be found in [260, 464, 466]. This method, however, runs into technical difficulties as the proof of the dynamic programming principle is very complicated. In this section we will present a more general method of showing existence of viscosity solutions based on finite-dimensional approximations. This method can be thought of as a Galerkin type approximation for PDEs in infinitely many variables. It was first introduced in [141] for first-order equations and later generalized to second-order equations in [537, 538]. We will present the proofs for the initial value problems.

Let A be a linear, densely defined, maximal dissipative operator in H. Let B be a bounded, strictly positive, self-adjoint, compact operator on H such that A^*B is bounded. For $N > 1$ let H_N be the finite-dimensional space spanned by the eigenvectors of B corresponding to the eigenvalues which are greater than or equal to $1/N$. Let P_N, Q_N be defined as in Sect. 3.2. We see that B commutes with P_N and Q_N, and P_N, Q_N are now also orthogonal projections in H.

We need to change slightly the structure conditions on the Hamiltonian F.

Hypothesis 3.77 There exists a modulus ω such that

$$F\left(t, x, \frac{B(x-y)}{\varepsilon}, X\right) - F\left(t, y, \frac{B(x-y)}{\varepsilon}, Y\right)$$

$$\geq -\omega\left(|x-y|\left(1 + \frac{|x-y|_{-1}}{\varepsilon}\right)\right)$$

for all $(t, x, y) \in (0, T) \times H \times H, \varepsilon > 0$, and $X, Y \in S(H)$, $X = P_N X P_N$, $Y = P_N Y P_N$ for some N and such that (3.52) holds.

For a bounded, strictly positive, self-adjoint, operator C on H we will use the notation

$$|x|_C := |C^{1/2} x|.$$

Hypothesis 3.78 Let C be a bounded, strictly positive, self-adjoint operator on H. We say that Hypothesis 3.78-C is satisfied if there exists a modulus ω_1 such that

$$F(t, x, c_1 C(x - y), X) - F(t, y, c_1 C(x - y), Y)$$

$$\geq -\omega_1 \left(|x - y|_C (1 + (c_1 + c_2 + c_3)|x - y|_C)\right)$$

for all $(t, x, y) \in (0, T) \times H \times H$, and $X, Y \in S(H)$, $X = P_N X P_N$, $Y = P_N Y P_N$ for some N and such that

$$- c_2 \begin{pmatrix} I & 0 \\ 0 & I \end{pmatrix} \leq \begin{pmatrix} X & 0 \\ 0 & -Y \end{pmatrix} \leq c_3 \begin{pmatrix} C & -C \\ -C & C \end{pmatrix}, \tag{3.191}$$

for some $c_1, c_2, c_3 \geq 0$.

Hypothesis 3.79 There exists an $h \in C^2(H)$, radial, nondecreasing, nonnegative, $h(x) \to \infty$ as $|x| \to \infty$, Dh, $D^2 h$ are bounded, and

$$F(t, x, p + \alpha Dh(x), X + \alpha D^2 h(x)) \geq F(t, x, p, X) - \sigma(\alpha, \|p\| + \|X\|) \tag{3.192}$$

$\forall x, p, X, \forall \alpha \geq 0$, where σ is a local modulus.

Hypotheses 3.77–3.79 will sometimes be applied to Hamiltonians F defined on finite-dimensional spaces, i.e. when $F : (0, T) \times H_{N_0} \times H_{N_0} \times S(H_{N_0}) \to \mathbb{R}$ for some N_0. In such cases it will be understood that N in Hypotheses 3.77, 3.78 will always be equal to N_0 and that every $X \in S(H_{N_0})$ is naturally extended to an operator in $S(H)$ by taking $P_{N_0} X P_{N_0}$.

We first show continuity estimates for viscosity solutions of finite-dimensional problems.

Lemma 3.80 Let $\delta > 0, l > 0$, and let ω be a modulus. Then there exist a nondecreasing, concave, C^2 function φ_δ on $[0, +\infty)$ such that $\varphi_\delta(0) < \delta$ and

$$\omega(|\varphi_\delta''(r)|r^2 + \varphi_\delta'(r)r + r) \leq \varphi_\delta(r) \quad \text{for } 0 \leq r \leq l. \tag{3.193}$$

Proof For $\varepsilon \in (0, l], 0 \leq r \leq l + 1, 0 < \gamma \leq 1$, thanks to the subadditivity of ω, we have

$$\omega(r) \leq \omega(\varepsilon) + \frac{\omega(\varepsilon)}{\varepsilon} r \leq \omega(\varepsilon) + \frac{\omega(\varepsilon)}{\varepsilon} (1 + l)^{1-\gamma} r^\gamma. \tag{3.194}$$

Let ε be such that $\omega(\varepsilon) < \delta/4$. Define

$$g_\gamma(r) = 2\omega(\varepsilon) + 2\frac{\omega(\varepsilon)}{\varepsilon}(1+l)^{1-\gamma}r^\gamma.$$

An elementary calculation and (3.194) give

$$g_\gamma(r) - \omega(|g_\gamma''(r)|r^2 + g_\gamma'(r)r + r) \geq \frac{\omega(\varepsilon)}{\varepsilon}(1+l)^{1-\gamma}r^\gamma\left(1 - 2\gamma(2-\gamma)\frac{\omega(\varepsilon)}{\varepsilon}(1+l)\right) \geq 0$$

if γ is small enough. We choose such γ_0 and set

$$\varphi_\delta(r) = g_{\gamma_0}(r + r_0),$$

where $0 < r_0 < 1$ is such that $g_{\gamma_0}(r_0) < \delta$. The function φ_δ has the required properties. $\qquad\square$

Proposition 3.81 *Let C be a bounded, strictly positive, self-adjoint, operator on $\mathbb{R}^k$. Let $u \in USC([0, T) \times \mathbb{R}^k)$, $v \in LSC([0, T) \times \mathbb{R}^k)$ be respectively a viscosity subsolution and a viscosity supersolution of*

$$\begin{cases} u_t + F(t, x, Du, D^2u) = 0 & \text{for } t \in (0, T), x \in \mathbb{R}^k, \\ u(0, x) = \psi(x) & \text{for } x \in \mathbb{R}^k, \end{cases} \tag{3.195}$$

where $F : (0, T) \times \mathbb{R}^k \times \mathbb{R}^k \times S(\mathbb{R}^k) \to \mathbb{R}$ is continuous, degenerate elliptic (Hypothesis 3.46) and satisfies Hypotheses 3.78 (with $H = H_N = \mathbb{R}^k$) and 3.79, and $\psi \in UC_b(\mathbb{R}^k)$.

(i) If $u, -v \leq M$, then there is a modulus of continuity m, depending only on M, T, ω_1, and a modulus of continuity of ψ in the $|\cdot|_C$ norm, such that

$$u(t, x) - v(t, y) \leq m(|x - y|_C) \tag{3.196}$$

for all $t \in [0, T)$ and $x, y \in \mathbb{R}^k$.

(ii) If

$$\sup_{x \in \mathbb{R}^k, t \in (0,T)} |F(t, x, 0, 0)| = K < +\infty, \tag{3.197}$$

then there exists a unique bounded viscosity solution u of (3.195). The norm $\|u\|_0$ only depends on $\|\psi\|_0$ and K.

Proof (i) Let m_1 be a modulus of continuity of ψ in the $|\cdot|_C$ norm. Given $\mu > 0$, set

$$u_1(t, x) = u(t, x) - \frac{\mu}{T - t} \tag{3.198}$$

$$v_1(t, x) = v(t, x) + \frac{\mu}{T - t}. \tag{3.199}$$

Let $\kappa = 3(T+1)(1+2\|C\|)$. Lemma 3.80 applied with the modulus $m_2(r) = m_1(r) + \kappa\omega(r) + (2M+1)r$ and $l = 2$ gives us for every $\delta > 0$ a function $\varphi_\delta \in C^2([0, \infty))$, nondecreasing, concave, such that

$$\varphi_\delta(0) < \delta, \quad \varphi_\delta(1) \geq 2M + 1 \tag{3.200}$$

and

$$\varphi_\delta(r) - m_2(|\varphi_\delta''(r)|r^2 + \varphi_\delta'(r)r + r) \geq 0 \tag{3.201}$$

for $0 \leq r \leq 2$.

We are going to show that for every $\delta > 0$

$$u_1(t, x) - v_1(t, y) \leq \varphi_\delta(|x - y|_C)(1 + t)$$

for $t \in [0, T]$ and $\{|x - y|_C < 1\} = \Delta$. Let for $\gamma > 0$

$$\varphi(t, x, y) = \varphi_\delta(|x - y|_C^2 + \gamma)^{\frac{1}{2}}(1 + t).$$

Suppose that

$$\sup_{(x,y)\in\Delta, t\in[0,T]} (u_1(t, x) - v_1(t, y) - \varphi(t, x, y)) > 0$$

(if not we are done). Then, for small $\alpha > 0$, using h from Hypothesis 3.79,

$$\sup_{(x,y)\in\Delta, t\in[0,T]} (u_1(t, x) - v_1(t, y) - \varphi(t, x, y) - \alpha h(x) - \alpha h(y)) > 0$$

and is attained at a point $(\bar{t}, \bar{x}, \bar{y})$. Moreover, (3.200) and (3.201) imply that $(\bar{x}, \bar{y}) \in \Delta$ and $0 < \bar{t} < T$.

We compute

$$D_x\varphi(\bar{t}, \bar{x}, \bar{y}) = \varphi_\delta'\left((|\bar{x} - \bar{y}|_C^2 + \gamma)^{\frac{1}{2}}\right) \frac{C(\bar{x} - \bar{y})}{(|\bar{x} - \bar{y}|_C^2 + \gamma)^{\frac{1}{2}}}(\bar{t} + 1), \tag{3.202}$$

$$D_{xx}^2\varphi(\bar{t}, \bar{x}, \bar{y}) = \varphi_\delta''((|\bar{x} - \bar{y}|_C^2 + \gamma)^{\frac{1}{2}}) \frac{C(\bar{x} - \bar{y}) \otimes C(\bar{x} - \bar{y})}{|\bar{x} - \bar{y}|_C^2 + \gamma}(\bar{t} + 1)$$

$$+ \varphi_\delta'((|\bar{x} - \bar{y}|_C^2 + \gamma)^{\frac{1}{2}}) \frac{C}{(|\bar{x} - \bar{y}|_C^2 + \gamma)^{\frac{1}{2}}}(\bar{t} + 1) \tag{3.203}$$

$$- \varphi_\delta'((|\bar{x} - \bar{y}|_C^2 + \gamma)^{\frac{1}{2}}) \frac{C(\bar{x} - \bar{y}) \otimes C(\bar{x} - \bar{y})}{(|\bar{x} - \bar{y}|_C^2 + \gamma)^{\frac{3}{2}}}(\bar{t} + 1).$$

We may rewrite (3.203) as $D_{xx}^2 \varphi(\bar{x}, \bar{y}) = B_1 + B_2 + B_3$, where B_1, B_2, B_3 are the three terms appearing in (3.203). Since φ_δ is nondecreasing and concave, $B_2 \geq 0$ and $B_1, B_3 \leq 0$. Using this notation we have

$$D^2\varphi(\bar{t}, \bar{x}, \bar{y}) = \begin{pmatrix} B_2 & -B_2 \\ -B_2 & B_2 \end{pmatrix} + \begin{pmatrix} B_1 + B_3 & -B_1 - B_3 \\ -B_1 - B_3 & B_1 + B_3 \end{pmatrix}. \tag{3.204}$$

If we denote the two matrices in (3.204) by D_1 and $-D_2$ respectively, we obtain $D^2\varphi(\bar{t}, \bar{x}, \bar{y}) = D = D_1 - D_2$, where $D_1, D_2 \geq 0$.

Applying Theorem E.11 with $\varepsilon = 1/(\|D_1\| + \|D_2\|)$, there exist $b_1, b_2 \in \mathbb{R}$ and matrices $X, Y \in S(\mathbb{R}^k)$ such that

$$\left(b_1, \varphi_\delta' \left((|\bar{x} - \bar{y}|_C^2 + \gamma)^{\frac{1}{2}} \right) \frac{C(\bar{x} - \bar{y})}{(|\bar{x} - \bar{y}|_C^2 + \gamma)^{\frac{1}{2}}} (1 + \bar{t}), X \right) \in \bar{\mathcal{P}}^{2,+}(u_1 - \alpha h)(\bar{t}, \bar{x}),$$

$$\left(b_2, \varphi_\delta' \left((|\bar{x} - \bar{y}|_C^2 + \gamma)^{\frac{1}{2}} \right) \frac{C(\bar{x} - \bar{y})}{(|\bar{x} - \bar{y}|_C^2 + \gamma)^{\frac{1}{2}}} (1 + \bar{t}), Y \right)$$

$$\in \bar{\mathcal{P}}^{2,-}(v_1 + \alpha h)(\bar{t}, \bar{y}), \tag{3.205}$$

$$b_1 - b_2 = \varphi_\delta \left((|\bar{x} - \bar{y}|_C^2 + \gamma)^{\frac{1}{2}} \right), \tag{3.206}$$

and

$$-2(\|D_1\| + \|D_2\|) \begin{pmatrix} I & 0 \\ 0 & I \end{pmatrix} \leq \begin{pmatrix} X & 0 \\ 0 & -Y \end{pmatrix}$$

$$\leq D + \frac{1}{\|D_1\| + \|D_2\|} D^2 \leq 2D_1,$$

where in the last line we used $D^2 \leq (\|D_1\| + \|D_2\|)(D_1 + D_2)$. Computing the norms, we thus have obtained

$$-2\|C\|(1 + T) \left[2|\varphi_\delta'' \left((|\bar{x} - \bar{y}|_C^2 + \gamma)^{\frac{1}{2}} \right)| + \frac{3\varphi_\delta' \left((|\bar{x} - \bar{y}|_C^2 + \gamma)^{\frac{1}{2}} \right)}{(|\bar{x} - \bar{y}|_C^2 + \gamma)^{\frac{1}{2}}} \right] \begin{pmatrix} I & 0 \\ 0 & I \end{pmatrix}$$

$$\leq \begin{pmatrix} X & 0 \\ 0 & -Y \end{pmatrix} \leq \frac{2\varphi_\delta' \left((|\bar{x} - \bar{y}|_C^2 + \gamma)^{\frac{1}{2}} \right)}{(|\bar{x} - \bar{y}|_C^2 + \gamma)^{\frac{1}{2}}} (1 + T) \begin{pmatrix} C & -C \\ -C & C \end{pmatrix}. \tag{3.207}$$

We set $\bar{r} = (|\hat{x} - \hat{y}|_C^2 + \gamma)^{\frac{1}{2}}$ and

$$d = \sup \left\{ |\varphi_\delta''(r)| + \varphi_\delta'(r) : 0 \leq r \leq 2 \right\}.$$

Using the Eqs. (3.205)–(3.207), and Hypothesis 3.79 we now have for $\gamma < 1$ and small α

$$
\varphi_\delta(\bar{r}) + \frac{2\mu}{T^2} \leq F\left(\bar{t}, \bar{y}, \frac{(1+\bar{t})\varphi_\delta'(\bar{r})}{\bar{r}} C(\bar{x} - \bar{y}), Y\right)
$$
$$
- F\left(\bar{t}, \bar{x}, \frac{(1+\bar{t})\varphi_\delta'(\bar{r})}{\bar{r}} C(\bar{x} - \bar{y}), X\right) + 2\sigma\left(\alpha, \frac{6d(T+1)\|C\|}{\gamma^{\frac{1}{2}}} + d\|C\|^{\frac{1}{2}}(T+1)+1\right).
$$

It thus follows from Hypothesis 3.78 that

$$
\varphi_\delta(\bar{r}) + \frac{2\mu}{T^2}
$$
$$
\leq \omega_1\left(|\hat{x} - \hat{y}|_C\left(1 + 3(T+1)(1+2\|C\|)\frac{\varphi_\delta'(\bar{r})}{\bar{r}} + 4(T+1)\|C\|\varphi_\delta''(\bar{r})|\right)|\hat{x} - \hat{y}|_C\right)
$$
$$
+ 2\sigma\left(\alpha, \frac{6d(T+1)\|C\|}{\gamma^{\frac{1}{2}}} + d\|C\|^{\frac{1}{2}}(T+1)+1\right)
$$

for some local modulus σ_1. Thus, since ω_1 is concave, we get

$$
\varphi_\delta(\bar{r}) + \frac{2\mu}{T^2} \leq 3(T+1)(1+2\|C\|)\omega_1\left(|\varphi_\delta''(\bar{r})|\bar{r}^2 + \varphi_\delta'(\bar{r})\bar{r} + \bar{r}\right)
$$
$$
+ 2\sigma\left(\alpha, \frac{6d(T+1)\|C\|}{\gamma^{\frac{1}{2}}} + d\|C\|^{\frac{1}{2}}(T+1)+1\right).
$$

Therefore we obtain a contradiction if we let $\alpha \to 0$. This implies

$$
u_1(t, x) - v_1(t, y) \leq \varphi_\delta(|x - y|_C)(1 + T) + 2M|x - y|_C
$$

for all $x, y \in \mathbb{R}^k$ and $t \in [0, T)$. The claim now follows by letting $\mu \to 0$.

(ii) We remark that part (i) in particular guarantees that the comparison principle holds for Eq. (3.195). It is standard to notice that under our assumptions one can construct a bounded viscosity subsolution $\underline{u}$ and a bounded viscosity supersolution $\bar{u}$ such that $\underline{u}(0, x) = \psi(x) = \bar{u}(0, x)$ and $\underline{u} \leq \bar{u}$ (see Proposition 3.94 for a similar construction). We can thus use Perron's method (see Theorem E.12) to obtain a bounded viscosity solution which is unique by (i). $\qquad\square$

The above existence and uniqueness result for finite-dimensional HJB equations will be an important tool in constructing viscosity solutions of HJB equations in Hilbert spaces by finite-dimensional approximations. We begin with the case when the strong B-condition for A is satisfied.

Proposition 3.82 *Let B be compact and satisfy the strong B-condition for A as in Definition 3.10. Let u, v be respectively a viscosity subsolution and a viscosity supersolution of*

$$\begin{cases} u_t - \langle Ax, Du \rangle + F(t, x, Du, D^2 u) = 0 \quad \text{for } t \in (0, T), x \in H, \\ u(0, x) = \psi(x) \quad \text{for } x \in H, \end{cases} \tag{3.208}$$

where $F : (0, T) \times H \times H \times S(H) \to \mathbb{R}$ satisfies (for $U = H$) Hypotheses 3.44, 3.46, 3.47, 3.77, and 3.79. Let $\psi \in UC_b(H)$. Let u, $-v \le M$ and be such that

$$|u(t, x) - u(t, y)| + |v(t, x) - v(t, y)| \le m(|x - y|) \tag{3.209}$$

for all $t \in [0, T)$ and $x, y \in H$, for some modulus m. Assume, moreover, that

$$\lim_{t \to 0} \tilde{\rho}(t) = 0, \tag{3.210}$$

where

$$\tilde{\rho}(t) = \sup_{x \in H} \left[(u(t, x) - \psi(e^{tA}x))^+ + (v(t, x) - \psi(e^{tA}x))^- \right].$$

Then for every $0 < \tau < T$ there exists a modulus m_τ, depending only on $\tau, m, \omega, \tilde{\rho}, T, M$, the constant c_0 in Definition 3.10 and the modulus of continuity of ψ, such that

$$u(t, x) - v(t, y) \le m_\tau(|x - y|) \quad \text{for all } x, y \in H, t \in [\tau, T). \tag{3.211}$$

Proof We will first show that there exists constants $C_\varepsilon > 0$, depending only on $\varepsilon, m, c_0, \omega$, such that

$$\lim_{\varepsilon \to 0} C_\varepsilon = 0, \tag{3.212}$$

and for every $0 < \tau < T$,

$$\sup_{t \in [\tau, T)} a_{\varepsilon, C_\varepsilon}(t) = a_{\varepsilon, C_\varepsilon}(\tau), \tag{3.213}$$

where

$$a_{\varepsilon, C}(t) = \sup_{x, y \in H} \left\{ u(t, x) - v(t, y) - \frac{|x - y|_{-1}^2}{2\varepsilon} - Ct \right\}.$$

For $\mu > 0, \alpha > 0, \beta > 0$ we consider the function

$$\Psi(t, s, x, y) = u(t, x) - \frac{\mu}{T - t} - v(s, y) - \frac{\mu}{T - s} - \frac{|x - y|_{-1}^2}{2\varepsilon}$$
$$- \alpha h(x) - \alpha h(y) - \frac{(t - s)^2}{2\beta} - Ct,$$

where h is the function from Hypothesis 3.79. Since B is compact, B-upper semi continuity is equivalent to weak sequential upper semicontinuity, so Ψ attains a maximum at some point $(\bar{t}, \bar{s}, \bar{x}, \bar{y})$. Moreover, as always we have

$$\lim_{\beta \to 0} \frac{(\bar{t} - \bar{s})^2}{2\beta} = 0 \quad \text{for fixed } \varepsilon, \alpha. \tag{3.214}$$

Therefore, using the weak sequential upper semicontinuity of the above function, it is easy to see that if $\sup_{t \in [\tau, T)} a_{\varepsilon, C}(t) > a_{\varepsilon, C}(\tau)$, then for small $\mu > 0, \alpha > 0, \beta > 0$, we must have $\tau < \bar{t}, \bar{s} < T$.

We can now argue as in the proof of Theorem 3.50 (from (3.87) to (3.89)) to obtain that for $N > 2$ there exist $X_N, Y_N \in S(H)$ satisfying (3.52) and such that

$$\frac{\bar{t} - \bar{s}}{\beta} + \frac{\mu}{(T - \bar{t})^2} + C - \left\langle \bar{x}, A^* \left(\frac{B(\bar{x} - \bar{y})}{\varepsilon} \right) \right\rangle$$
$$+ F\left(\bar{t}, \bar{x}, \frac{B(\bar{x} - \bar{y})}{\varepsilon} + \alpha Dh(\bar{x}), X_N + \frac{2BQ_N}{\varepsilon} + \alpha D^2 h(\bar{x}) \right) \leq 0 \tag{3.215}$$

and

$$\frac{\bar{t} - \bar{s}}{\beta} - \frac{\mu}{(T - \bar{s})^2} + \left\langle \bar{y}, A^* \left(\frac{B(\bar{x} - \bar{y})}{\varepsilon} \right) \right\rangle$$
$$+ F\left(\bar{s}, \bar{y}, \frac{B(\bar{x} - \bar{y})}{\varepsilon} - \alpha Dh(\bar{x}), Y_N - \frac{2BQ_N}{\varepsilon} - \alpha D^2 h(\bar{x}) \right) \geq 0. \tag{3.216}$$

Since $u, -v$ are bounded from below it is obvious that

$$\frac{|B(\bar{x} - \bar{y})|}{\varepsilon} \leq R_\varepsilon \tag{3.217}$$

for some R_ε, possibly depending[3] on $u, -v$. Also, since $\Psi(\bar{t}, \bar{s}, \bar{x}, \bar{x}) + \Psi(\bar{t}, \bar{s}, \bar{y}, \bar{y}) \leq 2\Psi(\bar{t}, \bar{s}, \bar{x}, \bar{y})$, we get

$$\frac{|\bar{x} - \bar{y}|^2_{-1}}{2\varepsilon} \leq m(|\bar{x} - \bar{y}|). \tag{3.218}$$

---

[3] In fact, using uniform continuity of u, since for every $w \in H$, $|w| = 1$ we have $\Psi(\bar{t}, \bar{s}, \bar{x} + w, \bar{y}) \leq \Psi(\bar{t}, \bar{s}, \bar{x}, \bar{y})$, we can obtain for $\alpha < 1$

$$\frac{|B(\bar{x} - \bar{y})|}{\varepsilon} = \sup_{|w|=1} \frac{\langle B(\bar{x} - \bar{y}), w \rangle}{\varepsilon} \leq \frac{\langle Bw, w \rangle}{2\varepsilon} + u(\bar{t}, \bar{x}) - u(\bar{t}, \bar{x} - w)$$

$$+ \alpha(h(\bar{x} - w) - h(\bar{x})) \leq \frac{\|B\|}{2\varepsilon} + m(1) + L,$$

where L is the Lipschitz constant of h.

Thus, subtracting (3.216) from (3.215), and using Hypotheses 3.44, 3.47, 3.77, 3.79, and (3.3), (3.214), (3.217), yields

$$C + \frac{2\mu}{T^2} \leq c_0 \frac{|\bar{x} - \bar{y}|^2_{-1}}{\varepsilon} - \frac{|\bar{x} - \bar{y}|^2}{\varepsilon} + w\left(|\bar{x} - \bar{y}|\left(1 + \frac{|\bar{x} - \bar{y}|_{-1}}{\varepsilon}\right)\right) + \sigma_1(\varepsilon; \alpha, \beta, N),$$
(3.219)

where $\lim_{\alpha \to 0} \lim \sup_{\beta \to 0} \lim \sup_{N \to +\infty} \sigma_1(\varepsilon; \alpha, \beta, N) = 0$. Let

$$\gamma(\varepsilon) = \sup_{x,y \in H}\left\{c_0 \frac{|\bar{x} - \bar{y}|^2_{-1}}{\varepsilon} - \frac{|\bar{x} - \bar{y}|^2}{\varepsilon} + w\left(|\bar{x} - \bar{y}|\left(1 + \frac{|\bar{x} - \bar{y}|_{-1}}{\varepsilon}\right)\right)\right\}.$$

Using (3.218) we have

$$\gamma(\varepsilon) \leq \sup_{r \geq 0}\left\{2c_0 m(r) - \frac{r^2}{\varepsilon} + w\left(r\left(1 + \frac{(2m(r))^{\frac{1}{2}}}{\varepsilon^{\frac{1}{2}}}\right)\right)\right\}.$$
(3.220)

This expression can be estimated from above by

$$C_1\left(1 + r + \frac{r}{\varepsilon^{\frac{1}{2}}} + \frac{r^{\frac{3}{2}}}{\varepsilon^{\frac{1}{2}}}\right) - \frac{r^2}{\varepsilon},$$
(3.221)

where C_1 only depends on w, m and c_0. It is easily seen that (3.221) is positive only if $r \leq C_2 \varepsilon^{\frac{1}{2}}$ for $\varepsilon \leq 1$, where C_2 only depends on C_1. But then it easily follows from (3.220) that $\lim_{\varepsilon \to 0} \gamma(\varepsilon) = 0$. Thus, if $C = C_\varepsilon := 2\gamma(\varepsilon)$, we obtain a contradiction after we take $\lim \sup_{\alpha \to 0} \lim \sup_{\beta \to 0} \lim \sup_{N \to +\infty}$ in (3.219). Hence (3.213) must be true with this choice of C_ε.

A consequence of (3.213) is that for all $t \in [\tau, T)$, $x, y \in H$

$$u(t, x) - v(t, y) - \frac{|x - y|^2_{-1}}{2\varepsilon} - C_\varepsilon t \leq \sup_{x,y \in H}\left\{u(\tau, x) - v(\tau, y) - \frac{|x - y|^2_{-1}}{2\varepsilon}\right\}.$$
(3.222)

Let m_ψ be the modulus of continuity of ψ. Then, by (3.20), (3.212) and (3.210), we obtain from (3.222)

$$u(t, x) - v(t, y) \leq \frac{|x - y|^2_{-1}}{2\varepsilon} + C_\varepsilon t + \sup_{x,y \in H}\left\{u(\tau, x) - v(\tau, y) - \frac{|x - y|^2_{-1}}{2\varepsilon}\right\}$$

$$\leq \frac{|x - y|^2_{-1}}{2\varepsilon} + C_\varepsilon T + 2\tilde{\rho}(\tau) + \sup_{x,y \in H}\left\{|\psi(e^{\tau A}x) - \psi(e^{\tau A}y)| - \frac{|x - y|^2_{-1}}{2\varepsilon}\right\}$$

$$\leq \frac{|x-y|^2_{-1}}{2\varepsilon} + C_\varepsilon T + 2\tilde\rho(\tau) + \sup_{x,y\in H} \left\{ m_\psi \left(\frac{e^{c_0 T}|x-y|_{-1}}{2\tau^{\frac{1}{2}}} \right) - \frac{|x-y|^2_{-1}}{2\varepsilon} \right\}$$

$$\leq \frac{|x-y|^2_{-1}}{2\varepsilon} + \rho_\tau(\varepsilon), \quad (3.223)$$

where ρ_τ depends only on C_ε, T, τ, m_ψ, $\tilde\rho$, and $\lim_{\varepsilon\to 0} \rho_\tau(\varepsilon) = 0$. Thus for every $\varepsilon > 0$

$$u(t,x) - v(t,y) \leq \min\left\{ 2M|x-y|_{-1}, \frac{|x-y|^2_{-1}}{2\varepsilon} + \rho_\tau(\varepsilon) \right\},$$

which implies (3.211).   □

Proposition 3.82 in particular implies the comparison principle for bounded and uniformly continuous viscosity sub- and supersolutions of (3.208). However we want to mention that comparison also holds without the requirement of uniform continuity of u and v, with almost the same proof.

We will be using the following operators to approximate the operator A. For $N \geq 1$ we define

$$A_N = (P_N A^* P_N)^*.$$

The A_N are bounded, dissipative, operators in H and it is easy to see that

$$A_N P_N = A_N = P_N A_N, \quad (3.224)$$

and thus it follows that

$$e^{tA_N} P_N = P_N e^{tA_N}. \quad (3.225)$$

Moreover, we have

$$-A_N^* B + c_0 B \geq P_N. \quad (3.226)$$

We alert the readers that in the lemma below we will use x_N to denote a sequence in H, not $P_N x$ as we have done in previous sections.

Lemma 3.83 *Let B be a positive, self-adjoint, compact operator satisfying the strong B-condition as in Definition 3.10. Then:*

(i) Let $x, x_N \in D(A^*)$, $x_N \to x$, and $A^* x_N \to A^* x$. Then $A_N^* x_N \to A^* x$.
(ii) For every $x \in H$, $T > 0$

$$e^{tA_N} x \to e^{tA} x \quad (3.227)$$

uniformly on $[0, T]$.

Proof (i) We know from Lemma 3.17(ii) that the operator $S = -A^* B + c_0 B$ is invertible, $S^{-1} \in \mathcal{L}(H)$, and $D(A^*) = D(B^{-1})$. We have

$$A^* = -SB^{-1} + c_0 I, \quad A_N^* = -P_N S B^{-1} P_N + c_0 P_N.$$

Since $B^{-1} = S^{-1}(-A^* + c_0 I)$ we thus obtain

$$B^{-1} x_N \to B^{-1} x.$$

Therefore

$$A^* x_N - A_N^* x_N = -Q_N S B^{-1} x_N - P_N S Q_N B^{-1} x_N + c_0 Q_N x_N \to 0$$

since P_N converges strongly to I and Q_N converges strongly to 0. This proves the claim.

(ii) We see that $e^{t A_N^*}$ and $e^{t A_N}$ are semigroups of contractions. Using (3.3) we have

$$|e^{tA}x|_{-1}^2 + 2 \int_0^t \langle e^{sA}x, S e^{sA}x \rangle ds - 2c_0 \int_0^t |e^{sA}x|_{-1}^2 ds = |x|_{-1}^2,$$

$$|e^{tA_N}x|_{-1}^2 + 2 \int_0^t \langle e^{sA_N}x, S_N e^{sA_N}x \rangle ds - 2c_0 \int_0^t |e^{sA_N}x|_{-1}^2 ds = |x|_{-1}^2,$$

$$(3.228)$$

where

$$S_N = -A_N^* B + c_0 B = -P_N A^* P_N B + c_0 B = P_N(-A^* B + c_0 B) P_N + c_0 Q_N B.$$

By the Trotter–Kato theorem (see Theorem B.46), for every $x \in H$, $e^{t A_N^*} x \to e^{t A^*} x$ uniformly on $[0, T]$. Thus, taking adjoints, it follows that

$$e^{t A_N} x \to e^{tA} x \qquad \text{for every } x \in H, t \geq 0. \tag{3.229}$$

Since B is compact, this implies

$$e^{t A_N} x \to e^{tA} x \qquad \text{in } H_{-1}. \tag{3.230}$$

Thus, passing to the limit as $N \to +\infty$ in (3.228) and using (3.230), we obtain

$$\int_0^t \langle e^{sA_N}x, S_N e^{sA_N}x \rangle ds \to \int_0^t \langle e^{sA}x, S e^{sA}x \rangle ds,$$

which, upon observing that $\|Q_N B\| \to 0$ and $P_N x \to x$, yields

$$\int_0^t \langle e^{sA_N}x, S e^{sA_N}x \rangle ds \to \int_0^t \langle e^{sA}x, S e^{sA}x \rangle ds. \tag{3.231}$$

Let $y = e^{sA}x$, $y_N = e^{sA_N}x$. Then

$$0 \leq |y - y_N|^2 \leq \langle y - y_N, S(y - y_N) \rangle = \langle y, Sy \rangle - \langle y_N, Sy \rangle - \langle y, Sy_N \rangle + \langle y_N, Sy_N \rangle.$$
$$(3.232)$$

Using (3.229) it thus follows that

$$0 \leq \liminf_{N \to +\infty} \langle y_N, Sy_N \rangle - \langle y, Sy \rangle.$$

This, together with Fatou's lemma and (3.231), implies that

$$\lim_{N \to +\infty} \langle e^{sA_N} x, Se^{sA_N} x \rangle = \langle e^{sA} x, Se^{sA} x \rangle$$

for a.e. s. We then get from (3.232) that

$$\lim_{N \to +\infty} |e^{sA} x - e^{sA_N} x|^2 = 0 \qquad \text{for a.e. } s. \qquad (3.233)$$

The uniform convergence on $[0, T]$ follows from standard arguments using the integral representation of the resolvent (see e.g. (3.2) of [479]) and Theorem 4.2. of [479]. □

Theorem 3.84 *Let B be compact and satisfy the strong B-condition for A as in Definition 3.10. Let $F : (0, T) \times H \times H \times S(H) \to \mathbb{R}$ satisfy (for $U = H$) Hypotheses 3.44, 3.46, Hypothesis 3.47 with $B = I$, and Hypotheses 3.77, 3.78-1, and 3.79. Let $\psi \in UC_b(H)$ and let for every $R > 0$*

$$F_R := \sup\{|F(t, x, p, X)| : t \in (0, T), x \in H, |p| + \|X\| \leq R\} < +\infty. \quad (3.234)$$

Then there exists a unique bounded viscosity solution $u \in UC_b^x([0, T) \times H) \cap UC_b^x([\tau, T) \times H_{-1})$ for $0 < \tau < T$, of

$$\begin{cases} u_t - \langle Ax, Du \rangle + F(t, x, Du, D^2 u) = 0 & \text{for } t \in (0, T), x \in H, \\ u(0, x) = \psi(x) & \text{for } x \in H, \end{cases} \quad (3.235)$$

satisfying

$$\limsup_{t \to 0}_{x \in H} |u(t, x) - \psi(e^{tA} x)| = 0. \quad (3.236)$$

Moreover, there is a modulus ρ such that

$$|u(t, x) - u(s, e^{(t-s)A} x)| \leq \rho(t - s) \quad \text{for all } 0 \leq s \leq t < T, x \in H. \quad (3.237)$$

Proof We consider two approximating equations.

$$\begin{cases} (u_N)_t - \langle A_N x, Du_N \rangle + F(t, P_N x, P_N Du_N, P_N D^2 u_N P_N) = 0 & \text{in } (0, T) \times H \\ u_N(0, x) = \psi(P_N x) & \text{in } H, \end{cases}$$
$$(3.238)$$

and

$$\begin{cases} (v_N)_t - \langle A_N x, D v_N \rangle + F(t, x, D v_N, P_N D^2 v_N P_N) = 0 & \text{in } (0, T) \times H_N \\ v_N(0, x) = \psi(x) & \text{in } H_N. \end{cases}$$

(3.239)

We notice that, since A_N is dissipative, the Hamiltonian $\tilde{F}_N : (0, T) \times H_N \times H_N \times S(H_N) \to \mathbb{R}$ defined by

$$\tilde{F}_N(t, x, p, X) = \langle A_N x, p \rangle + F(t, x, p, P_N X P_N)$$

satisfies all the assumptions of Proposition 3.81 with $C = I$, uniformly in N. Therefore, by Proposition 3.81, there is a unique bounded viscosity solution v_N of (3.239), $M \geq 0$, and a modulus m such that for all N

$$\begin{cases} \|v_N\|_0 \leq M \\ |v_N(t, x) - v_N(t, y)| \leq m(|x - y|) & \text{for all } t \in [0, T), \ x, y \in H_N \end{cases}$$

(3.240)

and so (3.240) is also satisfied by u_N on H.

We remark that the monotonicity of A_N guarantees that v_N is also a viscosity solution in the sense of Definition 3.34 on the finite-dimensional Hilbert space H_N.

We now extend v_N to H by setting $u_N(x) = v_N(P_N x)$. We claim that u_N is a viscosity solution of (3.238). Again, since all the terms are bounded and A_N is monotone it is enough to show it in the classical sense of (the parabolic counterpart of) Definition 3.40. To prove that u_N is a viscosity subsolution, suppose that $u_N(t, x) - \varphi(t, x)$ has a maximum at $(\hat{t}, \hat{x})$ for a smooth test function φ. Then $v_N(t, z) - \varphi(t, z + Q_N \hat{x})$ has a maximum at $(\hat{t}, P_N \hat{x})$ in $(0, T) \times H_N$. Therefore, using the fact that v_N is a subsolution of (3.239), we get

$$\varphi_t(\hat{t}, \hat{x}) - \langle A_N P_N \hat{x}, P_N D\varphi(\hat{t}, \hat{x}) \rangle + F(\hat{t}, P_N \hat{x}, P_N D\varphi(\hat{t}, \hat{x}), P_N D^2 \varphi(\hat{t}, \hat{x}) P_N) \leq 0$$

and the claim follows by (3.224). The supersolution case is similar.

We now show that there is a modulus ρ, depending only on m and the function F_R, such that

$$|v_N(t, x) - v_N(s, e^{-(t-s)A_N} x)| \leq \rho(t - s)$$

(3.241)

for $x \in H_N, 0 \leq s \leq t < T$. Because of (3.240) it is enough to show (3.241) for $s = 0$ since then the estimate can be reapplied at any later time. To do this we begin with $\psi \in C_b^{1,1}(H)$. We denote the Lipschitz constant of $D\psi$ by $L_{D\psi}$. We use the fact that $w(t, x) = \psi(e^{tA_N} x)$ is a classical (and viscosity) solution of

$$w_t - \langle A_N x, Dw \rangle = 0 \quad \text{in } (0, T) \times H_N, \quad u(0, x) = \psi(x) \text{ in } H_N,$$

which implies that

$$w + t F_{L_{D\psi} + \|D\psi\|_0}, \quad w - t F_{L_{D\psi} + \|D\psi\|_0}$$

are respectively a viscosity supersolution and a subsolution of (3.239). Comparison then gives

$$|v_N(t, x) - \psi(e^{tA_N} x)| \leq t F_{L_{D\psi} + \|D\psi\|_0}. \quad (3.242)$$

For $\psi \in UC_b(H)$ we can approximate it by its inf-sup convolutions $\overline{\psi}_\varepsilon \in C_b^{1,1}(H)$ (see Proposition D.26). This approximation is such that $c_\varepsilon = \|\psi - \overline{\psi}_\varepsilon\|_0$ and $K_\varepsilon = L_{D\overline{\psi}_\varepsilon} + \|D\overline{\psi}_\varepsilon\|_0$ only depend on the modulus of continuity of ψ, and moreover $\lim_{\varepsilon \to 0} c_\varepsilon = 0$. Let v_N^ε be the viscosity solution of (3.239) with initial condition $\overline{\psi}_\varepsilon$. It follows from comparison guaranteed by Proposition 3.81 that

$$\|v_N - v_N^\varepsilon\|_0 \leq \|\psi - \overline{\psi}_\varepsilon\|_0 = c_\varepsilon.$$

Using this and (3.242) we thus have

$$|v_N(t, x) - \psi(e^{tA_N} x)| \leq |v_N(t, x) - v_N^\varepsilon(t, x)| + |v_N^\varepsilon(t, x) - \overline{\psi}_\varepsilon(e^{tA_N} x)|$$
$$+ |\overline{\psi}_\varepsilon(e^{tA_N} x) - \psi(e^{tA_N} x)| \leq 2\|\psi - \overline{\psi}_\varepsilon\|_0 + t F_{L_{D\overline{\psi}_\varepsilon} + \|D\overline{\psi}_\varepsilon\|_0} = 2c_\varepsilon + t K_\varepsilon.$$

Therefore

$$|v_N(t, x) - \psi(e^{tA_N} x)| \leq \rho(t) = \inf_{\varepsilon > 0} \{2c_\varepsilon + t K_\varepsilon\},$$

which completes the proof of (3.241). We also conclude, by (3.225), that for $0 \leq s \leq t < T, x \in H$,

$$|u_N(t, x) - u_N(s, e^{(t-s)A_N} x)| = |v_N(t, P_N x) - v_N(s, P_N e^{(t-s)A_N} x)|$$
$$= |v_N(t, P_N x) - v_N(s, e^{(t-s)A_N} P_N x)| \leq \rho(t). \quad (3.243)$$

We will now show that for every $0 < \tau < T$ there exists a modulus m_τ such that

$$|u_N(t, x) - u_N(t, y)| \leq m_\tau(|x - y|_{-1}) \quad \text{for all } x, y \in H, t \in [\tau, T). \quad (3.244)$$

We notice that (3.226) implies that B restricted to H_N (i.e. $B_N = B P_N$) satisfies the strong condition for A_N on H_N with the same constant c_0. Therefore (3.244) follows from (3.240), (3.241) and Proposition 3.82 applied on spaces $H = H_N$, since all assumptions are independent of N. (In fact, we do not need the full force of Proposition 3.82 since we are dealing with bounded equations on finite-dimensional spaces.)

Since B also satisfies the weak B-condition for A_N with constant c_0, we notice that, by (3.19), for every N, $|e^{tA_N} x - x|_{-1} \leq C(R)\sqrt{t}$ for $|x| \leq R$, where $C(R)$ is independent of N. Thus for $0 < \tau \leq s < t < T, R > 0$, using (3.243), (3.244), we obtain for $|x| < R$

$$|u_N(t, x) - u_N(s, x)| \le |u_N(t, x) - u_N(s, e^{(t-s)A_N} x)| + |u_N(s, e^{(t-s)A_N} x) - u_N(s, x)|$$

$$\le \rho(|t - s|) + m_\tau(|e^{(t-s)A_N} x - x|) \le \rho(|t - s|) + m_\tau(C(R)\sqrt{|t - s|}) =: \rho_{\tau,R}(|t - s|).$$

Combining this with (3.244) we have

$$|u_N(t, x) - u_N(s, x)| \le m_\tau(|x - y|_{-1}) + \rho_{\tau,R}(|t - s|), \quad N \ge 1, \tau \le t, s < T, |x|, |y| \le R. \tag{3.245}$$

Therefore (extending u_N to $t = T$), the family $\{u_N\}$ is equicontinuous in the topology of $\mathbb{R} \times H_{-1}$ on sets $[\tau, T] \times \{|x| \le R\}$ for $\tau > 0$. But since B is compact such sets are compact in $\mathbb{R} \times H_{-1}$. Therefore, by the Arzela–Ascoli theorem there is a subsequence of u_N, still denoted by u_N, and a function u, such that $u_N \to u$ uniformly on bounded subsets of $[\tau, T] \times H$ for $\tau > 0$. Obviously u satisfies (3.237), (3.240), (3.244) and (3.245). The conclusion that u is a viscosity solution of (3.235) will follow from Theorem 3.41 (reformulated for the initial value problem), Lemmas 3.83(i) and 3.85 below applied with $\tilde{F}(X) := F(t, x, p, X)$ for some fixed $(t, x, p) \in (0, T) \times H \times H$. Uniqueness is a consequence of Proposition 3.82. $\quad\square$

Lemma 3.85 *If $\tilde{F} : S(H) \to \mathbb{R}$ is locally uniformly continuous and satisfies Hypotheses 3.46 and 3.47 with $B = I$, then for every $X \in S(H)$*

$$\tilde{F}(P_N X P_N) \to \tilde{F}(X) \quad as \quad N \to \infty.$$

Proof For every $\varepsilon > 0$ we have

$$P_N(X - \varepsilon X^2)P_N - \left(\|X\| + \frac{1}{\varepsilon}\right) Q_N \le X \le P_N(X + \varepsilon X^2)P_N + \left(\|X\| + \frac{1}{\varepsilon}\right) Q_N.$$

Therefore, Hypotheses 3.46 and 3.47 imply

$$\tilde{F}(P_N(X + \varepsilon X^2)P_N) - \sigma_1(N, \varepsilon) \le \tilde{F}(X) \le \tilde{F}(P_N(X - \varepsilon X^2)P_N) + \sigma_1(N, \varepsilon),$$

where σ_1 is a local modulus. Using uniform continuity of $\tilde{F}$ we thus obtain

$$\tilde{F}(P_N X P_N) - \sigma_1(N, \varepsilon) - \sigma_2(\varepsilon) \le \tilde{F}(X) \le \tilde{F}(P_N X P_N) + \sigma_1(N, \varepsilon) + \sigma_2(\varepsilon),$$

for some modulus σ_2. Thus

$$|\tilde{F}(X) - \tilde{F}(P_N X P_N)| \le \{\sigma_1(N, \varepsilon) + \sigma_2(\varepsilon)\} \to \sigma_2(\varepsilon) \quad as \quad N \to +\infty$$

and the claim follows thanks to the arbitrariness of ε. $\quad\square$

We now study the case when B satisfies the weak B-condition for A, i.e. when $-A^*B + c_0 B \ge 0$. In this case we do not have an analogue of Lemma 3.83 so we will have to add another layer of approximations of A. We will first replace A

by its Yosida approximation A_λ and then approximate A_λ by $A_{\lambda,N} = P_N A_\lambda P_N$. The operators A_λ and $A_{\lambda,N}$ are bounded and dissipative. We first notice that B also satisfies a weak B-condition for A_λ and $A_{\lambda,N}$ with a different constant. Indeed, since for every $y \in D(A)$,

$$(1 - \lambda c_0)\langle By, y \rangle \leq \langle B(I - \lambda A)y, y \rangle,$$

taking $y = (I - \lambda A)^{-1}x$, we get

$$(1 - \lambda c_0)|B^{\frac{1}{2}}(I - \lambda A)^{-1}x|^2 \leq |B^{\frac{1}{2}}x||B^{\frac{1}{2}}(I - \lambda A)^{-1}x|,$$

which yields

$$|B^{\frac{1}{2}}(I - \lambda A)^{-1}x| \leq \frac{|B^{\frac{1}{2}}x|}{1 - \lambda c_0}.$$

It thus follows that for every $x \in H$,

$$\langle B(I - \lambda A)^{-1}x, x \rangle \leq \frac{1}{1 - \lambda c_0}\langle Bx, x \rangle.$$

Therefore,

$$-BA_\lambda + \frac{c_0}{1 - \lambda c_0}B = \frac{1}{\lambda}\left(\frac{1}{1 - \lambda c_0}B - B(I - \lambda A)^{-1}\right) \geq 0$$

and we conclude that

$$- A_\lambda^* B + \frac{c_0}{1 - \lambda c_0}B \geq 0. \tag{3.246}$$

Thus B satisfies the weak B-condition for A_λ with constant $2c_0$ for $\lambda < 1/(2c_0)$. Obviously (3.246) is also satisfied if A_λ is replaced by $A_{\lambda,N}$.

Theorem 3.86 *Let B be compact and satisfy the weak B-condition for A as in Definition 3.9. Let $F : (0, T) \times H \times H \times S(H) \to \mathbb{R}$ satisfy (for $U = H$) Hypotheses 3.44, 3.46, Hypothesis 3.47 with $B = I$, and Hypotheses 3.78-B, and 3.79. Let $\psi \in UC_b(H_{-1})$ and let*

$$\sup\{|F(t, x, 0, 0)| : t \in (0, T), x \in H\} = K < +\infty. \tag{3.247}$$

Then there exists a unique bounded viscosity solution $u \in UC_b^x([0, T) \times H_{-1})$ of (3.235). Moreover, for every $R > 0$, there is a modulus ρ_R such that

$$|u(t, x) - u(s, x)| \leq \rho_R(|t - s|) \quad \text{for all } 0 \leq s, t < T, |x| \leq R. \tag{3.248}$$

Proof We first solve, for $N > 2, 0 < \lambda < 1/(2c_0)$, the equations

$$\begin{cases} (v_{\lambda,N})_t + \langle A_{\lambda,N}x, Dv_{\lambda,N} \rangle + F(t, x, Dv_{\lambda,N}, P_N D^2 v_{\lambda,N} P_N) = 0 & \text{in } (0, T) \times H_N \\ v_{\lambda,N}(0, x) = \psi(x) & \text{in } H_N. \end{cases}$$
(3.249)

To do this we notice that, since $A_{\lambda,N}$ is dissipative and the weak B condition holds with constant $1/(2c_0)$, the Hamiltonian $\tilde{F}_{\lambda,N} : (0, T) \times H_N \times H_N \times S(H_N) \to \mathbb{R}$ defined by

$$\tilde{F}_{\lambda,N}(t, x, p, X) = \langle A_{\lambda,N}x, p \rangle + F(t, x, p, P_N X P_N)$$

satisfies all the assumptions of Proposition 3.81 with $C = B$, uniformly in N. (Again we identify $X \in S(H_N)$ with $P_N X P_N \in S(H)$.) In fact, we have

$$\tilde{F}_{\lambda,N}(t, x, c_1 B(x - y), X) - \tilde{F}_{\lambda,N}(t, y, c_1 B(x - y), Y)$$

$$\geq -\omega_1 (|x - y|_{-1}(1 + (c_1 + c_2 + c_3)|x - y|_{-1})) - \frac{c_1}{2c_0}|x - y|_{-1}^2$$

in Hypothesis 3.78-B now. Therefore, there exists a unique viscosity solution of $v_{\lambda,N}$ of (3.249), $M \geq 0$, and a modulus m such that for all λ, N

$$\begin{cases} \|v_{\lambda,N}\|_0 \leq M \\ |v_{\lambda,N}(t, x) - v_{\lambda,N}(t, y)| \leq m(|x - y|_{-1}) & \text{for all } t \in [0, T), x, y \in H_N. \end{cases}$$
(3.250)

As in the proof of Theorem 3.84, the functions $u_{\lambda,N}(t, x) = v_{\lambda,N}(t, P_N x)$ are viscosity solutions of

$$\begin{cases} (u_{\lambda,N})_t - \langle A_{\lambda,N}x, Du_{\lambda,N} \rangle + F(t, P_N x, P_N Du_{\lambda,N}, P_N D^2 u_{\lambda,N} P_N) = 0 \\ u_{\lambda,N}(0, x) = \psi(P_N x) & \text{in } H, \end{cases}$$
(3.251)

and they also satisfy (3.250).

We will now show that for every $R > 0$ there exists a modulus ρ_R such that for all $N > 2, 0 < \lambda < 1/(2c_0)$,

$$|v_{\lambda,N}(t, x) - v_{\lambda,N}(s, x)| \leq \rho_R(|t - s|) \quad \text{for } 0 \leq t, s < T, |x| \leq R. \quad (3.252)$$

It is obvious that the function

$$\overline{w}(t, x) = Kt + M$$

is a classical and viscosity supersolution of (3.249) for every λ, N. For every $\varepsilon > 0, x \in H$ there exists a C_ε, depending only on m, such that

$$\psi(y) \leq \psi(x) + \varepsilon + C_\varepsilon |x - y|_{-1}^2.$$

Set

$$R_\varepsilon = \left(\frac{KT + 2M}{\varepsilon} \right)^{\frac{1}{2}}.$$

We notice that

$$\psi_{x,\varepsilon}(y) = \psi(x) + \varepsilon + C_\varepsilon |x - y|_{-1}^2 + \varepsilon |y|^2 > KT + M \quad \text{for } |y| \geq R_\varepsilon.$$

Now let $|x| \leq R$. Since $\|A_{\lambda,N}^* B\| \leq \|A^* B\|$, we have

$$|\langle A_{\lambda,N} y, 2 C_\varepsilon B(y - x) \rangle| \leq 2 C_\varepsilon \|A^* B\| R_\varepsilon (R_\varepsilon + R) \quad \text{for } |y| \leq R_\varepsilon.$$

Thus if we set

$$F_{R,\varepsilon} = \sup\{|F(t, y, p, X)| : t \in (0, T), |y| \leq R_\varepsilon,$$
$$|p| + \|X\| \leq 2[\varepsilon(R_\varepsilon + 1) + C_\varepsilon \|B\|(R + R_\varepsilon + 1)]\} + 2 C_\varepsilon \|A^* B\| R_\varepsilon (R_\varepsilon + R),$$

it is easy to see that for every ε, N the function

$$\eta_{x,\varepsilon}(t, y) = F_{R,\varepsilon} t + \psi_{x,\varepsilon}(y)$$

is a viscosity supersolution of (3.249) in $[0, T) \times \{|y| < R_\varepsilon\}$. Therefore, for every ε, λ, N, the function

$$\overline{w}_{x,\varepsilon} = \min\{\overline{w}, \eta_{x,\varepsilon}\}$$

is a bounded viscosity supersolution of (3.249) in $[0, T) \times H_N$. By comparison we have $v_{\lambda,N} \leq \overline{w}_{x,\varepsilon}$. In particular,

$$v_{\lambda,N}(t, x) - \psi(x) \leq \overline{w}_{x,\varepsilon}(t, x) - \psi(x) \leq \varepsilon + \varepsilon R^2 + F_{R,\varepsilon} t.$$

Taking the infimum over $\varepsilon > 0$ we obtain a modulus ρ_R such that

$$v_{\lambda,N}(t, x) - \psi(x) \leq \rho_R(t) \quad \text{for } t \geq 0, |x| \leq R.$$

A similar construction for subsolutions provides the same bound from below. Since the construction only depended on M and m in (3.250) it can be applied for any starting point $0 \leq s < T$, which yields (3.252), which is obviously also true for $u_{\lambda,N}$.

We can finish as in the proof of Theorem 3.84. By (3.250) and (3.252), the family $u_{\lambda,N}$ is equibounded and equicontinuous in the topology of $\mathbb{R} \times H_{-1}$ on bounded sets of $[0, T] \times H$ and thus, by the Arzela–Ascoli theorem, for every λ there is a subsequence of $u_{\lambda,N}$, still denoted by $u_{\lambda,N}$, and a function u_λ, such that $u_{\lambda,N} \to u_\lambda$ uniformly on bounded subsets of $[0, T] \times H$. Obviously u_λ satisfies (3.250) and (3.252). The fact that u_λ is a viscosity solution of (3.235) with A replaced by A_λ

is standard and follows from Theorem 3.41 and Lemma 3.85 since all the terms are bounded. We then again use the Arzela–Ascoli theorem to obtain that, up to a subsequence, u_λ converges uniformly on bounded subsets of $[0, T] \times H$ to a function u which satisfies (3.250) and (3.252). Using again Theorem 3.41, Lemma 3.85, and a well-known analogue of Lemma 3.83(i) for Yosida approximations we finally conclude that u is a viscosity solution of (3.235).

The proof of uniqueness is similar to the proof of Proposition 3.82 and it will be omitted. Alternatively it can be deduced from the proof of Theorem 3.50 where we now have to first let $\delta \to 0$ and then $\varepsilon \to 0$ there. $\qquad\square$

3.8 Singular Perturbations

Passing to limits with viscosity solutions for equations in infinite-dimensional spaces was discussed in Sect. 3.4. Despite its ease, some finite-dimensional techniques cannot be applied due to the lack of local compactness, and we need to know a priori that solutions converge locally uniformly. When A is more coercive a version of the method of half-relaxed limits will be discussed in Sect. 3.9. In this section we look at a classical "vanishing viscosity" limit in which one tries to establish convergence of viscosity solutions of singularly perturbed equations. Such problems arise, for instance, in large deviation considerations and we will focus on equations having such origins.

Suppose we have a sequence of SDEs

$$\begin{cases} dX_n(s) = (AX_n(s) + b(s, X_n(s)))ds + \frac{1}{\sqrt{n}}\sigma(s, X_n(s))dW_Q(s) & \text{for } s > t, \\ X(t) = x \in H \end{cases}$$
(3.253)

in a real, separable Hilbert space H, where A is a linear, densely defined, maximal dissipative operator in H, $Q \in \mathcal{L}^+(H)$ and W_Q is a Q-Wiener process defined on some reference probability space. We want to investigate the large deviation principle for the processes X_n. One of the key components in the study of large deviations is establishing the existence of the so-called Laplace limit, i.e.

$$\lim_{n \to +\infty} \frac{1}{n} \log \mathbb{E}\left[e^{-ng(X_n(T))} : X_n(t) = x\right]$$

for a given continuous and bounded function g, where $T > t$. Defining

$$v_n(t, x) := -\frac{1}{n} \log \mathbb{E}\left[e^{-ng(X_n(T))}\right],$$

by formally applying Itô's formula, the function v_n should be a viscosity solution of the second-order equation

$$\begin{cases} (v_n)_t + \frac{1}{2n}\text{Tr}\left((\sigma(t,x)Q^{1/2})(\sigma(t,x)Q^{1/2})^*D^2v_n\right) - \frac{1}{2}|(\sigma(t,x)Q^{\frac{1}{2}})^*Dv_n|^2 \\ \qquad\qquad\qquad\qquad\qquad +\langle Ax + b(t,x), Dv_n\rangle = 0, \\ v_n(T,x) = g(x) \qquad \text{in } (0,T) \times H. \end{cases}$$

$$(3.254)$$

Sending $n \to +\infty$ in (3.254) we obtain the limiting first-order PDE

$$\begin{cases} v_t + \langle Ax + b(t,x), Dv\rangle - \frac{1}{2}|(\sigma(t,x)Q^{\frac{1}{2}})^*Dv|^2 = 0, \\ v(T,x) = g(x) \qquad \text{in } (0,T) \times H. \end{cases}$$

$$(3.255)$$

This is the HJB equation associated to the deterministic optimal control problem characterized by the state equation

$$\begin{cases} \frac{d}{ds}X(s) = AX(s) + b(s,X(s)) + \sigma(s,X(s))Q^{\frac{1}{2}}z(s) \quad s > t, \\ X(t) = x, \end{cases}$$

$$(3.256)$$

where we minimize the cost functional

$$J(t,x;z(\cdot)) = \int_t^T \frac{1}{2}|z(s)|^2 ds + g(X(T)) \qquad (3.257)$$

over all controls $z(\cdot) \in L^2(t,T;H)$. The value function of the problem should be the unique viscosity solution of (3.255). Thus we can show the existence of the Laplace limit and identify it if we can prove that solutions v_n of the PDE (3.254) converge to the viscosity solution v of the limiting PDE (3.255). This is a classical singular perturbation limit problem which can be solved using the theory of viscosity solutions presented in this book. The details of the above program (which is based on a general PDE approach to large deviations developed in [250], see also [246–248]) and a further study of this large deviation problem are in [541]. Here we will only show how the convergence of the v_n can be established using the techniques from the proof of the comparison principle. We also point out that Eqs. (3.254) and (3.255) have a quadratic gradient term which makes them more difficult. In particular, they do not satisfy the assumptions of Sect. 3.5.

Let $T > 0$. Let B be an operator satisfying the weak B-condition (3.2) for A. We make the following assumptions.

Hypothesis 3.87 The functions $b : [0,T] \times H \to H, \sigma : [0,T] \times H \to \mathcal{L}_2(\Xi_0, H)$ are uniformly continuous on bounded sets and there exist constants L, M such that

$$|b(t,x) - b(t,y)| \le L|x-y|, \quad t \in [0,T], \ x,y \in H, \qquad (3.258)$$

$$\langle b(t,x) - b(t,y), B(x-y)\rangle \le L|x-y|_{-1}^2, \quad t \in [0,T], \ x,y \in H, \qquad (3.259)$$

$$\|\sigma(t, x) - \sigma(t, y)\|_{\mathcal{L}_2(\Xi_0, H)} \le L|x - y|_{-1}, \quad t \in [0, T], \ x, y \in H, \qquad (3.260)$$

$$\|\sigma(t, x)\|_{\mathcal{L}_2(\Xi_0, H)} \le M, \quad t \in [0, T], \ x \in H. \qquad (3.261)$$

The function $g : H \to \mathbb{R}$ is bounded and

$$|g(x) - g(y)| \le L|x - y|_{-1}, \quad x, y \in H. \qquad (3.262)$$

It was shown in [541] that the functions v_n are unique viscosity solutions of (3.254). The assumptions in [541] were slightly different from Hypothesis 3.87 and some additional restrictions were placed on test functions to deal with exponential moments, however the proof of existence follows the standard arguments and the test function restrictions can be circumvented by localization using Itô's formulas with stopping times and (for instance) Theorem 3.70. The uniqueness part is more difficult. Moreover, it was shown in [541] that the value function v of the deterministic control problem satisfies

$$|v(t, x) - v(s, y)| \le C_1|x - y|_{-1} + C_2(R)|t - s|^{\frac{1}{2}}$$

for all $x, y \in H, |x|, |y| \le R, R > 0$ and $t, s \in [0, T]$.

The following theorem addresses the convergence problem. It is a general statement about a singular perturbation problem. The theorem could be stated for more general HJB equations, however the main difficulty here is the quadratic gradient term.

Theorem 3.88 *Let Hypothesis 3.87 hold. Let v_n be a bounded viscosity solution of (3.254), and v be a bounded viscosity solution of (3.255) such that*

$$\lim_{t \to T}\{|v_n(t, x) - g(x)| + |v(t, x) - g(x)|\} = 0, \quad \text{uniformly on bounded sets} \quad (3.263)$$

and

$$|v(t, x) - v(t, y)| \le L|x - y|_{-1}, \quad 0 \le t \le T, x, y \in H. \qquad (3.264)$$

Then there exists a constant C independent of n such that

$$\|v_n - v\|_0 \le \frac{C}{\sqrt{n}}. \qquad (3.265)$$

Proof Set

$$u_n := v + \frac{C}{\sqrt{n}}(T - t + 1).$$

Then u_n is a viscosity solution of

$$(u_n)_t + \langle Ax + b(t, x), Du_n \rangle - \frac{1}{2}|(\sigma(t, x)Q^{\frac{1}{2}})^* Du_n|^2 = -\frac{C}{\sqrt{n}}. \qquad (3.266)$$

We will show that there exists a C independent of n such that $v_n \leq u_n$. If $v_n \not\leq u_n$, then for $\mu, \delta, \beta > 0, m \in \mathbb{N}$, there exist $p_m, q_m \in H, a_m, b_m \in \mathbb{R}$ such that $|p_m|, |q_m|, |a_m|, |b_m| \leq 1/m$, and

$$\Psi(t, s, x, y) := v_n(t, x) - u_n(s, y) - \frac{\mu}{t} - \frac{\mu}{s} - \frac{\sqrt{n}}{2}|x - y|_{-1}^2 - \delta(|x|^2 + |y|^2)$$

$$-\frac{(t - s)^2}{2\beta} + \langle Bp_m, x \rangle + \langle Bq_m, y \rangle + a_m t + b_m s \qquad (3.267)$$

has a global maximum over $(0, T] \times H \times (0, T] \times H$ at some points $\bar{t}, \bar{s}, \bar{x}, \bar{y}$, where $\Psi(\bar{t}, \bar{s}, \bar{x}, \bar{y}) \geq \eta_n > 0$ for small μ, δ and large m. Similarly to the proof of Theorem 3.50 we have

$$\limsup_{\beta \to 0} \limsup_{m \to \infty} \frac{(\bar{t} - \bar{s})^2}{2\beta} = 0 \quad \text{for fixed } \mu, \varepsilon, \delta, \qquad (3.268)$$

$$\limsup_{\delta \to 0} \limsup_{\beta \to 0} \limsup_{m \to \infty} \delta(|\bar{x}|^2 + |\bar{y}|^2) = 0 \text{ for fixed } \mu. \qquad (3.269)$$

Since $\Psi(\bar{t}, \bar{s}, \bar{x}, \bar{x}) \leq \Psi(\bar{t}, \bar{s}, \bar{x}, \bar{y})$, it follows from (3.264) that

$$\frac{\sqrt{n}}{2}|\bar{x} - \bar{y}|_{-1}^2 \leq u_n(\bar{s}, \bar{x}) - u_n(\bar{s}, \bar{y}) + \delta|x|^2 + \langle Bq_m, \bar{y} - \bar{x} \rangle$$

$$\leq \left(L + \frac{\|B^{1/2}\|}{m}\right)|\bar{x} - \bar{y}|_{-1} + \delta|x|^2.$$

Therefore

$$\limsup_{\delta \to 0} \limsup_{\beta \to 0} \limsup_{m \to \infty} |\bar{x} - \bar{y}|_{-1} \leq \frac{2L}{\sqrt{n}}. \qquad (3.270)$$

If either $\bar{s}$ or $\bar{t}$ is equal to T, we thus obtain from (3.263), (3.264), (3.268), (3.269) and (3.270) that

$$\eta_n \leq \limsup_{\delta \to 0} \limsup_{\beta \to 0} \limsup_{m \to \infty} \Psi(\bar{t}, \bar{s}, \bar{x}, \bar{y})$$

$$\leq \limsup_{\delta \to 0} \limsup_{\beta \to 0} \limsup_{m \to \infty} \left(L|\bar{x} - \bar{y}|_{-1} - \frac{C}{\sqrt{n}}\right) \leq \frac{2L^2 - C}{\sqrt{n}}.$$

Thus if $C \geq 2L^2$ we must have $0 < \bar{t}, \bar{s} < T$.

We now use that v_n is a viscosity subsolution of (3.254) to obtain

$$
-\frac{\mu}{\bar{t}^2} - a_m + \frac{\bar{t} - \bar{s}}{\beta} + \frac{1}{2n} \operatorname{Tr}\left((\sigma(\bar{t}, \bar{x})Q^{1/2})(\sigma(\bar{t}, \bar{x})Q^{1/2})^*(\sqrt{n}B + 2\delta I)\right)
$$
$$
- \frac{1}{2}|(\sigma(\bar{t}, \bar{x})Q^{\frac{1}{2}})^*(\sqrt{n}B(\bar{x} - \bar{y}) + 2\delta\bar{x} - Bp_m)|^2
$$
$$
+ \langle \bar{x}, A^*[\sqrt{n}B(\bar{x} - \bar{y}) - Bp_m]\rangle + \langle b(\bar{t}, \bar{x}), \sqrt{n}B(\bar{x} - \bar{y}) + 2\delta\bar{x} - Bp_m\rangle \geq 0.
$$
$$
(3.271)
$$

Moreover, since u_n is a viscosity supersolution of (3.266), we get

$$
\frac{\mu}{\bar{s}^2} + b_m + \frac{\bar{t} - \bar{s}}{\beta} - \frac{1}{2}|(\sigma(\bar{s}, \bar{y})Q^{\frac{1}{2}})^*(\sqrt{n}B(\bar{x} - \bar{y}) - 2\delta\bar{y} + Bq_m)|^2
$$
$$
+ \langle \bar{y}, A^*[\sqrt{n}B(\bar{x} - \bar{y}) + Bq_m]\rangle + \langle b(\bar{s}, \bar{y}), \sqrt{n}B(\bar{x} - \bar{y}) - 2\delta\bar{y} + Bq_m\rangle
$$
$$
\leq -\frac{C}{\sqrt{n}}.
$$
$$
(3.272)
$$

Subtracting (3.272) from (3.271) and using (3.2), (3.258), (3.259), (3.260), (3.261), (3.268), (3.269) and (3.270) give us

$$
2\frac{\mu}{T^2} \leq \frac{n}{2}|(\sigma(\bar{t}, \bar{y})Q^{\frac{1}{2}})^*B(\bar{x} - \bar{y})|^2 - \frac{n}{2}|(\sigma(\bar{t}, \bar{x})Q^{\frac{1}{2}})^*B(\bar{x} - \bar{y})|^2
$$
$$
+ \frac{1}{2\sqrt{n}}M^2\|B\| + c_1\sqrt{n}|\bar{x} - \bar{y}|^2_{-1} - \frac{C}{\sqrt{n}} + \gamma(\delta, \beta, m), \qquad (3.273)
$$

where c_1 is some constant depending only on L and c_0 in (3.2), and γ is a function such that $\limsup_{\delta \to 0} \limsup_{\beta \to 0} \limsup_{m \to \infty} \gamma(\delta, \beta, m) = 0$. Now

$$
|(\sigma(\bar{t}, \bar{y})Q^{\frac{1}{2}})^*B(\bar{x} - \bar{y})|^2 - |(\sigma(\bar{t}, \bar{x})Q^{\frac{1}{2}})^*B(\bar{x} - \bar{y})|^2
$$
$$
= \operatorname{Tr}\left((\sigma(\bar{t}, \bar{y})Q^{1/2})(\sigma(\bar{t}, \bar{y})Q^{1/2})^*B(\bar{x} - \bar{y}) \otimes B(\bar{x} - \bar{y})\right)
$$
$$
- \operatorname{Tr}\left((\sigma(\bar{t}, \bar{x})Q^{1/2})(\sigma(\bar{t}, \bar{x})Q^{1/2})^*B(\bar{x} - \bar{y}) \otimes B(\bar{x} - \bar{y})\right)
$$
$$
\leq c_2|\bar{x} - \bar{y}|^3_{-1},
$$

where c_2 is some constant depending only on $L, M, \|B^{1/2}\|$. Plugging this inequality into (3.273) and invoking (3.270) we thus obtain

$$
2\frac{\mu}{T^2} \leq \limsup_{\delta \to 0} \limsup_{\beta \to 0} \limsup_{m \to \infty} \left(c_1\sqrt{n}|\bar{x} - \bar{y}|^2_{-1} + c_2 n|\bar{x} - \bar{y}|^3_{-1}\right) + \frac{1}{2\sqrt{n}}M^2\|B\| - \frac{C}{\sqrt{n}}
$$
$$
\leq \frac{1}{\sqrt{n}}\left(4L^2c_1 + 8L^3c_2 + \frac{1}{2}M^2\|B\|\right) - \frac{C}{\sqrt{n}}.
$$

This yields a contradiction if $C \geq 4L^2c_1 + 8L^3c_2 + \frac{1}{2}M^2\|B\|$. Thus we must have

$$v_n \leq v + \frac{C}{\sqrt{n}}(T - t + 1) \leq v + \frac{C(T+1)}{\sqrt{n}}.$$

Similar arguments give us

$$v - \frac{C}{\sqrt{n}}(T - t + 1) \leq v_n$$

and thus the result follows. $\square$

The rate of convergence provided by Theorem 3.88 is the same as the rate for finite-dimensional problems.

Remark 3.89 It is obvious from the proof that Theorem 3.88 remains the same if the term $\langle b(t, x), Du \rangle$ in (3.254) and (3.255) is replaced by a general Hamiltonian $F(t, x, Du)$, where $F : [0, T] \times H \times H \to \mathbb{R}$ is uniformly continuous on bounded sets and, for instance, satisfies

$$|F(t, x, p) - F(t, x, q)| \leq C|p - q|(1 + |x|),$$

$$F\left(t, x, \frac{B(x-y)}{\varepsilon}\right) - F\left(t, y, \frac{B(x-y)}{\varepsilon}\right) \leq C|x - y|^2_{-1},$$

for all $t \in [0, T], x, y, p, q \in H$. Such equations arise in risk sensitive optimal control problems. We refer to [112, 113, 365, 462, 463, 465, 466, 540] for such problems in infinite-dimensional spaces and to [203, 263] for more on risk sensitive control problems. A result similar to Theorem 3.88 has been proved in [466] for a risk sensitive control problem using representation formulas and probabilistic methods. ■

3.9 Perron's Method and Half-Relaxed Limits

Perron's method is one of the main techniques for producing viscosity solutions of PDEs in finite-dimensional spaces (see [139, 358] and Appendix E.4). It is based on the principle that the supremum of the family of all viscosity subsolutions which are less than or equal to a viscosity supersolution of an equation is a (possibly discontinuous) viscosity solution. Thus to construct a viscosity solution, all we need is to produce one subsolution u_0 and one supersolution v_0 that both satisfy the boundary and initial conditions and such that $u_0 \leq v_0$. If we have a comparison theorem, the viscosity solution produced by Perron's method can then be proved to be continuous. Perron's method has a rather trivial extension to infinite-dimensional bounded equations (3.66), see [412]. Perron's method was also used to prove the existence of viscosity solutions using Ishii's definitions of viscosity solutions [360, 361]. However, it is not known if a version of Perron's method can be implemented for B-continuous viscosity solutions of (3.56) and (3.62), even if the equations are of first order. The reason for this is that B-continuous viscosity sub-/supersolutions are semicontinuous in a weaker topology and this makes the problem difficult. However, it was shown in [376] how to adapt Perron's method to B-continuous viscosity solutions of (3.56) and (3.62) when the operator A is more coercive. We only discuss the initial value problems

$$\begin{cases} u_t - \langle Ax, Du \rangle + F(t, x, u, Du, D^2u) = 0 & (t, x) \in (0, T) \times H \\ u(0, x) = g(x), \end{cases} \tag{3.274}$$

where H is a real, separable Hilbert space and A is a linear, densely defined, maximal dissipative operator in H. The presentation here is based on [376] and we refer to this paper for further results and more details.

In order to develop Perron's method we need to introduce a notion of a discontinuous viscosity solution. Let B be an operator satisfying (3.2). For a function u we will write $u^{*,-1}$ and $u_{*,-1}$ to denote the upper- and lower-semicontinuous envelopes of u in the $|\cdot| \times |\cdot|_{-1}$ norm, i.e.

$$u^{*,-1}(t, x) = \lim \sup \{u(s, y) : s \to t, |y - x|_{-1} \to 0\},$$

$$u_{*,-1}(t, x) = \lim \inf \{u(s, y) : s \to t, |y - x|_{-1} \to 0\}.$$

Observe that $u^{*,-1}$ is upper semicontinuous in the $|\cdot| \times |\cdot|_{-1}$ norm and thus, thanks to Lemma 3.6(ii), it is B-upper semicontinuous.

We assume that $F : (0, T) \times H \times \mathbb{R} \times H \times S(H) \to \mathbb{R}$ satisfies Hypotheses 3.44–3.46. We also impose the following coercivity condition on A.

$$-\langle A^*x, x \rangle \geq \lambda |x|_1^2, \qquad \text{for } x \in D(A^*) \tag{3.275}$$

for some $\lambda > 0$.

The above implies, in particular, that $D(A^*) \subset H_1$. Assumption (3.275) is satisfied, for instance, for self-adjoint invertible operators A if $B = (-A)^{-1}$.

Definition 3.90 A locally bounded function u is a discontinuous viscosity subsolution of (3.274) if $u(0, y) \leq g(y)$ on H, and whenever $(u - h)^{*,-1} - \varphi$ has a local maximum in the topology of $|\cdot| \times |\cdot|_{-1}$ at a point (t, x) for a test function $\psi(s, y) = \varphi(s, y) + h(s, |y|)$ from Definition 3.32 such that φ is B-continuous, $h_r(t, r) > 0, r \in (0, +\infty), t \in (0, T)$ and

$$u(s, y) - h(s, |y|) \to -\infty \quad \text{as } |y| \to \infty \text{ locally uniformly in } s \tag{3.276}$$

then

$$x \in H_1$$

and

$$\psi_t(t, x) + \lambda |x|_1^2 \frac{h_r(t, |x|)}{|x|} - \langle x, A^* D\varphi(t, x) \rangle$$
$$+ F(t, x, (u - h)^{*, -1}(t, x) + h(t, |x|), D\psi(t, x), D^2\psi(t, x)) \le 0,$$

where h_r is the partial derivative of h with respect to the second variable.

A locally bounded function u is a discontinuous viscosity supersolution of (3.274) if $u(0, y) \ge g(y)$ on H, and whenever $(u + h)_{*, -1} + \varphi$ has a local minimum in the topology of $|\cdot| \times |\cdot|_{-1}$ at a point (t, x) for a test function $\psi(s, y) = \varphi(s, y) + h(s, |y|)$ from Definition 3.32 such that φ is B-continuous, $h_r(t, r) > 0, r \in (0, +\infty), t \in (0, T)$ and

$$u(s, y) + h(s, |y|) \to +\infty \quad \text{as } |y| \to \infty \quad \text{locally uniformly in } s \qquad (3.277)$$

then

$$x \in H_1$$

and

$$-\psi_t(t, x) - \lambda |x|_1^2 \frac{h_r(|x|)}{|x|} + \langle x, A^* D\varphi(t, x) \rangle$$
$$+ F(t, x, (u + h)_{*, -1}(t, x) - h(t, |x|), -D\psi(t, x), -D^2\psi(t, x)) \ge 0.$$

A discontinuous viscosity solution of (3.274) is a function which is both a discontinuous viscosity subsolution and a discontinuous viscosity supersolution.

The maxima and minima in Definition 3.90 can be assumed to be global and strict in the $|\cdot| \times |\cdot|_{-1}$ norm. Compared to Definition 3.34, apart from discontinuity of sub/supersolutions, the main difference here is that we require that $x \in H_1$ and the term $\langle x, A^* Dh(t, x) \rangle$ is not dropped entirely. We notice that if $x \in A^*$ then $-\langle x, A^* Dh(t, x) \rangle = -\frac{h_r(t, |x|)}{|x|} \langle x, A^* x \rangle \ge \lambda |x|_1^2 \frac{h_r(t, |x|)}{|x|}$ and this term is well defined and is left in the definition. If $x = 0$ the term $|x|_1^2 \frac{h_r(t, |x|)}{|x|}$ by definition is equal to 0. We also remark that if u is B-upper semicontinuous then $(u - h)^{*, -1} = u - h$ and if u is B-lower semicontinuous then $(u + h)_{*, -1} = u + h$. Definitions of discontinuous viscosity solutions were first used in [360, 361]. Definitions requiring that points where maxima/minima occur belong to better spaces appeared in [97, 144] and have been successfully employed for some second-order equations which are discussed in this book in Sects. 3.11–3.13 (see also [318, 322, 323]).

For simplicity we restrict ourselves to F not depending on u. We will often say that u is a viscosity sub-/supersolution in an open set V. This will mean that we disregard the initial condition and the conditions of Definition 3.90 must be satisfied only if $(t, x) \in V$. However, all functions involved must be defined on $(0, T) \times H$ and the maxima/minima in Definition 3.90 are local in the topology of $|\cdot| \times |\cdot|_{-1}$ in the whole $(0, T) \times H$.

Lemma 3.91 *Let Hypotheses 3.44, 3.46 and condition (3.275) hold and let V be an open subset of $(0, T) \times H$. Let $\psi = \varphi + h$ be a test function from Definition 3.32 such that $w = -\psi$ (respectively, $w = \psi$) satisfies*

$$w_t(t, x) - \langle x, A^* Dw(t, x) \rangle + F(t, x, Dw(t, x), D^2 w(t, x)) \le 0 \quad x \in D(A^*) \cap V, t \in (0, T)$$

(respectively,

$$w_t(t, x) - \langle x, A^* Dw(t, x) \rangle + F(t, x, Dw(t, x), D^2 w(t, x)) \ge 0 \quad x \in D(A^*) \cap V, t \in (0, T).)$$

Then the function w is a viscosity subsolution (respectively, supersolution) of (3.274).

Proof We will only prove the lemma in the subsolution case. We see that $w = -\psi$ is B-upper semicontinuous. Suppose that $w(s, y) - \tilde{\varphi}(s, y) - \tilde{h}(s, |y|)$ has a local maximum at (t, x) for a test function $\tilde{\psi} = \tilde{\varphi} + \tilde{h}$. Then

$$w_t(t, x) = \tilde{\psi}_t(t, x), \quad -D\varphi(t, x) - \frac{h_r(t, |x|)}{|x|} x - D\tilde{\varphi}(t, x) - \frac{\tilde{h}_r(t, |x|)}{|x|} x = 0$$

and

$$D^2 w(t, x) \le D^2(\tilde{\varphi} + \tilde{h})(t, x).$$

Therefore, either $x = 0$ or

$$\left(\frac{h_r(t, |x|)}{|x|} + \frac{\tilde{h}_r(t, |x|)}{|x|} \right) x = -D\varphi(t, x) - D\tilde{\varphi}(t, x) \in D(A^*),$$

i.e. $x \in D(A^*)$. Thus, using (3.275) and Hypothesis 3.46, we obtain

$$\tilde{\psi}_t(t, x) + \lambda |x|_1^2 \frac{\tilde{h}_r(t, |x|)}{|x|} - \langle x, A^* D\tilde{\varphi}(t, x) \rangle + F(t, x, D\psi(t, x), D^2 \psi(t, x))$$
$$\le w_t(t, x) - \frac{\tilde{h}_r(t, |x|)}{|x|} \langle x, A^* x \rangle - \langle x, A^* D\tilde{\varphi}(t, x) \rangle + F(t, x, Dw(t, x), D^2 w(t, x))$$
$$= w_t(t, x) - \langle x, A^* Dw(t, x) \rangle + F(t, x, Dw(t, x), D^2 w(t, x)) \le 0$$

and the claim is proved. $\qquad\square$

Proposition 3.92 *Let Hypotheses 3.44, 3.46 and condition (3.275) be satisfied. Let $\mathcal{A}$ be a family of viscosity subsolutions of (3.274) in the sense of Definition 3.90. Suppose that the function*

$$u(x) = \sup \{w(x) : w \in \mathcal{A}\} \tag{3.278}$$

is locally bounded. Then u is a viscosity subsolution of (3.274) in the sense of Definition 3.90.

Proof Suppose that $(u-h)^{*,-1}-\varphi$ has a strict in $|\cdot|\times|\cdot|_{-1}$ norm global maximum at a point (t, x) for a test function $\psi = \varphi + h$. (We can assume that $(u-h)^{*,-1}(s, y) - \varphi(s, y) \le -|y|$ as $|y| \to \infty$.) Perturbed optimization (see Corollary 3.26) and Definition 3.90 yield that there exist $w_n \in \mathcal{A}$, $x_n \in H_1$, t_n, and $a_n \in \mathbb{R}$, $p_n \in H$, $|a_n| + |p_n| \le 1/n$ such that

$$t_n \to t, \quad B^{\frac{1}{2}}x_n \to B^{\frac{1}{2}}x, \quad x_n \rightharpoonup x \text{ in } H \text{ as } n \to \infty, \qquad (3.279)$$

$$(w_n - h)^{*,-1}(s, y) - \varphi(s, y) + \langle Bp_n, y \rangle + a_n s$$

has a strict in $|\cdot|\times|\cdot|_{-1}$ norm global maximum at (t_n, x_n), and

$$(w_n - h)^{*,-1}(t_n, x_n) \to (u - h)^{*,-1}(t, x) \text{ as } n \to \infty. \qquad (3.280)$$

Therefore,

$$\psi_t(t_n, x_n) - a_n + \lambda|x_n|_1^2 \frac{h_r(t_n, |x_n|)}{|x_n|} - \langle x_n, A^*(D\varphi(x_n) - Bp_n) \rangle$$
$$+ F(t_n, x_n, D\psi(t_n, x_n) - Bp_n, D^2\psi(t_n, x_n)) \le 0. \quad (3.281)$$

Since the x_n are bounded, using the local boundedness of F we thus obtain that either $x_n \to 0 = x$ or, up to a subsequence, $|x_n| > c > 0$ which leads to

$$|x_n|_1^2 \le C$$

for some constant C which, together with (3.279), implies that $x \in H_1$, and $B^{-\frac{1}{2}}x_n \rightharpoonup B^{-\frac{1}{2}}x$ as $n \to \infty$. Therefore, by (3.279),

$$|x_n - x|^2 = \langle B^{-\frac{1}{2}}(x_n - x), B^{\frac{1}{2}}(x_n - x) \rangle \to 0 \text{ as } n \to \infty,$$

i.e. $x_n \to x$ in H. Using this, the continuity of F, and the lower semicontinuity of $|\cdot|_1$ in H, we can now pass to the lim inf as $n \to \infty$ in (3.281) to obtain

$$\psi_t(t, x) + \lambda|x|_1^2 \frac{h_r(t, |x|)}{|x|} - \langle x, A^*D\varphi(t, x) \rangle + F(t, x, D\psi(t, x), D^2\psi(t, x)) \le 0,$$

which completes the proof. $\square$

Theorem 3.93 *Let Hypotheses 3.44, 3.46 and condition (3.275) be satisfied. Let u_0, v_0 be respectively a viscosity subsolution and a viscosity supersolution of (3.274) in the sense of Definition 3.90 such that $u_0 \le v_0$ and $u_0(0, x) = v_0(0, x) = g(x)$, $x \in H$. Then the function*

$$u(t, x) = \sup\{v(t, x) : u_0 \leq v \leq v_0, v \text{ is a viscosity subsolution}$$
$$\text{of (3.274) in the sense of Definition 3.90}\} \tag{3.282}$$

is a viscosity solution of (3.274) in the sense of Definition 3.90.

Proof It follows from Proposition 3.92 that u is a viscosity subsolution. Suppose now that $(u + h)_{*, -1} + \varphi$ has a strict in $|\cdot| \times |\cdot|_{-1}$ norm global minimum at a point (t, x) for a test function $\psi = \varphi + h$ satisfying (3.277). First we observe that if

$$(u + h)_{*, -1}(t, x) = (v_0 + h)_{*, -1}(t, x)$$

then $(v_0 + h)_{*, -1} + \varphi$ has a global minimum at (t, x) and so we are done since v_0 is a viscosity supersolution. Therefore we only need to consider the case

$$(u + h)_{*, -1}(t, x) < (v_0 + h)_{*, -1}(t, x).$$

It then follows from the above inequality, the B-continuity of φ and the weak sequential lower semi-continuity of $|\cdot|$ that there is an $\varepsilon_0 > 0$ such that for every $R > 0$ there exists an $\eta_0 > 0$ such that

$$\varepsilon + (u + h)_{*, -1}(t, x) + \varphi(t, x) - \varphi(s, y) - h(s, |y|)$$
$$< (v_0 + h)_{*, -1}(s, y) - h(s, |y|) \leq v_0(s, y) \tag{3.283}$$

for $(s, y) \in (t - \eta_0, t + \eta_0) \times (B_{H_{-1}}(x, \eta_0) \cap B(x, R))$, $0 < \varepsilon < \varepsilon_0$. Let

$$w(y) = \varepsilon + (u + h)_{*, -1}(t, x) + \varphi(t, x) - \varphi(s, y) - h(s, |y|). \tag{3.284}$$

By further modifying h for large values of $|y|$ and $s \notin (t - \eta_0, t + \eta_0)$ if necessary, we can also assume that there is an $R_0 > 0$ such that

$$w(s, y) \leq u(s, y) - 1 \quad y \notin B(x, R_0), \ s \in (0, T). \tag{3.285}$$

Moreover, if R_0 is big enough, there exist $s_n \to t$, $y_n \in B(x, R_0)$, $y_n \to x$ in H_{-1} such that

$$u(s_n, y_n) + h(s_n, |y_n|) + \varphi(s_n, y_n) \to (u + h)_{*, -1}(t, x) + \varphi(t, x),$$

which means that for every $\eta > 0$ there exist points $(s, y) \in (t - \eta, t + \eta) \times (B_{H_{-1}}(x, \eta) \cap B(x, R_0))$ for which

$$u(s, y) < w(s, y). \tag{3.286}$$

If the condition for u being a viscosity supersolution of (3.274) is violated at (t, x) for the test function ψ then one of the following must hold:

(i) $x \notin H_1$.

(ii) $x \in H_1$ but

$$-\psi_t(t, x) - \lambda |x|_1^2 \frac{h_r(t, |x|)}{|x|} + \langle x, A^* D\varphi(t, x) \rangle$$
$$+ F(t, x, -D\psi(t, x), -D^2\psi(t, x)) < -\nu < 0$$
$$(3.287)$$

for some $\nu > 0$.

If (i) is satisfied then we must have

$$\liminf_{\substack{y \to x \text{ in } H_{-1} \\ y \in H_1}} |y|_1 = +\infty. \qquad (3.288)$$

Otherwise we would have a sequence y_n such that $B^{\frac{1}{2}} y_n \to B^{\frac{1}{2}} x$ and $|B^{-\frac{1}{2}} y_n| \leq C$. Then for some subsequence (still denoted by y_n) $B^{-\frac{1}{2}} y_n \rightharpoonup z$ for some $z \in H$, which would imply $x \in H_1$ and $z = B^{-\frac{1}{2}} x$. Using the local boundedness of F, condition (3.288) now implies that for every $R > 0$

$$w_t(s, y) - \lambda |y|_1^2 \frac{h_r(s, |y|)}{|y|} + \langle y, A^* D\varphi(s, y) \rangle + F(s, y, Dw(s, y), D^2 w(s, y)) < -\frac{\nu}{2},$$
$$(3.289)$$

for $(s, y) \in (t - \eta_1, t + \eta_1) \times (B_{H_{-1}}(x, \eta_1) \cap B(x, R) \cap H_1)$, for some $\eta_1 > 0$.

Suppose that (ii) is true. We will show that for every $R > 0$ (3.289) holds for $(s, y) \in (t - \eta_1, t + \eta_1) \times (B_{H_{-1}}(x, \eta_1) \cap B(x, R) \cap H_1)$ for some $\eta_1 > 0$. If not there exist sequences $t_n \to t$, $x_n \to x$ in H_{-1}, $|x_n| \leq R$ such that

$$w_t(t_n, x_n) - \lambda |x_n|_1^2 \frac{h_r(t_n, |x_n|)}{|x_n|} + \langle x_n, A^* D\varphi(t_n, x_n) \rangle$$
$$+ F(t_n, x_n, Dw(t_n, x_n), D^2 w(t_n, x_n)) \geq -\frac{\nu}{2}. \qquad (3.290)$$

If $x_n \to 0 = x$ then letting $n \to +\infty$ in (3.290) would contradict (3.287). If $x_n \nrightarrow 0$ then for some subsequence (still denoted by x_n) we would have $h_r(t_n, |x_n|)/|x_n| \geq \gamma > 0$ for some γ and this would imply $|x_n|_1 \leq C$ for some constant C, as otherwise (3.290) would be violated. But then we must have $B^{-\frac{1}{2}} x_n \rightharpoonup B^{-\frac{1}{2}} x$ in H and thus we obtain $x_n \to x$. However then (3.287), (3.290) and the lower semi-continuity of $\| \cdot \|_1$ again imply

$$-\frac{\nu}{2} \leq \limsup_{n\to\infty} \left(w_t(t_n, x_n) - \lambda |x_n|_1^2 \frac{h_r(t_n, |x_n|)}{|x_n|} + \langle x_n, A^*D\varphi(t_n, x_n)\rangle \right.$$

$$\left. + F(t_n, x_n, Dw(t_n, x_n), D^2w(t_n, x_n)) \right) < -\nu,$$

which gives a contradiction.

Thus we have proved that in both cases (i) and (ii), for every $R > 0$ (3.289) holds for $(s, y) \in (t - \eta_1, t + \eta_1) \times (B_{H_{-1}}(x, \eta_1) \cap B(x, R) \cap H_1)$ for some $\eta_1 > 0$.

Recall now the definition of w given in (3.284). Since $(u + h)_{*,-1} + \varphi$ has a global minimum at (t, x), strict in $|\cdot| \times |\cdot|_{-1}$ norm, given $\eta > 0$ and ε small enough (depending on η) there exists a constant $\mu_\eta > 0$ such that

$$w(s, y) < (u + h)_{*,-1}(s, y) - h(s, |y|) - \mu_\eta \leq u(s, y) - \mu_\eta \qquad (3.291)$$

for $y \notin B_{H_{-1}}(x, \eta), s \in (t - \eta, t + \eta)$.

Using (3.283), (3.285), (3.291), and (3.289) we can therefore conclude that there exist numbers $R, \eta, \varepsilon, \mu > 0$ such that

$$w \leq v_0 \quad \text{in } [0, T) \times H, \qquad (3.292)$$

$$w(s, y) < u(s, y) - \mu \quad \text{for } (s, y) \notin (t - \eta, t + \eta) \times (B_{H_{-1}}(x, \eta) \cap B(x, R)), \quad (3.293)$$

and such that (3.289) is satisfied for $(s, y) \in (t - 2\eta, t + 2\eta) \times (B_{H_{-1}}(x, 2\eta) \cap B(x, 2R) \cap H_1)$.

We now claim that the function w is a viscosity subsolution of (3.274) in the interior of $(t - 2\eta, t + 2\eta) \times (B_{-1}(x, 2\eta) \cap B(x, 2R))$. This follows from Lemma 3.91 upon noticing that by (3.289) and (3.275) we have

$$w_t(s, y) - \langle y, A^*Dw(s, y)\rangle + F(s, y, Dw(s, y), D^2w(s, y))$$
$$\leq w_t(s, y) - \lambda|y|_1^2 \frac{h_r(s, |y|)}{|y|} + \langle y, A^*D\varphi(s, y)\rangle + F(s, y, Dw(s, y), D^2w(s, y)) < 0$$

for $(s, y) \in (t - 2\eta, t + 2\eta) \times (B_{H_{-1}}(x, 2\eta) \cap B(x, 2R) \cap D(A^*))$.

It remains to show that the function

$$u_1 = \max(w, u) \qquad (3.294)$$

is a viscosity subsolution in the sense of Definition 3.90. It follows from the definition that u_1 is a viscosity subsolution in the interior of $(t - 2\eta, t + 2\eta) \times (B_{-1}(x, 2\eta) \cap B(x, 2R))$. If $(s, y) \notin (t - 3/2\eta, t + 3/2\eta) \times (B_{-1}(x, 3/2\eta) \cap B(x, 3/2R))$ and $(u_1 - \tilde{h})^{*,-1} - \tilde{\varphi}$ has a maximum at (s, y), and

$$(u_1 - \tilde{h})^{*,-1}(s, y) = \lim_{n\to+\infty} (u_1(s_n, y_n) - \tilde{h}(s_n, y_n)),$$

where $|s_n - s| + |y_n - y|_{-1} \to 0$ and $y_n \rightharpoonup y$, then since $|y| \leq \liminf_{n \to +\infty} |y_n|$, we obtain $(s_n, y_n) \notin (t - \eta, t + \eta) \times (B_{-1}(x, \eta) \cap B(x, R))$ for large n. Thus by (3.293) $u_1(s_n, y_n) = u(s_n, y_n)$, which implies $(u_1 - \tilde{h})^{*,-1}(s, y) = (u - \tilde{h})^{*,-1}(s, y)$. Therefore the subsolution condition is satisfied for u_1 at (s, y) for the test function $\tilde{\psi} = \tilde{h} + \tilde{\varphi}$, and hence u_1 is a discontinuous viscosity subsolution of (3.274).

By the definition of u_1 and (3.292) we know that $u_0 \leq u_1 \leq v_0$. Thus, by (3.282), we should have $u_1 \leq u$, but this contradicts (3.286).   □

The comparison theorem in the whole space can be proved under the same assumptions as those of Theorem 3.50. The proof is almost exactly the same. The reader can also check the proof of Theorem 4.1 in [376] for a proof in a simpler time-independent case. The comparison theorem in particular implies that a discontinuous viscosity solution (if it exists) is in fact B-continuous. Thus, in particular, if the comparison theorem holds, a viscosity solution in the sense of Definition 3.90 is the usual viscosity solution in the sense of Definition 3.34 (with the additional requirements that test functions φ are B-continuous and $h_r(t, r) > 0, r > 0$). However, we now have a very convenient way to prove the existence of a solution by Perron's method. The remaining question is how to construct a sub- and a supersolution u_0 and v_0 as in Theorem 3.93 that in addition attain the initial condition locally uniformly so that we can later use the comparison theorem.

Proposition 3.94 *Let Hypotheses 3.44, 3.46 and condition (3.275) hold and let g be locally uniformly B-continuous and such that $|g(x)| \leq \mu(1 + |x|)$ for $x \in H$ for some constant μ. Then there are a viscosity subsolution u_0 and viscosity supersolution v_0 of Eq. (3.274) in the sense of Definition 3.90 such that*

$$\lim_{t \downarrow 0} (|u_0(t, x) - g(x)| + |v_0(t, x) - g(x)|) = 0$$

uniformly on bounded sets of H.

Proof We will only show how to construct v_0.
   Define

$$C(r) = \sup\{|F(t, x, p, X)| : x \in H, t \in [0, T], |p| \leq r, \|X\| \leq r\}.$$

Let $v(t, x) = \alpha t + 2\mu\sqrt{1 + |x|^2}$. Notice that $v(0, x) \geq g(x), x \in H$. By Lemma 3.91, v is a viscosity supersolution of (3.274) if

$$\alpha + F(t, x, Dv(t, x), D^2v(t, x)) \geq 0$$

for all $(t, x) \in (0, T) \times H$. Since $Dv(t, x)$ and $D^2v(t, x)$ are bounded we can therefore select α, depending only on μ, such that the above condition is satisfied.
   Let $z \in H, \varepsilon > 0$. We first choose a constant $R = R(|z|) \geq |z|$ such that $((|x| - |z|)^+)^4 \geq 2v(t, x)$ for $|x| \geq R, t \in (0, T)$. We then find $M = M(|z|, \varepsilon)$ such that

$$\bar{w}_{z,\varepsilon}(x) := g(z) + \varepsilon + M|x - z|_{-1}^2 + ((|x| - |z|)^+)^4 \geq g(x)$$

for $|x| \leq R$. Let now $\gamma = \sup\{|D\bar{w}_{z,\varepsilon}(x)| + \|D^2\bar{w}_{z,\varepsilon}(x)\| : |x| \leq R\}$. Using again Lemma 3.91, in order for $w_{z,\varepsilon}(t, x) := \beta t + \bar{w}_{z,\varepsilon}(x)$ to be a viscosity supersolution of (3.274) in the interior of $(0, T) \times B(0, R)$ we need

$$\beta + 2M\langle x, A^*B(x - z)\rangle + F(t, x, Dw_{z,\varepsilon}(t, x), D^2 w_{z,\varepsilon}(t, x)) \geq 0$$

in this set. This can be achieved by taking $\beta = 2RM(R + |z|)\|A^*B\| + C(\gamma)$. Since $w_{z,\varepsilon}(t, x) > v(t, x)$ if $t \in (0, T)$, $|x| \geq R$, it thus follows that

$$\hat{\omega}_{z,\varepsilon}(t, x) := \min\{w_{z,\varepsilon}(t, x), v(t, x)\}$$

is a B-lower semicontinuous viscosity supersolution of (3.274) in $[0, T) \times H$. It is now clear from the construction of the $\hat{\omega}_{z,\varepsilon}$ and Proposition 3.92 for supersolutions that the function $v_0(t, x) := \inf_{z,\varepsilon} \hat{\omega}_{z,\varepsilon}(t, x)$ is a viscosity supersolution of (3.274) in the sense of Definition 3.90 such that $\lim_{t\downarrow 0} |v_0(t, x) - g(x)| = 0$ uniformly on bounded sets of H. $\qquad\square$

In the last part of this section we show how the method of half-relaxed limits of Barles–Perthame (see [139]) can be generalized to infinite-dimensional spaces. This method improves the general consistency result of Sect. 3.4. Suppose that we have equations

$$u_t - \langle A_n x, Du\rangle + F_n(t, x, u, Du, D^2 u) = 0 \quad (t, x) \in (0, T) \times H, \qquad (3.295)$$

where $F_n : [0, T] \times H \times \mathbb{R} \times H \times S(H) \to \mathbb{R}$, and $A_n, n = 1, 2, \ldots$, are linear, densely defined maximal dissipative operators in H such that $D(A^*) \subset D(A_n^*)$. Let F^+, F_- be defined as in Theorem 3.41. We define

$$u^+(x) = \lim_{i \to \infty} \sup \left\{ u_n(y) : n \geq i, |x - y| \leq \frac{1}{i} \right\},$$

$$u_-(x) = \lim_{i \to \infty} \inf \left\{ u_n(y) : n \geq i, |x - y| \leq \frac{1}{i} \right\}.$$

Theorem 3.95 *Let the operator B satisfying (3.2) be compact. Let A_n be as above, let $A, A_n, n = 1, 2, \ldots$, satisfy (3.275), let (3.68) hold, and let for every test function φ, the family $A_n^* D\varphi, n = 1, 2, \ldots$, be locally uniformly bounded. Suppose that $F_n, n = 1, 2, \ldots$, are continuous, locally bounded uniformly in n, and satisfy Hypotheses 3.45 and 3.46. Let u_n be locally bounded, uniformly in n, B-upper semicontinuous (respectively, B-lower semicontinuous) viscosity subsolutions, (respectively, supersolutions) of*

$$(u_n)_t - \langle A_n x, Du_n\rangle + F_n(t, x, u_n, Du_n, D^2 u_n) = 0 \quad in \ (0, T) \times H \qquad (3.296)$$

in the sense of Definition 3.90. Then the function u^+ (respectively, u_-) is a viscosity subsolution (respectively, supersolution) of

$$(u^+)_t - \langle Ax, Du^+ \rangle + F_-(t, x, u^+, Du^+, D^2 u^+) = 0 \quad in \ (0, T) \times H$$

(respectively,

$$(u_-)_t - \langle Ax, Du_- \rangle + F^+(t, x, u_-, Du_-, D^2 u_-) = 0 \quad in \ (0, T) \times H)$$

in the sense of Definition 3.90.

Proof Let $(u^+ - h)^{*,-1} - \varphi$ have a local maximum (equal to 0) at (t, x) for some test function $\psi = \varphi + h$. In light of local uniform boundedness of the u_n we can assume that the maximum is global, strict in the $|\cdot| \times |\cdot|_{-1}$ norm, and such that

$$u^+(y) - h(s, |y|) \to -\infty, \quad (u^+ - h)^{*,-1}(y) - \varphi(s, y) \to -\infty,$$

and

$$u_n(s, y) - h(s, |y|) - \varphi(s, y) \to -\infty$$

as $|y| \to +\infty$, uniformly in n and $s \in (0, T)$, and as $s \to 0$ and $s \to T$, uniformly in n and y in bounded sets. Then there must exist sequences t_n, x_n such that $|t_n - t| + |x_n - x|_{-1} \to 0$, $|x_n| \leq C$, and

$$u^+(t_n, x_n) - h(t_n, |x_n|) - \varphi(t_n, x_n) \geq -\frac{1}{n}.$$

Therefore there exist τ_n, y_n and i_n such that

$$u_{i_n}(\tau_n, y_n) - h(\tau_n, |y_n|) - \varphi(\tau_n, y_n) \geq -\frac{2}{n}. \tag{3.297}$$

Let (s_n, z_n) be a global maximum of

$$u_{i_n}(s, y) - h(s, |y|) - \varphi(s, y).$$

It exists because of the decay of this function at infinity and around $0, T$, and the fact that, because B is compact, B-upper semicontinuity is equivalent to weak sequential upper semicontinuity. Obviously $|z_n| \leq C_1$ and we also have

$$\psi_t(s_n, z_n) + \lambda |z_n|_1^2 \frac{h_r(s_n, |z_n|)}{|z_n|} - \langle z_n, A_{i_n}^* D\varphi(s_n, z_n) \rangle$$

$$+ F_{i_n}(s_n, z_n, u_{i_n}(s_n, z_n), D\psi(s_n, z_n), D^2\psi(s_n, z_n)) \leq 0. \tag{3.298}$$

We can assume that $s_n \to s$. Now either $z_n \to 0$ or for a subsequence (still denoted by z_n) $|z_n| \geq c_1 > 0, n = 1, 2, \dots$, which implies $h_r(s_n, |z_n|)/|z_n| > c_2 > 0$,

$n = 1, 2, ...$ It then follows from the local uniform boundedness of the F_n and $A_{i_n}^* D\varphi$, that $|z_n|_1 \leq C_2$, which implies $z_n \rightharpoonup z$ in H_1 for some $z \in H_1$ and thus, since B is compact, $z_n \to z$ in H.

Therefore $u^+(s, z) \geq \lim \sup_{n \to \infty} u_{i_n}(s_n, z_n)$ which, together with (3.297), gives

$$0 \geq (u^+ - h)^{*,-1}(s, z) - \varphi(s, z) \geq u^+(s, z) - h(s, z) - \varphi(s, z)$$
$$\geq \lim_{n \to \infty} \sup(u_{i_n}(s_n, z_n) - h(s_n, |z_n|) - \varphi(s_n, z_n)) \geq 0.$$

Thus $(s, z) = (t, x)$ and moreover

$$(u^+ - h)^{*,-1}(t, x) + h(t, x) = \lim_{n \to \infty} \sup u_{i_n}(s_n, z_n).$$

It now remains to pass to $\lim \inf_{n \to +\infty}$ in (3.298) and use (3.68) to conclude the proof.   □

If $F^+ = F_-$ and comparison holds for the limiting equation one can obtain the convergence of the u_n to the unique viscosity solution of the limiting equation. Moreover, the limiting Hamiltonians F^+ and F^- may be of first order so the above theorem can be applied to singular perturbation problems discussed in Sect. 3.8. Other applications related to the convergence of finite-dimensional approximations (like those in Sect. 3.7) when condition (3.275) is satisfied by the operators A_n only on a family of finite-dimensional spaces can be found in [376].

3.10 The Infinite-Dimensional Black–Scholes–Barenblatt Equation

In this section we show how the theory of viscosity solutions and the results of previous sections can be used to deal with the infinite-dimensional Black–Scholes–Barenblatt equation (2.140) introduced in Sect. 2.6.7. We refer the reader to Sect. 2.6.7 for details about the financial meaning of the equation and the associated optimal control problem.

Let H be the Sobolev space $H^1([0, +\infty))$ and let A be the maximal dissipative operator

$$\begin{cases} D(A) := H^2([0, +\infty)) \\ A(x)(\sigma) := \dfrac{dx}{d\sigma}(\sigma). \end{cases}$$

The operator A generates, by Theorem B.45, a C_0-semigroup of contractions e^{tA} in H. Let B be a bounded, self-adjoint, strictly positive operator satisfying (3.2). We introduce the space

$$\mathfrak{V} = \left\{ x \in H \; : \; \sigma \to \sqrt{\sigma} x(\sigma), \; \sigma \to \sqrt{\sigma}\frac{dx}{d\sigma}(\sigma) \in L^2(0, \infty) \right\},$$

equipped with the norm

$$|x|_{\mathfrak{V}}^2 = \int_0^\infty (1+\sigma)\left(x^2(\sigma) + \left(\frac{dx}{d\sigma}(\sigma)\right)^2 \right) d\sigma,$$

and we denote by Λ a fixed bounded and closed subset of $\mathfrak{V}^d$. The space H will be the state space and Λ will be the control space.

The set of admissible controls $\mathcal{U}_t$ is defined as in Sect. 2.1.2, where the W in the reference probability spaces ν there are d-dimensional standard Brownian motions.

Lemma 3.96 *The function* $b : \mathfrak{V}^d \to H$ *defined by*

$$b(x)(\sigma) = \sum_{k=1}^d x_k(\sigma) \int_0^\sigma x_k(\mu) d\mu$$

is locally Lipschitz. Here $x = (x_1, ..., x_k)$.

Proof Let $R > 0$ and $x, y \in \mathfrak{V}^d$ be such that $|x_k|_{\mathfrak{V}}, |y_k|_{\mathfrak{V}} \le R$, for $k = 1, ..., d$. Then

$$|b(x)(\sigma) - b(y)(\sigma)|^2 \le \sum_{k=1}^d 2d \left(\sigma |x_k(\sigma) - y_k(\sigma)|^2 \int_0^\sigma x_k^2(\mu) d\mu \right.$$
$$\left. + \sigma y_k^2(\sigma) \int_0^\sigma |x_k(\mu) - y_k(\mu)|^2 d\mu \right).$$

Integrating we have

$$\int_0^{+\infty} |b(x)(\sigma) - b(y)(\sigma)|^2 d\sigma \le \sum_{k=1}^d 2d R^2 \int_0^{+\infty} (1+\sigma)|(x_k - y_k)(\sigma)|^2 d\sigma.$$

Similarly, we obtain

$$|(b(x))'(\sigma) - (b(y))'(\sigma)|^2 \le 3d \sum_{k=1}^d \left(4R^2 |x_k(\sigma) - y_k(\sigma)|^2 \right.$$
$$+ \sigma \left((y_k)'\right)^2 (\sigma) \int_0^\sigma |x_k(\mu) - y_k(\mu)|^2 d\mu$$
$$\left. + \sigma \left|\left((x_k)'\right)(\sigma) - \left((y_k)'\right)(\sigma)\right|^2 \int_0^\sigma x_k^2(\mu) d\mu \right),$$

which after integration yields

$$\int_0^{+\infty} |(b(x))'(\sigma) - (b(y))'(\sigma)|^2 d\sigma \leq 3d(4R^2 + 1) \sum_{k=1}^{d} \left(\int_0^{+\infty} |(x_k - y_k)(\sigma)|^2 d\sigma \right.$$

$$\left. + \int_0^{+\infty} \sigma \left| ((x_k)') (\sigma) - ((y_k)') (\sigma) \right|^2 d\sigma \right).$$

The claim now follows easily.   □

The previous lemma implies, in particular, that for $\tau \in \mathcal{U}_t$, the process $b(\tau(s))$ is progressively measurable and bounded. Therefore the state equation for the problem

$$\begin{cases} dr(s) = (Ar(s) + b(\tau(s)))ds + \tau(s) \cdot dW(s), \ s \in (t, T] \\ r(t) \ \ = x \end{cases} \tag{3.299}$$

is well posed in H for any reference probability space $\nu = (\Omega, \mathscr{F}, \mathscr{F}_s^t, \mathbb{P}, W)$ and any $\tau(\cdot) \in \mathcal{U}_t$ (see Theorem 1.127). We denote its unique mild solution by $r(\cdot)$.

Our control problem consists in maximizing the cost functional

$$\mathbb{E} \left(e^{-\int_t^T r^+(s,0)ds} g(r(T)) \right) \tag{3.300}$$

over all controls $\tau(\cdot) \in \mathcal{U}_t$. (We used $r^+(s, 0)$ to denote $r^+(s)(0)$.) This defines the value function

$$V(t, x) := \sup_{\tau(\cdot) \in \mathcal{U}_t} \mathbb{E} \left(e^{-\int_t^T r^+(s,0)ds} g(r_T) \right).$$

We assume the function g satisfies the following hypothesis.

Hypothesis 3.97 The function g is locally uniformly B-continuous and

$$|g(x)| \leq C(1 + |x|^m) \quad \text{for all } x \in H,$$

for some $C, m \geq 0$.

We can now apply the results of the previous sections to the HJB of the problem (2.140). Observe first that, if we define

$$c(x) = x^+(0),$$

for x in H, then c is weakly sequentially continuous on H and so it is uniformly continuous in the $| \cdot |_{-1}$ norm on bounded sets of H. Moreover, it is easy to see that c has at most linear growth at infinity and for instance

$$x^+(0) \leq 2|x|, \tag{3.301}$$

so the hypotheses needed to prove the "existence" part of Theorem 3.66 are satisfied. It guarantees the existence of a local modulus ω such that

$$|V(t, x) - V(s, y)| \le \omega\big(|t - s| + |x - y|_{-1}; R\big) \tag{3.302}$$

for all $0 \le t, s \le T, x, y \in B(0, R)$. It also ensures that V is a viscosity solution of the Hamilton–Jacobi–Bellman equation (the BSB equation) (2.140).

As regards the uniqueness of viscosity solutions of the BSB equation (2.140) we observe that Hypotheses 3.44–3.46, 3.48, 3.49 with $\gamma = 0$ are satisfied. To guarantee Hypothesis 3.47 we need an additional assumption.

We suppose that Λ is a compact subset of H_{-1}^d. It is then obvious that

$$\sup_{\tau \in \Lambda} \sum_{i=1}^{d} |Q_N \tau_i|_{-1}^2 \to 0 \quad \text{as } N \to \infty,$$

where Q_N is defined as in Sect. 3.5. This implies that Hypothesis 3.47 holds. Therefore, by Theorem 3.50, comparison holds for (2.140) and thus we have the following result.

Theorem 3.98 *Let Hypothesis 3.97 hold and let Λ be a bounded and closed subset of $\mathfrak{V}^d$ which is also a compact subset of H_{-1}^d. Then the value function V satisfies (3.302) and is the unique viscosity solution of the BSB equation (2.140) among functions satisfying (3.73) with $\gamma = 0$ and*

$$\lim_{t \to T} |u(t, x) - g(x))| = 0 \tag{3.303}$$

uniformly on bounded sets.

If g is bounded and weakly sequentially continuous, it can be shown that the value function can be approximated by viscosity solutions of finite-dimensional approximations of the BSB equation (2.140). This assumption holds in many interesting cases, for example if g is given by (2.136). We refer to [375] for further details.

A non-local BSB equation related to the HJMM model with Lévy noise was studied in [545]. The BSB equation in [545] was considered in the space $H = H^{1,\gamma}(\mathbb{R}^+)$ which is a weighted $H^1(\mathbb{R}^+)$ space with the weight $e^{\gamma\sigma}$ for some $\gamma > 0$. The results of this section could also be obtained in such a space.

3.11 The HJB Equation for Control of the Duncan–Mortensen–Zakai Equation

This section is a continuation of Sect. 2.6.6, which the reader should be familiar with. We have seen in Sect. 2.6.6 how the Duncan–Mortensen–Zakai (DMZ) equation arises in control problems with partial observation. In the so-called "separated" problem the DMZ equation is the state equation for the unnormalized conditional probability density of the state process with respect to the observation process. This gives rise to an optimal control problem for the DMZ equation which is fully observable. In this section we discuss how the HJB techniques can be applied to this problem. We first present basic results about variational solutions of SPDEs.

3.11.1 Variational Solutions

In this section we make the following assumptions. Let V, H be real separable Hilbert spaces. We identify H with its dual. Suppose that V is continuously and densely embedded in H. We then have the continuous and dense embeddings

$$V \subset H \subset V^*$$

and V^* is also separable, where V^* is the dual of V. We denote the norms in V, H, V^* by $|\cdot|_V, |\cdot|, |\cdot|_{V^*}$, respectively. The inner product in H is denoted by $\langle \cdot, \cdot \rangle$. The duality pairing between V^* and V is denoted by $\langle \cdot, \cdot \rangle_{\langle V^*, V \rangle}$. The duality pairing agrees with the inner product on H, i.e. for every $x \in H$, $v \in V$, $\langle x, v \rangle = \langle x, v \rangle_{\langle V^*, V \rangle}$. The triple (V, H, V^*) with the above properties is called a Gelfand triple.

Let Ξ be a real separable Hilbert space, Λ be a Polish space, $Q \in \mathcal{L}_1^+(\Xi)$ and $T \in (0, +\infty)$. Let $\mu = \left(\Omega, \mathcal{F}, \{\mathcal{F}_s\}_{s\in[0,T]}, \mathbb{P}, W_Q\right)$ be a generalized reference probability space. Let $a(\cdot) \in \mathcal{U}^\mu := \mathcal{U}_0^\mu$ (see (2.1)). We assume the following hypothesis.

Hypothesis 3.99 The following conditions are satisfied:

(i) The linear operators $A(t, a) : V \to V'$ are closed with a common domain $D(A)$ for $(t, a) \in [0, T] \times \Lambda$, and for every $t \in [0, T]$, the map $\tilde{A} : [0, T] \times \Lambda \times V \to V^*$, $\tilde{A}(s, a, v) = A(s, a)v$, restricted to $[0, t] \times \Lambda \times V$, is $\mathcal{B}([0, t]) \otimes \mathcal{B}(\Lambda) \otimes \mathcal{B}(V)/\mathcal{B}(V^*)$-measurable. Moreover, there exist C, γ, and $\beta > 0$ such that for all $u, v \in V$,

$$|\langle A(s, a)u, v \rangle_{\langle V^*, V \rangle}| \le C|u|_V|v|_V, \quad (s, a) \in [0, T] \times \Lambda, \tag{3.304}$$

$$\langle A(s, a)v, v \rangle_{\langle V^*, V \rangle} \le -\beta|v|_V^2 + \gamma|v|^2, \quad (s, a) \in [0, T] \times \Lambda. \tag{3.305}$$

(ii) The functions $b : [0, T] \times V \times \Lambda \to H$ and $\sigma : [0, T] \times V \times \Lambda \to \mathcal{L}_2(\Xi_0, H)$ are such that for every $t \in [0, T]$ their restrictions to $[0, t] \times V \times \Lambda$ are respectively $\mathcal{B}([0, t]) \otimes \mathcal{B}(V) \otimes \mathcal{B}(\Lambda)/\mathcal{B}(H)$- and $\mathcal{B}([0, t]) \otimes \mathcal{B}(V) \otimes \mathcal{B}(\Lambda)/\mathcal{B}(\mathcal{L}_2(\Xi_0, H))$-measurable.

(iii) There exists a C such that for all $u, v \in V$

$$|b(s, v, a)| + \|\sigma(s, v, a)\|_{\mathcal{L}_2(\Xi_0, H)} \le C(1 + |v|_V), \tag{3.306}$$

$$|b(s, u, a) - b(s, v, a)| + \|\sigma(s, u, a) - \sigma(s, v, a)\|_{\mathcal{L}_2(\Xi_0, H)} \le C|u - v|_V,$$
(3.307)

for all $(s, a) \in [0, T] \times \Lambda$.

(iv) There exist C, γ_1, and $\beta_1 > 0$ such that for all $v \in V$,

$$\langle A(s, a)v, v \rangle_{\langle V^*, V \rangle} + \|\sigma(s, v, a)\|^2_{\mathcal{L}_2(\Xi_0, H)} \le -\beta_1 |v|^2_V + \gamma_1 |v|^2 + C, \quad (3.308)$$

for all $(s, a) \in [0, T] \times \Lambda$.

(v) There exists a δ such that for all $u, v \in V$,

$$2\langle A(s, a)(u - v), u - v \rangle_{\langle V^*, V \rangle} + 2\langle b(s, u, a) - b(s, v, a), u - v \rangle$$
$$+ \|\sigma(s, u, a) - \sigma(s, v, a)\|^2_{\mathcal{L}_2(\Xi_0, H)} \le \delta |u - v|^2, \quad (3.309)$$

for all $(s, a) \in [0, T] \times \Lambda$.

It is now easy to see that the maps $A(s, a(s))v, b(s, v, a(s)), \sigma(s, v, a(s))$ defined on $[0, T] \times \Omega \times V$ are such that for every $t \in [0, T]$ their restrictions to $[0, t] \times \Omega \times V$ are respectively $\mathcal{B}([0, t]) \otimes \mathscr{F}_t \otimes \mathcal{B}(V)/\mathcal{B}(V^*)$, $\mathcal{B}([0, t]) \otimes \mathscr{F}_t \otimes \mathcal{B}(V)/\mathcal{B}(H)$ and $\mathcal{B}([0, t]) \otimes \mathscr{F}_t \otimes \mathcal{B}(V)/\mathcal{B}(\mathcal{L}_2(\Xi_0, H))$-measurable, and they satisfy (3.304)–(3.309) (with a replaced by $a(s)$) for a.e. $(s, \omega) \in [0, T] \times \Omega$.

We consider the following stochastic PDE

$$\begin{cases} dX(s) = (A(s, a(s))X(s) + b(s, X(s), a(s)))ds + \sigma(s, X(s), a(s))dW_Q(s) \\ X(0) = \xi. \end{cases}$$
(3.310)

Definition 3.100 (*Variational solution of* (3.310)) A process $X(\cdot) \in M^2_\mu(0, T; H)$ is called a *variational solution* of (3.310) if

$$\mathbb{E}\left[\int_0^T |X(r)|^2_V dr\right] < +\infty$$

and for every $\phi \in V$ we have

$$\langle X(s), \phi \rangle = \langle \xi, \phi \rangle + \int_0^s \langle A(r, a(r))X(r), \phi \rangle_{\langle V^*, V \rangle} dr + \int_0^s \langle b(r, X(r), a(r)), \phi \rangle dr$$
$$+ \int_0^s \langle \sigma(r, X(r), a(r))dW_Q(r), \phi \rangle \quad \text{for each } s \in [0, T], \ \mathbb{P} - \text{a.e.}$$
(3.311)

We remark that the integrand $A(r, a(r))X(r)$ above is evaluated at a V-valued progressively measurable equivalent version of $X(\cdot)$, and the process $\mathbf{1}_{X(s) \in V} X(s)$ is equivalent to the process $X(s)$ and, by Lemma 1.17-(iii), belongs to $M^2_\mu(0, T; V)$. Moreover, (3.311) is equivalent to the equality

$$X(s) = \xi + \int_0^s (A(r, a(r))X(r) + b(r, X(r), a(r)))dr + \int_0^s \sigma(r, X(r), a(r))dW_Q(r)$$

as elements of V^*.

The following result is taken from [386], Theorem I.3.1, in the version from [294], Theorem 4.3, p. 165 (see also [491], Theorem 4.2.5, and [519]).

Theorem 3.101 *Let μ be a generalized reference probability space, ξ be an $\mathscr{F}_0$-measurable H-valued random variable such that $\mathbb{E}^\mu[|\xi|^2] < +\infty$, and let $Y(\cdot) \in M_\mu^2(0, T; V^*)$, $Z(\cdot) \in \mathcal{N}_Q^2(0, T; H)$. We define the continuous V^*-valued process*

$$X(s) = \xi + \int_0^s Y(r)dr + \int_0^s Z(r)dW_Q(r), \quad s \in [0, T].$$

If $X(\cdot)$ has an equivalent version $\tilde{X}(\cdot) \in M_\mu^2(0, T; V)$, then $X(\cdot) \in M_\mu^2(0, T; H) \cap L^2(\Omega; C([0, T], H))$,

$$\mathbb{E}\left[\sup_{0 \leq s \leq T} |X(s)|^2\right] \leq +\infty,$$

and the following Itô's formula holds $\mathbb{P}$-a.e.

$$|X(s)|^2 = |\xi|^2 + \int_0^s \left(2\langle Y(r), \tilde{X}(r)\rangle_{\langle V^*, V\rangle} + \|Z(r)\|_{\mathcal{L}_2(U_0, H)}^2\right) dr$$
$$+2\int_0^s \langle Z(r)dW_Q(r), X(r)\rangle \quad s \in [0, T]. \tag{3.312}$$

Theorem 3.102 *Let μ be a generalized reference probability space, ξ be $\mathscr{F}_0$-measurable H-valued random variable such that $\mathbb{E}^\mu[|\xi|^2] < +\infty$, and $a(\cdot) \in \mathcal{U}^\mu$. Then:*

(i) *There exists a unique variational solution of (3.310) $X(\cdot) \in L^2(\Omega; C([0, T], H))$, and the energy equality holds $\mathbb{P}$-a.e.*

$$|X(s)|^2 = |\xi|^2 + 2\int_0^s \langle A(r, a(r))X(r), X(r)\rangle_{\langle V^*, V\rangle}dr + 2\int_0^s \langle b(r, X(r), a(r)), X(r)\rangle dr$$
$$+2\int_0^s \langle \sigma(r, X(r), a(r))dW_Q(r), X(r)\rangle + \int_0^s \|\sigma(r, X(r), a(r))\|_{\mathcal{L}_2(\Xi_0, H)}^2 dr \quad s \in [0, T]. \tag{3.313}$$

(ii) *If μ_1 is another generalized reference probability space, ξ_1 is an $\mathscr{F}_0^{\mu_1}$-measurable H-valued random variable such that $\mathbb{E}^{\mu_1}[|\xi_1|^2] < +\infty$, $a_1(\cdot) \in \mathcal{U}^\mu$, and*

$$\mathcal{L}_{\mathbb{P}_1}(\xi_1, a_1(\cdot), W_{Q,1}(\cdot)) = \mathcal{L}_{\mathbb{P}}(\xi, a(\cdot), W_Q(\cdot)),$$

then

$$\mathcal{L}_{\mathbb{P}_1}(a_1(\cdot), X_1(\cdot)) = \mathcal{L}_{\mathbb{P}}(a(\cdot), X(\cdot)), \tag{3.314}$$

where $X_1(\cdot)$ is the variational solution of (3.310) in μ_1 with control $a_1(\cdot)$ and initial condition ξ_1.

Proof We sketch the proof. The complete proof of the first part of the theorem can be found in [124], pp. 168–183 or [294, 386, 388, 491, 519]. Let $\{v_1, v_2, ...\}$ be an orthonormal basis of H composed of elements of V. We set $H_n := \mathrm{span}\{v_1, ..., v_n\}$, and define

$$P_n w := \sum_{k=1}^{n} \langle w, v_k \rangle_{\langle V^*, V \rangle} v_k, \quad w \in V^*.$$

If $w \in H$ we have

$$P_n w = \sum_{k=1}^{n} \langle w, v_k \rangle v_k$$

so P_n is an extension to V^* of the orthogonal projection in H onto H_n. We set

$$\begin{cases} A^n(s, \alpha)v := P_n A(s, \alpha)v, \ b^n(s, v, \alpha) := P_n b(s, v, \alpha), \ \sigma^n(s, v, \alpha) := P_n \sigma(s, v, \alpha) \\ \xi^n := P_n \xi. \end{cases}$$

(One can also project the Wiener process on a finite-dimensional subspace but it is not necessary.) Since the above functions are Lipschitz continuous in v on H_n, standard theory guarantees that there exists a unique strong solution (in the sense of Definition 1.118) $X^n(\cdot)$ of

$$\begin{cases} dX^n(s) = (A^n(s, a(s))X^n(s) + b^n(s, X^n(s), a(s)))ds + \sigma^n(s, X^n(s), a(s))dW_Q(s) \\ X^n(0) = \xi^n \end{cases}$$

$$(3.315)$$

which satisfies $X^n(\cdot) \in L^2(\Omega; C([0, T], H)) \cap M_\mu^2(0, T; V)$. Moreover, Itô's formula gives $\mathbb{P}$-a.e.

$$|X^n(s)|^2 = |\xi^n|^2 + 2\int_0^s \langle A^n(r, a(r))X^n(r), X^n(r) \rangle dr$$

$$+ 2\int_0^s \langle b^n(r, X^n(r), a(r)), X^n(r) \rangle dr + 2\int_0^s \langle \sigma^n(r, X^n(r), a(r))dW_Q(r), X^n(r) \rangle$$

$$+ \int_0^s \|\sigma^n(r, X^n(r), a(r))\|^2_{\mathcal{L}_2(\Xi_0, H)} dr, \quad s \in [0, T],$$

$$(3.316)$$

and using (3.316) and the assumptions one shows that

$$\mathbb{E}\left[\sup_{0 \le s \le T} |X^n(s)|^2 + \int_0^T |X^n(s)|^2_V ds \right] \le M \quad \text{for all } n.$$

Therefore, up to subsequences still denoted by X^n, b^n, σ^n, we obtain that there exist $X(\cdot)$, $\tilde{b}(\cdot) \in M_\mu^2(0, T; H)$, $\tilde{X}(\cdot) \in M_\mu^2(0, T; V)$, $\tilde{\sigma}(\cdot) \in \mathcal{N}_Q^2(0, T; H)$, and a $\mathscr{F}_T$-measurable random variable $\eta \in L^2(\Omega; H)$ such that

$$X^n(\cdot) \rightharpoonup \tilde{X}(\cdot) \text{ in } M_\mu^2(0, T; V), \quad X^n(T) \rightharpoonup \eta \text{ in } L^2(\Omega; H),$$

$$X^n(\cdot) \rightharpoonup X(\cdot), b^n(\cdot, X^n(\cdot), a(\cdot)) \rightharpoonup \tilde{b}(\cdot) \text{ in } M_\mu^2(0, T; H),$$

$$\sigma^n(\cdot, X^n(\cdot), a(\cdot)) \rightharpoonup \tilde{\sigma}(\cdot) \text{ in } \mathcal{N}_Q^2(0, T; H).$$

Obviously $\tilde{X}(\cdot)$ is an equivalent version of $X(\cdot)$. Passing to the limit as $n \to +\infty$ one obtains that $X(\cdot)$ is a variational solution of the linear equation

$$\begin{cases} dX(s) = (A(s, a(s))\tilde{X}(s) + \tilde{b}(s)ds + \tilde{\sigma}(s)dW_Q(s) \\ X(0) = \xi, \end{cases}$$

$X(T) = \eta$, and, by Theorem 3.101, $X(\cdot) \in L^2(\Omega; C([0, T], H))$ and it satisfies $\mathbb{P}$-a.e.

$$|X(s)|^2 = |\xi|^2 + 2 \int_0^s \langle A(r, a(r))X(r), X(r)\rangle_{\langle V^*, V\rangle} dr + 2 \int_0^s \langle \tilde{b}(r), X(r)\rangle dr$$
$$+ 2 \int_0^s \langle \tilde{\sigma}(r)dW_Q(r), X(r)\rangle + \int_0^s \|\tilde{\sigma}(r)\|_{\mathcal{L}_2(U_0, H)}^2 dr, \quad s \in [0, T].$$

One then uses monotonicity arguments to prove that

$$\tilde{b}(r) = b(r, X(r), a(r)), \quad \tilde{\sigma}(r) = \sigma(r, X(r), a(r)) \quad dt \otimes \mathbb{P}\text{-a.e.}$$

and hence $X(\cdot)$ is a variational solution of (3.310) and (3.313) holds. Moreover, it also follows from these arguments (see, for instance, [386] for details) that $\mathbb{E}[|X^n(T)|^2] \to \mathbb{E}[|X(T)|^2]$ and so $X^n(T) \to X(T)$ in $L^2(\Omega; H)$. Replacing interval $[0, T]$ by another interval $[0, t], 0 < t < T$, the same arguments give $X^n(t) \to X(t)$ in $L^2(\Omega; H)$ for all $0 \le t \le T$.

The uniqueness of variational solution in the generalized reference probability space μ follows from Theorem 3.101, elementary estimates using the assumptions on the coefficients, and Gronwall's lemma.

Finally, if $X_1(\cdot)$ is the variational solution in the generalized reference probability space μ_1 and $X_1^n(\cdot)$ are the solutions of the approximating problems (3.315) in this space then we have

$$\mathcal{L}_{\mathbb{P}_1}(a_1(\cdot), X_1^n(\cdot)) = \mathcal{L}_{\mathbb{P}}(a(\cdot), X^n(\cdot)),$$

and thus (3.314) follows since $X_1^n(t) \to X_1(t)$ in $L^2(\Omega_1; H)$ and $X^n(t) \to X(t)$ in $L^2(\Omega; H)$ for every $t \in [0, T]$. $\square$

3.11.2   Weighted Sobolev Spaces

We denote the norm and the inner product in $\mathbb{R}^d$ by $|\cdot|_{\mathbb{R}^d}$ and $\langle \cdot, \cdot \rangle_{\mathbb{R}^d}$, respectively. Given $k \in \mathbb{N}$ we denote by $H^k := H^k(\mathbb{R}^d)$ the standard Sobolev space on $\mathbb{R}^d$. We recall that $H^0 = L^2(\mathbb{R}^d)$, and the inner product in H^0 will be denoted by $\langle \cdot, \cdot \rangle_0$. Let $B := (-\Delta + I)^{-1}$, where $D(\Delta) = H^2$. We equip H^k with the inner product

$$\langle x, y \rangle_k := \langle B^{-k/2}x, B^{-k/2}y \rangle_0.$$

This inner product gives the norm $|x|_k = |B^{-k/2}x|$, which is equivalent to the standard norm in H^k given by

$$\left(\sum_{|\alpha| \leq k} \int_{\mathbb{R}^d} |\partial^\alpha x(\xi)|^2 d\xi \right)^{\frac{1}{2}}.$$

The topological dual space of H^k is denoted by H^{-k}. Except when explicitly stated we always identify H^0 with its dual. The space H^{-k} can be identified with the completion of H^0 under the norm

$$|x|_{-k} := |B^{k/2}x| = \langle B^k x, x \rangle_0^{1/2}$$

and then $B^{1/2}$ (after a natural extension) is an isometry between H^k and H^{k+1}, $k \in \mathbb{Z}$. The space H^{-k}, $k \in \mathbb{N}$, is a Hilbert space equipped with the inner product

$$\langle x, y \rangle_{-k} := \langle B^{k/2}x, B^{k/2}y \rangle_0.$$

The duality pairing between H^{-k} and H^k, $k \in \mathbb{N}$, is denoted by $\langle \cdot, \cdot \rangle_{\langle H^{-k}, H^k \rangle}$. We have $\langle a, b \rangle_{\langle H^{-k}, H^k \rangle} = \langle B^{k/2}a, B^{-k/2}b \rangle_0$. Observe also that, for $k \in \mathbb{Z}$, the adjoint of the operator $B^{1/2} : H^k \to H^{k+1}$ is $B^{1/2} : H^{-k-1} \to H^{-k}$.

Let $k = 0, 1, 2$. Given a strictly positive real-valued function $\rho \in C^2(\mathbb{R}^d)$, we define the *weighted Sobolev space* $H_\rho^k(\mathbb{R}^d)$ (or simply H_ρ^k) to be the completion of $C_c^\infty(\mathbb{R}^d)$ with respect to the *weighted norm*

$$|x|_{k,\rho} := |\rho x|_k.$$

The space H_ρ^k can also be defined as the space of all measurable functions $x : \mathbb{R}^d \to \mathbb{R}$ such that $\rho(\cdot)x(\cdot) \in H^k$. We recall that the norm $|x|_{k,\rho}$ is equivalent to the norm given by

$$\left(\sum_{|\alpha| \le k} \int_{\mathbb{R}^d} |\partial^\alpha [\rho(\xi) x(\xi)]|^2 \, d\xi \right)^{1/2}.$$

We denote by C_ρ the isometry $C_\rho : H_\rho^k \to H^k$ defined as $(C_\rho x)(\xi) = \rho(\xi) x(\xi)$, and by $C_{1/\rho} = C_\rho^{-1} : H^k \to H_\rho^k$ its inverse: $(C_{1/\rho} x)(\xi) = (\rho(\xi))^{-1} x(\xi)$. We observe that H_ρ^k is a Hilbert space with the inner product $\langle x, y \rangle_{k,\rho} = \langle C_\rho x, C_\rho y \rangle_k$. We denote the topological dual space of H_ρ^k by H_ρ^{-k} and, identifying H_ρ^0 with its dual, we have

$$H_\rho^k \subset H_\rho^0 = [H_\rho^0]' \subset [H_\rho^k]' = H_\rho^{-k}, \quad k \ge 0. \tag{3.317}$$

We always use this identification, except when explicitly stated.

The adjoint C_ρ^* of C_ρ is an isometry $C_\rho^* : H^{-k} \longmapsto H_\rho^{-k}$. Observe that C_ρ^* can be identified with $C_{1/\rho}$.

To simplify the notation we write $X_k := H_\rho^k$.

Let $B_\rho := C_{1/\rho} [(-\Delta + I)^{-1}] C_\rho = C_{1/\rho} B C_\rho$. Similarly to the case of non-weighted spaces, X_{-k} can be identified with the completion of X_0 under the norm $|x|_{-k,\rho}^2 := \langle B_\rho^k x, x \rangle_{0,\rho} = \langle B^k C_\rho x, C_\rho x \rangle_0$ and then $B_\rho^{1/2}$ is an isometry between X_{-2}, X_{-1}, X_0, X_1 and X_{-1}, X_0, X_1, X_2, respectively. We remark that $B_\rho^{-1} = C_{1/\rho} B^{-1} C_\rho$, $B_\rho^{1/2} = C_{1/\rho} B^{1/2} C_\rho$, $B_\rho^{-1/2} = C_{1/\rho} B^{-1/2} C_\rho$. Thus $|x|_{-k,\rho} = |B_\rho^{k/2} x|_{0,\rho}$ and $|x|_{k,\rho} = |B_\rho^{-k/2} x|_{0,\rho}$. The duality pairing between X_{-k} and X_k is denoted by $\langle \cdot, \cdot \rangle_{(X_{-k}, X_k)}$. We have

$$\langle a, b \rangle_{(X^{-k}, X^k)} = \langle B_\rho^{k/2} a, B_\rho^{-k/2} b \rangle_{0,\rho} = \langle C_\rho a, C_\rho b \rangle_{(H^{-k}, H^k)}.$$

In what follows we consider weight functions ρ of the form

$$\rho_\beta(\xi) = 1 + |\xi|_{\mathbb{R}^d}^\beta, \qquad \beta > 2. \tag{3.318}$$

With such a choice of ρ it can be shown that $X_k \subset H^k$, and if $\beta > d/2$ then $X_0 \subset L^1(\mathbb{R}^d)$ and $X_k \subset W^{k,1}(\mathbb{R}^d)$.

3.11.3 Optimal Control of the Duncan–Mortensen–Zakai Equation

We now study the optimal control problem for the DMZ equation derived in Sect. 2.6.6. We study it in the weighted space X_0 using the formalism of abstract control problems. Let $T > 0$. The control set Λ was originally a subset of $\mathbb{R}^n$ but we will consider Λ to be a more general Polish space. For every $0 \le t \le T$, the reference and generalized reference probability spaces are defined by Definitions 2.7 and 1.100 where W_Q is now just a standard Wiener process in $\mathbb{R}^m$ (i.e. $\Xi = \mathbb{R}^m$, $Q = I$) which is denoted by W. The classes of admissible controls with respect to all reference and generalized reference probability spaces are defined, as always, by $\mathcal{U}_t$ and $\overline{\mathcal{U}}_t$. We remark that for a reference probability space $\mu = \left(\Omega, \mathscr{F}, \{\mathscr{F}_s^t\}_{s \in [t,T]}, \mathbb{P}, W \right)$, $\mathbb{P}$ now corresponds to $\bar{\mathbb{P}}$ in Sect. 2.6.6, $\mathscr{F}_s^t$ to $\mathscr{F}_s^{y_1,t}$, and W to y_1. Without loss of generality we will always assume that the Q-Wiener processes in the reference probability spaces have everywhere continuous paths.

Recall from Sect. 2.6.6 that for every $a \in \Lambda$ we have the differential operators A_a and S_a^k $(k = 1, \ldots, m)$

$$(A_a x)(\xi) = \sum_{i,j=1}^{d} \partial_i \left[a_{i,j}(\xi, a) \partial_j x(\xi) \right] + \sum_{i=1}^{d} \partial_i \left[b_i(\xi, a) x(\xi) \right], \tag{3.319}$$

$$\left(S_a^k x \right)(\xi) = \sum_{i=1}^{d} d_{ik}(\xi, a) \partial_i x(\xi) + e_k(\xi, a) x(\xi); \quad k = 1, \ldots, m. \tag{3.320}$$

Typically we set $D(A_a) = X_2$, $D(S_a^k) = X_1$, $k = 1, \ldots, m$, however the operators will be considered with different domains in different Gelfand triples. Having in mind the original Hypothesis 2.48, we assume the following hypothesis.

Hypothesis 3.103

(i) Λ is a compact metric space.
(ii) The coefficients

$$\left(a_{ij} \right)_{i,j=1,\ldots,d}, (b_i)_{i=1,\ldots,d}, c, (d_{ik})_{i=1,\ldots,d; \; k=1,\ldots,m}, (e_k)_{k=1,\ldots,m} : \mathbb{R}^d \times \Lambda \to \mathbb{R}$$

are continuous in (ξ, a) and, as functions of ξ, are in $C_b^2 \left(\mathbb{R}^d \right)$ for every a, with their norms in $C_b^2 \left(\mathbb{R}^d \right)$ bounded uniformly in $a \in \Lambda$. Moreover, there exists a constant $\lambda > 0$ such that

$$\sum_{i,j=1}^{d} \left(a_{i,j}(\xi, a) - \frac{1}{2} \sum_{k=1}^{m} d_{ik}(\xi, a) d_{jk}(\xi, a) \right) z_i z_j \geq \lambda |z|^2 \tag{3.321}$$

for every $a \in \Lambda$ and $\xi, z \in \mathbb{R}^d$.
(iii) The weight ρ is of the form (3.318).

For every $a(\cdot) \in \mathcal{U}_t$ the DMZ equation is considered in X_0:

$$\begin{cases} dY(s) = A_{a(s)} Y(s) ds + \sum_{k=1}^{m} S_{a(s)}^k Y(s) dW_k(s), & s > t \\ Y(t) = x \in X_0. \end{cases} \tag{3.322}$$

We consider the cost functional

$$J(t, x; a(\cdot)) = \mathbb{E}\left\{\int_t^T l(Y(s), a(s))ds + g(Y(T))\right\}$$

and we make the following assumptions about the cost functions l and g.

Hypothesis 3.104

(i) $l : X_0 \times \Lambda \to \mathbb{R}$ and $g : X_0 \to \mathbb{R}$ are continuous and there exist $C > 0$ and $\gamma < 2$ such that

$$|l(x, a)|, |g(x)| \leq C\left(1 + |x|_{0,\rho}^\gamma\right)$$

for every $(x, a) \in X_0 \times \Lambda$;

(ii) for every $R > 0$ there exists a modulus ω_R such that

$$|l(x, a) - l(y, a)| \leq \omega_R\left(|x - y|_{0,\rho}\right), \quad |g(x) - g(y)| \leq \omega_R\left(|x - y|_{-1,\rho}\right)$$
$$(3.323)$$

for every $x, y \in B_{X_0}(0, R), a \in \Lambda$.

The optimal control problem in the weak formulation we study consists in minimizing the cost $J(t, x; a(\cdot))$ over all admissible controls $a(\cdot) \in \mathcal{U}_t$.

The associated HJB equation in X_0 has the form

$$\begin{cases} v_t + \inf_{a \in \Lambda}\left\{\frac{1}{2}\sum_{k=1}^m \langle D^2 v S_a^k x, S_a^k x\rangle_{\rho,0} + \langle A_a x, Dv\rangle_{\rho,0} + l(x, a)\right\} = 0, \\ \qquad\qquad\qquad\qquad\qquad\qquad\qquad\qquad\qquad\qquad \text{in } (0, T) \times X_0, \\ v(T, x) = g(x), \end{cases}$$
$$(3.324)$$

and the value function is

$$V(t, x) = \inf_{a(\cdot) \in \mathcal{U}_t} J(t, x; a(\cdot)). \qquad (3.325)$$

Let us now describe which restrictions may be placed on the original separated problem of Sect. 2.6.6 so that the current assumptions are satisfied.

First the law of the initial datum η of Eq. (2.124) must have density x in $L_\rho^2(\mathbb{R}^d)$. To guarantee that the density is also in $L^1(\mathbb{R}^d)$ we should assume $\beta > d/2$. The density x is polynomially decreasing when $|\xi|_{\mathbb{R}^d} \to +\infty$. This is of course a further restriction with respect to assuming only $x \in L^1(\mathbb{R}^d)$ but it is satisfied in many practical cases, for instance when the starting distribution is normal. One can consult e.g. [46], pp. 36, 204, for the use of x being Gaussian or [46], pp. 82, 167, for other integrability assumptions on x (see also [597], [472] on this). Regarding the cost functional (2.129) we have (recall that we now use $\mathbb{E}$ in place of $\bar{\mathbb{E}}$)

$$J(t, x; a(\cdot)) = \mathbb{E}\left\{\int_t^T \langle l_1(\cdot, a(s)), Y(s)\rangle_0 \, ds + \langle g_1(\cdot), Y(T)\rangle_0\right\}$$

$$= \mathbb{E}\left\{\int_t^T \langle (1/\rho^2)l_1(\cdot, a(s)), Y(s)\rangle_{0,\rho}\, ds + \langle (1/\rho^2)g_1(\cdot), Y(T)\rangle_{0,\rho}\right\}$$

$$= \mathbb{E}\left\{\int_t^T l(Y(s), a(s))ds + g(Y(T))\right\},$$

where we set

$$l(x, a) = \langle l_1(\cdot, a), x\rangle_0 = \left\langle \frac{1}{\rho^2}l_1(\cdot, a), x\right\rangle_{0,\rho},$$

$$g(x) = \langle g_1(\cdot), x\rangle_0 = \left\langle \frac{1}{\rho^2}g_1(\cdot), x\right\rangle_{0,\rho}. \tag{3.326}$$

It is easy to see that Hypothesis 3.104 is satisfied if the functions $l_1 : \mathbb{R}^d \times \mathbb{R}^n \to \mathbb{R}$ and $g_1 : \mathbb{R}^d \to \mathbb{R}$ are continuous and $\sup_{a\in\Lambda} |\frac{1}{\rho}l_1(\cdot, a)|_0 + |\frac{1}{\rho}g_1(\cdot)|_0 < +\infty$, since in this case the function g is weakly sequentially continuous in X_0. For instance, if

$$l_1(\xi, \alpha) = \langle M\xi, \xi\rangle_{\mathbb{R}^d} + \langle N\alpha, \alpha\rangle_{\mathbb{R}^n}, \quad g_1(\xi) = \langle G\xi, \xi\rangle_{\mathbb{R}^d},$$

where M, N and G are suitable non-negative definite matrices, $\Lambda = B_{\mathbb{R}^n}(0, R)$, Hypothesis 3.104 is satisfied if $\beta > 2 + d/2$. This is the main advantage of using weighted spaces. When the initial density is, say, polynomially decreasing at infinity, we can deal with polynomially growing cost functions. This would not be possible if we took $\rho = 1$. Finally, we mention that in the absence of density the separated problem has to be studied in the space of measures.

3.11.4 Estimates for the DMZ Equation

Lemma 3.105 *Let Hypothesis 3.103 hold. Then:*

(i) *The DMZ equation satisfies the assumptions of Hypothesis 3.99 for the Gelfand triple (X_1, X_0, X_{-1}) and also for (X_2, X_1, X_0).*
(ii) *There exist constants $\bar{\lambda} > 0$, $K \geq 0$ such that for all $a \in \Lambda$*

$$\langle A_a x, x\rangle_{\langle X_{-1}, X_1\rangle} + \frac{1}{2}\sum_{k=1}^m \langle S_a^k x, S_a^k x\rangle_{0,\rho}$$

$$\leq -\bar{\lambda}|x|_{1,\rho}^2 + K|x|_{0,\rho}^2, \quad x \in X_1, \tag{3.327}$$

$$\langle A_a x, B_\rho^{-1} x \rangle_{0,\rho} + \frac{1}{2} \sum_{k=1}^{m} \langle B_\rho^{-1} S_a^k x, S_a^k x \rangle_{\langle X_{-1}, X_1 \rangle}$$

$$\leq -\bar{\lambda} |x|_{2,\rho}^2 + K |x|_{1,\rho}^2, \quad x \in X_2, \tag{3.328}$$

$$\langle A_a x, B_\rho x \rangle_{\langle X_{-2}, X_2 \rangle} + \frac{1}{2} \sum_{k=1}^{m} \langle B_\rho S_a^k x, S_a^k x \rangle_{\langle X_1, X_{-1} \rangle}$$

$$\leq -\bar{\lambda} |x|_{0,\rho}^2 + K |x|_{-1,\rho}^2, \quad x \in X_0. \tag{3.329}$$

Proof Part (i) follows from direct computations and estimates in (ii). Part (ii) is proved in [323], Lemma 3.3. □

We also record for future use that

$$\sup_{a \in \Lambda, k=1,\dots,m} \left(\| B_\rho A_a \|_{\mathcal{L}(X_0)} + \| B_\rho^{1/2} S_a^k \|_{\mathcal{L}(X_0)} \right) \leq C. \tag{3.330}$$

Proposition 3.106 *Assume that Hypothesis 3.103 holds. Let $0 \leq t \leq T$, let μ be a generalized reference probability space, $a(\cdot) \in \mathcal{U}_t^\mu$ and $x \in L^2(\Omega; X_0)$ be an $\mathcal{F}_t$-measurable random variable. Then there exists a unique variational solution $Y(s) := Y(\cdot; t, x, a(\cdot)) \in L^2(\Omega; C([t, T], H))$ of the state equation (3.322). Moreover, we have:*

- $\mathbb{P}$-*a.s.*

$$|Y(s)|_{0,\rho}^2 = |x|_{0,\rho}^2 + 2 \int_t^s \langle A_{a(r)} Y(r), Y(r) \rangle_{\langle X_{-1}, X_1 \rangle} \, dr$$

$$+ \sum_{k=1}^{m} \int_t^s \langle S_{a(r)}^k Y(r), Y(r) \rangle_{0,\rho} \, dW_k(r) \tag{3.331}$$

$$+ \sum_{k=1}^{m} \int_t^s \langle S_{a(r)}^k Y(r), S_{a(r)}^k Y(r) \rangle_{0,\rho} \, dr, \quad s \in [t, T].$$

In particular,

$$\mathbb{E} |Y(s)|_{0,\rho}^2 = \mathbb{E} |x|_{0,\rho}^2 + 2\mathbb{E} \int_t^s \langle A_{a(r)} Y(r), Y(r) \rangle_{\langle X_{-1}, X_1 \rangle} \, dr$$

$$+ \sum_{k=1}^{m} \mathbb{E} \int_t^s \langle S_{a(r)}^k Y(r), S_{a(r)}^k Y(r) \rangle_{0,\rho} \, dr, \quad s \in [t, T]. \tag{3.332}$$

• There exists a constant $C > 0$ independent of μ, $a(\cdot) \in \mathcal{U}_t^\mu$ and x such that

$$\mathbb{E}\,|Y(s)|^2_{0,\rho} \leq \mathbb{E}\,|x|^2_{0,\rho}\,(1 + C(s-t)), \quad s \in [t, T], \tag{3.333}$$

$$\mathbb{E}\int_t^T |Y(s)|^2_{1,\rho}\,ds \leq C\mathbb{E}\,|x|^2_{0,\rho}, \quad s \in [t, T]. \tag{3.334}$$

• The conclusion of Theorem 3.102-(ii) (with 0 replaced by t) is satisfied.

Proof The results follow from Theorem 3.102.   □

The following proposition collects various estimates for solutions of (3.322).

Proposition 3.107 Assume that Hypothesis 3.103 holds and let $0 \leq t \leq T$. Let $a(\cdot) \in \mathcal{U}_t$ and $x \in X_0$. Then:

(i) There exists a constant $C > 0$ independent of $a(\cdot) \in \mathcal{U}_t$ and x such that, for all $s \in [t, T]$,

$$\mathbb{E}\,|Y(s)|^2_{-1,\rho} \leq |x|^2_{-1,\rho}\,(1 + C(s-t)), \tag{3.335}$$

$$\mathbb{E}\int_t^T |Y(s)|^2_{0,\rho}\,ds \leq C\,|x|^2_{-1,\rho}. \tag{3.336}$$

$$\mathbb{E}\,|Y(s) - x|^2_{-1,\rho} \leq C\,(s-t)\,|x|^2_{0,\rho}, \tag{3.337}$$

$$\mathbb{E}\int_t^s |Y(r) - x|^2_{0,\rho}\,dr \leq C\,(s-t)\,|x|^2_{0,\rho}. \tag{3.338}$$

There is a modulus σ_x, independent of $a(\cdot) \in \mathcal{U}_t$, such that

$$\mathbb{E}\,|Y(s) - x|^2_{0,\rho} \leq \sigma_x(s-t), \quad s \in [t, T]. \tag{3.339}$$

(ii) If in addition $x \in X_1$, then $Y(\cdot)$ is a strong solution and there exists a constant $C > 0$ independent of $a(\cdot) \in \mathcal{U}_t$ and x such that, for all $s \in [t, T]$,

$$\mathbb{E}\,|Y(s)|^2_{1,\rho} \leq |x|^2_{1,\rho}\,(1 + C(s-t)), \tag{3.340}$$

$$\mathbb{E}\int_t^T |Y(s)|^2_{2,\rho}\,ds \leq C\,|x|^2_{1,\rho}, \tag{3.341}$$

$$\mathbb{E}\,|Y(s) - x|^2_{0,\rho} \leq C\,|x|^2_{1,\rho}\,(s-t), \tag{3.342}$$

$$\mathbb{E}\int_t^s |Y(r) - x|^2_{1,\rho}\,dr \leq C(s-t)\,|x|^2_{1,\rho}. \tag{3.343}$$

There is a modulus σ_x, independent of $a(\cdot) \in \mathcal{U}_t$, such that

$$\mathbb{E}\,|Y(s) - x|^2_{1,\rho} \le \sigma_x(s - t), \qquad s \in [t, T]. \tag{3.344}$$

Proof (i). By Itô's formula we have

$$\begin{aligned}
\mathbb{E}\,|Y(s)|^2_{-1,\rho} &= \mathbb{E}\,\big|B_\rho^{1/2}Y(s)\big|^2_{0,\rho} \\
&= \mathbb{E}\,\big|B_\rho^{1/2}x\big|^2_{0,\rho} + 2\mathbb{E}\int_t^s \big(B_\rho^{1/2}A_{a(r)}Y(r),\,B_\rho^{1/2}Y(r)\big)_{0,\rho}\,dr \\
&\quad + \sum_{k=1}^m \mathbb{E}\int_t^s \big(B_\rho^{1/2}S_{a(r)}^k Y(r),\,B_\rho^{1/2}S_{a(r)}^k Y(r)\big)_{0,\rho}\,dr. \tag{3.345}
\end{aligned}$$

Since

$$\big(B_\rho^{1/2}A_{a(r)}Y(r),\,B_\rho^{1/2}Y(r)\big)_{0,\rho} = \big(A_{a(r)}Y(r),\,B_\rho Y(r)\big)_{\langle X_{-2},X_2\rangle}$$

and

$$\big(B_\rho^{1/2}S_{a(r)}^k Y(r),\,B_\rho^{1/2}S_{a(r)}^k Y(r)\big)_{0,\rho} = \big(B_\rho S_{a(r)}^k Y(r),\,S_{a(r)}^k Y(r)\big)_{\langle X_1,X_{-1}\rangle}$$

we have, thanks to (3.329),

$$\mathbb{E}\,|Y(s)|^2_{-1,\rho} + 2\bar{\lambda}\int_t^s \mathbb{E}\,|Y(r)|^2_{0,\rho}\,dr \le |x|^2_{-1,\rho} + 2K\int_t^s \mathbb{E}\,|Y(r)|^2_{-1,\rho}\,dr.$$

Estimates (3.335)–(3.336) follow by applying Gronwall's inequality.

To show (3.337), we have

$$\mathbb{E}\,|Y(s) - x|^2_{-1,\rho} = \mathbb{E}\,|Y(s)|^2_{-1,\rho} + |x|^2_{-1,\rho} - 2\mathbb{E}\,\big(Y(s),\,B_\rho x\big)_{0,\rho},$$

which gives, by (3.345) and by the definition of variational solution,

$$\begin{aligned}
&\mathbb{E}\,|Y(s) - x|^2_{-1,\rho} \\
&= 2\mathbb{E}\int_t^s \Bigg[\big(A_{a(r)}Y(r),\,B_\rho Y(r)\big)_{\langle X_{-1},X_1\rangle} + \frac{1}{2}\sum_{k=1}^m \big(B_\rho S_{a(r)}^k Y(r),\,S_{a(r)}^k Y(r)\big)_{0,\rho}\Bigg]dr \\
&\quad - 2\mathbb{E}\int_t^s \big(B_\rho A_{a(r)}Y(r),\,x\big)_{0,\rho}.
\end{aligned}$$

Therefore, by (3.329),

$$\begin{aligned}
&\mathbb{E}\,|Y(s) - x|^2_{-1,\rho} + 2\bar{\lambda}\mathbb{E}\int_t^s |Y(r)|^2_{0,\rho}\,dr \\
&\le 2K\mathbb{E}\int_t^s |Y(r)|^2_{-1,\rho}\,dr + 2\mathbb{E}\int_t^s |x|_{0,\rho}\,\big|B_\rho A_{a(r)}Y(r)\big|_{0,\rho}\,dr,
\end{aligned}$$

which, upon using (3.330), (3.333) and straightforward calculations, yields

$$\mathbb{E}\,|Y(s) - x|^2_{-1,\rho} + \bar{\lambda}\mathbb{E}\int_t^s |Y(r)|^2_{0,\rho}\,dr \le C\,(s-t)\left[|x|^2_{-1,\rho} + |x|^2_{0,\rho}\right]$$

for some constant $C > 0$. This proves (3.337). Estimate (3.338) follows upon observing that

$$\mathbb{E}\int_t^s |Y(r) - x|^2_{0,\rho}\,dr \le 2\mathbb{E}\int_t^s \left(|Y(r)|^2_{0,\rho} + |x|^2_{0,\rho}\right)dr \le C\,(s-t)\,|x|^2_{0,\rho}\,.$$

To prove (3.339), we assume by contradiction that it is not satisfied. In this case there are $a_n(\cdot) \in \mathcal{U}_t$ (which we can assume to be $\mathscr{F}_s^{t,0}$-predictable) and $t_n \to t$ such that $\mathbb{E}|Y_n(t_n) - x|^2_{0,\rho} \not\to 0$ as $n \to +\infty$. Because of Corollary 2.21 and Proposition 3.106 we can assume that all $a_n(\cdot)$ are defined on the same reference probability space. Since $\mathbb{E}\,|Y(t_n)|^2_{0,\rho}$ is bounded, there exists a subsequence, still denoted by t_n, $t_n \to 0$ as $n \to +\infty$, and an element $\bar{Y}$ of $L^2(\Omega; X_0)$ such that, as $n \to +\infty$

$$Y(t_n) \rightharpoonup \bar{Y}, \quad \text{weakly in } L^2(\Omega; X_0)$$

and hence also weakly in $L^2(\Omega; X_{-1})$. Since by (3.337)

$$Y(t_n) \to x, \quad \text{strongly in } L^2(\Omega; X_{-1})$$

as $n \to +\infty$, we obtain $\bar{Y} = x$. This, plus the fact that $\mathbb{E}\,|Y(t_n)|^2_{0,\rho} \to |x|^2_{0,\rho}$ as $n \to +\infty$ provided by (3.333), implies that $Y(t_n) \to x$ strongly in $L^2(\Omega; X_0)$, which gives a contradiction.

(ii). The existence of the strong solution is known (see [385, 386, 388]) and can be obtained similarly to Proposition 3.106 by applying it to the Gelfand triple (X_2, X_1, X_0). One now obtains

$$\mathbb{E}\,|Y(s)|^2_{1,\rho} = \mathbb{E}\,|x|^2_{1,\rho} + 2\mathbb{E}\int_t^s \langle A_{a(r)}Y(r), Y(r)\rangle_{\langle X_0, X_2\rangle}\,dr$$
$$+ \sum_{k=1}^m \mathbb{E}\int_t^s \langle S^k_{a(r)}Y(r), S^k_{a(r)}Y(r)\rangle_{1,\rho}\,dr,$$

which, upon using (3.328) and applying the same arguments as those in the proof of (i), yields (3.340) and (3.341). The proof of the final three estimates is analogous to the similar ones proved in (i). $\square$

3.11.5 Viscosity Solutions

The definition of viscosity solution for (3.324) is similar to the general definition given in Sect. 3.3. However, here we have unbounded first- and second-order terms so the equation is different. We also make use of the coercivity of the operators A_a.

Definition 3.108 A function ψ is a test function if $\psi = \varphi + \delta(t)|x|_{0,\rho}^2$, where:

(i) $\varphi \in C^{1,2}((0,T) \times X_0)$ is B_ρ-lower semicontinuous, and $\varphi_t \in UC((0,T) \times X_0)$, $D\varphi \in UC((0,T) \times X_0, X_2)$, $D^2\varphi \in UC_b((0,T) \times X_0, \mathcal{L}(X_{-1}, X_1))$.

(ii) $\delta \in C([0,T]) \cap C^1((0,T))$ is such that $\delta > 0$.

Definition 3.109 A locally bounded B_ρ-upper (respectively, lower) semicontinuous function $u : (0,T] \times X_0 \to \mathbb{R}$ is a viscosity subsolution (respectively, supersolution) of (3.324) if $u(T,x) \le g(x)$ (respectively, $u(T,x) \ge g(x)$) on X_0 and for every test function ψ, if $u - \psi$ (respectively, $u + \psi$) has a global maximum (respectively, minimum) at $(t,x) \in (0,T) \times X_0$, then $x \in X_{1,\rho}$ and

$$\psi_t(t,x) + \inf_{a \in \Lambda} \left\{ \frac{1}{2} \sum_{k=1}^{m} \left\langle D^2\psi(t,x)S_a^k x, S_a^k x \right\rangle_{0,\rho} + \langle A_a x, D\psi(t,x) \rangle_{\langle X_{-1}, X_1 \rangle} + f(x,a) \right\} \ge 0,$$

(respectively,

$$-\psi_t(t,x) + \inf_{a \in \Lambda} \left\{ \frac{1}{2} \sum_{k=1}^{m} \langle -D^2\psi(t,x)S_a^k x, S_a^k x \rangle_{0,\rho} \right.$$

$$\left. + \langle A_a x, -D\psi(t,x) \rangle_{\langle X_{-1}, X_1 \rangle} + f(x,a) \right\} \le 0).$$

A function is a viscosity solution if it is both a viscosity subsolution and a viscosity supersolution.

The main difference between this and the definition in Sect. 3.3 is that we require that the point x where the maximum/minimum occurs belongs to a smaller subspace X_1 (of more regular functions). This is possible because of the coercivity of the operators A_a. In this way all terms appearing in the equation are well defined and there is no need to discard any of them. Such definitions originated for first-order equations in [97, 144]. Compared to Definition 3.32 we have put more conditions on φ and restricted the class of radial functions. The role of A^* in Definition 3.32 is now played by B_ρ^{-1}. The radial test functions are quadratic since we only consider solutions with smaller growth rate at infinity, which is a reasonable assumption for value functions coming from separated problems (see (3.326)). We remark that the definition of viscosity solution here is different from the definition given in [323]) where it was required that $\varphi \in UC^{1,2}((0,T) \times X_{-1})$. Both allow us to prove the same results. We decided to change the definition to make it more in line with the presentation of the material in this book.

If u has less than quadratic growth in x as $|x|_{0,\rho} \to +\infty$ then the maxima and minima in the definition of solution can be assumed to be strict: if $u - (\varphi + \delta(t) |x|_{0,\rho}^2)$ has a global maximum at $(\hat{t}, \hat{x})$ and $\lambda \in C^2([0, +\infty))$ is such that $\lambda > 0$, $\lambda(r) = r^4$ if $r \leq 1$ and $\lambda(r) = 1$ if $r \geq 2$, then it is easy to see that $u - (\varphi + \delta(t) |x|_{0,\rho}^2) - \lambda(|x - \hat{x}|_{-1,\rho}) - (t - \hat{t})^2$ has a strict global maximum at $(\hat{t}, \hat{x})$.

It is easy to see that $\langle A_a x, D\varphi(t, x) \rangle_{\langle X_{-1}, X_1 \rangle} = \langle B_\rho A_a x, B_\rho^{-1} D\varphi(t, x) \rangle_{-,\rho}$ and $\langle D^2 \varphi(t, x) S_a^k x, S_a^k x \rangle_{0,\rho} = \langle B_\rho^{-1/2} D^2 \varphi(t, x) B_\rho^{-1/2} B_\rho^{1/2} S_a^k x, B_\rho^{1/2} S_a^k x \rangle_{0,\rho}, k = 1, ..., m$. Moreover,

$$B_\rho^{-1} D\varphi \in UC((0, T) \times X_0, X_0), \quad B_\rho^{-1/2} D^2 \varphi B_\rho^{-1/2} \in UC_b((0, T) \times X_0, \mathcal{L}(X_0)).$$
$$(3.346)$$

We also remark that Itô's formula holds for the test functions. For the radial part of a test function this follows from (3.331) and Itô's formula. As regards φ, the easiest way to see it is to use the fact that if $Y(s) := Y(\cdot; t, x, a(\cdot))$ is the solution of Eq. (3.322) on $[t, T]$, $x \in X_0$, then $Y(\cdot)$ is in fact a strong solution on any interval $[s, T]$, $s > t$. Thus if φ is a test function as above and $a(\cdot) \in \mathcal{U}_t$, then by the usual Itô's formula we have for $t < s < \eta$

$$\mathbb{E}\varphi(\eta, Y(\eta)) = \mathbb{E}\varphi(s, Y(s)) + \mathbb{E} \int_s^\eta \left[\varphi_t(r, Y(r)) + \langle A_{a(r)} Y(r), D\varphi(r, Y(r)) \rangle_{0,\rho} \right.$$
$$\left. + \frac{1}{2} \sum_{k=1}^m \langle D^2 \varphi(r, Y(r)) S_{a(r)}^k Y(r), S_{a(s)}^k Y(r) \rangle_{0,\rho} \right] dr$$
$$\to \varphi(t, x) + \mathbb{E} \int_t^\eta \left[\varphi_t(r, Y(r)) + \langle A_{a(r)} Y(r), D\varphi(r, Y(r)) \rangle_{\langle X_{-1}, X_1 \rangle} \right.$$
$$\left. + \frac{1}{2} \sum_{k=1}^m \langle D^2 \varphi(r, Y(r)) S_{a(r)}^k Y(r), S_{a(r)}^k Y(r) \rangle_{0,\rho} \right] dr$$

as $s \to t$ using, for instance, (3.333) and (3.339), since by (3.330)

$$|\varphi_t(r, Y(r))| + \left| \langle A_{a(r)} Y(r), D\varphi(r, Y(r)) \rangle_{\langle X_{-1}, X_1 \rangle} \right| \leq C \left(1 + |Y(r)|_{0,\rho}^2 \right),$$

$$\left| \langle D^2 \varphi(r, Y(r)) S_{a(r)}^k Y(r), S_{a(r)}^k Y(r) \rangle_{0,\rho} \right| \leq C |Y(r)|_{0,\rho}^2, \quad k = 1, ..., m.$$

Itô's formulas for test functions from [323] are proved in [467].

We prove the comparison principle. We use the notation of Sect. 3.2: $\{e_n\}_{n=1}^\infty \subset X_0$ is now an orthonormal basis in X^{-1}, $X^N = \text{span}\{e_1, ..., e_N\}$, P_N is the orthogonal projection from X_{-1} onto X^N, and $Q_N = I - P_N$, and $Y^N = Q_N X_{-1}$. We have an orthogonal decomposition $X_{-1} = X^N \times Y^N$, and for $x \in X_{-1}$ we write $x = (P_N x, Q_N x)$.

Theorem 3.110 *Let Hypotheses 3.103 and 3.104 hold. Let $u, v : X_0 \to \mathbb{R}$ be respectively a viscosity subsolution, and a viscosity supersolution of (3.324) (as defined in Definition 3.109). Let*

$$\limsup_{|x|_{0,\rho} \to \infty} \frac{u(t, x)}{|x|_{0,\rho}^2} = 0, \quad \limsup_{|x|_{0,\rho} \to \infty} \frac{-v(t, x)}{|x|_{0,\rho}^2} = 0, \tag{3.347}$$

uniformly for $t \in [0, T]$, and

$$\begin{cases} (i) & \lim_{t \uparrow T} (u(t, x) - g(x))^+ = 0 \\ (ii) & \lim_{t \uparrow T} (v(t, x) - g(x))^- = 0 \end{cases} \tag{3.348}$$

uniformly on bounded subsets of X_0. Then $u \le v$.

Proof Without loss of generality we can assume that u and $-v$ are bounded from above and such that

$$\lim_{|x|_{0,\rho} \to \infty} u(t, x) = -\infty, \quad \lim_{|x|_{0,\rho} \to \infty} v(t, x) = +\infty. \tag{3.349}$$

To see this we observe that if K is the constant from (3.327) then for every $\eta > 0$

$$u_\eta(t, x) = u(t, x) - \eta e^{2K(T-t)}|x|_{0,\rho}^2, \quad v_\eta(t, x) = v(t, x) + \eta e^{2K(T-t)}|x|_{0,\rho}^2$$

are respectively viscosity sub- and supersolutions of (3.324) and satisfy (3.348). This follows from (3.327) since, with $h(t, x) = \eta e^{2K(T-t)}|x|_{0,\rho}^2$, we have

$$h_t + \sup_{a \in \Lambda} \left\{ \frac{1}{2} \sum_{k=1}^{m} \langle D^2 h S_a^k x, S_a^k x \rangle_{0,\rho} + \langle A_a x, Dh \rangle_{\langle X_{-1}, X_1 \rangle} \right\} \le -\eta K e^{2K(T-t)}|x|_{0,\rho}^2 \le 0$$

on $(0, T) \times X_0$. The functions $u_\eta, -v_\eta$ satisfy (3.349). Therefore, if we can prove that $u_\eta \le v_\eta$ for every $\eta > 0$, we recover $u \le v$ by letting $\eta \to 0$.

The proof basically follows the lines of the proof of Theorem 3.50. Suppose that $u \not\le v$. Let for $\mu, \varepsilon, \delta, \beta > 0$,

$$\Psi(t, s, x, y) := u(t, x) - v(s, y) - \frac{\mu}{t} - \frac{\mu}{s} - \frac{|x - y|_{-1,\rho}^2}{2\varepsilon} - \delta(|x|_{0,\rho}^2 + |y|_{0,\rho}^2) - \frac{(t - s)^2}{2\beta}.$$

For every $n \in \mathbb{N}$ there exist $p_n, q_n \in X_0, a_n, b_n \in \mathbb{R}$ such that $|p_n|_{0,\rho}, |q_n|_{0,\rho}, |a_n|, |b_n| \le 1/n$, and

$$\Psi(t, s, x, y) + \langle B_\rho p_n, x \rangle_{0,\rho} + \langle B_\rho q_n, y \rangle_{0,\rho} + a_n t + b_n s$$

304 3 Viscosity Solutions

has a strict global maximum at $(\bar{t}, \bar{s}, \bar{x}, \bar{y})$ over $(0, T] \times (0, T] \times X_0 \times X_0$. Arguing as in the proof of Theorem 3.50 we obtain

$$\limsup_{\beta \to 0} \limsup_{n \to \infty} \frac{(\bar{t} - \bar{s})^2}{2\beta} = 0 \quad \text{for fixed } \mu, \varepsilon, \delta, \tag{3.350}$$

$$|\bar{x}|_{0,\rho} + |\bar{y}|_{0,\rho} \leq R \quad \text{for some } R, \text{ independently of } \mu, \varepsilon, \delta, \beta, n, \tag{3.351}$$

and

$$\limsup_{\varepsilon \to 0} \limsup_{\delta \to 0} \limsup_{\beta \to 0} \limsup_{n \to \infty} \frac{|\bar{x} - \bar{y}|^2_{-1,\rho}}{2\varepsilon} = 0 \quad \text{for fixed } \mu. \tag{3.352}$$

Therefore, it follows from (3.348)–(3.350) that $0 < \bar{t}$, and $\bar{s} < T$. We now fix $N \in \mathbb{N}$. Defining

$$u_1(t, x) = u(t, x) - \frac{\mu}{t} - \frac{\langle B_\rho Q_N(\bar{x} - \bar{y}), x \rangle_{0,\rho}}{\varepsilon} - \frac{|Q_N(x - \bar{x})|^2_{-1,\rho}}{\varepsilon}$$
$$+ \frac{|Q_N(\bar{x} - \bar{y})|^2_{-1,\rho}}{2\varepsilon} - \delta|x|^2_{0,\rho} + a_n t + \langle B p_n, x \rangle_{0,\rho}$$

and

$$v_1(s, y) = v(s, y) + \frac{\mu}{s} - \frac{\langle B_\rho Q_N(\bar{x} - \bar{y}), y \rangle_{0,\rho}}{\varepsilon} + \frac{|Q_N(y - \bar{y})|^2_{-1,\rho}}{\varepsilon}$$
$$+ \delta|y|^2_{0,\rho} - b_n s - \langle B q_n, y \rangle_{0,\rho},$$

we see that

$$u_1(t, x) - v_1(s, y) - \frac{1}{2\varepsilon}|P_N(x - y)|^2_{-1,\rho} - \frac{1}{2\beta}|t - s|^2$$

has a strict global maximum at $(\bar{t}, \bar{s}, \bar{x}, \bar{y})$. It now follows from Corollary 3.29 and the proof of Theorem 3.27 that for every $\nu > 1$ there exist test functions φ_i, and ψ_i, $i = 1, 2, ...,$ such that

$$u_1(t, x) - \varphi_i(t, x)$$

has a global maximum at some point (t_i, x_i),

$$v_1(s, y) - \psi_i(s, y)$$

has a global minimum at some point (s_i, y_i), and

$$\left(t_i, x_i, u_1(t_i, x_i), (\varphi_i)_t(t_i, x_i), D\varphi_i(t_i, x_i), D^2\varphi_i(t_i, x_i)\right)$$

$$\xrightarrow{k\to\infty} \left(\bar{t}, \bar{x}, u_1(\bar{t}, \bar{x}), \frac{\bar{t}-\bar{s}}{\beta}, \frac{B_\rho(\bar{x}_N - \bar{y}_N)}{\varepsilon}, L_N\right)$$

$$\text{in } \mathbb{R} \times X_0 \times \mathbb{R} \times \mathbb{R} \times X_2 \times \mathcal{L}(X_{-1}, X_1),$$
$$(3.353)$$

$$\left(s_i, y_i, v_1(s_i, y_i), (\psi_i)_t(s_i, y_i), D\psi_i(s_i, y_i), D^2\psi_i(s_i, y_i)\right)$$

$$\xrightarrow{k\to\infty} \left(\bar{s}, \bar{y}, v_1(\bar{s}, \bar{y}), \frac{\bar{t}-\bar{s}}{\beta}, \frac{B_\rho(\bar{x}_N - \bar{y}_N)}{\varepsilon}, M_N\right)$$

$$\text{in } \mathbb{R} \times X_0 \times \mathbb{R} \times \mathbb{R} \times X_2 \times \mathcal{L}(X_{-1}, X_1),$$
$$(3.354)$$

where $L_N = P_N^* L_N P_N$, $M_N = P_N^* M_N P_N$ and

$$\begin{pmatrix} L_N & 0 \\ 0 & -M_N \end{pmatrix} \leq \frac{\nu}{\varepsilon} \begin{pmatrix} B_\rho P_N & -B_\rho P_N \\ -B_\rho P_N & B_\rho P_N \end{pmatrix}. \tag{3.355}$$

Using the definition of a viscosity subsolution we thus obtain

$$\inf_{a\in\Lambda} \left\{ \frac{1}{2} \sum_{k=1}^m \langle (D^2\varphi_i(t_i, x_i) + \frac{2}{\varepsilon} B_\rho Q_N + 2\delta I) S_a^k x_i, S_a^k x_i \rangle_{0,\rho} \right.$$

$$+ \langle A_a x_i, D\varphi_i(t_i, x_i) + \frac{B_\rho Q_N(\bar{x} - \bar{y})}{\varepsilon} + \frac{2B_\rho Q_N(x_i - \bar{x})}{\varepsilon} + 2\delta x_i - B_\rho P_n \rangle_{\langle X_{-1}, X_1 \rangle}$$

$$\left. + f(x_i, a) \right\} - a_n + (\varphi_i)_t(t_i, x_i) \geq \frac{\mu}{T^2}. \tag{3.356}$$

(We remark that the function $\frac{\mu}{t}$ can be modified around 0 so that it is part of a test function.)

We now pass to the limit in (3.356) as $i \to \infty$. We see that by (3.327), for every $a \in \Lambda$

$$\sum_{k=1}^m \langle S_a^k x_i, S_a^k x_i \rangle_0 + 2\langle A_a x_i, x_i \rangle_{\langle X_{-1}, X_1 \rangle} \leq 2\delta K |x_i|_{0,\rho}^2 \to 2\delta K |\bar{x}|_{0,\rho}^2$$

as $i \to \infty$. (In fact one can prove $x_i \to \bar{x}$ in X_1.) Moreover we observe that by (3.346), $B_\rho^{-1/2} D^2\varphi_i(t_i, x_i) B_\rho^{-1/2} \to B_\rho^{-1/2} L_N B_\rho^{-1/2}$ in $\mathcal{L}(X_0)$. Using these, (3.353), and (3.330), i.e. that $\|B_\rho A_a\|_{\mathcal{L}(X_0)}, \|B_\rho^{\frac{1}{2}} S_a^k\|_{\mathcal{L}(X_0)} \leq C$ independently of $a \in \Lambda$, $1 \leq k \leq m$, we obtain upon passing to $\limsup_{i\to\infty}$ in (3.356) that

$$-a_n + \frac{\bar{t} - \bar{s}}{\beta} + \inf_{a \in \Lambda} \left\{ \frac{1}{2} \sum_{k=1}^{m} \langle (L_N + \frac{2}{\varepsilon} B_\rho Q_N) S_a^k \bar{x}, S_a^k \bar{x} \rangle_{0,\rho} \right.$$

$$\left. + \langle A_a \bar{x}, \frac{B_\rho(\bar{x} - \bar{y})}{\varepsilon} \rangle - B_\rho p_n \rangle_{\langle X_{-1}, X_1 \rangle} + f(\bar{x}, a) \right\} + 2\delta K |\bar{x}|_{0,\rho}^2 \geq \frac{\mu}{T^2}. \tag{3.357}$$

We obtain similarly for the supersolution v

$$b_n + \frac{\bar{t} - \bar{s}}{\beta} + \inf_{a \in \Lambda} \left\{ \frac{1}{2} \sum_{k=1}^{m} \langle (M_N - \frac{2}{\varepsilon} B_\rho Q_N) S_a^k \bar{y}, S_a^k \bar{y} \rangle_{0,\rho} \right.$$

$$\left. + \langle A_a \bar{y}, \frac{B_\rho(\bar{x} - \bar{y})}{\varepsilon} \rangle + B_\rho q_n \rangle_{\langle X_{-1}, X_1 \rangle} + f(\bar{y}, a) \right\} - 2\delta K |\bar{y}|_{0,\rho}^2 \leq -\frac{\mu}{T^2}. \tag{3.358}$$

By Hypothesis 3.103 the closures of the sets $\{S_a^k \bar{x} : a \in \Lambda, 1 \leq k \leq m\}$ and $\{S_a^k \bar{y} : a \in \Lambda, 1 \leq k \leq m\}$ are compact in X_0, and hence in X_{-1}. This yields

$$\sup\{|B_\rho^{1/2} Q_N S_a^k \bar{x}|_{0,\rho} + |B_\rho^{1/2} Q_N S_a^k \bar{y}|_{0,\rho} : a \in \Lambda, 1 \leq k \leq m\} \to 0 \tag{3.359}$$

as $N \to \infty$. Moreover, (3.355) implies that

$$\langle L_N S_a^k \bar{x}, S_a^k \bar{x} \rangle_{0,\rho} - \langle M_N S_a^k \bar{y}, S_a^k \bar{y} \rangle_{0,\rho} \leq \frac{\nu}{2\varepsilon} \langle B_\rho S_a^k (\bar{x} - \bar{y}), S_a^k (\bar{x} - \bar{y}) \rangle_{0,\rho}. \tag{3.360}$$

Therefore, subtracting (3.357) from (3.358) and using (3.323), (3.359), and (3.360), we have

$$\inf_{a \in \Lambda} \left\{ -\frac{\nu}{2\varepsilon} \sum_{k=1}^{m} \langle B_\rho S_a^k (\bar{x} - \bar{y}), S_a^k (\bar{x} - \bar{y}) \rangle_{0,\rho} - \frac{1}{\varepsilon} \langle A_a (\bar{x} - \bar{y}), B_\rho(\bar{x} - \bar{y}) \rangle_{\langle X_{-1}, X_1 \rangle} \right\}$$

$$+ a_n + b_n - \omega_R(|\bar{x} - \bar{y}|_0) - 2\delta K (|\bar{x}|_{0,\rho}^2 + |\bar{y}|_{0,\rho}^2) - \sigma(1/N, n) \leq -\frac{2\mu}{T^2}$$

for some local modulus σ. Now, if ν is close to 1, it follows from (3.329) that

$$a_n + b_n + \frac{\bar{\lambda}}{2\varepsilon} |\bar{x} - \bar{y}|_{0,\rho}^2 - \frac{K}{\varepsilon} |\bar{x} - \bar{y}|_{-1,\rho}^2$$

$$-\omega_R(|\bar{x} - \bar{y}|_{0,\rho}) - 2\delta K (|\bar{x}|_{0,\rho}^2 + |\bar{y}|_{0,\rho}^2) - \sigma(1/N, n) \leq -\frac{2\mu}{T^2}. \tag{3.361}$$

Since ω_R is a modulus we have

$$\liminf_{\varepsilon \to 0 \, r \geq 0} \left(\frac{\bar{\lambda}}{2\varepsilon} r^2 - \omega_R(r) \right) = 0. \tag{3.362}$$

Therefore we obtain a contradiction in (3.361) after sending $N \to \infty$, $n \to \infty$, $\beta \to 0$, $\delta \to 0$, $\varepsilon \to 0$ in the above order, and using (3.351), (3.352), and (3.362).   □

3.11.6 The Value Function and Existence of Solutions

In this subsection we show that the value function V defined by (3.325) is the unique viscosity solution of (3.324).

Proposition 3.111 *Assume that Hypotheses 3.103 and 3.104 are satisfied. Then for every $R > 0$ there exists a modulus σ_R such that*

$$|V(t, x) - V(s, y)| \leq \sigma_R(|t - s| + |x - y|_{-1,\rho}), \quad t, s \in [0, T], |x|_{0,\rho}, |y|_{0,\rho} \leq R,$$
(3.363)

and there is $C > 0$ such that

$$|V(t, x)| \leq C \left(1 + |x|_{0,\rho}^{\gamma}\right), \quad t \in [0, T], x \in X_0.$$
(3.364)

Moreover, the dynamic programming principle (2.23) is satisfied.

Proof We only sketch the proof since it is very similar to the proof of Proposition 3.61. Using similar arguments it follows from (3.333), (3.335), (3.336), linearity of the DMZ equation (3.322), and Hypothesis 3.104 that (3.364) holds and there are moduli σ_R^1 such that

$$|J(t, x; a(\cdot)) - J(t, y; a(\cdot))| \leq \sigma_R^1(|x - y|_{-1,\rho}), \quad t \in [0, T], |x|_{0,\rho}, |y|_{0,\rho} \leq R, a(\cdot) \in \mathcal{U}_t,$$
(3.365)

and thus the same inequality is satisfied by V. We now claim that the dynamic programming principle holds. To do this we need to check that the assumptions of Hypothesis 2.12 are satisfied. Parts $(A0)$ and $(A2)$ follow from the definition of a variational solution, properties of stochastic integrals and standard manipulations. Part $(A1)$ was proved in Proposition 3.106 (i.e. Theorem 3.102-(ii)). Part $(A3)$ can be proved similarly to the proof of Proposition 2.16. It thus follows from Theorem 2.24 that for every $x \in X_0, 0 \leq t \leq \eta \leq T$,

$$V(t, x) = \inf_{a(\cdot)\in\mathcal{U}_t} \mathbb{E}\left[\int_t^{\eta} l(Y(s), a(s))ds + V(\eta, Y(\eta))\right].$$
(3.366)

Using (3.366) we again argue as in the proof of Proposition 3.61 using (3.333), (3.337), (3.364), and (3.365) to obtain that there exist moduli σ_R^2 such that

$$|V(t, x) - V(s, x)| \leq \sigma_R^2(|t - s|), \quad t, s \in [0, T], |x|_{0,\rho} \leq R.$$
(3.367)

Obviously (3.365) and (3.367) produce (3.363).   □

Theorem 3.112 *Assume that Hypotheses 3.103 and 3.104 are true. Then the value function V is the unique viscosity solution of the HJB equation (3.324) among functions satisfying*

$$\limsup_{|x|_{0,\rho} \to \infty} \frac{|u(t,x)|}{|x|_{0,\rho}^2} = 0 \quad \text{uniformly for } t \in [0,T],$$

and

$$\lim_{t \uparrow T} |u(t,x) - g(x)| = 0 \quad \text{uniformly on bounded subsets of } X_0.$$

Proof The uniqueness is a consequence of Theorem 3.110 and Proposition 3.111. Therefore it remains to show that V is a viscosity solution of (3.324).

We only consider the supersolution property as the subsolution part is easier. Suppose that $V + (\varphi + \delta(t) |x|_{0,\rho}^2)$ has a global minimum at $(t_0, x_0) \in (0,T) \times X_0$. We need to prove that $x_0 \in X_1$. For every $(t,x) \in (0,T) \times X_0$

$$V(t,x) - V(t_0, x_0) \geq -(\varphi(t,x) - \varphi(t_0, x_0)) - \left(\delta(t) |x|_{0,\rho}^2 - \delta(t_0) |x_0|_{0,\rho}^2\right). \quad (3.368)$$

By the dynamic programming principle, for every $\varepsilon > 0$ there exists an $a_\varepsilon(\cdot) \in \mathcal{U}_{t_0}$ such that, writing $Y_\varepsilon(s)$ for $Y(s; t_0, x_0, a_\varepsilon(\cdot))$, $s \in [t_0, t_0 + \varepsilon]$, we have

$$V(t_0, x_0) + \varepsilon^2 > \mathbb{E}\left[\int_{t_0}^{t_0 + \varepsilon} l(Y_\varepsilon(s), a_\varepsilon(s)) \, ds + V(t_0 + \varepsilon, Y_\varepsilon(t_0 + \varepsilon))\right].$$

In light of Corollary 2.21 and Proposition 3.106, without loss of generality we can assume that all $a_\varepsilon(\cdot)$ are defined on the same reference probability space. We have, by (3.368),

$$\varepsilon^2 - \mathbb{E}\int_{t_0}^{t_0 + \varepsilon} l(Y_\varepsilon(s), a_\varepsilon(s)) \, ds \geq \mathbb{E}\left[V(t_0 + \varepsilon, Y_\varepsilon(t_0 + \varepsilon)) - V(t_0, x_0)\right]$$

$$\geq -\mathbb{E}[\varphi(t_0 + \varepsilon, Y_\varepsilon(t_0 + \varepsilon)) - \varphi(t_0, x_0)] - \mathbb{E}\left[\delta(t_0 + \varepsilon) |Y_\varepsilon(t_0 + \varepsilon)|_{0,\rho}^2 - \delta(t_0) |x_0|_{0,\rho}^2\right]$$

and, by Itô's formula

$$\varepsilon - \mathbb{E}\frac{1}{\varepsilon}\int_{t_0}^{t_0 + \varepsilon} l(Y_\varepsilon(s), \alpha_\varepsilon(s)) \, ds$$

$$\geq -\mathbb{E}\frac{1}{\varepsilon}\int_{t_0}^{t_0 + \varepsilon}\left[\varphi_t(s, Y_\varepsilon(s)) + \left\langle A_{a_\varepsilon(s)}Y_\varepsilon(s), D\varphi(s, Y_\varepsilon(s))\right\rangle_{\langle X_{-1}, X_1\rangle}\right.$$

$$\left. + \frac{1}{2}\sum_{k=1}^{m}\left\langle D^2\varphi(s, Y_\varepsilon(s)) S_{a_\varepsilon(s)}^k Y(s), S_{a_\varepsilon(s)}^k Y_\varepsilon(s)\right\rangle_{0,\rho}\right] ds \quad (3.369)$$

$$-\mathbb{E}\frac{1}{\varepsilon}\int_{t_0}^{t_0+\varepsilon}\left[\delta'(s)\,|Y_\varepsilon(s)|^2_{0,\rho}+2\delta(s)\left[\langle A_{a_\varepsilon(s)}Y_\varepsilon(s)\,,\,Y_\varepsilon(s)\rangle_{\langle X_{-1},X_1\rangle}\right.\right.$$
$$\left.\left.+\frac{1}{2}\sum_{k=1}^{m}\langle S^k_{a_\varepsilon(s)}Y_\varepsilon(s)\,,\,S^k_{a_\varepsilon(s)}Y_\varepsilon(s)\rangle_{0,\rho}\right]\right]ds.$$

By (3.327) we have

$$-2\delta(s)\left[\langle A_{a_\varepsilon(s)}Y_\varepsilon(s)\,,\,Y_\varepsilon(s)\rangle_{\langle X_{-1},X_1\rangle}+\frac{1}{2}\sum_{k=1}^{m}\langle S^k_{a_\varepsilon(s)}Y_\varepsilon(s)\,,\,S^k_{a_\varepsilon(s)}Y_\varepsilon(s)\rangle_{0,\rho}\right]$$
$$\geq 2\delta(s)\left[\bar\lambda\,|Y_\varepsilon(s)|^2_{1,\rho}-K\,|Y_\varepsilon(s)|^2_{0,\rho}\right].$$

Moreover,

$$|\varphi_t(s,Y_\varepsilon(s))|+\left|\langle A_{a_\varepsilon(s)}Y_\varepsilon(s)\,,\,D\varphi(s,Y_\varepsilon(s))\rangle_{\langle X_{-1},X_1\rangle}\right|\leq C_1(1+|Y_\varepsilon(s)|^2_{0,\rho}\,,$$

$$\left|\sum_{k=1}^{m}\langle D^2\varphi(s,Y_\varepsilon(s))\,S^k_{a_\varepsilon(s)}Y_\varepsilon(s)\,,\,S^k_{a_\varepsilon(s)}Y_\varepsilon(s)\rangle_{0,\rho}\right|\leq C_2\,|Y_\varepsilon(s)|^2_0\,.$$

Therefore, using Hypothesis 3.104, (3.333), and $\delta(s)\geq\gamma>0$ for s close to t_0 for some $\gamma>0$, we obtain

$$2\bar\lambda\gamma\mathbb{E}\frac{1}{\varepsilon}\int_{t_0}^{t_0+\varepsilon}|Y_\varepsilon(s)|^2_{1,\rho}\,ds\leq C_3\left[1+\mathbb{E}\frac{1}{\varepsilon}\int_{t_0}^{t_0+\varepsilon}|Y_\varepsilon(s)|^2_{0,\rho}\,ds\right]\leq C_4.$$

Take now $\varepsilon=1/n$ and set $Y_n(s):=Y(s;t_0,x_0,a_{1/n}(\cdot))$. The above inequality yields

$$n\int_{t_0}^{t_0+1/n}\mathbb{E}\,|Y_n(s)|^2_1\,ds\leq C_5$$

so that, along a sequence $t_n\in(t_0,t_0+1/n)$,

$$\mathbb{E}\,|Y_n(t_n)|^2_{1,\rho}\leq C_5$$

and thus, along a subsequence, still denoted by t_n, we have

$$Y_n(t_n)\rightharpoonup\bar Y$$

weakly in $L^2(\Omega;X_1)$ for some $\bar Y\in L^2(\Omega;X_1)$. This also clearly implies weak convergence in $L^2(\Omega;X_0)$. However, by (3.339), $Y_n(t_n)\to x_0$ strongly (and weakly) in $L^2(\Omega;X_0)$. Thus $\bar Y=x_0\in X_1$.

Having established that $x_0\in X_1$, we now go back to (3.369), use the properties of test functions and estimates of Proposition 3.107 to obtain

$$\varepsilon \geq -\mathbb{E}\frac{1}{\varepsilon}\int_{t_0}^{t_0+\varepsilon}\left[\varphi_t\left(t_0, x_0\right) + \left\langle A_{a_\varepsilon(s)}x_0, D\varphi\left(t_0, x_0\right)\right\rangle_{\langle X_{-1}, X_1\rangle}\right.$$

$$\left. + \frac{1}{2}\sum_{k=1}^{m}\left\langle D^2\varphi\left(t_0, x_0\right) S_{a_\varepsilon(s)}^k x_0, S_{a_\varepsilon(s)}^k x_0\right\rangle_{0,\rho}\right]ds$$

$$-\mathbb{E}\frac{1}{\varepsilon}\int_{t_0}^{t_0+\varepsilon}\left[\delta'\left(t_0\right)|x_0|_{0,\rho}^2 + 2\delta(t_0)\left[\left\langle A_{a_\varepsilon(s)}x_0, x_0\right\rangle_{\langle X_{-1}, X_1\rangle}\right.\right.$$

$$\left.\left. + \frac{1}{2}\sum_{k=1}^{m}\left\langle S_{a_\varepsilon(s)}^k x_0, S_{a_\varepsilon(s)}^k x_0\right\rangle_{0,\rho}\right] - l\left(x_0, \alpha_\varepsilon(s)\right)\right]ds - \gamma(\varepsilon),$$

$$\geq \mathbb{E}\frac{1}{\varepsilon}\int_{t_0}^{t_0+\varepsilon}\left[-\psi_t(t, x) + \inf_{a\in\Lambda}\left\{\frac{1}{2}\sum_{k=1}^{m}\left\langle -D^2\psi(t, x)S_a^k x, S_a^k x\right\rangle_{0,\rho}\right.\right.$$

$$\left.\left. + \left\langle A_a x, -D\psi(t, x)\right\rangle_{\langle X_{-1}, X_1\rangle} + f(x, a)\right\}\right]ds - \gamma(\varepsilon),$$

where $\lim_{\varepsilon\to 0}\gamma(\varepsilon) = 0$. It remains to let $\varepsilon \to 0$. We refer the reader to the proof of Theorem 5.4 in [323] for more details. $\qquad\square$

3.12 HJB Equations for Boundary Control Problems

In this section we discuss how the theory of viscosity solutions can be applied to solve HJB equations coming from the stochastic boundary control problems discussed in Sect. 2.6.2. We will only consider time-independent problems. Suppose that H is a real, separable Hilbert space and A is an operator in H satisfying the following hypothesis.

Hypothesis 3.113 $A : D(A) \subset H \to H$ is a (densely defined) self-adjoint operator, there exists $a > 0$ such that $\langle Ax, x\rangle \leq -a|x|^2$ for all $x \in D(A)$ and A^{-1} is compact.

Hypothesis 3.113 implies in particular that A is the infinitesimal generator of an analytic semigroup with compact resolvent satisfying $\|e^{tA}\| \leq e^{-at}$ for all $t \geq 0$ and that there is an orthonormal basis of H composed of eigenvectors of A such that the corresponding sequence of (negative) eigenvalues diverges to $-\infty$ as $n \to \infty$. Moreover, the fractional powers $(-A)^\gamma, \gamma > 0$, are well defined, and if $\gamma \in (0, 1]$ and $\alpha \in (0, \gamma)$, a well-known interpolation inequality (see e.g. [479], pp. 73–74) gives us that for every $\sigma > 0$ there exists a $C_\sigma > 0$ such that

$$|(-A)^\alpha x| \leq \sigma|(-A)^\gamma x| + C_\sigma|x|, \qquad \text{for every } x \in D((-A)^\gamma). \tag{3.370}$$

The HJB equations introduced in Sect. 2.6.2 (see (2.99)) have the form

$$\lambda v - \langle Ax, Dv\rangle + F(x, Dv, D^2v) = 0, \quad x \in H, \tag{3.371}$$

where $F : Z \subset H \times H \times S(H) \to \mathbb{R}$, $\lambda > 0$. In particular, $F(x, p, X)$ may only be defined if $p \in D((-A)^{\beta})$ for some $\beta > 0$ and may be undefined if X is not of trace class. The unboundedness in the first-order terms comes from the boundary control term rewritten as a distributed control term, and the unboundedness in the second derivative terms comes from noise with non-nuclear covariance in the control problem. Such second-order unboundedness has not been discussed so far in this book and indeed it is not easy to handle by the viscosity solution methods. Here we suggest one way to do it. The idea is the following. To deal with the unboundedness in the first and second derivatives we introduce a change of variables $x = (-A)^{\frac{\beta}{2}} y, \beta > 0$. Then the function $u(y) := v((-A)^{\frac{\beta}{2}} y)$ should formally solve

$$\lambda u - \langle Ay, Dv \rangle + F((-A)^{\frac{\beta}{2}} y, (-A)^{-\frac{\beta}{2}} Dv, (-A)^{-\frac{\beta}{2}} D^2 v (-A)^{-\frac{\beta}{2}}) = 0. \quad (3.372)$$

This equation contains fewer unbounded terms and is easier to handle in spite of the additional difficulty created by the presence of the new unbounded term $(-A)^{\frac{\beta}{2}} y$. We will define a viscosity solution of (3.371) to be a function v such that $u(\cdot) \overset{def}{=} v((-A)^{\frac{\beta}{2}} \cdot)$ is a viscosity solution of (3.372). We will make this idea rigorous in the next section. The definition is meaningful, indeed, when (3.371) comes from a stochastic boundary control problem, v and u can be respectively characterized as the value functions of their control problems.

3.12.1 Definition of a Viscosity Solution

We first consider the following HJB equation

$$\lambda u - \langle Ay, Du \rangle + G(y, Du, D^2 u) = 0, \qquad y \in H, \quad (3.373)$$

where $G : D((-A)^{\frac{\beta}{2}}) \times D((-A)^{\frac{\beta}{2}}) \times S(H) \to \mathbb{R}$.

Definition 3.114 We say that a function ψ is a test function if $\psi(x) = \varphi(x) + \delta|x|^2$, where $\delta > 0$ and

 (i) $\varphi \in C^2(H)$ and is weakly sequentially lower semicontinuous on H.
 (ii) $D\varphi \in UC(H, H) \cap UC \left(D((-A)^{\frac{1}{2}-\varepsilon}), D((-A)^{\frac{1}{2}}) \right)$ for some $\varepsilon = \varepsilon(\varphi) > 0$.
 (iii) $D^2\varphi \in UC_b(H, S(H))$.

Definition 3.115 We say that a function $w : H \to \mathbb{R}$ is a viscosity subsolution of (3.373) if w is weakly sequentially upper semicontinuous on H, and whenever $w - \psi$ has a local maximum at x for a test function ψ, then

$$x \in D((-A)^{\frac{1}{2}})$$

and

$$\lambda w(x) + \langle (-A)^{\frac{1}{2}} x, (-A)^{\frac{1}{2}} D\varphi(x) \rangle + 2\delta |(-A)^{\frac{1}{2}} x|^2 + G\left(x, D\psi(x), D^2\psi(x) \right) \le 0.$$

We say that w is a viscosity supersolution of (3.373) if w is weakly sequentially lower semicontinuous on H, and whenever $w + \psi$ has a local minimum at x for a test function ψ, then

$$x \in D((-A)^{\frac{1}{2}})$$

and

$$\lambda w(x) - \langle (-A)^{\frac{1}{2}} y, (-A)^{\frac{1}{2}} D\varphi(x) \rangle - 2\delta |(-A)^{\frac{1}{2}} x|^2 + G\left(x, -D\psi(x), -D^2\psi(x) \right) \ge 0.$$

We say that w is a viscosity solution of (3.373) if it is both a viscosity subsolution and a supersolution.

Suppose now that F from (3.371) is such that $F : H \times D((-A)^{\beta}) \times (-A)^{-\frac{\beta}{2}} S(H)$ $(-A)^{-\frac{\beta}{2}} \to \mathbb{R}$. We define

$$G_F(z, p, S) \overset{def}{=} F\left((-A)^{\frac{\beta}{2}} z, (-A)^{-\frac{\beta}{2}} p, (-A)^{-\frac{\beta}{2}} S(-A)^{-\frac{\beta}{2}} \right). \tag{3.374}$$

Definition 3.116 A bounded continuous function $v : H \to \mathbb{R}$ is said to be a viscosity solution of Eq. (3.371) if the function

$$u(y) \overset{def}{=} v((-A)^{\frac{\beta}{2}} y)$$

is a viscosity solution of the equation

$$\lambda u - \langle Ay, Du \rangle + G_F(y, Du, D^2 u) = 0, \qquad y \in H. \tag{3.375}$$

Similarly we define a viscosity subsolution and a supersolution of (3.371).

We remark that the function v is uniquely determined once u has been characterized on $D((-A)^{\frac{\beta}{2}})$.

3.12.2   Comparison and Existence Theorem

For $\gamma > 0$ we denote by $H_{-\gamma}$ the completion of H in the norm $|x|_{-\gamma} = |(-A)^{-\frac{\gamma}{2}} x|$, and $D((-A)^{\frac{\gamma}{2}})$ is equipped with the norm $|x|_{\gamma} = |(-A)^{\frac{\gamma}{2}} x|$. For $N > 2$ let H_N be finite-dimensional subspaces of H generated by eigenvectors of $(-A)^{-1}$ corresponding to the eigenvalues which are greater than or equal to $1/N$. Denote by P_N the orthogonal projection in H_{-1} onto H_N, $Q_N = I - P_N$, and $H_N^{\perp} = Q_N H$. P_N and

Q_N are also orthogonal projections in H. We then have an orthogonal decomposition $H = H_N \times H_N^\perp$ and we will write $x = (x_N, x_N^\perp) = (P_N x, Q_N x)$.

We assume:

Hypothesis 3.117

(i) There exists a $\beta \in (0, 1)$ such that the function $G : D((-A)^{\frac{\beta}{2}}) \times D((-A)^{\frac{\beta}{2}}) \times S(H) \to \mathbb{R}$ is uniformly continuous (in the topology of $D((-A)^{\frac{\beta}{2}}) \times D((-A)^{\frac{\beta}{2}}) \times S(H)$) on bounded sets of $D((-A)^{\frac{\beta}{2}}) \times D((-A)^{\frac{\beta}{2}}) \times S(H)$.

(ii) $G(y, p, S_1) \le G(y, p, S_2)$ if $S_1 \ge S_2$, for all $y, p \in D((-A)^{\frac{\beta}{2}})$.

(iii) There exists a modulus ρ such that

$$|G(y, p, S_1) - G(y, q, S_2)|$$
$$\le \rho\left((1 + |(-A)^{\frac{\beta}{2}} y|)|(-A)^{\frac{\beta}{2}}(p - q)| + (1 + |(-A)^{\frac{\beta}{2}} y|^2)\|S_1 - S_2\| \right)$$

for all $y, p, q \in D((-A)^{\frac{\beta}{2}})$ and $S_1, S_2 \in S(H)$.

(iv) There exist $0 < \eta < 1 - \beta$ and a modulus ω such that, for all $N > 2, \varepsilon > 0$,

$$G\left(x, \frac{(-A)^{-\eta}(x - y)}{\varepsilon}, Z \right) - G\left(y, \frac{(-A)^{-\eta}(x - y)}{\varepsilon}, Y \right)$$
$$\ge -\omega\left(|(-A)^{\frac{\beta}{2}}(x - y)|\left(1 + \frac{|(-A)^{\frac{\beta}{2}}(x - y)|}{\varepsilon} \right) \right)$$

for all $x, y \in D((-A)^{\frac{\beta}{2}})$ and $Z, Y \in S(H), Z = P_N Z P_N, Y = P_N Y P_N$ such that

$$\begin{pmatrix} Z & 0 \\ 0 & -Y \end{pmatrix} \le \frac{3}{\varepsilon} \begin{pmatrix} (-A)^{-\eta} P_N & -(-A)^{-\eta} P_N \\ -(-A)^{-\eta} P_N & (-A)^{-\eta} P_N \end{pmatrix}. \tag{3.376}$$

(v) For every $R < +\infty, |\lambda| \le R, p, x \in D((-A)^{\frac{\beta}{2}})$

$$\sup\left\{ |G(x, p, S + \lambda Q_N) - G(x, p, S)| : \|S\| \le R, S = P_N S P_N \right\} \to 0. \tag{3.377}$$

as $N \to \infty$.

Some of the conditions of Hypothesis 3.117 can be weakened. By the properties of moduli, Hypothesis 3.117-(iii) guarantees that there exists a constant C such that, for every y, p, S,

$$|G(y, p, S)| \le C\left(1 + (1 + |(-A)^{\frac{\beta}{2}} y|)|(-A)^{\frac{\beta}{2}} p| + (1 + |(-A)^{\frac{\beta}{2}} y|^2)\|S\| \right) + |G(y, 0, 0)|, \tag{3.378}$$

and conditions (i), (iv) of Hypothesis 3.117 imply that there is C_1 such that

$$|G(y, 0, 0)| \leq C_1 \left(1 + \left|(-A)^{\frac{\beta}{2}} y\right|\right). \tag{3.379}$$

Theorem 3.118 *Let Hypotheses 3.113 and 3.117 be satisfied. Then:*
Comparison: Let u, $-v \leq M$ for some constant M. If u is a viscosity subsolution of (3.373) and v is a viscosity supersolution of (3.373) then $u \leq v$ on H. Moreover, if u is a viscosity solution then

$$|u(x) - u(y)| \leq m(|(-A)^{-\frac{\eta}{2}}(x - y)|) \tag{3.380}$$

for all $x, y \in H$ and some modulus m, where η is the constant in (iv).
Existence: If

$$\sup_{x \in D(A^{\frac{\beta}{2}})} |G(x, 0, 0)| = K < \infty, \tag{3.381}$$

then there exists a unique viscosity solution $u \in UC_b(H_{-\eta})$ of (3.373).

Proof Comparison. Let $\varepsilon, \delta > 0$. We set

$$\Phi(x, y) = u(x) - v(y) - \frac{|(-A)^{-\frac{\eta}{2}}(x - y)|^2}{2\varepsilon} - \frac{\delta}{2}|x|^2 - \frac{\delta}{2}|y|^2.$$

Since $u - v$ is bounded from above and weakly sequentially upper-semicontinuous in $H \times H$, Φ must attain its maximum at some point $(\bar{x}, \bar{y}) \in D((-A)^{\frac{1}{2}}) \times D((-A)^{\frac{1}{2}})$ (which can be assumed to be strict by subtracting, for instance, $\mu(|(-A)^{-1}(x - \bar{x})|^2 + |(-A)^{-1}(y - \bar{y})|^2)$ and then letting $\mu \to 0$). Moreover, arguing similarly as in the proof of Theorem 3.56, we have

$$\lim_{\delta \to 0} \left(\delta|\bar{x}|^2 + \delta|\bar{y}|^2\right) = 0 \quad \text{for every fixed } \varepsilon > 0, \tag{3.382}$$

$$\lim_{\varepsilon \to 0} \limsup_{\delta \to 0} \left(\frac{|(-A)^{-\frac{\eta}{2}}(\bar{x} - \bar{y})|^2}{\varepsilon}\right) = 0. \tag{3.383}$$

Then (see the proof of Theorem 3.50), defining

$$u_1(x) = u(x) - \frac{\langle x, Q_N(-A)^{-\eta} Q_N(\bar{x} - \bar{y})\rangle}{\varepsilon} + \frac{\langle Q_N(-A)^{-\eta} Q_N(\bar{x} - \bar{y}), \bar{x} - \bar{y}\rangle}{2\varepsilon}$$
$$- \frac{|(-A)^{-\frac{\eta}{2}} Q_N(x - \bar{x})|^2}{\varepsilon} - \frac{\delta}{2}|x|^2,$$

$$v_1(y) = v(y) - \frac{\langle y, Q_N(-A)^{-\eta} Q_N(\bar{x} - \bar{y})\rangle}{\varepsilon} + \frac{|(-A)^{-\frac{\eta}{2}} Q_N(y - \bar{y})|^2}{\varepsilon} + \frac{\delta}{2}|y|^2,$$

it follows that the function

$$\widetilde{\Phi}(x, y) \overset{def}{=} u_1(x) - v_1(y) - \frac{|(-A)^{-\frac{\eta}{2}} P_N(x - y)|^2}{2\varepsilon}$$

always satisfies $\widetilde{\Phi} \leq \Phi$ and $\widetilde{\Phi}$ attains a strict global maximum at $(\bar{x}, \bar{y})$, where $\widetilde{\Phi}(\bar{x}, \bar{y}) = \Phi(\bar{x}, \bar{y})$. Using Corollary 3.28 (with $B = (-A)^{-\eta}$) we thus obtain functions $\varphi_n, -\psi_n$ satisfying conditions (i)–(iii) of Definition 3.114 such that

$$u_1(x) - \varphi_n(x)$$

has a global maximum at some point x^n,

$$v_1(y) + \psi_n(y)$$

has a global minimum at some point y^n, and

$$\left(x^n, u_1(x^n), D\varphi_n(x^n), D^2\varphi_n(x^n)\right) \xrightarrow{n \to \infty} \left(\bar{x}, u_1(\bar{x}), \frac{(-A)^{-\eta} P_N(\bar{x} - \bar{y})}{\varepsilon}, Z_N\right)$$

$$\text{in } H \times \mathbb{R} \times D(-A) \times \mathcal{L}(H, H), \quad (3.384)$$

$$\left(y^n, v_1(y^n), -D\psi_n(y^n), -D^2\psi_n(y^n)\right) \xrightarrow{n \to \infty} \left(\bar{y}, v(\bar{y}), \frac{(-A)^{-\eta} P_N(\bar{x} - \bar{y})}{\varepsilon}, Y_N\right)$$

$$\text{in } H \times \mathbb{R} \times D(-A) \times \mathcal{L}(H, H),$$
$$(3.385)$$

for some $Z_N, Y_N \in S(H)$ such that $Z_N = P_N Z_N P_N$, $Y_N = P_N Y_N P_N$, they satisfy (3.376) and $\|Z_N\| + \|Y_N\| \leq C_\varepsilon$ for some constant C_ε.

Therefore, by the definition of viscosity subsolution, $x^n \in D((-A)^{\frac{1}{2}})$ and

$$\lambda u(x^n) + \left\langle (-A)^{\frac{1}{2}} x^n, (-A)^{\frac{1}{2}} D\varphi_n(x^n) + \frac{(-A)^{\frac{1}{2}-\eta} Q_N(\bar{x} - \bar{y})}{\varepsilon} \right.$$

$$\left. + \frac{2(-A)^{\frac{1}{2}-\eta} Q_N(x^n - \bar{x})}{\varepsilon} \right\rangle + \delta |(-A)^{\frac{1}{2}} x^n|^2$$

$$+ G\left(x^n, D\varphi_n(x^n) + \frac{(-A)^{-\eta} Q_N(\bar{x} - \bar{y})}{\varepsilon} + \frac{2(-A)^{-\eta} Q_N(x^n - \bar{x})}{\varepsilon} + \delta x^n, \right.$$

$$\left. D^2\varphi_n(x^n) + \frac{2\|(-A)^{-\eta}\| Q_N}{\varepsilon} + \delta I \right) \leq 0. \qquad (3.386)$$

Thus, using (3.384), (3.378), (3.379), and (3.370), it follows from (3.386) that $|(-A)^{\frac{1}{2}} x^n|$ are bounded independently of n which implies, thanks to (3.384), that

$$(-A)^{\frac{1}{2}}x^n \rightharpoonup (-A)^{\frac{1}{2}}\bar{x} \quad \text{as } n \to +\infty. \tag{3.387}$$

Since $(-A)^{\frac{\beta-1}{2}}$ and $(-A)^{-\frac{\eta}{2}}$ are compact we conclude that, as $n \to +\infty$,

$$(-A)^{\frac{\beta}{2}}x^n = (-A)^{\frac{\beta-1}{2}}((-A)^{\frac{1}{2}}x^n) \to (-A)^{\frac{\beta}{2}}x \quad \text{and} \quad (-A)^{\frac{1-\eta}{2}}x^n \to (-A)^{\frac{1-\eta}{2}}\bar{x}. \tag{3.388}$$

Using (3.384), (3.387), (3.388), and the weak sequential lower semicontinuity of the norm we thus obtain

$$\left\langle (-A)^{\frac{1-\eta}{2}}\bar{x}, \frac{(-A)^{\frac{1-\eta}{2}}(\bar{x}-\bar{y})}{\varepsilon} \right\rangle + \delta|(-A)^{\frac{1}{2}}\bar{x}|^2$$

$$\leq \liminf_{n\to\infty}\left[\left\langle (-A)^{\frac{1}{2}}x^n, (-A)^{\frac{1}{2}}D\varphi_n(x^n) + \frac{(-A)^{\frac{1}{2}-\eta}Q_N(\bar{x}-\bar{y})}{\varepsilon} \right. \right.$$

$$\left. \left. + \frac{2(-A)^{\frac{1}{2}-\eta}Q_N(x^n-\bar{x})}{\varepsilon} \right\rangle + \delta|(-A)^{\frac{1}{2}}x^n|^2 \right]$$

and then letting $n \to \infty$ in (3.386) yields

$$\lambda u(\bar{x}) + \left\langle (-A)^{\frac{1-\eta}{2}}\bar{x}, \frac{(-A)^{\frac{1-\eta}{2}}(\bar{x}-\bar{y})}{\varepsilon} \right\rangle + \delta|(-A)^{\frac{1}{2}}\bar{x}|^2$$

$$+G\left(\bar{x}, \frac{(-A)^{-\eta}(\bar{x}-\bar{y})}{\varepsilon} + \delta\bar{x}, Z_N + \frac{2\|(-A)^{-\eta}\|Q_N}{\varepsilon} + \delta I \right) \leq 0. \tag{3.389}$$

Using Hypothesis 3.117-(iii) we have

$$G\left(\bar{x}, \frac{(-A)^{-\eta}(\bar{x}-\bar{y})}{\varepsilon}, Z_N + \frac{2\|(-A)^{-\eta}\|Q_N}{\varepsilon} \right) - \rho\left(c\delta(1+|(-A)^{\frac{\beta}{2}}\bar{x}|^2) \right)$$

$$\leq G\left(\bar{x}, \frac{(-A)^{-\eta}(\bar{x}-\bar{y})}{\varepsilon} + \delta\bar{x}, Z_N + \frac{2\|(-A)^{-\eta}\|Q_N}{\varepsilon} + \delta I \right) \tag{3.390}$$

for some constant $c > 0$. Now, given $\tau > 0$, let K_τ be such that $\rho(s) \leq \tau + K_\tau s$. Applying (3.370) with $\alpha = \beta/2$ and $\gamma = 1/2$ we obtain

$$\rho\left(c\delta(1+|(-A)^{\frac{\beta}{2}}\bar{x}|^2) \right) \leq \delta|(-A)^{\frac{1}{2}}\bar{x}|^2 + \delta C_\tau|\bar{x}|^2 + \tau + K_\tau c\delta$$

for some constant $C_\tau > 0$ independent of δ and ε. It then follows from (3.382) that

$$\limsup_{\delta\to 0}\left(\rho\left(c\delta(1+|(-A)^{\frac{\beta}{2}}\bar{x}|^2) \right) - \delta|(-A)^{\frac{1}{2}}\bar{x}|^2 \right) \leq 0. \tag{3.391}$$

Using (3.390), (3.391) and (3.377) in (3.389) we thus obtain

$$
\lambda u(\bar{x}) + \left\langle (-A)^{\frac{1-\eta}{2}} \bar{x}, \frac{(-A)^{\frac{1-\eta}{2}}(\bar{x} - \bar{y})}{\varepsilon} \right\rangle + G\left(\bar{x}, \frac{(-A)^{-\eta}(\bar{x} - \bar{y})}{\varepsilon}, Z_N \right) \tag{3.392}
$$
$$
\leq \omega_1(\varepsilon, \delta; N) + \omega_2(\varepsilon; \delta),
$$

where $\lim_{N \to \infty} \omega_1(\varepsilon, \delta; N) = 0$, $\lim_{\delta \to 0} \omega_2(\varepsilon; \delta) = 0$. Similarly we obtain

$$
\lambda v(\bar{y}) + \left\langle (-A)^{\frac{1-\eta}{2}} \bar{y}, \frac{(-A)^{\frac{1-\eta}{2}}(\bar{x} - \bar{y})}{\varepsilon} \right\rangle + G\left(\bar{y}, \frac{(-A)^{-\eta}(\bar{x} - \bar{y})}{\varepsilon}, Y_N \right) \tag{3.393}
$$
$$
\geq -\omega_1(\varepsilon, \delta; N) - \omega_2(\varepsilon; \delta).
$$

We subtract (3.393) from (3.392), use Hypothesis 3.117-(iv), and let $N \to +\infty$ to conclude that

$$
\lambda(u(\bar{x}) - v(\bar{y})) \leq \omega\left(|(-A)^{\frac{\beta}{2}}(\bar{x} - \bar{y})| \left(1 + \frac{|(-A)^{\frac{\beta}{2}}(\bar{x} - \bar{y})|}{\varepsilon} \right) \right)
$$
$$
- \frac{|(-A)^{\frac{1-\eta}{2}}(\bar{x} - \bar{y})|^2}{\varepsilon} + 2\omega_2(\varepsilon; \delta).
$$

By (3.370), for every $\sigma > 0$

$$
|(-A)^{\frac{\beta}{2}}(\bar{x} - \bar{y})| \leq \sigma |(-A)^{\frac{1-\eta}{2}}(\bar{x} - \bar{y})| + C_\sigma |(-A)^{-\frac{\eta}{2}}(\bar{x} - \bar{y})|. \tag{3.394}
$$

Since for every $\alpha > 0$, $\omega(s) \leq \alpha/2 + K_\alpha s$, if σ is sufficiently small, we obtain after elementary calculations

$$
\lambda(u(\bar{x}) - v(\bar{y})) \leq \alpha + \tilde{K}_\alpha \frac{|(-A)^{-\frac{\eta}{2}}(\bar{x} - \bar{y})|^2}{\varepsilon} + 2\omega_2(\varepsilon; \delta). \tag{3.395}
$$

By (3.383) this implies

$$
\limsup_{\varepsilon \to 0} \limsup_{\delta \to 0} (u(\bar{x}) - v(\bar{y})) \leq \frac{\alpha}{\lambda}
$$

for all $\alpha > 0$, which gives $u \leq v$ in H, since for all $x \in H$ we have

$$
\Phi(x, x) \leq \Phi(\bar{x}, \bar{y}) \leq u(\bar{x}) - v(\bar{y}).
$$

If u is a solution, we can set $u = v$ in the proof to obtain that for all $x, y \in H$

$$u(x) - u(y) - \frac{|(-A)^{-\frac{\eta}{2}}(x - y)|^2}{2\varepsilon} = \lim_{\delta \to 0} \Phi(x, y) \leq \lim_{\delta \to 0} \sup (u(\bar{x}) - u(\bar{y})) \leq \rho_1(\varepsilon)$$

for some modulus ρ_1 in light of (3.395). This proves (3.380).

Existence. The existence of a viscosity solution will be proved by the method of finite-dimensional approximations similar to that of Sect. 3.7. We consider for $N > 2$ the approximating equations

$$\lambda u_N - \langle Ax, Du_N \rangle + G(x, Du_N, D^2 u_N) = 0 \quad \text{in } H_N. \tag{3.396}$$

We see that for every $\gamma > 0$, $(-A)^\gamma x = P_N(-A)^\gamma x$, $(-A)^{-\gamma} x = P_N(-A)^{-\gamma} x$ for $x \in H_N$, and thus (3.396) satisfies Hypotheses 3.113 and 3.117 with constants and moduli independent of N. Since $\underline{u}(x) = -K/\lambda$ is a viscosity subsolution and $\bar{u}(x) = K/\lambda$ is a viscosity supersolution of (3.396), it follows from the finite-dimensional Perron's method that (3.396) has a (unique) bounded viscosity solution u_N such that $\|u_N\|_0 \leq K/\lambda$.

We will prove that there exists a modulus $\tilde{\sigma}_\eta$ independent of N such that

$$|u_N(x) - u_N(y)| \leq \tilde{\sigma}_\eta(|x - y|_{-\eta})$$

for all $x, y \in H_N$. To do this we adapt the technique of Sect. 3.7.

For every $\varepsilon > 0$ let K_ε be such that $\omega(r) \leq \lambda \varepsilon/2 + K_\varepsilon r$. For $L > K/\lambda + 1$ we set

$$\psi_L(r) = 2Lr^{\frac{1}{2L}}.$$

The function $\psi_L \in C^2(0, \infty)$ is increasing, concave, $\psi_L'(r) \geq 1$ for $0 < r \leq 1$, $\psi_L(0) = 0$, $\psi_L(1) > 2(K/\lambda + 1)$, and

$$\psi_L(r) > L\left(\psi_L'(r)r + r\right) \quad \text{for } 0 \leq r \leq 1. \tag{3.397}$$

We will show that for every $\varepsilon > 0$ there exists an $L = L_\varepsilon$ such that

$$u_N(x) - u_N(y) \leq \psi_L(|(-A)^{-\frac{\eta}{2}}(x - y)|) + \varepsilon \quad \text{for every } x, y \in H_N. \tag{3.398}$$

Set $\Delta = \{(x, y) \in H \times H : |(-A)^{-\frac{\eta}{2}}(x - y)| < 1\}$. It is clear from the properties of ψ_L that, for $(x, y) \notin \Delta$, (3.398) is always satisfied independently of L. Assume now by contradiction that (3.398) is false. Then, for any $L > \frac{K}{\lambda} + 1$ we have, for small $\delta > 0$,

$$\sup_{(x,y) \in H_N \times H_N} \left(u_N(x) - u_N(y) - \psi_L(|(-A)^{-\frac{\eta}{2}}(x - y)|) - \varepsilon - \frac{\delta}{2}|x|^2 - \frac{\delta}{2}|y|^2 \right) > 0 \tag{3.399}$$

and is attained at $(\bar{x}, \bar{y}) \in \Delta$ such that $\bar{x} \neq \bar{y}$. Denote $s = |(-A)^{-\frac{\eta}{2}}(\bar{x} - \bar{y})|$.

Repeating the arguments from the proof of Proposition 3.81 that led to (3.207) and then the arguments from the just finished proof of comparison we obtain that there exist $Z, Y \in S(H_N)$ such that

$$\begin{pmatrix} Z & 0 \\ 0 & -Y \end{pmatrix} \le \frac{2\psi'_L(s)}{s} \begin{pmatrix} (-A)^{-\eta}P_N & -(-A)^{-\eta}P_N \\ -(-A)^{-\eta}P_N & (-A)^{-\eta}P_N \end{pmatrix}$$

and

$$\lambda(u_N(\bar{x}) - u_N(\bar{y})) \le -\frac{\psi'_L(s)}{s}|(-A)^{\frac{1-\eta}{2}}(\bar{x}-\bar{y})|^2 + G(\bar{y}, \frac{\psi'_L(s)}{s}(-A)^{-\eta}(\bar{x}-\bar{y}), Y)$$

$$-G(\bar{x}, \frac{\psi'_L(s)}{s}(-A)^{-\eta}(\bar{x}-\bar{y}), Z) + \rho(L;\delta)$$

$$\le -\frac{\psi'_L(s)}{s}|(-A)^{\frac{1-\eta}{2}}(\bar{x}-\bar{y})|^2 + \frac{\lambda\varepsilon}{4}$$

$$+K_\varepsilon\left(|(-A)^{\frac{\beta}{2}}(\bar{x}-\bar{y})|\left(1 + \frac{\psi'_L(s)}{s}|(-A)^{\frac{\beta}{2}}(\bar{x}-\bar{y})|\right)\right) + \rho(L;\delta),$$

where $\lim_{\delta\to 0}\rho(L;\delta) = 0$. Therefore using (3.394) with sufficiently small σ it follows that

$$\lambda(u_N(\bar{x}) - u_N(\bar{y})) \le \frac{\lambda\varepsilon}{2} + C_\varepsilon(\psi'_L(s)s + s) + \rho(L;\delta),$$

where C_ε only depends on K_ε and the interpolation constant but not on L. Choosing $L = C_\varepsilon/\lambda$, using (3.397), and letting $\delta \to 0$ we arrive at

$$u_N(\bar{x}) - u_N(\bar{y}) \le \frac{\varepsilon}{2} + \psi_L(s).$$

This is a contradiction since we obviously have by (3.399)

$$\psi_L(s) + \varepsilon \le u_N(\bar{x}) - u_N(\bar{y}).$$

Hence we obtain the existence of the required modulus of continuity $\tilde{\sigma}_\eta$.

Now set $v_N(x) = u_N(P_N x)$. Since $(-A)^{-\frac{\eta}{2}}$ is compact we are in a position to apply the Arzela–Ascoli theorem to find a subsequence (still denoted by v_N) converging uniformly on bounded sets of H to a function u that obviously satisfies the same estimates as the u_N's. It remains to show that u solves the limiting equation (3.373). To this end let $u - \psi$ have a maximum at $\hat{x}$ (which we may assume to be strict) for some test function $\psi(x) = \varphi(x) + \delta|x|^2$. It follows from the local uniform convergence of the v_N and the strictness of the maximum at $\hat{x}$ that there exists a sequence $\hat{x}_N = P_N\hat{x}_N \to \hat{x}$ as $N \to \infty$ such that, for every $x \in H_N$,

$$v_N(x) - \varphi(x) - \delta|x|^2 \le v_N(\hat{x}_N) - \varphi(\hat{x}_N) - \delta|\hat{x}_N|^2.$$

Therefore, since $AP_N = P_N A$,

$$
\lambda u_N(\hat{x}_N) + \langle (-A)^{\frac{1}{2}}\hat{x}_N, (-A)^{\frac{1}{2}} D\varphi(\hat{x}_N)\rangle + 2\delta |(-A)^{\frac{1}{2}}\hat{x}_N|^2
$$
$$
+G\left(\hat{x}_N, P_N D\varphi(\hat{x}_N) + 2\delta\hat{x}_N, P_N(D^2\varphi(\hat{x}_N) + 2\delta I)P_N\right) \le 0. \tag{3.400}
$$

Since φ is a test function we have

$$
|(-A)^{\frac{1}{2}} D\varphi(\hat{x}_N)| \le C_1 + C_2|(-A)^{\frac{1}{2}-\varepsilon}\hat{x}_N| \tag{3.401}
$$

for some independent constants C_1, C_2. Also, by (3.370), (3.378), (3.381) and (3.401),

$$
|G\left(\hat{x}_N, P_N D\varphi(\hat{x}_N) + 2\delta\hat{x}_N, P_N(D^2\varphi(\hat{x}_N) + 2\delta I)P_N\right)|
$$
$$
\le C_3\left(1 + |(-A)^{\frac{\beta}{2}}\hat{x}_N|^2 + |(-A)^{\frac{1}{2}-\varepsilon}\hat{x}_N|^2\right) \le C_4 + \frac{\delta}{2}|(-A)^{\frac{1}{2}}\hat{x}_N|^2.
$$

Using this, (3.401) and (3.370), we therefore obtain from (3.400) that

$$
|(-A)^{\frac{1}{2}}\hat{x}_N| \le C_5
$$

for some constant C_5 independent of N. Thus $(-A)^{\frac{1}{2}}\hat{x}_N \rightharpoonup (-A)^{\frac{1}{2}}\hat{x}$ (so $\hat{x} \in D((-A)^{\frac{1}{2}})$) and hence

$$
(-A)^{\frac{\beta}{2}}\hat{x}_N \to (-A)^{\frac{\beta}{2}}\hat{x}, \quad \text{and} \quad (-A)^{\frac{1}{2}} D\varphi(\hat{x}_N) \to (-A)^{\frac{1}{2}} D\varphi(\hat{x}).
$$

These convergences and Lemma 3.85 allow us to pass to the limit in (3.400) as $N \to \infty$ to conclude that

$$
\lambda u(\hat{x}) + \langle (-A)^{\frac{1}{2}}\hat{x}, (-A)^{\frac{1}{2}} D\varphi(\hat{x})\rangle + 2\delta|(-A)^{\frac{1}{2}}\hat{x}|^2 + G\left(\hat{x}, D\varphi(\hat{x}) + \delta\hat{x}, D^2\varphi(\hat{x}) + \delta I\right) \le 0.
$$

The proof of the supersolution property is analogous.   □

3.12.3 A Stochastic Control Problem

We present an application of the results of the previous section to an abstract infinite horizon stochastic optimal control problem which includes the class of problems discussed in Sect. 2.6.2 and may come from a boundary control problem of Dirichlet type with distributed controls. We take the usual setup. Let H, Ξ be real separable Hilbert spaces, and $Q \in \mathcal{L}^+(\Xi)$. Let $\Lambda = \Lambda_1 \times \tilde{\Lambda}_2$, where Λ_1, Λ_2 are real separable Hilbert spaces, and $\tilde{\Lambda}_2$ is a closed bounded subset of Λ_2. We set $R := \sup_{a_2 \in \tilde{\Lambda}_2} |a_2|_{\Lambda_2}$.

Given a reference probability space $\nu = (\Omega, \mathscr{F}, \{\mathscr{F}_s\}_{s \geq 0}, \mathbb{P}, W_Q)$ we have the set of admissible controls

$$\mathcal{U}^\nu := \{a(\cdot) = (a_1(\cdot), a_2(\cdot)) : [0, +\infty) \times \Omega \to \Lambda :$$
$$a_1(\cdot), a_2(\cdot) \text{ are } \mathscr{F}_s - \text{progressively measurable}\}, \qquad (3.402)$$

and we define $\mathcal{U} := \bigcup_\nu \mathcal{U}^\nu$ to be the set of all admissible controls.

We control the state given by the SDE

$$\begin{cases} dX(t) = \left[AX(t) + b(X(t), a_1(t)) + (-A)^\beta Ca_2(t) \right] dt \\ \qquad\qquad\qquad + \sigma(X(t), a_1(t)) dW_Q(t), \quad t > 0 \\ X(0) = x_0 \in H, \end{cases}$$

$$(3.403)$$

i.e.

$$X(t) = e^{tA} x_0 + \int_0^t e^{(t-s)A} b(X(s), a_1(s)) ds + (-A)^\beta \int_0^t e^{(t-s)A} Ca_2(s) ds$$
$$+ \int_0^t (-A)^{\frac{\beta}{2}} e^{(t-s)A} (-A)^{-\frac{\beta}{2}} \sigma(X(s), a_1(s)) dW_Q(s),$$

$$(3.404)$$

and try to minimize the cost functional

$$J(x_0; a(\cdot)) = \mathbb{E} \int_0^{+\infty} e^{-\lambda t} l(X(t; x_0, a(\cdot)), a(t)) dt, \qquad (3.405)$$

over all admissible controls $a(\cdot) \in \mathcal{U}$. We denote by v the value function for this problem. We assume that A satisfies Hypothesis 3.113 and $\lambda > 0$. We also make the following assumptions.

Hypothesis 3.119

(i) The function b is continuous from $H \times \Lambda_1$ to H and there exists a constant $c_0 > 0$ such that

$$|b(x, a_1)| \leq c_0(1 + |x|) \text{ for all } x \in H, a_1 \in \Lambda_1,$$
$$|b(x_1, a_1) - b(x_2, a_1)| \leq c_0|x_1 - x_2| \text{ for all } x_1, x_2 \in H, a_1 \in \Lambda_1.$$

(ii) $C \in \mathcal{L}(\Lambda_2, H)$ and $\beta \in \left(\frac{3}{4}, 1\right)$.

(iii) $\sigma : H \times \Lambda_1 \to \mathcal{L}(\Xi_0, H)$, the map $(-A)^{-\frac{\beta}{2}}\sigma : H \times \Lambda_1 \to \mathcal{L}_2(\Xi_0, H)$ is continuous and moreover there exists a constant $K_1 > 0$ such that

$$\|(-A)^{-\frac{\beta}{2}}\sigma(x, a_1)\|_{\mathcal{L}_2(\Xi_0, H)} \leq K_1(1 + |x|)$$

for all $x \in H$, $a_1 \in \Lambda_1$, and

$$\|(-A)^{-\frac{\beta}{2}}[\sigma(x_1, a_1) - \sigma(x_2, a_1)]\|_{\mathcal{L}_2(\Xi_0, H)} \leq K_1 |x_1 - x_2|$$

for all $x_1, x_2 \in H$, $a_1 \in \Lambda_1$.

(iv) For all $x \in H$

$$\lim_{N \to +\infty} \sup_{a_1 \in \Lambda_1} \|Q_N (-A)^{-\frac{\beta}{2}} \sigma(x, a_1)\|_{\mathcal{L}_2(\Xi_0, H)} = 0.$$

(v) $l \in C(H \times \Lambda)$ and

$$|l(x, a)| \leq C_l, \quad \text{for all } (x, a) \in H \times \Lambda,$$

$$|l(x_1, a) - l(x_2, a)| \leq \omega_l(|x_1 - x_2|), \quad \text{for all } a \in \Lambda, \ x_1, x_2 \in H,$$

for some positive constant C_l and modulus ω_l.

Remark 3.120

(1) Hypotheses 3.119-*(iii),(iv)* are satisfied if we assume, for example, that there exists a constant $K_2 > 0$ such that

$$\|\sigma(x, a_1)\|_{\mathcal{L}(\Xi_0, H)} \leq K_2(1 + |x|)$$

for all $x \in H$, $a_1 \in \Lambda_1$

$$\|\sigma(x_1, a_1) - \sigma(x_2, a_1)\|_{\mathcal{L}(\Xi_0, H)} \leq K_2 |x_1 - x_2|$$

for all $x_1, x_2 \in H$, $a_1 \in \Lambda_1$, and if the operator $(-A)^{-\beta}$ is trace class.

(2) Hypothesis 3.119-*(iv)* is satisfied if, for instance, for every $x \in H$ there exists an $\eta \in (0, \beta/2)$ such that $(-A)^{-\eta} \sigma(x, a_1)$ is bounded in $\mathcal{L}_2(\Xi_0, H)$ independently of $a_1 \in \Lambda_1$. ∎

It is a consequence of Theorem 1.141 that for every generalized reference probability space $\mu = (\Omega, \mathcal{F}, \{\mathcal{F}_s\}_{s \geq 0}, \mathbb{P}, W_Q)$, $T > 0$, $a(\cdot) \in \mathcal{U}^\mu$, and $\xi \in L^2(\Omega, \mathcal{F}_0, \mathbb{P})$, Eq. (3.403) with $X(0) = \xi$ has a unique mild solution in $\mathcal{H}_2^\mu(0, T; H)$ with continuous trajectories. Following the strategy described in Remark 2.17 and using Proposition 1.142 one can then argue that all the assumptions needed to prove the DPP for the problem are satisfied. However, we will not look directly into this, since we need to study the transformed HJB equation (3.375) and the optimal control problem associated with it.

Let us first see how the state equation is transformed by the change of variables. If $X(\cdot; x_0, a(\cdot))$ satisfies (3.403) then $Y(\cdot) = Y(\cdot; y_0, a(\cdot)) := (-A)^{-\frac{\beta}{2}} X(\cdot; x_0, a(\cdot))$ satisfies the equation

$$\begin{cases} dY(t) = \left[AY(t) + (-A)^{-\frac{\beta}{2}}b((-A)^{\frac{\beta}{2}}Y(t), a_1(t)) + (-A)^{\frac{\beta}{2}}Ca_2(t) \right] dt \\ \qquad\qquad + (-A)^{-\frac{\beta}{2}}\sigma((-A)^{\frac{\beta}{2}}Y(t), a_1(t)) dW_Q(t) \\ \\ Y(0) = y_0 = (-A)^{-\frac{\beta}{2}}x_0 \in H, \end{cases}$$

(3.406)

which is understood in its mild form

$$Y(t) = e^{tA}y_0 + \int_0^t e^{(t-s)A}(-A)^{-\frac{\beta}{2}}b((-A)^{\frac{\beta}{2}}Y(s), a_1(s))ds$$

$$+ (-A)^{\frac{\beta}{2}}\int_0^t e^{(t-s)A}Ca_2(s)ds$$

$$+ \int_0^t e^{(t-s)A}(-A)^{-\frac{\beta}{2}}\sigma((-A)^{\frac{\beta}{2}}Y(s), a_1(s))dW_Q(s), \qquad t \geq 0. \quad (3.407)$$

We are now minimizing the cost functional

$$\tilde{J}(y_0; a(\cdot)) = \mathbb{E}\int_0^{+\infty} e^{-\lambda t} l\left((-A)^{\frac{\beta}{2}}Y(t; y_0, a(\cdot)), a(t)\right) dt, \qquad (3.408)$$

over all admissible controls and we denote by u the value function for this problem. The HJB equation associated with this new control problem is of the form (3.375) with $G : D((-A)^{\frac{\beta}{2}}) \times D((-A)^{\frac{\beta}{2}}) \times S(H) \to \mathbb{R}$ given by

$$G(y, q, S)$$
$$= \sup_{a\in\Lambda}\left\{ -\frac{1}{2}\mathrm{Tr}\left[((-A)^{-\frac{\beta}{2}}\sigma((-A)^{\frac{\beta}{2}}y, a_1)Q^{\frac{1}{2}})((-A)^{-\frac{\beta}{2}}\sigma((-A)^{\frac{\beta}{2}}y, a_1)Q^{\frac{1}{2}})^*S \right] \right.$$
$$\left. - \left\langle b((-A)^{\frac{\beta}{2}}y, a_1), (-A)^{-\frac{\beta}{2}}q \right\rangle - \left\langle Ca_2, (-A)^{\frac{\beta}{2}}p \right\rangle - l((-A)^{\frac{\beta}{2}}y, a) \right\}.$$

(3.409)

We will see that the value functions v and u are linked by the relation $v(x) = u((-A)^{-\frac{\beta}{2}}x)$ for $x \in H$. Thus u should correspond to an HJB equation (3.371) with a Hamiltonian $F : H \times D((-A)^{\beta}) \times (-A)^{-\frac{\beta}{2}}S(H)(-A)^{-\frac{\beta}{2}} \to \mathbb{R}$ such that $G = G_F$, where G_F is given by (3.374). An easy calculation shows that this is true if

$$F(x, p, S) = \sup_{a\in\Lambda}\left\{ -\frac{1}{2}\mathrm{Tr}\left[(\sigma(x, a_1)Q^{\frac{1}{2}})^*S(\sigma(x, a_1)Q^{\frac{1}{2}}) \right] \right.$$

$$\left. - \langle b(x, a_1), q \rangle - \left\langle Ca_2, (-A)^{\beta}p \right\rangle - l(x, a) \right\}.$$

(3.410)

This is just the formal Hamiltonian corresponding to the original control problem, however notice that the second-order terms in F are written in a slightly different form since we do not know that $\sigma(x, a_1)Q^{\frac{1}{2}} \in \mathcal{L}_2(U, H)$. We remark that if either

$(\sigma(x, a_1)Q^{\frac{1}{2}})^* S$ or $S(\sigma(x, a_1)Q^{\frac{1}{2}})$ is trace class then (see Appendix B.3)

$$\mathrm{Tr}\left[(\sigma(x, a_1)Q^{\frac{1}{2}})^* S(\sigma(x, a_1)Q^{\frac{1}{2}})\right] = \mathrm{Tr}\left[(\sigma(x, a_1)Q^{\frac{1}{2}})(\sigma(x, a_1)Q^{\frac{1}{2}})^* S\right].$$

Proposition 3.121 *Assume that Hypotheses 3.113 and 3.119 hold. Let* $\mu = (\Omega, \mathscr{F}, \{\mathscr{F}_s\}_{s \geq 0}, \mathbb{P}, W_Q)$ *be a generalized reference probability space,* $T > 0$, $a(\cdot) \in \mathcal{U}^\mu$, *and* $\xi \in L^2(\Omega, \mathscr{F}_0, \mathbb{P})$. *Then Eq. (3.406) with* $Y(0) = \xi$ *has a unique mild solution* $Y(\cdot) = Y(\cdot; \xi, a(\cdot))$ *among all processes which have* $dt \otimes \mathbb{P}$ *equivalent versions in* $M^2_\mu(0, T; D((-A)^{\frac{\beta}{2}}))$. *The solution has continuous trajectories in* H.

Proof The proof of existence and uniqueness will follow from the contraction mapping principle. Assume first that $\xi \in L^p(\Omega, \mathscr{F}_0, \mathbb{P})$, $2 \leq p < 2/\beta$. For $Z \in M^p_\mu(0, T; D((-A)^{\frac{\beta}{2}}))$ we define a map $\mathcal{K}$ on $M^p_\mu(0, T; D((-A)^{\frac{\beta}{2}})$ by

$$\mathcal{K}(Z)(t) = e^{tA}\xi + \int_0^t e^{(t-s)A}(-A)^{-\frac{\beta}{2}}b((-A)^{\frac{\beta}{2}}Z(s), a_1(s))ds$$

$$+ (-A)^{\frac{\beta}{2}} \int_0^t e^{(t-s)A} C a_2(s) ds$$

$$+ \int_0^t e^{(t-s)A}(-A)^{-\frac{\beta}{2}}\sigma((-A)^{\frac{\beta}{2}}Z(s), a_1(s))dW_Q(s), \qquad t \in [0, T]. \quad (3.411)$$

We see that $\mathcal{K}(Z)(\cdot)$ is progressively measurable as a process with values in H by arguments similar to those in Remark 1.123. Moreover, thanks to Hypothesis 3.119-(i) and (B.18), for suitable constants $C_1, C_2 > 0$,

$$|(-A)^{\frac{\beta}{2}}\mathcal{K}(Z)(t)| \leq |(-A)^{\frac{\beta}{2}}e^{tA}\xi| + C_1 \int_0^t e^{-a(t-s)}[1 + |(-A)^{\frac{\beta}{2}}Z(s)|]ds$$

$$+ C_2 R \int_0^t \frac{e^{-a(t-s)}}{(t-s)^\beta}ds + \left|\int_0^t (-A)^{\frac{\beta}{2}}e^{(t-s)A}(-A)^{-\frac{\beta}{2}}\sigma((-A)^{\frac{\beta}{2}}Z(s), a_1(s))dW_Q(s)\right|.$$

Then, taking the expectation of the p-th power of the terms of this last inequality and using (1.111) and Hypothesis 3.119-(iii) we get

$$\mathbb{E}\left|(-A)^{\frac{\beta}{2}}\mathcal{K}(Z)(t)\right|^p \leq C_3\left[\frac{1}{t^{\frac{p\beta}{2}}}\mathbb{E}|\xi|^p + 1\right.$$

$$\left. + \int_0^t \mathbb{E}|(-A)^{\frac{\beta}{2}}Z(s)|^p ds + \int_0^t \frac{1}{(t-s)^{\frac{p\beta}{2}}}[1 + \mathbb{E}|(-A)^{\frac{\beta}{2}}Z(s)|^p]ds\right]. \quad (3.412)$$

Therefore

$$
\begin{aligned}
|\mathcal{K}(Z)|^p_{M^p_\mu(0,T;D((-A)^{\frac{\beta}{2}}))} &= \int_0^T \mathbb{E}\left|(-A)^{\frac{\beta}{2}}\mathcal{K}(Z)(t)\right|^p dt \\
&\le C_4(T)\left[\mathbb{E}|\xi|^p + 1 + \int_0^T\int_0^t\left[1 + \frac{1}{(t-s)^{\frac{p\beta}{2}}}\right]\mathbb{E}|(-A)^{\frac{\beta}{2}}Z(s)|^p ds\, dt\right] \\
&\le C_4(T)\left[\mathbb{E}|\xi|^p + 1 + \int_0^T\mathbb{E}|(-A)^{\frac{\beta}{2}}Z(s)|^p\int_s^T\left[1 + \frac{1}{(t-s)^{\frac{p\beta}{2}}}\right]dt\, ds\right] \\
&\le C_5(T)\left[\mathbb{E}|\xi|^p + 1 + |Z|^p_{M^p_\mu(0,T;D((-A)^{\frac{\beta}{2}}))}\right].
\end{aligned}
$$

Thus $\mathcal{K}(Z)(\cdot) \in M^p_\mu(0,T;D((-A)^{\frac{\beta}{2}}))$. We now prove that $\mathcal{K}$ is a contraction on $M^p_\mu(0,T;D((-A)^{\frac{\beta}{2}}))$ if T is sufficiently small. Let $Z_1(\cdot), Z_2(\cdot) \in M^p_\mu(0,T; D((-A)^{\frac{\beta}{2}}))$. Then, arguing as above, we have

$$
\begin{aligned}
\mathbb{E}&\left|(-A)^{\frac{\beta}{2}}\left(\mathcal{K}(Z_1)(t) - \mathcal{K}(Z_2)(t)\right)\right|^p \\
&\le C_6\int_0^t\left(1 + \frac{1}{(t-s)^{\frac{p\beta}{2}}}\right)\mathbb{E}\left|(-A)^{\frac{\beta}{2}}(Z_1(s) - Z_2(s))\right|^p ds
\end{aligned}
$$

for some constant $C_6 > 0$, which implies

$$
|\mathcal{K}(Z_1) - \mathcal{K}(Z_2)|_{M^p_\mu(0,T;D((-A)^{\frac{\beta}{2}}))} \le C_7\left(T + T^{1-\frac{p\beta}{2}}\right)^{\frac{1}{2}}|Z_1 - Z_2|_{M^p_\mu(0,T;D((-A)^{\frac{\beta}{2}}))}.
$$

Thus $\mathcal{K}$ is a contraction on $M^p_\mu(0,T;D((-A)^{\frac{\beta}{2}}))$ for small $T > 0$ and hence it has a fixed point $Y(\cdot;\xi,a(\cdot))$. The second and third terms of the right-hand side of (3.411) have continuous trajectories in H by Lemma 1.115 whereas the stochastic integral there has continuous trajectories if $p > 2$ by Proposition 1.112. Thus $Y(\cdot;\xi,a(\cdot))$ has a $dt \otimes \mathbb{P}$-equivalent version which has continuous trajectories in H if $p > 2$. To prove that $Y(\cdot;\xi,a(\cdot))$ has continuous trajectories if $\xi \in L^2(\Omega,\mathscr{F}_0,\mathbb{P})$ we argue as in the proof of Theorem 1.141. We approximate ξ by random variables

$$
\xi_n = \begin{cases} \xi & \text{if } |\xi| \le n \\ 0 & \text{if } |\xi| > n. \end{cases}
$$

The solutions $Y(\cdot;\xi_n,a(\cdot))$ have continuous trajectories in H, $\mathbb{P}$-a.s., and since the solutions are obtained by fixed point, one can show that $Y(\cdot;\xi,a(\cdot)) = Y(\cdot;\xi_n,a(\cdot))$, $\mathbb{P}$-a.s. on $\{\omega : |\xi(\omega)| \le n\}$.

The existence of a unique solution in $M^p_\mu(0,T;D((-A)^{\frac{\beta}{2}}))$ for any $T > 0$ follows by repeating the argument a finite number of times. $\square$

Proposition 3.122 *Assume that Hypotheses 3.113 and 3.119 hold. Let $T > 0$, $y_0 \in H$. Then there exists a constant $C(T, |y_0|) \geq 0$ such that, for all $t \in (0, T]$ and $a(\cdot) \in \mathcal{U}$,*

$$\mathbb{E}|(-A)^{\frac{\beta}{2}} Y(t; y_0, a(\cdot))|^2 \leq C(T, |y_0|) \frac{1}{t^\beta}. \tag{3.413}$$

Moreover, for every $\gamma \in (0, 1 - \beta)$, there exists a constant $C_\gamma(T, |y_0|) \geq 0$ such that, for all $t \in (0, T]$ and $a(\cdot) \in \mathcal{U}$,

$$\int_0^t \frac{1}{(t-s)^{\beta+\gamma}} \mathbb{E}|(-A)^{\frac{\beta}{2}} Y(s; y_0, a(\cdot))|^2 ds \leq C_\gamma(T, |y_0|) \frac{1}{t^\beta}. \tag{3.414}$$

Proof Estimate (3.412) for $p = 2$ applied to the solution $Y(\cdot) = Y(\cdot; y_0, a(\cdot))$ implies

$$\mathbb{E}\left|(-A)^{\frac{\beta}{2}} Y(t)\right|^2 \leq C(T)\left[\frac{1}{t^\beta}(|y_0|^2 + 1) + \int_0^t \frac{1}{(t-s)^\beta} \mathbb{E}|(-A)^{\frac{\beta}{2}} Y(s)|^2 ds\right].$$

Estimate (3.413) thus follows from Proposition D.30. Now

$$\int_0^t \frac{1}{(t-s)^{\beta+\gamma}} \mathbb{E}|(-A)^{\frac{\beta}{2}} Y(s)|^2 ds \leq C(T, |y_0|) \int_0^t \frac{1}{(t-s)^{\beta+\gamma}} \frac{1}{s^\beta} ds \leq C_\gamma(T, |y_0|) \frac{1}{t^\beta}$$

since

$$\int_0^t \frac{1}{(t-s)^{\beta+\gamma}} \frac{1}{s^\beta} ds = t^{1-(2\beta+\gamma)} \int_0^1 \frac{1}{(1-s)^{\beta+\gamma}} \frac{1}{s^\beta} ds$$

and this last integral is bounded and $1 - (2\beta + \gamma) > -\beta$. $\square$

Theorem 3.123 *Assume that Hypotheses 3.113, 3.119 hold. Then the value function v is the unique $UC_b(H_{-\eta})$ viscosity solution (for every $\eta \in (0, 1)$) of the HJB equation (3.371) with the Hamiltonian F given by (3.410). Moreover, the dynamic programming principle holds for u and v, i.e. for $x \in H$ and all $T > 0$,*

$$v(x) = \inf_{a(\cdot) \in \mathcal{U}} \mathbb{E}\left\{\int_0^T e^{-\lambda t} l(X(t; x, a(\cdot)), a(t)) dt + e^{-\lambda T} v(X(T; x, a(\cdot)))\right\}$$

and

$$u(y) = \inf_{a(\cdot) \in \mathcal{U}} \mathbb{E}\left\{\int_0^T e^{-\lambda t} l((-A)^{\frac{\beta}{2}} Y(t; y, a(\cdot)), a(t)) dt + e^{-\lambda T} u(Y(T; y, a(\cdot)))\right\}.$$

We will prove this theorem by the approximation argument used in the proof of existence of Theorem 3.118. We consider for $N \geq 1$ the following SDE approximating the state Eq. (3.406).

$$
\begin{cases}
dY_N(t) = \left[P_N A Y_N(t) + (-A)^{-\frac{\beta}{2}} P_N b((-A)^{\frac{\beta}{2}} Y(t), a_1(t)) + (-A)^{\frac{\beta}{2}} P_N C a_2(t) \right] dt \\
\qquad + (-A)^{-\frac{\beta}{2}} P_N \sigma((-A)^{\frac{\beta}{2}} Y(t), a_1(t)) dW_Q(t) \\
Y_N(0) = P_N y_0 \in H_N.
\end{cases}
$$

$$(3.415)$$

These are finite-dimensional SDEs (even though the noise is infinite-dimensional) in the spaces H_N which have unique strong solutions $Y_N(\cdot) = Y_N(\cdot; y_0, a(\cdot))$ and good continuous dependence estimates like those in Sects. 1.4.3 and 3.1.2 with respect to the norm in H_N. The solutions $Y_N(\cdot)$ can be also written in the mild form

$$
Y_N(t) = e^{tA} P_N y_0 + \int_0^t e^{(t-s)A} (-A)^{-\frac{\beta}{2}} P_N b((-A)^{\frac{\beta}{2}} Y_N(s), a_1(s)) ds
$$

$$
+ (-A)^{\frac{\beta}{2}} \int_0^t e^{(t-s)A} P_N C a_2(s) ds
$$

$$
+ \int_0^t e^{(t-s)A} (-A)^{-\frac{\beta}{2}} P_N \sigma((-A)^{\frac{\beta}{2}} Y_N(s), a_1(s)) dW_Q(s), \qquad t \geq 0.
$$

Lemma 3.124 *Let Hypotheses 3.113, 3.119 hold, $y_0 \in H$, and $T > 0$. Then*

$$
\lim_{N \to +\infty} \sup_{a(\cdot) \in \mathcal{U}} |Y_N(\cdot; y_0, a(\cdot)) - Y(\cdot; y_0, a(\cdot))|_{M^2(0,T;D((-A)^{\frac{\beta}{2}}))} = 0.
$$

Proof We define $Y(\cdot) = Y(\cdot; y_0, a(\cdot))$, $Y_N(\cdot) = Y_N(\cdot; y_0, a(\cdot))$ and fix $\gamma \in (0, 1 - \beta)$. Recall that P_N, Q_N commute with $-A$, its fractional powers and e^{tA}. We have

$$
(-A)^{\frac{\beta}{2}} (Y_N(t) - Y(t)) = -(-A)^{\frac{\beta}{2}} e^{tA} Q_N y_0
$$

$$
- Q_N (-A)^{-\frac{\gamma}{2}} \int_0^t (-A)^{\frac{\gamma}{2}} e^{(t-s)A} b((-A)^{\frac{\beta}{2}} Y(s), a_1(s)) ds
$$

$$
- Q_N (-A)^{-\frac{\gamma}{2}} \int_0^t (-A)^{\beta + \frac{\gamma}{2}} e^{(t-s)A} C a_2(s) ds
$$

$$
- Q_N (-A)^{-\frac{\gamma}{2}} \int_0^t (-A)^{\frac{\beta+\gamma}{2}} e^{(t-s)A} (-A)^{\frac{\beta}{2}} \sigma((-A)^{\frac{\beta}{2}} Y(s), a_1(s)) dW(s)
$$

$$
+ \int_0^t e^{(t-s)A} P_N [b((-A)^{\frac{\beta}{2}} Y_N(s), a_1(s)) - b((-A)^{\frac{\beta}{2}} Y(s), a_1(s))] ds
$$

$$
+ \int_0^t (-A)^{\frac{\beta}{2}} e^{(t-s)A} (-A)^{\frac{\beta}{2}} P_N [\sigma((-A)^{\frac{\beta}{2}} Y_N(s), a_1(s)) - \sigma((-A)^{\frac{\beta}{2}} Y(s), a_1(s))] dW_Q(s),
$$

which yields, for a suitable $C_\gamma(T) > 0$,

$$
\mathbb{E}|(-A)^{\frac{\beta}{2}} (Y_N(t) - Y(t))|^2
$$

$$\leq C_\gamma(T)\left[\frac{1}{t^\beta}|Q_N y_0|^2 + \|Q_N A^{-\frac{\gamma}{2}}\|^2\left(1+\int_0^t\left(1+\frac{1}{(t-s)^{\beta+\gamma}}\right)\mathbb{E}|(-A)^{\frac{\beta}{2}}Y(s)|^2 ds\right)\right.$$

$$\left.+\int_0^t\left(1+\frac{1}{(t-s)^\beta}\right)\mathbb{E}|(-A)^{\frac{\beta}{2}}(Y_N(s)-Y(s))|^2 ds\right].$$

Since $A^{-\gamma/2}$ is compact, $\|Q_N A^{-\gamma/2}\| \to 0$ as $N \to +\infty$, and by using (3.414) we thus deduce that

$$\mathbb{E}|(-A)^{\frac{\beta}{2}}(Y_N(t)-Y(t))|^2 \leq C_{\gamma,T,y_0}(N)\left(1+\frac{1}{t^\beta}\right)$$

$$+\tilde{C}_{\gamma,T}\int_0^t \frac{1}{(t-s)^\beta}\mathbb{E}|(-A)^{\frac{\beta}{2}}(Y_N(s)-Y(s))|^2 ds,$$

where $C_{\gamma,T,y_0}(N) \to 0$ as $N \to +\infty$. Using Proposition D.30 we thus obtain

$$\mathbb{E}|(-A)^{\frac{\beta}{2}}(Y_N(t)-Y(t))|^2 \leq C_{\gamma,T,y_0}(N)M\frac{1}{t^\beta} \tag{3.416}$$

for some constant M independent of N. This implies the claim. $\square$

Proof of Theorem 3.123 We notice that under our assumptions Eq. (3.375) with G given by (3.409) has a unique viscosity solution in $UC_b(H_{-\eta})$ for every $\eta \in (0, 1-\beta)$. To verify that the value function u is the solution we consider the approximating problems

$$\lambda u_N - \langle Ax, Du_N\rangle + G(x, Du_N, D^2 u_N) = 0 \quad \text{in} \quad H_N. \tag{3.417}$$

Equation (3.417) is the one used in the proof of Theorem 3.118 and it is easy to see that it is the equation in H_N corresponding to the control problem with evolution given by (3.415). Therefore, by the results of Sect. 3.6.3, the function

$$u_N(y_0) = \inf_{a(\cdot)\in\mathcal{U}} \mathbb{E}\int_0^{+\infty} e^{-\lambda t}l((-A)^{\frac{\beta}{2}}Y_N(t; y_0, a(\cdot)), a(t))dt \tag{3.418}$$

belongs to $UC_b(H_N)$, it satisfies the dynamic programming principle, i.e. for every $y_0 \in H_N, T \geq 0$

$$u_N(y_0) = \inf_{a(\cdot)\in\mathcal{U}} \mathbb{E}\left\{\int_0^T e^{-\lambda t}l((-A)^{\frac{\beta}{2}}Y_N(t; y_0, a(\cdot)), a(t))dt\right.$$

$$\left. + e^{-\lambda T}u_N(Y_N(T; y_0, a(\cdot)))\right\}, \tag{3.419}$$

and u_N is the unique viscosity solution of (3.417) in $UC_b(H_N)$.

Since for every $y_0 \in H$, $Y_N(t; y_0, a(\cdot)) = Y_N(t; P_N y_0, a(\cdot))$, extending u_N to H by putting $u_N(y) = u_N(P_N y)$ we obtain (3.418) and (3.417) for every $y_0 \in H$. Moreover, from the proof of existence of Theorem 3.118, we know that for every $\eta \in (0, 1 - \beta)$ and $N \geq 1$

$$\|u_N\|_0 \leq \frac{C_l}{\lambda}, \qquad |u_N(x) - u_N(y)| \leq \tilde{\sigma}_\eta(|x - y|_{-\eta}) \tag{3.420}$$

for some modulus $\tilde{\sigma}_\eta$ and $u_N \to \bar{u}$ uniformly on bounded sets, where $\bar{u}$ is the unique viscosity solution of (3.375) in $UC_b(H_{-\eta})$, $\eta \in (0, 1 - \beta)$.

We need to show that $u = \bar{u}$. We will prove that u_N converges pointwise to u as $N \to \infty$. Let $y_0 \in H$. For every $T > 0$,

$|u_N(y_0) - u(y_0)|$

$$\leq \sup_{a(\cdot) \in \mathcal{U}} \int_0^T e^{-\lambda t} \mathbb{E} \omega_l(|(-A)^{\frac{\beta}{2}}(Y_N(t; y_0, a(\cdot)) - Y(t; y_0, a(\cdot)))|) dt + 2C_l \frac{e^{-\lambda T}}{\lambda}.$$

Let $\varepsilon > 0$ and $T_\varepsilon > 0$ be such that such that $2C_l e^{-\lambda T_\varepsilon}/\lambda \leq \varepsilon$. If $\omega_l(s) \leq \varepsilon + K_\varepsilon s$, $s \geq 0$, we obtain by the Cauchy–Schwarz inequality

$$\int_0^{T_\varepsilon} e^{-\lambda t} \mathbb{E} \omega_l(|(-A)^{\frac{\beta}{2}}(Y_N(t; y_0, a(\cdot)) - Y(t; y_0, a(\cdot)))|) dt$$

$$\leq \frac{\varepsilon}{\lambda} + \frac{K_\varepsilon}{\lambda} |Y_N(\cdot; y_0, a(\cdot)) - Y(\cdot; y_0, a(\cdot))|_{M^2(0, T_\varepsilon; D((-A)^{\frac{\beta}{2}}))}$$

for all $N \geq 1$ and all $a(\cdot) \in \mathcal{U}$. The conclusion thus follows by letting $N \to +\infty$ and using Lemma 3.124, since ε is arbitrary.

It remains to show the dynamic programming principle for u. By (3.419), we have

$$\left| u_N(y_0) - \inf_{a(\cdot) \in \mathcal{U}} \mathbb{E} \left\{ \int_0^T e^{-\lambda t} l((-A)^{\frac{\beta}{2}} Y(t; y_0, a(\cdot)), a(t)) dt + e^{-\lambda T} u(Y(T; y_0, a(\cdot))) \right\} \right|$$

$$\leq \sup_{a(\cdot) \in \mathcal{U}} \mathbb{E} \int_0^T e^{-\lambda t} \omega_l(|(-A)^{\frac{\beta}{2}}(Y_N(t; y_0, a(\cdot)) - Y(t; y_0, a(\cdot)))|) dt$$

$$+ e^{-\lambda T} \sup_{a(\cdot) \in \mathcal{U}} \mathbb{E} |u_N(Y_N(T; y_0, a(\cdot))) - u(y(T; y_0, a(\cdot)))|.$$

The first term of the right-hand side converges to 0 when N goes to infinity by the same argument as in the previous paragraph. For the second term, we proceed as follows:

$$\mathbb{E}|u_N(Y_N(T; y_0, a(\cdot))) - u(Y(T; y_0, a(\cdot)))|$$
$$\leq \mathbb{E}|u_N(Y_N(T; y_0, a(\cdot))) - u_N(Y(T; y_0, a(\cdot)))|$$
$$+ \mathbb{E}|u_N(Y(T; y_0, a(\cdot))) - u(Y(T; y_0, a(\cdot)))|.$$

The first term of the right-hand side converges to 0 uniformly in $a(\cdot) \in \mathcal{U}$ when N goes to infinity by (3.416) and (3.420). It remains to prove that

$$\sup_{a(\cdot) \in \mathcal{U}} \mathbb{E}|u_N(Y(T; y_0, a(\cdot))) - u(Y(T; y_0, a(\cdot)))|$$

goes to 0 when N goes to infinity. By Proposition 3.122, estimate (3.413), $\mathbb{E}|Y(T; y_0, a(\cdot))|^2$ is bounded by a constant $\tilde{C}(T, |y_0|) > 0$ which does not depend on $a(\cdot) \in \mathcal{U}$. Hence, for all $R > 0$,

$$\mathbb{P}\{|Y(T; y_0, a(\cdot))| > R\} \leq \frac{\tilde{C}(T, |y_0|)}{R^2}.$$

Let $\varepsilon > 0$ and choose $R_\varepsilon > 0$ sufficiently large so that this probability is smaller than ε. Then

$$\sup_{a(\cdot) \in \mathcal{U}} \mathbb{E}|u_N(Y(T; y_0, a(\cdot))) - u(Y(T; y_0, a(\cdot)))| \leq \frac{2C_l}{\lambda}\varepsilon + \sup_{|y| \leq R_\varepsilon} |u_N(y) - u(y)|.$$

We conclude by letting $N \to +\infty$ since ε was arbitrary.

Finally, we observe that for $x_0 \in H$ and $y_0 = (-A)^{-\frac{\beta}{2}}x_0$, the mild solutions $X(\cdot)$ of (3.404) and $Y(\cdot)$ of (3.407) are related by $Y(\cdot) = (-A)^{-\frac{\beta}{2}}X(\cdot)$. Therefore we have $v(x_0) = u((-A)^{-\frac{\beta}{2}}x_0)$. Thus the dynamic programming principle also holds for v, and by definition v is the unique viscosity solution in $UC_b(H_{-\eta})$, $\eta \in (0, 1)$, of the HJB equation (3.371) with F given by (3.410). □

Remark 3.125 We would obtain the same results if instead of the change of variables $y = (-A)^{-\frac{\beta}{2}}x$ we applied the change of variables $y = (-A)^{-\frac{\gamma}{2}}x$ for $\beta \leq \gamma < 1$. This may be beneficial for a boundary control problem with the Neumann boundary condition, where we have $\beta < 1/2$, as it may help make the second-order terms satisfy Hypothesis 3.119-(iii), (iv), which would then have $(-A)^{-\frac{\beta}{2}}$ replaced by $(-A)^{-\frac{\gamma}{2}}$ there (see also the example below). ∎

We now discuss a specific example of a stochastic boundary control problem with Dirichlet boundary conditions. It is more general than the one from Sect. 2.6.2 since it also contains distributed controls and allows multiplicative noise. Good examples of deterministic boundary control problems can be found in [400]. The results of this section would apply to suitable stochastic perturbations of examples belonging to the "first abstract class" in [400].

Let $\mathcal{O} \subset \mathbb{R}^N$ be an open, connected and bounded set with smooth boundary. Consider, as in Sect. 2.6.2, the following stochastic controlled PDE

$$\begin{cases} \dfrac{\partial x}{\partial t}(t,\xi) = \Delta_\xi x(t,\xi) + f_1\left(x(t,\xi),\alpha_1(t,\xi)\right) \\ \qquad\qquad\qquad + f_2\left(x(t,\xi),\alpha_1(t,\xi)\right)\dot{W}_Q(t,\xi) \text{ in } (0,\infty)\times\mathcal{O} \\ x(0,\xi) = x_0(\xi) \qquad\qquad\qquad\qquad\qquad\qquad \text{on } \mathcal{O} \\ x(t,\xi) = \alpha_2(t,\xi) \qquad\qquad\qquad\qquad\qquad\quad \text{on } (0,\infty)\times\partial\mathcal{O}, \end{cases} \tag{3.421}$$

where W_Q is a Q-Wiener process, $Q \in \mathcal{L}^+(L^2(\mathcal{O}))$, $x_0 \in L^2(\mathcal{O})$, and f_1, f_2 : $\mathbb{R}^2 \to \mathbb{R}$. We take $H = L^2(\mathcal{O})$, $\Lambda_1 = L^2(\mathcal{O})$ and Λ_2 to be the closed ball centered at 0 with radius R in $L^2(\partial\mathcal{O})$, and assume that the control $a(t) = (a_1(t), a_2(t)) := (\alpha_1(t,\cdot), \alpha_2(t,\cdot))$ belongs to $\mathcal{U}$ as defined in (3.402). As was discussed in Sect. 2.6.2, (3.421) can be rewritten as an abstract stochastic evolution equation (3.403) and (3.404), where A is the Laplace operator with zero Dirichlet boundary conditions, C is the Dirichlet operator, and

$$b(x,a_1)(\xi) = f_1(x(\xi),a_1(\xi)), \quad [\sigma(x,a_1)y](\xi) = f_2(x(\xi),a_1(\xi))y(\xi).$$

Suppose that f_1, f_2 satisfy for $i = 1, 2$

$$|f_i(r,s)| \le c_1(1 + |r|) \text{ for all } r, s \in \mathbb{R},$$
$$|f_i(r_1,s) - f_i(r_2,s)| \le c_1|r_1 - r_2| \text{ for all } r_1, r_2, s \in \mathbb{R}.$$

It is then easy to see that b satisfies Hypothesis 3.119-(i). As regards σ, suppose that $Q = I$ and $N = 1$. Let $\{e_k\}$ be the orthonormal basis of eigenvectors of A. In this case $-Ae_k = ck^2 e_k$, where $c > 0$. Moreover, the e_k are bounded in $L^\infty(\mathcal{O})$, uniformly in k. Therefore we obtain

$$\|(-A)^{-\frac{\beta}{2}}\sigma(x,a_1)\|^2_{\mathcal{L}_2(H)} = \sum_{k=1}^{+\infty}|(-A)^{-\frac{\beta}{2}}\sigma(x,a_1)e_k|^2$$
$$= c^{-\beta}\sum_{k=1}^{+\infty}\sum_{h=1}^{+\infty}h^{-2\beta}\langle\sigma(x,a_1)e_k,e_h\rangle^2 = c^{-\beta}\sum_{h=1}^{+\infty}h^{-2\beta}|\sigma(x,a_1)e_h|^2$$
$$= c^{-\beta}\sum_{h=1}^{+\infty}h^{-2\beta}\int_{\mathcal{O}}|f_2(x(\xi),a_1(\xi))e_h(\xi)|^2 d\xi \le C_1(1 + |x|^2),$$

where we used that $\langle\sigma(x,a_1)e_k,e_h\rangle = \langle e_k,\sigma(x,a_1)e_h\rangle$ to justify the third equality above, and the fact that $\beta > 1/2$. However, the above computation does not work for $N \ge 2$, where stronger assumptions either on f_2 or Q need to be imposed. In addition the basis $\{e_k\}$ may not be bounded in $L^\infty(\mathcal{O})$ if $N \ge 2$, however it is bounded when $\mathcal{O}$ is a rectangular parallelepiped. The above computation also does not work in the Neumann case when $\beta < 1/2$, even if $f_2 \equiv 1$, and this is why it may be beneficial to use the change of variables $y = (-A)^{-\frac{\gamma}{2}}x$ for $\beta \le \gamma < 1$, as discussed in Remark 3.125. The other conditions of Hypothesis 3.119-(iii), (iv) are checked similarly. If f_2 is constant (and thus so is σ), Hypothesis 3.119-(iii) (iv) holds if we assume that there is an orthonormal basis $\{e_k\}$ of H such that

$$Ae_k = -\lambda_k e_k, \quad Qe_k = \beta_k e_k, \quad k \in \mathbb{N},$$

where (λ_k) is a sequence of positive numbers increasing to $+\infty$ while (β_k) is a bounded sequence of nonnegative real numbers and

$$\sum_{k=1}^{\infty} \frac{\beta_k}{\lambda_k^{\beta}} < +\infty.$$

Since for the Laplace operator A we have $\lambda_k \approx k^{\frac{2}{N}}$ as $k \to +\infty$, this condition is fulfilled if for some $\varepsilon > 0$, $\beta_k \leq Ck^{\frac{2\beta}{N}-1-\varepsilon}$. When Q is invertible this is possible only for $N = 1$. However Q can have finite rank.

If the original cost functional was given by

$$\mathbb{E} \int_0^{+\infty} e^{-\lambda t} \int_{\mathcal{O}} f_3(x(t,\xi), \alpha_1(t,\xi)) d\xi dt,$$

where $\lambda > 0$ and $f_3 : \mathbb{R}^2 \to \mathbb{R}$, then the cost functional for the abstract evolution system (3.403) and (3.404) is given by (3.405), where

$$l(x, a_1) := \int_{\mathcal{O}} f_3(x(\xi), a_1(\xi)) d\xi.$$

If $f_3 \in UC_b(\mathbb{R}^2)$ then l satisfies Hypothesis 3.119-(v). The original cost functional can be more general and depend explicitly on the boundary control α_2.

3.13 HJB Equations for Control of Stochastic Navier–Stokes Equations

In this section we present another special class of equations which can be studied by viscosity solution methods, which however require modifications of the general definition of viscosity solution from Sect. 3.3 and the techniques of Sects. 3.5 and 3.6. We will study second-order HJB equations that arise in problems of optimal control of stochastic Navier–Stokes equations. Not much is known about equations of this type. Kolmogorov equations for stochastic Navier–Stokes equations have been studied by Komech and Vishik (see [567] and the references therein) and more recently in [33, 34, 255, 512] for two-dimensional stochastic Navier–Stokes equations and by Da Prato and Debussche [161] for the three-dimensional case. Only existence of strict and mild solutions has been proved in [161]. A semilinear equation associated to a special optimal control problem has been investigated by Da Prato and Debussche in [158] from the point of view of mild solutions. Some of these results have been generalized to the three-dimensional case in [424]. The mild solution approach of [158] is discussed in Sect. 4.9. The viscosity solution approach is more general in the sense that it can handle more complicated cost functionals and applies to stochastic optimal control problems with the associated HJB equations that are fully nonlinear in the gradient variable. On the other hand the covariance operator of the Wiener process here must be of trace class and thus the viscosity solution approach cannot cover non-degenerate cases studied in [158], where regular solutions were obtained and a formula for optimal feedback was derived.

We will consider an optimal control problem for the two-dimensional stochastic Navier–Stokes equations with periodic boundary conditions in the setting of an abstract stochastic evolution equation for the velocity vector field discussed in Sect. 2.6.5. Let $\mathcal{O} = [0, L] \times [0, L]$, and let $\nu > 0$. We define the spaces

$$V = \left\{ x \in H_p^1 \left(\mathcal{O}; \mathbb{R}^2 \right), \operatorname{div} x = 0, \int_{\mathcal{O}} x = 0 \right\},$$

$$H = \quad \text{the closure of } V \text{ in } L^2 \left(\mathcal{O}; \mathbb{R}^2 \right),$$

where for an integer $k \geq 1$, $H_p^k \left(\mathcal{O}; \mathbb{R}^2 \right)$ is the space of $\mathbb{R}^2$-valued functions x that are in $H_{\text{loc}}^k \left(\mathbb{R}^2; \mathbb{R}^2 \right)$ and such that $x(y + Le_i) = x(y)$ for every $y \in \mathbb{R}^2$ and $i = 1, 2$. We will denote by $\langle \cdot, \cdot \rangle$ and $| \cdot |$, respectively, the inner product and the norm in $L^2 \left(\mathcal{O}; \mathbb{R}^2 \right)$. The space H inherits the same inner product and norm, and V has the norm inherited from $H_p^1 \left(\mathcal{O}; \mathbb{R}^2 \right)$. Let P_H be the orthogonal projection in $L^2 \left(\mathcal{O}; \mathbb{R}^2 \right)$ onto H. Define $Ax = P_H \Delta x$ with the domain $D(A) = H_p^2 \left(\mathcal{O}; \mathbb{R}^2 \right) \cap V$, and $B(x, y) = P_H[(x \cdot \nabla)y]$ for $x, y \in V$. The operator A is maximal dissipative, self-adjoint, and $(-A)^{-1}$ is compact. For $\gamma = 1, 2$ we define $V_\gamma := D((-A)^{\frac{\gamma}{2}})$, equipped with the norm

$$|x|_\gamma := |(-A)^{\frac{\gamma}{2}} x|. \tag{3.422}$$

The space V_1 coincides with V. Recall that

$$\int_{\mathcal{O}} |\operatorname{curl} x(\xi)|^2 d\xi = \int_{\mathcal{O}} |\nabla x(\xi)|^2 d\xi, \quad \text{for } x \in V.$$

Hence the $|x|_1$-norm is equivalent to

$$\left(\int_{\mathcal{O}} |\operatorname{curl} x(\xi)|^2 d\xi \right)^{1/2}.$$

The dual space V^* of V can be identified with the space V_{-1}, which is the completion of H with respect to the norm

$$|x|_{-1} := |(-A)^{-\frac{1}{2}} x|.$$

The duality is then given by

$$\langle x, y\rangle_{\langle V^*, V\rangle} = \langle (-A)^{-\frac{1}{2}}x, (-A)^{\frac{1}{2}}y\rangle.$$

Let $T > 0$ and Λ be a complete separable metric space. For every $0 \le t < T$, reference probability space $\nu = (\Omega, \mathscr{F}, \mathscr{F}_s^t, \mathbb{P}, W_Q)$, where W_Q is an H-valued Q-Wiener process with $Q \in \mathcal{L}_1^+(H)$, and $a(\cdot) \in \mathcal{U}_t^\mu$, the abstract controlled stochastic Navier–Stokes (SNS) equations describe the evolution of the velocity vector field $X : [t, T] \times \mathcal{O} \times \Omega \to \mathbb{R}^2$ that satisfies the stochastic evolution equation

$$\begin{cases} dX(s) = (AX(s) - B(X(s), X(s)) + f(s, a(s)))\, ds + dW_Q(s) & \text{in } (t, T] \times H, \\ X(t) = x \in H, \end{cases}$$
(3.423)

where $f : [0, T] \times \Lambda \to V$. (We remark that without loss of generality we set the viscosity coefficient in front of A to be 1.) The optimal control problem consists in the minimization, over all controls $a(\cdot) \in \mathcal{U}_t$, of a cost functional

$$J(t, x; a(\cdot)) = \mathbb{E}\left\{\int_t^T l(s, X(s), a(s))ds + g(X(T))\right\}.$$

The value function
$$v(t, x) = \inf_{a(\cdot)\in\mathcal{U}_t} J(t, x; a(\cdot)),$$
(3.424)

and the associated Hamilton–Jacobi–Bellman equation is

$$\begin{cases} u_t + \frac{1}{2}\text{Tr}\left(QD^2u\right) + \langle Ax - B(x, x), Du\rangle + F(t, x, Du) = 0, \\ u(T, x) = g(x) \qquad \text{for } (t, x) \in (0, T) \times H, \end{cases}$$
(3.425)

where the Hamiltonian function F is defined by

$$F(t, x, p) := \inf_{a\in\Lambda}\left\{\langle f(t, a), p\rangle + l(t, x, a)\right\}.$$
(3.426)

It is convenient to introduce the trilinear form $b(\cdot, \cdot, \cdot) : V \times V \times V \to \mathbb{R}$, defined as

$$b(x, y, z) = \int_{\mathcal{O}} z(\xi) \cdot (x(\xi) \cdot \nabla_\xi)y(\xi)\, d\xi = \langle B(x, y), z\rangle.$$

It is a continuous operator on $V \times V \times V$ but it can also be extended to a continuous map in different topologies, for instance it is also continuous on $V \times V_2 \times H$ (see [555] and (3.431) below.) The incompressibility condition gives the standard orthogonality relations

$$b(x, y, z) = -b(x, z, y), \quad b(x, y, y) = 0. \tag{3.427}$$

Also, because of the periodic boundary conditions (see for instance [555]),

$$b(x, x, Ax) = 0 \quad \text{for } x \in V_2. \tag{3.428}$$

We will be using the following inequalities. If $x, y, z \in V$ then

$$|b(x, y, z)| \leq C |x|^{1/2} |x|_1^{1/2} |y|_1 |z|^{1/2} |z|_1^{1/2}, \tag{3.429}$$

which gives when $z = x$

$$|b(x, y, x)| \leq C |x| |x|_1 |y|_1. \tag{3.430}$$

Also, if $x \in V, y \in V_2, z \in H$, then

$$|b(x, y, z)| \leq C |x|_1 |y|_2 |z|. \tag{3.431}$$

We will assume the following hypothesis throughout the rest of this section

Hypothesis 3.126

(i) $(-A)^{\frac{1}{2}} Q^{\frac{1}{2}} \in \mathcal{L}_2(H)$.
(ii) The function $f : [0, T] \times \Lambda \to V$ is continuous and there is $R \geq 0$ such that

$$|f(t, a)|_1 \leq R \quad \text{for all } t \in [0, T], a \in \Lambda. \tag{3.432}$$

We remark that Hypothesis 3.126-(i) is equivalent to the requirement that $\text{Tr}(Q_1) < +\infty$, where $Q_1 := (-A)^{\frac{1}{2}} Q(-A)^{\frac{1}{2}}$. By this we mean that $(-A)^{\frac{1}{2}} Q(-A)^{\frac{1}{2}}$ is densely defined and it extends to a bounded operator, still denoted by Q_1, belonging to $\mathcal{L}_1^+(H)$.

3.13.1 Estimates for Controlled SNS Equations

We will be using the notions of variational and strong solutions of the SNS equations (3.423). The definition of a variational solution is the same as that in Sect. 3.11.1, however since the generic equation (3.310) there is slightly different from (3.423), we repeat the definition below.

Definition 3.127 Let $0 \leq t < T$. Let $\mu = \left(\Omega, \mathcal{F}, \{\mathcal{F}_s^t\}_{s \in [t, T]}, \mathbb{P}, W_Q \right)$ be a generalized reference probability space, let Hypothesis 3.126 be satisfied. Let ξ be an $\mathcal{F}_t^t$-measurable H-valued random variable such that $\mathbb{E}^\mu[|\xi|^2] < +\infty$, and let $a(\cdot) \in \mathcal{U}_t^\mu$.

- A process $X(\cdot) \in M_\mu^2(t, T; H)$ is called a *variational solution* of (3.423) with initial condition $X(t) = \xi$ if

$$\mathbb{E}\left[\int_t^T |X(r)|_V^2 dr\right] < +\infty$$

and for every $\phi \in V$ we have

$$\langle X(s), \phi \rangle = \langle \xi, \phi \rangle + \int_t^s \langle AX(r) - B(X(r), X(r)) + f(r, a(r)), \phi \rangle_{\langle V^*, V \rangle} dr$$

$$+ \int_t^s \langle dW_Q(r), \phi \rangle \quad \text{for each } s \in [t, T], \ \mathbb{P} - \text{a.e.}$$

- A process $X(\cdot) \in M_\mu^2(t, T; H)$ is called a *strong solution* of (3.423) with initial condition $X(t) = \xi$ if

$$\mathbb{E}\left[\int_t^T |X(r)|_{V_2}^2 dr\right] < +\infty$$

and we have

$$X(s) = \xi + \int_t^s (AX(r) - B(X(r), X(r)) + f(r, a(r)))dr + \int_t^s dW_Q(r)$$

for each $s \in [t, T]$, $\mathbb{P}$-a.e.

Proposition 3.128 *Let* $0 \leq t < T$ *and* $p \geq 2$. *Let* $\mu = \left(\Omega, \mathscr{F}, \{\mathscr{F}_s^t\}_{s \in [t,T]}, \mathbb{P}, W_Q\right)$ *be a generalized reference probability space, and let Hypothesis 3.126 be satisfied. Let* ξ *be an* $\mathscr{F}_t^t$-*measurable* H-*valued random variable such that* $\mathbb{E}^\mu[|\xi|^p] < +\infty$, *and let* $a(\cdot) \in \mathcal{U}_t^\mu$. *Then:*

(i) *There exists a unique variational solution* $X(\cdot) = X(\cdot; t, \xi, a(\cdot))$ *of (3.423) with initial condition* $X(t) = \xi$. *The solution has continuous trajectories and satisfies, for* $t \leq s \leq T$,

$$\mathbb{E}|X(s)|^p + \mathbb{E}\int_t^s |X(\tau)|_1^2 |X(\tau)|^{p-2} d\tau \leq \mathbb{E}|\xi|^p + C(p, R, Q)(s-t) \quad (3.433)$$

and

$$\mathbb{E}\left[\sup_{t \leq s \leq T} |X(s)|^p\right] \leq C(p, T, R, Q)\left(1 + \mathbb{E}|\xi|^p\right). \quad (3.434)$$

(ii) *If* $\mathbb{E}|\xi|_1^p < +\infty$, *then the variational solution* $X(\cdot) = X(\cdot; t, \xi, a(\cdot))$ *is a strong solution with trajectories continuous in* V. *Moreover, we have for* $t \leq s \leq T$

$$\mathbb{E}|X(s)|_1^p + \mathbb{E}\int_t^s |X(\tau)|_2^2 |X(\tau)|_1^{p-2} d\tau \le \mathbb{E}|\xi|_1^p + C(p, R, Q_1)(s - t)$$

(3.435)

and

$$\mathbb{E}\left[\sup_{t \le s \le T} |X(s)|_1^p\right] \le C(p, T, R, Q_1)\left(1 + \mathbb{E}|\xi|_1^p\right).$$

(3.436)

(iii) If μ_1 is another generalized reference probability space, ξ_1 is an $\mathscr{F}_t^{t,\mu_1}$-measurable H-valued random variable such that $\mathbb{E}^{\mu_1}[|\xi_1|^p] < +\infty$, $a_1(\cdot) \in \mathcal{U}_t^{\mu_1}$, and

$$\mathcal{L}_{\mathbb{P}_1}(\xi_1, a_1(\cdot), W_{Q,1}(\cdot)) = \mathcal{L}_{\mathbb{P}}(\xi, a(\cdot), W_Q(\cdot)),$$

then

$$\mathcal{L}_{\mathbb{P}_1}(a_1(\cdot), X_1(\cdot)) = \mathcal{L}_{\mathbb{P}}(a(\cdot), X(\cdot)),$$

(3.437)

where $X_1(\cdot) = X_1(\cdot; t, \xi_1, a_1(\cdot))$ is the variational solution of (3.423) in μ_1 with control $a_1(\cdot)$ and initial condition ξ_1.

Proof (i) The general strategy of the proof of part (i) is similar to the proof of Theorem 3.102. More precisely, part (i) is proved in [444], Proposition 3.3 (see also [124] for a similar proof and [177, 567] for related results and estimates). We sketch the main points of the proof since we will need them to explain parts (ii) and (iii).

Let $\{e_1, e_2, ...\}$ be the orthonormal basis of H composed of eigenvectors of A, $H_n := \text{span}\{e_1, ..., e_n\}$, and P_n be the orthogonal projection in H onto H_n. In this case P_n extends to the orthogonal projection in V^* onto H_n. Also we have $P_n A = A P_n$. Let $X^n(\cdot)$ be the unique strong solution of

$$\begin{cases} dX^n(s) = (P_n A X^n(s) - P_n B(X^n(s), X^n(s)) + P_n f(s, a(s)))ds + P_n dW_Q(s) \\ X^n(t) = P_n \xi. \end{cases}$$

(3.438)

We first assume that $p \ge 8$. It follows, using Itô's formula (see also [444]), that we have

$$\mathbb{E}\left[\sup_{t \le s \le T} |X^n(s)|^p + \int_t^T |X^n(s)|_V^2 (1 + |X^n(s)|^{p-2}) ds\right] \le M \quad \text{for all } n.$$

It can also be deduced from the estimates for $X^n(\cdot)$ obtained from Itô's formula, and Itô's isometry, that the norms of $X^n(\cdot)$ in $L^8(\Omega; L^4((t, T) \times \mathcal{O}))$ are bounded uniformly in n. Therefore, there exists a process $X(\cdot) \in M_\mu^2(t, T; V) \cap L^p(\Omega; L^\infty(t, T; H))$, and a $\mathscr{F}_T$-measurable random variable $\eta \in L^2(\Omega; H)$ such that (up to a subsequence and identifying $X(\cdot)$ with its versions)

$$X^n(\cdot) \rightharpoonup X(\cdot) \text{ in } M_\mu^2(t, T; V), \quad X^n(T) \rightharpoonup \eta \text{ in } L^2(\Omega; H),$$

$$X^n(\cdot) \to X(\cdot) \text{ weak star in } L^p(\Omega; L^\infty(t, T; H)).$$

We can also assume that $X(\cdot) \in L^8(\Omega; L^4((t, T) \times \mathcal{O}))$. Passing to the limit as $n \to +\infty$ we obtain that there is a process $F_0(\cdot) \in M_\mu^2(t, T; V^*)$ such that, up to a subsequence,

$$P_n A X^n(\cdot) - P_n B(X^n(\cdot), X^n(\cdot)) \rightharpoonup F_0(\cdot) \text{ in } M_\mu^2(t, T; V^*),$$

$X(\cdot)$ is a variational solution of

$$\begin{cases} dX(s) = (F_0(s) + f(s, a(s)))ds + dW_Q(s) \\ X(t) = \xi, \end{cases}$$

$X(T) = \eta$, and, by Theorem 3.101, $X(\cdot) \in L^p(\Omega; C([t, T], H))$ and $\mathbb{P}$-a.e.

$$|X(s)|^2 = |\xi|^2 + 2\int_t^s \langle F_0(r) + f(r, a(r)), X(r) \rangle_{\langle V^*, V \rangle} dr$$
$$+ 2\int_t^s \langle dW_Q(r), X(r) \rangle + \mathrm{Tr}(Q)(s - t).$$

One then uses an argument based on the monotonicity of the operator $-Ax + B(x, x)$ on balls in $L^4(\mathcal{O})$ (see [444]) to show that $F_0(\cdot) = AX(\cdot) - B(X(\cdot), X(\cdot))$, i.e. $X(\cdot)$ is a variational solution of (3.423), and thus using (3.427), we have

$$|X(s)|^2 = |\xi|^2 - 2\int_t^s |X(r)|_1^2 dr + 2\int_t^s \langle f(r, a(r)), X(r) \rangle dr$$
$$+ 2\int_t^s \langle dW_Q(r), X(r) \rangle + \mathrm{Tr}(Q)(s - t). \tag{3.439}$$

(The monotonicity argument uses the fact that $X(\cdot) \in L^8(\Omega; L^4((t, T) \times \mathcal{O}))$.) We also have a similar identity as (3.439) for $|X^n(s)|^2$. Taking expectation in both of them for $s = T$, passing to the limit as $n \to +\infty$, and recalling that $X^n(T) \rightharpoonup X(T)$, we deduce $\mathbb{E}|X^n(T)|^2 \to \mathbb{E}|X(T)|^2$, which gives $\mathbb{E}|X^n(T) - X(T)|^2 \to 0$ as $n \to +\infty$. Replacing T by $s \in (t, T)$, the same arguments give $\mathbb{E}|X^n(s) - X(s)|^2 \to 0$.

Estimates (3.433) and (3.434) can now be proved by applying Itô's formula to the function $\varphi(r) = r^{p/2}$ and using identity (3.439).

Uniqueness of variational solutions for any $p \geq 2$ follows from Proposition 3.129-(i).

Let now $2 \leq p < 8$. For $n \geq 1$ we define $\Omega^n := \{\omega \in \Omega : |\xi(\omega)| \leq n\}$ and $\xi^n := \xi 1_{\Omega^n}$. Then $X(\cdot; t, \xi^n, a(\cdot)) = X(\cdot; t, \xi^m, a(\cdot))$ on Ω^n if $n \leq m$, and estimates (3.433) and (3.434) are true for the processes $X(\cdot; t, \xi^n, a(\cdot)), n \geq 1$. Therefore, the process $X(s) := \lim_{n \to +\infty} X(\cdot; t, \xi^n, a(\cdot))$ is well defined, (3.433) and (3.434) for $X(\cdot; t, \xi^n, a(\cdot))$ follow from Fatou's lemma, and it is easy to see that $X(\cdot)$ is a variational solution of (3.423).

(ii) Let $p \geq 8$ and $X^n(\cdot)$ be the processes from part (i). It follows from Itô's formula, Hypothesis 3.126 and (3.428), that there is an $M \geq 0$ such that the processes $X^n(\cdot)$ satisfy in this case

$$\mathbb{E}\left[\sup_{t \leq s \leq T} |X^n(s)|_1^p + \int_t^T |X^n(s)|_2^2 (1 + |X^n(s)|_1^{p-2})ds\right] \leq M \quad \text{for all } n.$$

Therefore by passing to a weak limit we obtain that the variational solution from part (i) satisfies $X(\cdot) \in L^p(\Omega; L^\infty(t, T; V)) \cap M_\mu^2(t, T; V_2)$, and thus it is a strong solution. Moreover, $(-A)^{\frac{1}{2}} (AX(r) - B(X(r), X(r)) + f(r, a(r))) \in M_\mu^2(t, T; V_{-1})$, the process

$$(-A)^{\frac{1}{2}} X(s) = (-A)^{\frac{1}{2}} \xi$$
$$+ \int_t^s (-A)^{\frac{1}{2}} (AX(r) - B(X(r), X(r)) + f(r, a(r))) \, dr + \int_t^s (-A)^{\frac{1}{2}} dW_Q(r)$$

is a continuous process with values in V_{-1}, and $(-A)^{\frac{1}{2}} X(\cdot) \in M_\mu^2(t, T; V)$. Thus, by Theorem 3.101, $X(\cdot) \in L^p(\Omega; C([t, T], V))$ and $\mathbb{P}$-a.e.

$$|X(s)|_1^2 = |\xi|_1^2 - 2 \int_t^s \left(|AX(r)|^2 + \langle f(r, a(r)), AX(r) \rangle\right) dr$$
$$- 2 \int_t^s \langle dW_Q(r), AX(r) \rangle + \mathrm{Tr}(Q_1)(s - t). \tag{3.440}$$

Estimates (3.435) and (3.436) now follow by standard arguments applying Itô's formula to the function $\varphi(r) = r^{p/2}$ and using identity (3.440). For $2 \leq p < 8$ we proceed as in the proof of part (i).

(iii) Similarly to the proof of part (ii) of Theorem 3.102, if $p \geq 8$ and $X_1(\cdot)$ is the variational solution in the generalized reference probability space μ_1 and $X_1^n(\cdot)$ are the solutions of the approximating problems (3.438) in this space, then

$$\mathcal{L}_{\mathbb{P}_1}(a_1(\cdot), X_1^n(\cdot)) = \mathcal{L}_{\mathbb{P}}(a(\cdot), X^n(\cdot)),$$

and thus (3.437) follows since $X_1^n(s) \to X_1(s)$ in $L^2(\Omega_1; H)$ and $X^n(s) \to X(s)$ in $L^2(\Omega; H)$ for every $s \in [t, T]$. For $2 \leq p < 8$ we have

$$\mathcal{L}_{\mathbb{P}_1}(a_1(\cdot), X_1(\cdot; t, \xi_1^n, a_1(\cdot))) = \mathcal{L}_{\mathbb{P}}(a(\cdot), X(\cdot; t, \xi^n, a(\cdot))),$$

which gives the claim in the limit as $n \to +\infty$. $\qquad\square$

Without loss of generality we will always assume from now on that the Q-Wiener processes in the reference probability spaces have everywhere continuous paths.

Proposition 3.129 *Let $0 \le t < T$ and $p \ge 2$. Let $\nu = \left(\Omega, \mathscr{F}, \{\mathscr{F}_s^t\}_{s\in[t,T]}, \mathbb{P}, W_Q\right)$ be a reference probability space, and let Hypothesis 3.126 be satisfied. Let ξ, η be $\mathscr{F}_t^t$-measurable H-valued random variables such that $\mathbb{E}^\nu[|\xi|^p + |\eta|^p] < +\infty$, and let $a(\cdot) \in \mathcal{U}_t^\nu$. Then:*

(i) *There exists a constant C independent of $t, \xi, \eta, a(\cdot)$ and μ, such that a.s. on Ω*

$$|X(s) - Y(s)|^2 + \int_t^s |X(\tau) - Y(\tau)|_1^2 d\tau \le |\xi - \eta|^2 \exp\left\{\int_t^s C|X(\tau)|_1^2 d\tau\right\}$$

(3.441)

for all $s \in [t, T]$, where $X(\cdot) = X(\cdot; t, \xi, a(\cdot)), Y(\cdot) = Y(\cdot; t, \eta, a(\cdot))$ are solutions of (3.423) with initial conditions $X(t) = \xi$ and $Y(t) = \eta$.

(ii) *If $|x|_1 \le R_1$ then there exists a constant $C = C(p, T, R, R_1, Q)$ such that*

$$\mathbb{E}|X(s) - x|^p \le C(p, T, R, R_1, Q)(s - t), \quad \text{for all } s \in [t, T], \quad (3.442)$$

where $X(\cdot) = X(\cdot; t, x, a(\cdot))$.

(iii) *For every initial condition $x \in V$ there exists a modulus ω, independent of the reference probability spaces ν and controls $a(\cdot) \in \mathcal{U}_t^\nu$, such that*

$$\mathbb{E}|X(s) - x|_1^2 \le \omega_x(s - t), \quad \text{for all } s \in [t, T], \quad (3.443)$$

where $X(\cdot) = X(\cdot; t, x, a(\cdot))$.

Proof (i) Let $Z(\cdot) = X(\cdot) - Y(\cdot)$. Then $Z(\cdot)$ satisfies, for $s \in [t, T]$,

$$Z(s) = \xi - \eta + \int_t^s AZ(\tau)d\tau + \int_t^s [B(Y(\tau), Y(\tau)) - B(X(\tau), X(\tau))]d\tau.$$

Hence, using (3.427) and (3.430), we obtain

$$\begin{aligned}|Z(s)|^2 &= |\xi - \eta|^2 - 2\int_t^s |Z(\tau)|_1^2 d\tau - \int_t^s b(Z(\tau), X(\tau), Z(\tau))d\tau \\ &\le |\xi - \eta|^2 - 2\nu \int_t^s |Z(\tau)|_1^2 d\tau + \int_t^s C|Z(\tau)|_1 |X(\tau)|_1 |Z(\tau)| d\tau \\ &\le |\xi - \eta|^2 - \nu \int_t^s |Z(\tau)|_1^2 d\tau + C\int_t^s |X(\tau)|_1^2 |Z(\tau)|^2 d\tau. \quad (3.444)\end{aligned}$$

Here we have used Young's inequality. Then it follows from Gronwall's lemma that

$$|Z(s)|^2 \le |\xi - \eta|^2 \exp\{\int_t^s C|X(\tau)|_1^2 d\tau\} \quad \mathbb{P}\text{-a.s.}$$

Plugging this back into (3.444) yields (3.441) with another constant C.

(ii) Let $Y(\cdot) = X(\cdot) - x$. Then, for $s \in [t, T]$,

$$Y(s) = \int_t^s (AX(\tau) - B(X(\tau), X(\tau)) + f(\tau, a(\tau))) \, d\tau + \int_t^s dW(\tau).$$

Therefore, applying Itô's formula, taking expectation, and using (3.427) and the Cauchy–Schwarz inequality, we obtain

$$\begin{aligned}
\mathbb{E}|Y(s)|^p &\le \mathbb{E} \int_t^s p\langle AX(\tau) - B(X(\tau), X(\tau)) + f(\tau, a(\tau)), Y(\tau)\rangle |Y(\tau)|^{p-2} d\tau \\
&\quad + \mathbb{E} \int_t^s \frac{p(p-1)}{2} \mathrm{tr}(Q)|Y(\tau)|^{p-2} d\tau \\
&\le -\frac{p}{2}\mathbb{E} \int_t^s |X(\tau)|_1^2 |Y(\tau)|^{p-2} d\tau + C_p \mathbb{E} \int_t^s |x|_1^2 |Y(\tau)|^{p-2} d\tau \\
&\quad + C(p, R, R_1, Q)\mathbb{E} \int_t^s (|Y(\tau)|^{p-1} + |Y(\tau)|^{p-2}) d\tau \\
&\quad + p\mathbb{E} \int_t^s |b(X(\tau), X(\tau), x)||Y(\tau)|^{p-2} d\tau.
\end{aligned}$$

Since

$$|b(X(\tau), X(\tau), x)| \le C|X(\tau)|_1 |X(\tau)||x|_1 \le \frac{1}{2}|X(\tau)|_1^2 + \frac{C^2}{2}|X(\tau)|^2 |x|_1^2,$$

plugging this into the previous inequality and using (3.434) finally yields

$$\begin{aligned}
\mathbb{E}|Y(s)|^p &\le C(p, R, R_1, Q)\mathbb{E} \int_t^s (|Y(\tau)|^{p-1} + |Y(\tau)|^{p-2} + |X(\tau)|^2 |Y(\tau)|^{p-2}) d\tau \\
&\le C(p, T, R, R_1, Q)(s - t).
\end{aligned}$$

(iii) If (3.443) is not satisfied then there are $\varepsilon > 0$, $a_n(\cdot) \in \mathcal{U}_t$ (which we can assume to be $\mathscr{F}^{t,0}$-predictable) and $s_n \to t$ such that $\mathbb{E}|X_n(s_n) - x|_1^2 \ge \varepsilon$ for all $n \ge 1$, where $X_n(\cdot) = X(\cdot; t, x, a_n(\cdot))$. By Corollary 2.21 and Proposition 3.128-(iii), we can assume that all $a_n(\cdot)$ are defined on the same reference probability space.

However, it follows from (3.442) and (3.435) that, up to a subsequence, we have

$$X_n(s_n) \to x \text{ strongly in } L^2(\Omega; H) \text{ and weakly in } L^2(\Omega; V).$$

Since the weak sequential convergence in $L^2(\Omega; V)$ implies

$$|x|_1^2 \le \liminf_{n \to \infty} \mathbb{E}|X(s_n)|_1^2,$$

this, together with (3.435), implies $|x|_1^2 = \lim_{n \to \infty} \mathbb{E}|X(s_n)|_1^2$. Therefore $X(s_n) \to x$ strongly in $L^2(\Omega; V)$, contrary to our assumption. $\qquad\square$

3.13.2 The Value Function

In this section we show continuity properties of the value function of the stochastic optimal control problem and the dynamic programming principle.

To minimize non-essential technical difficulties we will assume that the running cost function l is independent of t. The case of l depending on t is a straightforward extension of the methods presented here. The continuity of the value function is not entirely trivial since continuous dependence estimates in the mean for solutions of the stochastic Navier–Stokes equations depend on exponential moments of solutions (3.441) and these seem to be bounded only for a short time (see Corollary XI.3.1 in [567], also [541]). We make the following assumptions about the cost functions l and g.

Hypothesis 3.130 The functions $l : V \times \Lambda \to \mathbb{R}$, and $g : H \to \mathbb{R}$ are continuous and there exist $k \geq 0$ and for every $r > 0$ a modulus σ_r such that

$$|l(x, a)|, |g(x)| \leq C(1 + |x|_1^k) \quad \text{for all } x \in V, a \in \Lambda, \tag{3.445}$$

$$|l(x, a) - l(y, a)| \leq \sigma_r (|x - y|_1) \quad \text{if } |x|_1, |y|_1 \leq r, a \in \Lambda \tag{3.446}$$

$$|g(x) - g(y)| \leq \sigma_r (|x - y|) \quad \text{if } |x|_1, |y|_1 \leq r. \tag{3.447}$$

Proposition 3.131 *Let Hypotheses 3.126 and 3.130 be satisfied. Then:*

(i) *For every $r > 0$ there exists a modulus ω_r such that for every $t \in [0, T]$, $a(\cdot) \in \mathcal{U}_t$*

$$|J(t, x; a(\cdot)) - J(t, y; a(\cdot))| \leq \omega_r (|x - y|) \quad \text{if } |x|_1, |y|_1 \leq r. \tag{3.448}$$

(ii) *The value function v satisfies the dynamic programming principle, i.e. for every $0 \leq t \leq \eta \leq T$ and $x \in V$,*

$$v(t, x) = \inf_{a(\cdot) \in \mathcal{U}_t} \mathbb{E} \left\{ \int_t^\eta l(X(s; t, x, a(\cdot)), a(s))ds + v(\eta, X(\eta; t, x, a(\cdot))) \right\}. \tag{3.449}$$

(iii) *For every $r > 0$ there exists a modulus ω_r such that*

$$|v(t_1, x) - v(t_2, y)| \leq \omega_r (|t_1 - t_2| + |x - y|) \tag{3.450}$$

for all $t_1, t_2 \in [0, T]$ and $|x|_1, |y|_1 \leq r$, and there exists a $C \geq 0$ such that

$$|v(t, x)| \leq C(1 + |x|_1^k) \tag{3.451}$$

for all $t \in [0, T]$ and $x \in V$.

Proof (i) Let $x, y \in V, t \in [0, T]$ and $a(\cdot) \in \mathcal{U}_t$. For every $m > 0$ let D_m be a constant such that $\sigma_m(s) \le \frac{1}{m} + D_m s$. Define, for $s \in [t, T]$, $X(s) = X(s; t, x, a(\cdot))$, $Y(s) = Y(s; t, y, a(\cdot))$, and $A_m = \{\omega \in \Omega : \max_{t \le s \le T} |X(s)|_1 \le m\}$, $B_m = \{\omega \in \Omega : \max_{t \le s \le T} |Y(s)|_1 \le m\}$. Then, using (3.436), (3.441), (3.445) and (3.446), we obtain

$$
\mathbb{E} \int_t^T |l(X(s), a(s)) - l(Y(s), a(s))| ds \le \frac{T}{m} + \mathbb{E} \int_t^T D_m |X(s) - Y(s)|_1 \mathbf{1}_{A_m \cap B_m} ds
$$

$$
+ \mathbb{E} \int_t^T C(2 + |X(s)|_1^k + |Y(s)|_1^k) \mathbf{1}_{\Omega \setminus (A_m \cap B_m)} ds
$$

$$
\le \frac{T}{m} + D_m |x - y| \mathbb{E} \int_t^T \exp \left\{ C \int_t^s |X(\tau)|_1^2 d\tau \right\} \mathbf{1}_{A_m} ds
$$

$$
+ \int_t^T C \left(2 + (\mathbb{E}|X(s)|_1^{2k})^{\frac{1}{2}} + (\mathbb{E}|Y(s)|_1^{2k})^{\frac{1}{2}} \right) \left((\mathbb{P}(\Omega \setminus A_m))^{\frac{1}{2}} + (\mathbb{P}(\Omega \setminus B_m))^{\frac{1}{2}} \right) ds
$$

$$
\le \frac{T}{m} + D_m T |x - y| e^{CTm^2} + C_1(p, T, R, Q_1)(1 + |x|_1^k + |y|_1^k) \frac{1 + |x|_1 + |y|_1}{m}.
$$

Applying the same process to estimate $|g(X(T)) - g(Y(T))|$ we therefore obtain that for every $r, m > 0$ there exist constants c_m, d_r such that, for every $t \in [0, T]$ and $a(\cdot) \in \mathcal{U}_t$,

$$
|J(t, x; a(\cdot)) - J(t, y; a(\cdot))| \le \frac{d_r}{m} + c_m |x - y| \quad \text{if } |x|_1, |y|_1 \le r.
$$

Estimate (3.448) now follows by taking the infimum over all $m > 0$.

(ii) We need to show that the problem satisfies the assumptions of Hypothesis 2.12. However, here the statement of the DPP is restricted to points in V but the filtrations are still generated by Q-Wiener processes with values in H. Thus to proceed with the proof of the DPP described in Sect. 2.3 it is enough to assume in Hypothesis 2.12 that the random variable ξ there satisfies $\mathbb{E}^\mu |\xi|_1^2 < +\infty$. We recall that in this case we have strong solutions, so conditions $(A0)$ and $(A2)$ follow from the definition of solution and standard arguments. In particular, $(A0)$ follows from (3.441) and $(A2)$ from an obvious generalization of (3.441). Condition $(A1)$ follows from Proposition 3.128-(iii). We point out that if $\mathcal{L}_{\mathbb{P}_1}(X_1(\cdot; t_1, x, a_1(\cdot)), a_1(\cdot)) = \mathcal{L}_{\mathbb{P}_2}(X_2(\cdot; t_1, x, a_2(\cdot)), a_2(\cdot))$ as processes with values in $H \times \Lambda$ then, by Lemma 1.17-(i), they have the same laws as processes with values in $V \times \Lambda$. The proof of condition $(A3)$ starts in the same way as in its proof in Proposition 2.16, however the arguments are now obvious since here the stochastic integral is just $W_{t_1}(s)$.

(iii) We notice that (3.448) implies

$$
|v(t, x) - v(t, y)| \le \omega_r (|x - y|) \tag{3.452}
$$

for all $t \in [0, T]$ and $|x|_1, |y|_1 \le r$. Moreover, (3.451) is a direct consequence of (3.436) and (3.445).

Let now $0 \le t_1 < t_2 \le T, x \in V, |x|_1 \le r$. We will define, for $s \in [t, T]$, $X(s) = X(s; t_1, x, a(\cdot))$. Using (3.436), (3.442), (3.449), (3.451), (3.452), we obtain for $m > r$,

$$
\begin{aligned}
|v(t_1, x) - v(t_2, x)| &\le \sup_{a(\cdot) \in \mathcal{U}_{t_1}} \mathbb{E} \int_{t_1}^{t_2} (1 + |X(s)|_1^k) ds \\
&\quad + \sup_{a(\cdot) \in \mathcal{U}_{t_1}} \mathbb{E} |v(t_2, X(t_2)) - v(t_2, x)| \\
&\le C(R, T, Q_1, r)(t_2 - t_1) + \sup_{a(\cdot) \in \mathcal{U}_{t_1}} \left\{ \mathbb{E} \left(C(1 + |X(t_2)|_1^k + |x|_1^k) \mathbf{1}_{\{|X(t_2)|_1 > m\}} \right) \right\} \\
&\quad + \sup_{a(\cdot) \in \mathcal{U}_{t_1}} \mathbb{E} \sigma_m \left(|X(t_2) - x| \right) \\
&\le C(R, T, Q_1, r)(t_2 - t_1) + C(R, T, Q_1)(1 + |x|_1^k) \frac{1 + |x|_1}{m} \\
&\quad + \sigma_m \left(C(R, Q, r)(t_2 - t_1)^{\frac{1}{2}} \right).
\end{aligned}
$$

(We also used that σ_m above can be assumed to be concave.) The result now follows by taking the infimum over $m > r$. $\square$

3.13.3 Viscosity Solutions and the Comparison Theorem

Since we only have continuity of the value function on $[0, T] \times V$, the definition of a viscosity solution has to be restricted to this space. From the point of view of the HJB equation it might be better to set it up in this space, however because of the associated control problem, we want to keep H as our reference space. We achieve it by a proper choice of test functions. By using a special radial function of $|\cdot|_1$ as test function we first restrict the points where maxima or minima occur in the definition of viscosity sub/solution to be in $(0, T) \times V$. Then we require that the points where the maxima/minima occur belong to $(0, T) \times V_2$. Having this property we can interpret all terms appearing in the HJB equation. In this way we gain some coercive terms which had to be discarded in the generic definition given in Sect. 3.3, which are very useful in the proof of the comparison principle. The definition is meaningful as we are able to show, using properties of the Navier–Stokes equations and the coercivity of the operator $-A$, that the value function is a viscosity solution. The definition of viscosity solution here is thus similar to the one used in Sects. 3.11 and 3.12, however we use a radial test function of a different type. If different continuity requirements were imposed in Hypothesis 3.130, we would have different continuity properties of the value function, and then we could work with a definition of viscosity solution which more closely resembles the definition from Sect. 3.11, as was done for first-order equations in [321].

Definition 3.132 A function ψ is a test function for Eq. (3.425) if $\psi = \varphi + \delta(t)$ $(1 + |x|_1^2)^m$, where

(i) $\varphi \in C^{1,2}((0, T) \times H)$, and is such that $\varphi_t, D\varphi, D^2\varphi$ are uniformly continuous on $[\varepsilon, T - \varepsilon] \times H$ for every $\varepsilon > 0$.
(ii) $\delta \in C^1((0, T))$ is such that $\delta > 0$ on $(0, T)$, and $m \geq 1$.

The function $h(t, x) = \delta(t)(1 + |x|_1^2)^m$ is not Fréchet differentiable in H. Therefore the terms involving Dh and D^2h, in particular $\langle Ax - B(x, x), Dh(t, x)\rangle$ and $\mathrm{Tr}(QD^2h(t, x))$ have to be understood properly. We define

$$Dh(t, x) := -\delta(t)\left(2m(1 + |x|_1^2)^{m-1}Ax\right),$$

and we will write

$$D\psi := D\varphi + Dh$$

even though this is a slight abuse of notation. Then, if $(t, x) \in (0, T) \times V_2$, $D\psi(t, x)$ makes sense, and so does the term $\langle Ax - B(x, x), D\psi(t, x)\rangle$. As regards the term $\mathrm{Tr}(QD^2\psi(t, x))$, without defining $D^2h(t, x)$, we interpret it by defining

$$\mathrm{Tr}(QD^2\psi(t, x)) := \mathrm{Tr}(QD^2\varphi(t, x)) + \delta(t)\left(2m(1 + |x|_1^2)^{m-1}\mathrm{Tr}(Q_1)\right.$$
$$\left. + 4m(m - 1)(1 + |x|_1^2)^{m-2}|Q^{\frac{1}{2}}Ax|^2\right).$$

It will be seen in the next section that the above interpretations appear as direct consequences of Itô's formula applied to h.

We give a definition of viscosity solution for a general Eq. (3.425) where the Hamiltonian function F is not necessarily given by (3.426). Thus we assume in this section that $F : [0, T] \times V \times H \to \mathbb{R}$ is any function.

Definition 3.133 A weakly sequentially upper-semicontinuous (respectively, lower-semicontinuous) function $u : (0, T] \times V \to \mathbb{R}$ is called a viscosity subsolution (respectively, supersolution) of (3.425) if $u(T, y) \leq h(y)$ (respectively, $u(T, y) \geq h(y)$) for all $y \in V$ and if, for every test function ψ, whenever $u - \psi$ has a global maximum (respectively $u + \psi$ has a global minimum) over $(0, T) \times V$ at (t, x), then $x \in V_2$ and

$$\psi_t(t, x) + \frac{1}{2}\mathrm{Tr}(QD^2\psi(t, x)) + \langle Ax - B(x, x), D\psi(t, x)\rangle + F(t, x, D\psi(t, x)) \geq 0$$

(respectively,

$$-\psi_t(t, x) - \frac{1}{2}\mathrm{Tr}(QD^2\psi(t, x)) - \langle Ax - B(x, x), D\psi(t, x)\rangle + F(t, x, -D\psi(t, x)) \leq 0.)$$

A function u is a viscosity solution of (3.425) if it is both a viscosity subsolution and a viscosity supersolution of (3.425).

Hypothesis 3.134 $F : [0, T] \times V \times H \to \mathbb{R}$ and there exist a modulus of continuity w, and moduli w_r such that for every $r > 0$ we have

$$|F(t, x, p) - F(t, y, p)| \le w_r (|x - y|_1) + w (|x - y|_1 |p|), \quad \text{if } |x|_1, |y|_1 \le r, \tag{3.453}$$

$$|F(t, x, p) - F(t, x, q)| \le w ((1 + |x|_1)|p - q|), \tag{3.454}$$

$$|F(t, x, p) - F(s, x, p)| \le w_r(|t - s|), \quad \text{if } |x|_1, |p|_1 \le r, \tag{3.455}$$

$$|g(x) - g(y)| \le w_r(|x - y|), \quad \text{if } |x|_1, |y|_1 \le r. \tag{3.456}$$

Theorem 3.135 *Let Hypothesis 3.134 hold. Let $u, v : (0, T] \times V \to \mathbb{R}$ be, respectively, a viscosity subsolution, and a viscosity supersolution of (3.425). Let*

$$u(t, x), \ -v(t, x), \ |g(x)| \le C(1 + |x|_1^k) \tag{3.457}$$

for some $k \ge 0$. Then $u \le v$ on $(0, T] \times V$.

Proof We observe that weak sequential upper-semicontinuity of u and weak sequential lower-semicontinuity of v imply that

$$\begin{cases} \lim_{t \uparrow T} (u(t, x) - g(x))^+ = 0 \\ \lim_{t \uparrow T} (v(t, x) - g(x))^- = 0 \end{cases} \tag{3.458}$$

uniformly on bounded subsets of V. We define for $\mu > 0$,

$$u_\mu(t, x) = u(t, x) - \frac{\mu}{t}, \quad v_\mu(t, x) = v(t, x) + \frac{\mu}{t}.$$

Then u_μ and v_μ are, respectively, a viscosity subsolution, and a viscosity supersolution of

$$(u_\mu)_t + \frac{1}{2}\mathrm{Tr}(QD^2u_\mu) + \langle Ax - B(x, x), Du_\mu \rangle + F(t, x, Du_\mu) = \frac{\mu}{T^2}$$

and

$$(v_\mu)_t + \frac{1}{2}\mathrm{Tr}(QD^2u_\mu) + \langle Ax - B(x, x), Dv_\mu \rangle + F(t, x, Dv_\mu) = -\frac{\mu}{T^2}.$$

Let m be a number such that $m \ge 1$ and $2m \ge k + 1$. For $0 < \varepsilon, \delta, \beta \le 1$, we consider the function

$$\Phi(t, s, x, y) = u_\mu(t, x) - v_\mu(s, y) - \frac{|x - y|^2}{2\varepsilon}$$

$$-\delta e^{K_\mu(T-t)}(1 + |x|_1^2)^m - \delta e^{K_\mu(T-s)}(1 + |y|_1^2)^m - \frac{(t - s)^2}{2\beta}$$

and set

$$\Phi(t, s, x, y) = -\infty \quad \text{if } x, y \notin V.$$

The constant K_μ will be chosen later. Obviously $\Phi \to -\infty$ as $\max(|x|_1, |y|_1) \to +\infty$. We claim that Φ is weakly sequentially upper-semicontinuous on $(0, T] \times (0, T] \times H \times H$.

It is well known that functions $x \to (1 + |x|_1^2)^m$, $y \to (1 + |y|_1^2)^m$ and $|x - y|^2$ are weakly sequentially lower-semicontinuous, respectively, in H and $H \times H$. To show that, say,

$$u_\mu(t, x) - \delta e^{K_\mu(T-t)}(1 + |x|_1^2)^m$$

is weakly sequentially upper-semicontinuous on $(0, T] \times H$, we suppose that this is not the case, i.e. there exist sequences $t_n \to t \in (0, T]$, $x_n \rightharpoonup x \in H$ such that

$$\limsup_{n \to \infty} \left(u_\mu(t_n, x_n) - \delta e^{K_\mu(T-t_n)}(1 + |x_n|_1^2)^m \right) > u_\mu(t, x) - \delta e^{K_\mu(T-t)}(1 + |x|_1^2)^m.$$

If $\liminf_{n \to \infty} |x_n|_1 = +\infty$, this is impossible by (3.457). So there must exist a subsequence (still denoted by (t_n, x_n)) such that $\limsup_{n \to \infty} |x_n|_1 < +\infty$. But then we have $x_n \rightharpoonup x$ in V, which contradicts the weak sequential upper-semicontinuity of u_μ.

Therefore Φ has a global maximum over $(0, T] \times (0, T] \times H \times H$ at some point $(\bar{t}, \bar{s}, \bar{x}, \bar{y}) \in (0, T] \times (0, T] \times V \times V$, where $|\bar{x}|_1, |\bar{y}|_1$ are bounded independently of ε, β for a fixed δ. We can assume that the maximum is strict. By the definition of viscosity solution, $\bar{x}, \bar{y} \in V_2$. Moreover, it is standard to observe that

$$\lim_{\beta \to 0} \frac{(\bar{t} - \bar{s})^2}{2\beta} = 0 \quad \text{for fixed } \delta, \varepsilon, \tag{3.459}$$

and

$$\lim_{\varepsilon \to 0} \limsup_{\beta \to 0} \frac{|\bar{x} - \bar{y}|^2}{2\varepsilon} = 0 \quad \text{for fixed } \delta. \tag{3.460}$$

If $u \nleq v$ it then follows from (3.460), (3.459), (3.456) and (3.458) that for small μ and δ, we have $\bar{t}, \bar{s} < T$ if β and ε are sufficiently small.

We use the projections from the proof of Proposition 3.128. Let $\{e_1, e_2, ...\}$ be the orthonormal basis of H composed of eigenvectors of A, $H_N := \text{span}\{e_1, ..., e_N\}$, P_N be the orthogonal projection in H onto H_N, and $Q_N = I - P_N$ for $N \geq 2$. We define

$$\hat{u}(t, x) = u_\mu(t, x) - \frac{\langle x, Q_N(\bar{x} - \bar{y}) \rangle}{\varepsilon} + \frac{|Q_N(\bar{x} - \bar{y})|^2}{2\varepsilon}$$

$$- \frac{|Q_N(x - \bar{x})|^2}{\varepsilon} - \delta e^{K_\mu(T-t)}(1 + |x|_1^2)^m,$$

$$\hat{v}(s, y) = v_\mu(s, y) - \frac{\langle y, Q_N(\bar{x} - \bar{y})\rangle}{\varepsilon} + \frac{|Q_N(y - \bar{y})|^2}{\varepsilon} + \delta e^{K_\mu(T-s)}(1 + |y|_1^2)^m,$$

and we set $\hat{u}(t, x) = -\infty$, $\hat{v}(s, y) = +\infty$ if $x, y \notin V$. Then $\hat{u}$, $\hat{v}$ are respectively weakly sequentially upper- and lower-semicontinuous on $(0, T] \times H$ and it is easy to see (as in the proof of Theorem 3.50) that

$$\hat{u}(t, x) - \hat{v}(s, y) - \frac{|P_N(x - y)|^2}{2\varepsilon} - \frac{(t - s)^2}{2\beta}$$

attains a strict global maximum over $(0, T] \times (0, T] \times H \times H$ at $(\bar{t}, \bar{s}, \bar{x}, \bar{y})$. Moreover, the functions $\hat{u}, -\hat{v}$ satisfy (3.53). Therefore the assumptions of Corollary 3.29 are satisfied for $B = I$ there. Therefore there exist functions $\varphi_k, \psi_k \in C^2((0, T) \times H)$ for $k = 1, 2, \ldots$ such that $\varphi_k, (\varphi_k)_t, D\varphi_k, D^2\varphi_k, \psi_k, (\psi_k)_t, D\psi_k, D^2\psi_k$ are bounded and uniformly continuous, and such that

$$\hat{u}(t, x) - \varphi_k(t, x)$$

has a global maximum at some point $(t_k, x_k) \in (0, T) \times V$,

$$\hat{v}(s, y) - \psi_k(s, y)$$

has a global minimum at some point $(s_k, y_k) \in (0, T) \times V$, and

$$\left(t_k, x_k, \hat{u}(t_k, x_k), (\varphi_k)_t(t_k, x_k), D\varphi_k(t_k, x_k), D^2\varphi_k(t_k, x_k)\right)$$
$$\xrightarrow[\mathbb{R}\times H\times\mathbb{R}\times\mathbb{R}\times H\times\mathcal{L}(H)]{k\to\infty} \left(\bar{t}, \bar{x}, \hat{u}(\bar{t}, \bar{x}), \frac{\bar{t} - \bar{s}}{\beta}, \frac{P_N(\bar{x} - \bar{y})}{\varepsilon}, X_N\right) \tag{3.461}$$

$$\left(s_k, y_k, \hat{v}(s_k, y_k), (\psi_k)_t(s_k, y_k), D\psi_k(s_k, y_k), D^2\psi_k(s_k, y_k)\right)$$
$$\xrightarrow[\mathbb{R}\times H\times\mathbb{R}\times\mathbb{R}\times H\times\mathcal{L}(H)]{k\to\infty} \left(\bar{s}, \bar{y}, \hat{v}(\bar{s}, \bar{y}), \frac{\bar{t} - \bar{s}}{\beta}, \frac{P_N(\bar{x} - \bar{y})}{\varepsilon}, Y_N\right), \tag{3.462}$$

where $X_N = P_N X_N P_N, Y_N = P_N Y_N P_N$ and $X_N \le Y_N$. Moreover, since the functions $\hat{u}, -\hat{v}$ are weakly sequentially upper-semicontinuous, it follows from the proof of Corollary 3.29 (see the proof of Theorem 3.27) that $\varphi_k(t, x) = \varphi_k(t, P_N x)$, $\psi_k(s, y) = \psi_k(s, P_N y)$, and thus in particular we have

$$D\varphi_k(t_k, x_k) \to \frac{P_N(\bar{x} - \bar{y})}{\varepsilon}, \quad D\psi_k(s_k, y_k) \to \frac{P_N(\bar{x} - \bar{y})}{\varepsilon} \quad \text{in } H_1. \tag{3.463}$$

In addition, it is easy to see that we must have $|x_k|_1, |y_k|_1 \leq C$ for some C. Therefore $x_k \rightharpoonup \bar{x}$, $y_k \rightharpoonup \bar{y}$ in V. This, together with (3.461), (3.462), and the fact that $(\bar{t}, \bar{s}, \bar{x}, \bar{y})$ is the maximum point of Φ, implies

$$|x_k|_1 \to |\bar{x}|_1, \quad |y_k|_1 \to |\bar{y}|_1,$$

which in turns gives

$$x_k \to \bar{x}, \ y_k \to \bar{y} \ \text{in } V. \tag{3.464}$$

By the definition of viscosity solution, we have $x_k, y_k \in V_2$, and

$$
\begin{aligned}
&- \delta K_\mu e^{K_\mu(T - t_k)}(1 + |x_k|_1^2)^m + (\varphi_k)_t(t_k, x_k) \\
&+ \frac{\delta}{2} e^{K_\mu(T - t_k)}\left(2m\,\mathrm{Tr}(Q_1)(1 + |x_k|_1^2)^{m-1} + 4m(m-1)|Q^{\frac{1}{2}}Ax_k|^2(1 + |x_k|_1^2)^{m-2}\right) \\
&+ \frac{1}{2}\mathrm{Tr}\left(QD^2\varphi_k(t_k, x_k) + 2QQ_N\right) \\
&+ \Big\langle Ax_k, D\varphi_k(t_k, x_k) + \frac{Q_N(\bar{x} - \bar{y})}{\varepsilon} + \frac{2Q_N(x_k - \bar{x})}{\varepsilon} \\
&\qquad\qquad - 2m\delta e^{K_\mu(T - t_k)}(1 + |x_k|_1^2)^{m-1}Ax_k\Big\rangle \\
&- b\left(x_k, x_k, D\varphi_k(t_k, x_k) + \frac{Q_N(\bar{x} - \bar{y})}{\varepsilon} + \frac{2Q_N(x_k - \bar{x})}{\varepsilon}\right) \\
&+ F\Big(t_k, x_k, D\varphi_k(t_k, x_k) + \frac{Q_N(\bar{x} - \bar{y})}{\varepsilon} + \frac{2Q_N(x_k - \bar{x})}{\varepsilon} \\
&\qquad\qquad - 2m\delta e^{K_\mu(T - t_k)}(1 + |x_k|_1^2)^{m-1}Ax_k\Big) \geq \frac{\mu}{T^2}. \tag{3.465}
\end{aligned}
$$

Above we have used (3.428) to get $b\,(x_k, x_k, Ax_k) = 0$. We now want to pass to the limit as $k \to \infty$. Let C_μ be a constant such that

$$\omega(s) \leq \frac{\mu}{2T^2} + C_\mu s.$$

It then follows from (3.454) that

$$
\begin{aligned}
&\left| F\Big(t_k, x_k, D\varphi_k(t_k, x_k) + \frac{Q_N(\bar{x} - \bar{y})}{\varepsilon} + \frac{2Q_N(x_k - \bar{x})}{\varepsilon}\right. \\
&\qquad\qquad - 2m\delta e^{K_\mu(T - t_k)}(1 + |x_k|_1^2)^{m-1}Ax_k\Big) \\
&\left. - F\Big(t_k, x_k, D\varphi_k(t_k, x_k) + \frac{Q_N(\bar{x} - \bar{y})}{\varepsilon} + \frac{2Q_N(x_k - \bar{x})}{\varepsilon}\Big)\right| \\
&\leq \frac{\mu}{2T^2} + C_\mu(1 + |x_k|_1)2m\delta e^{K_\mu(T - t_k)}(1 + |x_k|_1^2)^{m-1}|Ax_k|.
\end{aligned}
$$

Moreover,

$$
C_\mu(1 + |x_k|_1)2m\delta e^{K_\mu(T - t_k)}(1 + |x_k|_1^2)^{m-1}|Ax_k|
$$
$$
+ \frac{\delta}{2}e^{K_\mu(T - t_k)}\left(2m\mathrm{Tr}(Q_1)(1 + |x_k|_1^2)^{m-1} + 4m(m-1)|Q^{\frac{1}{2}}Ax_k|^2(1 + |x_k|_1^2)^{m-2}\right)
$$
$$
\leq 2m\delta C_\mu^2 e^{K_\mu(T - t_k)}(1 + |x_k|_1^2)^m + m\delta e^{K_\mu(T - t_k)}|Ax_k|^2(1 + |x_k|_1^2)^{m-1}
$$
$$
+ \delta e^{K_\mu(T - t_k)}m(2m-1)\mathrm{Tr}(Q_1)(1 + |x_k|_1^2)^m
$$
$$
\leq m\delta e^{K_\mu(T - t_k)}|Ax_k|^2(1 + |x_k|_1^2)^{m-1}
$$
$$
+ \delta e^{K_\mu(T - t_k)}\left(2mC_\mu^2 + m(2m-1)\mathrm{Tr}(Q_1)\right)(1 + |x_k|_1^2)^m. \tag{3.466}
$$

Therefore, choosing $K_\mu = 1 + 2(2mC_\mu^2 + m(2m-1)\mathrm{Tr}(Q_1))$ we obtain from (3.465) and (3.466) that

$$
-\frac{\delta}{2}K_\mu e^{K_\mu(T - t_k)}(1 + |x_k|_1^2)^m + (\varphi_k)_t(t_k, x_k)
$$
$$
+ \frac{1}{2}\mathrm{Tr}\left(QD^2\varphi_k(t_k, x_k) + 2QQ_N\right)
$$
$$
+ \left\langle Ax_k, D\varphi_k(t_k, x_k) + \frac{Q_N(\bar{x} - \bar{y})}{\varepsilon} + \frac{2Q_N(x_k - \bar{x})}{\varepsilon}\right.
$$
$$
\left. -m\delta e^{K_\mu(T - t_k)}Ax_k(1 + |x_k|_1^2)^{m-1}\right\rangle
$$
$$
- b\left(x_k, x_k, D\varphi_k(t_k, x_k) + \frac{Q_N(\bar{x} - \bar{y})}{\varepsilon} + \frac{2Q_N(x_k - \bar{x})}{\varepsilon}\right)
$$
$$
+ F\left(t_k, x_k, D\varphi_k(t_k, x_k) + \frac{Q_N(\bar{x} - \bar{y})}{\varepsilon} + \frac{2Q_N(x_k - \bar{x})}{\varepsilon}\right) \geq \frac{\mu}{2T^2}. \tag{3.467}
$$

Using (3.461), (3.463), (3.464), (3.453)–(3.455), and the continuity of b on $V \times V \times V$, we obtain from (3.467) that the norms $|Ax_k|$ are bounded and therefore $x_k \rightharpoonup \bar{x}$ in V_2. Therefore, using the above again, we can pass to the lim sup as $n \to \infty$ in (3.467) to get

$$
-\frac{\delta}{2}K_\mu e^{K_\mu(T - \bar{t})}(1 + |\bar{x}|_1^2)^m + \frac{\bar{t} - \bar{s}}{\beta} + \frac{1}{2}\mathrm{Tr}\left(QX_N + 2QQ_N\right)
$$
$$
+ \langle A\bar{x}, \frac{\bar{x} - \bar{y}}{\varepsilon}\rangle - b\left(\bar{x}, \bar{x}, \frac{\bar{x} - \bar{y}}{\varepsilon}\right) + F\left(\bar{t}, \bar{x}, \frac{\bar{x} - \bar{y}}{\varepsilon}\right) \geq \frac{\mu}{2T^2}. \tag{3.468}
$$

Similarly, we obtain

$$
\frac{\delta}{2}K_\mu e^{K_\mu(T - \bar{s})}(1 + |\bar{y}|_1^2)^m + \frac{\bar{t} - \bar{s}}{\beta} + \frac{1}{2}\mathrm{Tr}\left(QY_N - 2QQ_N\right)
$$
$$
+ \langle A\bar{y}, \frac{\bar{x} - \bar{y}}{\varepsilon}\rangle - b\left(\bar{y}, \bar{y}, \frac{\bar{x} - \bar{y}}{\varepsilon}\right) + F\left(\bar{s}, \bar{y}, \frac{\bar{x} - \bar{y}}{\varepsilon}\right) \leq -\frac{\mu}{2T^2}. \tag{3.469}
$$

Combining (3.468) and (3.469), using $X_N \leq Y_N$, and then sending $N \to \infty$ yields

$$\frac{\delta}{2}\left((1+|\bar{x}|_1^2)^m + (1+|\bar{y}|_1^2)^m\right) + \frac{|\bar{x}-\bar{y}|_1^2}{\varepsilon}$$
$$+b\left(\bar{x},\bar{x},\frac{\bar{x}-\bar{y}}{\varepsilon}\right) - b\left(\bar{y},\bar{y},\frac{\bar{x}-\bar{y}}{\varepsilon}\right)$$
$$+F\left(\bar{t},\bar{x},\frac{\bar{x}-\bar{y}}{\varepsilon}\right) - F\left(\bar{s},\bar{y},\frac{\bar{x}-\bar{y}}{\varepsilon}\right) \leq -\frac{\mu}{T^2}. \tag{3.470}$$

To estimate the trilinear form terms we use (3.427), (3.430), and then (3.460) to produce

$$\left|b\left(\bar{x},\bar{x},\frac{\bar{x}-\bar{y}}{\varepsilon}\right) - b\left(\bar{y},\bar{y},\frac{\bar{x}-\bar{y}}{\varepsilon}\right)\right|$$
$$= \frac{1}{\varepsilon}|b(\bar{x}-\bar{y},\bar{x},\bar{x}-\bar{y})| \leq \frac{C}{\varepsilon}|\bar{x}|_1|\bar{x}-\bar{y}||\bar{x}-\bar{y}|_1 \tag{3.471}$$
$$\leq \frac{\delta}{2}|\bar{x}|_1^2 + C_\delta \frac{|\bar{x}-\bar{y}|^2}{\varepsilon}\frac{|\bar{x}-\bar{y}|_1^2}{\varepsilon} \leq \frac{\delta}{2}(1+|\bar{x}|_1^2)^m + \sigma_2(\beta,\varepsilon;\delta,\mu)\frac{|\bar{x}-\bar{y}|_1^2}{\varepsilon},$$

where, for fixed μ, δ, $\lim_{\varepsilon\to 0}\limsup_{\beta\to 0}\sigma_2(\beta,\varepsilon;\delta,\mu) = 0$.

Finally, we need to estimate the terms containing F. We know that for μ and δ fixed, $|\bar{x}|_1, |\bar{y}|_1 \leq R_\delta$ for some $R_\delta > 0$. Let $D_{\mu,\delta}$ be a constant such that

$$\omega_{R_\delta}(s) \leq \frac{\mu}{4T^2} + D_{\mu,\delta}s,$$

and define $R_{\delta,\varepsilon} := 2R_\delta/\varepsilon$. Then (3.453), (3.455), (3.459) and (3.460) imply

$$\left|F\left(\bar{t},\bar{x},\frac{\bar{x}-\bar{y}}{\varepsilon}\right) - F\left(\bar{s},\bar{y},\frac{\bar{x}-\bar{y}}{\varepsilon}\right)\right|$$
$$\leq \omega_{R_{\delta,\varepsilon}}(|\bar{t}-\bar{s}|) + \omega_{R_\delta}(|\bar{x}-\bar{y}|_1) + \omega\left(|\bar{x}-\bar{y}|_1\frac{|\bar{x}-\bar{y}|}{\varepsilon}\right)$$
$$\leq \omega_{R_{\delta,\varepsilon}}(|\bar{t}-\bar{s}|) + \frac{3\mu}{4T^2} + D_{\mu,\delta}|\bar{x}-\bar{y}|_1 + C_\mu|\bar{x}-\bar{y}|_1\frac{|\bar{x}-\bar{y}|}{\varepsilon}$$
$$\leq \omega_{R_{\delta,\varepsilon}}(|\bar{t}-\bar{s}|) + \frac{3\mu}{4T^2} + D_{\mu,\delta}|\bar{x}-\bar{y}|_1 + \frac{|\bar{x}-\bar{y}|_1^2}{2\varepsilon} + 2C_\mu^2\frac{|\bar{x}-\bar{y}|^2}{\varepsilon}$$
$$\leq \frac{3\mu}{4T^2} + \sigma_3(\beta,\varepsilon;\delta,\mu) + D_{\mu,\delta}|\bar{x}-\bar{y}|_1 + \frac{|\bar{x}-\bar{y}|_1^2}{2\varepsilon}, \tag{3.472}$$

where, for fixed μ, δ, $\lim_{\varepsilon\to 0}\limsup_{\beta\to 0}\sigma_3(\beta,\varepsilon;\delta,\mu) = 0$.

Therefore, using (3.471) and (3.472) in (3.470), we obtain

$$\left(\frac{1}{2} - \sigma_2(\beta,\varepsilon;\delta,\mu)\right)\frac{|\bar{x}-\bar{y}|_1^2}{\varepsilon} - D_{\mu,\delta}|\bar{x}-\bar{y}|_1 \leq -\frac{\mu}{4T^2} + \sigma_3(\beta,\varepsilon,\delta;\mu). \tag{3.473}$$

We now see that if ε and β are small, then $\frac{1}{2} - \sigma_2(\beta, \varepsilon; \delta, \mu) > \frac{1}{4}$ and that

$$\liminf_{\varepsilon \to 0 \, r > 0} \left(\frac{r^2}{4\varepsilon} - D_{\mu,\delta} r \right) = 0.$$

Therefore, it remains to take $\liminf_{\varepsilon \to 0} \liminf_{\beta \to 0}$ in (3.473) to obtain a contradiction, which proves that we must have $u \leq v$. $\qquad\square$

3.13.4 Existence of Viscosity Solutions

We go back to the HJB equation (3.425) with the Hamiltonian function F defined by (3.426) and show that the value function of the associated stochastic optimal control problem is its viscosity solution.

Theorem 3.136 *Let Hypotheses 3.126 and 3.130 be satisfied, and let in addition $f : [0, T] \times \Lambda \to V$ be such that $f(\cdot, a)$ is uniformly continuous, uniformly for $a \in \Lambda$. Then the value function v defined by (3.424) is the unique viscosity solution of the HJB equations (3.425)–(3.426) within the class of viscosity solutions u satisfying*

$$|u(t, x)| \leq C(1 + |x|_1^k), \quad (t, x) \in (0, T] \times V,$$

for some $k \geq 0$.

Proof First of all we see that under our assumptions, the Hamiltonian F in (3.426) satisfies Hypothesis 3.134. Moreover, by Proposition 3.131, the value function v satisfies (3.450), (3.451), and the dynamic programming principle (3.449). In particular, v is weakly sequentially continuous on $(0, T] \times V$. Therefore, if v is a viscosity solution of (3.425)–(3.426), the uniqueness part is a direct consequence of Theorem 3.135. We will only show that the value function is a viscosity supersolution. The proof that v is a viscosity subsolution is easier and uses the same techniques. To this end, let $\psi(t, x) = \varphi(t, x) + \delta(t)(1 + |x|_1^2)^m$ be a test function and let $v + \psi$ have a global minimum at $(t_0, x_0) \in (0, T) \times V$.

Step 1. We need to show that $x_0 \in V$. By (3.449), for every $\varepsilon > 0$ there exists an $a_\varepsilon(\cdot) \in \mathcal{U}_{t_0}$ such that, writing $X_\varepsilon(\cdot)$ for $X(\cdot; t_0, x_0, a_\varepsilon(\cdot))$, we have

$$v(t_0, x_0) + \varepsilon^2 > \mathbb{E}\left\{ \int_{t_0}^{t_0+\varepsilon} l(X_\varepsilon(s), a_\varepsilon(s)) \, ds + v(t_0 + \varepsilon, X_\varepsilon(t_0 + \varepsilon)) \right\}.$$

We can assume that a_ε is $\mathscr{F}_s^{t_0,0}$-predictable and thus, by Corollary 2.21 and Proposition 3.128-(iii), we can assume that all $a_\varepsilon(\cdot)$ are defined on the same reference probability space ν, i.e. $a_\varepsilon(\cdot) \in \mathcal{U}_{t_0}^\nu$. Since for every $(t, x) \in (0, T) \times V$

$$v(t, x) - v(t_0, x_0) \geq -\varphi(t, x) + \varphi(t_0, x_0) - \delta(t)(1 + |x|_1^2)^m + \delta(t_0)(1 + |x_0|_1^2)^m,$$

we have

$$\varepsilon^2 - \mathbb{E}\int_{t_0}^{t_0+\varepsilon} l\left(X_\varepsilon(s), a_\varepsilon(s)\right)ds \geq \mathbb{E}\left[v\left(t_0+\varepsilon, X_\varepsilon(t_0+\varepsilon)\right) - v(t_0, x_0)\right]$$

$$\geq \mathbb{E}\Big[-\varphi\left(t_0+\varepsilon, X_\varepsilon(t_0+\varepsilon)\right) + \varphi(t_0, x_0)$$
$$-\delta(t_0+\varepsilon)\left(1 + |X_\varepsilon(t_0+\varepsilon)|_1^2\right)^m + \delta(t_0)\left(1 + |x_0|_1^2\right)^m \Big].$$

Set $\lambda = \inf_{t \in [t_0, t_0+\varepsilon_0]} \delta(t)$ for some fixed $\varepsilon_0 > 0$, and take $\varepsilon < \varepsilon_0$. Applying Itô's formula to $\varphi(s, X_\varepsilon(s))$ and $\delta(s)\left(1 + |X_\varepsilon(s)|_1^2\right)^m$, together with identity (3.440), in the inequality above, and then dividing both sides by ε, we obtain

$$\varepsilon - \frac{1}{\varepsilon}\mathbb{E}\int_{t_0}^{t_0+\varepsilon} l\left(X_\varepsilon(s), a_\varepsilon(s)\right)ds$$

$$\geq -\frac{1}{\varepsilon}\mathbb{E}\Bigg[\int_{t_0}^{t_0+\varepsilon}\bigg(\varphi_t\left(s, X_\varepsilon(s)\right) + \langle AX_\varepsilon(s) - B\left(X_\varepsilon(s), X_\varepsilon(s)\right), D\varphi\left(s, X_\varepsilon(s)\right)\rangle$$

$$+ \langle f\left(s, a_\varepsilon(s)\right), D\varphi\left(s, X_\varepsilon(s)\right)\rangle + \frac{1}{2}\mathrm{Tr}\left(QD^2\varphi\left(s, X_\varepsilon(s)\right)\right)\bigg)ds\Bigg]$$

$$-\frac{1}{\varepsilon}\mathbb{E}\Bigg[\int_{t_0}^{t_0+\varepsilon}\bigg(\delta'(s)\left(1 + |X_\varepsilon(s)|_1^2\right)^m + m\mathrm{Tr}(Q_1)\left(1 + |X_\varepsilon(s)|_1^2\right)^{m-1}$$

$$-2m\delta(s)\left(|AX_\varepsilon(s)|^2 + \langle f\left(s, a_\varepsilon(s)\right), AX_\varepsilon(s)\rangle\right)\left(1 + |X_\varepsilon(s)|_1^2\right)^{m-1}$$

$$+2m(m-1)|Q^{\frac{1}{2}}AX_\varepsilon(s)|^2\left(1 + |X_\varepsilon(s)|_1^2\right)^{m-2}\bigg)ds\Bigg]. \tag{3.474}$$

By the definition of λ it then follows that

$$\frac{2m\lambda}{\varepsilon}\mathbb{E}\int_{t_0}^{t_0+\varepsilon}|X_\varepsilon(s)|_2^2\left(1 + |X_\varepsilon(s)|_1^2\right)^{m-1}ds$$

$$\leq \varepsilon + \frac{1}{\varepsilon}\mathbb{E}\Bigg[\int_{t_0}^{t_0+\varepsilon}\bigg(-l\left(X_\varepsilon(s), a_\varepsilon(s)\right) + \varphi_t\left(s, X_\varepsilon(s)\right)$$

$$+ \langle AX_\varepsilon(s) - B\left(X_\varepsilon(s), X_\varepsilon(s)\right), D\varphi\left(s, X_\varepsilon(s)\right)\rangle$$

$$+ \langle f\left(s, a_\varepsilon(s)\right), D\varphi\left(s, X_\varepsilon(s)\right)\rangle + \frac{1}{2}\mathrm{Tr}\left(QD^2\varphi\left(s, X_\varepsilon(s)\right)\right)\bigg)ds\Bigg]$$

$$+\frac{1}{\varepsilon}\mathbb{E}\Bigg[\int_{t_0}^{t_0+\varepsilon}\bigg(\delta'(s)\left(1 + |X_\varepsilon(s)|_1^2\right)^m + m\mathrm{Tr}(Q_1)\left(1 + |X_\varepsilon(s)|_1^2\right)^{m-1}$$

$$-2m\delta(s)\langle f\left(s, a_\varepsilon(s)\right), AX_\varepsilon(s)\rangle\left(1 + |X_\varepsilon(s)|_1^2\right)^{m-1}$$

$$+2m(m-1)|Q^{\frac{1}{2}}AX_\varepsilon(s)|^2\left(1 + |X_\varepsilon(s)|_1^2\right)^{m-2}\bigg)ds\Bigg]. \tag{3.475}$$

We now have
$$|l\left(X_\varepsilon\left(s\right),a_\varepsilon\left(s\right)\right)| \le C\left(1 + |X_\varepsilon\left(s\right)|_1^k\right),$$

$$|\varphi_t\left(s,X_\varepsilon\left(s\right)\right)| \le C(1 + |X_\varepsilon\left(s\right)|),$$

$$|\langle AX_\varepsilon\left(s\right),D\varphi\left(s,X_\varepsilon\left(s\right)\right)\rangle| \le \frac{\lambda}{2}|X_\varepsilon\left(s\right)|_2^2 + C\left(1 + |X_\varepsilon\left(s\right)|^2\right),$$

$$|\langle B\left(X_\varepsilon\left(s\right),X_\varepsilon\left(s\right)\right),D\varphi\left(s,X_\varepsilon\left(s\right)\right)\rangle| = |b\left(X_\varepsilon\left(s\right),X_\varepsilon\left(s\right),D\varphi\left(s,X_\varepsilon\left(s\right)\right)\right)|$$
$$\le C|X_\varepsilon\left(s\right)|_1|X_\varepsilon\left(s\right)|_2\left(1 + |X_\varepsilon\left(s\right)|\right) \le \frac{\lambda}{2}|X_\varepsilon\left(s\right)|_2^2 + C\left(1 + |X_\varepsilon\left(s\right)|_1^4\right),$$

$$|\text{Tr}\left(QD^2\varphi\left(s,X_\varepsilon\left(s\right)\right)\right)|, |\langle f\left(s,a_\varepsilon\left(s\right)\right),D\varphi\left(s,X_\varepsilon\left(s\right)\right)\rangle| \le C(1 + |X_\varepsilon\left(s\right)|),$$

$$|\langle f\left(s,a_\varepsilon\left(s\right)\right),AX_\varepsilon\left(s\right)\rangle|\left(1 + |X_\varepsilon\left(s\right)|_1^2\right)^{m-1} \le C\left(1 + |X_\varepsilon\left(s\right)|_1^2\right)^m,$$

and
$$|Q^{\frac{1}{2}}AX_\varepsilon\left(s\right)|^2\left(1 + |X_\varepsilon\left(s\right)|_1^2\right)^{m-2} \le C\left(1 + |X_\varepsilon\left(s\right)|_1^2\right)^{m-1}.$$

Employing the above estimates in (3.475) and then using (3.436) yields

$$\frac{\lambda}{\varepsilon}\int_{t_0}^{t_0+\varepsilon} \mathbb{E}|X_\varepsilon\left(s\right)|_2^2\left(1 + |X_\varepsilon\left(s\right)|_1^2\right)^{m-1} ds \le C \qquad (3.476)$$

for some constant C independent of ε. Therefore there exist sequences $\varepsilon_n \to 0$ and $t_n \in (t_0, t_0 + \varepsilon_n)$ such that
$$\mathbb{E}|X_{\varepsilon_n}\left(t_n\right)|_2^2 \le C,$$

and thus there exist subsequences, still denoted by ε_n, t_n, such that

$$X_{\varepsilon_n}\left(t_n\right) \rightharpoonup \bar{x} \quad \text{weakly in } L^2(\Omega^\nu; V_2)$$

for some $\bar{x} \in L^2(\Omega^\nu; V_2)$ (and thus also weakly in $L^2(\Omega^\nu; H)$). However, by (3.442), $X_{\varepsilon_n}\left(t_n\right) \to x_0$ strongly in $L^2(\Omega^\nu; H)$. Therefore, by the uniqueness of the weak limit in $L^2(\Omega^\nu; H)$, it follows that $x_0 = \bar{x} \in V_2$.

Step 2. We now prove the supersolution inequality. We need to "pass to the limit" as $\varepsilon \to 0$ in (3.474), at least along a subsequence. This operation is rather standard for most of the terms, more precisely for those that only use convergence in the norms of H and V. To explain how we deal with the easy terms, let us consider the cost term.

Let $r \geq |x_0|_1$. Then, using (3.436), (3.443), (3.445) and (3.446), we have

$$\left| \frac{1}{\varepsilon} \mathbb{E} \int_{t_0}^{t_0+\varepsilon} \left[l\left(X_\varepsilon(s), a_\varepsilon(s)\right) ds - l\left(x_0, a_\varepsilon(s)\right) \right] ds \right|$$

$$\leq \frac{1}{\varepsilon} \mathbb{E} \int_{t_0}^{t_0+\varepsilon} \sigma_r \left(|X_\varepsilon(s) - x_0|_1 \right) ds$$

$$+ \frac{1}{\varepsilon} \mathbb{E} \int_{t_0}^{t_0+\varepsilon} C \left(1 + |X_\varepsilon(s)|_1^k + |x_0|_1^k \right) \mathbf{1}_{\{|X_\varepsilon(s)|_1 > r\}} ds$$

$$\leq \sigma_r \left(\frac{1}{\varepsilon} \mathbb{E} \int_{t_0}^{t_0+\varepsilon} |X_\varepsilon(s) - x_0|_1 ds \right) + C \left(1 + |x_0|_1^k \right) \frac{1 + |x_0|_1}{r}$$

$$\leq \sigma_r \left(\frac{1}{\varepsilon} \mathbb{E} \int_{t_0}^{t_0+\varepsilon} \sqrt{\omega_{x_0}(\varepsilon)} ds \right) + C \left(1 + |x_0|_1^k \right) \frac{1 + |x_0|_1}{r}. \tag{3.477}$$

The above implies that

$$\left| \frac{1}{\varepsilon} \mathbb{E} \int_{t_0}^{t_0+\varepsilon} \left[l\left(X_\varepsilon(s), a_\varepsilon(s)\right) ds - l\left(x_0, a_\varepsilon(s)\right) \right] ds \right| \leq \gamma(\varepsilon), \tag{3.478}$$

where $\lim_{\varepsilon \to 0} \gamma(\varepsilon) = 0$. Arguing like in (3.477), and using that $(-A)^{\frac{1}{2}} f$ is bounded in H, $(-A)^{\frac{1}{2}} f(\cdot, a)$ is uniformly continuous with values in H, uniformly for $a \in \Lambda$, and $Q^{\frac{1}{2}}(-A)^{\frac{1}{2}}$ extends to a bounded operator in H, we can deal with all the terms in (3.474), except the terms

$$- \frac{1}{\varepsilon} \mathbb{E} \int_{t_0}^{t_0+\varepsilon} \langle AX_\varepsilon(s), D\varphi(s, X_\varepsilon(s)) \rangle, \tag{3.479}$$

$$\frac{1}{\varepsilon} \mathbb{E} \int_{t_0}^{t_0+\varepsilon} \langle B\left(X_\varepsilon(s), X_\varepsilon(s)\right), D\varphi(s, X_\varepsilon(s)) \rangle, \tag{3.480}$$

$$\frac{1}{\varepsilon} \mathbb{E} \int_{t_0}^{t_0+\varepsilon} 2m\delta(s) |AX_\varepsilon(s)|^2 \left(1 + |X_\varepsilon(s)|_1^2 \right)^{m-1}, \tag{3.481}$$

which require special consideration.

We first notice that

$$\mathbb{E} \left| \frac{1}{\varepsilon} \int_{t_0}^{t_0+\varepsilon} \sqrt{\delta(s)} AX_\varepsilon(s) \left(1 + |X_\varepsilon(s)|_1^2 \right)^{\frac{m-1}{2}} ds \right|^2$$

$$\leq \mathbb{E} \frac{1}{\varepsilon} \int_{t_0}^{t_0+\varepsilon} \delta(s) |X_\varepsilon(s)|_2^2 \left(1 + |X_\varepsilon(s)|_1^2 \right)^{m-1} ds \leq C \tag{3.482}$$

by (3.476). Therefore, there exists a sequence $\varepsilon_n \to 0$ and $Y \in L^2(\Omega^\nu, H)$ such that

$$Y_n := \frac{1}{\varepsilon_n} \int_{t_0}^{t_0+\varepsilon_n} \sqrt{\delta(s)} A X_{\varepsilon_n}(s) \left(1 + |X_{\varepsilon_n}(s)|_1^2\right)^{\frac{m-1}{2}} ds \rightharpoonup Y \quad \text{in } L^2(\Omega^\nu, H)$$

as $n \to \infty$. However, using arguments similar to those in (3.477), it is easy to see that

$$A^{-1} Y_n = \frac{1}{\varepsilon_n} \int_{t_0}^{t_0+\varepsilon_n} \sqrt{\delta(s)} X_{\varepsilon_n}(s) \left(1 + |X_{\varepsilon_n}(s)|_1^2\right)^{\frac{m-1}{2}} ds \to \sqrt{\delta(t_0)} x_0 \left(1 + |x_0|_1^2\right)^{\frac{m-1}{2}}$$

strongly in $L^2(\Omega^\nu, H)$. Therefore it follows that

$$Y = \sqrt{\delta(t_0)} A x_0 \left(1 + |x_0|_1^2\right)^{\frac{m-1}{2}}.$$

Then, using the first inequality of (3.482), we get

$$\liminf_{n \to \infty} \mathbb{E} \frac{1}{\varepsilon_n} \int_{t_0}^{t_0+\varepsilon_n} \delta(s) |A X_{\varepsilon_n}(s)|^2 \left(1 + |X_{\varepsilon_n}(s)|_1^2\right)^{m-1} ds \tag{3.483}$$
$$\geq \delta(t_0) |A x_0|^2 \left(1 + |x_0|_1^2\right)^{m-1}.$$

This takes care of the term (3.481). The same argument also shows that we can assume that

$$\frac{1}{\varepsilon_n} \int_{t_0}^{t_0+\varepsilon_n} A X_{\varepsilon_n}(s) \, ds \rightharpoonup A x_0 \quad \text{in } L^2(\Omega^\nu, H) \text{ as } n \to \infty. \tag{3.484}$$

As regards (3.479), denoting by ω_φ a modulus of continuity of $D\varphi$, we have by (3.476), (3.442), and (3.484)

$$\left| \frac{1}{\varepsilon_n} \mathbb{E}_{t_0} \int_{t_0}^{t_0+\varepsilon_n} \left\langle A X_{\varepsilon_n}(s), D\varphi\left(s, X_{\varepsilon_n}(s)\right)\right\rangle ds - \left\langle A x_0, D\varphi\left(t_0, x_0\right)\right\rangle \right|$$
$$\leq \frac{1}{\varepsilon_n} \int_{t_0}^{t_0+\varepsilon_n} \left(\mathbb{E}|A X_{\varepsilon_n}(s)|^2\right)^{\frac{1}{2}} \left(\mathbb{E}\left(\omega_\varphi\left(\varepsilon_n + |X_{\varepsilon_n}(s) - x_0|\right)\right)^2\right)^{\frac{1}{2}} ds \tag{3.485}$$
$$+ \left| \mathbb{E}\left\langle \frac{1}{\varepsilon_n} \int_{t_0}^{t_0+\varepsilon_n} A X_{\varepsilon_n}(s) \, ds - A x_0, D\varphi\left(t_0, x_0\right)\right\rangle \right| \to 0 \quad \text{as } n \to \infty.$$

Finally, for (3.480), using (3.431), (3.476), (3.436), (3.442), (3.443), and (3.484),

$$\left| \frac{1}{\varepsilon_n} \mathbb{E} \int_{t_0}^{t_0+\varepsilon_n} b\left(X_{\varepsilon_n}(s), X_{\varepsilon_n}(s), D\varphi\left(s, X_{\varepsilon_n}(s)\right)\right) ds - b\left(x_0, x_0, D\varphi\left(t_0, x_0\right)\right) \right|$$
$$\leq \frac{1}{\varepsilon_n} \mathbb{E} \int_{t_0}^{t_0+\varepsilon_n} |X_{\varepsilon_n}(s)|_1 |X_{\varepsilon_n}(s)|_2 \omega_\varphi\left(\varepsilon_n + |X_{\varepsilon_n}(s) - x_0|\right) ds$$

$$+\frac{1}{\varepsilon_n}\mathbb{E}\int_{t_0}^{t_0+\varepsilon_n}|X_{\varepsilon_n}(s)-x_0|_1|X_{\varepsilon_n}(s)|_2|D\varphi(t_0,x_0)|ds$$

$$+\left|\frac{1}{\varepsilon_n}\mathbb{E}\int_{t_0}^{t_0+\varepsilon_n}b\left(x_0,X_{\varepsilon_n}(s)-x_0,D\varphi(t_0,x_0)\right)ds\right|$$

$$\leq\frac{1}{\varepsilon_n}\int_{t_0}^{t_0+\varepsilon_n}\left(\mathbb{E}|X_{\varepsilon_n}(s)|_2^2\right)^{\frac{1}{2}}\left(\mathbb{E}|X_{\varepsilon_n}(s)|_1^4\right)^{\frac{1}{4}}\left(\mathbb{E}\left(\omega_\varphi\left(\varepsilon_n+|X_{\varepsilon_n}(s)-x_0|\right)\right)^4\right)^{\frac{1}{4}}ds$$

$$+C\frac{1}{\varepsilon_n}\int_{t_0}^{t_0+\varepsilon_n}\left(\mathbb{E}|X_{\varepsilon_n}(s)|_2^2\right)^{\frac{1}{2}}\left(\mathbb{E}|X_{\varepsilon_n}(s)-x_0|_1^2\right)^{\frac{1}{2}}ds \qquad (3.486)$$

$$+\left|\mathbb{E}b\left(x_0,\frac{1}{\varepsilon_n}\int_{t_0}^{t_0+\varepsilon_n}X_{\varepsilon_n}(s)\,ds-x_0,D\varphi(t_0,X_0)\right)\right|\to 0 \quad\text{as } n\to\infty.$$

In particular, the last term goes to zero since, by (3.484),

$$\frac{1}{\varepsilon_n}\int_{t_0}^{t_0+\varepsilon_n}X_{\varepsilon_n}(s)\,ds\rightharpoonup x_0 \quad\text{in } L^2(\Omega^\nu,V_2)\text{ as } n\to\infty$$

and

$$Z\to b(x_0,Z,D\varphi(t_0,x_0))$$

is a bounded linear functional on $L^2(\Omega^\nu,V_2)$.

Therefore, using (3.478) (and similar estimates for other standard terms), (3.483), (3.485), and (3.486) in (3.474), we obtain for small ε_n that

$$-\psi_t(t_0,x_0)-\frac{1}{2}\text{Tr}\left(QD^2\psi(t_0,x_0)\right)-\langle Ax_0-B(x_0,x_0),D\psi(t,x_0)\rangle$$

$$+\frac{1}{\varepsilon_n}\mathbb{E}\int_{t_0}^{t_0+\varepsilon_n}[\langle f(t_0,a_\varepsilon(s)),-D\psi(t_0,x_0)\rangle+l(x_0,a_\varepsilon(s))]ds\leq\omega_1(\varepsilon_n)$$

for some modulus ω_1. It now remains to take the infimum over $a\in\Lambda$ inside the integral and then send $n\to\infty$. □

Example 3.137 The following example satisfies the assumptions of Theorem 3.136. Let

$$l(x,a)=|\text{curl }x|^2+\frac{1}{2}|a|^2,$$

$$g(x)=|x|^2,$$

$$f(t,a)=Ka,$$

where $K\in\mathcal{L}(H,V)$ and $\Lambda=B_H(0,R)\subset H$. Such a control and the singular kernel of K can be approximately realized by a suitable Lorentz force distribution in electrically conducting fluids such as liquid metals and salt water. The Hamiltonian function is then

$$F(x, p) = |\text{curl } x|^2 + h(K^*p),$$

where $h(\cdot) : H \to R$ is given by

$$h(z) := \inf_{a \in \Lambda} \left\{ \langle a, z \rangle_H + \frac{1}{2}|a|^2 \right\}$$

and K^* is the adjoint of K considered as an operator from H to H. We can in fact explicitly obtain h as

$$h(z) = \begin{cases} -\dfrac{1}{2}|z|^2 & \text{for } |z| \leq R \\ \\ -R|z| + \frac{1}{2}R^2 & \text{for } |z| > R. \end{cases}$$

We also remark that the optimal feedback control here is given formally as

$$\tilde{a}(t) = \Upsilon(K^*Du(t, x(t))),$$

where

$$\Upsilon(z) := Dh(z) = \begin{cases} -z & \text{for } |z| \leq R, \\ \\ -z\dfrac{R}{|z|} & \text{for } |z| > R. \end{cases}$$

Under additional conditions on Q, optimal feedback controls for this example are discussed in Sect. 4.9.1.2 using mild solutions. ∎

3.14 Bibliographical Notes

The material of Sect. 3.1.1 on B-continuity is based on [141, 142, 506]. The formulation and the proof of the exponential moment estimates of Proposition 3.18 is taken from [541] while the rest of Sect. 3.1.2 mostly follows [142, 374]. More general formulations of Theorem 3.25 are in [403, 535]. Corollary 3.26 was first introduced in [142]. Other smooth or partially smooth perturbed optimization principles can be found in [67, 185, 381].

The first version of a maximum principle for semicontinuous functions in Hilbert spaces appeared in [412]. It was an infinite-dimensional version of a maximum principle in domains of $\mathbb{R}^n$ (see e.g. [139]) and was applicable to a class of bounded second-order equations (3.66). By a reduction to a finite-dimensional case and the use of the finite-dimensional maximum principle it provided test functions whose second-order derivatives satisfied proper inequalities on finite-dimensional subspaces, and with the remaining parts of second-order derivatives becoming negligible for the class of equations considered as the dimension of the finite-dimensional subspaces increased to $+\infty$. A corrected and simplified proof of this result, which also included its time-dependent version, based on the use of so-called partial sup-convolutions, appeared in [140]. These maximum principles have been adapted to unbounded equations first in [361, 537, 538] and later in various settings in [318, 322, 323, 374, 376]. The version stated in Theorem 3.27 is general and new as we tried to formulate it in a way that would be more directly applicable to various classes of equations. Its proof draws on the collective body of work from the above cited papers. A scaling reduction to obtain Corollary 3.29 was introduced in [361]. A different type of time-dependent maximum principle, similar in the spirit to its finite-dimensional version in [139], is in [140], see Remark 3.30 for more on this.

A definition of viscosity solution similar to the one presented in Sect. 3.3 was introduced in [537, 538]. It was based on the notion of a B-continuous viscosity solution developed by Crandall and P.L. Lions in [141, 142]. An earlier paper [411] dealing with a specific second-order HJB equation for an optimal control of a Zakai equation also used some ideas of the B-continuous viscosity solution. The material of Sect. 3.3 is mostly based on [374, 537, 538], however the definitions of viscosity solutions are more general. Lemma 3.37 is taken from [376]. A different definition of viscosity solution for second-order equations in Hilbert spaces was proposed by Ishii in [361]. It was related to the definition for first-order equations in [360]. Ishii's viscosity sub/supersolutions are allowed to be discontinuous and the definition uses a special (convex) function to deal with the unboundedness in the equation. The function is related to the equation and can be thought of as an energy function for the controlled deterministic/stochastic PDE related to the HJB equation. The advantage of this definition is that viscosity solutions can be relatively easily obtained by Perron's method and the unbounded operator A (together with other terms) can be nonlinear. However, it seems to be difficult to apply this definition to control problems and no attempts have been made in this direction. The idea of using special functions as part of test functions in the definition of viscosity solution to exploit the coercivity of the operator A also appeared in [97, 144] (see also [249, 321]) and later for second-order equations in [318, 322, 323, 376], see Sects. 3.9, 3.11–3.13.

We only briefly mentioned bounded equations in Sect. 3.3.1. The definition of viscosity solution in Sect. 3.3.1 is taken from [412] where the theory of such equations was developed for equations satisfying Hypothesis 3.47. In this paper an equivalent definition using second-order jets is also discussed. For equations that do not satisfy Hypothesis 3.47 a stronger definition of viscosity solution was introduced in [410]. It allowed for more general test functions which are not necessarily twice Fréchet differentiable. Both papers contain comparison and existence results. Uniqueness of solutions is obtained in [410] by a combination of stochastic and analytic techniques. Perron's method is discussed in [412] while connections with stochastic optimal control are discussed in [410]. In particular, [410] contains proofs of sub- and super-optimality inequalities of dynamic programming. Regularity results for bounded equations and their obstacle problems have been obtained in [410, 542]. Existence and uniqueness results for bounded equations can also be found in [378], in particular one can find there proofs of comparison principles using a parabolic

maximum principle from [140], which allows one to relax the way viscosity subsolutions and supersolutions attain the initial/terminal values. A Dirichlet boundary
value problem for a linear equation was investigated in [374] and a risk-sensitive
control problem in [540]. An obstacle problem related to optimal stopping and pricing of American options was studied in [293]. Classical results for bounded linear
equations can be found in [179].

The first comparison theorems for B-continuous viscosity sub/supersolutions of
equations discussed in Sect. 3.5 were proved in [537, 538]. These works dealt with
the case of a compact operator B and the proofs of comparison in most part relied on
a combination of techniques developed for first-order equations in [141, 142] and the
maximum principle arguments of [140, 412]. Slightly different techniques, but also
based on the maximum principle of [412], were used to prove the comparison principle with Ishii's definition of solution in [361]. Some of Ishii's methods were later
used in other comparison proofs. Comparison results for the equations of Sect. 3.5
without the compactness assumption on B are in [374] and the proofs of comparison
in various special cases are contained in [318, 322, 323, 376, 380]. The comparison
theorem for equations with quadratic gradient terms is in [541]. The papers [361, 376]
show comparison for discontinuous viscosity sub- and supersolutions. The material
of Sect. 3.5 is to some extent new and incorporates formulations and techniques of
[361, 374, 537, 538]. The statements of Theorems 3.50, 3.54, 3.56, 3.58 are new and
include general growth conditions for viscosity sub- and supersolutions. The proofs
of the above comparison theorems are also to some extent new. Comparison theorems with general growth conditions for B-continuous viscosity sub/supersolutions
of first-order equations can be found in [403].

Direct proofs that value functions of stochastic optimal control problems in Hilbert
spaces are viscosity solutions of their HJB equations can be found, in various cases, in
[323, 410, 411]. An early attempt in this direction was also made in [461]. The general
finite time horizon optimal control problem (3.119) and (3.120) and its connection
to B-continuous viscosity solutions of Eq. (3.122) and (3.123) was studied in [374].
Our presentation expands and generalizes [374]. Some results which were part of the
folklore of the theory are stated in Sect. 3.6 for the first time. We presented continuity
properties of the value functions in both finite and infinite horizon cases and under
both weak and strong B-conditions. Only the weak B-condition case was discussed in
[374]. The use of the dynamic programming principle is also fully explained and the
proofs that value functions are viscosity solutions of the associated HJB equations
are given in all cases. We tried to include all the details. The proof of a stronger
version of the dynamic programming principle in the stopping time formulation in
Sect. 3.6.2 uses some arguments from [452].

The material of Sect. 3.7 on finite-dimensional approximations is based on the
results of [537, 538], however it contains some improvements of the results and
their proofs. The method of finite-dimensional approximations provides a way to
construct B-continuous viscosity solutions for equations which may not be HJB
equations related to optimal control problems, for instance for Isaacs equations. It
requires, however, that the operator B be compact. The method, together with its
basic techniques, was introduced in [141] for first-order equations, and was later

generalized to second-order equations in [537, 538]. A version of this method was also used in [318]. Lemma 3.83-(ii) was proved in [141] by viscosity solution arguments. Our proof uses direct functional analytic arguments. Other proofs of existence employ Perron's method (see the comments in this section in the paragraph on Perron's method). For Isaacs equations, probabilistic representation formulas can be obtained [260, 464, 466, 539].

Section 3.8 on singular perturbations is based on [541]. Singular perturbation problems in finite dimensional spaces have been studied extensively by viscosity solution methods and the reader can consult [40, 263] for results and references. The problems have not yet been widely investigated in Hilbert or other infinite-dimensional spaces. Nisio studied such a problem in [466] in connection with a risk-sensitive control problem. Also a singular limit problem related to a risk-sensitive control problem with bounded evolution in a Hilbert space was studied in [540]. In [541] convergence of viscosity solutions of singularly perturbed HJB equations was used to investigate large deviation problems for stochastic PDEs perturbed by small noise. The case of integro-PDEs was studied in [543]. Both papers [541, 543] use a general PDE approach to large deviations developed in [250] (see also [246–248]).

Perron's method for viscosity solutions of PDEs in finite-dimensional spaces was introduced by Ishii in [358] (see also [139]). It was extended to bounded equations in Hilbert spaces in [412]. Perron's method provides another way to obtain existence of viscosity solutions for equations that may not necessarily be of the HJB type, for instance for Isaacs equations. For unbounded first and second-order equations it was shown to work with Ishii's definitions of viscosity solution [360, 361] and with the Tataru–Crandall–Lions definition of viscosity solution [143]. For B-continuous viscosity solutions Perron's method was introduced in [376] under an assumption that the unbounded operator A has some coercivity properties. Perron's method requires the notion of a discontinuous viscosity solution, so in [376] a more general definition of a discontinuous viscosity solution using B-semicontinuous envelopes was introduced. This definition borrowed an idea from the definitions in [360, 361] of combining the upper and lower-semicontinuous envelopes with the radial test functions. The method of half-relaxed limits of Barles–Perthame requires compactness. The fact that it may not work in infinite-dimensional spaces was noticed in [15, 540] (see Example 3.43 in this book). A version of half-relaxed limits presented in Sect. 3.9 was developed in [376] where more results on Perron's method and half-relaxed limits can be found.

The material of Sect. 3.10 is based on [375]. The infinite-dimensional Black–Scholes equation was analyzed in [302], where the existence of smooth solutions was proved for smooth data, and an obstacle problem for the Black–Scholes equation was studied in [583] from the point of view of Bellman's inclusions. A similar obstacle problem for a related model was studied in [293]. A non-local Black–Scholes–Barenblatt equation associated with the HJMM model driven by a Lévy type noise was investigated in [545]. A Kolmogorov equation related to the problem of hedging of a derivative of a risky asset whose volatility as well as the claim may depend on the past history of the asset was studied in [517], where $C^{1+\alpha}$ regularity of viscosity solutions was obtained on special finite-dimensional subspaces.

Section 3.11 follows [323]. The first result about viscosity solutions of the HJB equation for control of the DMZ equation appeared in [411], where the equation was studied in a standard L^2 space and the operators S_a^k were bounded multiplication operators. The paper [411] used a combination of probabilistic and analytic techniques to deal with the uniqueness of the viscosity solution of the HJB equation. In [344] it was shown that the value function is a viscosity solution in a very weak sense when the HJB equation is considered in the space of measures. (A regularity result for a related equation in the space of measures was obtained in [346].) Another paper on the subject is [22]. The approach of [323] used the theory of B-continuous viscosity solutions of [537, 538] together with an idea that originated in [97, 144] (also [360, 361] had related ideas) to use a special radial function and the coercivity of operators in the equation to "improve" the points where the maxima/minima occur in the definition of a viscosity solution. This idea of using a special energy function related to the underlying controlled state equation as a part of the test functions was also used in many cases for first- and second-order equations [249, 318, 321, 322, 376]. The viscosity solution approach of [323] is also presented in [467], where a different proof of the dynamic programming principle is given. The book [467] discusses in detail, in Chaps. 5 and 6, a partially observed optimal control problem, the separated problem, the optimal control of the DMZ equation, and other related material. It complements the material in Sects. 2.6.6 and 3.11 and gives a slightly different perspective. Our short introduction to variational solutions in Sect. 3.11.1 is based on [124, 294, 386, 388, 491]. Semi-linear stochastic parabolic equations and the DMZ equation are also discussed in [467], where Itô's formulas are proved for the original test functions used in [323], which are similar to but different from the test functions in Sect. 3.11.5. Our presentation of the various energy and continuous dependence estimates for the DMZ equation follows, with small changes, [323]. The reader can also find similar results in [467]. The material on viscosity solutions and the value function of Sects. 3.11.5 and 3.11.6 has some differences from [323] as we merged it into the presentation of the book and made some improvements and corrections.

Section 3.12 is based on [318] and fills in some missing details there. The HJB equation in this section has second-order coefficients which are not trace class, so the equation is also unbounded in the second-order terms. Since the equation is fully nonlinear it cannot be dealt with by the techniques of mild solutions. A change of variables is done to convert the equation to one with bounded second-order terms. This is a rather ad hoc technique. Viscosity solutions of HJB equations with unbounded second-order terms coming from control problems with state equations driven by cylindrical Wiener processes have not yet been studied systematically. The definition of viscosity solution for the converted equation is similar to that in [97] and uses a special radial function to guarantee that the points where the maxima/minima occur belong to a better space (see the previous paragraph for the discussion on the origins of this definition). A Cauchy problem for equations similar to the "converted" equation was also studied in [560] using the techniques of [318]. Apart from [318], boundary control problems and their associated HJB equations have been studied via viscosity solution techniques in [577] for the stochastic case (with noise at the boundary) and in [94, 96, 97, 221, 222] for the deterministic case. Second-order HJB equations and stochastic boundary control/noise problems have been investigated via mild solutions and Backward SDEs in [181, 189, 225, 310, 437, 574, 591, 592], some of them also in connection with stochastic delay equations.

The material of Sect. 3.13 follows [322]. The definition of viscosity solution is similar to those of [318, 323] (see the previous comments on the origins of these definitions) however it uses a different energy function (a radial function of the $| \cdot |_1$ norm) which reduces the equation to a subspace of the Hilbert space H. In this respect, the definition is similar to the definitions in [360, 361]. A stationary equation similar to (3.425) was also investigated in [559]. Viscosity solution approaches to first-order HJB equations associated to optimal control of deterministic Navier–Stokes equations are in [321, 526] (see also [534] for earlier attempts). A PDE-viscosity solution approach to large deviations of stochastic two-dimensional Navier–Stokes equations with small noise intensities is considered in [541], where convergence of viscosity solutions of singularly perturbed HJB equations is studied. For results on Kolmogorov and HJB equations by other approaches [33, 34, 158, 161, 255, 424, 512, 567] we refer to Sect. 4.9.1, and the short discussion at the beginning of Sect. 3.13.

In this book we have not explicitly discussed Isaacs equations in Hilbert spaces which are associated to zero-sum two-player stochastic differential games. For Isaacs equations, one can prove existence of viscosity solutions by showing directly that the associated upper/lower value function of the game is a viscosity solution of the upper/lower Isaacs equation. Such results can be found in [260, 464, 466, 539]. This, however, is not easy since the proof of the dynamic programming principle is very complicated and thus only limited results are available. Related results on risk-sensitive stochastic control and differential games can be found in [462, 463, 465, 466, 540].

Other types of equations can be studied using the theory of viscosity solutions presented in this chapter, for instance obstacle problems for HJB equations related to optimal stopping problems, HJB equations for ergodic control problems. Comparison proofs easily extend to the case of obstacle problems. Explicit literature however is limited. Obstacle problems for bounded equations have been studied in [293, 410, 542]. HJB equations for ergodic control have not been investigated by viscosity solutions. In [301] they have been studied by the perturbation approach to mild solutions and in [277] by BSDEs. Likewise HJB equations for singular control problems and for state constraints problems have not been studied in the infinite-dimensional stochastic case in the viscosity solution framework. A singular stochastic control problem with delay is studied by other methods in [6] and [242]. Concerning infinite-dimensional state constraint control problems and viscosity solutions of the associated HJB equations, the readers may check [235, 236] for the stochastic case and [93, 238, 239, 244, 379] for the deterministic case. A stochastic viability problem for a subset of a Hilbert space was studied by viscosity solutions in [86].

The viscosity solution approach to HJB equations for optimal control of stochastic delay equations has also not been fully explored. Some results on the subject are in [235, 236, 517], see also [238, 239, 244, 589, 590] for the deterministic case and first-order HJB equations. For other methods applicable to HJB equations for control of stochastic delay equations we refer the reader to Chap. 5 (in particular Sects. 5.5 and 5.6) and Chap. 6 (in particular Sect. 6.5).

Another unexplored area is viscosity solutions of HJB equations with unbounded second-order terms which come from control problems with state equations driven by Q-Wiener processes with $\mathrm{Tr}(Q) = +\infty$, i.e. such that we may have $\mathrm{Tr}[(\sigma(t, x, a) Q^{\frac{1}{2}})(\sigma(t, x, a)Q^{\frac{1}{2}})^*] = +\infty$. So far [318] has been the only paper on the subject in a specific case. Up to now viscosity solution theory handles well fully nonlinear but "degenerate" equations while the theory of mild solutions handles well semilinear but "nondegenerate" equations. One would expect that the theory of viscosity solutions can be extended to the fully nonlinear "nondegenerate" HJB equations (see also the comments about [134] in the paragraph below discussing path-dependent PDEs).

There are also very few explicit results on viscosity solutions of boundary value problems in Hilbert spaces. Only some Dirichlet boundary value problems have been studied. There exist comparison theorems (see Sect. 3.5), however equations studied by viscosity solutions are "degenerate" and hence construction of barriers at the boundary is not easy. Thus value functions may not be continuous up to the boundary unless some conditions are imposed on the drift. A Dirichlet boundary value problem for a bounded linear equation was investigated in [374] and a boundary value problem for a bounded HJB equation related to a risk-sensitive control problem was studied in [540]. Some results about value functions in bounded sets are sketched in [410]. A related paper for first-order HJB equations is [93]. Results using approaches of mild and L^2 solutions are limited to linear equations. We refer the reader to [36, 37, 165–168, 179, 497, 498, 546] and the references there for more. In [36, 37, 168] Neumann boundary value problems are considered.

An interesting direction in the evolution of the notion of a viscosity solution in infinite-dimensional spaces may come from the concept of path-dependent PDEs. Path-dependent PDEs come from the study of problems driven by path-dependent SDEs. In the finite dimensional spaces the notion of a path-dependent viscosity solution was introduced in [205] and this notion was extended to infinite-dimensional spaces in [134]. In the Markovian case this approach gives an alternative way to treat the HJB equations studied in this book. Its advantage is that it avoids the use of the maximum principle and thus it can be applied to "non-degenerate" equations, the continuity assumptions in the $|\cdot|_{-1}$ norm can be dropped, and the operator A does not need to be maximal dissipative. It is not clear if this method can be applied to fully nonlinear equations.

An emerging area of development for second-order equations in infinite-dimensional spaces seems to be related to PDEs in spaces of probability measures, in particular in the Wasserstein space. A second-order HJB equation in the space of probability measures was studied in [346] in connection with partially observed control and regularity of solutions was proved. Similar results were obtained for first-order equations in [345]. Following the program described in [250], first-order HJB equations can be used to study large deviations for empirical measures of stochastic particle systems. Results in this direction are in [249–252] (see also the references therein). Equations in the space probability measures also appear in context of Mean Field Control and Mean Field Games [48–50, 84, 98–100, 114, 289, 405]. In particular, the so-called Master Equations of Mean Field Games have attracted a lot of attention. These are non-local equations, which in the case of second-order or stochastic Mean Field Games are of second-order. So far only limited results about existence and in some cases uniqueness of classical and strong solutions of first and second-order Master Equations of Mean Field Games have been obtained in [55, 84, 100, 114, 289]. An interesting approach proposed by P.L. Lions [98, 405] allows one to convert an equation in the Wasserstein space to an equation in the Hilbert space L^2, where measures with finite second moments become random variables in L^2 with given laws.

Another emerging direction is HJB integro-PDEs in Hilbert spaces which are associated to optimal control problems with state equations driven by Lévy processes or random measures. Viscosity solutions have been introduced for such non-local equations in [543–545]. Comparison theorems have been proved in [544] and existence of viscosity solutions and optimal control problems have been studied in [545]. Some linear non-local PDEs and properties of transition semigroups for processes with jumps have been studied by other methods in [14, 402, 485, 500–502].

Chapter 4
Mild Solutions in Spaces of Continuous Functions

In this chapter we present the theory of regular solutions (i.e. at least C^1 in the space variable in a suitable sense) for a class of HJB equations in Hilbert spaces through a *perturbation approach* which was first introduced in [147, 340] and then improved and developed in various subsequent papers like [89, 90, 306, 307, 317] and later [105, 107, 301, 309, 310, 431–434]. Similar results, but using a different method based on a convex regularization procedure, were obtained in earlier papers [28–30] in the special case of convex data and quadratic Hamiltonians.

The type of solutions we study here are called *mild solutions*, in the sense that they solve the HJB equation in a suitable integral form (see (4.5) and (4.9)), where only the first derivative appears. Such kinds of solutions also appear in subsequent chapters: Chap. 5, where they are used in a weaker sense (i.e. in spaces of integrable functions with respect to a suitably chosen measure m), and Chap. 6, where mild solutions are, like here, studied in spaces of continuous functions, but with a completely different method, based on the study of an associated Backward SDE.

As explained in the preface, the method presented in this chapter, in contrast to the viscosity solution method, works only for a special class of semilinear HJB equations featuring suitable smoothing properties of the semigroup associated to the linear part of the equation: these are the key tool to solve the equation and for this reason this method is called here a "smoothing method". The good thing is that this method allows us to find very powerful results on existence, uniqueness and regularity of solutions and to apply them to prove verification theorems and existence of optimal feedback controls in a satisfactory way, an outcome which, up to now, has not been achieved in the context of viscosity solutions.

We cover both the parabolic and the elliptic cases (associated, respectively, to finite and infinite horizon optimal stochastic control problems). After an introduction (Sect. 4.1), where we explain the main setup, and Sect. 4.2 with preliminaries, where we present some basic material on G-derivatives and on weighted spaces, we divide the core of the chapter into eight sections:

© Springer International Publishing AG 2017
G. Fabbri et al., *Stochastic Optimal Control in Infinite Dimension*,
Probability Theory and Stochastic Modelling 82,
DOI 10.1007/978-3-319-53067-3_4

- Section 4.3 is devoted to smoothing properties of transition semigroups, which is a key tool used to solve the HJB equations.
- Section 4.4 contains general results on existence and uniqueness of mild solutions.
- Section 4.5 explains how mild solutions can be seen as strong solutions, i.e. limits of classical solutions.
- Sections 4.6 and 4.7 contain more powerful results obtained when the underlying transition semigroup is of Ornstein–Uhlenbeck type.
- Section 4.8 contains applications of the results of Sects. 4.4 and 4.5 (plus the special cases of Sects. 4.6 and 4.7) to a class of optimal stochastic control problems without control in the diffusion coefficient.
- Section 4.9 is devoted to special cases which can be treated by variants of the same methods. For each case we present the results on existence and uniqueness of regular solutions for the HJB equation and applications to stochastic optimal control. Here proofs are not provided but precise references are given, together with some ideas of the proofs in a few cases.
- Section 4.10 is devoted to cases where an explicit representation of the solutions can be found allowing us to solve the associated optimal control problem.

We conclude the chapter with bibliographical notes. The setting we use here is partly borrowed from [179, 309, 310, 431–434, 582].

4.1   The Setting and an Introduction to the Methods

We present the class of HJB equations studied in this chapter and ideas about the methods used to prove existence, uniqueness and regularity of their solutions. Our main goal in this chapter is to develop a theory of such HJB equations which can be used to solve the associated optimal control problems, i.e. to prove the analogues of the results of Sect. 2.5, in particular the verification theorem (like Theorem 2.36) and the existence of optimal feedbacks (like Corollary 2.38). Such results are contained in Sect. 4.8, see in particular Sects. 4.8.1.5, 4.8.1.6, 4.8.2.4, 4.8.2.5.

Let H be a real separable Hilbert space with the inner product $\langle \cdot, \cdot \rangle_H$[1] and the norm $| \cdot |_H$. We consider the following two types of second-order HJB equations[2] in H: the parabolic HJB equation (for a given $T > 0$)

$$\begin{cases} v_t + \dfrac{1}{2} \operatorname{Tr} [\Sigma(t, x) D^2 v] + \langle Ax + b(t, x), Dv \rangle + F(t, x, v, Dv) = 0, \\ \qquad\qquad\qquad\qquad\qquad\qquad\qquad\qquad\qquad t \in [0, T), \ x \in H, \\ v(T, x) = \varphi(x), \ x \in H, \end{cases} \tag{4.1}$$

and, for $\lambda > 0$, the elliptic HJB equation

---

[1] We omit the subscript H when it is clear from the context.

[2] In the literature, such equations are sometimes called *semilinear Kolmogorov equations* (see e.g. [284]).

$$\lambda v - \frac{1}{2} \operatorname{Tr} \left[\Sigma(x) D^2 v\right] - \langle Ax + b(x), Dv \rangle - F(x, v, Dv) = 0, \qquad x \in H. \quad (4.2)$$

In both cases the linear operator $A : D(A) \subset H \to H$ is the infinitesimal generator of a strongly continuous semigroup $\{e^{tA}\}_{t \geq 0}$ while the functions $\Sigma : [0, T] \times H \to \mathcal{L}^+(H)$, $b : [0, T] \times H \to H$, $F : [0, T] \times H \times \mathbb{R} \times H \to H$,[3] $\varphi : H \to \mathbb{R}$, are Borel measurable functions (possibly unbounded in the sense that they may be defined on smaller dense subsets). Precise assumptions on all such data will be given later (see Sect. 4.4).

Since we have in mind applications to stochastic optimal control, we consider here *terminal value problems* for the parabolic case. Clearly also *initial value problems* of the same type can be studied using the same techniques.

The main idea for treating such equations here is to use a perturbation method that we briefly outline in the following subsections, distinguishing the parabolic and the elliptic cases. We mainly consider the case when the underlying basic space is $C_m(H)$ (or, in some cases, $B_m(H)$ or $UC_m(H)$) for $m \geq 0$.[4] In most of the literature the basic space is $C_b(H)$ (or $B_b(H)$ or $UC_b(H)$). We choose to work with functions with polynomial growth as in most applied examples arising in optimal control this is a natural requirement for the data while boundedness is usually too restrictive (see e.g. Sect. 2.6 and, in particular, Sects. 2.6.1 and 2.6.4). The case when the data are in $C_b(H)$ (or $B_b(H)$ or $UC_b(H)$) will then be a special case of the one presented here.

4.1.1 The Method in the Parabolic Case

We consider the linear operator corresponding to the linear part of Eq. (4.1) and which is formally given by:

$$\mathcal{A}(t) : D(\mathcal{A}(t)) \subset C_m(H) \to C_m(H);$$

$$\mathcal{A}(t)\phi = \frac{1}{2} \operatorname{Tr} \left[\Sigma(t, x) D^2 \phi\right] + \langle Ax + b(t, x), D\phi \rangle,$$

where $D(\mathcal{A}(t))$ has to be suitably defined. Suppose that $b : [0, T] \times H \to H$ and $\sigma : [0, T] \times H \to \mathcal{L}(\Xi, H)$ are given functions, where Ξ is another real separable Hilbert space (possibly equal to H), and that[5]

$$\Sigma(t, x) = \sigma(t, x)\sigma^*(t, x).$$

---

[3] In the elliptic case we have $\Sigma : H \to \mathcal{L}(H)$, $b : H \to H$ and $F : H \times \mathbb{R} \times H \to H$.

[4] Recall that when $m = 0$ we use the notation $C_b(H)$ and not $C_0(H)$ to denote the Banach space of bounded and continuous functions.

[5] This can always be done by choosing $\Xi = H$ and $\sigma(t, x) = \sqrt{\Sigma(t, x)}$ for every $(t, x) \in [0, T] \times H$, see e.g. Theorem 12.33 of [521].

Then the operator $\mathcal{A}(t)$ is, formally, the generator of the (two-parameter) transition semigroup $P_{t,s}$ associated to the H-valued diffusion process $X(\cdot)$ which solves the following SDE in H:

$$dX(s) = [AX(s) + b(s, X(s))]ds + \sigma(s, X(s))dW(s), \quad s \in [t, T], \quad X(t) = x,$$
$$(4.3)$$

where, given a filtered probability space $\left(\Omega, \mathscr{F}, \{\mathscr{F}_s\}_{s \in [0,T]}, \mathbb{P}\right)$, $W = W_I$ is a cylindrical Wiener process in Ξ, see Remark 1.89. Assuming well-posedness of (4.3) and denoting by $X(\cdot; t, x)$ its unique solution, the semigroup $P_{t,s}$ is formally defined, for any $\phi \in B_m(H)$ $(m \geq 0)$, as

$$P_{t,s}[\phi](x) := \mathbb{E}[\phi(X(s; t, x)], \qquad 0 \leq t \leq s \leq T,$$

(see Sect. 1.6, Eq. (1.95)). Using the operator $\mathcal{A}(t)$, the Eq. (4.1) can be rewritten as

$$\begin{cases} v_t + \mathcal{A}(t)v + F(t, x, v, Dv) = 0, & t \in [0, T), \ x \in H, \\ v(T, x) = \varphi(x), \ x \in H. \end{cases} \qquad (4.4)$$

Hence, using the semigroup $P_{t,s}$ and the formula of variation of constants, we deduce the following integral form (usually called the *mild form*) of Eq. (4.1):

$$v(t, x) = P_{t,T}[\varphi](x) + \int_t^T P_{t,s}\left[F(s, \cdot, v(s, \cdot), Dv(s, \cdot))\right](x)ds, \ x \in H. \quad (4.5)$$

Such form of our equation is weaker than the classical one in the sense that it requires less regularity. We only need one derivative of the unknown function v instead of two. Moreover, apart from φ and F, this equation depends on the other data only through the operators $P_{t,s}$. Thus (4.5) is the equation studied in this chapter for the parabolic case and the theory of existence and uniqueness of solutions is developed making the assumptions, beyond those on φ and F, directly on the operators $P_{t,s}$. Hence the theory may be applicable to more general cases, e.g. when the underlying process $X(\cdot; t, x)$ is not a diffusion.

It is important to note that in many papers in the literature the authors study initial value problems. To cover such problems we can simply reverse time, defining $u(t, x) := v(T - t, x)$. In this way the terminal value problem (4.1) becomes an initial value problem and the mild form is changed accordingly. Here we will keep the terminal value problem as it is the natural form in which the HJB equations are formulated when they are associated to optimal control problems.

To solve Eq. (4.5) we use a fixed point argument in a suitable space. Setting aside technicalities, this is typically possible for any initial datum $\varphi \in C_m(H)$, $m \geq 0$, if the semigroup $P_{t,s}$ possesses the following smoothing property:

- the function $x \rightarrow P_{t,s}[\varphi](x)$ is differentiable (Gâteaux or Fréchet) for $s > t$ and there exists an integrable map $\gamma : (0, T] \rightarrow (0, +\infty)$ such that
$$\|DP_{t,s}[\varphi]\|_{C_m} \leq \gamma(s - t)\|\varphi\|_{C_m} \quad \forall \varphi \in C_m(H), \; s \in (t, T], \; x \in H.$$
(4.6)

Of course, if we want to take $\varphi \in B_m(H)$, the above property should be true for all such functions.

In the literature, in many cases, the function $\gamma(s - t)$ is substituted, for simplicity, with $C(s - t)^{-\alpha}$ for some $C > 0$ and $\alpha \in (0, 1)$: occasionally we will also do this here. When the Hamiltonian F is globally Lipschitz continuous in the last two arguments (see Hypothesis 4.72) the Contraction Mapping Principle applied in a suitable space allows us to find directly a global solution of (4.5) which is (Gâteaux or Fréchet) differentiable. This case is treated in Sect. 4.4.1.

In the case when F is only locally Lipschitz (see Hypothesis 4.169) things are more complicated: we need to take more regular final datum φ (at least globally Lipschitz) and to assume more regularity on F (namely a certain local Lipschitz continuity of the derivative). With these assumptions we find local solutions by the Contraction Mapping Principle and we prove that such solutions are global, using suitable a priori estimates (see Sect. 4.7).

The solution of (4.5) that we will find with the method described above will be called a *mild solution* of Eq. (4.1) (see Definition 4.70).

To perform the optimal synthesis in the case when the HJB equation (4.1) arises from a stochastic optimal control problem, we need to extend to mild solutions some results presented in Sect. 2.5.1 when the solutions are taken in a classical sense (see e.g. Definition 2.32).

To do this we prove that the *mild solution* of (4.1) is a *strong solution*, i.e. it is the limit (in some sense which involves the so-called $\mathcal{K}$-convergence or π-convergence, see Appendix B.5.1) of *classical solutions* (defined similarly to Definition 2.32) of suitably chosen approximating equations (see Definitions 4.129 and 4.132).

In performing applications to stochastic optimal control it often happens that the "standard" derivative Dv is not the right one needed to find the optimal feedback control. This happens, for example, in the boundary control case (see Sect. 2.6.2) or in the cases treated in Sects. 2.6.4 and 2.6.8.1 (when "the noise enters the system with the control", see Chap. 6 and also [431, 432]). In such cases the Hamiltonian F depends on the differential Dv "through an operator G". When G is a constant (possibly unbounded) operator on H (we refer to Sect. 4.2.1 or to [432] for more general and precise definitions) this roughly means that $F(t, x, v, Dv) = F_0(t, x, v, G^*Dv)$ for a suitable function F_0. In such cases the method described above can be applied with a different smoothing assumption (which we call G-smoothing) on the semigroup $P_{t,s}$, namely that, denoting G^*D formally by D^G,

- the function $x \rightarrow D^G P_{t,s}[\varphi](x)$ is well defined (in some sense) for $s > t$ and there exists an integrable map $\gamma_G : (0, T] \rightarrow (0, +\infty)$ such that
$$\|D^G P_{t,s}[\varphi]\|_{C_m} \leq \gamma_G(s - t)\|\varphi\|_{C_m} \quad \forall \varphi \in C_m(H), \; s \in (t, T], \; x \in H.$$
(4.7)

The most common choice in the literature for the function γ_G is $\gamma_G(s) = Cs^{-\alpha}$ for some $C > 0$ and $\alpha \in (0, 1)$ (see e.g. [431, 432]).

This G-smoothing property holds under different assumptions than those guaranteeing the smoothing property (4.6). For example, in the boundary control case (see Sect. 2.6.2) $G = (-A)^\beta$ (where A is the Laplace operator) is unbounded and so the G-smoothing is somehow stronger and more difficult to obtain. In the cases of Sects. 2.6.4 and 2.6.8, G is bounded and with "narrow" image (finite-dimensional in the delay case) so the G-smoothing is a weaker property and easier to obtain.

We will use this setting (which includes the previous one as the particular case when $G = I$) in the rest of the chapter.

4.1.2   The Method in the Elliptic Case

The stationary HJB equation (4.2) is treated by a similar method. We first consider, formally, as for the parabolic case, the operator

$$\mathcal{A} : D(\mathcal{A}) \subset C_m(H) \to C_m(H); \qquad \mathcal{A}\phi = \frac{1}{2} \operatorname{Tr} [\Sigma(x)D^2\phi] + \langle Ax + b(x), D\phi \rangle,$$

where $D(\mathcal{A})$ has to be properly defined. Suppose that $b : H \to H$ and $\sigma : H \to \mathcal{L}(\Xi, H)$ are given functions for some real separable Hilbert space Ξ and $\Sigma(x) = \sigma(x)\sigma^*(x)$. Then the operator $\mathcal{A}$ is, again formally, the generator of the (one-parameter) transition semigroup P_s associated to the H-valued diffusion process X which solves the SDE

$$dX(s) = [AX(s) + b(X(s))]ds + \sigma(X(s))dW(s), \quad s \in [0, +\infty), \ X(0) = x \in H. \tag{4.8}$$

As in the previous section, $W = W_I$ is a cylindrical Wiener process in Ξ, on a given filtered probability space $(\Omega, \mathscr{F}, \{\mathscr{F}_s\}_{s \geq 0}, \mathbb{P})$. Assuming well-posedness of (4.8) and denoting by $X(\cdot; 0, x) = X(\cdot; x)$ its unique solution, the semigroup P_s is formally defined, for any $\phi \in B_m(H)$ $(m \geq 0)$, as

$$P_s[\phi](x) := \mathbb{E}[\phi(X(s; x))], \qquad 0 \leq s < +\infty$$

(see Sect. 1.6, in particular Eq. (1.99)). Similarly to what is done in the parabolic case we can then rewrite the HJB equation (4.2) as

$$[(\lambda I - \mathcal{A})u](x) = F(x, u(x), Du(x))$$

and so, by expressing formally the resolvent as the Laplace transform of the semigroup (see e.g. [479], proof of Theorem 3.1, p. 8–9), as

$$u(x) = (\lambda I - \mathcal{A})^{-1}[F(\cdot, u, Du)](x) = \int_0^{+\infty} e^{-\lambda s} P_s[F(\cdot, u, Du)](x)ds. \quad (4.9)$$

We call (4.9) the *mild form* of Eq. (4.2) and solutions of it are called *mild solutions* of Eq. (4.2) (see Definition 4.102). Similarly to the parabolic case, this equation only depends on the operators P_s and the function F and thus the theory may be applicable to more general cases, e.g. when the underlying process $X(\cdot; t, x)$ is not a diffusion.

As for the parabolic case, this equation is solved by a fixed point argument, finding a unique (Gâteaux or Fréchet) differentiable solution, if the semigroup P_s has the following smoothing property, completely analogous to the one required for the parabolic case (except for the exponential term which is needed here to control the growth at infinity):

- the function $x \to P_s[\varphi](x)$ is differentiable (Gâteaux or Fréchet) for $s > 0$ and there exist $a \geq 0$ and a map $\gamma : (0, +\infty) \to (0, +\infty)$ integrable on (0,T) for each $T > 0$ and bounded in a neighborhood of $+\infty$, such that
$$\|DP_s[\varphi]\|_{C_m} \leq \gamma(s)e^{as}\|\varphi\|_{C_m} \quad \forall \varphi \in C_m(H), \; s \in (0, +\infty), \; x \in H.$$
$$(4.10)$$

In contrast to the parabolic case, an existence/uniqueness theorem (see Sect. 4.4.2) holds in general only if the number λ is big enough; this is a standard fact which also arises with other techniques, see e.g. Hypothesis 3.72 in Chap. 3 on viscosity solutions. Under suitable additional hypotheses and using monotone operator techniques, one can extend such a result to any $\lambda > 0$, see Sects. 4.6.2 and 6.7. As for the parabolic case, to perform the optimal synthesis in applications to stochastic optimal control, one needs to prove that the mild solution is indeed a *strong solution* (see Definition 4.140), i.e. the limit (in a suitable sense involving π or $\mathcal{K}$-convergence) of classical solutions (see Definition 4.139).

Finally, again as in Sect. 4.1.1, to cover important families of applied examples, here it is more convenient to study directly the case when the Hamiltonian F is, for a given possibly unbounded operator G, of the form

$$F(x, v, Dv) = F_0(x, v, D^G v),$$

where $D^G v$ (roughly equal to $G^* Dv$) is the G-derivative, as explained heuristically in the previous subsection and, more precisely, in Sect. 4.2.1. This setting will be used in the rest of the chapter.

4.2 Preliminaries

4.2.1 G-Derivatives

We start by recalling the notion of G-derivative as in [286] (see also [431], Sect. 4 and [432]).

Definition 4.1 Let X, Y and Z be three real Banach spaces. Let $G : X \to \mathcal{L}(Z, X)$ and consider a mapping $f : X \to Y$.

- The G-directional derivative $\nabla^G f\,(x; h)$ at a point $x \in X$ in the direction $h \in Z$ is defined as

$$\nabla^G f\,(x; h) := \lim_{s \to 0} \frac{f\,(x + sG\,(x)\,h) - f\,(x)}{s}, \quad s \in \mathbb{R}, \qquad (4.11)$$

  where the limit above is taken in the norm of Y.
- We say that f is G-Gâteaux differentiable at a point $x \in X$ if f admits the G-directional derivative in every direction $h \in Z$ and there exists a bounded linear operator, the G-Gâteaux derivative $\nabla^G f(x) \in \mathcal{L}(Z, Y)$, such that $\nabla^G f(x; h) = \nabla^G f(x)h$ for all $h \in Z$. We say that f is G-Gâteaux differentiable on X if it is G-Gâteaux differentiable at every point $x \in X$.
- We say that f is G-Fréchet differentiable (or simply G-differentiable) at a point $x \in X$ if it is G-Gâteaux differentiable and if the limit in (4.11) is uniform for h in the unit ball of Z. In this case we call $D^G f(x)$ the G-Fréchet derivative (or simply the G-derivative) of f at x. We say that f is G-Fréchet differentiable on X if it is G-Fréchet differentiable at every point $x \in X$.

Note that, in the definition of the G-derivative, one considers only the directions in X selected by the image of $G(x)$. This is similar to what is done in many papers in the theory of abstract Wiener spaces, considering the K-derivative, where K is a subspace of X, see e.g. [329]. Similar concepts are also used in [147, 492] and in Sect. 3.3.1 of [179]. A generalized notion of G-derivative in spaces $L^p(H, m)$ is considered in relation to Dirichlet forms, see e.g. [422] (or also [154], Chap. 3, where it is called a Malliavin derivative). The same notion is used in Chap. 5 but with a slightly different notation (see Definition 5.11 and the subsequent remark).

If f is Gâteaux (Fréchet) differentiable on X then, given any G as in the definition above, f is G-Gâteaux (Fréchet) differentiable on X and

$$\nabla^G f\,(x)\,h = \nabla f\,(x)\,(G\,(x)\,h), \qquad \left(D^G f\,(x)\,h = Df\,(x)\,(G\,(x)\,h)\right), \qquad (4.12)$$

i.e. the G-directional derivative in the direction $h \in Z$ is just the usual directional derivative at a point $x \in X$ in the direction $G\,(x)\,h \in X$. However, the notion of the G-derivative allows us to deal with functions which are not Gâteaux differentiable.

Example 4.2 Consider $f : \mathbb{R}^2 \to \mathbb{R}$ so that in the previous notation $X = \mathbb{R}^2, Y = \mathbb{R}$. The function $f(x) = |x_1|\,x_2$ does not admit the directional derivative in the direction $h = (1, 0)$ at $x = (0, x_2)$ for $x_2 \neq 0$. However, taking $Z = \mathbb{R}$ and $G(x_1, x_2) \equiv (0, 1)$, we see that f admits the G-Fréchet derivative at every $x \in \mathbb{R}^2$. ∎

Notation 4.3 If $Y = \mathbb{R}$ then the G-derivative (Gâteaux or Fréchet) takes values in $\mathcal{L}(Z, \mathbb{R}) = Z^*$. If Z is a Hilbert space we will identify Z with its dual, so we will have $\nabla^G f : X \to Z$ and we will write $\left\langle \nabla^G f(x), h \right\rangle_Z$ for $\nabla^G f(x)\,h$. Similar comments apply to the G-Fréchet derivative.

In the same spirit, when $Y = \mathbb{R}$ and both X and Z are Hilbert spaces, we identify the spaces X and Z with their duals. Hence, whenever f is Gâteaux (Fréchet) differentiable at $x \in X$, identity (4.12) becomes

$$\langle \nabla^G f(x), h \rangle_Z = \langle \nabla f(x), G(x) h \rangle_X = \langle G(x)^* \nabla f(x), h \rangle_Z ,$$

and similarly for the Fréchet derivatives.   ∎

We denote by $\mathcal{L}_u(Z, X)$ the set of linear closed operators (possibly unbounded) with dense domain from Z to X. We extend the concept of the G-derivative to the case when $G : X \to \mathcal{L}_u(Z, X)$.

Definition 4.4 Let X, Y and Z be three real Banach spaces, let $f : X \to Y$ and let $G : X \to \mathcal{L}_u(Z, X)$.

- The G-directional derivative $\nabla^G f(x; h)$ at a point $x \in X$ in the direction $h \in D(G(x))$ is defined exactly as in (4.11).
- We say that f is G-Gâteaux differentiable at a point $x \in X$ if f admits the G-directional derivative in every direction $h \in D(G(x))$ and there exists a *bounded linear operator*, the G-Gâteaux derivative $\nabla^G f(x) \in \mathcal{L}(Z, Y)$, such that $\nabla^G f$ $(x; h) = \nabla^G f(x) h$ for $x \in X$ and $h \in D(G(x))$. We say that f is G-Gâteaux differentiable on X if it is G-Gâteaux differentiable at every point $x \in X$.
- We say that f is G-Fréchet differentiable (or simply G-differentiable) at a point $x \in X$ if it is G-Gâteaux differentiable and if the limit in (4.11) is uniform for h in the unit ball of Z intersected with $D(G(x))$. In this case we call $D^G f(x)$ the G-Fréchet derivative (or simply the G-derivative) of f at x. We say that f is G-Fréchet differentiable on X if it is G-Fréchet differentiable at every point $x \in X$.

Remark 4.5 Even if f is Fréchet differentiable at $x \in X$, the G-derivative may not exist at x when G is unbounded. Indeed, consider the following case. Let X be a Hilbert space and $G : D(G) \subset X \to X$ be a closed linear operator on X with dense domain and with unbounded adjoint G^* on X whose domain is $D(G^*)$. Let $f : X \to \mathbb{R}$ be Fréchet differentiable on X. By the definition of G-directional derivative we have, for every $x \in X$ and $h \in D(G)$,

$$\nabla^G f(x; h) = \langle Df(x), Gh \rangle_X .$$

On the other hand, if the G-derivative of f exists at $x \in X$ then we should have $D^G f(x) \in \mathcal{L}(X, \mathbb{R}) = X^*$ (which we identify with X). Hence, if f was G-differentiable on X this would imply that, for any $x \in X$,

$$\nabla^G f(x; h) = \langle D^G f(x), h \rangle_X ,$$

from which we get

$$|\nabla^G f(x; h)| = \left| \langle D^G f(x), h \rangle_X \right| \le c|h|, \quad \forall h \in D(G).$$

This would mean that $Df(x) \in D(G^*)$ for all $x \in X$. This may not be true, e.g. when $f(x) = |x|^2$.   ∎

Remark 4.6 Let X be a Hilbert space. Observe that if $G : D(G) \subset X \rightarrow X$ is an element of $\mathcal{L}_u(X, X)$ then we can consider it as a constant function $X \rightarrow \mathcal{L}(D(G), X)$ and, taking $Z = D(G)$ (with the usual Hilbert structure on $D(G)$, where the inner product is given by $\langle x, y \rangle + \langle Gx, Gy \rangle$), we fall into the setting of Definition 4.1. In this case, denoting by $\bar{G}^*$ the adjoint of G as an operator from $D(G)$ to X, if $f : X \rightarrow \mathbb{R}$ is Gâteaux (Fréchet) differentiable at $x \in X$ we have, for $h \in D(G)$,

$$\nabla^G f(x) h = \langle \nabla f(x), Gh \rangle_X = \langle \bar{G}^* \nabla f(x), h \rangle_Z ,$$

and similarly for the Fréchet derivatives. It should be clear to the reader that this way of seeing the G-derivative for unbounded operators G is weaker than that of Definition 4.4 due to the weaker continuity requirement on the G-derivative. For our purposes (in particular in treating boundary control problems) we will be using the stronger requirement of Definition 4.4.   ∎

We now define, following [284, 431, 432], relevant classes of spaces.

Definition 4.7 Let X, Y and Z be three real Banach spaces and let X_0 be a Borel subset of X. For $m \geq 0$, we define the linear space $B_m^s(X_0, \mathcal{L}(Z, Y))$ (respectively, $C_m^s(X_0, \mathcal{L}(Z, Y))$) to be the space of the mappings $L : X_0 \rightarrow \mathcal{L}(Z, Y)$ such that, for every $z \in Z$, $L(\cdot)z \in B_m(X_0, Y)$,[6] (respectively, $L(\cdot)z \in C_m(X_0, Y)$).[7] These spaces are equipped with the norm (which is finite by the Banach–Steinhaus theorem)

$$\|L\|_{B_m^s(X_0, \mathcal{L}(Z,Y))} := \sup_{x \in X_0} \frac{\|L(x)\|_{\mathcal{L}(Z,Y)}}{1 + |x|^m}. \tag{4.13}$$

When $L \in C_m^s(X_0, \mathcal{L}(Z, Y))$ we write $\|L\|_{C_m^s(X_0, \mathcal{L}(Z,Y))}$. When it is clear from the context we will simply write $\|L\|_{B_m^s}$ and, for elements of $C_m^s(X_0, \mathcal{L}(Z, Y))$, $\|L\|_{C_m^s}$. When $Y = \mathbb{R}$ and Z is a Hilbert space, we identify Z with its dual and so $\mathcal{L}(Z, Y)$ with Z, hence we will write $C_m^s(X_0, Z)$. In this case the strong continuity of a map $f : X_0 \rightarrow Z$ means that the map $\langle f(\cdot), z \rangle_Z$ is continuous for every $z \in Z$. When $m = 0$, we will use the notation $B_b^s(X_0, \mathcal{L}(Z, Y))$ and $C_b^s(X_0, \mathcal{L}(Z, Y))$.

Proposition 4.8 *Let $m \geq 0$ and X_0, X, Y and Z be as in Definition 4.7. The spaces $B_m^s(X_0, \mathcal{L}(Z, Y))$ and $C_m^s(X_0, \mathcal{L}(Z, Y))$, endowed with the norm (4.13), are Banach spaces.*

---

[6]This property is usually called strong measurability of L. Note that, thanks to the Pettis measurability theorem (see Lemma 1.17-(iv)), when Z is separable and $Y = \mathbb{R}^n$, strong measurability of L is equivalent to its measurability since the space $\mathcal{L}(Z, \mathbb{R}^n)$ is separable. This is not the case, in general, when Y is infinite-dimensional.

[7]This property is usually called strong continuity of L.

Proof We prove the claim for $C_m^s(X_0, \mathcal{L}(Z, Y))$. The proof for $B_m^s(X_0, \mathcal{L}(Z, Y))$ is the same. Let $(f_n)_{n \in \mathbb{N}}$ be a Cauchy sequence in $C_m^s(X_0, \mathcal{L}(Z, Y))$. Then, for each $x \in X_0$, by completeness of $\mathcal{L}(Z, Y)$, $f_n(x) \to f(x)$ in $\mathcal{L}(Z, Y)$ for some $f(x) \in \mathcal{L}(Z, Y)$. On the other hand, by completeness of $C_m(X_0, Y)$, we also have, for each $z \in Z$, $f_n(\cdot)z \to f_z(\cdot)$ in $C_m(X_0, Y)$ for some $f_z \in C_m(X_0, Y)$. By uniqueness of the limit we have $f(x)z = f_z(x)$ for each $z \in Z$ and $x \in X_0$. Hence, $f \in C_m^s(X_0, \mathcal{L}(Z, Y))$. It remains to show that $f_n \to f$ in the $\|\cdot\|_{C_m^s(X_0, \mathcal{L}(Z,Y))}$ norm, that is

$$\sup_{x \in X_0} \sup_{|z|_Z = 1} \frac{|(f_n(x) - f(x))z|_Y}{1 + |x|^m} \to 0.$$

To this end, for every $z \in Z$, $|z|_Z = 1$ and $n \in N$, we have

$$\frac{|(f_n(x) - f(x))z|_Y}{1 + |x|^m} = \lim_{k \to \infty} \frac{|(f_n(x) - f_k(x))z|_Y}{1 + |x|^m} \leq \limsup_{k \to \infty} \|f_n - f_k\|_{C_m^s(X_0, \mathcal{L}(Z,Y))}.$$

The result follows since $(f_n)_{n \in \mathbb{N}}$ is a Cauchy sequence in $C_m^s(X_0, \mathcal{L}(Z, Y))$. $\qquad \square$

Definition 4.9 Let X, Y and Z be three real Banach spaces and let $G : X \to \mathcal{L}_u(Z, X)$. Let $m \geq 0$.

A mapping $f : X \to Y$ belongs to the class $B_m^{1,G}(X, Y)$ if $f \in B_m(X, Y)$, f is G-Gâteaux differentiable on X, and $\nabla^G f \in B_m^s(X, \mathcal{L}(Z, Y))$.[8] We write

$$B_m^{1,G}(X, Y) := \left\{ f \in B_m(X, Y) : \nabla^G f \in B_m^s(X, \mathcal{L}(Z, Y)) \right\}. \tag{4.14}$$

Moreover, a mapping $f : X \to Y$ belongs to the class $\mathcal{G}_m^{1,G}(X, Y)$ if $f \in B_m^{1,G}(X, Y)$, f is continuous, and its G-Gâteaux derivative $\nabla^G f(\cdot)$ is strongly continuous, namely

$$\mathcal{G}_m^{1,G}(X, Y) := \left\{ f \in C_m(X, Y) : \nabla^G f \in C_m^s(X, \mathcal{L}(Z, Y)) \right\}. \tag{4.15}$$

Similarly we define

$$C_m^{1,G}(X, Y) := \left\{ f \in C_m(X, Y) : D^G f \in C_m(X, \mathcal{L}(Z, Y)) \right\}, \tag{4.16}$$

$$UC_m^{1,G}(X, Y) := \left\{ f \in UC_m(X, Y) : D^G f \in UC_m(X, \mathcal{L}(Z, Y)) \right\}. \tag{4.17}$$

Note that in the last two cases we require G-Fréchet differentiability.[9] When $m = 0$ we employ, as usual, the notation $B_b^{1,G}(X, Y)$, $\mathcal{G}_b^{1,G}(X, Y)$, $C_b^{1,G}(X, Y)$, $UC_b^{1,G}(X, Y)$. Moreover, when $Y = \mathbb{R}$ we omit it in the notation.

---

[8]Recall that, as noted above, when Z is separable and $Y = \mathbb{R}^n$, we have $B_m^s(X, \mathcal{L}(Z, Y)) = B_m(X, \mathcal{L}(Z, Y))$ thanks to the Pettis measurability theorem.

[9]One may think that, in such cases, G-Fréchet differentiability follows from the continuity of the Gâteaux differential, however this is not obvious without further assumptions, see Remark 4.19.

We endow $B_m^{1,G}(X, Y)$ and its subspaces $\mathcal{G}_m^{1,G}(X, Y), C_m^{1,G}(X, Y), UC_m^{1,G}(X, Y)$ with the norm

$$\|f\|_{B_m^{1,G}(X,Y)} := \sup_{x \in X} \frac{|f(x)|_Y}{1 + |x|^m} + \sup_{x \in X} \frac{|\nabla^G f(x)|_{\mathcal{L}(Z,Y)}}{1 + |x|^m}, \tag{4.18}$$

which we denote by $\|f\|_{\mathcal{G}_m^{1,G}(X,Y)}$ (respectively, $\|f\|_{C_m^{1,G}(X,Y)}$ or $\|f\|_{UC_m^{1,G}(X,Y)}$) when $f \in \mathcal{G}_m^{1,G}(X, Y)$ (respectively, $f \in C_m^{1,G}(X, Y)$ or $f \in UC_m^{1,G}(X, Y)$). If it is clear from the context, we will often simply write $\|f\|_{B_m^{1,G}}$ and $\|f\|_{\mathcal{G}_m^{1,G}}, \|f\|_{C_m^{1,G}}, \|f\|_{UC_m^{1,G}}$.

Remark 4.10 In Appendix A the function spaces $C_m^1(X, Y)$ and $UC_m^1(X, Y)$, are defined. It is immediate from the definitions that, when $Z = X$ and $G = I$,

$$C_m^{1,I}(X, Y) = C_m^1(X, Y), \quad UC_m^{1,I}(X, Y) = UC_m^1(X, Y).$$

See Remark A.1 for more on the definition of spaces with polynomial growth. ∎

We will need to perform the G-differentiation under the integral sign, in particular when we apply the Contraction Mapping Principle to find the mild solutions of (4.1) and (4.2). We will discuss two ways (Corollary 4.14 and Proposition 4.16) to do this task, however we will only use Proposition 4.16 in this book. We present both to make the reader aware of the difficulties arising here. The setting and the results given in the remainder of this subsection are mainly taken from [241].

We need the following assumption.

Hypothesis 4.11 Let X and Z be real separable Hilbert spaces. The map $G : X \to \mathcal{L}_u(Z, X)$ satisfies the following.

(i) $D(G(x)) = D(G(y))$ for every $x, y \in X$; we denote by $D(G) \subset Z$ the common domain.
(ii) $R(G(x)) = R(G(y))$ for every $x, y \in X$; we denote by $R(G) \subset X$ the common range.
(iii) Let $G(x)^{-1} : R(G) \to D(G)$ be the pseudo-inverse of $G(x)$ according to Definition B.1. For each $k \in R(G)$ the map $X \to Z, x \to G(x)^{-1}k$ is bounded on compact sets.

We will also use the following hypothesis.

Hypothesis 4.12 Hypothesis 4.11 holds true with *bounded* replaced by *continuous* in (iii).

The first way to perform differentiation under the integral sign is by using the closedness of the G-derivative operator.

Proposition 4.13 *Let Hypothesis 4.12 hold and let Y be a real separable Banach space. The spaces $\mathcal{G}_m^{1,G}(X, Y), C_m^{1,G}(X, Y),$ and $UC_m^{1,G}(X, Y)$ are Banach spaces when endowed with the norm (4.18).*

Proof We give the proof for $\mathcal{G}_m^{1,G}$. The proofs for $C_m^{1,G}$ and $UC_m^{1,G}$ are analogous.

Let $(\Phi_n)_{n\in\mathbb{N}}$ be a Cauchy sequence in $\mathcal{G}_m^{1,G}(X, Y)$. In particular, $(\Phi_n)_{n\in\mathbb{N}}$ is a Cauchy sequence in $C_m(X, Y)$, so that Φ_n converges to a function $\Phi \in C_m(X, Y)$.

Now, for all $x \in X$, $\left(\nabla^G \Phi_n(x)\right)_{n\in\mathbb{N}}$ is a Cauchy sequence of linear bounded operators in $\mathcal{L}(Z, Y)$, so that $\nabla^G \Phi_n(x)$ converges to a linear bounded operator $A(x)$. On the other hand, for all $z \in Z$, the sequence $\left(\nabla^G \Phi_n(\cdot)z\right)_{n\in\mathbb{N}}$ is a Cauchy sequence in $C_m(X, Y)$ so that $\nabla^G \Phi_n(\cdot)z$ converges to a function $A_z \in C_m(X, Y)$. Hence, we have $A_z(x) = A(x)z$, which yields $A \in C_m^s(X, \mathcal{L}(Z, Y))$.

Now notice that, by the definition of $\nabla^G \Phi_n(x)$, we have $\nabla^G \Phi_n(x)h = 0$ whenever $h \in \ker(G(x))$. It follows that

$$\ker(G(x)) \subset \ker(A(x)), \quad \forall x \in X. \tag{4.19}$$

We are now going to prove that $\Phi \in \mathcal{G}_m^{1,G}(X, Y)$ and $A = \nabla^G \Phi$. Let $x \in X$, $h \in D(G)$. Set, for $r \in \mathbb{R}$, $y(r) := x + rG(x)h$ and $\varphi_n(r) := \Phi_n(x + rG(x)h)$. Then,

$$
\begin{aligned}
\varphi_n'(r) &= \lim_{\tau \to 0} \frac{\Phi_n(x + (r+\tau)G(x)h) - \Phi_n(x + rG(x)h)}{\tau} \\
&= \lim_{\tau \to 0} \frac{\Phi_n(y(r) + \tau G(x)h) - \Phi_n(y(r))}{\tau} \\
&= \lim_{\tau \to 0} \frac{\Phi_n(y(r) + \tau G(y(r))G(y(r))^{-1}G(x)h) - \Phi_n(y(r))}{\tau} \\
&= \nabla^G \Phi_n(y(r))G(y(r))^{-1}G(x)h, \quad \forall r \in \mathbb{R}.
\end{aligned}
$$

Notice that, as $\nabla^G \Phi_n \in C_m^s(X, \mathcal{L}(Z, Y))$, by the Banach–Steinhaus Theorem the family $\{\nabla^G \Phi_n(y(r))\}_{r\in[-1,1]}$ is a family of uniformly bounded operators in $\mathcal{L}(X, Y)$. We have, for every $r, r_0 \in [-1, 1]$,

$$
\begin{aligned}
|\varphi_n'(r) - \varphi_n'(r_0)|_Y &\leq |\left(\nabla^G \Phi_n(y(r))G(y(r))^{-1} - \nabla^G \Phi_n(y(r_0)G(y(r_0))^{-1}\right)G(x)h|_Y \\
&\leq |\nabla^G \Phi_n(y(r))|_{\mathcal{L}(Z,Y)}|(G(y(r))^{-1} - G(y(r_0))^{-1})G(x)h|_Z \\
&\quad + |\left(\nabla^G \Phi_n(y(r)) - \nabla^G \Phi_n(y(r_0))\right)G(y(r_0))^{-1}G(x)h|_Z \\
&\leq \left(\sup_{s\in[-1,1]} |\nabla^G \Phi_n(y(s))|_{\mathcal{L}(Z,Y)}\right)|(G(y(r))^{-1} - G(y(r_0))^{-1})G(x)h|_Z \\
&\quad + |(\nabla^G \Phi_n(y(r)) - \nabla^G \Phi(y(r_0)))G(y(r_0))^{-1}G(x)h|_Z.
\end{aligned}
$$

Thus, by Hypothesis 4.12 and again by the fact that $\nabla^G \Phi_n \in C_m^s(X, \mathcal{L}(Z, Y))$, we see that $\varphi_n'(r) \to \varphi_n'(r_0)$ as $r \to r_0$. Hence $\varphi_n' \in C([-1, 1], Y)$. We can then apply Theorem D.20 obtaining for all $s \in [-1, 1] \setminus \{0\}$

$$\frac{\Phi_n(x + sG(x)h) - \Phi_n(x)}{s} = \frac{1}{s}\int_0^s \varphi_n'(r)dr = \frac{1}{s}\int_0^s \nabla^G \Phi_n(y(r))G(y(r))^{-1}G(x)h\,dr. \tag{4.20}$$

Now, as $n \to \infty$, we have the convergences

$$\Phi_n(x + sG(x)h) \to \Phi(x + sG(x)h),$$
$$\Phi_n(x) \to \Phi(x),$$
$$\nabla^G \Phi_n(y(r))G(y(r))^{-1}G(x)h \to A(y(r))G(y(r))^{-1}G(x)h, \quad r \in \mathbb{R}.$$

Thus, from (4.20), we get

$$\frac{\Phi(x + sG(x)h) - \Phi(x)}{s} = \frac{1}{s} \int_0^s A(y(r))G(y(r))^{-1}G(x)h\,dr. \tag{4.21}$$

Now, since $A \in C_m^s(X, \mathcal{L}(Z, Y))$, arguing as we did above for $\nabla^G \Phi_n$ and using Hypothesis 4.12 we see that the function $[-1, 1] \to \mathbb{R}$, $r \mapsto A(y(r))G(y(r))^{-1}G(x)h$ is continuous. Therefore, it follows from (4.21) that

$$\lim_{s \to 0} \frac{\Phi(x + sG(x)h) - \Phi(x)}{s} = A(x)G(x)^{-1}G(x)h. \tag{4.22}$$

Let $k := G(x)^{-1}G(x)h$. We observe that $G(x)h = G(x)k$, i.e. $h - k \in \ker(G(x))$. We now obtain from (4.22) and (4.19) that $\nabla^G \Phi(x)$ exists and it coincides with $A(x)$. The convergence of Φ_n to Φ in the norm $\|\cdot\|_{\mathcal{G}_m^{1,G}}$ then follows as in the proof of Proposition 4.8, completing the proof. $\square$

A straightforward consequence of the above result is the following corollary on differentiation under the integral sign.

Corollary 4.14 *Let Hypothesis 4.12 hold, let Y be a real separable Banach space and let $m \geq 0$. Let $T \in (0, +\infty]$, where as always, when $T = +\infty$, $[0, T]$ means $[0, +\infty)$.*

(i) The unbounded operators

$$\nabla^G : \mathcal{G}_m^{1,G}(X, Y) \subset C_m(X, Y) \to C_m^s(X, \mathcal{L}(Z, Y)),$$

$$D^G : C_m^{1,G}(X, Y) \subset C_m(X, Y) \to C_m(X, \mathcal{L}(Z, Y)),$$

$$D^G : UC_m^{1,G}(X, Y) \subset UC_m(X, Y) \to UC_m(X, \mathcal{L}(Z, Y)),$$

are closed.

(ii) Let $f : [0, T] \to \mathcal{G}_m^{1,G}(X, Y)$ be measurable and such that

$$\int_0^T \left[\|f(t)\|_{C_m} + \|\nabla^G f(t)\|_{C_m^s} \right] dt < +\infty.$$

Then

$$\int_0^T f(t)dt \in \mathcal{G}_m^{1,G}(X, Y) \quad and \quad \nabla^G \int_0^T f(t)dt = \int_0^T \nabla^G f(t)dt.$$

The same holds if we replace $\mathcal{G}_m^{1,G}(X, Y)$ with $C_m^{1,G}(X, Y)$ or $UC_m^{1,G}(X, Y)$, substituting ∇^G with D^G.

Proof We only consider the space $\mathcal{G}_m^{1,G}(X, Y)$ as the other cases are completely similar. The first part is a straightforward consequence of Lemma 4.13. The second part is a consequence of the first part, Remark 1.31 and the assumptions on f. □

Remark 4.15 The result of Corollary 4.14-(ii) is not sufficient for our purposes. First of all the proof of Proposition 4.13 does not work for the space $B_m^{1,G}(X, Y)$ as in the last part we need to use the continuity of the map $r \rightarrow A(y(r))(G(y(r))^{-1}G(x)h)$, which is not guaranteed in that case. This prevents the extension of it and of Corollary 4.14 to $B_m^{1,G}(X, Y)$.

Secondly, the result of Corollary 4.14-(ii) requires that the function $f : [0, T] \rightarrow \mathcal{G}_m^{1,G}(X, Y)$ be measurable. This assumption is in general not true for the maps f arising when we construct solutions of equations (4.1) or (4.2) in the mild form through the contraction mapping principle.

Indeed, in such cases we may have, for example, $f(t) = P_t[\psi]$, where ψ is a given function in $C_m(H)$ and P_t is a suitable transition semigroup, e.g. the Ornstein–Uhlenbeck semigroup defined in Sect. 4.3.1. As recalled in Proposition B.89, such transition semigroups are not strongly continuous in $C_m(H)$, hence the function $\mathbb{R}^+ \rightarrow C_m(H), t \rightarrow P_t[\psi]$ is not measurable in general. This problem can be overcome by performing "pointwise" differentiation, as will be shown in Proposition 4.16, see also [241] for more on this. Another approach to resolving this problem may involve a change of the topology used in such spaces, see [243]. ■

The next result shows how we can differentiate "pointwise" under the integral sign for functions with values in the spaces $B_m^{1,G}$, and consequently also in $\mathcal{G}_m^{1,G}$, $C_m^{1,G}, UC_m^{1,G}$.

Proposition 4.16 *Let Hypothesis 4.11 hold and let Y be a real separable Banach space. Let $T \in (0, +\infty]$, where, when $T = +\infty$, $[0, T]$ means $[0, +\infty)$. Let $m \geq 0$, $f : [0, T] \times X \rightarrow Y$ and assume that:*

- *f is jointly measurable and there exists a $g \in L^1(0, T; \mathbb{R}^+)$ such that*

$$|f(t, x)|_Y \leq g(t)(1 + |x|^m) \quad for \ a.e. \ t \in [0, T], \forall x \in X. \quad (4.23)$$

- *$f(t, \cdot) \in B_m^{1,G}(X, Y)$ (respectively, $\mathcal{G}_m^{1,G}(X, Y)$, $C_m^{1,G}(X, Y)$), for a.e. $t \in [0, T]$, and $\nabla^G f$ is jointly strongly measurable. Moreover, we have*

$$|\nabla^G f(t, x)|_{\mathcal{L}(Z, Y)} \leq g(t)(1 + |x|^m) \quad for \ a.e. \ t \in [0, T], \forall x \in X, \quad (4.24)$$

where g is from (4.23).

Then the function $L : X \to Y, L(x) := \int_0^T f(t, x)\, dt$, belongs to $B_m^{1,G}(X, Y)$ (respectively, $\mathcal{G}_m^{1,G}(X, Y)$, $C_m^{1,G}(X, Y))$ and

$$\nabla^G L(x) h = \int_0^T \nabla^G f(t, x) h\, dt, \quad \forall h \in Z. \tag{4.25}$$

When L belongs to $C_m^{1,G}(X, Y)$ this last formula also holds with D^G in place of ∇^G.

If we also assume that, for a.e. $t \in [0, T]$, we have $f(t, \cdot) \in UC_m^{1,G}(X, Y)$ and, for all $x, y \in X$,

$$|f(t, x) - f(t, y)|_Y + |\nabla^G f(t, x) - \nabla^G f(t, y)|_{\mathcal{L}(Z,Y)} \leq g(t)\rho(|x - y|),$$

where ρ is a suitable modulus and g is as above, then $L \in UC_m^{1,G}(X, Y)$.

Proof We first prove the claim for $B_m^{1,G}$. The function L is well defined, measurable and it belongs to $B_m(X, Y)$ thanks to (4.23) and Theorem 1.33.

Let now $x \in X, h \in D(G)$ and consider the limit

$$\lim_{s \to 0} \frac{L(x + sG(x)h) - L(x)}{s} = \lim_{s \to 0} \int_0^T \frac{1}{s}[f(t, x + sG(x)h) - f(t, x)]dt. \tag{4.26}$$

By our assumptions the integrand of the right-hand side converges, for a.e. t, to $\nabla^G f(t, x)h$. We will show that the integrand is bounded by an integrable function, uniformly for s in a neighborhood of 0. To do this we set

$$y : \mathbb{R} \to X, \quad y(r) := x + rG(x)h; \quad \varphi : [0, T] \times \mathbb{R} \to Y, \quad \varphi(t, r) := f(t, y(r)).$$

Then, using Hypothesis 4.11-(i) and (ii), we obtain for a.e. $t \in [0, T]$ and every $r \in \mathbb{R}$

$$
\begin{aligned}
\frac{\partial}{\partial r}\varphi(t, r) &= \lim_{\eta \to 0} \frac{f(t, y(r) + \eta G(x)h) - f(t, y(r))}{\eta} \\
&= \lim_{\eta \to 0} \frac{f(t, y(r) + \eta G(y(r))G(y(r))^{-1}G(x)h) - f(t, y(r))}{\eta} \\
&= \nabla^G f(t, y(r))(G(y(r))^{-1}G(x)h).
\end{aligned}
$$

Hence $\frac{\partial}{\partial r}\varphi(t, r)$ exists for every r and moreover, by (4.24), we have

$$\left|\frac{\partial}{\partial r}\varphi(t, r)\right| \leq g(t)(1 + |y(r)|^m)|G(y(r))^{-1}G(x)h|_Y.$$

Now, by Hypothesis 4.11-(iii), there exists a $C > 0$, depending on $G(x)h$ but independent of t, such that

$$|G(y(r))^{-1}G(x)h|_Y \leq C, \quad \text{for every } r \in [-1, 1].$$

The last two estimates give, for a.e. $t \in [0, T]$ and every $r \in [-1, 1]$,

$$\left| \frac{\partial}{\partial r} \varphi(t, r) \right| \le Cg(t) \left[1 + (|x| + |G(x)h|)^m \right]. \qquad (4.27)$$

The integrand in the right-hand side of (4.26) is equal to $s^{-1}[\varphi(t, s) - \varphi(t, 0)]$. Since, for a.e. $t \in [0, T]$, $\varphi(t, \cdot)$ is everywhere differentiable, then it is also continuous. Hence using the Mean Value Theorem for functions with values in Y (see Theorem D.20 or [586], Proposition 3.5, p. 76), and (4.27), we obtain, for a.e. $t \in [0, T]$,

$$s^{-1}[\varphi(t, s) - \varphi(t, 0)] \le Cg(t) \left[1 + (|x| + |G(x)h|)^m \right], \qquad \forall s \in (0, 1).$$

This gives the required uniform bound for $s \in (0, 1)$. Applying the dominated convergence theorem to (4.26) we now conclude

$$\nabla^G L(x; h) = \int_0^T \nabla^G f(t, x)h \, dt,$$

which implies that, for every $x \in X$ and $h \in D(G)$,

$$|\nabla^G L(x; h)|_Y \le \left(\int_0^T \|\nabla^G f(t, x)\|_{\mathcal{L}(Z,Y)} dt \right) |h| \le \left(\int_0^T g(t)(1 + |x|^m) dt \right) |h|.$$

This proves the required G-Gâteaux differentiability and (4.25). Theorem 1.33 and (4.25) imply that $\nabla^G L$ is strongly measurable and it belongs to $B_m^s(X, \mathcal{L}(Z, Y))$.

To prove the claim for $\mathcal{G}_m^{1,G}(X, Y)$ (respectively, $C_m^{1,G}(X, Y)$) it is enough to show that, if $f(t, \cdot)$ belongs to $\mathcal{G}_m^{1,G}(X, Y)$ for a.e. $t \in [0, T]$, then $L \in C_m(X, Y)$ and $\nabla^G L \in C_m^s(X, \mathcal{L}(Z, Y))$ (respectively, $C_m(X, \mathcal{L}(Z, Y))$). We do this in the first case as the second one is completely analogous. Take any $x \in X$ and any sequence $x_n \to x$. Without loss of generality we can assume $|x_n| \le 2|x|$. Then, for a.e. $t \in [0, T]$ and for all $h \in Z$ we have, as $n \to +\infty$,

$$f(t, x_n) \to f(t, x), \quad \text{and} \quad \nabla^G f(t, x_n)h \to \nabla^G f(t, x)h.$$

Thanks to (4.23) and (4.24) we get the claim by the dominated convergence theorem.

To prove the last statement we observe that, under the assumption there, for $x, y \in X$,

$$|L(x) - L(y)|_Y \le \int_0^T |f(t, x) - f(t, y)|_Y dt \le \int_0^T g(t)\rho(|x - y|) dt$$

and similarly we can estimate $|D^G L(x) - D^G L(y)|_{\mathcal{L}(Z,Y)}$. $\qquad \square$

We finally define analogues of the spaces introduced in Definition 4.9 when the functions are defined on $[0, T] \times X$ for some $T > 0$ (see e.g. [283], Sect. 2).

Definition 4.17 Let I be an interval in $\mathbb{R}$ and let X, Y, Z be real separable Banach spaces. Let $G : I \times X \to \mathcal{L}_u(Z, X)$. Let $m \geq 0$. We say that a mapping $f : I \times X \to Y$ belongs to the space $B_m^{0,1,G}(I \times X, Y)$ (respectively, $f \in \mathcal{G}_m^{0,1,G}(I \times X, Y)$, $C_m^{0,1,G}(I \times X, Y)$, $UC_m^{0,1,G}(I \times X, Y)$) if:

- $f \in B_m(I \times X, Y)$ (respectively, $C_m(I \times X, Y)$ in the first two cases and $UC_m^x(I \times X, Y)$ in the last one);
- For every $t \in I$, $f(t, \cdot)$ belongs to $B_m^{1,G(t,\cdot)}(X, Y)$ (respectively, $\mathcal{G}_m^{1,G(t,\cdot)}(X, Y)$, $C_m^{1,G(t,\cdot)}(X, Y)$, $UC_m^{1,G(t,\cdot)}(X, Y)$);
- The map $(t, x) \to \nabla^{G(t,x)} f(t, x)$ belongs to $B_m^s(I \times X, \mathcal{L}(Z, Y))$ (respectively $C_m^s(I \times X, \mathcal{L}(Z, Y))$, $C_m(I \times X, \mathcal{L}(Z, Y))$, $UC_m^x(I \times X, \mathcal{L}(Z, Y))$).

When the image space $Y = \mathbb{R}$, it will be dropped from the notation in all cases above.

The G-Gâteaux (Fréchet) derivative of f with respect to x in such spaces will always be denoted by $\nabla^G f$ ($D^G f$) or, if we want to underline the time dependence, $\nabla^{G(t,\cdot)} f(t, \cdot)$ ($D^{G(t,\cdot)} f(t, \cdot)$).

We endow $B_m^{0,1,G}(I \times X, Y)$ and its subspaces $\mathcal{G}_m^{0,1,G}(I \times X, Y)$, $C_m^{0,1,G}(I \times X, Y)$, $UC_m^{0,1,G}(I \times X, Y)$ with the norm[10]

$$\|f\|_{B_m^{0,1,G}(I\times X,Y)} := \sup_{(t,x)\in I\times X} \frac{|f(t,x)|_Y}{1+|x|^m} + \sup_{(t,x)\in I\times X} \frac{\left|\nabla^{G(t,\cdot)}f(t,x)\right|_{\mathcal{L}(Z,Y)}}{1+|x|^m}, \quad (4.28)$$

which we denote by $\|f\|_{\mathcal{G}_m^{0,1,G}(I\times X,Y)}$ (respectively, $\|f\|_{C_m^{0,1,G}(I\times X,Y)}$, $\|f\|_{UC_m^{0,1,G}(I\times X,Y)}$) when $f \in \mathcal{G}_m^{0,1,G}(I \times X, Y)$ (respectively, $f \in C_m^{0,1,G}(I \times X, Y)$, $f \in UC_m^{0,1,G}(I \times X, Y)$). If it is clear from the context, we will often simply write $\|f\|_{B_m^{0,1,G}}$ and $\|f\|_{\mathcal{G}_m^{0,1,G}}$, $\|f\|_{C_m^{0,1,G}}$, $\|f\|_{UC_m^{0,1,G}}$.

Notation 4.18 When $Z = X$ and $G = I$ we drop the superscript G in the notation for derivatives and all the spaces introduced in this subsection, writing for example $\mathcal{G}_m^1(X, Y)$ for $\mathcal{G}_m^{1,I}(X, Y)$. ∎

Remark 4.19 We point out that, dealing with G-gradients, other properties, beyond the already discussed exchange of differentiation and integration, are also not obvious. For instance, consider the following standard property (see Prop.4.8(c), p. 137 in [586]): if $f : X \to Y$ is Gâteaux differentiable and $\nabla f : X \to \mathcal{L}(X, Y)$ is continuous at $x \in X$, then f is Fréchet differentiable at x and $Df(x) = \nabla f(x)$. If we want to extend this property to G-gradients, we find problems similar to the ones in Proposition 4.13. A way to prove it is to strengthen Assumption 4.12, requiring that

$$\lim_{y\to x} \sup_{h\in D(G)\cap(\ker G)^\perp} \frac{|G^{-1}(y)G(x)h - h|}{|h|} = 0.$$

Without an assumption of this kind the conclusion does not seem guaranteed.

---

[10]Arguing as in Proposition 4.13 one can prove that the three subspaces are Banach spaces with this norm when Hypothesis 4.12 holds for $G(t, \cdot)$, for every $t \in I$. We will not need this fact.

Another important remark is that, in Proposition 4.16 we do not require any measurability of the map G. Indeed, we do not need it explicitly since we directly require the measurability of $\nabla^G f$ (which in order to be true usually needs some measurability of G). However, keeping the assumption on $\nabla^G f$ seems a bit sharper as it allows, for example, "bad" behavior of G in directions where f is constant. ∎

4.2.2 Weighted Spaces

We introduce suitable weighted spaces: these are Banach spaces of continuous functions in time and space $((t, x) \in (0, T] \times X$ for given $T > 0$ and a Banach space $X)$ blowing up at $t = 0$ at a prescribed rate. They will be used to apply the contraction mapping principle to solve the HJB equations (4.1) and (4.2). We also note that the setting used here is slightly different from the one introduced in part of the previous literature (see e.g. [89, 306]). Our setting is more general (close to that of [189, 309, 431, 432]).

We first define two classes of weights $\mathcal{I}_1$ and $\mathcal{I}_2$ as follows:

$$\mathcal{I}_1 := \left\{ \eta : (0, +\infty) \to (0, +\infty) \text{ decreasing and } \eta \in L^1(0, T), \forall T > 0 \right\}, \quad (4.29)$$

$$\mathcal{I}_2 := \left\{ \eta \in \mathcal{I}_1 : \exists \lim_{t \searrow 0^+} \frac{1}{\eta(t)} \int_0^t \eta(s)\eta(t - s)ds = 0 \right\}. \quad (4.30)$$

Remark 4.20 The two classes above are those that the weight γ_G, introduced in (4.7), must belong to in order for us to be able to solve the HJB equations (4.1) and (4.2). The class $\mathcal{I}_2$ is for the parabolic HJB equation (4.1) and $\mathcal{I}_1$ for the elliptic HJB equation (4.2). It is not clear if the two classes coincide or not. Clearly the function $\eta(t) = t^{-\theta}$ for $\theta \in (0, 1)$ belongs to $\mathcal{I}_2$. Moreover, for any $\beta > 1$ the function $\eta(t) = t^{-1} |\ln t|^{-\beta}$ also belongs to $\mathcal{I}_2$.

If $\eta_1 \le \eta_2$ and $\eta_2 \in \mathcal{I}_1$, then also $\eta_1 \in \mathcal{I}_1$ while the same is not clear for $\mathcal{I}_2$. Similarly if $\eta_1, \eta_2 \in \mathcal{I}_1$, also $\eta_1 \vee \eta_2 \in \mathcal{I}_1$ but the same is not clear for $\mathcal{I}_2$. ∎

We have the following.

Proposition 4.21 *Let $\eta \in \mathcal{I}_1$. Then:*

(i) $\liminf_{t \to 0^+} \frac{1}{\eta(t)} \int_0^t \eta(s)\eta(t - s)ds = 0$.

(ii) $\lim_{t \to 0^+} t\eta(t) = 0$.

(iii) *Let $\eta_0 \in \mathcal{I}_1$. Then for all $t > 0$ we have*

$$\int_0^t \eta(s)\eta_0(t - s)ds \le \eta(t/2) \int_0^{t/2} \eta_0(s)ds + \eta_0(t/2) \int_0^{t/2} \eta(s)ds.$$

Hence, for every $T > 0$,

$$\int_0^t \eta(s)\eta_0(t-s)ds \leq \eta_1(t) := \eta(t/2)\int_0^T \eta_0(s)ds + \eta_0(t/2)$$

$$\int_0^T \eta(s)ds, \quad \forall t \in (0, T]$$

and the right-hand side belongs to $\mathcal{I}_1$ if we set $\eta_1(t) = \eta_1(T)$ for $t > T$.

(iv) For $\eta \in \mathcal{I}_1$ we have, for every $T > 0$,

$$\sup_{t \in (0,T)} \left\{ \int_t^T \eta(T-s)e^{-\beta(s-t)}ds \right\} \longrightarrow 0$$

as $\beta \to +\infty$.

(v) For $\eta \in \mathcal{I}_2$ and $\beta \geq 0$ define the function

$$f_\beta : (0, +\infty) \to (0, +\infty), \quad f_\beta(t) := \frac{1}{\eta(t)}\int_0^t \eta(s)\eta(t-s)e^{-\beta s}ds.$$

Then, as $\beta \to +\infty$, f_β converges to 0 uniformly on $(0, T)$ for all $T > 0$. Moreover, for all $T > 0$ there exists a constant $C(T) > 0$ such that, for all $\beta \geq 0$

$$\sup_{t \in (0,T)} f_\beta(t) \leq C(T).$$

Exactly the same claims hold for the function

$$\bar{f}_\beta : (0, +\infty) \to (0, +\infty), \quad \bar{f}_\beta(t) := \frac{1}{\eta(t)}\int_0^t \eta(s)e^{-\beta s}ds.$$

Proof We first show (i). Let $\eta \in \mathcal{I}_1$ and define $\bar{\eta}(t) := \int_0^t \eta(s)ds$ and $\eta^{*2}(t) = \int_0^t \eta(s)\eta(t-s)ds$. Observe first that, by exchanging the integrals

$$\int_0^t \eta^{*2}(s)ds = \int_0^t \int_0^s \eta(r)\eta(s-r)drds = \int_0^t \eta(r)\int_r^t \eta(s-r)dsdr = \int_0^t \eta(r)\bar{\eta}(t-r)dr.$$

Since $\lim_{t \to 0^+} \bar{\eta}(t) = 0$, we obtain that for all $\varepsilon > 0$ there exists a $t_\varepsilon > 0$ such that, if $t \leq t_\varepsilon$ then

$$\int_0^t \eta^{*2}(s) - \varepsilon\eta(s)ds < 0.$$

If

$$\liminf_{t \to 0^+} \frac{\eta^{*2}(t)}{\eta(t)} = L > 0$$

then for $s > 0$ sufficiently small we must have $\eta^{*2}(s) \geq \frac{L}{2}\eta(s)$; hence, for sufficiently small $t > 0$ we get

$$\int_0^t \eta^{*2}(s) - \frac{L}{2}\eta(s)ds \geq 0,$$

which is a contradiction. Note that here we did not use the fact that η is decreasing.

We now prove (ii). We suppose by contradiction that (ii) is not true for a given $\eta \in \mathcal{I}_1$. Then there exists a sequence $(t_n)_{n \in \mathbb{N}}$ such that $t_n \searrow 0$ and $\lim_{n \to +\infty} t_n \eta(t_n) = L > 0$. By refining such a sequence we can assume that $t_{n+1} \leq \frac{1}{2}t_n$ for all $n \in \mathbb{N}$. This means that there exists an $\bar{n}$ such that, for all $n \geq \bar{n}$, $t_n \eta(t_n) > L/2$, i.e. $\eta(t_n) > L/(2t_n)$. Since η is decreasing we get

$$\int_0^1 \eta(t)dt \geq \sum_{n \in \mathbb{N}} \eta(t_n)(t_n - t_{n+1}) \geq \sum_{n > \bar{n}} \eta(t_n)(t_n - t_{n+1})$$

$$\geq \sum_{n > \bar{n}} \frac{L}{2t_n}(t_n - t_{n+1}) = \frac{L}{2}\sum_{n > \bar{n}} \frac{t_n - t_{n+1}}{t_n}.$$

Since $t_{n+1} \leq \frac{1}{2}t_n$ for all $n \in \mathbb{N}$,

$$\sum_{n > \bar{n}} \frac{t_n - t_{n+1}}{t_n} \geq \sum_{n > \bar{n}} \frac{1}{2} = +\infty.$$

This contradicts the integrability of η, so (ii) is shown.

The first part of claim (iii) follows by writing

$$\int_0^t \eta(s)\eta_0(t-s)ds = \int_0^{t/2} \eta(s)\eta_0(t-s)ds + \int_{t/2}^t \eta(s)\eta_0(t-s)ds$$

$$\leq \eta_0(t/2)\int_0^{t/2} \eta(s)ds + \eta(t/2)\int_0^{t/2} \eta_0(s)ds.$$

The second part is an immediate consequence of the first and the fact that for $\alpha > 0$ the functions $t \to \eta(\alpha t), t \to \eta_0(\alpha t)$ belong to $\mathcal{I}_1$.

To prove (iv) we observe that, for $\eta \in \mathcal{I}_1$ and $0 < t < T$,

$$\int_t^T \eta(T-s)e^{-\beta(s-t)}ds = \int_0^{T-t} \eta(T-t-s)e^{-\beta s}ds \leq \int_0^{T-t} \eta(T-t-s)ds$$

$$= \int_0^{T-t} \eta(s)ds.$$

Now for arbitrary $\varepsilon > 0$ take t_ε such that, for $t \in [t_\varepsilon, T)$

$$\int_0^{T-t} \eta(s)ds \le \varepsilon/2.$$

Then observe that, for $t \in (0, t_\varepsilon)$, since η is decreasing,

$$\int_0^{T-t} \eta(T - t - s)e^{-\beta s}ds \le \int_0^{T-t_\varepsilon} \eta(T - t_\varepsilon - s)ds + \int_{T-t_\varepsilon}^{T-t} \eta(T - t - s)e^{-\beta s}ds$$

$$\le \varepsilon/2 + e^{-\beta(T-t_\varepsilon)} \int_0^T \eta(s)ds,$$

from which the claim follows, since ε is arbitrary.

We prove (v) only for f_β as the proof for $\bar{f}_\beta$ is exactly the same and even simpler. Observe first that for every $\beta \ge 0$ we have, for all $t > 0$,

$$f_\beta(t) \le \frac{1}{\eta(t)} \int_0^t \eta(s)\eta(t - s)ds.$$

Now let $T > 0$. Since $\eta \in \mathcal{I}_2$, fixing any $\varepsilon > 0$, we can take $t_\varepsilon \in (0, T)$ such that, for all $t \in (0, t_\varepsilon]$, $f_\beta(t) < \varepsilon/2$. For $t \in (t_\varepsilon, T)$, thanks to the monotonicity of η, we have the following estimate

$$f_\beta(t) \le \frac{1}{\eta(T)} \left[\int_0^{t_\varepsilon} \eta(s)\eta(t - s)e^{-\beta s}ds + \int_{t_\varepsilon}^t \eta(s)\eta(t - s)e^{-\beta s}ds \right]$$

$$\le \frac{1}{\eta(T)} \int_0^{t_\varepsilon} \eta(s)\eta(t_\varepsilon - s)e^{-\beta s}ds + \frac{1}{\eta(T)} \int_{t_\varepsilon}^t \eta(t_\varepsilon)\eta(t - s)e^{-\beta s}ds$$

$$\le \frac{1}{\eta(T)} \int_0^{t_\varepsilon} \eta(s)\eta(t_\varepsilon - s)e^{-\beta s}ds + \frac{\eta(t_\varepsilon)}{\eta(T)}e^{-\beta t_\varepsilon} \int_0^T \eta(s)ds.$$

Hence (using the dominated convergence theorem for the first term), there exists a β_ε such that $\sup_{t \in (0,T)} f_\beta(t) \le \varepsilon$ for all $\beta > \beta_\varepsilon$ and the claim of convergence follows. Concerning the uniform boundedness it is enough to show it for $\beta = 0$, which immediately follows from the estimates used to prove the convergence, for, say, $\varepsilon = 1$. $\square$

We now present the list of the weighted spaces we use. Below we fix $T > 0$, $\eta \in \mathcal{I}_1$ and X, Y and Z are real Banach spaces.

$B_{m,\eta}((0, T] \times X, Y)$ **and its subspaces.** Let $m \ge 0$. Define:

$$B_{m,\eta}((0,T] \times X, Y) := \{w \colon (0,T] \times X \to Y \text{ measurable } :$$
$$w \in B_m([\tau, T] \times X, Y) \; \forall \tau \in (0,T) \text{ and } \eta^{-1}w \in B_m((0,T] \times X, Y)\} \tag{4.31}$$

and its subspaces

$$C_{m,\eta}((0,T] \times X, Y) :=$$
$$\{w \in B_{m,\eta}((0,T] \times X, Y) \; : \; w \in C_m([\tau,T] \times X, Y) \; \forall \tau \in (0,T)\}, \tag{4.32}$$

$$UC^x_{m,\eta}((0,T] \times X, Y) :=$$
$$\{w \in C_{m,\eta}((0,T] \times X, Y) \; : \; w \in UC^x_m([\tau,T] \times X, Y) \; \forall \tau \in (0,T)\}. \tag{4.33}$$

The three spaces above are Banach spaces when endowed with the norm (see e.g. [306, 310] for the case $m = 0$ and [102] for the case $m > 0$)

$$\|w\|_{B_{m,\eta}((0,T]\times X,Y)} := \sup_{(t,x)\in(0,T]\times X} \eta(t)^{-1}(1 + |x|^m)^{-1} |w(t,x)|_Y.$$

Often we will simply write $\|w\|_{B_{m,\eta}}$ and, for the subspaces, $\|w\|_{C_{m,\eta}}$ or $\|w\|_{UC^x_{m,\eta}}$. When $\eta(t) = t^{-\theta}$ for $\theta \in (0,1)$ we will use the notation $B_{m,\theta}((0,T] \times X, Y)$, $C_{m,\theta}((0,T] \times X, Y)$, $UC^x_{m,\theta}((0,T] \times X, Y)$. When $Y = \mathbb{R}$ we will omit it in the notation, as usual.

Also, when $m = 0$, we will write $B_{b,\eta}$ instead of $B_{0,\eta}$ and similarly for the other spaces.

$B^s_{m,\eta}((0,T] \times X, \mathcal{L}(Z,Y))$ **and its subspaces.** Given $m \geq 0$ we denote by $B^s_{m,\eta}$ $((0,T] \times X, \mathcal{L}(Z,Y))$ the linear space of the mappings $L : (0,T] \times X \to \mathcal{L}(Z,Y)$ such that for every $z \in Z$, $L(\cdot,\cdot)z \in B_{m,\eta}((0,T] \times X)$. The space $B^s_{m,\eta}((0,T] \times X, \mathcal{L}(Z,Y))$ is a Banach space if it is endowed with the norm[11]

$$\|L\|_{B^s_{m,\eta}((0,T]\times X,\mathcal{L}(Z,Y))} := \sup_{(t,x)\in(0,T]\times X} \eta(t)^{-1}(1 + |x|^m)^{-1} \|L(t,x)\|_{\mathcal{L}(Z,Y)}.$$

The subspace $C^s_{m,\eta}((0,T] \times X, \mathcal{L}(Z,Y))$ (respectively, $UC^{x,s}_{m,\eta}((0,T] \times X, \mathcal{L}(Z,Y))$) is the space of all elements of $B^s_{m,\eta}((0,T] \times X, \mathcal{L}(Z,Y))$ such that for every $z \in Z$, $L(\cdot,\cdot)z \in C_{m,\eta}((0,T] \times X)$ (respectively, $L(\cdot,\cdot)z \in UC^x_{m,\eta}((0,T] \times X)$). When $Y = \mathbb{R}$ we have $\mathcal{L}(Z,\mathbb{R}) = Z^*$. If Z is a Hilbert space we will identify Z^* with Z and so we will write Z in place of $\mathcal{L}(Z,\mathbb{R})$ in the above notation. As before, when $m = 0$ we use the subscript b instead of 0.

---

[11] The proof is similar to the proof of Proposition 4.8. Note also that $B^s_{m,\eta}((0,T] \times X, \mathcal{L}(Z,Y))$ can be identified with the space of operators $\mathcal{L}\left(Z, B_{m,\eta}((0,T] \times X, Y)\right)$.

Remark 4.22 If $f \in B_{m,\eta}((0, T] \times X, Y)$, then for every $t \in (0, T]$ the function $f(t, \cdot)$ belongs to $B_m(X, Y)$ (using Lemma 1.8-(iv)). However, see the discussion before Lemma 1.21, it is not true in general that the function

$$(0, T] \to B_m(X, Y), \qquad t \to f(t, \cdot)$$

is Borel measurable. On the other hand, by Lemma 1.21, if X is separable, then the map

$$\rho_1 : (0, T] \to \mathbb{R}, \qquad t \to \|f(t, \cdot)\|_{B_m(X,Y)} \tag{4.34}$$

is always Lebesgue measurable.

Thus, when X is separable, asking that there exists an $\eta \in \mathcal{I}_1$ such that the function f belongs to $B_{m,\eta}((0, T] \times X, Y)$ clearly implies that

$$\int_0^T \|f(t, \cdot)\|_{B_m} dt < +\infty. \tag{4.35}$$

Conversely, if a function $f : (0, T] \times X \to Y$ is jointly Borel measurable and satisfies (4.35) then, setting $\eta(t) = \|f(t, \cdot)\|_{B_m}$, we have $\eta \in L^1(0, T; \mathbb{R}^+)$ and, if also $\eta \in \mathcal{I}_1$, $f \in B_{m,\eta}((0, T] \times X, Y)$. ∎

Let now, as in Definition 4.17, $G : [0, T] \times X \to \mathcal{L}_u(Z, X)$. For $m \geq 0$ and $\eta \in \mathcal{I}_1$, we introduce the linear space[12]

$$B_{m,\eta}^{0,1,G}([0, T] \times X) := \left\{ v \in B_m([0, T] \times X) : \nabla^G v \in B_{m,\eta}((0, T] \times X, Z^*) \right\} \tag{4.36}$$

and its subspaces

$$\mathcal{G}_{m,\eta}^{0,1,G}([0, T] \times X) := \left\{ v \in C_m([0, T] \times X) : \nabla^G v \in C_{m,\eta}^s((0, T] \times X, Z^*) \right\}, \tag{4.37}$$

$$C_{m,\eta}^{0,1,G}([0, T] \times X) := \left\{ v \in C_m([0, T] \times X) : D^G v \in C_{m,\eta}((0, T] \times X, Z^*) \right\}, \tag{4.38}$$

$$UC_{m,\eta}^{0,1,G}([0, T] \times X) := \left\{ v \in UC_m^x([0, T] \times X) : D^G v \in UC_{m,\eta}^x((0, T] \times X, Z^*) \right\}. \tag{4.39}$$

We endow such spaces with the norm

$$\|v\|_{B_{m,\eta}^{0,1,G}([0,T]\times X)} := \|v\|_{B_m([0,T]\times X)} + \|\eta^{-1}\nabla^G v\|_{B_m((0,T]\times X, Z^*)}. \tag{4.40}$$

---

[12]Recall that here, since $Y = \mathbb{R}$, strong measurability of $\nabla^G v$ is equivalent to its measurability, see the previous footnotes.

We will often just write $\|v\|_{B^{0,1,G}_{m,\eta}}$ and, for the subspaces, $\|v\|_{\mathcal{G}^{0,1,G}_{m,\eta}}$, $\|v\|_{C^{0,1,G}_{m,\eta}}$ or $\|v\|_{UC^{0,1,G}_{m,\eta}}$. When Z is Hilbert, we identify it with its dual and thus write Z in place of Z^* in the above notation. When $G = I$ we drop it from the function space notation according to the convention of Notation 4.18. Using the same method as employed in the proof of Proposition 4.13, one can prove that the three spaces in (4.37)–(4.39) equipped with the norm (4.40) are Banach spaces when Hypothesis 4.12 holds for $G(t, \cdot)$, for every $t \in [0, T]$.

We also define, for $m \geq 0$, analogous spaces related to higher derivatives in the case when $X = Z$ is a Hilbert space and $G = I$:

$$B^{0,2,s}_{m,\eta}([0, T]\times X) := \{v \in B_m([0, T] \times X) : \nabla v \in B_m([0, T] \times X, X),$$
$$\nabla^2 v \in B^s_{m,\eta}((0, T] \times X, \mathcal{L}(X))\} \quad (4.41)$$

and its subspaces

$$C^{0,2,s}_{m,\eta}([0, T]\times X) := \{v \in C_m([0, T] \times X) : Dv \in C_m([0, T] \times X, X),$$
$$D^2 v \in C^s_{m,\eta}((0, T] \times X, \mathcal{L}(X))\}, \quad (4.42)$$

$$UC^{0,2,s}_{m,\eta}([0, T]\times X) := \{v \in UC^x_m([0, T] \times X) : Dv \in UC^x_m([0, T] \times X, X),$$
$$D^2 v \in UC^{x,s}_{m,\eta}((0, T] \times X, \mathcal{L}(X))\}. \quad (4.43)$$

All of them are Banach spaces if we use the norm

$$\|v\|_{B^{0,2,s}_{m,\eta}([0,T]\times X)} := \|v\|_{B_m([0,T]\times X)}$$
$$+ \|\nabla v\|_{B_m([0,T]\times X,X)} + \|\eta(T - \cdot)\nabla^2 v\|_{B_m((0,T]\times X,\mathcal{L}(X))}.$$

As always, to make the notation less cumbersome we will often drop the spaces from the notation by simply writing $\|v\|_{B^{0,2,s}_{m,\eta}}$ and similarly for all the terms appearing in its definition. On the subspaces (substituting ∇ with D) the norms will be denoted by $\|v\|_{C^{0,2,s}_{m,\eta}}$ and $\|v\|_{UC^{0,2,s}_{m,\eta}}$.

When $\gamma(t) = t^\theta$ for some $\theta \geq 0$ we will write $B^{0,1,G}_{m,\theta}([0, T] \times X)$ for $B^{0,1,G}_{m,\eta}$ $([0, T] \times X)$ and $B^{0,2,s}_{m,\theta}([0, T] \times X)$ for $B^{0,2,s}_{m,\eta}([0, T] \times X)$. Similar convention will be used for other spaces.

We finally observe that all the spaces presented in this subsection can also be defined if we replace $[0, T]$ by a closed bounded interval $I \subset \mathbb{R}$.

Remark 4.23 In some of the spaces defined in this and the previous subsection we require continuity in the so-called *strong sense*, i.e. continuity of the maps $L(\cdot)z : X \to Y$ or $L(\cdot, \cdot)z : [0, T) \times X \to Y$ for all $z \in Z$.

One reason for doing this is that sometimes (see some examples in Sect. 4.3.3, Remark 4.63) the G-Gâteaux derivatives of the solution of the HJB equation (even in the linear case) may neither be G-Fréchet nor continuous, hence it is not reasonable to require more than strong continuity from them.

Another reason is that, even when first space derivatives of the solution of the HJB equation are taken in the classical Fréchet sense, the second derivatives may fail to be continuous or even measurable. As an example, take $v(t, x) = \langle e^{tA}x, x \rangle /2$, where A is the generator of a strongly continuous semigroup which is also self-adjoint. Then $D^2 v(t, x) = e^{tA}$, which is not even measurable in general when it is considered as a map $[0, T] \times X \to \mathcal{L}(X)$, while for every fixed $\xi \in X$, the map $D^2 v(\cdot, \cdot)\xi : [0, T] \times X \to X$ is continuous (see e.g. [180], Sect. 1.2 for more on this).   ∎

We finally define weighted spaces with singularities at the right end of the time interval.

Definition 4.24 Let $\eta \in \mathcal{I}_1$. We define $\overline{B}_{m,\eta} ([0, T) \times X, Y)$ to be the Banach space of all functions $\psi : [0, T) \times X \to Y$ such that $\psi(T - \cdot, \cdot) \in B_{m,\eta} ((0, T] \times X, Y)$, with the norm $\|\psi\|_{\overline{B}_{m,\eta}([0,T) \times X,Y)} := \|\psi(T - \cdot, \cdot)\|_{B_{m,\eta}((0,T] \times X,Y)}$. In the same way we define its subspaces $\overline{C}_{m,\eta} ([0, T) \times X, Y)$ and $\overline{UC}^x_{m,\eta} ([0, T) \times X, Y)$.

In exactly the same way we define:

- The spaces $\overline{B}^s_{m,\eta}([0, T) \times X, \mathcal{L}(Z, Y))$, $\overline{C}^s_{m,\eta}([0, T) \times X, \mathcal{L}(Z, Y))$ and $\overline{UC}^{x,s}_{m,\eta}([0, T) \times X, \mathcal{L}(Z, Y))$.
- For $G : [0, T] \times X \to \mathcal{L}_u(Z, X)$, the spaces $\overline{B}^{0,1,G}_{m,\eta} ([0, T] \times H), \overline{\mathcal{G}}^{0,1,G}_{m,\eta} ([0, T] \times H), \overline{C}^{0,1,G}_{m,\eta} ([0, T] \times H), \overline{UC}^{0,1,G}_{m,\eta} ([0, T] \times H)$.
- The spaces $\overline{B}^{0,2,s}_{m,\eta} ([0, T] \times H), \overline{C}^{0,2,s}_{m,\eta} ([0, T] \times H), \overline{UC}^{0,2,s}_{m,\eta} ([0, T] \times H)$.

The bar will always indicate a space with a "singularity" at T. As usual, when $G = I$ we will drop it in the notation of such spaces.

4.3 Smoothing Properties of Transition Semigroups

In this section we recall some known results about smoothing properties of transition semigroups associated to SDEs of the form (4.3) and (4.8). Such results, as explained in Sect. 4.1, are a key ingredient to proving existence and uniqueness of mild solutions of the HJB equations (4.1) and (4.2). We divide the section into three parts corresponding to the methods used in the proofs. Section 4.3.1 is the main one and is equipped with full proofs. It deals with the Ornstein–Uhlenbeck case, where the smoothing property is proved using a change of measure based on the Cameron–Martin theorem (Theorem 1.60). Section 4.3.2 considers the case of a perturbed Ornstein–Uhlenbeck semigroup, where the smoothing property is proved by a suitable integration by parts performed using Malliavin calculus. Section 4.3.3

considers the case when one can apply the so-called Bismut–Elworthy–Li formula, which again is a way of performing integration by parts. In the last two subsections most of the results are presented without proofs, for which references to the literature are given.

4.3.1 The Case of the Ornstein–Uhlenbeck Semigroup

4.3.1.1 The Equation

Let H and Ξ be two real separable Hilbert spaces. Let X be the Ornstein–Uhlenbeck process in H, i.e. the solution of the SDE

$$\begin{cases} dX\,(s) = AX\,(s)\,ds + \sigma dW\,(s)\,,\, s \in [t,\,T] \\ X\,(t) = x, \end{cases} \tag{4.44}$$

where $W(\cdot)$ is a cylindrical Wiener process in Ξ, with identity covariance, on a given filtered probability space $(\Omega,\,\mathscr{F},\,\{\mathscr{F}_s\}_{s\geq 0},\,\mathbb{P})$ and A and σ satisfy Hypothesis 4.25 below. Equation (4.44) is (4.3) with $b = 0$ and $\sigma\,(t,x) = \sigma \in \mathcal{L}(\Xi,\,H)$. For the study of the properties of this process one can see, for example, Chaps. 5 and 9 of [180] and Chaps. 6 and 10 of [179].

Hypothesis 4.25 H and Ξ are real separable Hilbert spaces.

(i) The linear operator A is the generator of a strongly continuous semigroup $\{e^{tA},\,t \geq 0\}$ in the Hilbert space H and, for some $M \geq 1,\,\omega \in \mathbb{R}$,

$$\|e^{tA}\| \leq Me^{\omega t}, \quad t \geq 0. \tag{4.45}$$

(ii) $\sigma \in \mathcal{L}(\Xi,\,H)$, $e^{sA}\sigma\sigma^*e^{sA^*} \in \mathcal{L}_1(H)$ for all $s > 0$ and, for all $t \geq 0$,

$$\int_0^t \mathrm{Tr}\left[e^{sA}\sigma\sigma^*e^{sA^*}\right]ds < +\infty,$$

so the symmetric positive operator

$$Q_t : H \to H, \quad Q_t := \int_0^t e^{sA}\sigma\sigma^*e^{sA^*}ds, \tag{4.46}$$

is of trace class for every $t \geq 0$.

Under these assumptions (see Theorem 1.152) Eq. (4.44) has a unique mild solution $X(\cdot;t,x)$ (or simply $X(\cdot)$) written in the mild form as

$$X(s) = e^{(s-t)A}x + \int_t^s e^{(s-r)A}\sigma dW(r).$$

We recall that, under Hypothesis 4.25, continuity of the above process is not guaranteed (only mean square continuity holds, see Proposition 1.144 or also Theorem 5.2-(i) of [180]). If one adds the assumption that, for some $\theta \in (0, 1)$ and $T > 0$

$$\int_0^T s^{-\theta} \mathrm{Tr} \left[e^{sA} \sigma \sigma^* e^{sA^*} \right] ds < +\infty, \tag{4.47}$$

then the trajectories of $X(\cdot)$ are continuous (see Theorem 1.152 and also Theorem 5.11 of [180]). Without this additional assumption continuity of trajectories may fail to hold, see e.g. [357].

Remark 4.26 In cases of noise on the boundary (see e.g. [177], Chap. 13 or Sect. 2.6.3 for related control problems) the operator σ is unbounded but the process $X(\cdot)$ may still be well defined. We do not consider such cases here as they give rise to functions γ_G in (4.7) which are not integrable (see the introduction of [181]) and so the method used in this chapter to solve the related HJB equations cannot be used. ■

Remark 4.27 In some literature (see e.g. [153], p.117), one takes $\Xi = H$, and, given a bounded symmetric and positive operator Σ in H, the Ornstein–Uhlenbeck process is the solution to the equation

$$\begin{cases} dX(s) = AX(s) ds + \sqrt{\Sigma} dW(s), s \in [t, T], \\ X(t) = x. \end{cases} \tag{4.48}$$

If we take $\Sigma = \sigma \sigma^*$ (where σ is the operator introduced in (4.44)) the transition semigroup generated by the process (4.48) is the same as the one generated by the process (4.44). So, if one is only concerned about the transition semigroup, one may refer to (4.48) instead of (4.44). In what follows we always denote by Σ the operator $\sigma \sigma^*$. ■

4.3.1.2   The Semigroup and the Associated Kolmogorov Equation

The process X is clearly time-homogeneous so the associated transition semigroup is a one-parameter semigroup. Since it is a special one we will denote it by R_t instead of P_t, which is the notation used in the general case. It is well known that for every $\phi \in B_m(H)$, $m \geq 0$, we have,[13] for $t \geq 0$,

$$R_t[\phi](x) := \mathbb{E} \left[\phi(X(t; 0, x)) \right] = \mathbb{E} \left[\phi(e^{tA} x + W^A(t)) \right], \tag{4.49}$$

where $W^A(t) = \int_0^t e^{(t-s)A} \sigma dW(s)$ as in Sect. 1.5.2. Using Gaussian measures we can write

$$R_t[\phi](x) = \int_H \phi(y) \mathcal{N} \left(e^{tA} x, Q_t \right) (dy) = \int_H \phi(e^{tA} x + y) \mathcal{N}_{Q_t}(dy), \tag{4.50}$$

---

[13] See e.g. Sect. 6.3 of [179] for $m = 0$, [102, 300] for $m > 0$.

where, for given $a \in H$ and $Q \in \mathcal{L}^+(H)$, $\mathcal{N}(a, Q)(dy)$ (or $\mathcal{N}_Q(dy)$ when $a = 0$) denotes the Gaussian measure in H with mean a and covariance operator Q (see Definition 1.58 and also, for example, Chap. 1 of [179] for the related theory).

Since Hypotheses 1.143 and 1.145 are satisfied here, we can apply Theorem 1.157, which guarantees the semigroup property of R_t, and Theorem 1.162, which ensures that R_t has the Feller property (see Definition 1.159 and Lemma 1.160). Other continuity properties of R_t are discussed in Proposition 4.50.

The semigroup R_t is not strongly continuous on $C_b(H)$ (nor in $UC_b(H)$, $B_b(H)$, $C_m(H)$, $UC_m(H)$, $B_m(H)$ for $m > 0$), see e.g. [101, 492] and also Proposition B.89. It is a π-continuous and a $\mathcal{K}$-continuous semigroup, see the precise definitions in Appendix B.5.2, and its generator is the operator $\mathcal{A}$, which can be formally written as

$$\mathcal{A}f(x) = \frac{1}{2} \mathrm{Tr}\left(\Sigma D^2 f(x)\right) + \langle Ax, Df(x) \rangle.$$

It is well known (see Theorem 6.1.2 of [179]) that for any function $\phi \in UC_b^2(H)$ such that $BB^*D^2\phi \in UC_b(H, \mathcal{L}_1(H))$, the function

$$u(t, x) = R_t[\phi](x) \tag{4.51}$$

is a *strict solution* of the Kolmogorov equation with terminal value

$$u_t + \frac{1}{2} \mathrm{Tr}\left(\Sigma D^2 u\right) + \langle Ax, Du \rangle = 0, \quad u(T, x) = \phi(x), \ x \in H, \tag{4.52}$$

in the following sense.

Definition 4.28 A function $u(t, x)$, $t \in [0, T]$, $x \in H$, is said to be a *strict solution* to Eq. (4.52) if:

(i) u is continuous on $[0, T] \times H$ and $u(T, \cdot) = \phi$,
(ii) $u(t, \cdot) \in UC_b^2(H)$ for all $t \in [0, T]$, and $\Sigma D^2 u(t, x) \in \mathcal{L}_1(H)$ for all $x \in H$ and $t \in [0, T]$,
(iii) for any $x \in D(A)$, $u(\cdot, x)$ is continuously differentiable on $[0, T]$ and (4.52) is satisfied pointwise on $[0, T) \times D(A)$.

If $\phi \in C_b(H)$ (or $B_b(H)$, or also $B_m(H)$ for some $m > 0$) then the function u defined by (4.51) is called (see e.g. Sect. 6.2, p. 103 of [179]) the *generalized solution* (or sometimes also *the mild solution*) of (4.52). For more on the relationship between the semigroup R_t and the above Kolmogorov equation, see Appendix B.7.

4.3.1.3   Smoothing Properties of R_t: Assumptions and Null Controllability

We now provide conditions that guarantee the smoothing properties (4.6) and, more generally, (4.7). These smoothing properties will immediately yield regularity of generalized solutions of (4.52).

We start with the following hypothesis, introduced first in [579] in the case $U = H$ and $G = I$, which will guarantee the differentiability of $R_t[\phi]$.

Hypothesis 4.29 Let U be a real separable Hilbert space and $G : U \to H$ be a closed linear operator (possibly unbounded) with dense domain $D(G)$. Assume, whenever G is bounded, that

$$e^{tA}G(U) \subset Q_t^{1/2}(H), \qquad \forall t > 0 \tag{4.53}$$

and, whenever G is unbounded, that for all $t > 0$ the operator $e^{tA}G : D(G) \subset U \to H$ extends to a bounded operator, which we still denote by $e^{tA}G : U \to H$, such that

$$e^{tA}G(U) \subset Q_t^{1/2}(H), \qquad \forall t > 0. \tag{4.54}$$

Remark 4.30 When $U = H$ and $G = I$, Hypothesis 4.29 is equivalent to asking that, for every $t > 0$, the deterministic control system

$$z' = Az + \sigma a, \quad z(0) = x, \tag{4.55}$$

is *null controllable*, with controls in $L^2(0, t; \Xi)$, from every initial datum $x \in H$. This means that for every $t > 0$ and $x \in H$, there is a control strategy $a(\cdot) \in L^2(0, t; \Xi)$ such that $z(t; 0, x, a(\cdot)) = 0$. In terms of operators, since

$$z(t) = e^{tA}x + \int_0^t e^{(t-s)A}\sigma a(s)ds,$$

denoting by $\mathcal{L}_t$ the operator

$$\mathcal{L}_t : L^2(0, t; \Xi) \to H, \qquad \mathcal{L}_t a(\cdot) = \int_0^t e^{(t-s)A}\sigma a(s)ds, \tag{4.56}$$

the null controllability for an initial datum $x \in H$ means that

$$e^{tA}x \in \mathcal{L}_t(L^2(0, t; \Xi)).$$

The equivalence mentioned above then follows from the fact that, after easy computations, we get

$$|\mathcal{L}_t^* x|^2 = \langle Q_t x, x \rangle = |Q_t^{1/2}x|^2, \quad x \in H, \tag{4.57}$$

and so, by Proposition B.2,

$$\mathcal{L}_t(L^2(0,t;\Xi)) = Q_t^{1/2}(H) \tag{4.58}$$

(see also [180], Corollary B.7).

When G is bounded, in view of what was said above, Hypothesis 4.29 is equivalent to asking that system (4.55) is null controllable for every initial datum $x \in G(U) \subset H$ (see also Sect. 3.1 of [432]).

When G is unbounded we may consider Hypothesis 4.29 as a null controllability assumption for the extension of system (4.55) to a suitable extrapolation space (see e.g. [217], Sect. 2.5 for a definition).

Finally, we observe that when $U = H$ and $G = I$, if (4.53) holds for a given $t_0 > 0$, it must hold for all $t > t_0$ thanks to the fact that the images of e^{tA} decrease with t (by the semigroup property) while those of $Q_t^{1/2}$ increase with t (by (4.58)).[14] This is not ensured when $G \neq I$. Since in many cases we will be interested in the smoothing property of R_t on finite intervals $(0, T]$ for a given $T > 0$, in such cases (4.53) or (4.54) may be required to hold only for $t \in (0, T]$. ∎

We note that if (4.53) or (4.54) holds then the operator $\Gamma_G(t) : U \to H$,

$$\Gamma_G(t) := Q_t^{-1/2} e^{tA} G \tag{4.59}$$

(where $Q_t^{-1/2}$ is the pseudoinverse of $Q_t^{1/2}$, see Definition B.1), is bounded by the closed graph theorem, so it belongs to $\mathcal{L}(U, H)$. When $U = H$ and $G = I$ we will often simply write $\Gamma(t) := Q_t^{-1/2} e^{tA}$.

Remark 4.31

(i) If Hypothesis 4.29 holds, then, by Proposition B.2, for any $t > 0$ there exists a constant $c_t > 0$ such that

$$|(e^{tA}G)^* x|_U^2 \leq c_t \langle Q_t x, x \rangle_H, \qquad \forall x \in H, \tag{4.60}$$

or equivalently (in the sense of the ordering of positive operators),

$$e^{tA} G (e^{tA} G)^* \leq c_t Q_t. \tag{4.61}$$

The smallest c_t with such property is exactly $\|\Gamma_G(t)\|^2$. Since Q_t is of trace class, this implies that $e^{tA} G \in \mathcal{L}_2(H)$ (and so it is compact) for all $t > 0$.

(ii) Since the images $Q_t^{1/2}(H)$ increase as t increases, we clearly have that if Hypothesis 4.29 holds then also

$$e^{sA} G(H) \subset Q_t^{1/2}(H), \ 0 < s < t. \tag{4.62}$$

Since we can write, for $0 < s < t$,

---

[14]It is indeed constant when $U = H$ and $G = I$ and (4.53) holds, see e.g. [179], Theorem B.2.2.

$$\Gamma_G(t) = Q_t^{-1/2} e^{(t-s)A} e^{sA} G, \ 0 < s < t, \tag{4.63}$$

then, when $U = H$ and $G = I$, also $\Gamma(t) \in \mathcal{L}_2(H)$. This may not be true when $G \neq I$, in general.

For the proofs of all these facts the reader can see Appendix B of [179]. There the proofs are given for $U = H$ and $G = I$ but the generalizations to our case are completely straightforward. Such facts can be employed to get conditions for the regularity of the generalized solution u from (4.51), see [179], Sect. 6.2.   ∎

We now introduce the second hypothesis which, recalling (4.7), guarantees the existence of an integrable weight γ_G in this case.

Hypothesis 4.32 Let U be a real separable Hilbert space and $G : U \to H$ be a closed linear operator (possibly unbounded) with dense domain $D(G)$. Assume that Hypothesis 4.29 holds and that

the function $t \mapsto \|\Gamma_G(t)\|$ is integrable in a right neighborhood of 0.   (4.64)

Remark 4.33 In many cases in the literature assumption (4.64) is substituted, for simplicity, by the requirement that, for suitable $C > 0$ and $\theta \in (0, 1)$, one has

$$\|\Gamma_G(t)\| \le Ct^{-\theta}, \text{ for } t \in (0, T] \tag{4.65}$$

in the parabolic case with finite horizon T and

$$\|\Gamma_G(t)\| \le C(1 \vee t^{-\theta}), \text{ for } t > 0, \tag{4.66}$$

in the elliptic case. We will occasionally do this here in Sects. 4.6–4.8.   ∎

Remark 4.34 Define, for any $t > 0$ and $x \in H$, the *minimal energy* to steer, in time t, the deterministic control system (4.55) from x to 0, as

$$\mathcal{E}(t, x) := \inf \left\{ \left(\int_0^t |a(s)|^2 ds \right)^{1/2} : \ a(\cdot) \in L^2(0, t; \Xi), \ \ z(t; 0, x, a(\cdot)) = 0 \right\}, \tag{4.67}$$

with the agreement that the infimum of the empty set is $+\infty$. When $U = H$ and $G = I$ we have (see e.g. Appendix B, Remark B.9 of [180]) $\mathcal{E}(t, x) = |\Gamma(t)x|$ and so Hypothesis 4.32 is equivalent to asking that the function

$$t \to \sup_{x \in H, |x| \le 1} \mathcal{E}(t, x) \quad \text{is integrable in a right neighborhood of 0.}$$

In the case when G is bounded (see e.g. [432], Sect. 3.1, Proposition 3.9) we have $\mathcal{E}(t, Gk) = |\Gamma_G(t)k|$ and so Hypothesis 4.32 is equivalent to asking that the function

$$t \to \sup_{k \in U, |k| \le 1} \mathcal{E}(t, Gk) \quad \text{is integrable in a right neighborhood of 0.}$$

When G is unbounded we may view Hypothesis 4.32 as an estimate of the minimal energy for the extension of system (4.55) to a suitable extrapolation space.

Using this interpretation it is not difficult to prove that the function $t \to \|\Gamma_G(t)\|$ is decreasing. This fact is proved, in the case when $G = I$, e.g. in [312], Theorem 3.7-(ii). In Lemma 4.35 we give a general proof. Finally, we also mention that $t \to \Gamma_G(t)$ is strongly continuous (see [312], Theorem 3.7-(iii) when $G = I$) while the continuity may fail in general. ∎

Lemma 4.35 *Let Hypotheses 4.25 and 4.29 hold. Then for each $k \in U$ the map $t \to |\Gamma_G(t)k|$ is monotonically decreasing. Consequently the map $t \to \|\Gamma_G(t)\|_{\mathcal{L}(U,H)}$ is also monotonically decreasing.*

Proof Let first $k \in D(G)$. Consider the deterministic control system (4.55) with initial datum $x = Gk$. Using the operator $\mathcal{L}_t$ defined in (4.56) we can write the solution of (4.55) as

$$z(t; k, a(\cdot)) = e^{tA}Gk + \mathcal{L}_t a(\cdot)$$

for any $a(\cdot) \in L^2(0, t; \Xi)$. If $k \in U$ the last equation still makes sense, recalling that the operator $e^{tA}G$ extends to all of U by Hypothesis 4.29.

Now given any control $a(\cdot) \in L^2(0, t; \Xi)$, we have $z(t) = 0$ if and only if $\mathcal{L}_t a(\cdot) = -e^{tA}Gk$, i.e. if and only if $a(\cdot)$ belongs to the inverse image (through $\mathcal{L}_t$) of $-e^{tA}Gk$. Among all such controls, the one of minimum norm is, by the definition of the pseudoinverse, $\bar{a}_t(\cdot) := \mathcal{L}_t^{-1}\left(-e^{tA}Gk\right)$ for all $t \geq 0$. Its norm is clearly given by $|\mathcal{L}_t^{-1}\left(-e^{tA}Gk\right)|_{L^2(0,t;\Xi)}$. Now, from (4.57) and Proposition B.2-(ii), we get

$$|\mathcal{L}_t^{-1}\left(-e^{tA}Gk\right)|_{L^2(0,t;\Xi)} = |Q_t^{-1/2}\left(-e^{tA}Gk\right)|_H = |\Gamma_G(t)k|.$$

Thus it is enough to show that the norm $|\bar{a}_t(\cdot)|_{L^2(0,t;\Xi)}$ is decreasing in t. Indeed, let $t_1 > t_2$ and consider the control $a_1(\cdot) \in L^2(0, t_1; \Xi)$ defined as

$$a_1(s) = \begin{cases} \bar{a}_{t_2}(s) & s \in [0, t_2], \\ 0 & s \in (t_2, t_1]. \end{cases}$$

Then we have $z(t_1; 0, k, a_1(\cdot)) = 0$ since, by the definition of $a_1(\cdot)$,

$$e^{t_1 A}Gk + \mathcal{L}_{t_1}a_1 = e^{(t_1-t_2)A}\left[e^{t_2 A}Gk + \int_0^{t_2} e^{(t_2-s)A}\sigma\bar{a}_{t_2}(s)\right] = 0$$

(here we used the fact that $e^{t_1 A}Gk = e^{(t_1-t_2)A}e^{t_2 A}Gk$, which is obvious for $k \in D(G)$ and follows by density for all $k \in U$). It thus follows, by the minimality of $|\bar{a}_{t_1}|_{L^2(0,t_1;\Xi)}$ and the definition of $a_1(\cdot)$, that

$$|\bar{a}_{t_1}(\cdot)|_{L^2(0,t_1;\Xi)} \leq |a_1(\cdot)|_{L^2(0,t_1;\Xi)} = |\bar{a}_{t_2}(\cdot)|_{L^2(0,t_2;\Xi)},$$

which gives the claim. □

Remark 4.36 The two key Hypotheses 4.29 and 4.32 can also be formulated if G depends on $(s, x) \in [0, T] \times H$ for given $T > 0$. In this case we have to ask that (4.53) (or (4.54)) holds for all $(s, x) \in [0, T] \times H$, so that the bounded linear operator from U to H,

$$\Gamma_G(t, s, x) := Q_t^{-1/2} e^{tA} G(s, x),$$

is bounded for all $(t, s, x) \in (0, T] \times [0, T] \times H$. Moreover, assumption (4.64) would then require that the function

$$t \to \sup_{(s,x)\in[0,T]\times H} \|\Gamma_G(t, s, x)\|$$

(which is still decreasing thanks to the above remarks) is integrable in a right neighborhood of 0.   ∎

4.3.1.4   Smoothing Properties of R_t: Results and Estimates

We begin by recalling a well known result for the case $U = H$ and $G = I$ when Hypothesis 4.29 holds (see Theorem 6.2.2 and Exercise 6.3.3 of [179]).

Theorem 4.37 *Assume that Hypotheses 4.25 and 4.29 for $U = H$ and $G = I$ hold. Then for any $\phi \in B_b(H)$ and any $t > 0$ we have $R_t[\phi] \in UC_b^\infty(H)$.*
In particular, for any $k, h, x \in H$, we have

$$\langle DR_t[\phi](x), k \rangle = \int_H \langle \Gamma(t)k, Q_t^{-1/2}y \rangle \phi(e^{tA}x + y) \mathcal{N}_{Q_t}(dy) \qquad (4.68)$$

and

$$\langle D^2 R_t[\phi](x)k, h \rangle$$

$$= \int_H [\langle \Gamma(t)k, Q_t^{-1/2}y \rangle \langle \Gamma(t)h, Q_t^{-1/2}y \rangle - \langle \Gamma(t)k, \Gamma(t)h \rangle] \phi(e^{tA}x + y) \mathcal{N}_{Q_t}(dy).$$
$$(4.69)$$

Moreover, the following estimates hold:

$$|DR_t[\phi](x)| \leq \|\Gamma(t)\| \, \|\phi\|_0, \quad t > 0, \; x \in H, \qquad (4.70)$$

$$\|D^2 R_t[\phi](x)\| \leq \sqrt{2} \, \|\Gamma(t)\|^2 \, \|\phi\|_0, \quad t > 0, \; x \in H. \qquad (4.71)$$

More generally, for every $n \in \mathbb{N}$ there exists a constant $C_n > 0$ such that

$$\|D^n R_t[\phi](x)\| \leq C_n \|\Gamma(t)\|^n \|\phi\|_0, \quad n \in \mathbb{N}, \; t > 0, \; x \in H. \qquad (4.72)$$

If, conversely, Hypothesis 4.25 holds with $U = H$ and $G = I$ and $R_t[\phi] \in C_b(H)$ for any $\phi \in B_b(H)$ and any $t > 0$, then (4.53) is satisfied (for $U = H$ and $G = I$).

Remark 4.38 The result above is generalized to the case when the datum $\phi \in UC_m(H)$ ($m \in \mathbb{N}$) in [102] and the arguments used there can easily be applied to obtain the same result when $\phi \in B_m(H)$ or, with some changes, to $L^2(H, \mathcal{N}(0, Q_\infty))$ (see [179], Propositions 10.3.1 and 10.3.5). ∎

We now prove an analogous result for the case of G-derivatives (G possibly unbounded) which generalizes Lemma 3.4 of [432] and which can be found, in a similar form, in [189] and, in a slightly more general form, in [316].

We need two lemmas. In the following, the symbol $[t]$ denotes the greatest integer part of $t \in [0, +\infty)$.

Lemma 4.39 *Let Hypothesis 4.25 hold and let $M \geq 1$, $\omega \in \mathbb{R}$ be as in (4.45). Then*

$$\mathrm{Tr}[Q_t] \leq \mathrm{Tr}[Q_1]M^2 \frac{e^{2\omega([t]+1)} - 1}{e^{2\omega} - 1}, \quad \forall t \geq 0,$$

with the agreement

$$\frac{e^{2\omega([t]+1)} - 1}{e^{2\omega} - 1} := [t] + 1, \quad \text{if } \omega = 0.$$

Proof Note that

$$Q_t = Q_{t-1} + \int_{t-1}^t e^{sA}\Sigma e^{sA^*}\,ds = Q_{t-1} + e^{(t-1)A}Q_1 e^{(t-1)A^*}, \quad \forall t \geq 1.$$

Now, recall that, see Proposition B.28-(i), if $T \in \mathcal{L}_1(H)$ and $S \in \mathcal{L}(H)$, then $TS \in \mathcal{L}_1(H)$ and $|TS|_{\mathcal{L}_1(H)} \leq |T|_{\mathcal{L}_1(H)}|S|_{\mathcal{L}(H)}$ and that the trace is additive. Thus, setting $a_n := \mathrm{Tr}\,[Q_n]$, $n \in \mathbb{N}$, and $q := \mathrm{Tr}\,[Q_1]$, we get

$$a_0 = 0, \quad a_n \leq a_{n-1} + qM^2 e^{2\omega(n-1)}, \quad \forall n \in \mathbb{N} \setminus \{0\}.$$

This implies

$$a_n \leq qM^2 \sum_{k=1}^n e^{2\omega(k-1)} = qM^2 \frac{e^{2\omega n} - 1}{e^{2\omega} - 1}$$

(with the agreement specified in the statement when $\omega = 0$). The claim follows simply by observing that $t \leq [t] + 1$. □

The following lemma is an extension of Proposition 2.19 of [179] and of Lemma 3.1 of [102].

Lemma 4.40 *Let Hypothesis 4.25 hold. Then for every $\alpha \geq 0$ there exists a $K_1(\alpha) \geq 1$ such that for all $t \geq 0$ we have, for all $x \in H$,*

$$\int_H |y + e^{tA}x|^\alpha \mathcal{N}_{Q_t}(dy) \le K_1(\alpha)(1 + |x|^\alpha)e^{\alpha(\omega\vee 0)t}, \qquad \text{if } \omega \ne 0, \qquad (4.73)$$

$$\int_H |y + e^{tA}x|^\alpha \mathcal{N}_{Q_t}(dy) \le K_1(\alpha)(1 + |x|^\alpha)(1 + t^\alpha), \qquad \text{if } \omega = 0. \qquad (4.74)$$

Proof The case $\alpha = 0$ is obvious. Let $\alpha > 0$. We have

$$|y + e^{tA}x|^\alpha \le (1 \vee 2^{\alpha-1})(|y|^\alpha + |e^{tA}x|^\alpha),$$

so

$$\int_H |y + e^{tA}x|^\alpha \mathcal{N}_{Q_t}(dy) \le (1 \vee 2^{\alpha-1}) \left(|e^{tA}x|^\alpha + \int_H |y|^\alpha \mathcal{N}_{Q_t}(dy) \right).$$

By Proposition 1.59 we have

$$\int_H |y|^\alpha \mathcal{N}_{Q_t}(dy) \le K(\alpha/2)(\mathrm{Tr}[Q_t])^{\alpha/2},$$

where the constant $K(\alpha/2)$ is the one from (1.10), which is independent of t. We now use Lemma 4.39 for $\omega \ne 0$ getting

$$\mathrm{Tr}[Q_t]^{\alpha/2} \le \mathrm{Tr}[Q_1]^{\alpha/2}M^\alpha \left[\frac{e^{2\omega([t]+1)} - 1}{e^{2\omega} - 1} \right]^{\alpha/2}.$$

We thus obtain, using (4.45),

$$\int_H |y + e^{tA}x|^\alpha \mathcal{N}_{Q_t}(dy)$$

$$\le (1 \vee 2^{\alpha-1}) \left(M^\alpha e^{\alpha\omega t}|x|^\alpha + K(\alpha/2)\mathrm{Tr}[Q_1]^{\alpha/2}M^\alpha \left[\frac{e^{2\omega([t]+1)} - 1}{e^{2\omega} - 1} \right]^{\alpha/2} \right).$$

The claim for $\omega \ne 0$ follows by suitably choosing $K_1(\alpha)$. For $\omega = 0$ we have, by Lemma 4.39,

$$\mathrm{Tr}[Q_t]^{\alpha/2} \le \mathrm{Tr}[Q_1]^{\alpha/2}M^\alpha([t] + 1)^{\alpha/2},$$

so

$$\int_H |y + e^{tA}x|^\alpha \mathcal{N}_{Q_t}(dy)$$

$$\le (1 \vee 2^{\alpha-1}) \left(M^\alpha|x|^\alpha + K(\alpha/2)\mathrm{Tr}[Q_1]^{\alpha/2}M^\alpha([t] + 1)^{\alpha/2} \right)$$

and the claim follows by properly choosing $K_1(\alpha)$. $\qquad\qquad\square$

Theorem 4.41 *Let Hypotheses 4.25 and 4.29 hold true and let $m \geq 0$. Then we have the following:*

(i) *For every $\phi \in B_m(H)$ and $t > 0$, $R_t[\phi]$ is G-Fréchet differentiable in H and, for every $x \in H$, $k \in U$ we have the formula*

$$\langle D^G R_t[\phi](x), k \rangle_U = \int_H \phi\left(y + e^{tA}x\right) \left\langle \Gamma_G(t)k, Q_t^{-1/2}y \right\rangle \mathcal{N}_{Q_t}(dy) \quad (4.75)$$

and the estimates (here $K_1(\cdot)$ is the constant from Lemma 4.40)

$$|R_t[\phi](x)|_{\mathbb{R}} \leq \|\phi\|_{B_m}(1 + |x|^m)2K_1(m)e^{m(\omega \vee 0)t}, \quad (4.76)$$

$$|D^G R_t[\phi](x)|_U \leq \|\Gamma_G(t)\| \, \|\phi\|_{B_m}(1 + |x|^m)2[K_1(2m)]^{1/2}e^{m(\omega \vee 0)t}, \quad (4.77)$$

where $\omega \neq 0$ is as in (4.45). If $\omega = 0$ the estimates above hold substituting $e^{m(\omega \vee 0)t}$ with $1 + t^m$.

(ii) *Moreover, if $\phi \in C_m(H)$, then also $D^G R_t[\phi] \in C_m(H)$ for all $t > 0$.*

(iii) *Finally, if $\phi \in C_m^1(H)$, then for all $t > 0$*

$$\langle D^G R_t[\phi](x), k \rangle_U = \int_H \langle D\phi\left(y + e^{tA}x\right), e^{tA}Gk \rangle \mathcal{N}_{Q_t}(dy). \quad (4.78)$$

Proof
Proof of (i). We first compute, for $k \in D(G) \subset U$, the limit

$$\lim_{s \to 0} \frac{1}{s}\left[R_t[\phi](x + sGk) - R_t[\phi](x)\right].$$

Using (4.50) we have

$$\frac{1}{s}\left[R_t[\phi](x + sGk) - R_t[\phi](x)\right]$$

$$= \frac{1}{s}\left[\int_H \phi\left(y + e^{tA}(x + sGk)\right)\mathcal{N}_{Q_t}(dy) - \int_H \phi\left(y + e^{tA}x\right)\mathcal{N}_{Q_t}(dy)\right]$$

$$= \frac{1}{s}\left[\int_H \phi\left(y + e^{tA}x\right)\mathcal{N}\left(se^{tA}Gk, Q_t\right)(dy) - \int_H \phi\left(y + e^{tA}x\right)\mathcal{N}_{Q_t}(dy)\right].$$

The Gaussian measures $\mathcal{N}\left(se^{tA}Gk, Q_t\right)$ and $\mathcal{N}_{Q_t}$ are equivalent since, by Hypothesis 4.29, $se^{tA}Gk \in Q_t^{1/2}(H)$. Applying the Cameron–Martin formula (see Theorem 1.60 or, e.g., Theorem 1.3.6 of [179]), we define

$$d(t, x, sGk, y) = \frac{d\mathcal{N}\left(e^{tA}sGk, Q_t\right)}{d\mathcal{N}_{Q_t}}(y)$$

$$= \exp\left\{\left\langle sQ_t^{-1/2}e^{tA}Gk, Q_t^{-1/2}y\right\rangle - \frac{1}{2}s^2\left|Q_t^{-1/2}e^{tA}Gk\right|^2\right\}$$

and we get bringing, formally, the limit inside the integral,

$$\lim_{s\to 0}\frac{1}{s}[R_t[\phi](x + sGk) - R_t[\phi](x)]$$

$$= \lim_{s\to 0}\int_H \phi\left(y + e^{tA}x\right)\frac{(d(t, x, sGk, y) - 1)}{s}\mathcal{N}_{Q_t}(dy)$$

$$= \int_H \phi\left(y + e^{tA}x\right)\lim_{s\to 0}\frac{(d(t, x, sGk, y) - 1)}{s}\mathcal{N}_{Q_t}(dy)$$

$$= \int_H \phi\left(y + e^{tA}x\right)\left\langle Q_t^{-1/2}e^{tA}Gk, Q_t^{-1/2}y\right\rangle\mathcal{N}_{Q_t}(dy).$$

We now justify the limit above and get the required estimate. First note that, by the definition of the pseudoinverse (Definition B.1), $R(Q_t^{-1/2}) = \left[\ker\left(Q_t^{1/2}\right)\right]^{\perp}$, hence $Q_t^{-1/2}e^{tA}Gk \in \overline{R(Q_t^{1/2})}$. So, from Proposition 1.59 (see also [179] Sect. 1.2.4), we see that the function $y \to \left\langle Q_t^{-1/2}e^{tA}Gk, Q_t^{-1/2}y\right\rangle$ is well defined and square-integrable with respect to the measure $\mathcal{N}_{Q_t}$. Moreover, using the Cauchy–Schwarz inequality we get

$$\left|\int_H \phi\left(y + e^{tA}x\right)\left\langle Q_t^{-1/2}e^{tA}Gk, Q_t^{-1/2}y\right\rangle\mathcal{N}_{Q_t}(dy)\right|$$

$$\leq \|\phi\|_{B_m}\int_H (1 + |y + e^{tA}x|^m)\left|\left\langle Q_t^{-1/2}e^{tA}Gk, Q_t^{-1/2}y\right\rangle\right|\mathcal{N}_{Q_t}(dy)$$

$$\leq \|\phi\|_{B_m}\left(\int_H (1 + |y + e^{tA}x|^m)^2\mathcal{N}_{Q_t}(dy)\right)^{1/2}$$

$$\times \left(\int_H \left|\left\langle Q_t^{-1/2}e^{tA}Gk, Q_t^{-1/2}y\right\rangle\right|^2\mathcal{N}_{Q_t}(dy)\right)^{1/2}$$

$$\leq \|\phi\|_{B_m}\left(\int_H 2(1 + |y + e^{tA}x|^{2m})\mathcal{N}_{Q_t}(dy)\right)^{1/2}\left\|Q_t^{-1/2}e^{tA}G\right\||k|$$

$$\leq \|\phi\|_{B_m}\left[2 + 2K_1(2m)(1 + |x|^{2m})e^{2m(\omega\vee 0)t}\right]^{1/2}\left\|Q_t^{-1/2}e^{tA}G\right\||k|$$

$$\leq \|\phi\|_{B_m}2[K_1(2m)]^{1/2}(1 + |x|^m)e^{m(\omega\vee 0)t}\left\|Q_t^{-1/2}e^{tA}G\right\||k|. \qquad (4.79)$$

In the last three lines we used (1.8), Lemma 4.40 for $\omega \neq 0$, and the fact that $(1 + |x|^{2m})^{1/2} \leq 1 + |x|^m$. If $\omega = 0$ we substitute the term $e^{m(\omega\vee 0)t}$ with $1 + t^m$. From this last estimate we can easily see that the limit is uniform for k in the unit ball of U intersected with $D(G)$. So we can extend all the above computations to $k \in U$ and, using the definition of $\Gamma_G(t)$, we conclude that $R_t[\phi]$ is G-Fréchet differentiable and

$$D^G R_t[\phi](x)k = \int_H \phi\left(y + e^{tA}x\right)\left\langle\Gamma_G(t)k, Q_t^{-1/2}y\right\rangle \mathcal{N}_{Q_t}(dy).$$

Estimate (4.77) follows easily from (4.79). Estimate (4.76) follows observing that

$$|R_t[\phi](x)|_{\mathbb{R}} = \left|\int_H \phi\left(y + e^{tA}x\right) \mathcal{N}_{Q_t}(dy)\right| \le \|\phi\|_{B_m} \int_H \left(1 + |y + e^{tA}x|^m\right) \mathcal{N}_{Q_t}(dy)$$

and then using Lemma 4.40.

Proof of (ii). Let now $\phi \in C_m(H)$, $(t, x) \in (0, T] \times H$ and take any sequence $x_n \to x$ in H. Arguing similarly to (4.79) we get,

$$|D^G R_t[\phi](x_n) - D^G R_t[\phi](x)| = \sup_{|k|_U=1} \left\langle D^G R_t[\phi](x_n) - D^G R_t[\phi](x), k\right\rangle_U$$

$$= \sup_{|k|_U=1} \int_H \left[\phi\left(y + e^{tA}x_n\right) - \phi\left(y + e^{tA}x\right)\right]\left\langle\Gamma_G(t)k, Q_t^{-1/2}y\right\rangle_H \mathcal{N}_{Q_t}(dy)$$

$$\le \left(\int_H \left|\phi\left(y + e^{tA}x_n\right) - \phi\left(y + e^{tA}x\right)\right|^2 \mathcal{N}_{Q_t}(dy)\right)^{1/2} \|\Gamma_G(t)\|.$$

Hence the claim follows by the dominated convergence theorem.

Proof of (iii). Taking $\phi \in C_m^1(H)$, $t > 0$ and $k \in D(G) \subset U$, we have, using (4.49) and the dominated convergence theorem,

$$\left\langle D^G R_t[\phi](x), k\right\rangle_U = \lim_{s\to 0} \frac{1}{s}[R_t[\phi](x + sGk) - R_t[\phi](x)]$$

$$= \lim_{s\to 0} \int_H \frac{1}{s}\left[\phi\left(y + e^{tA}(x + sGk)\right) - \phi\left(y + e^{tA}x\right)\right]\mathcal{N}_{Q_t}(dy)$$

$$= \int_H \left\langle D\phi\left(y + e^{tA}x\right), e^{tA}Gk\right\rangle\mathcal{N}_{Q_t}(dy).$$

This is the claim when $k \in D(G) \subset U$. The claim when $k \in U$ simply follows using the density of $D(G)$ in U and the fact that, by Hypothesis 4.29, the operator $e^{tA}G$ extends to a bounded operator defined on the whole U. □

Remark 4.42

(i) Under the assumptions of Theorem 4.41 it is possible to prove, as was done in Theorem 4.37, the existence of the second (and higher) G-derivative with a formula like (4.69) and an estimate like (4.71) (or (4.72) for higher G-derivatives) with Γ_G in place of Γ. We do not do this here as we will not be using such regularity properties in this chapter.

(ii) In Theorem 4.41, in contrast to Theorem 4.37, we cannot say in general that the semigroup R_t is strongly Feller (Definition 1.159). Indeed the G-Fréchet differentiability of $R_t[\phi]$ may not imply, when ϕ is not continuous, its continuity (taking, for example, G as in Example 4.2 and suitably choosing ϕ and Σ). Similarly we cannot say that $D^G R_t \phi$ is continuous if ϕ is not continuous.

(iii) Note that Theorem 4.37 applies only when ϕ is bounded while in Theorem 4.41 polynomial growth of ϕ is also allowed (similarly to what is done in a special case in [102], Theorem 4.2).

(iv) By a straightforward modification of the proof, Theorem 4.41 can be generalized to the case when, for a given $T > 0$, G is a map from $[0, T] \times H$ to the set of linear closed operators from U to H satisfying Hypothesis 4.11 for $G(t, \cdot)$, for every $t \in [0, T]$. To obtain the result, due to measurability problems, one has to assume a bit more than the modification of Hypothesis 4.29 stated in Remark 4.36. The following would be a reasonable requirement.
For every $(t, x) \in (0, T] \times H, s \in [0, T]$ *the linear operator* $e^{tA} G(s, x)$ *can be extended to a bounded operator, which we still denote by* $e^{tA} G(s, x) : U \to H$, *such that the map*

$$
\begin{aligned}
(0, T] \times [0, T] \times H &\;\to\; \mathcal{L}(U, H), \\
(t, s, x) &\;\to\; e^{tA} G(s, x)
\end{aligned}
$$

is strongly measurable and for every $(t, s, x) \in (0, T] \times [0, T] \times H$

$$
e^{tA} G(s, x)(U) \subset Q_t^{1/2}(H).
$$

∎

4.3.1.5 Smoothing Properties of R_t: Examples with Diagonal Operators

We consider here the case when $U = \Xi = H$ and the operators A, $\Sigma := \sigma\sigma^*$ and G are all diagonal with respect to the same orthonormal basis. We assume the following.

Hypothesis 4.43 Let $U = \Xi = H$ and let $\{e_k\}$ be an orthonormal basis of H. We assume that $\sigma \in \mathcal{L}(H)$ is constant and A, $\Sigma = \sigma\sigma^*$ and G satisfy the following:

$$
Ae_k = -\alpha_k e_k, \quad \Sigma e_k = q_k e_k, \quad Ge_k = g_k e_k, \quad k \in \mathbb{N}, \tag{4.80}
$$

where for all $k \in \mathbb{N}$ we have $\alpha_k \geq 0$, $q_k > 0$ and $g_k \in \mathbb{R}$. Moreover, $\alpha_k \nearrow +\infty$.

Due to the assumptions, the set $\mathbb{N}_0 := \{k \in \mathbb{N} : \alpha_k = 0\}$ is finite. Set $\mathbb{N}_1 := \mathbb{N} - \mathbb{N}_0$. We have the following result (see [180], Proposition 9.44).

Proposition 4.44 *Let Hypothesis 4.43 hold. Then, for any* $t > 0$, *we have the following.*

(i) *The condition*

$$\sum_{k\in\mathbb{N}_1} \frac{q_k}{\alpha_k} < +\infty \tag{4.81}$$

is equivalent to requiring that $e^{sA}\Sigma e^{sA^*} \in \mathcal{L}_1^+(H)$ *for all* $s > 0$ *and* $\int_0^t \mathrm{Tr}[e^{sA}\Sigma e^{sA^*}]ds < +\infty$ *for all* $t > 0$, *i.e. Hypothesis 4.25-(ii). In this case the operator* $Q_t = \int_0^t e^{sA}\Sigma e^{sA^*}ds$ *is diagonal and nuclear and*

$$Q_t e_k = \frac{q_k}{2\alpha_k}(1 - e^{-2\alpha_k t})e_k$$

under the agreement that $\frac{1-e^{-2\alpha_k t}}{2\alpha_k} =: t$ *for* $k \in \mathbb{N}_0$. *Moreover, (4.47) is satisfied if*

$$\sum_{k\in\mathbb{N}_1} \frac{q_k}{\alpha_k^{1-\theta}} < +\infty. \tag{4.82}$$

(ii) *The operator* $e^{tA}G : D(G) \to H$ *extends to a bounded operator in* $\mathcal{L}(H)$ *(which we still denote by* $e^{tA}G$) *if and only if*

$$\sup_{k\in\mathbb{N}} e^{-t\alpha_k} g_k < +\infty. \tag{4.83}$$

Moreover, we have[15] $D(A) \subset D(G)$ *and* $|Gz| \le c|Az|$ *for all* $z \in D(A)$ *for some* $c > 0$, *if and only if* $g_k = 0$ *for* $k \in \mathbb{N}_0$ *and*

$$\sup_{k\in\mathbb{N}_1} \frac{g_k}{\alpha_k} < +\infty, \tag{4.84}$$

which also implies (4.83).

(iii) *The operator* $\Gamma_G(t)$ *is well defined on the elements of* $\{e_k\}$,

$$\Gamma_G(t)e_k = \sqrt{\frac{2\alpha_k}{e^{2t\alpha_k} - 1} \cdot \frac{g_k^2}{q_k}}\, e_k$$

and Hypotheses 4.29 and 4.32 hold if and only if there exists a function $\gamma_G \in \mathcal{I}_1$ *such that*

$$\|\Gamma_G(t)\| = \sup_{k\in\mathbb{N}} \sqrt{\frac{2\alpha_k}{e^{2t\alpha_k} - 1} \cdot \frac{g_k^2}{q_k}} \le \gamma_G(t), \qquad t > 0. \tag{4.85}$$

Proof Concerning point (i) observe that, when (4.81) holds, we clearly have

---

[15]This condition will be used to obtain strong solutions of HJB equations, see Proposition 4.148 and Theorems 4.150–4.158.

$$\sum_{k \in \mathbb{N}} q_k e^{-2s\alpha_k} = \sum_{k \in \mathbb{N}} \frac{q_k}{\alpha_k} e^{-2s\alpha_k} \alpha_k < +\infty \quad \text{for all } s > 0,$$

since for each $s > 0$, $e^{-2s\alpha_k} \alpha_k \leq (2es)^{-1}$. Moreover, since (recall the convention when $\alpha_k = 0$)

$$\int_0^t e^{-2s\alpha_k} q_k ds = \frac{q_k}{2\alpha_k}(1 - e^{-2\alpha_k t}),$$

we can apply Fubini's Theorem 1.33-(ii) to get $\int_0^t \text{Tr}[e^{sA} \Sigma e^{sA^*}] ds < +\infty$ for all $t > 0$.

In the other direction, since we always have $Q_t e_k = \frac{q_k}{2\alpha_k}(1 - e^{-2\alpha_k t}) e_k$, it is immediate that if Q_t is nuclear, then (4.81) must hold.

The last part of point (i) follows observing that, by a simple change of variable,

$$\int_0^t s^{-\theta} e^{-2s\alpha_k} ds \leq \alpha_k^{\theta-1} \int_0^{+\infty} s^{-\theta} e^{-2s} ds,$$

and using again Fubini's Theorem 1.33-(ii).

Concerning point (ii), the first part is immediate. The second part of (ii) can be easily seen since $Gz = \sum_{k=1}^{+\infty} g_k \langle z, e_k \rangle e_k$ and $Az = \sum_{k=1}^{+\infty} \alpha_k \langle z, e_k \rangle e_k$ on elements of their domains.

Regarding (iii), on the elements of the basis $\{e_k\}$, we have

$$\Gamma_G(t)e_k = Q_t^{-1/2} e^{tA} G e_k = \sqrt{\frac{2\alpha_k}{(1 - e^{-2t\alpha_k})q_k}} \, e^{-t\alpha_k} g_k e_k = \sqrt{\frac{2\alpha_k}{e^{2t\alpha_k} - 1} \cdot \frac{g_k^2}{q_k}} \, e_k.$$

Thus the claim follows since the boundedness of $\Gamma_G(t)$ is equivalent to $\sup_{k \in \mathbb{N}} |\Gamma_G(t)e_k| < +\infty$. □

The following corollary lists three common cases in which Hypotheses 4.25 and 4.32 are satisfied when Hypothesis 4.43 holds. We will denote by γ_G any function in $\mathcal{I}_1$ which dominates $\|\Gamma_G(\cdot)\|$.

Corollary 4.45 *Let Hypothesis 4.43 be satisfied. Then we have the following.*

(i) *Let $G = (-A)^\delta$ and $\Sigma = (-A)^{-\beta}$ for some $\delta, \beta \in [0, 1)$. In this case*

$$\|\Gamma_G(t)\|^2 = \sup_{k \in \mathbb{N}} \frac{2\alpha_k^{1+\beta+2\delta}}{e^{2t\alpha_k} - 1} \leq \frac{C}{t^{1+\beta+2\delta}},$$

where $C = \sup_{s>0} \frac{2s^{1+\beta+2\delta}}{e^{2s} - 1}$. Here (4.85) holds when $\beta + 2\delta < 1$ with $\gamma_G(t) = C^{1/2} t^{-(1+\beta+2\delta)/2}$. Note that, in this case, at least formally, $\beta < 0$ would also be fine.[16] Moreover, in this case (4.81) (respectively, (4.82)) is satisfied if and only

---

[16]This can be understood by looking at the control-theoretic interpretation of Hypotheses 4.29 and 4.32 given in Remarks 4.30 and 4.34. Indeed, it corresponds to the intuitive fact that if $\beta < 0$

if

$$\sum_{k\in\mathbb{N}_1} \alpha_k^{-\beta-1} < +\infty, \qquad \left(\text{respectively, } \sum_{k\in\mathbb{N}_1} \alpha_k^{\theta-\beta-1} < +\infty\right).$$

Both are true (the second for sufficiently small θ) e.g. if $\alpha_k \sim k^\eta$ for $\eta(1+\beta) > 1$.

(ii) *Let $G = \sqrt{\Sigma}$ (which includes the case $G = \Sigma = I$). In this case*

$$\|\Gamma_G(t)\|^2 = \sup_{k\in\mathbb{N}} \frac{2\alpha_k}{e^{2t\alpha_k} - 1} \le \frac{C}{t},$$

where $C = \sup_{s>0} \frac{2s}{e^{2s}-1}$. Thus (4.85) holds with $\gamma_G(t) = C^{1/2}t^{-1/2}$. Conditions (4.81) and (4.82) are unchanged.

(iii) *Let $G = (-A)^\beta\sqrt{\Sigma}$ for some $\beta \in (0, 1/2)$. In this case we have $g_k^2 = \alpha_k^{2\beta}q_k$ for every $k \in \mathbb{N}$, hence*

$$\|\Gamma_G(t)\|^2 = \sup_{k\in\mathbb{N}} \frac{2\alpha_k^{1+2\beta}}{e^{2t\alpha_k} - 1} \le \frac{C}{t^{1+2\beta}},$$

where $C = \sup_{s>0} \frac{2s^{1+2\beta}}{e^{2s}-1}$ and so (4.85) holds with $\gamma_G(t) = Ct^{-(1/2+\beta)}$. Also in this case (4.81) and (4.82) are unchanged.

Proof The result easily follows from Proposition 4.44. □

The above examples can be easily extended to cases when some (possibly infinitely many) of the q_k are zero provided that the corresponding g_k are zero too.

We now present a concrete example where Hypothesis 4.43 is satisfied (see e.g. Example 6.3 of [306], Example 13.1.2 of [179]) and [240, 241].

Example 4.46 Let $C_d = (0, \pi)^d$ and $H = L^2(C_d), d \in \mathbb{N}$. Take (see Proposition C.3 and Remark C.9) the Laplace operator with Dirichlet boundary condition

$$D(A_D) = H^2(C_d) \cap H_0^1(C_d), \quad A_D x = \Delta x, \quad \text{for } x \in D(A_D)$$

and (see Proposition C.7 and Remark C.9) the Laplace operator with Neumann boundary condition

$$D(A_N) = \left\{x \in H^2(C_d) : \frac{\partial x}{\partial n} = 0 \text{ on } \partial C_d\right\}, \quad A_N x = \Delta x, \quad \text{for } x \in D(A_N),$$

(Footnote 16 continued)
then the image of $(-A)^{-\beta}$ (in a suitable extrapolation space) is bigger and it is easier to steer any point $x \in H$ to zero. This case should not be confused with the case of boundary noise, where Σ is unbounded but with a very narrow image, see Sect. 2.6.3 and also Sects. C.4 and C.5. For example, in the case of Neumann boundary noise described in Sect. 2.6.3 we have $\Sigma := (\lambda I - A)N_\lambda : \mathbb{R}^2 \to H^{-1}(0, \pi)$, hence the image of Σ is only two-dimensional.

where $\frac{\partial x}{\partial n}$ is the derivative in the normal direction. Both operators A_D and A_N satisfy Hypothesis 4.25-(i) and generate analytic semigroups of compact operators. Moreover, they are both diagonal. For A_D the orthonormal basis of eigenvectors is

$$e^D_{n_1,..,n_d}(\xi) = \left(\frac{2}{\pi}\right)^{\frac{d}{2}} \sin(n_1\xi_1) \cdots \sin(n_d\xi_d), \quad n_i = 1, 2, ..., i = 1, ..., d,$$

with the eigenvalues $-\alpha^D_{n_1,...,n_d}$ where

$$\alpha^D_{n_1,...,n_d} = n_1^2 + \cdots + n_d^2. \tag{4.86}$$

For A_N the orthonormal basis of eigenvectors is

$$e^N_{n_1,..,n_d}(\xi) = C_{n_1,..,n_d} \cos(n_1\xi_1) \cdots \cos(n_d\xi_d), \quad n_i = 0, 1, 2, ..., i = 1, ..., d,$$

for appropriate constants $C_{n_1,...,n_d}$ with the eigenvalues $-\alpha^N_{n_1,...,n_d}$ where

$$\alpha^N_{n_1,...,n_d} = n_1^2 + \cdots + n_d^2. \tag{4.87}$$

In both cases, ordering the eigenvalues with a single index k, we have

$$\alpha_k \approx k^{\frac{2}{d}} \text{ as } k \to +\infty. \tag{4.88}$$

Note that, in the case of Neumann boundary condition there is a zero eigenvalue and so, in defining fractional powers, we have to take the operator $\lambda I - A_N$ for some $\lambda > 0$ (which has eigenvalues $\lambda + \alpha^N_{n_1,...,n_d} > 0$ and the same eigenvectors as A_N), and take, in Corollary 4.45-(i) and (iii), $(\lambda I - A_N)^\beta$ in place of $(-A_N)^\beta$.

Let us look at the three cases of Corollary 4.45 for the operator $A = A_D$ or $A = A_N$. We discuss when the key conditions (4.85) and (4.81) are satisfied.

(i) $Ge_k = \alpha_k^\delta e_k$ and $\Sigma e_k = \alpha_k^{-\beta} e_k$, $\delta, \beta \in [0, 1)$. Here (4.85) holds, when $\beta + 2\delta < 1$. Moreover, Q_t is nuclear if and only if $(1 + \beta)\frac{2}{d} > 1$, i.e. $d < 2(1 + \beta)$. Hence if $d = 1$ (respectively, $d = 2$) one can take $\beta = 0$ (respectively, $\beta > 0$ arbitrarily small) and $\delta < 1/2$, allowing us to cover the case of Neumann boundary control when $\Sigma = I$ (respectively, $\Sigma = (-A)^{-\beta}$ for $\beta \in (0, 1/2)$), see e.g. [240, 241] and Sects. 2.6.2 and 4.8. Recall that for Neumann boundary control (see Sect. C.4) one has to take $\delta = 1/4 + \varepsilon$ for arbitrarily small $\varepsilon > 0$. The case $d = 1$ with $G = I$ is also treated in [90].
    If $d = 3$ one has to take $\beta > 1/2$ and so one can treat the case when $G = I$ but not the case of Neumann boundary control since in this case $\delta = 1/4 + \varepsilon$, hence the condition $\beta + 2\delta < 1$ cannot be satisfied.
(ii) $G = \Sigma^{1/2}$. Here any $d \in \mathbb{N}$ can be considered by suitably choosing Σ. For example, taking $\Sigma e_k = 0$ for all but a finite set of eigenvectors, Q_t is nuclear for every $d \in \mathbb{N}$. Taking $\Sigma e_k = \alpha_k^{-\beta}$, for $\beta > 0$ we can take all d smaller than $2(1 + \beta)$.

(iii) $G = (-A)^{\beta} \Sigma^{1/2}$ for $\beta \in (0, 1/2)$. Also here any $d \in \mathbb{N}$ can be considered for a suitable choice of Σ.

∎

4.3.1.6 Smoothing Properties of R_t: Other Examples

Example 4.47 This example is taken from Example 3.10 in [432]. Let $U = \Xi = H$, $\Sigma \in \mathcal{L}^{+}(H)$, $G = \sigma = \Sigma^{1/2}$. Suppose that, for $t > 0$,

$$e^{tA} \Sigma^{1/2}(H) \subset \Sigma^{1/2}(H),$$

i.e. that $\Sigma^{1/2}(H)$ is invariant for the semigroup e^{tA}. By the closed graph theorem this implies that $\Sigma^{-1/2} e^{tA} \Sigma^{1/2}$ is a bounded operator so we can write

$$e^{tA} \Sigma^{1/2} x = \frac{1}{t} \int_0^t e^{(t-s)A} \Sigma^{1/2} \Sigma^{-1/2} e^{sA} \Sigma^{1/2} x \, ds.$$

Now we use the minimum energy formulation of Hypothesis 4.32 (see Remarks 4.30 and 4.34). Defining for any $x \in H$, $u_{t,x}(s) = -\frac{1}{t} \Sigma^{-1/2} e^{sA} \Sigma^{1/2} x$, we see that, setting

$$z(r) = e^{rA} \Sigma^{1/2} x + \int_0^r e^{(r-s)A} \Sigma^{1/2} u_{t,x}(s) ds, \quad r \in [0, t],$$

we have $z(t) = 0$. This means that $R(e^{tA} \Sigma^{1/2}) \subset R(\mathcal{L}_t)$. If we assume further that the function

$$[0, +\infty) \to \mathbb{R}, \quad s \to \|\Sigma^{-1/2} e^{sA} \Sigma^{1/2}\|$$

is locally p-summable for some $p \in (2, +\infty]$, then, for some $C > 0$,

$$\left(\int_0^t |u_{t,x}(s)|^2 ds \right)^{1/2} = \frac{1}{t} \left(\int_0^t |\Sigma^{-1/2} e^{sA} \Sigma^{1/2} x|^2 ds \right)^{1/2} \le \frac{C|x|}{t^{\frac{1}{2} + \frac{1}{p}}}.$$

In this case, (4.66) is satisfied with $\theta = \frac{1}{2} + \frac{1}{p}$, with the agreement that $1/p = 0$ when $p = +\infty$. ∎

Remark 4.48 If $G = \sigma$ and Ξ, H are finite-dimensional spaces (take $H = \mathbb{R}^n$ and $U = \Xi = \mathbb{R}^m$), Hypotheses 4.29 and 4.32 (with $\theta = 1/2$ in (4.65)) are always satisfied. Indeed, in finite dimension the system (4.55), even when it is not null controllable, has the property that every x in the subspace $\sigma(H)$ can be driven to 0 in any time $t > 0$ with minimal energy smaller than $Ct^{-\frac{1}{2}}|x|$, for a suitable constant C independent of t and x, see [525], Sect. 4.

When H is infinite-dimensional this fact is false in general. It is true in Example 4.47 and it is also true in the case when (4.44) is a linear stochastic wave equation with additive noise (see on this e.g. [432], Sect. 6.1, see also Example 3.69 for a more general wave equation). We give an example where this is false. Consider, on a given filtered probability space $\left(\Omega, \mathscr{F}, \{\mathscr{F}_s\}_{s \geq 0}, \mathbb{P}\right)$, the following 1-dimensional stochastic delay equation with pointwise delay $r > 0$, starting at time $t \geq 0$,

$$dy(s) = \beta_1 y(s - r)ds + dW_0(s), \qquad y(t) = x_0, \ y(t + \xi) = x_1(\xi), \ \xi \in [-r, 0),$$

driven by the 1-dimensional Wiener process W_0. In the setting introduced in Sect. 2.6.8.1 such equation can be rewritten as an SDE in the Hilbert space $H := \mathbb{R} \times L^2(-r, 0; \mathbb{R})$ for the new state variable $X(t) = (X_0(t), X_1(t)(\cdot)) = (y(t), y(t + \cdot))$, $t \geq 0$. The noise space is $\Xi = \mathbb{R}$ and the SDE is

$$dX(t) = AX(t) + \sigma dW_0(t),$$

where

$$\sigma w_0 = (w_0, 0)$$

and

$$\begin{cases} D(A) = \left\{(x_0, x_1(\cdot)) \in \mathbb{R} \times W^{1,2}(-r, 0; \mathbb{R}), \ x_0 = x_1(0)\right\} \\ A(x_0, x_1(\cdot)) = \left(\beta_1 x_1(-r), x_1'(\cdot)\right). \end{cases}$$

The operator A generates a strongly continuous semigroup e^{tA} on H. In the easiest case, when $\beta_1 = 0$, by simple computations we have, for $x = (x_0, x_1) \in H$ and $0 \leq t \leq r$,

$$e^{tA}x = (x_0, x_0 \mathbf{1}_{[-t,0]}(\cdot) + x_1(t + \cdot)\mathbf{1}_{[-d,-t]}(\cdot)). \tag{4.89}$$

Now we consider, for $0 < t \leq r$, the associated deterministic control system (here the control $u(\cdot)$ belongs to $L^2(0, t; \Xi)$)

$$z'(s) = Az(s) + \sigma u(s), \quad z(0) = x \in H,$$

whose mild solution is

$$z(t; 0, x, u(\cdot)) = e^{tA}x + \int_0^t e^{(t-s)A}\sigma u(s)ds.$$

Hypothesis 4.29 in this case is equivalent to requiring that for every $k \in \Xi = \mathbb{R}$ there exists a control $u_{t,k}(\cdot) \in L^2(0, t; \Xi)$ such that $z(t; 0, \sigma k, u_{t,k}(\cdot)) = 0$. We show that this is impossible when $k \neq 0$. Indeed, by (4.89) we have, for any $k \in \mathbb{R}$,

$$e^{tA}\sigma k = e^{tA}(k, 0) = (k, k\mathbf{1}_{[-t,0]}(\cdot))$$

and, for any $u(\cdot) \in L^2(0, t; \Xi)$,

$$\int_0^t e^{(t-s)A} \sigma u(s) ds = \left(\int_0^t u(s) ds, \int_0^t u(s) \mathbf{1}_{[-(t-s),0]}(\cdot) ds \right) = \left(\int_0^t u(s) ds, \int_0^{t+\cdot} u(s) ds \right).$$

To have null controllability we must be able to choose $u(\cdot)$ such that both components of $z(t; 0, \sigma k, u(\cdot))$ become 0. This means

$$k + \int_0^t u(s) ds = 0 \quad and \quad k + \int_0^{t+\xi} u(s) ds = 0 \quad \forall \xi \in [-t, 0),$$

which is clearly impossible, unless $k = 0$.   ∎

4.3.1.7   Joint Space–Time Regularity of R_t

Measurability and continuity properties of the function $(t, x) \to R_t[\phi](x)$ and of related convolutions are useful to establish the regularity of the solutions of our HJB equations in the parabolic case, see Sect. 4.4.1 and, in the specific Ornstein–Uhlenbeck case, Sects. 4.6 and 4.7.

We begin with a lemma about compactness properties of a particular family of Gaussian measures.

Lemma 4.49 *Let H be a real separable Hilbert space and let Hypothesis 4.25 be satisfied. Given any $T > 0$ and any compact set $U \in H$, the family $\{\mathcal{N}_{Q_t} : t \in [0, T]\}$ of probability measures on H is tight.*[17]
Moreover, if $t_n \to t$ as $n \to +\infty$, then $\|Q_{t_n} - Q_t\|_{\mathcal{L}_1(H)} \to 0$ and $\mathcal{N}_{Q_{t_n}}$ converges weakly to $\mathcal{N}_{Q_t}$.

Proof Regarding the first part we use Theorem 1.63 (see also [101], proof of Lemma 6.3, for a similar result). First of all, by the definition of Q_t we immediately have that, for $0 < s < t$,

$$Q_t - Q_s = e^{sA} Q_{t-s} e^{sA^*}, \tag{4.90}$$

which implies that $Q_s \leq Q_t$.

Now, taking a complete orthonormal system $\{e_i\}_i$ in H, we have, for every $t \in [0, T]$, see (1.8),

$$\int_H \sum_{i=N}^{+\infty} \langle x, e_i \rangle^2 \mathcal{N}_{Q_t}(dx) = \sum_{i=N}^{+\infty} \langle Q_t e_i, e_i \rangle \leq \sum_{i=N}^{+\infty} \langle Q_T e_i, e_i \rangle \longrightarrow 0 \quad as \quad N \to +\infty.$$

Hence, by Theorem 1.63, the family $\{\mathcal{N}_{Q_t} : t \in [0, T]\}$ is relatively compact and then tight by Theorem 1.62.

To prove the second part, in view of Proposition 1.64, it is enough to show that $\|Q_{t_n} - Q_t\|_{\mathcal{L}_1(H)} \to 0$ as $n \to +\infty$. By (4.90) and Proposition B.28-(i) it is enough

---

[17] See Definition 1.61-(iii).

to prove that, if $t_n \searrow 0$, then $\mathrm{Tr}[Q_{t_n}] \to 0$. Since, by the monotone convergence theorem, we have for every $t \in [0, T]$

$$\mathrm{Tr}[Q_t] = \int_0^t \mathrm{Tr}\left[e^{sA}\Sigma e^{sA^*}\right] ds$$

and, by Hypothesis 4.25, the map $s \to \mathrm{Tr}\left[e^{sA}\Sigma e^{sA^*}\right]$ is integrable, the claim follows. $\qquad\square$

The next result deals with the joint measurability and continuity of R_t.

Proposition 4.50 *Let $T > 0$ and $m \geq 0$. Suppose that Hypothesis 4.25 is satisfied. Then we have the following.*

(i) *For every $\phi \in B_m(H)$ (respectively $\phi \in C_m(H)$) the function*

$$\phi_R^0 : [0, T] \times H \to \mathbb{R}, \qquad (t, x) \to R_t[\phi](x)$$

belongs to $B_m([0, T] \times H)$ (respectively $C_m([0, T] \times H)$). If $\phi \in C_m^1(H)$ then $D\phi_R^0 \in C_m([0, T] \times H, H)$.

(ii) *Given any $\eta \in \mathcal{I}_1$ and $\psi \in B_{m,\eta}((0, T] \times H)$, defining $I_0 := \{(s, t) : 0 < s \leq t \leq T\}$, the function*

$$\bar{\psi}_R^0 : I_0 \times H \to \mathbb{R}, \qquad (t, s, x) \to R_{t-s}[\psi(s, \cdot)](x)$$

is measurable and the function[18]

$$\psi_R^0 : [0, T] \times H \to \mathbb{R}, \qquad (t, x) \to \int_0^t R_{t-s}[\psi(s, \cdot)](x)ds$$

belongs to $B_m([0, T] \times H)$. If also $\psi(s, \cdot) \in C_m(H)$ for all $s \in (0, T]$, then $\psi_R^0 \in C_m([0, T] \times H)$.

(iii) *Let $G : D(G) \subset U \to H$ be a closed linear operator such that Hypothesis 4.29 holds for it. Then, for every $\phi \in B_m(H)$, the function*

$$\phi_R^1 : (0, T] \times H \to U, \qquad (t, x) \to D^G R_t[\phi](x)$$

belongs to $B_{m,\gamma_G}((0, T] \times H, U)$, where $\gamma_G(t) := \|\Gamma_G(t)\|$. Moreover, if $\phi \in C_m(H)$ then ϕ_R^1 is continuous in x.

(iv) *Let G as in point (iii) above be such that Hypotheses 4.29 and 4.32 hold. Given any $\eta \in \mathcal{I}_1$ and $\psi \in B_{m,\eta}((0, T] \times H)$, defining $I_1 := \{(s, t) : 0 < s < t \leq T\}$, the function*

---

[18]For $t = 0$ such function is defined to be 0 by continuity.

$$\tilde{\psi}_R^1 : I_1 \times H \to U, \qquad (t, s, x) \to D^G R_{t-s}[\psi(s, \cdot)](x)$$

is measurable, and

$$\psi_R^1 : (0, T] \times H \to U, \qquad (t, x) \to \int_0^t D^G R_{t-s}[\psi(s, \cdot)](x)ds$$

belongs to $B_{m,\eta_1}((0, T] \times H, U)$, where $\eta_1 \in \mathcal{I}_1$ is defined in Proposition 4.21-(iii) for $\eta_0 = \Gamma_G$. Moreover, if $\psi(s, \cdot) \in C_m(H)$ for all $s \in (0, T]$, then ψ_R^1 is continuous in x.

Proof
Proof of (i). Let first $\phi \in B_m(H)$. The joint measurability of ϕ_R^0 simply follows from (4.49), while its polynomial growth follows by observing that, by (4.49),

$$\frac{|\phi_R^0(t, x)|}{1 + |x|^m} \leq \|\phi\|_{C_m} \int_H \frac{1 + |y + e^{tA}x|^m}{1 + |x|^m} \mathcal{N}_{Q_t}(dy)$$

and then using Lemma 4.40.

Let $\phi \in C_m(H)$. Let $(t_n, x_n) \to (t, x) \in [0, T] \times H$, as $n \to +\infty$. We evaluate, using (4.49),

$$\phi_R^0(t_n, x_n) - \phi_R^0(t, x) = \int_H \phi(e^{t_n A}x_n + y)\mathcal{N}_{Q_{t_n}}(dy) - \int_H \phi(e^{tA}x + y)\mathcal{N}_{Q_t}(dy)$$

$$= \int_H \left[\phi(e^{t_n A}x_n + y) - \phi(e^{tA}x + y)\right]\mathcal{N}_{Q_{t_n}}(dy)$$

$$+ \left(\int_H \phi(e^{tA}x + y)\mathcal{N}_{Q_{t_n}}(dy) - \int_H \phi(e^{tA}x + y)\mathcal{N}_{Q_t}(dy)\right) =: I_1 + I_2. \quad (4.91)$$

We estimate each term separately. We first observe that, by Lemma 4.49, the family of measures $\mathcal{N}_{Q_t}, t \in [0, T]$ is tight. Hence, for any $\varepsilon > 0$, we can choose a compact set K_ε in H such that $\mathcal{N}_{Q_t}(H - K_\varepsilon) < \varepsilon$ for any $t \in [0, T]$. Then we have

$$|I_1| \leq \|\phi\|_{C_m} \int_{H-K_\varepsilon} \left[(1 + |y + e^{t_n A}x_n|^m) + (1 + |y + e^{tA}x|^m)\right]\mathcal{N}_{Q_{t_n}}(dy)$$

$$+ \int_{K_\varepsilon} |\phi(e^{t_n A}x_n + y) - \phi(e^{tA}x + y)|\mathcal{N}_{Q_{t_n}}(dy).$$

Standard computations yield that, for some constant $C(m) > 0$,

$$\int_{H-K_\varepsilon} \left[(1 + |y + e^{t_n A}x_n|^m) + (1 + |y + e^{tA}x|^m)\right]\mathcal{N}_{Q_{t_n}}(dy)$$

$$\leq \varepsilon C(m)(1 + |x|^m) + C(m)\int_{H-K_\varepsilon} |y^m|\mathcal{N}_{Q_{t_n}}(dy) \leq \varepsilon C(m)(1 + |x|^m) + \rho_m(\varepsilon)$$

for some modulus ρ_m. Moreover, since K_ε is compact it is easy to see that

$$\sup_{y \in K_\varepsilon} |\phi(e^{t_n A} x_n + y) - \phi(e^{t A} x + y)| \le \rho_\varepsilon(1/n) = 0$$

for some modulus ρ_ε. Thus we obtain

$$|I_1| \le \|\phi\|_{C_m} [\varepsilon C(m)(1 + |x|^m) + \rho_m(\varepsilon)] + \rho_\varepsilon(1/n).$$

We now look at I_2. By Lemma 4.49 we know that, as $n \to +\infty$, $\|Q_{t_n} - Q_t\|_{\mathcal{L}_1(H)} \to 0$ and the measures $\mathcal{N}_{Q_{t_n}}$ weakly converge to $\mathcal{N}_{Q_t}$, i.e. for every $f \in C_b(H)$, $\int_H f(y)\mathcal{N}_{Q_{t_n}}(dy) \to \int_H f(y)\mathcal{N}_{Q_t}(dy)$. Define, for $M > 0$, $f_\infty(y) := \phi(e^{t A} x + y)$ and $f_M(y) := \phi(e^{t A} x + y) \wedge M$. We have

$$I_2 = \left(\int_H f_M(y)\mathcal{N}_{Q_{t_n}}(dy) - \int_H f_M(y)\mathcal{N}_{Q_t}(dy) \right)$$
$$+ \left(\int_H (f_\infty - f_M)(y)\mathcal{N}_{Q_{t_n}}(dy) - \int_H (f_\infty - f_M)(y)\mathcal{N}_{Q_t}(dy) \right). \quad (4.92)$$

We fix $\varepsilon > 0$ and choose $M > 0$ such that $f_M(y) = f_\infty(y)$ for all $y \in K_\varepsilon$. This implies that the second term of the right-hand side of (4.92) can be estimated, similarly as it was done for I_1, by $2\|\phi\|_{C_m} [\varepsilon C(m)(1 + |x|^m) + \rho_m(\varepsilon)]$. Moreover, when $t_n \to t$ the first term goes to 0 since f_M is bounded.

Therefore we obtained that, for all $\varepsilon > 0$,

$$\lim_{n \to +\infty} |\phi_R^0(t_n, x_n) - \phi_R^0(t, x)| \le 3\|\phi\|_{C_m} [\varepsilon C(m)(1 + |x|^m) + \rho_m(\varepsilon)],$$

which gives the joint continuity by the arbitrariness of ε.

Let now $\phi \in C_m^1(H)$. We use (4.50) and differentiate with respect to x under the integral sign (using the closedness of the derivative operator, see Corollary 4.14, and Remark 1.31) obtaining, for $x, h \in H$,

$$\langle D\phi_R^0(t, x), h \rangle = \int_H \langle D\phi(y + e^{t A} x), e^{t A} h \rangle \mathcal{N}_{Q_t}(dy).$$

We now take $(t_n, x_n) \to (t, x)$ and evaluate

$$|D\phi_R^0(t_n, x_n) - D\phi_R^0(t, x)| = \sup_{|h|=1} \langle D\phi_R^0(t_n, x_n) - D\phi_R^0(t, x), h \rangle$$

$$= \sup_{|h|=1} \left[\int_H \langle e^{t_n A^*} D\phi(y + e^{t_n A} x_n), h \rangle \mathcal{N}_{Q_{t_n}}(dy) - \int_H \langle e^{t A^*} D\phi(y + e^{t A} x), h \rangle \mathcal{N}_{Q_t}(dy) \right].$$

When h is fixed, the fact that $\langle D\phi_R^0(t_n, x_n) - D\phi_R^0(t, x), h \rangle$ goes to 0, as $n \to +\infty$, follows the same arguments as these we used above for $\phi_R^0(t_n, x_n) - \phi_R^0(t, x)$. To get the claim we need to show that in the above procedure the limit is uniform in h with $|h| = 1$. It is not difficult to see that the continuity of $D\phi$ and the strong continuity of e^{tA} imply the required uniformity in the estimates for the term analogous to I_1 defined in (4.91). Concerning the term analogous to I_2, the second term of the right-hand side of (4.92) is estimated uniformly as before. For the first term we need uniformity in the weak convergence of measures. This is guaranteed by Theorem 1.65, which then allows us to conclude.

Proof of (ii). Let $\psi \in B_{m,\eta}((0, T] \times H)$. The measurability of $\bar{\psi}_R^0$ follows from the measurability of ψ and from the definition of the semigroup R_t in (4.49). Defining $\psi_0(s, x) := \eta(s)^{-1}\psi(s, x)$, by the assumptions on ψ, we have that $\frac{\psi_0(s,x)}{1+|x|^m}$ is bounded. Moreover, $\bar{\psi}_R^0(t, s, x) := \eta(s)R_{t-s}[\psi_0(s, \cdot)](x)$.

The joint measurability of ψ_R^0 immediately follows from the Fubini theorem (Theorem 1.33-(i)) since

$$\psi_R^0(t, x) = \int_0^t \eta(s)R_{t-s}[\psi_0(s, \cdot)](x)ds. \tag{4.93}$$

Moreover, $\psi_R^0 \in B_m([0, T] \times H)$ since, for $(t, x) \in [0, T] \times H$,

$$|\psi_R^0(t, x)| \le \int_0^t \eta(s) |R_{t-s}[\psi_0(s, \cdot)](x)| \, ds$$

$$\le C(m)(1 + |x|^m)\|\psi\|_{B_{m,\eta}((0,T]\times H)} \int_0^t \eta(s)ds. \tag{4.94}$$

We now add the assumption that $\psi(s, \cdot) \in C_m(H)$ for all $s \in (0, T]$, and prove that $\psi_R^0 \in C_m([0, T] \times H)$.

First we observe that in such a case the function $\bar{\psi}_R^0$ is continuous in the variables (t, x). When $\psi(s, \cdot) \in C_m(H)$ for all $s \in (0, T]$, we observe that $\bar{\psi}_R^0(t, s, x) = (w_1 \circ g)(t, s, x)$, where $g : I_0 \to [0, T] \times (0, T] \times H, g(t, s, x) = (t - s, s, x)$ and $w_1 : [0, T] \times (0, T] \times H \to \mathbb{R}, w_1(t_1, s, x) = R_{t_1}[\psi(s, \cdot)](x)$. The continuity follows since g is clearly continuous and since, for all $s \in (0, T]$, $w_1(\cdot, s, \cdot)$ is continuous thanks to part (i).

From the above we thus have that the integrand in (4.93) is continuous in (t, x), integrable in s, and dominated in s uniformly for $(t, x) \in (0, T] \times H_0$ for every bounded subset H_0 of H. This immediately gives the required continuity by a straightforward application of dominated convergence.

Proof of (iii). Let $\phi \in B_m(H)$. We observe first that by (4.75) we have, for $(t, x) \in (0, T] \times H, k \in U$,

$$\langle D^G R_t[\phi](x), k \rangle_U = \int_H \phi\left(y + e^{tA}x\right) \left\langle \Gamma_G(t)k, Q_t^{-1/2}y \right\rangle_H \mathcal{N}_{Q_t}(dy).$$

This formula immediately implies the measurability of ϕ_R^1. Moreover, by using the Cauchy–Schwarz inequality exactly as in (4.79), we get that $\phi_R^1 \in B_{m,\gamma_G}((0,T] \times H, U)$.

When $\phi \in C_m(H)$ the continuity of ϕ_R^1 in x follows directly from Theorem 4.41-(ii).

Proof of (iv). We observe first that, by (4.75), taking ψ_0 as in (4.93) we have

$$\left\langle D^G R_{t-s}[\psi_0(s,\cdot)](x), k \right\rangle_U = \int_H \psi_0\left(s, y + e^{(t-s)A}x\right) \left\langle \Gamma_G(t-s)k, Q_{t-s}^{-1/2}y \right\rangle \mathcal{N}_{Q_{t-s}}(dy).$$

Clearly the function above is measurable in (t, s, x). Moreover, using the Cauchy–Schwarz inequality exactly as in (4.79), we get

$$\left| \left\langle D^G R_{t-s}[\psi_0(s,\cdot)](x), k \right\rangle_U \right| \le C(m)(1 + |x|^m) \|\psi_0\|_{B_m((0,T]\times H)} |\Gamma_G(t-s)k|.$$

We then deduce that the integral

$$\psi_R^1(t,x) := \int_0^t D^G R_{t-s}[\psi(s,\cdot)](x)ds = \int_0^t \eta(s) D^G R_{t-s}[\psi_0(s,\cdot)](x)ds$$

is well defined for all $(t, x) \in (0, T] \times H$ since the integrand is measurable in s and the norm of the integrand is estimated from above by

$$C(m)(1 + |x|^m)\eta(s)\|\Gamma_G(t-s)\| \, \|\psi\|_{B_{m,\eta}([0,T)\times H)}, \quad s \in (0,t),$$

which is integrable on $[0, t]$ for $t \in (0, T]$. The measurability of ψ_R^1 is immediate from the measurability of the integrand while the boundedness of $\frac{\psi_R^1(t,x)}{\eta_1(t)(1+|x|^m)}$ on $(0, T] \times H$ follows observing that, for (t, x) in this set, arguing as in (4.79),

$$\frac{|\psi_R^1(t,x)|}{1 + |x|^m} \le C(m)\|\psi\|_{B_{m,\eta}((0,T]\times H)} \int_0^t \eta(s)\|\Gamma_G(t-s)\| \, ds$$

for a suitable constant $C(m) > 0$. Thus the claim follows from Proposition 4.21-(iii).

Assume now that $\psi(s, \cdot)$ is continuous in x for all $s \in (0, T]$. The continuity of ψ_R^1 in x then follows, applying the dominated convergence theorem as in part (iii). □

The following result holds under stronger assumptions which are satisfied, for instance, in examples with boundary control (see Sect. 2.6.2).

Proposition 4.51 *Let $T > 0$ and $m \ge 0$. Suppose that Hypothesis 4.25 is satisfied, Hypothesis 4.29 holds with $U = H$ and $G = I$ and define $\gamma_I(t) := \|\Gamma(t)\|$. Then, in addition to the claims of Proposition 4.50, we have the following.*

(i) *For every $\phi \in B_m(H)$ we have $\phi_R^0 \in C_m((0, T] \times H)$.*

(ii) *Given any $\eta \in \mathcal{I}_1$ and $\psi \in B_{m,\eta}((0, T] \times H)$, we have $\psi_R^0 \in C_m([0, T] \times H)$.*

(iii) *Let $G : D(G){\subset}U \to H$ be a closed linear operator such that Hypothesis 4.29 holds for it. Then, for every $\phi \in B_m(H)$, we have $\phi_R^1 \in C_{m,\gamma_G}((0, T] \times H, U)$. Moreover, $D\phi_R^0 \in C_{m,\gamma_l}((0, T] \times H, H)$.*

(iv) *Let G be as in part (iii) above and such that Hypotheses 4.29 and 4.32 hold. Then for every $\eta \in \mathcal{I}_1$ and $\psi \in B_{m,\eta}((0, T] \times H)$, we have $\psi_R^1 \in C_{m,\eta_1}((0, T] \times H, U)$ with η_1 as in Proposition 4.50-(iv) (for $\eta_0 = \gamma_G$).*

Proof

Proof of (i). If Hypothesis 4.29 holds with $U = H$ and $G = I$, then the semigroup R_t is strongly Feller (see Definition 1.159) by Theorem 4.41. Hence, given $\phi \in B_m(H)$, for all $\varepsilon > 0$ we have $R_\varepsilon[\phi] \in C_m(H)$. Thus, writing for $t > \varepsilon$, $\phi_R^0(t, x) = R_{t-\varepsilon}[R_\varepsilon[\phi]](x)$, by the previous part we get $\phi_R^0 \in C_m([\varepsilon, T] \times H)$. The claim follows from the arbitrariness of ε.

Proof of (ii). Consider the function w_0 defined in the first lines of the proof of part (ii) of Proposition 4.50. Thanks to part (i) of this proposition this function is continuous in the variables (t, x). Thus the integrand in (4.93) is continuous in (t, x), integrable in s and dominated, uniformly for $(t, x) \in (0, T] \times H_0$ for every bounded subset H_0 of H. This immediately gives the required continuity by a straightforward application of dominated convergence.

Proof of (iii). Let $\phi \in B_m(H)$. Since Hypothesis 4.29 holds with $U = H$ and $G = I$ we apply Theorem 4.37 and Remark 4.38 (see also Remark 4.42) obtaining that, for all $\varepsilon > 0$, the function $\phi_\varepsilon := R_\varepsilon[\phi]$ belongs to $C_m^1(H)$. By Theorem 4.41-(iii) we then have, for $x \in H, k \in U$ and $t > \varepsilon$,

$$\left\langle D^G R_t[\phi](x), k \right\rangle_U = \left\langle D^G R_{t-\varepsilon}[\phi_\varepsilon](x), k \right\rangle_U$$

$$= \int_H \left\langle D\phi_\varepsilon(y + e^{(t-\varepsilon)A}x), e^{(t-\varepsilon)A}Gk \right\rangle \mathcal{N}_{Q_{t-\varepsilon}}(dy).$$

Now the joint continuity of the ϕ_R^1 in $(\varepsilon, T] \times H$ is proved exactly as the joint continuity of $D\phi_R^0$ when $\phi \in C_m^1(H)$, once we know that the map $t \to e^{tA}G$ is strongly continuous for $t > 0$. This is guaranteed since, by the semigroup property, we have for $t_1, t_2 > 0$

$$e^{(t_1+t_2)A}Gk = e^{t_1 A}e^{t_2 A}Gk.$$

The above is easily proved first for $k \in D(G)$ and then, by density, for all $k \in U$. The claim then follows by the arbitrariness of $\varepsilon > 0$.

The fact that $D\phi_0^R \in C_{m,\gamma_l}((0, T] \times H, H)$ follows using claim (iii) of Proposition 4.50 when $G = I$ and, for the joint continuity, the arguments above when $G = I$.

Proof of (iv). By the proof of part (iii), for all $s \in (0, T)$ the function $(t, x) \to D^G R_{t-s}[\psi(s, \cdot)](x)$ is continuous in $(s, T] \times H$. Thus, if $(t_n, x_n) \to (t, x) \in (0, T] \times H$, the convergence $\psi_R^1(t_n, x_n) \to \psi_R^1(t, x)$ follows easily by the dominated convergence theorem.   □

Remark 4.52

(i) One may expect that, in part (iii) of Proposition 4.50, the continuity of ϕ would imply the joint continuity of ϕ_R^1 in (t, x). This may indeed be true but, unfortunately, we are not able to prove it. A natural way to show such joint continuity, used, for example, in Lemma 4.8 of [306], would be to prove uniform continuity of the first G-derivative $D^G R_t[\phi]$ by showing that its (standard) derivative $D\left(D^G R_t[\phi]\right)$ is well defined and bounded. This can clearly be done if, as in Proposition 4.51-(iii), we also require that Hypothesis 4.29 holds for $U = H$ and $G = I$, as it happens, for instance, in our example with Neumann boundary control (Example 4.225). Similar remarks apply to part (iv) of Propositions 4.50 and 4.51.

(ii) Since the Ornstein–Uhlenbeck process $X(\cdot)$ is mean square continuous (see Proposition 1.144 or also, e.g., Theorem 5.2-(i) of [180]), the proof of the first statement of Proposition 4.50-(i) can be done directly using this fact, as is done in the proof of Proposition 4.67-(i). Here we presented a different proof, based on the Gaussian law of the process.

■

Remark 4.53   The claims of Propositions 4.50 and 4.51 can be generalized or adapted by suitably modifying the proofs in the following directions.

(i) Let $\phi \in UC_m(H)$ $(m \geq 0)$. Then ϕ_R^0 in Proposition 4.50-(i) belongs to $UC_m^x([0, T] \times H)$.

Moreover, if $\phi \in UC_m(H)$ $(m \geq 0)$, then ϕ_R^1 in Proposition 4.51-(iii) belongs to $UC_{m,\gamma_G}^x((0, T] \times H, U)$ and $D\phi_0^R \in UC_{m,\gamma_I}^x((0, T] \times H, H)$. See for both statements Proposition 3.3-(ii) of [306] in the case $m = 0$; the case $m > 0$ can be proved with straightforward modifications along the lines of what is done in Theorem 3.3 in [102].

Similar claims can be proved for the functions ψ_R^0 in part (ii) and ψ_R^1 in part (iv).

(ii) If G depends on $(s, x) \in [0, T] \times H$ then statements (iii) and (iv) of Propositions 4.50 and 4.51 remain true under suitable measurability and continuity assumptions about the maps $(t, s, x) \to e^{tA} G(s, x)$ and $(t, s, x) \to \Gamma_G(t, s, x)$. For example, a sufficient condition would be to ask that both maps be strongly continuous, the first be bounded on $(0, T] \times [0, T] \times H$ and the second be such that

$$t \to \sup_{(s,x)\in[0,T]\times H} \|\Gamma_G(t, s, x)\|$$

is integrable in a right neighborhood of 0.

■

4.3.2 The Case of a Perturbed Ornstein–Uhlenbeck Semigroup

We consider here the case, studied in [64, 271, 272, 311, 431, 434], when the transition semigroup is associated to an SDE of Ornstein–Uhlenbeck type with a Lipschitz perturbation of the drift. The proofs (which we do not provide) are based mainly on the Girsanov Theorem (see Theorem 6.34) and Malliavin calculus (see Sect. 6.2.2 for some basic material on it).

Let H and Ξ be two real separable Hilbert spaces. Given a filtered probability space $\left(\Omega, \mathscr{F}, \{\mathscr{F}_s\}_{s \in [0,T]}, \mathbb{P}\right)$, and a cylindrical Wiener process W on Ξ, we consider the time homogeneous SDE in H

$$\begin{cases} dX(s) = AX(s)ds + \sigma R(X(s))ds + \sigma dW(s), & s \in [t, T], \\ X(t) = x \in H, \end{cases} \qquad (4.95)$$

where $T > 0, 0 \le t < T$ and we assume that A and σ satisfy Hypotheses 4.25 and 4.29 for $U = H$ and $G = I$. Moreover, regarding the nonlinear term $R : H \to \Xi$, we assume the following hypothesis.

Hypothesis 4.54 The map R is globally Lipschitz continuous, i.e. for all $x, y \in H$, we have

$$|R(x) - R(y)|_{\Xi} \le \|R\|_{0,1} |x - y|.$$

Under the above assumptions Eq. (4.95) admits a unique mild solution $X(\cdot; t, x)$ (e.g. by Proposition 1.147). We write $X(\cdot; x)$ for $X(\cdot; 0, x)$. In mild form we have

$$X(s; x) = e^{sA}x + \int_0^s e^{(s-r)A}\sigma R(X(s; x))ds + W^A(s), \qquad (4.96)$$

where, as in (1.64),

$$W^A(s) = \int_0^s e^{(s-r)A}dW(s). \qquad (4.97)$$

We denote by P_t the associated transition semigroup (see Sect. 1.6).

We will also use a stronger assumption about R.

Hypothesis 4.55 R is Fréchet differentiable and $DR \in C_b(H, \mathcal{L}(H, \Xi))$.

We have the following result, proved first in [271] and then, with a simplified method, in [64].

Theorem 4.56 *Assume that Hypotheses 4.25, 4.29 (for $U = H$ and $G = I$) and 4.55 hold and that (4.47) is satisfied. Let $\phi \in B_b(H)$ and $t > 0$. Then $P_t[\phi]$ is Fréchet differentiable on H and for all $x, h \in H$, we have*[19]

---

[19]In the formula below the operator $Q_t^{-1/2}e^{(t-s)A}Q_s^{1/2}$ is always well defined for $0 \le s \le t$ since, thanks to Lemma 2.3 of [271] (see also Lemma 5 of [64]), we always have $Range\left(e^{(t-s)A}Q_s^{1/2}\right) \subset Range\left(Q_t^{1/2}\right)$ and $\|Q_t^{-1/2}e^{(t-s)A}Q_s^{1/2}\| \le 1$.

$$\langle DP_t[\phi](x), h \rangle = \mathbb{E}\left[\phi(X(t;x))\left\langle W_A(t), Q_t^{-1/2}\Gamma(t)h \right\rangle\right]$$
$$+\mathbb{E}\left[\phi(X(t;x))\int_0^t \left\langle DR(X(s;x))\left(e^{sA}h - Q_s^{1/2}\left(Q_t^{-1/2}e^{(t-s)A}Q_s^{1/2}\right)^*\Gamma(t)h\right), dW(s)\right\rangle\right].$$
(4.98)

The following estimate holds:

$$|DP_t[\phi](x)| \le C_t \|\phi\|_0, \qquad x \in H,$$

where

$$C_t = \|\Gamma(t)\| + \|R\|_{0,1}\left(\int_0^t \|e^{sA}\|^2 ds\right)^{1/2} + \|\Gamma(t)\|\|R\|_{0,1}\left(\int_0^t \|Q_s\|^2 ds\right)^{1/2}.$$
(4.99)

Hence, for some $C > 0$

$$C_t \le C\left(e^{(\omega \vee 0)t} + \sqrt{t}\right)(1 + \|\Gamma(t)\|).$$

Proof See the proofs of Theorem 2.5 and Corollary 2.7 in [271]. See also the proof of Theorem 8 in [64]. □

Remark 4.57 If we remove the regularity assumption of Hypothesis 4.55, the only known results are these contained in Theorem 2.6 of [271] or in Theorem 8 of [64], i.e. that, for every $t > 0$, $\phi \in B_b(H)$, $x, y \in H$

$$|P_t[\phi](x) - P_t[\phi](y)| \le C_t \|\phi\|_0 |x - y|, \tag{4.100}$$

where C_t is the constant from (4.99). It may be possible, with a careful use of approximation procedures, to improve such results, indeed obtaining Fréchet differentiability of $P_t[\phi]$ in such case. In [64, 271] this was not done as the main focus of the authors (as in [486] in a different context) was to prove the strong Feller property for which estimate (4.100) is enough.

Another way of removing Hypothesis 4.55 in Theorem 4.56 is to use Theorem 3.7 of [311]. Indeed, this theorem states that, under Hypotheses 4.25, 4.29, 4.32 (both for $U = H$ and $G = I$) and (4.47), the semigroup P_t transforms Lipschitz continuous functions into Fréchet differentiable functions with the estimate (for any given $T > 0$),

$$\|DP_t[\phi]\|_0 \le C_T \|\phi\|_{0,1}, \qquad \forall t \in (0, T]$$

for a suitable constant $C_T > 0$. Hence adding Hypothesis 4.32 (for $U = H$ and $G = I$) and using the semigroup property of P_t one can prove the smoothing property of Theorem 4.56 without assuming Hypothesis 4.55.

Finally, it is worth mentioning here the papers [272, 434] where cases of non-Lipschitz R have been studied. In [236] the author uses an ad hoc technique proving a formula similar to (4.98) and the consequent smoothing property. There $\Xi = H$, $\Sigma = \sigma\sigma^*$ and $R(x) = -\Sigma^{1/2}DU(x)$ where $U : H \to \mathbb{R}$ is a suitable smooth map with unbounded derivative and having special properties. In [434] the author exploits the dissipativity of R to prove a smoothing property in Banach spaces. ∎

Remark 4.58 One may ask why the smoothing properties of the perturbed Ornstein–Uhlenbeck semigroup have only been studied in the literature when the nonlinearity is of the type σR, where σ is the noise operator. One reason is that a generic Lipschitz perturbation may destroy the smoothing property of the non-perturbed semigroup. Consider the following simple two-dimensional example. Let $H = \mathbb{R}^2$, $\Xi = \mathbb{R}$,

$$A = \begin{pmatrix} 0 & 1 \\ 0 & 0 \end{pmatrix}, \qquad \sigma = \begin{pmatrix} 0 \\ 1 \end{pmatrix}.$$

Moreover, take the linear perturbation term Bx with $B = -A$. It is immediate to see that the original Ornstein–Uhlenbeck semigroup satisfies Hypothesis 4.29 with $U = H$ and $G = I$ while the perturbed one does not.

If we add the assumption that the original Ornstein–Uhlenbeck semigroup also satisfies Hypothesis 4.32 with $U = H$ and $G = I$ then the above example does not work and we do not know if there are any counterexamples. ∎

Remark 4.59 We finally make some remarks about possible extensions of the results of Theorem 4.56.

(i) In Sect. 5.4 in [431] it is suggested that Theorem 4.56 may also hold when R is time-dependent, i.e. when, for a given $T > 0$, $R : [0, T] \times H \to \Xi$ is measurable and, for some constant $C_R > 0$, we have

$$|R(t, x) - R(t, y)|_\Xi \le C_R |x - y|, \qquad |R(t, x)|_\Xi \le C_R(1 + |x|).$$

(ii) It is also suggested in [431] (Sect. 5.4) that the results of Theorem 4.56 can possibly be extended to the case when we look for the G-Fréchet derivative when $G = \sigma$, under Hypothesis 4.29 for such G.
(iii) Theorem 4.56 should extend to the case when the datum ϕ belongs to $B_m(H)$, for some $m > 0$.

∎

We conclude this section noting that the analogues of Propositions 4.50 and 4.51 (see also Proposition 4.67) can be proved. We do not do this here since we will not deal with this case in our main examples.

4.3.3 The Case of an Invertible Diffusion Coefficient

A useful method to prove the smoothing property stated in (4.6) (and consequently the one in (4.7), at least for bounded G) is to apply the so-called Bismut–Elworthy–Li formula, introduced in [60] and, later, revisited in [216] (see also [486] Lemma 2.4,

[179] Lemma 7.7.3, [283] for the version used here, and, for a generalization to the nonlinear superquadratic case, [439]).

The idea behind this method is similar to that described in Sect. 4.3.2: to exploit the tools of Malliavin calculus (see Sect. 6.2.2 for basic material on this) and to perform integration by parts, moving the derivative operation on the process, hence on the data of the SDE defining the semigroup.

As far as we know the Bismut–Elworthy–Li formula has been used to prove smoothing properties of transition semigroups in three main cases:

- stochastic Burgers and Navier Stokes equations (see [155, 158]);
- stochastic reaction-diffusion equations (see [103] and [106], Chaps. 6 and 7);
- SDEs with invertible diffusion coefficient (see e.g. [486] and, later, [283, 439] in more general cases).

Here we present the third case. The first two cases concern specific models and are postponed to Sect. 4.9. As in the previous sections we skip the proofs, giving a few comments on some generalizations and sending the reader to the paper [283] for the details. Moreover here, for simplicity, we only present estimates for the gradient $\nabla P_{t,s}$ of the semigroup on bounded time intervals, which do take into account what happens when $s - t$ approaches infinity, and hence can be used only to study finite horizon control problems. Estimates on an infinite time interval can be proved under further assumptions, as is done in Proposition 4.6-(ii) of [285], allowing then applications to infinite horizon control problems.

Let $T > 0$ and let H and Ξ be two real separable Hilbert spaces. Given a filtered probability space $\left(\Omega, \mathscr{F}, \{\mathscr{F}_s\}_{s\in[0,T]}, \mathbb{P}\right)$, and a cylindrical Wiener process W on Ξ with identity covariance, we consider the SDE in H

$$\begin{cases} dX(s) = AX(s)\,ds + b(s, X(s))\,ds + \sigma(s, X(s))\,dW(s)\,, s \in [t, T]\,, \\ X(t) = x \in H, \end{cases}$$

$$(4.101)$$

where we assume that the coefficients satisfy Hypothesis 1.149.

By Theorem 1.152 we know that, under such assumptions, there exists a unique mild solution $X(\cdot; t, x)$ of (4.101) for every initial data $(t, x) \in [0, T] \times H$. Such solution has continuous trajectories and satisfies, for every $p \in (0, \infty)$, the estimate $\mathbb{E}\sup_{\tau\in[t,T]} |X(\tau; t, x)|^p < C_p (1 + |x|^p)$, for some constant $C_p > 0$.

As discussed in Sect. 1.6, Theorems 1.157 and 1.162, the (two-parameter) transition semigroup $P_{t,s}$ associated to the process $X(\cdot; t, x)$ is well defined, for $\phi \in B_m(H)$, by the formula

$$P_{t,s}[\phi](x) = \mathbb{E}\phi(X(s; t, x))$$

$$(4.102)$$

and it has the Feller property.

To establish the smoothing property (4.6) through the Bismut–Elworthy–Li formula we first need to look at the derivative of the process $X(\cdot; t, x)$ with respect to the initial datum x. We need the following assumption.

Hypothesis 4.60 Let $T > 0$ and let A and σ be as in Hypothesis 1.149 (i) and (iv). Assume, moreover, the following.

(i) $b : [0, T] \times H \to H$ is measurable and such that

$$|b(t, x)| \le L(1 + |x|), \qquad \text{for all } t \in [0, T], \ x \in H, \qquad (4.103)$$

$$|b(t, x) - b(t, y)| \le L|x - y|, \qquad \text{for all } t \in [0, T], \ x, y \in H, \qquad (4.104)$$

where the constant L is that of Hypothesis 1.149-(iv).

(ii) For all $t \in [0, T]$ and $x \in H$ we have for the same constant L above,

$$|\sigma(t, x)|_{\mathcal{L}(\Xi, H)} \le L.$$

(iii) For every $t \in [0, T]$ and $s \in (0, T]$, the mapping $b(t, \cdot) : H \to H$ belongs to $\mathcal{G}^1(H, H)$ while $e^{sA}\sigma(t, \cdot) : H \to \mathcal{L}_2(\Xi, H)$ belongs to $\mathcal{G}^1(H, \mathcal{L}_2(\Xi, H))$.

Under the above assumptions we have the following result, which is Proposition 6.10, and which we repeat here for the reader's convenience.

Proposition 4.61 *Let $T > 0$. Let Hypotheses 1.149 and 4.60 be satisfied. Then for every $p \in [2, +\infty)$ the following hold.*[20]

(i) The map $(t, x) \to X(\cdot; t, x)$ belongs to the space

$$\mathcal{G}^{0,1}([0, T] \times H, L_{\mathcal{P}}^p(\Omega, C([0, T], H))).$$

(ii) Denoting by ∇_x the partial Gâteaux derivative in x, for every direction $h \in H$, the directional derivative process $\nabla_x X(s; t, x)h$, $s \in [0, T]$ solves, for all $h \in H$, $\mathbb{P}$-a.s., the equation

$$\nabla_x X(s; t, x)h = e^{(s-t)A}h + \int_t^s e^{(s-\tau)A}\nabla_x b(\tau, X(\tau; t, x))\nabla_x X(s; t, x)h\,d\tau$$

$$+ \int_t^s \nabla_x \left(e^{(s-\tau)A}\sigma(\tau, X(\tau; t, x))\right) \nabla_x X(s; t, x)h\,dW(r).$$

(iii) We have

$$|\nabla_x X(s; t, x)h|_{L_{\mathcal{P}}^p(\Omega, C([0,T],H))} \le C|h|$$

for some constant $C > 0$ depending only on L, γ_2 from Hypothesis 1.149, T, p, and $M_0 := \sup_{s \in [0,T]} \|e^{sA}\|$.

Remark 4.62 Proposition 4.61 belongs to a class of results concerning the regularity properties of the solutions of SDEs like (4.101) with respect to the initial

---

[20]The space $L_{\mathcal{P}}^p(\Omega, C([0, T], H))$ is defined in Sect. 6.1.1.

conditions. Such results are usually proved by means of the so-called parameter-dependent contraction mapping principle (see e.g. [180], Lemma 9.2 or [179], Theorems 7.1.2 and 7.1.3). With this method one needs to differentiate the map $\mathcal{K}$ given in (1.81) with respect to the process $Y \in \mathcal{H}_p(0, T_0; H)$. The differentiability results of [180], Sect. 9.1.1 or [179], Sect. 7.3, (like the ones quoted above) are formulated for Gâteaux derivatives. For analogous results proving continuous Fréchet differentiability see e.g. [177], Theorem 5.4.1 and [154], Sect. 8.3, [106], Sects. 4.2 and 6.3, [294], Theorem 3.9. We finally mention [269], Chap. 4, for an analysis of the first-order differentiability in both the Gâteaux and Fréchet cases and the recent working paper [516] where results on Gâteaux and Fréchet derivatives are given, together with a generalization of the above quoted results. ∎

Remark 4.63 The assumptions of the Proposition 4.61 are also designed to cover the cases when, as for reaction-diffusion equations (see e.g. [177], Chap. 11 or, our Chap. 2, Sect. 2.6.1 and, in particular, Eqs. (2.79) and (2.83)), where $\Xi = H$ and the operators b and σ are nonlinear Nemytskii type operators. Indeed, in such cases it is well known that, when the underlying space is $L^2(\mathcal{O})$ ($\mathcal{O} \subset \mathbb{R}^n$, open), the operator $b(t, \cdot)$ is never Fréchet differentiable while its Gâteaux differentiability holds under a differentiability condition for the underlying real-valued function defining the Nemytskii operator (see e.g. [10], Sect. 1.2).

Similarly, in such cases, the operator $\sigma(t, \cdot) : H \to \mathcal{L}(H)$ is never Lipschitz continuous (see Remark 1.150) while the operator $e^{sA}\sigma(t, \cdot) : H \to \mathcal{L}_2(H)$ can be proved to be Gâteaux differentiable (Fréchet differentiability fails in general) when the underlying real-valued function defining the Nemytskii operator is differentiable. ∎

To obtain the Bismut–Elworthy–Li formula and, consequently, the required smoothing property, we also need the following assumption (see [486] for the autonomous case and [283] for the non-autonomous case).

Hypothesis 4.64 Let A, b and σ be as in Hypothesis 4.60. Let $T > 0$. For every $t \in [0, T]$ and $x \in H$, $\sigma(t, x)$ is invertible and there is a constant $L > 0$ (for simplicity assumed to be the same as that in Hypothesis 4.60) such that

$$\left|\sigma^{-1}(t, x)\right|_{\mathcal{L}(H, \Xi)} \leq L$$

for all $t \in [0, T]$, $x \in H$.

We have the following result (see [283], Theorem 4.2).

Theorem 4.65 *Let $T > 0$ and let Hypotheses 4.60 and 4.64 be satisfied. Let $\phi \in C_m(H)$ ($m \geq 0$). Then for every $0 < s \leq T$ the function $(t, x) \to P_{t,s}[\phi](x)$ belongs to $\mathcal{G}^{0,1}([0, s) \times H)$. Moreover, there exists a constant $C(m) > 0$ (possibly depending on T) such that, for all $\phi \in C_m(H)$ and all $0 \leq t < s \leq T$,*

$$|\nabla P_{t,s}[\phi](x)| \leq \frac{C(m)}{(s-t)^{1/2}} \|\phi\|_{C_m}(1 + |x|^m), \quad x \in H, \qquad (4.105)$$

and we have the representation formula (Bismut–Elworthy–Li formula)

$$\langle \nabla P_{t,s}[\phi](x), h \rangle = \mathbb{E}\left[\phi(X(s;t,x))U^h(s,t,x)\right], \quad x, h \in H, \qquad (4.106)$$

where (denoting by ∇_x, as in Proposition 4.61, the partial Gâteaux derivative with respect to x),

$$U^h(s,t,x) := \frac{1}{s-t} \int_t^s \langle \sigma(\tau, X(\tau;t,x))^{-1} \nabla_x X(\tau;t,x)h, dW(\tau) \rangle. \qquad (4.107)$$

Proof The result is a special case of Theorem 4.2 of [283]. □

Remark 4.66 Similarly to what happens in the case of a perturbed Ornstein–Uhlenbeck semigroup treated in the previous subsection (Theorem 4.56), the smoothing result above can be generalized to the case when the initial datum ϕ belongs to $B_m(H)$ for $m \geq 0$. Indeed, one can first prove that the strong Feller property holds by using an approximation procedure like the one used to prove Theorem 1.2 of [486] (see also Theorem 7.7.1 in [179]) and then simply apply the semigroup property.

Moreover, requiring more regularity of the data, it is also possible to prove the Fréchet differentiability in Theorem 4.65. We do not do this here as, in view of Remark 4.63, we prefer to keep more reasonable assumptions on the data. For Fréchet differentiability one can see the references given in Remark 4.62.

Finally, observe that, using the exponential estimate given in Proposition 4.208 (see also [285], Proposition 4.6), we can extend the above result to the case when $T = +\infty$. More precisely, we can show that (4.105) holds for all $0 \leq t \leq s$ with $C(m)$ substituted by $C_1(m)e^{C_2(m)(s-t)}$, for some $C_1(m), C_2(m) > 0$ depending on m and independent of t, s. ∎

We now prove a useful result about the joint continuity of the two-parameter Markov semigroup $P_{t,s}$.

Proposition 4.67 *Let $T > 0$ and $m \geq 0$. Suppose that Hypotheses 4.60 and 4.64 are satisfied. Then we have the following.*

(i) *For every $\phi \in B_m(H)$ the function*

$$\phi_P^0 : [0, T] \times H \to \mathbb{R}, \qquad (t, x) \to P_{t,T}[\phi](x),$$

belongs to $C_m([0, T) \times H)$. If $\phi \in C_m(H)$ then $\phi_P^0 \in C_m([0, T] \times H)$.
(ii) *Given any $\eta \in \mathcal{I}_1$ and $\psi \in \overline{B}_{m,\eta}([0, T) \times H)$, defining $I_0 := \{(t, s) : 0 \leq t \leq s < T\}$, the function*

$$\bar{\psi}_P^0 : I_0 \times H \to \mathbb{R}, \qquad (t, s, x) \to P_{t,s}[\psi(s, \cdot)](x),$$

is measurable, continuous in (t, x) and

$$\psi_P^0 : [0, T] \times H \to \mathbb{R}, \qquad (t, x) \to \int_t^T P_{t,s}[\psi(s, \cdot)](x)ds,$$

belongs to $C_m([0, T] \times H)$.

(iii) For every $\phi \in B_m(H)$ the function

$$\phi_P^1 : [0, T) \times H \to H, \qquad (t, x) \to \nabla P_{t,T}[\phi](x),$$

belongs to $\overline{B}_{m,1/2}([0, T) \times H, H)$. Moreover, ϕ_P^1 is strongly continuous.

(iv) Let η, ψ be as in part (ii) above. Defining $I_1 := \{(t, s) : 0 \le t < s < T\}$, the function

$$\bar{\psi}_P^1 : I_1 \times H \to H, \qquad (t, s, x) \to \nabla P_{t,s}[\psi(s, \cdot)](x),$$

is measurable, continuous in (t, x) and

$$\psi_P^1 : [0, T] \times H \to H, \qquad (t, x) \to \int_t^T \nabla P_{t,s}[\psi(s, \cdot)](x)ds,$$

belongs to $\overline{B}_{m,\eta_1}([0, T) \times H, H)$, where $\eta_1 \in \mathcal{I}_1$ is defined in Proposition 4.21-(iii) for $\eta_0(t) = t^{-1/2}$. Moreover, ψ_P^1 is strongly continuous.

Proof

Proof of (i). Part (i) substantially follows from the statement of Theorem 4.65 proved in [283], however we give the proof for completeness. We first take $\phi \in C_m(H)$ and $(t_n, x_n) \to (t, x) \in [0, T] \times H$, as $n \to +\infty$. Then, by using estimate (1.83) and the fact that we can assume $|x_n| \le |x| + 1/2$, we have

$$\mathbb{E}[|X(T; t_n, x_n) - X(T; t, x)|^2]$$
$$\le 2C_2 \left[|x_n - x|^2 + (1 + |x|^2)|t_n - t|^{\gamma_3} + |e^{(t_n - t)A}x - x|^2 \right].$$

This means that, as $n \to +\infty$, we have $X(T; t_n, x_n) \to X(T; t, x)$ in $L^2(\Omega)$, and hence in probability. Using (1.80) and Lemma 1.51 we thus obtain $\phi_P^0(t_n, x_n) \to \phi_P^0(t, x)$. Moreover, since $\phi \in C_m(H)$, we have

$$|\phi_P^0(t, x)| \le \|\phi\|_{C_m}\mathbb{E}[1 + |X(T; t, x)|^m].$$

Using (1.80) we then easily get the claim.

Let now $\phi \in B_m(H)$. By the strong Feller property (see Remark 4.66) we have, for all $T > \varepsilon > 0$, $\phi_\varepsilon := P_{T-\varepsilon,T}[\phi] \in C_m(H)$. Since $\phi_P^0(t, x) = P_{t,T-\varepsilon}[\phi_\varepsilon](x)$, the claim follows by the first part of the proof and by the arbitrariness of ε.

Proof of (ii). Define $\psi_0(s, x) := \eta(T - s)^{-1}\psi(s, x)$. To prove that $\bar{\psi}_0$ is continuous in (t, x) we argue exactly as in part (i), simply freezing s. The measurability in s for each fixed (t, x) is a consequence of the measurability in (s, ω) of the process

$\psi_0(s, X(s; t, x))$ and Theorem 1.33-(i). Then the measurability in (t, s, x) follows from Lemma 1.18.

From the assumptions on ψ, it immediately follows that $\frac{\bar{\psi}_P^0(t,s,x)}{1+|x|^m}$ is bounded. We now have

$$\psi_P^0(t, x) = \int_t^T \eta(T - s) P_{t,s}[\psi_0(s, \cdot)](x) ds.$$

Since the integrand is continuous in (t, x) we can apply the dominated convergence theorem to get that ψ_P^0 is continuous. Moreover,

$$|\psi_P^0(t, x)| \leq \int_t^T \eta(T - s)|P_{t,s}[\psi_0(s, \cdot)](x)|ds$$

$$\leq C(m)(1 + |x|^m)\|\psi\|_{B_{m,\eta}((0,T]\times H)} \int_t^T \eta(T - s)ds. \qquad (4.108)$$

Proof of (iii). When $\phi \in C_m(H)$ joint strong continuity is already contained in the claim of Theorem 4.65. This implies joint measurability thanks to the Pettis measurability Theorem (see Lemma 1.17-(iv)). When $\phi \in B_m(H)$ we use Remark 4.66 and apply the semigroup property. Finally, the estimate about the singularity when t approaches T follows from (4.105).

Proof of (iv). We immediately deduce from Theorem 4.65 that the function $\bar{\psi}_P^0$ is strongly continuous in (t, x). Moreover, for every fixed $(t, x) \in [0, T] \times H$, it is measurable in s. To see this we use the representation formula (4.106) and (4.107). Indeed, the process $\phi(X(s; t, x))$ is clearly measurable in (s, ω) while the process $U^h(s, t, x)$ defined in (4.107) is also measurable in (s, ω), for instance by Lemma 1.73, since it is mean square continuous and hence stochastically continuous. Mean square continuity follows since, by Hypothesis 4.64 and Proposition 4.61-(iii), defining $\bar{U}^h(s, t, x) := (s - t)U^h(s, t, x)$, we get $\mathbb{E}[|\bar{U}^h(s, t, x)|^2] \leq C_1|h|^2$ and, for $t \leq s_1 \leq s_2 \leq T$,

$$\mathbb{E}[|\bar{U}^h(s_2, t, x) - \bar{U}^h(s_1, t, x)|^2] \leq \int_{s_1}^{s_2} |\sigma(\tau, X(\tau; t, x))^{-1}\nabla_x X(\tau; t, x)h|^2 d\tau$$

$$\leq C_1(s_2 - s_1)|h|^2.$$

The measurability in (t, s, x) then follows from Lemma 1.18.

Moreover, we also have, by (4.105),

$$|\nabla P_{t,s}[\psi_0(s, \cdot)](x)| \leq \|\psi_0\|_{B_m}(s - t)^{-1/2}C(m)(1 + |x|^m).$$

We then get that the integral

$$\psi_P^1(t, x) := \int_t^T \nabla P_{t,s}[\psi(s, \cdot)](x)ds = \int_t^T \eta(T - s)\nabla P_{t,s}[\psi_0(s, \cdot)](x)ds$$

is well defined for all $(t, x) \in [0, T) \times H$ since the integrand is measurable in s and the norm of the integrand is estimated from above by

$$\eta(T - s)(s - t)^{-1/2} \|\psi\|_{B_{m,\eta}((0,T] \times H)} C(m)(1 + |x|^m), \quad s \in (t, T),$$

which is integrable on (t, T) for $t \in [0, T)$. Measurability of ψ_P^1 is immediate from the measurability of the integrand. Strong continuity of ψ_P^1 follows by dominated convergence while boundedness of $\frac{\psi_P^1(T-t,x)}{\eta_1(t)(1+|x|^m)}$ on $(0, T) \times H$ follows, observing that, for (t, x) in this set,

$$|\psi_P^1(t, x)| \leq \|\psi\|_{B_{m,\eta}((0,T] \times H)} C(m)(1 + |x|^m) \int_t^T \eta(T - s)(s - t)^{-1/2} \, ds$$

and noticing that, by Proposition 4.21-(iii),

$$\int_t^T \eta(T - s)(s - t)^{-1/2} \leq \eta_1(T - t).$$

$\square$

Remark 4.68 In contrast to Proposition 4.50, the results in parts (iii) and (iv) give strong continuity in (t, x). This follows from the properties given in Proposition 4.61 and Theorem 4.65, which come from Hypotheses 4.60 and 4.64, which are stronger, in the Ornstein–Uhlenbeck case, than Hypothesis 4.25 as they imply, in particular, that $G = I$, hence Proposition 4.51 applies. Similarly to what was said in Remark 4.52-(i), it is not clear, at this stage, if such joint strong continuity would hold in the Ornstein–Uhlenbeck case under Hypotheses 4.25, 4.29 and 4.32.

The arguments of the proof of part (i) of Proposition 4.67 can also be used to prove part (i) of Proposition 4.50. ∎

Remark 4.69 Proposition 4.67 can be generalized or adapted, by suitably modifying the proof, to the case when $\phi \in UC_m(H)$ $(m \geq 0)$ or when $\psi \in \overline{UC}_{m,\eta}^x((0, T] \times H)$.

In such a case the statement (i) holds with $UC_m^x([0, T] \times H)$ in place of $C_m([0, T) \times H)$. To prove it for $m = 0$ the argument is straightforward. Let $\phi \in UC_b(H)$. We estimate, for $t \in [0, T]$ and $x_1, x_2 \in H$,

$$|P_{t,T}[\phi](x_1) - P_{t,T}[\phi](x_2)| \leq \mathbb{E}\rho_\phi(|X(T; t, x_1) - X(T; t, x_2)|),$$

where ρ_ϕ is a modulus of continuity of ϕ. Since this modulus can be chosen to be concave (see Sect. D.1) then we have, by Jensen's inequality,

$$|P_{t,T}[\phi](x_1) - P_{t,T}[\phi](x_2)| \leq \rho_\phi(\mathbb{E}|X(T; t, x_1) - X(T; t, x_2)|)$$

$$\leq \rho_\phi(\mathbb{E}|X(T; t, x_1) - X(T; t, x_2)|^2) \leq \rho_\phi(C_2|x_1 - x_2|^2),$$

where in the last inequality we used (1.83). The claim then follows. The proof when $m > 0$ is more complicated. Denoting by ρ_ϕ the modulus of continuity of $\phi(x)/(1 + |x|^m)$, we estimate

$$\left| \frac{P_{t,T}[\phi](x_1)}{1 + |x_1|^m} - \frac{P_{t,T}[\phi](x_2)}{1 + |x_2|^m} \right|$$

$$= \left| \mathbb{E}\left[\frac{\phi(X(T; t, x_1))}{1 + |X(T; t, x_1)|^m} \cdot \frac{1 + |X(T; t, x_1)|^m}{1 + |x_1|^m} \right] - \mathbb{E}\left[\frac{\phi(X(T; t, x_2))}{1 + |X(T; t, x_2)|^m} \cdot \frac{1 + |X(T; t, x_2)|^m}{1 + |x_2|^m} \right] \right|$$

$$\leq \left| \mathbb{E}\left[\rho_\phi(|X(T; t, x_1) - X(T; t, x_2)|) \frac{1 + |X(T; t, x_1)|^m}{1 + |x_1|^m} \right] \right|$$

$$+ \|\phi\|_{UC_m} \left| \mathbb{E}\left[\frac{1 + |X(T; t, x_1)|^m}{1 + |x_1|^m} - \frac{1 + |X(T; t, x_2)|^m}{1 + |x_2|^m} \right] \right|,$$

and then apply estimates (1.80) and (1.83).

Similar statements also hold for parts (ii)–(iii)–(iv).   ∎

4.4   Mild Solutions of HJB Equations

In this section we prove two general theorems on the existence and uniqueness of regular (mild) solutions for the HJB equations (4.1) and (4.2). Such theorems are proved following the methods explained in Sect. 4.1 and they use, as the fundamental assumption, the smoothing property of the linear transition semigroup $P_{t,s}$ described, for the parabolic case, in (4.6) (or in (4.7)). Here we will assume that this smoothing property, together with other basic assumptions, holds without making any connections between $P_{t,s}$ and the data of the underlying SDE (Eq. (4.3) in the parabolic case and (4.8) in the elliptic one). In Sect. 4.5, to prove that mild solutions can be approximated by classical solutions, we will require such a connection. In Sects. 4.6 and 4.7, we will then study more deeply the case when the linear part of the HJB equations (4.1) and (4.2) is of Ornstein–Uhlenbeck type, obtaining stronger regularity results.

4.4.1   The Parabolic Case

Let H be a real separable Hilbert space and $T > 0$. The parabolic HJB equation in $[0, T] \times H$ we consider is slightly different from (4.1) since, as announced in Sect. 4.1.1, the dependence on Dv in the Hamiltonian function is through an operator function G (which may reduce to the identity) which maps $[0, T] \times H$ into the set of closed linear operators with dense domain from U to H where U is another real separable Hilbert space. Our Hamiltonian thus has the form $F_0(t, x, v, D^G v)$ instead of $F(t, x, v, Dv)$. We will identify U with its topological dual U^*, so $D^G v$ will always take values in U and $F_0 : [0, T] \times H \times \mathbb{R} \times U \to \mathbb{R}$ is Borel measurable. The HJB equation is the following.

$$
\begin{cases}
v_t + \dfrac{1}{2} \operatorname{Tr} [\Sigma(t, x) D^2 v] + \langle Ax + b(t, x), Dv \rangle + F_0(t, x, v, D^G v) = 0, \\
\qquad\qquad\qquad t \in [0, T), \ x \in H, \\
v(T, x) = \varphi(x), \ x \in H.
\end{cases} \tag{4.109}
$$

Here, as in (4.1), the linear operator $A : D(A) \subset H \to H$ is the infinitesimal generator of a strongly continuous semigroup $\{e^{tA}\}_{t \geq 0}$ while the functions $\Sigma : [0, T] \times H \to \mathcal{L}^+(H)$, $b : [0, T] \times H \to H$, $\varphi : H \to \mathbb{R}$, are Borel measurable (possibly unbounded in the sense that they may be defined on smaller dense subsets).

Note that to denote the derivatives in the HJB equations we always use the symbols D and D^G, unless it is explicitly stated that the solution only possesses Gâteaux derivatives. We do not study Eq. (4.109) directly. We first give a formal argument to introduce its mild form and then we study such a mild form in an abstract way.

4.4.1.1 Formal Derivation of the Mild Form

Given $T > 0$ we take a generalized reference probability space $(\Omega, \mathscr{F}, \{\mathscr{F}_s\}_{s \in [0, T]}, \mathbb{P}, W)$ (where W is a cylindrical Wiener process in a real separable Hilbert space Ξ) and fix $(t, x) \in [0, T] \times H$. Consider, formally, the SDE related to the linear part of (4.109), where[21] $\sigma(t, x) = \sqrt{\Sigma(t, x)}$,

$$
\begin{cases}
dX(s) = [AX(s) + b(s, X(s))] \, ds + \sigma(s, X(s)) \, dW(s), \ s \in [t, T], \\
X(t) = x, \qquad\qquad\qquad\qquad\qquad\qquad\qquad\qquad\quad x \in H.
\end{cases} \tag{4.110}
$$

Again, formally, we define for every $\phi \in B_m(H)$ ($m \geq 0$), the (two-parameter) transition semigroup associated to such an SDE

$$
P_{t,s}[\phi](x) = \mathbb{E}[\phi(X(s; t, x))], \qquad 0 \leq t \leq s \leq T, \tag{4.111}
$$

where $X(\cdot; t, x)$ is the solution of (4.110). Still, formally, the generator of the semigroup $P_{t,s}$ will be the operator

$$
\mathcal{A}(t)\phi = \frac{1}{2} \operatorname{Tr} [\Sigma(t, x) D^2 \phi] + \langle Ax + b(t, x), D\phi \rangle
$$

with domain $D(\mathcal{A}(t))$ that has to be suitably defined, and the function

$$
u(t, x) = P_{t,T}[\varphi](x)
$$

---

[21] Recall that such square root always exists, see e.g. Theorem 12.33 of [521].

should be the solution of (4.109) when $F_0 = 0$. So (4.109) can be formally written as

$$v_t(t, x) + (\mathcal{A}(t)v)(t, x) + F_0(t, x, v(t, x), D^{G(t, \cdot)}v(t, x)) = 0, \qquad v(T, x) = \varphi(x)$$

and applying the variation of constants formula we get, for $t \in [0, T]$, $x \in H$,

$$v(t, x) = P_{t,T}\left[\varphi\right](x) + \int_t^T P_{t,s}\left[F_0(s, \cdot, v(s, \cdot), D^{G(s, \cdot)}v(s, \cdot)\right](x)\, ds. \quad (4.112)$$

This is the equation we want to study, which we call the *mild form* of (4.109). As we mentioned before, we will always use the symbols D^G in such an equation unless it is explicitly stated that the solution only possesses Gâteaux derivatives. Note that, to deal with Eq. (4.112), we only need to consider the data φ, F_0, G and the operators $P_{t,s}$, regardless of their origins. Indeed, the family of operators $P_{t,s}$ could be defined by (4.111) but also, without using the SDE (4.110), directly from the solution of the linear Kolmogorov equation (4.109) with $F_0 = 0$ (see e.g. Chap. 3 of [536] and [161] for results in this direction).

Moreover, with an appropriate choice of the operators $P_{t,s}$ the integral equation (4.112) could be seen as the mild form of a different, possibly more general, semi-linear parabolic equation, e.g. when $P_{t,s}$ is associated with jump diffusions driven by Lévy processes and hence to non-local generators $\mathcal{A}(t)$. Also in these cases the results for (4.112) will only depend, besides the data φ, F_0 G, on the properties of the family $P_{t,s}$.

Hence, to provide results that can be used in different or more general contexts, we will formulate all assumptions in terms of φ, F_0, G and the operators $P_{t,s}$, without specifying their relation to the SDE (4.110) or the operators $\mathcal{A}(t)$. In Sect. 4.4.1.9 we will explain the validity of the assumptions in various interesting cases discussed in Sect. 4.3.

4.4.1.2   Definition of Mild Solution

We now introduce the notion of a mild solution of the HJB equation (4.109).

Definition 4.70 We say that a function $u : [0, T] \times H \to \mathbb{R}$ is a mild solution of the HJB equation (4.109) if, for some $m \geq 0$, the following are satisfied:

(i) There exists an $\eta \in \mathcal{I}_1$ such that $u \in \overline{B}_{m,\eta}^{0,1,G}([0, T] \times H)$.
(ii) Equality (4.112) holds.

It might be better and more precise to say that Definition 4.70 defines a solution of the integral equation (4.112) since, as we discussed before, it is more general than (4.109). However, in the following sections we only study Eq. (4.109), so we decided to keep the description "a mild solution of Eq. (4.109)".

Remark 4.71 Concerning the above definition we observe the following.

(i) The space $\overline{B}_{m,\eta}^{0,1,G}([0, T] \times H)$ is, in some sense, the minimal one where the solution can exist, hence the one where it is easier to find solutions with a minimal set of assumptions. However, if the solutions only exist in this space we are not able to approximate them with classical solutions (see Sect. 4.5.1). Hence, after proving a result about existence and uniqueness in such a space, we will prove that, in many cases, the mild solution u is more regular, e.g. u belongs to $\overline{\mathcal{G}}_{m,\eta}^{0,1,G}([0, T] \times H)$, $\overline{C}_{m,\eta}^{0,1,G}([0, T] \times H)$ or $\overline{UC}_{m,\eta}^{0,1,G}([0, T] \times H)$.

(ii) We point out that several things are implicitly implied by the definition of a mild solution. Firstly, $P_{t,T}[\varphi](x)$ must be well defined for every $x \in H$ and $t \in [0, T]$ and secondly the function $s \to P_{t,s}[F_0(s, \cdot, u(s, \cdot), D^{G(s,\cdot)}u(s, \cdot))](x)$ must be integrable on $[t, T]$ for every $x \in H$ and $t \in [0, T]$. This requires appropriate conditions on the family $P_{t,s}$ and the functions F_0, φ. Moreover, it also suggests that, in Part (i) of Definition 4.70, we could require less as long as all terms on both sides of (4.112) are well defined and measurable.

∎

4.4.1.3 Existence and Uniqueness in B_m Spaces: Assumptions

The first assumption ensures enough regularity of the data F_0 and φ to apply a fixed point argument.

Hypothesis 4.72 The functions $F_0 : [0, T] \times H \times \mathbb{R} \times U \to \mathbb{R}$ and $\varphi : H \to \mathbb{R}$ satisfy the following, for given constants $L, L' > 0$ and $m \geq 0$.

(i) For every $t \in [0, T]$, $x \in H$, $y_1, y_2 \in \mathbb{R}$, $z_1, z_2 \in U$.

$$|F_0(t, x, y_1, z_1) - F_0(t, x, y_2, z_2)| \leq L(|y_1 - y_2| + |z_1 - z_2|_U).$$

(ii) For every $t \in [0, T]$, $x \in H$, $y \in \mathbb{R}$, $z \in U$.

$$|F_0(t, x, y, z)| \leq L'(1 + |x|^m + |y| + |z|_U).$$

(iii) F_0 is Borel measurable.
(iv) $\varphi \in B_m(H)$.

It is clear that, if Hypothesis 4.72 holds for a given $m \geq 0$, then it also holds for all $m_1 > m$.

Remark 4.73

(i) In principle it is also possible to study, with the techniques presented in this chapter, the case when the Hamiltonian F_0 is locally Lipschitz in the last variable. This case is very interesting for applications but the procedure is long and technical and has been investigated only in some special cases (see [105, 307, 438, 442]). We will treat this case in Sect. 4.7 in the special case where the underlying transition semigroup is of Ornstein–Uhlenbeck type (the case studied in [307, 438]) and in Sect. 4.9.2, only presenting the results.

(ii) If the Hamiltonian F_0 is more regular it is possible to obtain better regularity of mild solutions (i.e. C^2 regularity). A result of this type, up to now, has only been obtained in the case when the underlying transition semigroup is of Ornstein–Uhlenbeck type in [306] and is presented in Sect. 4.6.1.1. ∎

We now give the assumptions for the operators $P_{t,s}$. We divide them into three parts. The first establishes the semigroup property and a basic estimate.

Hypothesis 4.74 Let $m \geq 0$ be from Hypothesis 4.72. For every $0 \leq t \leq s \leq T$, $P_{t,s} \in \mathcal{L}(B_m(H))$. The family of operators $P_{t,s}$ satisfies $P_{t,t} = I$ for all $t \in [0, T]$ and the semigroup property

$$P_{t,r}P_{r,s} = P_{t,s}, \qquad \forall 0 \leq t \leq r \leq s \leq T.$$

Moreover, there exists a $C(m) > 0$ such that, for $\phi \in B_m(H)$ and $0 \leq t \leq s \leq T$,

$$\left| P_{t,s}[\phi](x) \right| \leq C(m) \|\phi\|_{B_m} (1 + |x|^m), \quad x \in H. \tag{4.113}$$

Remark 4.75 Assume that the SDE (4.110) has a unique mild solution $X(\cdot; t, x)$ for all $(t, x) \in [0, T] \times H$, and set $P_{t,s}[\phi](x) = \mathbb{E}[\phi(X(s; t, x))]$, for $0 \leq t \leq s \leq T$, $x \in H$. Then, when $m = 0$, the transition semigroup $P_{t,s}$ is a semigroup of contractions and (4.113) is immediately true with $C(0) = 1/2$. When $m > 0$ instead we know that

$$\left| P_{t,s}[\phi](x) \right| \leq \|\phi\|_{B_m} \mathbb{E}\left[1 + |X(s; t, x)|^m \right], \quad x \in H, \ 0 \leq t \leq s \leq T,$$

and so the validity of (4.113) depends on the estimates of the moments of the solutions of (4.110) like the one in (1.80). ∎

The second hypothesis is the key one and the most restrictive in applications. It is needed to ensure the smoothing property of the transition semigroup $P_{t,s}$.

Hypothesis 4.76 Let $m \geq 0$ be from Hypothesis 4.72. Given another real separable Hilbert space U (possibly equal to H), the function G maps $[0, T] \times H$ into the set of linear closed operators (possibly unbounded) from U to H. Moreover, $G(t, \cdot)$ satisfies, for all $t \in [0, T]$, Hypothesis 4.11, and, for all $\phi \in B_m(H)$, we have the following.

(i) The function $P_{t,s}[\phi](\cdot)$ is $G(t,\cdot)$-Gâteaux differentiable for every $0 \le t < s \le T$.

(ii) There exists a $\gamma_G \in \mathcal{I}_2$ (possibly depending on m and T) such that for $0 \le t < s \le T$,

$$\left|\nabla^{G(t,\cdot)} P_{t,s}[\phi](x)\right|_U \le \gamma_G(s-t)\|\phi\|_{B_m}(1+|x|^m), \quad x \in H. \tag{4.114}$$

The third hypothesis is needed to guarantee the joint measurability properties required to meet the definition of a mild solution and to apply Proposition 4.16.

Hypothesis 4.77 Let $m \ge 0$ be from Hypothesis 4.72 and γ_G be from Hypothesis 4.76.

(i) For all $\phi \in B_m(H)$ the map

$$\phi_P^0 : [0,T] \times H \to \mathbb{R}, \quad (t,x) \to P_{t,T}[\phi](x) \tag{4.115}$$

is measurable. Let $I_0 = \{(t,s) : 0 \le t \le s < T\}$. For every $\psi \in \overline{B}_{m,\gamma_G}([0,T) \times H)$ the function

$$\bar{\psi}_P^0 : I_0 \times H \to \mathbb{R}, \quad (t,s,x) \to P_{t,s}[\psi(s,\cdot)](x), \tag{4.116}$$

is measurable.

(ii) Let Hypothesis 4.76 hold. For all $\phi \in B_m(H)$ the map

$$\phi_P^1 : [0,T) \times H \to U, \quad (t,x) \to \nabla^{G(t,\cdot)} P_{t,T}[\phi](x) \tag{4.117}$$

is measurable. Moreover, define $I_1 = \{(t,s) : 0 \le t < s < T\}$ and let ψ be as in point (i) above. The function

$$\bar{\psi}_P^1 : I_1 \times H \to U, \quad (t,s,x) \to \nabla^{G(t,\cdot)} P_{t,s}[\psi(s,\cdot)](x), \tag{4.118}$$

is measurable.

Remark 4.78 Hypothesis 4.77-(i) is a substitute of strong continuity of the semigroup $P_{t,s}$, i.e. continuity of the map

$$[t,T] \to B_m(H), \quad s \to P_{t,s}[\phi] \tag{4.119}$$

for every $\phi \in B_m(H)$, which is not true. Indeed, even in the simple case when $P_{t,s}$ is the Ornstein–Uhlenbeck semigroup with $H = \mathbb{R}$, strong continuity fails, not only in $B_m(H)$ but also in $C_m(H)$ and $UC_m(H)$ (see Proposition B.89-(ii)). Consequently measurability of the map in (4.119) also fails (see Proposition B.89-(iii)). However Hypothesis 4.77 is satisfied in all the main examples we study and it allows us to apply Proposition 4.16. ∎

We will need the following simple lemma.

Lemma 4.79 *Let Hypotheses 4.74, 4.76 and 4.77 hold. Let* $\psi \in \overline{B}_{m,\gamma_G}([0, T) \times H)$
We then have the following.

(i) *The maps*

$$\psi_P^0 : [0, T] \times H \to \mathbb{R}, \quad (t, x) \to \int_t^T P_{t,s}[\psi(s, \cdot)](x)ds,$$

$$\psi_P^1 : [0, T) \times H \to U, \quad (t, x) \to \int_t^T \nabla^{G(t,\cdot)} P_{t,s}[\psi(s, \cdot)](x)ds,$$

are measurable.
(ii) *If the map* $\bar{\psi}_P^0$ *in Hypothesis 4.77-(i) is continuous in* (t, x), *then* ψ_P^0 *is continuous.*

If the map $\bar{\psi}_P^1$ *in Hypothesis 4.77-(ii) is continuous (respectively, strongly continuous) in* (t, x), *then* ψ_P^1 *is continuous (respectively, strongly continuous).*

Proof Part (i) is a direct consequence of Theorem 1.33-(i).

Part (ii) is a consequence of the continuity (respectively, strong continuity), the estimates (4.113) and (4.114) and the dominated convergence theorem. Here we only observe that, to prove the strong continuity of ψ_P^1, i.e. that, for any $h \in U$, the function

$$[0, T) \times H \to \mathbb{R}, \quad (t, x) \to \langle \psi_P^1(t, x), h \rangle_U$$

is continuous, we compute

$$\langle \psi_P^1(t, x), h \rangle_U = \left\langle \int_t^T \nabla^{G(t,\cdot)} P_{t,s}[\psi(s, \cdot)](x)ds, h \right\rangle_U$$

$$= \int_t^T \langle \nabla^{G(t,\cdot)} P_{t,s}[\psi(s, \cdot)](x), h \rangle_U ds,$$

where in the last equality we used Corollary 1.30 applied to the linear functional $T_h = \langle \cdot, h \rangle$. The claim now follows from the dominated convergence theorem. □

4.4.1.4 Existence and Uniqueness of Mild Solutions in B_m Spaces

The following theorem on the existence and uniqueness of mild solutions of Eq. (4.109) is the main result of this section.

Theorem 4.80 *Let* $m \geq 0$ *be such that Hypotheses 4.72, 4.74, 4.76, and 4.77 are satisfied. We have the following.*

(i) *Equation (4.109) has a mild solution* u *(in the sense of Definition 4.70) with* $u \in \overline{B}_{m,\gamma_G}^{0,1,G}([0, T] \times H)$. *Any mild solution* $u^* \in \overline{B}_{m,\gamma_G}^{0,1,G}([0, T] \times H)$ *is equal to* u.

(ii) If $\varphi \in B_m^{1,G(T,\cdot)}(H)$ and the map $(t,x) \to \nabla^{G(t,\cdot)} P_{t,T}[\varphi](x)$ belongs to B_m $([0,T] \times H, U)$, then $\nabla^G u \in B_m([0,T] \times H, U)$.

Proof In both parts we use the contraction mapping principle in a suitable Banach space.

Proof of (i). We consider the product space $B_m ([0,T] \times H) \times \overline{B}_{m,\gamma_G} ([0,T) \times H, U)$ (see Definition 4.24), endowed with the product norm given by the sum of the norms of the factor spaces. In this space we define the operator $\Upsilon = (\Upsilon_1, \Upsilon_2)$:

$$\Upsilon_1 [u, v] (t, x) = P_{t,T}[\varphi](x) + \int_t^T P_{t,s} [F_0 (s, \cdot, u(s, \cdot), v(s, \cdot))] (x)ds, \quad (4.120)$$

$$\Upsilon_2 [u, v] (t, x) = \nabla^{G(t,\cdot)} P_{t,T}[\varphi](x) + \int_t^T \nabla^{G(t,\cdot)} P_{t,s} [F_0 (s, \cdot, u(s, \cdot), v(s, \cdot))] (x)ds. \quad (4.121)$$

The proof will be accomplished in three steps.

Step 1. *The map Υ is well defined.*

We first prove that Υ is well defined on $B_m ([0,T] \times H) \times \overline{B}_{m,\gamma_G} ([0,T) \times H, U)$, with values in itself. Let $(u, v) \in B_m ([0,T] \times H) \times \overline{B}_{m,\gamma_G} ([0,T) \times H, U)$.

Concerning $\Upsilon_1[u, v]$ the first term is in $B_m ([0,T] \times H)$ thanks to Hypothesis 4.77-(i) and (4.113). For the second term we define

$$\psi(s, x) := F_0 (s, x, u(s, x), v(s, x)) \quad (4.122)$$

and prove that such ψ satisfies the requirements of Hypothesis 4.77-(i). Indeed, ψ is Borel measurable since F_0 is measurable (Hypothesis 4.72-(iii)). Moreover, the function

$$(s, x) \to \frac{1}{\gamma_G(T - s)} \frac{\psi(s, x)}{1 + |x|^m}$$

is bounded since, by Hypothesis 4.72-(ii), (writing from now on $\|v\|_{\overline{B}_{m,\gamma_G}}$ for $\|v\|_{\overline{B}_{m,\gamma_G}([0,T) \times H, U)}$)

$$|F_0 (s, x, u(s, x), v(s, x))| \le L' \left(1 + |x|^m + |u(s, x)| + |v(s, x)|_U\right)$$
$$\le L'(1 + |x|^m) \left(1 + \|u\|_{B_m} + \gamma_G(T - s)\|v\|_{\overline{B}_{m,\gamma_G}}\right). \quad (4.123)$$

The above implies, by Hypothesis 4.77-(i), that the map $\bar{\psi}_P^0$, associated to ψ defined in (4.122), is measurable. Hence, by Lemma 4.79-(i), the associated map ψ_P^0, which is equal to the second term of Υ_1, is also measurable. Finally, $\psi_P^0 \in B_m([0,T] \times H)$ by estimate (4.113).

Concerning $\Upsilon_2[u, v]$, the first term belongs to $\overline{B}_{m,\gamma_G} ([0,T) \times H, U)$ thanks to Hypothesis 4.77-(ii) and (4.114). The second term is measurable by Hypothesis 4.77-(ii) (using the properties of ψ discussed above) and Lemma 4.79-(i). To show that the second term belongs to $\overline{B}_{m,\gamma_G}([0, T) \times H, U)$ we estimate, using (4.114) and (4.123),

$$
\frac{1}{\gamma_G(T - t)(1 + |x|^m)} \left| \int_t^T \nabla^{G(t, \cdot)} P_{t,s} \left[F_0(s, \cdot, u(s, \cdot), v(s, \cdot)) \right](x) ds \right|
$$

$$
\leq \frac{L'}{\gamma_G(T - t)} \int_t^T \gamma_G(s - t) \left(1 + \|u\|_{B_m} + \gamma_G(T - s) \|v\|_{\overline{B}_{m,\gamma_G}} \right) ds
$$

$$
\leq \frac{L'}{\gamma_G(T - t)} \left(1 + \|u\|_{B_m} \right) \int_t^T \gamma_G(s - t) ds
$$

$$
+ \|v\|_{\overline{B}_{m,\gamma_G}} \frac{L'}{\gamma_G(T - t)} \int_t^T \gamma_G(s - t) \gamma_G(T - s) ds. \tag{4.124}
$$

The claim now follows from Proposition 4.21-(v) applied for $\beta = 0$.

Step 2. Υ *is a contraction on* $B_m([0, T] \times H) \times \overline{B}_{m,\gamma_G}([0, T) \times H, U)$ *endowed with a suitable equivalent norm.*
We define on $B_m([0, T] \times H)$ the equivalent norm

$$
\|f\|_{\beta, B_m} := \sup_{(t,x) \in [0,T] \times H} \exp(-\beta(T - t)) \frac{|f(t, x)|}{1 + |x|^m},
$$

and on $\overline{B}_{m,\gamma_G}([0, T) \times H, U)$ the equivalent norm

$$
\|f\|_{\beta, \overline{B}_{m,\gamma_G}} := \sup_{(t,x) \in [0,T) \times H} \exp(-\beta(T - t)) \frac{1}{\gamma_G(T - t)} \frac{|f(t, x)|_U}{1 + |x|^m},
$$

where β is a positive constant to be fixed later in the proof. We are going to prove that, for a suitable $\beta > 0$, the map $\Upsilon = (\Upsilon_1, \Upsilon_2)$ is a contraction on $(B_m([0, T] \times H)$, $\|\cdot\|_{\beta, B_m}) \times \left(\overline{B}_{m,\gamma_G}([0, T) \times H, U), \|\cdot\|_{\beta, \overline{B}_{m,\gamma_G}} \right)$ endowed with the product norm given by the sum of the norms of the factor spaces. We start with estimates on Υ_1 (defined in (4.120)). Taking any elements (u_1, v_1) and (u_2, v_2) of $B_m([0, T] \times H) \times B_{m,\gamma_G}([0, T) \times H, U)$ we have, using (4.113), Hypothesis 4.72-(i) and the definition of the equivalent norms,

$$
|\Upsilon_1[u_1, v_1](t, x) - \Upsilon_1[u_2, v_2](t, x)|
$$

$$
= \left| \int_t^T P_{t,s} \left[F_0(s, \cdot, u_1(s, \cdot), v_1(s, \cdot)) - F_0(s, \cdot, u_2(s, \cdot), v_2(s, \cdot)) \right](x) ds \right|
$$

$$
\leq C(m)(1 + |x|^m) \int_t^T \| F_0(s, \cdot, u_1(s, \cdot), v_1(s, \cdot)) - F_0(s, \cdot, u_2(s, \cdot), v_2(s, \cdot)) \|_{B_m} ds
$$

$$
\leq C(m)(1 + |x|^m) \int_t^T L \left[\|u_1(s, \cdot) - u_2(s, \cdot)\|_{B_m} + \|v_1(s, \cdot) - v_2(s, \cdot)\|_{B_m} \right] ds
$$

$$\leq C(m)L(1+|x|^m)\int_t^T \left[e^{\beta(T-s)}\|u_1-u_2\|_{\beta,B_m} + \gamma_G(T-s)e^{\beta(T-s)}\|v_1-v_2\|_{\beta,B_{m,\gamma_G}} \right]ds$$

$$\leq C(m)L(1+|x|^m)\left[\|u_1-u_2\|_{\beta,B_m} + \|v_1-v_2\|_{\beta,B_{m,\gamma_G}} \right]$$

$$\times \left[\left(\int_t^T e^{\beta(T-s)}ds \right) \vee \left(\int_t^T \gamma_G(T-s)e^{\beta(T-s)}ds \right) \right].$$

(Note that in the lines above we had to use Lemma 1.21 to ensure the measurability of functions like $s \rightarrow \|u_1(s,\cdot)-u_2(s,\cdot)\|_{B_m}$.) It follows that

$$\|\Upsilon_1[u_1,v_1]-\Upsilon_1[u_2,v_2]\|_{\beta,B_m}$$
$$= \sup_{t\in[0,T]} \left\{ e^{-\beta(T-t)}\|\Upsilon_1[u_1,v_1](t,\cdot)-\Upsilon_1[u_2,v_2](t,\cdot)\|_{B_m} \right\}$$
$$\leq C(m)LC_1(\beta)\left[\|u_1-u_2\|_{\beta,B_m} + \|v_1-v_2\|_{\beta,\overline{B}_{m,\gamma_G}} \right],$$

where

$$C_1(\beta) := \sup_{t\in[0,T]} \left\{ e^{-\beta(T-t)}\left[\left(\int_t^T e^{\beta(T-s)}ds \right) \vee \left(\int_t^T \gamma_G(T-s)e^{\beta(T-s)}ds \right) \right] \right\}$$
$$= \sup_{t\in[0,T]} \left\{ \frac{1-e^{-\beta(T-t)}}{\beta} \vee \int_t^T \gamma_G(T-s)e^{-\beta(s-t)}ds \right\}$$
$$\leq \frac{1}{\beta} \vee \sup_{t\in[0,T]} \left\{ \int_t^T \gamma_G(T-s)e^{-\beta(s-t)}ds \right\}.$$

Thanks to Proposition 4.21-(iv) we have $C_1(\beta) \rightarrow 0$ as $\beta \rightarrow +\infty$.

We now look at Υ_2. By using (4.114) and Hypothesis 4.72-(i) we get, for any (u_1,v_1) and (u_2,v_2) in $B_m([0,T]\times H) \times \overline{B}_{m,\gamma_G}([0,T)\times H,U)$ (still using Lemma 1.21 to ensure the measurability of functions like $s\rightarrow \|u_1(s,\cdot)-u_2(s,\cdot)\|_{B_m}$),

$$|\Upsilon_2[u_1,v_1](t,x)-\Upsilon_2[u_2,v_2](t,x)|$$
$$= \left| \int_t^T \nabla^{G(t,\cdot)}P_{t,s}\left[F_0(s,\cdot,u_1(s,\cdot),v_1(s,\cdot)) - F_0(s,\cdot,u_2(s,\cdot),v_2(s,\cdot)) \right](x)ds \right|$$
$$\leq (1+|x|^m)\int_t^T \gamma_G(s-t)\|F_0(s,\cdot,u_1(s,\cdot),v_1(s,\cdot)) - F_0(s,\cdot,u_2(s,\cdot),v_2(s,\cdot))\|_{B_m}ds$$
$$\leq L(1+|x|^m)\int_t^T \gamma_G(s-t)\left[\|u_1(s,\cdot)-u_2(s,\cdot)\|_{B_m} + \|v_1(s,\cdot)-v_2(s,\cdot)\|_{B_m} \right]ds$$
$$\leq L(1+|x|^m)\int_t^T \gamma_G(s-t)\Big[e^{\beta(T-s)}\|u_1-u_2\|_{\beta,B_m}$$
$$+ \gamma_G(T-s)e^{\beta(T-s)}\|v_1-v_2\|_{\beta,\overline{B}_{m,\gamma_G}} \Big]ds$$

$$\le L(1+|x|^m)\overline{C}_2(t,\beta)\Big[\|u_1-u_2\|_{\beta,B_m}+\|v_1-v_2\|_{\beta,\overline{B}_{m,\gamma_G}}\Big],$$

where we set

$$\overline{C}_2(t,\beta):=\Big[\Big(\int_t^T\gamma_G(s-t)e^{\beta(T-s)}ds\Big)\vee\Big(\int_t^T\gamma_G(s-t)\gamma_G(T-s)e^{\beta(T-s)}ds\Big)\Big].$$

It follows that

$$\|\Upsilon_2[u_1,v_1](t,x)-\Upsilon_2[u_2,v_2](t,x)\|_{\beta,\overline{B}_{m,\gamma_G}}$$
$$\le LC_2(\beta)\Big[\|u_1-u_2\|_{\beta,B_m}+\|v_1-v_2\|_{\beta,\overline{B}_{m,\gamma_G}}\Big],$$

where, changing variables in the integrals, we have

$$C_2(\beta):=\sup_{t\in[0,T)}\Big\{e^{-\beta(T-t)}\frac{1}{\gamma_G(T-t)}\overline{C}_2(t,\beta)\Big\}$$
$$=\sup_{t\in[0,T)}\Big\{\Big(\frac{1}{\gamma_G(T-t)}\int_0^{T-t}\gamma_G(s)e^{-\beta s}ds\Big)$$
$$\vee\Big(\frac{1}{\gamma_G(T-t)}\int_0^{T-t}\gamma_G(s)\gamma_G(T-t-s)e^{-\beta s}ds\Big)\Big\}.$$

Since $\gamma_G\in\mathcal{I}_2$, by Proposition 4.21-(iv) and (v), we obtain $C_2(\beta)\to 0$ as $\beta\to+\infty$.
We conclude that there exists a $\beta_0>0$ such that, for $\beta\ge\beta_0$

$$\|\Upsilon_1(u_1,v_1)-\Upsilon_1(u_2,v_2)\|_{\beta,B_m}+\|\Upsilon_2(u_1,v_1)-\Upsilon_2(u_2,v_2)\|_{\beta,\overline{B}_{m,\gamma_G}}$$
$$\le\frac{1}{2}\Big[\|u_1-u_2\|_{\beta,B_m}+\|v_1-v_2\|_{\beta,\overline{B}_{m,\gamma_G}}\Big]$$

so Υ is a contraction and thus it has a unique fixed point.

Step 3. *The first component of the fixed point of Υ is the unique mild solution of* (4.109).

We first observe that we do not know that the G-derivative is a closed operator (see Remark 4.15) and that we do not know if the integrands of the second terms of Υ_1 and Υ_2 are measurable as functions of s with values in $B_m(H)$ and $B_m(H,H)$, respectively, as required by Corollary 4.14. Hence we have to apply Proposition 4.16. The required assumptions are satisfied thanks to Hypothesis 4.76 and 4.77. We then obtain that, for all $(u,v)\in B_m([0,T]\times H)\times\overline{B}_{m,\gamma_G}([0,T)\times H,U)$, $(t,x)\in[0,T]\times H$, $h\in U$ $\Upsilon_1[u,v](t,x)$ is G-Gâteaux differentiable and $\langle\Upsilon_2[u,v](t,x),h\rangle_U=\langle\nabla^{G(t,\cdot)}\Upsilon_1[u,v](t,x),h\rangle_U$.

Let now $[\overline{u},\overline{v}]$ be the fixed point of Υ, so $\Upsilon[\overline{u},\overline{v}]=(\overline{u},\overline{v})$. It follows that $\overline{v}(t,x)=\nabla^{G(t,\cdot)}\overline{u}(t,x)$. So the first component $\overline{u}$ of the unique fixed point of Υ satisfies the following:

- $\overline{u}(t,x) = P_{t,T}\left[\varphi\right](x) + \int_t^T P_{t,s}\left[F_0\left(s,\cdot,\overline{u}\left(s,\cdot\right),\nabla^{G(s,\cdot)}\overline{u}\left(s,\cdot\right)\right)\right](x)\,ds;$
- $\overline{u} \in B_m\left([0,T] \times H\right);$
- $\overline{u}$ is G-Gâteaux differentiable and $\nabla^G\overline{u} \in \overline{B}_{m,\gamma_G}\left([0,T) \times H, U\right).$

The above imply that $\overline{u}$ is a mild solution of (4.109).

The required uniqueness is immediate since any other solution $u^* \in \overline{B}_{m,\gamma_G}^{0,1,G}$ $([0,T] \times H)$ must be, thanks to Proposition 4.16, equal to the first component of the fixed point of Υ in $B_m([0,T] \times H) \times \overline{B}_{m,\gamma_G}([0,T) \times H, U)$, hence it must be equal to $\overline{u}$.

Proof of (ii). If $\varphi \in B_m^{1,G(T,\cdot)}(H)$ and the map $(t,x) \to \nabla^{G(t,\cdot)}P_{t,T}[\varphi](x)$ belongs to $B_m([0,T] \times H, U)$, then we can perform the fixed point argument in the space $B_m([0,T] \times H) \times B_m([0,T] \times H, U)$, where the second space now has the norm

$$\|f\|_{\beta,B_m} = \sup_{(t,x) \in [0,T] \times H} \exp\left(-\beta\left(T-t\right)\right) \frac{|f(t,x)|_U}{1+|x|^m}.$$

The same proof works in this product space. Indeed, it is easier and it also holds when $\gamma_G \in \mathcal{I}_1$. □

Remark 4.81 The uniqueness statement of Theorem 4.80 can be generalized in the following way. Let u be the mild solution found in Theorem 4.80 and let u^* be another mild solution of Eq. (4.109) such that $u^* \in \overline{B}_{m,\eta}^{0,1,G}([0,T] \times H)$ for $\eta \in \mathcal{I}_1$ and such that $\gamma_1 := \gamma_G \vee \eta$ belongs to $\mathcal{I}_2$. We then must have $u^* \in \overline{B}_{m,\gamma_1}^{0,1,G}([0,T] \times H)$. Moreover,

$$u^*(t,x) = P_{t,T}[\varphi](x) + \int_t^T P_{t,s}\left[F_0\left(s,\cdot,u^*(s,\cdot),\nabla^{G(s,\cdot)}u^*(s,\cdot)\right)\right](x)\,ds,$$

and using again Proposition 4.16,

$$\nabla^{G(t,\cdot)}u^*(t,x) = \nabla^{G(t,\cdot)}P_{t,T}[\varphi](x) + \int_t^T \nabla^{G(t,\cdot)}P_{t,s}\left[F_0\left(s,\cdot,u^*(s,\cdot),\nabla^{G(s,\cdot)}u^*(s,\cdot)\right)\right](x)\,ds.$$

This implies that $\left(u^*, \nabla^G u^*\right)$ is a fixed point of Υ in the space $B_m([0,T) \times H) \times \overline{B}_{m,\gamma_1}([0,T) \times H, U)$. Arguing as in the case of the weight γ_G, we get that also in $B_m([0,T) \times H, U) \times \overline{B}_{m,\gamma_1}([0,T) \times H, U)$ there exists a unique fixed point of Υ. But this point must be equal to $(u, \nabla^G u)$, hence $u^* = u$. In particular, if $\gamma_G = Ct^{-\theta}$ for some $C > 0$ and $\theta \in (0,1)$, this implies that for any $\beta \in (0,1)$ any solution $u^* \in \overline{B}_{m,\eta}^{0,1,G}([0,T] \times H)$ with $\eta = C_1 t^{-\beta}$, for some $C_1 > 0$, is equal to u. ■

4.4.1.5 Existence and Uniqueness of Mild Solutions in $\mathcal{G}_m$ Spaces

In applications to optimal control it is important to show that the mild solution found in Theorem 4.80 has more regularity properties. This is true under stronger

4.4 Mild Solutions of HJB Equations

assumptions. We start with a result where we show that the mild solution belongs to $\overline{\mathcal{G}}_{m,\gamma_G}^{0,1,G}([0,T] \times H)$.

We first introduce new assumptions about the data F_0 and φ.

Hypothesis 4.82 Let $m \geq 0$ be fixed. The following are satisfied.

(i) $F_0 : [0,T] \times H \times \mathbb{R} \times U \to \mathbb{R}$ satisfies, for m fixed here and for given constants $L, L' > 0$, parts (i) and (ii) of Hypothesis 4.72.

(ii) Denote by (U, τ_U^w) the space U endowed with the weak topology. F_0 is sequentially continuous as a function from $[0,T] \times H \times \mathbb{R} \times (U, \tau_U^w)$ to $\mathbb{R}$.

(iii) $\varphi \in C_m(H)$.

Remark 4.83 The reason why in Hypothesis 4.82-(ii) we require more than the continuity of F is the following. To prove that the mild solution from Theorem 4.80 belongs to $\overline{\mathcal{G}}_{m,\gamma_G}^{0,1,G}([0,T] \times H)$ we apply the fixed point theorem in a space of more regular functions, i.e. $C_m([0,T] \times H) \times \overline{C}_{m,\gamma_G}^s([0,T) \times H, U)$ (see Definition 4.24). To do this we need to know that the integral term in (4.120) is continuous in (t,x) when v there (which is the G-Gâteaux derivative of the solution) is only strongly continuous. However, in such a case the composition of F_0 with v may not be continuous, for example when F_0 depends on the norm of v and U is infinite-dimensional, a case frequent in applications, see e.g. Sects. 2.6.1 and 2.6.4. Thus to have the continuity of the integral term in (4.120), we would need either a strong Feller property of the semigroup $P_{t,s}$ (a case which will be dealt with in Sect. 4.4.1.8), or a stronger continuity property of F_0 like the one used in Hypothesis 4.82-(ii). Indeed, under these assumptions it is easy to see that F_0 is continuous (with respect to the standard topologies) and has the following property: for every $(t,y) \in [0,T] \times \mathbb{R}$, and every measurable map $w : [0,T) \times H \to U$ which is strongly continuous in the x variable, the function $x \to F_0(t,x,y,w(t,x))$ is continuous in H. This is exactly what is needed to apply the fixed point argument, see the proof of Theorem 4.85.

We finally observe that Hypothesis 4.82, even if it is too restrictive to be applied in the cases discussed above, is equivalent to the assumption of "standard" continuity of F_0 when U is finite-dimensional, a case which arises, for example, in many control problems driven by delay equations (see Sect. 2.6.8) where the Hamiltonian function only depends on a finite-dimensional projection of the gradient (see on this [236, 238, 239, 244, 313, 316]). ∎

We now state the assumptions about the semigroup which are variations of Hypotheses 4.74, 4.76 and 4.77.

Hypothesis 4.84 Let $m \geq 0$ be from Hypothesis 4.82. We assume the following.

(i) Hypothesis 4.74 is satisfied substituting everywhere $C_m(H)$ in place of $B_m(H)$.

(ii) Hypothesis 4.76 is satisfied substituting everywhere $C_m(H)$ in place of $B_m(H)$.

(iii) Hypothesis 4.77 is satisfied in the following form.

(a) For all $\phi \in C_m(H)$ the map ϕ_P^0 in (4.115) is continuous. For every $\psi \in \overline{C}_{m,\gamma_G}([0,T) \times H)$, the map ψ_P^0 in (4.116) is measurable in s and continuous in (t,x).

(b) For all $\phi \in C_m(H)$ the map ϕ_P^1 in (4.117) is strongly continuous. Moreover, for every ψ as in point (a) above, the map $\tilde{\psi}_P^1$ in (4.118) is measurable in s and strongly continuous in (t, x).

We have the following result.

Theorem 4.85 *Let $m \geq 0$ be such that Hypotheses 4.82 and 4.84 are satisfied. Then the following are true.*

(i) *Equation (4.109) admits a mild solution u in the sense of Definition 4.70 with $u \in \overline{\mathcal{G}}_{m,\gamma_G}^{0,1,G}([0, T] \times H)$. Any mild solution $u^* \in \overline{\mathcal{G}}_{m,\gamma_G}^{0,1,G}([0, T] \times H)$ is equal to u.*

(ii) *If $\varphi \in \mathcal{G}_m^{1,G(T,\cdot)}(H)$ and the map $(t, x) \to \nabla^{G(t,\cdot)} P_{t,T}[\varphi](x)$ belongs to C_m^s $([0, T] \times H, U)$, then $\nabla^G u \in C_m^s([0, T] \times H, U)$.*

Proof The proof is analogous to the proof of Theorem 4.80. We only explain the changes needed here.

Proof of (i). We perform the fixed point argument in the product space C_m $([0, T] \times H) \times \overline{C}_{m,\gamma_G}^s$ $([0, T) \times H, U)$, endowed with the product norm given by the sum of the norms of the factor spaces. In this space we consider the operator $\Upsilon = (\Upsilon_1, \Upsilon_2)$ defined by (4.120) and (4.121). Once we show that Υ maps C_m $([0, T] \times H) \times \overline{C}_{m,\gamma_G}^s$ $([0, T) \times H, U)$ into itself, the rest of the proof is exactly the same as the proof of part (i) of Theorem 4.80, and will be omitted.

Let $(u, v) \in C_m$ $([0, T] \times H) \times \overline{C}_{m,\gamma_G}^s$ $([0, T) \times H, U)$.

The first term of $\Upsilon_1[u, v]$ is in C_m $([0, T] \times H)$ by Hypothesis 4.84-(iii)-(a) and (4.113). Regarding the second term we define

$$\psi(s, x) := F_0(s, x, u(s, x), v(s, x))$$

and observe that ψ is continuous. Indeed, if $(s_n, x_n) \to (s, x) \in [0, T) \times H$ we have, by the strong continuity of v, that $v(s_n, x_n)$ converges weakly in U to $v(s, x)$. Hence, thanks to Hypothesis 4.82-(i), $\psi(s_n, x_n) \to \psi(s, x)$. Moreover, using Hypothesis 4.82-(ii) and arguing as in (4.123), the function

$$(s, x) \to \frac{1}{\gamma_G(T - s)} \frac{\psi(s, x)}{1 + |x|^m}$$

is bounded. Thus the second term of $\Upsilon_1[u, v]$ belongs to C_m $([0, T] \times H)$ thanks to Hypothesis 4.84-(iii)-(a) and Lemma 4.79-(ii).

Concerning $\Upsilon_2[u, v]$, its first term belongs to $\overline{C}_{m,\gamma_G}^s$ $([0, T) \times H, U)$ thanks to Hypothesis 4.84-(iii)-(b) and (4.114). The second term is strongly continuous by Hypothesis 4.84-(iii)-(b) and Lemma 4.79-(ii) (using the continuity of ψ explained above). To show that the second term belongs to $\overline{C}_{m,\gamma_G}^s$ $([0, T) \times H, U)$ we argue exactly as in the proof of Theorem 4.80, using (4.124).

Proof of (ii). We consider the map Υ in the space $C_m([0, T] \times H) \times C_m^s([0, T] \times H, U)$. The proof that Υ maps this space into itself is completely similar to what is done in part (i) above. The proof that it is a contraction is the same as the proof of part (ii) of Theorem 4.80.   □

Remark 4.86 Hypothesis 4.82 is stronger than Hypothesis 4.72. On the other hand Hypothesis 4.84 does not imply Hypotheses 4.74, 4.76 and 4.77. However, in most examples we consider in this chapter, if Hypothesis 4.84 is satisfied then so are Hypotheses 4.74, 4.76 and 4.77. In such cases Theorem 4.85 shows that, when the data F_0 and φ are more regular, the mild solution u found in Theorem 4.80 is also more regular.   ■

4.4.1.6 Existence and Uniqueness of Mild Solutions in C_m Spaces

We study when the mild solution belongs to $\overline{C}_{m,\gamma_G}^{0,1,G}([0,T] \times H)$. We need to modify the assumptions about the data F_0 and φ and modify Hypotheses 4.74, 4.76 and 4.77.

Hypothesis 4.87 Let $m \geq 0$ be fixed. The following are satisfied.

(i) $F_0 : [0,T] \times H \times \mathbb{R} \times U \to \mathbb{R}$ satisfies, for m fixed here and for given constants $L, L' > 0$, parts (i) and (ii) of Hypothesis 4.72.
(ii) F_0 is continuous.
(iii) $\varphi \in C_m(H)$.

Remark 4.88 Note that the assumption about F_0 here is weaker than the one in Hypothesis 4.82 needed for working in the spaces $\mathcal{G}_m$.   ■

The assumptions about the semigroup are very similar to Hypothesis 4.84.

Hypothesis 4.89 Let $m \geq 0$ be from Hypothesis 4.87. We assume the following.

(i) Hypothesis 4.74 is satisfied substituting everywhere $C_m(H)$ in place of $B_m(H)$.
(ii) Hypothesis 4.76 is satisfied substituting everywhere $C_m(H)$ in place of $B_m(H)$ and *the G-Fréchet derivative* D^G *in place of the G-Gâteaux derivative* ∇^G.
(iii) Hypothesis 4.77 is satisfied in the following form.

(a) For all $\phi \in C_m(H)$ the map ϕ_P^0 in (4.115) is continuous. For every $\psi \in \overline{C}_{m,\gamma_G}([0,T] \times H)$, the map ψ_P^0 in (4.116) is measurable in s and continuous in (t, x).
(b) For all $\phi \in C_m(H)$ the map ϕ_P^1 in (4.117) is *continuous*. Moreover, for ψ as in point (a) above the map ψ_P^1 in (4.118) is measurable in s and *continuous* in (t, x).

We have the following result.

Theorem 4.90 *Let $m \geq 0$ be such that Hypotheses 4.87 and 4.89 are satisfied. Then the following are true.*

(i) *Equation (4.109) admits a mild solution u in the sense of Definition 4.70 with $u \in \overline{C}_{m,\gamma_G}^{0,1,G}([0,T] \times H)$. Any mild solution $u^* \in \overline{C}_{m,\gamma_G}^{0,1,G}([0,T] \times H)$ is equal to u.*

(ii) If $\varphi \in C_m^{1,G(T,\cdot)}(H)$ and the map $(t, x) \to D^{G(t,\cdot)} P_{t,T}[\varphi](x)$ belongs to C_m $([0, T] \times H, U)$, then $D^G u \in C_m([0, T] \times H, U)$.

Proof Similarly to the case of Theorem 4.85, the proof here is analogous to the proof of Theorem 4.80 and we only explain the changes needed here.

Proof of (i). We consider the product space $C_m ([0, T] \times H) \times \overline{C}_{m,\gamma_G} ([0, T) \times H, U)$ endowed with the product norm given by the sum of the norms of the factor spaces. In this space we consider the operator $\Upsilon = (\Upsilon_1, \Upsilon_2)$ defined by (4.120) and (4.121). We only show that Υ maps $C_m ([0, T] \times H) \times \overline{C}_{m,\gamma_G} ([0, T) \times H, U)$ into itself as the rest of the proof is exactly the same as in the proof of part (i) of Theorem 4.80.

Let $(u, v) \in C_m ([0, T] \times H) \times \overline{C}_{m,\gamma_G} ([0, T) \times H, U)$.

The proof that $\Upsilon_1[u, v] \in C_m ([0, T] \times H)$ is the same as the proof of this statement in the proof of Theorem 4.85 once we observe that

$$\psi(s, x) := F_0 (s, x, u(s, x), v(s, x))$$

is obviously continuous thanks to Hypothesis 4.87-(ii).

Concerning $\Upsilon_2[u, v]$, its first term belongs to $\overline{C}_{m,\gamma_G} ([0, T) \times H, U)$ by Hypothesis 4.89-(iii)-(b) and (4.114). The second term is continuous thanks to Hypothesis 4.89-(iii)-(b) and Lemma 4.79-(ii) (using the continuity of ψ). To prove that the second term belongs to $\overline{C}_{m,\gamma_G} ([0, T) \times H, U)$ we argue exactly as in the proof of Theorem 4.80, using (4.124).

Proof of (ii). We consider the map Υ in the space $C_m([0, T] \times H) \times C_m([0, T] \times H, U)$. The proof that Υ maps this space into itself is completely similar to what is done in the proof of point (i). The proof that it is a contraction follows the proof of point (ii) of Theorem 4.80. □

Similarly to what has been observed in Remark 4.86, Theorem 4.90 also applies to most examples studied in this chapter and provides additional regularity of mild solutions when the data are more regular.

Remark 4.91 In the case when $P_{t,s} = R_{s-t}$ is the Ornstein–Uhlenbeck semigroup described in Sect. 4.3.1, it is known that when $\phi \in C_m(H)$, the function $D^G R_s \phi(x)$ is continuous in x (Proposition 4.50-(iii)). However it is not clear, unless additional assumptions are made (see Proposition 4.51-(iii) and Remark 4.52-(i)), if such a function is jointly continuous in (s, x). Hence, in such cases, if we avoid the additional assumptions, Theorem 4.90 cannot be applied. Moreover, Theorem 4.85 cannot be applied as joint strong continuity of $D^G R_s \phi(x)$ is required there. So only Theorem 4.80 applies.

Nevertheless in such cases one can prove more. Indeed, one can modify Hypotheses 4.87 and 4.89 as follows:

• In Hypothesis 4.87-(ii) we require only that F is continuous in x.

- In Hypothesis 4.89-(iii)-(b) we require only that the map $\bar{\phi}_1^P$ is jointly measurable and continuous in x, and that the map $\bar{\psi}_1^P$ is measurable in (s, t) and continuous in x.

With this change of the assumptions it is possible to prove, by a straightforward modification of the proof, that Theorem 4.90 holds with the following variants:

- In Part (i) the mild solution u belongs to $C_m([0, T] \times H)$ while $Du \in \overline{B}_{m,\gamma_G}$ $([0, T) \times H, U)$ and is continuous in x. Uniqueness holds among functions of the same type.
- In Part (ii) the function $D^G u$ belongs to $B_m([0, T] \times H)$ and is continuous in x.

Such a result is similar to what is proved in [105] (see Sect. 4.9.2 and Remark 4.98). ∎

4.4.1.7 Existence and Uniqueness Mild Solutions in UC_m Spaces

This case was the first to be studied in the literature. However, it requires more assumptions and it does not give real advantages with respect to the results of the previous subsection. Indeed, in the case of HJB equations arising from optimal control problems, the results of Theorem 4.90, even if they provide a little less regularity than Theorem 4.94, already allow us to obtain verification theorems for the associated optimal control problem. Moreover, the assumptions are a little more complicated, as one can see in Hypotheses 4.92-(ii) and 4.93-(iii). For these reasons the results of the current subsection (and the related ones in this chapter, see e.g. Remark 4.53-(i) and Sect. 4.6) are presented without going into all the details.

The new assumptions about the data F_0, φ and the semigroup are the following.

Hypothesis 4.92 Let $m \geq 0$ be fixed. The following are satisfied.

(i) $F_0 : [0, T] \times H \times \mathbb{R} \times U \to \mathbb{R}$ satisfies, for m fixed here and for given constants $L, L' > 0$, parts (i) and (ii) of Hypothesis 4.72.

(ii) F_0 is continuous and the function $(s, x, v, w) \to F_0(s, x, v, w)/(1 + |x|^m + |v| + |w|)$ is uniformly continuous in the last three variables, uniformly with respect to the first.

(iii) $\varphi \in UC_m(H)$.

Hypothesis 4.93 Let $m \geq 0$ be from Hypothesis 4.92. We assume the following.

(i) Hypothesis 4.74 is satisfied substituting everywhere $UC_m(H)$ in place of $B_m(H)$.

(ii) Hypothesis 4.76 is satisfied substituting everywhere $UC_m(H)$ in place of $B_m(H)$ and the G-Fréchet derivative D^G in place of the G-Gâteaux derivative ∇^G.

(iii) Hypothesis 4.77 is satisfied in the following form.

(a) For all $\phi \in UC_m(H)$ the map ϕ_P^0 in (4.115) belongs to $UC_m^x([0, T] \times H)$. For every $\psi \in \overline{UC}_{m,\gamma_G}^x([0, T) \times H)$, the map

$$(t, s, x) \to (\gamma_G(T - s))^{-1}\bar{\psi}_P^0(t, s, x)$$

(see (4.116)) belongs to $UC_m^x(I_0 \times H)$, where $I_0 = \{(t, s) : 0 \le t \le s < T\}$.

(b) For every $\phi \in UC_m(H)$ the map ϕ_P^1 in (4.117) belongs to $\overline{UC}_{m,\gamma_G}^x([0, T) \times H)$. Moreover, for ψ as in point (a) above, the map

$$(t, s, x) \to (\gamma_G(T - s))^{-1}(\gamma_G(s - t))^{-1}\bar{\psi}_P^1(t, s, x)$$

(see (4.118)) belongs to $UC_m^x(I_1 \times H)$, where $I_1 = \{(t, s) : 0 \le t < s < T\}$.

We have the following result.

Theorem 4.94 *Let $m \ge 0$ be such that Hypotheses 4.92 and 4.93 are satisfied. Then the following are true.*

(i) *Equation (4.109) admits a mild solution u in the sense of Definition 4.70 with $u \in \overline{UC}_{m,\gamma_G}^{0,1,G}([0, T] \times H)$. Any mild solution $u^* \in \overline{UC}_{m,\gamma_G}^{0,1,G}([0, T] \times H)$ is equal to u.*

(ii) *If $\varphi \in UC_m^{1,G(T,\cdot)}(H)$ and the map $(t, x) \to D^{G(t,\cdot)}P_{t,T}[\varphi](x)$ belongs to $UC_m([0, T] \times H, U)$, then $D^G u \in UC_m^x([0, T] \times H, U)$.*

Proof Similarly to Theorems 4.85 and 4.90, the proof is similar to the proof of Theorem 4.80. We only explain the major changes needed here.

Proof of (i). We consider the product space $UC_m^x([0, T] \times H) \times \overline{UC}_{m,\gamma_G}^x([0, T) \times H, U)$ endowed with the product norm given by the sum of the norms of the factor spaces and we consider the operator $\Upsilon = (\Upsilon_1, \Upsilon_2)$ defined by (4.120) and (4.121). We only argue that Υ maps $UC_m^x([0, T] \times H) \times \overline{UC}_{m,\gamma_G}^x([0, T) \times H, U)$ into itself as the rest of the proof is exactly the same as the proof of part (i) of Theorem 4.80.

Let $(u, v) \in UC_m^x([0, T] \times H) \times \overline{UC}_{m,\gamma_G}^x([0, T) \times H, U)$.

We look first at $\Upsilon_1[u, v]$. The first term of it belongs to $UC_m^x([0, T] \times H)$ by Hypothesis 4.93-(iii)-(a). For the second term we consider (as in the proofs of Theorems 4.80, 4.85, 4.90), the function

$$\psi(s, x) := F_0(s, x, u(s, x), v(s, x)).$$

Straightforward computations, together with Hypothesis 4.92, imply that $\psi \in \overline{UC}_{m,\gamma_G}^x([0, T) \times H)$. Now, the second term of $\Upsilon_1[u, v]$ can be rewritten as, see (4.116),

$$\int_t^T \gamma_G(T - s)\left[(\gamma_G(T - s))^{-1}\bar{\psi}_P^0(t, s, x)\right] ds.$$

Since, by Hypothesis 4.93-(iii)-(a), the term in the square brackets above belongs to $UC_m^x (I_0 \times H)$, the above integral belongs to $UC_m^x ([0, T] \times H)$.

Concerning $\Upsilon_2[u, v]$, the first term belongs to $\overline{UC}_{m,\gamma_G} ([0, T) \times H, U)$ thanks to Hypothesis 4.93-(iii)-(b). The second term can be rewritten as, see (4.118),

$$\int_t^T \gamma_G(T - s)\gamma_G(s - t) \left[(\gamma_G(T - s))^{-1}(\gamma_G(s - t))^{-1} \bar{\psi}_P^1(t, s, x) \right] ds.$$

Since, by Hypothesis 4.93-(iii)-(b), the term in the square brackets above belongs to $UC_m^x (I_1 \times H)$ and $\gamma_G \in \mathcal{I}_2$, the above integral belongs to $\overline{UC}_{m,\gamma_G}^x ([0, T) \times H)$.

Proof of (ii). We consider the map Υ in the space $UC_m^x([0, T] \times H) \times UC_m^x ([0, T] \times H, U)$. The proof that Υ maps this space into itself is completely similar to the proof of part (i). The proof that it is a contraction follows the proof of part (ii) of Theorem 4.80. $\qquad\square$

Similar comments to what was observed in Remark 4.86 also apply to Theorem 4.94.

4.4.1.8 Existence and Uniqueness in the "Strong Feller" Case

In this section we add to the hypotheses of Sect. 4.4.1.2 the strong Feller property of the semigroup $P_{t,s}$, a property which is satisfied in many applied problems. For instance, the strong Feller property holds when $U = H$ and $G = I$ in Hypothesis 4.76. Moreover, under reasonable assumptions, the strong Feller property of $P_{t,s}$ also holds in examples discussed in Sects. 2.6.1 (see e.g. [103, 104] and [106], Chaps. 6-7), 2.6.2 (e.g. when $b = 0$, using the results of Sect. 4.3.1 for the case when $P_{t,s}$ is of Ornstein–Uhlenbeck type), 2.6.4 (see [155, 157]), and 2.6.5 (see [158]).

Hypothesis 4.95 Let m be from Hypothesis 4.72. The family $P_{t,s}, 0 \le t \le s \le T$, is strongly Feller (see Definition 1.159 and Lemma 1.160) in the sense that

$$P_{t,s}(B_m(H)) \subset C_m(H), \qquad \forall 0 \le t < s \le T.$$

We have the following result.

Theorem 4.96 *Let $m \ge 0$ be from Hypothesis 4.72. Let Hypotheses 4.72, 4.74, 4.76, 4.77 and 4.95 hold. We have the following.*

(i) *The mild solution u obtained in Theorem 4.80-(i) (or (ii)) of Eq. (4.109) is continuous in the variable $x \in H$ for all $t \in [0, T)$.*

(ii) *Let $\varphi \in C_m(H)$ and assume that Hypothesis 4.84 holds. Assume in addition that, for $\psi \in \overline{B}_{m,\gamma_G} ([0, T) \times H)$, the function*

$$(t, x) \to \int_t^T P_{t,s}[\psi(s, \cdot)](x) ds, \quad (respectively, (t, x) \to$$

$$\int_t^T \nabla^{G(t,\cdot)} P_{t,s}[\psi(s, \cdot)](x)) ds)$$

is continuous in $[0, T] \times H$ (respectively, strongly continuous in $[0, T) \times H$).
Then $u \in \overline{\mathcal{G}}_{m,\gamma_G}^{0,1,G}([0, T] \times H)$.

(iii) Assume, in addition to the hypotheses of part (ii), that $\varphi \in \mathcal{G}_m^{1,G}(H)$ and that the
map $(t, x) \to \nabla^{G(t,\cdot)} P_{t,T}[\varphi](x)$ belongs to $C_m^s([0, T] \times H, U)$. Then $\nabla^G u \in C_m^s([0, T] \times H, U)$.

Proof The proof of (i) easily follows from the strong Feller property, estimate (4.113)
and the dominated convergence theorem.

The proofs of (ii) and (iii) are the same (and even easier, thanks to the strong
assumptions made here) as those of Theorem 4.85 (parts (i) and (ii)).   □

Remark 4.97 Observe that the joint continuity (strong continuity) required in Theorem 4.96-(ii) is usually satisfied when the strong Feller property is true. This holds,
for example, in the three cases presented in Sect. 4.3.

Observe also that in Theorem 4.96, by appropriately changing the assumptions
and the claims, one could prove statements similar to (ii) and (iii) in C_m spaces. ■

Remark 4.98 In [105, 179] the authors prove the existence and uniqueness of a mild
solution applying the fixed point theorem in the space Z_T, which is the subspace of
$B_b^{1,I}([0, T] \times H)$ whose elements are continuous and bounded functions such that,
for all $t \in [0, T]$, $u(t, \cdot) \in UC_b^1(H)$. It is possible to prove a version of Theorem
4.80 which establishes existence and uniqueness of solutions in such a space if
one requires continuity of φ and F_0 in x, see Remark 4.91 for a similar result.
Similarly, it is also possible to prove a version of Theorem 4.96 in the space Z_T.
Such results may be useful when joint continuity of the maps ϕ_P^1 in (4.117) and ψ_P^1
in (4.118) are difficult to obtain, like in the Ornstein–Uhlenbeck case discussed in
Remark 4.91.   ■

4.4.1.9 Examples

Example 4.99 Assume that $P_{t,s} = R_{s-t}$, where R_t is the Ornstein–Uhlenbeck semigroup considered in Sect. 4.3.1. Assume that Hypotheses 4.25, 4.29 and 4.32 are
satisfied. Then, Hypothesis 4.74 (together with its variants in Hypotheses 4.84-(i),
4.89-(i), 4.93-(i)) is clearly satisfied as a consequence of the definition of R_t. Moreover, Hypothesis 4.76 (together with its variants in Hypotheses 4.84-(ii), 4.89-(ii),
4.93-(ii)) is satisfied thanks to Theorem 4.41.

Hypothesis 4.77 (together with its variant discussed in Remark 4.91) is satisfied
thanks to Proposition 4.50. The variants of Hypothesis 4.77 given in Hypotheses
4.84-(iii), 4.89-(iii), 4.93-(iii) are satisfied if we require that Hypothesis 4.29 also
holds for $U = H$ and $G = I$. This follows from Propositions 4.50, 4.51, and Remark
4.53-(i).

Hence, in this Ornstein–Uhlenbeck case, assuming Hypothesis 4.72, Theorem 4.80 applies, and also its version discussed in Remark 4.91.

To apply the other theorems we need to require that Hypothesis 4.29 also holds for $U = H$ and $G = I$. In this case, assuming, in place of Hypothesis 4.72, Hypothesis 4.82 (respectively Hypothesis 4.87 or 4.92), then Theorem 4.85, (respectively 4.90, 4.94) applies, too. Moreover, since Hypothesis 4.29 also holds for $U = H$ and $G = I$, then also Theorem 4.96 applies. ∎

Example 4.100 Assume that $P_{t,s} = P_{s-t}$, where P_t is the perturbed Ornstein–Uhlenbeck semigroup studied in Sect. 4.3.2. Assume that $U = H$ and $G = I$ and that for such G, Hypotheses 4.25, 4.29, 4.32, 4.54 and 4.55 are satisfied. Then Hypothesis 4.74 (together with its variants in Hypotheses 4.84-(i), 4.89-(i), 4.93-(i)) is satisfied by the definition of P_t. Hypothesis 4.76 (together with its variants in Hypotheses 4.84-(ii), 4.89-(ii), 4.93-(ii)) is satisfied thanks to Theorem 4.56 while the fact that Hypothesis 4.77 (together with its variants in Hypotheses 4.84-(iii), 4.89-(iii), 4.93-(iii)) is satisfied can be proved arguing similarly as in the proofs for the standard Ornstein–Uhlenbeck case in Propositions 4.50, 4.51 and in Remark 4.53.

Hence, for the perturbed Ornstein–Uhlenbeck case, assuming Hypothesis 4.72 (also, when needed, Hypotheses 4.82, 4.87, 4.92), the statements of Theorems 4.80, 4.85, 4.90, 4.94 hold. Since here the strong Feller property holds the same is true for Theorem 4.96.

Finally, arguing as in Remark 4.57, the above might be proved avoiding Hypothesis 4.55. ∎

Example 4.101 Assume that $P_{t,s}$ is the semigroup described in Sect. 4.3.3 and that Hypotheses 4.60 and 4.64 are satisfied. Then Hypothesis 4.74 (together with its variants in Hypotheses 4.84-(i), 4.89-(i), 4.93-(i)) is true thanks to estimate (1.80). Moreover, using Theorem 4.65 and Remark 4.66, we see that Hypothesis 4.76 (together with its variants in Hypotheses 4.84-(ii), 4.89-(ii), 4.93-(ii)) also holds with $U = H, G = I$ and $\gamma_G(t) = Ct^{-1/2}$ for some $C > 0$. Finally, by Proposition 4.67 and Remark 4.69, we see that Hypothesis 4.77 (together with its variants in Hypotheses 4.84-(iii), 4.89-(iii), 4.93-(iii)) is satisfied.

Thus, under Hypotheses 4.60 and 4.64 (also, when needed, Hypothesis 4.82), the statements of Theorems 4.80 and 4.85 hold. Theorems 4.90 and 4.94 do not apply since we do not have the continuity of the derivative. However, since here $U = H, G = I$ and the strong Feller property holds, Theorem 4.96 applies giving us, in particular, that the solution is in $\mathcal{G}_m^{0,1,G}([0, T] \times H)$ when the initial datum is continuous.

This result is in line with what is obtained using the BSDE approach in Chap. 6, Theorem 6.32 (see also Theorem 4.2 of [283]). ∎

4.4.2   The Elliptic Case with a Big Discount Factor

Similarly to the parabolic case we take a real separable Hilbert space H and we consider the following elliptic HJB equation in H which is slightly different from (4.2) since in the Hamiltonian function the dependence on Dv is through an operator function $G(x)$ (which may reduce to the identity), hence the Hamiltonian is $F_0(x, v, D^G v)$ instead of $F(x, v, Dv)$.

$$\lambda v - \frac{1}{2} \operatorname{Tr} [\Sigma(x) D^2 v] - \langle Ax + b(x), Dv \rangle - F_0(x, v, D^G v) = 0, \quad x \in H.$$
(4.125)

Recall that, to denote the derivatives in the HJB equations, we always use the symbols D and D^G unless it is precisely stated that the solution only possesses Gâteaux derivatives. We introduce a mild form of (4.125) and then we study the mild form of the equation. As in Sects. 3.6.3 and 6.9, we provide results for λ sufficiently big. A sharper result, which holds for all $\lambda > 0$, will be given in Sect. 4.6.2.2 for a specific Ornstein–Uhlenbeck case.

4.4.2.1   Formal Derivation of the Mild Form

Let us fix $x \in H$ and take a generalized reference probability space $(\Omega, \mathscr{F}, \{\mathscr{F}_s\}_{s \in [0,+\infty)}, \mathbb{P}, W)$ (where W is a cylindrical Wiener process in a real separable Hilbert space Ξ). We describe here a formal argument to obtain the mild form of (4.125). Consider the SDE related to the linear part of (4.125), where $\sigma(x) = \sqrt{\Sigma(x)}$

$$\begin{cases} dX(s) = [AX(s) + b(X(s))] ds + \sigma(X(s)) dW(s), & s \in [0, +\infty), \\ X(0) = x, & x \in H, \end{cases}$$
(4.126)

and denote by $X(\cdot; x)$ its solution (we omit the initial time since it is always equal to 0 in this subsection). For every $\phi \in B_m(H)$, $m \geq 0$, the (one parameter) transition semigroup associated to such SDE is

$$P_s[\phi](x) = \mathbb{E}[\phi(X(s; x))], \quad s \geq 0.$$
(4.127)

The generator of the semigroup P_s is, formally, defined by

$$\mathcal{A}\phi = \frac{1}{2} \operatorname{Tr} [\Sigma(x) D^2 \phi] + \langle Ax + b(x), D\phi \rangle$$

and then, for any $g \in B_m(H)$, the function

$$u(x) = (\lambda I - \mathcal{A})^{-1}[g](x) = \int_0^{+\infty} e^{-\lambda s} P_s[g](x) ds$$
(4.128)

(here we use the standard expression for the resolvent, see e.g. [479], proof of Theorem 3.1, pp. 8–9) is, still formally, the solution of the linear equation $\lambda u(x) - (\mathcal{A}u)(x) = g(x)$. Hence, taking $g(x) = F_0(x, v(x), D^G v(x))$, (4.125) can be rewritten as

$$\lambda v(x) - (\mathcal{A}v)(x) = F_0(x, v(x), D^G v(x))$$

and, applying the formula (4.128) for the resolvent, as

$$v(x) = \int_0^{+\infty} e^{-\lambda s} P_s \left[F_0(\cdot, v(\cdot), D^G v(\cdot)) \right](x)ds, \quad x \in H. \tag{4.129}$$

We call this equation the *mild form* of (4.125). According to our convention, to denote the derivatives in this mild version of the HJB equation, we always use the symbols D and D^G unless it is stated that the solution only possesses Gâteaux derivatives. As in the parabolic case, Eq. (4.129) is completely determined by F_0 and the operators P_s. Thus the assumptions will be formulated only in terms of F_0 and the family $\{P_s, \ s \geq 0\}$. In this way our results, once the required assumptions are satisfied, can be applied to cases where the integral equation (4.129) is associated to other types of SDE, e.g. when P_s is associated to jump diffusions driven by Lévy processes and hence to non-local generators $\mathcal{A}$. Examples when the assumptions are satisfied will be discussed in Sect. 4.4.2.9.

4.4.2.2 Definition of Mild Solution

Here is the notion of a mild solution of the HJB equation (4.125).

Definition 4.102 We say that a function $u : H \to \mathbb{R}$ is a mild solution of the HJB equation (4.125) if, for some $m \geq 0$, the following are satisfied:

(i) $u \in B_m^{1,G}(H)$.
(ii) Equality (4.129) holds.

Definition 4.102 introduces a solution of the integral equation (4.129) which, as we discussed before, is more general than (4.125). Since in the following sections we only study Eq. (4.125), we keep the description "a mild solution of Eq. (4.125)".

Remark 4.103

(i) Similarly to what we observed in the parabolic case, the space $B_m^{1,G}(H)$ is in a sense the minimal one where the solution can exist and so the one where it is easier to find solutions under a minimal set of assumptions. However, it may not be possible to approximate mild solutions in this space by classical solutions (see Sect. 4.5.2). Hence we will prove that in many cases solutions are more regular, e.g. they belong to $\mathcal{G}_m^{1,G}(H)$, $C_m^{1,G}(H)$ or $UC_m^{1,G}(H)$.

(ii) The definition of a mild solution implicitly implies that the function $s \rightarrow$ $e^{-\lambda s} P_s [F_0 \left(\cdot, u(\cdot), D^G u(\cdot) \right)](x)$ is integrable and its integral over $[0, +\infty)$ is measurable in H. Moreover, Definition 4.102 would still make sense without Part (i) as long as the right-hand side of (4.129) makes sense and is measurable.

∎

4.4.2.3   Existence and Uniqueness in B_m Spaces: Assumptions

The assumptions here are very similar to the assumptions in the parabolic case. The main differences are the independence of the data of the time variable t and the need for suitable exponential estimates for big t.

The first assumption prescribes conditions on F_0 needed to apply our fixed point argument.

Hypothesis 4.104 The function $F_0 : H \times \mathbb{R} \times U \rightarrow \mathbb{R}$ satisfies the following.

(i) There exists a constant $L > 0$ such that

$$|F_0(x, y_1, z_1) - F_0(x, y_2, z_2)| \leq L \left(|y_1 - y_2| + |z_1 - z_2|_U \right)$$

for every $x \in H$, $y_1, y_2 \in \mathbb{R}$, $z_1, z_2 \in U$.

(ii) There exist $L' > 0$ and $m \geq 0$ such that

$$|F_0(x, y, z)| \leq L' \left(1 + |x|^m + |y| + |z|_U \right)$$

for every $x \in H$, $y \in \mathbb{R}$, $z \in U$.

(iii) F_0 is Borel measurable.

Remark 4.105

(i) Using fixed point techniques similar to those employed in this section and suitable a priori estimates, it is also possible to study the case when F_0 is locally Lipschitz in the last variable. This case is very interesting in applications, however, up to now, it has only been studied in a special case (see [106, 107]) which is briefly presented in Sect. 4.9.2.

(iii) If F_0 is more regular, C^2 regularity of mild solutions can be obtained. Such a result has only been proved in the case when the underlying transition semigroup is of Ornstein–Uhlenbeck type in [317]. It is explained in Sect. 4.6.2.1.

∎

We now give the assumptions for the operators P_s. They are divided into three parts. The first establishes the semigroup property and a basic estimate.

Hypothesis 4.106 Let $m \geq 0$ be from Hypothesis 4.104. For every $s \geq 0$ $P_s \in$ $\mathcal{L}(B_m(H))$. The family of operators P_s satisfies $P_0 = I$ and the semigroup property

$$P_t P_s = P_{t+s}, \qquad \forall s, t \geq 0.$$

Moreover, there exist $C(m) > 0$ and $a(m) \in \mathbb{R}$ such that for every $\phi \in B_m(H)$ and $s \geq 0$,

$$|P_s[\phi](x)| \leq C(m) e^{a(m)s} \|\phi\|_{B_m} (1 + |x|^m), \quad x \in H. \tag{4.130}$$

Remark 4.107 Assume that the SDE (4.126) has a unique mild solution $X(\cdot; x)$ for all $x \in H$ and set $P_s[\phi](x) = \mathbb{E}[\phi(X(s; x))]$, for $s \geq 0$, $x \in H$. Then, when $m = 0$, (4.130) is immediately true with $C(0) = 1/2$ and $a(0) = 0$. When $m > 0$, we have

$$|P_s[\phi](x)| \leq \|\phi\|_{B_m} \mathbb{E}\left[(1 + |X(s; x)|^m)\right], \quad x \in H, \ s \geq 0.$$

Thus the validity of (4.130) depends on the estimates of the moments of the solution of (4.126), which must hold for all $s \geq 0$ with an exponential growth. A result of this kind can be found, for example, in Proposition 4.6 in [285] under suitable assumptions about the coefficients. We observe here that such an estimate can be deduced[22] from an analogous finite horizon estimate. We briefly describe the argument for the case when the coefficients of the SDE (4.126) satisfy Hypothesis 1.149 without the dependence on s and a. From Theorem 1.152 we know that, for $\xi \in L^m(\Omega, \mathscr{F}_0, \mathbb{P}; H), m \geq 2$, the SDE (4.126) has a unique solution $X(\cdot; \xi) = X(\cdot; 0, \xi)$. The same holds if the initial time is $t \geq 0$, in this case we denote the mild solution by $X(\cdot; t, \xi)$. By uniqueness we have $X(s; \xi) = X(s; t, X(t; \xi))$ for $s \geq t \geq 0$. Moreover, by Theorem 1.152 we have the estimate (see (1.79))

$$\mathbb{E}[|X(s; \xi)|^m] \leq c_m(1 + \mathbb{E}[|\xi|^m]), \qquad \forall s \in [0, 1]. \tag{4.131}$$

Hence, using uniqueness we have

$$\mathbb{E}\left[|X(s; \xi)|^m\right] = \mathbb{E}\left[|X(s; 1, X(1; \xi))|^m\right], \ \forall s \in (1, 2].$$

Now we use Theorem 1.152 and estimate (1.79) when the initial time is $t = 1$, yielding

$$\mathbb{E}\left[|X(s; 1, X(1; \xi))|^m\right] \leq c_m(1 + \mathbb{E}[|X(1; \xi)|^m]).$$

The last two equations and (4.131) then give

$$\mathbb{E}\left[|X(s; \xi)|^m\right] \leq c_m + c_m^2(1 + \mathbb{E}[|\xi|^m])), \ \forall s \in (1, 2].$$

Arguing by induction we get

---

[22]This was suggested to us by Mauro Rosestolato.

$$\mathbb{E}\left[|X(s;\xi)|^m\right] \le (c_m + \ldots + c_m^{n+1}) + c_m^{n+1}\mathbb{E}[|\xi|^m]$$
$$\le n(c_m \vee 1)^{n+1} + c_m^{n+1}\mathbb{E}[|\xi|^m]$$
$$= ne^{(n+1)\log(c_m\vee 1)} + e^{(n+1)\log c_m}\mathbb{E}[|\xi|^m], \qquad \forall s \in (n, n+1], \ \forall n \in \mathbb{N}.$$

Then, by suitably defining $C(m)$ and $a(m)$, we obtain

$$\mathbb{E}\left[|X(s;\xi)|^m\right] \le C(m)e^{a(m)s}(1 + \mathbb{E}[|\xi|^m])$$

for all $s \ge 0$, which is the required estimate.   ∎

The second hypothesis ensures the smoothing property of the transition semigroup P_s.

Hypothesis 4.108 Let $m \ge 0$ be from Hypothesis 4.104. Given another real separable Hilbert space U (possibly equal to H), the function G maps H into the set of closed linear operators (possibly unbounded) from U to H and satisfies Hypothesis 4.11. Moreover, for all $\phi \in B_m(H)$, we have the following.

(i) The function $P_s[\phi](\cdot)$ is G-Gâteaux differentiable for every $s > 0$.
(ii) There exist $\gamma_G \in \mathcal{I}_1$ (possibly depending on m) and $a(m) \in \mathbb{R}$ such that for every $s > 0$,

$$\left|\nabla^G P_s[\phi](x)\right|_U \le \gamma_G(s)e^{a(m)s}\|\phi\|_{B_m}(1 + |x|^m), \quad x \in H. \qquad (4.132)$$

To avoid unnecessary complications we assume that the constants $a(m)$ in Hypothesis 4.106 and here are the same.

Remark 4.109 In contrast to Hypotheses 4.74 and 4.76 in the parabolic case, here we need to require (in (4.130) and (4.132)) a prescribed exponential growth at infinity for both $P_s[\phi]$ and $D^G P_s[\phi]$. For the case discussed in Remark 4.107, (4.130) corresponds to exponential estimates for the moments of the solutions of (4.126). These exponential growth estimates are needed to guarantee the integrability over $\mathbb{R}^+$ of the right-hand side of (4.129).

Regarding the growth estimates in Hypotheses 4.106 and 4.108 we note that we generically require $a(m) \in \mathbb{R}$. In fact, for transition semigroups (see 1.6), which are the ones we will be dealing with from the next section on, we always have $a(m) \ge 0$ in (4.130) since, by (1.95), constant functions are invariant for such semigroups. Nevertheless, requiring $a(m) \ge 0$ is not necessary for Theorem 4.112 below. We keep the (a priori weaker) assumption $a(m) \in \mathbb{R}$, which may be satisfied in other cases (e.g. for semigroups P_s which are not transition semigroups and which have negative type; a case arising when the operator $\mathcal{A}$ is strictly dissipative), and may give rise to a sharper result in Theorem 4.112, whose conclusion may (possibly) also hold for negative λ.

Finally, in the elliptic case we do not need $\gamma_G \in \mathcal{I}_2$ but only to be in $\mathcal{I}_1$, since here the contraction estimates are simpler. Indeed, we may also avoid requiring γ_G to be decreasing with no change in the results. We keep this requirement since in all our examples γ_G is decreasing. ∎

The third hypothesis guarantees the joint measurability properties required in the definition of a mild solution and needed to apply Proposition 4.16.

Hypothesis 4.110 Let $m \geq 0$ be from Hypothesis 4.104 and γ_G be from Hypothesis 4.108.

(i) For every $\phi \in B_m(H)$ the function

$$\bar{\phi}_P^0 : [0, +\infty) \times H \to \mathbb{R}, \qquad (s, x) \to P_s[\phi](x) \qquad (4.133)$$

is measurable.

(ii) For every $\phi \in B_m(H)$ the function

$$\bar{\phi}_P^1 : (0, +\infty) \times H \to U, \qquad (s, x) \to \nabla^G P_s\lfloor\phi\rfloor(x) \qquad (4.134)$$

is measurable.

The observations made in Remark 4.78 also apply (with obvious changes) to Hypothesis 4.110-(i).

The following lemma is the infinite horizon analogue of Lemma 4.79.

Lemma 4.111 *Let Hypotheses 4.106, 4.108 and 4.110 hold. Let $\phi \in B_m(H)$ and $\lambda > a(m)$, where $a(m)$ is the constant from (4.130) and (4.132). We then have the following.*

(i) The functions

$$\phi_P^0 : H \to \mathbb{R}, \qquad x \to \int_0^{+\infty} e^{-\lambda s} P_s[\psi](x)ds,$$

$$\phi_P^1 : H \to U, \qquad x \to \int_0^{+\infty} e^{-\lambda s} \nabla^G P_s[\psi](x)ds,$$

are measurable.

(ii) If the function $\bar{\phi}_P^0$ in Hypothesis 4.110-(i) is also continuous in x, then ϕ_P^0 is continuous.

If the function $\bar{\phi}_P^1$ in Hypothesis 4.110-(ii) is also continuous (respectively, strongly continuous) in x, then ψ_P^1 is continuous (respectively, strongly continuous).

Proof The proof is the same (and even simpler) as the proof of Lemma 4.79. □

4.4.2.4 Existence and Uniqueness of Mild Solutions in B_m Spaces

In this section we prove the existence and uniqueness theorem for mild solutions of Eq. (4.125).

Theorem 4.112 *Let $m \geq 0$ be such that Hypotheses 4.104, 4.106, 4.108 and 4.110 are satisfied. Then there exists a $\lambda_0 \geq 0$ such that for every $\lambda > \lambda_0$, Eq. (4.125) admits a unique mild solution u in the sense of Definition 4.102.*

Proof The proof is similar to the proof of Theorem 4.80. However, we provide its full details since the result is important.

We fix $m \geq 0$ and consider the product (Banach) space $B_m(H) \times B_m(H, U)$ endowed with the product norm given by the sum of the norms of the factor spaces. We define the operator $\Upsilon = (\Upsilon_1, \Upsilon_2)$ on $B_m(H) \times B_m(H, U)$ as

$$\Upsilon_1 [u, v] (x) = \int_0^{+\infty} e^{-\lambda s} P_s [F_0(\cdot, u(\cdot), v(\cdot))] (x) ds, \qquad (4.135)$$

$$\Upsilon_2 [u, v] (x) = \int_0^{+\infty} e^{-\lambda s} \nabla^G P_s [F_0(\cdot, u(\cdot), v(\cdot))] (x) ds. \qquad (4.136)$$

The proof is accomplished in three steps.

Step 1. *The map Υ is well defined for $\lambda > a(m)$.*

Let $\lambda > a(m)$ and $(u, v) \in B_m(H) \times B_m(H, U)$. We prove that $\Upsilon[u, v]$ is well defined and belongs to $B_m(H) \times B_m(H, U)$. Concerning $\Upsilon_1(u, v)$ we define $\psi(x) := F_0 (x, u(x), v(x))$ and observe that ψ is Borel measurable thanks to Hypothesis 4.104-(iii). Moreover, the function

$$x \rightarrow \frac{\psi(x)}{1 + |x|^m}$$

is bounded since, by Hypothesis 4.104-(ii),

$$F_0 (x, u(x), v(x)) \leq L' \left(1 + |x|^m + |u(x)| + |v(x)|_U\right)$$
$$\leq L'(1 + |x|^m) \left(1 + \|u\|_{B_m} + \|v\|_{B_m}\right). \qquad (4.137)$$

Thus, by Hypothesis 4.110-(i), the function $(s, x) \rightarrow P_s [F_0(\cdot, u(\cdot), v(\cdot))] (x)$ is measurable. Hence, by Lemma 4.111-(i) and estimate (4.130), $\Upsilon_1[u, v]$ is well defined for $\lambda > a(m)$ and it belongs to $B_m(H)$.

Concerning $\Upsilon_2[u, v]$, in view of the fact that $\psi \in B_m(H)$, the integral is well defined for $\lambda > a(m)$ and belongs to $B_m(H, U)$ by estimate (4.132), Hypothesis 4.110-(ii) and Lemma 4.111-(i). In particular, using first (4.132) and then (4.137), we get

$$\frac{1}{1+|x|^m} \left| \int_0^{+\infty} e^{-\lambda s} \nabla^G P_s[\psi](x)ds \right| \leq \int_0^{+\infty} \gamma_G(s) e^{(\lambda - a(m))s} \|\psi\|_{B_m} ds$$

$$\leq L'\left(1 + \|u\|_{B_m} + \|v\|_{B_m}\right) \int_0^{+\infty} \gamma_G(s) e^{-(\lambda - a(m))s} ds.$$

Step 2. Υ *is a contraction in* $B_m(H) \times B_m(H, U)$ *for sufficiently big* λ.
We begin with the estimate for Υ_1. Taking any elements (u_1, v_1) and (u_2, v_2) of $B_m(H) \times B_m(H, U)$ we have, using (4.130) and then Hypothesis 4.104-(i),

$$|\Upsilon_1[u_1, v_1](x) - \Upsilon_1[u_2, v_2](x)|$$

$$= \left| \int_0^{+\infty} e^{-\lambda s} P_s \left[F_0(\cdot, u_1(\cdot), v_1(\cdot)) - F_0(\cdot, u_2(\cdot), v_2(\cdot)) \right](x)ds \right|$$

$$\leq \int_0^{+\infty} C(m) e^{-(\lambda - a(m))s} (1 + |x|^m) \| F_0(\cdot, u_1(\cdot), v_1(\cdot)) - F_0(\cdot, u_2(\cdot), v_2(\cdot)) \|_{B_m} ds$$

$$\leq \frac{1}{\lambda - a(m)} C(m)(1 + |x|^m) L \left(\|u_1 - u_2\|_{B_m} + \|v_1 - v_2\|_{B_m} \right).$$

It follows that

$$\|\Upsilon_1[u_1, v_1] - \Upsilon_1[u_2, v_2]\|_{B_m} \leq \frac{C(m)L}{\lambda - a(m)} \left(\|u_1 - u_2\|_{B_m} + \|v_1 - v_2\|_{B_m} \right).$$

We now look at Υ_2. By using (4.132) and Hypothesis 4.104-(i) we get, for any (u_1, v_1) and (u_2, v_2) in $B_m(H) \times B_m(H, U)$,

$$|\Upsilon_2[u_1, v_1](x) - \Upsilon_2[u_2, v_2](x)|$$

$$= \left| \int_0^{+\infty} e^{\lambda s} \nabla^G P_s \left[F_0(\cdot, u_1(\cdot), v_1(\cdot)) - F_0(\cdot, u_2(\cdot), v_2(\cdot)) \right](x)ds \right|$$

$$\leq \int_0^{+\infty} e^{-(\lambda - a(m))s} \gamma_G(s)(1 + |x|^m) \| F_0(\cdot, u_1(\cdot), v_1(\cdot)) - F_0(\cdot, u_2(\cdot), v_2(\cdot)) \|_{B_m} ds$$

$$\leq L(1 + |x|^m) \left(\|u_1 - u_2\|_{B_m} + \|v_1 - v_2\|_{B_m} \right) \int_0^{+\infty} e^{-(\lambda - a(m))s} \gamma_G(s)ds.$$

It follows that

$$\|\Upsilon_2[u_1, v_1] - \Upsilon_2[u_2, v_2]\|_{B_m} \leq L \left(\|u_1 - u_2\|_{B_m} + \|v_1 - v_2\|_{B_m} \right) \int_0^{+\infty} e^{-(\lambda - a(m))s} \gamma_G(s)ds.$$

Therefore we conclude that

$$\|\Upsilon_1[u_1, v_1] - \Upsilon_1[u_2, v_2]\|_{B_m} + \|\Upsilon_2[u_1, v_1] - \Upsilon_2[u_2, v_2]\|_{B_m}$$

$$\leq L \left[\frac{C(m)}{\lambda - a(m)} + \int_0^{+\infty} e^{-(\lambda - a(m))s} \gamma_G(s)ds \right] \left[\|u_1 - u_2\|_{B_m} + \|v_1 - v_2\|_{B_m} \right].$$

Defining

$$C(\lambda) := L\left[\frac{C(m)}{\lambda - a(m)} + \int_0^{+\infty} e^{-(\lambda-a(m))s}\gamma_G(s)ds\right]$$

it is easy to see that $C(\lambda)$ is strictly decreasing and continuous in $\lambda \in (a(m), +\infty)$, $\lim_{\lambda \to a(m)} C(\lambda) = +\infty$ and $\lim_{\lambda \to +\infty} C(\lambda) = 0$. Hence, if λ_0 is the unique number for which $C(\lambda_0) = 1$, we obtain that for all $\lambda > \lambda_0$ the mapping Υ is a contraction and it admits a unique fixed point in $B_m(H) \times B_m(H, U)$.

Step 3. *The first component of the fixed point of Υ is the unique mild solution of* (4.125).

We first observe that, even if we do not know that the G-derivative is a closed operator (see Remark 4.15), we can still apply Proposition 4.16, obtaining that, for all $(u, v) \in B_m(H) \times B_m(H, U)$, $x \in H$, $h \in U$, $\Upsilon_1[u, v](x)$ is G-Gâteaux differentiable and $\langle \Upsilon_2[u, v](x), h\rangle = \langle \nabla^G \Upsilon_1[u, v], h\rangle$.

Let now $[\bar{u}, \bar{v}]$ be the fixed point of Υ, i.e. $\Upsilon[\bar{u}, \bar{v}] = (\bar{u}, \bar{v})$. It follows from the above observation that $\bar{u}$ is G-Gâteaux differentiable and $\nabla^G\bar{u}(x) = \bar{v}(x)$, hence $\nabla^G\bar{u} \in B_m(H, U)$. Substituting it into the definition of Υ_1, we get

$$\bar{u}(x) = \int_0^{+\infty} e^{-\lambda s} P_s\left[F_0(\cdot, \bar{u}(\cdot), \nabla^G\bar{u}(\cdot))\right](x)ds.$$

The above implies that $\bar{u}$ is a mild solution of (4.125) according to Definition 4.102.

To prove uniqueness, let u^* be another mild solution of Eq. (4.125). Then u^* is G-Gâteaux differentiable with $\nabla^G u^* \in B_m(H, U)$. Hence, setting $v^* := \nabla^G u^*$, we easily see that (u^*, v^*) is a fixed point of Υ and so $u^* = \bar{u}$ and $v^* = \bar{v}$. This completes the proof. □

Remark 4.113 Differently from the parabolic case here we can take $\gamma_G \in \mathcal{I}_1$ (see the end of Remark 4.109) since we do not need to deal with the integral $\int_0^t \gamma_G(s)\gamma_G(t - s)ds$ in the contraction estimates, as the solution does not depend on t. ∎

4.4.2.5 Existence and Uniqueness of Mild Solutions in $\mathcal{G}_m$ Spaces

We now study cases where the mild solution found in Theorem 4.112 is more regular. In this section we discuss when the mild solution belongs to $\mathcal{G}_m^{1,G}(H)$.

We need new assumptions which are variations of Hypotheses 4.104, 4.106, 4.108 and 4.110.

Hypothesis 4.114 Let $m \geq 0$ be fixed. The following are satisfied.

(i) $F_0 : [0, T] \times H \times \mathbb{R} \times U \to \mathbb{R}$ satisfies, for m fixed here and for given constants $L, L' > 0$, parts (i) and (ii) of Hypothesis 4.104.
(ii) Denote by (U, τ_U^w) the space U endowed with the weak topology. F_0 is sequentially continuous as a function from $H \times \mathbb{R} \times (U, \tau_U^w)$ to $\mathbb{R}$.

The observations made in Remark 4.83 apply (with obvious changes) to Hypothesis 4.114.

Hypothesis 4.115 Let $m \geq 0$ be from Hypothesis 4.114. We assume the following.

(i) Hypothesis 4.106 is satisfied substituting everywhere $C_m(H)$ in place of $B_m(H)$.
(ii) Hypothesis 4.108 is satisfied substituting everywhere $C_m(H)$ in place of $B_m(H)$.
(iii) Hypothesis 4.110 is satisfied in the following form.

    (a) For all $\phi \in C_m(H)$ the function $\bar{\phi}_P^0$ in (4.133) is measurable in s and continuous in x.
    (b) For all $\phi \in C_m(H)$ the function $\bar{\phi}_P^1$ in (4.134) is measurable in s and strongly continuous in x.

We have the following result.

Theorem 4.116 *Let $m \geq 0$ be such that Hypotheses 4.114 and 4.115 are satisfied. Let λ_0 be from Theorem 4.112. Then for every $\lambda > \lambda_0$, Eq. (4.125) admits a unique mild solution u in $\mathcal{G}_m^{1,G}(H)$.*

Proof The proof is similar to the proof of Theorem 4.112. We only explain the changes needed here.

We consider the product space $C_m(H) \times C_m^s(H, U)$, endowed with the product norm given by the sum of the norms of the factor spaces. In this space we consider the operator $\Upsilon = (\Upsilon_1, \Upsilon_2)$ defined by (4.135) and (4.136). Once we show that Υ maps $C_m(H) \times C_m^s(H, U)$ into itself, the rest of the proof is exactly the same as the proof of Theorem 4.112, and will be omitted.

Let $(u, v) \in C_m(H) \times C_m^s(H, U)$. Observe that the function

$$\psi : H \to H, \qquad \psi(x) := F(x, u(x), v(x))$$

is continuous. Indeed, if $x_n \to x \in H$ we have, by the strong continuity of v, that $v(x_n)$ converges weakly in U to $v(x)$. Hence, thanks to Hypothesis 4.114-(ii), $\psi(x_n) \to \psi(x)$.

By Hypothesis 4.115-(iii)-(a) and Lemma 1.18, the integrand in $\Upsilon_1[u, v]$ is jointly measurable and is continuous in x. Hence, by estimate (4.130) and Lemma 4.111-(ii), the integral is well defined and $\Upsilon_1[u, v] \in C_m(H)$.

Concerning $\Upsilon_2[u, v]$, by Hypothesis 4.115-(iii)-(b), Lemma 1.18 and Lemma 1.17, the integrand in $\Upsilon_2[u, v]$ is jointly measurable and is strongly continuous in x. Hence, by estimate (4.132) and Lemma 4.111-(ii), the integral is well defined and $\Upsilon_2[u, v] \in C_m^s(H, U)$.

The rest follows the proof of Theorem 4.112. $\square$

The observations made in Remark 4.86 apply (with obvious changes) to Theorem 4.116.

4.4.2.6   Existence and Uniqueness of Mild Solutions in C_m Spaces

In this section we investigate mild solutions in $C_m^{1,G}(H)$. The modifications of Hypotheses 4.104, 4.106, 4.108 and 4.110 are the following.

Hypothesis 4.117 Let $m \geq 0$ be fixed. The following are satisfied.

(i) $F_0 : H \times \mathbb{R} \times U \to \mathbb{R}$ satisfies, for the m here and for given constants $L, L' > 0$, parts (i) and (ii) of Hypothesis 4.104.
(ii) F_0 is continuous.

Remark 4.118 Note that the assumption about F_0 here is weaker than the one in Hypothesis 4.104, needed for working in the spaces $\mathcal{G}_m$.   ∎

Hypothesis 4.119 Let $m \geq 0$ be from Hypothesis 4.117. We assume the following.

(i) Hypothesis 4.106 is satisfied substituting everywhere $C_m(H)$ in place of $B_m(H)$.
(ii) Hypothesis 4.108 is satisfied substituting everywhere $C_m(H)$ in place of $B_m(H)$ and *the G-Fréchet derivative D^G in place of the G-Gâteaux derivative ∇^G*.
(iii) Hypothesis 4.110 is satisfied in the following form.

   (a) For all $\phi \in C_m(H)$ the function $\bar{\phi}_P^0$ in (4.133) is measurable in s and continuous in x.
   (b) For all $\phi \in C_m(H)$ the function $\bar{\phi}_P^1$ in (4.134) is measurable in s and *continuous in x*.

We have the following result. It applies to most examples studied in this chapter and provides additional regularity of mild solutions when the data are more regular.

Theorem 4.120 *Let $m \geq 0$ be such that Hypotheses 4.117 and 4.119 are satisfied. Let λ_0 be from Theorem 4.112. Then for every $\lambda > \lambda_0$, Eq. (4.125) admits a unique mild solution u in $C_m^{1,G}(H)$.*

Proof Similarly to Theorem 4.90, the proof is analogous to the proof of Theorem 4.112 and we only explain the changes which have to be implemented here.

We consider the product space $C_m(H) \times C_m(H, U)$, endowed with the product norm given by the sum of the norms of the factor spaces. In this space we consider the operator $\Upsilon = (\Upsilon_1, \Upsilon_2)$ defined by (4.135) and (4.136). We only show that Υ maps $C_m(H) \times C_m(H, U)$ into itself as the rest of the proof is exactly the same as in the proof of Theorem 4.112.

Let $(u, v) \in C_m(H) \times C_m(H, U)$. The proof that $\Upsilon_1[u, v] \in C_m(H)$ is the same as the proof of this statement in the proof of Theorem 4.116 once we observe that

$$\psi(x) := F_0(x, u(x), v(x))$$

is obviously continuous, thanks to Hypothesis 4.117-(ii).

Concerning $\Upsilon_2[u, v]$, by Hypothesis 4.119-(iii)-(b) and Lemma 1.18, the integrand in $\Upsilon_2[u, v]$ is jointly measurable and is continuous in x. Hence, by estimate (4.132) and Lemma 4.111-(ii), the integral is well defined and $\Upsilon_2[u, v] \in C_m(H, U)$. $\qquad\qquad\qquad\qquad\qquad\qquad\qquad\qquad\qquad\qquad\qquad\qquad\qquad\qquad\qquad\quad\square$

4.4.2.7   Existence and Uniqueness of Mild Solutions in UC_m Spaces

As for the parabolic equation (4.109), this case was the first to be studied in the literature but it requires more assumptions and it does not provide real advantages compared to the results of the previous subsection. For HJB equations arising from optimal control problems, Theorem 4.120 already guarantees enough regularity to obtain verification theorems for the associated optimal control problem. Moreover, the assumptions of Theorem 4.123 are a little more complicated. Hence the results of this subsection (and related results, see e.g. Sect. 4.6) are not presented with all the details.

The new assumptions about F_0 and the semigroup are the following.

Hypothesis 4.121 Let $m \geq 0$ be fixed. The following are satisfied.

(i) $F_0 : H \times \mathbb{R} \times U \to \mathbb{R}$ satisfies, for m fixed here and for given constants $L, L' > 0$, parts (i) and (ii) of Hypothesis 4.104.
(ii) The function $(x, y, z) \to F_0(x, v, w)/(1 + |x|^m + |y| + |z|_U)$ is uniformly continuous.

Hypothesis 4.122 Let $m \geq 0$ be from Hypothesis 4.121. We assume the following.

(i)   Hypothesis 4.106 is satisfied substituting everywhere $UC_m(H)$ in place of $B_m(H)$.
(ii) Hypothesis 4.108 is satisfied substituting everywhere $UC_m(H)$ in place of $B_m(H)$ and the G-Fréchet derivative D^G in place of the G-Gâteaux derivative ∇^G.
(iii) Hypothesis 4.110 is satisfied in the following form.

   (a)   For every $\phi \in UC_m(H)$, the function $\bar{\phi}_P^0$ in (4.133) multiplied by $e^{-a(m)s}$ is measurable in s and uniformly continuous in x, uniformly with respect to s.
   (b)   For every $\phi \in UC_m(H)$, the function $\bar{\phi}_P^1$ in (4.134) multiplied by $e^{-a(m)s}$ is measurable in s and uniformly continuous in x, uniformly with respect to s.

We have the following result.

Theorem 4.123 *Let* $m \geq 0$ *be such that Hypotheses 4.121 and 4.122 are satisfied. Let* λ_0 *be from Theorem 4.112. Then, for every* $\lambda > \lambda_0$, *Eq. (4.125) admits a unique mild solution* u *in* $UC_m^{1,G}(H)$.

4 Mild Solutions in Spaces of Continuous Functions

Proof The proof is similar to the proof of Theorem 4.112. We consider the product space $UC_m(H) \times UC_m(H, U)$, endowed with the product norm given by the sum of the norms of the factor spaces, and we consider the operator $\Upsilon = (\Upsilon_1, \Upsilon_2)$ defined by (4.135) and (4.136). We only argue that Υ maps $UC_m(H) \times UC_m(H, U)$ into itself as the rest of the proof is exactly the same as the proof of Theorem 4.112.

Let $(u, v) \in UC_m(H) \times UC_m(H, U)$ and define

$$\psi(x) := F_0(x, u(x), v(x)).$$

Straightforward computations, together with Hypothesis 4.121, imply that $\psi \in UC_m(H)$. Hence, arguing as in the proof of Theorem 4.120 we obtain that, when $\lambda > a(m)$, $\Upsilon_1[u, v] \in C_m(H)$ and $\Upsilon_2[u, v] \in C_m(H, U)$. To prove uniform continuity of $\Upsilon_1[u, v]$ we use Hypothesis 4.122-(iii)-(a), since for $x_1, x_2 \in H$,

$$|\Upsilon_1[u, v](x_1) - \Upsilon_1[u, v](x_2)|$$
$$\leq \int_0^{+\infty} e^{-(\lambda - a(m))s} \left| e^{-a(m)s} P_s[\psi](x_1) - e^{-a(m)s} P_s[\psi](x_2) \right| ds$$
$$\leq \int_0^{+\infty} e^{-(\lambda - a(m))s} \rho(|x_1 - x_2|) ds.$$

Arguing in the same way and using Hypothesis 4.122-(iii)-(b) we obtain $\Upsilon_2[u, v] \in UC_m(H, U)$ when $\lambda > a(m)$. $\qquad\square$

Similar comments to what was observed in Remark 4.86 also apply to Theorem 4.123.

4.4.2.8 Existence and Uniqueness in the "Strong Feller" Case

In this section we proceed as in Sect. 4.4.1.8, i.e. we add to the hypotheses of Sect. 4.4.2.2 the strong Feller property of the semigroup P_s (see the beginning of Sect. 4.4.1.8 for references about the strong Feller property).

Hypothesis 4.124 Let $m \geq 0$ be from Hypothesis 4.104. The family P_s, $s \geq 0$, is strongly Feller (see Definition 1.159 and Lemma 1.160) in the sense that

$$P_s(B_m(H)) \subset C_m(H), \qquad \forall s > 0.$$

We have the following result.

Theorem 4.125 *Let $m \geq 0$ be from Hypothesis 4.104. Let Hypotheses 4.104, 4.106, 4.108, 4.110 and 4.124 hold true. Let λ_0 be from Theorem 4.112 and let $\lambda > \lambda_0$ in Eq. (4.125). Then we have the following.*

(i) *The unique mild solution u of Eq. (4.125) (obtained in Theorem 4.112) is continuous.*

(ii) *Assume, moreover, that*

$$P_s(B_m(H)) \subset \mathcal{G}_m^{1,G}(H), \qquad \forall s > 0.$$

Then $u \in \mathcal{G}_m^{1,G}(H)$.

(iii) *Assume, moreover, that*

$$P_s(B_m(H)) \subset C_m^{1,G}(H), \qquad \forall s > 0.$$

Then $u \in C_m^{1,G}(H)$.

Proof Concerning claim (i), it is enough to observe that u solves the integral equation (4.129) and the right-hand side is continuous by the strong Feller property, estimate (4.130) and the dominated convergence theorem.

The proof of claim (ii) (respectively, (iii)) is exactly the same once we observe that from (4.129) we have (see the proof of Theorem 4.112)

$$\nabla^G u(x) = \int_0^{+\infty} e^{-\lambda s} \nabla^G P_s \left[F_0(\cdot, u(\cdot), \nabla^G u(\cdot)) \right](x) ds. \qquad (4.138)$$

By assumption the integrand is jointly measurable and strongly continuous (respectively, continuous) in x. Hence the claim follows by (4.132) and the dominated convergence theorem. $\square$

Remark 4.126 We remark that the additional assumption in (ii) of Theorem 4.125 is satisfied in the case presented in Sect. 4.3.3 while the one in (iii) is satisfied in the case presented in Sect. 4.3.1.   ∎

4.4.2.9 Examples

We briefly discuss how to apply the results of the present subsection (Theorems 4.112 4.116, 4.120, 4.123 and 4.125) to Examples 4.99–4.101 presented in the parabolic case. Most of the discussion is exactly the same, hence we only focus on the main differences.

The main difference is that, to apply Theorems 4.112, 4.116, 4.120, 4.123 and 4.125, one needs first to prove that estimates (4.130) and (4.132) hold. These bounds with exponential growth in time are guaranteed in Examples 4.99 and 4.100 for which such estimates are proved in Theorems 4.41 and 4.56.[23] For Example 4.101

---

[23] In Theorem 4.56 only the case $m = 0$ is considered. However, as noted in Remark 4.59-(iii), it does not seem to be difficult to extend such a result to the case $m > 0$.

such estimates, as recalled at the beginning of Sect. 4.3.3, are proved under additional assumptions[24] which the reader can find in Proposition 4.6-(ii) of [285].

Moreover, concerning Example 4.99 we observe that, differently from the parabolic case, Hypothesis 4.110 and all its variants given in Hypotheses 4.115-(iii), 4.119-(iii), 4.122-(iii), are satisfied without requiring that Hypothesis 4.29 also holds for $U = H$ and $G = I$. This follows from Proposition 4.50. Hence, to apply Theorems 4.116, 4.120 and 4.123 we do not need such an extra assumption. On the other hand, to apply Theorem 4.125 we need the strong Feller property, hence we need to require that Hypothesis 4.29 also holds for $U = H$ and $G = I$.

4.5 Approximation of Mild Solutions: Strong Solutions

In applications to optimal control (see Sect. 4.8) it is desirable to know that the mild solutions obtained in the previous section are (or can be characterized as) the limits, in a sense to be made precise, of very regular (enough to apply Itô's formula) solutions that we call *classical solutions*. We cannot hope for uniform convergence for reasons recalled in Sects. B.6 and D.3. Hence we use the so-called π-*convergence* and $\mathcal{K}$-*convergence*. We refer the reader to Appendix B.5 for basic definitions and results about these notions of convergence.

Since solutions that are obtained as limits of solutions of approximating problems are usually called *strong solutions*, we will use here the terminology π-*strong* and $\mathcal{K}$-*strong solutions* (see Appendix B.7 for some basic results about mild and strong solutions for linear Cauchy problems for Kolmogorov equations).

4.5.1 The Parabolic Case

We first define the classical solutions[25] of (4.109). To do this, in contrast to the previous section, we connect Eq. (4.109), and thus the coefficients A, b, Σ, with the SDE (4.110).

Let $T > 0$ and let $\left(\Omega, \mathscr{F}, \{\mathscr{F}_s^0\}_{s \in [0, T]}, \mathbb{P}, W\right)$, where W is a cylindrical Wiener process in a real separable Hilbert space Ξ, be a generalized reference probability space. For $(t, x) \in [0, T] \times H$ we consider the SDE

$$\begin{cases} dX(s) = [AX(s) + b(s, X(s))]\, ds + \sigma(s, X(s))\, dW(s), \ s \in [t, T], \\ X(t) = x, \qquad\qquad\qquad\qquad\qquad\qquad\qquad\qquad\qquad\ x \in H. \end{cases}$$

$$(4.139)$$

---

[24]Using the argument of Remark 4.107, one can prove estimate (4.130) without such additional assumptions. It is likely that a similar argument might also be applied to prove (4.132).

[25]This definition is similar to Definition B.82 and is a bit more restrictive than the one used in Sect. 6.2 of [179], see Remark B.83-(1) for explanations.

We assume the following.

Hypothesis 4.127 Let $T > 0$.

   (i) The functions F_0 and φ satisfy Hypothesis 4.72 for some $m \geq 0$ and $L, L' > 0$.
  (ii) The operator A and the functions b and σ in (4.139) satisfy, in $[0, T] \times H$, either Hypothesis 1.149 with b and σ independent of a, or Hypotheses 1.143 and 1.145 without dependence on a_1 and with $a_2(\cdot) = 0$. We have $\Sigma(t, x) := \sigma(t, x)\sigma^*(t, x)$.
 (iii) There exist $r \in \mathbb{R}, r \in \varrho(A)$ such that $(rI - A)^{-1}b \colon [0, T] \times H \to H$ is continuous and for every $y, z \in H$ the function $\langle \Sigma y, z \rangle \colon [0, T] \times H \to \mathbb{R}$ is continuous. Moreover, there exists a $C > 0$ such that, for all $(t, x) \in [0, T] \times H$,

$$\begin{cases} |(rI - A)^{-1}b(t, x)| \leq C(1 + |x|) \\ \|\sigma(t, x)\|_{\mathcal{L}(\Xi, H)} \leq C(1 + |x|). \end{cases} \tag{4.140}$$

 (iv) For all $t \leq s \leq T$, $\phi \in B_m(H)$, where $m \geq 0$ is from point (i) above, we have $P_{t,s}[\phi](x) = \mathbb{E}[\phi(X(s; t, x))]$, where $X(\cdot; t, x)$ is the mild solution of SDE (4.139) (which exists and is unique by Theorem 1.152 or Proposition 1.147).

Remark 4.128 The assumptions of Hypothesis 4.127-(ii) imply that the family $P_{t,s}$ defined in (iv) has the semigroup property. This follows from Theorem 1.157 and Corollary 1.158, and is used in Theorem 4.135 to guarantee that Hypothesis 4.74 is satisfied. Moreover, if Hypothesis 1.149 is satisfied in Hypothesis 4.127-(ii), using moment and continuous dependence estimates for solutions of (4.139) and the continuity of the trajectories of their solutions, it follows that the functions ϕ_P^0 and $\bar{\psi}_P^0$ in Hypothesis 4.77 are continuous in (t, x) and (t, s, x), respectively, if ϕ and ψ are continuous there. In particular, Hypothesis 4.89-(iii)(a) is satisfied. The same is true if Hypotheses 1.143 and 1.145 hold in Hypothesis 4.127-(ii) if we use moment and continuous dependence estimates for solutions of (4.139) for this case and mean square continuity of the solutions. On the other hand one can employ Lemma 1.18 and the argument outlined in Sect. 1.6 before Theorem 1.157 to show that Hypothesis 4.77-(i) is also satisfied.

Finally, observe that Hypothesis 4.127-(iii) contains the conditions in the assumptions of Proposition 1.168.   ∎

Similarly to what is done in Appendix B.7 (see also Sect. 4.3 of [306] or Sect. 6.2 of [179]), we define the operator $\mathcal{A}_1(t)$ as follows:

$$\begin{cases} D(\mathcal{A}_1(t)) = \left\{ \phi \in UC_b^2(H) \ : \ A^*D\phi \in UC_b(H, H), D^2\phi \in UC_b(H, \mathcal{L}_1(H)) \right\} \\[2mm] \mathcal{A}_1(t)[\phi](x) = \frac{1}{2} \operatorname{Tr} [\Sigma(t, x)D^2\phi(x)] + \langle x, A^*D\phi(x) \rangle \\[2mm] \qquad\qquad + \left\langle (rI - A)^{-1}b(t, x), (rI - A)^*D\phi(x) \right\rangle. \end{cases}$$

$$\tag{4.141}$$

Above $r \in \mathbb{R}$ belongs to the resolvent set $\varrho(A)$. Since $D(\mathcal{A}_1(t))$ is independent of t we denote it simply by $D(\mathcal{A}_1)$ from now on. By Hypothesis 4.127 it is clear that for all $t \in [0, T]$

$$\mathcal{A}_1(t) : D(\mathcal{A}_1) \subset C_m(H) \longrightarrow C_m(H), \qquad \text{for } m \geq 2.$$

We endow the space $D(\mathcal{A}_1)$ with the norm

$$\|\phi\|_{D(\mathcal{A}_1)} := \|\phi\|_0 + \|D\phi\|_0 + \|A^*D\phi\|_0 + \sup_{x \in H} \|D^2\phi(x)\|_{\mathcal{L}_1(H)}. \tag{4.142}$$

The space $D(\mathcal{A}_1)$ with this norm is a Banach space (see Appendix B.7.1 for a similar result). We now define the notion of a classical solution.

Definition 4.129 A function $u : [0, T] \times H \to \mathbb{R}$ is a classical solution of the equation

$$\begin{cases} v_t + \dfrac{1}{2} \operatorname{Tr} \left[\Sigma(t, x) D^2 v \right] + \langle Ax + b(t, x), Dv \rangle + F_0(t, x, v, D^G v) = g(t, x), \\ \qquad\qquad t \in [0, T), \ x \in H, \\[4pt] v(T, x) = \varphi(x), \ x \in H, \end{cases} \tag{4.143}$$

(where g is a given Borel measurable function and the other data are as in Hypothesis 4.127) if u has the regularity properties

$$\begin{cases} u(\cdot, x) \in C^1([0, T]), \ \forall x \in H; \\ u(t, \cdot) \in D(\mathcal{A}_1), \ \forall t \in [0, T], \quad \text{and} \quad \sup_{t \in [0,T]} \|u(t, \cdot)\|_{D(\mathcal{A}_1)} < +\infty; \\ u \in C_b([0, T] \times H), \ Du, A^*Du \in C_b([0, T] \times H, H); \\ D^2 u \in C_b([0, T] \times H, \mathcal{L}_1(H)), \ D^G u \in C_b([0, T] \times H, U), \end{cases} \tag{4.144}$$

satisfies $u(T, \cdot) = \varphi$ and, for some $r \in \mathbb{R} \cap \varrho(A)$,

$$u_t(t, x) + \frac{1}{2}\operatorname{Tr}[\Sigma(t, x)D^2 u(t, x)] + \langle x, A^*Du(t, x) \rangle$$
$$+ \left\langle (rI - A)^{-1}b(t, x), (rI - A)^*Du(t, x) \right\rangle + F_0(t, x, u(t, x), D^G u(t, x)) = g(t, x) \tag{4.145}$$

for all $(t, x) \in [0, T) \times H$.

Remark 4.130 The above definitions of the operator $\mathcal{A}_1(t)$ and of a classical solution are independent of the choice of $r \in \mathbb{R} \cap \varrho(A)$. Indeed, let $r_1, r_2 \in \varrho(A) \cap \mathbb{R}$ and observe that, for $(t, x) \in [0, T) \times H$, using the resolvent identity (B.1),

$$\left\langle (r_1 I - A)^{-1}b(t, x), (r_1 I - A)^*Du \right\rangle - \left\langle (r_2 I - A)^{-1}b(t, x), (r_2 I - A)^*Du \right\rangle$$
$$= \left\langle \left[(r_1 I - A)^{-1} - (r_2 I - A)^{-1} \right] b(t, x), (r_1 I - A)^*Du \right\rangle$$
$$+ \left\langle (r_2 I - A)^{-1}b(t, x), \left[(r_1 I - A)^* - (r_2 I - A)^* \right] Du \right\rangle$$

$$= (r_2 - r_1)\big\langle \big[(r_1 I - A)^{-1}(r_2 I - A)^{-1}\big] b(t,x), (r_1 I - A)^* Du\big\rangle$$
$$+ \big\langle (r_2 I - A)^{-1} b(t,x), (r_1 - r_2)Du\big\rangle = 0.$$

Finally, note that in the definition above we do not require the continuity of u_t. This will be guaranteed if the datum F_0 is continuous. ∎

Before defining strong solutions we give a variant of the definitions of π-convergence and $\mathcal{K}$-convergence (see Definitions B.55 and B.56).

Definition 4.131 Let $m \geq 0$, $\eta \in \mathcal{I}_1$ and let Z be a real separable Hilbert space. We say that a sequence $(f_n)_{n\in\mathbb{N}} \subset B_{m,\eta}((0,T] \times H, Z)$ π-converges to $f \in B_{m,\eta}((0,T] \times H, Z)$ if

$$\begin{cases} \sup_{n\in\mathbb{N}} \|f_n\|_{B_{m,\eta}((0,T]\times H,Z)} < +\infty, \\ \lim_{n\to+\infty} \frac{1}{\eta(t)}|f_n(t,x) - f(t,x)| = 0 \quad \forall x \in H \text{ and for a.e. } t \in [0,T]. \end{cases} \tag{4.146}$$

In such a case we write $\pi - \lim_{n\to+\infty} f_n = f$ in $B_{m,\eta}([0,T) \times H, Z)$. Moreover, we say that $(f_n)_{n\in\mathbb{N}} \subset B_{m,\eta}((0,T] \times H, Z)$ $\mathcal{K}$-converges to $f \in B_{m,\eta}((0,T] \times H, Z)$ if

$$\begin{cases} \sup_{n\in\mathbb{N}} \|f_n\|_{B_{m,\eta}((0,T]\times H,Z)} < +\infty, \\ \lim_{n\to+\infty} \sup_{(t,x)\in(0,T]\times K} \frac{1}{\eta(t)}|f_n(t,x) - f(t,x)| = 0, \end{cases} \tag{4.147}$$

for every compact set $K \subset H$. In such a case we write $\mathcal{K} - \lim_{n\to+\infty} f_n = f$ in $B_{m,\eta}((0,T] \times H, Z)$ or, if all functions are also continuous, in $C_{m,\eta}((0,T] \times H, Z)$. Recalling the notation introduced in Definition 4.24, we say that $(f_n)_{n\in\mathbb{N}} \subset \overline{B}_{m,\eta}([0,T) \times H, Z)$ π-converges ($\mathcal{K}$-converges) to $f \in \overline{B}_{m,\eta}([0,T) \times H, Z)$ if $(f_n(T - \cdot, \cdot))_{n\in\mathbb{N}} \subset B_{m,\eta}((0,T] \times H, Z)$ π-converges ($\mathcal{K}$-converges) to $f(T - \cdot, \cdot) \in B_{m,\eta}((0,T] \times H, Z)$. The same definition applies to the π-convergence ($\mathcal{K}$-convergence) in $\overline{C}_{m,\eta}([0,T) \times H, Z)$.

Definition 4.132 We say that a function $u : [0,T] \times H \to \mathbb{R}$ is a π-strong solution (respectively a $\mathcal{K}$-strong solution) of Eq. (4.109) if there exist $m \geq 0$ and $\eta \in \mathcal{I}_1$ such that $u \in \overline{B}_{m,\eta}^{0,1,G}([0,T] \times H)^{26}$ and there exist three sequences $(\varphi_n) \subset D(\mathcal{A}_1)$, $(u_n) \subset C_b([0,T] \times H)$ and $(g_n) \subset \overline{B}_{m,\eta}([0,T) \times H)$ such that, for every $n \in \mathbb{N}$, u_n is a classical solution of the Cauchy problem

$$\begin{cases} w_t + \mathcal{A}_1(t)w + F_0(t,x,w,D^G w) = g_n(t,x) \\ w(T,x) = \varphi_n(x), \end{cases} \tag{4.148}$$

and moreover as $n \to +\infty$, using the notation of Definition B.55,

---

[26]This means that u satisfies (i) of Definition 4.70.

$$\begin{cases} \pi - \lim_{n \to +\infty} \varphi_n = \varphi & in \ B_m(H) \\ \pi - \lim_{n \to +\infty} u_n = u & in \ B_m([0, T] \times H), \end{cases} \tag{4.149}$$

and, using the notation of Definition 4.131,

$$\begin{cases} \pi - \lim_{n \to +\infty} g_n = 0 & in \ \overline{B}_{m,\eta}([0, T] \times H) \\ \pi - \lim_{n \to +\infty} D^G u_n = \nabla^G u & in \ \overline{B}_{m,\eta}([0, T] \times H, U) \end{cases} \tag{4.150}$$

(respectively, when all four convergences above hold in the $\mathcal{K}$ sense, using the notation of Definitions B.56 and 4.131).

The result we would like to prove is that any mild solution in the sense of Definition 4.70 is also a strong solution in the sense of Definition 4.132 (and possibly also the opposite). To prove such a result one needs to define properly the approximating sequence u_n. The choice of such u_n clearly depends on the properties of the transition semigroup $P_{t,s}$. Heuristically speaking there are two main cases:

- When the data of the SDE (4.110) are very regular (e.g. the Ornstein–Uhlenbeck case of Sect. 4.3.1) and the semigroup maps $D(\mathcal{A}_1)$ into itself. In this case it is enough to approximate only φ and F_0.
- When the data of the SDE are not regular enough (e.g. the case of invertible diffusion coefficient of Sect. 4.3.3 and the case of reaction-diffusion equations of Sect. 4.9.2) and the semigroup does not map $D(\mathcal{A}_1)$ into itself. In this case, in addition to φ and F_0, one also needs to approximate the semigroup $P_{t,s}$.

Here we present a result which only deals with the first case. The second case is more difficult and strongly dependent on a specific problem. Up to now it has only been studied in the specific case of reaction-diffusion equations (see [103, 105, 107] and also [106], Chaps. 9 and 10) which is presented briefly in Sect. 4.9.2.

We will need additional assumptions about the transition semigroup.

Hypothesis 4.133 Let $T > 0$ and let $m \geq 0$ be from Hypothesis 4.72 (and hence also Hypothesis 4.127).

(i) For all $0 \leq t \leq s \leq T$, and all $\varphi \in \mathcal{F}C_0^{\infty,A^*}(H)$, $f \in \mathcal{F}C_0^{\infty,A^*}([0, T] \times H)$, the function

$$(t, x) \to P_{t,T}[\varphi](x) + \int_t^T P_{t,s}[f(s, \cdot)](x)ds$$

satisfies the last three lines of (4.144).

(ii) If $\varphi_n \xrightarrow[n \to \infty]{\pi} \varphi$ in $B_m(H)$ then

$$D^G P_{t,T}[\varphi_n] \xrightarrow[n \to \infty]{\pi} \nabla^G P_{t,T}[\varphi] \quad in \ \overline{B}_{m,\gamma_G}([0, T] \times H, U).$$

(iii) If $f_n \xrightarrow[n\to\infty]{\pi} f$ in $\overline{B}_{m,\gamma_G}([0,T) \times H)$

then

$$\int_t^T D^G P_{t,s}[f_n]ds \xrightarrow[n\to\infty]{\pi} \int_t^T \nabla^G P_{t,s}[f]ds \quad in \quad \overline{B}_{m,\gamma_G}([0,T) \times H, U).$$

Observe that point (i) above will guarantee the regularity of the approximating solutions while points (ii) and (iii) will guarantee the convergence of their derivatives.

Before we state the main result, Theorem 4.135, we show a lemma which will be used in its proof.

Lemma 4.134 *Let the assumptions of Hypothesis 4.127-(ii)-(iii) be satisfied. Let $\varphi \in C([0,T] \times H)$ be such that $D\varphi, A^*D\varphi \in C([0,T] \times H, H), D^2\varphi \in C([0,T] \times H, \mathcal{L}_1(H))$. Then the function*

$$\frac{1}{2}\mathrm{Tr}[\Sigma(t,x)D^2\varphi(t,x)] + \langle x, A^*D\varphi(t,x)\rangle + \langle (rI-A)^{-1}b(t,x), (rI-A)^*D\varphi(t,x)\rangle$$

is continuous on $[0,T] \times H$.

Proof We only need to show the continuity of $F(t,x) := \frac{1}{2}\mathrm{Tr}[\Sigma(t,x)D^2\varphi(t,x)]$ as the continuity of the other terms is obvious. If F is not continuous then there are $(t_0,x_0) \in [0,T] \times H$, a sequence $(t_n,x_n) \in [0,T] \times H$ converging to (t_0,x_0) and $\varepsilon > 0$ such that

$$|F(t_0,x_0) - F(t_n,x_n)| \geq \varepsilon \quad \text{for all} \quad n = 1,2,\ldots.$$

Let $\{e_i\}_{i\in\mathbb{N}}$ be an orthonormal basis of H composed of eigenvectors of $D^2\varphi(t_0,x_0)$ and denote by P_N the orthogonal projection in H onto $\mathrm{span}\{e_1,\ldots e_N\}$. Then, since $D^2\varphi \in C([0,T] \times H, \mathcal{L}_1(H))$, it is easy to see that there exists an $N \in \mathbb{N}$ such that

$$\left|\frac{1}{2}\mathrm{Tr}[\Sigma(t_0,x_0)D^2\varphi(t_0,x_0)P_N] - \frac{1}{2}\mathrm{Tr}[\Sigma(t_n,x_n)D^2\varphi(t_0,x_0)P_N]\right| \geq \frac{\varepsilon}{2} \quad (4.151)$$

for sufficiently large N. But

$$\mathrm{Tr}[\Sigma(t_n,x_n)D^2\varphi(t_0,x_0)P_N] = \sum_{i=1}^N \langle \Sigma(t_n,x_n)D^2\varphi(t_0,x_0)e_i, e_i\rangle$$

$$\to \sum_{i=1}^N \langle \Sigma(t_0,x_0)D^2\varphi(t_0,x_0)e_i, e_i\rangle = \mathrm{Tr}[\Sigma(t_0,x_0)D^2\varphi(t_0,x_0)P_N]$$

by Hypothesis 4.127-(iii). This contradicts (4.151). $\qquad\square$

Theorem 4.135 *Let $T > 0$. Let Hypotheses 4.76, 4.77, 4.127 and 4.133 be satisfied. Let $m \geq 0$ be from Hypothesis 4.127. Let u be the unique mild solution of Eq. (4.109)* in $\overline{B}_{m,\gamma_G}^{0,1,G}([0, T] \times H)$. *Assume that u is continuous on* $[0, T] \times H$ *and also that the function* $f(t, x) := F_0(t, x, u(t, x), \nabla^G u(t, x))$ *is continuous on* $[0, T) \times H$. *Then u is also the unique π-strong solution, in* $\overline{B}_{m,\gamma_G}^{0,1,G}([0, T] \times H)$, *of the Cauchy problem* (4.109).

If Hypothesis 4.133 holds substituting everywhere π-convergence by $\mathcal{K}$-convergence, then u is the unique $\mathcal{K}$-strong solution in $\overline{B}_{m,\gamma_G}^{0,1,G}([0, T] \times H)$ *of* (4.109).

Proof We observe first that Hypothesis 4.74 is automatically satisfied thanks to Hypothesis 4.127-(ii) and estimate (1.80) (or (1.69)). This, together with the other assumptions, ensures the existence and uniqueness, in $\overline{B}_{m,\gamma_G}^{0,1,G}([0, T) \times H)$, of the mild solution u of (4.109) by Theorem 4.80-(i). We then have, for $(t, x) \in [0, T] \times H$,

$$u(t, x) = P_{t,T}[\varphi](x) + \int_t^T P_{t,s}\left[f(s, \cdot)\right](x)ds. \tag{4.152}$$

It follows from the assumptions that $f \in \overline{C}_{m,\gamma_G}([0, T) \times H)$. To find the family u_n we first define suitable approximations φ_n of φ and f_n of f, then we define

$$u_n(t, x) = P_{t,T}[\varphi_n](x) + \int_t^T P_{t,s}\left[f_n(s, \cdot)\right](x)ds \tag{4.153}$$

and show that u_n satisfies the required properties. This is done in three steps.

Step 1. The approximating functions u_n satisfy the last three lines of (4.144).

Let φ_n be a sequence provided by Lemma B.78 for $B = A^*$ there (i.e. taking an orthonormal basis contained in $D(A^*)$). We then have $\varphi_n \in \mathcal{F}C_0^{\infty,A^*}(H)$ and $\varphi_n \xrightarrow[n\to\infty]{\mathcal{K}} \varphi$.

We now show that $\mathcal{F}C_0^{\infty,A^*}(H) \subset D(\mathcal{A}_1)$ since Definition 4.132 requires $(\varphi_n) \subset D(\mathcal{A}_1)$. Let $\psi \in \mathcal{F}C_0^{\infty,A^*}(H)$. It is clear that $\psi \in UC_b^2(H)$. Moreover, by its definition, $\psi(x) = \psi(Px)$ for some finite-dimensional orthogonal projection P with $P(H) \subset D(A^*)$. Thus $D\psi(x) \in D(A^*)$ and the map $x \to A^*D\psi(x)$ belongs to $UC_b(H, H)$. Finally, since ψ depends on a finite number of coordinates, it is easy to see that the function $x \to D^2\psi(x)$ takes values in $\mathcal{L}_1(H)$ and is uniformly continuous and bounded there.

Since f may not belong to $C_m([0, T) \times H)$ due to the singularity at $t = T$, we have to modify a little the approximation argument of Lemma B.78. For each sufficiently big $k \in \mathbb{N}$ we define, for $t \in [0, T) \times H$, $\hat{f}_h(t, x) := \chi_k(t)f(t, x)$, where $\chi_k : [0, T) \to [0, 1]$ is a smooth function such that $\chi_k(t) = 1$ for $t \in [0, T - 2/k]$ and $\chi_k(t) = 0$ for $t \in [T - 1/k, T)$. Then, for every sufficiently big k, we take a sequence $(f_{k,n})_n$ from Lemma B.78, which $\mathcal{K}$-converges to $\hat{f}_k$. Then the diagonal sequence $(f_n)_n := (f_{n,n})_n$ has elements in $\mathcal{F}C_0^{\infty,A^*}([0, T] \times H)$ and can be proved to $\mathcal{K}$-converge to f in $\overline{B}_{m,\gamma_G}([0, T) \times H)$ exactly as in the last part of Lemma B.78.

With this choice of φ_n and f_n the last three lines of (4.144) immediately follow from Hypothesis 4.133-(i).

Step 2. u_n satisfies the first regularity property of (4.144) *and is a classical solution of* (4.148) *for a suitable g_n.*

The proof that u_n is a classical solution is based on the proof of Theorem 7.5.1 of [179] (see also Theorem 9.25 [180]) however here we also have to deal with the non-homogeneous term f. The proof is similar to that of Proposition B.91-(i) where we consider the case when $P_{t,s}$ is of Ornstein–Uhlenbeck type.

Let $X(s; t, x)$ be the mild solution of (4.110) at time s. We apply Dynkin's formula of Proposition 1.168 or Proposition 1.169 (whose assumptions are satisfied thanks to Hypothesis 4.127) to the process $\varphi_n(X(s; t, x))$ on the interval $s \in [t, T]$, obtaining

$$
\mathbb{E}\varphi_n(X(T; t, x)) = \varphi_n(x) + \mathbb{E} \int_t^T \langle X(s; t, x), A^* D\varphi_n(X(s; t, x))\rangle\, ds
$$

$$
+ \mathbb{E} \int_t^T \Big[\langle (rI - A)^{-1} b(s, X(s; t, x)), (rI - A)^* D\varphi_n(X(s; t, x))\rangle\, ds
$$

$$
+ \frac{1}{2}\mathrm{Tr}\left(\Sigma(s, X(s; t, x)) D^2\varphi_n(X(s; t, x))\right) \Big]\, ds. \tag{4.154}
$$

We now compute the left derivative of u_n at $t = T$. We observe that by the definition of u_n,

$$
\frac{u_n(T, x) - u_n(T - h, x)}{h}
$$

$$
= \frac{\varphi_n(x) - P_{T-h,T}[\varphi_n](x)}{h} - \frac{1}{h}\int_{T-h}^T P_{T-h,s}[f_n(s, \cdot)](x)\, ds. \tag{4.155}
$$

We see that by Lemma 4.134, the function

$$
F(s, y) := \langle y, A^* D\varphi_n(y)\rangle + \langle (rI - A)^{-1} b(s, y), (rI - A)^* D\varphi_n(y)\rangle + \frac{1}{2}\mathrm{Tr}\left(\Sigma(s, y) D^2\varphi_n(y)\right)
$$

is continuous. Moreover, by (4.140), $|F(s, y)| \le C_1(1 + |y|^2)$, $(s, y) \in [0, T] \times H$ for some constant C_1.

In the case when A, b and σ satisfy Hypothesis 1.149, by Theorem 1.152, (1.80) is satisfied where the supremum is taken over $s \in [t, T]$ and, by (1.83), $\mathbb{E}|X(s; t, x) - x|^2 \le \rho(s - t)$ for some modulus ρ independent of t. When A, b and σ satisfy Hypotheses 1.143 and 1.145, by (1.69) we have $\sup_{s \in [t,T]} \mathbb{E}|X(s; t, x)|^p \le C_p$ for some C_p independent of t. Moreover, it is also easy to see that we have $\mathbb{E}|X(s; t, x) - x|^2 \le \rho(s - t)$ for some modulus ρ independent of t. Thus, using continuity and growth of F and arguing similarly as in the proof of Theorem 3.66 we can conclude that when $h \searrow 0$, the first term of the right-hand side of (4.155) converges to

$$
-\langle x, A^* D\varphi_n(x)\rangle - \langle (rI - A)^{-1} b(T, x), (rI - A)^* D\varphi_n(x)\rangle - \frac{1}{2}\mathrm{Tr}[\Sigma(T, x) D^2\varphi_n(x)].
$$

Concerning the second term of the right-hand side of (4.155), we observe first that $P_{T-h,s}[f_n(s,\cdot)](x) = \mathbb{E}[f_n(s, X(s; T-h, x))]$. Thus, using the mean square continuity estimates above we obtain that the second term of the right-hand side of (4.155) converges to $f_n(T, x)$ as $h \searrow 0$.

Thus, denoting by D_t^- the left time derivative, we have

$$
\begin{aligned}
D_t^- u_n(T, x) &= \lim_{h \searrow 0} \frac{u_n(T, x) - u_n(T-h, x)}{h} \\
&= -\langle x, A^* D\varphi_n(x)\rangle - \langle (rI - A)^{-1}b(T, x), (rI - A)^* D\varphi_n(x)\rangle \\
&\quad - \frac{1}{2}\text{Tr}[\Sigma(T, x)D^2\varphi_n(x)] + f_n(T, x),
\end{aligned}
$$

hence the equation is satisfied for $t = T$.

For $t < T$ we observe that by the semigroup property (see Theorem 1.157)

$$
u_n(t-h, x) = P_{t-h,t}[u_n(t, \cdot)](x) + \int_{t-h}^t P_{t-h,s}[f_n(s, \cdot)](x)ds,
$$

hence we have

$$
\frac{u_n(t, x) - u_n(t-h, x)}{h} = \frac{u_n(t, x) - P_{t-h,t}[u_n(t, \cdot)](x)}{h} - \frac{1}{h}\int_{t-h}^t P_{t-h,s}[f_n(s, \cdot)](x)ds
$$

and, arguing as for the case $t = T$ with $u(t, \cdot)$ in place of φ, we get

$$
\begin{aligned}
D_t^- u_n(t, x) &= -\langle x, A^* Du_n(t, x)\rangle - \langle (rI - A)^{-1}b(t, x), (rI - A)^* Du_n(t, x)\rangle \\
&\quad - \frac{1}{2}\text{Tr}[\Sigma(t, x)D^2 u_n(t, x)] - f_n(t, x).
\end{aligned}
$$

Since, by Lemma 4.134, the right-hand side of the above identity is a continuous function on $[0, T] \times H$, it follows from Lemma D.19 that $u_n(\cdot, x)$ is continuously differentiable. Thus we obtain that $(u_n)_t$ is continuous and u_n solves the equation

$$
(u_n)_t + \mathcal{A}_1(t)u_n + f_n = 0, \qquad u_n(T, \cdot) = \varphi_n.
$$

The claim now follows setting

$$
g_n(t, x) := F_0(t, x, u_n(t, x), D^{G(t,\cdot)}u_n(t, x)) - f_n(t, x).
$$

Step 3. Proof of (4.149) and (4.150) and of uniqueness.

By construction and by the beginning of Step 1 we have that $(\varphi_n) \subset D(\mathcal{A}_1)$ and $\varphi_n \xrightarrow[n\to\infty]{\mathcal{K}} \varphi$ in $C_m(H)$ which also gives the π-convergence of φ_n to φ. The π-convergence of u_n to u follows applying the dominated convergence theorem. The required π-convergence of $D^G u_n$ to $\nabla^G u$ follows from Proposition 4.16 (which allows us to move D^G or ∇^G inside the integral) and Hypothesis 4.133-(ii) (iii). The π-convergence of g_n to 0 now follows from the π-convergence of f_n to f and the continuity of F_0 in the last two arguments.

Concerning uniqueness in $\overline{B}_{m,\gamma_G}^{0,1,G}([0,T] \times H)$, we observe that any π-strong solution u^* must possess an approximating sequence (u_n) satisfying (4.153) for some sequences (φ_n) and (f_n), $\mathcal{K}$-converging (hence also π-converging) to φ and to $f^*(t,x) = F_0(t,x,u^*(t,x), \nabla^G u^*(t,x))$, respectively. Then, by the dominated convergence theorem we get that u^* must be a mild solution and so it is unique in $\overline{B}_{m,\gamma_G}^{0,1,G}([0,T] \times H)$ by Theorem 4.80.

The proof of the final statement is the same if we take into account that the sequences φ_n and f_n are already $\mathcal{K}$-convergent.   □

Remark 4.136

(i) The assumption of continuity in x of the function f in Theorem 4.135 seems essential to perform the above proof. On the other hand the continuity of f in t is less essential and could be avoided by refining the approximation result of Lemma B.78, allowing for f only measurable in t. In such a case only π-convergence would hold, since $\mathcal{K}$-convergence would also imply continuity in t.

(ii) The continuity of u and f in Theorem 4.135 is guaranteed, for example, if Hypothesis 4.87 (or 4.82) holds. However, all points of Hypothesis 4.133 are nontrivial to check. In the next example we will see that they are satisfied in the case of the Ornstein–Uhlenbeck semigroup.

(iii) In Theorem 4.135 the uniqueness of strong solutions is a consequence of the uniqueness of mild solutions provided by Theorem 4.80, which holds in the space $\overline{B}_{m,\gamma_G}^{0,1,G}([0,T] \times H)$. Since, as noted in Remark 4.81, uniqueness of mild solutions holds in a more general framework, the same applies to strong solutions.

(iv) In some applications to control problems, it may be useful to extend Theorem 4.135 to the case when the data are not continuous in x, hence the mild solution u only belongs to $\overline{B}_{m,\gamma_G}^{0,1,G}([0,T] \times H)$. This may be possible, in principle, if the functions φ and f in (4.152) can be approximated, using π-convergence, by a sequence (or a multisequence, as in [500], Sect. 5.2) of smooth functions, similarly to what is done in Lemma B.78. This cannot be achieved for all Borel measurable functions (see on this e.g. [563], Chap. 5), however it might be possible to do it for suitable classes of functions arising in applications.

(v) If the final datum φ is in $C_m^1(H)$ then, using the result of Theorem 4.80 (or Theorem 4.90), by a straightforward modification of the above proof, we can show that the convergences of g_n and $D^G u_n$ in Theorem 4.135 hold in $B_m([0,T] \times H)$. This is used later in the proof of Proposition 4.174.

(vi) It is clear from the proof that the claim of Theorem 4.135 still holds if we substitute Hypotheses 4.76 and 4.77 with Hypothesis 4.89-(ii)-(iii) and in addition we require that φ and F_0 are continuous. In this case u is the unique mild solution in $\overline{C}_{m,\gamma_G}^{0,1,G}([0, T] \times H)$. Thus u and f are automatically continuous.

∎

Example 4.137 Take $P_{t,s} = R_{s-t}$, where R is the Ornstein–Uhlenbeck semigroup studied in Sect. 4.3.1. Assume that Hypotheses 4.25, 4.29 and 4.32 are satisfied, requiring also that, as noted in Remark 4.52, Hypotheses 4.29 and 4.32 hold for $U = H$ and $G = I$.

Assuming Hypothesis 4.82 (or 4.87) we can apply Theorem 4.85 (or 4.90). This gives the continuity of u and f required in the statement of Theorem 4.135.

Hypothesis 4.127 is immediately true in this case as Hypotheses 1.143 and 1.145 are satisfied.

Concerning the validity of Hypothesis 4.133, parts (ii) and (iii) directly follow (also for $\mathcal{K}$-convergence) from formula (4.75) and the dominated convergence theorem. Part (i) is a consequence of Proposition B.91-(i) when $U = H$ and $G = I$, but the same proof also works when $G \neq I$ if G is bounded or if $D(A^*) \subset D(G^*)$ and, for some $c > 0$, $|G^*z| \leq c|A^*z|$ for all $z \in D(A^*)$. See Proposition 4.148-(ii) for more on this.

Thus, under the above assumptions, we can apply Theorem 4.135. See also Theorem 4.150 for a more specific statement in this case.

It would be interesting to study concrete examples (e.g. control problems with delay, see Sect. 2.6.8 and also [316]) to show that in some cases mild solutions are also strong solutions without the additional requirement that Hypotheses 4.29 and 4.32 also hold for $U = H$ and $G = I$. In such a case, using Remark 4.91, one would have a mild solution u with u jointly continuous but $D^G u$ continuous only in x. Then, if F_0 is continuous in x, the function f in the statement of Theorem 4.135 would only be continuous in x. It seems possible, by refining the argument of the proof of Lemma B.78, that the approximation result given there also holds, with π-convergence, for functions f which are continuous in x but only measurable in t. If this can be done, then one can also prove a variant of Theorem 4.135 that applies to this case.

Finally, Theorem 4.135 can be applied to the cases of Examples 4.100 and 4.101 if one is able to prove that Hypothesis 4.133 is satisfied. We are not aware of any references regarding this but it seems possible under suitable regularity assumptions about b and σ.

∎

4.5.2   The Elliptic Case

Similarly to the parabolic case, to define a classical solution[27] of (4.125), we connect Eq. (4.125) to the SDE (4.126).

---

[27] This definition is similar to the definition of a *strict* solution used in Sect. 6.4 of [179].

Let $x \in H$ and let $(\Omega, \mathscr{F}, \{\mathscr{F}_s\}_{s \geq 0}, \mathbb{P}, W)$ be a generalized reference probability space, where W is a cylindrical Wiener process in a real separable Hilbert space Ξ. Consider the SDE

$$\begin{cases} dX(s) = [AX(s) + b(X(s))]\,ds + \sigma(X(s))\,dW(s), & s \geq 0, \\ X(0) = x, & x \in H. \end{cases} \quad (4.156)$$

We assume the following.

Hypothesis 4.138

(i) The function F_0 satisfies Hypothesis 4.104 for some $m \geq 0$ and $L, L' > 0$.

(ii) For some $T > 0$ (e.g. $T = 1$) the operator A and the functions b and σ in (4.156) satisfy, in H, either Hypothesis 1.149 with b and σ independent of t and a, or Hypotheses 1.143 and 1.145 without dependence on s, a_1 and with $a_2(\cdot) = 0$. We have $\Sigma(x) := \sigma(x)\sigma^*(x)$.

(iii) There exists $r \in \mathbb{R}$, $r \in \varrho(A)$ such that $(rI - A)^{-1}b: H \to H$ is continuous and for every $y, z \in H$ the function $\langle \Sigma y, z \rangle : H \to \mathbb{R}$ is continuous. Moreover, there exists a $C > 0$ such that, for all $x \in H$,

$$\begin{cases} |(rI - A)^{-1}b(x)| \leq C(1 + |x|) \\ \|\sigma(x)\|_{\mathcal{L}(\Xi, H)} \leq C(1 + |x|). \end{cases} \quad (4.157)$$

(iv) For all $s \geq 0$, $\phi \in B_m(H)$, where $m \geq 0$ is from point (i) above, we have $P_s[\phi](x) = \mathbb{E}[\phi(X(s; x))]$, where $X(\cdot; x)$ is the mild solution of the SDE (4.156) (which exists and is unique by Theorem 1.152 or Proposition 1.147).

Similar comments to those in Remark 4.128 apply here.

As in the parabolic case we consider the operator $\mathcal{A}_1$ which is time-independent here (see the operator $\mathcal{A}_0$ in Appendix B.7.1):

$$\begin{cases} D(\mathcal{A}_1) = \left\{ \phi \in UC_b^2(H) : A^*D\phi \in UC_b(H, H), \ D^2\phi \in UC_b(H, \mathcal{L}_1(H)) \right\} \\ \mathcal{A}_1\phi(x) = \tfrac{1}{2}\,\mathrm{Tr}\,[\Sigma(x)D^2\phi(x)] + \langle x, A^*D\phi(x) \rangle + \langle (rI - A)^{-1}b(x), (rI - A)^*D\phi(x) \rangle. \end{cases} \quad (4.158)$$

Above $r \in \mathbb{R}$ belongs to the resolvent set $\varrho(A)$. Notice that the space $D(\mathcal{A}_1)$ is the same as in (4.141), hence it is a Banach space with the norm defined in (4.142).

Definition 4.139 A function $u : H \to \mathbb{R}$ is a classical solution of equation

$$\lambda v - \frac{1}{2}\mathrm{Tr}[\Sigma(x)D^2v] - \langle Ax + b(x), Dv \rangle - F_0(x, v, D^G v) = g(x), \quad x \in H, \quad (4.159)$$

(where g is a given Borel measurable function and the other data are as in Hypothesis 4.138) if $u \in D(\mathcal{A}_1)$, $D^G u \in C_b(H)$ and, for some $r \in \mathbb{R} \cap \varrho(A)$,

$$\lambda v(x) - \frac{1}{2} \operatorname{Tr} [\Sigma(x) D^2 v(x)] - \langle x, A^* D v(x) \rangle - \Big\langle (rI - A)^{-1} b(x), (rI - A)^* D v(x) \Big\rangle$$
$$- F_0(x, v(x), D^G v(x)) = g(x) \quad (4.160)$$

for every $x \in H$.

Observe that, as proved in Remark 4.130, the above definitions of the operator $\mathcal{A}_1$ and of a classical solution are independent of $r \in \mathbb{R} \cap \varrho(A)$.

Now we explain the notions of π- and $\mathcal{K}$-strong solutions.

Definition 4.140 We say that a function $u : H \to \mathbb{R}$ is a π-strong solution (respectively, a $\mathcal{K}$-strong solution) of Eq. (4.125) if there exists an $m \geq 0$ such that $u \in B_m^{1,G}(H)$ and there exist two sequences, $(g_n) \subset B_m(H)$ and $(u_n) \subset D(\mathcal{A}_1)$, such that for every $n \in \mathbb{N}$, u_n is a classical solution of the equation

$$\lambda w - \mathcal{A}_1 w - F_0(x, w, D^G w) = g_n(x) \quad (4.161)$$

and moreover, as $n \to +\infty$, using the notation of Definition B.55,

$$\begin{cases} \pi - \lim_{n \to +\infty} u_n = u & in \ B_m(H) \\ \pi - \lim_{n \to +\infty} D^G u_n = \nabla^G u & in \ B_m(H, U) \\ \pi - \lim_{n \to +\infty} g_n = g & in \ B_m(H) \end{cases} \quad (4.162)$$

(respectively, when all three convergences above hold in the $\mathcal{K}$ sense, using the notation of Definition B.56).

As in the parabolic case, to establish that any mild solution in the sense of Definition 4.102 is also a strong solution in the sense of Definition 4.140 (and, possibly, also the opposite), we need to define properly the approximating sequence u_n. Here too we have the same two main cases:

- When the data of the SDE (4.156) are very regular (e.g. the Ornstein–Uhlenbeck case of Sect. 4.3.1), and the semigroup P_s maps $D(\mathcal{A}_1)$ into itself. In this case it is enough to approximate only F_0.
- When the data of the SDE (4.156) are not regular (e.g. the case of invertible diffusion coefficient of Sect. 4.3.3 and the case of reaction-diffusion equations of Sect. 4.9.2) and the semigroup P_s does not map $D(\mathcal{A}_1)$ into itself. In this case, in addition to F_0, we also need to approximate the semigroup P_s.

Similarly to the parabolic case we only present a result which deals with the first case. The second case, up to now, has only been studied in the context of reaction-diffusion equations in [107] (see also [106], Chap. 10) and is briefly presented in Sect. 4.9.2.

Hypothesis 4.141 Let $m \geq 0$ be from Hypothesis 4.104 (and hence also Hypothesis 4.138).

(i) For all $s \geq 0$ and all $\phi \in \mathcal{F}C_0^{\infty,A^*}(H)$, the function

$$(s,x) \to P_s[\phi](x)$$

satisfies the last three lines of (4.144) for every $T > 0$. Moreover, there exists $C_1 > 0$ and $a_1 \geq 0$ such that, for every $\phi \in \mathcal{F}C_0^{\infty,A^*}(H)$,

$$\|P_s[\phi]\|_{D(\mathcal{A}_1)} \leq C_1\|\phi\|_{D(\mathcal{A}_1)}e^{a_1 s}, \qquad \forall s \geq 0. \tag{4.163}$$

(ii) If $\phi_n \xrightarrow[n\to\infty]{\pi} \phi$ in $B_m(H)$ then, for every $s > 0$,

$$D^G P_s[\phi_n] \xrightarrow[n\to\infty]{\pi} \nabla^G P_s[\phi] \quad in \quad B_m(H,U).$$

Remark 4.142 The semigroup assumptions of Hypothesis 4.141 are similar to those of Hypothesis 4.133 and address the preservation of regularity (i) and the convergence of derivatives (ii). What is new here is the exponential estimate for $\|P_s[\phi]\|_{D(\mathcal{A}_1)}$ which (like the exponential estimates in Sect. 4.4.2) is needed to ensure convergence of the integrals on infinite time intervals. ∎

Theorem 4.143 *Let Hypotheses 4.108, 4.110, 4.138 and 4.141 be satisfied. Let $m \geq 0$ be from Hypothesis 4.138. Let $\lambda > \lambda_0 \vee a_1$, where λ_0 is from Theorem 4.112 and a_1 from (4.163). Let u be the unique mild solution, in $B_m^{1,G}(H)$, of Eq. (4.125). Assume that u and the function $f(x) := F_0(x, u(x), \nabla^G u(x))$ are continuous in H. Then u is also the unique π-strong solution of (4.125) in $B_m^{1,G}(H)$.*

If Hypothesis 4.141 holds substituting everywhere π-convergence by $\mathcal{K}$-convergence, then u is the unique $\mathcal{K}$-strong solution of (4.125) in $B_m^{1,G}(H)$.

Proof The proof is very similar to the proof of Theorem 4.135 except for the proof that the approximations are classical solutions of the approximating equations.

We see that Hypothesis 4.106 is satisfied thanks to Hypothesis 4.138-(ii), estimate (1.80) (or (1.69)) and Remark 4.107. This, together with the other assumptions, ensures the existence and uniqueness, in $B_m^{1,G}(H)$, of the mild solution u of (4.125) thanks to Theorem 4.112-(i). We then have, for $x \in H$,

$$u(x) = \int_0^{+\infty} e^{-\lambda s} P_s[f](x)ds. \tag{4.164}$$

We know, by assumption, that $f \in C_m(H)$. We approximate f by the sequence (f_n) from Lemma B.78 with $B = A^*$ there (i.e. for an orthonormal basis contained in $D(A^*)$). We then have $f_n \in \mathcal{F}C_0^{\infty,A^*}(H) \subset D(\mathcal{A}_1)$ (see the proof of Theorem 4.135 Step 1) and $f_n \xrightarrow[n\to\infty]{\mathcal{K}} f$. We define

$$u_n(x) = \int_0^{+\infty} e^{-\lambda s} P_s[f_n](x)ds \tag{4.165}$$

and show that u_n satisfies the required properties.

Step 1: u_n is a classical solution of (4.125) with f_n in place of f.
Consider the function

$$w_n : [0, +\infty) \times H \to \mathbb{R}, \qquad w_n(t, x) = P_t[f_n](x).$$

Using Hypothesis 4.141-(i) and arguing as in the proof of Theorem 4.135 (see also Proposition B.91-(i), Theorem 7.5.1 of [179] and Theorem 9.25 of [180]) we know that w_n is a classical solution of the equation

$$w_t = \mathcal{A}_1 w, \qquad w(0) = f_n. \tag{4.166}$$

Now, by the last three lines of (4.144), we have $w_n(s, \cdot) \in D(\mathcal{A}_1)$ for every $s \geq 0$, and the function $(s, x) \to \mathcal{A}_1 w_n(s, x)$ is continuous. Since estimate (4.163) also holds and $\lambda > a_1$, the function

$$H \to \mathbb{R}, \qquad x \to \int_0^{+\infty} e^{-\lambda s} \mathcal{A}_1 [w_n(s, \cdot)](x) ds$$

is well defined and continuous in H. The facts above allow us to compute $\mathcal{A}_1 u_n$ from (4.165), differentiating under the integral sign thanks to Proposition 4.16,[28] obtaining that $u_n \in D(\mathcal{A}_1)$ and

$$\mathcal{A}_1 u_n(x) = \int_0^{+\infty} e^{-\lambda s} \mathcal{A}_1 [w_n(s, \cdot)](x) ds. \tag{4.167}$$

Now, for any $T > 0$ and $x \in H$, we compute, integrating by parts on $[0, T]$,

$$\int_0^T e^{-\lambda s} w_n(s, x) ds = -\frac{1}{\lambda} [e^{-\lambda T} w_n(T, x) - w_n(0, x)] + \frac{1}{\lambda} \int_0^T e^{-\lambda s} (w_n)_s(s, x) ds.$$

Hence, using (4.166),

$$\int_0^T e^{-\lambda s} w_n(s, x) ds = \frac{1}{\lambda} [-e^{-\lambda T} w_n(T, x) + f_n(x)] + \frac{1}{\lambda} \int_0^T e^{-\lambda s} \mathcal{A}_1 [w_n(s, \cdot)](x) ds. \tag{4.168}$$

We know by (4.130) that

$$\lim_{T \to +\infty} e^{-\lambda T} w_n(T, x) = 0.$$

Moreover, since $\lambda > a_1$, the function $s \to e^{-\lambda s} \|w_n(s, \cdot)\|_{D(\mathcal{A}_1)}$ is integrable on $[0, +\infty)$, hence we get

---

[28] As in other cases we cannot apply Corollary 4.14 here since the map $\mathbb{R}^+ \to C_b(H), s \to P_s[f_n]$, may not be measurable, see Remark 4.15.

$$\lim_{T \to +\infty} \int_0^T e^{-\lambda s} \mathcal{A}_1[w_n(s, \cdot)](x)ds = \int_0^{+\infty} e^{-\lambda s} \mathcal{A}_1[w_n(s, \cdot)](x)ds.$$

Thus passing to the limit as $T \to +\infty$ in (4.168) and using (4.165) and (4.167) we obtain

$$\lambda u_n(x) = f_n(x) + \mathcal{A}_1 u_n(x).$$

The claim follows setting

$$g_n(x) := f_n(x) - F_0(x, u_n(x), D^G u_n(x)).$$

Step 3. Proof of (4.162) *and the uniqueness.*

Concerning the convergences, by construction we have $f_n \xrightarrow[n \to \infty]{\mathcal{K}} f$ in $C_m(H)$. The π-convergence of u_n to u follows applying the dominated convergence theorem. The π-convergence of $D^G u_n$ to $\nabla^G u$ follows by Hypothesis 4.141-(ii). The π-convergence of g_n to 0 follows by the convergence of f_n to f and the continuity of F_0 in the last two arguments.

Concerning uniqueness, we observe that any π-strong solution u^* belongs to $B_m^{1,G}(H)$ by definition and must possess an approximating sequence (u_n) satisfying (4.165) for some sequence (f_n), π-converging to $f^*(x) = F_0(x, u^*(x), \nabla^G u^*(x))$. Then, by dominated convergence we obtain that u^* must be a mild solution in $B_m^{1,G}(H)$ and so it is unique by Theorem 4.112.

The proof of the final statement is immediate, after noticing that the sequences φ_n and f_n are already $\mathcal{K}$-convergent. $\square$

Remark 4.144

(i) Similarly to the parabolic case, the assumption of continuity of the functions u and f in Theorem 4.143 seems essential to perform the above proof. Extensions to cases where u and f are not continuous might be possible following the ideas explained in Remark 4.136-(iv).

(ii) The continuity of u and f are guaranteed, for example, if Hypothesis 4.117 (or 4.114) holds. However, Hypothesis 4.141 is not easy to check. From the discussion in Example 4.137 we see that points (i) and (ii) are satisfied in the case of the Ornstein–Uhlenbeck semigroup, except for estimate (4.163). This estimate holds by Lemma B.90-(i).

(iii) Similar observations as those in points (iii), (iv) and (vi) of Remark 4.136 apply here. In particular, the claim of Theorem 4.143 still holds if we substitute Hypotheses 4.108 and 4.110 with Hypothesis 4.119-(ii)-(iii) and in addition we require that F_0 is continuous. In this case u is the unique mild solution in $C_m^{1,G}(H)$. Also in this case the functions u and f are automatically continuous.

     ■

We conclude remarking that, as observed for the parabolic case in Example 4.137, one can apply Theorem 4.143 to the Ornstein–Uhlenbeck case mentioned there. Moreover, Theorem 4.143 could also be applied to the infinite horizon version of Examples 4.100 and 4.101 if one could prove that Hypothesis 4.141 is satisfied. This has not been studied but it seems possible under proper regularity assumptions about b and σ.

4.6 HJB Equations of Ornstein–Uhlenbeck Type: Lipschitz Hamiltonian

In Sects. 4.6 and 4.7 we consider special cases of HJB equations (4.109) and (4.125) where the linear parts of these HJB equations are associated to a stochastic process of Ornstein–Uhlenbeck type, namely a process of the type described in Sect. 4.3.1 (see also Sect. B.7.2).

Such equations were the first to be studied in the literature and better results can be proved for them. In this section we consider the case when the Hamiltonian F_0 is (as in Sect. 4.4) Lipschitz continuous in the last two variables while in the next section we consider the case when F_0 is only locally Lipschitz continuous. We always use Fréchet derivatives in these two sections.

As always H is a real separable Hilbert space. The parabolic and elliptic equations we study are the following (compare with (4.109) and (4.125)).

$$
\begin{cases}
v_t + \dfrac{1}{2}\, \mathrm{Tr}\,[\Sigma D^2 v] + \langle Ax, Dv \rangle + F_0(t, x, v, D^G v) = 0, & t \in [0, T), \ x \in D(A), \\
v(T, x) = \varphi(x), \ x \in H,
\end{cases}
$$

$$ (4.169) $$

$$
\lambda v - \frac{1}{2}\, \mathrm{Tr}\,[\Sigma D^2 v] - \langle Ax, Dv \rangle - F_0(x, v, D^G v) = 0, \qquad x \in D(A), \quad (4.170)
$$

where $\lambda > 0$. We make the following assumption in this section.

Hypothesis 4.145

(i) The linear operator $A : D(A) \subset H \to H$ is the infinitesimal generator of a strongly continuous semigroup $\{e^{tA}\}_{t \geq 0}$ on H so there exist $M \geq 1$ and $\omega \in \mathbb{R}$ such that

$$
\|e^{tA}\| \leq M e^{\omega t} \qquad \forall t \geq 0.
$$

(ii) $\Sigma \in \mathcal{L}^+(H)$ and for any $s > 0$, $e^{sA} \Sigma e^{sA^*} \in \mathcal{L}_1(H)$. Moreover, for all $t \geq 0$,

$$
\int_0^t \mathrm{Tr}\,\left[e^{sA} \Sigma e^{sA^*} \right] ds < +\infty, \qquad (4.171)
$$

so $\mathrm{Tr}\,[Q_t] < +\infty$, where $Q_t = \int_0^t e^{sA} \Sigma e^{sA^*} ds$.

(iii) The operator $G : D(G) \subset U \to H$ is linear, closed (possibly, but not necessarily, unbounded), where U is another real separable Hilbert space, possibly equal to H.

(iv) For every $t > 0$ the operator $e^{tA}G$ can be extended to a bounded operator, which we still denote by $e^{tA}G : U \to H$, and $e^{tA}G(U) \subset Q_t^{1/2}(H)$.

(v) Setting, for $t > 0$, $\Gamma_G(t) = Q_t^{-1/2}e^{tA}G$, the map $t \to \|\Gamma_G(t)\|$ belongs to $\mathcal{I}_1$.

We recall that we denote by γ_G a function in $\mathcal{I}_2$ (respectively $\mathcal{I}_1$) satisfying Hypothesis 4.76 (respectively, 4.108). Hence in this section, in light of Hypothesis 4.145 and Theorem 4.41, γ_G will be a properly chosen function in $\mathcal{I}_2$ (or $\mathcal{I}_1$) such that $\gamma_G(t) \geq c\|\Gamma_G(t)\|$, for given $c > 0$ and for all $t > 0$.

Remark 4.146

- A typical case where the operator G is not bounded is when we consider stochastic optimal control problems with the control acting only on the boundary or in a subdomain (see e.g. Sect. 2.6.2 for an explanation of the model and [189, 241, 310] for results in this case using the approach of mild solutions). Hypothesis 4.145 above is satisfied, for example, in the case when A is the Laplace operator with Neumann boundary conditions and one controls the normal derivative of the state on the boundary, (see Chap. 2, Sect. 2.6.2), taking in this case $G = (-A)^\beta$ for some $\beta \in (1/4, 1/2)$. If A is the Laplace operator with Dirichlet boundary conditions and we control the value of the state on the boundary, then taking $G = (-A)^\beta$ for some $\beta \in (3/4, 1)$, Hypothesis 4.145 holds except for point (v). Some ideas to remove this difficulty are in the recent paper [315].

- Another case where the operator G is unbounded comes from problems with pointwise delay in the control (see Sect. 2.6.8) which are studied in [316] by a variation of the approach of this chapter.

- In stochastic optimal control problems with noise at the boundary (see Sect. 2.6.3) both operators Σ and G are unbounded. The methods used in this chapter have not yet been applied to such a case. As is explained in the introduction of [181], in this case it is difficult to obtain the validity of Hypothesis 4.145-(v) since, even in the simplest cases, the minimal energy $\mathcal{E}(t, x)$ (which is equal to $|\Gamma_G(t)x|$, see Remark 4.34) needed to steer a given state x to 0 in time t blows up in a non-integrable way (indeed, in some cases, with a rate bigger than $e^{-1/t}$) as $t \to 0^+$. This case can be studied using the BSDE approach, see Chap. 6. We do not consider it here but we refer to the recent paper [315] where some new ideas to apply the approach of this chapter to such boundary noise/control problems are introduced.

- We finally recall that, choosing G to be the identity operator, our setting reduces to the cases investigated in [89, 90, 306, 307]. ∎

The transition semigroup R_t formally associated to the linear part of (4.169) and (4.170), and called the Ornstein–Uhlenbeck semigroup, is given (see (4.49) and (4.50)) by

$$R_t[\phi](x) = \int_H \phi(y + e^{tA}x)\mathcal{N}_{Q_t}(dy)$$

for any $\phi \in B_m(H)$. It is associated to the SDE (4.44) if $\sigma\sigma^* = \Sigma$. Using the results of Sect. 4.3.1 we easily prove that Hypothesis 4.145 implies the regularity properties of R_t which are needed to solve the HJB equations (4.169) and (4.170). More precisely, we have the following results, the first concerning the assumptions for the theorems about mild solutions, the second concerning additional assumptions for the theorems about strong solutions.

Proposition 4.147 *Let Hypothesis 4.145 be satisfied. Consider the transition semigroup* $P_{t,s} = R_{s-t}$. *Then for any* $m \geq 0$ *we have the following (recall that* $K_1(m)$ *below is the constant from Theorem 4.41).*

(A) *Parabolic case.*

   (i) *Hypothesis 4.74 is satisfied with* $C(m) = 2K_1(m)e^{m(\omega\vee 0)T}$ *when* $\omega \neq 0$, *and with* $C(m) = 2K_1(m)(1 + T^m)$ *when* $\omega = 0$.
   (ii) *If the function* $t \to \|\Gamma_G(t)\|$ *belongs to* $\mathcal{I}_2$, *then Hypothesis 4.76 is satisfied for all* $\phi \in B_m(H)$, *where in addition we have the continuous G-Fréchet differentiability of* $R_{s-t}[\phi]$, *for* $s > t \geq 0$. *Moreover, we can choose* $\gamma_G(s) = 2K_1(m/2)^{1/2}e^{m(\omega\vee 0)T}\|\Gamma_G(s)\|$ *when* $\omega \neq 0$ *and* $\gamma_G(s) = 2K_1(m/2)^{1/2}(1 + T^m)\|\Gamma_G(s)\|$ *when* $\omega = 0$.
   (iii) *Parts (i) and (ii) of Hypotheses 4.84, 4.89, 4.93 are satisfied.*
   (iv) *Hypothesis 4.77 is satisfied, together with its variant given in Remark 4.91.*
   (v) *Assume, moreover, that Hypothesis 4.145-(iv) also holds for* $U = H$ *and* $G = I$. *Then the variants of Hypothesis 4.77 described in parts (iii) of Hypotheses 4.84, 4.89, 4.93 are also satisfied.*

(B) *Elliptic case.*

   (i) *Hypothesis 4.106 is satisfied with* $C(m) = 2K_1(m)$ *and* $a(m) = \varepsilon + m(\omega \vee 0)$ *(for any* $\varepsilon > 0$ *which can be chosen to be 0 if* $\omega \neq 0$*).*
   (ii) *Hypothesis 4.108 is satisfied for all* $\phi \in B_m(H)$, *where in addition we have the continuous G-Fréchet differentiability of* $R_s[\phi]$, *for* $s > 0$. *Moreover, we can choose* $\gamma_G(s) = 2K_1(2m)^{1/2}\|\Gamma_G(s)\|$ *and* $a(m) = \varepsilon + m(\omega \vee 0)$ *(for any* $\varepsilon > 0$, *which can be chosen to be 0 if* $\omega \neq 0$*).*
   (iii) *Parts (i) and (ii) of Hypotheses 4.115, 4.119, 4.122 are satisfied.*
   (iv) *Hypothesis 4.110 is satisfied, together with its variants described in parts (iii) of Hypotheses 4.115, 4.119, 4.122.*

Proof Parabolic case.

   (i) Concerning Hypothesis 4.74, the semigroup property follows from Corollary 1.158 while estimate (4.113) follows from (4.76).
   (ii) Hypothesis 4.76 follows from Theorem 4.41.
   (iii) Parts (i) and (ii) of Hypotheses 4.84, 4.89, 4.93 follow from points (i) and (ii) of the parabolic case of this proposition.

(iv) Hypothesis 4.77 and its variant given in Remark 4.91 follows from Proposition 4.50.
 (v) When Hypothesis 4.145-(iv) also holds for $U = H$ and $G = I$, the variants of Hypothesis 4.77 described in (iii) of Hypotheses 4.84, 4.89, 4.93 follow from Proposition 4.51 and Remark 4.53-(i).

Elliptic case.

 (i) The semigroup property in Hypothesis 4.106 follows from Corollary 1.158 and (4.130) follows from (4.76).
 (ii) Hypothesis 4.108 follows from Theorem 4.41.
(iii) Parts (i) and (ii) of Hypotheses 4.84, 4.89, 4.93 follow from points (i) and (ii) of the elliptic case of this proposition.
(iv) Hypothesis 4.110 and its variants described in parts (iii) of Hypotheses 4.115, 4.119, 4.122, follow from Proposition 4.50 and Remark 4.53-(i). Note that, in Proposition 4.50 and Remark 4.53-(i), T is arbitrary.

$\square$

Proposition 4.148 *Let Hypothesis 4.145 be satisfied. Consider the transition semigroup $P_{t,s} = R_{s-t}$. Then we have the following.*

(A) Parabolic case.

  (i) Hypothesis 4.127-(ii)-(iii)-(iv) is satisfied.
 (ii) Hypothesis 4.133 is satisfied (both with π- and $\mathcal{K}$-convergence) when G is bounded or when $D(A^*) \subset D(G^*)$ and, for some $c > 0$, $|G^*z| \leq c|A^*z|$ for all $z \in D(A^*)$.

(B) Elliptic case.

  (i) Hypothesis 4.138-(ii)-(iii)-(iv) is satisfied.
 (ii) Hypothesis 4.141 is satisfied (both with π- and $\mathcal{K}$-convergence) when G is bounded or when $D(A^*) \subset D(G^*)$ and, for some $c > 0$, $|G^*z| \leq c|A^*z|$ for all $z \in D(A^*)$.

Proof Parabolic case.

 (i) Since here $b = 0$ and we can take $\sigma = \sqrt{\Sigma}$, which is constant, parts (iii) and (iv) of Hypothesis 4.127 are immediately satisfied. Hypothesis 4.127-(ii) follows from Hypothesis 4.145, in particular from (4.171).
 (ii) Concerning Hypothesis 4.133, if $U = H$ and $G = I$, part (i) is a direct consequence of Proposition B.91. When $G \neq I$ the only thing which remains to be proved is that the function in point (i) of Hypothesis 4.133 (which we call u) admits a G-derivative with respect to x and that $D^G u \in C_b([0, T] \times H)$. When G is bounded this follows since $Du \in C_b([0, T] \times H)$. When G is unbounded and $D(A^*) \subset D(G^*)$ and, for some $c > 0$, $|G^*z| \leq c|A^*z|$ for all $z \in D(A^*)$, this follows since $A^*Du \in C_b([0, T] \times H)$. Parts (ii) and (iii) directly follow from formula (4.75) and the dominated convergence theorem.

Elliptic case.

(i) Parts (ii), (iii) and (iv) of Hypothesis 4.127 are true for the same reasons as in the parabolic case.
(ii) Concerning Hypothesis 4.141, the regularity in part (i) follows by Proposition B.91 exactly as in the parabolic case. The exponential estimate (4.163) is a consequence of Lemma B.90-(i). Part (ii) directly follows from formula (4.75) and the dominated convergence theorem.

$$\square$$

To be able to use Gronwall's lemma estimate of Proposition D.30, we will sometimes work with the stronger assumption (used in most of the literature, see e.g. [89, 90, 306, 307]) that, for every $t \geq 0$,

$$\|\Gamma_G(t)\| \leq C(1 \vee t^{-\alpha}) \qquad \text{for some } C > 0 \text{ and } \alpha \in (0, 1). \tag{4.172}$$

Recalling the operator $\mathcal{A}_0$ introduced in Sect. B.7.1, we can rewrite the two HJB equations (4.169) and (4.170) in a slightly stronger form (since we require $Dv \in D(A^*)$):

$$\begin{cases} v_t + \mathcal{A}_0 v + F_0(t, x, v, D^G v) = 0, & t \in [0, T), \ x \in H, \\ v(T, x) = \varphi(x), & x \in H, \end{cases} \tag{4.173}$$

$$\lambda v - \mathcal{A}_0 v - F_0(x, v, D^G v) = 0, \qquad x \in H. \tag{4.174}$$

The same can be done substituting the operator $\mathcal{A}_0$ with $\hat{\mathcal{A}}_0$, which is also introduced in Sect. B.7.1.

Since (see Proposition B.92) the operators $\mathcal{A}_0$ (respectively $\hat{\mathcal{A}}_0$) are $\mathcal{K}$-closable in $C_m(H)$ for $m \geq 1$ (respectively for $m \geq 0$) and their $\mathcal{K}$-closure $\mathcal{A}^m$ is the generator, in the sense of $\mathcal{K}$-semigroups,[29] of the Ornstein–Uhlenbeck semigroup R_t in $C_m(H)$, we can write the above two equations in the weaker forms

$$\begin{cases} v_t + \mathcal{A}^m v + F_0(t, x, v, D^G v) = 0, & t \in [0, T), \ x \in H, \\ v(T, x) = \varphi(x), & x \in H, \end{cases} \tag{4.175}$$

$$\lambda v - \mathcal{A}^m v - F_0(x, v, D^G v) = 0, \qquad x \in H. \tag{4.176}$$

The mild forms of the equation are thus the following:

$$v(t, x) = R_{T-t}[\varphi](x) + \int_t^T R_{s-t}\Big[F_0(s, \cdot, v(s, \cdot), D^G v(s, \cdot))\Big](x)ds, \ t \in [0, T], \ x \in H, \tag{4.177}$$

---

[29]This result can also be proved using the π-semigroups framework, see [493], Part III.

$$v(x) = \int_0^{+\infty} e^{-\lambda s} R_s \left[F_0(\cdot, v(\cdot), D^G v(\cdot)) \right](x) ds, \quad x \in H. \qquad (4.178)$$

We analyze separately the parabolic and the elliptic equations.

4.6.1 The Parabolic Case

The first result is a straightforward corollary of Theorems 4.80, 4.96 and Proposition 4.147. Recall that the concept of a mild solution we use here is the one introduced in Definition 4.70.

Theorem 4.149 *Let Hypothesis 4.145 hold with the function* $t \to \|\Gamma_G(t)\|$ *belonging to* $\mathcal{I}_2$ *and assume Hypothesis 4.72 is satisfied for a given* $m \geq 0$. *We have the following.*

  (i) *Equation (4.169) has a mild solution* u *which is unique in* $\overline{B}_{m,\gamma_G}^{0,1,G}([0,T] \times H)$ *for* γ_G *from Proposition 4.147. Moreover, if* F_0 *is also continuous in* x *and* φ *is continuous, then* u *and* $D^G u$ *are also continuous in* x. *If* $\varphi \in C_m^1(H)$ *then* $D^G u \in B_m([0,T] \times H, U)$.
 (ii) *If Hypothesis 4.145 also holds with* $U = H$ *and* $G = I$ *in point (iv), then, even if* $\varphi \in B_m(H)$, *the functions* u *and* $D^G u$ *are jointly continuous in* $(t,x) \in [0,T) \times H$. *If in addition Hypothesis 4.145-(v) also holds with* $U = H$ *and* $G = I$ *then also* Du *is jointly continuous in* $(t,x) \in [0,T) \times H$.
(iii) *Let Hypothesis 4.145 also hold with* $U = H$ *and* $G = I$ *in point (iv). If* $\varphi \in C_m(H)$ *and* F_0 *is continuous then the mild solution* u *belongs to* $\overline{C}_{m,\gamma_G}^{0,1,G}([0,T] \times H)$ *for* γ_G *as in point (i). If* $\varphi \in C_m^1(H)$ *then* $D^G u \in C_m([0,T] \times H, U)$.
 (iv) *Let Hypothesis 4.145 also hold with* $U = H$ *and* $G = I$ *in point (iv). Let* $\varphi \in UC_m(H)$ *and the function* $(s,x,y,z) \to F_0((s,x,y,z)/(1 + |x|^m + |y| + |z|)$ *be uniformly continuous in the last three variables, uniformly with respect to the first. Then the mild solution* u *belongs to* $\overline{UC}_{m,\gamma_G}^{0,1,G}([0,T] \times H)$ *for* γ_G *as in point (i). If* $\varphi \in UC_m^1(H)$ *then* $D^G u \in UC_m^x([0,T] \times H, U)$.

Proof The first part of (i) follows from Theorem 4.80 and Proposition 4.147-(A). Continuity in x follows from Remark 4.91. When $\varphi \in C_m^1(H)$, the claim is a consequence of Theorem 4.41-(iii).

Point (ii), when Hypothesis 4.145-(iv) (respectively, Hypothesis 4.145-(iv)-(v)) also holds with $U = H$ and $G = I$, follows since u satisfies (4.177) whose right-hand side is continuous, together with its derivative D^G (respectively, D), thanks to Proposition 4.51.

Point (iii) follows from Theorem 4.90, Proposition 4.147-(A) and, when $\varphi \in C_m^1(H)$, from Theorem 4.41-(iii).

Point (iv) follows from Theorem 4.94, Proposition 4.147-(A) and, when $\varphi \in UC_m^1(H)$, from Theorem 4.41-(iii). □

We now prove two results. The first is relatively simple and deals with the convergence of approximating sequences to the mild solution. The second concerns the C^2 space regularity of the mild solutions. Since it is more complex it will be the subject of a separate subsection.

Concerning strong solutions we show that Theorem 4.135 holds here with a stronger type of convergence.

Theorem 4.150 *Let $T > 0$ and let Hypothesis 4.145 be satisfied with the following additions.*

(i) The map $t \mapsto \|\Gamma_G(t)\|$ belongs to $\mathcal{I}_2$.
(ii) Either G is bounded or $D(A^*) \subset D(G^*)$ and, for some $c > 0$, $|G^*z| \leq c|A^*z|$ for all $z \in D(A^*)$.

Let Hypothesis 4.72 be satisfied for some $m \geq 0$ and assume, moreover, that φ is continuous. Let u be the mild solution of Eq. (4.169) from Theorem 4.149 and let the function $F_0(\cdot, \cdot, u(\cdot, \cdot), \nabla^G u(\cdot, \cdot))$ be continuous on $[0, T) \times H$. Then the function u is the unique $\mathcal{K}$-strong solution of Eq. (4.169) among solutions in $\overline{B}_{m,\gamma_G}^{0,1,G}([0, T] \times H)$, for γ_G from Proposition 4.147.

Assume moreover that Hypothesis 4.145 also holds with $U = H$ and $G = I$ in point (iv). Then the sequence u_n approximating u can be chosen so that its convergence to u is uniform on $[0, T - \varepsilon] \times H_0$ for all $\varepsilon \in (0, T)$ and all bounded subsets H_0 of H.

Proof We know from Propositions 4.147-(A) and 4.148-(A) that the assumptions of Theorem 4.135 are satisfied. Hence Theorem 4.135 still holds in this case and u is also the unique $\mathcal{K}$-strong solution as claimed in the statement. The improvement of the convergence follows from the compactness of the operators e^{tA}, $t > 0$, which follows from Remark 4.31-(i) taking there $U = H$ and $G = I$. Recall that, setting $f(t, x) = F_0(t, x, u(t, x), D^G u(t, x))$, the mild solution u is written as

$$u(t, x) = R_{T-t}[\varphi](x) + \int_t^T R_{s-t}[f(s, \cdot)](x)ds$$

and its approximations u_n, as in Theorem 4.135, are

$$u_n(t, x) = R_{T-t}[\varphi_n](x) + \int_t^T R_{s-t}[f_n(s, \cdot)](x)ds$$

for suitable sequences (φ_n), (f_n) which $\mathcal{K}$-converge to φ in $C_m(H)$ and to f in $\overline{B}_{m,\gamma_G}([0, T] \times H)$, respectively.

We first prove the required convergence for the first terms of the above formulae. Let φ_n be $\mathcal{K}$-convergent to φ and observe that, for $s \geq \varepsilon > 0$, $x \in H$,

$$R_s[\varphi_n](x) - R_s[\varphi](x) = R_{s-\varepsilon} R_\varepsilon[\varphi_n - \varphi](x).$$

Now, defining $z_n^\varepsilon(x) := R_\varepsilon[\varphi_n - \varphi](x)$, we see that for any bounded subset H_0 of H,

$$\sup_{x\in H_0} |z_n^\varepsilon(x)| \le \int_H [\sup_{x\in H_0} |(\varphi_n - \varphi)(y + e^{\varepsilon A}x)|]\mathcal{N}_{Q_\varepsilon}(dy).$$

By compactness of the operator $e^{\varepsilon A}$ we immediately conclude that the right-hand side converges to 0. Now, for fixed $0 < \varepsilon < T$, we have

$$\sup_{s\in[\varepsilon,T],x\in H_0} |R_s[\varphi_n](x) - R_s[\varphi](x)| = \sup_{s\in[\varepsilon,T],x\in H_0} |R_{s-\varepsilon}[z_n^\varepsilon](x)|$$

$$\le \int_H [\sup_{s\in[\varepsilon,T],x\in H_0} |z_n^\varepsilon(y + e^{(s-\varepsilon)A}x)|]\mathcal{N}_{Q_{s-\varepsilon}}(dy).$$

Since z_n^ε converges to 0 uniformly on bounded subsets of H we obtain that the last term converges to 0 as $n \to +\infty$.

We now look at the second terms. We have, for $(t,x) \in [0,T] \times H$, $n \in \mathbb{N}$,

$$\left| \int_t^T R_{s-t}[f_n(s,\cdot)](x)ds - \int_t^T R_{s-t}[f(s,\cdot)](x)ds \right|$$

$$\le \int_0^{T-t} |R_s[(f_n - f)(s+t,\cdot)](x)|\, ds \le \int_0^T |R_s[(f_n - f)(s,\cdot)](x)|\, ds.$$

Let H_0 be a bounded subset of H. We have, for $n \in \mathbb{N}$,

$$\sup_{x\in H_0} \int_0^T |R_s[(f_n - f)(s,\cdot)](x)|\, ds \le \int_0^T \int_H \sup_{x\in H_0} \left|(f_n - f)\left(s, y + e^{sA}x\right)\right| \mathcal{N}_{Q_s}(dy)ds.$$

Now, by the compactness of e^{sA}, we have for any $s \in (0,T]$ and $y \in H$, $\sup_{x\in H_0} \left|(f_n - f)\left(s, y + e^{sA}x\right)\right| \to 0$ as $n \to +\infty$. Moreover, for each $s \in (0,T]$, $n \in \mathbb{N}$, and $z \in H$, using Definition 4.131 of $\mathcal{K}$-convergence in $\overline{B}_{m,\gamma_G}([0,T) \times H)$ and the fact that the approximating functions (f_n) satisfy (B.31), we have

$$|(f_n - f)(s,z)| \le C_1\gamma_G(s)\|f\|_{\overline{B}_{m,\gamma_G}}(1 + |z|^m)$$

for some $C_1 > 0$. Hence, by the boundedness of H_0, we obtain for $n \in \mathbb{N}$,

$$\sup_{x\in H_0} \left|(f_n - f)\left(s, y + e^{sA}x\right)\right|$$

$$\le C_1\gamma_G(s)\|f\|_{\overline{B}_{m,\gamma_G}} \sup_{x\in H_0} (1 + |y + e^{sA}x|^m) \le C_2\gamma_G(s)(1 + |y|^m)$$

for some $C_2 > 0$. Then, by the dominated convergence theorem, for any $s \in (0,T]$,

$$\overline{f}_n(s) := \int_H \sup_{x\in H_0} \left|(f_n - f)\left(s, y + e^{sA}x\right)\right| \mathcal{N}_{Q_s}(dy) \to 0, \qquad \text{as } n \to +\infty.$$

Since for $n \in \mathbb{N}$, $\overline{f}_n(s) \leq C_3 \gamma_G(s)$ for some $C_3 > 0$, we can apply the dominated convergence theorem again to get the claim. $\square$

4.6.1.1 C^2 Regularity of Mild Solutions

We prove, under additional assumptions, C^2 regularity in x of the mild solution, following and generalizing [306]. Even if some generalizations seem possible, we only consider the case when $U = H$, $G = I$ and $m = 0$. In this subsection (as was done, for example, in Sect. 4.3.1), we set $\Gamma(t) := \Gamma_I(t)$ (see (4.59)) and $\gamma(t) := \|\Gamma(t)\|$.

Hypothesis 4.151 Let $U = H$ and $T > 0$. Assume that F_0 satisfies Hypothesis 4.72 for $m = 0$ and is continuous. Assume, moreover, that the map $(x, v, p) \to F_0$ (t, x, v, p) is Fréchet differentiable with continuous Fréchet derivatives $D_x F_0$, $D_v F_0$, $D_p F_0$. Suppose also that for some constant $C_1 \geq (L \vee L') > 0$, we have, for all $t \in [0, T]$, $x \in H$, $v_1, v_2 \in \mathbb{R}$, $p_1, p_2 \in H$,

$$|D_x F_0(t, x, v_1, p_1) - D_x F_0(t, x, v_2, p_2)| + |D_v F_0(t, x, v_1, p_1) - D_v F_0(t, x, v_2, p_2)|$$
$$+ |D_p F_0(t, x, v_1, p_1) - D_p F_0(t, x, v_2, p_2)| \leq C_1(|v_1 - v_2| + |p_1 - p_2|) \quad (4.179)$$

and[30]

$$|D_x F_0(t, x, v, p)| \leq C_1(1 + |v| + |p|),$$
$$|D_v F_0(t, x, v, p)| + |D_p F_0(t, x, v, p)| \leq C_1, \quad (4.180)$$

for all $t \in [0, T]$, $x \in H$, $v \in \mathbb{R}$, $p \in H$.

The first theorem deals with the case when the initial datum belongs to $C_b^1(H)$.

Theorem 4.152 Let Hypothesis 4.145 hold with $U = H$, $G = I$ and $\gamma \in \mathcal{I}_2$ and assume that Hypothesis 4.151 is satisfied. Let $T > 0$, $\varphi \in C_b^1(H)$, and let $u \in C_b^{0,1}([0, T] \times H)$ be the mild solution of (4.169) from Theorem 4.149. Then u is twice Fréchet differentiable with respect to x on $[0, T) \times H$ and $D^2 u \in \overline{C}_{b,\gamma}^s([0, T) \times H, \mathcal{L}(H))$.

Proof The mild solution u of (4.169) obtained in Theorem 4.149 belongs to $C_b^{0,1}([0, T] \times H)$. To show the higher regularity we consider the map

$$\Upsilon[v](t, x) = R_{T-t}[\varphi](x) + \int_t^T R_{s-t}[F_0(s, \cdot, v(s, \cdot), Dv(s, \cdot))](x)ds. \quad (4.181)$$

We know from Theorem 4.149 that u is a fixed point of Υ in the space $C_b^{0,1}$ $([0, T] \times H)$. We now consider Υ in the Banach space of more regular functions (depending on T_0 to be fixed later)

---

[30]The second inequality in (4.180) follows from Hypothesis 4.72-(i) and the fact that $C_1 \geq L$, but we repeat it here for the reader's convenience.

$$\overline{\mathcal{H}}_{T_0,T} = \{v \in C_b ([T_0, T] \times H) : Dv \in C_b ([T_0, T] \times H, H),$$
$$D^2 v \in \overline{C}^s_{b,\gamma} ([T_0, T) \times H, \mathcal{L}(H))\} \tag{4.182}$$

with the norm

$$\|v\|_{\overline{\mathcal{H}}_{T_0,T}} = \sup_{t \in [T_0, T)} \left[\|v(t, \cdot)\|_0 + \|Dv(t, \cdot)\|_0 + \gamma(T - t)^{-1} \|D^2 v(t, \cdot)\|_0 \right],$$

where $\overline{C}^s_{b,\gamma} ([T_0, T) \times H, \mathcal{L}(H))$ is as in Definition 4.24.
We look at the equation $v = \Upsilon[v]$ in this space.

Step 1. $\Upsilon : \overline{\mathcal{H}}_{T_0,T} \to \overline{\mathcal{H}}_{T_0,T}$.
Setting, for $v \in \overline{\mathcal{H}}_{T_0,T}$, $(s, x) \in [T_0, T) \times H$,

$$\psi^v(s, x) = F_0(s, x, v(s, x), Dv(s, x)) \tag{4.183}$$

we have $\psi^v \in C^{0,1}_{b,\gamma}([T_0, T) \times H)$ and, for $(s, x) \in [T_0, T) \times H, h \in H$,

$$\langle D\psi^v(s, x), h \rangle = \langle D_x F_0(s, x, v(s, x), Dv(s, x)), h \rangle$$
$$+ D_v F_0(s, x, v(s, x), Dv(s, x)) \langle Dv(s, x), h \rangle \tag{4.184}$$
$$+ \langle D_p F_0(s, x, v(s, x), Dv(s, x)), D^2 v(s, x)h \rangle.$$

By (4.181) and Propositions 4.50 and 4.51, it follows that $\Upsilon[v] \in C_b([T_0, T] \times H)$ for $v \in \overline{\mathcal{H}}_{T_0,T}$. Moreover, by Proposition 4.16, differentiating (4.181) we obtain for $(t, x) \in [T_0, T) \times H$,

$$D\Upsilon[v](t, x) = DR_{T-t}[\varphi](x) + \int_t^T DR_{s-t} \left[\psi^v(s, \cdot) \right](x)ds. \tag{4.185}$$

Since we can write, for $(t, x) \in [T_0, T) \times H$,

$$\Upsilon[v](t, x) = \int_H \varphi(y + e^{tA}x) \mathcal{N}_{Q_t}(dy) + \int_t^T \int_H \psi^v(s, y + e^{(s-t)A}x) \mathcal{N}_{Q_{s-t}}(dy)ds, \tag{4.186}$$

then, for $(t, x) \in [T_0, T) \times H, h \in H$, we have, using Theorem 4.41-(iii) (see also the proof of Lemma B.90),

$$\langle D\Upsilon[v](t, x), h \rangle = \int_H \langle D\varphi(y + e^{tA}x), e^{tA}h \rangle \mathcal{N}_{Q_t}(dy)$$
$$+ \int_t^T \int_H \langle D\psi^v(s, y + e^{(s-t)A}x), e^{(s-t)A}h \rangle \mathcal{N}_{Q_{s-t}}(dy)ds, \tag{4.187}$$

i.e.

$$\langle D\Upsilon[v](t,x), h\rangle = R_{T-t}[\langle D\varphi, e^{tA}h\rangle](x) + \int_t^T R_{s-t}\left[\langle D\psi^v(s,\cdot), e^{(s-t)A}h\rangle\right](x)ds$$

$$(4.188)$$

so, again from Propositions 4.50 and 4.51 and from the strong continuity of the semigroup e^{tA}, we get $D\Upsilon[v] \in C_b([T_0, T] \times H, H)$. Finally, for $(t, x) \in [T_0, T) \times H$, $h \in H$, we get, similarly to (4.185),

$$D^2\Upsilon[v](t,x)h = DR_{T-t}[\langle D\varphi, e^{tA}h\rangle](x) + \int_t^T DR_{s-t}\left[\langle D\psi^v(s,\cdot), e^{(s-t)A}h\rangle\right](x)ds.$$

$$(4.189)$$

Since in the right-hand side of (4.189) we have first derivatives applied to the semi-group R_t, we can use Theorem 4.41 and Propositions 4.50–4.51, together with the strong continuity of the semigroup e^{tA}, to see that $D^2\Upsilon[v] \in \overline{C}_{b,\gamma}^s([T_0, T) \times H, \mathcal{L}(H))$.

Step 2. Local existence.
Let $T_0 \in [0, T]$. Using (4.181), (4.188), (4.189) and standard computations we have, for $v \in \overline{\mathcal{H}}_{T_0, T}$ and $t \in [T_0, T]$ (note that integrability of the norms below follows from Lemma 1.21)

$$\|\Upsilon[v](t, \cdot)\|_0 \le \|\varphi\|_0 + \int_t^T \|\psi^v(s, \cdot)\|_0 ds$$

$$\|D\Upsilon[v](t, \cdot)\|_0 \le Me^{(\omega \vee 0)T}\left[\|D\varphi\|_0 + \int_t^T \|D\psi^v(s, \cdot)\|_0 ds\right]$$

$$\|D^2\Upsilon[v](t, \cdot)\|_0 \le Me^{(\omega \vee 0)T}\left[\gamma(T-t)\|D\varphi\|_0 + \int_t^T \gamma(s-t)\|D\psi^v(s, \cdot)\|_0 ds\right] \quad (4.190)$$

(using (4.70) in the last inequality) so that

$$\gamma(T-t)^{-1}\|D^2\Upsilon[v](t, \cdot)\|_0$$

$$\le Me^{(\omega \vee 0)T}\left[\|D\varphi\|_0 + \gamma(T-t)^{-1}\int_t^T \gamma(s-t)\|D\psi^v(s, \cdot)\|_0 ds\right],$$

$$(4.191)$$

which gives

$$\|\Upsilon[v]\|_{\overline{\mathcal{H}}_{T_0, T}} \le \|\varphi\|_0 + 2Me^{(\omega \vee 0)T}\|D\varphi\|_0 + (T - T_0)\|\psi^v\|_0$$

$$+ Me^{(\omega \vee 0)T}\int_t^T \left(\gamma(T-s) + \frac{\gamma(s-t)\gamma(T-s)}{\gamma(T-t)}\right)\gamma(T-s)^{-1}\|D\psi^v(s, \cdot)\|_0 ds$$

$$\leq Me^{(\omega \vee 0)T}\|\varphi\|_1 + \rho_1(T-T_0)\left[\|\psi^v\|_0 + \sup_{s\in[T_0,T)}[\gamma(T-s)^{-1}\|D\psi^v(s,\cdot)\|_0]\right], \quad (4.192)$$

where

$$\rho_1(T-T_0) := Me^{(\omega\vee 0)T}\left((T-T_0)\vee \sup_{t\in[T_0,T)}\int_t^T\left(\gamma(T-s)+\frac{\gamma(s-t)\gamma(T-s)}{\gamma(T-t)}\right)ds\right).$$
$$(4.193)$$

Moreover, by (4.183) and Hypothesis 4.72-(ii),

$$\|\psi^v\|_0 \leq C_1[2+\|v\|_0+\|Dv\|_0]$$

and, by (4.184) and (4.180),

$$\|D\psi^v(t,\cdot)\|_0 \leq 2C_1\left[1+\|v(t,\cdot)\|_0+\|Dv(t,\cdot)\|_0+\|D^2v(t,\cdot)\|_0\right]. \quad (4.194)$$

Hence, for $C_2 = 2C_1(\gamma(T)^{-1}\vee 1)$, we have

$$\sup_{t\in[T_0,T)}\gamma(T-t)^{-1}\|D\psi^v(t,\cdot)\|_0$$

$$\leq C_2\left[1+\|v\|_0+\|Dv\|_0+\sup_{t\in[T_0,T)}\gamma(T-t)^{-1}\|D^2v(t,\cdot)\|_0\right]$$
$$(4.195)$$

and then estimate (4.192) becomes

$$\|\Upsilon[v]\|_{\overline{\mathcal{H}}_{T_0,T}} \leq Me^{(\omega\vee 0)T}\|\varphi\|_1$$

$$+(2C_1+C_2)\rho_1(T-T_0)\left[1+\|v\|_0+\|Dv\|_0+\sup_{t\in[T_0,T)}\gamma(T-t)^{-1}\|D^2v(t,\cdot)\|_0\right],$$

which gives, taking $\rho_2(T-T_0) := (2C_1+C_2)\rho_1(T-T_0)$ and $C_3 := Me^{(\omega\vee 0)T}$,

$$\|\Upsilon[v]\|_{\overline{\mathcal{H}}_{T_0,T}} \leq C_3\|\varphi\|_1 + \rho_2(T-T_0)\left[1+\|v\|_{\overline{\mathcal{H}}_{T_0,T}}\right]. \quad (4.196)$$

Now take $w_1, w_2 \in \overline{\mathcal{H}}_{T_0,T}$. We have

$$\Upsilon[w_1](t,x) - \Upsilon[w_2](t,x) = \int_t^T R_{s-t}\left[\psi^{w_1}(s,\cdot) - \psi^{w_2}(s,\cdot)\right](x)ds.$$

We estimate the right-hand side exactly as in (4.192) taking $\varphi = 0$ and $\psi^{w_1} - \psi^{w_2}$ in place of ψ^v. We then get

$$\|\Upsilon[w_1] - \Upsilon[w_2]\|_{\overline{\mathcal{H}}_{T_0,T}}$$

$$\leq \rho_1(T - T_0)\left[\|\psi^{w_1} - \psi^{w_2}\|_0 + \sup_{t \in [T_0,T)}\left[\gamma(T-t)^{-1}\|D\psi^{w_1}(t,\cdot) - D\psi^{w_2}(t,\cdot)\|_0\right]\right].$$
(4.197)

Observe now that, by (4.183), (4.184) and Hypotheses 4.72-(i) and 4.151, we have

$$\|\psi^{w_1}(t,\cdot) - \psi^{w_2}(t,\cdot)\|_0 \leq C_1\left[\|w_1(t,\cdot) - w_2(t,\cdot)\|_0 + \|Dw_1(t,\cdot) - Dw_2(t,\cdot)\|_0\right]$$

and

$$\|D\psi^{w_1}(t,\cdot) - D\psi^{w_2}(t,\cdot)\|_0 \leq C_1\left[\|D(w_1-w_2)(t,\cdot)\|_0 + \|D^2(w_1-w_2)(t,\cdot)\|_0\right] +$$
$$+ C_1\left[\|(w_1-w_2)(t,\cdot)\|_0 + \|D(w_1-w_2)(t,\cdot)\|_0\right]\left[1 + \|Dw_1(t,\cdot)\|_0 + \|D^2w_1(t,\cdot)\|_0\right].$$
(4.198)

We then obtain, by standard calculations,

$$\|\Upsilon[w_1] - \Upsilon[w_2]\|_{\overline{\mathcal{H}}_{T_0,T}}$$
$$\leq C_4\rho_1(T - T_0)\left[\|w_1 - w_2\|_{\overline{\mathcal{H}}_{T_0,T}} + \|w_1\|_{\overline{\mathcal{H}}_{T_0,T}}\|w_1 - w_2\|_{C_b^1([T_0,T]\times H)}\right]$$
(4.199)

for a suitable $C_4 > 0$.

We now define a subset of $\overline{\mathcal{H}}_{T_0,T}$ where it is possible to apply the contraction mapping principle. Let $u \in C_b^{0,1}([0,T] \times H)$ be the mild solution of (4.169) and set $\|u\|_{C_b^{0,1}} := \|u\|_{C_b^{0,1}([0,T]\times H)}$, so that

$$\|\varphi\|_1 \leq \|u\|_{C_b^{0,1}}.$$

We then take

$$R > C_3\|u\|_{C_b^{0,1}}$$
(4.200)

and choose $T_0 \in [0,T)$ such that (see (4.196))

$$C_3\|u\|_{C_b^{0,1}} + \rho_2(T - T_0)(1 + R) < R$$
(4.201)

and (see (4.199))

$$C_4\rho_1(T - T_0)(1 + R) < \frac{1}{2}.$$
(4.202)

This is possible since $\lim_{t\to 0^+}\rho_1(t) = 0$ due to $\gamma \in \mathcal{I}_2$ (see (4.30)). Let $B_{T_0}(0, R)$ be the closed ball centered at 0 with radius R in the space $\overline{\mathcal{H}}_{T_0,T}$. Then, applying (4.201) and (4.202) to (4.196) and (4.199), it follows that Υ is a contraction in $B_{T_0}(0, R)$. Thus the contraction mapping principle guarantees that there exists a unique solution of $v = \Upsilon[v]$ in $B_{T_0}(0, R)$. This solution must coincide with the restriction of the mild solution u to $[T_0, T] \times H$ since we have a unique mild solution in $C_b^{0,1}([T_0, T] \times H)$.

Step 3. Global existence. We repeat the argument of Step 2 in the space $\overline{\mathcal{H}}_{(3T_0-T)/2,(T+T_0)/2}$. (If $3T_0 < T$ we use the space $\overline{\mathcal{H}}_{0,(T+T_0)/2}$.) Now the final datum at $(T + T_0)/2$ is $u((T + T_0)/2, \cdot) \in C_b^1(H)$. Hence if R is as in $(4.200)^{31}$ then (4.201) and (4.202) still hold with T replaced by $(T + T_0)/2$ and T_0 replaced by $(3T_0 - T)/2$ and so Υ has a unique fixed point in the closed ball of radius R centered at 0 of $\overline{\mathcal{H}}_{(3T_0-T)/2,(T+T_0)/2}$. This fixed point must coincide with the restriction of the mild solution u to $[(3T_0 - T)/2, (T + T_0)/2] \times H$ since we have a unique mild solution in $C_b^{0,1}([(3T_0 - T)/2, (T + T_0)/2] \times H)$. This shows that $D^2u \in C_b^s([(3T_0 - T)/2, (T + T_0)/2] \times H, \mathcal{L}(H))$. Finally, we can get an explicit estimate. If R is from (4.200) then for $t \in [T_0, T)$

$$\|D^2u(t, \cdot)\| \le R\gamma(T - t)$$

and, since γ is decreasing, we have for $t \in [(3T_0 - T)/2, T_0]$,

$$\|D^2u(t, \cdot)\| \le R\gamma((T + T_0)/2 - t) \le R\gamma((T - T_0)/2).$$

Hence, for $t \in [(3T_0 - T)/2, T)$,

$$\|D^2u(t, \cdot)\| \le R[\gamma(T - t) \vee \gamma((T - T_0)/2)].$$

This implies that $D^2u \in \overline{C}_{b,\gamma}^s([(3T_0 - T)/2, T) \times H, \mathcal{L}(H))$.

If $(3T_0 - T)/2 > 0$ we repeat the above argument a finite number of times to obtain the claim.   □

Remark 4.153

(i) The above proof does not work when $m > 0$ since, estimating the difference $D\psi^{w_1} - D\psi^{w_2}$ as in (4.198), the product of two functions in C_m arises and we do not know if such a product is still in C_m unless $m = 0$.

(ii) In principle, the above scheme of proof may also work in the case when $G \neq I$, clearly adjusting the assumptions and finding second-order G-derivatives.

(iii) If we also assume that (4.172) holds, then one can use Gronwall's inequality of Proposition D.30 to estimate more precisely the norm of the second derivative. Indeed, using that $u = \Upsilon[u]$ on $[0, T] \times H$, we get from (4.190) and (4.195)

$$\|D^2u(t, \cdot)\|_0 \le Me^{(\omega \vee 0)T}\left[\gamma(T - t)\|D\varphi\|_0 + 2C_1\int_t^T \gamma(s - t)(1 + \|u(s, \cdot)\|_2)\,ds\right]$$

$$\le C_5\left[\gamma(T - t)\|\varphi\|_1 + (1 + \|u\|_{C_b^{0,1}}) + \int_t^T \gamma(s - t)\|D^2u(s, \cdot)\|_0\,ds\right]. \tag{4.203}$$

---

[31] Indeed, in (4.200) it would have been enough to choose $R > C_3\|\varphi\|_1$. The choice $R > C_3\|u\|_{C_b^{0,1}}$ allows us to repeat the argument in the subsequent intervals.

for some $C_5 > 0$. Using (4.172) and Gronwall's inequality of Proposition D.30 we get (also using $\|\varphi\|_1 \le \|u\|_{C_b^{0,1}}$)

$$\|D^2 u(t, \cdot)\|_0 \le (T - t)^{-\alpha} K_1 (1 + \|u\|_{C_b^{0,1}}) \tag{4.204}$$

for some constant K_1.

∎

We now consider the case when φ is only measurable and bounded. We need the following lemma.

Lemma 4.154 *Let Hypothesis 4.145 hold with $U = H$, $G = I$ and $\gamma \in \mathcal{I}_2$ and assume that Hypothesis 4.151 is satisfied. Let $T > 0$, $\varphi \in B_b(H)$ and let u be the mild solution of (4.169) from Theorem 4.149. Take $\varepsilon \in (0, T)$ and define $\varphi^\varepsilon(x) = u(T - \varepsilon, x)$ for every $x \in H$. Consider the integral equation (for $t \in [0, T]$, $x \in H$)*

$$w(t, x) = R_{T-\varepsilon-t}[\varphi^\varepsilon](x) + \int_t^{T-\varepsilon} R_{s-t}[F_0(s, \cdot, w(s, \cdot), Dw(s, \cdot))](x) ds. \tag{4.205}$$

Then the function u, restricted to $[0, T - \varepsilon] \times H$, is the unique mild solution of (4.205) in $C_b^{0,1}([0, T - \varepsilon] \times H)$.

Proof Uniqueness of the mild solution of (4.205) comes from Theorem 4.149. To get the claim it is then enough to prove that u solves (4.205). Observe that, since u solves (4.177), we have

$$\varphi^\varepsilon(x) = R_\varepsilon[\varphi](x) + \int_{T-\varepsilon}^T R_{s-(T-\varepsilon)}[F_0(s, \cdot, u(s, \cdot), Du(s, \cdot))](x) ds.$$

Moreover, using the semigroup property of R_t, for every $0 \le t \le T - \varepsilon$ we have

$$u(t, x) = R_{T-\varepsilon-t}[R_\varepsilon \varphi](x) + \int_t^{T-\varepsilon} R_{s-t}[F_0(s, \cdot, u(s, \cdot), Du(s, \cdot))](x) ds$$

$$+ \int_{T-\varepsilon}^T R_{T-\varepsilon-t}[R_{s-(T-\varepsilon)}[F_0(s, \cdot, u(s, \cdot), Du(s, \cdot))]](x) ds$$

$$= R_{T-\varepsilon-t}[\varphi^\varepsilon](x) + \int_t^{T-\varepsilon} R_{s-t}[F_0(s, \cdot, u(s, \cdot), Du(s, \cdot))](x) ds$$

(in the last equality we use Fubini Theorem 1.33-(i)), which gives the claim. □

Theorem 4.155 *Let Hypothesis 4.145 hold with $U = H$, $G = I$ and $\gamma \in \mathcal{I}_2$ and assume that Hypothesis 4.151 is satisfied. Let $T > 0$, $\varphi \in B_b(H)$ and let u be* *the mild solution of (4.169) from Theorem 4.149. Then u is twice Fréchet differentiable with respect to x on $[0, T) \times H$ and, for every $\varepsilon \in (0, T)$, $D^2 u \in C_b^s([0, T - \varepsilon] \times H, \mathcal{L}(H))$.*

Proof Let u be the mild solution of (4.169) from Theorem 4.149-(i). Set, for every $\varepsilon > 0$,

$$\varphi^\varepsilon = u(T - \varepsilon, \cdot).$$

By Theorem 4.149-(i) we have $\varphi^\varepsilon \in C_b^1(H)$ for every $\varepsilon > 0$. Consider the following map defined on $C_b^{0,1}([0, T - \varepsilon] \times H)$,

$$\Upsilon[w](t, x) = R_{T-\varepsilon-t}[\varphi^\varepsilon](x) + \int_t^{T-\varepsilon} R_{s-t}\left[F_0(s, \cdot, w(s, \cdot), Dw(s, \cdot))\right](x)ds.$$

We already know, using Lemma 4.154 and Theorem 4.149, that u is a fixed point of Υ in the space $C_b^{0,1}([0, T - \varepsilon] \times H)$. Moreover, by Theorem 4.152 applied on the interval $[0, T - \varepsilon]$, we immediately get that u is twice Fréchet differentiable with respect to x on $[0, T - \varepsilon) \times H$ and $D^2 u \in C_b^s([0, T - \varepsilon) \times H, \mathcal{L}(H))$. The result thus follows by the arbitrariness of $\varepsilon > 0$. $\qquad\square$

Remark 4.156 Assume that γ satisfies (4.172). Then, using $D^2 u \in C_b^s([0, T - \varepsilon] \times H, \mathcal{L}(H))$ and estimate (4.204) (which is valid only when the initial datum is C^1), we obtain, for a suitable constant $C > 0$,

$$\|D^2 u(t, \cdot)\|_0 \leq C(T - t)^{-2\alpha}, \quad \forall t \in [0, T).$$

Indeed using (4.204) we have, fixing any $t \in [0, T)$ and taking $\varepsilon = (T - t)/2$,

$$\|D^2 u(t, \cdot)\|_0 \leq K_1(1 + \|u\|_{C_b^{0,1}([0,T-\varepsilon]\times H)})(T - \varepsilon - t)^{-\alpha}$$

and since by Theorem 4.149, $u \in C_{b,\alpha}^{0,1}([0, T] \times H)$, we obtain for some $K_2 > 0$ (which may depend on u)

$$\|D^2 u(t, \cdot)\|_0 \leq K_2\varepsilon^{-\alpha}(T - \varepsilon - t)^{-\alpha} = K_2 2^{2\alpha}(T - t)^{-2\alpha}.$$

We finally observe that the procedure used in this section to prove C^2 regularity can be iterated (of course under stronger and stronger assumptions) to obtain higher regularity and similar estimates for the singularities of higher derivatives. $\qquad\blacksquare$

4.6.2 The Elliptic Case

Concerning the elliptic HJB equation (4.170) and its mild form (4.178), we will always assume that Hypotheses 4.104 and 4.145 hold. We begin with a result which is an easy corollary of Theorems 4.112, 4.125 and Proposition 4.147.

Theorem 4.157 *Let Hypothesis 4.145 hold. Let Hypothesis 4.104 be satisfied for some $m \geq 0$ and let λ_0 be from Theorem 4.112. Then we have the following.*

(i) *For every $\lambda > \lambda_0$ Eq. (4.170) admits a unique mild solution u, in the sense of Definition 4.102. If Hypothesis 4.145 holds also with $U = H$ and $G = I$ in point (iv), then u and $D^G u$ are continuous in H. If in addition point (v) of Hypothesis 4.145 holds with $U = H$ and $G = I$ then also Du is continuous in H.*

(ii) *If F_0 is continuous then the mild solution u belongs to $C_m^{1,G}(H)$.*

(iii) *If the function $(x, v, w) \rightarrow F_0(x, v, w)/(1 + |x|^m + |v| + |w|)$ is uniformly continuous, then the mild solution u belongs to $U C_m^{1,G}(H)$.*

Proof The first part of (i) follows from Theorem 4.112 and Proposition 4.147-(B). The second part of (i), when Hypothesis 4.145 also holds with $U = H$ and $G = I$ in point (iv) (respectively, (iv)-(v)), follows since u satisfies (4.178) whose right-hand side is continuous, together with its derivative D^G (respectively, D), thanks to Proposition 4.51.[32]

Part (ii) follows from Theorem 4.120 and Proposition 4.147-(B).

Part (iii) follows from Theorem 4.123 and Proposition 4.147-(B).   □

Concerning strong solutions, as in Theorem 4.150 for the parabolic case, we show here that Theorem 4.143 holds with a stronger convergence.

Theorem 4.158 *Let Hypothesis 4.145 be satisfied with the following addition.*

(i) *Either G is bounded or $D(A^*) \subset D(G^*)$ and, for some $c > 0$, $|G^*z| \leq c|A^*z|$ for all $z \in D(A^*)$.*

Let Hypothesis 4.104 be satisfied for some $m \geq 0$. Let $\lambda > \lambda_0 \vee a_1$ where λ_0 is from Theorem 4.112 and a_1 is from (4.163).[33] Let $u \in B_m^{1,G}(H)$ be the mild solution of Eq. (4.170) and assume, moreover, that u and $F_0(\cdot, u(\cdot), \nabla^G u(\cdot))$ are continuous.

Then the function u is the unique $\mathcal{K}$-strong solution of (4.170) in $B_m^{1,G}(H)$. Assume that Hypothesis 4.145 also holds with $U = H$ and $G = I$ in point (iv). Then the sequence u_n approximating u can be chosen so that the convergence is uniform on bounded subsets of H.

Proof We already know, from Propositions 4.147-(B) and 4.148-(B), that the assumptions of Theorem 4.143 hold, so u is the unique $\mathcal{K}$-strong solution in $B_m^{1,G}(H)$. For the proof of convergence we argue exactly as we did for the convergence of the integral terms in the last part of the proof of Theorem 4.150.   □

In the remaining part of this subsection we prove, under suitable assumptions, the following results:

- C^2 regularity of mild solutions.
- Existence of mild solutions for all $\lambda > 0$.

---

[32] In Proposition B.92 the time horizon $T < +\infty$ but the arguments are the same as when $T = +\infty$ once, as in this case, integrability is guaranteed.

[33] From Lemma B.90-(i) we easily see that, in this case, $a_1 = 2(\omega \vee 0)$.

4.6.2.1 C^2 Regularity of Mild Solutions

As in Sect. 4.6.1.1 we consider the case $U = H, G = I, m = 0$ and set $\Gamma(t) := \Gamma_I(t)$.

Hypothesis 4.159 Let $U = H$. Assume that $F_0 : H \times \mathbb{R} \times H \to \mathbb{R}$ satisfies Hypothesis 4.104 for $m = 0$, it is continuously Fréchet differentiable and, for some constant $C_1 \geq L > 0$ (where L is the constant from Hypothesis 4.104-(i)), we have for all $x \in H$, $v_1, v_2 \in \mathbb{R}$, $p_1, p_2 \in H$,

$$|D_x F_0(x, v_1, p_1) - D_x F_0(x, v_2, p_2)| + |D_v F_0(x, v_1, p_1) - D_v F_0(x, v_2, p_2)|$$
$$+ |D_p F_0(x, v_1, p_1) - D_p F_0(x, v_2, p_2)| \leq C_1(|v_1 - v_2| + |p_1 - p_2|) \quad (4.206)$$

and[34]

$$|D_x F_0(x, v, p)| \leq C_1(1 + |v| + |p|),$$
$$|D_v F_0(x, v, p)| + |D_p F_0(x, v, p)| \leq C_1, \quad (4.207)$$

for all $x \in H$, $v \in \mathbb{R}$, $p \in H$.

Theorem 4.160 *Let Hypothesis 4.145 hold with $U = H$ and $G = I$ and assume that Hypothesis 4.159 is satisfied. Let λ_0 be from Theorem 4.112. Let $\lambda > \lambda_0$ and let $u \in C_b^1(H)$ be the mild solution of (4.170) from Theorem 4.157. Then there exists a $\lambda_1 \geq \lambda_0$ such that, for $\lambda > \lambda_1$, u is twice Fréchet differentiable and $D^2 u \in C_b(H, \mathcal{L}(H))$.*[35]

Proof The unique mild solution u of (4.170) obtained, for $\lambda > \lambda_0$, in Theorem 4.157 belongs to $C_b^1(H)$. To show the required regularity we consider, for $\lambda > \lambda_0$, the map

$$\Upsilon[v](x) = \int_0^{+\infty} e^{-\lambda t} R_t \left[F_0(\cdot, v(\cdot), Dv(\cdot)) \right] (x) dt. \quad (4.208)$$

We know from Theorem 4.157 that u is a fixed point of Υ in the space $C_b^1(H)$. We now consider Υ in the Banach space

$$\overline{\mathcal{H}} = C_b^2(H) = \left\{ v \in C_b^1(H) : D^2 v \in C_b(H, \mathcal{L}(H)) \right\}$$

with the norm

$$\|v\|_{\overline{\mathcal{H}}} = \sup_{x \in H} \left[|v(x)| + |Dv(x)|_H + \|D^2 v(x)\|_{\mathcal{L}(H)} \right]$$

---

[34]The second inequality in (4.206) follows from Hypothesis 4.104-(i) and the fact that $C_1 \geq L$, but we repeat it here for the reader's convenience.

[35]In [317], Theorem 3.3, twice differentiability of the mild solution is proved, in a special case, using a different method, based on interpolation spaces and a bootstrap argument. This method may also be applied here, under a weak additional assumption about the singularity of $\|\Gamma(t)\|$ at $t = 0$, to show that u is twice differentiable under weaker assumptions about F_0.

and we look at the equation $v = \Upsilon[v]$ in $\overline{\mathcal{H}}$.

Step 1. $\Upsilon : \overline{\mathcal{H}} \to \overline{\mathcal{H}}$.
Define, for $v \in \overline{\mathcal{H}}$,

$$\psi^v(x) = F_0(x, v(x), Dv(x)). \tag{4.209}$$

We have $\psi^v \in C_b^1(H)$ and, for $h \in H$,

$$\langle D\psi^v(x), h \rangle = \langle D_x F_0(x, v(x), Dv(x)), h \rangle + D_v F_0(x, v(x), Dv(x)) \langle Dv(x), h \rangle$$
$$+ \left\langle D_p F_0(x, v(x), Dv(x)), D^2 v(x)h \right\rangle. \tag{4.210}$$

Since we can write

$$\Upsilon[v](x) = \int_0^{+\infty} e^{-\lambda t} \int_H \psi^v(y + e^{tA}x) \mathcal{N}_{Q_t}(dy)dt, \tag{4.211}$$

we can differentiate using Proposition 4.16, as in the proof of Theorem 4.155, obtaining for $h \in H$ (recall that $\lambda_0 \geq \omega$)

$$\langle D\Upsilon[v](x), h \rangle = \int_0^{+\infty} e^{-\lambda t} \int_H \left\langle D\psi^v(y + e^{tA}x), e^{tA}h \right\rangle \mathcal{N}_{Q_t}(dy)dt, \tag{4.212}$$

i.e.

$$\langle D\Upsilon[v](x), h \rangle = \int_0^{+\infty} e^{-\lambda t} R_t \left[\langle D\psi^v(\cdot), e^{tA}h \rangle \right](x)dt. \tag{4.213}$$

Similarly, for $h_1, h_2 \in H$,

$$\left\langle D^2\Upsilon[v](x)h_1, h_2 \right\rangle = \int_0^{+\infty} e^{-\lambda t} \left\langle DR_t \left[\langle D\psi^v(\cdot), e^{tA}h_1 \rangle \right](x), h_2 \right\rangle dt$$
$$= \int_0^{+\infty} e^{-\lambda t} \int_H \left\langle D\psi^v(y + e^{tA}x), e^{tA}h_1 \right\rangle \left\langle \Gamma(t)h_2, Q_t^{-1/2}y \right\rangle \mathcal{N}_{Q_t}(dy)dt. \tag{4.214}$$

Hence, given a sequence $x_n \to x$ as $n \to +\infty$, we have

$$|D^2\Upsilon[v](x_n) - D^2\Upsilon[v](x)|_{\mathcal{L}(H)} = \sup_{|h_1|=|h_2|=1} \left\langle \left(D^2\Upsilon[v](x_n) - D^2\Upsilon[v](x) \right) h_1, h_2 \right\rangle$$

$$= \sup_{|h_1|=|h_2|=1} \int_0^{+\infty} e^{-\lambda t}$$

$$\int_H \left\langle D\psi^v(y + e^{tA}x_n) - D\psi^v(y + e^{tA}x), e^{tA}h_1 \right\rangle \left\langle \Gamma(t)h_2, Q_t^{-1/2}y \right\rangle \mathcal{N}_{Q_t}(dy)dt$$

$$\leq \sup_{|h_1|=1} \int_0^{+\infty} e^{-\lambda t} \|\Gamma(t)\| \left[\int_H \left\langle D\psi^v(y + e^{tA}x_n) - D\psi^v(y + e^{tA}x), e^{tA}h_1 \right\rangle^2 \mathcal{N}_{Q_t}(dy) \right]^{1/2} dt,$$

where in the last line we used Schwarz's inequality as in (4.79). The continuity of $D\psi^v$ now implies the continuity of $D^2\Upsilon[v]$.

Step 2. Contraction estimates and conclusion.
By standard computations using (4.208), (4.212) and (4.214), we have for $v \in \overline{\mathcal{H}}$

$$\|\Upsilon[v]\|_0 \le \int_0^\infty e^{-\lambda t} \|\psi^v\|_0 dt,$$

$$\|D\Upsilon[v]\|_0 \le M \int_0^{+\infty} e^{-(\lambda-\omega)t} \|D\psi^v\|_0 dt,$$

$$\|D^2\Upsilon[v]\|_0 \le M \int_0^{+\infty} e^{-(\lambda-\omega)t} \|\Gamma(t)\| \|D\psi^v\|_0 dt, \qquad (4.215)$$

which gives

$$\|\Upsilon[v]\|_{\overline{\mathcal{H}}} \le \frac{1}{\lambda} \|\psi^v\|_0 + \rho_1(\lambda) \|D\psi^v\|_0, \qquad (4.216)$$

where

$$\rho_1(\lambda) := M \int_0^{+\infty} e^{-(\lambda-\omega)t} (1 + \|\Gamma(t)\|) dt.$$

Moreover, by (4.209) and Hypothesis 4.104-(ii),

$$\|\psi^v\|_0 \le C_1 [2 + \|v\|_0 + \|Dv\|_0]$$

and, by (4.210) and (4.207),

$$\|D\psi^v\|_0 \le 2C_1 \left[1 + \|v\|_0 + \|Dv\|_0 + \|D^2v\|_0 \right]. \qquad (4.217)$$

Hence estimate (4.216) becomes

$$\|\Upsilon[v]\|_{\overline{\mathcal{H}}} \le \rho_2(\lambda) \left[2 + \|v\|_0 + \|Dv\|_0 + \|D^2v\|_0 \right],$$

where $\rho_2(\lambda) := \frac{1}{\lambda}C_1 + 2C_1\rho_1(\lambda)$. Thus, for $R > 0$, if

$$\rho_2(\lambda)(2 + R) \le R \iff \rho_2(\lambda) \le \frac{R}{2 + R},$$

then $\Upsilon : B(0, R) \to B(0, R)$ in $\overline{\mathcal{H}}$.
   Now take $w_1, w_2 \in \overline{\mathcal{H}}$. We have

$$\Upsilon[w_1](x) - \Upsilon[w_2](x) = \int_0^{+\infty} e^{-\lambda t} R_t \left[\psi^{w_1}(\cdot) - \psi^{w_2}(\cdot) \right](x) ds.$$

We estimate the right-hand side exactly as in (4.216), taking $\psi^{w_1} - \psi^{w_2}$ in place of ψ^v. We get

$$\|\Upsilon[w_1] - \Upsilon[w_2]\|_{\overline{\mathcal{H}}} \leq \frac{1}{\lambda} \left[\|\psi^{w_1} - \psi^{w_2}\|_0 + \rho_1(\lambda) \|D\psi^{w_1} - D\psi^{w_2}\|_0 \right]. \quad (4.218)$$

We observe that, by (4.209) and Hypothesis 4.104-(i),

$$\|\psi^{w_1} - \psi^{w_2}\|_0 \leq C_1 \left[\|w_1 - w_2\|_0 + \|D(w_1 - w_2)\|_0 \right]$$

and, by Hypothesis 4.159 and (4.210),

$$\|D\psi^{w_1} - D\psi^{w_2}\|_0 \leq C_1 \left[\|D(w_1 - w_2)\|_0 + \|D^2(w_1 - w_2)\|_0 \right] +$$
$$+ C_1 \left[\|(w_1 - w_2)\|_0 + \|D(w_1 - w_2)\|_0 \right] \left[1 + \|Dw_1\|_0 + \|D^2 w_1\|_0 \right]. \quad (4.219)$$

Thus we obtain

$$\|\Upsilon[w_1] - \Upsilon[w_2]\|_{\overline{\mathcal{H}}} \leq \frac{C_1}{\lambda} \|w_1 - w_2\|_{C_b^1(H)} +$$

$$+ C_1 \rho_1(\lambda) \left[\|w_1 - w_2\|_{\overline{\mathcal{H}}} + \|w_1 - w_2\|_{C_b^1(H)} \left(1 + \|w_1\|_{\overline{\mathcal{H}}} \right) \right]. \quad (4.220)$$

If $w_1, w_2 \in B(0, R)$ then we have

$$\|\Upsilon[w_1] - \Upsilon[w_2]\|_{\overline{\mathcal{H}}} \leq \rho_2(\lambda) \|w_1 - w_2\|_{\overline{\mathcal{H}}} (1 + R). \quad (4.221)$$

So, choosing $R > 1$ and $\lambda_1 \geq \lambda_0$ such that, for all $\lambda > \lambda_1$, we have

$$\rho_2(\lambda)(1 + R) < \frac{1}{2},$$

we get that $\Upsilon : B(0, R) \to B(0, R)$ is a contraction. Then the contraction mapping principle guarantees that there exists a unique solution of $v = \Upsilon[v]$ in $B(0, R)$. This solution must coincide with the mild solution u, so there is a unique mild solution in $C_b^1(H)$. $\qquad \square$

Remark 4.161

(i) As in the parabolic case the above proof does not work when $m > 0$ since, in estimating the difference $D\psi^{w_1} - D\psi^{w_2}$ as in (4.198), the product of two functions in C_m arises and we do not know if such a product is still in C_m unless $m = 0$.

(ii) If, in Hypothesis 4.159, we add the requirement that F_0 has uniformly continuous Fréchet derivatives then Theorem 4.160 holds true with the solution u belonging to $UC_b^2(H)$. The proof repeats the same arguments (with obvious modifications) in the space $\overline{\mathcal{H}} = UC_b^2(H)$.

(iii) The general scheme of proof may also work in the case when $G \neq I$.

■

4.6.2.2   Existence of Mild Solutions for All $\lambda > 0$

In this subsection we take $U = H$, $G = I$, $m = 0$ and set $\Gamma(t) := \Gamma_I(t)$, $\gamma(t) := \|\Gamma(t)\|$.

We assume, in addition to Hypothesis 4.145 for $U = H$ and $G = I$, the following.

Hypothesis 4.162 (i) The semigroup e^{tA}, $t \geq 0$, is of negative type, i.e. there exist $M \geq 1$, $\omega < 0$ such that $\|e^{tA}\| \leq Me^{\omega t}$ for $t \geq 0$.
(ii) The Hamiltonian $F_0 : H \times H \to \mathbb{R}$ is of the form

$$F_0(x, p) = \langle b_1(x), p \rangle + F_1(p) + \psi(x),$$

where $b_1 \in C_b^{0,1}(H, H)$, $\psi \in UC_b(H)$, $F_1 \in C^{0,1}(H)$. Moreover, F_1 is concave.

Remark 4.163 (i) The uniform continuity of data is needed here to prove the crucial estimates of Lemma 4.166. Thus in this subsection we work in the space $UC_b(H)$ instead of the more typical $C_b(H)$.
(ii) Hypothesis 4.162-(ii) implies, in particular, that Hypothesis 4.104 is satisfied with $m = 0$ and that F_0 is continuous.
(iii) Without loss of generality (replacing ψ by $\psi + F_1(0)$ if necessary) we can assume that $F_1(0) = 0$.

■

Equation (4.170) is rewritten in our case as

$$\lambda v - \frac{1}{2} \operatorname{Tr}[\Sigma D^2 v] - \langle Ax, Dv \rangle - \langle b_1(x), Dv \rangle - F_1(Dv) - \psi(x) = 0, \qquad x \in D(A), \tag{4.222}$$

and its mild form is

$$v(x) = \int_0^{+\infty} e^{-\lambda t} R_t[\langle b_1, Dv \rangle + F_1(Dv) + \psi](x)dt, \qquad x \in H. \tag{4.223}$$

We know by Theorem 4.157-(iii) that, under Hypotheses 4.145 and 4.162, the mild solution u of (4.222) exists and is unique in $UC_b^1(H)$ for all $\lambda > \lambda_0$, where λ_0 is from Theorem 4.112. Moreover, thanks to Hypothesis 4.162-(ii), we know that, if $b_1 \in C_b^1(H, H)$, $F_1 \in C_b^{1,1}(H)$ and $\psi \in C_b^1(H)$ then Hypothesis 4.159 is satisfied and so, by Theorem 4.160 if $\lambda > \lambda_1$, u is twice differentiable and $D^2 u \in C_b(H, \mathcal{L}(H))$.

To get the existence of solutions for all $\lambda > 0$ we use the theory of m-dissipative operators, see e.g. [26, 146].

The idea is the following. First, similarly to (4.174), Eq. (4.222) can be rewritten as

$$\lambda v - \hat{A}_0 v - \langle b_1, Dv \rangle - F_1(Dv) - \psi = 0, \quad x \in H,$$

where the operator $\hat{A}_0 : D(\hat{A}_0) \subset UC_b(H) \to UC_b(H)$ is defined in Appendix B, Sect. B.7.1 (note that the analogous operator $\mathcal{A}_0$ there does not take values in $UC_b(H)$ and this is why here we prefer to use $\hat{A}_0$, see Remark B.97). It is proved in Proposition B.92-(ii) that the operator $\hat{A}_0$ is $\mathcal{K}$-closable in $UC_b(H)$ and its $\mathcal{K}$-closure is the operator[36] $\mathcal{A}$ which is the generator of the $\mathcal{K}$-continuous semigroup R_t in $UC_b(H)$. Such a generator, see Remark B.72, is defined through the resolvent operator which is given, for all $\lambda > 0$ and $\phi \in UC_b(H)$, by

$$(\lambda I - \mathcal{A})^{-1} \phi(x) := \int_0^{+\infty} e^{-\lambda t} R_t[\phi](x) dt, \quad x \in H.$$

It is immediate to see that the mild form (4.223) can then be seen as

$$v(x) = (\lambda I - \mathcal{A})^{-1} (\langle b_1, Dv \rangle + F_1(Dv) + \psi)(x).$$

Moreover, as proved in Proposition 4.165 (see the beginning of its proof), when $\phi \in UC_b(H)$ we must have $(\lambda I - \mathcal{A})^{-1} \phi \in UC_b^1(H)$ so $D(\mathcal{A}) \subset UC_b^1(H)$. Hence we can consider the nonlinear operator $\mathcal{B} : D(\mathcal{B}) \subset UC_b(H) \to UC_b(H)$ defined as

$$D(\mathcal{B}) = D(\mathcal{A}) \subset UC_b^1(H), \qquad \mathcal{B}(u) = \mathcal{A}u + \langle b_1, Du \rangle + F_1(Du). \qquad (4.224)$$

Using this operator our Eq. (4.222) can be written, at least formally, as

$$\lambda v(x) - \mathcal{B}(v)(x) = \psi(x)$$

or applying the resolvent of $\mathcal{B}$ as

$$v = (\lambda I - \mathcal{B})^{-1}(\psi),$$

which can be seen as another way of writing the mild form (4.223). It is then clear that the m-dissipativity (see Sect. B.1) of the operator $\mathcal{B}$, which in particular implies that $(\lambda I - \mathcal{B})^{-1}$ is well defined for all $\lambda \in (0, +\infty)$, is the key property to solving (4.223) for all $\lambda > 0$.

In the literature there are two ways of studying the m-dissipativity of $\mathcal{B}$. One, used e.g. in [106, 179], looks directly at the operator $\mathcal{B}$ defined above; the other, used e.g. in [146], studies the m-dissipativity of $\mathcal{B}$ through the family of the resolvent operators.

The first approach does not require concavity of F_1 but it does require its differentiability. Since here our main interest is in equations coming from optimal control problems, where F_1 is always concave, we follow the second approach, generalizing the results of [317].

---

[36]In Proposition B.92 it is denoted by $\mathcal{A}^m$. Here $m = 0$ and we write $\mathcal{A}$ for simplicity.

Notation 4.164 When $\lambda > \lambda_0$ (from Theorem 4.157-(iii)) and $\psi \in UC_b(H)$, we denote by $u_{\lambda,\psi}$ the unique solution of (4.223). We also define, for $\lambda > \lambda_0$, the map $R(\lambda) : UC_b(H) \to UC_b^1(H)$ as $R(\lambda)(\psi) := u_{\lambda,\psi}$.

Moreover, for any $\lambda > 0$, we denote by T_λ the resolvent of $\mathcal{A}$, i.e. the linear operator on $UC_b(H)$ defined by

$$T_\lambda \psi(x) := (\lambda I - \mathcal{A})^{-1}\psi(x) = \int_0^{+\infty} e^{-\lambda t} R_t[\psi](x)dt, \quad \forall x \in H, \qquad (4.225)$$

and by T_λ^ψ the nonlinear operator on $UC_b^1(H)$ defined by

$$T_\lambda^\psi(u)(x) := T_\lambda[\psi + \langle b_1, Du \rangle + F_1(Du)](x), \quad \forall x \in H \text{ and } \psi \in UC_b(H).$$

Note that T_λ, and thus also T_λ^ψ, are well defined for all $\lambda > 0$ since $(R_t)_{t>0}$ is a contraction semigroup on $UC_b(H)$.   ∎

Proposition 4.165 *Assume that Hypotheses 4.145 for $U = H$ and $G = I$ and 4.162 hold. Let λ_0 be from Theorem 4.157-(iii). Then we have the following.*

(i) For all $\lambda, \mu \geq \lambda_0$, the following so-called identity of the resolvents holds:

$$R(\lambda) = R(\mu) \circ (I + (\mu - \lambda)R(\lambda)). \qquad (4.226)$$

(ii) For all $\lambda \geq \lambda_0$ the map $R(\lambda)$ is injective.
(iii) For all $\lambda \geq \lambda_0$ and all $\psi, \varphi \in UC_b(X)$, we have

$$\|R(\lambda)(\psi) - R(\lambda)(\varphi)\|_0 \leq \frac{1}{\lambda}\|\psi - \varphi\|_0. \qquad (4.227)$$

Proof We first observe that, for all $\lambda > 0$, T_λ maps $UC_b(H)$ into $UC_b^1(H)$. Indeed, using Hypothesis 4.145 for $U = H$ and $G = I$, we have from Proposition B.92-(iii) and Remark 4.53-(i) that the function $(t, x) \to DR_t[\psi](x)$ is in $UC_{m,\gamma}^x((0, T] \times H, H)$ for every $T > 0$. Moreover, from (4.70) we have the estimate

$$|DR_t[\psi](x)| \leq \|\Gamma(t)\|\|\psi\|_0, \quad \forall t > 0, \ x \in H.$$

Then, by Proposition 4.16,

$$DT_\lambda\psi(x) = \int_0^{+\infty} e^{-\lambda t} DR_t[\psi](x)dt, \quad \forall x \in H. \qquad (4.228)$$

Hence, by (4.75) (here ρ_ψ is the modulus of continuity of ψ),

$$|DT_\lambda\psi(x_1) - DT_\lambda\psi(x_2)| \leq \int_0^{+\infty} e^{-\lambda t}\|\Gamma(t)\|\rho_\psi(|e^{tA}(x_1 - x_2)|)dt, \quad \forall x_1, x_2 \in H,$$

$$(4.229)$$

so the uniform continuity of $DT_\lambda\psi$ follows by straightforward computations.

We also observe that (4.223) is equivalent to $u = T_\lambda^\psi(u)$, so that $u \in UC_b^1(H)$ is a mild solution of (4.223) if and only if it is a fixed point of T_λ^ψ.

Proof of (i). Let $\lambda, \mu > 0, \psi \in UC_b(H)$. We first observe that the resolvent identity holds for the family T_λ. Indeed, using (4.49) we have, for $\lambda > 0$,

$$R_t[T_\lambda\psi](x) = \mathbb{E}\left[T_\lambda\psi(e^{tA}x + W^A(t))\right] = \mathbb{E}\left[\left(\int_0^{+\infty} e^{-\lambda s} R_s[\psi](e^{tA}x + W^A(t))ds\right)\right]$$

and so, by the Fubini Theorem 1.33 and the semigroup property of R_t,

$$R_t[T_\lambda\psi](x) = \int_0^{+\infty} e^{-\lambda s}\mathbb{E}\left[R_s[\psi](e^{tA}x + W^A(t))\right]ds = \int_0^{+\infty} e^{-\lambda s} R_{t+s}[\psi](x)ds.$$

When $\mu, \lambda > 0, \mu \neq \lambda$, we then obtain, first changing variables and then integrating by parts,

$$T_\mu(T_\lambda\psi)(x) = \int_0^{+\infty} e^{-\mu t}\int_0^{+\infty} e^{-\lambda s} R_{t+s}[\psi](x)ds\,dt$$

$$= \int_0^{+\infty} e^{-\mu t}\int_t^{+\infty} e^{-\lambda(r-t)} R_r[\psi](x)dr\,dt = \int_0^{+\infty} e^{-(\mu-\lambda)t}\int_t^{+\infty} e^{-\lambda r} R_r[\psi](x)dr\,dt$$

$$= \frac{1}{\lambda-\mu}\left(-\int_0^{+\infty} e^{-\lambda r} R_r[\psi](x)ds + \int_0^{+\infty} e^{-\mu r} R_r[\psi](x)ds\right)$$

$$= \frac{1}{\lambda-\mu}\left(T_\mu\psi(x) - T_\lambda\psi(x)\right).$$

Hence the family T_λ satisfies, for $\mu, \lambda > 0, \mu \neq \lambda$,

$$T_\lambda = T_\mu\circ[I + (\mu - \lambda)T_\lambda]. \tag{4.230}$$

Now, take $\lambda, \mu > \lambda_0$ and observe that if $u = u_{\lambda,\psi}$, (4.226) is equivalent to

$$T_\mu^{\psi+(\mu-\lambda)u}(u) = u.$$

However, using (4.230), we have

$$T_\mu^{\psi+(\mu-\lambda)u}(u) = T_\mu\left[\psi + (\mu - \lambda)u + \langle b_1, Du\rangle + F_1(Du)\right]$$
$$= T_\mu\left[\psi + \langle b_1, Du\rangle + F_1(Du) + (\mu - \lambda)T_\lambda\left[\psi + \langle b_1, Du\rangle + F_1(Du)\right]\right]$$
$$= T_\lambda\left[\psi + \langle b_1, Du\rangle + F_1(Du)\right] = u$$

and the proof of (4.226) is complete.

Proof of (ii). Let $\varphi, \psi \in UC_b(H)$, $\lambda > \lambda_0$, and assume that $R(\lambda)(\varphi) = R(\lambda)(\psi) =: u$. By the definition of $R(\lambda)$, (4.223) and (4.225) we must have

$$T_\lambda(\langle b_1, Du\rangle + F_1(Du) + \varphi) = T_\lambda(\langle b_1, Du\rangle + F_1(Du) + \psi),$$

which gives $T_\lambda \varphi = T_\lambda \psi$. On the other hand, by the resolvent identity, for any $\mu > \lambda > \lambda_0$,

$$R(\mu)(\varphi + (\mu - \lambda)u) = R(\mu)(\psi + (\mu - \lambda)u),$$

which, arguing as above, gives $T_\mu \varphi = T_\mu \psi$. We can now use the injectivity of the Laplace transform and the continuity of the map $t \to R_t[\varphi - \psi](x)$ to deduce that for all $x \in H$, $R_t[\varphi](x) = R_t[\psi](x)$ for all $t > 0$. Now, since the semigroup R_t is $\mathcal{K}$-continuous, for all $x \in H$ $\lim_{t \to 0^+} R_t[\varphi](x) = \varphi(x)$ and $\lim_{t \to 0^+} R_t[\psi](x) = \psi(x)$. Thus $\varphi(x) = \psi(x)$ for all $x \in H$.

Proof of (iii). Let $\varphi, \psi \in UC_b(H)$. Let $\lambda \geq \lambda_0$. We set $u = R(\lambda)(\psi)$ and $v = R(\lambda)(\varphi)$. If we try to compute $\|u - v\|_0$ directly, we cannot obtain the desired estimate since T_λ^ψ and T_λ^φ are nonlinear. The next lemma shows how to approximate the nonlinear term.

Lemma 4.166 *There exists a family of operators* $(N_\varepsilon)_{\varepsilon \geq 0}$ *which satisfies:*

(i) For all $w_1, w_2 \in UC_b(H)$,

$$\|N_\varepsilon(w_1) - N_\varepsilon(w_2)\|_0 \leq \|w_1 - w_2\|_0, \quad \forall \varepsilon \geq 0. \tag{4.231}$$

(ii) For all $w \in UC_b^1(H)$,

$$\lim_{\varepsilon \to 0} \left\| \frac{N_\varepsilon(w) - w}{\varepsilon} - \langle b_1, Dw \rangle - F_1(Dw) \right\|_0 = 0. \tag{4.232}$$

Proof of Lemma 4.166. Roughly speaking, $(N_\varepsilon)_{\varepsilon \geq 0}$ has to be the nonlinear semigroup associated to the equation

$$\begin{cases} z_t - \langle b_1(x), Dz \rangle + F_1(Dz) = 0 \\ z(0) = w, \end{cases}$$

which is a time-dependent first-order HJB equation whose solution "should be" the value function of the following optimal control problem. The dynamic of the system is described by the equation

$$\begin{cases} y'(s) = b_1(y(s)) + \alpha(s), \quad s > 0, \\ y(0) = x, \end{cases}$$

whose unique solution at time s is denoted by $y(s; x, \alpha(\cdot))$ or, when clear from the context, simply by $y(s)$. The control $\alpha(\cdot)$ belongs to the set

$$\Lambda_M = \left\{ \alpha(\cdot) \in L^\infty(0, +\infty; H) : |\alpha(s)| \leq M \text{ for a.e. } s \in [0+, \infty) \right\},$$

where $M > 0$ is a constant to be chosen later. The cost functional is given by

$$J(t, x; \alpha(\cdot)) = \int_0^t g(\alpha(s))ds + w(y(t)),$$

where the function g on H is defined by

$$g(\alpha) = \sup_{|p| \leq M} \{-\langle \alpha, p \rangle + F_1(p)\}.$$

The value function is

$$z(t, x) = \inf_{\alpha(\cdot) \in \Lambda_M} J(x, t; \alpha(\cdot)). \tag{4.233}$$

Now, for $w \in UC_b(H)$, we set $N_\varepsilon(w) = z(\varepsilon, \cdot)$, where z is the value function just defined in (4.233). We will prove that such defined operators N_ε satisfy the properties listed in Lemma 4.166.

Let $w_1, w_2 \in UC_b(H)$. For all $\varepsilon > 0$ and for all $x \in H$, we have

$$|N_\varepsilon(w_1)(x) - N_\varepsilon(w_2)(x)| \leq \sup_{\alpha(\cdot) \in \Lambda_M} |w_1(y(\varepsilon; x, \alpha(\cdot))) - w_2(y(\varepsilon; x, \alpha(\cdot)))|$$
$$\leq \|w_1 - w_2\|_0,$$

which shows (4.231). The proof of (4.232) is more complicated and we do it in three steps.

Step 1. If $M \geq [F_1]_{0,1}$ then, for all $p \in H$ such that $|p| \leq M$, we have

$$F_1(p) = \inf_{|\alpha| \leq M} \{\langle \alpha, p \rangle + g(\alpha)\}.$$

Let G be defined on H by

$$G(p) = \inf_{|\alpha| \leq M} \{\langle \alpha, p \rangle + g(\alpha)\}$$

and let $p_0 \in H$ be such that $|p_0| \leq M$. Then

$$G(p_0) = \inf_{|\alpha| \leq M} \sup_{|p| \leq M} \{\langle \alpha, p_0 - p \rangle + F_1(p)\} \geq F_1(p_0)$$

by choosing $p = p_0$. Moreover,

$$G(p_0) = F_1(p_0) + \inf_{|\alpha| \leq M} \sup_{|p| \leq M} \{\langle \alpha, p_0 - p \rangle + F_1(p) - F_1(p_0)\}.$$

Since F_1 is concave, we have $F_1(p) - F_1(p_0) \leq \langle q_0, p - p_0 \rangle$ for every q_0 in the superdifferential of F_1 at p_0. Thus, if q_0 is in the superdifferential of F_1 at p_0,

$$G(p_0) \leq F_1(p_0) + \inf_{|\alpha| \leq M} \sup_{|p| \leq M} \{\langle \alpha, p_0 - p \rangle + \langle q_0, p - p_0 \rangle\}.$$

This yields

$$G(p_0) \leq F_1(p_0)$$

by choosing $\alpha = q_0$, which is possible since $|q_0| \leq [F_1]_{0,1} \leq M$.

Step 2. Estimate from above.

Let $u \in UC_b^1(H)$ and choose a constant $M \geq \max\{\|Du\|_0, [F_1]_{0,1}\}$. Let $\varepsilon > 0$ and $x \in H$ and set, for all $w \in UC_b^1(X)$,

$$\psi_\varepsilon(w) = \frac{N_\varepsilon(w) - w}{\varepsilon} - \langle b_1, Dw \rangle - F_1(Dw). \tag{4.234}$$

Using Step 1 and the fact that the value function is bounded by the infimum over all constant controls (which we denote by α) of the cost functional we get

$$\psi_\varepsilon(u)(x) \leq \inf_{|\alpha| \leq M}\left\{ g(\alpha) + \frac{u(y(\varepsilon; x, \alpha)) - u(x)}{\varepsilon} \right\} - \inf_{|\alpha| \leq M}\{\langle b_1(x) + \alpha, Du(x) \rangle + g(\alpha)\}$$

$$\leq \sup_{|\alpha| \leq M}\left\{ \frac{u(y(\varepsilon; x, \alpha)) - u(x)}{\varepsilon} - \langle b_1(x) + \alpha, Du(x) \rangle \right\}.$$

We have

$$u(y(\varepsilon; x, \alpha)) - u(x) = \int_0^\varepsilon \langle b_1(y(s; x, \alpha)) + \alpha, Du(y(s; x, \alpha)) \rangle\, ds, \tag{4.235}$$

which then yields

$$\psi_\varepsilon(u)(x) \leq \sup_{|\alpha| \leq M}\left\{ \frac{1}{\varepsilon} \int_0^\varepsilon [\langle b_1(y(s; x, \alpha)) + \alpha, Du(y(s; x, \alpha)) \rangle \right.$$

$$\left. - \langle b_1(x) + \alpha, Du(x) \rangle] ds \right\}.$$

Clearly the right-hand side converges to 0 as $\varepsilon \to 0^+$, thanks to the uniform continuity of b_1 and Du and the fact that $|y(s; x, \alpha) - x| \to 0$ as $s \to 0^+$, uniformly with respect to α for $|\alpha| \leq M$. Note that here uniform continuity of Du (at least on bounded sets) is really needed as the set $\{y(s; x, \alpha) : s \in [0, \varepsilon], |\alpha| \leq M\}$ is not compact in general. The same comment is relevant for the proof of Step 3 below.

Step 3. Estimate from below.

Let $u \in UC_b^1(H)$ and choose again a constant $M \geq \max\{\|Du\|_0, [F_1]_{0,1}\}$. Let $\varepsilon > 0$ and $x \in H$. By the dynamic programming principle we know that there exists an $\alpha_\varepsilon(\cdot) \in \Lambda_M$ (which in fact also depends on x) such that

$$N_\varepsilon(u)(x) \geq -\varepsilon^2 + \int_0^\varepsilon g(\alpha_\varepsilon(s))ds + u(y_\varepsilon(\varepsilon)),$$

where $y_\varepsilon(s) = y(s; x, \alpha_\varepsilon(\cdot))$. Therefore, using also the analogue of (4.235) in this case, we get

$$\psi_\varepsilon(u)(x) \geq -\varepsilon + \frac{1}{\varepsilon} \int_0^\varepsilon g(\alpha_\varepsilon(s)) ds + \frac{u(y_\varepsilon(\varepsilon)) - u(x)}{\varepsilon} - \langle b_1(x), Du(x)\rangle - F_1(Du(x))$$

$$= -\varepsilon + \frac{1}{\varepsilon} \int_0^\varepsilon [g(\alpha_\varepsilon(s)) + \langle b_1(y_\varepsilon(s)) + \alpha_\varepsilon(s), Du(y_\varepsilon(s))\rangle] ds - \langle b_1(x), Du(x)\rangle - F_1(Du(x))$$

$$= -\varepsilon + \frac{1}{\varepsilon} \int_0^\varepsilon [\langle b_1(y_\varepsilon(s)), Du(y_\varepsilon(s))\rangle - \langle b_1(x), Du(x)\rangle] ds$$

$$+ \frac{1}{\varepsilon} \int_0^\varepsilon [\langle \alpha_\varepsilon(s), Du(y_\varepsilon(s))\rangle + g(\alpha_\varepsilon(s)) - F_1(Du(x))] ds.$$

We now observe that the integrand of the second term of the right-hand side converges to 0 as $\varepsilon \to 0^+$, thanks to the uniform continuity of b_1 and Du and the fact that $|y_\varepsilon(s) - x| \to 0$ as $s \to 0^+$, uniformly with respect to ε. Moreover, thanks to Step 1, the integrand in the third term is greater than or equal to $F_1(Du(y_\varepsilon(s))) - F_1(Du(x))$. Arguing as before we can thus conclude that

$$\psi_\varepsilon(u)(x) \geq -\varepsilon + \rho(\varepsilon),$$

where $\rho(\varepsilon) \to 0$ as $\varepsilon \to 0^+$. *End of the proof of Lemma 4.166.*

Continuation of the proof of (iii).
Let $\varphi, \psi \in UC_b(H)$. Let $\lambda \geq \lambda_0$. We set $u = R(\lambda)(\psi)$ and $v = R(\lambda)(\varphi)$ and observe that, by Hypothesis 4.162 and Theorem 4.112-(iii), both u and v belong to $UC_b^1(H)$. Let $\varepsilon > 0$ and set $\mu = \lambda + 1/\varepsilon$ in (4.226). We get

$$u = R\left(\lambda + \frac{1}{\varepsilon}\right)\left(\psi + \frac{1}{\varepsilon} u\right) = T_{\lambda + \frac{1}{\varepsilon}}\left[\psi + \frac{1}{\varepsilon} u + \langle b_1, Du\rangle + F_1(Du)\right]$$

and an equivalent identity for v. This yields, using the definition of ψ_ε in (4.234),

$$u - v = T_{\lambda + \frac{1}{\varepsilon}}\left[\psi - \varphi + \frac{1}{\varepsilon}(N_\varepsilon(u) - N_\varepsilon(v)) - \psi_\varepsilon(u) + \psi_\varepsilon(v)\right]$$

and thus, by using (4.231),

$$\|u - v\|_0 \leq \frac{1}{\lambda + \frac{1}{\varepsilon}}\left(\|\psi - \varphi\|_0 + \frac{1}{\varepsilon}\|u - v\|_0 + \|\psi_\varepsilon(u)\|_0 + \|\psi_\varepsilon(v)\|_0\right).$$

This implies

$$\lambda\|u - v\|_0 \leq \|\psi - \varphi\|_0 + \|\psi_\varepsilon(u)\|_0 + \|\psi_\varepsilon(v)\|_0.$$

We now obtain (4.232) by letting $\varepsilon \to 0$ and using (4.232).   □
    The next theorem is the main result of this subsection.

Theorem 4.167 *Assume that Hypotheses 4.145 for $U = H$ and $G = I$ and 4.162 hold. Then, for all $\lambda > 0$ and $\psi \in UC_b(H)$, there exists a unique mild solution $u \in UC_b^1(H)$ of (4.222). Moreover, the operator $\mathcal{B}$ defined in (4.224) is m-dissipative and we have, for all $\lambda > 0$,*

$$u = (\lambda I - \mathcal{B})^{-1}(\psi).$$

Finally, if $\psi \in UC_b^1(H)$, $F \in C_b^{1,1}(H)$, $b_1 \in C_b^1(H, H)$, then $u \in UC_b^2(H)$.[37]

Proof By (i) and (ii) of Proposition 4.165 and Theorem 4.112-(iii) we know that there exists a $\lambda_0 > 0$ such that the operator $R(\lambda)$, defined for all $\lambda \geq \lambda_0$, takes its values in $UC_b^1(X)$, is injective and satisfies the resolvent identity (4.226). Thus, thanks to Proposition B.19 (see also Proposition I.3.3 of [146]) there exists a unique operator

$$\mathcal{B}_1 : D(\mathcal{B}_1) \subset UC_b^1(H) \to UC_b(H)$$

such that $R(\lambda) = (\lambda I - \mathcal{B}_1)^{-1}$ for all $\lambda \geq \lambda_0$. Moreover, since $R(\lambda)$ satisfies (4.227) we know, again by Proposition B.19 (see also Proposition II.9.6 of [146]), that $\mathcal{B}_1$ is m-dissipative.

The m-dissipativity of $\mathcal{B}_1$ then yields that $R(\lambda) = (\lambda I - \mathcal{B}_1)^{-1}$ is well defined for all $\lambda \in (0, +\infty)$ and satisfies (4.226) and (4.227) for any $\lambda, \mu > 0$ (again by Proposition B.19, see also Propositions I.3.2 and II.9.1 of [146]).

The proof is completed in three steps.

Step 1. For all $\lambda > 0$ and $\psi \in UC_b(X)$, $(\lambda I - \mathcal{B}_1)^{-1}(\psi)$ is a fixed point of T_λ^ψ. Conversely, if $u \in UC_b^1(X)$ is a fixed point of T_λ^ψ, then $u = (\lambda I - \mathcal{B}_1)^{-1}(\psi)$.

Recall that, for all $\lambda > \lambda_0$, $R(\lambda)(\psi) = (\lambda I - \mathcal{B}_1)^{-1}(\psi)$ is the unique fixed point of T_λ^ψ. Let $\lambda > 0$ and $u = (\lambda I - \mathcal{B}_1)^{-1}(\psi)$. Since (4.226) holds for $(\lambda I - \mathcal{B}_1)^{-1}$ for all $\lambda > 0$, we have $u = ((\lambda + \lambda_0)I - \mathcal{B}_1)^{-1}(\psi + \lambda_0 u)$, which is the unique fixed point of $T_{\lambda+\lambda_0}^{\psi+\lambda_0 u}$. Thus, by (4.230) we get

$$u = T_{\lambda+\lambda_0}[\psi + \lambda_0 u + \langle b_1, Du \rangle + F_1(Du)]$$
$$= T_\lambda[\psi + \lambda_0 u + \langle b_1, Du \rangle + F_1(Du) + (\lambda - (\lambda + \lambda_0))u] = T_\lambda^\psi(u)$$

and u is a fixed point of T_λ^ψ.

Conversely, if u is a fixed point of T_λ^ψ for $\lambda > 0$, then using the same equality above, we get that we must have $u = T_{\lambda+\lambda_0}^{\psi+\lambda_0 u} u$ so

$$u = ((\lambda + \lambda_0)I - \mathcal{B}_1)^{-1}(\psi + \lambda_0 u) = (\lambda I - \mathcal{B}_1)^{-1}(\psi).$$

Step 2. $\mathcal{B}_1 = \mathcal{B}$ as defined in (4.224).

---

[37]The last claim of Theorem 4.167 is proved in Theorem 3.3 of [317] using interpolation spaces and a bootstrap argument, assuming only $b_1 \in C_b^{0,1}(H, H)$, $F_1 \in C_b^{0,1}(H)$ and $\psi \in C_b^1(H)$ and a weak additional assumption about the singularity of $\|\Gamma(t)\|$ at $t = 0$.

First we recall that one can define $D(\mathcal{A})$ as the set of all functions $T_\lambda \psi$ for $\psi \in UC_b(X)$ and for an arbitrary $\lambda > 0$, so that $D(\mathcal{A})$ is a subset of $UC_b^1(X)$. Now let $u \in D(\mathcal{B}_1)$ and $\psi = \lambda u - \mathcal{B}_1(u)$. Then, by Step 1 above,

$$u = T_\lambda[\psi + \langle b_1, Du \rangle + F_1(Du)]$$

and thus $u \in D(\mathcal{A})$.

Conversely, let $u \in D(\mathcal{A})$ and $u = T_\lambda \psi$ for some $\lambda > 0$ and $\psi \in UC_b(H)$. Then

$$u = T_\lambda[\psi - \langle b_1, Du \rangle - F_1(Du) + \langle b_1, Du \rangle + F_1(Du)] = T_\lambda^{\psi - \langle b_1, Du \rangle - F_1(Du)}(u)$$

and thus $u \in D(\mathcal{B}_1)$. Hence $D(\mathcal{B}_1) = D(\mathcal{A}) = D(\mathcal{B})$.

Now let $u \in D(\mathcal{B}) = D(\mathcal{B}_1)$ and let $\psi = \mathcal{B}_1(u)$. We have, for $\lambda > 0$, $\lambda u - \psi = \lambda u - \mathcal{B}_1(u)$, i.e.

$$u = T_\lambda^{\lambda u - \psi} u = T_\lambda[\lambda u - \psi + \langle b_1, Du \rangle + F_1(Du)].$$

Thus, since $T_\lambda = (\lambda I - \mathcal{A})^{-1}$,

$$\lambda u - \mathcal{A}u = \lambda u - \psi + \langle b_1, Du \rangle + F_1(Du),$$

i.e.

$$\psi = \mathcal{A}u + \langle b_1, Du \rangle + F_1(Du),$$

which concludes the proof.

Step 3. If $\psi \in UC_b^1(H)$, then $u \in UC_b^2(H)$.

When $\lambda > \lambda_1$, where λ_1 is from Theorem 4.160, the statement is a direct consequence of Hypothesis 4.162 and Remark 4.161. When $\lambda > 0$ we use the fact that, as seen above, the mild solution $u \in UC_b^1(H)$ satisfies

$$u = T_{\lambda + \lambda_1}[\psi + \lambda_1 u + \langle b_1, Du \rangle + F_1(Du)],$$

i.e. u is the mild solution of (4.222) with λ substituted by $\lambda + \lambda_1$ and ψ substituted by $\psi + \lambda_1 u$. The claim thus follows.   □

We now show that the mild solution of (4.222) is also a $\mathcal{K}$-strong solution, i.e. it can be approximated by classical solutions for all $\lambda > 0$. The definitions of a classical solution and a $\mathcal{K}$-strong solution are given by Definitions 4.139 and 4.140 with the operator $\mathcal{A}_1$ in (4.158) replaced by the operator $\mathcal{A}_0$ defined in (B.36). It could also be substituted by the operator $\hat{\mathcal{A}}_0$ defined in (B.37) without any big difficulties. We have the following result.

Theorem 4.168 *Assume that Hypotheses 4.145 with $U = H$ and $G = I$ and 4.162 are satisfied. Let $\lambda > 0$. Then we have the following.*

(i) *If u is a classical solution of (4.222) then it is also a mild solution.*

(ii) If u is a mild solution of (4.222) and $u \in D(\mathcal{A}_0)$ then u is also a classical solution.

(iii) u is a mild solution of (4.222) if and only if it is a $\mathcal{K}$-strong solution.

Proof Statement (i) follows immediately from the definitions, since for a classical solution $u \in D(\mathcal{A}_0)$, we immediately have $u \in D(\mathcal{B})$ and $\lambda u - \mathcal{B}(u) = \psi$, hence $u = T_\lambda^\psi u$.

Concerning statement (ii) we observe that, if u is a mild solution, we have $\lambda u - \mathcal{B}(u) = \lambda u - \mathcal{A}u - \langle b_1, Du \rangle - F_1(Du) = \psi$. If $u \in D(\mathcal{A}_0)$ then $\mathcal{A}_0 u = \mathcal{A}u$, hence u satisfies

$$\lambda u - \mathcal{A}u - \langle b_1, Du \rangle - F_1(Du) = \psi$$

and then it is a classical solution according to Definition 4.139.

We now prove (iii) starting with the "only if" part.

Let $\psi \in UC_b(H)$. Let $u = (\lambda I - \mathcal{B})^{-1}(\psi)$ be the mild solution of (4.222). Then $u \in D(\mathcal{B}) = D(\mathcal{A}) \subset UC_b^1(H)$. We now approximate u by a sequence $(u_n)_{n \in \mathbb{N}} \subset \mathcal{F}C_0^\infty(H) \subset D(\mathcal{A}_0)$ given in Lemma B.78. Since both the operators D (the Fréchet derivative) and $\mathcal{A}$ are $\mathcal{K}$-closed, the sequence u_n satisfies

$$u_n \xrightarrow{\mathcal{K}} u, \quad \text{in } UC_b(H), \qquad Du_n \xrightarrow{\mathcal{K}} Du, \quad \text{in } UC_b(H, H),$$

$$\mathcal{A}_0 u_n = \mathcal{A}u_n \xrightarrow{\mathcal{K}} \mathcal{A}u, \quad \text{in } UC_1(H).$$

Hence, setting

$$\psi_n = \lambda u_n - \mathcal{A}_0 u_n - \langle b_1, Du_n \rangle - F_1(Du_n),$$

we have $\psi_n \in UC_1(H)$ and $\psi_n \xrightarrow{\mathcal{K}} \psi$ in $UC_1(H)$. This concludes the proof of the "only if" part.

To prove the "if" part, let u be a $\mathcal{K}$-strong solution and let $(u_n)_{n \in \mathbb{N}}$ be the approximating sequence as in Definition 4.140. Then for every $n \in \mathbb{N}$, u_n satisfies

$$\mathcal{A}_0 u_n = \lambda u_n - \langle b_1, Du_n \rangle - F_1(Du_n) - \psi_n.$$

Hence, by Definition 4.140 the right-hand side $\mathcal{K}$-converges to $\lambda u - \langle b_1, Du \rangle - F_1(Du) - \psi$. Since the $\mathcal{K}$-closure of $\mathcal{A}_0$ is $\mathcal{A}$ (see Proposition B.92-(i)) it follows that $u \in D(\mathcal{A})$ and

$$\mathcal{A}u = \lambda u - \langle b_1, Du \rangle - F_1(Du) - \psi \iff \lambda u - \mathcal{B}(u) = \psi,$$

which gives the claim. $\qquad\qquad\qquad\qquad\qquad\qquad\qquad\qquad\qquad\qquad\qquad\square$

4.7   HJB Equations of Ornstein–Uhlenbeck Type: Locally Lipschitz Hamiltonian

In this section we again consider the HJB equations (4.169) and (4.170), however now in the case when, in addition to Hypothesis 4.145 with $U = H$ and $G = I$, we only assume that the Hamiltonian F_0 is Lipschitz continuous on bounded subsets in the variable $p = Dv$. For simplicity we only consider the Hamiltonian F_0, which is independent of v. We analyze the parabolic case, which is mainly taken from [307]. Elliptic equations, up to now, have only been studied for problems related to reaction diffusion equations (see [106, 107], Sect. 9.5.2) and this material is briefly presented in Sect. 4.9.2.2.

In this section we take $U = H$, $G = I$, $m = 0$ and set $\Gamma(t) := \Gamma_I(t)$, $\gamma(t) := \|\Gamma(t)\|$.

4.7.1   The Parabolic Case

We replace Hypothesis 4.72 by the following set of assumptions.

Hypothesis 4.169

(i) $F_0 : [0, T] \times H \times H \to \mathbb{R}$ is continuous and for every $R > 0$ there exists a constant $C_{1,F_0}(R)$ such that for all $t \in [0, T]$, $x \in H$, $p, p_1, p_2 \in H$, $|p|, |p_1|, |p_2| \le R$,

$$|F_0(t, x, p_1) - F_0(t, x, p_2)| \le C_{1,F_0}(R)|p_1 - p_2|$$

$$|F_0(t, x, p)| \le C_{1,F_0}(R).$$

(ii) $F_0(t, \cdot, \cdot)$ is Fréchet differentiable with $D_x F$ and $D_p F$ continuous and for every $R > 0$ there exists a constant $C_{2,F_0}(R)$ such that for all $t \in [0, T], x, x_1, x_2 \in H$, $p, p_1, p_2 \in H$, $|p|, |p_1|, |p_2| \le R$,

$$|D_x F_0(t, x, p_1) - D_x F_0(t, x, p_2)| \le C_{2,F_0}(R)|p_1 - p_2|$$

$$|D_p F_0(t, x_1, p_1) - D_p F_0(t, x_2, p_2)| \le C_{2,F_0}(R)[|x_1 - x_2| + |p_1 - p_2|]$$

$$|D_x F_0(t, x, p)| \le C_{2,F_0}(R).$$

(iii) $\varphi \in C_b^1(H)$.

Remark 4.170   Hypothesis 4.169 is obviously satisfied in the case (studied in [307]) when $F_0(t, x, p) = \langle b_1(x), p \rangle + F_1(p) + g(x)$, where $b_1 \in UC_b^1(H, H)$, $g \in UC_b^1(H)$ and $F_1 \in C_b^{1,1}(B(0, R))$ for all $R > 0$.   ∎

In the next three subsections we prove the following results.

- The existence and uniqueness of a local mild solution. A local mild solution is a mild solution on a short time interval.
- An a priori estimate for the local solution that allows us to extend it to a global mild solution.
- The mild solution is a $\mathcal{K}$-strong solution.

4.7.1.1 Local Existence and Uniqueness of Mild Solutions

We prove the existence and uniqueness of a local mild solution of (4.169) in a suitable weighted Banach space by applying the contraction mapping principle. The proof generalizes the ideas from Theorem 4.5 of [307] (see also [90, 107]).

Theorem 4.171 *Let Hypothesis 4.145 hold with $U = H$ and $G = I$ and let $\gamma \in \mathcal{I}_2$. Assume that Hypothesis 4.169 holds. Then there exists a $T_0 \in [0, T]$ such that the Cauchy problem (4.169) has a local mild solution $u \in \overline{C}_{b,\gamma}^{0,2,s}([T_0, T] \times H)$. The mild solution u is unique among all mild solutions in $C_b^{0,1}([T_0, T] \times H)$.*

Proof This proof is similar to the first two steps of the proof of Theorem 4.152. We first prove that the map Υ defined in (4.181) is a contraction in the space $\overline{\mathcal{H}}_{T_0,T}$ (defined in (4.182)) for a suitably chosen $T_0 \in [0, T)$. Set

$$\psi^v(s, x) = F_0(s, x, Dv(s, x)). \tag{4.236}$$

Arguing exactly as in Step 2 of the proof of Theorem 4.152, we get the estimate

$$\|\Upsilon[v]\|_{\overline{\mathcal{H}}_{T_0,T}} \le Me^{(\omega \vee 0)T} \|\varphi\|_1 + \rho_1(T - T_0)\left[\|\psi^v\|_0 + \sup_{t \in [T_0,T)} \gamma(T - t)^{-1}\|D\psi^v(t, \cdot)\|_0\right], \tag{4.237}$$

where ρ_1 is given as in (4.193). Moreover, by (4.236) and Hypothesis 4.169, we have

$$\|\psi^v\|_0 \le C_{1,F_0}(\|Dv\|_0),$$

$$\|D\psi^v(t, \cdot)\|_0 \le C_{2,F_0}(\|Dv\|_0) + C_{1,F_0}(\|Dv\|_0)\|D^2v(t, \cdot)\|_0$$

so that (using that γ is decreasing) estimate (4.237) becomes

$$\|\Upsilon[v]\|_{\overline{\mathcal{H}}_{T_0,T}} \le Me^{(\omega \vee 0)T} \|\varphi\|_1 + \rho_1(T - T_0)\Big[C_{1,F_0}(\|Dv\|_0)+$$

$$\gamma(T)^{-1}C_{2,F_0}(\|Dv\|_0) + C_{1,F_0}(\|Dv\|_0) \sup_{t \in [T_0,T)} \gamma(T - t)^{-1}\|D^2v(t, \cdot)\|_0\Big],$$

i.e.

$$\|\Upsilon[v]\|_{\overline{\mathcal{H}}_{T_0,T}} \le Me^{(\omega \vee 0)T} \|\varphi\|_1 + \rho_2(T - T_0, \|Dv\|_0)(1 + \|v\|_{\overline{\mathcal{H}}_{T_0,T}}), \tag{4.238}$$

where

$$\rho_2(T - T_0, R) := \rho_1(T - T_0)\left[C_{1,F_0}(R) + [\gamma(T)^{-1} \vee 1]C_{2,F_0}(R)\right].$$

We now take $w_1, w_2 \in \overline{\mathcal{H}}_{T_0,T}$ and estimate $\Upsilon[w_1] - \Upsilon[w_2]$ exactly as in (4.197) to obtain

$$\|\Upsilon w_1 - \Upsilon w_2\|_{\overline{\mathcal{H}}_{T_0,T}}$$

$$\leq \rho_1(T - T_0)\left[\|\psi^{w_1} - \psi^{w_2}\|_0 + \sup_{t \in [T_0,T)} \gamma(T - t)^{-1}\|D\left(\psi^{w_1} - \psi^{w_2}\right)(t, \cdot)\|_0\right].$$
$$(4.239)$$

Defining $R(w_1, w_2) := \|Dw_1\|_0 \vee \|Dw_2\|_0$ we get, by (4.236) and Hypothesis 4.169,

$$\|\psi^{w_1}(t, \cdot) - \psi^{w_2}(t, \cdot)\|_0 \leq C_{1,F_0}(R(w_1, w_2))\|Dw_1(t, \cdot) - Dw_2(t, \cdot)\|_0$$

and

$$\|D\psi^{w_1}(t, \cdot) - D\psi^{w_2}(t, \cdot)\|_0 \leq C_{2,F_0}(R(w_1, w_2))\|Dw_1(t, \cdot) - Dw_2(t, \cdot)\|_0$$
$$+ C_{2,F_0}(R(w_1, w_2))\|Dw_1(t, \cdot) - Dw_2(t, \cdot)\|_0\|D^2w_1(t, \cdot)\|_0$$
$$+ C_{1,F_0}(R(w_1, w_2))\|D^2w_1(t, \cdot) - D^2w_2(t, \cdot)\|_0.$$

Combining the above with (4.239), we obtain by straightforward calculations

$$\|\Upsilon[w_1] - \Upsilon[w_2]\|_{\overline{\mathcal{H}}_{T_0,T}} \leq \rho_1(T - T_0)\|Dw_1 - Dw_2\|_0 \times$$

$$\times \left[C_{1,F_0}(R(w_1, w_2)) + C_{2,F_0}(R(w_1, w_2))\left(\gamma(T)^{-1} + \sup_{t \in [T_0,T)} \gamma(T - t)^{-1}\|D^2w_1(t, \cdot)\|_0\right)\right]$$
$$+ \rho_1(T - T_0)C_{1,F_0}(R(w_1, w_2)) \sup_{t \in [T_0,T)} \gamma(T - t)^{-1}\|D^2w_1(t, \cdot) - D^2w_2(t, \cdot)\|_0$$

which gives

$$\|\Upsilon[w_1] - \Upsilon[w_2]\|_{\overline{\mathcal{H}}_{T_0,T}} \leq \rho_2(T - T_0, R(w_1, w_2))(1 + \|w_1\|_{\overline{\mathcal{H}}_{T_0,T}})\|w_1 - w_2\|_{\overline{\mathcal{H}}_{T_0,T}}.$$
$$(4.240)$$

We now define a subset of $\overline{\mathcal{H}}_{T_0,T}$ where it is possible to apply the contraction mapping principle. We take

$$R \geq 2Me^{(\omega \vee 0)T}\|\varphi\|_1 \tag{4.241}$$

and choose $T_0 \in [0, T)$ such that (see (4.238))

$$R/2 + \rho_2(T - T_0, R)(1 + R) < R \tag{4.242}$$

and (see (4.240))

$$\rho_2(T - T_0, R)(1 + R) < \frac{1}{2}. \tag{4.243}$$

Let $B_{T_0}(0, R)$ be the closed ball centered at 0 with radius R in $\overline{\mathcal{H}}_{T_0,T}$. Applying (4.242) and (4.243) to (4.239) and (4.240), it follows that Υ is a contraction on $B_{T_0}(0, R)$. The contraction mapping principle now guarantees that there exists a unique solution u of $v = \Upsilon[v]$ in $B_{T_0}(0, R)$. The solution u is also unique in the whole space $C_b^{0,1}([T_0, T] \times H)$. Indeed, let u_1 be another mild solution in $C_b^{0,1}([T_0, T] \times H)$. We set $R_1 := \|Du\|_0 \vee \|Du_1\|_0$. Since $u_1 = \Upsilon[u_1]$, we have, for $(t, x) \in [T_0, T] \times H$, using that R_t is a contraction semigroup in $C_b(H)$,

$$|u(t, x) - u_1(t, x)| = |\Upsilon[u](t, x) - \Upsilon[u_1](t, x)|$$

$$\leq \int_t^T |R_{s-t}[F_0(s, \cdot, Du(s, \cdot)) - F_0(s, \cdot, Du_1(s, \cdot))](x)|\, ds$$

$$\leq C_{1,F_0}(R_1) \int_t^T \|Du(s, \cdot) - Du_1(s, \cdot)\|_0 ds$$

and, using (4.70),

$$|Du(t, x) - Du_1(t, x)| \leq \int_t^T |DR_{s-t}[F_0(t, \cdot, Du(t, \cdot)) - F_0(t, \cdot, Du_1(t, \cdot))](x)|\, ds$$

$$\leq C_{1,F_0}(R_1) \int_t^T \gamma(s - t)\|Du_1(s, \cdot) - Du_2(s, \cdot)\|_0 ds.$$

Thus

$$\|u(t, \cdot) - u_1(t, \cdot)\|_1 \leq C_{1,F_0}(R_1) \int_t^T [1 + \gamma(s - t)]\|u(s, \cdot) - u_1(s, \cdot)\|_1 ds,$$

which gives the required uniqueness thanks to Gronwall's Lemma D.29.

Clearly the argument above cannot be iterated on a time interval of the same length without an a priori estimate for the $C_b^{0,1}$ norm of the solution since the choice of R and then of T_0 depend on $\|\varphi\|_1$. $\qquad\square$

4.7.1.2 A Priori Estimates

In this subsection we let the assumptions of Theorem 4.171 hold. We denote by u the local mild solution of Eq. (4.1). We need the following additional assumption.

Hypothesis 4.172 (i) For every $t \geq 0$ we have $\|e^{tA}\| \leq e^{\omega t}$, i.e. $M = 1$ in Hypothesis 4.145.

(ii) For every $t \in [0, T]$, $x \in H$, $p \in H$ we have

$$\langle D_x F_0(t, x, p), p \rangle \leq C_{3, F_0} |p|^2.$$

Remark 4.173 Thanks to the Lumer–Phillips Theorem B.45 the above assumption (i) is satisfied if and only if $A^* - \omega I$ is maximal dissipative. So, in particular, for all $p \in D(A^*)$ we must have

$$\langle p, A^* p \rangle \leq \omega |p|^2. \tag{4.244}$$

Moreover, assumption (ii) is satisfied if we take $F_0(t, x, p) = \langle b_1(t, x), p \rangle + F_1(p) + g$ with bounded $D_x b_1$ and Dg (which is the case studied in [307]) or more generally if, for some $C > 0$

$$\langle D_x b_1(t, x) p, p \rangle \leq C |p|^2, \qquad \forall (t, x, p) \in [0, T] \times H \times H$$

and Dg is bounded.   ∎

For simplicity, from now on we will use the notation $g(t, x) = F_0(t, x, 0)$ and $F_2(t, x, p) = F_0(t, x, p) - F_0(t, x, 0)$ so that we will substitute $F_0(t, x, p)$ by $F_2(t, x, p) + g(t, x)$. Hence our HJB equation is now written as

$$\begin{cases} v_t + \dfrac{1}{2} \operatorname{Tr}[\Sigma D^2 v] + \langle Ax, Dv \rangle + F_2(t, x, v, D^G v) + g(t, x) = 0, & t \in [0, T), \ x \in D(A), \\ v(T, x) = \varphi(x), & x \in H. \end{cases}$$
$$\tag{4.245}$$

We prove an estimate for $\|u\|_{C_b^{0,1}([T_0, T] \times H)}$ when $\varphi \in C_b^1(H)$, from which the global existence will follow easily.

Proposition 4.174 *Let Hypothesis 4.145 hold with $U = H$ and $G = I$ and let $\gamma \in \mathcal{I}_2$. Assume that Hypotheses 4.169 and 4.172 hold. Let $\varphi \in C_b^1(H)$ and let $u \in \overline{C}_{b,\gamma}^{0,2,s}([0, T - T_0] \times H)$ be the local mild solution of (4.245) from Theorem 4.171. Then, setting $C = 2(\omega + C_{3, F_0} + 1)$, we have*

$$\|u\|_{C_b^{0,1}([T_0, T] \times H)} \leq e^{C(T - T_0)} \left[\|\varphi\|_1 + (T - T_0) \|g\|_{C_b^{0,1}([T_0, T] \times H)} \right]. \tag{4.246}$$

Proof We first prove an estimate for $\|u\|_0$. For $(t, x) \in [T_0, T] \times H$ we define

$$G_0(t, x) := \int_0^1 D_p F_2(t, x, \lambda Du(t, x)) d\lambda. \tag{4.247}$$

By Hypothesis 4.169 and the regularity of u we deduce that G_0 is continuous and satisfies

$$|G_0(t, x)| \leq C_{1, F_0}(\|Du\|_0), \qquad \forall (t, x) \in [T_0, T] \times H, \tag{4.248}$$

and

$$|G_0(t, x_1) - G_0(t, x_2)| \leq C_{2, F_0}(\|Du\|_0)|[|x_1 - x_2| + |Du(t, x_1) - Du(t, x_2)|]$$

$$\leq C_{2, F_0}(\|Du\|_0) \left[1 + \gamma(t)\|u\|_{\overline{C}_{b,\gamma}^{0,2,s}([T_0, T] \times H)} \right] |x_1 - x_2| \quad (4.249)$$

for all $t \in [T_0, T]$, $x_1, x_2 \in H$. Hence, G_0 satisfies Hypothesis 1.145-(i) for b (when b_0 does not depend on a_1 and $a_2 = 0$) on $[T_0, T] \times H$. Thus, by Proposition 1.147, the SDE

$$dY(s) = [AY(s) + G_0(s, Y(s))]ds + \sqrt{\Sigma}dW(s), \quad s \in [t, T], \qquad Y(t) = x, \quad (4.250)$$

$t \in [T_0, T]$, $x \in H$, has a unique solution $Y(s; t, x)$. For $\phi \in B_b(H)$ we define $P_{t,s}[\phi](x) := \mathbb{E}[\phi(Y(s; t, x))]$ to be the two-parameter transition semigroup associated to the above SDE. We claim that, for $(t, x) \in [T_0, T] \times H$,

$$u(t, x) = P_{t,T}[\varphi](x) + \int_t^T P_{t,s}[g(s, \cdot)](x)ds. \quad (4.251)$$

Indeed, by Theorem 4.150, we know that u is a $\mathcal{K}$-strong solution of (4.245).[38] Using Remark 4.136-(v), this means (by Definition 4.132 where $\eta \equiv 1$, $m = 0$, and spaces of Borel measurable functions are substituted by spaces of continuous functions since here φ, g, u, Du are continuous) that there exist three sequences $(\varphi_n), (g_n), (u_n)$, such that u_n is a classical solution of the Eq. (4.245) with data φ_n and g_n and we have

$$\varphi_n \xrightarrow{\mathcal{K}} \varphi, \quad \text{in } C_b(H), \qquad Du_n \xrightarrow{\mathcal{K}} Du \quad \text{in } C_b([0, T] \times H, H),$$

$$g_n \xrightarrow{\mathcal{K}} g, \quad u_n \xrightarrow{\mathcal{K}} u, \quad \text{in } C_b([0, T] \times H).$$

Letting $Y(s) = Y(s; t, x)$ we can then apply Dynkin's formula of Proposition 1.169 to the process $u_n(s, Y(s))$ obtaining

$$\mathbb{E}\varphi_n(Y(T)) - u_n(t, x)$$

$$= \mathbb{E} \int_t^T \left[(u_n)_t(s, Y(s)) + \mathcal{A}_0 u_n(s, Y(s)) + \langle G_0(s, Y(s)), Du_n(s, Y(s)) \rangle \right] ds.$$

Since u_n is a classical solution of (4.245) with data φ_n and g_n, we get

---

[38] To be precise, in this case Hypothesis 4.72 is not satisfied. However, looking at the proof of Theorem 4.150 (which uses Theorem 4.135), we see that this assumption is only used there to guarantee the existence of a mild solution. Since here we already have a mild solution, the proof proceeds in exactly the same way.

$$\mathbb{E}\varphi_n(Y(T)) - u_n(t, x) = \mathbb{E}\int_t^T \Big[\langle G_0(s, Y(s)), Du_n(s, Y(s)) \rangle$$

$$- F_2(s, Y(s), Du_n(s, Y(s))) - g_n(s, Y(s)) \Big] ds.$$

Letting $n \to +\infty$ and using the dominated convergence theorem we thus obtain

$$\mathbb{E}\varphi(Y(T)) - u(t, x) = \mathbb{E}\int_t^T \Big[\langle G_0(s, Y(s)), Du(s, Y(s)) \rangle$$

$$- F_2(s, Y(s), Du(s, Y(s))) - g(s, Y(s)) \Big] ds = -\mathbb{E}\int_t^T g(s, Y(s)) ds,$$

where to get the last equality we used the definition of G_0 and the fact that $F_2(t, x, 0) = 0$. This shows (4.251) and we immediately get the estimate

$$\|u(t, \cdot)\|_0 \leq \|\varphi\|_0 + (T - t)\|g\|_0. \tag{4.252}$$

The proof of the estimate for $\|Du\|_0$ is more complicated. We use the so-called Bernstein's method, consisting in finding the equation satisfied by $\|Du\|_0^2$ and using it to get the required estimate. We set

$$f(t, x) = F_0(t, x, Du(t, x)) + g(t, x).$$

By Hypothesis 4.169 and the regularity of u we have $f \in \overline{C}_{b,\gamma}^{0,1}([T_0, T] \times H)$. By Theorem B.95 (which generalizes the approximation scheme used in [108, 308, 494]) we can find two sequences $\varphi_n \in \mathcal{F}C_0^{\infty, A^*}(H)$ and $f_n \in \mathcal{F}C_0^{\infty, A^*}([T_0, T] \times H)$ such that

$$\begin{aligned}
\mathcal{K} - \lim_{n \to \infty} \varphi_n = \varphi && \text{in } C_b(H) \\
\mathcal{K} - \lim_{n \to \infty} D\varphi_n = D\varphi && \text{in } C_b(H, H)
\end{aligned}$$

$$\tag{4.253}$$

$$\begin{aligned}
\mathcal{K} - \lim_{n \to \infty} f_n = f && \text{in } C_b([T_0, T] \times H) \\
\mathcal{K} - \lim_{n \to \infty} Df_n = Df && \text{in } \overline{C}_{b,\gamma}([T_0, T) \times H, H)
\end{aligned}$$

(see Definition 4.131 for the last convergence). If we consider the sequence

$$u_n(t, x) = R_t[\varphi_n](x) + \int_0^t R_{t-s}[f_n(s, \cdot)](x) ds$$

then, by Theorem B.95 it follows that, for all $n \in \mathbb{N}$, u_n satisfies (4.144) on $[T_0, T]$. Moreover, simply differentiating the above formula and using the regularity properties of R_t, φ_n and f_n we get that, for every $n \in \mathbb{N}$, $t \in [T_0, T]$, the function u_n satisfies

$$u_n(t, \cdot) \in UC_b^\infty(H)$$

$$D^3 u_n(t, \cdot) D u_n(t, \cdot) \in UC_b(X, \mathcal{L}_1(H)) \tag{4.254}$$

$$A^* D^2 u_n(t, \cdot) D u_n(t, \cdot) \in UC_b(H)$$

and, by (4.253),

$$\mathcal{K} - \lim_{n \to \infty} u_n = u \qquad \qquad \text{in } C_b([T_0, T] \times H)$$

$$\mathcal{K} - \lim_{n \to \infty} D u_n = D u \qquad \qquad \text{in } C_b([T_0, T] \times H, H) \tag{4.255}$$

$$\mathcal{K} - \lim_{n \to \infty} D^2 u_n h = D^2 u h \qquad \text{in } \overline{C}_{b, \gamma}([T_0, T) \times H, H), \quad \forall h \in H.$$

Furthermore, still according to Theorem B.95, every u_n is a classical solution of the approximating problem

$$\begin{cases} w_t + \dfrac{1}{2} \operatorname{Tr} [\Sigma D^2 w] + \langle Ax, Dw \rangle + F_2(t, x, Dw) + g_n = 0, \quad t \in [T_0, T), \ x \in H \\[2mm] w(T, x) = \varphi_n(x), \quad x \in H, \end{cases} \tag{4.256}$$

where

$$g_n(t, x) = g(t, x) + [f_n(t, x) - f(t, x)] + [F_2(t, x, Du(t, x)) - F_2(t, x, Du_n(t, x))].$$

We now set for $(t, x) \in [T_0, T] \times H$,

$$z_n(t, x) = \frac{1}{2} |D u_n(t, x)|^2, \qquad z(t, x) = \frac{1}{2} |D u(t, x)|^2. \tag{4.257}$$

We prove that, for every $n \in \mathbb{N}$, the function z_n is a classical solution of a certain Kolmogorov equation. The required estimate will follow by a suitable representation of the solution as in the first part of the proof. In the proof we will be using that $D^3 u_n(t, x)$, when it is identified with an element of $\mathcal{L}^3(H)$, is symmetric (see Appendix D.2).

Formally, we have

$$(z_n)_t(t, x) = \langle D_{tx} u_n(t, x), D u_n(t, x) \rangle. \tag{4.258}$$

Since by (4.256)

$$(u_n)_t(t, x) = -\frac{1}{2} \operatorname{Tr} [\Sigma D^2 u_n(t, x)] - \langle x, A^* D u_n(t, x) \rangle$$

$$- F_2(t, x, D u_n(t, x)) - g_n(t, x), \quad t \in [T_0, T), \ x \in H,$$

then, by (4.254), we can deduce that the map $(u_n)_t$ is continuously differentiable with respect to x. Indeed, by the definition of the trace, for a given orthonormal basis $\{e_j\}$ of H and any $h \in H$,

$$\left\langle D\mathrm{Tr}\,[\Sigma D^2 u_n], h \right\rangle = \left\langle D\left[\sum_{j=1}^{+\infty}\left\langle \Sigma D^2 u_n e_j, e_j \right\rangle\right], h \right\rangle = \sum_{j=1}^{+\infty}\left\langle D\left\langle D^2 u_n e_j, \Sigma e_j \right\rangle, h \right\rangle$$

$$= \sum_{j=1}^{+\infty}\left\langle [D^3 u_n e_j]\Sigma e_j, h \right\rangle = \sum_{j=1}^{+\infty}\left\langle [D^3 u_n h]e_j, \Sigma e_j \right\rangle = \sum_{j=1}^{+\infty}\left\langle \Sigma[D^3 u_n h]e_j, e_j \right\rangle$$

$$\tag{4.259}$$

(in the last line we used that $D^3 u_n$ is symmetric) which makes classical sense thanks to (4.254). Moreover,

$$\left\langle D\left[\langle x, A^* Du_n(t, x)\rangle + F_2(t, x, Du_n(t, x)) + g_n(t, x)\right], h \right\rangle$$

$$= \left\langle x, A^* D^2 u_n(t, x)h \right\rangle + \left\langle h, A^* Du_n(t, x) \right\rangle + \left\langle D_x F_2(t, x, Du_n(t, x)), h \right\rangle \quad (4.260)$$

$$+ \left\langle D_p F_2(t, x, Du_n(t, x)), D^2 u_n(t, x)h \right\rangle + \left\langle Dg_n(t, x), h \right\rangle,$$

which again makes classical sense by (4.254). Hence $D_{tx}u_n$ is well defined. This implies that $z_n(\cdot, x)$ is differentiable for every $x \in H$ and that (4.258) is true. Then we can write

$$-(z_n)_t(t, x) = -\left\langle D_{tx}u_n(t, x), Du_n(t, x) \right\rangle$$

$$= \left\langle D\left[\tfrac{1}{2}\mathrm{Tr}\,[\Sigma D^2 u_n(t, x)] + \langle x, A^* Du_n(t, x)\rangle\right. \right. \quad (4.261)$$

$$\left.\left. + F_2(t, x, Du_n(t, x)) + g_n(t, x)\right], Du_n(t, x) \right\rangle$$

which yields, by (4.259) and (4.260),

$$-(z_n)_t(t, x) = \frac{1}{2}\sum_{j=1}^{+\infty}\left\langle \Sigma[D^3 u_n(t, x)(Du_n(t, x))]e_j, e_j \right\rangle + \left\langle x, A^* D^2 u_n(t, x)Du_n(t, x) \right\rangle$$

$$+ \left\langle Du_n(t, x), A^* Du_n(t, x) \right\rangle + \left\langle D_x F_2(t, x, Du_n(t, x)), Du_n(t, x) \right\rangle$$

$$+ \left\langle D_p F_2(t, x, Du_n(t, x)), D^2 u_n(t, x)Du_n(t, x) \right\rangle + \left\langle Dg_n(t, x), Du_n(t, x) \right\rangle.$$

$$\tag{4.262}$$

This, in particular, implies that $(z_n)_t \in C_b([T_0, T] \times H)$, which is a part of the definition of a classical solution (see Definition 4.129).

Now, by (4.254), $z_n(t, \cdot) \in UC_b^2(H)$ for every $t \in [T_0, T]$ and differentiating with respect to x we get

$$\langle Dz_n(t,x), h \rangle = \langle D^2 u_n(t,x)h, Du_n(t,x) \rangle \qquad \text{and}$$

$$\langle D^2 z_n(t,x)h, k \rangle = \langle [D^3 u_n(t,x)h]k, Du_n(t,x) \rangle + \langle D^2 u_n(t,x)h, D^2 u_n(t,x)k \rangle \tag{4.263}$$

for all $h, k \in H$. Observe now that, for $h, k \in H$, since $D^3 u_n$ is symmetric,

$$\langle [D^3 u_n(t,x)h]\Sigma k, Du_n(t,x) \rangle = \langle [D^3 u_n(t,x)Du_n(t,x)]h, \Sigma k \rangle$$
$$= \langle \Sigma[D^3 u_n(t,x)Du_n(t,x)]h, k \rangle.$$

Hence

$$\langle \Sigma D^2 z_n(t,x)h, k \rangle = \langle D^2 z_n(t,x)h, \Sigma k \rangle$$
$$= \langle [D^3 u_n(t,x)h]\Sigma k, Du_n(t,x) \rangle + \langle D^2 u_n(t,x)h, D^2 u_n(t,x)\Sigma k \rangle$$
$$= \langle \Sigma[D^3 u_n(t,x)Du_n(t,x)]h, k \rangle + \langle \Sigma D^2 u_n(t,x)D^2 u_n(t,x)h, k \rangle.$$

Now (4.254) and (4.263) imply (recall that the operator $\mathcal{A}_0$ is defined in (B.37)) that $\|z_n(t,\cdot)\|_{D(\mathcal{A}_0)}$ is bounded on $[T_0, T]$. Moreover,

$$\mathcal{A}_0 z_n(t,x) = \tfrac{1}{2}\mathrm{Tr}\,[\Sigma D^2 z_n(t,x)] + \langle x, A^* Dz_n(t,x) \rangle$$
$$= \frac{1}{2}\sum_{j=1}^{+\infty}\left(\langle \Sigma[D^3 u_n(t,x)Du_n(t,x)]e_j, e_j \rangle + \langle \Sigma D^2 u_n(t,x)D^2 u_n(t,x)e_j, e_j \rangle\right)$$
$$+ \langle x, A^* D^2 u_n(t,x)Du_n(t,x) \rangle, \tag{4.264}$$

so that $\mathcal{A}_0 z_n \in C_1([T_0, T] \times H)$ and, putting (4.263) and (4.264) in (4.262) we obtain the following equation for z_n

$$-(z_n)_t(t,x) = \mathcal{A}_0 z_n(t,x) - \frac{1}{2}\mathrm{Tr}\,[\Sigma D^2 u_n(t,x)D^2 u_n(t,x)]$$
$$+ \langle Du_n(t,x), A^* Du_n(t,x) + D_x F_2(t,x, Du_n(t,x)) \rangle \tag{4.265}$$
$$+ \langle D_p F_2(t,x, Du_n(t,x)), Dz_n(t,x) \rangle + \langle Dg_n(t,x), Du_n(t,x) \rangle.$$

We observe that

$$\langle Dg_n(t,x), Du_n(t,x) \rangle \le |Dg_n(t,x)||Du_n(t,x)|$$
$$\le \frac{|Dg_n(t,x)|^2}{2} + \frac{|Du_n(t,x)|^2}{2} = \frac{|Dg_n(t,x)|^2}{2} + z_n(t,x)$$

and, by Hypothesis 4.172,

$$\langle Du_n(t,x), A^* Du_n(t,x) \rangle \le \omega|Du_n(t,x)|^2 \le 2\omega z_n(t,x)$$

and

$$\langle Du_n(t, x), D_x F_2(t, x, Du_n(t, x)) \rangle$$
$$= \langle Du_n(t, x), D_x F_0(t, x, Du_n(t, x)) \rangle - \langle Du_n(t, x), Dg_n(t, x) \rangle$$
$$\leq C_{3,F_0}|Du_n(t, x)|^2 + \frac{|Dg_n(t, x)|^2}{2} + z_n(t, x) \leq [2C_{3,F_0} + 1]z_n + \frac{|Dg_n(t, x)|^2}{2}.$$

Then, setting

$$L_n(t, x) = \frac{1}{2}\mathrm{Tr}[\Sigma D^2 u_n(t, x) D^2 u_n(t, x)] + \left([2\omega + 2C_{3,F_0} + 1]z_n(t, x) \right.$$

$$+ \frac{|Dg_n(t, x)|^2}{2} - \left\langle Du_n(t, x), A^* Du_n(t, x) + D_x F_2(t, x, Du_n(t, x)) \right\rangle \right) \qquad (4.266)$$

$$+ \left(\frac{|Dg_n(t, x)|^2}{2} + z_n(t, x) - \langle Dg_n(t, x), Du_n(t, x) \rangle \right),$$

we have that $L_n \in C_b([T_0, T] \times H)$, $L_n \geq 0$ and

$$-(z_n)_t(t, x) = \mathcal{A}_0 z_n(t, x) + \left\langle D_p F_2(t, x, Du_n(t, x)), Dz_n(t, x) \right\rangle$$
$$+ |Dg_n|^2 + 2[\omega + C_{3,F_0} + 1]z_n(t, x) - L_n(t, x). \qquad (4.267)$$

Setting

$$G_1(t, x) = D_p F_2(t, x, Du(t, x)) \text{ and } C = 2(\omega + C_{3,F_0} + 1),$$

(4.267) can be written as

$$-(z_n)_t(t, x) = \mathcal{A}_0 z_n(t, x) + \langle G_1(t, x), Dz_n(t, x) \rangle$$

$$+ \left\langle D_p F_2(t, x, Du_n(t, x)) - D_p F_2(t, x, Du(t, x)), Dz_n(t, x) \right\rangle \qquad (4.268)$$

$$+ Cz_n(t, x) + |Dg_n(t, x)|^2 - L_n(t, x).$$

We now argue exactly as in the first part of the proof but now for the function $e^{-Ct}z_n(t, x)$.

By Hypothesis 4.169 and the regularity of u we deduce that G_1 is continuous and satisfies

$$|G_1(t, x)| \leq C_{1,F_0}(\|Du\|_0), \quad \forall(t, x) \in [T_0, T] \times H, \qquad (4.269)$$

and

$$|G_1(t, x_1) - G_1(t, x_2)| \leq C_{2,F_0}(\|Du\|_0)[|x_1 - x_2| + |Du(t, x_1) - Du(t, x_2)|]$$

$$\leq C_{2,F_0}(\|Du\|_0)\left[1 + \gamma(t)\|u\|_{\overline{C}_{b,\gamma}^{0,2,s}([T_0,T] \times H)}\right] |x_1 - x_2| \quad (4.270)$$

for all $t \in [T_0, T]$, $x_1, x_2 \in H$. Hence, for every $(t, x) \in [T_0, T] \times H$, by Proposition 1.147 the SDE

$$dY(s) = [AY(s) + G_1(s, Y(s))]ds + \sqrt{\Sigma}dW(s), \quad s \in [t, T], \qquad Y(t) = x, \quad (4.271)$$

has a unique solution $Y_1(s; t, x)$. Define $P_{t,s}^1[\phi](x) := \mathbb{E}[\phi(Y_1(s; t, x))]$, for $\phi \in B_b(H)$, to be the two-parameter transition semigroup associated to the SDE (4.271). We fix $(t, x) \in [T_0, T] \times H$. We use Dynkin's formula of Proposition 1.169 applied to the function $e^{-C(T-s)}z_n(s, Y_1(s; t, x))$, $s \in [t, T]$ and, arguing similarly as in the first part of this proof, we obtain

$$z_n(t, x) = \frac{1}{2}e^{C(T-t)}P_{t,T}^1\left[|D\varphi_n|^2\right](x) + \int_t^T e^{C(s-t)}P_{t,s}^1\left[|Dg_n|^2(s, \cdot) - L_n(s, \cdot)\right](x)ds$$

$$+ \int_t^T e^{C(s-t)}P_{t,s}^1\left[\langle D_p F_2(s, \cdot, Du_n(s, \cdot)) - D_p F_2(s, \cdot, Du(s, \cdot)), Dz_n(s, \cdot)\rangle\right](x)ds,$$

so that, since $L_n \geq 0$,

$$z_n(t, x) \leq \frac{1}{2}e^{C(T-t)}P_{t,T}^1\left[|D\varphi_n|^2\right](x) + \int_t^T e^{C(s-t)}P_{t,s}^1\left[|Dg_n|^2(s, \cdot)\right](x)ds$$

$$+ \int_t^T e^{C(s-t)}\left|P_{t,s}^1\left[\langle D_p F_2(s, \cdot, Du_n(s, \cdot)) - D_p F_2(s, \cdot, Du(s, \cdot)), Dz_n(s, \cdot)\rangle\right](x)\right|ds.$$

At this point we observe that, by (4.255), we have, in $C_b([T_0, T] \times H)$,

$$\mathcal{K} - \lim_{n \to \infty} D_p F_2(\cdot, \cdot, Du_n(\cdot, \cdot)) = D_p F_2(\cdot, \cdot, Du(\cdot, \cdot)), \qquad \mathcal{K} - \lim_{n \to \infty} z_n = z$$

and, in $\overline{C}_{b,\gamma}([T_0, T] \times H, H)$,

$$\mathcal{K} - \lim_{n \to \infty} Dz_n = Dz.$$

Similarly, by (4.253),

$$\mathcal{K} - \lim_{n \to \infty} D\varphi_n = D\varphi \qquad \text{in} \quad C_b(H)$$

and

$$\mathcal{K} - \lim_{n \to \infty} Dg_n = Dg \qquad \text{in} \quad \overline{C}_{b,\gamma}([T_0, T) \times H, H),$$

so that, letting $n \to \infty$ and applying the dominated convergence theorem, we get

$$z(t, x) \leq \frac{1}{2} e^{C(T-t)} P_{t,T}^1 \left[|D\varphi|^2 \right](x) + \int_t^T e^{C(s-t)} P_{t,s}^1 \left[|Dg|^2(s, \cdot) \right](x) ds, \quad (4.272)$$

which yields

$$z(t, x) \leq \frac{e^{C(T-t)}}{2} \|D\varphi\|_0^2 + \int_t^T e^{C(s-t)} |Dg(s, \cdot)|_0^2 ds$$

and the claim easily follows.   □

4.7.1.3   Global Existence of Mild and Strong Solutions

The following is the main result of this subsection.

Theorem 4.175 *Let Hypothesis 4.145 hold with $U = H$ and $G = I$ and let $\gamma \in \mathcal{I}_2$. Assume that Hypotheses 4.169 and 4.172 hold. Then there exists a mild solution u of (4.169) in $\overline{C}_{b,\gamma}^{0,2,s}([0, T] \times H)$, and u is unique in $C_b^{0,1}([0, T] \times H)$. Assume, moreover, that $\gamma(t) \leq \bar{C} t^{-\alpha}$, $t \in (0, T]$, for some $\bar{C} > 0$ and $\alpha \in (0, 1)$. If, in Hypothesis 4.169, we only require $\varphi \in C_b^{0,1}(H)$, then there exists a mild solution $u \in \overline{C}_{b,\alpha}^{0,1}([0, T] \times H)$ of problem (4.169), which is unique in $\overline{C}_{b,\alpha}^{0,1}([0, T] \times H)$.*

Proof Step 1. The case $\varphi \in C_b^1(H)$.

The required global existence follows by the local existence (Theorem 4.171) and by the a priori estimate for the $C_b^{0,1}$ norm of the solution proved in Proposition 4.174). We iterate a finite number of times the procedure used in the proof of Theorem 4.171, as is done in Step 3 of the proof of Theorem 4.152. To clarify this we look closely at the second iteration. We consider the map Υ in the space $\overline{\mathcal{H}}_{T_0-t_1, T_0+t_1}$, where $t_1 \in (0, T - T_0)$ will be defined later. The final datum at $T_0 + t_1$ is $u(T_0 + t_1, \cdot) \in C_b^1(H)$, where u is the solution given by Theorem 4.171. From Proposition 4.174, formula (4.246), we know that, setting $C = 2(\omega + C_{3, F_0} + 1)$ and recalling that $g(t, x) = F_0(t, x, 0)$, we have, for all $t \in [T_0, T]$,

$$\|u(t, \cdot)\|_1 \leq e^{C(T-T_0)} \left[\|\varphi\|_1 + (T - T_0) \|g\|_{C_b^{0,1}([T_0,T] \times H)} \right] \quad (4.273)$$

$$\leq e^{CT} \left[\|\varphi\|_1 + T \|g\|_{C_b^{0,1}([0,T] \times H)} \right] =: K(T). \quad (4.274)$$

Similarly to (4.241), we choose $R_0 := 2M e^{(\omega \vee 0)T} K(T) \geq R$, where R is from (4.241), and then $\tilde{t}_1 \in [0, T - T_0]$ such that, as in (4.242) and (4.243),

$$R_0/2 + \rho_2(2t_1, R_0)(1 + R_0) < R_0$$

and

$$\rho_2(2t_1, R_0)(1 + R_0) < \frac{1}{2}.$$

The argument of the proof of Theorem 4.171 shows that Υ has a unique fixed point in the closed ball centered at 0 with radius R_0 in $\overline{\mathcal{H}}_{T_0-t_1,T_0+t_1}$, and this fixed point must coincide with the local mild solution u in $[T_0, T_0 + t_1] \times H$ since $R_0 \geq R$. This proves the existence of a mild solution defined in $[T_0 - t_1, T) \times H$. By construction we have $u \in C_b^{0,1}([T_0 - t_1, T] \times H)$ and, thanks to Proposition 4.174, $\|u\|_{C_b^{0,1}([T_0-t_1,T]\times H)} \leq K(T)$. Moreover, for all $t \in [T_0, T)$

$$\|D^2 u(t, \cdot)\| \leq R\gamma(T - t)$$

and, since γ is decreasing, we have for all $t \in [T_0 - t_1, T_0]$,

$$\|D^2 u(t, \cdot)\| \leq R_0\gamma(T_0 + t_1 - t) \leq R_0\gamma(t_1).$$

Hence, for $t \in [T_1, T)$,

$$\|D^2 u(t, \cdot)\| \leq [R\gamma(T - t)] \vee [R_0\gamma(t_1)].$$

This implies that $D^2 u \in \overline{C}_{b,\gamma}^s([T_0 - t_1, T) \times H, \mathcal{L}(H))$.

If $T_0 - t_1 > 0$ we can now repeat the above argument a finite number of times to get the claim. Indeed, thanks to the a priori estimate (4.274), the choice of R_0 remains the same and thus also the choice of t_1.

The uniqueness is proved exactly as in the last part of the proof of Theorem 4.171 since the argument works in the same way on the interval $[0, T]$ instead of $[T_0, T]$.

Step 2. The case $\varphi \in C_b^{0,1}(H)$.

We consider a sequence of functions $(\varphi_n)_{n\in\mathbb{N}} \subset C_b^1(H)$ converging uniformly to φ and such that $\|\varphi_n\|_1 \leq \|\varphi\|_{0,1}$ (this is always possible using inf-sup convolutions, see Definition D.24 and Proposition D.26). We then associate to φ_n the unique mild solution $u_n \in C_b^{0,1}([0, T] \times H)$ of Eq. (4.169) with initial datum φ_n so that

$$u_n(t, x) = R_{T-t}[\varphi_n](x) + \int_t^T R_{s-t}[F_2(s, \cdot, Du_n(s, \cdot)) + g(s, \cdot)](x)ds.$$

It follows from Step 1 that

$$\|u_n\|_{C_b^{0,1}([0,T]\times H)} \leq e^{CT}\left[\|\varphi\|_{0,1} + T\|g\|_{C_b^{0,1}([0,T]\times H)}\right] =: K(T).$$

Subtracting we then get

$$u_n(t, x) - u_m(t, x) = R_{T-t}[\varphi_n - \varphi_m](x)$$
$$+ \int_t^T R_{s-t}[F_2(s, \cdot, Du_n(s, \cdot)) - F_2(s, \cdot, Du_m(s, \cdot))](x)ds$$

$$(4.275)$$

so that

$$\|(Du_n - Du_m)(t, \cdot)\|_0 \leq \gamma(T - t)\|\varphi_n - \varphi_m\|_0$$
$$+ C_{1,F_0}(K(T)) \int_t^T \gamma(s - t)\|(Du_n - Du_m)(s, \cdot)\|_0 ds.$$

Using $\gamma(t) \leq Ct^{-\alpha}$ and the second Gronwall's inequality of Proposition D.30 we get, for some constant $C_4 := C_4\left(T, \|\varphi\|_{0,1}, \|g\|_{C_b^{0,1}([0,T]\times H)}\right)$,

$$\|(Du_n - Du_m)(t, \cdot)\|_0 \leq C_4(T - t)^{-\alpha}\|\varphi_n - \varphi_m\|_0.$$

Using this estimate in (4.275) we get, for a suitable C_5,

$$\|u_n - u_m\|_{\overline{C}_{b,\alpha}^{0,1}([0,T]\times H)} \leq C_5\|\varphi_n - \varphi_m\|_0,$$

which implies that u_n converges in $\overline{C}_{b,\alpha}^{0,1}([0, T] \times H)$ to a function u satisfying the integral equation (4.177).

To prove uniqueness, if u_1 is another mild solution in $\overline{C}_{b,\alpha}^{0,1}([0, T] \times H)$, similarly to (4.275), defining $R_1 = \|Du\|_0 \vee \|Du_1\|_0$, we get

$$u(t, x) - u_1(t, x) = \int_t^T R_{s-t}[F_2(s, \cdot, Du(s, \cdot)) - F_2(s, \cdot, Du_1(s, \cdot))](x)ds \tag{4.276}$$

and so

$$\|(Du - Du_1)(t, \cdot)\|_0 \leq C_{1,F_0}(R_1) \int_t^T \gamma(s - t)\|(Du - Du_1)(s, \cdot)\|_0 ds.$$

Using Gronwall's inequality of Proposition D.29 we then get $Du_1 = Du_2$ and so the claim follows by (4.276). □

The analogue of Theorem 4.150 also holds in this case.

Theorem 4.176 *Let $T > 0$. Assume that Hypothesis 4.145 is satisfied with $U = H$, $G = I$ and $\|\Gamma(t)\| \leq \bar{C}t^{-\alpha}$, $t \in (0, T]$, for some $\bar{C} > 0$ and $\alpha \in (0, 1)$. Let also Hypotheses 4.169 and 4.172 be satisfied with $\varphi \in C^{0,1}(H)$. Let $u \in \overline{C}_{b,\alpha}^{0,1}([0, T] \times H)$ be the mild solution of Eq. (4.169). Then the function u is the unique $\mathcal{K}$-strong solution of the Cauchy problem (4.169) in $\overline{C}_{b,\alpha}^{0,1}([0, T] \times H)$. Moreover, the approximating sequence u_n to u can be chosen so that the convergence is uniform on $[0, T - \varepsilon] \times H_0$ for every $\varepsilon \in (0, T)$ and every bounded subset H_0 of H.*

Proof The proof is completely similar to that of Theorem 4.150 as the different assumptions on F_0 do not affect the approximation procedure. □

4.8 Stochastic Control: Verification Theorems and Optimal Feedbacks

In this section we study a class of stochastic optimal control problems to which we apply the results of the previous sections. The aim is to replicate, as much as possible, the results on optimal synthesis presented in Sect. 2.5.1. More precisely, we want to prove the following.

- The value function of a control problem coincides with the unique mild/strong solution of the associated HJB equation.
- A verification theorem, i.e. a sufficient (and in some cases necessary) condition for optimality.
- Existence, and in some cases uniqueness, of optimal feedback controls, where the feedback formula is written in terms of the space derivative of the value function.

We present separately the finite horizon case (leading to a parabolic HJB equation like (4.109)) and the infinite horizon case (leading to an elliptic HJB equation like (4.125)).

4.8.1 The Finite Horizon Case

The plan of this section is the following. We first collect results about HJB equations associated with our optimal control problems. Then we discuss the optimal control problems and finally prove verification theorems and existence of optimal feedback controls. We choose, differently from what seems more intuitive, to start with the results on HJB equations, to have a more linear presentation of the material, and to avoid confusion with various choices of the needed generalized reference probability spaces.

4.8.1.1 The HJB Equation

Let $T > 0$ and H be a real separable Hilbert space. We consider HJB equations of the form (4.1) (which are special cases of (2.42))

$$
\begin{cases}
v_t + \dfrac{1}{2}\, \mathrm{Tr}\, [\Sigma(t,x)D^2 v] + \langle Ax + b(t,x), Dv \rangle + F(t,x,Dv) = 0 \\
\qquad\qquad t \in [0,T), \; x \in H, \\
v(T,x) = g(x), \quad x \in H,
\end{cases}
\tag{4.277}
$$

where $\Sigma(t,x) = \sigma(t,x)\sigma^*(t,x)$ and the Hamiltonian F is given by

$$F(t, x, p) = \inf_{a \in \Lambda} F_{CV}(t, x, p, a) = \inf_{a \in \Lambda} \left\{ \langle G(t, x) R(t, x, a), p \rangle_H + l(t, x, a) \right\}.$$

$$\tag{4.278}$$

The connections of these equations to specific control problems are discussed in Sect. 4.8.1.4. Note that, differently from the definition of Hamiltonians (2.43) and (2.44) in Sect. 2.5.1, here we drop the term $\langle b(t, x), Dv \rangle$ from F_{CV} since it does not depend on the control variable. However, sometimes in the next subsections it will be included in the Hamiltonian, see e.g. (4.292) and (4.293).

To be able to study more general cases where $G(t, x) R(t, x, a)$ may not be well defined, as was done in Sect. 4.4.1, we introduce the modified Hamiltonian F_0 as follows

$$F_0(t, x, q) = \inf_{a \in \Lambda} F_{0,CV}(t, x, q, a) = \inf_{a \in \Lambda} \left\{ \langle R(t, x, a), q \rangle_U + l(t, x, a) \right\} \quad (4.279)$$

and observe that, for $p \in D(G(t, x)^*)$,

$$F(t, x, p) = F_0(t, x, G(t, x)^* p),$$

so the term $F(t, x, Dv)$ in (4.277) can be formally rewritten (see Sect. 4.2.1) as $F_0(t, x, D^G v)$.

We consider two different sets of assumptions. The first is the following.

Hypothesis 4.177

(i) A, b, σ satisfy Hypotheses 1.149 and 4.127-(iii).
(ii) For a given real separable Hilbert space U, G is a family of (possibly unbounded with dense domains) closed linear operators $G(t, x) : D(G(t, x)) \subset U \to H$, $(t, x) \in [0, T] \times H$, such that for every $t \in [0, T]$, $G(t, \cdot)$ satisfies Hypothesis 4.11.
(iii) For every $(s, t, x) \in (0, T] \times [0, T] \times H$, $e^{sA} G(t, x)$ extends to an operator in $\mathcal{L}(U, H)$, which we still denote by $e^{sA} G(t, x)$, the function

$$(0, T] \times [0, T] \times H \to \mathcal{L}(U, H), \qquad (s, t, x) \to e^{sA} G(t, x)$$

is strongly measurable and, for all $s \in (0, T]$, $(t, x) \in [0, T] \times H$,

$$\| e^{sA} G(t, x) \|_{\mathcal{L}(U,H)} \le f_G(s)(1 + |x|),$$

where $f_G(s) = Ls^{-\beta}$ for some $L > 0$ and $\beta \in [0, 1)$. Moreover, there exists an $r \in \mathbb{R} \cap \varrho(A)$ such that for every $(t, x) \in [0, T] \times H$, $(rI - A)^{-1} G(t, x)$ extends to an operator in $\mathcal{L}(U, H)$, which we still denote by $(rI - A)^{-1} G(t, x)$, and the function

$$[0, T] \times H \to \mathcal{L}(U, H), \qquad (t, x) \to (rI - A)^{-1} G(t, x)$$

is strongly measurable and there exists a $C > 0$ such that

$$\|(rI - A)^{-1}G(t, x)\|_{\mathcal{L}(U,H)} \leq C(1 + |x|), \quad \forall (t, x) \in [0, T] \times H.$$

(iv) Λ is a Polish space.
(v) $R \in C_b([0, T] \times H \times \Lambda, U)$. Moreover, for all $(s, t) \in (0, T] \times [0, T], a \in \Lambda$ and $x_1, x_2 \in H$,

$$|e^{sA}G(t, x_1)R(t, x_1, a) - e^{sA}G(t, x_2)R(t, x_2, a)| \leq f_G(s)|x_1 - x_2|,$$

where f_G is as in (iii) above.

Here is the second set of assumptions.

Hypothesis 4.178

(i) A, σ satisfy Hypothesis 1.143 with $Q = I$. Moreover, b satisfies Hypothesis 1.145 in the case when b_0 is independent of a_1, and $a_2(\cdot) = 0$.
(ii) For a given real separable Hilbert space U, $G(t, x) \equiv G$, where $G : D(G) \subset U \to H$ is a closed linear operator (possibly unbounded with dense domain).
(iii) For every $s \in (0, T]$, $e^{sA}G$ extends to an operator in $\mathcal{L}(U, H)$, which we still denote by $e^{sA}G$, and we have

$$\|e^{sA}G\|_{\mathcal{L}(U,H)} \leq Ls^{-\beta}$$

for some $L > 0$ and $\beta \in [0, 1)$. Moreover, there exists an $r \in \mathbb{R} \cap \varrho(A)$ such that $(rI - A)^{-1}G$ extends to an operator in $\mathcal{L}(U, H)$.
(iv) Λ is a subset of a real separable Banach space E.
(v) The function R is independent of x, it belongs to $C([0, T] \times \Lambda, U)$ and is bounded on bounded sets.

Observe that $(rI - A)^{-1}G$ extends to an operator in $\mathcal{L}(U, H)$ if $D(A^*) \subset D(G^*)$.
We remark that the results about the HJB equation (4.277) in Sect. 4.8.1.3 will be stated without the need to refer to conditions (iii), (iv) and (v) of the above hypotheses. These conditions will be used later, in Sect. 4.8.1.4, to have the well-posedness of the state equation (4.294), and in Sect. 4.8.1.5, to prove the verification theorem (Theorem 4.197). The assumptions about l and g will be specified later in Sect. 4.8.1.3 since they involve growth conditions related to the properties of the transition semigroups $P_{t,s}$ there.

Remark 4.179 Recall that Λ is a control set. The space U, where the map R has its values, is introduced to give more flexibility to the class of problems we study. In most cases U will be either equal to the state space H, or to the control space Λ (or, more generally, Λ will be a subset of U). The choice of U depends on a specific problem. For example, in the distributed control case of Sect. 2.6.1 it is natural to take $U = H$, while in the problems of Sect. 2.6.8 it is natural to take $U = \Lambda$. ∎

Remark 4.180 If we consider HJB equations for optimal control of the heat equation (Sects. 2.6.1, 2.6.2), we see that the above assumptions include both the case of distributed control and the case of boundary control (or the one with both distributed and boundary controls).

A typical case where Hypothesis 4.177 is true is the distributed control case of (2.84), where we can take $U = E = H$, $R(t, x, a) = a$ and $G(t, x) = I$ and assume that Λ is a bounded subset of H. In the boundary control case (see (2.93) and (2.94) for the Dirichlet case and (2.96) and (2.97) for the Neumann case) we can take $U = H$, $R(t, x, a) = Ba$ with $B \in \mathcal{L}(E, H)$ and $G = (-A)^\beta$ ($\beta \in (3/4, 1)$ in the Dirichlet case and $\beta \in (1/4, 1/2)$ in the Neumann case). Again we have to assume that Λ is a bounded subset of H.

Hypothesis 4.178 is introduced since in some examples it is interesting to treat cases where R is unbounded on $[0, T] \times \Lambda$. For example, for the distributed control case (see (2.84)) we can take $U = E = \Lambda = H$, $R(t, a) = a$ and $G = I$. In the boundary control case (see (2.93) and (2.94) for the Dirichlet case and (2.96) and (2.97) for the Neumann case) we can take $U = H$, $R(t, a) = Ba$ with $B \in \mathcal{L}(E, H)$ and $G = (-A)^\beta$ ($\beta \in (3/4, 1)$ in the Dirichlet case and $\beta \in (1/4, 1/2)$ in the Neumann case).

HJB equations for distributed control problems have been studied e.g. in [89, 90, 105, 306, 307], while, for the case of boundary control, see [189, 310].

We also note that one may assume something different about b, G and R (e.g. dissipativity with respect to the variable x). We avoid generalizations here, referring the reader to Sect. 4.9, where some special cases are discussed. ∎

4.8.1.2 Properties of the Hamiltonian F_0

We prove three results about the properties of F_0.

Proposition 4.181 *Let $m \geq 0$. Let Λ be a Polish space and $R \in C_b([0, T] \times H \times \Lambda, U)$. Let $l \in C([0, T] \times H \times \Lambda)$ be such that*

$$|l(t, x, a)| \leq C(1 + |x|^m) \quad \forall (t, x, a) \in [0, T] \times H \times \Lambda. \tag{4.280}$$

Then Hypothesis 4.72-(i)-(ii)-(iii) is satisfied. If R and l are continuous in (t, x), uniformly with respect to $a \in \Lambda$, then F_0 is continuous.

Proof In this case, for $(t, x) \in [0, T] \times H$ and $q, q_1, q_2 \in U$,

$$|F_0(t, x, q_1) - F_0(t, x, q_2)| \leq \sup_{a \in \Lambda}\{\langle R(t, x, a), q_1 - q_2\rangle_U\} \leq \|R\|_0 |q_1 - q_2|$$

and

$$|F_0(t, x, q)| \leq \|R\|_0 |q| + C(1 + |x|^m).$$

Hence Hypothesis 4.72-(i)-(ii) holds with $L = \|R\|_0$ and $L' = C \vee \|R\|_0$. Measurability of F_0 is immediate from the separability of Λ. Concerning the continuity of F_0 we simply observe that, if $(t_n, x_n, q_n) \to (t, x, q)$, we have

$$|F_0(t_n, x_n, q_n) - F_0(t, x, q)| \leq \|R\|_0 |q_n - q|$$
$$+ \sup_{a \in \Lambda} |\langle R(t_n, x_n, a), q \rangle_U + l(t_n, x_n, a) - \langle R(t, x, a), q \rangle_U + l(t, x, a)|$$

and the claim immediately follows since R and l are continuous in (t, x), uniformly with respect to $a \in \Lambda$. $\qquad\square$

Proposition 4.182 *Let $m \geq 0$. Let E be a real separable Hilbert space and let $\Lambda \subset E$ be closed and unbounded. Let $R(t, x, a) = Ba$, where $B \in \mathcal{L}(E, U)$.*

(i) *Let $l \in C([0, T] \times H \times \Lambda)$ and let $C : \mathbb{R}^+ \to \mathbb{R}^+$ be an increasing function such that*

$$|l(t, x, a)| \leq C(|a|_E)(1 + |x|^m), \quad \forall (t, x, a) \in [0, T] \times H \times \Lambda \quad (4.281)$$

and

$$\frac{l(t, x, a)}{|a|_E} \to +\infty \quad as \quad |a|_E \to +\infty, \ a \in \Lambda, \quad (4.282)$$

uniformly for $(t, x) \in [0, T] \times H$. Then there exists a $C_1 > 0$ and an increasing function $C_2 : \mathbb{R}^+ \to \mathbb{R}^+$ such that

$$C_1(1 + |x|^m + |q|) \geq F_0(t, x, q) \geq -C_2(|q|_U)\left[1 + |x|^m\right]$$

and

$$|F_0(t, x, q_1) - F_0(t, x, q_2)| \leq C_2(|q_1|_U \vee |q_2|_U)\|B\||q_1 - q_2|. \quad (4.283)$$

If l is continuous in (t, x) uniformly with respect to $a \in \Lambda$, then F_0 is continuous.

(ii) *Let $\Lambda = E$ and $l(t, x, a) = l_0(t, x) + l_1(a)$, where $l_0 \in C_m([0, T] \times H)$ and $l_1 : \Lambda \to \mathbb{R}$ is continuous and such that*

$$\frac{l_1(a)}{|a|_E} \to +\infty \quad as \quad |a|_E \to +\infty. \quad (4.284)$$

Then F_0 is finite and we have

$$F_0(t, x, q) = l_0(t, x) + \bar{l}_1(B^*q),$$

where for $z \in E$,

$$\bar{l}_1(z) := \inf_{a \in E}\{\langle a, z \rangle_E + l_1(a)\} = -l_1^*(-z). \quad (4.285)$$

$(l_1^*(w) := \sup_{a \in E} \{\langle a, w \rangle_E - l_1(a)\}$ is the Légendre transform of l_1.)

(iii) *Let the assumptions of (ii) be satisfied and let l_1 be also strictly convex. Then $\bar{l}_1$ is Gâteaux differentiable and $\nabla \bar{l}_1$ is bounded on bounded sets. If also $l_1 \in C^1(E)$ with Dl_1 invertible and $(Dl_1)^{-1}$ Lipschitz continuous on bounded sets, then $D\bar{l}_1$ is Lipschitz continuous on bounded sets.*

Proof

Proof of (i). Let $a_0 \in \Lambda$ be fixed. We notice first that if $(t, x, q) \in [0, T] \times H \times U$ then

$$F_0(t, x, q) \le \langle Ba_0, q \rangle_U + l(t, x, a_0) \le |Ba_0||q| + l(t, x, a_0). \tag{4.286}$$

On the other hand, (4.282) implies that for every $M > 0$ there exists a $\bar{C}_2(M) > 0$ such that, for $|q| \le M$,

$$F_0(t, x, q) = \inf_{a \in \Lambda, |a|_E \le \bar{C}_2(M)} \{\langle Ba, q \rangle + l(t, x, a)\}, \quad \forall (t, x) \in [0, T] \times H.$$

Hence

$$F_0(t, x, q) \ge -\bar{C}_2(|q|_U) \|B\| |q|_U + \inf_{a \in \Lambda, |a|_E \le \bar{C}_2(|q|_U)} l(t, x, a),$$

which, together with (4.281) and (4.286), gives (4.283). Moreover,

$$|F_0(t, x, q_1) - F_0(t, x, q_2)| \le \sup_{|a|_E \le \bar{C}(|q_1|_U) \vee \bar{C}(|q_2|_U)} |\langle Ba, q_1 - q_2 \rangle_U|,$$

which yields (4.283) after redefining the function C_2. Concerning the continuity of F_0 we observe that, for $(t_n, x_n, q_n) \to (t, x, q)$, we have, using (4.283),

$$|F_0(t_n, x_n, q_n) - F_0(t, x, q)| \le C_2(|q|_U)|q_n - q| + \sup_{a \in \Lambda} |l(t_n, x_n, a) - l(t, x, a)|.$$

Hence, if l is continuous in (t, x), uniformly with respect to $a \in \Lambda$, then F_0 is continuous.

Proofs of (ii) and (iii). The claims are straightforward consequences of the properties of the Légendre transform in Hilbert spaces. In particular Gâteaux differentiability follows from Corollary 18.12 of [43]. Moreover, since l_1 is continuous and strictly convex, there exists a unique minimum point $\bar{a}(z)$ in (4.285) and, thanks to (4.284), $\bar{a}(z)$ is bounded on bounded sets. Since we have $\bar{a}(z) = \nabla \bar{l}_1(z)$ (see e.g. Proposition 16.9 of [43]), then $\nabla \bar{l}_1$ is also bounded on bounded sets.

Finally, let $l_1 \in C^1(E)$ with Dl_1 invertible and $(Dl_1)^{-1}$ Lipschitz continuous on bounded sets. Then from (4.285) we have, at the minimum point $\bar{a} := \bar{a}(z)$,

$$z + Dl_1(\bar{a}) = 0 \iff \bar{a} = (Dl_1)^{-1}(-z),$$

and the claim follows.					□

Proposition 4.183 *Let $m \geq 0$. Let E be a real separable Hilbert space and let $\Lambda \subset E$ be closed, convex and bounded. Let $R(t, x, a) = Ba$, where $B \in \mathcal{L}(E, U)$. Assume also that $l(t, x, a) = l_0(t, x) + l_1(a)$, where $l_0 \in C_m([0, T] \times H)$ and $l_1 : \Lambda \to \mathbb{R}$ is continuous, strictly convex with always non-empty subdifferential $D^- l_1$. Then we have*

$$F_0(t, x, q) = l_0(t, x) + \bar{l}_1(B^*q),$$

where $\bar{l}_1 : E \to \mathbb{R}$ is given by (4.285). Moreover, $\bar{l}_1(p)$ is concave and Gâteaux differentiable with bounded derivative.

If the unique minimum point $\bar{a}(p)$ in (4.285) is a Lipschitz continuous function of $p \in E$, then $\bar{l}_1$ is Fréchet differentiable and $D\bar{l}_1$ is Lipschitz continuous.

Proof Since $\bar{l}_1 : E \to \mathbb{R}$ is clearly concave as an infimum of linear functions, and Lipschitz continuous since Λ is bounded, we only have to prove that it is Gâteaux differentiable. The proof follows from a standard adaptation of Corollary 18.12 of [43] but we provide it here for the reader's convenience. Since Λ is closed, convex and bounded, and l_1 is continuous and strictly convex, there exists, for each $p \in E$, a unique minimum point $\bar{a}(p)$ of the map

$$\Lambda \to \mathbb{R}, \qquad a \to \langle a, p \rangle_E + l_1(a).$$

We conclude the result in two steps. Define, for $a \in \Lambda$,

$$\bar{\bar{l}}_1(a) := \sup_{p \in H} \{ -\langle a, p \rangle + \bar{l}_1(p) \}.$$

Step 1. We have $\bar{\bar{l}}_1(a) = l_1(a)$ for all $a \in \Lambda$.
It is clear that $\bar{\bar{l}}_1$ is convex. Since, for all $a \in \Lambda$, $\bar{l}_1(p) \leq \langle a, p \rangle + l_1(a)$, we have

$$\bar{\bar{l}}_1(a) \leq \sup_{p \in H} \{ -\langle a, p \rangle + \langle a, p \rangle + l_1(a) \} = l_1(a).$$

To prove the opposite inequality we take any $\hat{a} \in \Lambda$ and $\hat{p} \in -D^- l_1(\hat{a})$ (it always exists since $D^- l_1(\hat{a})$ is non-empty). Then, taking $a_0 := \bar{a}(\hat{p})$, we have

$$\bar{l}_1(\hat{p}) = \langle a_0, \hat{p} \rangle + l_1(a_0) \tag{4.287}$$

and, by convexity of l_1 and (4.287),

$$l_1(\hat{a}) \leq l_1(a_0) + \langle a_0 - \hat{a}, \hat{p} \rangle = \bar{l}_1(\hat{p}) - \langle \hat{a}, \hat{p} \rangle \leq \bar{\bar{l}}_1(\hat{a}).$$

Step 2. Conclusion.
Let $\bar{p} \in H$ and $a \in D^+ \bar{l}_1(\bar{p})$. We recall that, since $\bar{l}_1$ is concave and Lipschitz continuous, $D^+ \bar{l}_1(\bar{p})$ is always non-empty, see e.g. (D.7). Then, by the previous step and the concavity of l_1, we have

$$l_1(a) = \bar{\bar{l}}_1(a) = \sup_{p \in H}\{-\langle a, p\rangle + \bar{l}_1(p)\} = -\langle a, \bar{p}\rangle + \bar{l}_1(\bar{p}),$$

i.e. $\bar{l}_1(\bar{p}) = l_1(a) + \langle a, \bar{p}\rangle$ which, by the uniqueness of the minimum point, implies $a = \bar{a}(\bar{p})$. This means that $D^+\bar{l}_1(\bar{p})$ is a singleton for each $\bar{p} \in H$. The claim now follows by the properties of the superdifferential (see e.g. [43], Proposition 17.26).

To prove the last statement we simply observe that, if the minimum point $\bar{a}(p)$ is a Lipschitz continuous function of $p \in E$, then the Gâteaux derivative $\nabla \bar{l}_1$ is Lipschitz continuous and so is the Fréchet derivative. □

Remark 4.184 (i) When R is bounded on Λ (Proposition 4.181 and 4.183), Hypothesis 4.72-(i)-(ii)-(iii), and also the continuity of F_0 are true under quite general assumptions about l. Hence in this case the theory of Sects. 4.4 and 4.5 can be applied if the transition semigroup associated to the linear part of the HJB equation satisfies the required properties.

(ii) In the case considered in Proposition 4.182, when R is linear and Λ (and hence R) is unbounded it is very unlikely that Hypothesis 4.72-(i)-(ii) is satisfied. Indeed, only local Lipschitz continuity estimates can be proved and superlinearity in $|q|_U$ also arises in simple cases: for example, when $\Lambda = E = U$ and $B = I$, we have

$$l_1^*(p)/|p| \longrightarrow -\infty \qquad \text{as} \quad |p| \to +\infty \tag{4.288}$$

(see e.g. [43], Proposition 16.17). Hence in this case the theory of Sects. 4.4 and 4.5 does not apply (see Remark 4.73-(ii)). One can apply the results of Sect. 4.7, adding suitable regularity assumptions about the data. For example, if one assumes that $l = l_0 + l_1$, where l_1 is as in point (iii) of Proposition 4.182 and $l_0 \in C_b^{0,1}([0, T] \times H)$ with Dl_0 Lipschitz continuous on bounded sets, then Hypothesis 4.169 (i)–(ii) and Hypothesis 4.172 are satisfied. ■

Remark 4.185 Cost functions $l(t, x, a) = l_0(t, x) + l_1(a)$ frequently occur in applications, see for example Eq. (2.99) in Sect. 2.6.2, Eqs. (2.146) and (2.150) in Sect. 2.6.8. These cases are studied e.g. in [89, 90, 306, 307].

A typical example when Proposition 4.182-(iii) applies is when $l_1(a) = |a|^\theta/\theta$, $\theta \in (1, 2]$, and $\Lambda = E$ so, denoting by $\theta' := \theta/(\theta - 1)$ the conjugate exponent of θ,

$$\bar{l}_1(p) = -\frac{|p|^{\theta'}}{\theta'}$$

with the minimum point $\bar{a}(p) = -|p|^{\frac{2-\theta}{\theta-1}} p$. Hence

$$F_0(t, x, q) = l_0(t, x) - \frac{|B^*q|^{\theta'}}{\theta'}.$$

On the other hand, Proposition 4.183 applies, for example, when $\Lambda = B_H(0, M)$, $l_1(a) = |a|^\theta/\theta$ $(\theta \in (1, 2])$, so we have

$$
\bar{l}_1(p) = \begin{cases} -\dfrac{|p|^{\theta'}}{\theta'} & \text{if } |p| \le M^{\theta-1} \\[2mm] -|p|M + \dfrac{M^\theta}{\theta} & \text{if } |p| > M^{\theta-1} \end{cases}
$$

with the minimum point $\bar{a}(p) = -|p|^{\frac{2-\theta}{\theta-1}} p$ when $|p| \le M^{\theta-1}$ and $\bar{a}(p) = -\frac{M}{|p|} p$ when $|p| > M^{\theta-1}$. Hence

$$
F_0(t, x, q) = l_0(t, x) + \begin{cases} \dfrac{|B^*q|^{\theta'}}{\theta'} & \text{if } |B^*q| \le M^{\theta-1}, \\[2mm] -|B^*q|M + \dfrac{M^\theta}{\theta} & \text{if } |B^*q| > M^{\theta-1}. \end{cases} \tag{4.289}
$$

∎

4.8.1.3 Results for the HJB Equation

We present three results where we apply the theory of Sects. 4.4–4.7 to our HJB equations. The first uses the general assumptions of Sect. 4.4.1.2 while the other two refer to the more specific cases treated in Sects. 4.6 and 4.7. Variations of such results and/or applications to specific models are possible: some specific cases are presented in Sects. 4.9 and 4.10, while some other cases are the subject of current research (see, for example, the recent work [316]).

In this subsection we fix a generalized reference probability space $\mu_0 := (\Omega, \mathscr{F}, \{\mathscr{F}_s\}_{s\in[0,T]}, \mathbb{P}, W)$, where W is a cylindrical Wiener process in a real separable Hilbert space Ξ. It is used to introduce a two-parameter transition semigroup $P_{t,s}$ and thus define a mild solution of the HJB equation.

For the first result we will use the SDE

$$
\begin{cases} dX(s) = [AX(s) + b(s, X(s))]\, ds + \sigma(s, X(s)) dW(s), \\ X(t) = x. \end{cases} \tag{4.290}
$$

It is clear that under the assumptions of either Hypothesis 4.177 or 4.178 there exists a unique (within a certain class) mild solution $X_0(\cdot; t, x)$ of the state equation (4.290). Indeed, Hypothesis 4.177 allows us to apply Theorem 1.152 while Hypothesis 4.178 allows us to apply Proposition 1.147. Moreover, $P_{t,s}[\phi](x) = \mathbb{E}[\phi(X_0(s; t, x))], 0 \le t \le s \le T$, defines a two-parameter transition semigroup on $B_m(H)$ for all $m \ge 0$. Uniqueness in law for (4.290) guarantees that the semigroup is independent of the choice of the generalized reference probability space μ_0.

To apply the results of Sects. 4.4 and 4.5 we need assumptions about the data F_0 and g and the transition semigroup $P_{t,s}$. We provide here two sets of assumptions, the first for the B_m space case, the second for the C_m space case.

Hypothesis 4.186 Fix $m \geq 0$ and assume that, for this m, the following hold.

(i) Hypothesis 4.72 is satisfied with F_0 defined by (4.279)[39] and $\varphi = g$.
(ii) The two-parameter transition semigroup $P_{t,s}$ satisfies Hypotheses 4.74, 4.76, 4.77.

Hypothesis 4.187 Fix $m \geq 0$ and assume that, for this m, the following hold.

(i) Hypothesis 4.87 is satisfied with F_0 defined by (4.279) and $\varphi = g$.
(ii) The two-parameter transition semigroup $P_{t,s}$ satisfies Hypothesis 4.89.

Remark 4.188 The results for the associated optimal control problems proved in Sects. 4.8.1.5 and 4.8.1.6 will only use Hypothesis 4.187, which is concerned with the C_m space case. To obtain results for the B_m space case one would need a result about strong solutions in this case, which might be proved along the lines suggested in Remark 4.136-(iv). In Corollaries 4.189–4.191 we keep for completeness the results about existence and uniqueness of mild solutions in B_m spaces, as they may be useful for extensions of the present theory. ∎

We have the following corollary.

Corollary 4.189 *(i) Assume that Hypothesis 4.186 is satisfied. Then the HJB equation (4.277) admits a unique mild solution v in $\overline{B}_{m,\gamma_G}^{0,1,G}([0, T] \times H)$.*

(ii) Assume that Hypothesis 4.187 is satisfied. Then the HJB equation (4.277) admits a unique mild solution v in $\overline{C}_{m,\gamma_G}^{0,1,G}([0, T] \times H)$.

(iii) Let Hypothesis 4.177 or Hypothesis 4.178 hold. Let Hypotheses 4.187 and 4.133 for $\mathcal{K}$-convergence be satisfied with the same U and G as in Hypothesis 4.177 or 4.178. Then the mild solution v is also the unique $\mathcal{K}$-strong solution in $\overline{C}_{m,\gamma_G}^{0,1,G}([0, T] \times H)$.

Proof Part (i) is an immediate consequence of Theorem 4.80.

Part (ii) follows from Theorem 4.90.

Part (iii) follows from Theorem 4.135 and Remark 4.136-(vi) once we observe that Hypothesis 4.127 is implied by both 4.177-(i) or 4.178-(i). □

In the second and third cases the process $X_0(s; t, x)$ is of Ornstein–Uhlenbeck type, i.e. it is the solution of the SDE

$$\begin{cases} dX(s) = AX(s)ds + \sigma dW(s), \\ X(t) = x. \end{cases} \tag{4.291}$$

---

[39] Note that Proposition 4.181 gives sufficient conditions on R and l for this. Similarly for Hypothesis 4.187-(i).

We now have $P_{t,s} = R_{s-t}$, where R_s is the Ornstein–Uhlenbeck semigroup associated to A and σ. We assume the following.

In the Hypothesis 4.190 and Corollaries 4.191 and 4.193, the weight function γ_G comes from an Ornstein–Uhlenbeck case and is given in Proposition 4.147, i.e. $\gamma_G(s) \geq c\|\Gamma_G(s)\|$ for some constant $c > 0$.

Hypothesis 4.190 (i) Hypothesis 4.145 is satisfied for A, $\Sigma = \sigma\sigma^*$ given in Eq. (4.291) and for a given operator G. Moreover, the map $t \to \gamma_G(t)$ belongs to $\mathcal{I}_2$.

(ii) The function b in (4.277) satisfies Hypothesis 1.145 in the case when b_0 is independent of a_1, and $a_2(\cdot) = 0$.[40] Moreover, b is such that $\langle b, p \rangle_H = \langle \bar{b}, G^*p \rangle_U$ for some $\bar{b} \in C_b([0, T] \times H, U)$.

(iii) Either G is bounded or $D(A^*) \subset D(G^*)$ and, for some $c > 0$, $|G^*z| \leq c|A^*z|$ for all $z \in D(A^*)$.

The Hamiltonian F_0 now becomes

$$\tilde{F}_0(t, x, q) := F_0(t, x, q) + \langle \bar{b}(t, x), q \rangle_U \tag{4.292}$$

and the current value Hamiltonian is

$$\tilde{F}_{0,CV}(t, x, q, a) := F_{0,CV}(t, x, q, a) + \langle \bar{b}(t, x), q \rangle_U. \tag{4.293}$$

The second result is also for a Lipschitz continuous (in the gradient variable) Hamiltonian F_0. It is an immediate consequence of Theorems 4.149 and 4.150.

Corollary 4.191 *Let Hypothesis 4.190 hold. Assume moreover that, for a given $m \geq 0$, we have $g \in B_m(H)$ and the function F_0 satisfies Hypothesis 4.72-(i)-(ii)-(iii) for such m. Then we have the following.*

(i) *The HJB equation (4.277) has a unique mild solution v in $\overline{B}_{m,\gamma_G}^{0,1,G}([0, T] \times H)$. If $g \in C_m(H)$ and F_0 is also continuous in x, then v is continuous in x.*

(ii) *If $g \in C_m(H)$, F_0 is continuous and Hypothesis 4.145-(iv) also holds with $U = H$ and $G = I$, then $v \in \overline{C}_{m,\gamma_G}^{0,1,G}([0, T] \times H)$ and v is also the unique $\mathcal{K}$-strong solution of (4.277) in $\overline{C}_{m,\gamma_G}^{0,1,G}([0, T] \times H)$. Assume moreover that Hypothesis 4.145 also holds with $U = H$ and $G = I$ in point (iv). Then the solutions of the approximating problems can be chosen to converge uniformly on $[0, T - \varepsilon] \times H_0$ for all $\varepsilon \in (0, T)$ and all bounded subsets H_0 of H.*

Proof The mild form of the HJB equation is now written with $P_{t,s} = R_{s-t}$. Since $\bar{b} \in C_b([0, T] \times H, U)$ and thanks to the assumptions on F_0, the new Hamiltonian $\tilde{F}_0$ given by (4.292) satisfies Hypothesis 4.72-(i)-(ii)-(iii). Hence point (i) follows from

---

[40]This first part of (ii) is not needed to prove Corollaries 4.191 and 4.193 but we keep it here since it is needed in further results on optimal control problems.

Theorem 4.149-(i). The mild solution $v \in \overline{C}_{m,\gamma_G}^{0,1,G}([0, T] \times H)$ by Theorem 4.149-(iii). The claims of (ii) about the strong solution follows from Theorem 4.150 thanks to Hypothesis 4.190.   □

The third result is concerned with the case of a locally Lipschitz Hamiltonian. We add the following assumption.

Hypothesis 4.192

(i) The operator A satisfies Hypothesis 4.172-(i).
(ii) The function F_0 introduced in (4.279) satisfies Hypotheses 4.169-(i)-(ii) and 4.172-(ii).
(iii) $g \in C^{0,1}(H)$.

The corollary below follows from Theorems 4.175 and 4.176.

Corollary 4.193 *Let $m = 0$ and let Hypothesis 4.190 hold with $U = H$, $G = I$, $\gamma(s) := \gamma_I(s) \leq \bar{C}s^{-\alpha}$ for some $\bar{C} > 0$ and $\alpha \in (0, 1)$, and b differentiable in x and such that $Db \in C_b^s([0, T] \times H, \mathcal{L}(H))$. Assume that Hypothesis 4.192 is satisfied. Then the HJB equation (4.277) admits a unique mild solution v in $\overline{C}_{b,\alpha}^{0,1}([0, T] \times H)$ which is also the unique $\mathcal{K}$-strong solution in $\overline{C}_{b,\alpha}^{0,1}([0, T] \times H)$. If $g \in C_b^1(H)$ then $v \in \overline{C}_{b,\alpha}^{0,2,s}([0, T] \times H)$.*

Proof As in Corollary 4.191 we write the mild form of the HJB equation using $P_{t,s} = R_{s-t}$. The assumptions about b imply that the required assumptions about the new Hamiltonian $\tilde{F}_0$ defined by (4.292) (notice that now $\tilde{F}_0(t, x, q) := F_0(t, x, q) + \langle b(t, x), q \rangle_H$) are satisfied so that we can apply Theorems 4.175 and 4.176.   □

Note that parts (ii) and (iii) of Proposition 4.182 provide reasonable conditions under which Hypothesis 4.169-(i) holds. Checking the validity of Hypotheses 4.169-(ii) and 4.172 depends on specific cases.

4.8.1.4   Optimal Control Problems

We use the setting of the *strong formulation* of a stochastic optimal control problem (see Sect. 2.1.1).

Let Ξ be a real separable Hilbert space from Sect. 4.8.1.3 and Λ be a Polish space (the control space). For every $t \in [0, T)$ we fix a generalized reference probability space $\mu := (\Omega, \mathscr{F}, \{\mathscr{F}_s^t\}_{s \in [t,T]}, \mathbb{P}, W)$ (possibly different from μ_0), where W is a cylindrical Wiener process in Ξ on $[t, T]$.

For an initial time $t \in [0, T)$ and an initial state $x \in H$, the state equation is the controlled SDE (where $s \in (t, T]$)

$$\begin{cases} dX(s) = [AX(s) + b(s, X(s)) + G(s, X(s))R(s, X(s), a(s))]\,ds \\ \qquad\qquad\qquad\qquad\qquad\qquad + \sigma(s, X(s))dW(s), \quad (4.294) \\ X(t) = x. \end{cases}$$

The set Λ and, consequently, the set of admissible controls, will be one of two types. The first type, which will be used together with Hypothesis 4.177, i.e. when R is bounded, is when Λ is a Polish space. In this case the set of admissible controls is, as in (2.1),

$$\mathcal{U}_t^\mu := \left\{ a(\cdot) : [t, T] \times \Omega \to \Lambda : a(\cdot) \text{ is } \mathscr{F}_s^t - \text{progressively measurable} \right\}.$$

The second type will be used together with Hypothesis 4.178, i.e. when R is unbounded and Λ is a closed but not bounded subset of a given real separable Banach space E (if Λ is bounded we are back to the previous case). In this case the set of admissible controls is

$$\mathcal{U}_t^{\infty,\mu} := \left\{ a(\cdot) \in M_\mu^\infty(t, T; E) : a(s) \in \Lambda, \mathbb{P}\text{-a.s.}, \forall s \in [0, T] \right\}, \tag{4.295}$$

where $M_\mu^\infty(t, T; E)$ is the space of bounded $\mathscr{F}_s^t$-progressively measurable processes.

We observe that if either Hypothesis 4.177 is satisfied and $a(\cdot) \in \mathcal{U}_t^\mu$ or if Hypothesis 4.178 is satisfied and $a(\cdot) \in \mathcal{U}_t^{\infty,\mu}$, there exists a unique mild solution $X(\cdot; t, x, a(\cdot))$ in $\mathcal{H}_p^\mu(t, T; H)$, for every $p \geq 2$, of the state equation (4.294). Indeed, Hypothesis 4.177 allows us to apply Theorem 1.152 while Hypothesis 4.178 allows us to apply Proposition pr2:exmildOUF0spsapp1.147. We remark that the term $G(s, X(s))R(s, X(s), a(s))$ may not be well defined and the term $e^{(s-r)A}G(r, X(r))R(r, X(r), a(r))$ in the definition of a mild solution is interpreted using the extensions of $e^{(s-r)A}G(r, X(r))$, which exist by Hypotheses 4.177 and 4.178. Note that for the moment we do not require continuity of the trajectories of the solution.

Remark 4.194 In many problems where the control set Λ is unbounded it may often be more desirable to consider larger sets of admissible controls which may include unbounded controls. A typical choice is

$$\mathcal{U}_t^{p,\mu} := \left\{ a(\cdot) \in M_\mu^p(t, T; E) : a(s) \in \Lambda, \mathbb{P}\text{-a.s.}, \forall s \in [0, T] \right\} \tag{4.296}$$

for some $p \geq 1$. For instance, if R in Hypothesis 4.178 satisfies

$$|R(t, a)|_U \leq C_R[1 + |a|_E]$$

we could take $\mathcal{U}_t^{p,\mu}$ with $p > 1/(1 - \beta)$ (see Proposition 1.147). However, if $a(\cdot) \in \mathcal{U}_t^{p,\mu}$, defining $a_n(s) := a(s)\mathbf{1}_{\{(s,\omega):|a(s,\omega)|\leq n\}}(s, \omega)$ for $n \in \mathbb{N}$, we have $a_n(\cdot) \in \mathcal{U}_t^{\infty,\mu}$, $|a(\cdot) - a_n(\cdot)|_{M_\mu^p(t,T;E)} \to 0$ and it is easy to see from the proof of Proposition 1.147 that $|X(\cdot; t, x, a(\cdot)) - X(\cdot; t, x, a(\cdot))|_{\mathcal{H}_p^\mu(t,T;H)} \to 0$ as $n \to +\infty$. Thus, under reasonable assumptions about l and g, the control problems using $\mathcal{U}_t^{p,\mu}$ and $\mathcal{U}_t^{\infty,\mu}$ as the sets of admissible controls would be equivalent. This is indeed the case in many concrete examples. Since the choice $p < +\infty$ increases the number of technical details we decided for the sake of presentation to work with bounded controls. This also allowed us to make Hypothesis 4.178-(iv) more general. However, the results presented in this section can be reproduced in the $\mathcal{U}_t^{p,\mu}$ set-up without much effort. ∎

The cost functional to minimize is

$$J^{\mu}(t, x; a(\cdot)) = \mathbb{E}\left\{\int_t^T l(s, X(s; t, x, a(\cdot)), a(s))ds + g(X(T; t, x, a(\cdot)))\right\}$$

$$(4.297)$$

over all controls $a(\cdot)$ in $\mathcal{U}_t^{\mu}$ or $\mathcal{U}_t^{\infty,\mu}$. Here $X(\cdot; t, x, a(\cdot))$ is the mild solution of (4.294) at time s, which will also be denoted by $X(\cdot)$ when the context is clear. The assumptions about l and g will be given later.

The value function for this problem, in the strong formulation, is defined as in (2.4) by

$$V_t^{\mu}(x) = \inf_{a(\cdot)\in\mathcal{U}_t^{\mu}} J^{\mu}(t, x; a(\cdot)) \quad \text{or} \quad V_t^{\mu}(x) = \inf_{a(\cdot)\in\mathcal{U}_t^{\infty,\mu}} J^{\mu}(t, x; a(\cdot)) \quad (4.298)$$

and optimal pairs are defined as in Definition 2.3.

We also consider the weak formulation from Sect. 2.1.2. In this case the generalized reference probability space μ varies with the controls. Then, as in (2.6), the set of admissible controls is either

$$\overline{\mathcal{U}}_t := \bigcup_{\mu} \mathcal{U}_t^{\mu} \quad \text{or} \quad \overline{\mathcal{U}}_t^{\infty} := \bigcup_{\mu} \mathcal{U}_t^{\infty,\mu}.$$

The value function is

$$\overline{V}(t, x) = \inf_{a(\cdot)\in\overline{\mathcal{U}}_t} J^{\mu}(t, x; a(\cdot)) \quad \text{or} \quad \overline{V}(t, x) = \inf_{a(\cdot)\in\overline{\mathcal{U}}_t^{\infty}} J^{\mu}(t, x; a(\cdot)). \quad (4.299)$$

To study more general cases when, given an admissible control strategy $a(\cdot)$, the existence and uniqueness of mild solutions of (4.294) are not guaranteed, we may use what we called the "extended weak formulation" introduced in Remark 2.6. This would allow us, for example, to treat the case when $G = \sigma$ and R is only Borel measurable and bounded (without assuming the Lipschitz continuity in Hypothesis 4.177-(iv)). In principle, the results of the present section can also be proved in this setting, but we do not study it here (see Sect. 6.5 for results in this case).

4.8.1.5 The Verification Theorem

We prove a verification theorem, i.e. a version of Theorem 2.36. Since the starting point of a verification theorem is a strong solution of the HJB equation, we rely on the three results from Sect. 4.8.1.3: Corollaries 4.189-(iii), 4.191-(ii) and 4.193. For each of these three cases the proof of our verification theorem is slightly different. We give the complete proof only for the "more general" result corresponding to Corollary 4.189-(iii). For the other two results, which correspond to cases when the transition semigroup $P_{t,s}$ is of Ornstein–Uhlenbeck type, we only explain the main changes which must be done in the proof.

We begin with the so-called fundamental identity (see (2.56)).

Hypothesis 4.195 The function $l \in C([0, T] \times H \times \Lambda)$ and for a given $m \geq 0$, we have the following: either, when Hypothesis 4.177 holds, there exists a $C > 0$ such that

$$|l(t, x, a)| \leq C(1 + |x|^m), \qquad \forall(t, x, a) \in [0, T] \times H \times \Lambda \qquad (4.300)$$

or, when Hypothesis 4.178 holds, there exist constants K_c for all $c > 0$ such that

$$|l(t, x, a)| \leq K_c(1 + |x|^m), \qquad \forall(t, x, a) \in [0, T] \times H \times \Lambda, \ |a|_E \leq c. \quad (4.301)$$

Lemma 4.196 *Let Hypothesis 4.177 hold[41] and let $m \geq 0$ be fixed. Let l satisfy for this m Hypothesis 4.195, formula (4.300). Let the other assumptions of Corollary 4.189-(iii) be satisfied for the same m, namely Hypotheses 4.187 and 4.133 for $\mathcal{K}$-convergence (for the same U and G as in Hypothesis 4.177). Let $v \in \overline{C}_{m,\gamma_G}^{0,1,G}([0, T] \times H)$ be the mild and strong solution of (4.277). Let $t \in [0, T]$, $x \in H$. Suppose that $a(\cdot) \in \mathcal{U}_t^\mu$. Then we have the identity*

$$v(t, x) = J^\mu(t, x; a(\cdot))$$

$$- \mathbb{E} \int_t^T \left[F_{0,CV}\left(s, X(s), D^G v(s, X(s), a(s))\right) - F_0\left(s, X(s), D^G v(s, X(s))\right) \right] ds,$$

$$(4.302)$$

where $X(\cdot) := X(\cdot; t, x, a(\cdot))$ is the mild solution of (4.294).

The same statement holds if we take the following sets of assumptions:

- *Hypothesis 4.178 holds, $a(\cdot) \in \mathcal{U}_t^{\infty,\mu}$ and l satisfies Hypothesis 4.195, formula (4.301) for a given $m \geq 0$ (respectively, for $m = 0$).*
- *The other assumptions of Corollary 4.191-(ii) (respectively, Corollary 4.193) are satisfied for the same m (respectively, for $m = 0$).*

Proof We know from Definition 4.132 and Theorem 4.135 that there exist three sequences $(g_n) \subset D(\mathcal{A}_1)$, $(h_n) \subset \overline{B}_{m,\gamma_G}([0, T) \times H)$ and $(v_n) \subset C_b([0, T] \times H)$, such that for every $n \in \mathbb{N}$, v_n is a classical solution (see Definition 4.129) of the Cauchy problem

$$\begin{cases} w_t(t, x) + \mathcal{A}_1(t)[w](t, x) + F_0(t, x, D^G w) = h_n(t, x) \\ w(0, x) = g_n(x) \end{cases} \qquad (4.303)$$

and moreover, as $n \to +\infty$ (using the notation of Definition B.56)

---

[41] Here we are in the framework of Corollary 4.189-(iii), hence we could also take Hypothesis 4.178. We avoid this here for simplicity.

$$\begin{cases} \mathcal{K}-\lim_{n\to+\infty} g_n = g & \text{in } B_m(H) \\ \mathcal{K}-\lim_{n\to+\infty} v_n = v & \text{in } B_m([0,T]\times H) \end{cases} \tag{4.304}$$

and (using the notation of Definition 4.131)

$$\begin{cases} \mathcal{K}-\lim_{n\to+\infty} h_n = 0 & \text{in } \overline{B}_{m,\gamma_G}([0,T)\times H, U) \\ \mathcal{K}-\lim_{n\to+\infty} D^G v_n = D^G v & \text{in } \overline{B}_{m,\gamma_G}([0,T)\times H, U). \end{cases} \tag{4.305}$$

We now show that (4.302) holds for v_n applying Dynkin's formula of Proposition 1.168[42] to the process $v_n(s, X(s)), s \in [t, T]$, where $X(s) = X(s; t, x, a(\cdot))$, obtaining

$$\mathbb{E}[v_n(T, X(T)) - v_n(t, x)]$$
$$= \mathbb{E}\int_t^T \left[(v_n)_t(s, X(s)) + \frac{1}{2}\mathrm{Tr}\Big[\Sigma(s, X(s))D^2 v_n(s, X(s))\Big] + \langle X(s), A^* Dv_n(s, X(s))\rangle \right] ds$$
$$+ \mathbb{E}\int_t^T \Big[\langle (rI - A)^{-1} b(s, X(s)), (rI - A^*)Dv_n(s, X(s))\rangle$$
$$+ \langle (rI - A)^{-1} G(s, X(s))R(s, X(s), a(s)), (rI - A^*)Dv_n(s, X(s))\rangle \Big] ds.$$

Since v_n is a classical solution of (4.303) and since Dv_n and $D^G v_n$ are both well defined, we have (see Notation 4.3)

$$\langle (rI - A)^{-1} G(s, X(s))R(s, X(s), a(s)), (rI - A^*)Dv_n(s, X(s))\rangle$$
$$= \langle R(s, X(s), a(s)), [(rI - A)^{-1} G(s, X(s))]^*(rI - A^*)Dv_n(s, X(s))\rangle$$
$$= \langle R(s, X(s), a(s)), D^G v_n(s, X(s))\rangle,$$

where in the last equality we used (see e.g. [521], Theorem 13.2) $[(rI - A)^{-1}G (s, X(s))]^* = G(s, X(s))^*[(rI - A)^{-1}]^*$. We then get

$$\mathbb{E}[g_n(X(T))] - v_n(t, x) = \mathbb{E}\int_t^T \left[-F_0\Big(s, X(s), D^G v_n(s, X(s))\Big) - h_n(s, X(s)) \right] ds$$
$$+ \mathbb{E}\int_t^T \langle R(s, X(s), a(s)), D^G v_n(s, X(s))\rangle ds.$$

Hence, adding and subtracting $\mathbb{E}\int_t^T l(s, X(s), a(s))ds + \mathbb{E}[g(X(T))]$ (such terms are well defined thanks to the assumptions on l and g) and rearranging the terms, we obtain

---

[42] This is similar to the proof of Theorem 2.36 but using a different Dynkin's formula since here we are dealing with the case $Q = I$ and σ possibly not belonging to $\mathcal{L}_2(\Xi, H)$.

$$v_n(t, x) = J^\mu(t, x; a(\cdot)) + \mathbb{E}[g_n(X(T)) - g(X(T))] + \mathbb{E}\int_t^T h_n(s, X(s))ds$$

$$- \mathbb{E}\int_t^T \left[F_{0,CV}\left(s, X(s), D^G v_n(s, X(s)), a(s)\right) - F_0\left(s, X(s), D^G v_n(s, X(s))\right) \right] ds.$$
$$(4.306)$$

We now pass to the limit as $n \to +\infty$ in (4.306) using the dominated convergence theorem and the convergences (4.304) and (4.305) to get the claim.

Now let Hypothesis 4.178 hold, $a(\cdot) \in \mathcal{U}_t^{\infty,\mu}$, let l satisfy Hypothesis 4.195, formula (4.301) and also let the assumptions of Corollary 4.191-(ii) (or Corollary 4.193) be satisfied. We then have (4.303) with F_0 replaced by $\tilde{F}_0$ from (4.292) and $\mathcal{A}_1(t)$ without the b term. However, since v_n is a classical solution, the new (4.303) can be rewritten in the old form too. We then apply Dynkin's formula of Proposition 1.169 where we set $a_2(s) = GR(s, a(s))$. Once this is done the rest of the proof is exactly the same, and even easier, since G is constant and R does not depend on x. The term $\langle b(s, X(s)), Dv_n(s, X(s)) \rangle$ is now converted into $\langle \bar{b}(s, X(s)), D^G v_n(s, X(s)) \rangle$. Finally, note that, in this case, the term $F_{0,CV} - F_0$ in (4.302) and (4.306) is equal to $\tilde{F}_{0,CV} - \tilde{F}_0$, where $\tilde{F}_{0,CV}$ is from (4.293). We keep $F_{0,CV} - F_0$, here and in the next statements, to make them simpler. $\qquad\square$

We can now prove our verification theorem.

Theorem 4.197 (Verification Theorem, Sufficient Condition) *Let Hypothesis 4.177 hold*[43] *and let $m \geq 0$ be fixed. Let l satisfy for this m Hypothesis 4.195, formula (4.300). Let the other assumptions of Corollary 4.189-(iii) be satisfied for the same m, namely Hypotheses 4.187 and 4.133 for $\mathcal{K}$-convergence (for the same U and G as in Hypothesis 4.177). Let $v \in \overline{C}_{m,\gamma_G}^{0,1,G}([0,T] \times H)$ be the mild and strong solution of (4.277). Let $t \in [0,T]$, $x \in H$ and let the set of admissible controls be $\mathcal{U}_t^\mu$. Then:*

(i) *For every generalized reference probability space μ on $[t, T]$ we have*

$$v(t, x) \leq \overline{V}(t, x) \leq V_t^\mu(x). \qquad (4.307)$$

(ii) *For a fixed generalized reference probability space $\hat{\mu}$ on $[t, T]$, let $a^*(\cdot) \in \mathcal{U}_t^{\hat{\mu}}$ be such that, with $X^*(s) = X^*(s; t, x, a^*(\cdot))$, we have*

$$a^*(s) \in \arg\min_{a \in \Lambda} F_{0,CV}(s, X^*(s), D^G v(s, X^*(s)), a), \qquad (4.308)$$

for almost every $s \in [t, T]$ and $\mathbb{P}$-a.s. Then the pair $(X^*(\cdot), a^*(\cdot))$ is $\hat{\mu}$-optimal at (t, x) (and thus is also optimal for the weak formulation) and $v(t, x) = \overline{V}(t, x) = V_t^{\hat{\mu}}(x)$.

The same statements hold if we take the following assumptions:

---

[43] Here we are in the framework of Corollary 4.189-(iii), hence we could also take Hypothesis 4.178.

- *Hypothesis 4.178 holds and l satisfies Hypothesis 4.195, formula (4.301) for a given $m \geq 0$ (respectively, for $m = 0$).*
- *The other assumptions of Corollary 4.191-(ii) (respectively, Corollary 4.193) are satisfied for the same m (respectively, for $m = 0$).*
- *The set of admissible controls is $\mathcal{U}_t^{\infty,\mu}$ and $a^*(\cdot) \in \mathcal{U}_t^{\infty,\hat{\mu}}$ in (ii).*

Proof By the definition, $F_{0,CV} - F_0 \geq 0$ everywhere, so, using Lemma 4.196 we get $v(t, x) \leq J^\mu(t, x; a(\cdot))$ for all generalized reference probability spaces μ and for all $a(\cdot) \in \mathcal{U}_t^\mu$. By taking the infimum over $a(\cdot) \in \overline{\mathcal{U}}_t$ in the right-hand side of (4.302) we obtain (i).

Regarding (ii), let $(X^*(\cdot), a^*(\cdot))$ be an admissible pair at (t, x) (for a given fixed $\hat{\mu}$) satisfying (4.308) for almost every $s \in [t, T]$ and $\mathbb{P}$-a.s. We then have

$$\mathbb{E} \int_t^T \left[F_{0,CV} \left(r, X^*(r), D^G v(r, X^*(r)), a^*(r) \right) \right.$$
$$\left. - F_0 \left(r, X^*(r), D^G v(r, X^*(r)) \right) \right] dr = 0.$$

Thus, by (4.302), we get

$$v(t, x) = J^{\hat{\mu}}(t, x; a^*(\cdot)) \tag{4.309}$$

which, together with (i), implies that $(X^*(\cdot), a^*(\cdot))$ is $\hat{\mu}$-optimal at (t, x) and $v(t, x) = \overline{V}(t, x) = V_t^{\hat{\mu}}(x)$.

The proofs under the other sets of assumptions are the same. $\qquad\square$

4.8.1.6 Optimal Feedbacks

To obtain the existence of optimal feedback controls we need to solve the *Closed Loop Equation* as was explained in Corollary 2.38. In our case, defining as in (2.61) the multivalued function

$$\begin{cases} \Phi: (0, T) \times H \to \mathcal{P}(\Lambda) \\ \Phi: (t, x) \to \arg\min_{a \in \Lambda} F_{0,CV}(t, x, D^G v(t, x), a), \end{cases} \tag{4.310}$$

the Closed Loop Equation associated with our problem and the mild solution v of the HJB equation (4.277) is

$$\begin{cases} dX(s) \in AX(s)ds + G(s, X(s))R(s, X(s), \Phi(s, X(s)))ds + \sigma(s, X(s))dW(s) \\ X(t) = x. \end{cases}$$
$$\tag{4.311}$$

Similarly to Corollary 2.38 we have here the following result.

Corollary 4.198 *Let Hypothesis 4.177 hold[44] and let $m \geq 0$ be fixed. Let l sat-isfy for this m Hypothesis 4.195, formula (4.300). Let the other assumptions of Corollary 4.189-(iii) be satisfied for the same m, namely Hypotheses 4.187 and 4.133 for $\mathcal{K}$-convergence (for the same U and G as in Hypothesis 4.177).[45] Let $v \in \overline{C}_{m,\gamma_G}^{0,1,G}([0,T] \times H)$ be the mild and strong solution of (4.277). Fix $(t,x) \in [0,T) \times H$. Assume moreover that, on $[t,T) \times H$, the feedback map Φ defined in (4.310) admits a measurable selection $\phi_t : [t,T) \times H \to \Lambda$ such that the Closed Loop Equation*

$$\begin{cases} dX(s) = [AX(s) + b(s, X(s)) + G(s, X(s))R(s, X(s), \phi_t(s, X(s)))] \, ds \\ \qquad\qquad + \sigma(s, X(s))dW(s) \\ X(t) = x, \end{cases}$$

(4.312)

has a weak mild solution (see Definition 1.121) $X_\phi(\cdot; t, x)$ in some generalized refer-ence probability space $\overline{\mu}$ on $[t,T]$. Define, for $s \in [t,T]$, $a_{\phi_t}(s) = \phi_t(s, X_{\phi_t}(s; t, x))$. Then the pair $(a_{\phi_t}(\cdot), X_{\phi_t}(\cdot; t, x))$, if it is admissible[46] is $\overline{\mu}$-strongly optimal at (t,x) and $v(t,x) = \overline{V}(t,x) = V_t^{\overline{\mu}}(x)$. If, finally, $\Phi(t,x)$ is always a singleton and the weak mild solution of (4.312) is unique in the generalized reference probability space $\overline{\mu}$, then $a_{\phi_t}(\cdot)$ is the unique $\overline{\mu}$-optimal control.

Proof The optimality of $(a_{\phi_t}(\cdot), X_{\phi_t}(\cdot; t, x))$ immediately follows from Theorem 4.197-(ii). We only need to discuss the part about the uniqueness of $\overline{\mu}$-optimal con-trols. To see this we observe that if $(\hat{a}(\cdot), \hat{X}(\cdot))$ is another $\overline{\mu}$-optimal pair at (t,x), we immediately have, by (4.302) and using $v(t,x) = \overline{V}(t,x) = V_t^{\overline{\mu}}(x)$, that, on $s \in [t,T]$,

$$\mathbb{E} \int_t^T \left[F_{0,CV}\left(s, \hat{X}(s), D^G v(s, \hat{X}(s)), \hat{a}(s)\right) - F_0\left(s, \hat{X}(s), D^G v(s, \hat{X}(s))\right) \right] ds = 0.$$

This implies that, for a.e. $s \in [t,T]$ and $\mathbb{P}$-a.s., we have $\hat{a}(s) = \phi_t(s, \hat{X}(s))$. The uniqueness of mild solutions of (4.312) in $\overline{\mu}$ thus gives the claim. $\qquad\square$

Providing conditions for the data that allow us to apply the above corollary is a difficult problem which has not yet been solved in many interesting cases. Here we

---

[44] Here we are in the framework of Corollary 4.189-(iii), hence we could also take Hypothesis 4.178.

[45] All assumptions up to this point can be substituted, as in Lemma 4.196, by the following:

- Hypothesis 4.178 holds and l satisfies Hypothesis 4.195, formula (4.301) for a given $m \geq 0$ (respectively, for $m = 0$).
- The other assumptions of Corollary 4.191-(ii) (respectively, Corollary 4.193) are satisfied for the same m (respectively, for $m = 0$).
- The set of admissible controls is $\mathcal{U}_t^{\infty,\mu}$.

The proof is exactly the same.

[46] This is always guaranteed when Hypothesis 4.177 holds. When Hypothesis 4.178 holds (see the previous footnote) we need to have $a_{\phi_t}(\cdot) \in \mathcal{U}_t^{\infty,\overline{\mu}}$. Similarly, if we use more general sets of admissible controls, like $\mathcal{U}_t^{p,\mu}$ with finite p, or like in cases with state constraints (see e.g. Sect. 4.10.2 for an example) then the admissibility is not guaranteed and must be checked.

state three results which can be applied to some of our examples of Sect. 2.6. The reader should be aware of the fact that these results are far from being sharp and there is a lot of room for possible improvements.

We start with a quite general result when Hypotheses 4.177 is satisfied and $G = \sigma$, similarly to the setting used in Chap. 6.

Proposition 4.199 *Let Hypothesis 4.177 hold and let $m \geq 0$ be fixed. Let l satisfy for this m Hypothesis 4.195, formula (4.300). Let the other assumptions of Corollary 4.189-(iii) be satisfied for the same m, namely Hypotheses 4.187 and 4.133 for $\mathcal{K}$-convergence (for the same U and G as in Hypothesis 4.177). Let $v \in \overline{C}_{m,\gamma_G}^{0,1,G}([0, T] \times H)$ be the mild and strong solution of (4.277). Assume that $G = \sigma$. Let $(t, x) \in [0, T] \times H$ and assume that, on $[t, T) \times H$, the feedback map Φ defined in (4.310) admits a measurable selection $\phi_t : [t, T) \times H \to \Lambda$.*

Then Eq. (4.312) admits a weak mild solution $X_{\phi_t}(\cdot; t, x)$ in some generalized reference probability space $\overline{\mu}$ on $[t, T]$. Moreover, defining, for $s \in [t, T]$, $a_{\phi_t}(s) = \phi_t(s, X_{\phi_t}(s; t, x))$, the pair $(a_{\phi_t}(\cdot), X_{\phi_t}(\cdot; t, x))$ is $\overline{\mu}$-optimal at (t, x) and $v(t, x) = \overline{V}(t, x) = V_t^{\overline{\mu}}(x)$.

Proof It is proved in the proof of Theorem 6.36 that, under our assumptions, there exists a weak mild solution of (4.312). Obviously $a_{\phi_t}(\cdot) \in \mathcal{U}_t^{\overline{\mu}}$. Thus the claim follows from Corollary 4.198. □

We now discuss two special cases arising in many applications, see e.g. Sects. 2.6.1, 2.6.2, 2.6.8. First we consider a special case of Corollary 4.191, where the Hamiltonian is as in Proposition 4.183. We assume the following.

Hypothesis 4.200 We assume that $U = H$ and Hypotheses 4.190 and 4.178 hold. Moreover, we assume the following.

(i) Hypothesis 4.145-(iv) also holds with $U = H$ and $G = I$.

(ii) E is a real separable Hilbert space and Λ is a closed, convex and bounded subset of E. Moreover, $R(t, x, a) = Ba$ for some $B \in \mathcal{L}(E, H)$ and $l(t, x, a) = l_0(t, x) + l_1(a)$ with $l_0 \in C_m([0, T] \times H)$ for some $m \geq 0$ and $l_1 : \Lambda \to \mathbb{R}$ continuous, strictly convex and with always non-empty subdifferential.

(iii) Defining

$$F_{2,CV}(z, a) := \langle a, z \rangle_E + l_1(a) \quad and \quad F_2(z) := \inf_{a \in \Lambda}\{\langle a, z \rangle_E + l_1(a)\},$$

the unique minimum point $\bar{a}(z)$ of the map

$$\Lambda \to \mathbb{R}, \qquad a \to F_{2,CV}(z, a)$$

is a Lipschitz continuous function of $z \in E$.

(iv) $g \in C_m(H)$ for the same m as in (ii).

Here is the result.

Theorem 4.201 *Let Hypothesis 4.200 hold. Let $v \in \overline{C}_{m,\gamma_G}^{0,1,G}([0,T] \times H)$ be the mild and strong solution of (4.277) and assume that*

$$|D^G v(t, x_1) - D^G v(t, x_2)| \leq \bar{f}_G(t)|x_1 - x_2|, \tag{4.313}$$

where $\bar{f}_G : [0,T) \to \mathbb{R}^+$ is bounded on $[0, T - \varepsilon]$ for all $\varepsilon \in (0,T)$. Then, for every fixed $t \in [0,T]$ and every generalized reference probability space μ on $[t,T]$, we have the following.

(i) *For every $x \in H$, $v(t, x) = V_t^\mu(x)$. Hence also $v(t, x) = \overline{V}(t, x)$.*
(ii) *For every $x \in H$, there exists a unique μ-optimal control $a^*(\cdot) \in \mathcal{U}_t^{\infty, \mu}$ which is related to the corresponding optimal state $X^*(\cdot)$ by the feedback formula*

$$a^*(s) = \arg\min_{a \in \Lambda} F_{2,CV}(B^* D^G v(s, X^*(s)), a) = DF_2\left(B^* D^G v(s, X^*(s))\right),$$

for almost every $s \in [t, T]$ and $\mathbb{P}$-a.s.

Proof We first observe that, by Hypothesis 4.200-(iii) and Proposition 4.183, the function F_2 is concave, belongs to $C^{1,1}(E)$ and DF_2 is bounded. Also, by Hypothesis 4.190-(i), we have $b = G\bar{b}$.

The Hamiltonian in this case, $\tilde{F}_0$, is defined by (4.292):

$$\tilde{F}_0(t, x, q) = \langle \bar{b}(t, x), q \rangle + F_0(t, x, q) = \langle \bar{b}(t, x), q \rangle + l_0(t, x) + F_2(B^* q)$$

and, again from the proof of Proposition 4.183, the minimum point of the function $a \to \tilde{F}_{0,CV}(t, x, q, a)$ from (4.293) is $\bar{a}(B^* q) = DF_2(B^* q)$.[47]

It thus follows that (4.308) in this case rewrites as

$$a^*(s) = DF_2(B^* D^G v(s, X^*(s))), \qquad \text{for a.e. } s \in [t, T], \mathbb{P} - \text{a.s.},$$

where $X^*(\cdot)$ is the state associated to $a^*(\cdot)$, i.e. $X^*(\cdot)$ is the solution of the closed loop equation

$$\begin{cases} dX(s) = \left[AX(s) + b(s, X(s)) + GBDF_2(B^* D^G v(s, X(s)))\right] ds \\ \qquad\qquad\qquad\qquad\qquad\qquad +\sigma dW(s), \quad s \in (t, T], \quad (4.314) \\ X(t) = x, \quad x \in H, \end{cases}$$

which can be written in the mild form as

$$X(s) = e^{(s-t)A}x + \int_t^s e^{(s-r)A}[b(r, X(r)) + GBDF_2(B^* D^G v(r, X(r)))]ds$$

$$+ W^A(t, s), \quad t \leq s \leq T,$$

---

[47]Note that in this proof we may use equivalently the Hamiltonians $F_{0,CV}$ or $\tilde{F}_{0,CV}$ as they have the same minimum points. We use $\tilde{F}_{0,CV}$ since $\tilde{F}_0$ is the actual Hamiltonian in the associated HJB equation in this case.

where $W^A(t, s) = \int_t^s e^{(s-r)A} dW(s)$ (as defined after Proposition 1.144). It remains to show that (4.314) has a unique mild solution.

Since G satisfies Hypothesis 4.178, Hypothesis 1.145 (where $a_2(\cdot) \equiv 0$ and the b_0 there does not depend on a_1 and is equal to $b + GBDF_2(B^* D^G v)$) is satisfied on every interval $[t, T - \varepsilon]$ for all $\varepsilon \in (0, T - t)$. Hence, by Proposition 1.147, there exists a unique mild solution $X^*(\cdot)$ of (4.314) in $\mathcal{H}_p^\mu(t, T - \varepsilon; H)$ for every $p \geq 1$. By the arbitrariness of ε we can then define a process $X^*(\cdot) : [t, T) \times \Omega \to H$ such that, for every $\varepsilon \in (0, T - t)$, $X^*(\cdot)|_{[t,T-\varepsilon]}$ is the unique mild solution of (4.314) in $\mathcal{H}_p^\mu(t, T - \varepsilon; H)$ for every $p \geq 1$. Moreover, exploiting the assumptions on b from Hypothesis 1.145-(i) and the boundedness of DF_2 we have, for some $C > 0$,

$$\sup_{s \in [t,T)} \mathbb{E}|X^*(s)|^p \leq C < +\infty$$

which implies that $X^*(\cdot) \in \mathcal{H}_p^\mu(t, T; H)$ for every $p \geq 1$. So $X^*(\cdot)$ is the unique mild solution of (4.314) in $\mathcal{H}_p^\mu(t, T; H)$. The claim then easily follows from Corollary 4.198.   □

Remark 4.202 A typical example where Hypothesis 4.200-(iii) holds is described in Remark 4.185. Indeed, in this case $\Lambda = B_H(0, M)$, which is closed, convex and bounded, and $l_1(a) = |a|^\theta/\theta$ with $\theta \in (1, 2]$. We see from Remark 4.185 that the minimum point of the Hamiltonian F_0 is

$$\bar{a}(q) = -|B^* q|^{\frac{2-\theta}{\theta-1}} B^* q \qquad \text{if} \qquad |B^* q| \leq M^{\theta-1}$$

and

$$\bar{a}(q) = -M \frac{B^* q}{|B^* q|} \qquad \text{if} \qquad |B^* q| > M^{\theta-1}.$$

Hence it follows from Proposition 4.183 and its proof that Hypothesis 4.200-(iii) is satisfied.

To obtain (4.313) we need to assume more about the data. In particular, if the data are such that Theorem 4.155 can be applied, then v would have bounded second derivatives on $[0, T - \varepsilon] \times H$ for all $\varepsilon \in (0, T)$ and hence (4.313) would be satisfied. Other ways to get such a regularity depends on specific cases, see e.g. Sect. 7 of [316] for an example in this direction.   ■

We now consider a special case of Corollary 4.193, where the Hamiltonian is as in Proposition 4.182-(iii). We need a new hypothesis.

Hypothesis 4.203 We assume that $U = H$, $G = I$, $\gamma(s) := \gamma_I(s) \leq \bar{C} s^{-\alpha}$ for some $\bar{C} > 0$ and $\alpha \in (0, 1)$, and Hypotheses 4.190 and 4.178[48] hold. Moreover, we assume the following.

(i) A generates a C_0-semigroup of pseudo-contractions, i.e. Hypothesis 4.172-(i) is satisfied.

---

[48] It is understood that in this case $\beta = 0$ there.

(ii) In Hypothesis 4.190, b is differentiable in x and $Db \in C_b^s([0, T] \times H, \mathcal{L}(H))$.
(iii) $\Lambda = E$ is a real separable Hilbert space and $R(t, x, a) \equiv Ba$ for some $B \in \mathcal{L}(E, H)$. Moreover, $l(t, x, a) = l_0(t, x) + l_1(a)$ with $l_0 \in C_b([0, T] \times H)$ and $l_1 \in C(H)$ strictly convex and such that

$$\lim_{|a|_E \to +\infty} \frac{l_1(a)}{|a|_E} = +\infty.$$

In addition, $l_1 \in C^1(E)$ and Dl_1 is invertible with $(Dl_1)^{-1}$ Lipschitz continuous on bounded sets.
(iv) $l_0 \in C_b^{0,1}([0, T] \times H)$ and $g \in C_b^1(H)$.

Theorem 4.204 *Assume that Hypothesis 4.203 holds. Then (4.277) has a mild solution $v \in \overline{C}_{b,\alpha}^{0,2,s}([0, T] \times H)$, which is unique in $\overline{C}_{b,\alpha}^{0,1}([0, T] \times H)$. The mild solution v is also a unique strong solution of (4.277) in $\overline{C}_{b,\alpha}^{0,1}([0, T] \times H)$. Let*[49]

$$F_{2,CV}(z, a) := \langle a, z \rangle_E + l_1(a) \quad and \quad F_2(z) := \inf_{a \in \Lambda} \{ \langle a, z \rangle_E + l_1(a) \}.$$

Then, for every fixed $t \in [0, T]$ and every generalized reference probability space μ on $[t, T]$, we have the following.

(i) *For every $x \in H$, $v(t, x) = V_t^\mu(x)$. Hence also $v(t, x) = \overline{V}(t, x)$.*
(ii) *For every $x \in H$ there exists a unique optimal control $a^*(\cdot) \in \mathcal{U}_t^{\infty,\mu}$ which is related to the corresponding optimal state $X^*(\cdot)$ by the feedback formula*

$$a^*(s) = \arg\min_{a \in E} F_{2,CV}(B^* Dv(s, X^*(s)), a) = DF_2\left(B^* Dv(s, X^*(s))\right),$$

for almost every $s \in [t, T]$ and $\mathbb{P}$-a.s.

Proof We observe that in this case the Hamiltonian is given by the function $\tilde{F}_0$ introduced in (4.292). Since $U = H$ and $G = I$, we have

$$\tilde{F}_0(t, x, q) = \langle b(t, x), q \rangle + l_0(t, x) + F_2(B^* q)$$

and it satisfies Hypotheses 4.169-(i)-(ii) and 4.172-(ii) (in particular, the function F_2 is concave, and DF_2 is bounded and Lipschitz continuous on bounded sets). Indeed, this follows from Proposition 4.182 and from the assumptions on b and l_0 in Hypothesis 4.203. Thus we can apply Corollary 4.193 which gives the existence and uniqueness of a mild and strong solution of (4.277) with the required regularity.

The Current Value Hamiltonian is

$$\tilde{F}_{0,CV}(t, x, q, a) = \langle b(t, x), q \rangle + l_0(t, x) + \langle Ba, q \rangle + l_1(a)$$
$$= \langle b(t, x), q \rangle + l_0(t, x) + F_{2,CV}(B^* q, a).$$

---

[49] See point (iii) of Hypothesis 4.200.

From Proposition 4.182-(iii) and its proof we know that the minimum point of $a \to \tilde{F}_{0,CV}(t, x, q, a)$ is $\bar{a}(B^*q) = DF_2(B^*q)$.

It follows that (4.308) holds when for a.e. $s \in [t, T]$, $\mathbb{P}$-a.s., we have

$$a^*(s) = DF_2(B^* Dv(s, X^*(s))),$$

where $X^*(\cdot)$ is the state associated to $a^*(\cdot)$, which should solve the closed loop equation

$$\begin{cases} dX(s) = [AX(s) + b(s, X(s)) + GBDF_2(B^* Dv(s, X(s)))] \, ds \\ \qquad\qquad\qquad\qquad\qquad\qquad\qquad +\sigma dW(s), \quad s \in (t, T], \\ X(t) = x, \quad x \in H. \end{cases}$$

$$(4.315)$$

Equation (4.315) can be solved similarly as in the proof of Theorem 4.201 if we take into account that Dv is bounded on $[0, T] \times H$ and $Dv(s, \cdot)$ are Lipschitz continuous on H, uniformly for $s \in [0, T - \varepsilon]$ for every $\varepsilon > 0$. The claim then easily follows from Corollary 4.198. □

Remark 4.205 A typical example where the function F_2 satisfies the assumptions of Theorem 4.204 is (as in Remark 4.185) when $\Lambda = E$ and $l_1(a) = |a|^\theta / \theta$ for some $\theta \in (1, 2]$. Indeed, from Remark 4.185 we see that

$$F_2(q) = -\frac{|B^* q|^{\theta'}}{\theta'}$$

and the minimum point of the Current Value Hamiltonian is

$$\bar{a}(q) = -|B^* q|^{\frac{2-\theta}{\theta-1}} B^* q.$$

■

4.8.2 The Infinite Horizon Case

Similarly to the finite horizon case we mainly use the setting of the *strong formulation* of stochastic optimal control problems (see Sect. 2.1.1, Hypothesis 2.1 and the discussion after Definition 2.3). Let H, Ξ be real separable Hilbert spaces (the state space and the noise space) and Λ be a Polish space (the control space). The initial time is now $t = 0$ and the horizon is $T = +\infty$. Since most of the results are completely similar to the finite horizon case we will present them without going into full details. However, for the reader's convenience, we will repeat the setting of the problem and the main assumptions, pointing out the differences with respect to the finite horizon case.

4.8.2.1   The State Equation

Let $\mu := \left(\Omega, \mathscr{F}, \{\mathscr{F}_s\}_{s\geq 0}, \mathbb{P}, W\right)$, where W is a cylindrical Wiener process in Ξ, be a generalized reference probability space. For any initial state $x \in H$ the state equation is the controlled SDE (where $s \in (0, +\infty)$)

$$\begin{cases} dX(s) = [AX(s) + b(X(s)) + G(X(s))R(X(s), a(s))]\,ds + \sigma(X(s))dW(s), \\ X(0) = x. \end{cases}$$

$$(4.316)$$

Differently from the finite horizon case we only consider one set of assumptions, which is similar to Hypothesis 4.177. For infinite horizon cases where unboundedness of R may arise (as in Hypothesis 4.178), see Remark 4.221.

Hypothesis 4.206

(i) A, b, σ satisfy Hypothesis 1.149 for $s \in (0, 1]$ with b and σ independent of t, a and with $f_1(s) = Ls^{-\gamma_1}$ and $f_2(s) = Ls^{-\gamma_2}$ for some $\gamma_1 \in (0, 1)$, $\gamma_2 \in (0, 1/2)$, and constant $L \geq 0$. Moreover, they also satisfy Hypothesis 4.138-(iii).

(ii) For a given real separable Hilbert space U, G is a family of closed linear operators (possibly unbounded with dense domain) $G(x) : D(G(x)){\subset}U \to H$, $x \in H$, satisfying Hypothesis 4.11.

(iii) For every $(s, x) \in (0, 1] \times H$, $e^{sA}G(x)$ extends to an operator in $\mathcal{L}(U, H)$, which we still denote by $e^{sA}G(x)$, the function

$$(0, T] \times H \to \mathcal{L}(U, H), \qquad (s, x) \to e^{sA}G(x)$$

is strongly measurable and, for every $(s, x) \in (0, 1] \times H$,

$$\|e^{sA}G(x)\|_{\mathcal{L}(U,H)} \leq f_G(s)(1 + |x|),$$

where $f_G(s) = L_1 s^{-\beta}$ for some $L_1 > 0$ and $\beta \in [0, 1)$. Moreover, there exists an $r \in \mathbb{R} \cap \varrho(A)$ such that for every $x \in H$, $(rI - A)^{-1}G(x)$ extends to an operator in $\mathcal{L}(U, H)$, which we still denote by $(rI - A)^{-1}G(x)$, and the function

$$H \to \mathcal{L}(U, H), \qquad x \to (rI - A)^{-1}G(x)$$

is strongly measurable and there exists a $C > 0$ such that

$$\|(rI - A)^{-1}G(x)\|_{\mathcal{L}(U,H)} \leq C(1 + |x|), \quad \forall x \in H.$$

(iv) Λ is a Polish space.

(v) $R \in C_b(H \times \Lambda, U)$. Moreover, for every $(s, x_1, x_2, a) \in (0, 1] \times H \times H \times \Lambda$,

$$|e^{sA}G(x_1)R(x_1, a) - e^{sA}G(x_2)R(x_2, a)| \leq f_G(s)|x_1 - x_2|,$$

where f_G is as in (iii) above.

Remark 4.207 Notice that if Hypothesis 4.206 is satisfied then, by the semigroup property of e^{tA}, Hypotheses 1.149 and 4.206 hold for all $s \in (0, +\infty)$, substituting the functions f_1, f_2 and f_G by $\bar{f}_1(s) := L_2(s^{-\gamma_1} \vee 1)e^{\omega s}$, $\bar{f}_2(s) := L_2(s^{-\gamma_2} \vee 1)e^{\omega s}$ and $\bar{f}_G(s) := L_2(s^{-\beta} \vee 1)e^{\omega s}$ for some $L_2 \geq 0$, where $\omega \in \mathbb{R}$ is such that $\|e^{sA}\| \leq Me^{\omega s}$, $s \geq 0$.   ∎

The set Λ in Part (iii) of Hypothesis 4.206 is the control space. Precise assumptions about admissible controls are given in the next subsection.

4.8.2.2 Optimal Control Problems and the HJB Equation

The set of admissible controls is of the same type as in the finite horizon case when Hypothesis 4.177 holds: indeed it is

$$\mathcal{U}_0^\mu := \{a : [0, +\infty) \to \Lambda : a(\cdot) \text{ is } \mathscr{F}_s\text{-progressively measurable}\}.$$

If Hypothesis 4.206 holds and $a(\cdot) \in \mathcal{U}_0^\mu$, the existence of a unique mild solution $X(\cdot; x, a(\cdot))$ of (4.316) follows from Theorem 1.152. The existence and uniqueness is in the spaces provided by this theorem. We will often write $X(s)$ for $X(s; x, a(\cdot))$, $s \in [t, T]$, when its meaning is clear from the context.

Proposition 4.208 *Let the assumptions of Hypothesis 4.206 be satisfied. Then for every $m \geq 0$ there exist constants $C_1(m)$, $\lambda_1(m) \geq 0$ such that for every $x \in H, a(\cdot) \in \mathcal{U}_0^\mu$, the mild solution $X(\cdot; x, a(\cdot))$ of (4.316) satisfies*

$$\mathbb{E}[|X(s; x, a(\cdot))|^m] \leq C_1(m)e^{\lambda_1(m)s}(1 + |x|^m), \quad \text{for all } s \geq 0.$$

Proof The proof repeats the strategy outlined in Remark 4.107. We first notice that, denoting by $X(\cdot; t, \xi, a(\cdot))$ the solution of (4.316) with initial condition $X(t) = \xi$, we have, by the uniqueness of the mild solution proved in Theorem 1.152[50] (notice that here R is bounded), the equality $X(s; x, a(\cdot)) = X(s; 0, x, a(\cdot)) = X(s; t, X(t; x, a(\cdot)), a(\cdot))$, $s \geq t \geq 0$. Then for every $n \in \mathbb{N}$, applying Theorem 1.152 with the initial time $t = n$, we get

$$\sup_{s \in [n, n+1]} \mathbb{E}[|X(s; x, a(\cdot))|^m] \leq C_0(m)(1 + \mathbb{E}[|X(n; x, a(\cdot))|^m]).$$

Therefore, iterating, we obtain that for $s \in [n, n+1]$,

$$\mathbb{E}[|X(s; x, a(\cdot))|^m] \leq (C_0(m) + \dots + C_0(m)^{n+1}) + C_0(m)^{n+1}(1 + |x|^m)$$

and we conclude as in Remark 4.107.   □

---

[50] Theorem 1.152 is stated with the initial time $t = 0$ but all claims there are valid for any initial time $t \geq 0$.

The cost functional to minimize is, for a given $\lambda > 0$,

$$J^\mu(x; a(\cdot)) = \mathbb{E}\left\{\int_0^{+\infty} e^{-\lambda s} l(X(s; x, a(\cdot)), a(s))ds\right\} \tag{4.317}$$

over all controls $a(\cdot)$ in $\mathcal{U}_t^\mu$. The assumptions about l and g will be given later.

The value function for this problem is defined as in (2.4)

$$V^\mu(x) = \inf_{a(\cdot)\in\mathcal{U}_0^\mu} J^\mu(x; a(\cdot)) \tag{4.318}$$

and optimal pairs are defined as in Definition 2.3.

We will also consider the weak formulation of the control problem from Sect. 2.1.2 for which the generalized reference probability space μ varies with the controls. Following (2.6) the set of admissible controls is then

$$\overline{\mathcal{U}}_0 := \bigcup_\mu \mathcal{U}_0^\mu.$$

The value function is

$$\overline{V}(x) = \inf_{a(\cdot)\in\overline{\mathcal{U}}_0} J^\mu(x; a(\cdot)). \tag{4.319}$$

To study more general cases we can also consider the so-called "extended weak formulation" introduced in Remark 2.6, see the paragraph after (4.299) for more remarks about this.

The HJB equation associated with this problem is (see Sect. 2.5.2)

$$\lambda v - \frac{1}{2}\,\mathrm{Tr}[\Sigma(x)D^2v] - \langle Ax + b(x), Dv\rangle - F(x, Dv) = 0, \quad x \in H, \tag{4.320}$$

where $\Sigma(x) = \sigma(x)\sigma^*(x)$ and the Hamiltonian F is given by

$$F(x, p) = \inf_{a\in\Lambda} F_{CV}(x, p, a) = \inf_{a\in\Lambda}\left\{\langle G(x)R(x, a), p\rangle_H + l(x, a)\right\}. \tag{4.321}$$

Note that, in contrast to the definition of Hamiltonians (2.64) and (2.65) in Sect. 2.5.2, here we drop the term $\langle b(x), Dv\rangle$ from F_{CV} since it does not depend on the control variable. However, sometimes it will be included in the Hamiltonian, see e.g. (4.325) and (4.326).

Exactly as in the parabolic case we introduce the modified Hamiltonian F_0

$$F_0(x, q) = \inf_{a\in\Lambda} F_{0,CV}(x, q, a) = \inf_{a\in\Lambda}\left\{\langle R(x, a), q\rangle_U + l(x, a)\right\} \tag{4.322}$$

and observe that, for $p \in D(G(x)^*)$,

$$F(x, p) = F_0(x, G(x)^* p),$$

so the term $F(x, Dv)$ in (4.320) can be formally rewritten (see Sect. 4.2.1) as $F_0(x, D^G v)$. It is clear that the results of Sect. 4.8.1.2, namely Propositions 4.181, 4.182 and 4.183, apply to this case.

We remark that similar observations as those in Remarks 4.179 and 4.180 are also valid here.

4.8.2.3 Results for the HJB Equation

We present two results where we apply the theory of Sects. 4.4.2, 4.5.2 and 4.6.2 to our HJB equation (4.320). The first uses general assumptions of Sect. 4.4.1.2 while the second refers to the more specific Ornstein–Uhlenbeck case of Sect. 4.6.2. We point out, as we already mentioned discussing the parabolic case, that variations of such results and/or applications to specific models are possible. Some cases are discussed in Sects. 4.9 and 4.10.

For our first result the transition semigroup is defined using the SDE

$$\begin{cases} dX(s) = [AX(s) + b(X(s))]ds + \sigma(X(s))dW(s), \\ X(0) = x. \end{cases} \tag{4.323}$$

We recall that, if Hypothesis 4.206 holds, this equation has a unique mild solution, which we denote by $X_0(\cdot; x)$, and $P_s[\phi](x) = \mathbb{E}[\phi(X_0(s; x))]$, $s \geq 0$, is a one-parameter transition semigroup on $B_m(H)$ for every $m \geq 0$. P_s is independent of the choice of a generalized reference probability space μ.

To apply the results of Sects. 4.4 and 4.5 we need assumptions about the datum F_0 and the transition semigroup P_s. We provide here two sets of assumptions, the first for the B_m space case, the second for the C_m space case.

Hypothesis 4.209 Let $m \geq 0$ and assume that, for this m, the following hold.

(i) The function F_0 introduced in (4.322) satisfies Hypothesis 4.104.[51]
(ii) The one-parameter transition semigroup P_s defined above satisfies Hypotheses 4.106, 4.108, 4.110.

Hypothesis 4.210 Let $m \geq 0$ and assume that, for this m, the following hold.

(i) The function F_0 introduced in (4.322) satisfies Hypothesis 4.117.
(ii) The one-parameter transition semigroup P_s defined above satisfies Hypothesis 4.119.

Similar observations as those in Remark 4.188 also apply here. We have the following corollary.

---

[51]Note that Proposition 4.181 provides sufficient conditions on R and l for this. Similarly for Hypothesis 4.210-(i).

Corollary 4.211

(i) *Assume that Hypothesis 4.209 is satisfied. Let λ_0 be from Theorem 4.112. Then for all $\lambda > \lambda_0$ the HJB equation (4.320) has a unique mild solution v in $B_m^{1,G}(H)$.*

(ii) *Assume that Hypothesis 4.210 is satisfied. Let λ_0 be from Theorem 4.112. Then for all $\lambda > \lambda_0$ the HJB equation (4.320) has a unique mild solution v in $C_m^{1,G}(H)$.*

(iii) *Let Hypothesis 4.206 hold. Let Hypotheses 4.210 and 4.141 for $\mathcal{K}$-convergence be satisfied with the same U and G as in Hypothesis 4.206. Let $\lambda > \lambda_0 \vee a_1$, where λ_0 is from Theorem 4.112 and a_1 from (4.163). Then the mild solution v is also the unique $\mathcal{K}$-strong solution in $C_m^{1,G}(H)$.*

Proof Part (i) is an immediate consequence of Theorem 4.112.

Part (ii) follows from Theorem 4.120.

Part (iii) follows from Theorem 4.143 and Remark 4.144-(iii) once we observe that Hypothesis 4.138 is implied by Hypothesis 4.206-(i) and Hypothesis 4.210. □

For the second result the process $X_0(\cdot; x)$ defining the transition semigroup is of Ornstein–Uhlenbeck type, i.e. it is the solution of the SDE

$$\begin{cases} dX(s) = AX(s)ds + \sigma dW(s), \\ X(0) = x. \end{cases} \tag{4.324}$$

We now have $P_s = R_s$, where R_s is the Ornstein–Uhlenbeck semigroup associated to A and σ. We assume the following.

Hypothesis 4.212

(i) Let Hypothesis 4.145 be satisfied for A, G and $\Sigma = \sigma\sigma^*$ given in Eq. (4.316). Moreover, (4.47) holds.[52]

(ii) The function b in (4.316) satisfies what is required for it in Hypothesis 4.206. Moreover, b is such that $\langle b, p \rangle_H = \langle \bar{b}, G^* p \rangle_U$, for some $\bar{b} \in C_b(H, U)$.

(iii) Either G is bounded or $D(A^*) \subset D(G^*)$ and, for some $c > 0$, $|G^* z| \leq c |A^* z|$ for all $z \in D(A^*)$.

The Hamiltonian F_0 now becomes

$$\tilde{F}_0(x, q) := F_0(x, q) + \langle \bar{b}(x), q \rangle_U, \tag{4.325}$$

and the corresponding current value Hamiltonian is

$$\tilde{F}_{0,CV}(x, q, a) := F_{0,CV}(x, q, a) + \langle \bar{b}(x), q \rangle_U. \tag{4.326}$$

The second result is also for a Lipschitz continuous (in the gradient variable) Hamiltonian F_0. It is an immediate consequence of Theorems 4.157 and 4.158.

---

[52]This is not needed in the proof of Corollary 4.213, however we included it so that this case is covered by Hypothesis 4.206.

Corollary 4.213 *Let Hypothesis 4.212-(i)-(ii) hold. Assume, moreover, that for a given $m \geq 0$, F_0 satisfies Hypothesis 4.104. Let λ_0 be from Theorem 4.112 for this case. Then we have the following.*

(i) *For all $\lambda > \lambda_0$ the HJB equation (4.320) has a unique mild solution v in $B_m^{1,G}(H)$.*

(ii) *If F_0 is continuous then, for all $\lambda > \lambda_0$, $v \in C_m^{1,G}(H)$. Let $\lambda > \lambda_0 \vee a_1$, where λ_0 is from Theorem 4.112 and a_1 from (4.163). Assume moreover that Hypothesis 4.145 also holds with $U = H$ and $G = I$ in point (iv). Then v is also the unique $\mathcal{K}$-strong solution in $C_m^{1,G}(H)$. The solutions of the approximating problems can be chosen to converge uniformly on bounded subsets of H.*

Proof The mild form of the HJB equation is now written with $P_s = R_s$. Since $\bar{b} \in C_b(H, U)$ and thanks to the assumptions about F_0, the new Hamiltonian $\tilde{F}_0$, given by (4.325), satisfies Hypothesis 4.104. Hence point (i) follows from Theorem 4.157-(i). The fact that the mild solution $v \in C_m^{1,G}(H)$ in point (ii) follows from Theorem 4.157-(ii). The claims of point (ii) concerning the strong solution follow from Theorem 4.158 thanks to Hypothesis 4.212.   □

Note that Proposition 4.183 gives reasonable conditions under which Hypotheses 4.209-(i) and 4.210-(i) hold.

4.8.2.4   The Verification Theorem

In this section we prove a verification theorem. We first prove the infinite horizon version of the so-called fundamental identity (see (2.73)). The proof is a little different from the one for the finite horizon case, hence we provide it.

Hypothesis 4.214 The function $l \in C(H \times \Lambda)$ and there exists a $C > 0$ such that

$$|l(x,a)| \leq C(1 + |x|^m), \qquad \forall (x,a) \in H \times \Lambda. \tag{4.327}$$

Lemma 4.215 *Let Hypothesis 4.206 hold and let $m \geq 0$ be fixed. Let l satisfy, for this m, Hypothesis 4.214. Let the other assumptions of Corollary 4.211-(iii) be satisfied for the same m, namely Hypotheses 4.210 and 4.141 for $\mathcal{K}$-convergence (for the same U and G as in Hypothesis 4.206). Let $\lambda > \lambda_0 \vee a_1 \vee \lambda_1(m)$, where λ_0 is from Theorem 4.112, a_1 is from (4.163), and $\lambda_1(m)$ is from Proposition 4.208.*

Let $v \in C_m^{1,G}(H)$ be the mild and strong solution of (4.320) and $x \in H$. Suppose that $a(\cdot) \in \mathcal{U}_0^\mu$. Then we have the identity

$$v(x) = J^\mu(x; a(\cdot))$$

$$- \mathbb{E} \int_0^{+\infty} e^{-\lambda s} \left[F_{0,CV}\left(X(s), D^G v(X(s), a(s))\right) - F_0\left(X(s), D^G v(X(s))\right) \right] ds,$$

$$\tag{4.328}$$

where $X(\cdot) = X(\cdot; x, a(\cdot))$ is the mild solution of (4.316).

The same statement holds if we take the following assumptions, for a given $m \geq 0$:

- *Hypothesis 4.206 holds, $a(\cdot) \in \mathcal{U}_0^\mu$, l satisfies Hypothesis 4.214 for this m and $\lambda > \lambda_0 \vee a_1 \vee \lambda_1(m)$ as above.*
- *The other assumptions of Corollary 4.213, including those from part (ii), are satisfied for this m.*

Proof We know from Definition 4.140 and Theorem 4.143 that there exist two sequences $(h_n) \subset B_m(H)$ and $(v_n) \subset D(\mathcal{A}_1)$ such that for every $n \in \mathbb{N}$, v_n is a classical solution (see 4.139) of the Cauchy problem

$$\lambda w(x) - (\mathcal{A}_1 w)(x) - F_0(x, D^G w(x)) = h_n(x) \tag{4.329}$$

and moreover, as $n \to +\infty$, (using the notation of Definition B.56)

$$\begin{cases} \mathcal{K}-\lim_{n \to +\infty} h_n = 0 & \text{in } B_m(H), \\ \mathcal{K}-\lim_{n \to +\infty} v_n = v & \text{in } B_m(H), \\ \mathcal{K}-\lim_{n \to +\infty} D^G v_n = D^G v & \text{in } B_m(H, U). \end{cases} \tag{4.330}$$

We first prove that (4.328) holds for v_n. We apply Dynkin's formula of Proposition 1.168[53] to the process $e^{-\lambda s} v_n(s, X(s))$, $s \in [t, T]$, where $X(s) = X(s; x, a(\cdot))$, obtaining for all $T > 0$,

$$e^{-\lambda T} \mathbb{E}[v_n(X(T))] - v_n(x)$$
$$= \mathbb{E} \int_0^T e^{-\lambda s} \left[-\lambda v_n(X(s)) + \frac{1}{2} \text{Tr} \left[\Sigma(X(s)) D^2 v_n(X(s)) \right] + \langle X(s), A^* D v_n(X(s)) \rangle \right] ds$$
$$+ \mathbb{E} \int_0^T e^{-\lambda s} \left[\langle (rI - A)^{-1} b(X(s)), (rI - A^*) D v_n(X(s)) \rangle \right.$$
$$\left. + \langle (rI - A)^{-1} G(X(s)) R(X(s), a(s)), (rI - A^*) D v_n(X(s)) \rangle \right] ds.$$

Since v_n is a classical solution of (4.329) and since $D v_n$ and $D^G v_n$ are both well defined, we have, for $s \geq t$ (see Notation 4.3),

$$\langle (rI - A)^{-1} G(X(s)) R(X(s), a(s)), (rI - A^*) D v_n(X(s)) \rangle$$
$$= \langle R(X(s), a(s)), [(rI - A)^{-1} G(X(s))]^* (rI - A^*) D v_n(X(s)) \rangle$$
$$= \langle R(X(s), a(s)), D^G v_n(X(s)) \rangle,$$

where in the last equality we used that $[(rI - A)^{-1} G(X(s))]^* = G(X(s))^* [(rI - A)^{-1}]^*$ (see e.g. [521], Theorem 13.2). We thus get

---

[53] This is similar to the proof of Theorem 2.42 but using a different Dynkin's formula since here we are dealing with the case when $Q = I$ and σ may not belong to $\mathcal{L}_2(\Xi, H)$.

$$e^{-\lambda T}\mathbb{E}[v_n(X(T))] - v_n(x) = \mathbb{E}\int_0^T e^{-\lambda s}\left[-F_0\left(X(s), D^G v_n(X(s))\right) - h_n(X(s))\right]ds$$

$$+ \mathbb{E}\int_t^T e^{-\lambda s}\left\langle R(X(s), a(s)), D^G v_n(X(s))\right\rangle ds.$$

Hence, adding and subtracting $\mathbb{E}\int_0^T e^{-\lambda s}l(X(s), a(s))ds$ and rearranging the terms, we obtain

$$v_n(x) = J^\mu(x; a(\cdot)) + e^{-\lambda T}\mathbb{E}[v_n(X(T))] - \mathbb{E}\int_T^{+\infty} e^{-\lambda s}l(X(s), a(s))ds$$

$$+ \mathbb{E}\int_0^T e^{-\lambda s}h_n(X(s))ds$$

$$- \mathbb{E}\int_0^T e^{-\lambda s}\left[F_{0,CV}\left(X(s), D^G v_n(X(s)), a(s)\right) - F_0\left(X(s), D^G v_n(X(s))\right)\right]ds.$$
$$\tag{4.331}$$

Since $\lambda > \lambda_1(m)$ from Proposition 4.208, all terms in (4.331) are well-defined and we can pass to the limit as $T \to +\infty$, getting

$$v_n(x) = J^\mu(x; a(\cdot)) + \mathbb{E}\int_0^{+\infty} e^{-\lambda s}h_n(X(s))ds$$

$$- \mathbb{E}\int_0^{+\infty} e^{-\lambda s}\left[F_{0,CV}\left(X(s), D^G v_n(X(s)), a(s)\right) - F_0\left(X(s), D^G v_n(X(s))\right)\right]ds.$$
$$\tag{4.332}$$

Finally, we let $n \to +\infty$ in (4.332) using the dominated convergence theorem and (4.330) to obtain the claim.

The proof when the assumptions of Corollary 4.213-(ii) hold is similar. $\square$

The verification theorem in this case is proved as in the finite horizon case, so we omit the proof.

Theorem 4.216 (Verification Theorem, Sufficient Condition) *Let Hypothesis 4.206 hold and let $m \geq 0$ be fixed. Let l satisfy, for this m, Hypothesis 4.214. Let the other assumptions of Corollary 4.211-(iii) be satisfied for the same m, namely Hypotheses 4.210 and 4.141 for $\mathcal{K}$-convergence (for the same U and G as in Hypothesis 4.206). Let $\lambda > \lambda_0 \vee a_1 \vee \lambda_1(m)$, where λ_0 is from Theorem 4.112, a_1 is from (4.163), and $\lambda_1(m)$ is from Proposition 4.208.*

Let $v \in C_m^{1,G}(H)$ be the mild and strong solution of (4.320). Then:

(i) For every generalized reference probability space μ we have

$$v(x) \leq \overline{V}(x) \leq V^\mu(x) \quad \text{for all } x \in H. \tag{4.333}$$

(ii) Let $x \in H$. For a fixed generalized reference probability space $\hat{\mu}$, let $a^*(\cdot) \in \mathcal{U}_0^{\hat{\mu}}$ be such that, with $X^*(s) = X(s; x, a^*(\cdot))$, $s \geq 0$, we have

$$a^*(s) \in \arg\min_{a \in \Lambda} F_{0,CV}(X^*(s), D^G v(s, X^*(s)), a), \qquad (4.334)$$

for almost every $s \in [0, +\infty)$ and $\mathbb{P}$-almost surely. Then the pair $(X^*(\cdot), a^*(\cdot))$ is $\hat{\mu}$-optimal at x (and thus it is also optimal for the weak formulation) and $v(x) = \overline{V}(x) = V^{\hat{\mu}}(x)$.

The same statements hold if we take the following assumptions, for a given $m \geq 0$:

- *Hypothesis 4.206 holds, l satisfies Hypothesis 4.214 for this m, $\lambda > \lambda_0 \vee a_1 \vee \lambda_1(m)$ as above and $a^*(\cdot) \in \mathcal{U}_0^{\mu}$ in (ii).*
- *The other assumptions of Corollary 4.213, including those from part (ii), are satisfied for this m.*

4.8.2.5 Optimal Feedbacks

Define, as in (2.61), the multivalued function

$$\begin{cases} \Phi: H \to \mathcal{P}(\Lambda) \\ \Phi: x \mapsto \arg\min_{a \in \Lambda} F_{0,CV}(x, D^G v(t, x), a). \end{cases} \qquad (4.335)$$

The Closed Loop Equation associated with our problem and to the mild solution v of the HJB equation (4.320) is

$$\begin{cases} dX(s) \in AX(s)ds + G(X(s))R(X(s), \Phi(X(s)))ds + \sigma(X(s))dW(s) \\ X(0) = x. \end{cases}$$
$$(4.336)$$

Similarly to Corollary 2.44 we have here the following result, whose proof is omitted as it is completely similar to that of Corollary 4.198.

Corollary 4.217 *Let the assumptions of Theorem 4.216 hold for a given $m \geq 0$. Let $v \in C_m^{1,G}(H)$ be the mild and strong solution of (4.320) and let $x \in H$ be fixed. Assume, moreover, that the feedback map Φ defined in (4.335) admits a measurable selection $\phi : H \to \Lambda$ such that the Closed Loop Equation*

$$\begin{cases} dX(s) = [AX(s) + G(X(s))R(X(s), \phi(X(s)))]\,ds + \sigma(X(s))dW(s) \\ X(t) = x \end{cases} \qquad (4.337)$$

has a weak mild solution (see Definition 1.121) $X_\phi(\cdot; x)$ in some generalized reference probability space $\overline{\mu}$. Define, for $s \geq 0$, $a_\phi(s) = \phi(X_\phi(s; x))$. Then the pair

$(a_\phi(\cdot), X_\phi(\cdot; x))$,[54] *is $\overline{\mu}$-strongly optimal at x and $v(x) = \overline{V}(x) = V^{\overline{\mu}}(x)$. If, finally, $\Phi(x)$ is always a singleton and the weak mild solution of (4.337) is unique in $\overline{\mu}$, then the optimal control is unique in $\overline{\mu}$.*

Providing conditions on the data that allow us to apply the above corollary is a difficult problem which has not yet been solved in many interesting cases.

Similarly to the parabolic case, we discuss two results which can be used for some of the examples of Sect. 2.6. These results are far from optimal and leave a lot of room for improvement.

We start with a result, completely analogous to Proposition 4.199 (hence we omit the proof, which is similar), where $G = \sigma$, similarly to the setting used in Chap. 6.

Proposition 4.218 *Let the same assumptions of Theorem 4.216 hold for a given $m \geq 0$. Let $v \in C_m^{1,G}(H)$ be the mild and strong solution of (4.320) and let $x \in H$ be fixed. Furthermore, assume that $G = \sigma$ and the feedback map Φ defined in (4.335) admits a measurable selection $\phi : H \to \Lambda$.*

Then Eq. (4.337) admits a weak mild solution $X_\phi(\cdot; x)$ in some generalized reference probability space $\overline{\mu}$. Moreover, defining for $s \in [0, +\infty)$, $a_\phi(s) = \phi(X_\phi(s; x))$, the pair $(a_\phi(\cdot), X_\phi(\cdot; x))$ is $\overline{\mu}$-strongly optimal at x and $v(x) = \overline{V}(x) = V^{\overline{\mu}}(x)$.

We now consider a case which covers some infinite horizon problems like the ones in Sects. 2.6.1, 2.6.2, 2.6.8. It is a special case of Corollary 4.213, where the Hamiltonian is as in Proposition 4.183. In contrast to the finite horizon case, we do not consider the case when the control set may be unbounded. The assumptions below are similar to Hypothesis 4.200.

Hypothesis 4.219 *Let $U = H$ and let Hypothesis 4.212 hold. Moreover, assume the following.*

(i) Λ is a closed, convex and bounded subset of a real separable Hilbert space E. Moreover, $R(t, x, a) \equiv Ba$ for given $B \in \mathcal{L}(E, H)$ and $l(x, a) = l_0(x) + l_1(a)$ with $l_0 \in C_m(H)$ for some $m \geq 0$, and $l_1 : \Lambda \to \mathbb{R}$ is continuous, strictly convex and with always non-empty subdifferential.

(ii) Defining

$$F_{2,CV}(z, a) := \langle a, z \rangle_E + l_1(a) \quad and \quad F_2(z) := \inf_{a \in \Lambda}\{\langle a, z \rangle_E + l_1(a)\},$$

the unique minimum point $\bar{a}(z)$ of the map

$$\Lambda \to \mathbb{R}, \qquad a \to F_{2,CV}(z, a),$$

is a Lipschitz continuous function of $z \in E$.

---

[54]Notice that here the pair $(a_\phi(\cdot), X_\phi(\cdot; x))$ is always admissible. In other cases, when the set of admissible controls allows for unbounded controls and/or state constraints, the admissibility of the pair would have to be checked.

The theorem below is similar to Theorem 4.201.

Theorem 4.220 *Let Hypothesis 4.219 hold. Let* $\lambda > \lambda_0 \vee a_1 \vee \lambda_1(m)$ *be from Lemma 4.215. Let* $v \in C_m^{1,G}(H)$ *be the mild and strong solution of* (4.320) *and assume that, for some* $C > 0$,

$$|D^G v(x_1) - D^G v(x_2)| \leq C|x_1 - x_2| \quad \text{for all } x_1, x_2 \in H. \tag{4.338}$$

Then, for every generalized reference probability space μ, *we have the following.*

(i) For all $x \in H$, $v(x) = V^\mu(x)$. *Hence also* $v(x) = \overline{V}(x)$.
(ii) For every $x \in H$ *there exists a unique* μ-*optimal control* $a^*(\cdot) \in \mathcal{U}_0^\mu$ *which is related to the corresponding optimal state* $X^*(\cdot)$ *by the feedback formula*

$$a^*(s) = \arg\min_{a \in \Lambda} F_{2,CV}(B^* D^G v(X^*(s)), a) = DF_2\left(B^* D^G v(X^*(s))\right),$$

for almost every $s \geq 0$ *and* $\mathbb{P}$-*almost surely.*

The same observations as those of Remark 4.202 apply here. Differently from the parabolic case, we do not consider here the case when the Hamiltonian is locally Lipschitz continuous (see Theorem 4.204) since results about HJB equations in this case are not available yet, except for a special case discussed in Sect. 4.9.2.

Remark 4.221 In many infinite horizon optimal control problems (e.g. the special case treated in Sect. 4.10.2) it is natural to require that the function R be unbounded and Λ be an unbounded subset of a real separable Hilbert space E. Moreover, in these cases, differently from what usually happens for finite horizon problems (see Remark 4.194), it is common that the optimal controls are unbounded on $[0, +\infty)$ (even if they are, usually, locally bounded, see again the case of Sect. 4.10.2). Hence in these cases a natural choice for the set of admissible controls may be

$$\widehat{\mathcal{U}}_0^{p,\mu} := \left\{a(\cdot) \in \mathcal{U}_0^\mu : a(\cdot) \in M_\mu^p(0, T; E) \; \forall T > 0\right\},$$

where $p \geq 1$ is chosen depending on the specific problem. The results proved in Sects. 4.8.2.4 and 4.8.2.5 for the bounded case may be extended, using similar ideas, to cases with unboundedness once suitable growth and integrability conditions are satisfied. Notice that usually in such cases the Hamiltonian F_0 is only locally Lipschitz continuous in the last variable (see e.g. Remark 4.185). Hence, to treat cases like these satisfactorily, one should extend the results proved in this chapter. Up to now the only result in this direction can be found in [107] which is presented in Sect. 4.9.2.2. ∎

4.8.3 Examples

In this subsection we present some examples of optimal control problems where the theory of this chapter can be applied, possibly at different stages (e.g. existence/

uniqueness of mild/strong solutions of the associated HJB equation, verification
theorem, existence/uniqueness of optimal feedback controls). The construction of
optimal feedback controls is clearly the ultimate goal, but this is possible only in a
few cases and often requires an ad hoc study of the specific case. We discuss examples
with diagonal operators and problems with invertible diffusion coefficients. The
theory used in this chapter is still developing so the reader should be aware that the
examples presented here are only a sample of what could be done in such a framework.
Other examples can be found in the recent literature, such as, for example, control of
the stochastic wave equation [432], Sect. 6.1, control of stochastic delay equations
[316], and boundary control of Dirichlet type [315]. We only present finite horizon
examples, but analogous infinite horizon cases can be treated similarly.

4.8.3.1 Diagonal Cases

We consider here the case when the underlying (two-parameter) transition semigroup
$P_{t,s}$ is the (one-parameter) Ornstein–Uhlenbeck semigroup R_t and the operators A,
σ and G are diagonal with respect to the same orthonormal basis, as in Sect. 4.3.1.5.

Example 4.222 We consider the setting of Example 4.46. We start with a problem
with distributed controls as in Sect. 2.6.1. The state space is $H := L^2((0, \pi)^d)$ for
$d \in \mathbb{N}$ and the noise space is $\Xi = H$. The control space is the closed ball $\Lambda_M :=
\{a \in L^2((0, \pi)^d) : |a|_{L^2} \le M\}$ for some $M > 0$. Fixing the initial time $t \in [0, T]$
and a generalized reference probability space μ, the control strategies belong to $\mathcal{U}_t^\mu$.
   The state equation is basically the same as (2.84), i.e.

$$\begin{cases} dX(s) = [AX(s) + b(X(s)) + a(s)]ds + \sigma dW(s), & s \in (t, T], \\ X(t) = x \in H. \end{cases} \quad (4.339)$$

Differently from (2.84), here W is a cylindrical Wiener process in H, hence $Q = I$
and the role of Q in (2.84) is played by $\Sigma := \sigma\sigma^*$. The operator A is the Laplace
operator with Dirichlet or Neumann boundary conditions (i.e. A_D or A_N). Let $\{e_k\}_{k\in\mathbb{N}}$
be an orthonormal basis of H such that $Ae_k = \alpha_k e_k, k = 1, 2, \ldots$. We assume that
$\Sigma \in \mathcal{L}^+(H)$ is diagonal with respect to $\{e_k\}$ and $b : H \to H$ is a Nemytskii type
operator as in Sect. 2.6.1.2. We consider the problem of minimizing, as in (2.85), the
functional

$$J_2(t, x; a(\cdot)) := \mathbb{E}\left[\int_t^T l(X(s), a(s))ds + g(x(T))\right] \quad (4.340)$$

over all controls $a(\cdot) \in \mathcal{U}_t^\mu$. Here l and g are the same functions as those in
Sect. 2.6.1.2.
   In this case $G = I$ and, as in point (i) in Example 4.46, we choose, for a suitable
$\bar\beta \in [0, 1)$, $\Sigma e_k = \alpha_k^{-\bar\beta} e_k$. As was explained in Example 4.46, Hypothesis 4.145
holds if we choose $\bar\beta$ such that $d < 2(1 + \bar\beta)$. Moreover, for such values of $\bar\beta$ (4.47)
also holds, see Corollary 4.45-(i). Hence we can take $\bar\beta = 0$ when $d = 1$, $\bar\beta \in (0, 1)$

when $d = 2$, $\bar{\beta} \in (1/2, 1)$ for $d = 3$. When $d = 4$ there is no value of $\bar{\beta}$ for which Hypothesis 4.145 can be satisfied. Note also that in these cases we have $\|\Gamma_G(t)\| \leq Ct^{1/2+\bar{\beta}/2}$, hence in the following we will choose $\gamma_G(t) = Ct^{1/2+\bar{\beta}/2}$. ∎

Regarding b, l and g we assume the following.

Hypothesis 4.223

(a) For all $x \in H$ and $\xi \in (0, \pi)^d$, $b(x(\xi)) = f(x(\xi))$, where $f : \mathbb{R} \to \mathbb{R}$ is Lipschitz continuous and bounded.
(b) For all $x \in H$ and $a \in \Lambda$,

$$l(x, a) = l_0(x) + l_1(a) = \int_{(0,\pi)^d} \beta_0(x(\xi))d\xi + \int_{(0,\pi)^d} \beta_1(a(\xi))d\xi,$$

where $\beta_0 \in C_2(\mathbb{R})$ and $\beta_1(r_1) := r_1^2/2$.
(c) For all $x \in H$, $g(x) = \int_{(0,\pi)^d} \gamma_0(x(\xi))d\xi$, where $\gamma_0 \in C_2(\mathbb{R})$.

Under all these assumptions the hypotheses of Corollary 4.191-(ii) (mild solution of the associated HJB equation), Theorem 4.197 (verification theorem) and Corollary 4.198[55] (optimal feedback when the closed loop equation has a solution) are satisfied with $m = 2$.

To perform optimal synthesis, we need to show that the corresponding closed loop equation (4.311) has a solution, at least, as required in Corollary 4.198, in a weak mild sense. To see what the map Φ in (4.310) is here, we look at the current value Hamiltonian which is (see (4.293))

$$\tilde{F}_{0,CV}(t, x, q, a) = \langle b(x) + a, q \rangle + l(x, a) = \langle b(x), q \rangle + l_0(x) + \langle a, q \rangle + l_1(a).$$

In this case there is always a unique minimum point of the function $a \to \tilde{F}_{0,CV}$ (t, x, q, a) (see Remark 4.185 with $\theta = 2$ there). Hence the map Φ is single-valued and, denoting by $\phi : H \to \Lambda$ the corresponding (selection) function, we have $\phi(q) = -q$ for $|q| \leq M$ and $\phi(q) = -Mq/|q|$ for $|q| > M$. The map ϕ is Lipschitz continuous. The closed loop equation is then (4.339) with $a(s), s \in [t, T]$, substituted by $\phi(Dv(s, X(s)))$.

Now, since we are in a case of a bounded control set and a globally Lipschitz Hamiltonian, we see if and when we can apply Proposition 4.199 or Theorem 4.201.

(i) To apply Proposition 4.199, since here $\sigma = G = I$, we need to take $d = 1$. In this case the assumptions of Proposition 4.199 are all satisfied (note, in particular, that Hypothesis 4.177 applies here since (4.47) holds). Hence in this case we can find optimal feedbacks as in the claim of Proposition 4.199.

(ii) Theorem 4.201 can also be applied in the case $d = 2$ or 3. However, apart from the other assumptions, which are all easily verified, we need the Lipschitz continuity of $D^G v$, i.e. Dv, in (4.313) for the mild solution v of the associated

---

[55] See the footnote there concerning the fact that the assumptions are satisfied.

HJB equation. One way to prove this is to use the regularity result of Theorem 4.155. However, in this case, to guarantee that Hypothesis 4.151 is satisfied we would need to add much stronger assumptions on the data, namely that $f = const.$,[56] $\beta_0 \in C_b^1(\mathbb{R})$ and $\gamma_0 \in C_b(\mathbb{R})$.

Remark 4.224 The above point (ii) exposes the major weakness of the theory based on the use of Fréchet derivatives. Even if the function f in Hypothesis 4.223 is smooth, the Nemytskii operator $b : L^2((0, \pi)^d) \to L^2((0, \pi)^d)$ is not Fréchet differentiable unless f is an affine function. Thus we cannot treat more general and realistic reaction terms f. To do this a theory using the framework of Gâteaux differentiable functions would have to be developed more, starting from the results of Sects. 4.4.1.5 and 4.4.1.8. In some cases other approaches are possible. For instance, approximations of Nemytskii operators in a Banach space of continuous functions were used in [105–107] (see the discussion of the results there in Sect. 4.9.2). Also the boundedness of β_0, γ, etc. required in (ii) above is a limitation of the theory. However, this restriction seems easier to overcome if the regularity results can be extended to the case of mild solutions and data with polynomial growth. ∎

It is also possible to consider an unbounded control set, taking e.g. $\Lambda = H$ in the above control problem. In this case, to obtain existence of regular mild solutions of the HJB equations (Corollary 4.193) we also have to require strong assumptions about the data, similar to and even stronger than these in point (ii) above. Under such conditions one can apply Theorem 4.204. The comments of Remark 4.224 also apply here. However, in the case when the Hamiltonian is exactly $-\frac{1}{2}|\Sigma Dv(t, x)|^2$, (like in the one-dimensional case, when $\sigma = I$), one can use a change of variable to reduce the HJB equation to a linear one (see on this Sect. 4.10.1 and also [178], Chap. 13 in [179] and Sect. 6 in [307]). In this case, see Theorem 4.262, one gets the existence of a unique mild solution in $\overline{C}_{b,1/2+\bar{\beta}/2}^{0,1}([0, T] \times H)$ when Hypothesis 4.223 holds with β_0 and γ_0 bounded. We finally note that by adapting the techniques used in Sect. 4.4, it is not difficult to extend the result of Theorem 4.262 to the case of data g and l with polynomial growth, hence removing the boundedness condition for β_0 and γ_0. ∎

Example 4.225 We now look at the problem with boundary control of Neumann type as in Sect. 2.6.2. The state and the noise space are $H = L^2((0, \pi)^d)$ and the control space is $\Lambda_M := \{a \in L^2(\partial(0, \pi)^d) : |a|_{L^2} \leq M\}$, as before. The state equation is a slight variation of (2.96)–(2.97), i.e.

$$\begin{cases} dX(s) = [AX(s) + b(X(s)) + (\lambda I - A)^{1/4+\varepsilon}B_\lambda a(s)]ds + \sigma dW(s), & s \in (t, T] \\ X(t) = x \in H, \end{cases}$$

(4.341)

where b, σ and W are as in (4.339), while $\lambda > 0$, $A = A_N$, $\varepsilon \in (0, 1/4)$) and B_λ is as in (2.97). We minimize the functional $J_2(t, x; a(\cdot))$ given in (4.340) over all $a(\cdot) \in \mathcal{U}_t^\mu$, for a given generalized reference probability space μ.

---

[56]Indeed, we need b to be Fréchet differentiable and this is satisfied, for Nemytskii operators in L^2 with bounded f, only if f is constant.

In such a case $G = (-A)^{1/4+\varepsilon}$ for $\varepsilon \in (0, 1/4)$). As in point (i) at the end of Example 4.46, we choose, for some $\bar\beta \in [0, 1)$, $\Sigma e_k = \alpha_k^{-\bar\beta} e_k$, where (α_k) is the increasing sequence of eigenvalues of A. As was explained in Example 4.46, Hypothesis 4.145 holds if we choose $\bar\beta$ such that $d < 2(1 + \bar\beta)$ and $\bar\beta + 1/2 + 2\varepsilon < 1$. Hence we can take $\bar\beta = 0$ when $d = 1$ and $\bar\beta \in (0, 1/2 - 2\varepsilon)$ when $d = 2$. When $d = 3$ this is not possible since $\bar\beta$ must be greater than $1/2$ to guarantee that the operator Q_t is nuclear.

We assume that Hypothesis 4.223 holds with the following change:

- in point (b) we assume that $l(x, a) = \int_{(0,\pi)^d} \beta_0(x(\xi))d\xi + \int_{\partial(0,\pi)^d} \beta_1(a(\xi))d\xi$, where $\beta_0 \in C_2(\mathbb{R})$ and $\beta_1(r_1) = r_1^2/2$.

Similarly to the distributed control case, under all these assumptions the hypotheses of Corollary 4.191-(ii) (mild solution of the associated HJB equation), Theorem 4.197 (verification theorem) and Corollary 4.198[57] (optimal feedback when the closed loop equation has a solution) are satisfied with $m = 2$.

Also here, to perform optimal synthesis, we need to show that the corresponding closed loop equation (4.311) has a solution, at least, as required in Corollary 4.198, in a weak mild sense. Here

$$\tilde F_{0,CV}(t, x, q, a) = \langle (\lambda I - A)^{-1/4-\varepsilon}b(x), q \rangle + l_0(x) + \langle B_\lambda a, q \rangle + l_1(a).$$

Hence the map $\phi : H \to \Lambda$ is again Lipschitz continuous and is given by $\phi(q) = -B_\lambda^* q$ for $|B_\lambda^* q| \le M$ and $\phi(q) = -M B_\lambda^* q/|B_\lambda^* q|$ for $|B_\lambda^* q| > M$. The closed loop equation is given by (4.339) where we substitute $a(s)$, $s \in [t, T]$, by $\phi(D^G v(s, X(s)))$.

In contrast to the distributed control case, we cannot apply Proposition 4.199 or Theorem 4.201. We briefly explain why.

(i) Proposition 4.199 cannot be applied since we do not have $G = \sigma$. This may be possible, in principle, if the noise operator σ is unbounded and equal to G (e.g. in the case of boundary noise, see Sect. 2.6.3).

(ii) To apply Theorem 4.201, as in previous point (ii), we need the Lipschitz continuity of $D^G v$ in (4.313) for the mild solution v of the associated HJB equation. However, we do not know if v satisfies (4.313), even if we add additional assumptions about β_0, f, γ_0 as in the distributed control case, since Theorem 4.155 was only proved for the case $G = I$. Generalizations to other cases, including unbounded operators G, seem possible but they have not been studied yet.

Some results about optimal feedbacks can be proved adapting the techniques of this chapter to specific cases (see e.g. [316]) or using different approximation techniques (see e.g. [240, 316]).

---

[57] See the footnote there concerning the fact that the assumptions are satisfied.

We finally observe that the case of an unbounded control set, e.g. $\Lambda = H$, would need variants of Theorems 4.175 and 4.176, when G is unbounded. Such results are not known at the present time. ∎

Remark 4.226 Some of the results mentioned in the previous two examples about distributed control or Neumann boundary control may possibly be adapted to the interesting case (briefly presented in Sect. 2.6.2, Remark 2.46), where in the state equation we have an additional boundary condition containing a term depending on the state. Some ideas in this direction are given in the paper [315]. ∎

Example 4.227 Consider the same distributed control problem of Example 4.222, where the state space is $H = L^2((0, \pi)^d)$. The only difference is that the operator A is now the iterated Laplace operator with Dirichlet conditions at the boundary defined as

$$D(A_i) = \left\{ x \in H^{2i}((0, \pi)^d), \ x, \ \Delta x, \ ..., \ \Delta^{i-1}x = 0 \ \text{ on } \partial(0, \pi)^d \right\}$$

$$A_i x = (-1)^{i-1}(\Delta)^m x, \quad \text{for } x \in D(A_i).$$

The operator A_i (which occurs in elasticity theory when $i = 2$) generates an analytic semigroup of compact operators of negative type. Moreover, A_i satisfies Hypothesis 4.43 as in the case $i = 1$ and we have

$$\alpha_k \approx k^{\frac{2i}{d}} \quad \text{as } k \to +\infty. \tag{4.342}$$

So, if we take $\Sigma = (-A)^{-\bar{\beta}}, \bar{\beta} \geq 0$, then, arguing as in point (i) at the end of Example 4.46, and using point (i) of Corollary 4.45, Hypothesis 4.145 is satisfied provided

$$d < 2i(1 + \bar{\beta}) \quad \text{and} \quad \bar{\beta} < 1, \tag{4.343}$$

which is possible for $d < 4i$. If $d < 2i$ then we can take $\bar{\beta} = 0$ (see [89]). All the results mentioned in Example 4.222 hold in this case for values of $\bar{\beta}, d$ and i satisfying (4.343). ∎

4.8.3.2   Invertible Diffusion Coefficients

Consider the state equation (4.294) under the assumptions of Hypothesis 4.177 with $U = H, G = I$ (i.e. $\beta = 0$ there) and where Λ is a bounded subset of a real separable Hilbert space E. Assume moreover that the functions b and σ satisfy Hypotheses 4.60 and 4.64.

In this case, see Sect. 4.3.3, the transition semigroup $P_{t,s}$ associated to the SDE (4.294) without the term GR satisfies Hypotheses 4.74, 4.76 and 4.77 and, similarly, when b and σ are time-independent, Hypotheses 4.106, 4.108 and 4.110. Moreover, $P_{t,s}$ also satisfies Hypothesis 4.84 and, when b and σ are time-independent, Hypothesis 4.115.

Consider the problem of minimizing the functional

$$J(t, x; a(\cdot)) := \mathbb{E}\left[\int_t^T l(X(s), a(s))ds + g(x(T))\right] \qquad (4.344)$$

over all controls $a(\cdot) \in \mathcal{U}_t^\mu$ for a given generalized reference probability space μ. Let $g \in C_2(H), l \in C(H \times \Lambda)$ and $|l(x, a)| \le C(1 + |x|^2)$ for all $(x, a) \in H \times \Lambda$. Then, by Proposition 4.181, the Hamiltonian F_0 satisfies Hypothesis 4.72 with $m = 2$.

If the above conditions are satisfied we can thus apply Theorem 4.96-(ii)[58] to obtain the existence and uniqueness of a mild solution v in $\overline{\mathcal{G}}_{2,1/2}^{0,1}([0, T] \times H)$ of the associated HJB equation.

Let us examine when all the above assumptions are satisfied for the example of Sect. 2.6.1 (Eq. (2.79) and cost function (2.81)). Recall that in this example $H = L^2(\mathcal{O})$ for a given bounded regular domain $\mathcal{O} \subset \mathbb{R}^N$, and Λ is a closed bounded subset of H. Moreover, the assumptions about the state equation are satisfied if:

- $f \in C^1(\mathbb{R})$ and has bounded derivative;
- $Q = I$;
- the additive noise term $dW(s)(\xi)$ is substituted by $\sigma_0(y(s, \xi))dW(s)(\xi)$, where $\sigma_0 \in C^1(\mathbb{R})$ has bounded derivative, and $0 < M_1 \le \sigma_0 \le M_2$ for some positive constants M_1, M_2 (see on this e.g. [283], pp. 460–463).

Finally, the assumptions about the cost function are satisfied, for example, if

- $\beta(y, \alpha) = \beta_0(y) + \beta_1(\alpha)$ with $\beta_0 \in C_2(\mathbb{R})$ and $\beta_1 \in C_2(\mathbb{R})$;
- $\gamma \in C_2(\mathbb{R})$.

Unfortunately we cannot say more. It may be possible to prove that the mild solution is a strong solution using Theorem 4.135, under suitable regularity assumptions about the coefficients b and σ of (4.294). A result of this kind, as explained in the discussion before Theorem 4.135, could possibly be done by approximating the semigroup, a procedure used, for example, in [105] in the special case presented in Sect. 4.9.2.

4.9 Mild Solutions of HJB for Two Special Problems

In this section we collect results obtained with the approach used in the previous sections in two special cases, where the theory does not apply as it is (a common feature of infinite-dimensional problems) but needs some nontrivial adaptations. Section 4.9.1 is devoted to the presentation of results for HJB equations associated with optimal control problems driven by stochastic Burgers and Navier–Stokes equations, recalling a series of results from [155–158]. Section 4.9.2 is concerned with

---

[58]This can be done avoiding Hypothesis 4.82-(ii) as, since in this case $G = I$, the strong Feller property holds, hence Theorem 4.96-(ii) applies.

HJB equations associated to optimal control problems driven by reaction-diffusion equations. The results we present there were obtained in [103, 105–107].

In both cases we will discuss existence and uniqueness results about the solutions of the HJB equations and, when available, the characterization of optimal feedbacks for the corresponding optimal control problems in terms of the solutions of the HJB equations. We omit the proofs giving precise references for all results.

4.9.1   Control of Stochastic Burgers and Navier–Stokes Equations

In this subsection we discuss some results on regular/mild solutions of HJB equations related to optimal control of stochastic Burgers and Navier–Stokes equations. Below we give a short description of each of the papers from which these results are taken.

- In [155] the authors proved a smoothing property (similar to (4.6) but with an exponential weight) for the one-dimensional Burgers equation case. They also considered the HJB equation associated to a corresponding control problem and proved the existence of a mild solution using a contraction mapping principle as in Sect. 4.4. Due to the presence of the exponential weight in the smoothing property, the fixed point argument used in [155] can only be applied to a very special class of Hamiltonians. No specific applications to the control problem are developed.
- In [156] the authors obtained the existence of a regular (C^1 in time and C^2 in space) solution v to an HJB equation with a quadratic Hamiltonian which is associated to an optimal control problem for the one-dimensional Burgers equation with an unbounded cost functional, see Sect. 2.6.4. The proof is done first showing (as in [178], see Sect. 4.10.1) that the equation for the Hopf transform $u = e^{-v}$ is linear and then proving, through a Galerkin approximating procedure, that such a linear equation has a regular (C^1 in time and C^2 in space) solution. This strong regularity allows us to find optimal feedback controls.
- In [157] the authors considered a control problem similar to that of [155] but with an unbounded cost functional and a more general Lipschitz continuous Hamiltonian. A fixed point argument is not applicable here due to the exponential weight used in the smoothing property proved in [155]. Hence the authors find the mild solution by combining the smoothing property and a Galerkin approximating procedure.
- In [158] the approach of [157] is generalized, adapting the same ideas, to the case of the HJB equation for the optimal control problem for two-dimensional stochastic Navier–Stokes equations. The authors proved an appropriate smoothing property. They considered derivatives in a weaker sense (similarly to what is done here in Sects. 4.4 and 4.6 with G-derivatives) and used a change of variable ($u = e^{-K|x|^2} v$ for a suitable $K > 0$). Similar techniques are used in [424] to study three-dimensional controlled stochastic Navier–Stokes equations.

We also recall that many of the results described above are related to results about Kolmogorov equations for the Burgers and Navier–Stokes equations (see e.g. [162,

425, 510, 511] for the Burgers equation case, [33, 34, 255, 512, 567] for the 2-D Navier–Stokes case, and [161] for the 3-D Navier–Stokes case).

We divide the presentation into two subsections, one for the Burgers equation and the other for the two-dimensional stochastic Navier–Stokes equations.

4.9.1.1 The Case of the Stochastic Burgers Equation

We present here the results from [155–157].

The state equation.
Let μ be a generalized reference probability space $\left(\Omega, \mathscr{F}, \{\mathscr{F}_s\}_{s\in[0,T]}, \mathbb{P}, W_Q\right)$, where $Q \in \mathcal{L}_1^+(H)$ and is strictly positive, where $H := L^2(0, 1)$. The state equation is the same as the one we introduced in Sect. 2.6.4:

$$
\begin{cases}
dy(s, \xi) = \left[\dfrac{\partial^2 y(s, \xi)}{\partial \xi^2} + \dfrac{1}{2}\dfrac{\partial}{\partial \xi}y^2(s, \xi) + G\alpha(s, \cdot)(\xi) \right] ds + dW_Q(s)(\xi), \\
\qquad\qquad\qquad\qquad\qquad\qquad\qquad\qquad\qquad\quad s \in (0, T], \xi \in (0, 1), \\[2mm]
y(0, \xi) = x(\xi), \qquad \xi \in [0, 1], \\[2mm]
y(s, 0) = y(s, 1) = 0, \qquad s \in [0, T],
\end{cases}
$$

$$(4.345)$$

where $\alpha : [0, T] \times \Omega \to \Lambda$ is a control process. We assume for simplicity that Λ is a closed subset of H and $G \in \mathcal{L}(H)$. In [155, 156] $G = \sqrt{Q}$ while in [157] $G = I$, but the results also cover more general cases. We consider an optimal control problem in the strong formulation for μ.

We refer the reader to the discussion after (2.107) for a description of the physical meaning of the variables and of the terms appearing in the equation and for related references.

As described in Sect. 2.6.4, (4.345) can be rewritten as an evolution equation in the Hilbert space H. Once we have defined A and B as in (2.109) and (2.110), the state equation can be reformulated as follows:

$$
\begin{cases}
dX(s) = (AX(s) + B\left(X(s)\right) + G a(s)) ds + dW_Q(s) \\
X(0) = x,
\end{cases}
$$

$$(4.346)$$

where $X(\cdot)$ and $a(\cdot)$ are respectively the state and the control variable in the $L^2(0, 1)$ setting. We use the notation

$$
W^A(s) := \int_0^s e^{(s-r)A} dW_Q(s).
$$

Since Q is nuclear, $W^A(\cdot)$ is a continuous process (see Proposition 1.112).

Definition 4.228 Let $x \in H$ and $a(\cdot) \in M_\mu^2(0, T; H)$. We say that an H-valued process X is a solution of (4.346) if $X(\cdot) = Z(\cdot) + W^A(\cdot)$, where $Z(\cdot)$ is a mild solution of the equation

$$\begin{cases} dZ(s) = \left(AZ(s) + B\left(Z(s) + W^A(s)\right) + Ga(s)\right) ds, \\ Z(0) = x. \end{cases} \tag{4.347}$$

Theorem 4.229 *For any $x \in H$ and any $a(\cdot) \in M_\mu^2(0, T; H)$, there exists a solution, $X(\cdot) = X(\cdot; x, a(\cdot))$ of Eq. (4.346), in the sense of Definition 4.228, which is unique among those with trajectories $\mathbb{P}$-a.s. in*

$$L^\infty(0, T; H) \cap L^2(0, T; H_0^1(0, 1)).$$

Proof See [156], Sect. 5, [157], p. 148 and Theorem 14.2.4, p. 260 of [177] (see also [163]). □

The optimal control problem and the HJB equation.
We consider the following functional to minimize:

$$J(x; a(\cdot)) = \mathbb{E}\left\{\int_0^T \left(g(X(s)) + \frac{1}{2}|a(s)|_H^2\right) ds + \varphi(X(T))\right\}, \tag{4.348}$$

where $\varphi : H \to \mathbb{R}$ is continuous while g is possibly unbounded and only defined on $D((-A)^{1/2})$. The set of admissible controls is $M_\mu^2(0, T; \Lambda)$ and, as in (2.4), we denote by $V^\mu(x)$ the value function.

The HJB equation associated with the optimal control problem (4.346)–(4.348) is

$$\begin{cases} v_t(t, x) + \frac{1}{2}\text{Tr}\left[QD^2v(t, x)\right] + \langle Dv(t, x), Ax + B(x)\rangle \\ \qquad\qquad + F_1(G^*Dv(t, x)) + g(x) = 0, \\ v(T, x) = \varphi(x), \end{cases} \tag{4.349}$$

where for $p \in H$ we set

$$F_1(p) := \inf_{a \in \Lambda}\left\{\langle a, p\rangle + \frac{1}{2}|a|^2\right\}.$$

The quadratic case through a Hopf change of variable.
We start with the special case treated in [156], where $G = \sqrt{Q}$, $\Lambda = H$, $g(x) = |(-A)^{1/2}x|^2$, $\varphi(x) = \frac{1}{2}|x|^2$. In this case, the HJB equation is

$$
\begin{cases}
v_t(t, x) + \dfrac{1}{2}\mathrm{Tr}\left[QD^2v(t, x)\right] + \langle Dv(t, x), Ax + B(x)\rangle \\
\qquad\qquad\qquad - \dfrac{1}{2}\left|\sqrt{Q}Dv(t, x)\right|^2 + \left|(-A)^{1/2}x\right|^2 = 0, \\
v(T, x) = \dfrac{1}{2}|x|^2.
\end{cases}
$$

$$(4.350)$$

Definition 4.230 A continuous function $v\colon [0, T] \times H \to \mathbb{R}$ is a strict solution of (4.350) if:

(i) v is a C^2 function with respect to x.
(ii) For any $x \in D(A)$, $t \to v(t, x)$ is a C^1 function.
(iii) (4.350) holds for any $(t, x) \in D(A) \times [0, T]$.

The main result proved in [156] using the Hopf transform is the following.

Theorem 4.231 *Consider the function $w\colon [0, T] \times H \to \mathbb{R}$ defined by*

$$
w(t, x) := \mathbb{E}\left[\exp\left(-\frac{1}{2}|Y(t)|^2 - \int_0^t |(-A)^{1/2}Y(r)|^2 dr\right)\right],
$$

where Y is the unique solution (in the sense of Definition 4.228) having trajectories $\mathbb{P}$-a.s. in $L^\infty(0, T; H) \cap L^2(0, T; H_0^1(0, 1))$ of the following uncontrolled Burgers equation

$$
\begin{cases}
dY(t) = [AY(t) + B(Y(t))]\,dt + dW_Q(t), \\
Y(0) = x.
\end{cases}
$$

Then the function $v(t, x) := -\ln(w(T - t, x))$ is a strict solution of (4.350). Moreover, $v(0, x) = V^\mu(x)$.

Finally, there exists a solution $X^*(s)$, in the sense of Definition 4.228, of the closed loop equation

$$
\begin{cases}
dX^*(s) = [AX^*(s) + B(X^*(s))]\,ds - QDv(s, X^*(s)) + dW_Q(s), \\
X^*(0) = x,
\end{cases}
$$

which is unique among those with trajectories $\mathbb{P}$-a.s. in $L^\infty(0, T; H) \cap L^2(0, T; H_0^1(0, 1))$, the process

$$
a^*(s) := -Q^{1/2}Dv(s, X^*(s))
$$

belongs to $M_\mu^2(0, T; H)$ and it is a μ-optimal control for the problem (4.346)–(4.348).

Proof See Theorem 2.4 and Sect. 5 of [156]. □

The general terminal cost case through smoothing.

Fix again a generalized reference probability space $\mu = \left(\Omega, \mathscr{F}, \{ \mathscr{F}_s \}_{s \in [0,T]}, \right.$ $\left. \mathbb{P}, W_Q \right)$ and consider the optimal control problem in the strong formulation characterized by the state equation (4.346), the functional (4.348) and the control space $\Lambda = B(0, R)$ in H for some $R > 0$. Hence the set of admissible controls is

$$\mathcal{U}^{\mu, R} := \{ a \colon [0, T] \times \Omega \to B(0, R), \text{ progressively measurable} \}. \tag{4.351}$$

The HJB equation associated to the problem is (4.349) with the Hamiltonian F_1 given by

$$F_1(p) := \begin{cases} -\frac{1}{2} |p|^2 & \text{if } |p| \leq R \\ -(R|p| - R^2/2) & \text{if } |p| \geq R. \end{cases} \tag{4.352}$$

In [155] the authors prove the following smoothing property. We denote by $X(\cdot; x)$ the solution of (4.346) when $a(\cdot) \equiv 0$.

Theorem 4.232 *Assume that* $D\left((-A)^{\beta/2} \right) \subset D \left(Q^{-1/2} \right)$ *and*

$$|Q^{-1/2} x| \leq C |(-A)^{\beta/2} x|, \quad \forall x \in D \left((-A)^{\beta/2} \right) \tag{4.353}$$

for some $C > 0$ *and* $\beta \in (1/2, 1)$. *Then, for every* $s > 0$, $x \to X(s; x)$ *is* $\mathbb{P}$-*a.s. Gâteaux differentiable. Moreover, for every* $s > 0$, $h \in H$ *and* $\phi \in C_b(H)$, *the function* $x \to P_s[\phi](x) := \mathbb{E}[\phi(X(s; x))]$ *is twice Gâteaux differentiable and*

$$\langle \nabla P_s[\phi](x), h \rangle = \frac{1}{s} \mathbb{E} \left[\phi(X(s; x)) \int_0^s \langle Q^{-1/2} \nabla_x X(r; x) h, dW(r) \rangle \right].$$

Finally, for any $T > 0$ *and* $\varepsilon > 0$, *there exists a* $C_{\varepsilon, T} > 0$ *such that*

$$|\nabla P_s[\phi](x)| \leq C_{\varepsilon, T} s^{-(1+\beta)/2} \|\phi\|_0 e^{\varepsilon |x|^2}, \ s \in (0, T],$$

and

$$|\nabla^2 P_s[\phi](x)| \leq C_{\varepsilon, T} s^{-1-\beta} \|\phi\|_0 e^{\varepsilon |x|^2}, \ s \in (0, T].$$

Proof See Proposition 4.1. in [155]. $\qquad\qquad\qquad\qquad\qquad\qquad\qquad\square$

Definition 4.233 $(C_\gamma^k(H))$ Let $\gamma \in \mathbb{R}^+$. We define $C_\gamma^0(H)$ to be the space of all functions ϕ of $C(H)$ such that the quantity

$$|\phi|_{C_\gamma^0(H)} := \sup_{r > 0} e^{-\gamma r^2} \left(\sup_{|x| \leq r} |\phi(x)| \right) < +\infty.$$

For $k \in \mathbb{N}$, we denote by $C_\gamma^k(H)$ the space of all functions ϕ in $C^k(H)$ which are in $C_\gamma^0(H)$ together with all their Fréchet derivatives up to the order k.

We set

$$\tilde{\gamma} := \frac{\pi^2}{2\|Q\|_{\mathcal{L}(H)}}.$$

Theorem 4.234 *Suppose that (4.353) and the following set of hypotheses are satisfied:*

(i) $G = I.$
(ii) $\varphi \in C_\gamma^0(H)$ *for some* $\gamma < \tilde{\gamma}$ *and it is bounded from below.*
(iii) g *is bounded below and it is of the form* $g = g_1 + g_2$ *where* $g_1 \in C_\gamma^1(H)$ *for some* $\gamma < \tilde{\gamma}$ *and* $g_2 \in C^2\left(D((-A)^{1/2})\right).$ *Moreover, there exists a* $c_g \in \mathbb{R}^+$ *such that* g_2 *satisfies the following estimates for any* $x, h \in D((-A)^{1/2})$:

$$|g_2(x)| \le c_g \left(1 + |x|_{D((-A)^{1/2})}\right),$$
$$|\langle Dg_2(x), h\rangle| \le c_g \left(1 + |x|_{D(-A)^{1/2}}\right) |h|_{D((-A)^{1/2})},$$
$$|\langle D^2 g_2(x)h, h\rangle| \le c_g |h|^2_{D((-A)^{1/2})}.$$

Then there exists a mild *solution* v *of (4.349) in the following sense:* $v \in C([0, T] \times H)$ *with* $v(t, \cdot) \in C_{\tilde{\gamma}}^1(H)$ *for any* $t \in [0, T)$ *and, for every* $t \in [0, T], x \in H,$

$$v(t, x) = P_{T-t}[g](x) + \int_t^T P_{s-t}[F_1(Dv(s; \cdot)) + g](x)\, ds,$$

where P_s *is defined in Theorem 4.232.*

Proof See Theorem 2.2, p. 149 of [157]. □

Theorem 4.235 *Let the assumptions of Theorem 4.234 be satisfied and* v *be a* mild *solution of (4.349). Let* $x \in H.$ *The following closed loop equation*

$$\begin{cases} dX^*(s) = (AX^*(s) + B\,(X^*(s)) + DF_1(Dv(t, X^*(t))))\, ds + dW_Q(s) \\ X(0) = x \end{cases}$$

$$(4.354)$$

has a solution $X^*(s)$ *in the sense of Definition 4.228 which is unique among those with trajectories* $\mathbb{P}$-a.s. *in* $L^\infty(0, T; H) \cap L^2(0, T; H_0^1(0, 1)).$ *The process*

$$a^*(s) := DF_1(Dv(s, X^*(s)))$$

belongs to $\mathcal{U}^{\mu, R}$ *and it is a* μ-*optimal control at* $x.$

Proof See Theorem 2.3, p. 149 of [157]. □

4.9.1.2   Two-Dimensional Stochastic Navier–Stokes Equations

In this section we discuss some results from [158] related to optimal control of two-dimensional Navier–Stokes equations.

The infinite-dimensional problem.
The problem is described in Sect. 2.6.5. The two-dimensional stochastic Navier–Stokes (state) equations and the functional to minimize are described respectively by (2.114) and (2.116). After rewriting the problem in the infinite-dimensional framework the state equation has the form

$$
\begin{cases}
dX(s) = (AX(s) + B\,(X(s)) + a(s))\,ds + P\,dW_Q(s) \\
X(0) = x
\end{cases}
\tag{4.355}
$$

in the state space H, where A, B, P, Q, W_Q are given in Sect. 2.6.5.2. We recall (see [557], Sect. 2I.2.1, in particular p. 107 or [555], Sect. 2.2) that there exists a sequence $0 > \lambda_1 \geq \lambda_2 \geq \dots$ with $\lambda_n \xrightarrow{n\to\infty} -\infty$ of eigenvalues of A and an orthonormal basis $\{e_n\}_{n\in\mathbb{N}}$ of H composed of eigenvectors of A, i.e. elements of $D(A)$ such that $Ae_n = \lambda_n e_n$, for all n. Partly following [253] we assume the following hypothesis.

Hypothesis 4.236 $Q \in \mathcal{L}_1^+(H)$ and is strictly positive. Moreover, $Qe_n = \theta_n e_n$ for some $\theta_n > 0$, for all $n = 1, 2, \dots$.

Lemma 4.237 *If Hypothesis 4.236 is satisfied then the trajectories of the stochastic convolution* $W^A(s) = \int_0^s e^{(s-r)A} dW_Q(s)$ *are $\mathbb{P}$-a.s. in $C([0, T], H) \cap L^2(0, T; D((-A)^{1/2}))$.*

Proof The continuity of the trajectories in H follows from Proposition 1.112. To conclude, observe that (A, Q and e^{sA} commute thanks to Hypothesis 4.236)

$$
\mathbb{E} \int_0^T \left| (-A)^{1/2} \int_0^s e^{(s-r)A} dW_Q(s) \right|^2 ds = \int_0^T \sum_{n=1}^{\infty} \int_0^s \langle Ae^{2rA} Qe_n, e_n \rangle\, dr\, ds
$$

$$
= \int_0^T \sum_{n=1}^{\infty} \int_0^s \theta_n \lambda_n e^{2\lambda_n r}\, dr\, ds = \int_0^T \sum_{n=1}^{\infty} \theta_n \left(\frac{e^{2\lambda_n s} - 1}{2} \right) ds
$$

$$
= \sum_{n=1}^{\infty} \theta_n \left(\frac{e^{2\lambda_n T} - 1 - 2\lambda_n T}{4\lambda_n} \right),
$$

which is finite because $\left(\frac{e^{2\lambda_n T} - 1 - 2\lambda_n T}{4\lambda_n} \right)$ is uniformly bounded in n and Q is trace class, so that $\sum_n \theta_n < +\infty$. □

We consider the functional (2.122) with $\bar{y} = 0$ so that

$$
J(x; a(\cdot)) = \mathbb{E} \left[\int_0^T |(-A)^{-1/2} X(s)|_H^2 + \frac{1}{2} |a(s)|_H^2 ds + |X(T)|_H^2 \right].
\tag{4.356}
$$

We fix a generalized reference probability space $\mu = \left(\Omega, \mathscr{F}, \{\mathscr{F}_s\}_{s\in[0,T]}, \mathbb{P}, W_Q\right)$ and we consider the problem in the strong formulation. We take the control space $\Lambda = B(0, R)$ in H for some $R > 0$. Hence the set of admissible controls is

$$\mathcal{U}^{\mu,R} := \{a\colon [0, T] \times \Omega \to B(0, R), \text{ progressively measurable}\}. \tag{4.357}$$

Definition 4.238 Let $x \in H$ and $a(\cdot) \in \mathcal{U}^{\mu,R}$. We say that an H-valued process X is a solution of (4.355) if $X(\cdot) = Z(\cdot) + W^A(\cdot)$, where $W^A(s) = \int_0^s e^{(s-r)A}dW_Q(s)$ and $Z(\cdot)$ is a mild solution of the equation

$$\begin{cases} dZ(s) = \left(AZ(s) + B\left(Z(s) + W^A(s)\right) + a(s)\right) ds, \\ Z(0) = x. \end{cases}$$

Theorem 4.239 *Let Hypothesis 4.236 be satisfied, let $x \in H$ and $a(\cdot) \in \mathcal{U}^{\mu,R}$. Then (4.355) has a unique solution X in the sense of Definition 4.238 among those having trajectories $\mathbb{P}$-a.s. in $L^\infty(0, T; H) \cap L^2(0, T; D((-A)^{1/2}))$.*

Proof Denote by $Ł^4$ the closure of $\mathscr{V}$ defined in (2.118) in $L^4(\mathcal{O}; \mathbb{R}^2)$. Theorem 15.3.1, p. 291 of [177] guarantees that there exists a unique solution X in the sense of Definition 4.238 among those having trajectories $\mathbb{P}$-a.s. in $L^4(0, T; Ł^4)$ while Proposition 15.1.1, p. 283 of [177] shows that $L^4(0, T; Ł^4)$ contains $L^\infty(0, T; H) \cap L^2(0, T; D((-A)^{1/2}))$. Thus to prove the statement we only have to know that the trajectories of X belong $\mathbb{P}$-a.s. to $L^\infty(0, T; H) \cap L^2(0, T; D((-A)^{1/2}))$. This fact is true because $\mathbb{P}$-a.s. the trajectories of $X - W^A$ belong to $L^\infty(0, T; H) \cap L^2(0, T; D((-A)^{1/2}))$ (see the proof of Lemma 15.2.4, p. 289, in [177]) and the trajectories of W^A belong $\mathbb{P}$-a.s. to $L^\infty(0, T; H) \cap L^2(0, T; D((-A)^{1/2}))$ by Lemma 4.237. $\qquad\square$

The Hamiltonian for the problem is given by

$$F_1(p) := \inf_{a\in\Lambda}\left\{\langle a, p\rangle + \frac{1}{2}|a|^2\right\} = \begin{cases} -\frac{1}{2}|p|^2 & \text{if } |p| \leq R \\ -|p|R + \frac{1}{2}R^2 & \text{if } |p| > R, \end{cases}$$

and the HJB equation is

$$\begin{cases} v_t + \frac{1}{2}\mathrm{Tr}\left[PQP^*D^2v\right] + \langle Dv, Ax + B(x)\rangle + F_1(Dv) + l(x) = 0 \\ v(T, x) = |x|^2, \end{cases} \tag{4.358}$$

where $l(x) = |(-A)^{-1/2}x|^2$.

Following [158] we introduce the change of variables

$$u(t, x) = e^{-K|x|^2}v(t, x).$$

Then, denoting for simplicity $\bar{Q} := PQP^*$, (4.358) transforms into

$$\begin{cases} u_t + \dfrac{1}{2}\mathrm{Tr}\left[\bar{Q}D^2 u\right] + \langle Du, Ax + B(x) - 2K|x|^2 u\rangle + \tilde{F}(x, u, Du) + \tilde{l}(x) = 0 \\ u(T, x) = \tilde{g}(x), \end{cases}$$

$$(4.359)$$

where

$$\tilde{F}(x, u, Du) = 2K\langle \bar{Q}x, Du\rangle + (2K^2|\bar{Q}^{1/2}x|^2 + 2K\mathrm{Tr}(\bar{Q}))u \\ - e^{-K|x|^2} F_1(e^{K|x|^2}(Du + 2Kxu))$$

and

$$\tilde{l}(x) = e^{-K|x|^2} l(x), \qquad \tilde{g}(x) = e^{-K|x|^2}|x|^2.$$

If the hypotheses of Theorem 4.239 are satisfied, for every $g \in B_b(H)$ we can define $R_t(g)$ as follows

$$R_t(g)(x) = \mathbb{E}\left(e^{-2K\int_0^t |X(s;x)|^2 ds} g(X(t;x))\right), \qquad (4.360)$$

where $X(\cdot; x)$ is the solution of (4.355) starting at x and with $a(\cdot) \equiv 0$. We use it to introduce the mild form of (4.359):

$$u(t, x) = R_{T-t}(\tilde{g})(x) + \int_t^T R_{s-t}\left[\tilde{F}(\cdot, u(s, \cdot), Du(s, \cdot)) + \tilde{l}\right](x)\, ds. \quad (4.361)$$

The smoothing property.
For $\gamma > 0$, $k, l \in \mathbb{N}$ and $\alpha \in [0, 1]$ we introduce the following function spaces:

$$C^{0,k,l+\alpha} = \left\{\psi: H \to \mathbb{R} \,:\, \psi \text{ is } l \text{ times differentiable and } |\psi|_{0,k,l+\alpha} < +\infty\right\},$$

where

$$|\psi|_{0,k,l+\alpha} = \sup_{x\in H}(1+|x|)^{-k}|\psi(x)| + \sup_{r>0}(1+r)^{-k} \sup_{|x|\le r,|y|\le r} \frac{\|D^l\psi(x) - D^l\psi(y)\|_{\mathcal{L}(H^l)}}{|x-y|^\alpha}$$

and

$$C^{\gamma,k,l+\alpha} = \left\{\psi: D((-A)^\gamma) \to \mathbb{R} \,:\, \psi((-A)^{-\gamma}\cdot) \in C^{0,k,l+\alpha}\right\}$$

endowed with the norm

$$|\psi|_{\gamma,k,l+\alpha} = |\psi((-A)^{-\gamma}\cdot)|_{0,k,l+\alpha}.$$

Proposition 4.240 *Let Hypothesis 4.236 be satisfied. For any $\gamma_2 > \gamma_1$, $l \in \{0, 1, 2\}$, $\alpha \in [0, 1]$ such that $l + \alpha \le 2$ and $k \ge 0$, there exist $c_0(\gamma_1, \gamma_2, k)$ and $c_1(\gamma_1, \gamma_2, k)$ such that, if $K \ge c_0(\gamma_1, \gamma_2, k)$ then, for any $\varphi \in C^{\gamma,k,\alpha}$ and any $t \in [0, T]$,*

$$|R_t\varphi|_{\gamma_2,k+2l,l+\alpha} \leq c_1(\gamma_1,\gamma_2,k)t^{-\alpha(1+\gamma_2-\gamma_1)}|\varphi|_{\gamma_2,k,\alpha}.$$

Proof See Proposition 3.3 and Remark 1, p. 469 of [158].   □

Solution of the HJB equation and optimal feedback.

Theorem 4.241 *Let Hypothesis 4.236 be satisfied. Assume that $\bar{Q} = PQP^*$ is trace class and that there exist $c_{\bar{Q}}$ and $\eta \in (0, \frac{1}{2})$ such that*

$$|\bar{Q}^{-1/2}x| \leq c_{\bar{Q}}|(-A)^{\frac{1}{2}+\eta}x|, \quad \text{for all } x \in D((-A)^{\frac{1}{2}+\eta}).$$

Then, if K is big enough, for some $\gamma \in (0, \frac{1}{2})$ and $k \geq 0$, we have the following.

  (i) *If we denote by $V^\mu(x)$ the value function of the problem (4.355)–(4.356)–(4.357) and we introduce $u(0, x) := V^\mu(x)e^{K|x|^2}$ then u belongs to $C([0, T], C^{\gamma,d,2})$ and it is a mild solution of Eq. (4.359).*
  (ii) *The following closed loop equation*

$$\begin{cases} dX^*(s) = \big(AX^*(s) + B\,(X^*(s)) - DF(DV_s^\mu(X^*(s)))\big)\,ds + PdW_Q(s) \\ X^*(0) = x \end{cases}$$

  has a unique mild solution X^*. If we define, for $s \in [0, T]$, $a^*(s) = -DF$ $(DV_s^\mu(X^*(s)))$ then $a^*(\cdot)$ belongs to $\mathcal{U}^{\mu,R}$ and it is a μ-optimal control at x for the problem (4.355)–(4.356)–(4.357).

Proof Part (i) is contained in [158], p. 473 and Theorem 2.1 of the same paper. Part (ii) is proved in Theorem 2.2 of [158].   □

Remark 4.242 To prove the statements described in Theorem 4.241 the authors of [158] use a Galerkin approximation technique which improves the one used in [157]. More precisely, they first consider the orthogonal projection P_m in H onto the linear span of the first m elements of the orthonormal basis $\{e_n\}$ composed of eigenvectors of A and, for $x \in H$, introduce $B_m(x) := P_m B(P_m x)$, $Q_m := P_m \bar{Q} P_m$ and

$$\tilde{F}_m(x, u^m, Du^m) = 2K\langle Q_m x, Du^m\rangle$$
$$+ (2K^2|Q_m^{1/2}x|^2 + 2K\text{Tr}(Q_m))u - e^{-K|x|^2}F(e^{K|x|^2}(Du^m + 2Kxu^m)).$$

They define, for bounded Borel measurable functions $g\colon P_m H \to \mathbb{R}$ and $x \in P_m H$,

$$R_t^m(g)(x) = \mathbb{E}\left(e^{-2K\int_0^t |Y_m(s)|^2 ds}g(Y_m(t))\right),$$

where Y_m is the solution of

$$\begin{cases} dY_m(t) = \big(AY_m(t) + B_m(Y_m(t))dt + dW_{Q_m}(t) \\ Y_m(0) = x. \end{cases} \tag{4.362}$$

They prove that the approximating equations

$$u^m(t)(\cdot) = R^m_{T-t}(\tilde{g})(\cdot) + \int_t^T R^m_{s-t}\left[\tilde{F}_m(\cdot, u^m(s, \cdot), Du^m(s, \cdot)) + \tilde{l}\right] ds \quad (4.363)$$

have unique (mild) solutions u^m belonging to $C([0, T], C^{\gamma,d,2})$ and converging (up to a subsequence) to some u which is then identified with the required transformation of the valued function.   ∎

Remark 4.243 A Kolmogorov equation associated to three-dimensional stochastic Navier–Stokes equations has been studied in [161] where the existence of special strict and mild solutions was obtained. Results for optimal control problems driven by the controlled three-dimensional stochastic Navier–Stokes equations are obtained in [424]. The approach is similar to the one used in [158] for the two-dimensional case presented above. Indeed, the main ingredient used in [424] is again a Galerkin finite-dimensional approximation. The limit of the approximating problems can be used again to characterize an optimal feedback (see Theorem 4.7 of [424]) for the problem in the weak formulation. However, more restrictive conditions, notably on the cost functional, are needed. In particular, the choice of a running cost including a term of the form $|\mathrm{curl}\, x|^2$, (which corresponds to the term $|A^{-1/2}x|^2$ appearing in the infinite-dimensional formulation for the two-dimensional problem), is not possible under the assumptions of [424].   ∎

4.9.2 Control of Reaction-Diffusion Equations

In this section we present a special case of results on the optimal control of stochastic reaction-diffusion systems contained in [103, 105, 107] and in Chaps. 9 and 10 of [106]. We begin by briefly recalling the content of these works:

- In [103] a smoothing property of the transition semigroup associated with the studied reaction-diffusion equation is proved (see also [104]) and then, as an application, an existence and uniqueness result (in a suitable function space) for mild solutions of a family of parabolic HJB equations is obtained. A contraction mapping argument similar to that used in Sect. 4.4 is used. The optimal synthesis is not studied.

- In [105] a wider family of parabolic HJB equations associated with the control of reaction-diffusion equations is investigated, allowing also for locally Lipschitz Hamiltonians. The equations are studied using a method similar to that of Sect. 4.7. The solutions of the HJB equations are characterized as the value functions of the corresponding optimal control problems and the optimal synthesis in the case of spatial dimension 1 is obtained.

- In [107] the infinite horizon analogue of [105] is studied. Again, using the smoothing properties of the transition semigroup, an existence and uniqueness result for mild solutions of a family of elliptic HJB equations is proved. The results are obtained in the cases of Lipschitz and locally Lipschitz Hamiltonians.

While the above mentioned papers deal with problems involving more general second-order uniformly elliptic differential operators, here we limit our attention to the case of a Laplacian with zero Dirichlet boundary conditions.

4.9.2.1   The Finite Horizon Problem

We consider a bounded domain $\mathcal{O} \subset \mathbb{R}^N$ for $N \leq 3$ whose boundary $\partial\mathcal{O}$ is regular and denote by H the Hilbert space $L^2(\mathcal{O})$ and by E the Banach space $C(\overline{\mathcal{O}})$.

Let $T > 0$ and $t \in [0, T)$. We introduce the following controlled stochastic reaction-diffusion equation on $[t, T]$,

$$
\begin{cases}
\partial y(s, \xi) = [\Delta y(s, \xi) + f(\xi, y(s, \xi)) + \alpha(s, \xi)] \, ds + dW_Q(s)(\xi), \\[2mm]
y(t, \xi) = x(\xi), \quad \xi \in \overline{\mathcal{O}}, \\[2mm]
y(s, \xi) = 0, \quad \xi \in \partial\mathcal{O}.
\end{cases}
\tag{4.364}
$$

The function $f : \overline{\mathcal{O}} \times \mathbb{R} \to \mathbb{R}$ and the process α are, respectively, the reaction term and the control. Specific assumptions about f are given below (see Hypotheses 4.245 and 4.246). As always W_Q is an H-valued Q-Wiener process defined on a filtered probability space $\left(\Omega, \mathscr{F}, \{\mathscr{F}_s\}_{s \in [0,T]}, \mathbb{P}\right)$, where $Q \in \mathcal{L}^+(H)$. We denote by μ the generalized reference probability space $\left(\Omega, \mathscr{F}, \{\mathscr{F}_s\}_{s \in [0,T]}, \mathbb{P}, W_Q\right)$.

Remark 4.244 Equation (4.364) is a particular case of a system considered in Sect. 4.1.1, p. 107 of [106]. Using the notation of [106] our case is characterized by $\mathcal{A} = \Delta$ and $\mathcal{B} = I$. ∎

Hypothesis 4.245 We assume that there exist two continuous functions $g : \overline{\mathcal{O}} \times \mathbb{R} \to \mathbb{R}$ and $h : \overline{\mathcal{O}} \times \mathbb{R} \to \mathbb{R}$ such that, for any $\xi \in \overline{\mathcal{O}}$ and $\sigma \in \mathbb{R}$, we have

$$
f(\xi, \sigma) = g(\xi, \sigma) + h(\xi, \sigma).
$$

The functions g and h, together with a natural number $l \geq 2$, have the following properties:

(1) For any $\xi \in \overline{\mathcal{O}}$ the function $h(\xi, \cdot)$ belongs to $C^l(\mathbb{R})$ and has bounded derivatives up to the l-th order, uniformly with respect to $\xi \in \overline{\mathcal{O}}$. Moreover, the mapping $D^j_\sigma h : \overline{\mathcal{O}} \times \mathbb{R} \to \mathcal{L}^j(\mathbb{R})$ is continuous, for $j = 1, .., l$.

(2) For any $\xi \in \overline{\mathcal{O}}$, the function $g(\xi, \cdot)$ belongs to $C^l(\mathbb{R})$ and there exists an $m \geq 0$ such that, for any $j = 1, .., l$,

$$\sup_{\xi \in \overline{\mathcal{O}}} \sup_{t \in \mathbb{R}} \frac{|D_t^j g(\xi, t)|}{1 + |t|^{2m+1-j}} < \infty.$$

Moreover, the mapping $D_t^j g : \overline{\mathcal{O}} \times \mathbb{R} \to \mathbb{R}$ is continuous for $j = 1, .., l$.
(3) If the constant m from (2) satisfies $m \geq 1$ then there exists a $c_1 \in \mathbb{R}$ such that

$$\sup_{t \in \mathbb{R}} \sup_{\xi \in \overline{\mathcal{O}}} D_t g(\xi, t) \leq c_1.$$

Hypothesis 4.246 Suppose that Hypothesis 4.245 is satisfied. If the constant m from Hypothesis 4.245-(2) satisfies $m \geq 1$ then there exist $a, \gamma > 0$ and $c_2 \in \mathbb{R}$ such that

$$\sup_{\xi \in \overline{\mathcal{O}}} \Big(g(\xi, t+s) - g(\xi, t)\Big)s \leq -as^{2m+2} + c_2(1 + |t|^\gamma)|s|$$

for all $s, t \in \mathbb{R}$.

The infinite-dimensional formulation.
To rewrite the state equation as a stochastic evolution equation in H we introduce, following [106] Sect. 4.1, the unbounded linear operator A on H as follows:

$$\begin{cases} D(A) = H^2(\mathcal{O}) \cap H_0^1(\mathcal{O}) \\ Ax = \Delta x - x. \end{cases}$$

It generates (see e.g. [479] Theorem 3.6, p. 215) an analytic semigroup on H and

$$\left\| e^{tA} \right\| \leq e^{-t} \qquad \text{for all } t \geq 0.$$

Remark 4.247 Observe that (see again [479], Theorem 3.6, p. 215) the constant ρ introduced on p. 109 of [106] can be chosen here to be 0 so our A corresponds to the operator A defined on p. 109 of [106]. Observe also that, in the case of a Laplacian with zero Dirichlet boundary condition, the operator $\mathcal{G}$ defined on p. 174 of [106] is trivial and then the operator C defined there equals $\Delta - I$ so that its realization C is exactly the operator A defined above.   ∎

We need the following assumptions about the operators A and Q.

Hypothesis 4.248

(1) There exists an orthonormal basis $\{e_k\}$ of H composed of elements of E which diagonalizes A and such that $\sup_{k \in \mathbb{N}} |e_k|_E < \infty$. If $(-\theta_k)$ are the corresponding eigenvalues then, for any $\delta > \frac{N}{2}$, we have

$$\sum_{k \in \mathbb{N}} \theta_k^{-\delta} < +\infty. \tag{4.365}$$

(2) $Q \in \mathcal{L}^+(H)$ is diagonal with respect to the orthonormal basis $\{e_k\}$ described in point (1) so that $Qe_k = \lambda_k e_k$ for some set of eigenvalues (λ_k). We suppose that

$$\sum_{k=1}^{\infty} \frac{\lambda_k^2}{\theta_k^{1-\gamma}} < +\infty$$

for some $\gamma \in (0, 1)$.

(3) There exists an $\varepsilon < 1$ such that

$$R((-A)^{-\varepsilon/2}) \subset R(Q^{1/2}).$$

It can be shown (see Remark 6.1.1 of [106]) that (4.365) is satisfied if the domain $\mathcal{O}$ satisfies suitable regularity conditions.

We introduce the Nemytskii operator b associated to the function $(\xi, \sigma) \to f(\xi, \sigma)$. It is defined as

$$b(x)(\xi) := f(\xi, x(\xi)) + x(\xi), \quad \xi \in \mathcal{O}. \tag{4.366}$$

We assume that the control processes $\alpha = (\alpha_1, \ldots, \alpha_r)$ belong to $M_\mu^2(t, T; H)$. We define $a(t) := \alpha(t, \cdot)$.

Using the above notation, the controlled equation (4.364) can be rewritten as the following infinite-dimensional SDE

$$\begin{cases} dX(s) = (Ax(s) + b(X(s)) + a(s)) \, ds + dW_Q(s), \\ X(t) = x, \end{cases} \tag{4.367}$$

for $0 \leq t < s \leq T$.

The next theorem is an existence and uniqueness result for (4.367) which is followed by a result about the differentiability of the solutions of (4.367) with respect to the initial condition when there is no control.

Theorem 4.249 *Let Hypotheses 4.245 and 4.248 be satisfied. Then*

(i) *For any $\mathscr{F}_s$-progressively measurable $a(\cdot) \in L^2(\Omega, L^p(t, T; H))$, with $p > 4/(4 - N)$, and $x \in E$, Eq. (4.367) admits a unique mild solution $X(\cdot; t, x, a(\cdot)) \in L^2(\Omega, C_b((t, T], E)))$ such that, on $s \in [t, T]$,*

$$|X(s; t, x, a(\cdot))|_E \leq c_T \left(|x|_E + |a(\cdot)|_{L^p(t,s;H)}^{2m+1} + \sup_{r \in [t,s]} |W^A(t, r)|_E^{2m+1} \right) \quad \mathbb{P} \text{-a.s.}, \tag{4.368}$$

where $W^A(t, r) = \int_t^r e^{(r-\theta)A} dW_Q(\theta)$ (see Proposition 1.144).

(ii) *For any $x \in H$ and any $a(\cdot) \in M_\mu^2(t, T; H)$, Eq. (4.367) admits a unique generalized solution $X(\cdot, t; x, a(\cdot)) \in L^2(\Omega, C([t, T], H))$ such that, for $s \in [t, T]$,*

$$|X(s; t, x, a(\cdot))|_H \leq c_T \left(|x|_H + |a(\cdot)|_{L^2(t,s;H)}^{2m+1} + \sup_{r \in [t,s]} |W^A(t, r)|_E^{2m+1} \right) \quad \mathbb{P} \text{-a.s.}$$

in the following sense: for any sequence $(x_n) \subset E$ converging to x in H and $(a_n(\cdot)) \subset L^2(\Omega, L^2(t, T; E))$, $\mathscr{F}_s$-progressively measurable, converging to $a(\cdot)$ in $L^2(\Omega, L^2(t, T; H))$, the corresponding sequence of mild solutions $(X(\cdot; t, x_n, a_n(\cdot)))$ converges to $X(\cdot; t, x, a(\cdot))$ in $C([t, T], H)$, $\mathbb{P}$-a.s.

Proof See Theorem 9.1.2, p. 240 of [106]. The result was originally stated, in a slightly different form, in Theorem 3.2 of [105]. □

Theorem 4.250 *Let Hypotheses 4.245, 4.246 and 4.248 be satisfied. Denote by $X(\cdot; t, x)$ the generalized solution of Eq. (4.367) starting from x at time t and with control $a(\cdot) \equiv 0$. Then:*

(i) For any $p \geq 1$ the function

$$\begin{cases} E \to \mathcal{H}_p^\mu(t, T; E) \\ x \to X(\cdot; t, x) \end{cases}$$

is l times Gâteaux differentiable.

(ii) Let $x, h \in H$ and let x_n and h_n be any two sequences in E converging respectively to x and h in H. Then the sequence $(\nabla_x X(\cdot; t, x_n)h_n)_{n\in\mathbb{N}}$ converges in $C([t, T], H)$ $\mathbb{P}$-a.s. to a process which we denote by $v(\cdot; t, x, h)$.

Proof See Theorem 6.3.3, p. 194 and Proposition 7.2.1, p. 211 of [106]. □

A smoothing result.
Given $\varphi \in B_b(H)$ we introduce the transition semigroup P_t associated to the uncontrolled equation (4.367) by setting, for any $x \in H$ and $s \geq 0$,

$$P_s[\varphi](x) = \mathbb{E}[\varphi(X(s; 0, x))], \quad s \geq 0,$$

where $X(\cdot; 0, x)$ is the solution of (4.367) starting from x at time $t = 0$ if we take the control $a(\cdot) \equiv 0$.

Theorem 4.251 *Suppose that Hypotheses 4.245, 4.246 and 4.248 are satisfied. Then P_s is a (not always strongly continuous) semigroup of contractions in $C_b(H)$ and, for any $s > 0$, it maps $B_b(H)$ into $C_b^1(H)$. If ε is the constant from Hypothesis 4.248-(3), P_s satisfies the following smoothing property: there exists a $c \geq 0$ such that*

$$\|P_s[\varphi]\|_1 \leq c(1 \wedge s)^{-\frac{1+\varepsilon}{2}} \|\varphi\|_0, \quad \text{for every } \varphi \in C_b(H), s > 0.$$

Moreover, for every $\varphi \in C_b(H)$ and $x, h \in H$, the following Bismut–Elworthy type formula (see also Section 4.3.3) holds,

$$\langle \nabla P_s[\varphi](x), h \rangle = \frac{1}{s}\mathbb{E}[\varphi(X(s; 0, x)) \int_0^s \langle Q^{-1}v(\tau; 0, x, h)dW_Q(\tau)\rangle, \quad s > 0,$$

where the process $v(\cdot; 0, x, h)$ is defined in Theorem 4.250-(ii).

Proof See Theorem 7.3.1, p. 217 of [106]. Given the structure of the Bismut–Elworthy type formula, we need in particular, for $s > 0$, $v(\tau; 0, x, h) \in D(Q^{-1})$, $\mathbb{P}$-a.s., which is proved, for instance, in Proposition 6.4.1, p. 197 of [106].   □

The HJB equation and its solution.
The smoothing result of Theorem 4.251 can be used to show, via a fixed point argument similar to the one used in Sect. 4.4, the existence and uniqueness of mild solutions for a class of HJB equations associated to the optimal control of the stochastic reaction-diffusion equation (4.364).

Consider the following infinite-dimensional Cauchy problem

$$\begin{cases} \dfrac{\partial v}{\partial t} + \dfrac{1}{2}\,\mathrm{Tr}\,[QD^2v] + \langle Ax + b(x), Dv \rangle + F_0(Dv) + l_1(x) = 0, \\ v(T, x) = g(x). \end{cases} \quad (4.369)$$

Hypothesis 4.252 The Hamiltonian $F_0 : H \to \mathbb{R}$ is Fréchet differentiable and locally Lipschitz continuous, together with its derivative. Moreover, $F_0(0) = 0$.

We define $\mathcal{V}_T^1$ to be the space of all bounded and continuous functions $u : [0, T] \times H \to \mathbb{R}$ such that $u(t, \cdot) \in C_b^1(H)$ for all $t \in [0, T)$ and the mapping

$$\begin{cases} [0, T) \times H \to H, \\ (t, x) \to Du(t, x) \end{cases}$$

is bounded and measurable. It is easy to check that $\mathcal{V}_T^1$, endowed with the norm

$$\|u\|_{\mathcal{V}_T^1} = \sup_{t \in [0,T]} \|u(t, \cdot)\|_0 + \sup_{t \in [0,T)} \|Du(t, \cdot)\|_0,$$

is a Banach space. Note that the space $\mathcal{V}_T^1$, which is the space where the mild solutions exist here, is a bit different from the spaces used in Sect. 4.4 (see Remark 4.91). Indeed, we have

$$C_b^{0,1}([0, T] \times H) \subset \mathcal{V}_T^1 \subset B_b^{0,1}([0, T] \times H).$$

The mild form of (4.369) is given, as in Sect. 4.4, by

$$v(t, x) = P_{T-t}[g](x) + \int_t^T P_{s-t}\,[F_0(Dv(s, \cdot)) + l_1]\,(x)\,ds. \quad (4.370)$$

The definition of a mild solution is the same as in Definition 4.70. The following result is proved using the same techniques as those employed to prove Theorem 4.175.

Theorem 4.253 *Let Hypotheses 4.245, 4.246, 4.248, and 4.252 hold. Assume that* $g, l_1 : H \to \mathbb{R}$ *are bounded and Lipschitz continuous functions. Then (4.369) admits a unique mild solution* u *in* $\mathcal{V}_T^1$.

Proof See Theorem 9.4.2, p. 255 of [106]. A previous version of the result, with some minor differences, is given in Theorem 6.3 of [105].   □

Application to optimal control problems.
As in the previous paragraphs we work with a fixed generalized reference probability space $\mu = \left(\Omega, \mathscr{F}, \{\mathscr{F}_s\}_{s \in [0,T]}, \mathbb{P}, W_Q \right)$. Consider an optimal control problem in the μ-strong formulation characterized by the state equation (4.367) and the cost functional

$$J^\mu(x; a(\cdot)) = \mathbb{E} \int_0^T (l_1(X(s)) + l_2(a(s))) \, ds + \mathbb{E} \, g(X(T)), \qquad (4.371)$$

where $X(\cdot) = X(\cdot; 0, x, a(\cdot))$ is the unique solution of (4.367). Our set of admissible controls will be $M_\mu^2(0, T; H)$ so the value function of the problem is, as in (2.5),

$$V_0^\mu(x) = \inf \left\{ J^\mu(x; a(\cdot)) \, : \, a(\cdot) \in M_\mu^2(0, T; H) \right\}.$$

We define the Hamiltonian F_0 by

$$F_0(p) = \inf_{a \in H} \{\langle a, p \rangle + l_2(a)\}, \qquad p \in H. \qquad (4.372)$$

Hypothesis 4.254 The function $l_2 : H \to \mathbb{R}$ is such that F_0 defined by (4.372) is well defined and satisfies Hypothesis 4.252.

Proposition 4.182-(iii) provides conditions under which Hypothesis 4.254 is satisfied.

The HJB equation associated with the above optimal control problem is then given by (4.369), where F_0 is defined by (4.372). If Hypotheses 4.245, 4.248 and 4.254 are satisfied, then the existence and uniqueness of a mild solution of (4.369) is guaranteed by Theorem 4.253. The connection between the mild solution and the value function of the control problem is provided by the following result.

Theorem 4.255 *Consider the HJB equation (4.369), where* F_0 *is defined by (4.372). Let Hypotheses 4.245, 4.246, 4.248, and 4.254 hold. Assume that* $g, l_1 : H \to \mathbb{R}$ *are bounded and Lipschitz continuous. Then* $V_0^\mu(x) = u(0, x)$ *for every* $x \in H$, *where* u *is the unique mild solution of (4.369).*

Proof See Theorem 10.1.2, p. 286 of [106]. An earlier version of the result is in Theorem 7.2 of [105].   □

We conclude with an optimal synthesis result in the one-dimensional case.

Theorem 4.256 *Let the hypotheses of Theorem 4.255 hold with $N = 1$ (one-dimensional reaction-diffusion state equation) and let the constant m in Hypothesis 4.245 be strictly smaller than 2. Suppose that, for $j = 0, 1, 2$,*

$$\sup_{\xi \in \overline{\mathcal{O}}} \sup_{\sigma \in \mathbb{R}^r} \frac{D_\sigma^j f(\xi, \sigma)}{|\sigma|^{3-j}} < +\infty.$$

Then for every bounded and Lipschitz continuous functions g and l_1 and for every $x \in H$ there exists a unique μ-optimal control for the problem (4.367)–(4.371).

Moreover, if u is the unique mild solution of (4.369), the closed loop equation

$$\begin{cases} dX(s) = (Ax(s) + b(X(s)) + DF_0(Du(s, X(s)))) \, ds + dW_Q(s), \\ X(0) = x \end{cases} \quad (4.373)$$

has a unique solution $X^*(\cdot)$,

$$a^*(s) = DF_0(Du(s, X^*(s))), \quad s \in [0, T],$$

is the unique μ-optimal control at x and $X^*(\cdot)$ is the corresponding optimal state trajectory.

Proof See Theorem 10.3.1, p. 297 of [106]. An earlier version of the result can be found in Theorem 7.3 of [105].   □

4.9.2.2   The Infinite Horizon Problem

We now consider a stationary version of the HJB equation studied in the previous subsection.

The state equation starting at time 0 is

$$\begin{cases} dX(s) = (AX(s) + b(X(s)) + a(s)) \, ds + dW_Q(s), \\ X(0) = x, \end{cases} \quad (4.374)$$

for $s \geq 0$. The spaces H and E as well as A, b, Q are as in Sect. 4.9.2.1. As before, the generalized reference probability space $\mu = (\Omega, \mathscr{F}, \{\mathscr{F}_s\}_{s \geq 0}, \mathbb{P}, W_Q)$ is fixed and the optimization problem is considered in the strong formulation. The control processes $a(\cdot)$ belong to $M_\mu^2(0, \infty; H)$.

The HJB equation and its solution.
We consider a family of HJB equations

$$\lambda v - \frac{1}{2} \operatorname{Tr} \left[QD^2v \right] - \langle Ax + b(x), Dv \rangle - F_0(Dv) - l_1(x) = 0. \quad (4.375)$$

The mild form of (4.375) is

$$v(x) = \int_0^{+\infty} e^{-\lambda t} P_t \left[l_1(\cdot) - F_0(Dv(\cdot))) \right](x) \, dt, \quad x \in H \qquad (4.376)$$

and a mild solution is defined in Definition 4.102.

We state two existence and uniqueness results for the mild solutions of (4.376) for the cases of Lipschitz and locally Lipschitz Hamiltonians.

Theorem 4.257 *Let Hypotheses 4.245, 4.246, 4.248 be satisfied. Suppose that Hypothesis 4.252 is satisfied and in addition F_0 and its derivative are Lipschitz continuous. Then, for any $\lambda > 0$ and $l_1 \in C_b(H)$, there exists a unique mild solution $u \in C_b^1(H)$ of (4.376).*

Proof See Theorem 9.5.9, p. 274 of [106]. An earlier version of the result can be found in [107], Theorem 4.9. □

Theorem 4.258 *Let Hypotheses 4.245, 4.246, 4.248 and 4.252 be satisfied. Then there exists a $\lambda_0 > 0$ such that, for any $\lambda > \lambda_0$ and for any $l_1 \in C_b^1(H)$, there exists a unique mild solution in $u \in C_b^1(H)$ of (4.376).*

Proof See Theorem 9.5.13, p. 279 of [106]. An earlier version of the result can be found in [107], Theorem 4.13. □

Remark 4.259 In [106] the existence and uniqueness results recalled in Theorems 4.257 and 4.258 are stated in a certain operator domain $D(L)$ rather than in $C_b^1(H)$, where, thanks to Lemma 9.5.1 of [106], $D(L) \subset C_b^1(H)$. The operator L is defined, similarly to what is done in [101] for the Ornstein–Uhlenbeck semigroup (see Remark B.72), through the family of resolvent operators

$$R(\lambda)[\phi](x) := \int_0^{+\infty} e^{-\lambda t} P_t [\phi](x) \, dt, \quad \phi \in C_b(H), x \in H. \qquad (4.377)$$

Hence, by construction, $R(\lambda)[\phi] = (\lambda I - L)^{-1}[\phi]$. Now any solution $u \in C_b^1(H)$ of (4.376) can be written as $u = (\lambda I - L)^{-1} [l_1 - F_0(Du)]$. Since $l_1 - F_0(Du) \in C_b(H)$ it follows that $u \in D(L)$, hence uniqueness in $D(L)$ implies uniqueness in $C_b^1(H)$. ∎

Application to optimal control problems.

We fix a generalized reference probability space $\mu = (\Omega, \mathscr{F}, \{\mathscr{F}_s\}_{s \geq 0}, \mathbb{P}, W_Q)$ and we consider the problem of minimizing the following functional

$$J^\mu(x, a(\cdot)) = \mathbb{E} \int_0^{+\infty} e^{-\lambda t} \left[l_1(X(t; x, a(\cdot))) + l_2(a(t)) \right] dt, \qquad (4.378)$$

where $X(\cdot; x, a(\cdot))$ is the solution of (4.374), over all controls $a(\cdot)$ in $M_\mu^2(0, \infty; H)$. The value function of the problem is given by

$$V^\mu(x) = \inf\left\{ J^\mu(x; a(\cdot)) \ : \ a(\cdot) \in M^2_\mu(0, \infty; H) \right\}. \tag{4.379}$$

The Hamiltonian F_0 associated to this optimal control problem is defined by (4.372).

Theorem 4.260 *Let Hypotheses 4.245, 4.246 and 4.248 be satisfied. Suppose that Hypothesis 4.254 is satisfied and in addition F_0 and its derivative are Lipschitz continuous. Then, for any $\lambda > 0$ and any bounded Lipschitz continuous $l_1 : H \to \mathbb{R}$, the unique mild solution u of the HJB equation (4.376) coincides with the value function V^μ defined in (4.379).*

Proof See Theorem 10.2.3, p. 295 of [106]. An earlier version of the result is in [107], Theorem 5.3. □

Theorem 4.261 *Let Hypotheses 4.245, 4.246, 4.248 and 4.254 be satisfied. Then there exists a $\lambda_0 > 0$ such that, for any $\lambda > \lambda_0$ and for any $l_1 \in C^1_b(H)$, the unique mild solution u of the HJB equation (4.376) coincides with the value function V^μ defined in (4.379).*

Proof See Theorem 10.2.2, p. 292 of [106]. An earlier version of the result is in [107], Theorem 5.2. □

4.10 Regular Solutions Through "Explicit" Representations

In this section we collect some results about explicit representations of the solutions of HJB equations. Such representations are seldom possible, however they are always coveted and are of interest in applications.

We devote Sect. 4.10.1 to the case of HJB equations with quadratic Hamiltonians that can be turned into linear equations by a Hopf change of variable. Section 4.10.2 presents a specific control problem in a form often arising in economic applications, where explicit solutions can be found.

4.10.1 Quadratic Hamiltonians

When the Hamiltonian F in Eq. (4.1) is a special quadratic function of Dv we can reduce the HJB equation to a linear equation for which an explicit representation formula is known. This well-known technique has been used in the infinite-dimensional context, for instance, in [178, 307] and, in the case of the control of the stochastic Burgers equation, in [156] (see Sect. 4.9.1). For linear equations in Hilbert spaces we refer to [179].

We first explain the heuristic argument. Consider the following parabolic HJB equation

$$
\begin{cases}
v_t + \dfrac{1}{2} \operatorname{Tr} \left[\Sigma(t,x) D^2 v \right] + \langle Ax + b(t,x), Dv(t,x) \rangle \\
\qquad\qquad - \dfrac{1}{2} | \sqrt{\Sigma(t,x)} Dv(t,x) |^2 + l(t,x) = 0 \quad t \in [0,T), \ x \in D(A), \\[2mm]
v(T,x) = g(x), \ x \in H.
\end{cases}
$$

$$(4.380)$$

We set, formally

$$
w(t,x) := e^{-v(t,x)},
$$

so that $v(t,x) = -\log w(t,x)$. If v is smooth, by straightforward computations we have

$$
w_t = -w v_t, \qquad Dw = -w Dv, \qquad D^2 w = w Dv \otimes Dv - w D^2 v.
$$

Defining

$$
\mathcal{A}_1 w(t,x) := \dfrac{1}{2} \operatorname{Tr} \left[\Sigma(t,x) D^2 w(t,x) \right] + \langle Ax + b(t,x), Dw(t,x) \rangle ,
$$

we have

$$
\mathcal{A}_1 w(t,x) = -w(t,x) \left(\mathcal{A}_1 v(t,x) - \dfrac{1}{2} | \sqrt{\Sigma(t,x)} Dv(t,x) |^2 \right).
$$

If v is a solution of (4.380), we thus obtain

$$
\mathcal{A}_1 w(t,x) = -w(t,x)(-v_t(t,x) - l(t,x)) = -w_t(t,x) + w(t,x) l(t,x).
$$

Thus w satisfies the linear equation

$$
\begin{cases}
w_t + \dfrac{1}{2} \operatorname{Tr} \left[\Sigma(t,x) D^2 w \right] + \langle Ax + b(t,x), Dw(t,x) \rangle - w(t,x) l(t,x) = 0, \\
\qquad\qquad\qquad\qquad\qquad\qquad\qquad\qquad\qquad\qquad t \in [0,T), \ x \in H, \\[2mm]
w(T,x) = e^{-g(x)}, \ x \in H.
\end{cases}
$$

$$(4.381)$$

A good candidate for a solution of (4.381) is given by the Feynman–Kac formula

$$
w(t,x) = \mathbb{E} \left(e^{-g(Y(T;t,x)) - \int_t^T l(s,Y(s;t,x))ds} \right),
$$

$$(4.382)$$

where $Y(s,x)$ is the mild solution of the problem

$$\begin{cases} dY(s) = [AY(s) + b(s, Y(s))]\,ds + \sqrt{\Sigma(s, Y(s))}dW(s), & t \le s \le T, \\ Y(t) = x \in H, \end{cases}$$

and W is a cylindrical Wiener process in H.

In many cases it can be proved that the function w in (4.382) is a classical solution of the Kolmogorov equation (4.381). Then $v := -\ln(w)$ is a classical solution of (4.380) and it can then be used to solve the associated control problem. This was done, for example, in [156] for a special case of an optimal control problem for the one-dimensional stochastic Burgers equation (see Theorem 4.231).

Another possibility is to prove that the function w in (4.382) is a mild solution of (4.381). This was done, under certain assumptions, in [178] and [307]. We mention here two results when the operators A and Q satisfy Hypothesis 4.145 used in the Ornstein–Uhlenbeck case.

Theorem 4.262 *Assume that Hypothesis 4.145 with $U = H$ and $G = I$ holds. Suppose also that the function $\Gamma(t) := \Gamma_I(t)$ defined in Hypothesis 4.145-(v) satisfies the estimate $\|\Gamma(t)\| \le Ct^{-\theta}$ for some $C > 0$ and $\theta \in (0, 1)$.*

Let $g \in UC_b(H), l \in UC_b(H)$ and $b \in UC_b(H, H)$. Then Eq.(4.381) has a unique mild solution $w \in C^{0,1}_{b,\theta}([0, T] \times H)$.

Proof See [307], Theorem 6.1, p. 441. □

Theorem 4.263 *Let the hypotheses of Theorem 4.262 be satisfied and suppose that b is Lipschitz continuous. Then the mild solution w of (4.381) is given by (4.382) and*

$$w(t, x) \ge e^{-\|g\|_0 - t\|l\|_0} \quad \forall (t, x) \in [0, T] \times H.$$

Moreover, (4.380) has a unique mild solution $v \in C^{0,1}_{b,\theta}([0, T] \times H)$ and $v = -\log w$ so that

$$|v(t, x)| \le \|g\|_0 + t\|l\|_0 \quad \forall (t, x) \in [0, T] \times H.$$

Proof See [307], Theorems 6.1 and 6.2, p. 442. □

4.10.2 Explicit Solutions in a Homogeneous Case

In a certain number of examples, when the state equation is linear and the cost functional is homogeneous in the state and the control, the value function can be proved to be a homogeneous function of the state and its expression can be explicitly found. We present here an example arising from an optimal delayed portfolio model studied in [57].[59] In this model an agent chooses her consumption and her portfolio strategies in order to maximize a certain inter-temporal expected utility (i.e. minimize

---

[59]In the same spirit, another example of an explicit solution is given, for instance, in a model driven by a stochastic neutral differential equation in [224].

the opposite of such a quantity in the formulation of this subsection). The problem
is studied in its strong formulation.

We fix an n-dimensional Brownian motion w defined on a filtered probability
space $(\Omega, \mathscr{F}, \{\mathscr{F}_t\}_{t\geq 0}, \mathbb{P})$ and we denote by μ the generalized reference probability
space $(\Omega, \mathscr{F}, \{\mathscr{F}_t\}_{t\geq 0}, \mathbb{P}, w)$. Let $\alpha, \sigma_y \in \mathbb{R}^n$, $\alpha_y \in \mathbb{R}$, $r > 0$ and σ be an $n \times n$
matrix such that $\sigma\sigma^\top$ is positive definite. We study the dynamics of the variables z
and y (the wealth and the labor income in [57]) given by

$$
\begin{cases}
dz(t) = \left[z(t)r + \theta^\top(t)(\alpha - r\mathbf{1}) + y(t) - c(t))\right] dt + \theta^\top(t)\sigma dw(t), \\
dy(t) = \left[y(t)\alpha_y + \int_{-d}^0 \phi(s)y(t+s)ds\right] dt + y(t)\sigma_y^\top dw(t), \\
z(0) = z^0, \\
y(0) = x_0^0, \quad y(s) = x_1^0(s) \text{ for } s \in [-d, 0).
\end{cases}
\tag{4.383}
$$

In the system above, $d > 0$, $\mathbf{1} = (1, \ldots, 1)^\top$ is the unitary vector in $\mathbb{R}^n$, $c(\cdot)$ and $\theta(\cdot)$
are two progressively measurable controls, respectively real and $\mathbb{R}^n$-valued (they
represent, in [57], the consumption and the portfolio strategies, further constraints
will be introduced below), ϕ is a certain fixed element of $L^2(-d, 0)$ while $z^0, x_0^0 \in \mathbb{R}$
and $x_1^0 \in L^2(-d, 0)$ are the initial data.

Remark 4.264 Since we are not interested here in the economic implications of the
model and we use it as an example, we consider a simplified version of the problem
studied in [57]. More precisely, using their notation, we take $\delta = 0$ in the problem
investigated in [57].   ∎

The infinite-dimensional formulation.
We denote by H the Hilbert space $\mathbb{R} \times \mathbb{R} \times L^2(-d, 0)$ and by $\langle \cdot, \cdot \rangle_{L^2}$ the inner
product in $L^2(-d, 0)$. Similarly to what we have seen in Sect. 2.6.8, Theorem 3.1 of
[118] ensures that the pair $\left(y(t), y(t+s)_{|s\in[-d,0)}\right)$, where $y(\cdot)$ is the solution of the
second equation in (4.383), can be identified with the $\mathbb{R} \times L^2(-d, 0)$-valued unique
solution $X = (X_0, X_1)$ of the evolution equation

$$
\begin{cases}
dX(t) = AX(t)dt + (BX(t))^\top dw(t), \\
X_0(0) = x_0^0, \\
X_1(0)(s) = x_1^0(s) \text{ for } s \in (-d, 0),
\end{cases}
\tag{4.384}
$$

where the operators A and B are defined as follows,

$$
\begin{cases}
D(A) := \{(x_0, x_1) \in \mathbb{R} \times L^2(-d, 0) : x_1(\cdot) \in W^{1,2}(-d, 0), x_0 = x_1(0)\} \\
A : D(A) \to \mathbb{R} \times L^2(-d, 0) \\
A(x_0, x_1) := \left(\alpha_y x_0 + \langle \phi, x_1 \rangle_{L^2}, x_1'\right),
\end{cases}
$$

and

$$
\begin{cases}
B : \mathbb{R} \times L^2(-d, 0) \to \mathbb{R}^n \times L^2(-d, 0) \\
B(x_0, x_1) := (x_0\sigma_y, 0).
\end{cases}
$$

The optimization problem and the HJB equation.
The evolution of the whole system is thus described by the following system of equations:

$$
\begin{cases}
dz(t) = \left[rz(t) + \theta^\top(t)(\alpha - r\mathbf{1}) + X_0(t) - c(t) \right] dt + \theta^\top(t)\sigma dw(t), \\
dX(t) = AX(t)dt + \left(BX(t) \right)^\top dw(t), \\
z(0) = z^0, \\
X(0) = x^0,
\end{cases}
\tag{4.385}
$$

whose unique solution (see Theorem 1.127) is denoted by $\left(z(\cdot\,; z^0, x^0, c, \theta), X(\cdot\,; x^0) \right)$ or simply by $(z(\cdot), X(\cdot))$.

We fix $\gamma \in (0, 1) \cup (1, +\infty)$. Given an initial datum $(z^0, x^0) \in H$, the agent chooses the nonnegative (consumption) process $c(\cdot)$ and the (portfolio) process $\theta(\cdot)$ in order to minimize the following functional

$$
J^\mu \left(z^0, x^0; c(\cdot), \theta(\cdot) \right) := \mathbb{E}\left(\int_0^{+\infty} -e^{-\lambda t} \frac{(c(t))^{1-\gamma}}{1-\gamma} dt \right),
\tag{4.386}
$$

being constrained by the following no-borrowing-without-repayment condition[60]

$$
z(t) + g_\infty X_0(t) + \langle h_\infty, X_1(t) \rangle_{L^2} \geq 0 \quad \text{for all } t,
\tag{4.387}
$$

where (g_∞, h_∞) are defined as

$$
g_\infty := \frac{1}{\beta - \beta_\infty},
$$
$$
h_\infty(s) := \frac{1}{\beta - \beta_\infty} \int_{-d}^{s} e^{-r(s-\tau)} \phi(\tau) d\tau,
\tag{4.388}
$$

and β and β_∞ are given by

$$
\beta := r - \alpha_y + \sigma_y^\top \kappa
\tag{4.389}
$$

with

$$
\kappa := (\sigma^\top)^{-1} (\mu - r\mathbf{1})
\tag{4.390}
$$

and

$$
\beta_\infty := \int_{-d}^{0} e^{r\tau} \phi(\tau) d\tau.
\tag{4.391}
$$

Defining for $(z, x) \in H$

$$
\Gamma_\infty = \Gamma_\infty(z, x) := z + g_\infty x_0 + \langle h_\infty, x_1 \rangle_{L^2},
\tag{4.392}
$$

---

[60]This condition has a precise economic interpretation which is explained in [57], Sect. 4.

the constraint (4.387) rewrites as

$$\Gamma_\infty(z(t), X(t)) \geq 0.$$

The set of admissible controls depends on the initial state (z^0, x^0) and is

$$\mathcal{U}_0^\mu(z^0, x^0) := \left\{ (c(\cdot), \theta(\cdot)) \in M_\mu^2(0, +\infty; \mathbb{R} \times \mathbb{R}^n) \; : \; c(t) \geq 0 \quad \text{and} \right.$$

$$\left. \Gamma_\infty\left(z(t; z^0, x^0, c, \theta), X_0(t; x^0)\right) \geq 0 \quad \forall t \in [0, +\infty), \; \mathbb{P}\text{-a.s.} \right\}.$$

(4.393)

The value function for the problem is given by

$$V^\mu(z^0, x^0) := \inf_{(c(\cdot), \theta(\cdot)) \in \mathcal{U}_0^\mu(z^0, x^0)} J^\mu\left(z^0, x^0; c(\cdot), \theta(\cdot)\right).$$

(4.394)

The problem is studied under the following assumptions.

Hypothesis 4.265 $z^0, x_0^0 > 0$, and $x_1^0(s) > 0$ for every $s \in (-d, 0)$.

Hypothesis 4.266 $\phi(s) > 0$ for every $s \in (-d, 0)$, and $\beta - \beta_\infty > 0$.

Hypothesis 4.267 $\lambda - (1 - \gamma)(r + \frac{\kappa^\top \kappa}{2\gamma}) > 0$.

The HJB equation and its solution.
For this problem we can identify explicitly a classical solution of the associated
HJB equation. We are only interested in positive initial data (see Hypothesis 4.265),
however the HJB equation is studied in a larger set H_0 which is identified by the
no-borrowing-without-repayment condition (4.387) and is an open half space of the
whole Hilbert space H. More precisely

$$H_0 := \{(z, x) \in H : z + g_\infty x_0 + \langle h_\infty, x_1 \rangle_{L^2} > 0\}.$$

(4.395)

We denote by

$$p = (p_1, p_2) = (p_1, p_{20}, p_{21})$$

(4.396)

any element of H, by $S(2)$ the set of all real symmetric 2×2 matrices, and by

$$P = \begin{pmatrix} P_{11} & P_{12} \\ P_{21} & P_{22} \end{pmatrix}$$

(4.397)

a matrix in $S(2)$. Given $(z, x) \in H_0$ $p \in H$ and $P \in S(2)$, the Hamiltonian F associated to the described optimization problem can be written as follows

$$F(z, x, p, P) := F_1(z, x, p, P) + F_{min}(z, x, p, P), \qquad (4.398)$$

where

$$F_1(z, x, p, P) := rzp_1 + x_0p_1 + \langle Ax, p_2 \rangle_{\mathbb{R} \times L^2} + \frac{1}{2}\sigma_y^\top \sigma_y x_0^2 P_{22}, \qquad (4.399)$$

and

$$F_{min}(z, x, p, P) := \inf_{(c, \theta) \in [0, +\infty) \times \mathbb{R}^n} F_{cv}(z, x, p, P; c, \theta), \qquad (4.400)$$

where

$$F_{cv}(z, x, p, P; c, \theta) := \frac{-(c^{1-\gamma})}{1-\gamma} + [\theta^\top(\alpha - r\mathbf{1}) - c]p_1 + \frac{1}{2}\theta^\top \sigma\sigma^\top\theta P_{11} + \theta^\top \sigma\sigma_y x_0 P_{12}.$$

It is clear that $F_{min} > -\infty$ when $p_1 < 0$ and $P_{11} > 0$.

The HJB equation associated to the optimization problem (4.385)–(4.386) can then be expressed as

$$\lambda v = F\left(z, x, (v_z, v_{x_0}, v_{x_1}), \begin{pmatrix} v_{zz} & v_{zx_0} \\ v_{x_0z} & v_{x_0x_0} \end{pmatrix}\right), \qquad (z, x) \in H_0. \qquad (4.401)$$

Definition 4.268 A function $v : H_0 \to \mathbb{R}$ is said to be a *classical solution* of the HJB equation (4.401) in H_0 if:

- v admits, on H_0, continuous first Fréchet derivatives with respect to $(z, x) = (z, x_0, x_1)$ and continuous second Fréchet derivatives with respect to (z, x_0).
- On H_0, $v_{x_0} < 0$ and $v_{x_0x_0} > 0$.
- $(v_{x_0}, v_{x_1}) \in C(H_0, D(A^*))$.
- v satisfies the HJB equation (4.401) at every point $(z, x) \in H_0$ in the following sense:

$$\lambda v(z, x) = rzv_z(z, x) + x_0v_z(z, x) + \left\langle x, A^*\left((v_{x_0}, v_{x_1})(z, x)\right)\right\rangle\Big|_{\mathbb{R} \times L^2}$$
$$+ \frac{1}{2}\sigma_y^\top \sigma_y x_0^2 v_{x_0x_0}(z, x) + F_{min}\left(z, x, (v_z, v_{x_0}, v_{x_1})(z, x), \begin{pmatrix} v_{zz} & v_{zx_0} \\ v_{x_0z} & v_{x_0x_0} \end{pmatrix}(z, x)\right).$$

Proposition 4.269 *Let Hypotheses 4.266 and 4.267 be satisfied. Then the function*

$$\bar{v}(z, x) := -\frac{f_\infty^\gamma\left(z + g_\infty x_0 + \langle h_\infty, x_1 \rangle_{L^2}\right)^{1-\gamma}}{1-\gamma}, \qquad (4.402)$$

where $f_\infty \in \mathbb{R}$ is given by

$$f_\infty := \frac{\gamma}{\lambda - (1 - \gamma)(r + \frac{\kappa^\top \kappa}{2\gamma})} > 0 \qquad (4.403)$$

and g_∞ and h_∞ are defined in (4.388), is a classical solution of (4.401) in H_0.

Proof See Proposition 4.5 of [57]. □

Properties of the value function and solution of the optimization problem.

Definition 4.270 Fix $(z^0, x^0) \in H_0$ and let $X(\cdot) := X(\cdot; x^0)$ be the unique mild solution of $dX(t) = AX(t)dt + (BX(t))^\top dw(t)$ with the initial condition $X(0) = x^0$. We say that a Borel measurable function $(c, \theta) : H_0 \to \mathbb{R}^+ \times \mathbb{R}^n$ is an *admissible closed loop strategy* at (z^0, x^0) if

$$\begin{cases} dz(t) = \left[z(t)r + \theta^\top(z(t), X(t))(\alpha - r\mathbf{1}) + X_0(t) - c(z(t), X(t))\right]dt \\ \qquad\qquad +\theta^\top(z(t), X(t))\sigma dw(t) \quad \forall t > 0, \\ z(0) = z^0, \end{cases}$$

has a unique strong (in probabilistic sense) solution[61] $z_{c,\theta}(\cdot)$ and $\left(c\big(z_{c,\theta}(\cdot), X(\cdot)\big)\right.$, $\left.\theta\big(z_{c,\theta}(\cdot), X(\cdot)\big)\right)$ belongs to $\mathcal{U}_0^\mu(z^0, x^0)$.

Definition 4.271 Fix $(z^0, x^0) \in H_0$. We say that an admissible closed loop strategy at (z^0, x^0), $(c, \theta) : H_0 \to \mathbb{R}^+ \times \mathbb{R}^n$, is an *optimal closed loop strategy* at (z^0, x^0) if

$$V^\mu(z^0, x^0) = \mathbb{E}\left\{\int_0^{+\infty} e^{-\lambda t}\left(\frac{\big(c(z_{c,\theta}(t), X(t))\big)^{1-\gamma}}{1 - \gamma}\right)dt\right\},$$

where $z_{c,\theta}(\cdot)$ and $X(\cdot)$ are as in Definition 4.270.

Theorem 4.272 *Let Hypotheses 4.266 and 4.267 be satisfied. Then, for any (z^0, x^0) $\in H_0$ satisfying Hypothesis 4.265,*

$$\bar{v}(z^0, x^0) = V^\mu(z^0, x^0),$$

where $\bar{v}$ and V^μ are defined, respectively, in (4.402) and (4.394).

Proof See Proposition 4.12 of [57]. □

Theorem 4.273 *Fix $(z^0, x^0) \in H_0$. Let Hypotheses 4.266 and 4.267 be satisfied. Then the following pair is an optimal closed loop strategy at (z^0, x^0):*

$$\hat{c}(z, x) := f_\infty^{-1}\Gamma_\infty(z, x),$$

$$\hat{\theta}(z, x) := (\sigma\sigma^\top)^{-1}(\alpha - r\mathbf{1})\frac{\Gamma_\infty(z, x)}{\gamma} - \sigma^{-1}\sigma_y g_\infty x_0. \qquad (4.404)$$

---

[61] As a one-dimensional stochastic differential equation. See e.g. [356] Sect. 4.1.

Proof See Proposition 4.12 of [57].   □

Finally, one can explicitly find optimal control processes.

Proposition 4.274 *Fix* $(z^0, x^0) \in H_0$. *Let Hypotheses 4.266 and 4.267 be satisfied. If we denote by* $\Gamma_\infty^*(t)$ *the solution of*

$$
\begin{cases}
d\Gamma_\infty^*(t) = \Gamma_\infty^*(t)\left(r + \frac{\kappa^\top\kappa}{\gamma} - f_\infty^{-1}\right)dt + \frac{\Gamma_\infty^*(t)}{\gamma}\kappa^\top dw(t), \\
\Gamma_\infty^*(0) = \Gamma_\infty(z^0, x^0),
\end{cases}
$$

then optimal controls are given by

$$
\begin{cases}
c^*(t) := f_\infty^{-1}\Gamma_\infty^*(t), \\
\theta^*(t) := (\sigma\sigma^\top)^{-1}(\alpha - r\mathbf{1})\frac{\Gamma_\infty^*(t)}{\gamma} - \sigma^{-1}\sigma_y g_\infty x_0.
\end{cases}
$$

Proof See Theorem 5.1 of [57].   □

4.11 Bibliographical Notes

The theory of mild solutions of infinite-dimensional second-order HJB equations in spaces of continuous functions presented in this chapter exploits smoothing properties of related transition semigroups to apply the Contraction Mapping Principle to the integral form of the equations to produce mild solutions. Once this is done, using suitable approximations it is possible to solve the associated stochastic optimal control problems by finding optimal feedback controls. This method (which we call "the smoothing method" in the remainder of this section) was first introduced in the papers of Da Prato [147] and Havarneanu [340], then developed by Cannarsa and Da Prato in [89, 90] and later studied by many authors. Such a method has been used before to treat semilinear parabolic equations in finite dimension. An account of the theory in this case can be found, for example, in Chap. 7 of [416].

4.11.1 The First Papers

Before analyzing the literature in detail it must be pointed out that some existence and uniqueness results for regular solutions of similar semilinear parabolic HJB equations have been obtained in [28, 30] (see also the book [29], Chap. 5). These results are proved using a different technique. First the nonlinear term F, which is simply equal to $\frac{1}{2}|Dv|^2$, is approximated using convex regularization techniques: indeed, $\frac{1}{\varepsilon}(v - v_\varepsilon)$ (where v_ε is the inf-convolution of v, see Definition D.22) converges to $F(Dv)$ in a suitable space of functions with polynomial growth. Then the approximated equation is written in an integral form and is solved through successive approximations finding a regular solution v^ε. Finally, it is proved that v^ε converges with its derivatives to a limit v which is the solution of the HJB equation.

To obtain such results, which were only proved for parabolic HJB equations, a strong regularity of the data needs to be assumed (e.g. the initial condition function must be taken convex and twice continuously differentiable), however one has the advantage of not requiring any smoothing property of the transition semigroup associated with the corresponding linear problem.

This method is a generalization of results that were obtained for first-order equations (associated to deterministic control problems), see e.g. [27, 35] and the book [29], Chaps. 1–4. It has not been developed further for second-order HJB equations in later years[62] even though it might be useful to treat specific equations when the data are regular and convex.

Coming back to the literature directly related to the material of this chapter, the paper [147] was the first where semilinear HJB equations like (4.1) and (4.2) were solved by studying the associated linear problem and the regularizing properties of the related transition semigroup. In contrast to what has been done in subsequent papers, in [147] the properties of the transition semigroup were not fully exploited as a full understanding of such properties was achieved only later (see e.g. [173]). Moreover, the existence/uniqueness of the solution was proved in [147] using the theory of m-dissipative operators, similarly to what we do here for the infinite horizon case in Sect. 4.6.2.2. Finally, most of the proofs, including the solution of the associated control problem, were achieved passing through finite-dimensional approximations, which calls for stronger assumptions. We also mention the paper [340], where a similar HJB equation, where the term $\langle Ax, Du \rangle$ was substituted by $\langle F(x), Du \rangle$ for a bounded, nonlinear, smooth function F, was solved using a smoothing property of the semigroup associated with the operator $\phi \to \mathrm{Tr}(BD^2\phi)$.

The papers [89, 90] were the first to develop, one for parabolic equations and the other for elliptic equations, the smoothing method as it is presented in this chapter. However, they considered a very special case (the linear part is of Ornstein–Uhlenbeck type, A is diagonal, $Q = G = I$ and the Hamiltonian is the simplest possible: quadratic in a ball and linear outside of it). They also solved the associated control problems. Since then, many papers have been devoted to various generalizations of this method and to applications to various types of stochastic optimal control problems.

4.11.2 Development of the Method

The next papers in the development of the smoothing method, and its applications in construction of optimal feedback maps for the associated control problems, were [306, 317] which treated (respectively, parabolic and elliptic) HJB equations whose

---

[62]Some extensions have been done for first-order equations, see e.g. [91, 186, 187, 231, 233, 234, 304, 305].

linear parts were of a general Ornstein–Uhlenbeck type and $G = I$. In these papers the authors assumed, on A and Q, conditions (4.53) and (4.65) and exploited the smoothing property given here in Theorem 4.37 (proved in [173]).

These papers were the first to employ the concept of $\mathcal{K}$-strong solutions (introduced first, together with the notion of $\mathcal{K}$-convergence and $\mathcal{K}$-semigroup, in [101, 108]) and to use systematically such an approximation scheme to prove the verification theorem and solve the associated control problem (see Sects. B.5–B.7 for more on this concept and its use). The related concepts of π-convergence,[63] π-semigroup and π-strong solutions were introduced and studied in [492, 493, 496] and, recently, in [500], but only in the context of linear Kolmogorov equations in infinite dimension. Results about the topology associated to π-convergence (respectively, $\mathcal{K}$-convergence) were obtained in [492, 493] (respectively, [300]). See Sect. B.5 for more on this.

After these two papers many other manuscripts were published, generalizing in various forms the smoothing method to other classes of problems. In [310] a first attempt was made, indeed incomplete, to study HJB equations for boundary control problems, while [301] generalized results of [317] to cover HJB equations associated with ergodic control problems.

The Hamiltonians in all the papers listed up to here were Lipschitz continuous. The first generalizations to locally Lipschitz Hamiltonians were done in [178] (see also [179], Sect. 13.3 and Sect. 4.10.1 of this chapter), for a quadratic Hamiltonian by a Hopf-type change of variable (later exploited also to solve the optimal control of the one-dimensional stochastic Burgers equation in a special case in [156] see Sect. 4.9.1.1), and in [307] for a more general case assuming C^1 regularity of the data to perform the needed a priori estimates. The results of [307], with some adjustments and slight generalizations to cover more general Hamiltonians, are presented in Sect. 4.7.

4.11.3   Beyond the Ornstein–Uhlenbeck Semigroup

In the literature discussed up to this point, the transition semigroup P_t used to define the mild form of the equations was always the Ornstein–Uhlenbeck semigroup with $G = I$ and the key smoothing property needed to perform a fixed point argument was that of Theorem 4.37. As soon as smoothing properties of other transition semigroups were obtained, the smoothing method was applied to other HJB equations.

The HJB equations associated to optimal control of stochastic Burgers and Navier–Stokes equations, in addition to the already mentioned paper [156], were studied by the smoothing method in [155, 157, 158, 424]. The results of these papers are summarized in Sect. 4.9.1. The presence of the unbounded nonlinear operator in these equations makes it more difficult to apply the standard version of the smoothing method. Indeed, the smoothing properties of the associated transition semigroups

---

[63]This concept had already been used before, see e.g. [219] (p. 11 and pp. 495–496), but without a connection to π-semigroups and π-strong solutions.

are weaker in these cases and include exponential terms. The contraction mapping principle is hard to use and hence the mild solutions are found through Galerkin approximations and a priori estimates.

The papers [105, 107], discussed in Sect. 4.9.2, studied optimal control of a family of stochastic reaction-diffusion equations and their HJB equations. The smoothing property of the transition semigroup used there was proved by Cerrai in [103, 104] exploiting the dissipativity of the reaction term (see on this also her book [106], Chap. 9). These two papers study both the cases of Lipschitz and locally Lipschitz Hamiltonians. Moreover, to handle the nonlinear term, the state equation is first considered in a Banach space of continuous functions and then, by a density argument, in a Hilbert space of square-integrable functions.

The Ph.D. thesis of Masiero [431] and her subsequent papers [432, 433] used an abstract approach to study mild solutions, where, as in this chapter, the transition semigroup was not directly related to an explicit SDE. These works were the first to use G-derivatives in the context of HJB equations, also with non-constant G. The paper [433] (which dealt with the elliptic case) was also the first to investigate mild solutions in C_m spaces with m possibly strictly positive. However, G was bounded and equal to the diffusion coefficient σ, hence preventing the use of the results in cases where they are different or one of them is unbounded (like boundary control, see Sect. 2.6.2, boundary noise, see Sect. 2.6.3, or delay in the control, see Sect. 2.6.8). In [431–433] the results for mild solutions are also applied to a class of optimal control problems but using a method different from the method of strong solutions used in Sect. 4.5. Instead the author exploited the fact that $G = \sigma$ to apply the verification method through BSDEs (see Chap. 6, Sects. 6.5 and 6.10) which requires stronger regularity assumptions on the data but does not need the additional Hypothesis 4.133 (or 4.141) about the semigroup.

The abstract approach of Masiero was also used in other papers [434, 438, 440, 442]. In [434] the setting is generalized, without application to a control problem, to a Banach space case where the state equation is set up in a Banach space E which is continuously and densely embedded in a Hilbert space H. In [438] Eq. (4.109) with a locally Lipschitz Hamiltonian was studied. There the operators A and Σ commute, $b = 0$ and F_0 only depends on the last variable $\nabla^G u$. The a priori estimates for the gradient were obtained using the BSDE representation of Theorem 6.32. In [442] the results of [438] were generalized using the BSDE approach, relying on an existence result for BSDE given in [441]. Moreover, the smoothing method was used to show that the results obtained with the BSDE approach (which needs differentiability of the data F_0 and φ) can be extended to cases where the data are less regular (Lipschitz continuous). Finally, the paper [440] used the same ideas as [442] to solve HJB equations associated to control problems with control and noise in a subdomain. This means that the function γ_G here is usually not integrable. Again the BSDE approach was used to deal with the case of regular data and then the smoothing method was employed to extend the results to Lipschitz continuous data.

The smoothing method was further improved in [316] where, for the first time, G was not equal to Σ (on the other hand G was constant there). In this paper Theorem 4.41 was proved in the case when G is possibly unbounded and the datum ϕ

is bounded, hence $m = 0$. Then [316] focused on an application to control problems with delay in the control, which needed a generalization of Theorem 4.41 and a delicate application of the contraction mapping principle. Indeed, the condition $e^{tA}G(K) \subset Q_t^{1/2}(H)$ in (4.54), which is not satisfied in the delay in the control case, was substituted by $Pe^{tA}G(K) \subset Q_t^{1/2}(H)$, where $P \in \mathcal{L}(H)$. This condition, when P is suitably chosen, is true for a larger class of cases and still allows us to prove a smoothing property for a suitable class of data ϕ. It holds in the case with delay in the control when P is the projection on the first component of the state space and allows us to solve the control problem for a reasonable class of data. Similar ideas are also used in the recent paper [315] which focuses on boundary control problems and generalizes Theorem 4.41 in a different direction. Results of [315, 316] are not discussed in this book.

Finally, papers [189, 241] contain the approach to the HJB equations (4.109) and (4.125) developed in Sect. 4.4. They introduced G derivatives for the case when G is non-constant (fixing some delicate gaps in the previous theory) and possibly unbounded, and proved existence and uniqueness theorems in the spaces $\mathcal{G}_m$ and C_m, the most suitable for applications to optimal control (Theorems 4.85, 4.90, 4.116, 4.120). The proofs in these papers are substantially the ones provided here. Regarding applications to the Ornstein–Uhlenbeck semigroup case in these papers, Theorem 4.41 is proved there exactly in the form given here.

We also mention that smoothing results have recently been proved (see [502], Sect. 4.4) for transition semigroups of Ornstein–Uhlenbeck type associated to Levy processes. Other properties of semigroups of this kind were studied in [500]. In principle, such results allow us to apply the theory presented in this chapter, but this has not been done yet. Paper [500] also studies a Kolmogorov equation with a non-local Ornstein–Uhlenbeck operator. We refer to the bibliographical notes in Sect. 3.14 for more about non-local equations in Hilbert spaces.

Other smoothing results have been proved for the case of the perturbed Ornstein–Uhlenbeck semigroup, as explained in Sect. 4.3.2 (see [64, 271, 272]), but they have not yet been applied to study HJB equations.

There exists an extensive literature about infinite-dimensional Kolmogorov equations, i.e. linear HJB equations of the type discussed in this chapter. Such equations have been widely studied, in particular due to their connections with the solutions of the associated SDE. A full account of their theory can be found in the book of Da Prato and Zabczyk [179] and also, partly, in their book [180, Chap. 9]. More recent results for various types of solutions of such equations which have some regularity can be found, for example, in [106, 133–136, 255, 257, 493, 496, 510–512, 516, 517].

In particular, we mention papers on infinite-dimensional Kolmogorov equations in domains, which may serve as a basis for the development of a theory for mild solutions of HJB equation in domains. See on this [179] (Chap. 8) and, e.g. [166–168, 497–499, 546].

Smoothing properties of transition semigroups are also useful for other purposes, for instance to study invariant measures associated with SDEs. A good reference for the theory in infinite dimension is the book of Da Prato and Zabczyk [177]. Other references on the subject can be found in the bibliographical notes of Chap. 5.

4.11.4 Explicit Solutions of HJB Equations

Explicit solutions of HJB equations, when available, can be very useful in applications to solve the associated control problem and to derive various properties of the optimal strategies. For this reason they have been extensively studied in the finite-dimensional case. In the infinite-dimensional case the only available results for nonlinear HJB equations we are aware of are those discussed in Sect. 4.10.

4.11.5 The Results and the Proofs of This Chapter
Compared with the Literature

The results of Sect. 4.2.1 generalize and fix some gaps in the approach of [431–433]. The proofs of the main results, Propositions 4.13, 4.16 and Corollary 4.14, are mainly taken from the recent papers [189, 241], even if here the arguments are also generalized to the B_m spaces.

The definitions of the weights in (4.29), (4.30) and Proposition 4.21 in Sect. 4.2.2 are taken from [189] and are introduced to extend the theory to cover parabolic equations when the function γ_G in Hypothesis 4.76 is not necessarily of power type.

The results of Sect. 4.3 contain the key tools which are needed to apply the general abstract theory of Sects. 4.4 and 4.5 to specific classes of transition semigroups. Most of Sect. 4.3 is devoted to the case of the Ornstein–Uhlenbeck semigroup, where the smoothing is obtained through some controllability assumptions which generalize the conditions introduced and developed in [193, 195, 578] (see also [584], Chapter IV-2). Here all details are given, see in particular Theorem 4.41 (including the lemmas before it) and Propositions 4.50 and B.92, whose proofs are generalizations (to include the B_m spaces) of the proofs in [189, 241]. In Sect. 4.3 we also provide the proof (taken from [189]) of the monotonicity of the map $t \to \|\Gamma_G(t)\|$ (see Lemma 4.35) which generalizes a well known result in the case $G = I$, see e.g. [312].

The results of Sect. 4.3.2 are stated without proofs as the smoothing method for HJB equations in this case has not yet been developed. The reader can find more in [64, 271, 272, 311, 431] on the material discussed there.

The material of Sect. 4.3.3 is an example of the smoothing results that can be obtained through the Bismut–Elworthy–Li formula. We only present the case of an invertible diffusion operator for which we supplement the results of [283] (proving Proposition 4.67) in order to apply the theory of Sect. 4.4.

The existence and uniqueness results of Sect. 4.4 follow [189] in the parabolic case and [241] in the elliptic case. The difference is that here we generalize the results to the case of measurable data F_0 and φ and we also show how to obtain solutions in the UC_m spaces. The parabolic case is more complicated and the contraction mapping principle is used by employing norms with exponential terms, as it was done in [431, 432].

The results about strong solutions in Sect. 4.5 connect the mild solutions with the original HJB partial differential equations, through approximations using π- (or $\mathcal{K}$-) convergence. This idea (employing $\mathcal{K}$-convergence) was first used in the paper [306] and then exploited in many subsequent papers, see e.g. [307, 317]. However, up to now, it was applied only when the underlying transition semigroup $P_{t,s}$ is of the Ornstein–Uhlenbeck type.

Proving that a mild solution is a strong solution is also a crucial step towards a proof of the verification theorem in applications to optimal control. To accomplish this step other approaches were used in the literature in the case where $P_{t,s}$ is not of the Ornstein–Uhlenbeck type. In [431–433, 438, 440, 442] the verification theorem is proved through BSDE techniques when the data are Gâteaux differentiable and then when the data are continuous, approximating them by Gâteaux differentiable functions in the uniform norm with the help of the inf-sup convolutions (see Sect. D.3).

We also mention [105–107] (see Sect. 4.9.2), where the author develops an approximation scheme for the mild solutions of the HJB equations which is similar to the one used in Sect. 4.5. This scheme is used to obtain results for the optimal control problem, but the concept of a strong solution is not used there.

Other papers concerning strong solutions are [188, 228]. There, extending some finite-dimensional results proved in [319, 320], the authors present a general procedure to prove verification theorems for stochastic optimal control problems once a strong solution v of the associated HJB equation is known. In these papers two types of strong solutions are used. The first is exactly the one used in Sect. 4.5, Definition 4.132, while the second is similar but it does not require the convergence of the derivatives of the approximating classical solutions.

The results of Sect. 4.6 are substantially generalizations of the results of [306, 317]. Here we use the same techniques of proofs with some improvements. In particular, in Sect. 4.6.2.2, the operator $\hat{\mathcal{A}}_0$ is used in place of $\mathcal{A}_0$ (which was used in both [306] and [317], to fix problems related to the unboundedness of the term $\langle x, A^*Dv \rangle$ in the definition of $D(\mathcal{A}_0)$, see Sect. B.7.1, and Remark B.97. Similarly the material of Sect. 4.7 slightly generalizes the results of [307].

The framework of Sect. 4.8 is designed to show how the results in the "regular case" of Sect. 2.5 can be partly replicated here. The setting includes that of [306, 307, 317] and also covers some more general cases, such as the case of unbounded control operators (e.g. boundary control of Neumann type, see Example 4.225), and is partly borrowed from the recent paper [240] where also an alternative approach is introduced. Since the results for optimal control problems strongly depend on the properties of the Hamiltonian F_0, we included in Sect. 4.8.1.2 three results which are only partially available in the literature, where useful properties of F_0 are obtained from the assumptions about the data of control problems.

Sections 4.9 and 4.10 present without proofs a few special interesting cases. Precise references to the related literature are given there.

Chapter 5
Mild Solutions in L^2 Spaces

This chapter is devoted to the presentation of the L^2 theory for the existence and uniqueness of mild solutions for a class of second-order infinite-dimensional HJB equations in Hilbert spaces through a perturbation approach. As in the previous chapter, the concept of mild solution concerns the HJB equation in an integral form that uses the transition semigroup associated to the linear part of the equation.

In the previous chapter the perturbation approach was used in Banach spaces of regular (at least differentiable in the x variable, in a suitable sense) real-valued functions defined on a Hilbert space H. The space where we seek the solutions here is a space of functions which are square-integrable (with their x derivative defined in a suitable sense) with respect to a suitable reference measure m on H.

One of the main reasons for the development of the L^2 theory is the need to study HJB equations without the smoothing Hypothesis 4.76 about the behavior of the transition semigroup, which was used in the previous chapter (see Sect. 4.1 for a discussion). Indeed, once the existence of the reference measure is postulated, the estimates that allow us to ensure, in the L^2 framework, the applicability of a fixed point argument, can be proved under weaker assumptions (see Sect. 5.1 for details).

As for the mild solutions in spaces of continuous functions, the L^2 theory can be applied to obtain optimal synthesis. The class of applicable infinite-dimensional stochastic optimal control problems includes cases which cannot be treated in the context presented in Chap. 4, like the stochastic delay differential equations and first-order SPDEs. On the other hand, specific hypotheses ensuring the existence of the reference measure m and the compatibility of the Hamiltonian with it, need to be satisfied. Moreover, the synthesis provided by the L^2 theory is less regular.

The approach we describe was mostly developed in [3, 4, 125, 298]. We will mainly follow [298].

The chapter is organized as follows:

- In Sect. 5.1 we describe the main ideas of the L^2 method.

© Springer International Publishing AG 2017

G. Fabbri et al., *Stochastic Optimal Control in Infinite Dimension*,
Probability Theory and Stochastic Modelling 82,
DOI 10.1007/978-3-319-53067-3_5

- In Sect. 5.2 we recall some classical results about invariant measures and other preliminary facts.
- Sections 5.3 and 5.4 are devoted to parabolic HJB equations. Section 5.3 contains existence and uniqueness results, while in Sect. 5.4 a result on approximation of mild solutions by classical solutions is provided.
- In Sect. 5.5 we apply the results of Sects. 5.3 and 5.4 to perform the optimal synthesis for stochastic optimal control problems, while in Sect. 5.6 we provide specific examples related to those of Chap. 2.
- In Sect. 5.7 we describe complementary results, mainly from [3, 4], which cover an additional class of problems. This section also contains existence and uniqueness results for a family of elliptic HJB equations without applications to control problems.
- Section 5.8 contains bibliographical notes.

5.1 Introduction to the Methods

We briefly sketch the main ideas of the method developed in the next sections. We consider a class of second-order infinite-dimensional HJB equations of the form

$$
\begin{cases}
v_t + \dfrac{1}{2}\, \mathrm{Tr}\, \left[Q D^2 v \right] + \langle Ax + b(x), Dv \rangle + F\,(t, x, Dv) + l(t, x) = 0, \\
\qquad\qquad\qquad\qquad\qquad\qquad\qquad\qquad t \in [0, T),\ x \in D(A) \qquad (5.1) \\
v(T, x) = g(x), \qquad x \in H,
\end{cases}
$$

and

$$
\lambda v - \frac{1}{2}\, \mathrm{Tr}\, [Q D^2 v] - \langle Ax, Dv \rangle - F(x, Dv) = g, \qquad x \in H, \qquad (5.2)
$$

where $T > 0$ is fixed, A is the generator of a C_0-semigroup on a real separable Hilbert space H, $Q \in \mathcal{L}^+(H)$, and $b \colon H \to \mathbb{R}$, $l \colon [0, T] \times H \to \mathbb{R}$, $g \colon H \to \mathbb{R}$, $F \colon [0, T] \times H \times H \to \mathbb{R}$ (or $F \colon H \times H \to \mathbb{R}$) are measurable functions. Further hypotheses on b, l, g and F will be introduced later.

Since the results available in the literature up to now are mainly oriented towards the evolutionary HJB equation (5.1), we devote most of the chapter to the theory in this case, limiting the treatment of the stationary equation (5.2) to Sect. 5.7.3.

Given a reference measure on H, the basic idea is to introduce mild and strong solutions of (5.1) and (5.2) in the space of real square-integrable functions on $[0, T] \times H$ (or on H). If H were a finite-dimensional space, the Lebesgue measure would be the natural choice for the reference measure but in infinite dimension the situation is more delicate. We consider the following stochastic evolution equation

$$\begin{cases} dX(s) = (AX(s) + b(X(s)))\, ds + dW_Q(s), \quad s \geq 0, \\[2mm] X(0) = x \in H, \end{cases} \tag{5.3}$$

we suppose it admits a mild solution and an invariant measure m and we work in the space $L^2(H, \overline{\mathcal{B}}, m)$ where $\overline{\mathcal{B}}$ is the completion of the Borel σ-field $\mathcal{B}(H)$ with respect to m.

Under suitable assumptions on the operators A and Q and on the function b (see, e.g., [180] Chap. 9), the solution w of the following Kolmogorov equation

$$\begin{cases} w_t = \dfrac{1}{2}\, \mathrm{Tr}\,[QD^2 w] + \langle Ax + b(x), Dw \rangle, \\[3mm] w(0, x) = \phi(x) \end{cases} \tag{5.4}$$

can be associated to the transition semigroup P_t of the solution $X(\cdot\,; x)$ of (5.3) as follows:

$$w(t, x) = P_t[\phi](x) = \mathbb{E}\phi(X(t, x)) \tag{5.5}$$

for any bounded continuous ϕ.

The semigroup P_t extends to a strongly continuous semigroup of contractions on $L^2(H, \overline{\mathcal{B}}, m)$ with generator $\mathcal{A}$, whose explicit expression on regular functions is

$$\mathcal{A}\phi(x) = \frac{1}{2}\, \mathrm{Tr}\,[QD^2\phi] + \langle Ax + b(x), D\phi \rangle\,; \tag{5.6}$$

this fact is recalled in Lemma 5.37.

The original HJB equation (5.1) can be seen as a perturbation of (5.4) and, by formally applying the variation of parameters formula, it can be written in the following integral (mild) form

$$u(t, x) = P_{T-t}[g](\cdot) + \int_t^T P_{s-t}\,[l(s, \cdot) + F\,(s, \cdot, Du(s, \cdot))]\,(x)ds. \tag{5.7}$$

To prove the existence and uniqueness of mild solutions in spaces of continuous functions we needed, as a key assumption, a smoothing property for the transition semigroup P_t of the following form[1]: there exist $C > 0$ and $\theta \in (0, 1)$ such that for every $\varphi \in B_b(X)$, $s > t$, $x \in H$,

$$|DP_{t-s}[\varphi](x)| \leq C(1 \vee (s - t)^{-\theta})\|\varphi\|_0$$

(or a similar hypothesis which uses an operator G and an integrable function γ, see Sect. 4.1.1 for details). This assumption was needed to prove the existence and

---

[1] See Hypothesis 4.76.

uniqueness of the solution using a fixed point theorem in a Banach space of continuous and differentiable functions (see e.g. Theorem 4.80).

In the L^2 setting, an important role is played by the space $W_Q^{1,2}(H, m)$ which is, formally, the Sobolev space of functions which admit a weak derivative in $L^2(H, \overline{\mathcal{B}}, m)$, endowed with the norm

$$|\phi|_{W_Q^{1,2}}^2 = \int_H |\phi|^2 dm + \int_H \left| Q^{1/2} D\phi \right|^2 dm.$$

In fact, the definition of such a space is more complicated (see Definition 5.11) due to the fact that the operator $Q^{1/2}D$ is not assumed to be closable in $L^2(H, \overline{\mathcal{B}}, m)$. We work in this framework because $Q^{1/2}D$ is not closable in some relevant cases, such as, for example, in the case of delay equations (see Sect. 5.6). The existence and uniqueness result is found by applying a fixed point argument in the space $L^2\left(0, T; W_Q^{1,2}(H, m)\right)$ (see Theorem 5.35). In this new context a milder smoothing property is required (see estimate (5.36) in Proposition 5.20) and, thanks to the properties of the invariant measure m, it can be verified without strong requirements on the data A, b and Q. This is the main reason why the L^2 theory developed in the present chapter allows us to deal with equations and control problems which cannot be treated by the techniques of Chap. 4.

More precisely:

(i) We do not need any smoothing properties of the Ornstein–Uhlenbeck semigroup associated with (A, Q) (see Remark 5.21). Therefore we do not impose any restrictions on Q: it is possible, for example, to take Q a one-dimensional projection.

(ii) $g, l \in L^2(H, \overline{\mathcal{B}}, m)$: they are not necessarily continuous, bounded or with polynomial growth.

This generality comes at a price. Similarly to Chap. 6 and differently from Chap. 4, we can only deal with a class of Hamiltonians of the form $F(t, x, p) = F_0\left(t, x, Q^{1/2}p\right)$. If we look at this restriction in terms of the optimal control problems we can study, it means that we are only able to deal with problems where the control appears in the state equation via a term of the form $Q^{1/2}R(t, x, a(t))$ (see (5.78)). This assumption may seem restrictive, but in fact it is quite natural in many control problems when the operator Q is degenerate. It implies that the system should be controlled by feedback taking values in the same space in which the noise disturbing the system is concentrated. Let us note that if $Q^{1/2} = 0$ then both the control and the noise disappear. A natural interpretation of this fact is that the uncontrolled system is in fact deterministic and the noise is brought into the system only by the control.

Another drawback is the fact that mild solutions found in the setting of this chapter possess weaker regularity properties due to the choice of the spaces. In particular, if Q is very degenerate (e.g. a finite-dimensional projection) the measure substantially ignores most of the space H. However, despite this weak regularity, when (5.7) is the HJB equation related to a stochastic optimal control problem, one can characterize its solution as the value function of the problem and use it to perform the optimal synthesis.

5.2   Preliminaries and the Linear Problem

5.2.1   Notation

As usual we denote by H a real separable Hilbert space with the norm $|\cdot|$ and the inner product $\langle \cdot, \cdot \rangle$ and by Q an element of $\mathcal{L}^+(H)$. $\mathcal{B}(H)$ is the Borel σ-field of H. The function spaces $C(H), UC(H), C_b(H), UC_b(H), C_b(H, H), C_b^k(H), C_0^k(\mathbb{R}^n)$, ... are defined in Appendix A.

5.2.2   The Reference Measure m and the Main Assumptions on the Linear Part

We will work under the following set of assumptions.

Hypothesis 5.1 (A) A is the generator of a strongly continuous semigroup $\{e^{tA}, \ t \geq 0\}$ on a real separable Hilbert space H. $M \geq 1$ and $\omega \in \mathbb{R}$ are two real constants such that

$$\left\| e^{tA} \right\| \leq M e^{\omega t}, \qquad \forall t \geq 0.$$

(B) $Q \in \mathcal{L}^+(H)$, and $\mu_0 = (\Omega, \mathscr{F}, \{\mathscr{F}_t\}_{t \geq 0}, \mathbb{P}, W_Q)$ is every generalized reference probability space (see Definition 1.100).

(C) $e^{sA} Q e^{sA^*} \in \mathcal{L}_1(H)$ for all $s > 0$. Moreover, for every $t \geq 0$,

$$\int_0^t \text{Tr} \left[e^{sA} Q e^{sA^*} \right] ds < +\infty,$$

so the symmetric positive operator

$$Q_t : H \to H, \qquad Q_t := \int_0^t e^{sA} Q e^{sA^*} ds,$$

is of trace class for every $t \geq 0$.

(D) The function $b : H \to H$ is continuous and Gâteaux differentiable, its Gâteaux differential ∇b is strongly continuous and

$$\|\nabla b\|_0 = \sup_{x \in H} \|\nabla b(x)\| \leq K < +\infty.$$

Proposition 5.2 *Let Hypothesis 5.1 be satisfied. Then:*

(i) The equation

$$\begin{cases} dX(s) = (AX(s) + b(X(s)))\,ds + dW_Q(s), & s \in [0, T], \\ X(0) = x \in H \end{cases} \tag{5.8}$$

has a unique mild solution $X(\cdot; x) \in \mathcal{H}_p^{\mu_0}(0, T; H)$ (see Definition 1.126) for all $p \geq 1$. We also have

$$\lim_{s \to 0} \mathbb{E}\,|X(s, x) - x|^2 = 0. \tag{5.9}$$

(ii) There exists a $\mathcal{B}([0, T]) \otimes \mathcal{B}(H) \otimes \mathscr{F}/\mathcal{B}(H)$-measurable function

$$\begin{cases} [0, T] \times H \times \Omega \to H \\ (s, x, \omega) \to \tilde{X}(s; x)(\omega) \end{cases}$$

such that, for every $x \in H$, $\tilde{X}(\cdot; x)$ is a version of the solution $X(\cdot; x)$. Thus in the future we will not make a distinction between $X(\cdot; x)$ and $\tilde{X}(\cdot; x)$.

Proof Part (i), except (5.9), is proved in Theorem 1.147 (observe that b is globally Lipschitz continuous thanks to Hypothesis 5.1-(D) and Theorem D.18). To prove (5.9) we can observe that, using Hypotheses 5.1-(A) and (D),

$$\mathbb{E}\,|X(s, x) - x|^2 \leq 3 \left| e^{sA}x - x \right|^2 + 3C \int_0^s \mathbb{E}\left(1 + |X(r)|^2\right) dr + 3\mathbb{E}\left| W^A(s) \right|^2, \quad s \in [0, T],$$

where C is a constant depending only on b. The first term converges to zero when $s \to 0$, the second goes to zero because $X(\cdot; x) \in \mathcal{H}_2^{\mu_0}(0, T; H)$ while the term concerning the stochastic convolution converges to zero thanks to its mean square continuity ensured by Proposition 1.144.

Part (ii) is proved in Proposition 5.44 for a more general controlled version of the equation (even though Proposition 5.44 is in a later section, its proof is independent). □

The *transition semigroup* P_s, $s \geq 0$, associated to (5.8) is defined for every $\phi \in C_b(H)$ as[2]

$$\begin{cases} P_s[\phi]: H \to \mathbb{R} \\ P_s[\phi]: x \to \mathbb{E}\phi(X(s; x)), \end{cases} \tag{5.10}$$

---

[2]In Sect. 1.6 we define the semigroup directly on all the functions of $B_b(H)$. The arguments of the present chapter are more transparent if we start by defining the semigroup only on $C_b(H)$. Since it will be extended (Proposition 5.9) to $L^p(H, \overline{\mathcal{B}}, m)$, and (Lemma 5.10), for any $\phi \in L^p(H, \overline{\mathcal{B}}, m)$, $P_t[\phi](x) = \mathbb{E}\phi(X(t; x))$, the two approaches are equivalent.

where $X(s; x)$ is the solution of (5.8) at time s. It follows from Proposition 1.147 that $P_s(C_b(H)) \subset C_b(H)$ (see Theorem 1.162) and P_s has the semigroup property in $C_b(H)$ as was remarked in Corollary 1.158. Moreover, P_s does not depend on μ_0 so the theory developed in this chapter is independent of the choice of μ_0.

In the setting described by Hypothesis 5.1, we can introduce the notion of an invariant measure.

Definition 5.3 (*Invariant measure*) Let P_t be the transition semigroup introduced in (5.10). A probability measure m on $(H, \mathcal{B}(H))$ is said to be an *invariant measure* for (5.8) if, for any $\phi \in C_b(H)$ and $t \geq 0$,

$$\int_H P_t[\phi](x)dm(x) = \int_H \phi(x)dm(x). \tag{5.11}$$

If Hypothesis 5.1 holds, we formulate the following assumption.

Hypothesis 5.4 There exists an invariant measure m for Eq. (5.8). Moreover,

$$\int_H |x|^2 \, dm(x) < \infty. \tag{5.12}$$

We denote by $\overline{\mathcal{B}}$ the completion (see Sect. 1.1.1) of the Borel σ-field $\mathcal{B}(H)$ with respect to the measure m.

Notation 5.5 L^p spaces have been introduced in Sect. 1.1.3. In order to distinguish the norms in $L^p(H, \overline{\mathcal{B}}, m)$ and $L^p(H, \overline{\mathcal{B}}, m; H)$ (i.e., the L^p norms computed using the measure m) from other L^p-norms that appear in this chapter, we will denote them by $|\cdot|_{L^p_m}$ and by $|\cdot|_{L^p_{m,H}}$.

We first recall some density results that we will use frequently.

Lemma 5.6 *Suppose that A satisfies Hypothesis 5.1 (A). Denote by $\mathcal{E}_A(H)$ the linear subspace of $UC_b(H)$ given by the linear span of the set of all real parts of the functions $e^{i\langle x,h\rangle}$ for some $h \in D(A^*)$. Then, for any $f \in UC_b(H)$ there exists a multi-sequence $\left(f_{n_1,n_2,n_3}\right)_{n_1,n_2,n_3\in\mathbb{N}}$ in $\mathcal{E}_A(H)$ such that*

$$\|f_{n_1,n_2,n_3}\|_0 \leq \|f\|_0, \quad \text{for any } n_1, n_2, n_3 \in \mathbb{N}$$

and

$$\lim_{n_1\to+\infty} \lim_{n_2\to+\infty} \lim_{n_3\to+\infty} f_{n_1,n_2,n_3}(x) = f(x), \quad \text{for any } x \in H.$$

Proof See Lemma 6.2.3, p. 112 in [179]. □

Lemma 5.7 *Given any bounded measure $\bar{m}$ defined on the Borel σ-field $\mathcal{B}(H)$ of H, denoting by $\overline{\mathcal{B}}_{\bar{m}}$ the completion of $\mathcal{B}(H)$ with respect to $\bar{m}$, we have the following density results:*

(i) $UC_b(H)$ and $UC_b^k(H)$, for any integer $k > 0$, are dense in $L^2(H, \overline{\mathcal{B}}_{\bar{m}}, \bar{m})$.

(ii) Let A be the generator of a C_0-semigroup on H and let A^* be its adjoint. Then $\mathcal{F}C_0^{k,A^*}(H)$, defined in (A.4), is dense in $L^2(H, \overline{\mathcal{B}}_{\bar{m}}, \bar{m})$ for any integer $k \geq 0$.

(iii) For every $\psi \in L^2(0, T; L^2(H, \overline{\mathcal{B}}_{\bar{m}}, \bar{m}))$ there exists a sequence $\psi_n \colon [0, T] \to \mathcal{F}C_0^{2,A^*}(H)$ such that

$$\begin{cases} \psi_n \in C([0,T], UC_b(H)), \\ D\psi_n, A^*D\psi_n \in C([0,T], UC_b(H,H)), \\ D^2\psi_n \in C([0,T], UC_b(H, \mathcal{L}_1(H))), \end{cases}$$

and

$$\psi_n \xrightarrow{n \to +\infty} \psi \quad \text{in } L^2(0, T; L^2(H, \overline{\mathcal{B}}_{\bar{m}}, \bar{m})).$$

Proof Part (i): $UC_b(H)$ is dense in $L^2(H, \overline{\mathcal{B}}_{\bar{m}}, \bar{m})$ thanks to Theorem 1.34. The density of $UC_b^k(H)$ in $L^2(H, \overline{\mathcal{B}}_{\bar{m}}, \bar{m})$ for $k > 0$ will be proved below.

Part (ii): Given $f \in L^2(H, \overline{\mathcal{B}}_{\bar{m}}, \bar{m})$ and any $n \in \mathbb{N}$ we need to find $\tilde{f}_n \in \mathcal{F}C_0^{k,A^*}(H)$ with $|f - \tilde{f}_n|_{L^2(H, \overline{\mathcal{B}}_{\bar{m}}, \bar{m})} \leq \frac{1}{n}$. Thanks to the already recalled density of $UC_b(H)$ in $L^2(H, \overline{\mathcal{B}}_{\bar{m}}, \bar{m})$ we can suppose that $f \in UC_b(H)$ and we can then consider an approximating multi-sequence $f_{n_1,n_2,n_3} \in \mathcal{E}_A(H)$ from Lemma 5.6. We define, for any $x \in H$, for $n_1 \in \mathbb{N}$, $f_{n_1}(x) := \lim_{n_2 \to +\infty} \lim_{n_3 \to +\infty} f_{n_1,n_2,n_3}(x)$ and, for $n_1, n_2 \in \mathbb{N}$, $f_{n_1,n_2}(x) := \lim_{n_3 \to +\infty} f_{n_1,n_2,n_3}(x)$ so that, pointwise, $f = \lim_{n_1 \to +\infty} f_{n_1}$. Using Egoroff's Theorem (Lemma 1.50-(iv)) we can find n_1 such that $|f - f_{n_1}|_{L^2(H, \overline{\mathcal{B}}_{\bar{m}}, \bar{m})} \leq \frac{1}{6n}$, then n_2 such that $|f_{n_1} - f_{n_1,n_2}|_{L^2(H, \overline{\mathcal{B}}_{\bar{m}}, \bar{m})} \leq \frac{1}{6n}$ and n_3 such that $|f_{n_1,n_2} - f_{n_1,n_2,n_3}|_{L^2(H, \overline{\mathcal{B}}_{\bar{m}}, \bar{m})} \leq \frac{1}{6n}$. We denote such an f_{n_1,n_2,n_3} by f_n and we have $|f - f_n|_{L^2(H, \overline{\mathcal{B}}_{\bar{m}}, \bar{m})} \leq \frac{1}{2n}$. The function f_n is a linear combination of real parts of functions $e^{i\langle x, h_i \rangle}$ for some $h_i \in D(A^*)$, $i = 1, \ldots, k_n$, so it does not belong to $\mathcal{F}C_0^{k,A^*}(H)$ and we need to modify it.

Let $\lambda \colon \mathbb{R} \to [0, 1]$ be a C^∞ function compactly supported in $(-2, 2)$ and identically equal to 1 in the interval $[-1, 1]$. We choose $\delta > 0$ and we replace the real part of each term $e^{i\langle x, h_i \rangle}$ in the linear combination by the real part of $e^{i\langle x, h_i \rangle} \lambda(\delta \langle x, h_i \rangle)$. We call the new function $\tilde{f}_n$. It belongs to $\mathcal{F}C_0^{k,A^*}(H)$ and if we choose δ small enough we have $|f_n - \tilde{f}_n|_{L^2(H, \overline{\mathcal{B}}_{\bar{m}}, \bar{m})} \leq \frac{1}{2n}$. It then follows that $|f - \tilde{f}_n|_{L^2(H, \overline{\mathcal{B}}_{\bar{m}}, \bar{m})} \leq \frac{1}{n}$.

The density of $UC_b^k(H)$ claimed in Part (i) now follows from Part (ii).

The proof of Part (iii) follows by applying the results of Part (ii) to the Hilbert space $\tilde{H} := \mathbb{R} \times H$, with the operator $\tilde{A} := \begin{pmatrix} 1 & 0 \\ 0 & A \end{pmatrix}$ (having domain $\mathbb{R} \times D(A)$) and the measure $\tilde{m} := \mathbf{1}_{[0,T]}dt \otimes \bar{m}$, where dt is the Lebesgue measure on $\mathbb{R}$. $\qquad\square$

Lemma 5.8 *The following results hold:*

(i) *If b satisfies Hypothesis 5.1-(D), there exists a sequence $(b_n) \subset C^2(H, H)$ such that*

$$\sup_n \|Db_n\|_0 \leq K < +\infty, \tag{5.13}$$

and for all $h, x \in H$ and for any sequence x_n of elements of H converging to x,

$$\lim_{n\to\infty} b_n(x_n) = b(x), \quad \lim_{n\to\infty} Db_n(x_n)(h) = \nabla b(x)(h).$$

(ii) *If b satisfies Hypothesis 5.1-(D) and $\|b\|_0 < +\infty$, then the sequence in Part (i) can be chosen such that*

$$\sup_n \|b_n\|_0 \le l < +\infty. \tag{5.14}$$

(iii) *Given $\phi \in C_b^1(H)$, there exists a sequence $(\phi_n) \subset UC_b^2(H)$ such that*

$$\sup_n \|\phi_n\|_0 \le l < +\infty, \quad \sup_n \|D\phi_n\|_0 \le l < +\infty, \tag{5.15}$$

and, for all $x \in H$,

$$\lim_{n\to\infty} \phi_n(x) = \phi(x), \quad \lim_{n\to\infty} D\phi_n(x) = D\phi(x).$$

Proof We only prove (i) since the proofs of (ii) and (iii) use the same arguments. The proof is based on a standard procedure of mollification over finite-dimensional subspaces (see e.g. the proof of Lemma 1.2, p. 164 of [486]). Take an orthonormal basis $\{e_n\}$ of H and, for $z \in H$, let $z = \sum_{i=1}^{\infty} z_i e_i$. For every $n \in \mathbb{N}$ let P_n be the orthogonal projection onto the n-dimensional subspace of H spanned by $\{e_1, \ldots e_n\}$. Define

$$\Pi_n : H \to \mathbb{R}^n, \qquad \Pi_n z = (z_1, \ldots, z_n),$$

$$Q_n : \mathbb{R}^n \to H, \qquad Q_n(z_1, \ldots, z_n) = z_1 e_1 + \cdots + z_n e_n,$$

and recall that $P_n = Q_n \circ \Pi_n$. Given a family of C^∞ mollifiers $\eta_n : \mathbb{R}^n \to \mathbb{R}$ with support in $B(0, 1/n)$, we define

$$b_n(z) = \int_{\mathbb{R}^n} b(Q_n y)\eta_n(\Pi_n z - y)dy = \int_{\mathbb{R}^n} b(P_n z - Q_n y)\eta_n(y)dy.$$

From the first equality above, we easily conclude that $b_n \in C^\infty(H, H)$. We have, in particular,

$$b_n(x_n) = \int_{\mathbb{R}^n} b(P_n x_n - Q_n y)\eta_n(y)dy.$$

From this equation, the fact that $P_n x_n \to x$ and the continuity of b we can conclude that

$$\lim_{n\to\infty} b_n(x_n) = b(x).$$

Fix $z \in H$. For any $h \in H$ with $|h| = 1$ and $\tau > 0$ we have

$$\frac{b_n(z+\tau h)-b_n(z)}{\tau} = \frac{1}{\tau}\int_{\mathbb{R}^n}\left[b(P_n(z+\tau h)-Q_n y)-b(P_n z-Q_n y)\right]\eta_n(y)dy$$

$$= \int_{\mathbb{R}^n}\left[\int_0^1 \nabla b(P_n(z+r\tau h)-Q_n y)(h)dr\right]\eta_n(y)dy,$$

(5.16)

where in last equality we used Theorem D.18. In particular,

$$\frac{b_n(x_n+\tau h)-b_n(x_n)}{\tau} = \int_{\mathbb{R}^n}\left[\int_0^1 \nabla b(P_n(x_n+r\tau h)-Q_n y)(h)dr\right]\eta_n(y)dy.$$

(5.17)

Since $b_n \in C^\infty(H,H)$ the left-hand side of the previous equality converges, when $\tau \to 0$, to $Db_n(x_n)(h)$ while, thanks to the strong continuity of ∇b, the right-hand side converges to $\int_{\mathbb{R}^n}\nabla b(P_n x_n - Q_n y)(h)\eta_n(y)dy$. Taking the limits of the two expressions when $n \to \infty$ we get (again thanks to the strong continuity of ∇b)

$$\lim_{n\to\infty} Db_n(x_n)(h) = \nabla b(x)(h).$$

Thanks to the last equality in (5.16), for any $z \in H$, we also have $\left|\frac{b_n(z+\tau h)-b_n(z)}{\tau}\right| \le \|\nabla b\|_0$ and then, letting $\tau \to 0$, we obtain

$$\sup_n \|Db_n\|_0 \le \|\nabla b\|_0.$$

$\square$

Proposition 5.9 Let $p \in [1,+\infty)$. Assume that Hypotheses 5.1 and 5.4 hold. Then P_t, defined on $C_b(H)$ by (5.10), extends to a strongly continuous semigroup of contractions on $L^p(H,\overline{\mathcal{B}},m)$. Moreover, for any $\phi \in L^p(H,\overline{\mathcal{B}},m)$ and $t \ge 0$, the relation (5.11) holds.

Proof We follow the proof of Theorem 10.1.5, p. 209 of [179], where the statement is proved for the Ornstein–Uhlenbeck case. Given $\phi \in C_b(H)$, for any $x \in H$, thanks to Jensen's inequality we have $|P_t[\phi](x)|^p \le |P_t[|\phi|^p](x)|$. Thus, since m is invariant,

$$\int_H |P_t[\phi](x)|^p dm(x) \le \int_H |P_t[|\phi|^p](x)|dm(x) = \int_H |\phi|^p(x)dm(x),$$

where the last expression is finite since ϕ is bounded and m is a finite measure. Thanks to the density of $C_b(H)$ in $L^p(H,\overline{\mathcal{B}},m)$ (Theorem 1.34), P_t extends to a contraction on $L^p(H,\overline{\mathcal{B}},m)$ for any $t \ge 0$.

To prove the strong continuity we observe first that it follows easily from the Lebesgue dominated convergence theorem and (5.9) that for every $\phi \in C_b(H)$ and $x \in H$, we have $\lim_{t\to 0^+} P_t[\phi](x) = \phi(x)$. Moreover, since $\|P_t[\phi]\|_0 \le \|\phi\|_0$, we then obtain, again using the Lebesgue dominated convergence theorem,

$$\lim_{t\to 0^+} P_t[\phi] = \phi \quad \text{in } L^p(H,\overline{\mathcal{B}},m).$$

Since P_t is a semigroup of contractions on $L^p(H, \overline{\mathcal{B}}, m)$ this implies strong continuity for every $\phi \in L^p(H, \overline{\mathcal{B}}, m)$.

To show the last claim, let $\phi \in L^p(H, \overline{\mathcal{B}}, m)$ and let $\phi_n \in C_b(H)$ be a sequence such that $\phi_n \to \phi$ in $L^p(H, \overline{\mathcal{B}}, m)$. We have, in particular, $\int_H \phi_n(x) \, dm(x) \to \int_H \phi(x) \, dm(x)$. Moreover, since for any $t \geq 0$, $P_t \in \mathcal{L}(L^p(H, \overline{\mathcal{B}}, m))$, $P_t[\phi_n] \to P_t[\phi]$ in $L^p(H, \overline{\mathcal{B}}, m)$ and, in particular, $\int_H P_t[\phi_n](x) \, dm(x) \to \int_H P_t[\phi](x) \, dm(x)$, so (5.11) follows letting $n \to \infty$ because it holds for the elements of $C_b(H)$. $\qquad \square$

In the previous proposition we extended, for any $t \geq 0$, the operator P_t to the whole space $L^p(H, \overline{\mathcal{B}}, m)$ by continuity. In other words, given $\phi \in L^p(H, \overline{\mathcal{B}}, m)$, $P_t[\phi]$ is defined as the limit in $L^p(H, \overline{\mathcal{B}}, m)$ of $P_t[\phi_n]$, where ϕ_n is a (any) sequence of elements of $C_b(H)$ converging to ϕ in $L^p(H, \overline{\mathcal{B}}, m)$. In the following lemma we show that this limit is indeed equal to $\mathbb{E}\phi(X(t; x))$ (which will be proved to be a well-defined expression) even for non-bounded and non-Borel measurable elements of $L^p(H, \overline{\mathcal{B}}, m)$.

Lemma 5.10 *Let $p \in [1, +\infty)$. Assume that Hypotheses 5.1 and 5.4 hold. Consider $\phi \in L^p(H, \overline{\mathcal{B}}, m)$ and $t \in [0, T]$. Then the function*

$$\begin{cases} H \times \Omega \to \mathbb{R} \\ (x, \omega) \to \phi(X(t; x)(\omega)) \end{cases}$$

is $\overline{\mathcal{B}(H) \otimes \mathcal{F}}/\mathcal{B}(\mathbb{R})$-measurable, where $\overline{\mathcal{B}(H) \otimes \mathcal{F}}$ is the completion of the σ-field $\mathcal{B}(H) \otimes \mathcal{F}$ w.r.t. the measure $m \otimes \mathbb{P}$. Moreover, $x \to \mathbb{E}\phi(X(t; x))$ is a $\overline{\mathcal{B}}/\mathcal{B}(\mathbb{R})$-measurable function and

$$P_t[\phi](x) = \mathbb{E}\phi(X(t; x)) \quad \text{for } m\text{-a.e. } x \in H. \tag{5.18}$$

Proof Suppose first that ϕ is Borel-measurable and $\phi \geq 0$. By Proposition 5.44 we can assume that $(t, x, \omega) \to \phi(X(t; x)(\omega))$ is a $\mathcal{B}[0, T] \otimes \mathcal{B}(H) \otimes \mathcal{F}/\mathcal{B}(H)$-measurable function and then (see Lemma 1.8(iv)), for any $t \in [0, T]$, $(x, \omega) \to \phi(X(t; x)(\omega))$ is $\mathcal{B}(H) \otimes \mathcal{F}/\mathcal{B}(H)$-measurable so that the function $(x, \omega) \to \phi(X(t; x)(\omega))$, being the composition of a $\mathcal{B}(H) \otimes \mathcal{F}/\mathcal{B}(H)$-measurable function and a $\mathcal{B}(H)/\mathcal{B}(\mathbb{R})$-measurable function, is $\mathcal{B}(H) \otimes \mathcal{F}/\mathcal{B}(\mathbb{R})$-measurable. The (Borel) measurability of $x \to \mathbb{E}\phi(X(t; x))$ then follows (see e.g. Lemma 1.26, p. 14 of [370]). Moreover, if we consider $\phi_n := \phi \wedge n$, thanks to the monotone convergence theorem, we have

$$\mathbb{E}\phi(X(t; x)) = \lim_{n \to \infty} \mathbb{E}\phi_n(X(t; x)) = \lim_{n \to \infty} P_t[\phi_n](x), \quad x \in H$$

(the limit can also be $+\infty$ for certain x). Since (again by monotone convergence) we have $\lim_{n \to \infty} \phi_n := \phi$ in $L^p(H, \overline{\mathcal{B}}, m)$, we also have

$$\lim_{n \to \infty} P_t[\phi_n] = P_t[\phi]$$

in $L^p(H, \overline{\mathcal{B}}, m)$ and then, extracting if necessary a subsequence, m-a.e. Thus we obtain (5.18).

As a second step we consider a positive $\phi \in L^p(H, \overline{\mathcal{B}}, m)$. By Lemma 1.16 we can find $\tilde{\phi} \in L^p(H, \mathcal{B}(H), m)$ and $V \in \mathcal{B}(H)$, $m(V) = 0$ such that $\phi(x) = \tilde{\phi}(x)$ for any $x \in H \setminus V$. Denoting by $\mathbf{1}_V$ the characteristic function of V we have

$$\int_H \mathbb{P}\{X(t; x)(\omega) \in V\} dm(x) = \int_H \mathbb{E}[\mathbf{1}_V(X(t; x))] dm(x)$$

$$= \int_H P_t[\mathbf{1}_V](x) dm(x) = \int_H \mathbf{1}_V(x) dm(x) = 0. \qquad (5.19)$$

So the functions $(x, \omega) \to \phi(X(t; x)(\omega))$ and $(x, \omega) \to \tilde{\phi}(X(t; x)(\omega))$ disagree only on a subset of $H \times \Omega$ which has $m \otimes \mathbb{P}$-measure 0 and thus, since we have already observed that $(x, \omega) \to \tilde{\phi}(X(t; x)(\omega))$ is $\mathscr{F} \otimes B/\mathcal{B}(\mathbb{R})$-measurable, $(x, \omega) \to \phi$ $(X(t; x)(\omega))$ is $\overline{\mathscr{F} \otimes B}/\mathcal{B}(\mathbb{R})$-measurable.

Therefore (see e.g. Theorem 2.39, p. 68 of [267]) $\mathbb{E}[\phi(X(t; x))] = \mathbb{E}\left[\tilde{\phi}(X(t; x))\right]$ is well defined for m-a.e. $x \in H$ and the function $x \to \mathbb{E}\left[\tilde{\phi}(X(t; x))\right]$ is $\overline{\mathcal{B}}/\mathcal{B}(\mathbb{R})$-measurable. However, for m-a.e. $x \in H$, $P_t[\phi](x) = P_t[\tilde{\phi}](x) = \mathbb{E}\left[\tilde{\phi}(X(t; x))\right] = \mathbb{E}[\phi(X(t; x))]$, which establishes (5.18).

The proof for a non-positive function follows by the previous arguments after decomposing the function into the sum of its positive and negative parts. $\square$

5.2.3 The Operator $\mathcal{A}$

From now on we fix the constant p of Proposition 5.9 and Lemma 5.10 equal to 2 and work in the space $L^2(H, \overline{\mathcal{B}}, m)$.

Let Hypotheses 5.1 and 5.4 be satisfied and let P_t be defined as in (5.10). We denote by $\mathcal{A}$ the generator of P_t as a strongly continuous semigroup on $L^2(H, \overline{\mathcal{B}}, m)$ (see Proposition 5.9). Its domain is denoted by $D(\mathcal{A}) \subset L^2(H, \overline{\mathcal{B}}, m)$.

We will often use the elements of the space $\mathcal{F}C_0^{2,A^*}(H)$ to approximate less regular functions and it will be useful to know how to calculate explicitly the operator $\mathcal{A}$ on them. Indeed, as proved in Lemma 5.37, $\mathcal{F}C_0^{2,A^*}(H) \subset D(\mathcal{A})$ and for any $\phi \in \mathcal{F}C_0^{2,A^*}(H)$ we have

$$\mathcal{A}\phi(x) = \frac{1}{2} \text{Tr}\left[QD^2\phi(x)\right] + \langle x, A^*D\phi(x)\rangle + \langle b(x), D\phi(x)\rangle. \qquad (5.20)$$

5.2.4   The Gradient Operator D_Q and the Space $W_Q^{1,2}(H, m)$

Let Q be an operator satisfying Hypothesis 5.1-(B). We then introduce the following operator D_Q.

Definition 5.11 (*The operator D_Q and the space $W_Q^{1,2}(H, m)$*) We define the operator
$$D_Q\phi := Q^{1/2}D\phi, \quad \phi \in C_b^1(H), \tag{5.21}$$

where $D\phi$ denotes the Fréchet derivative of ϕ.
  For $\phi \in C_b^1(H)$ we define the norm
$$|\phi|^2_{W_Q^{1,2}} = |\phi|^2_{L_m^2} + \left|D_Q\phi\right|^2_{L_{m,H}^2}.$$

The completion of $C_b^1(H)$ with respect to the norm $|\cdot|_{W_Q^{1,2}}$ will be denoted by $W_Q^{1,2}(H, m)$.

  The space $W_Q^{1,2}(H, m)$ may be identified with the subspace of $L^2(H, \overline{\mathcal{B}}, m) \times L^2(H, \overline{\mathcal{B}}, m; H)$ which consists of all pairs
$$(\psi, \Psi) \in L^2(H, \overline{\mathcal{B}}, m) \times L^2(H, \overline{\mathcal{B}}, m; H)$$

such that there exists a sequence $(\phi_n) \subset C_b^1(H)$ with the property
$$\phi_n \to \psi, \quad \text{in} \ \ L^2(H, \overline{\mathcal{B}}, m)$$

and
$$D_Q\phi_n \to \Psi, \quad \text{in} \ \ L^2(H, \overline{\mathcal{B}}, m; H).$$

  In the cases where the operator D_Q is closable (as an unbounded operator from its domain $C_b^1(H) \subset L^2(H, \overline{\mathcal{B}}, m)$ to $L^2(H, \overline{\mathcal{B}}, m; H)$), for any two pairs $(\psi_1, \Psi_1), (\psi_2, \Psi_2) \in W_Q^{1,2}(H, m)$ such that $\psi_1 = \psi_2$ in $L^2(H, \overline{\mathcal{B}}, m)$ we also have $\Psi_1 = \Psi_2$, so that $W_Q^{1,2}(H, m)$ is naturally embedded in $L^2(H, \overline{\mathcal{B}}, m)$.
  If D_Q is not closable then we can find a sequence $(\phi_n) \subset C_b^1(H)$ such that
$$\phi_n \to 0 \ \ \text{in} \ \ L^2(H, \overline{\mathcal{B}}, m) \quad \text{and} \quad D_Q\phi_n \to \Phi \neq 0, \quad \text{in} \ \ L^2(H, \overline{\mathcal{B}}, m; H).$$

Therefore, elements of $W_Q^{1,2}(H, m)$ cannot be identified, in general, with functions of $L^2(H, \overline{\mathcal{B}}, m)$ (e.g., the above element $(0, \Phi)$).[3] This means that the structure of

---

[3]For this reason, since we are interested in a definition that also works when the operator D_Q is non-closable, we do not work in the space $W^{1,2}(H, m)$ defined (see e.g. Chap. 9, p. 196 of [179]) as the linear space of all functions $\phi \in L^2(H, \overline{\mathcal{B}}, m)$ such that $D\phi \in L^2(H, \overline{\mathcal{B}}, m; H)$.

the Sobolev space changes significantly when we want to take into account the case of non-closable D_Q.

Observe that, in any case, even when D_Q is not closable, it can be extended to a well-defined continuous operator from $W_Q^{1,2}(H,m)$ (endowed with the norm described in Definition 5.11) to $L^2(H,\overline{B},m;H)$. Indeed, if $|\phi_n|^2_{W_Q^{1,2}} \to 0$ then $|D_Q\phi_n|^2_{L^2_{m,H}} \to 0$. We denote the continuous extension of D_Q from $W_Q^{1,2}(H,m)$ to $L^2(H,\overline{B},m;H)$ again by D_Q. When D_Q is not closable, considering the characterization of $W_Q^{1,2}(H,m)$ as a subspace of $L^2(H,\overline{B},m) \times L^2(H,\overline{B},m;H)$ and the notation described above, we have $D_Q(\psi,\Psi) = \Psi$.

The notation we use here is a little different from the one used in Chap. 4. Indeed, to be consistent with the notation of Chap. 4, we should write $D^{Q^{1/2}}$ instead of D_Q. We choose to use this notation for two reasons: it is simpler and, even if not very intuitive, it is fairly standard in the literature.

Sometimes in the literature the notation D_Q is used for different operators. We want to underline in particular the difference with respect to Chap. 9 of [179] where D_Q is used for the Malliavin derivative, which is again an operator of the form $Q^{\frac{1}{2}}D$ for some $Q \in \mathcal{L}_1^+(H)$. The difference is that, in our case Q is the covariance operator of the Wiener process, while in [179] it is the covariance operator of the (Gaussian) reference measure. When $b = 0$ and $\omega < 0$, the operator used in [179] is $Q_\infty = \int_0^{+\infty} e^{sA}Qe^{sA^*}ds$.

Remark 5.12 When (5.8) is linear (if $b = 0$) and $\omega < 0$, the problem of closability of D_Q can be approached using some characterizations that can be found in the literature. A negative result ensuring the non-closability of the operator is, for example, Theorem 3.5 of [299], which allows us to prove that D_Q is not closable, for example, in the two cases recalled in Sect. 5.6.

When the operator Q is injective, a characterization of closability is given by Theorem 6.1 of [299], which shows that the closability of the operator D_Q is equivalent to the closability of the operator $Z: D(Z) \subset H \to H$ given by

$$\begin{cases} D(Z) = Q_\infty^{1/2}(H) \\ Z\left(Q_\infty^{1/2}x\right) = Q^{\frac{1}{2}}x. \end{cases}$$

In the particular case considered, for example, in [3, 4, 125] (see also Example 4.46 and Sect. 4.8.3.1) the generator of the semigroup is

$$Ax = \sum_{n=1}^{+\infty} -\alpha_n \langle e_n, x \rangle e_n, \quad x \in D(A),$$

for some orthonormal basis $\{e_n\}$ and $0 < \alpha_1 \le \alpha_2 \le \alpha_3 \dots$. Moreover Q is given by

$$Qx = \sum_{i=n}^{+\infty} q_n \langle e_n, x \rangle e_n, \quad x \in H,$$

for a sequence of positive eigenvalues q_n. The expression for Z is given, for any $y = Q_\infty^{1/2} x$, by

$$Zy = Q^{\frac{1}{2}} Q_\infty^{-1/2} y = \sum_{n=1}^{+\infty} \sqrt{q_n} \sqrt{\frac{2\alpha_n}{q_n}} \langle e_n, y \rangle e_n$$

$$= \sum_{n=1}^{+\infty} -\sqrt{2\alpha_n} \langle e_n, y \rangle e_n = \sqrt{2}(-A)^{1/2} y.$$

Thus, since $Q_\infty^{1/2}(H) \subset D\left((-A)^{1/2}\right)$ and since $(-A)^{1/2}$ is closed (see Theorem B.53-(i)), Z admits a closed extension and so (see Theorem 5.4(a), p. 91 of [569]) it is closable. Therefore, thanks to Theorem 6.1 of [299], the operator D_Q is closable. ∎

5.2.5 The Operator $\mathcal{R}$

Let Q be an operator satisfying Hypothesis 5.1-(B) and let D_Q be defined as in Definition 5.11. We introduce and study here the properties of the operator $\mathcal{R}$ defined below (Definition 5.19).

We begin by studying the regularity of the solution $X(\cdot; x)$ of (5.8) with respect to the initial datum. We use Proposition 6.7. The following lemma specifies it in the particular case we are interested in.

Lemma 5.13 Let $\mathcal{H}_2^{\mu_0}(0, T; H)$ be the space defined in Definition 1.126. Let $\mathcal{K}: H \times \mathcal{H}_2^{\mu_0}(0, T; H) \to \mathcal{H}_2^{\mu_0}(0, T; H)$ be a continuous mapping satisfying, for some $\alpha \in [0, 1)$,

$$|\mathcal{K}(x, X) - \mathcal{K}(x, Y)|_{\mathcal{H}_2^{\mu_0}(0,T;H)} \le \alpha \, |\mathcal{K}(x, X) - \mathcal{K}(x, Y)|_{\mathcal{H}_2^{\mu_0}(0,T;H)} \tag{5.22}$$

for all $x \in H$ and $X, Y \in \mathcal{H}_2^{\mu_0}(0, T; H)$. Then:

(i) There exists a unique mapping $\varphi: H \to \mathcal{H}_2^{\mu_0}(0, T; H)$ such that

$$\varphi(x) = \mathcal{K}(x, \varphi(x)), \quad \text{for every } x \in H,$$

and it is continuous.

(ii) Suppose that, for any $(x, X) \in H \times \mathcal{H}_2^{\mu_0}(0, T; H)$ and for any $h \in H$ there exists the directional derivative of $\mathcal{K}$ with respect to x in the direction h and that, for any fixed h, the mapping

$$\begin{cases} H \times \mathcal{H}_2^{\mu_0}(0, T; H) \to \mathcal{H}_2^{\mu_0}(0, T; H) \\ (x, X) \to \nabla_x \mathcal{K}(x, X; h) \end{cases}$$

is continuous. Assume that, for any (x, X), $h \to \nabla_x \mathcal{K}(x, X; h)$ is continuous from H to $\mathcal{H}_2^{\mu_0}(0, T; H)$. Suppose also that for any $(x, X) \in H \times \mathcal{H}_2^{\mu_0}(0, T; H)$ and for any $Y \in \mathcal{H}_2^{\mu_0}(0, T; H)$ there exists the directional derivative of $\mathcal{K}$ with respect to X in the direction Y and that, for any fixed Y, the mapping

$$\begin{cases} H \times \mathcal{H}_2^{\mu_0}(0, T; H) \to \mathcal{H}_2^{\mu_0}(0, T; H) \\ (x, X) \to \nabla_X \mathcal{K}(x, X; Y) \end{cases}$$

is continuous. Assume that, for any (x, X), $Y \to \nabla_X \mathcal{K}(x, X; Y)$ is continuous from $\mathcal{H}_2^{\mu_0}(0, T; H)$ to $\mathcal{H}_2^{\mu_0}(0, T; H)$. Then, for any $x \in H$, there exists the Gâteaux derivative $\nabla \varphi(x)$. Moreover, $(x, h) \to \nabla \varphi(x)(h)$ is continuous as a mapping from $H \times H$ to $\mathcal{H}_2^{\mu_0}(0, T; H)$ and it satisfies the equation

$$\nabla \varphi(x)(h) = \nabla_x \mathcal{K}(x, \varphi(x); h) + \nabla_X \mathcal{K}(x, \varphi(x); \nabla \varphi(x)(h)), \qquad x, h \in H.$$

Proof This is a particular case of Proposition 6.7. In the claim of part (ii) we also made use of Lemma 6.4 (in a two-variable version) to verify the hypothesis "$F \in \mathcal{G}^{1,1}(X \times Y; X)$" of Proposition 6.7 for our spaces and of Lemma 6.3 to derive the continuity properties of $\nabla \varphi$.   □

Lemma 5.14 *Let Hypothesis 5.1 be satisfied and let $x, h \in H$. Denote by $X(\cdot; x)$ the solution of (5.8). Then:*

(i) *$X(\cdot; x)$ is Gâteaux differentiable as a mapping from H to $\mathcal{H}_2^{\mu_0}(0, T; H)$ and $x \to \nabla X(\cdot; x)$ is strongly continuous. For any $h \in H$ the (directional derivative) process $\zeta^{x,h}(\cdot) := \nabla X(\cdot; x)h$ is the unique mild solution in $\mathcal{H}_2^{\mu_0}(0, T; H)$ of the following equation*

$$\begin{cases} \frac{d\zeta^{x,h}(s)}{ds} = (A + \nabla b(X(s; x))) \zeta^{x,h}(s) \\ \zeta^{x,h}(0) = h \end{cases} \tag{5.23}$$

on $[0, T]$. The process $\zeta^{x,h}(\cdot)$ has $\mathbb{P}$-a.s. continuous trajectories.

(ii) *There exist universal constants $\alpha, a > 0$, α also depends on K, such that*

$$\left| \zeta^{x,h}(s) \right| \le a e^{\alpha s} |h|$$

for any $s \ge 0$. Therefore the solution to (5.23) defines, for any $x \in H$, $\omega \in \Omega$ and $s \ge 0$, a bounded operator $\zeta^x(s) : H \to H$, $\zeta^x(s)h = \zeta^{x,h}(s)$.

(iii) *For any $h \in H$ there exists a $\mathcal{B}([0, T]) \otimes \mathcal{B}(H) \otimes \mathscr{F}/\mathcal{B}(H)$-measurable function*

$$\begin{cases} [0, T] \times H \times \Omega \to H \\ (s, x, \omega) \to \tilde{\zeta}^{x,h}(s)(\omega) \end{cases} \tag{5.24}$$

such that, for every $x \in H$, $\tilde{\zeta}^{x,h}(\cdot)$ is a version of $\zeta^{x,h}(\cdot)$. Thus in the future we will not make a distinction between $\tilde{\zeta}^{x,h}(\cdot)$, $\zeta^{x,h}(\cdot)$, and $\nabla X(\cdot; x)h$.

Proof Since other similar results appearing in the book are proved for slightly different sets of hypotheses,[4] we provide the proofs.

To prove part (i), except for the $\mathbb{P}$-a.s. continuity of the trajectories of $\zeta^{x,h}(\cdot)$, we use Proposition 6.7 in the particular case stated in Lemma 5.13. The mapping $\mathcal{K}$ is defined as

$$\mathcal{K}(x, X)(s) = e^{sA}x + \int_0^s e^{(s-r)A}b(X(r))dr + W^A(s), \quad s \in [0, T],$$

where W^A is defined in (1.64). It is shown in the proof of Proposition 1.147 that if T is small enough then (5.22) is satisfied. The joint continuity of $\mathcal{K}$ is straightforward.

To verify the hypotheses of part (ii) of Lemma 5.13, we follow the arguments used in Sect. 9.1.1 of [180] (we repeat them because our hypotheses are a little different). The directional derivatives with respect to x are not a problem since one can easily see that $\nabla_x \mathcal{K}(x, X; h) = e^{\cdot A}h$ which is jointly continuous in all three variables.

As regards the directional derivative $\nabla_X \mathcal{K}(x, X; Y)$, we begin by showing that for any $X, Y \in \mathcal{H}_2^{\mu_0}(0, T; H)$ and any $x \in H$,

$$\nabla_X \mathcal{K}(x, X; Y)(s) = \int_0^s e^{(s-r)A}\nabla b(X(r))Y(r)dr, \quad s \in [0, T].$$

Indeed, we have

$$\sup_{s\in[0,T]} \mathbb{E}\left|\frac{1}{\varepsilon}(\mathcal{K}(x, X + \varepsilon Y) - \mathcal{K}(x, X))(s) - \int_0^s e^{(s-r)A}\nabla b(X(r))Y(r)dr\right|^2$$

$$= \sup_{s\in[0,T]} \mathbb{E}\left|\int_0^s e^{(s-r)A}\left[\frac{1}{\varepsilon}(b(X(r) + \varepsilon Y(r)) - b(X(r)) - \nabla b(X(r))Y(r))\right]dr\right|^2.$$

Using Theorem D.18 the last expression above becomes

$$\sup_{s\in[0,T]} \mathbb{E}\left|\int_0^s e^{(s-r)A}\left[\int_0^1 \nabla b(X(r) + \theta\varepsilon Y(r))Y(r) - \nabla b(X(r))Y(r)d\theta\right]dr\right|^2$$

$$\leq T\left(M\max\{e^{\omega T}, 1\}\right)^2 \mathbb{E}\int_0^T\left[\int_0^1 |\nabla b(X(r) + \theta\varepsilon Y(r))Y(r) - \nabla b(X(r))Y(r)|^2 d\theta\right]dr$$

which, thanks to the boundedness of ∇b and its strong continuity, converges to 0 when $\varepsilon \to 0$ by the Lebesgue dominated convergence theorem. We now prove the continuity properties of $\nabla_X \mathcal{K}(x, X; Y)$. We first fix (x, X) and we consider $Y_n \to Y$

---

[4]In particular, in Propositions 4.61 and 6.10 we work in $L_{\mathcal{P}}^p(\Omega; C([0, T], H))$, while here we use $\mathcal{H}_2^{\mu_0}(0, T; H)$. Indeed, in the mentioned propositions it is assumed that $\mathrm{Tr}\left[e^{sA}Qe^{sA^*}\right] \leq C_\beta s^{-2\beta}$ for some $\beta \in [0, 1/2)$ and $C_\beta > 0$.

in $\mathcal{H}_2^{\mu_0}(0, T; H)$. We have, using Hypothesis 5.1 and Hölder's inequality,

$$
|\nabla_X \mathcal{K}(x, X; Y_n) - \nabla_X \mathcal{K}(x, X; Y)|^2_{\mathcal{H}_2^{\mu_0}(0,T;H)}
$$

$$
= \sup_{s \in [0,T]} \mathbb{E} \left| \int_0^s e^{(s-r)A} \nabla b(X(r))(Y_n(r) - Y(r)) dr \right|^2
$$

$$
\leq K^2 \left(M \max\{e^{\omega T}, 1\} \right)^2 \sup_{s \in [0,T]} s \mathbb{E} \int_0^s |Y_n(r) - Y(r)|^2 dr \leq C|Y_n - Y|^2_{\mathcal{H}_2^{\mu_0}(0,T;H)} \to 0
$$

as $n \to +\infty$. To prove the strong continuity property we fix Y and suppose, to the contrary, that there are $\delta > 0$ and a sequence (x_n, X_n) such that $x_n \to x$ in H, $X_n \to X$ in $\mathcal{H}_2^{\mu_0}(0, T; H)$ but $|\nabla_X \mathcal{K}(x_n, X_n; Y) - \nabla_X \mathcal{K}(x, X; Y)|^2_{\mathcal{H}_2^{\mu_0}(0,T;H)} \geq \delta$ for any $n \in \mathbb{N}$. We have

$$
|\nabla_X \mathcal{K}(x_n, X_n; Y) - \nabla_X \mathcal{K}(x, X; Y)|^2_{\mathcal{H}_2^{\mu_0}(0,T;H)}
$$

$$
= \sup_{s \in [0,T]} \mathbb{E} \left| \int_0^s e^{(s-r)A} [\nabla b(X_n(r)) - \nabla b(X(r))] Y(r) dr \right|^2
$$

$$
\leq C \mathbb{E} \int_0^T |[\nabla b(X_n(r)) - \nabla b(X(r))] Y(r)|^2 dr,
$$

where C is a constant depending only on M, ω, T and K. For every $n \in \mathbb{N}$ the integrand in the last line above is dominated by $4K^2|Y(r)|^2$, moreover, since $X_n \to X$ in $\mathcal{H}_2^{\mu_0}(0, T; H)$ we can extract a subsequence X_{n_k} which converges to X, $dr \otimes \mathbb{P}$-a.e., and we can conclude using the Lebesgue dominated convergence theorem that $|\nabla_X \mathcal{K}(x_{n_k}, X_{n_k}; Y) - \nabla_X \mathcal{K}(x, X; Y)|^2_{\mathcal{H}_2^{\mu_0}(0,T;H)} \to 0$ as $k \to +\infty$, which contradicts our hypothesis.

Thus part (i) follows from Lemma 5.13. The continuity of the trajectories of $\zeta^{x,h}(\cdot)$ is a consequence of Lemma 1.115.

To prove part (ii) we observe that, thanks to Hypothesis 5.1 (A) and (D), we have, for all $s \in [0, T]$,

$$
\left| \zeta^{x,h}(s) \right| \leq \left| M \max\{e^{\omega T}, 1\} \right| |h| + M \max\{e^{\omega T}, 1\} \int_0^s K |\zeta^{x,h}(r)| dr
$$

and hence the conclusion follows from Gronwall's lemma (Proposition D.29).

To prove the claim of part (iii), we use the result of Proposition 5.44 (even though Proposition 5.44 is in a later section, its proof is independent). Let $(s, x, \omega) \to \tilde{X}(s; x)(\omega)$ be the $\mathcal{B}([s, T]) \otimes \mathcal{B}(H) \otimes \mathscr{F}/\mathcal{B}(H)$-measurable function found in Proposition 5.44 (we consider here the case when $t = 0$ and $R = 0$). Observe that, by construction, $\tilde{X}$ satisfies (1.70) for any $s \in [0, T]$, any $x, y \in H$ and any $\omega \in \Omega$ and in particular it is continuous in the variable x for any choice of $(s, \omega) \in [0, T] \times \Omega$.

We denote by $\tilde{\zeta}^{x,h}(\cdot)$ the unique solution of

$$\tilde{\zeta}^{x,h}(s) = e^{As}h + \int_0^s e^{(s-r)A} \nabla b(\tilde{X}(r;x))\tilde{\zeta}^{x,h}(r)\,dr, \qquad s \in [0,T].$$

We remark that $\tilde{\zeta}^{x,h}(s)$ is defined for every $(s,\omega) \in [0,T] \times \Omega$. Since $\tilde{X}(\cdot;x)$ is a version of $X(\cdot;x)$, $\tilde{\zeta}^{x,h}(\cdot)$ is a version of $\zeta^{x,h}(\cdot)$. Moreover, we claim that, for any choice of $(s,\omega) \in [0,T] \times \Omega$, $\tilde{\zeta}^{x,h}(s)$ is continuous in the variable x. To prove this we fix $\omega \in \Omega$ and consider $x \in H$ and any sequence x_n in H converging to x. We have

$$\left| \tilde{\zeta}^{x,h}(s)(\omega) - \tilde{\zeta}^{x_n,h}(s)(\omega) \right| \le I_1^n(s) + I_2^n(s)$$

$$:= \left| \int_0^s e^{(s-r)A} \left(\nabla b(\tilde{X}(r;x))(\omega) - \nabla b(\tilde{X}(r;x_n))(\omega) \right) \tilde{\zeta}^{x,h}(r)(\omega)\,dr \right|$$

$$+ \left| \int_0^s e^{(s-r)A} \nabla b(\tilde{X}(r;x_n))(\omega) \left(\tilde{\zeta}^{x,h}(r)(\omega) - \tilde{\zeta}^{x_n,h}(r)(\omega) \right) dr \right|, \qquad s \in [0,T].$$

$I_1^n(s)$ converges to zero, uniformly for $s \in [0,T]$, thanks to the Lebesgue dominated convergence theorem as Hypothesis 5.1 (A) and (D) and part (ii) give the uniform bound and the continuity of $x \to \tilde{X}(r;x)(\omega)$ gives the pointwise convergence. Thus the convergence (which is indeed uniform in s and thus even stronger than what we need) of $\left| \tilde{\zeta}^{x,h}(s)(\omega) - \tilde{\zeta}^{x_n,h}(s)(\omega) \right| \to 0$ follows from Gronwall's Lemma (using again Hypothesis 5.1 (A) and (D) which gives $|e^{(s-r)A} \nabla b(\tilde{X}(r;x_n))(\omega)| \le K\left(M \vee Me^{\omega T} \right)$ independently of s, r, n, ω).

Since $\tilde{\zeta}^{x,h}(\cdot)$ has continuous trajectories and is a version of $\zeta^{x,h}(\cdot) \in \mathcal{H}_2^{\mu_0}(0,T;H)$, it itself belongs to $\mathcal{H}_2^{\mu_0}(0,T;H)$. In particular, for every $x \in H$, $\tilde{\zeta}^{x,h}(\cdot)$ is $\mathcal{B}([0,T]) \otimes \mathscr{F}/\mathcal{B}(H)$-measurable as function of the variables s and ω. Moreover, we proved that, for any fixed $(s,\omega) \in [0,T] \times \Omega$, $\tilde{\zeta}^{x,h}(s)(\omega)$ is a continuous function of the variable x. It then follows from Lemma 1.18, that $\tilde{\zeta}^{x,h}(s)(\omega)$ is $\mathcal{B}([t,T]) \otimes \mathcal{B}(H) \otimes \mathscr{F}/\mathcal{B}(H)$-measurable. $\qquad \square$

Lemma 5.15 *Assume that Hypotheses 5.1 and 5.4 hold. Fix $t \in [0,T]$. Given $\phi \in C_b^1(H)$, $P_t[\phi] \in C_b(H)$, $P_t[\phi]$ is Gâteaux differentiable at any $x \in H$ and*

$$\langle \nabla P_t[\phi](x), h \rangle = \mathbb{E}\left(\left\langle \left(\zeta^x(t)\right)^* D\phi(X(t;x)), h \right\rangle \right), \qquad h \in H. \tag{5.25}$$

Moreover, $\nabla P_t[\phi]$ is strongly continuous and

$$\sup_{x \in H} |\nabla P_t[\phi](x)| < +\infty. \tag{5.26}$$

Proof The continuity of $P_t[\phi]$ follows from Theorem 1.162. Differentiating $P_t[\phi]$ and using its definition we obtain

$$\langle \nabla P_t[\phi](x), h \rangle = \mathbb{E} \langle D\phi(X(t;x)), \nabla(X(t;x))h \rangle$$

so (5.25) follows from Lemma 5.14. The strong continuity of the differential can be proved as follows. Given $h \in H$ and $t > 0$, consider a sequence x_n of elements of H converging to $x \in H$. We have

$$
\begin{aligned}
\langle \nabla P_t[\phi](x), h \rangle &- \langle \nabla P_t[\phi](x_n), h \rangle \\
&= \mathbb{E}\left(\langle D\phi(X(t;x)), \zeta^{x,h}(t) \rangle\right) - \mathbb{E}\left(\langle D\phi(X(t;x_n)), \zeta^{x_n,h}(t) \rangle\right) \\
&\leq \mathbb{E}\left|\langle D\phi(X(t;x)) - D\phi(X(t;x_n)), \zeta^{x,h}(t) \rangle\right| \\
&\quad + \mathbb{E}\left|\langle D\phi(X(t;x_n)), \zeta^{x_n,h}(t) - \zeta^{x,h}(t) \rangle\right| \\
&\leq I_1(n) + I_2(n) := ae^{\alpha t}|h|\, \mathbb{E}\left|D\phi(X(t;x)) - D\phi(X(t;x_n))\right| \\
&\quad + \|D\phi\|_0\, \mathbb{E}\left|\zeta^{x_n,h}(t) - \zeta^{x,h}(t)\right|,
\end{aligned}
$$

where $ae^{\alpha t}|h|$ is introduced in Lemma 5.14.

$I_1(n)$ converges to 0 when $n \to +\infty$ thanks to the dominated convergence theorem, the boundedness and the continuity of $D\phi$ and (1.70). Observe that, since $\{x_n\}_{n\in\mathbb{N}}$ is countable, we can find a subset of Ω of measure 1 where (1.70) holds for any n (with x_n and x as ξ_1 and ξ_2, respectively, moreover $f(r)$ appearing in (1.70) is, in our case, just a positive constant independent of r).

For $I_2(n)$ observe that

$$
\begin{aligned}
\mathbb{E}\left|\zeta^{x_n,h}(t) - \zeta^{x,h}(t)\right| \\
= \mathbb{E}\left|\int_0^t e^{(t-s)A}\left(\nabla b(X(s,x_n))\zeta^{x_n,h}(s) - \nabla b(X(s,x))\zeta^{x,h}(s)\right)ds\right| \\
\leq \mathbb{E}\left[C\int_0^t \left|(\nabla b(X(s,x)) - \nabla b(X(s,x_n)))\,\zeta^{x,h}(s)\right| ds \right. \\
\left. + C\int_0^t \left|\nabla b(X(s,x_n))\left(\zeta^{x,h}(s) - \zeta^{x_n,h}(s)\right)\right| ds\right] \\
\leq \mathbb{E}\left[C\int_0^t \left|(\nabla b(X(s,x)) - \nabla b(X(s,x_n)))\,\zeta^{x,h}(s)\right| ds\right] \\
+ CK\int_0^t \mathbb{E}\left|\left(\zeta^{x_n,h}(s)\right) - \left(\zeta^{x,h}(s)\right)\right| ds \quad (5.27)
\end{aligned}
$$

for some positive constant C coming from Hypothesis 5.1-(A) and with K from Hypothesis 5.1-(D). Thanks to the strong continuity of ∇b, the boundedness of $\|\nabla b\|_0$ and of $|\zeta^{x,h}(s)|$ (Hypothesis 5.1-(D) and Lemma 5.14), (1.70) (recall again that we can find a subset of Ω of measure 1 where (1.70) holds for any n) and the dominated convergence theorem, the term

$$
\mathbb{E}\left[C\int_0^t \left|(\nabla b(X(s,x)) - \nabla b(X(s,x_n)))\,\zeta^{x,h}(s)\right| ds\right]
$$

converges to 0 when $n \to \infty$. Thus we can apply Gronwall's Lemma to (5.27) and conclude that $I_2(n)$ converges to 0 when $n \to +\infty$. This concludes the proof of the strong continuity of $DP_t[\phi]$.

The bound (5.26) follows from the bound for the Gâteaux differential of X proved in Lemma 5.14 and the hypotheses on ϕ. $\qquad \square$

Corollary 5.16 *Assume that Hypotheses 5.1 and 5.4 hold. For any $\phi \in C_b^1(H)$, $P_t[\phi] \in W_Q^{1,2}(H, m)$. In particular, $D_Q P_t[\phi]$ is well defined and it equals $Q^{1/2} \nabla P_t[\phi]$.*

Proof Thanks to Lemma 5.15, $P_t[\phi]$ satisfies Hypothesis 5.1-(D) and it is bounded so we can apply to it Lemma 5.8-(i)(ii). The conclusion follows by the characterization of $W_Q^{1,2}(H, m)$ given after Definition 5.11. $\qquad \square$

Lemma 5.17 *Let Hypothesis 5.1 be satisfied, let b_n be as in Part (i) of Lemma 5.8, let $x \in H$ and $X(\cdot) = X(\cdot; x)$ be the solution of (5.8). The following hold:*

(i) *If, for some sequence x_n converging to x in H, we denote by $X_n(\cdot) = X_n(\cdot; x_n)$ the unique solution of the equation*

$$\begin{cases} dX_n(s) = (AX_n(s) + b_n(X_n(s))) \, dt + dW_Q(s), \\ X(0) = x_n, \end{cases} \tag{5.28}$$

then, for any $p > 1$,

$$\lim_{n \to \infty} \sup_{t \in [0,T]} \mathbb{E} |X_n(t; x_n) - X(t; x)|^p = 0. \tag{5.29}$$

(ii) *Let $X_n(\cdot)$, x_n be as in Part (i) above. Denote by $\zeta_n^{x_n, h}(\cdot)$ the solution of (5.23), where $X(\cdot)$ is replaced by $X_n(\cdot)$, b by b_n and x by x_n. Then, for any $p > 1$,*

$$\lim_{n \to \infty} \sup_{t \in [0,T]} \mathbb{E} \left(\sup_{|h| \le 1} \left| \zeta^{x,h}(t) - \zeta_n^{x_n, h}(t) \right| \right)^p = 0. \tag{5.30}$$

Proof For Part (i) we observe that for any $t \in [0, T]$

$$X_n(t; x_n) - X(t; x) = e^{tA}(x_n - x) + \int_0^t e^{(t-s)A}(b_n(X_n(s; x_n)) - b(X(s; x)))ds$$

$$= e^{tA}(x_n - x) + \int_0^t e^{(t-s)A}\big([b_n(X_n(s; x_n)) - b_n(X(s; x))]$$

$$+[b_n(X(s; x)) - b(X(s; x))]\big)ds$$

and thus

$$\mathbb{E}|X_n(t; x_n) - X(t; x)|^p$$

$$\leq C_T |x_n - x|^p + C_T \int_0^t l^p \mathbb{E}|X_n(s; x_n) - X(s; x)|^p ds$$

$$+ C_T \int_0^T \mathbb{E}|b_n(X(s; x)) - b(X(s; x))|^p ds$$

for a constant C_T depending on T. For any $s \in [0, T]$ the expression $\mathbb{E}|b_n(X(s; x)) - b(X(s; x))|^p$ converges to 0 thanks to Lemma 1.51 if we use Lemma 5.8 and the uniform moment estimates of (1.69). The claim thus follows by applying Gronwall's Lemma.

The argument for Part (ii) is similar. Indeed, for any $t \in [0, T]$,

$$(\zeta_n^{x_n, h}(t) - \zeta^{x, h}(t)) = \int_0^t e^{(t-s)A}[Db_n(X_n(s; x_n))\zeta_n^{x_n, h}(s) - \nabla b(X(s; x))\zeta^{x, h}(s)]ds$$

$$= \int_0^t e^{(t-s)A}[Db_n(X_n(s; x_n))(\zeta_n^{x_n, h}(s) - \zeta^{x, h}(s))$$

$$+ (Db_n(X_n(s; x_n)) - \nabla b(X(s; x)))\zeta^{x, h}(s)]ds.$$

So,

$$\sup_{|h| \leq 1} \left|\zeta_n^{x_n, h}(t) - \zeta^{x, h}(t)\right| \leq C_T \int_0^t \left[\|Db_n(X_n(s; x_n))\| \sup_{|h| \leq 1} \left|\zeta_n^{x_n, h}(s) - \zeta^{x, h}(s)\right| \right.$$

$$+ |(Db_n(X_n(s; x_n)) - \nabla b(X(s; x)))\zeta^{x, h}(s)| ds.$$

By taking the p-th powers and the expectations of the two sides and then using (5.13) we obtain, for a different constant C_T,

$$\mathbb{E}\left(\sup_{|h| \leq 1} \left|\zeta_n^{x_n, h}(t) - \zeta^{x, h}(t)\right|\right)^p \leq C_T \int_0^t K^p \mathbb{E}\left(\sup_{|h| \leq 1} \left|\zeta_n^{x_n, h}(s) - \zeta^{x, h}(s)\right|\right)^p ds + I_n,$$

where

$$I_n := C_T \int_0^T \mathbb{E}\left[|(Db_n(X_n(s; x_n)) - \nabla b(X(s; x)))\zeta^{x, h}(s)|^p\right]ds.$$

All we need to do now is to prove that I_n converges to 0. Then the claim will be a direct consequence of Gronwall's Lemma. To show this it is enough to show that for any subsequence I_{n_k} there exists a sub-subsequence converging to 0.

Let us then consider a subsequence of X_n (denoted again by X_n). Thanks to (5.29),

$$\int_0^T \mathbb{E}\left[|X_n(s, x_n) - X(s, x)|^p\right]ds \xrightarrow{n \to \infty} 0$$

and then we can extract a subsequence (denoted again by X_n) such that $X_n(\cdot, x_n)(\cdot)$ converges $(ds \otimes \mathbb{P})$-a.e. to $X(\cdot, x)(\cdot)$ (ds denotes the Lebesgue measure on $\mathbb{R}$). So, using Lemma 5.8-(i), $|(Db_n(X_n(\cdot; x_n)) - \nabla b(X(\cdot; x)))\zeta^{x,h}(\cdot)|^p$ converges to 0, $(ds \otimes \mathbb{P})$-a.e. Since, by (5.13) and the bound on $|\zeta^{x,h}|$ given by Lemma 5.14, these functions are bounded uniformly in n, we can thus conclude using the dominated convergence theorem that $I_n \to 0$. $\qquad\square$

Lemma 5.18 *Assume that Hypotheses 5.1 and 5.4 hold and $\phi \in C_b^1(H)$. Then, for any $t \in [0, T]$,*

$$\phi(X(t; x)) = P_t[\phi](x) + \int_0^t \langle \nabla P_{t-s}[\phi](X(s; x)), dW_Q(s)\rangle \qquad \mathbb{P} \ a.e. \quad (5.31)$$

Proof Step 1. The claim is proved for $b \in UC_b^2(H, H)$ and $\phi \in UC_b^2(H)$ in [582], Lemma 6.11, p. 181.

To extend the result to the general case, in the next step we will consider $\phi \in UC_b^2(H)$ and b which satisfies Hypothesis 5.1-(D), and in the third step we will prove the result in full generality.

Step 2. Consider $\phi \in UC_b^2(H)$ and b satisfying Hypothesis 5.1-(D). Let b_n be the sequence found in Part (i) of Lemma 5.8, $X_n(t; x)$ be the solution of (5.28) with $x_n = x$, and $P_t^n[\phi](x) = \mathbb{E}\phi(X_n(t; x))$, $t \geq 0$, be the corresponding transition semigroup.

Thanks to (5.29) (with $x_n = x$), up to extracting a subsequence,

$$\lim_{n \to +\infty} X_n(s; x) = X(s; x) \qquad \text{for } ds \otimes \mathbb{P}\text{-almost any } (s; \omega) \in [0, T] \times \Omega. \quad (5.32)$$

Observe now that, for any $x \in H$,

$$\lim_{n \to \infty} P_t^n[\phi](x) = P_t[\phi](x). \quad (5.33)$$

Indeed, we have

$$\left|P_t^n[\phi](x) - P_t[\phi](x)\right| = \mathbb{E}|\phi(X_n(t; x)) - \phi(X(t; x))| \leq C\mathbb{E}|X_n(t; x) - X(t; x)|^2$$

so the claim follows from (5.29). Observe also that by Lemma 5.15 we have

$$\nabla P_t[\phi](x) = \mathbb{E}\left((\zeta^x(t))^* D\phi(X(t; x))\right), \qquad \nabla P_t^n[\phi](x) = \mathbb{E}\left((\zeta_n^x(t))^* D\phi(X_n(t; x))\right).$$

Thus, using (5.29), (5.30), and a universal bound on $\|\zeta_n^x(t)\|$ given by Lemma 5.14, we easily obtain that

$$\sup_{n, x, t}(|\nabla P_t^n[\phi](x)| + |\nabla P_t[\phi](x)|) \leq C \quad (5.34)$$

for some constant C. We can then conclude using the dominated convergence theorem if we can show that for almost every $s \in [0, t]$, $\lim_{n \to \infty} \nabla P^n_{t-s}[\phi](X_n(s; x)$ $(\omega)) = \nabla P_{t-s}[\phi](X(s; x)(\omega))$ for $\mathbb{P}$-a.e. ω. In fact, we prove this convergence for any (s, ω) where (5.32) holds.

Given $(s, \omega) \in [0, t] \times \Omega$ where the convergence (5.32) holds, we rewrite it as $y_n := X_n(s, x)(\omega)$, $y_n \xrightarrow{n \to \infty} y := X(s, x)(\omega)$ in H. By Lemma 5.15,

$$
\begin{aligned}
\left| \nabla P^n_{t-s}[\phi](y_n) - \nabla P_{t-s}[\phi](y) \right| \\
= \sup_{h \in H,\, |h| \leq 1} \left| \langle \nabla P^n_{t-s}[\phi](y_n) - \nabla P_{t-s}[\phi](y), h \rangle \right| \\
\leq I^n_1 + I^n_2 := \sup_{h \in H,\, |h| \leq 1} \left| \mathbb{E} \langle D\phi(y_n), \zeta^{y_n,h}_n(t-s) - \zeta^{y,h}(t-s) \rangle \right| \\
+ \sup_{h \in H,\, |h| \leq 1} \left| \mathbb{E} \langle D\phi(y_n) - D\phi(y), \zeta^{y,h}(t-s) \rangle \right|.
\end{aligned}
$$

We have

$$
I^n_1 \leq \| D\phi \|_0 \left(\mathbb{E} \sup_{h \in H,\, |h| \leq 1} \left| \zeta^{y_n,h}_n(t-s) - \zeta^{y,h}(t-s) \right| \right),
$$

which converges to 0 by (5.30). Moreover, $I^n_2 \to 0$ thanks to the boundedness of $\zeta^y(t-s)$ given by Lemma 5.14 and the continuity of $D\phi$.

The result is thus true for any $\phi \in UC^2_b(H)$ and b satisfying Hypothesis 5.1-(D).

Step 3. Assume now that b satisfies Hypothesis 5.1-(D) and $\phi \in C^1_b(H)$. Let ϕ_n be the approximating sequence described in Part (iii) of Lemma 5.8. We have, for any $x \in H$,

$$
\lim_{n \to \infty} P_t[\phi_n](x) = P_t[\phi](x). \tag{5.35}
$$

Indeed,

$$
|P_t[\phi_n](x) - P_t[\phi](x)| = \mathbb{E} \, |\phi_n(X(t; x)) - \phi(X(t; x))|,
$$

which converges to 0 thanks to Lemma 5.8-(iii) and the dominated convergence theorem. Moreover, for any $x \in H$, by Lemma 5.15,

$$
\begin{aligned}
|\nabla P_{t-s}[\phi_n](x) - \nabla P_{t-s}[\phi](x)| \\
= \sup_{h \in H,\, |h| \leq 1} |\langle \nabla P_{t-s}[\phi_n](x) - \nabla P_{t-s}[\phi](x), h \rangle| \\
= \sup_{h \in H,\, |h| \leq 1} \left| \mathbb{E} \langle D\phi_n(x) - D\phi(x), \zeta^{x,h}(t-s) \rangle \right| \\
\leq |D\phi_n(x) - D\phi(x)| \sup_{h \in H,\, |h| \leq 1} \mathbb{E} \left| \zeta^{y,h}(t-s) \right|,
\end{aligned}
$$

which converges to 0 thanks to Lemmas 5.8-(iii) and 5.14. $\square$

We define the operator $\mathcal{R}$ as follows.

Definition 5.19 (*The operator* $\mathcal{R}$) Given $\phi \in C_b^1(H)$, we define for any $t \in [0, T]$,

$$(\mathcal{R}\phi)(t) := D_Q P_t[\phi].$$

The operator $\mathcal{R}$ is well defined thanks to Corollary 5.16.

The next proposition provides an identity which allows us to extend the operator $\mathcal{R}$ to the whole space $L^2(H, \overline{B}, m)$.

Proposition 5.20 *Assume that Hypotheses 5.1 and 5.4 hold. For every $\phi \in C_b^1(H)$*

$$\int_0^T \left| D_Q P_t[\phi] \right|_{L_m^2}^2 dt = |\phi|_{L_m^2}^2 - |P_T[\phi]|_{L_m^2}^2 . \tag{5.36}$$

Moreover, the operator $\mathcal{R}$ has a unique extension to a bounded operator

$$\mathcal{R} : L^2(H, \overline{B}, m) \to L^2\left(0, T; L^2(H, \overline{B}, m)\right),$$

with

$$|(\mathcal{R}\phi)|_{L^2(0,T;L^2(H,\overline{B},m))}^2 = \int_0^T |(\mathcal{R}\phi)(t)|_{L_m^2}^2 dt = |\phi|_{L_m^2}^2 - |P_T[\phi]|_{L_m^2}^2 \tag{5.37}$$

for any $\phi \in L^2(H, \overline{B}, m)$.

Proof Let $\phi \in C_b^1(H)$. Then (5.31) yields

$$\mathbb{E}[\phi^2(X(T,x))] = (P_T[\phi](x))^2 + \int_0^T \mathbb{E}\left| Q^{1/2} \nabla P_{T-t}[\phi](X(t,x)) \right|^2 dt.$$

Recall that, by Corollary 5.16, since $\phi \in C_b^1(H)$, we have $D_Q P_t[\phi] = Q^{1/2}\nabla P_t[\phi]$. Thus, integrating the previous identity with respect to m and rearranging the terms we get

$$\int_0^T \int_H \mathbb{E}\left| D_Q P_t[\phi](X(t,x)) \right|^2 dm(x)dt$$

$$= \int_0^T \int_H \mathbb{E}[\phi^2(X(T,x))dm(x)] - \int_0^T \int_H (P_T[\phi](x))^2 dm(x),$$

so, by using the invariant measure property (5.11), we obtain (5.36) for all $\phi \in C_b^1(H)$. The result follows thanks to the density of $C_b^1(H)$ in $L^2(H, \overline{B}, m)$ (Lemma 5.7-(i)). $\qquad\square$

Remark 5.21 In the particular case where $b = 0$, the operator $\mathcal{A}$ reduces to the Ornstein–Uhlenbeck operator and the semigroup P_t is called the Ornstein–Uhlenbeck semigroup. In particular, if $\|e^{At}\| \le Me^{-\omega t}$ with $M \in \mathbb{R}$ and $\omega > 0$ (the condition assumed in the whole remark), the invariant measure for P_t is the Gaussian measure $N(0, Q_\infty)$, where

$$Q_\infty := \int_0^{+\infty} e^{sA} Q e^{sA^*} ds.$$

In this case there are links between the closability of the operator D_Q, the smoothing properties of the semigroup P_t and the characteristics of certain controllability problems:

(1) If we consider the following linear controlled system,

$$\frac{dX(t)}{dt} = AX(t) + Q^{1/2}a(t), \quad X(0) = 0, \qquad (5.38)$$

the set of points of H that can be reached by the system in an infinite time using a control in the set $L^2(0, +\infty; H)$ is equal to $Q_\infty^{1/2}(H)$ (see [584], Theorem 2.3, page 210) and it can be proved (see [299], Theorem 6.1) that the closability of the operator D_Q is equivalent to the density of the set

$$\left\{ x \in H : Q^{1/2}x \in Q_\infty^{1/2}(H) \right\}$$

in H.

(2) Fix $t > 0$. The null-controllability in time t of the system

$$\frac{dX(t)}{dt} = AX(t) + Q^{1/2}a(t), \qquad X(0) = x,$$

is defined as the capability, by choosing a suitable control in $L^2(0, t; H)$, of reaching at time t the point 0, given any initial condition $x \in H$. The null-controllability of the described system (see [584], Theorem 2.3, p. 210) is equivalent to the condition

$$e^{tA}(H) \subset Q_t^{1/2}(H).$$

This condition is equivalent (see Theorem 2.23, p. 53 of [180]) to the fact that all the transition probabilities are mutually absolutely continuous and (see Theorem 9.26, p. 260 and Remark 9.29, p. 265 of [180]) to the fact that the semigroup P_t is strong Feller (see Definition 1.159).

By the results of Sect. 4.3.1, given $\phi \in L^2(H, \overline{B}, m)$, it can be seen that $\nabla P_t[\phi]$ is well defined for $t > 0$ if and only if (5.39) is satisfied (see Hypothesis 4.29, Remark 4.30 and Theorem 4.37). In this case (see Proposition 10.3.1, page 218 of [179]) the singularity of $|\nabla P_t[\phi]|_{L^2_{m,H}}$ at $t = 0^+$, similarly to the one of $|\nabla P_t[\phi]|_0$, is estimated from above by $\|\Gamma(t)\|$, where as in (4.59), $\Gamma(t) := Q_t^{-1/2} e^{tA}$. Similarly, $D_Q P_t[\phi]$ is well defined for $\phi \in L^2(H, \overline{B}, m)$ and $t > 0$ if and only if

$$e^{tA} Q^{1/2}(H) \subset Q_t^{1/2}(H), \qquad (5.39)$$

i.e. if and only if every point of $Q^{1/2}(H)$ is null controllable in time t (see again Hypothesis 4.29, Remark 4.30 and Theorem 4.41 when $G = Q^{1/2}$). In this case the singularity of $\left|D_Q P_t[\phi]\right|_{L^2_{m,H}}$ at 0^+ has the same behavior as the norm of the operator

$$\Gamma_{Q^{1/2}}(t) := Q_t^{-1/2} e^{tA} Q^{1/2}.$$

More on this subject can be found in [120], Sect. 10.3 of [179], Sect. 5.3 of [431, 432].

The observations of part (2) are useful to provide examples where the approach of the previous chapter cannot be applied while the theory of this chapter works. This is the case when the hypotheses of this chapter hold but (5.39) does not hold or when it holds but $\|\Gamma_{Q^{1/2}}(t)\|$ is not integrable at 0^+. Such examples are, for instance, delay equations (see Sect. 5.6.1), where the semigroup can never be strong Feller for t smaller than the delay appearing in the equation (r in Sect. 5.6.1) or certain classes of second-order SPDEs in the whole space, see Sect. 5.6.3. ∎

Remark 5.22 If D_Q is closable in $L^2(H, \overline{\mathcal{B}}, m)$ then $\mathcal{R}(\phi)(t) = \overline{D_Q P_t}[\phi](t)$ for all $t > 0$ and $\phi \in L^2(H, \overline{\mathcal{B}}, m)$. In this case (5.36) is easier to obtain and the whole study of the HJB equation (5.1) is simpler. This is true, in particular, when Q is boundedly invertible. ∎

5.2.6 Two Key Lemmas

Here we use Proposition 5.20 to provide two estimates that will be essential in the following. We begin with an estimate regarding the convolution of P_t.

Lemma 5.23 *Assume that Hypotheses 5.1 and 5.4 hold and let P_t be defined as in (5.10). Given $f \in L^2\left(0, T; L^2(H, \overline{\mathcal{B}}, m)\right)$ we define*

$$G_1 f(t) := \int_t^T P_{s-t}[f(s)]\, ds, \quad t \in [0, T],$$

and

$$G_2 f(t) := \int_t^T \mathcal{R}(f(s))(s-t)\, ds, \quad t \in [0, T].$$

Then

$$\int_0^T |G_1 f(t)|^2_{L^2_m}\, dt \le T^2 \int_0^T |f(t)|^2_{L^2_m}\, dt, \tag{5.40}$$

$G_2 f(t) \in L^2(H, \overline{\mathcal{B}}, m; H)$ *for almost every $t \in [0, T]$ and*

$$\int_0^T |G_2 f(t)|^2_{L^2_{m,H}}\, dt \le T \int_0^T |f(t)|^2_{L^2_m}\, dt. \tag{5.41}$$

Proof For the first estimate, observe that

$$\int_0^T |G_1 f(t)|^2_{L^2_m} \, dt = \int_0^T \left| \int_t^T P_{s-t}[f(s)] ds \right|^2_{L^2_m} dt$$

$$\leq \int_0^T \left(\int_t^T |P_{s-t}[f(s)]|_{L^2_m} \, ds \right)^2 dt \leq \int_0^T \left(\int_0^T |f(s)|_{L^2_m} \, ds \right)^2 dt$$

$$\leq \int_0^T T \int_0^T |f(s)|^2_{L^2_m} \, ds dt = T^2 \int_0^T |f(s)|^2_{L^2_m} \, ds.$$

We prove the second inequality. Assume first that $f \in C^1_b([0, T] \times H)$ and $f(t) \in \mathcal{F}C^1_0(H)$ (defined in Sect. A.2) for all $t \geq 0$. Then $D_Q P_{s-t}[f(s)]$ is well defined for $s \geq t$ and so is $D_Q G_1(t)$ for $t > 0$. Moreover,

$$\int_0^T |G_2 f(t)|^2_{L^2_{m,H}} \, dt \leq \int_0^T \left(\int_t^T |D_Q P_{s-t}[f(s)]|_{L^2_{m,H}} \, ds \right)^2 dt$$

$$\leq \int_0^T T \int_t^T |D_Q P_{s-t}[f(s)]|^2_{L^2_{m,H}} \, ds dt = T \int_0^T \int_0^s |D_Q P_r[f(s)]|^2_{L^2_{m,H}} \, dr ds$$

$$\leq \int_0^T T \int_0^T |D_Q P_r[f(s)]|^2_{L^2_{m,H}} \, dr dt.$$

Hence by (5.36),

$$\int_0^T |G_2(t)|^2_{L^2_{m,H}} \, dt \leq T \int_0^T |f(t)|^2_{L^2_m} dt.$$

If $f \in L^2\left(0, T; L^2(H, \overline{B}, m)\right)$ is arbitrary, then, thanks to Lemma 5.7 applied to the space $[0, T] \times H$, there exists a sequence $f_n \in C^1_b([0, T] \times H)$, with $f_n(t) \in \mathcal{F}C^1_0(H)$ for any $t \in [0, T]$, which converges to f in $L^2\left(0, T; L^2(H, \overline{B}, m)\right)$. Repeating the above arguments for

$$G_1^n(t) = \int_t^T P_{s-t}[f_n(s)] ds$$

we find that

$$\int_0^T |D_Q \left(G_1^n(t) - G_1^m(t)\right)|^2_{L^2_{m,H}} \, dt \leq T \int_0^T |f_n(t) - f_m(t)|^2_{L^2_m} \, dt.$$

Hence the sequence $D_Q G_1^n$ is convergent in $L^2\left(0, T; L^2(H, \overline{B}, m; H)\right)$. Moreover, by the Fubini Theorem,

$$\int_0^T \left| D_Q G_1^n(t) - G_2(t) \right|_{L^2_{m,H}}^2 dt$$

$$= \int_0^T \left| \int_t^T \left[D_Q P_{s-t}[f_n(s)]ds - \mathcal{R}\left(f(s)\right)(s-t) \right] ds \right|_{L^2_{m,H}}^2 dt$$

$$\leq T \int_0^T ds \int_0^T \left| D_Q P_t[f_n(s)] - \mathcal{R}\left(f(s)\right)(t) \right|_{L^2_{m,H}}^2 dt$$

$$= T \int_0^T ds \int_0^T \left| \mathcal{R}\left(f_n(s) - f(s)\right)(t) \right|_{L^2_{m,H}}^2 dt,$$

which gives, by Proposition 5.20,

$$\int_0^T \left| D_Q G_1^n(t) - G_2(t) \right|_{L^2_{m,H}}^2 dt$$

$$\leq T \int_0^T \left[|f_n(s) - f(s)|_{L^2_m}^2 - |P_T[f_n(s) - f(s)]|_{L^2_m}^2 \right] ds$$

$$\leq T \int_0^T |f_n(s) - f(s)|_{L^2_m}^2 ds, \tag{5.42}$$

so that $D_Q G_1^n$ is convergent in $L^2\left(0, T; L^2(H, \overline{\mathcal{B}}, m; H)\right)$ to G_2 and (5.41) holds. $\square$

The following corollary can be deduced from the proof of Lemma 5.23.

Corollary 5.24 *Assume that Hypotheses 5.1 and 5.4 hold. Let $f_n \to f$ be in $L^2\left(0, T; L^2(H, \overline{\mathcal{B}}, m)\right)$. Then, by (5.42), there exists a subsequence f_{n_k} such that for a.e. $(s, t) \in [0, T] \times [0, T]$ and $s \leq t$,*

$$D_Q P_{t-s}[f_{n_k}(s)] \to \mathcal{R}\left(f(s)\right)(t-s) \quad \text{in} \quad L^2(H, \overline{\mathcal{B}}, m; H).$$

This fact will be useful in Sect. 5.5.

We now extend the operator D_Q to all functions u that are mild solutions to suitable Cauchy problems.

Consider $g \in L^2\left(H, \overline{\mathcal{B}}, m\right)$ and $f \in L^2\left(0, T; L^2\left(H, \overline{\mathcal{B}}, m\right)\right)$. Consider the Cauchy problem:

$$\begin{cases} u_t(t) + \mathcal{A}u(t) + f(t) = 0 \ \ t \in [0, T), \\ \\ u(T, x) = g(x) \end{cases} \tag{5.43}$$

and define the mild solution of (5.43) as

$$u(t) = P_{T-t}[g] + \int_t^T P_{s-t}[f(s)]ds, \quad t \in [0, T]. \tag{5.44}$$

We denote by $\Upsilon_A(0, T)$ the set of all the functions in $L^2\left(0, T; L^2\left(H, \overline{B}, m\right)\right)$ that can be written in the form (5.44) for some f, g as above. The functions in $\Upsilon_A(0, T)$ belong to $C\left([0, T], L^2\left(H, \overline{B}, m\right)\right)$.

For the functions in $\Upsilon_A(0, T)$ we define the operator $\tilde{D}_Q$ by

$$(\tilde{D}_Q u)(t) := \mathcal{R}(g)(T - t) + \int_t^T \mathcal{R}(f(s))(s - t)\, ds, \quad t \in [0, T]. \tag{5.45}$$

Observe that $\tilde{D}_Q$ is well defined on $\Upsilon_A(0, T)$. Indeed, if we have $P_{T-t}[g_1] + \int_t^T P_{s-t}[f_1(s)]ds = P_{T-t}[g_2] + \int_t^T P_{s-t}[f_2(s)]ds$ then, taking $t = T$ we obtain $g_1 = g_2$ and then, $\int_t^T P_{s-t}[f_1(s)]ds = \int_t^T P_{s-t}[f_2(s)]ds$ so that $\int_t^T \mathcal{R}(f_1(s))(s - t)\, ds = \int_t^T \mathcal{R}(f_2(s))(s - t)\, ds$.

The following proposition gives a continuity result for $\tilde{D}_Q$.

Proposition 5.25 *Suppose that Hypotheses 5.1 and 5.4 hold. Consider two sequences $g_n \subset L^2\left(H, m\right)$ and $f_n \subset L^2\left(0, T; L^2\left(H, \overline{B}, m\right)\right)$ such that*

$$g_n \longrightarrow g \qquad \text{in } L^2\left(H, \overline{B}, m\right),$$

$$f_n \longrightarrow f \qquad \text{in } L^2\left(0, T; L^2\left(H, \overline{B}, m\right)\right).$$

Then, setting

$$u_n(t) = P_{T-t}[g_n] + \int_t^T P_{s-t}[f_n(s)]ds, \quad t \in [0, T], \tag{5.46}$$

and

$$\tilde{D}_Q u_n(t) = \mathcal{R}(g_n)(T - t) + \int_t^T \mathcal{R}(f_n(s))(s - t)\, ds, \quad t \in [0, T],$$

we have

$$u_n \longrightarrow u \qquad \text{in } C\left([0, T], L^2\left(H, \overline{B}, m\right)\right), \tag{5.47}$$

$$\tilde{D}_Q u_n \longrightarrow \tilde{D}_Q u \qquad \text{in } L^2\left(0, T; L^2\left(H, \overline{B}, m; H\right)\right). \tag{5.48}$$

Proof We start with the first claim. Subtracting (5.44) from (5.46) we get

$$u_n(t) - u(t) = P_{T-t}[g_n - g] + \int_t^T P_{s-t}[f_n(s) - f(s)]\, ds$$

so that, by the strong continuity of P_t,

$$|u_n(t) - u(t)|^2_{L^2_m} \leq C_T \left[|g_n - g|^2_{L^2_m} + \int_t^T |f_n(s) - f(s)|^2_{L^2_m} \, ds \right],$$

which gives (5.47) by taking the supremum over $[0, T]$. To prove (5.48) we observe that we have

$$\tilde{D}_Q (u_n(t) - u(t)) = \mathcal{R}(g_n - g)(T - t) + \int_t^T \mathcal{R}(f_n(s) - f(s))(s - t) \, ds$$

so that, by (5.37) and (5.41),

$$\int_0^T \left| \tilde{D}_Q u_n(t) - \tilde{D}_Q u(t) \right|^2_{L^2_{m,H}} \leq |g_n - g|^2_{L^2_m} + T \int_0^T |f_n(s) - f(s)|^2_{L^2_m} \, ds,$$

which shows (5.48). □

Remark 5.26 If g and f are differentiable functions, the operator D_Q is well defined on the functions u of the form (5.44). In (5.45) we define the operator $\tilde{D}_Q$ on all the functions of the form (5.44), where $g \in L^2\left(H, \overline{\mathcal{B}}, m\right)$ and $f \in L^2\left(0, T; L^2\left(H, \overline{\mathcal{B}}, m\right)\right)$. Thus Proposition 5.25 asserts that the operator $\tilde{D}_Q$ extends D_Q on $\Upsilon_A(0, T)$ "without closability problems" if the functions in the approximating sequence have the form (5.46). ■

5.3 The HJB Equation

In this section we study the existence and uniqueness of solutions to the HJB equation[5]

$$\begin{cases} u_t + \mathcal{A}u + F_0\left(t, x, D_Q u\right) + l(t, x) = 0, \\ u(T, x) = g(x) \end{cases} \tag{5.49}$$

with $g \in L^2(H, \overline{\mathcal{B}}, m)$. Observe that this corresponds to F in (5.1) having the form $F(t, x, p) = F_0\left(t, x, Q^{1/2}p\right)$. We assume that the following conditions are satisfied.

Hypothesis 5.27 (A) $F_0 : [0, T] \times H \times H \to \mathbb{R}$ is $Leb \otimes \overline{\mathcal{B}} \otimes \overline{\mathcal{B}}/\mathcal{B}(\mathbb{R})$-measurable (where Leb is the σ-field of Lebesgue measurable sets in $\mathbb{R}$) and there exists an $L \in \mathbb{R}$ such that

---

[5]Following the notation we use for HJB equations throughout the book, in the first line of (5.49) we only explicitly mention the dependence on t and x of the functions F_0 and l while we do not do so for u_t, $D_Q u$ and $\mathcal{A}u$.

$$|F_0(t, x, p) - F_0(t, x, q)| \leq L|p - q| \quad \text{and} \quad |F_0(t, x, p)| \leq L(1 + |p|)$$
$$(5.50)$$

for all $t \in [0, T]$ and $x, p, q \in H$.

(B) $l \in L^2\left(0, T; L^2(H, \overline{B}, m)\right)$ and $g \in L^2(H, \overline{B}, m)$.

Using the semigroup P_t defined in (5.10) and the variation of constants formula, as was done in Chap. 4, we can formally rewrite Eq. (5.49) in the following mild form:

$$u(t) = P_{T-t}[g] + \int_t^T P_{s-t}\left[F_0\left(s, \cdot, D_Q u(s)\right)\right] ds + \int_t^T P_{s-t}[l(s)] ds, \quad 0 \leq t \leq T,$$
$$(5.51)$$

where for simplicity we have written $D_Q u(s), l(s)$ for $D_Q u(s, \cdot), l(s, \cdot)$ and a similar convention is used later for other functions. We use this integral form to define a solution.

We will prove the existence of the solution of the HJB equation using a fixed point argument in the space $L^2\left(0, T; W_Q^{1,2}(H, m)\right)$. We can identify any element of $L^2\left(0, T; W_Q^{1,2}(H, m)\right)$ with an element (v, V) in $L^2\left(0, T; L^2(H, \overline{B}, m)\right) \times L^2\left(0, T; L^2(H, \overline{B}, m; H)\right)$. If $v(t) \in C_b^1(H)$ for almost every t, then $V(t) = D_Q v(t)$ for almost every t and the norm of $(v, V) = (v, D_Q v)$ in $L^2\left(0, T; W_Q^{1,2}(H, m)\right)$ can be written explicitly as follows

$$|(v, D_Q v)|^2_{L^2\left(0,T;W_Q^{1,2}\right)} = \int_0^T \left(|v(t)|^2_{L^2_m} + |D_Q v(t)|^2_{L^2_{m,H}}\right) dt.$$

To avoid any confusion in the notation we will always denote the elements of $L^2\left(0, T; W_Q^{1,2}(H, m)\right)$ as pairs.

Definition 5.28 By a *solution* of Eq. (5.51) (or *mild solution* of Eq. (5.49)), we mean a pair of functions

$$(u, U) \in L^2\left(0, T; W_Q^{1,2}(H, m)\right)$$
$$\subset L^2\left(0, T; L^2(H, \overline{B}, m)\right) \times L^2\left(0, T; L^2(H, \overline{B}, m; H)\right)$$

such that, for a.e. $t \in [0, T]$ and m-a.e.

$$u(t) = P_{T-t}[g] + \int_t^T P_{s-t}\left[F_0\left(s, \cdot, U(s)\right)\right] ds + \int_t^T P_{s-t}[l(s)] ds, \quad (5.52)$$

and

$$U(t) = \mathcal{R}(g)(T - t) + \int_t^T \mathcal{R}\left(F_0(s, \cdot, U(s))\right)(s - t)ds + \int_t^T \mathcal{R}\left(l(s)\right)(s - t)ds.$$
$$(5.53)$$

Remark 5.29 By the definition of $\tilde{D}_Q u$ in (5.45) and Definition 5.28 we immediately get $U = \tilde{D}_Q u$. ∎

Remark 5.30 If D_Q were closable, then it would be natural to define the solution of Eq. (5.51) as an element of $L^2\left(0, T; W_Q^{1,2}(H, m)\right)$ such that (5.51) is satisfied for a.e. $t \in [0, T]$ and m-a.e. But D_Q may not be closable, so elements of $W_Q^{1,2}(H, m)$ are not functions in general, but pairs of functions belonging to the product space $L^2(H, \overline{B}, m) \times L^2(H, \overline{B}, m; H)$.

Note that the second equation (5.53) is an obvious consequence of (5.52) if the operator D_Q is closable and, in this case, $U = D_Q u$. ∎

We will introduce a suitable nonlinear operator $\mathcal{M}$ which will allow us to use the fixed point argument. It will be defined in terms of a certain operator $\mathcal{M}_1$ and its derivative. Both of these operators will be initially defined on a subspace of $L^2\left(0, T; L^2(H, \overline{B}, m)\right)$ and then extended to $L^2\left(0, T; W_Q^{1,2}(H, m)\right)$. To make the distinction we will denote the extensions using the "overline": $\overline{\mathcal{M}_1}$ and $\overline{D_Q \mathcal{M}_1}$. As emphasized before, since the elements of $L^2\left(0, T; W_Q^{1,2}(H, m)\right)$ can be identified with a subspace of $L^2\left(0, T; L^2(H, \overline{B}, m)\right) \times L^2\left(0, T; L^2(H, \overline{B}, m; H)\right)$, we will use a one-argument notation for the non-extended operators (e.g. $\mathcal{M}_1(u)$) and a two-argument notation for the extended ones (e.g. $\overline{\mathcal{M}_1}(u, U)$).

Given g, l and F_0 satisfying Hypothesis 5.27, we define the operator $\mathcal{M}_1$ as follows:

$$
\begin{cases}
D(\mathcal{M}_1) = \left\{ v \in L^2\left(0, T; L^2(H, \overline{B}, m)\right) \right. \\
\qquad\qquad \left. : v(t) \in C_b^1(H) \quad \text{for a.e. } t \text{ and} \quad |(v, D_Q v)|_{L^2\left(0, T; W_Q^{1,2}\right)} < \infty \right\}, \\
\mathcal{M}_1 v(t) = P_{T-t}[g] + \displaystyle\int_t^T P_{s-t}\left[F_0\left(s, \cdot, D_Q v(s)\right)\right] ds + \int_t^T P_{s-t}[l(s)] ds, \quad t \leq T.
\end{cases}
$$

Remark 5.31 If g, l and F_0 are regular enough, then we can directly define $D_Q \mathcal{M}_1$. If $g \in L^2(H, \overline{B}, m)$, $l \in L^2(0, T; L^2(H, \overline{B}, m))$ and $F_0\left(s, x, D_Q v(s)\right) \in L^2(0, T; L^2(H, \overline{B}, m))$ we can use Lemma 5.23 to define $\tilde{D}_Q \mathcal{M}_1 v \in L^2\left(0, T; L^2\left(H, \overline{B}, m; H\right)\right)$ and it can be written as follows:

$$
\tilde{D}_Q \mathcal{M}_1 v(t) = \mathcal{R}(g)(T - t) + \int_t^T \mathcal{R}\left(F_0(s, \cdot, D_Q v(s))\right)(s - t)ds + \int_t^T \mathcal{R}\left(l(s)\right)(s - t)ds
$$

on $[0, T]$. In the following lemma we extend by continuity the operator $D_Q \mathcal{M}_1$ to $L^2\left(0, T; W_Q^{1,2}(H, m)\right)$ obtaining $\overline{D_Q \mathcal{M}_1}$. Since the definitions of $\tilde{D}_Q \mathcal{M}_1 v$ and $\overline{D_Q \mathcal{M}_1}$ coincide on $D(\mathcal{M}_1)$, they coincide once $\overline{D_Q \mathcal{M}_1}$ is extended to $L^2\left(0, T; W_Q^{1,2}(H, m)\right)$. ∎

Lemma 5.32 *Assume that Hypotheses 5.1, 5.4 and 5.27 hold. Then $\mathcal{M}_1$ extends to a Lipschitz mapping*

$$\overline{\mathcal{M}}_1 : L^2\left(0, T; W_Q^{1,2}(H, m)\right) \to L^2\left(0, T; L^2(H, \overline{\mathcal{B}}, m)\right)$$

with Lipschitz constant LT. The mapping $D_Q\mathcal{M}_1 : D(\mathcal{M}_1) \to L^2\left(0, T; L^2\right.$ $\left.\left(H, \overline{\mathcal{B}}, m; H\right)\right)$ extends to a Lipschitz mapping

$$\overline{D_Q\mathcal{M}_1} : L^2\left(0, T; W_Q^{1,2}(H, m)\right) \to L^2\left(0, T; L^2\left(H, \overline{\mathcal{B}}, m; H\right)\right)$$

with Lipschitz constant $LT^{1/2}$.

Proof Since $|F_0(t, x, p)| \le L(1 + |p|)$ for all $t \in [0, T]$ and $x, p \in H$, it follows from Lemma 5.23 that $\mathcal{M}_1 v \in L^2\left(0, T; L^2(H, \overline{\mathcal{B}}, m)\right)$ and $D_Q\mathcal{M}_1 v \in L^2 (0, T;$ $L^2\left(H, \overline{\mathcal{B}}, m\right))$ for every $v \in D(\mathcal{M}_1)$.

Given v_1 and v_2 in $D(\mathcal{M}_1)$, we have

$$\mathcal{M}_1\left(v_1 - v_2\right)(t) = \int_t^T P_{s-t}\left[F_0\left(s, \cdot, D_Q v_1(s)\right) - F_0\left(s, \cdot, D_Q v_2(s)\right)\right] ds, \quad t \in [0, T],$$

and therefore, since $\|P_t\| \le 1$ and by Hypothesis 5.27-(A),

$$\left|\mathcal{M}_1\left(v_1 - v_2\right)(t)\right|_{L_m^2} \le L \int_t^T \left|D_Q v_1(s) - D_Q v_2(s)\right|_{L_{m,H}^2} ds, \quad t \in [0, T].$$

Hence,

$$\int_0^T \left|\mathcal{M}_1\left(v_1 - v_2\right)(t)\right|_{L_m^2}^2 dt \le L^2 T^2 \int_0^T \left|D_Q v_1(t) - D_Q v_2(t)\right|_{L_{m,H}^2}^2 dt.$$

It follows that $\mathcal{M}_1$ may be extended to the whole space $L^2\left(0, T; W_Q^{1,2}(H, m)\right)$ by continuity and the resulting mapping is Lipschitz continuous with constant LT. Similarly, for v_1 and v_2 in $D(\mathcal{M}_1)$ and $t \in [0, T]$,

$$D_Q\mathcal{M}_1\left(v_1 - v_2\right)(t) = \int_t^T D_Q P_{s-t}\left[F_0\left(s, \cdot, D_Q v_1(s)\right) - F_0\left(s, \cdot, D_Q v_2(s)\right)\right] ds.$$

Using the notation introduced in Lemma 5.23 we obtain

$$\int_0^T \left|D_Q\mathcal{M}_1\left(v_1 - v_2\right)(t)\right|_{L_{m,H}^2}^2 dt$$

$$= \int_0^T \left|G_2\left(F_0\left(t, \cdot, D_Q v_1(t)\right) - F_0\left(t, \cdot, D_Q v_2(t)\right)\right)\right|_{L_{m,H}^2}^2 dt$$

$$\leq T \int_0^T \left| F_0\left(t, \cdot, D_Q v_1(t)\right) - F_0\left(t, \cdot, D_Q v_2(t)\right) \right|_{L_m^2}^2 dt$$

$$\leq L^2 T \int_0^T \left| D_Q \left(v_1(t) - v_2(t)\right) \right|_{L_{m,H}^2}^2 dt,$$

and therefore $D_Q \mathcal{M}_1$ extends to a Lipschitz continuous mapping on $L^2\left(0, T; W_Q^{1,2}(H, m)\right)$ with constant $LT^{1/2}$. $\qquad\square$

Remark 5.33 The operators $\overline{\mathcal{M}}_1$ and $\overline{D_Q \mathcal{M}}_1$ depend only on the second component of the elements of $L^2\left(0, T; W_Q^{1,2}(H, m)\right)$ but it is convenient for us to define them on $L^2\left(0, T; W_Q^{1,2}(H, m)\right)$ to apply the fixed point argument below. $\qquad\blacksquare$

Taking into account the extensions of the operators $\mathcal{M}_1$ and $D_Q \mathcal{M}_1$ provided by Lemma 5.32 we can define the operator

$$\begin{cases} \overline{\mathcal{M}} : L^2\left(0, T; W_Q^{1,2}(H, m)\right) \to L^2\left(0, T; W_Q^{1,2}(H, m)\right) \\ \overline{\mathcal{M}}(u, U) = (\overline{\mathcal{M}}_1(u, U), \overline{D_Q \mathcal{M}}_1(u, U)). \end{cases}$$

Remark 5.34 Using Proposition 5.20 and Lemma 5.23 we find that for a.e. $t \in [0, T]$,

$$\overline{\mathcal{M}}_1(u, U)(t) = P_{T-t}[g] + \int_t^T P_{s-t}\left[F_0(s, \cdot, U(s))\right] ds + \int_t^T P_{s-t}[l(s)] ds \tag{5.54}$$

and

$$\overline{D_Q \mathcal{M}}_1(u, U)(t)$$
$$= \mathcal{R}(g)(T - t) + \int_t^T \mathcal{R}\left(F_0(s, \cdot, U(s))\right)(s - t) ds + \int_t^T \mathcal{R}(l(s))(s - t) ds. \tag{5.55}$$

$\qquad\blacksquare$

Theorem 5.35 *Assume that Hypotheses 5.1, 5.4 and 5.27 hold. Then for every $g \in L^2(H, \overline{\mathcal{B}}, m)$ there exists a unique mild solution (u, U) to Eq. (5.49) in the sense of Definition 5.28. Moreover, $u \in C\left([0, T], L^2(H, \overline{\mathcal{B}}, m)\right)$ and $U = \tilde{D}_Q u$.*

Proof We apply the Banach Fixed Point Theorem to the mapping $\overline{\mathcal{M}}$ in the space $L^2\left(0, T; W_Q^{1,2}(H, m)\right)$ endowed with the norm $|\cdot|_{L^2\left(0,T;W_Q^{1,2}\right)}$ when T is sufficiently small. By Lemma 5.32, for any $(v_1, V_1), (v_2, V_2) \in L^2\left(0, T; W_Q^{1,2}(H, m)\right)$,

$$\int_0^T \left| \overline{\mathcal{M}}_1(v_1(t), V_1(t)) - \overline{\mathcal{M}}_1(v_2(t), V_2(t)) \right|_{L_m^2}^2 dt$$

$$\leq L^2 T^2 |(v_1, V_1) - (v_2, V_2)|_{L^2\left(0,T;W_Q^{1,2}\right)}^2 \quad (5.56)$$

and

$$\int_0^T \left| \overline{D_Q \mathcal{M}}_1(v_1(t), V_1(t)) - \overline{D_Q \mathcal{M}}_1(v_2(t), V_2(t)) \right|_{L_{m,H}^2}^2 dt$$

$$\leq L^2 T |(v_1, V_1) - (v_2, V_2)|_{L^2\left(0,T;W_Q^{1,2}\right)}^2. \quad (5.57)$$

From (5.56) and (5.57) we have

$$|\overline{\mathcal{M}}(v_1, V_1) - \overline{\mathcal{M}}(v_2, V_2)|_{L^2\left(0,T;W_Q^{1,2}\right)}$$

$$\leq L\sqrt{T(T+1)}|(v_1, V_1) - (v_2, V_2)|_{L^2\left(0,T;W_Q^{1,2}\right)}, \quad (5.58)$$

thus $\overline{\mathcal{M}}$ is a strict contraction for T sufficiently small. Thus we obtain a unique solution on a small time interval. The rest follows by standard iteration. Finally, denoting the solution by (u, U), since $F_0(s, \cdot, U(s)) \in L^2(0, T; L^2(H, \overline{B}, m))$ and P_t is a C_0-semigroup, we find that $u \in C([0, T], L^2(H, \overline{B}, m))$ thanks to (5.54).

The last statement is an immediate consequence of the definitions (see Remark 5.29).   □

Remark 5.36 Observe that the uniqueness of the solution stated in Theorem 5.35 has to be understood with respect to the reference measure m whose support can also be very thin. This is one of the drawbacks of the method. For results about existence of non-degenerate invariant measures, see Sect. 5.6 and the comments in the bibliographical notes.   ■

5.4   Approximation of Mild Solutions

We now show, following the approach of Chap. 4, that the mild solution of the HJB equation can be obtained as a limit of classical solutions. Thus we need to introduce the concept of a classical solution.

We introduce the operator $\mathcal{A}_1$ which is defined similarly to the operator $\mathcal{A}_1$ in (4.141):

$$\begin{cases} D(\mathcal{A}_1) = \left\{ \phi \in UC_b^2(H) \ : \ A^*D\phi \in UC_b(H, H) \text{ and } D^2\phi \in UC_b(H, \mathcal{L}_1(H)) \right\} \\[2mm] \mathcal{A}_1\phi = \frac{1}{2}\text{Tr}[QD^2\phi] + \langle x, A^*D\phi \rangle + \langle b(x), D\phi \rangle . \end{cases}$$

(5.59)

It is easy to see that $D(\mathcal{A}_1)$ endowed with the norm

$$\|\phi\|_{D(\mathcal{A}_1)} := \|\phi\|_0 + \|D\phi\|_0 + \|A^*D\phi\|_0 + \sup_{x \in H} \|D^2\phi(x)\|_{\mathcal{L}_1(H)} \qquad (5.60)$$

is a Banach space.

In Sect. 5.2.3 we introduced the operator $\mathcal{A}$ as the generator of the C_0-semigroup P_t on $L^2(H, \overline{\mathcal{B}}, m)$ (see Proposition 5.9). In the following lemma we study its relations with the operator $\mathcal{A}_1$.

Lemma 5.37 *Let Hypotheses 5.1 and 5.4 hold. Then:*

(i) $\mathcal{F}C_0^{2,A^*}(H) \subset D(\mathcal{A}_1)$.
(ii) $D(\mathcal{A}_1)$ *is embedded in* $D(\mathcal{A})$. *Moreover, for any* $\phi \in D(\mathcal{A}_1)$,

$$\mathcal{A}\phi(x) = \frac{1}{2}\text{Tr}\left[QD^2\phi(x)\right] + \langle x, A^*D\phi(x) \rangle + \langle b(x), D\phi(x) \rangle . \qquad (5.61)$$

(iii) *If we consider the Banach space structure on* $D(\mathcal{A}_1)$ *described above and the graph norm on* $D(\mathcal{A})$, *the embedding* $D(\mathcal{A}_1) \subset D(\mathcal{A})$ *is continuous.*

Proof Part (i) follows straightforwardly from the definitions of $\mathcal{F}C_0^{2,A^*}(H)$ and $D(\mathcal{A}_1)$.

Part (ii): We choose $\phi \in D(\mathcal{A}_1)$ and we start by showing that, for any $x \in H$,

$$\lim_{t \to 0} \frac{P_t[\phi](x) - \phi(x)}{t} = \frac{1}{2}\text{Tr}\left[QD^2\phi(x)\right] + \langle x, A^*D\phi(x) \rangle + \langle b(x), D\phi(x) \rangle . \qquad (5.62)$$

Indeed, applying Dynkin's formula (Proposition 1.169), we have

$$\frac{P_t[\phi](x) - \phi(x)}{t} = \frac{\mathbb{E}\phi(X(t; x)) - \phi(x)}{t}$$

$$= \frac{1}{t}\mathbb{E}\int_0^t \left[\frac{1}{2}\text{Tr}\left[QD^2\phi(X(s; x))\right] + \langle X(s; x), A^*D\phi(X(s; x)) \rangle \right.$$

$$\left. + \langle b(X(s; x)), D\phi(X(s; x)) \rangle \right] ds.$$

(5.63)

We need to show that every term in the right-hand side of (5.63) converges to the corresponding one in (5.62). Let us look at the middle term. We define

$$I_t^1(x) := \frac{1}{t}\int_0^t \left[\langle X(s; x), A^*D\phi(X(s; x)) \rangle - \langle x, A^*D\phi(x) \rangle\right] ds.$$

Let σ be a modulus of continuity of $A^*D\phi$ which we can assume to be concave. We have

$$
\begin{aligned}
|I_t^1(x)| &\leq \frac{1}{t}\int_0^t \mathbb{E}\Big[\,|\langle X(s;x)-x, A^*D\phi(X(s;x))\rangle| \\
&\qquad + |\langle x, A^*D\phi(X(s;x)) - A^*D\phi(x)\rangle|\,\Big]ds \\
&\leq \frac{1}{t}\int_0^t \big(\|A^*D\phi\|_0 \mathbb{E}\,|X(s;x)-x| + |x|\mathbb{E}\sigma\,(|X(s;x)-x|)\big)\,ds \\
&\leq \frac{1}{t}\int_0^t \big(\|A^*D\phi\|_0 \mathbb{E}\,|X(s;x)-x| + |x|\sigma\,(\mathbb{E}\,|X(s;x)-x|)\big)\,ds,
\end{aligned}
$$

where we used Jensen's inequality to obtain the last inequality. The last line above converges to 0 as $t \to 0$ by (5.9). The convergence of other terms in (5.62) is proved similarly.

We now need to show that the convergence takes place in $L^2(H, \overline{B}, m)$. We see that, thanks to (1.69) and since $\|A^*D\phi\|_0$ is finite, we have $\sup_{t\in(0,1]} \big(\mathbb{E}[I_t^1(x)]\big)^2 \leq g(x) = C_1 + C_2|x|^2$ for some positive constants C_1, C_2. Since $g \in L^1(H, \overline{B}, m)$ by (5.12), we can thus use the dominated convergence theorem to conclude that $\lim_{t\to 0}\mathbb{E}[I_t^1(\cdot)] = 0$ in $L^2(H, \overline{B}, m)$. We argue similarly to get the convergence of the other terms. Therefore $\phi \in D(\mathcal{A})$ and (5.61) holds.

Part (iii): Given $\phi \in D(\mathcal{A}_1)$ we have

$$
\begin{aligned}
|\phi|^2_{D(\mathcal{A})} &= |\phi|^2_{L^2_m} + |\mathcal{A}\phi|^2_{L^2_m} \\
&\leq |\phi|^2_{L^2_m} + 3\left|\frac{1}{2}\operatorname{Tr}\big[QD^2\phi\big]\right|^2_{L^2_m} + 3\,|\langle\cdot, A^*D\phi(\cdot)\rangle|^2_{L^2_m} + 3\,|\langle b(\cdot), D\phi(\cdot)\rangle|^2_{L^2_m} \\
&\leq \|\phi\|_0^2 + \frac{3}{4}\|Q\|^2_{\mathcal{L}(H)}\sup_{x\in H}\|D^2\phi(x)\|^2_{\mathcal{L}_1(H)} \\
&\qquad + 3\|A^*D\phi\|_0^2\int_H |x|^2\,dm(x) + 3\|D\phi\|_0^2\int_H (|b(0)| + K|x|)^2\,dm(x).
\end{aligned}
$$

Thanks to (5.12) there exists a constant C, depending only on m, b and Q such that the last expression is smaller than $C\|\phi\|^2_{D(\mathcal{A}_1)}$. This concludes the proof. $\square$

The concept of a classical solution of (5.49) is also similar to the one introduced in Definition 4.129, however here we limit our interest to functions belonging to $\Upsilon_A(0, T)$ to be able to define $\tilde{D}_Q$.

Definition 5.38 A function $u \in \Upsilon_A(0, T)$ is a classical solution of (5.49) if u has the following regularity properties

$$\begin{cases} u(\cdot, x) \in C^1([0, T]), \quad \forall x \in H \text{ and } u_t \in C_b([0, T] \times H), \\ u(t, \cdot) \in D(\mathcal{A}_1), \ \forall t \in [0, T] \quad and \quad \sup_{t \in [0,T]} \|u(t, \cdot)\|_{D(\mathcal{A}_1)} < +\infty, \\ u, \mathcal{A}_1 u \in C_b([0, T] \times H), \\ Du, A^* Du, \tilde{D}_Q u \in C_b([0, T] \times H, H), \\ D^2 u \in C_b([0, T] \times H, \mathcal{L}_1(H)), \end{cases}$$

(where $\tilde{D}_Q$ is defined in (5.45)) and satisfies

$$\begin{cases} u_t + \mathcal{A}_1 u + F_0 \left(t, x, \tilde{D}_Q u \right) + l(t, x) = 0, \quad t \in [0, T), \ for \ m - a.e. \ x \in H, \\ u(T, x) = g(x), \quad for \ m - a.e. \ x \in H. \end{cases}$$

$$(5.64)$$

Definition 5.39 A function $u \in \Upsilon_A(0, T)$ is a strong solution of Eq. (5.49) if $(u, \tilde{D}_Q u) \in L^2 \left(0, T; W_Q^{1,2}(H, m) \right)$ and there exist sequences $(u_n), (l_n) \subset L^2 \left(0, T; W_Q^{1,2}(H, m) \right)$ and $g_n \subset \mathcal{F}C_0^{2,A^*}(H)$ such that for every $n \in \mathbb{N}$, u_n is the classical solution of the Cauchy problem

$$\begin{cases} w_t + \mathcal{A}w + F_0(t, x, D_Q w) + l_n(t, x) = 0, \\ \\ w(T, x) = g_n(x), \end{cases}$$

$$(5.65)$$

and the following limits hold as $n \to +\infty$:

$$\begin{aligned} g_n &\longrightarrow g & \text{in } L^2 \left(H, \overline{\mathcal{B}}, m \right) \\ l_n &\longrightarrow l & \text{in } L^2 \left(0, T; L^2 \left(H, \overline{\mathcal{B}}, m \right) \right) \\ u_n &\longrightarrow u & \text{in } C \left([0, T], L^2 \left(H, \overline{\mathcal{B}}, m \right) \right) \\ \tilde{D}_Q u_n &\longrightarrow \tilde{D}_Q u & \text{in } L^2 \left(0, T; L^2 \left(H, \overline{\mathcal{B}}, m; H \right) \right). \end{aligned}$$

In principle we can have several strong solutions of Eq. (5.49), depending on the choice of the approximating sequences. Nevertheless we will see that in our case, if a strong solution exists, it is unique. See the discussion that follows the proof of Theorem 5.41 for more on this.

Theorem 5.40 *Assume that Hypotheses 5.1, 5.4 and 5.27 hold. If u is a strong solution of (5.49) then the pair $(u, U) := (u, \tilde{D}_Q u)$ is a mild solution of Eq. (5.49).*

Proof Let u_n, l_n, g_n be its approximating sequences as in Definition 5.39. Recalling that P_t is a strongly continuous semigroup on $L^2(H, \overline{\mathcal{B}}, m)$ (see Proposition 5.9), using Lemma 5.37 and the properties of classical solutions demanded in Definition 5.38, we can compute, for a fixed $t \in [0, T]$, the derivative in the variable s of $P_{s-t}[u_n(s)]$ (as a mapping from $[t, T]$ to $L^2(H, \overline{\mathcal{B}}, m)$). We get

$$\frac{d}{ds} P_{s-t}[u_n(s)] = P_{s-t}[\mathcal{A}u_n(s)] + P_{s-t}\left[\frac{d}{ds}u_n(s)\right]$$

$$= P_{s-t}[\mathcal{A}u_n(s)] + P_{s-t}\left[\left(-\mathcal{A}_1 u_n(s) - F_0\left(s,\cdot,\tilde{D}_Q u_n(s)\right) - l_n(s)\right)\right]$$

$$= P_{s-t}\left[\left(-F_0\left(s,\cdot,\tilde{D}_Q u_n(s)\right) - l_n(s)\right)\right], \quad s \in [t,T].$$

Integrating both sides of this expression over $[t,T]$, using that $u_n(T) = g_n$ and reordering the terms we obtain for every n

$$u_n(t) = P_{T-t}[g_n] + \int_t^T P_{s-t}\left[F_0(s,\cdot,\tilde{D}_Q u_n(s)) + l_n(s)\right] ds.$$

Setting $\psi_n(s) = F_0(s,\cdot,\tilde{D}_Q u_n(s)) + l_n(s)$, the last expression becomes

$$u_n(t) = P_{T-t}[g_n] + \int_t^T P_{s-t}[\psi_n(s)] ds,$$

where $g_n \in \mathcal{F}C_0^{2,A^*}(H), \psi_n \in L^2\left(0,T; L^2\left(H,\overline{\mathcal{B}},m\right)\right)$,

$$g_n \xrightarrow{n \to +\infty} g \quad \text{in } L^2(H,\overline{\mathcal{B}},m),$$

and, thanks to Hypothesis 5.27-(A),

$$\psi_n \xrightarrow{n \to +\infty} F_0(\cdot,\cdot,\tilde{D}_Q u) + l \quad \text{in } L^2\left(0,T; L^2\left(H,\overline{\mathcal{B}},m\right)\right).$$

We can now apply Proposition 5.25 and pass to the limit as $n \to +\infty$ to get the claim.
                                                                                        $\square$

Theorem 5.41 *Assume that Hypotheses 5.1, 5.4 and 5.27 hold and suppose $b = 0$. If the pair $(u,U) \in L^2\left(0,T; W_Q^{1,2}(H,m)\right)$ is a mild solution of Eq. (5.49) then $U = \tilde{D}_Q u$ and u is a strong solution of (5.49).*

Proof In the particular case $b = 0$ the semigroup P_t simplifies to the Ornstein–Uhlenbeck semigroup studied in Sect. B.7.2. The notation used in other parts of the book in this case is R_t but here, to be consistent with the general notation used in the chapter, we continue to denote the semigroup by P_t. Hypotheses 5.1-(A)-(B)-(C) imply Hypothesis B.79, needed in all the results of Sect. B.7 used in this proof. Observe that, if $b = 0$, the operator $\mathcal{A}_1$ defined in (5.59) reduces to the operator $\mathcal{A}_0$ defined in (B.36).

As argued in Remark 5.29 we immediately get $U = \tilde{D}_Q u$. Let g_n, ψ_n be two sequences such that

$$g_n \in \mathcal{F}C_0^{2,A^*}(H), \tag{5.66}$$

$$\psi_n \colon [0, T] \to \mathcal{F}C_0^{2, A^*}(H), \tag{5.67}$$

$$\psi_n \text{ and } \mathcal{A}_1\psi_n \text{ belong to } C\left([0, T], UC_b(H)\right), \tag{5.68}$$

$$g_n \overset{n\to+\infty}{\longrightarrow} g \quad \text{in } L^2(H, \overline{\mathcal{B}}, m) \tag{5.69}$$

and

$$\psi_n \overset{n\to+\infty}{\longrightarrow} F_0(\cdot, \cdot, \tilde{D}_Q u) + l \quad \text{in } L^2\left(0, T; L^2\left(H, \overline{\mathcal{B}}, m\right)\right). \tag{5.70}$$

These sequences exist thanks to Lemma 5.7.

Since $(u, U) = (u, \tilde{D}_Q u)$ is a mild solution of (5.49) we have

$$u(t) = P_{T-t}[g] + \int_t^T P_{s-t}\left[F_0(s, \cdot, \tilde{D}_Q u(s)) + l(s)\right]ds.$$

If we set

$$u_n(t, x) = P_{T-t}[g_n] + \int_t^T P_{s-t}\left[\psi_n(s)\right]ds, \tag{5.71}$$

by Proposition 5.25 we obtain that

$$u_n \overset{n\to+\infty}{\longrightarrow} u \quad \text{in } C\left([0, T], L^2\left(H, \overline{\mathcal{B}}, m\right)\right), \tag{5.72}$$

$$\tilde{D}_Q u_n \overset{n\to+\infty}{\longrightarrow} \tilde{D}_Q u \quad \text{in } L^2\left(0, T; L^2\left(H, \overline{\mathcal{B}}, m; H\right)\right). \tag{5.73}$$

The latter, thanks to Hypotheses 5.27-(A), implies in particular that

$$F_0(\cdot, \cdot, \tilde{D}_Q u_n) \overset{n\to+\infty}{\longrightarrow} F_0(\cdot, \cdot, \tilde{D}_Q u) \quad \text{in } L^2\left(0, T; L^2\left(H, \overline{\mathcal{B}}, m\right)\right).$$

So, thanks to (5.70), if we set

$$l_n = \psi_n - [F_0(\cdot, \cdot, \tilde{D}_Q u_n)], \tag{5.74}$$

we get

$$l_n \overset{n\to+\infty}{\longrightarrow} l \quad \text{in } L^2\left(0, T; L^2\left(H, \overline{\mathcal{B}}, m\right)\right). \tag{5.75}$$

We can now apply Proposition B.91-(ii). Observe that the existence of the function g_0 demanded in the hypotheses of this proposition can be easily found thanks to (5.68) and the constant C in (B.33) and (B.35) is here equal to zero. The time is reversed (t in Proposition B.91 corresponds to our $T - t$ for any $t \in [0, T]$). It thus follows that u_n satisfies in the classical sense the approximating HJB equation

$$\begin{cases} (u_n)_t + \mathcal{A}u_n + F_0(t, x, D_Q u_n) + l_n(t, x) = 0 \\ \\ u(T, x) = g_n(x). \end{cases} \tag{5.76}$$

Given the regularity of u_n, g_n and ψ_n, $\tilde{D}_Q u_n = D_Q u_n$ and then the fact that $\tilde{D}_Q u_n \in C_b([0, T] \times H, H)$, not directly stated in Proposition B.91, follows from $Du_n \in C_b([0, T] \times H, H)$ and the continuity of Q.

   This, together with the convergences (5.69), (5.72), (5.73) and (5.75), shows that u is a strong solution in the sense of Definition 5.39.   □

   Theorem 5.35 shows that, under Hypotheses 5.1, 5.4 and 5.27, there exists a unique mild solution (u, U) of Eq. (5.49). Theorem 5.40 ensures that, under the same hypotheses, any strong solution is also a mild solution so, in particular there exists at most one strong solution of (5.49) and, whenever it exists, it can be identified with the mild solution. Theorem 5.41 proves, under the additional assumption $b = 0$, the reverse implication, ensuring in particular the existence of a (unique) strong solution in this case. This result was stated in [298] (see in particular Proposition 4.3) without the assumption $b = 0$ but the proof of the regularity of the u_n in the general case was not complete.

   In Sect. 5.5, we work again under Hypotheses 5.1, 5.4 and 5.27 but we also suppose that a strong solution exists or, equivalently, that the mild solution of the equation is also strong. This is always the case if $b = 0$.

5.5 Application to Stochastic Optimal Control

We apply the results on abstract HJB equations from previous sections to study a family of optimal control problems.

5.5.1 The State Equation

We work, as usual, in a real separable Hilbert space H which will be both the state space and the noise space (see Sect. 1.2.4), that is we have $\Xi = H$. The control space Λ is a closed ball in a real separable Banach space E:

$$\Lambda = \overline{B_\varrho(0)}. \tag{5.77}$$

The linear operators A, Q and the function b satisfy Hypothesis 5.1. As in Chap. 2, the notation $\mu := \left(\Omega^\mu, \mathscr{F}^\mu, \{\mathscr{F}^t_{\mu,s}\}_{s \in [t,T]}, \mathbb{P}^\mu, W^\mu_Q \right)$ (or without the index μ if the context is clear) will be used to denote a generalized reference probability space (see Definition 1.100). We limit our attention here to the case where the σ-fields of the filtration $\mathscr{F}^t_{\mu,s}$ are countably generated up to sets of measure zero. This holds, for example, for filtrations generated by Wiener processes, see Lemma 1.94. We recall that the generalized reference probability spaces μ used in Sect. 5.5 may be different from μ_0 in Hypothesis 5.1.

We consider a stochastic controlled system governed by the state equation

$$\begin{cases} dX(s) = \Big(AX(s) + b(X(s)) + Q^{\frac{1}{2}} R(s, X(s), a(s))\Big) ds + dW_Q(s), \\ X(t) = x, \quad x \in H, \end{cases} \quad (5.78)$$

where R and a satisfy the following hypothesis.

Hypothesis 5.42 We assume that:

(i) $R: [0, T] \times H \times \Lambda \to H$ is Borel measurable and there exists an $M_R > 0$ such that

$$\sup_{(s,x,a) \in [0,T] \times H \times \Lambda} |R(s, x, a)| \le M_R < +\infty,$$

and, for all $s \in [0, T], a \in \Lambda, x, y \in H$,

$$|R(s, x, a) - R(s, y, a)| \le M_R |x - y|.$$

(ii) For every $t \in [0, T]$ and a generalized reference probability space μ on $[t, T]$, the σ-fields of the filtration $\{\mathscr{F}_{\mu,s}^t\}_{s \in [t,T]}$ are countably generated up to sets of measure zero. Λ is as in (5.77) and the control processes $a(\cdot) : [t, T] \times \Omega \to \Lambda$ belong to the set

$$\mathcal{U}_t^\mu := \big\{ a(\cdot) : [t, T] \times \Omega \to \Lambda \ : \ a(\cdot) \text{ is } \mathscr{F}_s^t - \text{progressively measurable} \big\}. \quad (5.79)$$

We recall that the control processes in $\mathcal{U}_t^\mu$ depend on the choice of the generalized reference probability space (Definition 1.100) μ because they are progressively measurable with respect to the filtration $\{\mathscr{F}_t^s\}_{s \in [t,T]}$ that depends on the choice of μ. See Sect. 2.1.1 for more on this.

Remark 5.43 The boundedness of R is imposed to be able to solve later, in Theorem 5.55, the closed loop equation using Girsanov's theorem. A similar approach is also used in Sect. 6.5. ∎

Proposition 5.44 *Let Hypotheses 5.1 and 5.42 be satisfied. Then, for any $t \in [0, T]$, $x \in H$, $a(\cdot) \in \mathcal{U}_t^\mu$, the state equation (5.78) has a unique solution $X(\cdot; t, x, a(\cdot)) \in \mathcal{H}_p^\mu(t, T; H)$ (see Definition 1.126) for all $p \ge 1$. In particular, $X(\cdot; t, x, a(\cdot)) \in M_\mu^p(t, T; H)$ (defined in (1.29)) for all $p \ge 1$.*

Moreover, there exists a $\mathcal{B}([t, T]) \otimes \mathcal{B}(H) \otimes \mathscr{F}/\mathcal{B}(H)$-measurable function

$$\begin{cases} [t, T] \times H \times \Omega \to H \\ (s, x, \omega) \to \tilde{X}(s; t, x, a(\cdot))(\omega) \end{cases} \quad (5.80)$$

such that, for every $x \in H$, $\tilde{X}(\cdot; t, x, a(\cdot))$ is a version of the solution $X(\cdot; t, x, a(\cdot))$. Thus in the future we will not make a distinction between $X(\cdot; t, x, a(\cdot))$ and $\tilde{X}(\cdot; t, x, a(\cdot))$.

Proof The result, except for the last claim, follows from Proposition 1.147. The whole term $\left[Q^{1/2}b(X(s)) + Q^{1/2}R(s, X(s), a(s))\right]$ corresponds to the term b_0 in Hypothesis 1.145, $a(\cdot)$ plays the role of $a_1(\cdot)$ and we have no $a_2(\cdot)$.

To prove the last claim, we consider a countable dense subset $S := \{x_n\}_{n \in \mathbb{N}}$ of H. Thanks to (1.70) we can find $\Omega_2 \subset \Omega$ with $\mathbb{P}(\Omega_2) = 1$ such that (1.70) holds with $\xi_1 = x_1$ and $\xi_2 = x_2$ for any $s \in [t, T]$ and $\omega \in \Omega_2$. Similarly, for every $N > 2$ we can find a subset $\Omega_N \subset \Omega$ with $\mathbb{P}(\Omega_N) = 1$ such that (1.70) is satisfied for any choice $\xi_1 = x_i, \xi_2 = x_j, i, j = 1, \dots, N$, for all $s \in [t, T]$ and $\omega \in \Omega_N$. If we define $\Omega_\infty = \bigcap_{n \geq 1} \Omega_n$ we have again $\mathbb{P}(\Omega_\infty) = 1$. Given $s \in [t, T]$ and $\omega \in \Omega_\infty$, we define, for any $x \in H$,

$$\tilde{X}(s; t, x, a(\cdot))(\omega) := \lim_{n \to \infty} X(s; t, y_n, a(\cdot))(\omega), \tag{5.81}$$

where y_n is a sequence of elements of S such that $y_n \to x$ (the limit exists and it does not depend on the chosen sequence y_n, again thanks to (1.70) and the choice of Ω_∞). We define $\tilde{X}(s; t, x, a(\cdot))(\omega) = 0$ for $(s, x, \omega) \in [t, T] \times H \times (\Omega \setminus \Omega_\infty)$. The pointwise convergence (5.81) and the progressive measurability (and thus the $\mathcal{B}([t, T]) \otimes \mathscr{F}/\mathcal{B}(H)$-measurability) of X ensures that (see Lemma 1.8(iii)), for any $x \in H$, the restriction of $\tilde{X}(\cdot; t, x, a(\cdot))(\cdot)$ to $[t, T] \times \Omega_\infty$ is $\mathcal{B}([t, T]) \otimes (\mathscr{F} \cap \Omega_\infty)/\mathcal{B}(H)$-measurable. This fact, the completeness of $\mathscr{F}$ and the fact that $\tilde{X}$ is constant on $[t, T] \times H \times (\Omega \setminus \Omega_\infty)$ give easily the $\mathcal{B}([t, T]) \otimes \mathscr{F}/\mathcal{B}(H)$-measurability of $\tilde{X}(\cdot; t, x, a(\cdot))(\cdot)$ on $[t, T] \times \Omega$. Moreover, by construction, for any $s \in [t, T]$ and $\omega \in \Omega$, $x \to \tilde{X}(s; t, x, a(\cdot))(\omega)$ is continuous so that (see Lemma 1.18) the function defined in (5.80) is $\mathcal{B}([t, T]) \otimes \mathcal{B}(H) \otimes \mathscr{F}/\mathcal{B}(H)$-measurable. □

5.5.2 The Optimal Control Problem and the HJB Equation

Let Hypotheses 5.1, 5.4 and 5.42 be satisfied. We study an optimal control problem in its *strong* formulation (see Sect. 2.1.1 for details) so that the generalized reference probability space μ is fixed. We consider the following cost functional

$$J^\mu(t, x; a(\cdot)) = \mathbb{E}\left\{\int_t^T l(s, X(s; t, x, a(\cdot))) + h_2(a(s))ds + g(X(T; t, x, a(\cdot)))\right\} \tag{5.82}$$

which we want to minimize over the control set $\mathcal{U}_t^\mu$. In this expression $X(s; t, x, a(\cdot))$ represents the mild solution of (5.78) at time s which, as always, we will often denote by $X(s)$. The functions l, h_2 and g satisfy the following hypothesis.

Hypothesis 5.45 $l : [0, T] \times H \to \mathbb{R}$ and $g : H \to \mathbb{R}$ satisfy Hypothesis 5.27-(B) while $h_2 : \Lambda \to \mathbb{R}$ is Borel measurable and bounded.

The value function of the problem depends on μ and it is defined as in (2.4):

$$V_t^\mu(x) = \inf_{a(\cdot) \in \mathcal{U}_t^\mu} J^\mu(t, x; a(\cdot)). \tag{5.83}$$

The HJB equation corresponding to the described optimal control problem is

$$\begin{cases} v_t + \mathcal{A}v + F_0(t, x, D_Q v) + l(t, x) = 0 \\ v(T, x) = g(x), \end{cases} \tag{5.84}$$

where the operator $\mathcal{A}$ is defined in Sect. 5.2.3 and the Hamiltonian F_0 is given by

$$F_0(t, x, p) = \inf_{a \in \Lambda} \{\langle R(t, x, a), p \rangle + h_2(a)\} =: \inf_{a \in \Lambda} F_{0,CV}(t, x, p, a). \tag{5.85}$$

We will suppose that F_0 satisfies Hypothesis 5.27-(A). Indeed, thanks to Hypotheses 5.42 and 5.45, the Lipschitz continuity and growth conditions (5.50) are always satisfied but the $Leb \otimes \overline{B} \otimes \overline{B}/\mathcal{B}(\mathbb{R})$ measurability may not always be ensured. However, when R does not depend on t and x, the Hamiltonian F_0 is just a function from H to $\mathbb{R}$ and Lemma 1.21 then guarantees that it is $\overline{B}/\mathcal{B}(\mathbb{R})$-measurable, so that Hypothesis 5.27-(A) is satisfied. Hypothesis 5.27-(A) is also always true if $R(t, x, \cdot)$ is continuous for every t and x due to the separability of Λ.

5.5.3 The Verification Theorem

We now show how to obtain a verification theorem and an explicit expression for optimal controls in feedback form.

Lemma 5.46 *Let* $t \in [0, T]$, $x \in H$, $\mu = \left(\Omega, \mathcal{F}, \{\mathcal{F}_s^t\}_{s \in [t, T]}, \mathbb{P}, W_Q\right)$ *be a generalized reference probability space on* $[t, T]$ *and let* $a(\cdot) \in \mathcal{U}_t^\mu$. *Assume that Hypotheses 5.1, 5.4, 5.42 and 5.45 hold. Define*

$$\rho_{a(\cdot)} = \exp\left(-\int_t^T \langle R(r, X(r; t, x, a(\cdot)), a(r)), dW_Q(r) \rangle\right.$$
$$\left. -\frac{1}{2} \int_t^T |R(r, X(r; t, x, a(\cdot)), a(r))|^2 \, dr\right).$$

Then:

(i) *The measure* $\tilde{\mathbb{P}}$ *on* $(\Omega, \mathcal{F})$ *defined by setting* $d\tilde{\mathbb{P}}(A) := \rho_{a(\cdot)}(T) d\mathbb{P}(\omega)$, *that is, for any* $A \in \mathcal{F}$,

$$\tilde{\mathbb{P}}(A) := \int_A \rho_{a(\cdot)}(\omega) \, d\mathbb{P}(\omega),$$

is a probability measure on Ω, in particular $\mathbb{E}[\rho_{a(\cdot)}] = 1$.
(ii) There exists a positive constant $\tilde{c} < +\infty$ such that

$$\mathbb{E}\left[\left(\rho_{a(\cdot)}\right)^{-1}\right] \le \tilde{c}, \quad \text{for any } x \in H \tag{5.86}$$

and we have

$$d\mathbb{P}(A) := \left(\rho_{a(\cdot)}\right)^{-1} d\tilde{\mathbb{P}}(\omega). \tag{5.87}$$

(iii) Denote by $X(\cdot; t, x)$ the solution of

$$\begin{cases} dX(s) = (AX(s) + b(X(s)))\,ds + dW_Q(s), \quad s \in [0, T], \\ X(t) = x \in H. \end{cases} \tag{5.88}$$

For any $s \in [t, T]$, $\mathcal{L}_{\mathbb{P}}(X(s; t, x)) = \mathcal{L}_{\tilde{\mathbb{P}}}(X(s; t, x, a(\cdot)))$.
(iv) For any nonnegative $w \in L^2(H, \mathcal{B}(H), m)$, for any $s \in [t, T]$,

$$\int_H \mathbb{E}w(X(s; t, x, a(\cdot)))dm(x) \le \sqrt{\tilde{c}} \left(\int_H \mathbb{E}w^2(X(s; t, x))dm(x)\right)^{1/2} = \sqrt{\tilde{c}}|w|_{L^2_m}, \tag{5.89}$$

where $\tilde{c}$ is the constant introduced in (5.86).

Proof Most of the statements of the lemma are corollaries of the Girsanov Theorem.
Part (i): Given the boundedness of R the claim follows from Proposition 10.17 and Theorem 10.14 of [180].
Part (ii): Observe first that if we replace $R(s, X(s; t, x, a(\cdot)), a(s))$ by $-R(s, X(s; t, x, a(\cdot)), a(s))$ we have again a bounded function so that the results of Part (i) hold: we get

$$\mathbb{E}\exp\left(\int_t^T \langle R(s, X(s; t, x, a(\cdot)), a(s)), Q^{-1/2}dW_Q(s)\rangle \right.$$
$$\left. -\frac{1}{2}\int_t^T |R(s, X(s; t, x, a(\cdot)), a(s))|^2\,ds\right) = 1. \tag{5.90}$$

Since by Hypothesis 5.42 there exists an $M_R \in \mathbb{R}$ such that $|R(s, X(s; t, x, a(\cdot)), a(s))| \le M_R$ for any choice of $s \in [t, T]$, $x \in H$ and any $a(\cdot)$,

$$\mathbb{E}\left[\left(\rho_{a(\cdot)}\right)^{-1}\right] = \mathbb{E}\exp\left(\int_t^T \langle R(r, X(r; t, x, a(\cdot)), a(r)), dW_Q(r)\rangle \right.$$
$$\left. +\frac{1}{2}\int_t^T |R(r, X(r; t, x, a(\cdot)), a(r))|^2\,dr\right) \le$$
$$e^{(T-t)M_R^2}\mathbb{E}\exp\left(\int_t^T \langle R(r, X(r; t, x, a(\cdot)), a(r)), dW_Q(r)\rangle\right.$$

$$-\frac{1}{2}\int_t^T |R(r, X(r; t, x, a(\cdot)), a(r))|^2\, dr\Bigg) = e^{(T-t)M^2} =: \tilde{c},$$

where in the last step we used (5.90).

The second claim follows by the strict positivity of $\rho_{a(\cdot)}$ as a corollary of the Radon–Nikodym Theorem (see [18], p. 64).

Part (iii): Thanks to Theorem 10.14 of [180] we know that the process defined by

$$\tilde{W}_Q(s) = W_Q(s) - W_Q(t) + \int_t^s Q^{\frac{1}{2}} R(r, X(r; t, x, a(\cdot)), a(r))dr, \qquad s \in [t, T],$$

is a Q-Wiener process in H with respect to $\{\mathscr{F}_s^t\}_{s\geq t}$ and the probability measure $\tilde{\mathbb{P}}$. We have

$$
\begin{aligned}
X(s; t, x, a(\cdot)) &= e^{(s-t)A}x + \int_t^s e^{(s-r)A}b(X(r; t, x, a(\cdot)))dr \\
&\quad + \int_t^s e^{(s-r)A}Q^{\frac{1}{2}}R(r, X(r; t, x, a(\cdot)), a(r))dr + \int_t^s e^{(s-r)A}dW_Q(r) \\
&= e^{(s-t)A}x + \int_t^s e^{(s-r)A}b(X(r; t, x, a(\cdot)))dr + \int_t^s e^{(s-r)A}Q^{\frac{1}{2}}R(r, X(r; t, x, a(\cdot)), a(r))dr \\
&\quad + \int_t^s e^{(s-r)A}d\tilde{W}_Q(r) - \int_t^s e^{(s-r)A}Q^{\frac{1}{2}}R(r, X(r; t, x, a(\cdot)), a(r))dr \\
&= e^{(s-t)A}x + \int_t^s e^{(s-r)A}b(X(r; t, x, a(\cdot)))dr + \int_t^s e^{(s-r)A}d\tilde{W}_Q(r), \quad s \in [t, T],
\end{aligned}
$$

so $X(\cdot; t, x, a(\cdot))$ solves the same equation as $X(\cdot; t, x)$. The claim thus follows thanks to Proposition 1.148-(ii).

Part (iv): For any $s \in [t, T]$, the joint measurability of the function $(x, \omega) \to w(X(s; t, x, a(\cdot))(\omega))$ follows by the Borel measurability of w and by the measurability of X stated in Proposition 5.44.

Using first (5.87) and then the Cauchy–Schwarz inequality we have, for $s \in [t, T]$,

$$
\begin{aligned}
\int_H \mathbb{E}w(X(s; t, x, a(\cdot)))dm(x) &= \int_H \int_\Omega w(X(s; t, x, a(\cdot))(\omega))d\mathbb{P}(\omega)dm(x) \\
&= \int_\Omega w(X(s; t, x, a(\cdot))(\omega))\left(\rho_{a(\cdot)}(\omega)\right)^{-1}d\tilde{\mathbb{P}}(\omega)dm(x) \\
&\leq \left(\int_H \int_\Omega \left(\rho_{a(\cdot)}(\omega)\right)^{-2}d\tilde{\mathbb{P}}(\omega)dm(x)\right)^{1/2}\left(\int_H \int_\Omega w^2(X(s; t, x, a(\cdot))(\omega))d\tilde{\mathbb{P}}(\omega)dm(x)\right)^{1/2} \\
&= \left(\int_H \int_\Omega \left(\rho_{a(\cdot)}(\omega)\right)^{-1}d\mathbb{P}(\omega)dm(x)\right)^{1/2}\left(\int_H \int_\Omega w^2(X(s; t, x)(\omega))d\mathbb{P}(\omega)dm(x)\right)^{1/2},
\end{aligned}
$$

where in the last step we used, in the two terms, respectively Part (i) and Part (iii). Therefore, by (5.86) and then using the definition of the transition semigroup, the fact that it does not depend on a generalized reference probability space, and the property of the invariant measure (observe that w^2 belongs to $L^1(H, \mathcal{B}(H), m)$ so we refer to Proposition 5.9 for $p = 1$), we obtain

$$\int_H \mathbb{E}w(X(s;t,x,a(\cdot)))dm(x) \leq \sqrt{\tilde{c}} \left(\int_H \mathbb{E}w^2(X(s;t,x))dm(x) \right)^{1/2}$$

$$= \sqrt{\tilde{c}} \left(\int_H P_{s-t} \left[|w(\cdot)|^2 \right](x)dm(x) \right)^{1/2} = \sqrt{\tilde{c}} \left(\int_H |w(x)|^2 dm(x) \right)^{1/2} = \sqrt{\tilde{c}}|w|_{L_m^2},$$

which gives the claim.   □

The result of Part (iv) of Lemma 5.46 will be extended to a general $w \in L^2(H, \overline{B}, m)$ in Corollary 5.48.

Lemma 5.47 *Let* $t \in [0, T]$, $\mu = \left(\Omega, \mathscr{F}, \{\mathscr{F}_s^t\}_{s \in [t,T]}, \mathbb{P}, W_Q \right)$ *be a generalized reference probability space on* $[t, T]$ *and* $a(\cdot) \in \mathcal{U}_t^\mu$. *Assume that Hypotheses 5.1, 5.4, 5.42 and 5.45 hold.*

Consider a $\overline{B}/B(\mathbb{R})$-*measurable function* $\phi: H \rightarrow \mathbb{R}$ *(respectively, a* $\overline{B}/B(H)$-*measurable function* $\phi: H \rightarrow H$*) and* $s \in [t, T]$. *Then the function*

$$\begin{cases} H \times \Omega \rightarrow \mathbb{R} \\ (x, \omega) \rightarrow \phi(X(s;t,x,a(\cdot)))(\omega) \end{cases}$$

is $\overline{B(H) \otimes \mathscr{F}}/B(\mathbb{R})$-*measurable (respectively,* $\overline{B(H) \otimes \mathscr{F}}/B(H)$-*measurable), where* $\overline{B(H) \otimes \mathscr{F}}$ *is the completion of the* σ-*field* $B(H) \otimes \mathscr{F}$ *w.r.t. the measure* $m \otimes \mathbb{P}$.

Similarly, given a $B([t, T]) \otimes \overline{B}/B(\mathbb{R})$-*measurable function* $\phi: [t, T] \times H \rightarrow \mathbb{R}$, *the function*

$$\begin{cases} [t, T] \times H \times \Omega \rightarrow \mathbb{R} \\ (s, x, \omega) \rightarrow \phi(s, X(s;t,x,a(\cdot)))(\omega) \end{cases}$$

is $\overline{B([t, T]) \otimes B(H) \otimes \mathscr{F}}/B(\mathbb{R})$-*measurable, where* $\overline{B([t, T]) \otimes B(H) \otimes \mathscr{F}}$ *is the completion of the* σ-*field* $B([t, T]) \otimes B(H) \otimes \mathscr{F}$ *w.r.t. the measure* $ds \otimes m \otimes \mathbb{P}$.

Proof The proof follows the same arguments as those used in the proof of Lemma 5.10. We give it for completeness.

If $\phi: H \rightarrow \mathbb{R}$ is Borel-measurable the statement follows from the measurability of the solutions of (5.78) stated in Proposition 5.44. If $\phi: H \rightarrow \mathbb{R}$ is $\overline{B}/B(\mathbb{R})$-measurable, let $\tilde{\phi}: H \rightarrow \mathbb{R}$ be a $B/B(\mathbb{R})$-measurable function and $V \in B(H)$, $m(V) = 0$ be such that $\phi(x) = \tilde{\phi}(x)$ for all $x \in H \setminus V$. Then

$$0 \leq \int_H \mathbb{P}\{X(s;t,x,a(\cdot))(\omega) \in V\} dm(x)$$

$$= \int_H \mathbb{E}[\mathbf{1}_V(X(s;t,x,a(\cdot)))] dm(x) \leq \sqrt{\tilde{c}} \left(\int_H \mathbb{E}[\mathbf{1}_V^2(X(s;t,x))] dm(x) \right)^{1/2}$$

$$= \sqrt{\tilde{c}} \left(\int_H |\mathbf{1}_V(x)|^2 dm(x) \right)^{1/2} = \sqrt{\tilde{c}} (m(V))^{1/2} = 0, \quad (5.91)$$

where we used (5.89) and then the property of the invariant measure. This fact shows that the functions $(x, \omega) \rightarrow \phi(X(t, x)(\omega)$ and $(x, \omega) \rightarrow \tilde{\phi}(X(t, x)(\omega)$ are $m \otimes \mathbb{P}$-e.e. equal on $H \times \Omega$. Thus, since $(x, \omega) \rightarrow \tilde{\phi}(X(t, x)(\omega)$ is $\mathscr{F} \otimes \mathcal{B}/\mathcal{B}(\mathbb{R})$-measurable, $(x, \omega) \rightarrow \phi(X(t, x)(\omega)$ is $\mathscr{F} \otimes \overline{\mathcal{B}}/\mathcal{B}(\mathbb{R})$-measurable.

The same proof applies if $\phi \colon H \to H$ is a $\overline{\mathcal{B}}/\mathcal{B}(H)$-measurable function.

Similarly, if $\phi \colon [t, T] \times H \to \mathbb{R}$ is $\mathcal{B}([t, T]) \otimes \overline{\mathcal{B}}/\mathcal{B}(\mathbb{R})$-measurable we can find (again by Lemma 1.16, recalling that $\mathcal{B}([t, T]) \otimes \overline{\mathcal{B}} \subset \overline{\mathcal{B}([t, T]) \otimes \mathcal{B}})$) a $\mathcal{B}([t, T]) \otimes \mathcal{B}/\mathcal{B}(\mathbb{R})$-measurable function $\tilde{\phi} \colon [t, T] \times H \to \mathbb{R}$ and $V \in \mathcal{B}([t, T]) \otimes \mathcal{B}(H)$ such that $(ds \otimes m)(V) = 0$ and $\phi(s, x) = \tilde{\phi}(s, x)$ for all $(s, x) \in [t, T] \times H \setminus V$. If we define $V_s := \{x \in H \ : \ (s, x) \in V\}$ then $V_s \in \mathcal{B}(H)$ and $m(V_s) = 0$ for almost every $s \in [0, T]$. Instead of (5.91) we now have

$$0 \leq \int_t^T \int_H \mathbb{P}\{(s, X(s; t, x, a(\cdot))(\omega)) \in V\} dm(x) ds$$

$$= \int_t^T \int_H \mathbb{P}\{X(s; t, x, a(\cdot))(\omega) \in V_s\} dm(x) ds$$

$$= \int_t^T \int_H \mathbb{E}\left[\mathbf{1}_{V_s}(X(s; t, x, a(\cdot)))\right] dm(x)$$

$$\leq \sqrt{\tilde{c}} \int_t^T \left(\int_H |\mathbf{1}_{V_s}(x)|^2 dm(x)\right)^{1/2} ds = 0$$

and the proof ends as before. □

Corollary 5.48 *Let* $t \in [0, T]$, $x \in H$, $\mu = \left(\Omega, \mathscr{F}, \{\mathscr{F}_s^t\}_{s \in [t, T]}, \mathbb{P}, W_Q\right)$ *be a generalized reference probability space on* $[t, T]$ *and let* $a(\cdot) \in \mathcal{U}_t^\mu$. *Assume that Hypotheses 5.1, 5.4, 5.42 and 5.45 hold. Then, for any* $w \in L^2(H, \overline{\mathcal{B}}, m)$, *the map* $x \rightarrow \mathbb{E}w(X(s; t, x, a(\cdot)))$ *belongs to* $L^1\left(H, \overline{\mathcal{B}}, m\right)$ *and for almost every* $s \in [t, T]$,

$$\int_H \mathbb{E}w(X(s; t, x, a(\cdot))) dm(x) \leq \sqrt{\tilde{c}} \left(\int_H \mathbb{E}w^2(X(s; t, x)) dm(x)\right)^{1/2} = \sqrt{\tilde{c}}|w|_{L_m^2},$$

where $\tilde{c}$ *is the constant introduced in (5.86).*

Proof The statements about the joint measurability proved in Lemma 5.47 allow us, in particular, to ensure the measurability in s and x of integrals with respect to ω and then to extend Lemma 5.46-(iv) to any $w \in L^2\left(H, \overline{\mathcal{B}}, m\right)$. □

Lemma 5.49 *Assume that Hypotheses 5.1, 5.4, 5.42, 5.45 hold and let* $a(\cdot) \in \mathcal{U}_t^\mu$. *Then, for every* $s \in [t, T]$ *and* $w \in L^2\left(H, \overline{\mathcal{B}}, m\right)$ *(respectively,* L^2 *$\left(H, \overline{\mathcal{B}}, m; H\right)$), the map* $x \rightarrow \mathbb{E}w(X(s; t, x, a(\cdot)))$ *belongs to* $L^1\left(H, \overline{\mathcal{B}}, m\right)$ *(respectively,* $L^1\left(H, \overline{\mathcal{B}}, m; H\right)$). *Moreover, given a sequence* w_n *converging to* w *in* $L^2\left(H, \overline{\mathcal{B}}, m\right)$ *(respectively in* $L^2\left(H, \overline{\mathcal{B}}, m; H\right)$), the sequence $\mathbb{E}w_n(X(s; t, x, a(\cdot)))$ converges to $\mathbb{E}w(X(s; t, x, a(\cdot)))$ in $L^1\left(H, \overline{\mathcal{B}}, m\right)$ (respectively, $L^1\left(H, \overline{\mathcal{B}}, m; H\right)$).*

Similarly, given $w \in L^2\left(t, T; L^2\left(H, \overline{B}, m\right)\right)$ *(respectively,* $L^2\left(t, T; L^2\left(H, \overline{B}, \right.\right.$
$\left.\left.m; H\right)\right)$*), the map* $(s, x) \rightarrow \mathbb{E}w\left(s, X(s; t, x, a(\cdot))\right)$ *belongs to* $L^1\left((t, T) \times \right.$
$\left. H, \overline{\mathcal{B}([t, T]) \otimes \mathcal{B}}, ds \otimes m\right)$ *(respectively,* $L^1\left((t, T) \times H, \overline{\mathcal{B}([t, T]) \otimes \mathcal{B}}, ds \otimes m; H\right)$*),*
where ds is the Lebesgue measure on $[t, T]$ *and* $\overline{\mathcal{B}([t, T]) \otimes \mathcal{B}}$ *is the completion of the*
σ-field $\mathcal{B}([t, T]) \otimes \mathcal{B}$ *w.r.t.* $ds \otimes m$*. Moreover, given a sequence* w_n *converging to* w
in $L^2\left(t, T; L^2\left(H, \overline{B}, m\right)\right)$ *(respectively,* $L^2\left(t, T; L^2\left(H, \overline{B}, m; H\right)\right)$*), the sequence*
$\mathbb{E}w_n\left(s, X(s; t, x, a(\cdot))\right)$ *converges to* $\mathbb{E}w\left(s, X(s; t, x, a(\cdot))\right)$ *in* $L^1\left((t, T) \times H, \right.$
$\left. \overline{\mathcal{B}([t, T]) \otimes \mathcal{B}}, ds \otimes m\right)$ *(respectively,* $L^1\left((t, T) \times H, \overline{\mathcal{B}([t, T]) \otimes \mathcal{B}}, ds \right.$
$\left. \otimes m; H\right)$*).*

Proof The statements about joint measurability of the various functions involved follow from Lemma 5.47, Corollary 5.48 or can be proved by similar arguments. Recall,
for the case when $w \in L^2\left(t, T; L^2\left(H, \overline{B}, m\right)\right)$, that there exists (see Theorem 11.47,
p. 427 of [8]) a $\tilde{w} \in L^2\left([t, T] \times H, \mathcal{B}([t, T]) \otimes \overline{B}, ds \otimes m\right)$, uniquely determined
up to a $ds \otimes m$-null set, such that, for a.e. $s \in [t, T]$, $\tilde{w}(s, \cdot) = w(s)(\cdot)$ m-a.e.

We only prove the remaining statements related to $w \in L^2\left(t, T; L^2\left(H, \overline{B}, m\right)\right)$,
the others being similar. Invoking Corollary 5.48 and Hölder's inequality, we obtain

$$\int_t^T \int_H \mathbb{E}\left|w\left(s, X(s; t, x, a(\cdot))\right)\right| dm(x)\, ds$$

$$\leq C_T \left(\int_t^T \int_H \mathbb{E}\left|w\left(s, X(s; t, x)\right)\right|^2 dm(x)\, ds\right)^{1/2}$$

$$= C_T \left(\int_t^T \int_H \left|w\left(s, \cdot\right)\right|^2(x)\, dm(x)\, ds\right)^{1/2} < +\infty$$

and the first claim follows. The statements about the convergence follow using the
same arguments as indeed we have

$$\int_t^T \int_H \left|\mathbb{E}w_n\left(s, X(s; t, x, a(\cdot))\right) - \mathbb{E}w\left(s, X(s; t, x, a(\cdot))\right)\right| dm(x)\, ds$$

$$\leq C_T \left(\int_t^T \int_H \left|w_n\left(s, \cdot\right) - w\left(s, \cdot\right)\right|^2(x)\, dm(x)\, ds\right)^{1/2} \xrightarrow{n \to \infty} 0.$$

Similar estimates give the other claims.   □

We are now ready to prove the fundamental identity.

Lemma 5.50 *Let* $t \in [0, T]$, $\mu = \left(\Omega, \mathscr{F}, \{\mathscr{F}_s^t\}_{s \in [t, T]}, \mathbb{P}, W_Q\right)$ *be a generalized*
reference probability space on $[t, T]$ *and let* $a(\cdot) \in \mathcal{U}_t^{\mu}$*. Assume that Hypotheses 5.1,*
5.4, 5.27, 5.42, 5.45 hold. Suppose that the mild solution $(u, U) \in$
$L^2\left(0, T; W_Q^{1,2}(H, m)\right)$ *of (5.84) is also a strong solution. Then the following identity*
holds for m-a.e. $x \in H$:

$$u(t, x) + \mathbb{E} \int_t^T F_{0,CV}\left(s, X(s), \tilde{D}_Q u(s, X(s)), a(s)\right) - F_0\left(s, X(s), \tilde{D}_Q u(s, X(s))\right) ds$$

$$= \mathbb{E}\left\{\int_t^T [l(s, X(s)) + h_2(a(s))]ds + g(X(T))\right\} = J^\mu(t, x; a(\cdot)), \qquad (5.92)$$

where $X(\cdot) := X(\cdot; t, x, a(\cdot))$ denotes the mild solution of (5.78).

Proof We denote by g_n and ψ_n the approximating sequences of g and $F_0 + l$ characterized in (5.66), (5.67), (5.68), (5.69) and (5.70). We set

$$u_n(t, x) = P_{T-t}[g_n] + \int_t^T P_{s-t}[\psi_n(s)]ds.$$

We know that u_n satisfies in the classical sense the approximating HJB equation

$$\begin{cases} (u_n)_t + \mathcal{A}u_n + F_0(t, x, \tilde{D}_Q u_n) + l_n(t, x) = 0 \\ u_n(T, x) = g_n(x), \ x \in H, \end{cases} \qquad (5.93)$$

where

$$l_n(t, x) := \psi_n(t, x) - F_0(t, x, \tilde{D}_Q u_n) \xrightarrow{n \to +\infty} l \quad \text{in } L^2\left(0, T; L^2(H, \overline{\mathcal{B}}, m; H)\right).$$

By Dynkin's formula (see Proposition 1.169) and (5.78) we obtain

$$\mathbb{E}u_n(T, X(T)) - u_n(t, x)$$

$$= \mathbb{E}\int_t^T \left[(u_n)_s(s, X(s)) + \langle X(s), A^* D u_n(s, X(s))\rangle + \frac{1}{2}\mathrm{Tr}\left[QD^2 u_n(s, X(s))\right]\right] ds$$

$$+ \mathbb{E}\int_t^T \left[\left\langle D u_n(s, X(s)), b(X(s)) + Q^{\frac{1}{2}} R(s, X(s), a(s))\right\rangle\right] ds. \qquad (5.94)$$

Then, using (5.93) and the notation $F_{0,CV}$ introduced in (5.85), we get

$$\mathbb{E}g_n(X(T)) - u_n(t, x) = \mathbb{E}\int_t^T \left[F_{0,CV}\left(s, X(s), \tilde{D}_Q u_n(s, X(s)), a(s)\right)\right.$$

$$\left. - F_0(s, X(s), \tilde{D}_Q u_n(s, X(s))) - l_n(s, X(s)) - h_2(a(s))\right] ds. \qquad (5.95)$$

We now pass to the limit as $n \to +\infty$ in (5.95). We use (5.69), (5.70) and the convergences of the sequences u_n and $\tilde{D}_Q u_n$ prescribed by Definition 5.39 (indeed they are proved explicitly in our context in (5.72) and (5.73)). Thanks to Lemma 5.49 it thus follows that, for m-a.e. $x \in H$,

$$\mathbb{E}g(X(T)) - u(t, x) = \mathbb{E} \int_t^T \Big[F_{0,CV}(s, X(s), \tilde{D}_Q u(s, X(s)), a(s))$$

$$- F_0(s, X(s), \tilde{D}_Q u(s, X(s))) - l(s, X(s)) - h_2(a(s)) \Big] ds,$$

which gives (5.92) after rearranging the terms.   □

Lemma 5.51 *Let $t \in [0, T]$, $\mu = \left(\Omega, \mathscr{F}, \{\mathscr{F}_s^t\}_{s \in [t,T]}, \mathbb{P}, W_Q \right)$ be a generalized reference probability space on $[t, T]$ satisfying Hypothesis 5.42 and let Λ be as in (5.77). For any $p \geq 1$ there exists a countable subset NU_t^μ of $\mathcal{U}_t^\mu$ dense in $\mathcal{U}_t^\mu$ endowed with the $M_\mu^p(t, T; E)$ norm.*

Proof A possible choice for NU_t^μ is a set of elementary processes (see Definition 1.96). Indeed, in the construction of Lemma 1.98 we can clearly limit the choice of the times t_i appearing in Definition 1.96 to those of a dense and countable subset of $[t, T]$ and the choice of the $\mathscr{F}_t^t$-random variables to those of a dense and countable subset $\{\xi_j^{t_i}\}_{j \in \mathbb{N}}$ of $L^p(\Omega, \mathscr{F}_{t_i}, \mathbb{P}; E)$ (this subset exists thanks to Lemma 1.25). Since we look for processes belonging to $\mathcal{U}_t^\mu$ (and thus having images in $\overline{B_\varrho(0)}$), instead of $\{\xi_j^{t_i}\}_{j \in \mathbb{N}}$ we consider the random variables $\tilde{\xi}_j^{t_i} := \left(\max \left\{ \frac{|\xi_j^{t_i}|}{\varrho}, 1 \right\} \right)^{-1} \xi_j^{t_i}$. They create a required dense set of $\overline{B_\varrho(0)}$-valued processes. This can be seen by observing that if $x, y \in E$, $|x|_E \leq \varrho$ and $|y|_E > \varrho$, if $\tilde{y} := \left(\max \left\{ \frac{|y|}{\varrho}, 1 \right\} \right)^{-1} y$, we have

$$|x - \tilde{y}|_E \leq |x - y|_E + |y - \tilde{y}|_E \leq 2|x - y|_E,$$

where the last inequality follows from the fact that $\tilde{y}$ is among the elements of $B_\varrho(0)$ nearest to y, so $|x - y|_E \geq |y - \tilde{y}|_E$.   □

In the following lemma we give a sufficient condition to ensure that the functional $J^\mu(t, x; \cdot)$ is continuous with respect to the $M_\mu^p(t, T; E)$ norm.

Lemma 5.52 *Suppose that Hypotheses 5.42-(i) and 5.45 hold and that R, l, g and h_2 satisfy the following additional conditions:*

(i) *There exists an $M_R > 0$ such that*

$$|R(s, x, a_1) - R(s, y, a_2)| \leq M_R(|x - y| + |a_1 - a_2|) \, \forall s \in [0, T], x, y \in H, a_1, a_2 \in \Lambda.$$

(ii) *For some $C, q > 0$,*

$$|l(t, x)| \leq C(1 + |x|^q), \quad \text{for all } t \in [0, T], x \in H,$$

$$|g(x)| \leq C(1 + |x|^q), \quad \text{for all } x \in H.$$

(iii) *$h_2 : \Lambda \to \mathbb{R}$ is continuous.*

Then for every $t \in [0, T]$, $x \in H$ and every generalized reference probability space μ on $[t, T]$, the functional $J^\mu(t, x; \cdot)$ is continuous with respect to the $M_\mu^p(t, T; E)$ norm, for any $p > q$. In other words, for any sequence of controls $a_n(\cdot)$ in $\mathcal{U}_t^\mu$ converging to $a(\cdot) \in \mathcal{U}_t^\mu$ such that

$$\lim_{n \to \infty} |a_n(\cdot) - a(\cdot)|_{M_\mu^p}^p = \lim_{n \to \infty} \mathbb{E} \int_t^T |a_n(s) - a(s)|_E^p ds = 0, \qquad (5.96)$$

we have

$$\lim_{n \to \infty} J(t, x; a_n(\cdot)) = J(t, x; a(\cdot)). \qquad (5.97)$$

Moreover,

$$\lim_{n \to \infty} \sup_{s \in [t, T]} \left[\mathbb{E}|X(s; t, x, a_n(\cdot)) - X(s; t, x, a(\cdot))|^p \right] = 0. \qquad (5.98)$$

Proof We denote, for $s \in [t, T]$, $X(s; t, x, a_n(\cdot))$ by $X_n(s)$ and $X(s; t, x, a(\cdot))$ by $X(s)$ and also denote by N a positive constant such that $\|e^{tA}\| \leq N$ for any $t \in [0, T]$ and $\sup_{x \in H} |\nabla b(x)| \leq N$. We have, for any $s \in [t, T]$,

$$|X_n(s) - X(s)| \leq \left| \int_t^s e^{(t-r)A}(b(X_n(r)) - b(X(r)))dr \right|$$

$$+ \left| \int_t^s e^{(t-r)A}(R(r, X_n(r), a_n(r)) - R(r, X(r), a(r)))dr \right|$$

$$\leq N^2 \int_t^s |X_n(r) - X(r)|dr + NM_R \int_t^s (|a_n(r) - a(r)| + |X_n(r) - X(r)|)dr$$

and then, for $s \in [t, T]$,

$$\mathbb{E}\left[|X_n(s) - X(s)|^p \right]$$

$$\leq 3^{p-1}(N^{2p} + N^p M_R^p)T^{1/p} \int_t^s \mathbb{E}|X_n(\tau) - X(\tau)|^p dr + 3^{p-1}T^{1/p}|a_n(\cdot) - a(\cdot)|_{M_\mu^p}^p$$

and we obtain (5.98) using (5.96) and Gronwall's Lemma (Proposition D.29).

It follows from (5.96) and an easy application of the Lebesgue dominated convergence theorem that $\mathbb{E} \int_t^T h_2(a_n(s))ds$ converges to $\mathbb{E} \int_t^T h_2(a(s))ds$.

So to show (5.97) it remains to prove the convergence of the term $\mathbb{E} \left\{ \int_t^T l(s, X_n(s))ds + g(X_n(T)) \right\}$. We define the following linear operators:

$$\begin{cases} S_n : L^2(H, \overline{B}, m) \times L^2(t, T; L^2(H, \overline{B}, m)) \to \mathbb{R} \\ S_n(g, l) := \mathbb{E} \left\{ \int_t^T l(s, X_n(s))ds + g(X_n(T)) \right\} \end{cases}$$

and

$$\begin{cases} S: L^2(H, \overline{\mathcal{B}}, m) \times L^2(t, T; L^2(H, \overline{\mathcal{B}}, m)) \to \mathbb{R} \\ S(g, l) := \mathbb{E}\left\{ \int_t^T l(s, X(s))ds + g(X(T)) \right\}. \end{cases}$$

Since the constant $\tilde{c}$ appearing in (5.89) only depends on M_R (introduced in Hypothesis 5.42(i)) we know from (5.89) that the family $\{S_n\}$ is equi-continuous. Using the Lebesgue dominated convergence theorem and (5.98) it is easy to see that, for any $(g, l) \in C_b(H) \times C_b([t, T] \times H)$, we have $S_n(g, l) \xrightarrow{n \to \infty} S(g, l)$. Since $C_b(H) \times C_b([t, T] \times H)$ is dense in $L^2(H, \overline{\mathcal{B}}, m) \times L^2(t, T; L^2(H, \overline{\mathcal{B}}, m))$ and $\{S_n\}$ is equi-continuous, we can conclude that $S(g, l) = \lim_{n \to \infty} S_n(g, l)$ for any $(g, l) \in L^2(H, \overline{\mathcal{B}}, m) \times L^2(t, T; L^2(H, \overline{\mathcal{B}}, m))$, which completes the proof of (5.97).

$\square$

Theorem 5.53 (Verification Theorem, Sufficient Condition) *Let $p \geq 1$ and let Hypotheses 5.1, 5.4, 5.27, 5.42, 5.45 hold. Suppose that the mild solution $(u, U) \in L^2\left(0, T; W_Q^{1,2}(H, m)\right)$ of (5.84) is also a strong solution. Then the following are true:*

(i) For any $t \in [0, T]$ and any generalized reference probability space $\mu = \left(\Omega^\mu, \mathscr{F}^\mu, \{\mathscr{F}_t^{\mu,s}\}_{s \in [t,T]}, \mathbb{P}^\mu\right)$ satisfying Hypothesis 5.42, if $J^\mu(t, x; \cdot)$ is continuous with respect to the $M_\mu^p(t, T; E)$ norm, then there exists a set Z_t^μ with $m(Z_t^\mu) = 1$ such that, for all $x \in Z_t^\mu$ and all $a(\cdot) \in \mathcal{U}_t^\mu$ we have

$$u(t, x) \leq V_t^\mu(x) \leq J^\mu(t, x; a(\cdot)). \tag{5.99}$$

(ii) Choose $t \in [0, T]$. Let $\hat{\mu}$ be a generalized reference probability space satisfying Hypothesis 5.42 such that $J^{\hat{\mu}}(t, x; \cdot)$ is continuous with respect to the $M_{\hat{\mu}}^p(t, T; E)$ norm. Let x be in $Z_t^{\hat{\mu}}$. Let $a^*(\cdot) \in \mathcal{U}_t^{\hat{\mu}}$ be such that, denoting by $X^*(\cdot)$ the corresponding state, we have

$$a^*(s) \in \arg\min_{a \in \Lambda} F_{0,CV}(s, X^*(s), \tilde{D}_Q u(s, X^*(s)), a), \tag{5.100}$$

for almost every $s \in [t, T]$ and $\mathbb{P}$-almost surely. Then, the pair $(a^*(\cdot), X^*(\cdot))$ is $\hat{\mu}$-optimal at (t, x) and $u(t, x) = V_t^{\hat{\mu}}(x) = J^{\hat{\mu}}(t, x; a^*(\cdot))$.

Proof Part (i): We fix $t \in [0, T]$. By definition, for every $a \in \Lambda$, $F_{0,CV}(\cdot, a) - F_0(\cdot) \geq 0$ everywhere so for any $a(\cdot) \in \mathcal{U}_t^\mu$, by (5.92), $v(t, x) \leq J^\mu(t, x; a(\cdot))$ for m-a.e. $x \in H$. Thanks to Lemma 5.51 we can then choose a countable subset NU_t^μ dense in $\mathcal{U}_t^\mu$ in the M_μ^p norm containing minimizing sequences for any $x \in H$ (observe that the set of the controls depends on t but it does not depend on the initial datum x). By taking the infimum over $a(\cdot)$ in NU_t^μ in the right-hand side of (5.92) we obtain (i).

Part (ii): Since

$$\mathbb{E} \int_t^T \left[F_{0,CV} \left(s, X^*(s), \tilde{D}_Q u(s, X^*(s)), a^*(s) \right) \right.$$

$$\left. - F_0 \left(s, X^*(s), \tilde{D}_Q u(s, X^*(s)) \right) \right] ds = 0,$$

by (5.92) we thus get

$$u(t, x) = J^{\hat{\mu}}(t, x; a^*(\cdot)). \tag{5.101}$$

Since (5.99) is satisfied at (t, x) because $x \in Z_t^{\hat{\mu}}$, it follows that $(a^*(\cdot), X^*(\cdot))$ is $\hat{\mu}$-optimal at (t, x) and $u(t, x) = V_t^{\hat{\mu}}(x)$. □

5.5.4 Optimal Feedbacks

Similarly to what we observed in Sect. 2.5.1 for the regular case and in Sect. 4.8 for mild solutions in spaces of continuous functions, we use the fundamental identity and the verification theorem to characterize optimal feedbacks in the L^2 framework.

We consider the hypotheses of Theorem 5.53 and we look at the, possibly multi-valued (and not always defined), function

$$\begin{cases} \Phi \colon (0, T) \times H \to \mathcal{P}(\Lambda) \\ \Phi \colon (s, x) \to \arg\min_{a \in \Lambda} F_{0,CV}(s, x, \tilde{D}_Q u(s, x), a), \end{cases} \tag{5.102}$$

where $(u, U) \in L^2 \left(0, T; W_Q^{1,2}(H, m) \right)$ is the mild solution of (5.84). The corresponding Closed Loop Equation is

$$\begin{cases} dX(s) \in \left(AX(s) + b(X(s)) + Q^{\frac{1}{2}} R(s, X(s), \Phi(s, X(s))) \right) ds + dW_Q(s), \\ X(t) = x, \quad x \in H. \end{cases}$$
$$\tag{5.103}$$

Similarly to Sect. 4.8 we have the following corollary of Theorem 5.53.

Corollary 5.54 *Let $p \geq 1$ and let Hypotheses 5.1, 5.4, 5.27, 5.42, 5.45 hold. Suppose that the mild solution $(u, U) \in L^2 \left(0, T; W_Q^{1,2}(H, m) \right)$ of (5.84) is also a strong solution.*

Choose $t \in [0, T]$ and $x \in H$. Assume that, on $[t, T) \times H$, the feedback map Φ defined in (5.103) admits a measurable selection $\phi_t : [t, T) \times H \to \Lambda$. Then:

(i) The Closed Loop Equation

$$\begin{cases} dX(s) = \left(AX(s) + b(X(s)) + Q^{\frac{1}{2}} R(s, X(s), \phi_t(s, X(s))) \right) ds + dW_Q(s), \\ X(t) = x, \end{cases}$$
$$\tag{5.104}$$

has a weak mild solution (see Definition 1.121) $X_{\phi_t}(\cdot; t, x)$ in a suitable generalized reference probability space $\bar{\mu}$ (and unique in such a space); the elements
of the filtration $\mathscr{F}^t_{\bar{\mu},s}$ are countably generated up to sets of measure zero.

(ii) Suppose that the generalized reference probability space $\bar{\mu}$ from part (i) is
     such that $J^{\bar{\mu}}(t, x; \cdot)$ is continuous with respect to the $M^p_{\bar{\mu}}(t, T; E)$ norm and
     that x in $Z^{\bar{\mu}}_t$. Define, for $s \in [t, T)$, $a_{\phi_t}(s) = \phi_t(s, X_{\phi_t}(s; t, x))$. Then the
     pair $(a_{\phi_t}(\cdot), X_{\phi_t}(\cdot; t, x))$ is $\bar{\mu}$-optimal at (t, x) and $u(t, x) = V^{\bar{\mu}}_t(x)$. If, finally,
     $\Phi(s, x)$ is a singleton for any $(s, x) \in (t, T) \times H$, then $a_{\phi_t}(\cdot)$ is the unique $\bar{\mu}$-
     optimal control.

Proof Part (i) follows from Theorem 6.36. We can always take the filtration to be the
one generated by the Wiener process to ensure that the elements of the filtration are
countably generated up to sets of measure zero.

   All the statements of part (ii) follow immediately from Theorem 5.53-(ii) except
for the uniqueness of optimal controls. If $(\hat{a}(\cdot), \hat{X}(\cdot))$ is another optimal pair at (t, x)
with generalized reference probability space $\bar{\mu}$, we immediately have, by Lemma 5.50
and the fact that $u(t, x) = V^{\bar{\mu}}_t(x)$,

$$\mathbb{E} \int_t^T \left[F_{0,CV}\left(s, \hat{X}(s), \tilde{D}_Q u(s, \hat{X}(s)), \hat{a}(s) \right) - F_0\left(s, \hat{X}(s), \tilde{D}_Q u(s, \hat{X}(s)) \right) \right] ds = 0.$$

This implies that, for a.e. $s \in [t, T]$ and $\mathbb{P}$-a.s., we have $\hat{a}(s) = \phi_t(s, \hat{X}(s))$. Uniqueness of solutions of (5.104) in $\bar{\mu}$ gives the claim.   □

   We conclude with a result in a specific case.

Theorem 5.55 *Let $p \geq 1$ and let Hypotheses 5.1, 5.4, 5.27, 5.42, 5.45 hold. Suppose
that the mild solution $(u, U) \in L^2\left(0, T; W^{1,2}_Q(H, m)\right)$ of (5.84) is also a strong
solution. Suppose also that:*

(i) $E = H$ and $R(t, x, a) \equiv a$, hence $F_{0,CV}$ does not depend on t and x and it is
    given by
$$F_{0,CV}(p, a) = \langle a, p \rangle + h_2(a).$$

(ii) $h_2 : \Lambda \to \mathbb{R}$ is strictly convex and lower semicontinuous.

(iii) $F_0(p) := \inf_{a \in \Lambda} (\langle a, p \rangle + h_2(a))$ is differentiable.

   *Then, for any $t \in [0, T]$ and $x \in H$, there exists a generalized reference prob-
ability space μ (where the elements of the filtration $\mathscr{F}^t_{\mu,s}$ are countably generated
up to sets of measure zero) and a control $a^*(\cdot) \in \mathcal{U}^\mu_t$ which satisfies, together with
the corresponding trajectory $X^*(\cdot) := X(\cdot; t, x, a^*(\cdot))$, the relation*

$$a^*(s) = D_p F_0(\tilde{D}_Q u(s, X^*(s))), \qquad s \in [t, T]. \tag{5.105}$$

*If $x \in Z^\mu_t$ and $J^\mu(t, x; \cdot)$ is continuous with respect to the $M^p_\mu(t, T; H)$ norm, then
the control $a^*(\cdot)$ is μ-optimal.*

Proof We extend the function $h_2 \colon \Lambda \to \mathbb{R}$ to a function $\tilde{h}_2 \colon H \to \mathbb{R} \cup \{+\infty\}$ by defining $\tilde{h}_2(a) = +\infty$ for any $a \notin \overline{B_\varrho(0)}$. One can easily see that $a \to \tilde{h}_2(a)$ is strictly convex and lower semicontinuous on H. Moreover (see e.g. Proposition 2.19, p. 77 of [39]), the function

$$\tilde{h}_2^* \colon H \to \mathbb{R}, \qquad \tilde{h}_2^*(p) := \sup_{a \in H} \Big(\langle a, p \rangle - \tilde{h}_2(a) \Big)$$

is convex and lower semicontinuous on H. Thanks to the way we extended h_2, we necessarily have $\sup_{a \in H} \big(\langle a, -p \rangle - \tilde{h}_2(a) \big) = \sup_{a \in \Lambda} \big(\langle a, -p \rangle - h_2(a) \big)$ and thus $\tilde{h}_2^*(-p) = -F_0(p)$ for any $p \in H$.

Let now $p \in H$. It follows from the lower semi-continuity of $\tilde{h}_2$, its convexity and the fact that its value is $+\infty$ on $H \setminus \overline{B_\varrho(0)}$, that $\arg\min_{a \in H} \big(\langle a, p \rangle + \tilde{h}_2(a) \big)$ is non-empty (Theorem 2.11 page 72 of [39]). Since $\tilde{h}_2$ is strictly convex it is single-valued (see p. 84 of [39]). Thanks to the way we extended h_2, this unique point a^* where the minimum is attained belongs to Λ so it is also the unique minimizer of the problem

$$\inf_{a \in \Lambda} \left(\langle a, p \rangle + h_2(a) \right).$$

Moreover (see [39], Proposition 2.33, p. 84), a^* must be in the sub-differential (Definition 2.30, p. 82 of [39]) of $\tilde{h}_2^*(\cdot)$ at $-p$ which is equal to the super-differential of $F_0(\cdot)$ at p. Since by hypothesis F_0 is differentiable, we must have $a^* = D_p F_0(p)$ (see Proposition 2.40, p. 87 of [39]).

We now define the feedback control by

$$a(t) = D_p F_0(\tilde{D}_Q u(t, X(t))). \tag{5.106}$$

Consider, for $s \in [t, T]$, the closed loop equation in the mild form

$$X(s) = e^{(s-t)A} x + \int_t^s e^{(s-r)A} \left[b(X(r)) + Q^{\frac{1}{2}} D_p F_0(\tilde{D}_Q u(s, X(s))) \right] dr$$

$$+ \int_t^s e^{(r-s)A} dW_Q(r). \tag{5.107}$$

There exists (Theorem 6.36, where the selection is given by (5.106)) a generalized reference probability space μ where this equation has a mild solution $X^*(\cdot)$. We then take

$$a^*(s) = D_p F_0(\tilde{D}_Q u(s, X^*(s))), \qquad s \in [t, T],$$

and we conclude thanks to Theorem 5.53. $\qquad\square$

5.5.5 Continuity of the Value Function and Non-degeneracy of the Invariant Measure

The results we have described so far show one of the intrinsic limitations of the L^2 approach. Indeed, they can only describe the behavior of the value function in the support of the invariant measure. Such a support can be, in principle, very small. Also the verification theorem and construction of optimal feedbacks hold only on sets of full measure which may change with the generalized reference probability space. To remedy this we are going to introduce a non-degeneracy hypothesis. The non-degeneracy hypothesis, coupled with some continuity assumptions, will help us refine previous results and prove a number of propositions concerning the weak formulation of the optimal control problem (see Sect. 2.1.2).

Hypothesis 5.56 The invariant measure in Hypothesis 5.4 is non-degenerate. In other words, for any non-void open set $O \subset H$, $m(O) > 0$.

Recall that in the weak formulation of the optimal control problem the generalized reference probability space μ varies with the controls so that the set of admissible controls becomes

$$\overline{\mathcal{U}}_t := \bigcup_\mu \mathcal{U}_t^\mu,$$

where $\mathcal{U}_t^\mu$ is the set of admissible controls for a given generalized reference probability space μ defined in (5.79). The value function for the optimal control problem in the weak formulation is then

$$\overline{V}(t, x) = \inf_{a(\cdot) \in \overline{\mathcal{U}}_t} J^\mu(t, x; a(\cdot)).$$

Corollary 5.57 *Let the hypotheses of Lemma 5.50 and Hypothesis 5.56 be satisfied. Suppose moreover that, for any choice of t, μ and $a(\cdot)$, the functions $u(t, x)$ and $J^\mu(t, x, a(\cdot))$ are continuous in the x variable. Then, for every $(t, x) \in [0, T] \times H$ and any generalized reference probability space μ on $[t, T]$, we have*

$$u(t, x) \leq \overline{V}(t, x) \leq V_t^\mu(x).$$

Proof Lemma 5.50 ensures that, for any choice of t, μ and $a(\cdot)$, $u(t, x) \leq J^\mu(t, x, a(\cdot))$ for m-almost every $x \in H$. For any $y \in H$ we consider the sequence of balls $B_{1/n}(y)$, where $n \in \mathbb{N}$. Given the non-degeneracy of m, $m\left(B_{1/n}(y)\right) > 0$ and then $u(t, \cdot)$ cannot be strictly bigger than $J^\mu(t, \cdot, a(\cdot))$ on $B_{1/n}(y)$. We can thus obtain a sequence y_n converging to y such that $u(t, y_n) \leq J^\mu(t, y_n, a(\cdot))$. By continuity we get $u(t, y) \leq J^\mu(t, y, a(\cdot))$. Taking the infimum over $a(\cdot)$ and μ we have the claim. $\qquad\square$

More precise results can be obtained under stronger continuity assumptions.

Corollary 5.58 *Let the assumptions of Corollary 5.57 be satisfied. Suppose that, for any choice of t, μ and $a \in \Lambda$, the functions $\tilde{D}_Q u(t, \cdot)$, $R(t, \cdot, a)$, $l(t, \cdot)$ and $F_0(t, \cdot, \cdot)$ are continuous. Suppose that there exist $C > 0$ and $N \in \mathbb{N}$ such that, for all (t, x), $|\tilde{D}_Q u(t, x)|$, $|l(t, x)| \le C(1 + |x|^N)$. Then the fundamental identity (5.92) holds for any $(t, x) \in [0, T] \times H$, any generalized reference probability space μ and any $a(\cdot) \in \mathcal{U}_t^\mu$.*

Proof Lemma 5.50 ensures that, for any choice of t, μ and $a(\cdot)$, we have, for m-a.e. $x \in H$,

$$u(t, x) + \mathbb{E} \int_t^T F_{0,CV}\left(s, X(s), \tilde{D}_Q u(s, X(s)), a(s)\right)$$
$$- F_0\left(s, X(s), \tilde{D}_Q u(s, X(s))\right) ds = J^\mu(t, x; a(\cdot)),$$
(5.108)

where $X(s) := X(s; t, x, a(\cdot))$, for $s \in [t, T]$, is the mild solution of (5.78). Thus, as we did in the proof of Lemma 5.57, thanks to the non-degeneracy of m, for every $x \in H$ we can find a sequence y_n converging to x in H such that

$$u(t, y_n) + \mathbb{E} \int_t^T F_{0,CV}\left(s, X(s; t, y_n, a(\cdot)), \tilde{D}_Q u(s, X(s; t, y_n, a(\cdot))), a(s)\right)$$
$$- F_0\left(s, X(s; t, y_n, a(\cdot)), \tilde{D}_Q u(s, X(s; t, y_n, a(\cdot)))\right) ds = J^\mu(t, y_n; a(\cdot)).$$
(5.109)

We need to show that, taking the limit $n \to \infty$, every term of (5.109) converges to the respective term in (5.108). The convergence of $J^\mu(t, y_n; a(\cdot))$ and $u(t, y_n)$ follows from their continuity in the x variable.

The terms inside the integral converge pointwise to the respective terms in (5.108) $\mathbb{P}$-a.s. and for almost any s thanks to (1.70) and the various continuity hypotheses. The convergence of the integral thus follows from Lemma 1.51, the uniform moment bounds from (1.69), the polynomial growth of $|\tilde{D}_Q u(t, \cdot)|$ and $l(t, \cdot)$, the boundedness of R and the bounds on the growth of b and F_0. □

Using this result we find the counterparts of Theorem 5.53, Corollary 5.54 and Theorem 5.55 as follows.

Theorem 5.59 (Verification Theorem, Sufficient Condition) *Let the assumptions of Corollary 5.58 be satisfied. Choose $(t, x) \in [0, T] \times H$ and denote by $\hat{\mu}$ a generalized reference probability space. Let $a^*(\cdot) \in \mathcal{U}_t^{\hat{\mu}}$ be such that, denoting by $X^*(\cdot)$ the corresponding state, we have*

$$a^*(s) \in \arg\min_{a \in \Lambda} F_{0,CV}(s, X^*(s), \tilde{D}_Q u(s, X^*(s)), a)$$
(5.110)

for almost every $s \in [t, T]$ and $\mathbb{P}$-almost surely. Then the pair $(a^*(\cdot), X^*(\cdot))$ is optimal at (t, x) for the weak formulation (and so in the $\hat{\mu}$-strong formulation) and $u(t, x) = \overline{V}(t, x) = V_t^{\hat{\mu}}(x) = J^{\hat{\mu}}(t, x; a^*(\cdot))$.

Proof The proof is identical to the proof of Theorem 4.197 if we use Corollary 5.58.

□

Corollary 5.60 *Let the assumptions of Corollary 5.58 be satisfied. Choose $(t, x) \in [0, T] \times H$. Assume, moreover, that on $[t, T) \times H$ the feedback map Φ defined in (5.102) admits a measurable selection $\phi_t : [t, T) \times H \to \Lambda$. Then:*

(i) The Closed Loop Equation

$$\begin{cases} dX(s) = \Big(AX(s) + b(X(s)) + Q^{\frac{1}{2}} R(s, X(s), \phi_t(s, X(s))) \Big) ds + dW_Q(s), \\ X(t) = x, \end{cases}$$

(5.111)

has a weak mild solution (see Definition 1.121) $X_{\phi_t}(\cdot; t, x)$ in a suitable generalized reference probability space $\bar{\mu}$ and it is unique in this space if (5.111) is considered as an equation with the control process $a_{\phi_t}(s) := \phi(s, X_{\phi_t}(s; t, x))$, $s \in [t, T)$.

(ii) The pair $(a_{\phi_t}(\cdot), X_{\phi_t}(\cdot; t, x))$ is optimal for the weak formulation (and a fortiori $\bar{\mu}$-optimal) at (t, x) and $u(t, x) = \overline{V}(t, x) = V_t^{\overline{\mu}}(x) = J^{\bar{\mu}}(t, x; a_{\phi_t}(\cdot))$. If, finally, $\Phi(s, x)$ a singleton for any $(s, x) \in (t, T) \times H$, then a_{ϕ_t} is the unique $\bar{\mu}$-optimal control.

Proof The proof is the same as that of Corollary 5.54 but we have to use Corollary 5.58 instead of Lemma 5.50. □

Observe that, in the above corollary, if the uniqueness of solutions of (5.111) is not guaranteed, the optimality of the pair $(a_{\phi_t}(\cdot), X_{\phi_t}(\cdot; t, x))$ needs to be understood in terms of the extended weak formulation introduced in Remark 2.6.

Theorem 5.61 *Let the assumptions of Corollary 5.58 be satisfied.*
Suppose also that:

(i) $E = H$ and $R(t, x, a) \equiv a$, hence $F_{0,CV}$ does not depend on t and x and it is given by

$$F_{0,CV}(p, a) = \langle a, p \rangle + h_2(a).$$

(ii) $h_2 : \Lambda \to \mathbb{R}$ is strictly convex and lower semicontinuous.
(iii) $F_0(p) := \inf_{a \in \Lambda} (\langle a, p \rangle + h_2(a))$ is differentiable.

Then, for any $t \in [0, T]$ and $x \in H$, there exists a generalized reference probability space μ (where the elements of the filtration $\mathscr{F}_{\mu,s}^t$ are countably generated up to sets of measure zero) and a control $a^*(\cdot) \in \mathcal{U}_t^\mu$ which satisfies, together with the corresponding trajectory $X^*(\cdot) := X(\cdot; t, x, a^*(\cdot))$, the relation

$$a^*(s) = D_p F_0(\tilde{D}_Q u(s, X^*(s))), \qquad s \in [t, T].$$

$a^*(\cdot)$ is an optimal control for the weak formulation at (t, x) and the unique $\bar{\mu}$-optimal control at (t, x). For any $t \in [0, T]$ and $x \in H$, $u(t, x)$ equals the value function $\overline{V}(t, x)$.

Proof The proof follows the same arguments as these used in the proof of Theorem 5.55. In the very last step we use Corollary 5.60 instead of Theorem 5.53. □

5.6 Examples

We show how the L^2-theory we have developed so far can be used to treat some specific optimal control problems.

5.6.1 Optimal Control of Delay Equations

Let us consider a simple controlled one-dimensional linear stochastic differential equation with a delay $r > 0$:

$$
\begin{cases}
dy(s) = (\beta_0 y(s) + \beta_1 y(s - r) + \alpha(s))\, ds + \sigma d W_0(s), \\
y(t) = x_0, \\
y(t + \theta) = x_1(\theta),\ \theta \in [-r, 0),
\end{cases}
\tag{5.112}
$$

where $\sigma > 0$, $\beta_0, \beta_1 \in \mathbb{R}$ are given constants; W_0 is a one-dimensional standard Brownian motion defined on a complete probability space $(\Omega, \mathscr{F}, \mathbb{P})$; and $\{\mathscr{F}_s^t\}_{s \in [t,T]}$ is the augmented filtration generated by W_0. The control $\alpha(\cdot)$ is an $\mathscr{F}_s^t$-progressively measurable process with values in the interval $\Lambda = [0, R]$ for some $R > 0$. We assume that $x_1(\cdot) \in L^2(-r, 0)$.

As recalled in Sect. 2.6.8, Eq. (5.112) can be rewritten as a linear evolution equation in the Hilbert space $H = \mathbb{R} \times L^2(-r, 0)$ of the following form:

$$
\begin{cases}
dX(s) = (A_1 X(s) + B_1 a(s))\, dt + G d W_0(s), \\
X(t) = \begin{pmatrix} x_0 \\ x_1 \end{pmatrix} := \begin{pmatrix} y_0 \\ y_1 \end{pmatrix} \in H,
\end{cases}
\tag{5.113}
$$

where $a(\cdot) = \alpha(\cdot)$, A_1 is a suitable generator of a C_0-semigroup on H; $B_1 : \mathbb{R} \to H$ and $G : \mathbb{R} \to H$ are continuous operators $B_1 w_0 = \begin{pmatrix} w_0 \\ 0 \end{pmatrix}$ and $G w_0 = \begin{pmatrix} \sigma w_0 \\ 0 \end{pmatrix}$ (further details can be found in Sect. 2.6.8). Finally, considering $Q \in \mathcal{L}^+(H) = \mathcal{L}^+(\mathbb{R} \times L^2(-\tau, 0))$ defined as $Q := \begin{pmatrix} \sigma^2 & 0 \\ 0 & 0 \end{pmatrix}$, we can rewrite the equation once more obtaining

$$
\begin{cases}
dX(s) = \left(A_1 X(s) + Q^{1/2} \frac{1}{\sigma} B_1 a(s)\right) ds + d W_Q(s), \\
X(t) = \begin{pmatrix} x_0 \\ x_1 \end{pmatrix} := \begin{pmatrix} y_0 \\ y_1 \end{pmatrix} \in H,
\end{cases}
\tag{5.114}
$$

which is the form required by (5.78).

Proposition 5.62 *Assume that $\sigma \neq 0$, that $\beta_0 < 1$ and denote by γ a real number in $(0, \pi)$ such that $\gamma \coth \gamma = \beta_0$. Assume that*

$$\beta_0 < -\beta_1 < \sqrt{\gamma^2 + \beta_0^2}. \qquad (5.115)$$

Then Eq.(5.113) and (5.114) have a unique invariant measure m which is non-degenerate.

Proof See Remark 10.2.6(i), Chap. 10 of [177]. □

Proposition 5.63 *Consider the operator $D_Q := Q^{1/2}D$ defined on $C_b^1(H) \subset L^2 (H, \overline{\mathcal{B}}, m)$. Then:*

(i) D_Q is not closable in $L^2(H, \overline{\mathcal{B}}, m)$.
(ii) Hypothesis 5.1 holds.

Proof Part (i) is proved in [299], Sect.7.2, pp. 15–16. The second statement can easily be verified. □

Thanks to Part (ii) of Proposition 5.63, the whole theory developed so far in this chapter can be applied even if the operator D_Q is not closable in the classical sense.

Remark 5.64 We considered a simple one-dimensional case of controlled stochastic delay equations for simplicity of presentation. In fact, this framework can be applied to more general cases like semilinear d-dimensional equations presented in Sect.2.6.8. Conditions to guarantee the existence of a nontrivial invariant measure for the multidimensional case can be found in Sect.10.3 of [177] (see, in particular, Theorem 10.2.5(i)). Using the same methodology, problems with cost functions f_0 and g_0 depending also on the history of the state y can be treated as well. ∎

5.6.2   Control of Stochastic PDEs of First Order

The second example is an optimal control problem driven by a first-order stochastic PDE similar to the one considered in Sect.2.6.7. This kind of equation is important in financial modeling since it provides a description of the time evolution of forward rates under the non-arbitrage assumption; we refer the reader to Sect.2.6.7 and [303].

Fix $\kappa > 0$. The state space H we consider here is given by the following weighted L^2 space of real-valued functions defined on $[0, +\infty)$:

$$H := \left\{ f : [0, +\infty) \to \mathbb{R} \text{ measurable } : \int_0^{+\infty} f^2(\xi) e^{-\kappa \xi} d\xi < +\infty \right\}.$$

In particular, if $\kappa = 0$, $H = L^2(\mathbb{R})$. The inner product on H is given by

$$\langle f, g \rangle_H := \int_0^{+\infty} f(\xi) g(\xi) e^{-\kappa \xi} d\xi$$

and the induced norm will be denoted by $| \cdot |_H$.

The following result can be easily proved.

Proposition 5.65 *The semigroup $S(t)$ defined as*

$$S(t) f(\xi) := f(t + \xi), \qquad \xi \geq 0$$

is a C_0-semigroup on H. Its generator is given by

$$\begin{cases} D(A) = H^1_\kappa(0, \infty) := \left\{ f \in L^2_\rho : \frac{df}{d\xi} \in L^2_\rho \right\} \\ A = \frac{d}{d\xi} \end{cases}$$

(where $\frac{df}{d\xi}$ denotes the distribution derivative of f here). Moreover,

$$\|S(t)\|_{\mathcal{L}(H)} \leq e^{-\kappa t}.$$

We consider the following equation, studied for instance in [303],

$$dX(t) = (AX(t) + b(X(t)) + Bh_1(a(t))) dt + \tau dW_0(t), \tag{5.116}$$

where W_0 is a one-dimensional Brownian motion; $\tau \in H \cap B_b([0, +\infty), \mathbb{R})$; $B \in \mathcal{L}(H)$ and $h_1 : \Lambda \to \mathbb{R}$; $a(t) = a(t, \cdot) \in H$ is a control process and b is an operator defined on H as follows

$$b(x)(\xi) = -\tau(x(\xi)) \int_0^\xi \frac{1}{1 + e^{x(r)}} \tau(r) dr - \frac{1}{2} |\tau(\xi)|^2 \frac{1}{1 + e^{x(\xi)}} \tau(\xi) \int_0^\xi \tau(r) dr.$$

In order to apply the L^2 theory we need to ensure the existence of an invariant measure for the uncontrolled version of (5.116). This is the content of the following lemma.

Lemma 5.66 *If*

$$\|\tau\|_0 + |\tau|_H |\tau e^{\kappa \cdot}|_H \leq \kappa,$$

then there exists a non-degenerate invariant measure m for

$$dX(t) = (AX(t) + b(X(t))) dt + \tau dW_0(t).$$

Proof See Proposition 3.2 in [303]. □

Observe that $\tau dW_0(t)$ is of the form $dW_Q(t)$ prescribed by Hypothesis 5.1-(B) if we consider, for instance, the operator $Qx = \tau \langle \tau, x \rangle$. In this case one can easily see that Hypothesis 5.1-(C) is satisfied as well. To verify Hypothesis 5.42 we need $Bh_1(a(t))$ to be of the form $Q^{1/2}R$ for some R satisfying Hypothesis 5.42-(i). This is the case if we take $B = Q^{\frac{1}{2}}$ and $h_1 \colon \Lambda \to H$ some bounded Borel measurable function. Λ needs to be specified, as in (5.77), as a closed ball of a real separable Banach space.

Remark 5.67 The operator $(D_Q, C_b^1(H))$ is not always closable in $L^2(H, \overline{B}, m)$ (see, e.g., Paragraph 7.1, pp. 13–14 of [299]).   ∎

5.6.3   Second-Order SPDEs in the Whole Space

The third example regards a stochastic controlled parabolic equation in the whole space (see Sects. 2.6.1 and 2.6.2 for stochastic controlled parabolic equations in bounded domains). We consider the problem using a weighted L^2 space as the underlying Hilbert space. For simplicity we limit our observations to the one-dimensional case.

We denote by H the weighted $L^2(\mathbb{R})$ space $L^2(\mathbb{R}, \rho_\kappa(\xi)dz)$, where the weight $\rho_\kappa(\xi) = e^{-\kappa|\xi|}$ with $\kappa > 0$.

The inner product and the norm in H are denoted by $\langle \cdot, \cdot \rangle_H$ and $|\cdot|_H$, respectively. Fix $\lambda > 0$ and define $A^{(0)} = \Delta - \lambda I$, where $\Delta \colon D(\Delta) \subset L^2(\mathbb{R}) \to L^2(\mathbb{R})$ is the Laplacian with domain $D(\Delta)$, which is the Sobolev space $H^2(\mathbb{R})$. Let $S^{(0)}(t)$ denote the C_0-semigroup on $L^2(\mathbb{R})$ generated by $A^{(0)}$. The semigroup $S^{(0)}(t)$ is self-adjoint on $L^2(\mathbb{R})$ and

$$\left\| S^{(0)}(t) \right\| \le e^{-\lambda t}. \tag{5.117}$$

Proposition 5.68 $\{S^{(0)}(t),\, t \ge 0\}$ *can be uniquely extended to a C_0-semigroup* $\{S^{(\kappa)}(t),\, t \ge 0\}$ *on H. Moreover,*

$$\left\| S^{(\kappa)}(t) \right\|_{\mathcal{L}(H)} \le e^{(\frac{1}{2}\kappa^2 - \lambda)t}, \quad t \ge 0. \tag{5.118}$$

Proof See Proposition 9.4.1, p. 187 of [177].   □

We denote by $A^{(\kappa)}$ the generator of $\{S^{(\kappa)}(t),\, t \ge 0\}$.
Consider the controlled equation

$$dX(t) = \left(AX^{(\kappa)}(t) + JR(X(t)) - Ja(t)\right)dt + JdW(t), \tag{5.119}$$

where W is a standard cylindrical Wiener process on $L^2(\mathbb{R})$; J is the embedding $L^2(\mathbb{R}) \hookrightarrow H$ and $a(\cdot)$ is a control process taking values in $L^2(\mathbb{R})$. Assume that the Lipschitz continuous map $R \colon L^2(\mathbb{R}) \to L^2(\mathbb{R})$ extends to a map $H \to H$ which satisfies Hypothesis 5.42-(i).

The following equation is the uncontrolled counterpart of (5.119)

$$dX(t) = A^{(\kappa)}X(t)dt + JdW(t). \tag{5.120}$$

Proposition 5.69 *For any $\kappa > 0$ and $\lambda > 0$ the solution of (5.120) is well defined in H and it admits a non-degenerate invariant measure m.*

Proof For the existence of the invariant measure, see Proposition 9.4.6, page 191 of [177]. In [119], Sect. 4.3, it is proved that the invariant measure can be chosen to be non-degenerate. □

It can be shown that the transition semigroup for this process is not strongly Feller, hence it violates the smoothing property required, for example, in Hypothesis 4.76. Thus the theory of the HJB equations developed in Chap. 4 does not apply in this case. Nevertheless, we can study the problem using the results of this chapter.

Remark 5.70 We observe that the family of optimal controls described by the state equation (5.78) needs to satisfy the *structural condition* described in Chap. 2: the image of the drift is always contained in the image of $Q^{1/2}$. The same kind of structure is also present in the state equation of the parabolic problem studied in [225] and described in (2.104). In that case the same operator B acts on the drift and on the diffusion but it is unbounded, so the theory described in this chapter cannot be used. Still, such a similarity in the structure suggests that some further development of the theory will probably be able to treat such a case. ∎

5.7 Results in Special Cases

In this section we present further results about existence and uniqueness of solutions of HJB equations when a certain "commutative assumption" for the operators A and Q is satisfied. We will indeed suppose (see Hypothesis 5.71-(D) for a more precise statement) that there exists an orthonormal basis of H made of eigenvectors of both A and Q.

The problem was studied in [3, 4, 123, 125] in this case. In this section we recall some results, mainly from [4, 123]. We omit the proofs. An element of interest of the approaches developed in [4, 125] is the use of variational solutions of the HJB equations. In this kind of approach the solution is defined via the duality pairing of the candidate solution with regular functions. Since the duality is obtained by extending an L^2 inner product on H, the use of this scheme is strictly linked to the identification of a reference measure on H.

5.7.1 Parabolic HJB Equations

We consider the following set of assumptions (similar to Hypothesis 5.1).

Hypothesis 5.71 (A) A is the generator of a strongly continuous semigroup $\{e^{tA}, t \geq 0\}$ on a real separable Hilbert space H and there exist constants $M \geq 1$ and $\omega > 0$ such that

$$\|e^{tA}\| \le Me^{-\omega t}, \qquad \forall t \ge 0.$$

(B) $Q \in \mathcal{L}^+(H)$, $T > 0$ and $\mu := \left(\Omega^\mu, \mathscr{F}^\mu, \{\mathscr{F}_{\mu,s}\}_{s\in[0,T]}, \mathbb{P}^\mu, W^\mu_Q\right)$ is a generalized reference probability space.
(C) $e^{sA}Qe^{sA^*} \in \mathcal{L}_1(H)$ for all $s > 0$. Moreover, for all $t \ge 0$,

$$\int_0^t \mathrm{Tr}\left[e^{sA}Qe^{sA^*}\right]ds < +\infty,$$

so the symmetric positive operator

$$Q_t : H \to H, \qquad Q_t := \int_0^t e^{sA}Qe^{sA^*}ds,$$

is of trace class for every $t \ge 0$.
(D) There exists an orthonormal basis $\{e_1, e_2, \ldots\}$ of H made of elements of $D(A)$
such that

$$Ax = \sum_{n=1}^{+\infty} -\alpha_n \langle e_n, x\rangle e_n, \quad x \in D(A),$$

for some eigenvalues $0 < \alpha_1 \le \alpha_2 \le \alpha_3 \ldots$ and

$$Qx = \sum_{i=n}^{+\infty} q_n \langle e_n, x\rangle e_n, \quad x \in H,$$

for a sequence of nonnegative eigenvalues q_n.

If Hypothesis 5.71 holds, the existence of an invariant measure m associated with
the following Ornstein–Uhlenbeck process

$$\begin{cases} dX(s) = AX(s)ds + dW_Q(s), & 0 \le s \le T, \\ X(0) = x \in H \end{cases} \tag{5.121}$$

is proved, for example, in [180], Theorem 11.30, page 325. Observe that, differently
from what we did in previous sections, here the reference measure is the invariant
measure of the homogeneous Cauchy problem (which coincides with that of previous
sections if $b = 0$ in (5.3)). For any $\phi \in C_b(H)$, the notation[6] $P_t[\phi](x)$ will be used

---

[6]In Chap. 4 and in Appendix B, when the transition semigroup reduces to the Ornstein–Uhlenbeck
case, the notation R_t is used. In this section, and in the proof of Theorem 5.41, we keep the notation
P_t even for the Ornstein–Uhlenbeck case because the semigroup plays exactly the same role, from
the perspective of the L^2 approach to the HJB equation, as the semigroup P_t in Sect. 5.3 and,
differently from Chap. 4 and Appendix B, the two semigroups never appear at the same time, so
there is no possibility of confusion.

to denote the transition semigroup P_t for (5.121):

$$P_t[\phi](x) = \mathbb{E}\phi(X(t, x)).$$

Denoting by $\overline{\mathcal{B}}$ the completion of the Borel σ-field $\mathcal{B}(H)$ with respect to m, P_t extends to a strongly continuous semigroup of contractions on $L^2(H, \overline{\mathcal{B}}, m)$ with the generator

$$\begin{cases} \mathcal{A}: D(\mathcal{A}) \subset L^2(H, \overline{\mathcal{B}}, m) \to L^2(H, \overline{\mathcal{B}}, m) \\ \\ \mathcal{A}: \phi \to \mathcal{A}\phi, \end{cases}$$

whose explicit expression on regular functions is

$$\mathcal{A}\phi(x) = \frac{1}{2} \operatorname{Tr} [QD^2\phi] + \langle Ax, D\phi \rangle. \tag{5.122}$$

When Hypothesis 5.71, and in particular its part (D), is satisfied, Remark 5.12 ensures that the operator D_Q introduced in Definition 5.11 is closable so that the closability problem we mentioned in Sect. 5.2.4 is no longer an issue. Therefore we work here with more conventional Sobolev spaces. We introduce them now together with some notations that will be useful in the variational approach to the solution of the HJB equation described below. Denote by $\mathcal{H}$ the space $L^2(H, \overline{\mathcal{B}}, m)$, by $\mathcal{V}$ the Sobolev space $W^{1,2}(H, m)$ made of all functions f of $L^2(H, \overline{\mathcal{B}}, m)$ such that $Df \in L^2(H, \overline{\mathcal{B}}, m)$, and by $\mathcal{V}^*$ its dual. Identifying $\mathcal{H}$ with its dual, one gets the following Gelfand triple

$$\mathcal{V} \subset \mathcal{H} \subset \mathcal{V}^*.$$

Given $T > 0$ we introduce

$$\mathcal{W}_T := \left\{ f \; : \; f \in L^2(0, T; \mathcal{V}), \frac{d}{dt} f \in L^2(0, T; \mathcal{V}^*) \right\}.$$

It follows, for instance, from Theorem 1.2.15 of [5] that $\mathcal{W}_T \subset C([0, T], \mathcal{H})$. In particular, given $f \in \mathcal{W}_T$, $f(T)$ is a well-defined element of $\mathcal{H}$ and thus an m-a.e. defined function from H to $\mathbb{R}$. We will use this fact in the following, in particular in the statements of Theorems 5.78 and 5.79.

Lemma 5.72 *Let Hypothesis 5.71 be satisfied. The operator* $\mathcal{A}: D(\mathcal{A}) \subset \mathcal{H} \to \mathcal{H}$ *extends uniquely to a linear operator* $\tilde{\mathcal{A}} \in \mathcal{L}(\mathcal{V}, \mathcal{V}^*)$ *such that, for any* $\phi, \psi \in \mathcal{V}$,

$$\left\langle \tilde{\mathcal{A}}\phi, \psi \right\rangle_{\langle \mathcal{V}^*, \mathcal{V} \rangle} = \left\langle \phi, \tilde{\mathcal{A}}\psi \right\rangle_{\langle \mathcal{V}, \mathcal{V}^* \rangle} = \frac{1}{2} \int_H \left\langle \sqrt{Q}D\phi, \sqrt{Q}D\psi \right\rangle_H dm(x).$$

Finally, $\tilde{\mathcal{A}}$ *satisfies the following coercivity estimate: there exist* $\alpha, \beta > 0$ *such that, for any* $\phi \in \mathcal{V}$,

$$-\left\langle \tilde{A}\phi, \phi \right\rangle_{\langle V^*, V \rangle} \geq \alpha|\phi|_V^2 - \beta|\phi|_\mathcal{H}^2 \tag{5.123}$$

and (5.123) holds in particular if one considers $\alpha = 1/2$ and $\beta = 1/2$.

Proof See [3], Lemma 4.2, p. 111. For the last statement, see [4], p. 503. □

Given a measurable map $G: V \to V^*$, a function $f \in L^2(0, T; V^*)$ and $g \in \mathcal{H}$ we consider the equation

$$\begin{cases} u_t + \mathcal{A}u + G(u) + f(t, x) = 0, \\ u(T, x) = g(x). \end{cases} \tag{5.124}$$

Definition 5.73 A function $u \in \mathcal{W}_T$ is a solution of (5.124) in the variational sense, if for any $\psi \in V$ and any $t \in [0, T]$,

$$\langle u(t), \psi \rangle = \langle g, \psi \rangle + \int_t^T \left\langle \tilde{A}u(s), \psi \right\rangle_{\langle V^*, V \rangle} ds + \int_t^T \langle Gu(s), \psi \rangle_{\langle V^*, V \rangle} ds$$
$$+ \int_t^T \langle f(s), \psi \rangle_{\langle V^*, V \rangle} ds. \tag{5.125}$$

Theorem 5.74 *Assume that Hypothesis 5.71 is satisfied. Assume that $G: V \to V^*$ and there exists a positive constant $K < \alpha$ (where α is the constant from (5.123)) such that:*

(G1) $|G(\xi)|_{V^*} \leq K(1 + |\xi|_V)$ *for all $\xi \in V$,*
(G2) $|G(\xi) - G(\eta)|_{V^*} \leq K|\xi - \eta|_V$ *for all $\xi, \eta \in V$.*

Then, for every $g \in \mathcal{H}$ and $f \in L^2(0, T; V^*)$ the evolution equation (5.124) has a unique solution in $\mathcal{W}_T$ in the sense of Definition 5.73.

Proof See Theorem 5.2 in [3]. □

One can remove the restriction $K < \alpha$ assuming a stronger regularity of the function G.

Theorem 5.75 *Assume that Hypothesis 5.71 is satisfied. Assume that $G: V \to \mathcal{H}$ and there exists a positive constant K such that:*

(G1) $|G(\xi)|_\mathcal{H} \leq K(1 + |\xi|_V)$ *for all $\xi \in V$,*
(G2) $|G(\xi) - G(\eta)|_\mathcal{H} \leq K|\xi - \eta|_V$ *for all $\xi, \eta \in V$.*

Then, for every $g \in \mathcal{H}$ and $f \in L^2(0, T; V^*)$ the evolution equation (5.124) has a unique solution in $\mathcal{W}_T$ in the sense of Definition 5.73.

Proof See Theorem 5.3 in [3]. □

5.7.2 Applications to Finite Horizon Optimal Control Problems

Let Hypothesis 5.71 be satisfied. We denote by Λ the closed ball $\overline{B}_\varrho(0)$ of radius ϱ in H. Given some generalized reference probability space $\mu := \left(\Omega^\mu, \mathscr{F}^\mu, \{\mathscr{F}_{\mu,s}\}_{s \in [0,T]}, \mathbb{P}^\mu, W_Q^\mu\right)$ we consider the class of admissible controls given by

$$\mathcal{U}_0^\mu = \{a(\cdot)\colon [0,T] \to \Lambda \ : \ a(\cdot) \text{ is } \mathscr{F}_{\mu,s} - \text{progressively measurable}\}. \quad (5.126)$$

We consider the optimal control problem, in the weak formulation, characterized by the state equation

$$\begin{cases} dX(s) = (AX(s) + b(X(s)) + B(X(s))a(s))\, ds + dW_Q^\mu(s), & 0 \le s \le T \\ X(0) = x, \quad x \in H, \end{cases}$$
$$(5.127)$$

and the target functional

$$J^\mu(x; a(\cdot)) = \mathbb{E}^\mu \left\{ \int_0^T [f(s, X(s; 0, x, a(\cdot))) + h(a(s))]\, ds + g(X(T; 0, x, a(\cdot))) \right\}.$$
$$(5.128)$$

The hypotheses on the functions $b\colon H \to H$, $B\colon H \to \mathcal{L}(H)$, f, h and g are specified below.

Since we are interested in the weak formulation of the problem, we let the generalized reference probability space μ vary and we consider the set of controls given by

$$\overline{\mathcal{U}}_0 := \bigcup_\mu \mathcal{U}_0^\mu, \quad (5.129)$$

where $\mathcal{U}_0^\mu$ is defined in (5.126). The value function of the problem is

$$\overline{V}_0(x) = \inf_{a(\cdot) \in \overline{\mathcal{U}}_0} J^\mu(x; a(\cdot)). \quad (5.130)$$

The corresponding HJB equation is

$$\begin{cases} v_t + Av + \langle b(x), Dv \rangle + F(x, Dv) + f(t, x) = 0, \\ v(T, x) = g(x), \quad x \in H, \end{cases}$$
$$(5.131)$$

where the Hamiltonian F is given by

$$F(x, p) = \inf_{a \in \Lambda} \{\langle B(x)a, p \rangle + h(a)\}. \quad (5.132)$$

If we introduce

$$G(v)(x) := \langle b(x), Dv(x) \rangle + F(x, Dv(x)),$$

equation (5.131) can be rewritten in the form (5.124),

$$\begin{cases} v_t + \mathcal{A}v + G(v) + f(t, x) = 0, \\ v(T, x) = g(x) \end{cases} \tag{5.133}$$

and Theorems 5.74 and 5.75 can be applied. One gets the following propositions, as corollaries.

Proposition 5.76 *Assume that Hypothesis 5.71 is satisfied. Suppose that b and $x \to B(x)a$, for any $a \in \Lambda$, are Borel measurable maps from H to H, have images in $\sqrt{Q}(H)$ and there exist two positive constants k_1 and k_2 such that*

$$|Q^{-1/2}b(x)| \leq k_1(1 + |x|) \quad \text{for all } x \in H \tag{5.134}$$

and

$$\|B^*(x)Q^{-1/2}\|_{\mathcal{L}(H)} \leq k_2(1 + |x|) \quad \text{for all } x \in H, \tag{5.135}$$

where $Q^{-1/2}$ denotes the pseudoinverse of $Q^{1/2}$. Moreover, assume that $h : \Lambda \to \mathbb{R}$ is measurable and bounded. Then, for any $g \in \mathcal{H}$ and $f \in L^2(0, T; \mathcal{V}^*)$, (5.133) has a unique solution $v \in \mathcal{W}_T$, provided that k_1 and k_2 are sufficiently small.

Proof See Corollary 4.3 in [4]. ☐

One can remove the restrictions on k_1 and k_2 if the regularity of b and B is stronger.

Proposition 5.77 *Assume that Hypothesis 5.71 is satisfied. Suppose that b and $x \to B(x)a$, for any $a \in \Lambda$, are Borel measurable maps from H to H, have images in $\sqrt{Q}(H)$, and that*

$$\sup_{x \in H} |Q^{-1/2}b(x)| < +\infty \tag{5.136}$$

and

$$\sup_{x \in H} \|B^*(x)Q^{-1/2}\|_{\mathcal{L}(H)} < +\infty, \tag{5.137}$$

where $Q^{-1/2}$ denotes the pseudoinverse of $Q^{1/2}$. Moreover, assume that $h : \Lambda \to \mathbb{R}$ is measurable and bounded. Then, for any $g \in \mathcal{H}$ and $f \in L^2(0, T; \mathcal{V}^*)$, (5.133) has a unique solution $v \in \mathcal{W}_T$.

Proof See Corollary 4.4 in [4]. ☐

We now state two results that ensure the existence of an optimal control and characterize the value function as the unique variational solution of the HJB equation.

Theorem 5.78 *Assume that the hypotheses of Proposition 5.76 are satisfied. Moreover, assume that:*

(i) $f \in L^2(0, T; \mathcal{V}^*)$.
(ii) $b: H \to H$ and $B: H \to \mathcal{L}(H)$ are Lipschitz-continuous.
(iii) $h : \Lambda \to \mathbb{R}$ is lower semicontinuous.

Then, for each initial datum $x \in H$, there exists an optimal control for the optimal control problem (5.127)–(5.129). Moreover, if $v \in \mathcal{W}_T \subset C([0, T], \mathcal{H})$ is the unique solution of (5.133) and $\overline{V}_0$ is the value function defined in (5.130), we have $v(0, x) = \overline{V}_0(x)$ for m-a.e. $x \in H$.

Proof See Theorem 5.4 in [4]. $\qquad\qquad\qquad\qquad\qquad\qquad\qquad\qquad\Box$

Theorem 5.79 *Assume that the hypotheses of Proposition 5.77 are satisfied. Moreover, assume that:*

(i) $f \in L^2(0, T; \mathcal{H})$.
(ii) $b: H \to H$ and $B: H \to \mathcal{L}(H)$ are Lipschitz-continuous.
(iii) $h : \Lambda \to \mathbb{R}$ is lower semicontinuous.

Then, for each initial datum $x \in H$, there exists an optimal control for the optimal control problem (5.127)–(5.129) and the unique solution of (5.133) is given by the value function defined in (5.130). Moreover, if $v \in \mathcal{W}_T \subset C([0, T], \mathcal{H})$ is the unique solution of (5.133) and $\overline{V}_0$ the value function defined in (5.130), we have $v(0, x) = \overline{V}_0(x)$ for m-a.e. $x \in H$.

Proof See Theorem 5.2 in [4]. $\qquad\qquad\qquad\qquad\qquad\qquad\qquad\qquad\Box$

Remark 5.80 We can compare the results and the assumptions of this last section with those obtained in the previous parts of the chapter. We observe that:

(i) In this section, differently from Sects. 5.2–5.4, the "commutative" Hypothesis 5.71-(D) is needed.
(ii) The Gâteaux differentiability of b, which was demanded in part (D) of Hypothesis 5.1 and then required in Sects. 5.2–5.4, is not needed here.
(iii) In the formulation of the state equation (5.78) we find $Q^{1/2}$ in front of the coefficient B. Even if in this respect the state equation (5.127) seems more general, the situation is not much different since Hypotheses (5.134)–(5.135) or (5.136)–(5.137) are needed.
(iv) While in Sect. 5.2 we consider the invariant measure m related to the nonhomogeneous Cauchy problem (5.3) (see Hypothesis 5.4), here m represents the invariant measure associated with the homogeneous stochastic equation (5.121). Still, as discussed after Theorem 5.41, in Sect. 5.4 the mild solution of the HJB equation can be characterized as a strong solution only if $b = 0$ and the properties of strong solutions are needed (see Sect. 5.5) to identify the solution of the HJB equation and the value function of the optimal control problem.
(v) The results in Sects. 5.2–5.4 refer to the case where the operator D_Q can be nonclosable. Conversely, as observed in Remark 5.12, Hypothesis 5.71, in particular Hypothesis 5.71-(D), implies the closability of the operator D_Q.

∎

5.7.3   Elliptic HJB Equations

In this section we present some results regarding the use of L^2 theory for the elliptic equation (5.2). They are mainly taken from [125] which, to the best of our knowledge, is the only article where an L^2-approach for HJB equations arising from optimal control problems with infinite horizon is developed. A variational solution of the HJB equation, different from the one given in Definition 5.73, is used. The identification of the solution with the value function is not provided.

We introduce the following set of assumptions.

Hypothesis 5.81 (A) A is the generator of a strongly continuous semigroup $\{e^{tA}, \ t \geq 0\}$ on a real separable Hilbert space H and there exist constants $M \geq 1$ and $\omega > 0$ such that

$$\left\| e^{tA} \right\| \leq M e^{-\omega t}, \qquad \forall t \geq 0.$$

Moreover, A is self-adjoint and $A^{-1} \in \mathcal{L}(H)$.
(B) $Q \in \mathcal{L}^+(H)$ and $\mathrm{Tr}[A^{-1}Q] < +\infty$.
(C) There exists a reflexive Banach space V with $D(A) \subset V \subset H$ having the following property: A extends to a continuous operator $A \colon V \to V^*$ (where V^* is the dual of V).
(D) $\mu := \left(\Omega, \mathscr{F}, \{\mathscr{F}_s\}_{s \in [0,+\infty)}, \mathbb{P}, W_Q \right)$ is a generalized reference probability space.
(E) There exists an orthonormal basis $\{e_1, e_2, \ldots\}$ of H made of elements of $D(A)$ such that

$$Ax = \sum_{n=1}^{+\infty} -\alpha_n \langle e_n, x \rangle e_n, \quad x \in D(A)$$

for some eigenvalues $0 < \alpha_1 < \alpha_2 < \alpha_3 \ldots$ and

$$Qx = \sum_{i=n}^{+\infty} q_n \langle e_n, x \rangle e_n, \quad x \in H,$$

for a sequence of nonnegative eigenvalues q_n.

We consider the following SDE

$$\begin{cases} dX(s) = AX(s)ds + dW_Q(s), & s > 0, \\ X(0) = x \in H \end{cases} \tag{5.138}$$

and denote by $X(\cdot\,; x)$ its mild solution at time t (the existence and the uniqueness of the solution are provided, for instance, by Theorem 1.147).

Proposition 5.82 *Suppose that Hypothesis 5.81 is satisfied. Then there exists a unique invariant measure m for (5.138). The measure m is a centered Gaussian measure supported in V with covariance operator* $\Gamma := -\frac{1}{2}A^{-1}Q$.

Proof See Theorem 6.2.1, p. 97 of [177].   □

We denote by $\overline{\mathcal{B}}$ the completion of the Borel σ-field $\mathcal{B}(H)$ with respect to m and by $\mathcal{H}$ the Hilbert space $L^2(H, \overline{\mathcal{B}}, m)$. We also denote by $P_t, t \geq 0$, the transition semigroup (indeed the Ornstein–Uhlenbeck semigroup) associated to (5.138). For any $\phi \in C_b(H)$ it is given by

$$P_t[\phi](x) = \mathbb{E}\phi(X(t, x)).$$

Proposition 5.83 *Suppose that Hypothesis 5.81 is satisfied. Then P_t extends to a strongly continuous semigroup of contractions on $L^2(H, \overline{\mathcal{B}}, m)$. Its generator $\mathcal{A}: D(\mathcal{A}) \subset L^2(H, \overline{\mathcal{B}}, m) \to L^2(H, \overline{\mathcal{B}}, m)$ is self-adjoint.*

Proof The first part of the proposition is a particular case of Proposition 5.9. The last claim is part of Lemma 2.4 of [125].   □

Notation 5.84 Denote by $\mathcal{I}$ the set of all sequences $\ell = (\ell_1, \ell_2, \ldots) \in \mathbb{N}^{\mathbb{N}}$ such that $\ell_i = 0$, except for a finite number of indices.   ∎

Definition 5.85 Let $\{e_n\}$ be the orthonormal basis of H introduced in Hypothesis 5.81-(E). For $j = 0, 1, 2 \ldots$, denote by h_j the standard j-th one-dimensional Hermite polynomials

$$h_j(\xi) := \frac{(-1)^j}{\sqrt{n!}} e^{\frac{\xi^2}{2}} \frac{d^j \left(e^{\frac{-\xi^2}{2}}\right)}{d\xi^j}, \quad \xi \in \mathbb{R}.$$

Given $\ell \in \mathcal{I}$ we define

$$K_\ell(x) := \prod_{i \in \mathbb{N}} h_{\ell_i}\left(\langle x, \Gamma^{-1/2} e_i \rangle_H\right), \quad x \in H,$$

the *Hermite polynomial on H of index* ℓ.

Proposition 5.86 *Suppose that Hypothesis 5.81 is satisfied. The set of the Hermite polynomials K_ℓ is an orthonormal basis in $L^2(H, \overline{\mathcal{B}}, m)$. Moreover, for any $\ell \in \mathcal{I}$, $K_\ell \in D(\mathcal{A})$ and*

$$\mathcal{A}(K_\ell) = \Lambda_\ell K_\ell,$$

where $\Lambda_\ell := -\sum_i \ell_i \alpha_i$ (it is a finite sum), and the α_i are from Hypothesis 5.81-(E).

Proof See Theorem 9.1.5, p. 191 of [179] and Lemma 2.2 of [125].   □

Definition 5.87 We define the following function spaces:

(i) The *Gauss–Sobolev space of order k*, for $k = 1, 2, \ldots$, is the space $\mathcal{H}_k$ defined by

$$\mathcal{H}_k := \left\{ \phi \in \mathcal{H} \; : \; \left(\sum_{\ell \in \mathcal{I}} (1 - \Lambda_\ell)^k \, \langle \phi, K_\ell \rangle_\mathcal{H}^2 \right)^{1/2} = |(I - \mathcal{A})^{k/2} \phi|_\mathcal{H} < +\infty \right\}$$

(observe that the expression is well defined since all α_i and ℓ_i are nonnegative and then $\Lambda_\ell \leq 0$).
(ii) We denote by $\mathcal{H}_k^*$ the dual of $\mathcal{H}_k$.
(iii) Given the weight $\rho_n(x) := \left(1 + |x|^2\right)^n$ for $x \in H$, we denote by $\mathcal{H}_{0,n}$ the space

$$\mathcal{H}_{0,n} := \left\{ f \in \mathcal{H} \; : \; \int_H f^2(x) \rho_n(x) dm(x) \right\}$$

endowed with the usual L^2-weighted Hilbert space structure.
(iv) Given $k = 1, 2, \ldots$ and $n = 0, 1, \ldots$, we denote by $\mathcal{H}_{k,n}$ the space

$$\mathcal{H}_{k,n} := \mathcal{H}_k \cap \mathcal{H}_{0,n},$$

and by $\mathcal{H}_{k,n}^*$ its dual.

Observe that, for any $\phi \in D(\mathcal{A})$, we have

$$\sum_{\ell \in \mathcal{I}} |\Lambda_\ell|^2 \, \langle \phi, K_\ell \rangle_\mathcal{H}^2 = \sum_{\ell \in \mathcal{I}} \langle \mathcal{A}\phi, K_\ell \rangle_\mathcal{H}^2 = |\mathcal{A}\phi|_\mathcal{H}^2 < +\infty$$

so one can easily see that $D(\mathcal{A}) \subset \mathcal{H}_1$. $\mathcal{A}$ can be extended to the whole space $\mathcal{H}_1$ as is shown in the next lemma.

Lemma 5.88 *Suppose that Hypothesis 5.81 is satisfied. Then $\mathcal{A}$ extends to a continuous linear operator from $\mathcal{H}_1$ to $\mathcal{H}_1^*$.*

Proof See Lemma 2.4 of [125]. □

Hypothesis 5.89 (i) Λ is a Polish space.
(ii) $\tilde{R} \colon V \times \Lambda \to H$ is Borel measurable and such that, for some $n \geq 0$ and $R_0 > 0$,
$$|\tilde{R}(x, a)| \leq R_0 (1 + |x|^2)^{n/2} \quad \text{for all } (x, a) \in V \times \Lambda.$$

We denote by $R \colon V \times \Lambda \to Q^{1/2}(H)$ the function $R := Q^{\frac{1}{2}} \tilde{R}$.
(iii) $\lambda \colon H \to \mathbb{R}^+$ is Borel measurable and there exist two real constants $\lambda_0, \lambda_1 > 0$ such that

$$\lambda_0 (1 + |x|^2)^n \leq \lambda(x) \leq \lambda_1 (1 + |x|^2)^n \quad \text{for all } x \in H.$$

(iv) $l: V \times \Lambda \to \mathbb{R}$ is Borel measurable and there exists a $c_0 > 0$ such that

$$|l(x, a)| \leq c_0(1 + |x|^2)^{n/2} \qquad \text{for all } (x, a) \in V \times \Lambda.$$

We are interested in studying the HJB equation

$$(\lambda(x)I - \mathcal{A}) v - F(v) = 0, \tag{5.139}$$

where

$$F(v)(x) := \inf_{a \in \Lambda} \{\langle R(x, a), Dv(x)\rangle + l(x, a)\}.$$

Remark 5.90 The HJB equation (5.139) is associated with the optimal control problem characterized by:

(i) The state equation

$$\begin{cases} dX(s) = (AX(s) + R(X(s), a(s))) \, ds + dW_Q(s), & s > 0, \\ X(0) = x, & x \in H. \end{cases}$$

(ii) The cost functional

$$\int_0^{+\infty} e^{\int_0^t -\lambda(X(s))ds} l(X(t), a(t)) dt.$$

(iii) The set of admissible controls

$$\mathcal{U}_0 = \{a(\cdot) \colon [0, +\infty) \to \Lambda \; : \; a(\cdot) \text{ is } \mathscr{F}_s\text{-progressively measurable}\}.$$

∎

In order to define and study the solution of (5.139) we introduce the nonlinear operator

$$\mathcal{M}(v) := (\lambda(x)I - \mathcal{A}) v - F(v)$$

which, thanks to Lemma 5.88, can be defined for any $v \in \mathcal{H}_{1,n}$. We have the following regularity result for $\mathcal{M}$.

Lemma 5.91 *Under Hypotheses 5.81 and 5.89 the operator $\mathcal{M}$ is locally bounded and Lipschitz continuous from $\mathcal{H}_{1,n}$ to $\mathcal{H}_{1,n}^*$. Moreover, if $\lambda_0 > R_0^2/2$, then there exists a $\delta > 0$ such that, for any $f, g \in \mathcal{H}_{1,n}$,*

$$\langle \mathcal{M}(f) - \mathcal{M}(g), f - g\rangle_{\langle \mathcal{H}_{1,n}^*, \mathcal{H}_{1,n}\rangle} \geq \delta |f - g|_{\mathcal{H}_{1,n}}^2.$$

Proof See Lemmas 4.1 and 4.2 of [125].

Definition 5.92 The function $v \in \mathcal{H}_{1,n}$ is a solution of (5.139) if

$$\langle \mathcal{M}(v), f \rangle_{\langle \mathcal{H}_{1,n}^*, \mathcal{H}_{1,n} \rangle} = 0$$

for any $f \in \mathcal{H}_{1,n}$.

Theorem 5.93 *If Hypotheses 5.81 and 5.89 are satisfied and $\lambda_0 > R_0^2/2$ then Eq. (5.139) has a unique solution v in the sense of the Definition 5.92. Moreover, $v \in \mathcal{H}_{2,n}$.*

Proof See Theorem 4.3 of [125]. ◻

5.8   Bibliographical Notes

In this chapter we focused our attention on HJB equations in L^2 spaces with respect to the invariant measure of an SDE with addictive noise and globally Lipschitz continuous drift independent of time. A number of existence results for various abstract classes of SDEs of this form can be found in the literature, for instance: for linear systems in [164, 354, 355], Sect. 6.2 of [177] and Sect. 11.5 of [180]; for the dissipative case in [164, 174, 426, 427, 533], Sects. 6.3 and 6.4 of [177] and Sect. 11.6 of [180]; for the case of a compact semigroup in [56, 164] and Sect. 11.7 of [180]; for equations with additive noise and weakly continuous drift in [120].[7]

Some approximation lemmas are presented in Sect. 5.2.2. Lemma 5.6 is a standard approximation result for uniformly continuous functions. Observe that in fact we do no need the approximating sequence to be in $\mathcal{E}_A(H)$, a weaker regularity would be enough for our purposes. The technique of mollification over finite-dimensional subspaces used to prove the pointwise convergences of Lemma 5.8 is well known (see e.g. Lemma 1.2, page 164 of [486] or [410]); we also use this kind of approach in the proof of Lemma B.78. The approximation result of Lemma 5.7 (especially its part (iii)) is *ad hoc* for the approximation of HJB equations in L^2 spaces. Even if we are not able to quote directly a specific published result, the proof uses completely standard arguments. Observe that the claim holds for any L^2 space on H w.r.t. any bounded measure, so the fact that we are working with an invariant measure of (5.8) plays no role. Obviously this specific measure is essential in Proposition 5.9. The claim of Proposition 5.9 is proved for the Ornstein–Uhlenbeck case (the proof is exactly the same), together with some characterization of the domain of the generator (the operator $\mathcal{A}$ defined at the beginning of Sect. 5.2.3), in [148, 149, 176], see also [121, 122, 152, 153, 184, 270, 297], Chap. 7 of [294] and Chap. 10 of [179]. We also mention, respectively, [417, 446] and [19] for the finite-dimensional and Banach space cases.

Lemma 5.37 provides a way to approximate elements of $D(\mathcal{A})$ even when its explicit characterization is missing. The space $\mathcal{F}C_0^{2,A^*}(H)$ is used because we can

---

[7]For uniqueness results the reader is referred to the review [443] and the references there.

explicitly compute the operator $\mathcal{A}$ in it (as well as other operators that will be defined later) and it is dense in $L^2(H, \overline{\mathcal{B}}, m)$. Other possible choices can be found in the literature, for example in Chap. 9 of [179] or in Chap. 8 of [153], the authors use, for the Gaussian case, a space of exponential functions. Using $\mathcal{F}C_0^{2,A^*}(H)$ is consistent with other similar approximations employed in the book, in particular in Chap. 4, see e.g. Hypotheses 4.133 and 4.141.

In Definition 5.11 we introduce a notion of Sobolev space for the case when the derivative operator $Q^{1/2}D$ is non-closable. Sobolev spaces in infinite dimension with respect to Gaussian measures are studied, for example, in [153, 484], Chap. 10 and [179], Chaps. 9 and 10. Sobolev spaces with respect to Gibbs measures are studied in [150, 151, 171, 172], Chap. 11 of [153] and Chap. 12 of [179]. In all of these cases the derivative operator is closable. Regarding the non-closable case needed here (see, in particular, Sect. 5.2.4) there is much less in the literature, the readers may consult [298, 299]. The closability of D_Q is related to the closability of the associated Dirichlet form, see [270, 509] for more on this and [422] for a general introduction to Dirichlet forms.

For some comments about the results of Lemma 5.14 and a discussion of the related literature, the readers may check the proof of Proposition 4.61 and Remark 4.62. The proofs of Lemmas 5.15 and 5.17 are standard but we could not find precise references. Results similar to Lemma 5.18 are often used in the literature as a step to prove Bismut–Elworthy–Li formulae, see for instance [486, 582] or [180], Sect. 9.4 (original results for the finite-dimensional case are, for example, in [60, 216]). In its proof, which expands the ideas contained in Step 1 of the proof of Proposition 2.4 of [298], the claim of Lemma 6.11 of [582], originally proved there for $b \in UC_b^2(H, H)$ and $\varphi \in UC_b^2(H)$, is extended. Results similar to Proposition 5.20 are given in [179] (they follow as corollaries of the proofs of Propositions 10.5.2 and 11.2.17) or in [184] (see p. 241); we follow here the arguments of [298]. More details and references about the claims of Remarks 5.21 are given in Sect. 4.3.1.3 and in the bibliographical notes of Chap. 4.

Sections 5.3 and 5.4 contain the main results of the chapter. We generalize the theorems contained in [298] to take into account Hamiltonians dependent on $x \in H$ and $t \in [0, T]$. In [298] only Hamiltonians of the form $F_0(D_Q u)$ were studied. Apart from this the setting is the same, beginning with Definition 5.28 of a mild solution. The main arguments used to prove the key result of Sect. 5.3, i.e. Theorem 5.35, are the same as those used in the proof of Theorem 3.7 of [298]. The proofs of Theorems 5.40 and 5.41 follow the lines of the proof of Proposition 4.3 of [298]. The literature on solutions of HJB equations in L^2 spaces is not very extensive and this chapter contains most of the published results (in Sects. 5.3, 5.4 and then in Sect. 5.7), so we cannot present a long genealogy of the results. However, many ideas and techniques have been used before to study HJB equations in spaces of regular functions discussed in Chap. 4. Thus we refer the reader to Sects. 4.4 and 4.5 and to the bibliographical notes of Chap. 4 for more.

The structure of Sect. 5.5 follows the structure of Sect. 4.8, starting from the proof of the fundamental identity (Lemma 5.50) and its use to obtain a verification theorem and optimal feedbacks (Theorem 5.53, Corollary 5.54, Theorem 5.55); the counterparts in Sect. 4.8 are Lemma 4.196, Theorem 4.197, Corollary 4.198 and Theorem 4.201. We refer the reader to the bibliographical notes of Chap. 4 for references on the subject. Compared to [298], the generalization of the Hamiltonian studied in Sects. 5.3 and 5.4 allows us to consider in Sect. 5.5 a more general optimal control problem, where the function R appearing in (5.78) also depends on s and $X(s)$ in addition to $a(s)$. Lemmas 5.46 and 5.49 are similar to results in [298], other proofs of the section are new. Proposition 5.44 is a standard existence and uniqueness result for solutions of stochastic evolution equations in Hilbert spaces, see the references mentioned in Chap. 1. Lemma 5.46 is a corollary of Girsanov's Theorem, the reader is referred, for example, to [44, 180, 382, 383, 448, 483, 580] for more on its Hilbert space formulations and various consequences. Because of the L^2 context, the result of Lemma 5.50 holds only m-almost everywhere. This is the main reason for introducing additional hypotheses (namely the boundedness of Λ used in Lemma 5.51 and the continuity of $J^{\mu}(t, x; \cdot)$) that we need in the proofs of Theorems 5.53 and 5.55. The formulations of the results of Sect. 5.5.5 are new even if the use of the non-degeneracy hypothesis, together with some continuity assumptions, was already suggested in Remark 3.10 of [298].

In Sect. 5.6 we show how some of the examples from Sect. 2.6 can be treated using the approach introduced in this chapter. We focus in particular on the existence of a (possibly non-degenerate) invariant measure, which is the key assumption needed here. For material on invariant measures for stochastic delay differential equations, besides Chap. 10 of [177, 299] which were already mentioned in Sect. 5.6.1, we refer the reader to [56, 338, 562]; for first-order stochastic equations, especially those connected to financial problems, results can be found in [299, 303, 430, 522, 553, 565] and Chap. 20 of [487].

The material of Sect. 5.7 essentially comes from [3, 4, 125]. More precisely, the results described in Sect. 5.7.1 (in particular Theorems 5.74 and 5.75) are proved in [3] (the two mentioned theorems correspond to Theorems 5.2 and 5.3 of [3]) while the content of Sect. 5.7.2 comes from [4]. Theorems 5.78 and 5.79 are Theorems 5.4 and 5.2, respectively, in [4]. Section 5.7.3 is based on the results obtained in [125] and the main theorem (Theorem 5.93) is Theorem 4.3 of [125]. In [123] the author uses a similar technique to deal with the Kolmogorov equation while in [125], Sect. 3, the authors study the related unbounded case. Even if we use in various parts of the book the variational solution of the state equation, this is the only section where we use the notion of a variational solution of the HJB equation (see Definitions 5.73 and 5.92). Indeed, it naturally needs some reference measure on the Hilbert state space and it is then linked to the study of HJB equations in the L^2 space. As far as we know, the above mentioned papers are the only ones that use this kind of notion of solution in the context of optimal control but, in the same spirit, a characterization of the value function for optimal stopping time problems, in terms of variational inequalities, is given in [38, 116], see also [125, 581, 583].

We also mention the recent paper [574] where the L^2 theory for HJB equations in Hilbert spaces, employing the ideas discussed in Sects. 5.1–5.5, is used to study an infinite horizon optimal control problem with boundary noise and boundary control.

The key prerequisite for the approach developed in this chapter is clearly the theory of invariant measures for infinite-dimensional PDEs. The results we use in this chapter concern invariant measures for SDEs with addictive noise, but the existing generalizations can be employed to develop applications to optimal control theory for other classes of stochastic partial differential equations in the spirit of the theory described here. In particular, the existence results for invariant measures for SPDEs with multiplicative noise (see, e.g., [218], Chap. 6 and Sect. 11.2 of [177] and Sect. 11.4 of [180]) and extensions to stochastic Burgers, Euler and Navier–Stokes equations (e.g. [7, 59, 81, 82, 159, 161, 253, 256, 336, 337, 389, 390, 515, 570, 571] and Chaps. 14 and 15 of [177]), stochastic reaction-diffusion equations (see for instance [109, 110]), stochastic porous media equations (as in [32, 169]) and stochastic nonlinear damped wave equations [31] can be a starting point in the study of optimal control problems driven by such state equations.

Results about invariant measures for transition semigroups for stochastic evolution equations in Banach spaces (such as those contained in [83, 292]) can be exploited to extend the techniques presented in this chapter to the Banach space case. Similarly the studies of SPDEs in domains/half-spaces and related invariant measures (see, e.g., [19, 165, 166, 494, 495, 497, 498, 546]) can be used as a first step to try to apply the methods to problems with state constraints. Another possible extension of the results presented here is the case of locally Lipschitz continuous Hamiltonians, following the results and the techniques introduced in [105, 307, 438].

Chapter 6
HJB Equations Through Backward Stochastic Differential Equations

Marco Fuhrman and Gianmario Tessitore

This last chapter of the book completes the picture of the main methods used to study second-order HJB equations in Hilbert spaces and related optimal control problems by presenting a survey of results that can be achieved with the techniques of Backward SDEs in infinite dimension.

The chapter has been written independently and autonomously. In order to maintain some coherence with the notation used in the Backward SDE literature, the notation used in this chapter is not always identical to that in the rest of the book. This is explained in Sects. 6.1.1 and 6.1.2.

The chapter has the following structure.

- Section 6.1 explains the basic notation and collects some useful results about generalized gradients and SDEs which are needed in the rest of the chapter.
- Section 6.2 provides results about regular dependence of solutions of SDEs on the data.
- Section 6.3 presents results about well-posedness and regular dependence on the data for Backward SDEs (BSDEs from now on) and Forward–Backward systems (FBSDEs) in Hilbert spaces.
- In Sect. 6.4 existence and uniqueness of mild solutions of HJB equations through FBSDEs are discussed.
- Section 6.5 gives applications of the results of Sect. 6.4 to optimal control problems. An example of a control problem with delay is studied in Sect. 6.6.

M. Fuhrman
Dipartimento di Matematica, Università degli studi di Milano, Milano, Italy

G. Tessitore (✉)
Dipartimento di Matematica e Applicazioni,Università degli studi di Milano-Bicocca, Milano, Italy
e-mail: gianmario.tessitore@unimib.it

© Springer International Publishing AG 2017
G. Fabbri et al., *Stochastic Optimal Control in Infinite Dimension*,
Probability Theory and Stochastic Modelling 82,
DOI 10.1007/978-3-319-53067-3_6

685

- Sections 6.7–6.10 develop the same program for elliptic HJB equations and infinite horizon control problems. An application to an infinite horizon optimal control problem driven by a heat equation with additive noise is discussed in Sect. 6.11.
- Results for elliptic HJB equations with non-constant second-order coefficients and some applications are collected in Sect. 6.12.

6.1 Complements on Forward Equations with Multiplicative Noise

6.1.1 Notation on Vector Spaces and Stochastic Processes

The notation for Banach spaces and linear operators between them is the same as that used in the other parts of the book, see, for instance, Appendix A.1.

In this chapter the letters Ξ, H, K will always denote Hilbert spaces. The scalar product is denoted, as usual, by $\langle \cdot, \cdot \rangle$, with a subscript to specify the space, if necessary. All Hilbert spaces are assumed to be real and separable.

We only consider stochastic differential equations driven by *cylindrical* Wiener processes W. By a cylindrical Wiener process with values in a Hilbert space Ξ, defined on a complete probability space $(\Omega, \mathscr{F}, \mathbb{P})$, we mean a family $W(t), t \geq 0$, of linear mappings $\Xi \to L^2(\Omega)$ such that

 (i) for every $u \in \Xi$, $\{W(t)u\}_{t \geq 0}$ is a real Wiener process (admitting a continuous modification);
 (ii) for every $u, v \in \Xi$ and $t, s \geq 0$, $\mathbb{E}\,(W(t)u \cdot W(s)v) = \min(t, s)\,\langle u, v \rangle_{\Xi}$.

Recall that, in this case, when the noise space Ξ has finite dimension d the Wiener process can be naturally identified with a d-dimensional standard Wiener process $(\beta_1, \ldots, \beta_d)$, where $\beta_i(t) = W(t)e_i$ and $(e_1, \ldots, e_d)$ denotes an orthonormal basis of Ξ. In other parts of the book Q-Wiener processes and in particular cylindrical Wiener processes are introduced in a slightly different (but equivalent) way, see Sect. 1.2.4 and in particular Remark 1.89.

Unless stated otherwise, $\{\mathscr{F}_t\}_{t \geq 0}$ will denote the natural filtration of W, augmented by the family $\mathcal{N}$ of $\mathbb{P}$-null sets of $\mathscr{F}$:

$$\mathscr{F}_t = \sigma(W(s)u \,:\, s \in [0, t],\ u \in \Xi) \vee \mathcal{N}.$$

The filtration $\mathscr{F}_t$ satisfies the usual conditions. All the concepts of measurability for stochastic processes (e.g. adaptedness, predictability etc.) refer to this filtration. By $\mathcal{P}$ we denote the predictable σ-field on $\Omega \times [0, \infty)$ or (by abuse of notation) its trace on $\Omega \times [0, T]$.

For $[a, b] \subset [0, T]$ we use the notation

$$\mathscr{F}_{[a,b]} = \sigma(W(s)u - W(a)u \,:\, s \in [a, b],\ u \in \Xi) \vee \mathcal{N}.$$

To denote the value of a process X at time s, sometimes instead of $X(s)$ the shortened notation X_s will be used, especially in proofs. The short-hand "a.a. (a.e.)" means "almost all (almost everywhere) with respect to the Lebesgue measure".

Next we define several classes of stochastic processes with values in a Hilbert space K.

- $L^2_{\mathcal{P}}(\Omega \times [0, T]; K)$ denotes the space of equivalence classes of processes $Y \in L^2(\Omega \times [0, T]; K)$, admitting a predictable version. $L^2_{\mathcal{P}}(\Omega \times [0, T]; K)$ is endowed with the norm

$$|Y|^2 = \mathbb{E} \int_0^T |Y(s)|^2 ds.$$

- $L^p_{\mathcal{P}}(\Omega; L^2([0, T]; K))$ denotes the space of equivalence classes of processes Y such that the norm

$$|Y|^p = \mathbb{E} \left(\int_0^T |Y(s)|^2 ds \right)^{p/2}$$

is finite, and Y admits a predictable version.

- $C_{\mathcal{P}}([0, T], L^2(\Omega; K))$ denotes the space of K-valued processes Y such that $Y : [0, T] \to L^2(\Omega; K)$ is continuous and Y has a predictable modification, endowed with the norm

$$|Y|^2 = \sup_{s \in [0,T]} \mathbb{E} |Y(s)|^2.$$

Elements of $C_{\mathcal{P}}([0, T], L^2(\Omega; K))$ are identified up to modification.

- $L^p_{\mathcal{P}}(\Omega; C([0, T], K))$ denotes the space of predictable processes Y with continuous paths in K, such that the norm

$$|Y|^p = \mathbb{E} \sup_{s \in [0,T]} |Y(s)|^p$$

is finite. Elements of $L^p_{\mathcal{P}}(\Omega; C([0, T], K))$ are identified up to indistinguishability.

Recall that, for a given element Ψ of $L^2_{\mathcal{P}}(\Omega \times [0, T]; \mathcal{L}_2(\Xi, K))$, the Itô stochastic integral $\int_0^t \Psi(s) \, dW(s)$, $t \in [0, T]$, is a K-valued martingale belonging to $L^2_{\mathcal{P}}(\Omega; C([0, T], K))$.

If Ψ belongs to $L^2_{\mathcal{P}}(\Omega \times [0, T]; \Xi)$, the real-valued Itô stochastic integral $\int_0^t \langle \Psi(s), dW(s) \rangle_\Xi$ is by definition the integral $\int_0^t \Psi(s)^* \, dW(s)$, where $\Psi(\omega, s)^* \in \Xi^*$ denotes the element corresponding to $\Psi(\omega, s) \in \Xi$ by the Riesz isometry (i.e., $\Psi(\omega, s)^* h = \langle \Psi(\omega, s), h \rangle_\Xi$, $h \in \Xi$).

The previous definitions have obvious extensions to processes defined on subintervals of $[0, T]$.

6.1.2   The Class $\mathcal{G}$

In this section we introduce a class of maps acting among Banach spaces, possessing suitable continuity and differentiability properties. Many assumptions in the following sections will be stated in terms of membership in this class.

The class we are going to introduce has several useful properties. First, membership in this class is often easy to verify: see Lemmas 6.4 and 6.6 below. Next, it is a well-behaved class as far as chain rules are concerned. Finally, it is sufficiently large to include operators commonly arising in applications to stochastic partial differential equations, such as Nemytskii (evaluation) operators; it is well known that the Nemytskii operators are not Fréchet differentiable except in trivial cases.

In this subsection, X, Y, Z, V denote Banach spaces. We recall that for a mapping $F : X \rightarrow V$ the directional derivative at a point $x \in X$ in the direction $h \in X$ is defined as

$$\nabla F(x; h) = \lim_{s \rightarrow 0} \frac{F(x + sh) - F(x)}{s},$$

whenever the limit exists in the topology of V. F is called Gâteaux differentiable at the point x if it has directional derivative in every direction at x and there exists an element of $\mathcal{L}(X, V)$, denoted $\nabla F(x)$ and called Gâteaux derivative, such that $\nabla F(x; h) = \nabla F(x)h$ for every $h \in X$.

Remark 6.1 When $V = \mathbb{R}$ the Gâteaux derivative $\nabla F(x)$ belongs to $\mathcal{L}(X, \mathbb{R}) = X^*$, the dual space of X. If, in addition, X is a Hilbert space then it can be identified canonically with X^* and the Gâteaux derivative of F at x can be thought of as an element of X that we denote by $DF(x)$. Thus, $DF(x)$ is the unique element of X such that $\nabla F(x; h) = \nabla F(x)h = \langle DF(x), h \rangle_X$ for every $h \in X$. Similarly, in the same circumstances, the second Gâteaux derivative will be identified with a (symmetric) element of $\mathcal{L}(X)$, denoted by $D^2 F(x)$. This convention is a little different from the rest of the book, where the notation $DF(x)$ is employed for the Fréchet derivative of F at x.   ∎

Definition 6.2 We say that a mapping $F : X \rightarrow V$ belongs to the class $\mathcal{G}^1(X, V)$ if it is continuous, Gâteaux differentiable on X, and $\nabla F : X \rightarrow \mathcal{L}(X, V)$ is strongly continuous.

The last requirement of the definition means that for every $h \in X$ the map $\nabla F(\cdot)h : X \rightarrow V$ is continuous. Note that $\nabla F : X \rightarrow \mathcal{L}(X, V)$ is not continuous in general if $\mathcal{L}(X, V)$ is endowed with the norm operator topology; clearly, if this happens then F is Fréchet differentiable on X. Some features of the class $\mathcal{G}^1(X, V)$ are collected below.

Lemma 6.3 *Suppose $F \in \mathcal{G}^1(X, V)$. Then*

(i) $(x, h) \rightarrow \nabla F(x)h$ is continuous from $X \times X$ to V;
(ii) if $G \in \mathcal{G}^1(V, Z)$ then $G(F) \in \mathcal{G}^1(X, Z)$ and $\nabla(G(F))(x) = \nabla G(F(x))\nabla F(x)$.

Proof (*i*) Let $x_n \to x$ and $h_n \to h$ in X. By the Banach–Steinhaus theorem we have $|\nabla F(x_n)|_{\mathcal{L}(X,V)} < L$ for every n and for a suitable constant L. Therefore

$$|\nabla F(x_n)h_n - \nabla F(x)h| \leq L|h - h_n| + |\nabla F(x_n)h - \nabla F(x)h|$$

and the claim follows immediately.

(*ii*) First we notice that for all $x, y \in H$:

$$F(x + y) = F(x) + \int_0^1 \nabla F(x + ry)y \, dr. \tag{6.1}$$

Therefore, given $x, h \in X, s \in (0, 1]$, repeated application of (6.1) yields

$$G(F(x + sh)) - G(F(x))$$
$$= \int_0^1 \left[\nabla G \left(F(x) + \sigma \int_0^1 \nabla F(x + srh)sh \, dr \right) \int_0^1 \nabla F(x + srh)sh \, dr \right] d\sigma.$$

Let $g(s) = \int_0^1 \nabla F(x + srh)h \, dr, K = \{\nabla F(x + rh)h : r \in [0, 1]\}$ and $\hat{K}$ be the closed convex hull of K. Clearly K, and hence $\hat{K}$, are compact subsets of V and $g(s) \in \hat{K}$ for all $s \in [0, 1]$. Moreover,

$$\{F(x) + \sigma s \int_0^1 \nabla F(x + srh)dr : \sigma \in [0, 1], s \in [0, 1]\}$$
$$\subset \hat{K}_1 := \{F(x) + \sigma k : \sigma \in [0, 1], k \in \hat{K}\},$$

which is itself compact. By the dominated convergence theorem $\lim_{s \to 0^+} g(s) = \nabla F(x)h$ and since, by the continuity of ∇G, $\sup_{z \in \hat{K}_1, k \in \hat{K}} |\nabla G(z)k| < +\infty$, applying again the dominated convergence theorem we can conclude that

$$\lim_{s \to 0^+} \frac{G(F(x + sh)) - G(F(x))}{s}$$
$$= \int_0^1 \lim_{s \to 0^+} [\nabla G(F(x) + \sigma sg(s))g(s)] d\sigma = \nabla G(F(x))\nabla F(x)h.$$

The proof that the map $x \to \nabla G(F(x))\nabla F(x)$ is strongly continuous is identical to the proof of point (*i*). $\qquad\square$

In addition to the ordinary chain rule in point (*ii*) above, a chain rule for the Malliavin derivative operator holds: see Sect. 6.2.2. Membership of a map in $\mathcal{G}^1(X, V)$ may be conveniently checked as shown in the following lemma.

Lemma 6.4 *A map* $F : X \to V$ *belongs to* $\mathcal{G}^1(X, V)$ *provided the following conditions hold:*

(i) *the directional derivatives $\nabla F(x; h)$ exist at every point $x \in X$ and in every direction $h \in X$;*
(ii) *for every h, the mapping $\nabla F(\cdot; h) : X \to V$ is continuous;*
(iii) *for every x, the mapping $h \to \nabla F(x; h)$ is continuous from X to V.*

Proof We have to show that F is continuous and the map $h \to \nabla F(x; h)$, where $\nabla F(x; h)$ denotes the directional derivative of F at a fixed point $x \in X$ in the direction $h \in X$, is linear. To start, we notice that a version of formula (6.1) still holds under the present assumptions, namely: $F(x + y) = F(x) + \int_0^1 \nabla F(x + ry; y) \, dr$ for all $x, y \in X$.

First we show linearity. By definition of the directional derivative it is obvious that for all $\rho \geq 0$ and all $x, h \in X$: $\nabla F(x, \rho h) = \rho \nabla F(x, h)$. Since, for fixed $h, k \in X$,

$$\frac{F(x + s(h + k)) - F(x)}{s} = \frac{F(x + s(h + k)) - F(x + sh)}{s} + \frac{F(x + sh) - F(x)}{s},$$

we have, by (6.1),

$$\nabla F(x; h + k) = \lim_{s \to 0^+} \int_0^1 \nabla F(x + sh + rsk; k) \, dr + \nabla F(x; h),$$

provided the limit exists. The continuity of $\nabla F(\cdot; k)$ implies that we can pass to the limit under the integral, by a dominated convergence argument, obtaining $\nabla F(x; h + k) = \nabla F(x; k) + \nabla F(x; h)$. It follows, in particular, that $\nabla F(x; -h) = -\nabla F(x; h)$ and so $\nabla F(x, \rho h) = \rho \nabla F(x, h)$ for all $\rho \in \mathbb{R}$ and all $x, h \in X$. Linearity is proved. From now on, we denote the directional derivative $\nabla F(x; k)$ by $\nabla F(x)k$.

Now we come to the continuity of F. Let $y_n \to 0$ in X and fix $x \in X$. By (6.1) we have: $F(x + y_n) - F(x) = \int_0^1 \nabla F(x + ry_n)y_n \, dr$. We see that the set $\{x + ry_n : r \in [0, 1], n \in \mathbb{N}\}$ is a compact subset of X. Therefore (using again the Banach–Steinhaus theorem) $\sup_{r \in [0,1], n \in \mathbb{N}} |\nabla F(x + ry_n)|_{\mathcal{L}(X,V)} < +\infty$ and we can apply the dominated convergence theorem to conclude that $F(x + y_n) - F(x) \to 0$. $\square$

We need to generalize these definitions to functions depending on several variables. For a function $F : X \times Y \to V$ the partial directional and Gâteaux derivatives with respect to the first argument, at point (x, y) and in the direction $h \in X$, are denoted $\nabla_x F(x, y; h)$ and $\nabla_x F(x, y)$, respectively, their definitions being obvious.

Definition 6.5 We say that a mapping $F : X \times Y \to V$ belongs to the class $\mathcal{G}^{1,0}(X \times Y, V)$ if it is continuous, Gâteaux differentiable with respect to x on $X \times Y$, and $\nabla_x F : X \times Y \to \mathcal{L}(X, V)$ is strongly continuous.

As in Lemma 6.3 one can prove that for $F \in \mathcal{G}^{1,0}(X \times Y, V)$ the mapping $(x, y, h) \to \nabla_x F(x, y)h$ is continuous from $X \times Y \times X$ to V, and analogues of the previously stated chain rules hold. The following result is proved in the same way as Lemma 6.4 (but note that continuity is explicitly required).

Lemma 6.6 *A continuous map $F : X \times Y \to V$ belongs to $\mathcal{G}^{1,0}(X \times Y, V)$ provided the following conditions hold:*

   (i) *the directional derivatives* $\nabla_x F(x, y; h)$ *exist at every point* $(x, y) \in X \times Y$
       and in every direction $h \in X$;
  (ii) *for every h, the mapping* $\nabla F(\cdot, \cdot; h) : X \times Y \to V$ *is continuous;*
 (iii) *for every* (x, y), *the mapping* $h \mapsto \nabla_x F(x, y; h)$ *is continuous from* X *to* V.

    The previous definitions and properties have obvious generalizations to slightly different situations, provided obvious changes are made. For instance, the space Y might be replaced by an interval $[0, T]$ or $[0, \infty)$. Another situation occurs when F depends on additional arguments: for instance, we say that $F : X \times Y \times Z \to V$ belongs to $\mathcal{G}^{1,1,0}(X \times Y \times Z, V)$ if it is continuous, Gâteaux differentiable with respect to x and y on $X \times Y \times Z$, and $\nabla_x F : X \times Y \times Z \to \mathcal{L}(X, V)$ and $\nabla_y F : X \times Y \times Z \to \mathcal{L}(Y, V)$ are strongly continuous.

    We will make systematic use of a parameter-dependent contraction principle, stated below as Proposition 6.7. It will be used to study regular dependence of solutions to stochastic equations on their initial data, which is crucial to the investigation of regularity properties of the nonlinear Kolmogorov equation which is the object of this Chapter. The first part of the following proposition is proved in [582], Theorems 10.1, 10.2 (see also [106] Appendix C). The second part is an immediate corollary.

Proposition 6.7 (Parameter-dependent contraction principle) *Let* $F : X \times Y \to X$ *be a continuous mapping satisfying*

$$|F(x_1, y) - F(x_2, y)| \le \alpha |x_1 - x_2|,$$

for some $\alpha \in [0, 1)$ *and every* $x_1, x_2 \in X$, $y \in Y$. *Let* $\phi(y)$ *denote the unique fixed point of the mapping* $F(\cdot, y) : X \to X$. *Then* $\phi : Y \to X$ *is continuous. If, in addition,* $F \in \mathcal{G}^{1,1}(X \times Y, X)$, *then* $\phi \in \mathcal{G}^1(Y, X)$ *and*

$$\nabla \phi(y) = \nabla_x F(\phi(y), y) \nabla \phi(y) + \nabla_y F(\phi(y), y), \qquad y \in Y.$$

More generally, let $F : X \times Y \times Z \to X$ *be a continuous mapping satisfying*

$$|F(x_1, y, z) - F(x_2, y, z)| \le \alpha |x_1 - x_2|,$$

for some $\alpha \in [0, 1)$ *and every* $x_1, x_2 \in X$, $y \in Y$, $z \in Z$. *Let* $\phi(y, z)$ *denote the unique fixed point of the mapping* $F(\cdot, y, z) : X \to X$. *Then* $\phi : Y \times Z \to X$ *is continuous. If, in addition,* $F \in \mathcal{G}^{1,1,0}(X \times Y \times Z, X)$, *then* $\phi \in \mathcal{G}^{1,0}(Y \times Z, X)$ *and*

$$\nabla_y \phi(y, z) = \nabla_x F(\phi(y, z), y, z) \nabla_y \phi(y, z) + \nabla_y F(\phi(y, z), y, z), \qquad y \in Y, \ z \in Z.$$
$$(6.2)$$

6.1.3 The Forward Equation: Existence, Uniqueness and Regularity

Let $W(t)$, $t \in [0, T]$, be a cylindrical Wiener process with values in a Hilbert space Ξ, defined on a probability space $(\Omega, \mathscr{F}, \mathbb{P})$. We fix an interval $[t, T] \subset [0, T]$ and we consider the Itô stochastic differential equation for an unknown process $X(s)$, $s \in [t, T]$, with values in a Hilbert space H:

$$\begin{cases} dX(s) = AX(s)\, ds + b(s, X(s))\, ds + \sigma(s, X(s))\, dW(s), & s \in [t, T], \\ X(t) = x \in H. \end{cases} \tag{6.3}$$

The precise notion of solution will be given next. For the moment we emphasize the fact that W will only denote a cylindrical Wiener process. Other cases can be reduced to this one by standard reformulations; for instance, the case of a finite-dimensional driving Brownian motion corresponds to the case where Ξ has finite dimension.

We assume the following:

Hypothesis 6.8 (i) The operator A is the generator of a strongly continuous semi-group e^{tA}, $t \geq 0$, in the Hilbert space H.

(ii) The mapping $b : [0, T] \times H \to H$ is measurable and satisfies, for some constant $L > 0$,

$$|b(t, x) - b(t, y)| \leq L\, |x - y|, \qquad t \in [0, T], \ x, y \in H.$$

(iii) σ is a mapping $[0, T] \times H \to \mathcal{L}(\Xi, H)$ such that for every $v \in \Xi$ the map $\sigma v : [0, T] \times H \to H$ is measurable, $e^{sA}\sigma(t, x) \in \mathcal{L}_2(\Xi, H)$ for every $s > 0$, $t \in [0, T]$ and $x \in H$. Moreover, for every $s > 0$, $t \in [0, T]$, $x, y \in H$,

$$\begin{aligned} |e^{sA}\sigma(t, x)|_{\mathcal{L}_2(\Xi, H)} &\leq L\, s^{-\gamma}(1 + |x|), \\ |e^{sA}\sigma(t, x) - e^{sA}\sigma(t, y)|_{\mathcal{L}_2(\Xi, H)} &\leq L\, s^{-\gamma}|x - y| \end{aligned} \tag{6.4}$$

and

$$|\sigma(t, x)|_{\mathcal{L}(\Xi, H)} \leq L\, (1 + |x|), \tag{6.5}$$

for some constants $L > 0$ and $\gamma \in [0, 1/2)$.

(iv) For every $s > 0$ and $t \in [0, T]$,

$$b(t, \cdot) \in \mathcal{G}^1(H, H), \qquad e^{sA}\sigma(t, \cdot) \in \mathcal{G}^1(H, \mathcal{L}_2(\Xi, H)).$$

By a solution to Eq. (6.3) we mean an $\mathscr{F}_t$-adapted process $X(s)$, $s \in [t, T]$, with continuous paths in H, such that, $\mathbb{P}$-a.s.,

$$X(s) = e^{(s-t)A}x + \int_t^s e^{(s-r)A}b(r, X(r))\, dr + \int_t^s e^{(s-r)A}\sigma(r, X(r))\, dW(r), \quad s \in [t, T]. \tag{6.6}$$

To shorten the notation slightly, we will often write X_s and W_s instead of $X(s)$, $W(s)$. We note that X is clearly a predictable process in H and that the measurability assumption in Hypothesis 6.8-(iii) is needed to ensure that the integrand process $e^{(s-r)A}G(r, X(r))$, $r \in [s, t]$, is a predictable process with values in $\mathcal{L}_2(\Xi, H)$ (endowed with the Borel σ-field). To stress dependence on the initial data we denote the solution by $X(s; t, x)$. Note that $X(s; t, x)$ is $\mathscr{F}_{[t,T]}$-measurable, hence independent of $\mathscr{F}_t$.

The inequality (6.5) and Hypothesis 6.8-(iv) are needed to have additional regularity for the process X, but they are not used in Proposition 6.9 below. It is a consequence of our assumptions that for every $s > 0$, $t \in [0, T]$, $x, h \in H$,

$$|\nabla_x b(t, x)h| \leq L\,|h|, \qquad |\nabla_x(e^{sA}\sigma(t, x))h|_{\mathcal{L}_2(\Xi, H)} \leq L\,s^{-\gamma}|h|. \qquad (6.7)$$

Proposition 6.9 *Under the assumptions of Hypothesis 6.8-(i)-(ii)-(iii), for every $p \in [2, \infty)$ there exists a unique process $X \in L^p_{\mathcal{P}}(\Omega; C([t, T], H))$ which is a solution to (6.6). Moreover,*

$$\mathbb{E} \sup_{s \in [t,T]} |X(s; t, x)|^p \leq C(1 + |x|)^p, \qquad (6.8)$$

for some constant C depending only on p, γ, T, L and $M := \sup_{s \in [0,T]} |e^{sA}|$.

Proof The result is well known, see e.g. [177], Theorem 5.3.1. We include the proof for completeness and because it will be useful in the following. We often write X_s for $X(s)$ and similar conventions are used for other stochastic processes. The argument is as follows: we define a mapping Φ from $L^p_{\mathcal{P}}(\Omega; C([t, T], H))$ to itself by the formula

$$\Phi(X)_s = e^{(s-t)A}x + \int_t^s e^{(s-r)A}b(r, X_r)\,dr + \int_t^s e^{(s-r)A}\sigma(r, X_r)\,dW_r, \qquad s \in [t, T],$$

and show that it is a contraction, under an equivalent norm. The unique fixed point is the required solution.

For simplicity, we set $t = 0$ and we treat only the case $b = 0$, the general case being handled in a similar way. Let us introduce the norm $\|X\|^p = \mathbb{E} \sup_{s \in [0,T]} e^{-\beta s p}|X_s|^p$, where $\beta > 0$ will be chosen later. In the space $L^p(\Omega; C([0, T], H))$ this norm is equivalent to the original one. We will use the so-called factorization method, see [177], Theorem 5.2.5. Let us take $p > 2$ and $\alpha \in (0, 1)$ such that

$$\frac{1}{p} < \alpha < \frac{1}{2} - \gamma, \qquad \text{and let} \qquad c_\alpha^{-1} = \int_r^s (s - u)^{\alpha-1}(u - r)^{-\alpha}du.$$

Then, by the stochastic Fubini theorem,

$$\Phi(X)_s = e^{sA}x + c_\alpha \int_0^s \int_r^s (s-u)^{\alpha-1}(u-r)^{-\alpha}e^{(s-u)A}e^{(u-r)A} \, du \, \sigma(r, X_r) \, dW_r$$
$$= e^{sA}x + c_\alpha \int_0^s (s-u)^{\alpha-1}e^{(s-u)A}Y_u \, du,$$

where

$$Y_u = \int_0^u (u-r)^{-\alpha}e^{(u-r)A}\sigma(r, X_r) \, dW_r.$$

By the Hölder inequality, setting $M = \sup_{s\in[0,T]} |e^{sA}|$, $p' = p/(p-1)$,

$$e^{-\beta s}\left| \int_0^s (s-u)^{\alpha-1}e^{(s-u)A}Y_u \, du \right| \leq \left(\int_0^s e^{-p'\beta(s-u)}(s-u)^{(\alpha-1)p'} \, ds \right)^{\frac{1}{p'}} \cdot$$
$$\cdot \left(\int_0^s e^{-p\beta u}|e^{(s-u)A}Y_u|^p \, du \right)^{\frac{1}{p}}$$
$$\leq M \left(\int_0^T e^{-p'\beta u}u^{(\alpha-1)p'} \, du \right)^{\frac{1}{p'}} \left(\int_0^T e^{-p\beta u}|Y_u|^p \, du \right)^{\frac{1}{p}}, \quad (6.9)$$

and we obtain

$$\|\Phi(X)\| \leq M|x| + Mc_\alpha \left(\int_0^T e^{-p'\beta u}u^{(\alpha-1)p'} \, du \right)^{\frac{1}{p'}} \left(\mathbb{E} \int_0^T e^{-p\beta u}|Y_u|^p \, du \right)^{\frac{1}{p}}.$$

By the Burkholder–Davis–Gundy inequalities, taking into account the assumption (6.4), we have, for some constant c_p depending only on p,

$$\mathbb{E}|Y_u|^p \leq c_p \mathbb{E} \left(\int_0^u (u-r)^{-2\alpha}|e^{(u-r)A}\sigma(r, X_r)|^2_{L_2(\Xi, H)} \, dr \right)^{\frac{p}{2}}$$
$$\leq L^p c_p \mathbb{E} \left(\int_0^u (u-r)^{-2\alpha-2\gamma}(1+|X_r|)^2 \, dr \right)^{\frac{p}{2}}$$
$$\leq L^p c_p \mathbb{E} \sup_{r\in[0,u]} [(1+|X_r|)^p e^{-p\beta r}] \left(\int_0^u (u-r)^{-2\alpha-2\gamma}e^{2\beta r} \, dr \right)^{\frac{p}{2}},$$

which implies

$$e^{-p\beta u}\mathbb{E}|Y_u|^p \leq L^p c_p (1+\|X\|^p) \left(\int_0^u (u-r)^{-2\alpha-2\gamma}e^{-2\beta(u-r)} \, dr \right)^{\frac{p}{2}}$$
$$\leq L^p c_p (1+\|X\|^p) \left(\int_0^T r^{-2\alpha-2\gamma}e^{-2\beta r} \, dr \right)^{\frac{p}{2}}.$$

We conclude that

$$\|\Phi(X)\| \le M|x| + MLc_\alpha \left(Tc_p(1 + \|X\|^p)\right)^{\frac{1}{p}} \cdot$$
$$\cdot \left(\int_0^T e^{-p'\beta u} u^{(\alpha-1)p'} \, du\right)^{\frac{1}{p'}} \left(\int_0^T r^{-2\alpha-2\gamma} e^{-2\beta r} \, dr\right)^{\frac{1}{2}}.$$

This shows that Φ is a well defined mapping on $L^p(\Omega; C([0, T], H))$. If X, X^1 are processes belonging to this space, similar passages show that

$$\|\Phi(X) - \Phi(X^1)\| \le MLc_\alpha \left(Tc_p\right)^{\frac{1}{p}} \|X - X^1\| \cdot$$
$$\cdot \left(\int_0^T e^{-p'\beta u} u^{(\alpha-1)p'} \, du\right)^{\frac{1}{p'}} \left(\int_0^T r^{-2\alpha-2\gamma} e^{-2\beta r} \, dr\right)^{\frac{1}{2}},$$

so that, for β sufficiently large, the mapping Φ is a contraction.

In particular, we obtain $\|X\| \le C(1 + |x|)$, which proves the estimate (6.8). $\quad\square$

6.2   Regular Dependence on Data

6.2.1   Differentiability

For further developments we need to investigate the dependence of the solution $X(s; t, x)$ on the initial data x and t. We first reformulate Eq. (6.6) as an equation on $[0, T]$. We set

$$S(s) = e^{sA} \quad \text{for} \quad s \ge 0, \qquad S(s) = I \quad \text{for} \quad s < 0, \qquad (6.10)$$

and we consider the equation

$$X(s) = S(s - t)x + \int_0^s 1_{[t,T]}(r)S(s - r)b(r, X(r)) \, dr$$
$$+ \int_0^s 1_{[t,T]}(r)S(s - r)\sigma(r, X(r)) \, dW(r), \quad (6.11)$$

for the unknown process $X(s)$, $s \in [0, T]$. Under the assumptions of Hypothesis 6.8, Eq. (6.11) has a unique solution $X \in L^p_\mathcal{P}(\Omega; C([0, T], H))$ for every $p \in [2, \infty)$. It clearly satisfies $X(s) = x$ for $s \in [0, t)$, and its restriction to the time interval $[t, T]$ is the unique solution to (6.6).

From now on we denote by $X(s; t, x)$, $s \in [0, T]$, the solution to (6.11).

Proposition 6.10 *Assume Hypothesis 6.8. Then, for every $p \in [2, \infty)$, the following hold.*

(i) *The map $(t, x) \rightarrow X(\cdot; t, x)$ belongs to $\mathcal{G}^{0,1}\left([0, T] \times H, L^p_\mathcal{P}(\Omega; C([0, T], H))\right)$.*

(ii) *Denoting by $\nabla_x X$ the partial Gâteaux derivative, for every direction $h \in H$ the directional derivative process $\nabla_x X(s; t, x)h$, $s \in [0, T]$, solves, $\mathbb{P}$-a.s., the equation:*

$$
\begin{cases}
\nabla_x X(s; t, x)h = e^{(s-t)A}h + \displaystyle\int_t^s e^{(s-r)A}\nabla_x b(r, X(r; t, x))\nabla_x X(r; t, x)h \, dr \\
\qquad + \displaystyle\int_t^s \nabla_x(e^{(s-r)A}\sigma(r, X(r; t, x)))\nabla_x X(r; t, x)h \, dW(r), \quad s \in [t, T], \\
\nabla_x X(s; t, x)h = h, \quad s \in [0, t).
\end{cases}
$$

$$\tag{6.12}$$

(iii) *Finally, $|\nabla_x X(\cdot; t, x)h|_{L_{\mathcal{P}}^p(\Omega; C([0,T],H))} \leq c \, |h|$ for some constant c.*

Proof Let us consider again the map Φ defined in the proof of Proposition 6.9. In our present notation, Φ can be seen as a mapping from $L_{\mathcal{P}}^p(\Omega; C([0, T], H)) \times [0, T] \times H$ to $L_{\mathcal{P}}^p(\Omega; C([0, T], H))$:

$$
\Phi(X, t, x)_s = S(s - t)x + \int_0^s 1_{[t,T]}(r)S(s - r)b(r, X_r) \, dr
$$
$$
+ \int_0^s 1_{[t,T]}(r)S(s - r)\sigma(r, X_r) \, dW_r,
$$

for $s \in [0, T]$. By the arguments of the proof of Proposition 6.9, $\Phi(\cdot, t, x)$ is a contraction in $L_{\mathcal{P}}^p(\Omega; C([0, T], H))$, under an equivalent norm, uniformly with respect to t, x. The process $X(\cdot; t, x)$ is the unique fixed point of $\Phi(\cdot, t, x)$. So, by the parameter-dependent contraction principle (Proposition 6.7), it suffices to show that

$$
\Phi \in \mathcal{G}^{1,0,1}\left(L_{\mathcal{P}}^p(\Omega; C([0, T], H)) \times [0, T] \times H, L_{\mathcal{P}}^p(\Omega; C([0, T], H))\right).
$$

By an obvious extension of Lemma 6.6, the proof is concluded by the following steps.

Step 1. Φ is continuous. We have already noticed that $\Phi(\cdot, t, x)$ is a contraction, uniformly with respect to $x \in H$ and $t \in [0, T]$, and so $\Phi(\cdot, t, x)$ is continuous, uniformly in t, x. Moreover, for fixed X it is easy to verify that $\Phi(X, \cdot, \cdot)$ is continuous from $[0, T] \times H$ to $L_{\mathcal{P}}^p(\Omega; C([0, T], H))$.

Step 2. The directional derivative $\nabla_X \Phi(X, t, x; N)$ in the direction $N \in L_{\mathcal{P}}^p(\Omega; C([0, T], H))$ is the process given by

$$
\nabla_X \Phi(X, t, x; N)_s = \int_t^s e^{(s-r)A}\nabla_x b(r, X_r)N_r \, dr
$$
$$
+ \int_t^s \nabla_x(e^{(s-r)A}\sigma(r, X_r))N_r \, dW_r, \quad s \in [t, T],
$$
$$
\nabla_X \Phi(X, t, x; N)_s = 0, \quad s \in [0, t);
$$

moreover, the mappings $(X, t, x) \to \nabla_X \Phi(X, t, x; N)$ and $N \to \nabla_X \Phi(X, t, x; N)$ are continuous.

We limit ourselves to proving this claim in the special case $b = 0$, the general case being a straightforward extension. For fixed $t \in [0, T]$ and $x \in H$, for all $s \in [t, T]$:

$$I_s^\varepsilon := \frac{1}{\varepsilon}\Phi(X + \varepsilon N, t, x)_s - \frac{1}{\varepsilon}\Phi(X, t, x)_s - \int_t^s \nabla_x(e^{(s-r)A}\sigma(r, X_r))N_r dW_r$$
$$= \int_t^s \left(\int_0^1 \left(\nabla_x(e^{(s-r)A}\sigma(r, X_r + \zeta\varepsilon N_r))N_r - \nabla_x(e^{(s-r)A}\sigma(r, X_r))N_r \right) d\zeta \right) dW_r.$$

Proceeding as in the proof of Proposition 6.9 (with $\beta = 0$) we get for $1/p < \alpha < 1/2 - \gamma$ and for a suitable constant c_p:

$$|I^\varepsilon|^p_{L^p_{\mathcal{P}}(\Omega;C([0,T],H))} \le c_p \mathbb{E} \int_t^T |Y_u^\varepsilon|^p du,$$

where

$$Y_u^\varepsilon = \int_t^u (u-r)^{-\alpha}\left(\int_0^1 \left(\nabla_x(e^{(u-r)A}\sigma(r, X_r + \zeta\varepsilon N_r))N_r \right.\right.$$
$$\left.\left. - \nabla_x(e^{(u-r)A}\sigma(r, X_r))N_r \right) d\zeta \right) dW_r.$$

Therefore

$$\mathbb{E}|Y_u^\varepsilon|^p \le c\mathbb{E}\left(\int_t^u (u-r)^{-2\alpha}\left| \int_0^1 \left(\nabla_x(e^{(u-r)A}\sigma(r, X_r + \zeta\varepsilon N_r))N_r \right.\right.\right.$$
$$\left.\left.\left. - \nabla_x(e^{(u-r)A}\sigma(r, X_r))N_r \right) d\zeta \right|^2_{L_2(\Xi,H)} dr \right)^{p/2}$$

for a suitable constant c. Since for all ε

$$\left| \int_0^1 \nabla_x(e^{(u-r)A}\sigma(r, X_r + \zeta\varepsilon N_r))N_r d\zeta \right|_{L_2(\Xi,H)} \le L(u-r)^{-\gamma}|N|_{C([0,T],H)}$$

and $\nabla_x(e^{sA}\sigma(t, x)v)$ is continuous in x then, by dominated convergence, we get $\mathbb{E} \int_t^T |Y_u^\varepsilon|^p du \to 0$ and the claim follows.

Continuity of the mappings $(X, t, x) \to \nabla_X\Phi(X, t, x; N)$ and $N \to \nabla_X\Phi(X, t, x; N)$ can be proved in a similar way.

Step 3. Finally, it is clear that the directional derivative $\nabla_x\Phi(X, t, x; h)$ in the direction $h \in H$ is the process given by

$$\nabla_x\Phi(X, t, x; h)_s = e^{(s-t)A}h, \quad s \in [t, T],$$
$$\nabla_x\Phi(X, t, x; h)_s = h, \quad s \in [0, t),$$

and that the mappings $(X, t, x) \to \nabla_x \Phi(X, t, x; h)$ and $h \to \nabla_x \Phi(X, t, x; h)$ are continuous.

To complete the proof we observe that the Eq. (6.12) is just a re-writing of (6.2) and that the estimate in (iii) is a trivial consequence of Eq. (6.12) and the fact that $|\nabla_X \Phi|$ is uniformly bounded by a constant <1, by the contraction property of Φ.   □

6.2.2 Differentiability in the Sense of Malliavin

In order to proceed further in the study of the properties of the solution to the forward equation we need to introduce basic notions and tools of the Malliavin calculus. We refer the reader to the book [468] for a detailed exposition; the paper [328] treats the extensions to Hilbert space-valued random variables and processes. We will report without proofs only the results that will be used in the sequel. This digression on the Malliavin calculus ends after Lemma 6.12, when we come back to the forward equation.

We also inform the reader that the aim of this entire section is just to prove Proposition 6.17, whose statement can be understood after reading a few introductory lines preceding it, and that no reference to the Malliavin calculus will be made in the sections that follow.

Our starting point will be a cylindrical Wiener process $\{W_t\}_{t \geq 0}$ on a real separable Hilbert space Ξ. For every (deterministic) function $h \in L^2([0, T]; \Xi)$ the integral $\int_0^T h(t)^* dW_t$ will be denoted by $W(h)$, where $h(t)^* \in \Xi^*$ denotes the image of $h(t) \in \Xi$ under the Riesz isometry. We will also use the notation $W(h) = \int_0^T \langle h(t), dW_t \rangle_\Xi$. Given a Hilbert space K, let S_K be the set of K-valued random variables F of the form

$$F = \sum_{j=1}^m f_j(W(h_1), \ldots, W(h_n)) e_j,$$

where $h_1, \ldots, h_n \in L^2([0, T]; \Xi)$, (e_j) is a basis of K and $f_1, \ldots f_m$ are infinitely differentiable functions $\mathbb{R}^n \to \mathbb{R}$ bounded together with all their derivatives. The Malliavin derivative $D^\mathcal{M} F$ of $F \in S_K$ is defined as the process $D_\eta^\mathcal{M} F$, $\eta \in [0, T]$,

$$D_\eta^\mathcal{M} F = \sum_{j=1}^m \sum_{k=1}^n \partial_k f_j(W(h_1), \ldots, W(h_n)) e_j \otimes h_k(\eta),$$

with values in $\mathcal{L}_2(\Xi, K)$; by ∂_k we denote the partial derivatives with respect to the k-th variable and by $e_j \otimes h_k(\eta)$ the operator $u \to e_j \langle h_k(\eta), u \rangle_\Xi$. It is known that the operator $D^\mathcal{M} : S_K \subset L^2(\Omega; K) \to L^2(\Omega \times [0, T]; \mathcal{L}_2(\Xi, K))$ is closable. We denote by $\mathbb{D}^{1,2}(K)$ the domain of its closure, and use the same letter to denote $D^\mathcal{M}$ and its closure:

$$D^{\mathcal{M}} : \mathbb{D}^{1,2}(K) \subset L^2(\Omega; K) \to L^2(\Omega \times [0, T]; \mathcal{L}_2(\Xi, K)).$$

The adjoint operator of $D^{\mathcal{M}}$,

$$\delta : \text{ dom } (\delta) \subset L^2(\Omega \times [0, T]; \mathcal{L}_2(\Xi, K)) \to L^2(\Omega; K),$$

is called the Skorohod integral. Thus, δ acts on a certain subset of square-integrable stochastic processes u_η, $\eta \in [0, T]$, with values in $\mathcal{L}_2(\Xi, K)$ (more precisely, on equivalence classes up to the product measure $\mathbb{P} \otimes d\eta$) and its value at u is a square-integrable random variable with values in K (more precisely, a $\mathbb{P}$-equivalence class), that will be denoted $\delta(u)$ or $\int_0^T u_\eta \, \hat{d} W_\eta$, because of its close connections with the Itô integral (see, for instance, Proposition 6.11 below). We also need to introduce the space $\mathbb{L}^{1,2}(\mathcal{L}_2(\Xi, K))$ of processes $u \in L^2(\Omega \times [0, T]; \mathcal{L}_2(\Xi, K))$ such that $u_r \in \mathbb{D}^{1,2}(\mathcal{L}_2(\Xi, K))$ for a.e. $r \in [0, T]$, and there exists a measurable version of $D_\eta^{\mathcal{M}} u_r$ satisfying

$$\|u\|^2_{\mathbb{L}^{1,2}(\mathcal{L}_2(\Xi,K))} = \|u\|^2_{L^2(\Omega \times [0,T]; \mathcal{L}_2(\Xi,K))} + \mathbb{E} \int_0^T \int_0^T \|D_\eta^{\mathcal{M}} u_r\|^2_{\mathcal{L}_2(\Xi, \mathcal{L}_2(\Xi,K))} \, dr \, d\eta < \infty.$$

The definition of $\mathbb{L}^{1,2}(K)$ for an arbitrary Hilbert space K (instead of $\mathcal{L}_2(\Xi, K)$) is entirely analogous.

In the following proposition we summarize all the properties that we need in the sequel concerning the objects introduced above. We omit the proofs, which can be found in [328] or, after appropriate reformulation, in [468] or [469]. In particular, point 4 is proved in [328], Proposition 3.4. Point 5 can be found in [469], Theorem 3.2, or [328], Proposition 2.11.

Proposition 6.11 *With the previous notation, the following holds.*

(1) If $F \in \mathbb{D}^{1,2}(K)$ is $\mathscr{F}_t$-adapted then $D^{\mathcal{M}} F = 0$ a.s. on $\Omega \times (t, T]$.
(2) If u is an (adapted) process belonging to $L^2_{\mathcal{P}}(\Omega \times [0, T]; \mathcal{L}_2(\Xi, K))$ then $u \in \text{dom}(\delta)$ and the Skorohod integral $\delta(u)$ coincides with the Itô integral, i.e.,

$$\int_0^T u_\eta \, \hat{d} W_\eta = \int_0^T u_\eta \, dW_\eta.$$

(3) If $u \in \mathbb{L}^{1,2}(\mathcal{L}_2(\Xi, K))$ then $u \in \text{dom}(\delta)$ and $\|\delta(u)\|^2_{L^2(\Omega;K)} \leq \|u\|^2_{\mathbb{L}^{1,2}(\mathcal{L}_2(\Xi,K))}$. In particular, the Skorohod integral δ is a continuous linear operator from $\mathbb{L}^{1,2}(\mathcal{L}_2(\Xi, K))$ to $L^2(\Omega; K)$.
(4) If $u \in \mathbb{L}^{1,2}(\mathcal{L}_2(\Xi, K))$, and for a.a. η the process $\{D_\eta^{\mathcal{M}} u_r\}_{r \in [0,T]}$ belongs to $\text{dom}(\delta)$, and the map $\eta \to \delta(D_\eta^{\mathcal{M}} u)$ belongs to $L^2(\Omega \times [0, T]; \mathcal{L}_2(\Xi, K))$, then $\delta(u) \in \mathbb{D}^{1,2}(K)$ and $D_\eta^{\mathcal{M}} \delta(u) = u_\eta + \delta(D_\eta^{\mathcal{M}} u)$, i.e.,

$$D_\eta^{\mathcal{M}} \int_0^T u_r \, \hat{d} W_r = u_\eta + \int_0^T D_\eta^{\mathcal{M}} u_r \, \hat{d} W_r.$$

(5) *If* $F \in \mathbb{D}^{1,2}(\mathbb{R})$, $u \in L^2(\Omega \times [0,T]; \mathcal{L}_2(\Xi, \mathbb{R})) \simeq L^2(\Omega \times [0,T]; \Xi^*)$ *belongs to* $dom(\delta)$ *and* $Fu \in L^2(\Omega \times [0,T]; \Xi^*)$, *then* $Fu \in dom(\delta)$ *and* $\delta(Fu) = F\delta(u) - \langle D^{\mathcal{M}}F, u \rangle$, *which means*

$$\int_0^T Fu_\eta \, \hat{d}W_\eta = F \int_0^T u_\eta \, \hat{d}W_\eta - \int_0^T \langle D_\eta^{\mathcal{M}} F, u_\eta \rangle_{\mathcal{L}_2(\Xi, K)} \, d\eta,$$

provided the right-hand side belongs to $L^2(\Omega; \mathbb{R})$.

In particular, if $0 \le a \le b \le T$, $\xi \in \Xi$, *and upon taking* $u_\eta = \xi^* 1_{[a,b]}(\eta)$, *we have* $F\xi^* 1_{[a,b]} \in dom(\delta)$ *and*

$$\int_a^b F \, \xi^* \hat{d}W_\eta = F \int_a^b \xi^* \hat{d}W_\eta - \int_a^b D_\eta^{\mathcal{M}} F\xi \, d\eta = F(W_b\xi - W_a\xi) - \int_a^b D_\eta^{\mathcal{M}} F\xi \, d\eta,$$

(6.13)

provided $F \in \mathbb{D}^{1,2}(\mathbb{R})$ *and the right-hand side of (6.13) belongs to* $L^2(\Omega; \mathbb{R})$.

Finally, we need to define the space $\mathbb{D}^{1,2}_{loc}(K)$. If $F \in \mathbb{D}^{1,2}(K)$ and $F = 0$ on a measurable subset $A \subset \Omega$ then $1_A D^{\mathcal{M}} F = 0$; this follows immediately from the corresponding result for $K = \mathbb{R}^d$ ([469], Lemma 2.6). Therefore the following definition is meaningful: we say that a random variable $F : \Omega \to K$ belongs to the space $\mathbb{D}^{1,2}_{loc}(K)$ if there exists an increasing sequence of measurable subsets $\Omega_k \subset \Omega$ and elements $F_k \in \mathbb{D}^{1,2}(K)$ such that $\cup_k \Omega_k = \Omega$ P-a.s. and $1_{\Omega_k} F = 1_{\Omega_k} F_k$. $D^{\mathcal{M}} F$: $\Omega \times [0,T] \to \mathcal{L}_2(\Xi, K)$ is then defined by requiring $1_{\Omega_k} D^{\mathcal{M}} F = 1_{\Omega_k} D^{\mathcal{M}} F_k$. The following chain rule holds; the proof consists in standard approximation arguments and is left to the reader.

Lemma 6.12 *Suppose* K, H *are Hilbert spaces,* $\psi \in \mathcal{G}^1(K, H)$ *and*

$$\sup_{|x| \le n} |\nabla \psi(x)|_{\mathcal{L}(K,H)} < \infty, \qquad n = 1, 2, \ldots. \tag{6.14}$$

(i) *If* $F \in \mathbb{D}^{1,2}_{loc}(K)$ *then* $\psi(F) \in \mathbb{D}^{1,2}_{loc}(H)$.
(ii) *If* $F \in \mathbb{D}^{1,2}(K)$ *and* $\sup_{x \in K} |\nabla \psi(x)|_{\mathcal{L}(K,H)} < \infty$ *then* $\psi(F) \in \mathbb{D}^{1,2}(H)$.
(iii) *More generally, if* $F \in \mathbb{D}^{1,2}(K)$, *(6.14) holds and*

$$\mathbb{E} |\psi(F)|_H^2 < \infty, \qquad \mathbb{E} \int_0^T |\nabla \psi(F) D_\eta^{\mathcal{M}} F|_{\mathcal{L}_2(K,H)}^2 d\eta < \infty,$$

then $\psi(F) \in \mathbb{D}^{1,2}(H)$.

In any of the cases (i)–(iii) we have $D^{\mathcal{M}} \psi(F) = \nabla \psi(F) D^{\mathcal{M}} F$.

After this digression on general Malliavin calculus we come back to the properties of the forward equation and consider again the solution $X = \{X(s; t, x)\}_{s \in [t,T]}$ to (6.6) with (t, x) fixed, denoted simply by (X_s). We set as before $X_s = x$, $s \in [0, t)$.

We will soon prove that X belongs to $\mathbb{L}^{1,2}(H)$. Then it is clear that the equality $D_\eta^{\mathcal{M}} X_s = 0$ $\mathbb{P}$-a.s. holds for a.a. η, t, s if $s < t$ or $\eta > s$.

Proposition 6.13 *Assume Hypothesis 6.8. Then the following properties hold.*

(i) $X \in \mathbb{L}^{1,2}(H)$.

(ii) *There exists a version of $D^{\mathcal{M}} X$ such that for every $\eta \in [0, T)$, $\{D_\eta^{\mathcal{M}} X_s\}_{s \in (s,T]}$ is a predictable process in $\mathcal{L}_2(\Xi, H)$ with continuous paths satisfying, for every $p \in [2, \infty)$,*

$$\sup_{\eta \in [0,T]} \mathbb{E}\left(\sup_{s \in (\eta, T]} (s - \eta)^{p\gamma} |D_\eta^{\mathcal{M}} X_s|_{\mathcal{L}_2(\Xi, H)}^p \right) \leq c, \tag{6.15}$$

where $c > 0$ depends only on p, L, T, γ and $M = \sup_{s \in [0,T]} |e^{sA}|$; moreover, $\mathbb{P}$-a.s.

$$\begin{aligned}
D_\eta^{\mathcal{M}} X_s &= e^{(s-\eta)A} \sigma(\eta, X_\eta) + \int_\eta^s e^{(s-r)A} \nabla_x b(r, X_r) D_\eta^{\mathcal{M}} X_r \, dr \\
&\quad + \int_\eta^s \nabla_x (e^{(s-r)A} \sigma(r, X_r)) D_\eta^{\mathcal{M}} X_r \, dW_r, \quad s \in (\eta, T].
\end{aligned} \tag{6.16}$$

Moreover, $X_s \in \mathbb{D}^{1,2}(H)$ for every $s \in [0, T]$.

(iii) *Given any element v of Ξ, the process $Q_{\eta s} = D_\eta^{\mathcal{M}} X_s v$ is a solution to the equation:*

$$\begin{aligned}
Q_{\eta s} &= e^{(s-\eta)A} \sigma(\eta, X_\eta) v + \int_\eta^s e^{(s-r)A} \nabla_x b(r, X_r) Q_{\eta r} \, dr \\
&\quad + \int_\eta^s \nabla_x (e^{(s-r)A} \sigma(r, X_r)) Q_{\eta r} \, dW_r, \quad \mathbb{P}\text{-a.s.}
\end{aligned} \tag{6.17}$$

for a.a. η, s with $t \leq \eta \leq s \leq T$. It is unique in the sense that if $\{Q_{\eta s}, t \leq \eta \leq s \leq T\}$ is another process with values in H such that $\{Q_{\eta s}\}_{s \in [\eta, T]}$ is predictable for every $\eta \in [t, T]$ and $\mathbb{E} \int_t^T \int_\eta^T |Q_{\eta s}|^2 ds d\eta < \infty$ then, for a.a. η, s, we have $Q_{\eta s} = D_\eta^{\mathcal{M}} X_s v$ $\mathbb{P}$-a.s.

In order to prove this proposition we need some preparation. We start with the following lemma.

Lemma 6.14 *If $X \in \mathbb{L}^{1,2}(H)$ then the random processes*

$$\int_0^s e^{(s-r)A} b(r, X_r) \, dr, \qquad \int_0^s e^{(s-r)A} \sigma(r, X_r) \, dW_r, \qquad s \in [0, T],$$

belong to $\mathbb{L}^{1,2}(H)$ and for a.a. η and s with $\eta < s$

$$D_\eta^{\mathcal{M}} \int_0^s e^{(s-r)A} b(r, X_r) \, dr = \int_\eta^s e^{(s-r)A} \nabla_x b(r, X_r) D_\eta^{\mathcal{M}} X_r \, dr,$$

$$D_\eta^{\mathcal{M}} \int_0^s e^{(s-r)A} \sigma(r, X_r) \, dW_r = e^{(s-\eta)A} \sigma(s, X_s) + \int_\eta^s \nabla_x (e^{(s-r)A} \sigma(r, X_r)) D_\eta^{\mathcal{M}} X_r \, dW_r.$$

$$(6.18)$$

Proof We will prove only (6.18). Recall that, by Proposition 6.11-4, if $u \in \mathbb{L}^{1,2}$ ($\mathcal{L}_2(\Xi, H)$), and for a.a. η the process $\{D_\eta^{\mathcal{M}} u_r\}_{r \in [0,T]}$ belongs to $\mathrm{dom}(\delta)$, and the map $\eta \to \delta(D_\eta^{\mathcal{M}} u)$ belongs to $L^2(\Omega \times [0, T]; \mathcal{L}_2(\Xi, H))$, then $\delta(u) \in \mathbb{D}^{1,2}(H)$ and $D_\eta^{\mathcal{M}} \delta(u) = u_\eta + \delta(D_\eta^{\mathcal{M}} u)$.

We fix s and we apply this result to the process $u_r = e^{(s-r)A} \sigma(r, X_r)$ (we set $u_r = 0$ for $r > s$). First notice that

$$\mathbb{E} \int_0^T |u_r|^2 \, dr = \mathbb{E} \int_0^s |e^{(s-r)A} \sigma(r, X_r)|_{\mathcal{L}_2(\Xi, H)}^2 \, dr$$
$$\leq L^2 \mathbb{E} \int_0^s (s-r)^{-2\gamma} (1 + |X_r|)^2 \, dr.$$

The right-hand side is finite for a.a. s; indeed, by exchanging the integrals we verify that

$$\int_0^T \left(\mathbb{E} \int_0^s (s-r)^{-2\gamma} (1 + |X_r|)^2 \, dr \right) ds$$
$$\leq \int_0^T r^{-2\gamma} \, dr \int_0^T \mathbb{E} (1 + |X_r|)^2 \, dr < \infty,$$

since $X \in \mathbb{L}^{1,2}(H) \subset L^2(\Omega \times [0, T]; H)$. Next, for every r, by the chain rule for the Malliavin derivative (Lemma 6.12-(ii)), $D_\eta^{\mathcal{M}} u_r = \nabla_x (e^{(s-r)A} \sigma(r, X_r)) D_\eta^{\mathcal{M}} X_r$ for a.a. $\eta < r$, whereas $D_\eta^{\mathcal{M}} u_r = 0$ for a.a. $\eta > r$, by adaptedness. Next, recalling (6.7),

$$\mathbb{E} \int_0^T |D_\eta^{\mathcal{M}} u_r|^2 \, dr = \mathbb{E} \int_\eta^s |\nabla_x (e^{(s-r)A} \sigma(r, X_r)) D_\eta^{\mathcal{M}} X_r|_{\mathcal{L}_2(\Xi, \mathcal{L}_2(\Xi, H))}^2 \, dr$$
$$\leq L^2 \mathbb{E} \int_\eta^s (s-r)^{-2\gamma} |D_\eta^{\mathcal{M}} X_r|_{\mathcal{L}_2(\Xi, H)}^2 \, dr,$$

so that

$$\mathbb{E} \int_0^T \int_0^T |D_\eta^{\mathcal{M}} u_r|^2 \, dr \, d\eta \leq L^2 \mathbb{E} \int_0^s \int_\eta^s (s-r)^{-2\gamma} |D_\eta^{\mathcal{M}} X_r|_{\mathcal{L}_2(\Xi, H)}^2 \, dr \, d\eta$$
$$= L^2 \int_0^s (s-r)^{-2\gamma} \int_0^r \mathbb{E} |D_\eta^{\mathcal{M}} X_r|_{\mathcal{L}_2(\Xi, H)}^2 \, d\eta \, dr.$$

The right-hand side is finite for a.a. s; indeed, by exchanging the integrals we verify that

$$\int_0^T \left(\int_0^s (s-r)^{-2\gamma} \int_0^r \mathbb{E}\,|D_\eta^{\mathcal{M}} X_r|^2_{\mathcal{L}_2(\Xi,H)}\, d\eta\, dr \right) ds$$

$$\leq \int_0^T r^{-2\gamma}\, dr \int_0^T \int_0^r \mathbb{E}\,|D_\eta^{\mathcal{M}} X_r|^2_{\mathcal{L}_2(\Xi,H)}\, d\eta\, dr$$

$$= \int_0^T r^{-2\gamma}\, dr\, |D^{\mathcal{M}} X|^2_{L^2(\Omega \times [0,T] \times [0,T];\mathcal{L}_2(\Xi,H))} < \infty,$$

since $X \in \mathbb{L}^{1,2}(H)$. Now we recall that the Skorohod and the Itô integral coincide for adapted integrands, so that

$$\int_0^T \mathbb{E}|\delta(D_\eta^{\mathcal{M}} u)|^2\, d\eta = \int_0^T \mathbb{E}\left|\int_0^T D_\eta^{\mathcal{M}} u_r\, dW_r\right|^2 d\eta = \mathbb{E}\int_0^T \int_0^T |D_\eta^{\mathcal{M}} u_r|^2\, dr\, d\eta < \infty.$$

So for a.a. s we can apply the result mentioned above and since

$$\delta(u) = \int_0^s e^{(s-r)A}\sigma(r,X_r)\, dW_r, \qquad \delta(D_\eta^{\mathcal{M}} u) = \int_\eta^s \nabla_x(e^{(s-r)A}\sigma(r,X_r))D_\eta^{\mathcal{M}} X_r\, dW_r,$$

formula (6.18) is proved. The estimate

$$\int_0^T \int_0^s \mathbb{E}\left|D_\eta^{\mathcal{M}} \int_0^s e^{(s-r)A}\sigma(r,X_r)\, dW_r\right|^2 d\eta\, ds$$

$$\leq 2\int_0^T \int_0^s \mathbb{E}|e^{(s-\eta)A}\sigma(\eta,X_\eta)|^2_{\mathcal{L}_2(\Xi,H)}\, d\eta\, ds$$

$$+2\int_0^T \int_0^s \mathbb{E}\int_\eta^s |\nabla_x(e^{(s-r)A}\sigma(r,X_r))D_\eta^{\mathcal{M}} X_r|^2_{\mathcal{L}_2(\Xi,\mathcal{L}_2(\Xi,H))}\, dr\, d\eta\, ds$$

$$\leq 2L^2 \int_0^T r^{-2\gamma}\, dr \int_0^T \mathbb{E}\,(1+|X_r|)^2\, dr$$

$$+2L^2 \int_0^T r^{-2\gamma}\, dr\, |D^{\mathcal{M}} X|^2_{L^2(\Omega \times [0,T] \times [0,T];\mathcal{L}_2(\Xi,H))} < \infty,$$

is a consequence of the previous passages, and shows that the process $\int_0^s e^{(s-r)A}$ $\sigma(r,X_r)\, dW_r, s \in [0,T]$, belongs to $\mathbb{L}^{1,2}(H)$. $\qquad\square$

For $\eta \in [0,T)$ and for arbitrary predictable processes $X_s, Q_s, s \in [\eta,T]$, with values in H and $\mathcal{L}_2(\Xi,H)$ respectively, we define, for $s \in [\eta,T]$,

$$\Gamma_1(X,Q)_{\eta s} = \int_\eta^s e^{(s-r)A}\nabla_x b(r,X_r)Q_r\, dr,$$

$$\Gamma_2(X,Q)_{\eta s} = \int_\eta^s \nabla_x(e^{(s-r)A}\sigma(r,X_r))Q_r\, dW_r.$$

The same notation will be used when $Q_s, s \in [\eta,T]$, is a process with values in H.

Proof of Proposition 6.13. We fix $t \in [0, T)$. Let us consider the sequence X^n defined as follows: $X^0 = 0$,

$$X_s^{n+1} = e^{(s-t)A}x + \int_t^s e^{(s-r)A}b(r, X_r^n)\, dr + \int_t^s e^{(s-r)A}\sigma(r, X_r^n)\, dW_r, \quad s \in [t, T],$$

and $X_s^n = x$ for $s < t$. It follows from the proof of Proposition 6.9 that X^n converges to the solution X of Eq. (6.6) in the space $L_\mathcal{P}^p(\Omega; C([0, T], H))$ hence, in particular, in the space $L^2(\Omega \times [0, T]; H)$. By Lemma 6.14, $X^n \in \mathbb{L}^{1,2}(H)$ and, for a.a. η and s with $\eta < s$,

$$\begin{aligned}
D_\eta^{\mathcal{M}} X_s^{n+1} &= e^{(s-\eta)A}\sigma(\eta, X_\eta^n) + \int_\eta^s e^{(s-r)A}\nabla_x b(r, X_r^n)D_\eta^{\mathcal{M}} X_r^n\, dr \\
&\quad + \int_\eta^s \nabla_x(e^{(s-r)A}\sigma(r, X_r^n))D_\eta^{\mathcal{M}} X_r^n\, dW_r.
\end{aligned} \tag{6.19}$$

Setting $I(X^n)_{\eta s} = e^{(s-\eta)A}\sigma(\eta, X_\eta^n)$ for $s > \eta$ and $I(X^n)_{\eta s} = 0$ for $s < \eta$, and recalling the operators introduced above, we may write equality (6.19) as

$$D^{\mathcal{M}} X^{n+1} = I(X^n) + \Gamma_1(X^n, D^{\mathcal{M}} X^n) + \Gamma_2(X^n, D^{\mathcal{M}} X^n).$$

We note that $I(X^n)$ is a bounded sequence in $L^2(\Omega \times [0, T] \times [0, T]; \mathcal{L}_2(\Xi, H))$, since

$$\begin{aligned}
\mathbb{E}\int_0^T \int_0^s |e^{(s-\eta)A}&\sigma(\eta, X_\eta^n)|_{\mathcal{L}_2(\Xi, H)}^2\, d\eta\, ds \\
&\leq L^2 \mathbb{E}\int_0^T \int_0^s (s-\eta)^{-2\gamma}(1+|X_\eta^n|)^2\, d\eta\, ds \\
&\leq L^2 \int_0^T s^{-2\gamma}\, ds \int_0^T \mathbb{E}\,(1+|X_\eta^n|)^2\, d\eta,
\end{aligned}$$

and X^n is a bounded sequence in $L^2(\Omega \times [0, T]; H)$. Next we show that there exists an equivalent norm $\|\cdot\|$ in $L^2(\Omega \times [0, T] \times [0, T]; \mathcal{L}_2(\Xi, H))$ such that

$$\|\Gamma_1(X^n, D^{\mathcal{M}} X^n)\| + \|\Gamma_2(X^n, D^{\mathcal{M}} X^n)\| \leq \alpha \|D^{\mathcal{M}} X^n\|, \tag{6.20}$$

for some $\alpha \in [0, 1)$ independent of n. For simplicity we only consider the operator Γ_2. For a process $(Z_{\eta s}) \in L^2(\Omega \times [0, T] \times [0, T]; \mathcal{L}_2(\Xi, H))$ we introduce the norm

$$\|Z\|^2 = \int_0^T \int_0^T \mathbb{E}\,|Z_{\eta s}|_{\mathcal{L}_2(\Xi, H)}^2 e^{-\beta(s-\eta)}\, ds\, d\eta,$$

where $\beta > 0$ will be chosen later. We have

$$\int_\eta^T \mathbb{E}|\Gamma_2(X^n, D^{\mathcal{M}} X^n)_{\eta s}|^2_{\mathcal{L}_2(\Xi, H)} e^{-\beta(s-\eta)}\, ds$$

$$= \int_\eta^T \int_\eta^s \mathbb{E}\, |\nabla_x(e^{(s-r)A}\sigma(r, X^n_r)) D^{\mathcal{M}}_\eta X^n_r|^2_{\mathcal{L}_2(\Xi, \mathcal{L}_2(\Xi, H))}\, dr\ e^{-\beta(s-\eta)}\, ds$$

$$\leq L^2 \int_\eta^T \int_\eta^s (s-r)^{-2\gamma} \mathbb{E}\, |D^{\mathcal{M}}_\eta X^n_r|^2_{\mathcal{L}_2(\Xi, H)}\, dr\ e^{-\beta(s-\eta)}\, ds$$

$$= L^2 \int_\eta^T e^{-\beta(r-\eta)} \mathbb{E}\, |D^{\mathcal{M}}_\eta X^n_r|^2_{\mathcal{L}_2(\Xi, H)} \int_r^T (s-r)^{-2\gamma}\ e^{-\beta(s-r)}\, ds\, dr$$

$$\leq L^2 \int_\eta^T e^{-\beta(r-\eta)} \mathbb{E}\, |D^{\mathcal{M}}_\eta X^n_r|^2_{\mathcal{L}_2(\Xi, H)}\, dr \left(\sup_{r\in[\eta,T]} \int_r^T (s-r)^{-2\gamma}\ e^{-\beta(s-r)}\, ds \right).$$

The supremum on the right-hand side can be estimated by $\int_0^T r^{-2\gamma}\, e^{-\beta r}\, dr$; so we obtain

$$\|\Gamma_2(X^n, D^{\mathcal{M}} X^n)\|^2 \leq L^2 \int_0^T r^{-2\gamma}\, e^{-\beta r}\, dr\, \|D^{\mathcal{M}} X^n\|^2.$$

Now to prove (6.20) it suffices to take β sufficiently large.

From (6.20) and from the fact that $I(X^n)$ is bounded in $L^2(\Omega \times [0, T] \times [0, T]; \mathcal{L}_2(\Xi, H))$, it follows easily that the sequence $D^{\mathcal{M}} X^n$ is also bounded in this space. Since, as mentioned before, X^n converges to X in $L^2(\Omega \times [0, T]; H)$, it follows from the closedness of the operator $D^{\mathcal{M}}$ that X belongs to $\mathbb{L}^{1,2}(H)$. Point (i) of Proposition 6.13 is now proved.

By Lemma 6.14, we can compute the Malliavin derivative of both sides of (6.6) and we obtain, for a.a. η and s with $\eta < s$,

$$D^{\mathcal{M}}_\eta X_s = I(X)_{\eta s} + \Gamma_1(X, D^{\mathcal{M}} X)_{\eta s} + \Gamma_2(X, D^{\mathcal{M}} X)_{\eta s}, \qquad \mathbb{P}\text{-a.s.,} \qquad (6.21)$$

where

$$I(X)_{\eta s} = e^{(s-\eta)A}\sigma(\eta, X_\eta). \qquad (6.22)$$

Let us introduce the space $\mathcal{K}$ of processes $Q_{\eta s}$, $0 \leq \eta < s \leq T$, such that for every $\eta \in [t, T)$, $\{Q_{\eta s}\}_{s\in(\eta,T]}$ is a predictable process in $\mathcal{L}_2(\Xi, H)$ with continuous paths, and such that

$$\sup_{\eta\in[0,T]} \mathbb{E}\left(\sup_{s\in(\eta,T]} e^{-\beta p(s-\eta)}(s-\eta)^{p\gamma} |Q_{\eta s}|^p_{\mathcal{L}_2(\Xi, H)} \right) < \infty. \qquad (6.23)$$

Here $p \in [2, \infty)$ is fixed and $\beta > 0$ is a parameter, to be chosen later. Let us consider the equation: for every $\eta \in [0, T)$, $\mathbb{P}$-a.s.,

$$Q_{\eta s} = I(X)_{\eta s} + \Gamma_1(X, Q)_{\eta s} + \Gamma_2(X, Q)_{\eta s}, \qquad s \in (\eta, T]. \qquad (6.24)$$

We are going to prove that there exists a unique solution $Q \in \mathcal{K}$ of this equation. Assume this for a moment. Then, subtracting (6.24) from (6.21), we obtain for a.a. η and s with $\eta < s$

$$D_\eta^\mathcal{M} X_s - Q_{\eta s} = \Gamma_1(X, D^\mathcal{M} X - Q)_{\eta s} + \Gamma_2(X, D^\mathcal{M} X - Q)_{\eta s}, \qquad \mathbb{P}\text{-a.s.}$$

Repeating the passages that led to (6.20) we obtain

$$\|\Gamma_1(X, D^\mathcal{M} X - Q)\| + \|\Gamma_2(X, D^\mathcal{M} X - Q)\| \le \alpha \|D^\mathcal{M} X - Q\|,$$

for some $\alpha \in [0, 1)$. This proves that Q is a version of $D^\mathcal{M} X$. Then equality (6.24) coincides with (6.16), and this proves point (ii) of the Proposition, except for the last assertion.

Now we prove unique solvability of (6.24) in the space $\mathcal{K}$. It suffices to show that $I(X) \in \mathcal{K}$ and that $\Gamma_1(X, \cdot) + \Gamma_2(X, \cdot)$ is a contraction in $\mathcal{K}$. Since, for $s > \eta$,

$$|e^{(s-\eta)A} \sigma(\eta, X_\eta)|_{\mathcal{L}_2(\Xi, H)} \le L(s - \eta)^{-\gamma} (1 + |X_\eta|),$$

we have

$$\sup_{\eta \in [0, T]} \mathbb{E} \sup_{s \in (\eta, T]} (s - \eta)^{p\gamma} |e^{(s-\eta)A} \sigma(\eta, X_\eta)|_{\mathcal{L}_2(\Xi, H)}^p \le L^p \sup_{\eta \in [0, T]} \mathbb{E} (1 + |X_\eta|)^p,$$

which is finite, since $X \in L_\mathcal{P}^p(\Omega; C([0, T], H))$. This shows that $I(X) \in \mathcal{K}$; the contraction property for $\Gamma_1(X, \cdot) + \Gamma_2(X, \cdot)$ requires a longer argument, and it is postponed to Lemma 6.15 below.

The last assertion of point (ii) is clear for $s \in [0, t]$, since $X_s = x$. For $s \in (t, T]$ we take a sequence $s_n \uparrow s$ such that $X_{s_n} \in \mathbb{D}^{1,2}(H)$ and we note that by (6.15) the sequence $\mathbb{E} \int_0^T |D_\eta^\mathcal{M} X_{s_n}|^2 d\eta$ is bounded by a constant independent of n; since $X_{s_n} \to X_s$ in $L^2(\Omega; H)$, it follows from the closedness of the operator $D^\mathcal{M}$ that $X_s \in \mathbb{D}^{1,2}(H)$.

Now we proceed to proving point (iii) of the Proposition. Let us fix $v \in \Xi$ and define the space $\mathcal{S}$ of processes $\{Q_{\eta s}, t \le \eta \le s \le T\}$, with values in H, such that $\{Q_{\eta s}\}_{s \in [\eta, T]}$ is predictable for every $\eta \in [t, T]$ and the norm

$$\|Q\|^2 = \int_t^T \int_\eta^T \mathbb{E} |Q_{\eta s}|_H^2 e^{-\beta(s-\eta)} ds \, d\eta$$

is finite, where $\beta > 0$ is a parameter to be chosen later. Since $I(X)$ (defined in (6.22)) belongs to the space $\mathcal{K}$ introduced above, $I(X)v$ belongs to $\mathcal{S}$ and the equality (6.17) is equivalent to the equality in the space $\mathcal{S}$:

$$Q = I(X)v + \Gamma_1(X, Q) + \Gamma_2(X, Q). \tag{6.25}$$

It turns out that this equation has a unique solution in $\mathcal{S}$: indeed, $\Gamma_1(X, \cdot) + \Gamma_2(X, \cdot)$ is a contraction in the space $\mathcal{S}$ if β is chosen sufficiently large, as it can be proved by passages almost identical to those leading to (6.20). Finally, $D^{\mathcal{M}} X v$ belongs to $\mathcal{S}$ since $D^{\mathcal{M}} X \in L^2(\Omega \times [0, T] \times [0, T]; \mathcal{L}_2(\Xi, H))$, and applying both sides of (6.16) to v we check that $D^{\mathcal{M}} X v = I(X)v + \Gamma_1(X, D^{\mathcal{M}} X v) + \Gamma_2(X, D^{\mathcal{M}} X v)$. Point (iii) of the proposition is now proved. $\qquad\square$

To complete the previous proof, it remains to state and prove the following lemma.

Lemma 6.15 *For $\eta \in [0, T)$, let X_s, $s \in [\eta, T]$, be a predictable process in H and let Q_s, $s \in (\eta, T]$, be an $\mathcal{L}_2(\Xi, H)$-valued continuous adapted process.*

For $p \in [2, \infty)$ sufficiently large and for every $\beta > 0$, the following estimate holds:

$$\mathbb{E}\left(\sup_{s \in [\eta, T]} (s - \eta)^{\gamma p} e^{-\beta p(s-\eta)} \left(|\Gamma_1(X, Q)_{\eta s}|^p_{\mathcal{L}_2(\Xi, H)} + |\Gamma_2(X, Q)_{\eta s}|^p_{\mathcal{L}_2(\Xi, H)} \right) \right)$$
$$\leq C(\beta) \mathbb{E}\left(\sup_{s \in [\eta, T]} (s - \eta)^{\gamma p} e^{-\beta p(s-\eta)} |Q_s|^p_{\mathcal{L}_2(\Xi, H)} \right),$$

where $C(\beta)$ depends on β, p, L, γ, T and $M = \sup_{s \in [0, T]} |e^{sA}|$, and is such that $C(\beta) \to 0$ as $\beta \to 0$.

Proof For simplicity, we only consider the operator Γ_2. Fixing $\eta \in [0, T)$ we introduce the space of $\mathcal{L}_2(\Xi, H)$-valued continuous adapted processes Q_s, $s \in (\eta, T]$ such that the norm

$$\|Q\|^p_\eta := \mathbb{E} \sup_{s \in [\eta, T]} (s - \eta)^{\gamma p} e^{-\beta p(s-\eta)} |Q_s|^p_{\mathcal{L}_2(\Xi, H)}$$

is finite. We use the factorization method, see [177], Theorem 5.2.5. Let us take $p > 2$ and $\alpha \in (0, 1)$ such that

$$\frac{1}{p} < \alpha < \frac{1}{2} - \gamma, \qquad \text{and let} \qquad c_\alpha^{-1} = \int_r^s (s - u)^{\alpha - 1}(u - r)^{-\alpha} du.$$

Then, by the stochastic Fubini theorem,

$$\Gamma_2(X, Q)_{\eta s} = c_\alpha \int_\eta^s \int_r^s (s - u)^{\alpha - 1}(u - r)^{-\alpha} e^{(s-u)A} \nabla_x (e^{(u-r)A} \sigma(r, X_r)) Q_r \, du \, dW_r$$
$$= c_\alpha \int_\eta^s (s - u)^{\alpha - 1} e^{(s-u)A} V_u \, du,$$

where

$$V_u = \int_\eta^u (u - r)^{-\alpha} \nabla_x (e^{(u-r)A} \sigma(r, X_r)) Q_r \, dW_r.$$

By the Hölder inequality, setting $M = \sup_{s \in [0, T]} |e^{sA}|$, $p' = p/(p - 1)$,

$$\left|\Gamma_2(X, Q)_{\eta s}\right| \leq c_\alpha M \int_\eta^s (s - u)^{\alpha-1} |V_u|\, du$$

$$\leq c_\alpha M \left(\int_\eta^s e^{-p\beta(u-\eta)} (u - \eta)^{\gamma p} |V_u|^p\, du \right)^{\frac{1}{p}}$$

$$\cdot \left(\int_\eta^s e^{p'\beta(u-\eta)} (u - \eta)^{-\gamma p'} (s - u)^{(\alpha-1)p'}\, du \right)^{\frac{1}{p'}}.$$

$$\|\Gamma_2(X, Q)\|_\eta^p \leq c_\alpha^p M^p \int_\eta^T e^{-p\beta(u-\eta)} (u - \eta)^{\gamma p} \mathbb{E}\, |V_u|^p\, du$$

$$\cdot \sup_{s \in (\eta, T]} (s - \eta)^{\gamma p} e^{-\beta p (s-\eta)} \left(\int_\eta^s e^{p'\beta(u-\eta)} (u - \eta)^{-\gamma p'} (s - u)^{(\alpha-1)p'}\, du \right)^{\frac{p}{p'}}.$$

Changing u into $(u - \eta)/(s - \eta)$, it is easily seen that the supremum on the right-hand side equals

$$\sup_{s \in (\eta, T]} (s - \eta)^{p\alpha - 1} e^{-\beta p (s-\eta)} \left(\int_0^1 e^{p'\beta u (s-\eta)} u^{-\gamma p'} (1 - u)^{(\alpha-1)p'}\, du \right)^{\frac{p}{p'}} \leq a(\beta)^p,$$

where we set

$$a(\beta) := \sup_{\lambda \in (0, T]} \lambda^{\alpha - \frac{1}{p}} e^{-\beta\lambda} \left(\int_0^1 e^{p'\beta u \lambda} u^{-\gamma p'} (1 - u)^{(\alpha-1)p'}\, du \right)^{\frac{1}{p'}}.$$

So we arrive at

$$\|\Gamma_2(X, Q)\|_\eta \leq c_\alpha M a(\beta) \left(\int_\eta^T e^{-p\beta(u-\eta)} (u - \eta)^{\gamma p} \mathbb{E}\, |V_u|^p\, du \right)^{\frac{1}{p}}.$$

By the Burkholder–Davis–Gundy inequalities, for some constant c_p depending only on p, we have

$$\mathbb{E}\, |V_u|^p \leq c_p \mathbb{E} \left(\int_\eta^u (u - r)^{-2\alpha} |\nabla_x (e^{(u-r)A} \sigma(r, X_r)) Q_r|^2_{\mathcal{L}_2(\Xi, \mathcal{L}_2(\Xi, H))}\, dr \right)^{\frac{p}{2}}$$

$$\leq L^p c_p \mathbb{E} \left(\int_\eta^u (u - r)^{-2\alpha - 2\gamma} |Q_r|^2_{\mathcal{L}_2(\Xi, H)}\, dr \right)^{\frac{p}{2}}$$

$$\leq L^p c_p \|Q\|_s^p \left(\int_\eta^u (u - r)^{-2\alpha - 2\gamma} (r - \eta)^{-2\gamma} e^{2\beta(r-\eta)}\, dr \right)^{\frac{p}{2}}.$$

Changing r into $(r - \eta)/(u - \eta)$ and taking into account that $\beta > 0$ and $\alpha + \gamma < 1/2$ we obtain

$$(u - \eta)^{\gamma p} e^{-p\beta(u-\eta)} \mathbb{E} |V_u|^p \leq L^p c_p \|Q\|_\eta^p (u - \eta)^{p(-\alpha-\gamma+1/2)}$$
$$\cdot \left(\int_0^1 (1 - r)^{-2\alpha-2\gamma} r^{-2\gamma} e^{-2\beta(1-r)(r-\eta)} \, dr \right)^{\frac{p}{2}}$$
$$\leq L^p c_p \|Q\|_\eta^p T^{p(\frac{1}{2}-\alpha-\gamma)} \left(\int_0^1 (1 - r)^{-2\alpha-2\gamma} r^{-2\gamma} \, dr \right)^{\frac{p}{2}}.$$

We conclude that

$$\|\Gamma_2(X, Q)\|_\eta \leq c_\alpha M L c_p^{\frac{1}{p}} a(\beta) T^{\frac{1}{2}-\alpha-\gamma+\frac{1}{p}} \left(\int_0^1 (1 - r)^{-2\alpha-2\gamma} r^{-2\gamma} \, dr \right)^{\frac{1}{2}} \|Q\|_\eta.$$

This inequality proves the lemma, since the property that $a(\beta) \to 0$ as $\beta \to +\infty$ follows easily from the definition of $a(\beta)$. $\qquad\square$

The following result relates the Malliavin derivative of the process X with $\nabla_x X(s; t, x)$, the partial Gâteaux derivative with respect to x (compare Proposition 6.10).

Proposition 6.16 *Assume Hypothesis 6.8. Then for a.a. η, s such that $t \leq \eta \leq s \leq T$ we have*

$$D_\eta^M X(s; t, x) = \nabla_x X(s; \eta, X(\eta; t, x))\sigma(\eta, X(\eta; t, x)), \qquad \mathbb{P}\text{-a.s.} \qquad (6.26)$$

Moreover, $D_\eta^M X(T; t, x) = \nabla_x X(T; \eta, X(\eta; t, x))\sigma(\eta, X(\eta; t, x))$, $\mathbb{P}$-a.s. *for a.a. η.*

Proof Proposition 6.10 states that for every $\eta \in [0, T]$ and every direction $h \in H$ the directional derivative process $\nabla_x X(s; \eta, x)h, s \in [\eta, T]$, solves the equation: $\mathbb{P}$-a.s.,

$$\nabla_x X(s; \eta, x)h = e^{(s-\eta)A}h + \int_\eta^s e^{(s-r)A}\nabla_x b(r, X(r; \eta, x))\nabla_x X(r; \eta, x)h \, dr$$
$$+ \int_\eta^s \nabla_x (e^{(s-r)A}\sigma(r, X(r; \eta, x))\nabla_x X(r; \eta, x)h \, dW_r, \quad s \in [\eta, T].$$

Given $v \in \Xi$ and $t \in [0, \eta]$, we can replace x by $X(\eta; t, x)$ and h by $\sigma(\eta, X(\eta; t, x))v$ in this equation, since $X(\eta; t, x)$ is $\mathscr{F}_\eta$-measurable. Next we note the equality: $\mathbb{P}$-a.s.,

$$X(r; \eta, X(\eta; t, x)) = X(r; t, x), \qquad r \in [\eta, T],$$

which is a consequence of the uniqueness of the solution to (6.6), and we obtain: $\mathbb{P}$-a.s.,

$$\nabla_x X(s; \eta, X(\eta; t, x))\sigma(\eta, X(\eta; t, x))v = e^{(s-\eta)A}\sigma(\eta, X(\eta; t, x))v$$

$$+ \int_\eta^s e^{(s-r)A} \nabla_x b(r, X(r; t, x)) \nabla_x X(r; \eta, X(\eta; t, x))\sigma(\eta, X(\eta; t, x))v \, dr$$

$$+ \int_\eta^s \nabla_x (e^{(s-r)A} \sigma(r, X(r; t, x)) \nabla_x X(r; \eta, X(\eta; t, x))\sigma(\eta, X(\eta; t, x))v \, dW_r, \quad s \in [\eta, T].$$

This shows that the process $\{\nabla_x X(s; t, X(\eta; t, x))\sigma(\eta, X(\eta; t, x))v : t \leq \eta \leq s \leq T\}$ is a solution to Eq. (6.17). Then (6.26) follows from the uniqueness property.

To prove the last assertion, it suffices to take a sequence $s_n \uparrow T$ such that (6.26) holds for s_n and let $n \to \infty$. The conclusion follows from the regularity properties of $D^{\mathcal{M}} X$ and $\nabla_x X$ stated above, as well as the closedness of the operator $D^{\mathcal{M}}$. □

Now, for $\xi \in \Xi$, recall that $W\xi = \{W(\tau)\xi\}_{\tau \geq 0}$ is a real Wiener process. Also fix $t \in [0, T]$ and $x \in H$ and set $X_\tau = X(\tau; t, x)$, $\tau \in [t, T]$, for simplicity. Given a function $u : [0, T] \times H \to \mathbb{R}$, we investigate the existence of the joint quadratic variation of the process $\{u(\tau, X_\tau)\}_{\tau \in [t,T]}$ with $W\xi$. As usual, this is defined for every $\tau \in [t, T]$ as the limit in probability of

$$\sum_{i=1}^n (u(\tau_i, X_{\tau_i}) - u(\tau_{i-1}, X_{\tau_{i-1}}))(W(\tau_i)\xi - W(\tau_{i-1})\xi),$$

where $\{\tau_i\}$, $t = \tau_0 < \tau_1 < \cdots < \tau_n = \tau$, is an arbitrary subdivision of $[t, \tau]$ whose mesh tends to 0. The existence of the joint quadratic variation is not trivial. Indeed, due to the occurrence of convolution type integrals in the definition of a mild solution, it is not obvious that the process X is a semimartingale. Moreover, even in this case, the process $u(\cdot, X.)$ might fail to be a semimartingale if u is not regular enough. Nevertheless, the following result holds true. Its proof could be deduced from the generalization of some results obtained in [469] to the infinite-dimensional case, but we prefer to give a simpler direct proof.

Proposition 6.17 *Assume Hypothesis 6.8, let u be a function in $\mathcal{G}^{0,1}([0, T] \times H, \mathbb{R})$ having polynomial growth together with its derivative $\nabla_x u$. Then the process $\{u(\tau, X_\tau)\}_{\tau \in [t,T]}$ admits a joint quadratic variation process V with $W\xi$, given by*

$$V_\tau = \int_t^\tau \nabla_x u(s, X_s) \, \sigma(s, X_s)\xi \, ds, \quad \tau \in [t, T].$$

Proof Let us write $\bar{u}_\tau = u(\tau, X_\tau)$, $\tau \in [t, T]$, for simplicity. By Proposition 6.13 and the assumptions on u we can apply the chain rule for the Malliavin derivative operator presented in Lemma 6.12 and conclude that, for every $\tau \in [t, T]$, we have $\bar{u}_\tau \in \mathbb{D}^{1,2}(\mathbb{R})$ and $D^{\mathcal{M}}\bar{u}_\tau = \nabla_x u(\tau, X_\tau)D^{\mathcal{M}}X_\tau$. Taking into account (6.26), for a.e. $s \in [0, \tau]$ we obtain

$$D_s^{\mathcal{M}}\bar{u}_\tau\xi = \nabla_x u(\tau, X_\tau) \, \nabla_x X(\tau; s, X_s) \, \sigma(s, X_s) \, \xi, \quad \mathbb{P}\text{-a.s.} \qquad (6.27)$$

whereas $D_s^\mathcal{M} \bar{u}_\tau \xi = 0$ $\mathbb{P}$-a.s., for a.e. $s \in (\tau, T]$.

Let us now compute the joint quadratic variation of $\bar{u}$ and $W\xi$. Let $t = \tau_0 < \tau_1 < \cdots < \tau_n = \tau$ be a subdivision of $[t, \tau] \subset [0, T]$. We use formula (6.13) in Proposition 6.11 with $[a, b] = [\tau_{i-1}, \tau_i]$ and $F = \bar{u}_{\tau_i} - \bar{u}_{\tau_{i-1}}$ and obtain

$$(\bar{u}_{\tau_i} - \bar{u}_{\tau_{i-1}})(W(\tau_i)\xi - W(\tau_{i-1})\xi) = \int_{\tau_{i-1}}^{\tau_i} (\bar{u}_{\tau_i} - \bar{u}_{\tau_{i-1}})\xi^* \, \hat{d}W_s$$

$$+ \int_{\tau_{i-1}}^{\tau_i} D_s^\mathcal{M}(\bar{u}_{\tau_i} - \bar{u}_{\tau_{i-1}})\xi \, ds,$$

where as usual we use the symbol $\hat{d}W$ to denote the Skorohod integral. We note that $D_s^\mathcal{M} \bar{u}_{\tau_{i-1}} = 0$ for $s > \tau_{i-1}$, so recalling (6.27) and setting $U_n(s) = \sum_{i=1}^n (\bar{u}_{\tau_i} - \bar{u}_{\tau_{i-1}}) \, 1_{(\tau_{i-1}, \tau_i]}(s)$ we obtain

$$\sum_{i=1}^n (\bar{u}_{\tau_i} - \bar{u}_{\tau_{i-1}})(W_{\tau_i}^\xi - W_{\tau_{i-1}}^\xi)$$

$$= \int_t^\tau U_n(s) \, \xi^* \, \hat{d}W_s + \sum_{i=1}^n \int_{\tau_{i-1}}^{\tau_i} \nabla_x u(\tau_i, X_{\tau_i}) \, \nabla_x X(\tau_i; s, X_s)\sigma(s, X_s)\xi \, ds.$$

By (6.27) and the continuity properties asserted in Proposition 6.10, it is easily verified that the maps $\tau \to \bar{u}_\tau$ and $\tau \to D^\mathcal{M} \bar{u}_\tau \xi$ are continuous on $[0, T]$ with values in $L^2(\Omega; \mathbb{R})$ and $L^2(\Omega \times [0, T]; \mathbb{R})$, respectively. In particular, $U_n \to 0$ in $\mathbb{L}^{1,2}(\mathbb{R})$, which implies that the Skorohod integral in the last equation tends to zero in $L^2(\Omega; \mathbb{R})$. Letting the mesh of the subdivision tend to 0 and using the continuity properties of $\nabla_x u$, X, $\nabla_x X$, we obtain

$$\sum_{i=1}^n (\bar{u}_{\tau_i} - \bar{u}_{\tau_{i-1}})(W(\tau_i)\xi - W(\tau_{i-1})\xi) \to V_\tau,$$

in probability, which finishes the proof of the proposition.   □

6.3   Backward Stochastic Differential Equations (BSDEs)

6.3.1   Well-Posedness

Some of the basic results on backward equations rely on the following well-known representation theorem (see e.g. [350]). Recall that $(\mathscr{F}_t)$ is the filtration generated by the cylindrical Wiener process W, augmented in the usual way. We denote by $\mathbb{E}^{\mathscr{F}_s}$ the conditional expectation with respect to $\mathscr{F}_s$.

Proposition 6.18 *Let K be a Hilbert space and T > 0. For arbitrary $\mathscr{F}_T$-measurable $\xi \in L^2(\Omega; K)$ there exists a $V \in L^2_\mathcal{P}(\Omega \times [0, T]; \mathcal{L}_2(\Xi, K))$ such that $\xi = \mathbb{E}\,\xi + \int_0^T V(r)\,dW(r)$, $\mathbb{P}$-a.s. Equivalently, for every $s \in [0, T]$,*

$$\mathbb{E}^{\mathscr{F}_s}\xi = \xi - \int_s^T V(r)\,dW(r), \qquad \mathbb{P}\text{-a.s.}$$

Lemma 6.19 *Assume $\eta \in L^2(\Omega; K)$ is $\mathscr{F}_T$-measurable and $f \in L^2_\mathcal{P}(\Omega \times [0, T]; K)$. Then there exists a unique pair of processes $Y(s), Z(s), s \in [0, T]$, such that*

(i) $Y \in L^2_\mathcal{P}(\Omega \times [0, T]; K)$, $Z \in L^2_\mathcal{P}(\Omega \times [0, T]; \mathcal{L}_2(\Xi, K))$;
(ii) for a.a. $s \in [0, T]$, $\mathbb{P}$-a.s.,

$$Y(s) + \int_s^T Z(r)\,dW(r) = \int_s^T f(r)\,dr + \eta. \tag{6.28}$$

Moreover, Y has a continuous version and for every $\beta \neq 0$,

$$\begin{aligned}
\mathbb{E}\int_0^T e^{2\beta r}|Z(r)|^2 dr &\leq \frac{4}{\beta}\,\mathbb{E}\int_0^T e^{2\beta r}|f(r)|^2 dr + 8\,e^{2\beta T}\mathbb{E}\,|\eta|^2,\\
\mathbb{E}\sup_{s\in[0,T]} e^{2\beta s}|Y(s)|^2 &\leq \frac{4}{\beta}\,\mathbb{E}\int_0^T e^{2\beta r}|f(r)|^2 dr + 8\,e^{2\beta T}\mathbb{E}\,|\eta|^2.
\end{aligned} \tag{6.29}$$

In particular, $Y \in C_\mathcal{P}([0, T], L^2(\Omega; K))$.

If, in addition, there exists a $p \in [2, \infty)$ such that

$$\mathbb{E}\left(\int_0^T |f(r)|^2 dr\right)^{p/2} < \infty, \qquad \mathbb{E}\,|\eta|^p < \infty,$$

then for every δ such that $0 \leq T - \delta < T$ we have

$$\mathbb{E}\sup_{s\in[T-\delta,T]}|Y(s)|^p + \mathbb{E}\left(\int_{T-\delta}^T |Z(r)|^2 dr\right)^{p/2} \leq c_p \delta^{p/2}\mathbb{E}\left(\int_{T-\delta}^T |f(r)|^2 dr\right)^{p/2} + c_p\mathbb{E}\,|\eta|^p, \tag{6.30}$$

where c_p is a positive constant, depending only on p.

Proof We modify the argument in [350]. We write Y_s instead of $Y(s)$ etc. to shorten notation.

Uniqueness. Assume that (6.28) holds. Then, taking conditional expectation with respect to $\mathscr{F}_s$ we obtain, for a.e. s,

$$Y_s = \mathbb{E}^{\mathscr{F}_s}\eta + \int_s^T \mathbb{E}^{\mathscr{F}_s} f_r\,dr. \tag{6.31}$$

If $\eta = 0$ and $f = 0$ this equality implies that $Y = 0$; from (6.28) it follows that $\int_s^T Z_r \, dW_r = 0$, which implies $Z = 0$ as well.

Existence. Define $\xi = \eta + \int_0^T f_r \, dr$. Since $\xi \in L^2(\Omega; K)$ is $\mathscr{F}_T$-measurable, by Proposition 6.18 there exists a $Z \in L^2_{\mathcal{P}}(\Omega \times [0, T]; \mathcal{L}_2(\Xi, K))$ such that

$$\mathbb{E}^{\mathscr{F}_s} \xi = \xi - \int_s^T Z_r \, dW_r,$$

for every $s \in [0, T]$. Now it suffices to define $Y_s = \mathbb{E}^{\mathscr{F}_s} \xi - \int_0^s f_r \, dr$ and Eq. (6.28) is satisfied. The existence of a continuous version is immediate, since (6.28) implies

$$Y_s - Y_0 = \int_0^s Z_r \, dW_r - \int_0^s f_r \, dr.$$

Estimates (6.29). Since $\eta \in L^2(\Omega; K)$ is $\mathscr{F}_T$-measurable, by Proposition 6.18 there exists an $L \in L^2_{\mathcal{P}}(\Omega \times [0, T]; \mathcal{L}_2(\Xi, K))$ such that

$$\mathbb{E}^{\mathscr{F}_s} \eta = \eta - \int_s^T L_\theta \, dW_\theta, \tag{6.32}$$

for every $s \in [0, T]$. Similarly, for a.a. r there exists a predictable process $\{K(\theta, r)\}_{\theta \in [0, r]}$ in $L^2_{\mathcal{P}}(\Omega \times [0, r]; \mathcal{L}_2(\Xi, K))$ such that

$$\mathbb{E}^{\mathscr{F}_s} f_r = f_r - \int_s^r K(\theta, r) \, dW_\theta, \tag{6.33}$$

for $s \in [0, r]$. We set $K(\theta, r) = 0$ for $\theta \in (r, T]$ and we can verify that the map $K : \Omega \times [0, T] \times [0, T] \to \mathcal{L}_2(\Xi, K)$ can be taken to be $\mathcal{P} \times \mathcal{B}([0, T])$-measurable, where $\mathcal{P}$ is the predictable σ-field on $\Omega \times [0, T]$ and $\mathcal{B}([0, T])$ denotes the Borel subsets of $[0, T]$; the existence of such a version of K can be proved by approximating f by simple processes and by a monotone class argument (or one can argue as in [350], proof of Lemma 2.1). Substituting into (6.31) and applying the stochastic Fubini theorem gives

$$Y_s = \eta - \int_s^T L_\theta \, dW_\theta + \int_s^T \left(f_r - \int_s^r K(\theta, r) \, dW_\theta \right) dr$$
$$= \eta + \int_s^T f_r \, dr - \int_s^T L_\theta \, dW_\theta - \int_s^T \left(\int_\theta^T K(\theta, r) \, dr \right) dW_\theta.$$

Comparing with the backward equation, we conclude by uniqueness that for a.a. θ,

$$Z_\theta = L_\theta + \int_\theta^T K(\theta, r) \, dr.$$

Now let $\beta \neq 0$.

From (6.32) we deduce that

$$\mathbb{E}\int_0^T e^{2\beta\theta}|L_\theta|^2\,d\theta \le e^{2\beta T}\mathbb{E}\left|\int_0^T L_\theta\,dW_\theta\right|^2 = e^{2\beta T}\mathbb{E}\left|\eta - \mathbb{E}^{\mathscr{F}_0}\eta\right|^2$$
$$\le 2e^{2\beta T}\mathbb{E}\,|\eta|^2 + 2e^{2\beta T}\mathbb{E}\,|\mathbb{E}^{\mathscr{F}_0}\eta|^2 \le 4e^{2\beta T}\mathbb{E}\,|\eta|^2.$$

Next note that

$$\left|\int_\theta^T K(\theta,r)\,dr\right|^2 \le \int_\theta^T e^{-2\beta r}\,dr\int_\theta^T e^{2\beta r}|K(\theta,r)|^2\,dr \le \frac{e^{-2\beta\theta}}{2\beta}\int_\theta^T e^{2\beta r}|K(\theta,r)|^2\,dr,$$

so that

$$\mathbb{E}\int_0^T e^{2\beta\theta}\left|\int_\theta^T K(\theta,r)\,dr\right|^2\,d\theta \le \frac{1}{2\beta}\mathbb{E}\int_0^T\int_\theta^T e^{2\beta r}|K(\theta,r)|^2\,dr\,d\theta$$
$$= \frac{1}{2\beta}\int_0^T e^{2\beta r}\mathbb{E}\int_0^r |K(\theta,r)|^2\,d\theta\,dr.$$

Since (6.33) yields

$$\mathbb{E}\int_0^r |K(\theta,r)|^2\,d\theta = \mathbb{E}\left|\int_0^r K(\theta,r)\,dW_\theta\right|^2 = \mathbb{E}\,|f_r - \mathbb{E}^{\mathscr{F}_0}f_r|^2$$
$$\le 2\mathbb{E}|f_r|^2 + 2\mathbb{E}\left|\mathbb{E}^{\mathscr{F}_s}f_r\right|^2 \le 4\mathbb{E}|f_r|^2,$$

the proof of the first inequality in (6.29) is finished. Now we prove the second one, estimating separately the two terms on the right-hand side of (6.31). By the Doob inequality for martingales,

$$\mathbb{E}\sup_{s\in[0,T]} e^{2\beta s}|\mathbb{E}^{\mathscr{F}_s}\eta|^2 \le e^{2\beta T}4\,\mathbb{E}\,|\eta|^2.$$

Next, since

$$\left(\int_s^T |f_r|\,dr\right)^2 \le \int_s^T e^{-2\beta r}\,dr\int_s^T e^{2\beta r}|f_r|^2\,dr \le \frac{e^{-2\beta s}}{2\beta}\int_s^T e^{2\beta r}|f_r|^2\,dr,$$

we obtain

$$e^{\beta s}\left|\int_s^T \mathbb{E}^{\mathscr{F}_s}f_r\,dr\right| \le \mathbb{E}^{\mathscr{F}_s}\left(e^{\beta s}\int_s^T |f_r|\,dr\right) \le \frac{1}{\sqrt{2\beta}}\mathbb{E}^{\mathscr{F}_s}\left(\int_s^T e^{2\beta r}|f_r|^2\,dr\right)^{1/2}$$

and by the Doob inequality,

$$\mathbb{E} \sup_{s \in [0,T]} e^{2\beta s} \left| \int_s^T \mathbb{E}^{\mathscr{F}_s} f_r \, dr \right|^2 \le \frac{4}{2\beta} \mathbb{E} \int_0^T e^{2\beta r} |f_r|^2 \, dr.$$

Estimates (6.30). Since, for $s \in [T - \delta, T]$,

$$\int_s^T |f_r| \, dr \le \left(\int_s^T |f_r|^2 \, dr \right)^{1/2} (T - s)^{1/2} \le \left(\int_s^T f_r \, dr \right)^{1/2} \delta^{1/2},$$

it follows from (6.31) that

$$\mathbb{E} \sup_{s \in [T-\delta,T]} |Y_s|^p \le c_p \mathbb{E} \sup_{s \in [T-\delta,T]} |\mathbb{E}^{\mathscr{F}_s} \eta|^p$$

$$+ c_p \delta^{p/2} \mathbb{E} \sup_{s \in [T-\delta,T]} \left| \mathbb{E}^{\mathscr{F}_s} \left(\int_s^T |f_r|^2 \, dr \right)^{1/2} \right|^p$$

$$\le c_p \mathbb{E} |\eta|^p + c_p \delta^{p/2} \mathbb{E} \left(\int_{T-\delta}^T |f_r|^2 \, dr \right)^{p/2},$$

which proves the desired inequality on the process Y. To obtain a similar estimate on Z we first set $Z_\theta^1 = \int_\theta^T K(\theta, r) \, dr$, so that $Z_\theta = L_\theta + Z_\theta^1$.

From (6.32) it follows that $\mathbb{E}^{\mathscr{F}_s} \eta - \mathbb{E}^{\mathscr{F}_{T-\delta}} \eta = \int_{T-\delta}^s L_\theta \, dW_\theta$, so by the Burkholder–Davis–Gundy and the Doob inequalities,

$$\mathbb{E} \left(\int_{T-\delta}^T |L_\theta|^2 d\theta \right)^{\frac{p}{2}} \le c_p \mathbb{E} \sup_{s \in [T-\delta,T]} \left| \int_{T-\delta}^s L_\theta \, dW_\theta \right|^p$$

$$= c_p \mathbb{E} \sup_{s \in [T-\delta,T]} |\mathbb{E}^{\mathscr{F}_s} \eta - \mathbb{E}^{\mathscr{F}_{T-\delta}} \eta|^p \le c_p \mathbb{E} |\eta|^p.$$

In order to prove a similar estimate for Z^1 we first note that, setting $Y_s^1 = \int_s^T \mathbb{E}^{\mathscr{F}_s} f_r \, dr$, the pair (Y^1, Z^1) is the solution corresponding to $\eta = 0$. Therefore

$$Y_s^1 - Y_{T-\delta}^1 = \int_{T-\delta}^s Z_r^1 \, dW_r - \int_{T-\delta}^s f_r \, dr.$$

So we obtain

$$\mathbb{E} \left(\int_{T-\delta}^T |Z_r^1|^2 dr \right)^{\frac{p}{2}} \le c_p \mathbb{E} \sup_{s \in [T-\delta,T]} \left| \int_{T-\delta}^s Z_r^1 \, dW_r \right|^p$$

$$\le c_p \mathbb{E} \sup_{s \in [T-\delta,T]} |Y_s^1|^p + c_p \mathbb{E} \left(\int_{T-\delta}^T |f_r| \, dr \right)^p.$$

For Y^1 we can use the estimate proved above with $\eta = 0$:

$$\mathbb{E} \sup_{s\in[T-\delta,T]} |\mathbb{E}^{\mathscr{F}_s} Y_s^1|^p \le c_p \delta^{p/2} \mathbb{E} \left(\int_{T-\delta}^T |f_r|^2 \, dr \right)^{p/2}.$$

Finally, the required estimate follows from

$$\int_{T-\delta}^T |f_r| \, dr \le \left(\int_{T-\delta}^T |f_r|^2 \, dr \right)^{1/2} \delta^{1/2}.$$

$\square$

Now we are concerned with the equation

$$Y_s + \int_s^T Z_r \, dW_r = \int_s^T f(r, Y_r, Z_r) \, dr + \eta. \tag{6.34}$$

In the following Proposition K is a Hilbert space, the mapping $f : \Omega \times [0, T] \times K \times \mathcal{L}_2(\Xi, K) \to K$ is assumed to be measurable with respect to $\mathcal{P} \times \mathcal{B}([0, T] \times K \times \mathcal{L}_2(\Xi, K))$ and $\mathcal{B}(K)$, respectively (we recall that by $\mathcal{P}$ we denote the predictable σ-field on $\Omega \times [0, T]$ and by $\mathcal{B}(\Lambda)$ the Borel σ-field of any topological space Λ). $\eta : \Omega \to K$ is assumed to be $\mathscr{F}_T$-measurable.

Proposition 6.20 *Assume that*

(i) there exists an $L > 0$ such that

$$|f(t, y_1, z_1) - f(t, y_2, z_2)| \le L(|y_1 - y_2| + |z_1 - z_2|),$$

$\mathbb{P}$-*a.s. for every* $t \in [0, T]$, $y_1, y_2 \in K$, $z_1, z_2 \in \mathcal{L}_2(\Xi, K)$;

(ii) $\mathbb{E} \int_0^T |f(r, 0, 0)|^2 dr < \infty, \qquad \mathbb{E} |\eta|^2 < \infty.$

Then there exists a unique pair of processes $Y(s), Z(s), s \in [0, T]$, such that

$$Y \in C_{\mathcal{P}}([0, T], L^2(\Omega; K)), \qquad Z \in L^2_{\mathcal{P}}(\Omega \times [0, T]; \mathcal{L}_2(\Xi, K))$$

and (6.34) holds for $s \in [0, T]$. Moreover, Y has a continuous version and $\mathbb{E} \sup_{s\in[0,T]} |Y(s)|^2 < \infty$.

If, in addition, there exists a $p \in [2, \infty)$ such that

$$\mathbb{E} \left(\int_0^T |f(r, 0, 0)|^2 dr \right)^{p/2} < \infty, \qquad \mathbb{E} |\eta|^p < \infty, \tag{6.35}$$

then we have $Y \in L^p_{\mathcal{P}}(\Omega; C([0, T], K))$, $Z \in L^p_{\mathcal{P}}(\Omega \times [0, T]; \mathcal{L}_2(\Xi, K))$ and

$$\mathbb{E} \sup_{s \in [0,T]} |Y(s)|^p + \mathbb{E} \left(\int_0^T |Z(r)|^2 dr \right)^{p/2} \leq c \, \mathbb{E} \left(\int_0^T |f(r,0,0)|^2 dr \right)^{p/2} + c \, \mathbb{E} \, |\eta|^p,$$

$$(6.36)$$

for some constant c > 0 depending only on p, L, T.

Finally assume that, for all λ *in a metric space* Λ, *a function* f_λ *is given satisfying (6.35) and assumption i) with L independent of* λ. *Also assume that, as* $\lambda \to \lambda_0$,

$$\mathbb{E} \left(\int_0^T |f_\lambda(r, Y, Z) - f_{\lambda_0}(r, Y, Z)|^2 dr \right)^{p/2} \to 0 \qquad (6.37)$$

for all $Y \in L_{\mathcal{P}}^p(\Omega; C([0,T], K))$, $Z \in L_{\mathcal{P}}^p(\Omega; L^2([0,T]; \mathcal{L}_2(\Xi, K)))$.

If we denote by $(Y(\lambda, \eta), Z(\lambda, \eta))$ *the solution to (6.34) corresponding to* $f = f_\lambda$ *and to the final data* $\eta \in L^p(\Omega; \mathbb{R})$ *then the map* $(\lambda, \eta) \to (Y(\lambda, \eta), Z(\lambda, \eta))$ *is continuous from* $\Lambda \times L^p(\Omega; \mathbb{R})$ *to* $L_{\mathcal{P}}^p(\Omega; C([0,T], K)) \times L_{\mathcal{P}}^p(\Omega; L^2([0,T]; \mathcal{L}_2(\Xi, K)))$.

Proof We let $\mathcal{K} = C_{\mathcal{P}}([0,T], L^2(\Omega; K)) \times L_{\mathcal{P}}^2(\Omega \times [0,T]; \mathcal{L}_2(\Xi, K))$ and we define a mapping $\Gamma : \mathcal{K} \to \mathcal{K}$ by setting $(Y, Z) = \Gamma(U, V)$ if (Y, Z) is the pair satisfying

$$Y_s + \int_s^T Z_r \, dW_r = \int_s^T f(r, U_r, V_r) \, dr + \eta, \qquad (6.38)$$

compare Lemma 6.19. The estimates (6.29) show that Γ is well defined, and it is a contraction if $\mathcal{K}$ is endowed with the norm

$$|(Y, Z)|_{\mathcal{K}}^2 = \mathbb{E} \int_0^T e^{2\beta r} \left(|Y_r|^2 + |Z_r|^2 \right) dr,$$

provided β is sufficiently large. For simplicity, we only verify the contraction property: if $(U^1, V^1) \in \mathcal{K}$, $(Y^1, Z^1) = \Gamma(U^1, V^1)$ and we let $\overline{Y} = Y - Y^1, \overline{Z} = Z - Z^1$, $\overline{U} = U - U^1, \overline{V} = V - V^1, \overline{f}_r = f(r, U_r, V_r) - f(r, U_r^1, V_r^1)$, we have

$$\overline{Y}_s + \int_s^T \overline{Z}_r \, dW_r = \int_s^T \overline{f}_r \, dW_r, \qquad (6.39)$$

so that by (6.29),

$$|(\overline{Y}, \overline{Z})|_{\mathcal{K}}^2 \leq T \, \mathbb{E} \sup_{s \in [0,T]} e^{2\beta s} |\overline{Y}_s|^2 + \mathbb{E} \int_0^T e^{2\beta r} |\overline{Z}_r|^2 \, dr \leq \frac{8(1+T)}{\beta} \mathbb{E} \int_0^T e^{2\beta r} |\overline{f}_r|^2 dr$$

$$\leq \frac{8(1+T)L^2}{\beta} \mathbb{E} \int_0^T e^{2\beta r} (|\overline{U}_r| + |\overline{V}_r|)^2 dr \leq \frac{16(1+T)L^2}{\beta} |(\overline{U}, \overline{V})|_{\mathcal{K}}^2.$$

Now we prove the estimate (6.36). We let $\mathcal{K}_{p,\delta} = L^p(\Omega; C([T - \delta, T], \mathbb{R})) \times L^p(\Omega; L^2([T - \delta, T]; \mathcal{L}_2(\Xi, \mathbb{R})))$ and define $\Gamma : \mathcal{K}_{p,\delta} \to \mathcal{K}_{p,\delta}$, setting $(Y, Z) = \Gamma(U, V)$ if (Y, Z) is the pair satisfying Eq. (6.38) for $s \in [T - \delta, T]$. It is easily verified that Γ is well defined and it is a contraction in $\mathcal{K}_{p,\delta}$, provided $\delta > 0$ is chosen sufficiently small; indeed, arguing as before, we deduce from (6.39) and from (6.30) the inequalities

$$
\begin{aligned}
|(\overline{Y}, \overline{Z})|_{\mathcal{K}}^p &= \mathbb{E} \sup_{s \in [T-\delta, T]} |\overline{Y}_s|^p + \mathbb{E} \left(\int_{T-\delta}^T |\overline{Z}_r|^2 \, dr \right)^{\frac{p}{2}} \\
&\leq c_p \delta^{p/2} L^p \, \mathbb{E} \left(\int_{T-\delta}^T (|\overline{U}_r| + |\overline{V}_r|)^2 dr \right)^{\frac{p}{2}} \\
&\leq c_p 2^{p/2} \delta^p L^p \delta \, \mathbb{E} \sup_{s \in [T-\delta, T]} |\overline{U}_s|^p + c_p (2\delta)^{p/2} L^p \, \mathbb{E} \left(\int_{T-\delta}^T |\overline{V}_r|^2 dr \right)^{\frac{p}{2}} \\
&\qquad\qquad\qquad\qquad \leq c_p (2\delta)^{p/2} L^p (1 + \delta^{p/2}) \, |(\overline{U}, \overline{V})|_{\mathcal{K}}^p,
\end{aligned}
$$

and the contraction property holds provided $c_p (2\delta)^{p/2} L^p (1 + \delta^{p/2}) < 1$. Repeating this argument on intervals $[T - \delta, T - 2\delta]$, $[T - 2\delta, T - 3\delta]$ etc. shows that $Y \in L^p(\Omega; C([0, T], \mathbb{R}))$ and $Z \in L^p(\Omega; L^2([0, T]; \mathcal{L}_2(\Xi, \mathbb{R})))$.

Next note that it follows from our assumptions that

$$
|f(r, x, y)| \leq |f(r, 0, 0)| + L(|x| + |y|).
$$

Applying the estimate (6.30) to Eq. (6.34) we obtain

$$
\begin{aligned}
\mathbb{E} \sup_{s \in [T-\delta, T]} |Y_s|^p &+ \mathbb{E} \left(\int_{T-\delta}^T |Z_r|^2 dr \right)^{p/2} \\
&\leq c_p \delta^{p/2} \mathbb{E} \left(\int_{T-\delta}^T |f(r, Y_r, Z_r)|^2 dr \right)^{p/2} + c_p \mathbb{E} |\eta|^p \\
&\leq c_p \mathbb{E} |\eta|^p + c_p 3^{p-1} \delta^{p/2} \mathbb{E} \left(\int_{T-\delta}^T |f(r, 0, 0)|^2 dr \right)^{p/2} \\
&+ c_p 3^{p-1} L^p \delta^{p/2} \mathbb{E} \left(\int_{T-\delta}^T |Y_r|^2 dr \right)^{p/2} + c_p 3^{p-1} L^p \delta^{p/2} \mathbb{E} \left(\int_{T-\delta}^T |Z_r|^2 dr \right)^{p/2} \\
&\leq c_p \mathbb{E} |\eta|^p + c_p 3^{p-1} \delta^{p/2} \mathbb{E} \left(\int_{T-\delta}^T |f(r, 0, 0)|^2 dr \right)^{p/2} \\
&+ c_p 3^{p-1} L^p \delta^p \mathbb{E} \sup_{s \in [T-\delta, T]} |Y_s|^p + c_p 3^{p-1} L^p \delta^{p/2} \mathbb{E} \left(\int_{T-\delta}^T |Z_r|^2 dr \right)^{p/2}.
\end{aligned}
$$

$$(6.40)$$

Choosing $\delta > 0$ so small that $\alpha := c_p 3^{p-1} L^p (\delta^p + \delta^{p/2}) < 1$ we obtain

$$\mathbb{E} \sup_{s \in [T-\delta,T]} |Y_s|^p + \mathbb{E} \left(\int_{T-\delta}^{T} |Z_r|^2 dr \right)^{p/2}$$

$$\leq c_p \mathbb{E} |\eta|^p + c_p 3^{p-1} \delta^{p/2} \mathbb{E} \left(\int_{T-\delta}^{T} |f(r,0,0)|^2 dr \right)^{p/2}$$

$$+ \alpha \left[\mathbb{E} \sup_{s \in [T-\delta,T]} |Y_s|^p + \mathbb{E} \left(\int_{T-\delta}^{T} |Z_r|^2 dr \right)^{p/2} \right],$$

$$(6.41)$$

and it follows that

$$\mathbb{E} \sup_{s \in [T-\delta,T]} |Y_s|^p + \mathbb{E} \left(\int_{T-\delta}^{T} |Z_r|^2 dr \right)^{p/2} \leq c \mathbb{E} |\eta|^p + c \mathbb{E} \left(\int_{T-\delta}^{T} |f(r,0,0)|^2 dr \right)^{p/2},$$

with c depending only on p and L. Next we note that for $s \leq T - \delta$,

$$Y_s + \int_{s}^{T-\delta} Z_r \, dW_r = \int_{s}^{T-\delta} f(r, Y_r, Z_r) \, dr + Y_{T-\delta},$$

and proceeding as before we obtain

$$\mathbb{E} \sup_{s \in [T-2\delta,T-\delta]} |Y_s|^p + \mathbb{E} \left(\int_{T-2\delta}^{T-\delta} |Z_r|^2 dr \right)^{p/2} \leq c \mathbb{E} |Y_{T-\delta}|^p + c \mathbb{E} \left(\int_{T-2\delta}^{T-\delta} |f(r,0,0)|^2 dr \right)^{p/2},$$

with the same choice of δ and the same value of c. After a finite number of steps we arrive at (6.36).

Finally, the proof of the last assertion can be done in a straightforward way, repeating the above argument. $\square$

Remark 6.21 The mapping Γ defined in the previous proof was shown to be a contraction in the space $\mathcal{K} = C_\mathcal{P}([0,T], L^2(\Omega; K)) \times L^2_\mathcal{P}(\Omega \times [0,T]; \mathcal{L}_2(\Xi, K))$. In a similar way, the estimates (6.29) allow us to show that Γ is well defined and it is a contraction in the space $L^2_\mathcal{P}(\Omega; C([0,T], K)) \times L^2_\mathcal{P}(\Omega \times [0,T]; \mathcal{L}_2(\Xi, K))$ as well as in the space $L^2_\mathcal{P}(\Omega \times [0,T]; K) \times L^2_\mathcal{P}(\Omega \times [0,T]; \mathcal{L}_2(\Xi, K))$. In particular, uniqueness holds for Eq. (6.34) in the latter space, too. ∎

6.3.2 Regular Dependence on Data

Now we are dealing with the backward equation

$$Y(s) + \int_{s}^{T} Z(r) \, dW(r) = \int_{s}^{T} F(r, X(r), Y(r), Z(r)) \, dr + \eta, \qquad (6.42)$$

on the time interval $[0, T]$, where η is a given $\mathscr{F}_T$-measurable real random variable and $X(s)$, $s \in [0, T]$, is a given predictable process. The mapping $F : [0, T] \times H \times K \times \mathcal{L}_2(\Xi, K) \to K$ is assumed to be Borel measurable. The solution we are looking for is a pair of predictable processes $Y(s)$, $Z(s)$, $s \in [0, T]$, with values in K and $\mathcal{L}_2(\Xi, K)$, respectively.

We fix the following assumptions on F.

Hypothesis 6.22 (i) There exists an $L > 0$ such that

$$|F(t, x, y_1, z_1) - F(t, x, y_2, z_2)| \le L(|y_1 - y_2| + |z_1 - z_2|),$$

for every $t \in [0, T]$, $x \in H$, $y_1, y_2 \in K$, $z_1, z_2 \in \mathcal{L}_2(\Xi, K)$.
(ii) For every $t \in [0, T]$, $F(t, \cdot, \cdot, \cdot) \in \mathcal{G}^{1,1,1}(H \times K \times \mathcal{L}_2(\Xi, K), K)$.
(iii) There exist $L > 0$ and $m \ge 0$ such that

$$|\nabla_x F(t, x, y, z)h| \le L|h|(1 + |z|)(1 + |x| + |y|)^m,$$

for every $t \in [0, T]$, $x, h \in H$, $y \in K$, $z \in \mathcal{L}_2(\Xi, K)$.
(iv) There exists an $L > 0$ such that $|F(t, 0, 0, 0)| \le L$ for every $t \in [0, T]$.

Conditions (i) and (ii) imply that the Gâteaux derivatives of F with respect to y and z are uniformly bounded: for every point (x, y, z) and all directions $k \in K$, $v \in \mathcal{L}_2(\Xi, K)$,

$$|\nabla_y F(t, x, y, z)k| \le L \, |k|, \quad |\nabla_z F(t, x, y, z)v| \le L \, |v|.$$

Moreover, conditions (i)–(iv) imply that

$$|F(t, x, y, z)| \le L(1 + |x|^{m+1} + |z| + |y|). \tag{6.43}$$

Finally, conditions (i) (ii) and (iii) imply

$$|F(t, x_1, y, z) - F(t, x_2, y, z)| \le L(1 + |z|)(1 + |x_1|^m + |x_2|^m + |y|^m)|x_2 - x_1|. \tag{6.44}$$

Remark 6.23 Instead of condition (iii), in some of the statements below we will assume that the stronger condition holds: there exists $L > 0$ such that

$$|\nabla_x F(t, x, y, z)h| \le L|h|, \quad t \in [0, T], \ x, h \in H, \ y \in K, \ z \in \mathcal{L}_2(\Xi, K). \tag{6.45}$$

Whenever (6.45) is assumed to hold, this will be explicitly mentioned.   ∎

To start we need the following general lemma that generalizes the classical result on continuity of evaluation operators, see e.g. [10].

Lemma 6.24 *Let K_1, K_2 and K_3 be Banach spaces and $\ell : [0, T] \times K_1 \times K_2 \to K_3$ be a measurable map such that, for all $t \in [0, T]$, $\ell(t, \cdot) : K_1 \times K_2 \to K_3$ is continuous*

(i) Suppose that for some $c > 0$ and $\mu \geq 1$,

$$|\ell(t, v_1, v_2)|_{K_3} \leq c(1 + |v_1|_{K_1}^{\mu})(1 + |v_2|_{K_2}), \qquad t \in [0, T], v_1 \in K_1, v_2 \in K_2.$$

For all $U \in L_{\mathcal{P}}^{r_1}(\Omega; C([0, T], K_1))$, $V \in L_{\mathcal{P}}^{r_2}(\Omega; L^2([0, T]; K_2))$ with $r_1, r_2 \geq 1$, let us define in the natural way the evaluation operator $\ell(U, V)(t, \omega) = \ell(t, U(t, \omega), V(t, \omega))$.
If $\mu/r_1 + 1/r_2 = 1/r_3$ and $r_1 \geq \mu$ then the evaluation operator is continuous from $L_{\mathcal{P}}^{r_1}(\Omega; C([0, T], K_1)) \times L_{\mathcal{P}}^{r_2}(\Omega; L^2([0, T]; K_2))$ to $L_{\mathcal{P}}^{r_3}(\Omega; L^2([0, T]; K_3))$.
(ii) Similarly, if

$$|\ell(t, v_1, v_2)|_{K_3} \leq c(1 + |v_1|_{K_1}^{\mu} + |v_2|_{K_2}), \qquad t \in [0, T], v_1 \in K_1, v_2 \in K_2,$$

and $r_2 = \mu r_1$ then the evaluation operator is continuous from $L_{\mathcal{P}}^{r_1}(\Omega; L^2([0, T]; K_2)) \times L_{\mathcal{P}}^{r_2}(\Omega; C([0, T], K_1))$ to $L_{\mathcal{P}}^{r_1}(\Omega; L^2([0, T]; K_3))$.

Proof We prove only (i), the proof of (ii) being identical.
Step 1. Firstly we consider only dependence on t. Define the evaluation operator (denoted again by ℓ by abuse of language): $\ell(\mathcal{U}, \mathcal{V})(t) = \ell(t, \mathcal{U}(t), \mathcal{V}(t))$ with $\mathcal{U} \in C([0, T], K_1)$, $\mathcal{V} \in L^2([0, T]; K_2)$. We claim that ℓ is continuous from $C([0, T], K_1) \times L^2([0, T]; K_2)$ to $L^2([0, T]; K_3)$. It is enough to prove that

$$\int_0^T |\ell(t, \mathcal{U}_n(t), \mathcal{V}_n(t)) - \ell(t, \mathcal{U}(t), \mathcal{V}(t))|^2 \, dt \to 0$$

for each pair of sequences $\mathcal{U}_n$, $\mathcal{V}_n$ with $\mathcal{U}_n \to \mathcal{U}$ in $C([0, T], K_1)$ and $\mathcal{V}_n \to \mathcal{V}$ in $L^2([0, T]; K_2)$. Extracting a subsequence, if necessary, we can always assume that $\sum_{n=1}^{\infty} |\mathcal{V}_n - \mathcal{V}|_{L^2([0,T];K_2)} < +\infty$ and $\mathcal{V}_n(t) \to \mathcal{V}(t)$ for a.a. $t \in [0, T]$. Let $\mathcal{V}^*(t) = \sum_{n=1}^{\infty} |\mathcal{V}_n(t) - \mathcal{V}(t)|_{K_2}$. By construction $\mathcal{V}^* \in L^2([0, T]; \mathbb{R})$ and $|\mathcal{V}_n(t)|_{K_2} \leq |\mathcal{V}(t)|_{K_2} + \mathcal{V}^*(t)$. Therefore

$$|\ell(t, \mathcal{U}_n(t), \mathcal{V}_n(t)) - \ell(t, \mathcal{U}(t), \mathcal{V}(t))|^2$$

$$\leq L \left(1 + \sup_n |\mathcal{U}_n|_{C([0,T],K_1)}^{\mu}\right)^2 \left(1 + |\mathcal{V}(t)|_{K_2} + \mathcal{V}^*(t)\right)^2,$$

for a suitable constant L. Since the right-hand term is a fixed summable function of $t \in [0, T]$ the claim follows from the dominated convergence theorem. Finally, we observe that

$$|\ell(\mathcal{U}, \mathcal{V})|_{L^2([0,T];K_3)} \leq L \left(1 + |\mathcal{U}|^{\mu}_{C([0,T],K_1)}\right)\left(1 + |\mathcal{V}|_{L^2([0,T];K_2)}\right)$$

for a suitable constant L.

Step 2. Now we consider dependence on ω. Let $\hat{\ell}$ be a continuous map $\hat{K}_1 \times \hat{K}_2 \to \hat{K}_3$, with $\hat{K}_i$ Banach spaces, $i = 1, 2, 3$, and $|\hat{\ell}(u, v)|_{\hat{K}_3} \leq L(1 + |u|^{\mu}_{\hat{K}_1})(1 + |v|_{\hat{K}_2})$. For $U \in L^{r_1}(\Omega; \hat{K}_1)$, $V \in L^{r_2}(\Omega; \hat{K}_2)$ with $\mu/r_1 + 1/r_2 = 1/r_3$, we define the evaluation operator $\hat{\ell}(U, V)(\omega) = \hat{\ell}(U(\omega), V(\omega))$ and claim that it is continuous from $L^{r_1}(\Omega; \hat{K}_1) \times L^{r_2}(\Omega; \hat{K}_2)$ to $L^{r_3}(\Omega; \hat{K}_3)$. Before proving the claim we notice that it completes the proof of Lemma 6.24: indeed, it suffices to apply it to $\hat{K}_1 = C([0, T], K_1)$, $\hat{K}_2 = L^2([0, T]; K_2)$, $\hat{K}_3 = L^2([0, T]; K_3)$ and to the evaluation operator introduced in Step 1.

The proof of the claim is similar to that of Step 1. It is enough to show that:

$$\mathbb{E}\left(\left|\hat{\ell}(U_n, V_n) - \hat{\ell}(U, V)\right|^{r_3}_{\hat{K}_3}\right) \to 0$$

for each pair of sequences U_n in $L^{r_1}(\Omega; \hat{K}_1)$ and V_n in $L^{r_2}(\Omega; \hat{K}_2)$ with $U_n \to U$ in $L^{r_1}(\Omega; \hat{K}_1)$ and $V_n \to V$ in $L^{r_2}(\Omega; \hat{K}_2)$. Extracting a subsequence, if necessary, we can assume that $U_n \to U$ and $V_n \to V$ $\mathbb{P}$-a.s., and

$$\sum_{n=1}^{\infty} |U_n - U|_{L^{r_1}(\Omega;\hat{K}_1)} < +\infty, \qquad \sum_{n=1}^{\infty} |V_n - V|_{L^{r_2}(\Omega;\hat{K}_2)} < +\infty.$$

Let:

$$U^* = \sum_{n=1}^{\infty} |U_n - U|_{\hat{K}_1}, \qquad V^* = \sum_{n=1}^{\infty} |V_n - V|_{\hat{K}_2}.$$

By construction $U^* \in L^{r_1}(\Omega; \mathbb{R})$ and $V^* \in L^{r_2}(\Omega; \mathbb{R})$. Moreover:

$$|U_n(\omega)|_{\hat{K}_1} \leq |U(\omega)|_{\hat{K}_1} + U^*(\omega), \qquad |V_n(\omega)|_{\hat{K}_2} \leq |V(\omega)|_{\hat{K}_2} + V^*(\omega), \qquad \mathbb{P}\text{-a.s.}$$

Therefore

$$\left|\hat{\ell}(U_n(\omega), V_n(\omega)) - \hat{\ell}(U(\omega), V(\omega))\right|^{r_3}_{\hat{K}_3} \leq L\left(1 + |U(\omega)|^{\mu r_3}_{\hat{K}_1} + (U^*(\omega))^{\mu r_3}\right)$$
$$\cdot \left(1 + |V(\omega)|^{r_3}_{\hat{K}_2} + (V^*(\omega))^{r_3}\right), \qquad \mathbb{P}\text{-a.s.,}$$

for a suitable constant L. Since $(\mu r_3)/r_1 + r_3/r_2 = 1$ the left-hand term has finite mean and the claim follows from the dominated convergence theorem. $\square$

We are now in a position to show the existence and uniqueness and regular dependence on data of the solution to Eq. (6.42). For $p \geq 2$ we define:

$$\mathcal{K}_p = L^p_{\mathcal{P}}(\Omega; C([0, T], K)) \times L^p_{\mathcal{P}}(\Omega; L^2([0, T]; \mathcal{L}_2(\Xi, K))),$$

endowed with the natural norm.

Proposition 6.25 *Assume Hypotheses 6.8 and 6.22.*

(i) *If $X \in L^\rho_{\mathcal{P}}(\Omega; C([0, T], H))$, $\eta \in L^r(\Omega; K)$ with $\rho = r(m + 1)$, $r \geq 2$ then there exists a unique solution in $\mathcal{K}_r$ of Eq. (6.42), which we will denote by $(Y(\cdot, X, \eta), Z(\cdot, X, \eta))$.*

(ii) *The following estimate holds:*

$$\mathbb{E} \sup_{s \in [0,T]} |Y(s, X, \eta)|^r + \left(\mathbb{E} \int_0^T |Z(s, X, \eta)|^2 ds \right)^{r/2}$$
$$\leq c \left(1 + |X|^\rho_{L^\rho_{\mathcal{P}}(\Omega; C([0,T], H))} \right) + c \mathbb{E} |\eta|^r$$

$$(6.46)$$

for a suitable constant c depending only on ρ, r and F.

(iii) *The map $(X, \eta) \to (Y(\cdot, X, \eta), Z(\cdot, X, \eta))$ is continuous from $L^\rho_{\mathcal{P}}(\Omega; C([0, T], H)) \times L^r(\Omega; K)$ to $\mathcal{K}_r$.*

(iv) *The map $(X, \eta) \to (Y(\cdot, X, \eta), Z(\cdot, X, \eta))$ is in $\mathcal{G}^{1,1}(L^\rho_{\mathcal{P}}(\Omega; C([0, T], H)) \times L^r(\Omega; \mathbb{R}), \mathcal{K}_p)$ with $r = (m + 2)p$, $p \geq 2$ (consequently $\rho = p(m + 1)(m + 2)$).*

Moreover, for all $X \in L^\rho_{\mathcal{P}}(\Omega; C([0, T], H))$, $\eta \in L^r(\Omega; K)$ the directional derivative in the direction (N, ζ) with $N \in L^\rho_{\mathcal{P}}(\Omega; C([0, T], H))$ and $\zeta \in L^r(\Omega; K)$, which we will denote by $(\nabla_{X,\eta} Y(\cdot, X, \eta)(N, \zeta), \nabla_{X,\eta} Z(\cdot, X, \eta)(N, \zeta))$, is the unique solution in $\mathcal{K}_p$ of:

$$\nabla_{X,\eta} Y(s, X, \eta)(N, \zeta) + \int_s^T \nabla_{X,\eta} Z(r, X, \eta)(N, \zeta) dW_r$$
$$= \int_s^T \nabla_x F(r, X_r, Y_r(X, \eta), Z_r(X, \eta)) N_r dr$$
$$+ \int_s^T \nabla_y F(r, X_r, Y_r(X, \eta), Z_r(X, \eta)) \nabla_{X,\eta} Y(r, X, \eta)(N, \zeta) dr$$
$$+ \int_s^T \nabla_z F(r, X_r, Y_r(X, \eta), Z_r(X, \eta)) \nabla_{X,\eta} Z(r, X, \eta)(N, \zeta) dr + \zeta.$$

(v) *Finally, the following estimate holds:*

$$\mathbb{E} \sup_{s \in [0,T]} |\nabla_{X,\eta} Y(s, X, \eta)(N, \zeta)|^p + \mathbb{E} \left(\int_0^T |\nabla_{X,\eta} Z(s, X, \eta)(N, \zeta)|^2 ds \right)^{p/2}$$
$$\leq c |N|^p_{L^\rho_{\mathcal{P}}(\Omega; C([0,T], H))} \left(1 + |X|^{(m+1)^2}_{L^\rho_{\mathcal{P}}(\Omega; C([0,T], H))} + |\eta|^{m+1}_{L^r(\Omega; K)} \right)^p + c |\zeta|^p_{L^p(\Omega; K)}.$$

$$(6.47)$$

(vi) If, in addition, there exists an $L > 0$ such that

$$|\nabla_x F(t, x, y, z)h| \leq L|h|, \qquad t \in [0, T], \ x, h \in H, \ y \in K, \ z \in \mathcal{L}_2(\Xi, K),$$

then the following estimate (stronger than (6.47)) holds:

$$\mathbb{E} \sup_{s \in [0,T]} |\nabla_{X,\eta} Y(s, X, \eta)(N, \zeta)|^p + \mathbb{E} \left(\int_0^T |\nabla_{X,\eta} Z(s, X, \eta)(N, \zeta)|^2 ds \right)^{p/2}$$
$$\leq c|N|^p_{L^p_{\mathcal{P}}(\Omega; C([0,T], H))} + c|\zeta|^p_{L^p(\Omega; K)}.$$

$$(6.48)$$

Proof Let $\Lambda = L^p_{\mathcal{P}}(\Omega; C([0, T], H))$ and, for every $X \in \Lambda$,

$$f_X(s, y, z) = F(s, X_s, y, z).$$

By (6.43) and Lemma 6.24-(ii) applied with $K_1 = H$, $K_2 = K \times \mathcal{L}_2(\Xi, K)$, $U = X$, $V = (Y, Z)$ we obtain that for all $(Y, Z) \in \mathcal{K}_r$ the map $X \to f_X(Y, Z)$ is continuous from Λ to $L^r_{\mathcal{P}}(\Omega; L^2([0, T]; K))$ and

$$\mathbb{E} \left(\int_0^T |f_X(s, 0, 0)|^2 ds \right)^{r/2} \leq c \left(1 + \mathbb{E} (\sup_{s \in [0,T]} |X_s|^{r(m+1)}) \right).$$

Therefore points (i)–(iii) of the claim follow immediately from Proposition 6.20.

To deal with point (iv) it is convenient now to introduce another backward stochastic equation; we will eventually show that it is satisfied by the derivatives of (Y, Z) with respect to X and η. For all $\zeta \in L^p(\Omega; K)$, $X, N \in L^r_{\mathcal{P}}(\Omega; C([0, T], H))$, $(Y, Z) \in \mathcal{K}_r$ we look for $(\widehat{Y}(X, N, Y, Z, \zeta), \widehat{Z}(X, N, Y, Z, \zeta)) \in \mathcal{K}_p$ solving:

$$\widehat{Y}_s + \int_s^T \widehat{Z}_r dW_r = \int_s^T \nabla_x F(r, X_r, Y_r, Z_r) N_r dr$$
$$\int_s^T \nabla_y F(r, X_r, Y_r, Z_r) \widehat{Y}_r dr + \int_s^T \nabla_z F(r, X_r, Y_r, Z_r) \widehat{Z}_r dr + \zeta.$$

$$(6.49)$$

By Hypothesis 6.22-(iii) we have

$$\mathbb{E} \left(\int_0^T |\nabla_x F(r, X_r, Y_r, Z_r) N_r|^2 dr \right)^{p/2}$$
$$\leq L|N|^p_{L^r_{\mathcal{P}}(\Omega; C([0,T], H))} \left(1 + |Z|_{L^r_{\mathcal{P}}(\Omega; L^2([0,T]; \mathcal{L}_2(\Xi, K)))} \right)^p$$
$$\cdot \left(1 + |X|^m_{L^r_{\mathcal{P}}(\Omega; C([0,T], H))} + |Y|^m_{L^r_{\mathcal{P}}(\Omega; C([0,T], H))} \right)^p$$

for a suitable constant L. Since $\nabla_y F$ and $\nabla_z F$ are bounded, by Proposition 6.20 the Eq. (6.49) admits a unique solution in $\mathcal{K}_p$. Moreover, by Lemma 6.24-(i), the map $(X, N, Y, Z) \to \nabla_x F(\cdot, X_{(\cdot)}, Y_{(\cdot)}, Z_{(\cdot)}) N_{(\cdot)}$ is continuous from the space

$$K^{\#} := L^r_{\mathcal{P}}(\Omega; C([0, T], H)) \times L^r_{\mathcal{P}}(\Omega; C([0, T], H)) \times \mathcal{K}_r$$

to $L^p_{\mathcal{P}}(\Omega; L^2([0, T]; K))$. Therefore, taking into account once more the boundedness of $\nabla_y F$ and $\nabla_z F$, we can apply the final statement of Proposition 6.20 with $\Lambda = K^{\#}$ and conclude that the map $(X, N, Y, Z, \zeta) \to (\widehat{Y}(X, N, Y, Z, \zeta), \widehat{Z}(X, N, Y, Z, \zeta))$ is continuous from $K^{\#} \times L^p(\Omega; K)$ to $\mathcal{K}_p$ and the estimate

$$\mathbb{E}(\sup_{s \in [0,T]} |\widehat{Y}_s|^p) + \mathbb{E}\left(\int_0^T |\widehat{Z}_r|^2 dr \right)^{p/2}$$
$$\leq c |N|^p_{L^r_{\mathcal{P}}(\Omega; C([0,T],H))} \left(1 + |Z|_{L^r_{\mathcal{P}}(\Omega; L^2([0,T]; \mathcal{L}_2(\Xi, K)))} \right)^p$$
$$\cdot \left(1 + |X|^m_{L^r_{\mathcal{P}}(\Omega; C([0,T],H))} + |Y|^m_{L^r_{\mathcal{P}}(\Omega; C([0,T],H))} \right)^p + c \mathbb{E}|\zeta|^p$$

$$(6.50)$$

holds for some constant $c > 0$.

It remains to prove that if $X, N \in L^p_{\mathcal{P}}(\Omega; C([0, T], H))$ and $\eta, \zeta \in L^r(\Omega; K)$ then the directional derivative of $(Y(X, \eta), Z(X, \eta))$ in the direction (N, ζ) is given by

$$(\widehat{Y}(X, N, Y(X, \eta), Z(X, \eta), \zeta), \widehat{Z}(X, N, Y(X, \eta), Z(X, \eta), \zeta)).$$

Let us define

$$\overline{Y}^\varepsilon := \frac{1}{\varepsilon} [Y(X + \varepsilon N, \eta + \varepsilon \zeta) - Y(X, \eta)] - \widehat{Y}(X, N, Y(X, \eta), Z(X, \eta), \zeta),$$

$$\overline{Z}^\varepsilon := \frac{1}{\varepsilon} [Z(X + \varepsilon N, \eta + \varepsilon \zeta) - Z(X, \eta)] - \widehat{Z}(X, N, Y(X, \eta), Z(X, \eta), \zeta).$$

For $\varepsilon \to 0$ we show that $\overline{Y}^\varepsilon \to 0$ in $L^p_{\mathcal{P}}(\Omega; C([0, T], K))$ and $\overline{Z}^\varepsilon \to 0$ in $L^p_{\mathcal{P}}(\Omega; L^2([0, T]; \mathcal{L}_2(\Xi, K)))$. For short we let $Y = Y(X, \eta)$, $Z = Z(X, \eta)$, $Y^\varepsilon = Y(X + \varepsilon N, \eta + \varepsilon \zeta)$, $Z^\varepsilon = Z(X + \varepsilon N, \eta + \varepsilon \zeta)$, $\widehat{Y} = \widehat{Y}(X, N, Y(X, \eta), Z(X, \eta), \zeta)$, and $\widehat{Z} = \widehat{Z}(X, N, Y(X, \eta), Z(X, \eta), \zeta)$.

The proof will be done by induction, dividing the interval $[0, T]$ into subintervals $[T - \delta, T], [T - 2\delta, T - \delta]$ and so on, for a suitable δ depending only on F and p. All the subintervals are treated in the same way (the proof for $[T - \delta, T]$ being even easier), so we concentrate on the second one, namely $[T - 2\delta, T - \delta]$. On such an interval we have:

$$\overline{Y}^\varepsilon_s + \int_s^{T-\delta} \overline{Z}^\varepsilon_r \, dr = \int_s^{T-\delta} \nu^\varepsilon(r) dr + \overline{Y}^\varepsilon_{T-\delta},$$

where $\nu^\varepsilon = \nu_1^\varepsilon + \nu_2^\varepsilon$ and:

$$\nu_1^\varepsilon(r) = \frac{1}{\varepsilon}\left[F(r, X_r + \varepsilon N_r, Y_r^\varepsilon, Z_r^\varepsilon) - F(r, X_r, Y_r^\varepsilon, Z_r^\varepsilon)\right] - \nabla_x F(r, X_r, Y_r, Z_r)N_r,$$

$$\nu_2^\varepsilon(r) = \frac{1}{\varepsilon}\left[F(r, X_r, Y_r^\varepsilon, Z_r^\varepsilon) - F(r, X_r, Y_r, Z_r)\right]$$
$$- \nabla_y F(r, X_r, Y_r, Z_r)\widehat{Y}_r - \nabla_z F(r, X_r, Y_r, Z_r)\widehat{Z}_r.$$

By Proposition 6.20 we have:

$$\mathbb{E}\sup_{s\in[T-2\delta, T-\delta]}|\overline{Y}_s^\varepsilon|^p + \mathbb{E}\left(\int_{T-2\delta}^{T-\delta}|\overline{Z}_r^\varepsilon|^2 dr\right)^{p/2}$$
$$\leq c_p\delta^{p/2}\sum_{i=1}^{2}\mathbb{E}\left(\int_{T-2\delta}^{T-\delta}|\nu_i^\varepsilon(r)|^2 dr\right)^{p/2} + c_p\mathbb{E}|\overline{Y}_{T-\delta}^\varepsilon|^p$$

and by the inductive assumption $\mathbb{E}|\overline{Y}_{T-\delta}^\varepsilon|^p \to 0$.

We start to evaluate the integral terms on the right. We can write

$$\nu_1^\varepsilon(r) = \int_0^1 \nabla_x F(r, X_r + \varepsilon\tau N_r, Y_r^\varepsilon, Z_r^\varepsilon)N_r d\tau - \int_0^1 \nabla_x F(r, X_r, Y_r, Z_r)N_r d\tau.$$

For all $x, g, n \in H$, $y \in K$, $z \in \mathcal{L}_2(\Xi, K)$ let $\chi(x, g, n, y, z) = \int_0^1 \nabla_x F(x + \tau g, y, z)n d\tau$, so that $\nu_1^\varepsilon(r) = \chi(X_r, \varepsilon N_r, N_r, Y_r^\varepsilon, Z_r^\varepsilon) - \chi(X_r, 0, N_r, Y_r, Z_r)$. Moreover, $|\chi(x, g, n, y, z)| \leq L|n|(1 + |z|)(1 + |x|^m + |g|^m + |y|^m)$ and χ is a continuous map. Applying Lemma 6.24-(i) with $K_1 = H^{\times 3} \times K$ $K_2 = \mathcal{L}_2(\Xi, K)$, $r_1 = r_2 = r$, $\mu = m + 1$ and taking into account that $(X, \varepsilon N, N, Y^\varepsilon) \to (X, 0, N, Y)$ in $L_\mathcal{P}^r(\Omega, C([T - 2\delta, T - \delta], K_1))$ and $Z^\varepsilon \to Z$ in $L_\mathcal{P}^r(\Omega, L^2([T - 2\delta, T - \delta], K_2))$ we immediately obtain $\mathbb{E}\left(\int_{T-2\delta}^{T-\delta}|\nu_1^\varepsilon(r)|^2 dr\right)^{p/2} \to 0$.

Dealing now with ν_2^ε we can rewrite $\nu_2^\varepsilon = \nu_{2.1}^\varepsilon + \nu_{2.2}^\varepsilon$ where:

$$\nu_{2.1}^\varepsilon(r) = \int_0^1\left(\nabla_y F(r, X_r, Y_r + \tau(Y_r^\varepsilon - Y_r), Z_r + \tau(Z_r^\varepsilon - Z_r))\widehat{Y}_r\right.$$
$$\left. - \nabla_y F(r, X_r, Y_r, Z_r)\widehat{Y}_r\right)d\tau$$
$$+ \int_0^1\left(\nabla_z F(r, X_r, Y_r + \tau(Y_r^\varepsilon - Y_r), Z_r + \tau(Z_r^\varepsilon - Z_r))\widehat{Z}_r\right.$$
$$\left. - \nabla_z F(r, X_r, Y_r, Z_r)\widehat{Z}_r\right)d\tau,$$

$$\nu_{2.2}^\varepsilon(r) = \int_0^1 \nabla_y F(r, X_r, Y_r + \tau(Y_r^\varepsilon - Y_r), Z_r + \tau(Z_r^\varepsilon - Z_r))\overline{Y}_r^\varepsilon d\tau$$
$$+ \int_0^1 \nabla_z F(r, X_r, Y_r + \tau(Y_r^\varepsilon - Y_r), Z_r + \tau(Z_r^\varepsilon - Z_r))\overline{Z}_r^\varepsilon d\tau.$$

Since $\nabla_y F$ and $\nabla_z F$ are bounded, by the dominated convergence theorem we immediately obtain $\mathbb{E}\left(\int_{T-2\delta}^{T-\delta} |\nu_{2.1}^\varepsilon(r)|^2 dr\right)^{p/2} \to 0$. Moreover,

$$\mathbb{E}\left(\int_{T-2\delta}^{T-\delta} |\nu_{2.2}^\varepsilon(r)|^2 dr\right)^{p/2} \le \overline{c}\left(\mathbb{E}\sup_{\tau\in[T-2\delta,T-\delta]} |\overline{Y}_\tau^\varepsilon|^p + \mathbb{E}\left(\int_{T-2\delta}^{T-\delta} |\overline{Z}_r^\varepsilon|^2 dr\right)^{p/2}\right)$$

for a suitable constant $\overline{c}$ depending only on $F, p\, T$. Choosing δ such that $c_p\overline{c}\delta^{p/2} < 1$ the claim follows immediately.

Finally, (6.47) follows plugging (6.46) into (6.50), and (6.48) is proved in the same way, taking into account the additional assumption. $\qquad\square$

6.3.3 Forward–Backward Systems

In this subsection we consider the system of stochastic differential equations

$$\begin{cases} X(s) = e^{(s-t)A}x + \int_t^s e^{(s-r)A}b(r, X(r))\, dr + \int_t^s e^{(s-r)A}\sigma(r, X(r))\, dW(r), \\ Y(s) + \int_s^T Z(r)dW(r) = \int_s^T F(r, X(r), Y(r), Z(r))dr + g(X(T)), \end{cases}$$
$$(6.51)$$

for s varying on the time interval $[t, T] \subset [0, T]$. As in Sect. 6.2 we extend the domain of the solution setting $X(s) = x$ for $s \in [0, t)$. We assume that $F : [0, T] \times H \times \mathbb{R} \times \mathcal{L}_2(\Xi, \mathbb{R}) \to \mathbb{R}$ satisfies Hypothesis 6.22 with $K = \mathbb{R}$. On the function $g : H \to \mathbb{R}$ we make the following assumptions:

Hypothesis 6.26 (i) $g \in \mathcal{G}^1(H, \mathbb{R})$;
(ii) There exist $L > 0$ and $m \ge 0$ such that, for every $x, h \in H$,

$$|\nabla g(x)h| \le L\,|h|\,(1 + |x|)^m.$$

For simplicity, and without any real loss of generality, we suppose that m is the same as in Hypothesis 6.22. Notice that Hypothesis 6.26 implies that

$$|g(x)| \le c(1 + |x|^{m+1}).$$

In some of the statements below we will assume the stronger condition: $|\nabla g(x)h| \le L|h|$, for every $x, h \in H$. Whenever this is the case, this requirement will be explicitly mentioned.

We note that the system (6.51) is decoupled, i.e., the first equation does not contain the solution (Y, Z) of the second one. Therefore, under the assumptions of Hypotheses 6.8, 6.22 and 6.26 by Propositions 9.9 and 6.25 there exists a unique solution to (6.51). We remark that the process X is $\mathscr{F}_{[t,T]}$-measurable, so that Y_t is measurable both with respect to $\mathscr{F}_{[t,T]}$ and $\mathscr{F}_t$; it follows that Y_t is indeed deterministic (see also [207]).

We denote the solution by $(X(s; t, x), Y(s; t, x), Z(s; t, x))$, $s \in [t, T]$, in order to stress dependence on the parameters $t \in [0, T]$ and $x \in H$.

For later use we notice two useful identities: for $t \le r \le T$ the equality: $\mathbb{P}$-a.s.,

$$X(s; r, X(r; t, x)) = X(s; t, x), \qquad s \in [r, T], \tag{6.52}$$

is a consequence of the uniqueness of the solution to (6.6). Since the solution to the backward equation is uniquely determined on an interval $[r, T]$ by the values of the process X on the same interval, for $t \le r \le T$ we have, $\mathbb{P}$-a.s.,

$$\begin{aligned}
Y(s; r, X(r; t, x)) &= Y(s; t, x), \quad \text{for } s \in [r, T], \\
Z(s; r, X(r; t, x)) &= Z(s; t, x) \quad \text{for a.a. } s \in [r, T].
\end{aligned} \tag{6.53}$$

Next we proceed to investigate regularity properties of the dependence on t and x. To this end we first notice that with the notation of Propositions 6.10 and 6.25:

$$Y(s; t, x) = Y(s; X(\cdot; t, x), g(X(T; t, x))),$$

$$Z(s; t, x) = Z(s; X(\cdot; t, x), g(X(T; t, x))).$$

Moreover, as a consequence of Hypothesis 6.26, it can be easily proved that the map $\eta \to g(\eta)$ belongs to the space $\mathcal{G}^1(L^p(\Omega; H), L^q(\Omega; \mathbb{R}))$, for every $p \in [2, \infty)$ and for all q sufficiently large (depending on p and m). The following Proposition is then an immediate consequence of Propositions 6.9, 6.10 and 6.25, and the chain rule for the class $\mathcal{G}$, stated in Lemma 6.3.

Proposition 6.27 *Assume Hypotheses 6.8, 6.22 and 6.26. Recall the notation:*

$$\mathcal{K}_p = L_{\mathcal{P}}^p(\Omega; C([0, T], \mathbb{R})) \times L_{\mathcal{P}}^p(\Omega; L^2([0, T]; \mathcal{L}_2(\Xi, \mathbb{R}))).$$

Then the map $(t, x) \to (Y(\cdot, t, x), Z(\cdot, t, x))$ belongs to $\mathcal{G}^{0,1}([0, T] \times H, \mathcal{K}_p)$ for all $p \in [2, \infty)$.

Denoting by $\nabla_x Y$, $\nabla_x Z$ the partial Gâteaux derivatives with respect to x, the directional derivative process in the direction $h \in H$, $\{(\nabla_x Y(s; t, x)h, \nabla_x Z(s; t, x)h)\}_{s \in [0, T]}$, solves the equation: $\mathbb{P}$-a.s.,

$$\nabla_x Y(s; t, x)h + \int_s^T \nabla_x Z(r; t, x)h \, dW_r$$

$$= \int_s^T \nabla_x F(r, X(r; t, x), Y(r; t, x), Z(r; t, x))\nabla_x X(r; t, x)h \, dr$$

$$\int_s^T \nabla_y F(r, X(r; t, x), Y(r; t, x), Z(r; t, x))\nabla_x Y(r; t, x)h \, dr \qquad (6.54)$$

$$\int_s^T \nabla_z F(r, X(r; t, x), Y(r; t, x), Z(r; t, x))\nabla_x Z(r; t, x)h \, dr$$

$$+\nabla g(X(T; t, x))\nabla_x X(T; t, x)h, \qquad s \in [0, T].$$

Finally, the following estimate holds:

$$\left[\mathbb{E} \sup_{s \in [0,T]} |\nabla_x Y(s; t, x)h|^p \right]^{\frac{1}{p}} + \left[\mathbb{E} \left(\int_0^T |\nabla_x Z(r; t, x)h|^2 dr \right)^{\frac{p}{2}} \right]^{\frac{1}{p}} \leq c|h|(1 + |x|^{(m+1)^2}).$$

$$(6.55)$$

If, in addition, there exists an $L > 0$ such that

$$|\nabla_x F(t, x, y, z)h| \leq L|h|, \qquad |\nabla g(x)h| \leq L|h|,$$

for every $t \in [0, T]$, $x, h \in H$, $y \in \mathbb{R}$, $z \in \mathcal{L}_2(\Xi, \mathbb{R})$, then the following stronger estimate holds:

$$\left[\mathbb{E} \sup_{s \in [0,T]} |\nabla_x Y(s; t, x)h|^p \right]^{1/p} + \left[\mathbb{E} \left(\int_0^T |\nabla_x Z(r; t, x)h|^2 dr \right)^{p/2} \right]^{1/p} \leq c|h|.$$

$$(6.56)$$

Proof We have already commented on the first two statements. The estimate (6.55) follows from (6.47) applied with

$$X = X(\cdot; t, x), \ N = \nabla_x X(\cdot; t, x)h, \ \eta = g(X(T; t, x)), \ \zeta = \nabla g(X(T; t, x))\nabla_x X(T; t, x)h,$$

taking into account that by Propositions 6.9 and 6.10 we have

$$|N|_{L_{\mathcal{P}}^p(\Omega; C([0,T],H))} \leq c|h|, \ |X|_{L_{\mathcal{P}}^p(\Omega; C([0,T],H))} \leq c(1 + |x|),$$

and, by Hypothesis 6.26, we also obtain $|\eta|_{L^r(\Omega)} \leq c(1 + |x|)^{m+1}$, $|\zeta|_{L^p(\Omega)} \leq c|h|$ $(1 + |x|)^m$ for a suitable constant c.

The estimate (6.56) is proved in a similar way, applying (6.48) instead of (6.47) and taking into account that under the additional assumption we have $|\eta|_{L^r(\Omega)} \leq c(1 + |x|)$, $|\zeta|_{L^p(\Omega)} \leq c|h|$ for a suitable constant c. $\qquad \square$

Proposition 6.28 *Assume Hypotheses 6.8, 6.22 and 6.26. Then the function $u(t, x) = Y(t, t, x)$ has the following properties:*

(i) $u \in \mathcal{G}^{0,1}([0, T] \times H, \mathbb{R})$;
(ii) *there exists a $C > 0$ such that $|\nabla_x u(t, x)h| \leq C|h|(1 + |x|^{(m+1)^2})$ for all $t \in$*
    $[0, T], x \in H, h \in H$;
(iii) *if, in addition,*

$$\sup_{t \in [0,T], x \in H} |F(t, x, 0, 0)| < \infty, \qquad \sup_{x \in H} |g(x)| < \infty,$$

then $\sup_{t \in [0,T], x \in H} |u(t, x)| < \infty$;
(iv) *similarly, if there exists an $L > 0$ such that*

$$|\nabla_x F(t, x, y, z)h| \leq L|h|, \qquad |\nabla g(x)h| \leq L|h|,$$

for every $t \in [0, T], x, h \in H, y \in \mathbb{R}, z \in \mathcal{L}_2(\Xi, \mathbb{R})$, then

$$|\nabla_x u(t, x)h| \leq c|h|$$

for a suitable constant c and all $x, h \in H$.

Proof (i) Since $Y(t; t, x)$ is deterministic, we have $u(t, x) = \mathbb{E}\, Y(t; t, x)$. So the map $(t, x) \to u(t, x)$ can be written as a composition, letting $u(t, x) = \Gamma_3(\Gamma_2(t, \Gamma_1(t, x)))$ with:

$$\Gamma_1 : [0, T] \times H \to L_{\mathcal{P}}^p(\Omega; C([0, T], \mathbb{R})), \qquad \Gamma_1(t, x) = Y(\cdot; t, x),$$

$$\Gamma_2 : [0, T] \times L_{\mathcal{P}}^p(\Omega; C([0, T], \mathbb{R})) \to L^p(\Omega; \mathbb{R}), \qquad \Gamma_2(t, U) = U(t),$$

$$\Gamma_3 : L^p(\Omega; \mathbb{R}) \to \mathbb{R}, \qquad \Gamma_3\zeta = \mathbb{E}\zeta.$$

By Proposition 6.27, $\Gamma_1 \in \mathcal{G}^{0,1}$. The inequality

$$|U(t) - V(s)|_{L^p(\Omega;\mathbb{R})} \leq |U(t) - U(s)|_{L^p(\Omega;\mathbb{R})} + |U - V|_{L_{\mathcal{P}}^p(\Omega;C([0,T],\mathbb{R}))}$$

shows that Γ_2 is continuous; moreover Γ_2 is clearly linear in the second variable. Finally, Γ_3 is a bounded linear operator. Then the assertion follows from the chain rule.

(ii) is an immediate consequence of the estimate in Proposition 6.27-(iii): indeed,

$$|u(t, x)|^2 = |Y(t; t, x)|^2 = \mathbb{E}\,|Y(t; t, x)|^2 \leq \sup_{s \in [t,T]} \mathbb{E}\,|Y(s; t, x)|^2.$$

(iii) Since (Y, Z) is a solution to the backward equation, the estimate in Proposition 6.20 yields

$$\sup_{s \in [t,T]} \mathbb{E}\,|Y(s; t, x)|^2 \leq c\,\mathbb{E}\int_0^T |F(r, X(r; t, x), 0, 0)|^2 dr + c\,\mathbb{E}\,|g(X(T; t, x))|^2 \leq c.$$

(iv) follows immediately from (6.56).   □

Corollary 6.29 *For every $t \in [0, T]$, $x \in H$ we have*

$$Y(s; t, x) = u(s, X(s; t, x)), \quad \text{for } s \in [u, T], \tag{6.57}$$
$$Z(s; t, x) = \nabla_x u(s, X(s; t, x))\, \sigma(s, X(s; t, x)), \quad \text{for a.a. } s \in [u, T]. \tag{6.58}$$

Proof Setting $s = r$ in the first equality of (6.53) we obtain (6.57).

To prove (6.58) we first write the backward equation in system (6.51) as

$$Y_s = Y_t + \int_t^s Z_r \, dW_r - \int_t^s F(r, X_r, Y_r, Z_r) \, dr, \quad s \in [t, T]$$

and by (6.57) this can be written

$$u(s, X(s; t, x)) = u(t, x) + \int_t^s Z_r \, dW_r - \int_t^s F(r, X_r, Y_r, Z_r) \, dr, \quad s \in [t, T]. \tag{6.59}$$

Now we fix an arbitrary $\xi \in \Xi$ and take the joint quadratic variation of both sides of (6.59) with the Wiener process $W\xi$. The joint quadratic variation of the left-hand side is

$$\int_t^s \nabla_x u(r, X(r; t, x))\sigma(r, X(r; t, x))\xi \, dr, \quad s \in [t, T], \tag{6.60}$$

by Proposition 6.17. Since the ordinary integral in (6.59) is a finite variation process, the joint quadratic variation of $W\xi$ and the right-hand side of (6.59) is

$$\int_t^s Z_r \xi \, dr, \quad s \in [t, T]. \tag{6.61}$$

Equating (6.60) and (6.61) we obtain (6.58).   □

6.4   BSDEs and Mild Solutions to HJB

We denote by $\mathcal{B}_p(H)$ the set of measurable functions $\phi : H \to \mathbb{R}$ with polynomial growth, i.e., such that $\sup_{x \in H} |\phi(x)|(1 + |x|^a)^{-1} < \infty$ for some $a > 0$.

Let $X(s; t, x)$, $s \in [t, T]$, denote the solution to the stochastic equation

$$X(s) = e^{(s-t)A}x + \int_t^s e^{(s-r)A}b(r, X(r)) \, dr + \int_t^s e^{(s-r)A}\sigma(r, X(r)) \, dW(r),$$

where A, b, σ, satisfy the assumptions in Hypothesis 6.8. The transition semigroup $P_{t,s}$ is defined for arbitrary $\phi \in \mathcal{B}_p(H)$ and for $0 \le t \le s \le T$ by the formula

$$P_{t,s}[\phi](x) = \mathbb{E}\,\phi(X(s;t,x)), \qquad x \in H.$$

The estimate $\mathbb{E}\sup_{s\in[t,T]} |X(s;t,x)|^p \le C(1+|x|)^p$, see (6.8), shows that $P_{t,s}$ is well defined as a linear operator $\mathcal{B}_p(H) \to \mathcal{B}_p(H)$; the semigroup property $P_{t,u}P_{u,s} = P_{t,s}$, $t \le u \le s$, is well known.

Let us denote by $\mathcal{A}(t)$ the (formal) generator of $P_{t,s}$:

$$\mathcal{A}(t)[\phi](x) = \frac{1}{2}\mathrm{Tr}\left(\sigma(t,x)\sigma(t,x)^* D^2\phi(x)\right) + \langle Ax + b(t,x), D\phi(x)\rangle,$$

where $D\phi$ and $D^2\phi$ are first and second Gâteaux derivatives of ϕ (here identified with elements of H and $\mathcal{L}(H)$, respectively). This definition is formal, since the domain of $\mathcal{A}(t)$ is not specified; however, if $g : H \to \mathbb{R}$ is a sufficiently regular function, the function $v(t,x) = P_{t,T}[g](x)$ is a classical solution to the backward Kolmogorov equation:

$$\begin{cases} \dfrac{\partial v(t,x)}{\partial t} + \mathcal{A}(t)[v(t,\cdot)](x) = 0, & t \in [0,T],\ x \in H, \\ v(T,x) = g(x). \end{cases}$$

We refer to [179, 180, 582] for a detailed exposition. When g is not regular, the function $v(t,x) = P_{t,T}[g](x)$ can be considered as a generalized solution to the backward Kolmogorov equation.

Here we are interested in a generalization of this equation, written formally as

$$\begin{cases} \dfrac{\partial u(t,x)}{\partial t} + \mathcal{A}(t)[u(t,\cdot)](x) + F(t,x,u(t,x),\nabla_x u(t,x)\sigma(t,x)) = 0, & t \in [0,T],\ x \in H, \\ u(T,x) = g(x). \end{cases}$$

$$(6.62)$$

We will refer to this equation as the nonlinear Kolmogorov equation. In the sequel we will be mostly concerned with the case when F is a Hamiltonian function related to an optimal control problem and in this case Eq. (6.62) is the Hamilton–Jacobi–Bellman equation for the corresponding value function. However, the results given in this section are more general, they do not rely on a control-theoretic interpretation and may be of independent interest.

In (6.62) $F : [0,T] \times H \times \mathbb{R} \times \Xi^* \to \mathbb{R}$ is a given function satisfying Hypothesis 6.22. Note that $\nabla_x u(t,x)$, the Gâteaux derivative of $u(t,x)$ with respect to x, is an element of H^*, so that the composition $\nabla_x u(t,x)\sigma(t,x)$ belongs to $\Xi^* = \mathcal{L}(\Xi,\mathbb{R}) = \mathcal{L}_2(\Xi,\mathbb{R})$. Thus, we are in the framework of Hypothesis 6.22 with $K = \mathbb{R}$.

Remark 6.30 A different formulation of Eq. (6.62) is possible, which differs only notationally. We could start with a real-valued function F defined on $[0,T] \times H \times \mathbb{R} \times \Xi$ and write the first equality in (6.62) as

$$\frac{\partial u(t, x)}{\partial t} + \mathcal{A}(t)[u(t, \cdot)](x) + F(t, x, u(t, x), \sigma(t, x)^* D_x u(t, x)) = 0,$$

where $\sigma(t, x)^* \in \mathcal{L}(H, \Xi)$ denotes the Hilbert space adjoint of $\sigma(t, x) \in \mathcal{L}(\Xi, H)$. We recall that D_x denotes the Gâteaux derivative identified with an element of H, so that $\nabla_x u(t, x)h = \langle D_x u(t, x), h \rangle_H$ for every $h \in H$. Of course, identifying Ξ with Ξ^* by the Riesz isometry, one checks immediately the equivalence of the two formulations.   ∎

Now we define the notion of solution to the nonlinear Kolmogorov equation. We consider the variation of constants formula for (6.62):

$$u(t, x) = \int_t^T P_{t,s}[F(s, \cdot, u(s, \cdot), \nabla_x u(s, \cdot)\sigma(s, \cdot))](x)\, ds + P_{t,T}[g](x), \quad (6.63)$$

for $t \in [0, T]$ and $x \in H$, and we see that formula (6.63) is meaningful, provided $F(t, \cdot, \cdot, \cdot)$, $u(t, \cdot)$ and $\nabla_x u(t, \cdot)$ have polynomial growth (and, of course, provided they satisfy appropriate measurability assumptions). We use this formula as a definition for the solution to (6.62):

Definition 6.31 We say that a function $u : [0, T] \times H \to \mathbb{R}$ is a mild solution to the nonlinear Kolmogorov equation (6.62) if the following conditions hold:

(i) $u \in \mathcal{G}^{0,1}([0, T] \times H, \mathbb{R})$;
(ii) there exist $C > 0$ and $d \in \mathbb{N}$ such that $|\nabla_x u(t, x)h| \leq C|h|(1 + |x|^d)$ for all $t \in [0, T]$, $x \in H$, $h \in H$;
(iii) equality (6.63) holds.

Note that the specific form of the operator $\mathcal{A}(t)$ plays no role in this definition. We are now ready to state the main result of this section.

Theorem 6.32 *Assume that Hypothesis 6.8 holds, and let F, g be functions satisfying the assumptions in Hypotheses 6.22 (with $K = \mathbb{R}$) and 6.26. Then there exists a unique mild solution to the nonlinear Kolmogorov equation (6.62).*

The solution u is given by the formula

$$u(t, x) = Y(t; t, x),$$

where (X, Y, Z) is the solution to the forward–backward system (6.51).

If, in addition, $\sup_{t \in [0,T], x \in H} |F(t, x, 0, 0)| < \infty$ and g is bounded then u is also bounded.

Similarly, if $|\nabla_x F|$ is uniformly bounded then $|\nabla_x u|$ is also uniformly bounded.

Proof Existence. By Proposition 6.28, the proposed solution u has the regularity properties stated in Definition 6.31 and the last two statements of the claim hold. It remains to verify that equality (6.63) holds. To this purpose we first fix $t \in [0, T]$ and $x \in H$ and write the backward equation of system (6.51) for $s = t$:

$$Y(t; t, x) + \int_t^T Z(s; t, x) \, dW_s$$
$$= \int_t^T F\left(s, X(s; t, x), Y(s; t, x), Z(s; t, x)\right) ds + g(X(T; t, x)).$$

Taking the expectation we obtain

$$u(t, x) = \mathbb{E} \int_t^T F\left(s, X(s; t, x), Y(s; t, x), Z(s; t, x)\right) ds + P_{t,T}[g](x).$$

By (6.57), (6.58) we have

$$u(t, x) = \mathbb{E} \int_t^T F\left(s, X(s; t, x), u(s, X(s; t, x)), \nabla_x u(s, X(s; t, x)) \, \sigma(s, X(s; t, x))\right) ds$$
$$+ P_{t,T}[g](x)$$

and equality (6.63) follows.

Uniqueness. Let u be a mild solution. We look for a convenient expression for the process $u(r, X(r; t, x)), r \in [t, T]$. By (6.63) and the definition of $P_{r,s}$, for every $r \in [t, T]$ and $x \in H$,

$$u(r, x) = \mathbb{E}[g(X(T; r, x))]$$
$$+ \mathbb{E}\left[\int_r^T F\left(s, X(s; r, x), u(s, X(s; r, x)), \nabla_x u(s, X(s; r, x)) \, \sigma(s, X(s; r, x))\right) ds\right].$$

Since $X(s; r, x)$ is $\mathscr{F}_r$-independent, we can replace the expectation by the conditional expectation given $\mathscr{F}_r$:

$$u(r, x) = \mathbb{E}^{\mathscr{F}_r}[g(X(T; r, x))]$$
$$+ \mathbb{E}^{\mathscr{F}_r}\left[\int_r^T F\left(s, X(s; r, x), u(s, X(s; r, x)), \nabla_x u(s, X(s; r, x)) \, \sigma(s, X(s; r, x))\right) ds\right].$$

For the same reason, we can replace x by $X(r; t, x)$ and use the equality: $\mathbb{P}$-a.s.

$$X(s; r, X(r; t, x)) = X(s; t, x), \qquad \text{for } s \in [r, T].$$

We arrive at

$$u(r, X(r; t, x)) = \mathbb{E}^{\mathscr{F}_r}[g(X(T; t, x))]$$
$$+ \mathbb{E}^{\mathscr{F}_r}\left[\int_r^T F\left(s, X(s; t, x), u(s, X(s; t, x)), \nabla_x u(s, X(s; t, x)) \sigma(s, X(s; t, x))\right) ds\right]$$
$$= \mathbb{E}^{\mathscr{F}_r}[\xi]$$
$$- \int_t^r F\left(s, X(s; t, x), u(s, X(s; t, x)), \nabla_x u(s, X(s; t, x)) \, \sigma(s, X(s; t, x))\right) ds,$$

where we have defined

$$\xi = g(X(T; t, x))$$
$$+ \int_t^T F\Big(s, X(s; t, x), u(s, X(s; t, x)) \nabla_x u(s, X(s; t, x)) \sigma(s, X(s; t, x))\Big) \, ds.$$

We note that $\mathbb{E}^{\mathscr{F}_t}[\xi] = u(t, x)$. Since $\xi \in L^2(\Omega; \mathbb{R})$ is $\mathscr{F}_T$-measurable, by the representation theorem recalled in Proposition 6.18, there exists a $\widetilde{Z} \in L^2_{\mathcal{P}}(\Omega \times [t, T]; \mathcal{L}_2(\Xi, \mathbb{R}))$ such that $\mathbb{E}^{\mathscr{F}_r}[\xi] = \int_t^r \widetilde{Z}_s \, dW_s + u(t, x)$. We conclude that the process $u(r, X(r; t, x)), r \in [t, T]$, is a (real) continuous semimartingale with canonical decomposition

$$u(r, X(r; t, x)) = \int_t^r \widetilde{Z}_s \, dW_s$$
$$+ u(t, x) - \int_t^r F\Big(s, X(s; t, x), u(s, X(s; t, x)), \nabla_x u(s, X(s; t, x)) \sigma(s, X(s; t, x))\Big) \, ds$$
$$(6.64)$$

into its continuous martingale part and continuous finite variation part. Let $\xi \in \Xi$. By Proposition 6.17, the joint quadratic variation process of $u(r, X(r; t, x))$ and $W(r)\xi$, $r \in [t, T]$, is

$$\int_t^r \nabla_x u(s, X(s; t, x)) \sigma(s, X(s; t, x)) \xi \, ds, \quad r \in [t, T]. \tag{6.65}$$

Taking into account the canonical decomposition (6.64), we note that the process (6.65) can also be obtained as the joint quadratic variation process between $W(r)\xi$, $r \in [t, T]$, and the process $\int_t^r \widetilde{Z}_s \, dW_s$. This yields the identity

$$\int_t^r \nabla_x u(s, X(s; t, x)) \sigma(s, X(s; t, x)) \xi \, ds = \int_t^s \widetilde{Z}_s \xi \, ds. \quad r \in [t, T].$$

Therefore, for a.a. $s \in [t, T]$, we have $\mathbb{P}$-a.s.

$$\nabla_x u(s, X(s; t, x)) \sigma(s, X(s; t, x)) = \widetilde{Z}_s.$$

Substituting into (6.64) we obtain

$$u(r, X(r; t, x)) = \int_t^r \nabla_x u(s, X(s; t, x)) \sigma(s, X(s; t, x)) \, dW_s + u(t, x)$$
$$+ \int_t^r F\Big(s, X(s; t, x), u(s, X(s; t, x)), \nabla_x u(s, X(s; t, x)) \sigma(s, X(s; t, x))\Big) \, ds,$$

for $r \in [t, T]$. Since $u(T, X(T; t, x)) = g(X(T; t, x))$, we also have

$$u(r, X(r; t, x)) + \int_r^T \nabla_x u(s, X(s; t, x)) \, \sigma(s, X(s; t, x)) \, dW_s = g(X(T; t, x))$$

$$+ \int_r^T F\Big(s, X(s; t, x), u(s, X(s; t, x)), \nabla_x u(s, X(s; t, x)) \sigma(s, X(s; t, x))\Big) \, ds,$$

for $r \in [t, T]$. Comparing with the backward equation in (6.51) we note that the pairs

$$\Big(Y(r; t, x), Z(r; t, x)\Big) \quad \text{and} \quad \Big(u(r, X(r; t, x)), \nabla_x u(r, X(r; t, x)) \sigma(r, X(r; t, x))\Big),$$

for $r \in [t, T]$, solve the same equation. By uniqueness, we have in particular $Y(r; t, x) = u(r, X(r; t, x))$, $r \in [t, T]$. Setting $r = t$ we obtain $Y(t; t, x) = u(t, x)$. □

6.5 Applications to Optimal Control Problems

We wish to apply the above results to perform the synthesis of the optimal control for a general nonlinear control system. We will see that this approach allows great generality, particularly with respect to degeneracy of the noise. To be able to use non-smooth feedbacks we settle the problem in the framework of optimal control problems formulated in the extended weak formulation, but we will present results on the extended strong formulation as well.

Let again H, Ξ, denote real separable Hilbert spaces (the state space and the noise space, respectively) and let Λ be a Polish space (the control space). For $t \in [0, T]$ a *generalized reference probability space* is given by $\mu = (\Omega, \mathscr{F}, \mathscr{F}_s^t, \mathbb{P}, W)$, where

- $(\Omega, \mathscr{F}, \mathbb{P})$ is a complete probability space;
- $\{\mathscr{F}_s^t\}_{s \geq t}$ is a filtration in it, satisfying the usual conditions;
- $(W(s))_{s \geq t}$ is a cylindrical $\mathbb{P}$-Wiener process in Ξ, with respect to the filtration $\mathscr{F}_s^t$, starting from $W(t) = 0$.

Given such μ, for every starting point $x \in H$ we will consider the following controlled state equation

$$\begin{cases} dX(s) = \Big(AX(s) + b(s, X(s)) + \sigma(s, X(s))R(s, X(s), a(s))\Big) \, ds \\ \qquad\qquad\qquad\qquad\qquad\qquad + \sigma(s, X(s)) \, dW(s), \ s \in [t, T], \\ X(t) = x \in H. \end{cases}$$

$$(6.66)$$

In (6.66), and below in this section, the equation is understood in the mild sense. $a(\cdot) : \Omega \times [t, T] \to \Lambda$ is the control process, which is always assumed to be progressively measurable with respect to $\{\mathscr{F}_s^t\}_{s \geq t}$. On the coefficients A, b, σ, R precise assumptions will be formulated in Hypothesis 6.33 below. In particular, to allow more generality, on the coefficient R we will only impose measurability and boundedness assumptions, so that, in particular, we cannot guarantee the existence or uniqueness of the solution to the state equation for an arbitrary control process $a(\cdot)$. Therefore the formulations of the control problems require some slight changes with respect to the previous sections and is given as follows (the word *extended* is used to distinguish such formulations, see Remark 2.6). We call $(a(\cdot), X(\cdot))$ an *admissible control pair* if $a(\cdot)$ is an $\mathscr{F}_s^t$-progressively measurable process with values in Λ and $X(\cdot)$ is a mild solution to (6.66) corresponding to $a(\cdot)$. To every admissible control pair we associate the cost:

$$J^\mu(t, x; a(\cdot), X(\cdot)) = \mathbb{E} \int_t^T l(s, X(s), a(s)) \, ds + \mathbb{E}\, g(X(T)),$$

where l, g are suitable real functions. The optimal control problem in the extended strong formulation consists in minimizing the functional $J^\mu(t, x; a(\cdot), X(\cdot))$ over all admissible control pairs $(a(\cdot), X(\cdot))$, and characterizing the value function

$$V_t^\mu(x) = \inf_{(a(\cdot), X(\cdot))} J^\mu(t, x; a(\cdot), X(\cdot)).$$

We will also address the optimal control problem in the extended weak formulation, which consists in further minimizing with respect to all generalized reference probability spaces, i.e., in characterizing the value function

$$\overline{V}(t, x) = \inf_\mu V_t^\mu(x).$$

Notice the occurrence of the operator σ in the control term of (6.66): this special structure of the state equation is imposed by our techniques and seems to be essential in different contexts as well (see [298]). The corresponding Hamiltonian function is defined for all $t \in [0, T]$, $x \in H$, $z \in \Xi^*$ setting

$$F_0(t, x, z) = \inf_{a \in \Lambda} \{ l(t, x, a) + z\, R(t, x, a) \}. \tag{6.67}$$

Note that this differs from the Hamiltonian as introduced in the previous chapters. In particular, the third argument z ranges over Ξ^* instead of H.

We make the following assumptions:

Hypothesis 6.33 The following holds:

(1) A, b and σ satisfy Hypothesis 6.8.
(2) $R : [0, T] \times H \times \Lambda \to \Xi$ is Borel measurable and $|R(t, x, a)|_\Xi \leq L$ for a suitable constant $L > 0$ and all $t \in [0, T]$, $x \in H$, $a \in \Lambda$.
(3) $l : [0, T] \times H \times \Lambda \to \mathbb{R}$ is continuous and $|l(t, x, a)| \leq L(1 + |x|^m)$ for suitable constants $L > 0$, $m \geq 0$ and all $t \in [0, T]$, $x \in H$, $a \in \Lambda$.
(4) g satisfies Hypothesis 6.26.
(5) Taking $K = \mathbb{R}$ (and noting that $\mathcal{L}_2(\Xi, \mathbb{R}) = \Xi^*$) the function $F_0 : [0, T] \times H \times \Xi^* \to \mathbb{R}$ satisfies Hypothesis 6.22.

(6) For all $t \in [0, T]$, $x \in H$ and $z \in \Xi^*$ we denote by $\Gamma(t, x, z) \subset \Lambda$ the set of elements $a \in \Lambda$ such that the infimum in (6.67) is attained and we assume that $\Gamma(t, x, z)$ is non-empty. We will denote by γ a measurable selection of Γ, i.e., a measurable function $\gamma : [0, T] \times H \times \Xi^* \to \Lambda$ such that $\gamma(t, x, z) \in \Gamma(t, x, z)$ for every $t \in [0, T]$, $x \in H$ and $z \in \Xi^*$. γ is not always assumed to exist.

The Hamilton–Jacobi–Bellman equation relative to the above stated problem is written formally:

$$\begin{cases} \dfrac{\partial v(t, x)}{\partial t} + \mathcal{A}(t)[v(t, \cdot)](x) + F_0(t, x, \nabla_x v(t, x) \sigma(t, x)) = 0, & t \in [0, T], \ x \in H, \\ v(T, x) = g(x). \end{cases}$$

(6.68)

Notice the special form of this equation where the nonlinear term depends on $\nabla_x v$ only via the composition $\nabla_x v \, \sigma$: this is consistent with the definition of F_0 given above.

The Hamilton–Jacobi–Bellman equation takes the form of a nonlinear Kolmogorov equation as considered in the previous sections. In particular, under our assumptions, it admits a unique mild solution in the sense specified by Theorem 6.32.

In the proof of our main results, Theorems 6.35 and 6.36 below, we will make use of a classical tool in stochastic analysis, namely the Girsanov Theorem. We recall its statement, in a form suitable for our purposes. Its infinite-dimensional version, which we are about to state, can be found, for example, in [180].

Theorem 6.34 Let $\mu = (\Omega, \mathcal{F}, \mathcal{F}_s^t, \mathbb{P}, W)$ be a generalized reference probability space, let $R(r)$, $r \in [t, T]$, be an $\mathcal{F}_s^t$-progressively measurable process with values in Ξ such that $\int_t^T |R(r)|_\Xi^2 dr < \infty$ $\mathbb{P}$-a.s., and define

$$\rho^t(s) = \exp\left(-\int_t^s \langle R(r), dW(r) \rangle_\Xi - \frac{1}{2} \int_t^s |R(r)|_\Xi^2 \, dr \right), \qquad s \in [t, T].$$

Then the following holds:

(1) $\rho^t(\cdot)$ is a $\mathbb{P}$-supermartingale;
(2) if

$$\mathbb{E}[\rho^t(T)] = 1 \tag{6.69}$$

then $\rho^t(\cdot)$ is a $\mathbb{P}$-martingale and we can define a probability $\widetilde{\mathbb{P}}$ setting $\widetilde{\mathbb{P}}(A) = \mathbb{E}[1_A \rho^t(T)]$, $A \in \mathcal{F}$;
(3) the process $\widetilde{W}$ defined by

$$\widetilde{W}(s) = W(s) - W(t) + \int_t^s R(r) \, dr, \qquad s \in [t, T], \tag{6.70}$$

is a cylindrical Wiener process in Ξ with respect to $\mathcal{F}_s^t$ and $\widetilde{\mathbb{P}}$;

(4) finally, if R is bounded in Ξ then (6.69) holds and for every $p \in [1, \infty)$ we have

$$\mathbb{E}\left[(\rho^t(T))^p\right] < \infty, \qquad \widetilde{\mathbb{E}}\left[(\rho^t(T))^{-p}\right] < \infty, \qquad (6.71)$$

where $\widetilde{\mathbb{E}}$ denotes expectation with respect to $\widetilde{\mathbb{P}}$.

Note that (6.70) does not make sense as it is written since W, being a cylindrical Wiener process, is not a genuine stochastic process taking values in Ξ. (6.70) should be understood as the equality $\widetilde{W}(s)h = W(s)h - W(t)h + \int_t^s \langle R(r), h \rangle_\Xi \, dr$ for any $h \in \Xi$. Nevertheless, in the following we will use a shortened notation as in (6.70).

We are in a position to prove the main results of this section:

Theorem 6.35 *Assume Hypothesis 6.33 and let $t \in [0, T]$, $x \in H$.*

(1) For all generalized reference probability spaces μ and all admissible control pairs (a, X) we have $J^\mu(t, x; a(\cdot), X(\cdot)) \geq v(t, x)$.
  It follows that $V_t^\mu(x) \geq v(t, x)$ for every μ, and so $\overline{V}(t, x) \geq v(t, x)$.

(2) For all μ and all admissible control pairs (a, X), the equality $J^\mu(t, x; a(\cdot), X(\cdot)) = v(t, x)$ holds if and only if the following feedback law is satisfied:

$$a(s) \in \Gamma(s, X(s), \nabla_x v(s, X(s)) \sigma(s, X(s))), \quad \mathbb{P}\text{-a.s. for a.a. } s \in [t, T]. \tag{6.72}$$

Therefore, (6.72) implies the optimality of an admissible control pair in the extended strong formulation with respect to a given generalized reference probability space μ. If such a control pair exists then $V_t^\mu(x) = v(t, x)$.

Proof For all $\mu = (\Omega, \mathcal{F}, \mathcal{F}_s^t, \mathbb{P}, W)$ and admissible control pairs $(a(\cdot), X(\cdot))$, using the boundedness of R, the Girsanov theorem ensures that there exists a probability measure $\widetilde{\mathbb{P}}$ on Ω such that

$$\widetilde{W}_s := W_s - W_t + \int_t^s R(r, X_r, a(r)) \, dr, \qquad s \in [t, T],$$

is a $\widetilde{\mathbb{P}}$-Wiener process (note that $\widetilde{\mathbb{P}}$ and $\widetilde{W}$ depend on (a, X), but we neglect this dependence in the notation). Equation (6.66) can be rewritten as:

$$\begin{cases} dX_s = AX_s \, ds + b(s, X_s) \, ds + \sigma(s, X_s) \, d\widetilde{W}_s, & s \in [t, T], \\ X_t = x \in H, \end{cases} \tag{6.73}$$

which, as usual, is to be understood in the mild sense. The process X turns out to be adapted to the filtration, denoted $(\widetilde{\mathcal{F}}_s^t)_{s \in [t, T]}$, generated by $\widetilde{W}$ and completed in the usual way by means of null sets. In the filtered probability space $(\Omega, \mathcal{F}, \widetilde{\mathcal{F}}_s^t, \widetilde{\mathbb{P}})$ we can consider the system of forward–backward equations on $[t, T]$:

$$
\begin{cases}
\widetilde{X}(s;t,x) = e^{(s-t)A}x + \int_t^s e^{(s-r)A} b(r,\widetilde{X}(r;t,x))\,dr + \int_t^s e^{(s-r)A}\sigma(r,\widetilde{X}(r;t,x))\,d\widetilde{W}_r, \\
\widetilde{Y}(s;t,x) + \int_s^T \widetilde{Z}(r;t,x)\,d\widetilde{W}_r = \int_s^T F_0(r,\widetilde{X}(r;t,x),\widetilde{Z}(r;t,x))\,dr + g(\widetilde{X}(T;t,x)).
\end{cases}
$$

$$(6.74)$$

We notice that $\widetilde{X}(s;t,x) = X_s$. Writing the backward equation in (6.74) for $s = t$ and with respect to the original process W we get:

$$
\widetilde{Y}(t;t,x) + \int_t^T \widetilde{Z}(r;t,x)\,dW_r
$$
$$
= \int_t^T \left[F_0(r,X_r,\widetilde{Z}(r;t,x)) - \widetilde{Z}(r;t,x)R(r,X_r,a(r)) \right] dr + g(X_T).
$$

$$(6.75)$$

We note that

$$
\mathbb{E}\left[\left(\int_t^T |\widetilde{Z}(r;t,x)|^2\,dr \right)^{1/2} \right] = \widetilde{\mathbb{E}}\left[(\rho^t(T))^{-1} \left(\int_t^T |\widetilde{Z}(r;t,x)|^2\,dr \right)^{1/2} \right]
$$
$$
\leq \left(\widetilde{\mathbb{E}}[(\rho^t(T))^{-2}] \right)^{1/2} \left(\widetilde{\mathbb{E}} \int_t^T |\widetilde{Z}(r;t,x)|^2\,dr \right)^{1/2} < \infty
$$

by (6.71). Therefore, by the Burkholder–Davis–Gundy inequalities, the stochastic integral $\int_t^T \widetilde{Z}(r;t,x)\,dW_r$ has finite $\mathbb{P}$-expectation, equal to zero. Now we recall the equalities (6.57) and (6.58) which imply in the present notation that $\widetilde{Y}(t;t,x) = v(t,x)$ and

$$
\widetilde{Z}(s;t,x) = \nabla_x v(s,\widetilde{X}(s;t,x))\,\sigma(s,\widetilde{X}(s;t,x)) = \nabla_x v(s,X_s)\,\sigma(s,X_s).
$$

Taking expectation with respect to the original probability $\mathbb{P}$ in (6.75) we obtain:

$$
\mathbb{E}\,g(X_T) - v(t,x) = -\mathbb{E}\int_t^T F_0(r,X_r,\nabla_x v(r,X_r)\sigma(r,X_r))\,dr
$$
$$
+ \mathbb{E}\int_t^T \nabla_x v(r,X_r)\sigma(r,X_r)R(r,X_r,a(r))\,dr.
$$

Adding and subtracting $\mathbb{E}\int_t^T l(r,X_r,a(r))\,dr$ we conclude that:

$$
J^\mu(t,x;a(\cdot),X(\cdot)) = v(t,x) + \mathbb{E}\int_t^T \Big[-F_0(r,X_r,\nabla_x v(r,X_r)\sigma(r,X_r))
$$
$$
+\nabla_x v(r,X_r)\sigma(r,X_r)R(r,X_r,a(r)) + l(r,X_r,a(r)) \Big]\,dr.
$$

$$(6.76)$$

The above equality is known as the *fundamental identity*. By the definition of F_0 and Γ it implies immediately that $v(t,x) \leq J^\mu(t,x;a(\cdot),X(\cdot))$ and that equality holds if and only if (6.72) holds. This proves all the conclusions of the theorem. $\qquad\square$

Theorem 6.36 *Assume Hypothesis 6.33, assume in addition that Γ admits a measurable selection γ, and let $t \in [0,T]$, $x \in H$. Then there exists at least one generalized reference probability space $\overline{\mu} = (\overline{\Omega}, \overline{\mathscr{F}}, \overline{\mathscr{F}}_s^t, \overline{\mathbb{P}}, \overline{W})$ and an admissible control pair $(\overline{a}(\cdot), \overline{X}(\cdot))$ for which the analogue of (6.72) holds. In particular, it follows that $V_t^{\overline{\mu}}(x) = v(t,x)$ and so $\overline{V}(t,x) = v(t,x)$. In the space $\overline{\mu}$ the process $\overline{X}$ is a mild solution to the closed loop equation:*

$$\begin{cases} d\overline{X}(s) = A\overline{X}(s)\,ds + \sigma(s,\overline{X}(s))\,R\Big(s,\overline{X}(s),\,\gamma(s,\overline{X}(s),\nabla_x v(s,\overline{X}(s))\sigma(s,\overline{X}(s)))\Big)\,ds \\ \qquad\qquad + b(s,\overline{X}(s))\,ds + \sigma(s,\overline{X}(s))\,d\overline{W}(s), \quad s \in [t,T], \\ \overline{X}(t) = x \in H, \end{cases}$$

$$(6.77)$$

the feedback law takes the form

$$\overline{a}(s) = \gamma(s,\overline{X}(s),\nabla_x v(s,\overline{X}(s))\sigma(s,\overline{X}(s))), \qquad \overline{\mathbb{P}}\text{-a.s. for a.a. } s \in [t,T],$$

and the pair $(\overline{a}(\cdot), \overline{X}(\cdot))$ is optimal for the control problem in the extended weak formulation.

Proof We start by showing the existence of a extended weak solution to Eq. (6.77), again by an application of the Girsanov theorem. We take an arbitrary generalized reference probability space $(\overline{\Omega}, \overline{\mathscr{F}}, \overline{\mathscr{F}}_s, \mathbb{P}, W)$ and denote by $\overline{X}$ the mild solution on $[t,T]$ of the (uncontrolled) equation

$$\begin{cases} d\overline{X}_s = A\overline{X}_s\,dt + b(s,\overline{X}_s)\,ds + \sigma(s,\overline{X}_s)\,dW_s, \\ \overline{X}_t = x. \end{cases}$$

Recalling the boundedness assumption on R, we see that the Girsanov Theorem provides a probability $\overline{\mathbb{P}}$ on $\overline{\Omega}$ under which the process

$$\overline{W}_s := -\int_t^s R(r,\overline{X}_r,\gamma(r,\overline{X}_r,\nabla_x v(r,\overline{X}_r)\sigma(r,\overline{X}_r)))\,dr + W_s - W_t, \quad s \in [t,T],$$

is a Wiener process. Then $\overline{X}$ is the mild solution to Eq. (6.77) relative to the generalized reference probability space $\overline{\mu} := (\overline{\Omega}, \overline{\mathscr{F}}, \overline{\mathscr{F}}_s^t, \overline{\mathbb{P}}, \overline{W})$. Setting $\overline{a}(s) := \gamma(s,\overline{X}_s,\nabla_x v(s,\overline{X}_s)\sigma(s,\overline{X}_s))$, the feedback inclusion (6.72) holds by definition of γ and all the required conclusions follow from Theorem 6.35. $\qquad\square$

Remark 6.37 Slight changes in the arguments of Theorem 6.35 allow us to prove an existence result for the control problem in the extended strong formulation, under additional assumptions. More precisely, assume Hypothesis 6.33 and, in addition, that the following holds:

(i) $|\nabla_x F_0(t, x, z)h| \le L|h|$ for a suitable constant L and all $t \in [0, T]$, $x, h \in H$ and $z \in \Xi^*$.

(ii) $\sup_{t \in [0,T], x \in H} |\sigma(t, x)|_{\mathcal{L}(\Xi, H)} < \infty$.

(iii) Γ admits a measurable selection γ; in addition the functions $R(t, \cdot, a) : H \to \Xi$, $\gamma(t, \cdot, \cdot) : H \times \Xi^* \to \Lambda$ and $\nabla_x v(t, \cdot) : H \to H$ are globally Lipschitz, uniformly with respect to $t \in [0, T]$, $a \in \Lambda$ (Lipschitzianity of γ is understood with respect to the metric defined in Λ).

Notice that, by the last statement in Theorem 6.32, (i) implies that $|\nabla_x v|$ is uniformly bounded.

Now, given $t \in [0, T]$ and $x \in H$, fix an *arbitrary* generalized reference probability space $\overline{\mu} = (\overline{\Omega}, \overline{\mathcal{F}}, \overline{\mathcal{F}}'_s, \overline{\mathbb{P}}, \overline{W})$. Then Eq. (6.77) admits a unique mild solution $\overline{X}$, since it has globally Lipschitz coefficients. If we define the control process $\overline{a}(s) = \gamma(s, \overline{X}(s), \nabla_x v(s, \overline{X}(s)) \sigma(s, \overline{X}(s)))$ we see that the pair $(\overline{a}(\cdot), \overline{X}(\cdot))$ is optimal for the control problem in the extended strong formulation corresponding to $\overline{\mu}$, namely

$$J^{\overline{\mu}}(t, x; \overline{a}(\cdot), \overline{X}(\cdot)) = v_t^{\overline{\mu}}(x).$$

Also note that under the additional assumptions the state equation admits a unique mild solution for an arbitrary control process, so the optimal control problem could also be formulated in a more standard way as in the previous chapters, i.e., as a minimization problem over a class of control processes. ∎

6.6 Application: Controlled Stochastic Equation with Delay

In this section we show how the previous results can be applied to perform the synthesis of an optimal control for a stochastic differential equation in $\mathbb{R}^n$ with unit delay:

$$\begin{cases} dx(s) = \left[\int_{-1}^0 x(s + \theta) \, \alpha(d\theta) + f(s, x(s)) + r(s, x(s), a(s)) \right] ds \\ \qquad\qquad\qquad\qquad\qquad\qquad + \sigma_0(s, x(s))dW(s), \quad s \in [t, T], \\ x(t) = y, \qquad x(t + \theta) = \beta(\theta), \quad \text{for } \theta \in (-1, 0), \end{cases}$$

(6.78)

and a cost functional of the form

$$J^\mu(t, y, \beta; a(\cdot), x(\cdot)) = \mathbb{E} \int_t^T h(s, x(s), a(s)) \, ds + \mathbb{E} \, k(x(T)).$$

Here $\mu = (\Omega, \mathcal{F}, \mathcal{F}_s^t, \mathbb{P}, W)$ denotes a generalized reference probability space as defined at the beginning of Sect. 6.5 and $(a(\cdot), x(\cdot))$ is an admissible control pair, i.e., the control process $a(\cdot)$ is $\{\mathcal{F}_s^t\}_{s \ge t}$ progressive with values in $\Lambda \subset \mathbb{R}^N$ and $x(\cdot)$

is a corresponding solution to Eq. (6.78). We will address the optimal control problem in the extended weak formulation, which consists in minimizing the functional $J^\mu(t, y, \beta; a(\cdot), x(\cdot))$ over all triples $(\mu, a(\cdot), X(\cdot))$, and characterizing the value function

$$\overline{V}(t, y, \beta) = \inf_{(\mu, a(\cdot), x(\cdot))} J^\mu(t, y, \beta; a(\cdot), x(\cdot)).$$

We assume the following (other assumptions are needed and will be stated below):

- $y \in \mathbb{R}^n$, $\beta \in L^2((-1, 0); \mathbb{R}^n)$;
- Λ is a Borel subset of $\mathbb{R}^N$;
- α is an $\mathcal{L}(\mathbb{R}^n, \mathbb{R}^n)$-valued finite measure on $[-1, 0]$;
- $f : [0, T] \times \mathbb{R}^n \to \mathbb{R}^n$ is measurable, $f(s, \cdot) \in C^1(\mathbb{R}^n)$ and there exists a constant $C > 0$ such that

$$|f(s, 0)| \leq C, \quad |\nabla_x f(s, x)| \leq C, \quad s \in [0, T], \, x \in \mathbb{R}^n;$$

- $\sigma_0 : [0, T] \times \mathbb{R}^n \to \mathcal{L}(\mathbb{R}^n, \mathbb{R}^n)$ is measurable and, for $t \in [0, T], x \in \mathbb{R}^n, \sigma_0(s, x)$ is invertible, we have $\sigma_0(s, \cdot) \in C^1(\mathbb{R}^n)$ and

$$|\sigma_0(s, 0)| \leq C, \quad |\nabla_x \sigma_0(s, x)| \leq C, \quad |\sigma_0^{-1}(s, x)| \leq C;$$

- $r : [0, T] \times \mathbb{R}^n \times \Lambda \to \mathbb{R}^n$ is measurable, $r(s, \cdot, a) \in C^1(\mathbb{R}^n)$ and, for some constant $m \geq 0$ and every $s \in [0, T], a \in \Lambda, x \in \mathbb{R}^n$,

$$|r(s, x, a)| \leq C, \quad |\nabla_x r(s, x, a)| \leq C(1 + |x|)^m.$$

- $h : [0, T] \times \mathbb{R}^n \times \Lambda \to \mathbb{R}$ is continuous, $h(s, \cdot, a) \in C^1(\mathbb{R}^n)$ and, for every $s \in [0, T], a \in \Lambda, x \in \mathbb{R}^n$,

$$|h(s, x, a)| + |\nabla_x h(s, x, a)| \leq C(1 + |x|)^m.$$

- $k : \mathbb{R}^n \to \mathbb{R}$ belongs to $C^1(\mathbb{R}^n)$ and satisfies

$$|\nabla_x k(x)| \leq C(1 + |x|)^m, \quad x \in \mathbb{R}^n.$$

We set $H = \mathbb{R}^n \times L^2((-1, 0); \mathbb{R}^n)$, $\Xi = \mathbb{R}^n$,

$$D(A) = \left\{ \begin{pmatrix} y \\ \beta \end{pmatrix} \in H : \beta \in W^{1,2}((-1, 0); \mathbb{R}^n) \text{ and } \beta(0) = y \right\},$$

$$A \begin{pmatrix} y \\ \beta \end{pmatrix} = \begin{pmatrix} \int_{-1}^0 \beta(\theta) a(d\theta) \\ \frac{d\beta}{d\theta} \end{pmatrix}.$$

Then A generates a strongly continuous semigroup in H. Moreover, if we set, for $t \in [0, T]$, $y \in \mathbb{R}^n$, $\beta \in L^2((-1, 0); \mathbb{R}^n)$, $a \in \Lambda$,

$$x = \begin{pmatrix} y \\ \beta \end{pmatrix}, \quad b\left(t, \begin{pmatrix} y \\ \beta \end{pmatrix}\right) = \begin{pmatrix} f(t, y) \\ 0 \end{pmatrix}, \quad \sigma\left(t, \begin{pmatrix} y \\ \beta \end{pmatrix}\right) = \begin{pmatrix} \sigma_0(t, y) \\ 0 \end{pmatrix},$$

$$R\left(t, \begin{pmatrix} y \\ \beta \end{pmatrix}, a\right) = \sigma_0^{-1}(t, y)r(t, y, a),$$

$$l\left(t, \begin{pmatrix} y \\ \beta \end{pmatrix}, a\right) = h(t, y, a), \quad g\begin{pmatrix} y \\ \beta \end{pmatrix} = k(y),$$

then Eq. (6.78) is reformulated as

$$\begin{cases} dX(s) = \Big(AX(s) + b(s, X(s)) + \sigma(s, X(s))R(s, X(s), a(s))\Big)\, ds \\ \qquad\qquad\qquad\qquad + \sigma(s, X(s))\, dW(s), \quad s \in [t, T], \\ X(t) = x. \end{cases}$$

Noting the product form of the state space H, we will write $X(s) = (x(s), x(s + \cdot))$ when we need to distinguish the two components of the solution process. The functional to be minimized can be rewritten as

$$\mathbb{E} \int_t^T l(s, X(s), a(s))\, ds + \mathbb{E}\, g(X(T)).$$

Remark 6.38 We see that the special form of the infinite-dimensional controlled equation (6.66) arises naturally from the finite-dimensional equation (6.78) of general form. ∎

Taking into account that Ξ is finite-dimensional, it is easy to check that the assumptions of Hypothesis 6.8 are satisfied. In particular, we may take $\gamma = 0$ in Hypothesis 6.8-(iii).

Next we define, for $s \in [0, T]$, $y \in \mathbb{R}^n$, $\beta \in L^2((-1, 0); \mathbb{R}^n)$, $z \in (\mathbb{R}^n)^*$ (this notation means that z is considered as a row vector),

$$F_0\left(s, \begin{pmatrix} y \\ \beta \end{pmatrix}, z\right) = F_{00}(s, y, z) := \inf_{a \in \Lambda} \left\{ h(s, y, a) + z\sigma_0^{-1}(s, y)r(s, y, a) \right\},$$
$$\tag{6.79}$$

$$\Gamma\left(s, \begin{pmatrix} y \\ \beta \end{pmatrix}, z\right) = \Gamma_0(s, y, z)$$
$$:= \{a \in \Lambda : F_{00}(s, y, a) = h(s, y, a) + z\sigma_0^{-1}(s, y)r(s, y, a)\}.$$
$$\tag{6.80}$$

We notice that F_0 and Γ only depend on the finite-dimensional coordinate in H.

The (linear) function $z \to z\sigma_0^{-1}(s, y)r(s, y, a)$ has a Lipschitz constant that only depends on the uniform bounds imposed on r and σ_0^{-1}. It follows that $F_{00}(s, \cdot, a)$ is Lipschitz on $\mathbb{R}^n$ with a Lipschitz constant that does not depend on (s, a).

Moreover, taking into account the growth conditions on the gradients of h, σ, r, it is easy to prove an estimate of the form

$$\left| \nabla_y \left[h(s, y, a) + z\sigma_0^{-1}(s, y)r(s, y, a) \right] \right| \le C(1 + |z|)(1 + |y|)^m,$$

which implies a local Lipschitz estimate on the function in square parentheses and hence on F_{00}:

$$|F_{00}(s, y, z) - F_{00}(s, y', z)| \le C(1 + |z|)(1 + |y| + |y'|)^m |y - y'|, \qquad (6.81)$$

for $s \in [0, T]$, $z \in (\mathbb{R}^n)^*$ and $y, y' \in \mathbb{R}^n$.

To proceed further we also need the following assumptions.

- F_{00} is Borel measurable and, for every $s \in [0, T]$, $F_{00}(s, \cdot, \cdot)$ is of class C^1.
- We assume that $\Gamma_0(s, y, z) \neq \emptyset$ and that there exists a measurable selection γ_0 of Γ_0, i.e., a measurable function $\gamma_0 : [0, T] \times \mathbb{R}^n \times (\mathbb{R}^n)^* \to \Lambda$ such that $\gamma_0(s, y, z) \in \Gamma_0(s, y, z)$ for every $s \in [0, T]$, $y \in \mathbb{R}^n$ and $z \in (\mathbb{R}^n)^*$. It follows that $\gamma(s, (y, \beta), z) := \gamma_0(s, y, z)$, defined on $[0, T] \times H \times (\mathbb{R}^n)^*$, is a measurable selection of Γ.

We note that the local Lipschitz estimate (6.81) implies

$$|\nabla_y F_{00}(s, y, z)| \le C(1 + |z|)(1 + |y|)^m$$

for $s \in [0, T]$, $z \in (\mathbb{R}^n)^*$ and $y \in \mathbb{R}^n$. Now it is easy to see that the conditions required in Hypothesis 6.22 (in the case $K = \mathbb{R}$) are all satisfied by F_0 and that Hypothesis 6.33 holds.

As a consequence of Theorem 6.36 we have the following result.

Theorem 6.39 *Under the previous assumptions there exists at least one generalized reference probability space* $\overline{\mu} = (\overline{\Omega}, \overline{\mathscr{F}}, \overline{\mathscr{F}}_s^t, \mathbb{P}, \overline{W})$ *and an admissible control pair* $(\overline{a}(\cdot), \overline{x}(\cdot))$ *for which*

$$\overline{V}(t, y, \beta) = J^{\overline{\mu}}(t, y, \beta; \overline{a}(\cdot), \overline{x}(\cdot)), \qquad t \in [0, T], y \in \mathbb{R}^n, \beta \in L^2((-1, 0); \mathbb{R}^n).$$

In particular, the triple $(\overline{\mu}, \overline{a}(\cdot), \overline{x}(\cdot))$ *is optimal.*

The value function $\overline{V}(t, y, \beta) = \overline{V}(t, x)$ *coincides with the function* $v(t, x)$ *which is the unique mild solution to the Hamilton–Jacobi–Bellman equation (6.68) in the sense specified by Theorem 6.32.*

In the space $\overline{\mu}$ *the process* $\overline{X}$ *given by* $\overline{X}(s) = (\overline{x}(s), \overline{x}(s + \cdot))$ *is a mild solution to the closed loop equation (6.77) and the optimal pair* $(\overline{a}(\cdot), \overline{X}(\cdot))$ *satisfies the feedback law equality*

$$\bar{a}(s) = \gamma(s, \overline{X}(s), \nabla_x v(s, \overline{X}(s)) \, \sigma(s, \overline{X}(s)))$$
$$= \gamma_0(s, \bar{x}(s), \nabla_\mu \overline{V}(s, \bar{x}(s), \bar{x}(s + \cdot)) \, \sigma_0(s, \bar{x}(s))) \quad \mathbb{P}\text{-}a.s. \text{ for a.a. } s \in [t, T].$$

6.7 Elliptic HJB Equation with Arbitrarily Growing Hamiltonian

In this section we address the solvability of the nonlinear stationary Kolmogorov equation:

$$\mathcal{A}u(x) - \lambda \, u(x) + F(x, u(x), Du(x) \, \sigma) = 0, \qquad x \in H. \tag{6.82}$$

We recall that, formally, the generator $\mathcal{A}$ of (P_t) is the operator

$$\mathcal{A}\phi(x) = \frac{1}{2} \mathrm{Tr}\left(\sigma\sigma^* D^2\phi(x)\right) + \langle Ax + b(x), D\phi(x)\rangle.$$

Our purpose is to extend the probabilistic techniques and BSDE representation to cover elliptic equations such as (6.82). We consider a general nonlinearity F that will only be assumed to be locally Lipschitz (with arbitrary growth) and no limitations are made on the size of λ. On the other hand, we assume that F is bounded with respect to x and that the noise is additive (that is, σ is independent of x).

We add the following standard piece of notation. If K is a Hilbert space, by $L^p_{\mathcal{P},\mathrm{loc}}(\Omega; L^2([0, \infty); K))$ we denote the space of processes $Y : \Omega \times [0, \infty) \to K$ such that Y restricted to $[0, T]$ is in $L^p_{\mathcal{P}}(\Omega; L^2([0, T]; K))$, $T > 0$.

An analogous definition is given for $L^p_{\mathcal{P},\mathrm{loc}}(\Omega, C([0, +\infty), H))$.

The standing assumptions will be (as far as the linear part of the HJB equation, or, equivalently the forward equation, is concerned):

Hypothesis 6.40 (i) The operator A is the generator of a strongly continuous semigroup e^{tA}, $t \geq 0$, in the Hilbert space H.
(ii) σ does not depend on x (that is, $\sigma \in \mathcal{L}(\Xi, H)$). Moreover, $|e^{tA}\sigma|_{\mathcal{L}_2(\Xi, H)} \leq Lt^{-\gamma}e^{at}$, for a suitable $\gamma \in [0, 1/2)$).
(iii) $b(\cdot) \in \mathcal{G}^1(H, H)$ and $|\nabla b(x)|_{\mathcal{L}(H)} \leq L$.
(iv) The operators $A + \nabla b(x)$ are dissipative (that is, $\langle Ay, y\rangle + \langle \nabla b(x)y, y\rangle \leq 0$ for all $x \in H$ and $y \in D(A)$).
(v) $\lambda > 0$,

and as far as the nonlinear part is concerned:

Hypothesis 6.41 (i) F is locally Lipschitz in z and y, that is, for all $R > 0$ there exists a K_R such that $|F(x, y, z) - F(x, y', z')| \leq K_R(|z - z'| + |y - y'|)$, $\forall x \in H, \forall y, y' \in H, \forall z, z' \in \Xi^*$ with $|z| \leq R, |z'| \leq R, |y| \leq R, |y'| \leq R$.
(ii) The map $x \to F(x, y, z)$ is continuous for all $z \in \Xi^*$, $y \in \mathbb{R}$.

(iii) $\sup_{x \in H} |F(x, 0, 0)| := M < +\infty$.

(iv) $F(\cdot, \cdot, \cdot) \in \mathcal{G}^1(H \times \mathbb{R} \times \Xi^*, \mathbb{R})$ and $|\nabla_x F(x, y, z)|_{H^*} \le c$, for a suitable constant $c > 0$ and all $x \in H, y \in \mathbb{R}, z \in \Xi^*$.

(v) F is dissipative with respect to y, that is, $\nabla_y F(x, y, z) \le 0$ for all $x \in H$, $y \in \mathbb{R}, z \in \Xi^*$

We will also need to add the following Lipschitzianity assumption, which we will eventually remove

Hypothesis 6.42 F is Lipschitz in z and y with constant κ:

$$|F(x, y, z) - F(x, y', z')| \le \kappa(|z - z'| + |y - y'|), \forall x \in H, \forall y, y' \in H, \forall z, z' \in \Xi^*.$$

6.8 The Associated Forward–Backward System

We start from a known result on bounded solutions of Lipschitz BSDEs on an infinite horizon, i.e., the following type of BSDE:

$$Y(\tau) = Y(T) + \int_\tau^T (f(\zeta, Y(\zeta), Z(\zeta)) - \lambda Y(\zeta))d\zeta - \int_\tau^T Z(\zeta)dW(\zeta), \quad 0 \le \tau \le T < \infty,$$
$$(6.83)$$

where $f : \Omega \times [0, \infty) \times \mathbb{R} \times \Xi^* \to \mathbb{R}$ is such that the process $(f(t, z))_{t \ge 0}$ is progressively measurable for all $z \in \Xi^*$. We suppose the following:

Hypothesis 6.43 (i) f is uniformly Lipschitz in z with Lipschitz constant K:

$$\forall t \ge 0, \forall y \in \mathbb{R}, \forall z, z' \in \Xi^*, \quad |f(t, y, z) - f(t, y, z')| \le K|z - z'|, \quad \mathbb{P}\text{-a.s.}$$

(ii) f is uniformly Lipschitz in y with Lipschitz constant k:

$$\forall t \ge 0, \forall y, y' \in \mathbb{R}, \forall z \in \Xi^*, \quad |f(t, y, z) - f(t, y', z)| \le k|y - y'|, \quad \mathbb{P}\text{-a.s.}$$

(iii) f is dissipative with respect to y that is

$$\forall t \ge 0, \forall y, y' \in \mathbb{R}, \forall z, z' \in \Xi^*, \quad (f(t, y, z) - f(t, y', z))(y - y') \le 0, \quad \mathbb{P}\text{-a.s.}$$

(iv) There exists a constant M such that $\forall t \ge 0, |f(t, 0, 0)| \le M$, $\mathbb{P}$-a.s.
We denote $\sup_{t \ge 0} |f(t, 0, 0)|$ by M.

We now turn to the existence and uniqueness of solution to (6.83) under Hypothesis 6.43.

Lemma 6.44 *Let us suppose that Hypothesis 6.43 holds. Then we have:*

(i) *There exists a solution (Y, Z) to the BSDE (6.83) such that Y is a continuous process bounded by $\frac{M}{\lambda}$, and $Z \in L^2_{\mathcal{P}, \text{loc}}(\Omega; L^2([0, \infty); \Xi))$ with $\mathbb{E} \int_0^\infty e^{-2\varepsilon s}$*

$|Z_s|^2 ds < \infty$ for all $\varepsilon > 0$. Moreover, the solution is unique in the class of processes (Y, Z) such that Y is continuous and uniformly bounded, and Z belongs to $L^2_{\mathcal{P},\text{loc}}(\Omega; L^2([0, \infty); \Xi))$.

(ii) Denoting by (Y^n, Z^n) the unique solution to the following finite horizon BSDE:

$$Y^n(\tau) = \int_\tau^n (f(\zeta, Y^n(\zeta), Z^n(\zeta)) - \lambda Y^n(\zeta))d\zeta - \int_\tau^n Z^n(\zeta)dW(\zeta), \quad t \in [0, T],$$
$$(6.84)$$

we have $|Y^n(\tau)| \leq \frac{M}{\lambda}$ and the following convergence rate holds:

$$|Y^n(\tau) - Y(\tau)| \leq \frac{M}{\lambda} \exp(-\lambda(n - \tau)). \qquad (6.85)$$

Moreover, $\forall \varepsilon > 0$

$$\mathbb{E} \int_0^{+\infty} e^{-2\varepsilon\zeta} |Z^n(\zeta) - Z(\zeta)|^2 d\zeta \to 0. \qquad (6.86)$$

Proof The result is contained in [79] and, under more general assumptions, in [518]. For the reader's convenience we report the proof here.

We start from a priori estimates. Fixing T, suppose that (Y, Z) with $Y \in L^2_{\mathcal{P}}(\Omega; C([0, T], K))$ and $Z \in L^2_{\mathcal{P}}(\Omega; L^2([0, T]; \Xi))$ satisfy

$$Y(\tau) = Y(T) + \int_\tau^T (f(\zeta, Y(\zeta), Z(\zeta)) - \lambda Y(\zeta))d\zeta - \int_\tau^T Z(\zeta)dW(\zeta), \quad 0 \leq \tau \leq T.$$
$$(6.87)$$

Applying Itô's rule to $e^{-\lambda(s-t)}Y_s$, $s \geq t$, we get

$$-d_s\left(e^{-\lambda(s-t)}Y_s\right) = e^{-\lambda(s-t)}f(s, Y_s, 0)ds - e^{-\lambda(s-t)}Z_s(-\theta_s ds + dW(s)),$$

where

$$\theta_s = [f(s, Y_s, Z_s) - f(s, Y_s, 0)] |Z_s|^{-2} Z_s^*$$

is a bounded process. Thus by Girsanov's Theorem there exists a probability $\tilde{\mathbb{P}}$ (mean value $\tilde{\mathbb{E}}$) under which $\tilde{W}(t) = -\int_t^s \theta_r dr + W(s)$ is an Ξ-valued Wiener process.

With respect to $(\tilde{W}(t))$ the above equation reads:

$$-d_s\left(e^{-\lambda(s-t)}Y_s\right) = e^{-\lambda(s-t)}f(s, Y_s, 0)ds - e^{-\lambda(s-t)}Z_s d\tilde{W}(s).$$

So applying Itô's rule to $\left(\varepsilon + e^{-2\lambda(s-t)}|Y_s|^s\right)^{1/2} := \mathcal{Y}_s$, $s \geq t$, we obtain

$$d_s\mathcal{Y}_s = \mathcal{Y}_s e^{-2\lambda(s-t)}\left[-\langle Y_s, f(s, 0, 0)\rangle - \langle Y_s, f(s, Y_s, 0) - f(s, 0, 0)\rangle\right] ds$$
$$+ \mathcal{Y}_s e^{-2\lambda(s-t)} \langle Y_s, Z_s\rangle d\tilde{W}(s)$$
$$+ \frac{1}{2}\mathcal{Y}_s e^{-2\lambda(s-t)} \left[|Z_s|^2 - \mathcal{Y}_s^{-2} e^{-2\lambda(s-t)} \langle Y_s, Z_s\rangle^2\right]. \quad (6.88)$$

Taking into account the dissipativity of f with respect to Y and the fact that, by construction, $\mathcal{Y}_s^{-1} e^{-\lambda(s-t)}|Y_s| \leq 1$ we obtain, integrating in $[t, T]$ and then computing the conditional expectation with respect to $\tilde{\mathbb{P}}$:

$$\sqrt{|Y_t|^2 + \varepsilon} \leq \tilde{\mathbb{E}}\left(\sqrt{e^{-2\lambda(T-t)}|Y_T|^2 + \varepsilon} \,\bigg|\, \mathscr{F}_t\right) + \tilde{\mathbb{E}}\left(\int_t^T e^{-\lambda(s-t)}|f(s, 0, 0)|ds \,\bigg|\, \mathscr{F}_t\right)$$

and by dominated convergence, recalling that $|f(s, 0, 0)| \leq M$:

$$|Y_t| \leq e^{-\lambda(T-t)}\tilde{\mathbb{E}}\left(|Y_T| \,\bigg|\, \mathscr{F}_t\right) + M/\lambda.$$

In particular, if (Y^n, Z^n) is a solution to (6.84) then $|Y_t^n| \leq M/\lambda$ for all $t \leq n$.

Moreover, if (Y, Z) is a solution in the whole $[0, \infty)$ with $Z \in L^2_{\mathcal{P}, \text{loc}}(\Omega; L^2([0, \infty); \Xi))$ and Y bounded then, letting $T \to \infty$, we get again: $|Y_t| \leq M/\lambda$.

If now $(Y^{(i)}, Z^{(i)})$, $i = 1, 2$, with $Y^{(i)} \in L^2_{\mathcal{P}}(\Omega; C([0, T], K))$ and $Z^{(i)} \in L^2_{\mathcal{P}}(\Omega; L^2([0, T]; \Xi))$ are both solutions to Eq. (6.87) then, by the above computations, applied this time to $(Y_t^{(2)} - Y_t^{(1)}, Z^{(2)} - Z^{(1)})$ we get

$$|Y_T^{(2)} - Y_T^{(1)}| \leq e^{-\lambda(T-t)}\tilde{\mathbb{E}}\left(|Y_T^{(2)} - Y_T^{(1)}| \,\bigg|\, \mathscr{F}_t\right), \quad \forall t \in [0, T].$$

Consequently, if $m > n$ and (Y^n, Z^n) and (Y^m, Z^m) satisfy Eq. (6.84) then

$$|Y_t^n - Y_t^m| \leq e^{-\lambda(n-t)}\tilde{\mathbb{E}}\left(|Y_n^m| \,\bigg|\, \mathscr{F}_t\right) \leq e^{-\lambda(n-t)}M/\lambda \quad \forall t \in [0, T]. \tag{6.89}$$

In the same way, if (Y, Z) is a solution of (6.83) on the whole $[0, \infty)$ with $Z \in L^2_{\mathcal{P}, \text{loc}}(\Omega; L^2([0, \infty); \Xi))$ and we know that (Y_s) is bounded, we get

$$|Y_t - Y_t^n| \leq e^{-\lambda(n-t)}M/\lambda.$$

We notice that the above relation immediately yields that if $(Y^{(i)}, Z^{(i)})$, $i = 1, 2$, are both solutions to Eq. (6.83) on the whole $[0, \infty)$ and we a priori know that both $(Y_t^{(1)})$ and $(Y_t^{(2)})$ are bounded, then $Y_t^{(1)} = Y_t^{(2)}$, $\mathbb{P}$-a.s., for all $t \in [0, T)$.

Concerning the estimate of the Z term we again fix T. If (Y, Z) with $Y \in L^2_{\mathcal{P}}(\Omega; C([0, T], K))$ and $Z \in L^2_{\mathcal{P}}(\Omega; L^2([0, T]; \Xi))$ satisfy (6.87) then, applying Itô's rule to $e^{-2\varepsilon}|Y_s|^2$, (with $0 < \varepsilon < \lambda$) and integrating between 0 and T, we get:

$$\int_0^T e^{-2\varepsilon s}|Z_s|^2 ds + |Y_0|^2 = 2e^{-2\varepsilon T}|Y_T|^2$$

$$+ 2\int_0^T e^{-2\varepsilon s}\left[\langle f(s, Y_s, Z_s), Y_s\rangle - (\lambda - \varepsilon)|Y_s|^2\right] ds$$

$$- \int_0^T e^{-2\varepsilon s}\langle Y_s, Z_s dW(s)\rangle.$$

Since $\mathbb{E}\left(\int_0^T e^{-2\varepsilon s}\langle Y_s, Z_s\rangle^2 ds\right)^{1/2} \leq \mathbb{E}\left[(\sup_{t\in[0,T]}|Y_t|)\left(\int_0^T |Z_s|^2 ds\right)^{1/2}\right] < \infty$, the stochastic integral in the above formula is a martingale. Thus, computing the expectation, taking into account the Lipschitzianity of f with respect to Z and its dissipativity with respect to Y, we get

$$\mathbb{E}\int_0^T e^{-2\varepsilon s}|Z_s|^2 ds \leq ce^{-2\varepsilon T}\mathbb{E}|Y_T|^2 + c\mathbb{E}\int_0^T e^{-2\varepsilon s}|Y_s|^2 ds + c\mathbb{E}\int_0^T e^{-2\varepsilon s}|f(s, 0, 0)|^2 ds,$$

where c is a constant depending only on f and ε.

In particular, if (Y, Z) is a solution on the whole $[0, \infty)$ with $Z \in L^2_{\mathcal{P},\mathrm{loc}}(\Omega; L^2([0, \infty); \Xi))$ and Y is bounded, then:

$$\mathbb{E}\int_0^\infty e^{-2\varepsilon s}|Z_s|^2 ds < +\infty.$$

Similarly, if $(Y^{(i)}, Z^{(i)})$, $i = 1, 2$, with $Y^{(i)} \in L^2_{\mathcal{P}}(\Omega; C([0, T], K))$ and $Z^{(i)} \in L^2_{\mathcal{P}}(\Omega; L^2([0, T]; \Xi))$ are solutions to Eq. (6.87), then:

$$\int_0^T e^{-2\varepsilon s}|Z_s^{(2)} - Z_s^{(1)}|^2 ds \leq ce^{-2\varepsilon T}\mathbb{E}|Y_T^{(2)} - Y_T^{(1)}|^2 + c\mathbb{E}\int_0^T e^{-2\varepsilon s}|Y_s^{(2)} - Y_s^{(1)}|^2 ds.$$

In particular, if $m > n$ and (Y^n, Z^n) and (Y^m, Z^m) satisfy Eq. (6.84) then, exploiting the estimates on Y^n and Y^m, we get, for all $T < n$

$$\mathbb{E}\int_0^T e^{-2\varepsilon t}|Z_t^n - Z_t^m| \leq ce^{-\lambda(n-T)},$$

and if (Y, Z) is a solution on the whole $[0, \infty)$ with Y bounded, then

$$\mathbb{E}\int_0^T e^{-2\varepsilon t}|Z_t^n - Z_t| \leq ce^{-\lambda(n-T)}. \tag{6.90}$$

Thus we have proved that, if a solution of Eq. (6.83) with (Y) bounded on the whole $[0, +\infty)$ exists, then it is unique and it satisfies estimates (6.85) and (6.86).

We now need to prove the existence of a bounded solution. By (6.89), fixing an arbitrary $T > 0$, the sequence of continuous functions $[0, T] \ni t \to Y_t^n$ is, $\mathbb{P}$ almost surely, a Cauchy sequence in $C([0, T])$. Thus there exists an adapted process with continuous trajectories such that, for any $T > 0$:

$$\sup_{t \in [0,T]} |Y_t^n - Y_t| \to 0, \qquad \mathbb{P}\text{-a.s.}$$

Notice that $|Y_t| \le M/\lambda$.

Moreover, by (6.90), for any $T > 0$, the sequence (Z^n) is Cauchy in $L_{\mathcal{P}}^2(\Omega; L^2([0, T]; \Xi))$, so there exists a $Z \in L_{\mathcal{P},\mathrm{loc}}^2(\Omega; L^2([0, \infty); \Xi))$ such that

$$\mathbb{E} \int_0^T |Z_t - Z_t^n|^2 dt \to 0.$$

To prove that (Y, Z) is the desired solution to Eq. (6.83) it is enough to observe that, for any fixed $0 < t < T < n$, we have

$$Y^n(\tau) = Y^n(T) + \int_\tau^T (f(\zeta, Y^n(\zeta), Z^n(\zeta)) - \lambda Y^n(\zeta))d\zeta - \int_\tau^T Z^n(\zeta)dW_\zeta.$$

The claim then follows just by letting $n \to \infty$ in the above formula. □

Now we come to the actual (Markovian) forward backward system. As far as the forward equation is concerned we consider the following special case of (6.6):

$$X(s; x) = e^{sA}x + \int_0^s e^{(s-\zeta)A}b(X(\zeta; x))d\zeta + \int_0^s e^{(s-\zeta)A}\sigma dW(\zeta), \quad s \ge 0. \quad (6.91)$$

We know that for every $p \in [2, \infty)$ and $T > 0$ there exists a unique process $X(\cdot; x) \in L_{\mathcal{P}}^p(\Omega; C([0, T], H))$ which is a solution to (6.91). Moreover, for all fixed $T > 0$, the map $x \to X(\cdot; x)$ is continuous from H to $L_{\mathcal{P}}^p(\Omega; C([0, T], H))$.

$$\mathbb{E} \sup_{\tau \in [0,T]} |X(\tau; x)|^p \le C(1 + |x|)^p, \qquad (6.92)$$

for some constant C depending only on T and pm.

We then consider the infinite horizon BSDE under the extra assumption (which will be removed later) that F is Lipschitz with respect to z. Namely, we deal with the equation (for $0 \le \tau \le T < \infty$)

$$Y(\tau; x) = Y(T; x) + \int_\tau^T (F(X(\zeta; x), Y(\zeta; x), Z(\zeta; x)) - \lambda Y(\zeta; x))d\zeta$$

$$- \int_\tau^T Z(\zeta; x)dW(\zeta). \qquad (6.93)$$

Here $X(\cdot; x)$ is the unique mild solution to (6.91) starting with $X(0; x) = x$.

Applying Lemma 6.44, we obtain:

Proposition 6.45 *Let us suppose that Hypotheses 6.40–6.42 hold. Then we have:*

(i) *For any $x \in H$, there exists a solution $(Y(\cdot; x), Z(\cdot; x))$ to the BSDE (6.93) such that $Y(\cdot; x)$ is a continuous process bounded by M/λ, and $Z \in L^2_{\mathcal{P}, loc}(\Omega; L^2 ([0, \infty); \Xi))$ with $\mathbb{E} \int_0^\infty e^{-2\lambda s} |Z(s; x)|^2 ds < \infty$. The solution is unique in the class of processes (Y, Z) such that Y is continuous and bounded, and Z belongs to $L^2_{\mathcal{P}, loc}(\Omega; L^2([0, \infty); \Xi))$.*

(ii) *Denoting by $(Y^n(\cdot; x), Z^n(\cdot; x))$ the unique solution of the following BSDE (with finite horizon):*

$$Y^n(\tau; x) = \int_\tau^n (F(X(\zeta; x), Y^n(\zeta; x), Z^n(\zeta; x)) - \lambda Y^n(\zeta; x)) d\zeta$$
$$- \int_\tau^n Z^n(\zeta; x) dW(\zeta), \qquad (6.94)$$

we have $|Y^n(\zeta; x)| \le \frac{M}{\lambda}$ and the following convergence rate holds:

$$|Y^n(\tau; x) - Y(\tau; x)| \le \frac{M}{\lambda} \exp(-\lambda(n - \tau)). \qquad (6.95)$$

Moreover,

$$\mathbb{E} \int_0^{+\infty} e^{-2\lambda\zeta} |Z^n(\zeta; x) - Z(\zeta; x)|^2 d\zeta \to 0. \qquad (6.96)$$

(iii) *For all $T > 0$ and $p \ge 1$, the map $x \to (Y(\cdot; x)|_{[0,T]}, Z(\cdot; x)|_{[0,T]})$ is continuous from H to the space $L^p_{\mathcal{P}}(\Omega; C([0, T], \mathbb{R})) \times L^p_{\mathcal{P}}(\Omega; L^2([0, T]; \Xi))$.*

Proof Statements (i) and (ii) are immediate consequences of Lemma 6.44. Let us prove (iii). If $x'_m \to x$ as $m \to +\infty$ then

$$|Y(T; x'_m) - Y(T; x)| \le |Y(T; x'_m) - Y^n(T; x'_m)| + |Y^n(T; x'_m) - Y^n(T; x)|$$
$$+ |Y^n(T; x) - Y(T; x)|$$
$$\le 2\frac{M}{\lambda} \exp(-\lambda(n - T)) + |Y^n(T; x'_m) - Y^n(T; x)|.$$

Moreover, for fixed n, $Y^n(\cdot; x'_m) \to Y^n(\cdot; x)$ in $L^p(\Omega, \mathscr{F}_T, \mathbb{P}; \mathbb{R})$ (see Proposition 6.27) and notice that we are now dealing with a finite horizon BSDE. Thus $Y(T; x'_m) \to Y(T; x)$ in $L^p(\Omega, \mathscr{F}_T, \mathbb{P}; \mathbb{R})$.

Now we can see that $(Y(\cdot; x)|_{[0,T]}, Z(\cdot; x)|_{[0,T]})$ is the unique solution of the following BSDE (with finite horizon):

$$Y(\tau; x) = Y(T; x) + \int_\tau^T (F(X(\zeta; x), Y(\zeta; x), Z(\zeta; x)) - \lambda Y(\zeta; x)) d\zeta$$

$$- \int_\tau^T Z(\zeta; x) dW(\zeta),$$

and the same holds for $(Y(\cdot; x_m')\big|_{[0,T]}, Z(\cdot; x_m')\big|_{[0,T]})$. So it is enough to apply again the continuity result in Proposition 6.27 to conclude that $(Y(\cdot; x_m')\big|_{[0,T]}, Z(\cdot; x_m')\big|_{[0,T]})$ converges to $(Y(\cdot; x)\big|_{[0,T]}, Z(\cdot; x)\big|_{[0,T]})$ in $L_{\mathcal{P}}^p(\Omega; C([0, T], \mathbb{R})) \times L_{\mathcal{P}}^p(\Omega; L^2([0, T]; \Xi))$. $\qquad\square$

Remark 6.46 We stress the fact that the uniform bound of Y does not depend on the Lipschitz constant κ of F with respect to y and z (provided that F is dissipative with respect to y). $\qquad\blacksquare$

6.8.1 Differentiability of the BSDE and a Priori Estimate on the Gradient

We need to study the regularity of $Y(\cdot, x)$. More precisely, we would like to show that $Y(0, x)$ belongs to $\mathcal{G}^1(H, \mathbb{R})$. Moreover, we will obtain a crucial a priori bound on the derivative $\nabla Y(0; x)$ independent of the Lipschitz constant of F with respect to z.

Lemma 6.47 *Under Hypothesis 6.40 the map $x \to X(\cdot, x)$ is Gâteaux differentiable (that is, it belongs to $\mathcal{G}(H, L_{\mathcal{P}}^p(\Omega, C([0, T], H)))$). Moreover, denoting by $\nabla X(\cdot, x)$ the partial Gâteaux derivative, for every direction $h \in H$, the directional derivative process $\nabla X(\cdot, x)h$, $\tau \in \mathbb{R}$, solves, $\mathbb{P}$-a.s., the equation*

$$\nabla X(\tau; x)h = e^{\tau A}h + \int_0^\tau e^{\zeta A}\nabla b(X(\zeta; x))\nabla X(\zeta; x)h\, d\zeta, \quad \tau \in \mathbb{R}^+. \qquad (6.97)$$

Finally, $\mathbb{P}$-a.s., $|\nabla X(\tau; x)h| \le |h|$, for all $\tau > 0$.

Proof The first assertion and relation (6.97) is a special case of Proposition 6.10. To prove the last assertion we proceed by a classical approximation argument (notice that the equation for ∇X has no stochastic integral term). Let $J_n := n(nI - A)^{-1}$ be the Yosida approximation for n large enough. As is well known (see also Appendix B.4.2) $J_n \in \mathcal{L}(H, D(A))$, $J_n x \to x$ for all $x \in H$. Let $L_t^n = J_n \nabla X(t; x)h$, then, for all $T > 0$, $L^n \in L_{\mathcal{P}}^p(\Omega; C([0, T], D(A)))$ and satisfies

$$(L_t^n)' = A L_t^n + J_n \nabla b(X(t; x))\nabla X(t; x)h.$$

Computing $\frac{d}{dt}|L_t^n|^2$, by Hypothesis 6.40 (iv) we get:

$$\frac{d}{dt}|L_t^n|^2 \le 2\langle L_t^n, \left(J_n\nabla b(X_t^x)\nabla X(t;x)h - \nabla b(X(t;x))J_n\nabla_x X(t;x)h\right)\rangle$$

and

$$|L_t^n|^2 \le |J_n h|^2 +$$
$$+2\int_0^t \langle L_s^n, (J_n\nabla b(X(s;x))\nabla X(s;x)h - \nabla b(X(s;x))J_n\nabla_x X(s;x)h)\rangle ds$$

and the claim follows by passing to the limit as $n \to \infty$. $\qquad\square$

The following is the main technical result of this section.

Theorem 6.48 *Under Hypotheses 6.40–6.42 the map $x \to Y(0; x)$ belongs to $\mathcal{G}^1(H, \mathbb{R})$. Moreover, $|Y(0; x)| + |\nabla Y(0; x)| \le c$, for a suitable constant c. We notice that the constant c does not depend on the Lipschitz constant κ of F with respect to y and z*

Proof The uniform bound on $|Y(0; x)|$ is an immediate consequence of Proposition 6.45.

Coming now to differentiability, fix $n \ge 1$, and let us consider the solution $(Y^n(\cdot; x), Z^n(\cdot; x))$ of (6.94). Then, see Proposition 6.27, the map $x \to (Y^n(\cdot; x), Z^n(\cdot; x))$ is Gâteaux differentiable from H to $L_{\mathcal{P}}^p(\Omega; C([0, T], \mathbb{R})) \times L_{\mathcal{P}}^p(\Omega; L^2([0, T]; \Xi^*))$, $\forall p \in [2, \infty)$. Denoting by $\nabla Y^n(\cdot; x)h, \nabla Z^n(\cdot; x)h$ the partial Gâteaux derivatives with respect to x in the direction $h \in H$, the processes

$$\{\nabla Y^n(\tau; x)h\}_{\tau \in [0,n]}, \qquad \{\nabla Z^n(\tau; x)h\}_{\tau \in [0,n]}$$

solve the following equation, $\mathbb{P}$-a.s.,

$$\nabla Y^n(\tau; x)h = \int_\tau^n \nabla_x F(X(\zeta; x), Y^n(\zeta; x), Z^n(\zeta; x))\nabla X(\zeta; x)h \, d\zeta$$
$$+ \int_\tau^n (-\lambda + \nabla_y F(X(\zeta; x), Y^n(\zeta; x), Z^n(\zeta; x)))\nabla Y^n(\zeta; x)h \, d\zeta$$
$$+ \int_\tau^n \nabla_z F(X(\zeta; x), Y^n(\zeta; x), Z^n(\zeta; x)\nabla Z^n(\zeta; x)h \, d\zeta \qquad (6.98)$$
$$- \int_\tau^n \nabla Z^n(\zeta; x)h \, dW(\zeta).$$

We see that in the above formula, we are considering that $Z^n(\cdot; x)$, $\nabla Z(\cdot; x)$ have values in Ξ^* and $\nabla_z F$ has values in Ξ^{**}. So if we identify Ξ^{**} and Ξ we can assume that $\nabla_z F$ has values in Ξ and Eq. (6.98) can be rewritten as:

$$\nabla Y^n(\tau; x)h = \int_\tau^n \nabla_x F(X(\zeta; x), Y^n(\zeta; x), Z^n(\zeta; x))\nabla X(\zeta; x)h \, d\zeta$$

$$+ \int_\tau^n (-\lambda + \nabla_y F(X(\zeta; x), Y^n(\zeta; x), Z^n(\zeta; x)))\nabla Y^n(\zeta; x)h \, d\zeta$$

$$+ \int_\tau^n (\nabla Z^n(\zeta; x)h) \left(\nabla_z F(X(\zeta; x), Y^n(\zeta; x), Z^n(\zeta; x)) \, d\zeta - dW(\zeta)\right).$$

By Hypotheses 6.41 and Lemma 6.47, we have that for all $x, h \in H$ the following holds $\mathbb{P}$-a.s. for all $n \in \mathbb{N}$ and all $\zeta \in [0, n]$:

$$\left|\nabla_x F(X(\zeta; x), Y^n(\zeta; x), Z^n(\zeta; x))\nabla_x X(\zeta; x)h\right| \le c|h|,$$

$$\nabla_y F(X(\zeta; x), Y^n(\zeta; x), Z^n(\zeta; x)) \le 0, \qquad \left|\nabla_z F(X(\zeta; x), Y^n(\zeta; x), Z^n(\zeta; x))\right|_\Xi \le \hat{c}.$$

Therefore, by Lemma 6.44, we obtain:

$$\sup_{\tau \in [0, n]} |\nabla Y^n(\tau; x)| \le C|h|, \quad \mathbb{P}\text{-a.s.,} \qquad (6.99)$$

where C does not depend on $\hat{c}$. Applying Itô's formula to $e^{-2\lambda t}|\nabla Y^n(\cdot; x)_t h|^2$, we get:

$$\mathbb{E} \int_0^\infty e^{-2\lambda t}(|\nabla Y^n(t; x)h|^2 + |\nabla_x Z^n(t; x)h|^2)dt \le C|h|^2. \qquad (6.100)$$

Let now $\mathcal{M}^{2,-2\lambda}$ be the Hilbert space of all pairs of $\{\mathscr{F}_t\}_{t \ge 0}$-adapted and measurable processes (y, z), where y has values in $\mathbb{R}$ and z in Ξ^*, such that

$$|(y, z)|^2_{\mathcal{M}^{2,-2\lambda}} := \mathbb{E} \int_0^\infty e^{-2\lambda t}(|y_t|^2 + |z_t|^2)dt < +\infty.$$

Fix $x, h \in H$, then there exists a subsequence of $\left(\nabla Y^n(\cdot; x)h, \nabla Z^n(\cdot; x)h, \nabla Y^n(0; x)h\right)_{n \in \mathbb{N}}$ which we still denote by itself, such that $(\nabla_x Y^n(\cdot; x)h, \nabla Z^n(\cdot; x)h)$ converges weakly to $(U^1(\cdot; x, h), V^1(\cdot; x, h))$ in $\mathcal{M}^{2,-2\lambda}$ and $\nabla_x Y^n(0; x)h$ converges to $\xi(x, h) \in \mathbb{R}$.

We define now

$$U^2(\tau; x, h) = \xi(x, h) - \int_0^\tau \nabla_x F(X(\zeta; x), Y(\zeta; x), Z(\zeta; x))\nabla X(\zeta; x) h d\zeta$$

$$- \int_0^\tau (-\lambda + \nabla_y F(X(\zeta; x), Y(\zeta; x), Z(\zeta; x)))U^1(\zeta; x, h)d\zeta$$

$$- \int_0^\tau \nabla_z F(X(\zeta; x), Y(\zeta; x), Z(\zeta; x))V^1(\zeta; x, h)d\zeta \qquad (6.101)$$

$$+ \int_0^\tau V^1(\zeta; x, h)dW(\zeta),$$

where $(Y(\cdot; x), Z(\cdot; x))$ is the unique bounded solution to the backward equation (6.93), see Proposition 6.45. Moreover, we rewrite (6.98) as follows:

$$\nabla Y^n(\tau; x)h = \nabla Y^n(0; x)h - \int_0^\tau \nabla_x F(X(\zeta; x), Y^n(\zeta; x), Z^n(\zeta; x))\nabla X(\zeta; x)hd\zeta$$

$$+ \int_0^\tau (\lambda - \nabla_y F(X(\zeta; x), Y^n(\zeta; x), Z^n(\zeta; x)))\nabla Y^n(\zeta; x)hd\zeta$$

$$- \int_0^\tau \nabla_z F(X(\zeta; x), Y^n(\zeta; x), Z^n(\zeta; x))\nabla Z^n(\zeta; x)hd\zeta \qquad (6.102)$$

$$+ \int_0^\tau \nabla Z^n(\zeta; x)hdW(\zeta).$$

Since, in particular, $(Y^n(\cdot; x), Z^n(\cdot; x)) \to (Y(\cdot; x), Z(\cdot; x))$ in measure $\mathbb{P} \times dt$; $\nabla_x F, \nabla_y F, \nabla_z F$ are bounded and finally $(\nabla Y^n(\cdot; x)h, \nabla Z^n(\cdot; x)h) \rightharpoonup (Y(\cdot; x), Z(\cdot; x))$ weakly in $\mathcal{M}^{2,-2\lambda}$, it is easy to show that $\nabla Y^n(\cdot; x)h$ converges to $U^2(\cdot; x, h)$ weakly in $L^2_{\mathcal{P}}(\Omega \times [0, T]; \mathbb{R})$ for all $T > 0$. Thus $U^2(t; x, h) = U^1(t; x, h)$, $\mathbb{P}$-a.s. for a.e. $t \in \mathbb{R}^+$ and $|U^2(t; x, h)| \leq c|h|$, $\mathbb{P}$-a.s. for all $t \in \mathbb{R}^+$ (this last assertion follows from continuity of the trajectories of $U^2(\cdot; x, h)$ and from the fact that $|U^1(t; x, h)| \leq c|h|$ $\mathbb{P}$-a.s. for almost every $t \in \mathbb{R}^+$). Therefore, coming back to Eq. (6.101), we have that $(U^2(\cdot; x, h), V^1(\cdot; x, h))$ is the unique bounded solution in $\mathbb{R}^+$ of the equation

$$U(\tau, x, h) = U(0, x, h) - \int_0^\tau \nabla_x F(X(\zeta; x), Y(\zeta; x), Z(\zeta; x))\nabla X(\zeta; x)hd\zeta$$

$$- \int_0^\tau (-\lambda + \nabla_y F(X(\zeta; x), Y(\zeta; x), Z(\zeta; x)))U(\tau, x, h)d\zeta$$

$$- \int_0^\tau \nabla_z F(X(\zeta; x), Y(\zeta; x), Z(\zeta; x))V(\zeta, x, h)d\zeta \qquad (6.103)$$

$$+ \int_0^\tau V(\zeta, x, h)dW(\zeta).$$

Notice that in particular $U(0, x, h) = \xi(x, h)$ is the limit of $\nabla Y^n(\cdot; x)_0 h$ (along the chosen subsequence). The uniqueness of the solution to (6.103) (see Lemma 6.44) implies that in reality $U(0, x, h) = \lim_{n\to\infty} \nabla Y^n(\cdot; x)_0 h$ along the original sequence.

Now let $x'_m \to x$. By (6.85), proceeding as in the proof of point (iii) in Proposition 6.45,

$$|U(0, x, h) - U(0, x'_m, h)| \leq \frac{2c}{\lambda}e^{-\lambda n}|h| + |U_n(0, x, h) - U_n(0, x'_m, h)|, \quad (6.104)$$

where $(U_n(\cdot, x, h), V_n(\cdot, x, h)) \in L^p_{\mathcal{P}}(\Omega; C([0, T], \mathbb{R})) \times L^p_{\mathcal{P}}(\Omega; L^2([0, T]; \Xi))$ is the unique solution of the finite horizon BSDE:

$$U_n(\tau, x, h) = \int_\tau^n \nabla_x F(X(\zeta; x), Y(\zeta; x), Z(\zeta; x)) \nabla X(\zeta; x) h d\zeta$$

$$+ \int_\tau^n (-\lambda + \nabla_y F(X(\zeta; x), Y(\zeta; x), Z(\zeta; x))) U_n(\tau, x, h) d\zeta$$

$$+ \int_\tau^n \nabla_z F(X(\zeta; x), Y(\zeta; x), Z(\zeta; x)) V_n(\zeta, x, h) d\zeta \qquad (6.105)$$

$$- \int_\tau^n V_n(\zeta, x, h) dW(\zeta),$$

and similarly for $(U_n(\cdot, x'_m, h), V_n(\cdot, x'_m, h))$. We now see that $\nabla_x F, \nabla_y F, \nabla_z F$ are, by assumptions, continuous and bounded. Moreover, the following statements on continuous dependence on x hold:

the maps $x \to X^x$, $x \to \nabla X^x h$ are continuous from H to $L_{\mathcal{P}}^p(\Omega; C([0, T], H))$ (see Proposition 6.10);

the map $x \to Y^x|_{[0,T]}$ is continuous from H to $L_{\mathcal{P}}^p(\Omega; C([0, T], \mathbb{R}))$ (see Proposition 6.45);

the map $x \to Z^x|_{[0,T]}$ is continuous from H to $L_{\mathcal{P}}^p(\Omega; L^2([0, T]; \Xi))$ (see Proposition 6.45).

We can therefore apply to (6.105) the continuous dependence on data result for finite horizon BSDEs (see Proposition 6.20) to obtain in particular that $U_n(0, x'_m, h) \to U_n(0, x, h)$ for all fixed n as $m \to \infty$. And by (6.104) we can conclude that $U(0, x'_m, h) \to U(0, x, h)$ as $m \to \infty$.

Summarizing, $U(0, x, h) = \lim_{n \to \infty} \nabla Y^n(\cdot; x)_0 h$ exists, moreover it is clearly linear in h and satisfies $|U(0, x, h)| \le C|h|$. Finally, it is continuous in x for every fixed h.

Lastly, for $t > 0$,

$$\lim_{t \searrow 0} \frac{1}{t} [Y(0; x + th) - Y(0; x)] = \lim_{t \searrow 0} \frac{1}{t} \lim_{n \to +\infty} [Y^n(0; x + th) - Y^n(0; x)]$$

$$= \lim_{t \searrow 0} \lim_{n \to +\infty} \int_0^1 \nabla Y^n(0; x + th) h d\theta$$

$$= \lim_{t \searrow 0} \int_0^1 U(0, x + \theta th) h d\theta = U(0, x) h$$

and the claim is proved. $\qquad\qquad\qquad\qquad\qquad\qquad\qquad\qquad\qquad \square$

6.9 Mild Solution of the Elliptic PDE

Assuming that Hypothesis 6.40 holds, we define in the usual way the transition semigroup $(P_t)_{t \ge 0}$ associated to the process X:

$$P_t[\phi](x) = \mathbb{E}\, \phi(X(t; 0, x)), \qquad x \in H, \qquad (6.106)$$

for every bounded measurable function $g : H \to \mathbb{R}$. Formally, the generator $\mathcal{A}$ of (P_t) is the operator

$$\mathcal{A}\phi(x) = \frac{1}{2}\mathrm{Tr}\left(\sigma\sigma^* D^2\phi(x)\right) + \langle Ax + b(x), D\phi(x)\rangle.$$

In this section we address the solvability of the nonlinear stationary Kolmogorov equation:

$$\mathcal{A}u(x) - \lambda\, u(x) + F(x, u(x), \nabla u(x)\,\sigma) = 0, \qquad x \in H. \qquad (6.107)$$

Definition 6.49 We say that a function $u : H \to \mathbb{R}$ is a mild solution of the nonlinear stationary Kolmogorov equation (6.107) if the following conditions hold:

(i) $u \in \mathcal{G}^1(H, \mathbb{R})$ and $\exists C > 0$ such that $|u(x)| \leq C$, $|\nabla u(x)h| \leq C\,|h|$, for all $x, h \in H$;

(ii) the following equality holds, for every $x \in H$ and $T \geq 0$:

$$u(x) = e^{-\lambda T}\, P_T[u](x) + \int_0^T e^{-\lambda \tau}\, P_\tau\left[F\left(\cdot, u(\cdot), \nabla u(\cdot)\,\sigma\right)\right](x)\, d\tau. \quad (6.108)$$

Remark 6.50 In order to motivate this definition one may consider the equation $\mathcal{A}u - \lambda u = -F$, where u, F are elements of a Banach space and $\mathcal{A}$ is a generator of a strongly continuous semigroup of bounded linear operators $(P_t)_{t\geq 0}$: if λ is sufficiently large, then

$$u = \int_0^\infty e^{-\lambda \tau} P_\tau F\, d\tau,$$

and, for arbitrary $T \geq 0$, by a change of variable,

$$e^{-\lambda T} P_T u = \int_T^\infty e^{-\lambda \tau} P_\tau F\, d\tau = u - \int_0^T e^{-\lambda \tau} P_\tau F\, d\tau.$$

∎

Theorem 6.51 *Assume that Hypothesis 6.40 and 6.41 hold, then Eq. (6.107) has a unique mild solution given by the formula*

$$u(x) = Y(0; x). \qquad (6.109)$$

Moreover, the following holds:

$$Y(\tau; x) = u(X(\tau; x)), \quad Z(\tau; x) = \nabla u(X(\tau; x))\, \sigma. \qquad (6.110)$$

Proof We initially assume that in addition F is Lipschitz with respect to z, uniformly in x and y.

We introduce the following equation, slightly more general than (6.91) since we consider a general initial time $t \geq 0$:

$$X(\tau) = e^{(\tau-t)A}x + \int_t^\tau e^{(\tau-\zeta)A}b(X(\zeta))\,d\zeta, + \int_t^\tau e^{(\tau-\zeta)A}\sigma\,dW(\zeta), \qquad (6.111)$$

for τ varying on an arbitrary time interval $[t, \infty) \subset [0, \infty)$. We set $X(\tau) = x$ for $\tau \in [0, t)$ and we denote by $\{X(\tau; t, x)\}_{\tau \geq 0}$ the solution, to indicate dependence on x and t. By an obvious extension of the results in the previous sections, we can solve the backward equation (6.93) with X given by (6.111); we denote the corresponding solution (Y, Z) by $\{(Y(\tau; t, x), Z(\tau; t, x))\}_{\tau \geq 0}$.

Thus, $\{(X(\tau; 0, x), Y(\tau; 0, x), Z(\tau; 0, x))\}_{\tau \geq 0}$ coincides with the process $\{X(\tau; x), Y(\tau; x), Z(\tau; x), \tau \geq 0\}$ occurring in relations (6.91) and (6.93). Note that, for bounded measurable $\phi : H \to \mathbb{R}$, we have

$$P_{\tau-t}[\phi](x) = \mathbb{E}\,\phi(X(\tau; t, x)), \qquad x \in H, \ 0 \leq t \leq \tau,$$

since the coefficients of Eq. (6.111) do not depend on time.

We first prove that u, given by (6.109), is a solution. The solutions of (6.111) satisfy the well-known property: for $0 \leq t \leq s$, $\mathbb{P}$-a.s.,

$$X(\tau; s, X(s; t, x)) = X(\tau; t, x), \qquad \text{for } \tau \in [s, \infty).$$

Since the solution of the backward equation is uniquely determined on an interval $[s, \infty)$ by the values of the process X on the same interval, for $0 \leq t \leq s$ we have, $\mathbb{P}$-a.s.,

$$\begin{aligned}
Y(\tau; s, X(s; t, x)) &= Y(\tau; t, x), \ \text{for } \tau \in [s, \infty), \\
Z(\tau; s, X(s; t, x)) &= Z(\tau; t, x) \ \text{for a.a.}\ \tau \in [s, \infty).
\end{aligned} \qquad (6.112)$$

In particular, for every $\tau \geq 0$,

$$Y(\tau; \tau, X(\tau; 0, x)) = Y(\tau; 0, x), \qquad \mathbb{P}\text{-a.s.} \qquad (6.113)$$

Since the coefficients of Eq. (6.111) do not depend on time, we have

$$X(\cdot; 0, x) \overset{(d)}{=} X(\cdot + t; t, x), \qquad t \geq 0,$$

where $\overset{(d)}{=}$ denotes equality in distribution (both sides of the equality are viewed as random elements with values in the space $C(\mathbb{R}^+, H)$). As a consequence we obtain

$$(Y(\cdot; 0, x), Z(\cdot; 0, x)) \overset{(d)}{=} (Y(\cdot + t; t, x), Z(\cdot + t; t, x)), \qquad t \geq 0,$$

where both sides of the equality are viewed as random elements with values in the space $C(\mathbb{R}^+, \mathbb{R}) \times L^2_{\text{loc}}(\mathbb{R}^+; \Xi^*)$. In particular, $Y(0; 0, x) \overset{(d)}{=} Y(t; t, x)$, and since they are both deterministic we have

$$u(x) = Y(0; 0, x) = Y(t; t, x), \qquad x \in H, \ t \geq 0.$$

Denoting for simplicity

$$(X(\tau), Y(\tau), Z(\tau)) = (X(\tau, 0, x), Y(\tau, 0, x), Z(\tau, 0, x)), \qquad \tau \geq 0,$$

it follows from (6.113) and path continuity that, $\mathbb{P}$-a.s.,

$$u(X(\tau)) = Y(\tau), \qquad \tau \geq 0.$$

It follows that, for all $0 < t < T$,

$$Y(t) = u(X(T)) - \int_t^T Z(\zeta) \, dW(\zeta) + \lambda \int_t^T Y(\zeta) \, d\zeta + \int_t^T F(X(\zeta), Y(\zeta), Z(\zeta)) \, d\zeta.$$
$$(6.114)$$

Thus by Corollary 6.29, considering the above equation as a BSDE on the finite horizon $[0, T]$ with final condition, it follows that, $\mathbb{P}$-a.s. for a.a. $\tau \geq 0$,

$$Z(\tau) = \nabla u(X(\tau)) \sigma.$$

We see that by Theorem 6.48 ∇u and consequently Z is bounded by a constant that does not depend on the Lipschitz constant of F with respect to z.

Applying the Itô formula to the equation solved by (Y, Z) we get

$$e^{-\lambda \tau} Y(\tau) - e^{-\lambda T} Y(T) + \int_\tau^T e^{-\lambda \zeta} Z(\zeta) \, dW(\zeta)$$
$$= \int_\tau^T e^{-\lambda \zeta} F(X(\zeta), Y(\zeta), Z(\zeta)) \, d\zeta, \ 0 \leq \tau \leq T < \infty,$$

and it follows that

$$\int_0^T e^{-\lambda \tau} P_\tau \Big[F\big(\cdot, u(\cdot), \nabla u(\cdot) \big) \Big](x) \, d\tau$$
$$= \mathbb{E} \int_0^T e^{-\lambda \tau} F(X(\tau), u(X(\tau)), \nabla u(X(\tau)) \sigma) \, d\tau$$
$$= \mathbb{E} \int_0^T e^{-\lambda \tau} F(X(\tau), Y(\tau), Z(\tau)) \, d\tau$$
$$= \mathbb{E} \Big[Y(0) - e^{-\lambda T} Y(T) + \int_0^T e^{-\lambda \tau} Z(\tau) \, dW(\tau) \Big]$$

$$= u(x) - e^{-\lambda T} \, \mathbb{E}\,[u(X_T)] = u(x) - e^{-\lambda T} \, P_T[u](x).$$

This completes the proof of the existence part.

Now we prove the uniqueness of the solution. Assume that u is a solution. For any $y \in H, 0 \le \tau \le T$ we have

$$u(y) = e^{-\lambda(T-\tau)} \, P_{T-\tau}[u](y) + \int_0^{T-\tau} e^{-\lambda t} \, P_t\Big[F\Big(\cdot, u(\cdot), \nabla u(\cdot)\,\sigma \Big)\Big](y)\, dt.$$

Set $y = X(\tau, 0, x)$, which we denote by $X(\tau)$ for simplicity. By the Markov property of X, denoting by $\mathbb{E}^{\mathscr{F}_\tau}$ the conditional expectation with respect to $\mathscr{F}_\tau$, we obtain

$$u(X(\tau)) = e^{-\lambda(T-\tau)} \, \mathbb{E}^{\mathscr{F}_\tau} u(X_T)$$
$$+ \int_0^{T-\tau} e^{-\lambda t} \, \mathbb{E}^{\mathscr{F}_\tau} F\Big(X(t + \tau), u(X(t + \tau)), \nabla u(X(t + \tau))\,\sigma \Big) dt$$

and, by a change of variable,

$$e^{-\lambda\tau} u(X(\tau)) = e^{-\lambda T} \, \mathbb{E}^{\mathscr{F}_\tau} u(X_T) + \int_\tau^T e^{-\lambda\zeta} \, \mathbb{E}^{\mathscr{F}_\tau} F\Big(X(\zeta), u(X(\zeta)), \nabla u(X(\zeta))\,\sigma \Big) d\zeta.$$

Now let $T > 0$ be fixed and let us define

$$F_\zeta = F(X(\zeta), u(X(\zeta)), \nabla u(X(\zeta))\,\sigma), \quad \zeta \in [0, T],$$

$$\xi = e^{-\lambda T} u(X_T) + \int_0^T e^{-\lambda\zeta} F_\zeta \, d\zeta.$$

Then we obtain

$$e^{-\lambda\tau} u(X(\tau)) = \mathbb{E}^{\mathscr{F}_\tau}\xi + \mathbb{E}^{\mathscr{F}_\tau} \int_0^\tau e^{-\lambda\zeta} F_\zeta \, d\zeta = \mathbb{E}^{\mathscr{F}_\tau}\xi + \int_0^\tau e^{-\lambda\zeta} F_\zeta \, d\zeta,$$

where the last equality holds since $\int_0^\tau e^{-\lambda\zeta} F_\zeta \, d\zeta$ is $\mathscr{F}_\tau$-adapted. Notice that ξ is square-integrable. Since $\mathscr{F}_t$ is generated by the Wiener process W, it follows that there exists a square-integrable, $\mathscr{F}_t$-predictable process $\widetilde{Z}(\tau), \tau \in [0, T]$, with values in Ξ^*, such that, $\mathbb{P}$-a.s.,

$$\mathbb{E}^{\mathscr{F}_\tau}\xi = \mathbb{E}\,\xi + \int_0^\tau \widetilde{Z}(\zeta)\, dW(\zeta), \qquad \tau \in [0, T].$$

An application of the Itô formula gives

$$u(X(\tau)) = \mathbb{E}\,\xi + \int_0^\tau e^{\lambda\zeta}\widetilde{Z}(\zeta)\,dW(\zeta) + \lambda\int_0^\tau u(X(\zeta))\,d\zeta + \int_0^\tau F(\zeta)\,d\zeta.$$

$$\text{(6.115)}$$

This shows that $u(X(\tau)), \tau \in [0, T]$, is a semimartingale. For $\xi \in \Xi$, we denote again by W^ξ the real Wiener process $W^\xi(\tau) := \langle \xi, W(\tau)\rangle, \tau \geq 0$. Let us consider the joint quadratic variation process of W^ξ with both sides of (6.115). Applying Proposition 6.17 (recall that u is by definition differentiable) we obtain, $\mathbb{P}$-a.s.,

$$\int_0^\tau \nabla u(X(\zeta))\,\sigma\xi\,d\zeta = \int_0^\tau e^{\lambda\zeta}\widetilde{Z}(\zeta)\xi\,d\zeta, \qquad \tau \in [0, T],\ \xi \in \Xi,$$

and we deduce that $\nabla u(X(\tau))\,\zeta = e^{\lambda\tau}\widetilde{Z}(\tau)$, $\mathbb{P}$-a.s. for almost all $\tau \in [0, T]$. Now setting

$$Y'(\tau) = u(X(\tau)), \qquad Z'(\tau) = e^{\lambda\tau}\nabla u(X(\tau))\sigma, \qquad \tau \geq 0,$$

it follows from (6.115) that, $\mathbb{P}$-a.s.,

$$Y(0) = Y'(\tau) + \int_0^\tau Z'(\zeta)\,dW(\zeta) + \lambda\int_0^\tau Y'(\zeta)\,d\zeta + \int_0^\tau F(X(\zeta), Y'(\zeta), Z'(\zeta))\,d\zeta,$$

for $\tau \in [0, T]$. Since T is arbitrary, we conclude that the process (Y', Z') is a solution of the backward equation, so that, by uniqueness, it must coincide with (Y, Z). In particular,

$$u(x) = u(X_0) = Y'(0) = Y(0).$$

This concludes the proof of the theorem. $\square$

6.10 Application to Optimal Control in an Infinite Horizon

We wish to apply the above results to perform the synthesis of the optimal control for a general nonlinear control system on an infinite time horizon. To be able to use non-smooth feedbacks we settle the problem in the framework of weak control problems.

As above, by H, Ξ we denote separable real Hilbert spaces.

Moreover, a *generalized reference probability space* is given by $\mu = (\Omega, \mathscr{F}, \mathscr{F}_s, \mathbb{P}, W)$, where

- $(\Omega, \mathscr{F}, \mathbb{P})$ is a complete probability space;
- $\{\mathscr{F}_s\}_{s\geq 0}$ is a filtration in it, satisfying the usual conditions;
- $(W(s))_{s\geq 0}$ is a cylindrical $\mathbb{P}$-Wiener process in Ξ, with respect to the filtration $\mathscr{F}_s$

(notice that, since our problem is homogeneous in time, we always choose the initial time $t = 0$).

Given such μ, we call an *admissible control pair* the pair $(a(\cdot), X(\cdot))$ of progressively measurable processes with respect to $\{\mathscr{F}_s\}_{s \geq 0}$ such that: a is defined on $\Omega \times [0, \infty)$ and takes its values in a fixed closed subset (not necessarily bounded) Λ of a Banach space E. Moreover, a is uniformly bounded, that is belongs to $L^\infty(\Omega \times [0, \infty), \mathbb{P} \otimes dt; E)$. Finally, X is the mild solution (on the whole $[0, \infty)$) of the following state equation:

$$\begin{cases} dX(\tau) = (AX(\tau) + b(X(\tau) + \sigma R(a(\tau)))) \, d\tau + \sigma \, dW(\tau), & \tau \geq 0, \\ X^a(0) = x \in H. \end{cases}$$

$$(6.116)$$

Notice that in the present case the assumptions on R will guarantee the existence and uniqueness of the mild solution X given and control a satisfying the above, so we work in the framework of the weak and strong formulations in the sense of Sect. 2.1.

To each admissible control pair we associate the cost:

$$J^\mu(x; a(\cdot), X(\cdot)) = \mathbb{E} \int_0^{+\infty} e^{-\lambda \zeta} [l(X(\zeta)) + |a(\zeta)|_E^2] \, d\zeta, \qquad (6.117)$$

where $l : H \times \Lambda \to \mathbb{R}$. As in the finite horizon case we minimize the functional $J^\mu(x; a(\cdot))$ over all admissible controls $a(\cdot)$ and characterize the value function

$$V^\mu(x) = \inf_a J^\mu(x; a(\cdot), X(\cdot)).$$

We will also address the optimal control problem in the weak formulation, which consists in further minimizing with respect to all generalized reference probability spaces, i.e., in characterizing the value function

$$\overline{V}(x) = \inf_\mu V^\mu(x).$$

Notice the occurrence of the operator σ in the control term: this special structure of the state equation is imposed by our techniques. Also notice that in contrast to what happens in the previous sections of this Chapter we now restrict ourselves to R that does not depend on x. This also ensures that for all $a(\cdot) \in \overline{\mathcal{U}}^\mu$ and $x \in H$ Eq. (6.116) admits a unique mild solution

We define in a classical way the Hamiltonian function relative to the above problem: for all $x \in H, z \in \Xi^*$,

$$\begin{aligned} F_0(x, z) &= l(x) + \inf\{|a|_E^2 + zR(a) : a \in \Lambda\} \\ \Gamma(z) &= \{a \in \Lambda : |a|_E^2 + zR(a) = \inf_{a \in \Lambda}\{|a|_E^2 + zR(a)\}\}. \end{aligned} \qquad (6.118)$$

We will work in the following general setting:

Hypothesis 6.52 The following holds:

(1) A, b, σ and satisfy Hypothesis 6.40.

(2) $R : \Lambda \to \Xi$ is Lipschitz.
(3) $l : H \to \mathbb{R}$ is uniformly Lipschitz, bounded and of class $\mathcal{G}^1(H, \mathbb{R})$.
(4) F_0 is of class $\mathcal{G}^1(\Xi^*, \mathbb{R})$.

Remark 6.53 Since R is Lipschitz $\inf\{|a|_E^2 + zR(a) : a \in \Lambda\}$ is always a real number. Moreover, there exists a constant c_R such that

$$\inf\{|a|_E^2 + zR(a) : a \in \Lambda\} = \inf\{|a|_E^2 + zR(a) : a \in \Lambda \cap B_E(0, c|z|)\}.$$

This immediately implies that $\Gamma(z) \subset B(0, c_R|z|)$ and that

$$\left| \inf\{|a|_E^2 + zR(a) : a \in \Lambda\} - \inf\{|a|_E^2 + z'R(a) : a \in \Lambda\} \right| \leq c_{1,R}(|z| + |z'|)|z - z'|.$$

So Hypothesis 6.41 holds true.   ∎

We see that for all $\lambda > 0$ the cost functional is well defined and $J^\mu(x; a(\cdot), X(\cdot)) < \infty$ for all $x \in H$, all admissible control systems μ and all admissible control pairs (a, X).

By Theorem 6.51, for all $\lambda > 0$ the stationary Hamilton–Jacobi–Bellman equation relative to the above stated problem, namely:

$$\mathcal{A}v(x) = \lambda v(x) - F_0(x, \nabla v(x)\sigma), \qquad x \in H, \tag{6.119}$$

admits a unique mild solution, in the sense of Definition 6.49.

We are in a position to prove the main result of this section:

Theorem 6.54 *Assume Hypothesis 6.52 and suppose that $\lambda > 0$. Then the following holds*

(1) For all generalized reference probability spaces and admissible pairs (a, X) we have $J^\mu(x; a(\cdot), X(\cdot)) \geq v(x)$. Therefore $V^\mu(x) \leq v(x)$.
(2) The equality $J^\mu(x; a(\cdot), X(\cdot)) = v(x)$ holds if and only if the following feedback law is satisfied:

$$a(\tau) \in \Gamma(\nabla v(X(\tau))\sigma), \qquad \mathbb{P}\text{-a.s. for a.e. } \tau \geq 0. \tag{6.120}$$

Notice that since ∇v is bounded, if (6.120) holds then the control a is uniformly bounded.

Proof Choose any generalized reference probability space μ and denote by $\rho(T)$ the Girsanov density

$$\rho(T) = \exp\left(-\int_0^T \langle R(a(\zeta)), dW(\zeta) \rangle_\Xi - \frac{1}{2} \int_0^T |R(a(\zeta))|_\Xi^2 \, d\zeta \right), \tag{6.121}$$

Let $\widetilde{\mathbb{P}}_T$ be the probability measure on $\mathscr{F}_T$ defined by $\widetilde{\mathbb{P}}_T = \rho(T)\,\mathbb{P}\big|_{\mathscr{F}_T}$ and let $\widetilde{\mathbb{E}}_T$ be the corresponding expectation. By Girsanov's Theorem (see Theorem 6.34) under $\widetilde{\mathbb{P}}_T$ the process

$$\widetilde{W}_\tau := \int_0^\tau R(a(\zeta))\,d\zeta + W_\tau, \qquad 0 \le \tau \le T, \tag{6.122}$$

is a cylindrical Wiener process. Equation (6.116) can be written:

$$\begin{cases} dX(\tau) = AX(\tau)\,d\tau + b(X(\tau))\,d\tau + \sigma\,d\widetilde{W}_\tau, & \tau \ge 0, \\ X_0 = x. \end{cases} \tag{6.123}$$

Let v be the unique mild solution of Eq. (6.119). Consider the following finite horizon Markovian forward–backward system (with respect to probability $\widetilde{\mathbb{P}}_T$ and to the filtration generated by $\{\widetilde{W}_\tau\}_{\tau \in [0,T]}$).

$$\begin{cases} \widetilde{X}(\tau; x) = e^{\tau A}x + \int_0^\tau e^{(\tau-\zeta)A}b(\widetilde{X}(\zeta; x))\,d\zeta + \int_0^\tau e^{(\tau-\zeta)A}\sigma\,d\widetilde{W}_\zeta, & \tau \ge 0, \\ \widetilde{Y}(\tau; x) - v(\widetilde{X}(T; x)) + \int_\tau^T \widetilde{Z}(\zeta; x)\,d\widetilde{W}_\zeta + \lambda\int_\tau^T \widetilde{Y}(\zeta; x)\,d\zeta \\ \qquad\qquad\qquad = \int_\tau^T F_0(\widetilde{X}(\zeta; x), \widetilde{Z}(\zeta; x))\,d\zeta, & 0 \le \tau \le T, \end{cases} \tag{6.124}$$

and let $(\widetilde{X}(x), \widetilde{Y}(x), \widetilde{Z}(x))$ be its unique solution with the three processes predictable relative to the filtration generated by $\{\widetilde{W}_\tau\}_{\tau \in [0,T]}$ and: $\widetilde{\mathbb{E}}_T \sup_{t \in [0,T]} |\widetilde{X}(t; x)|^2 < +\infty$, $\widetilde{Y}(x)$ bounded and continuous, $\widetilde{\mathbb{E}}_T \int_0^T |\widetilde{Z}(t; x)|^2 dt < +\infty$.

Moreover, Theorem 6.51 and uniqueness of the solution of system (6.124) together with Theorem 6.32 yields

$$\widetilde{Y}(\tau; x) = v(\widetilde{X}(\tau; x)), \qquad \widetilde{Z}(\tau; x) = \nabla v(\widetilde{X}(\tau; x))\sigma. \tag{6.125}$$

Comparing the forward equation in (6.124) with the state equation, rewritten as (6.123), we get $\widetilde{X}(t; x) = X_t$, $t \in [0, T]$, $\mathbb{P}$-a.s. Applying the Itô formula to $e^{-\lambda\tau}\widetilde{Y}(\tau; x)$, and restoring the original noise W, we get

$$\widetilde{Y}(0; x) + \int_0^T e^{-\lambda\zeta}\widetilde{Z}(\zeta; x)\,dW_\zeta$$
$$= \int_0^T e^{-\lambda\zeta}\big[F_0(X(\zeta), \widetilde{Z}(\zeta; x)) - \widetilde{Z}(\zeta; x)R(a(\zeta))\big]\,d\zeta + e^{-\lambda T}v(X(T)). \tag{6.126}$$

Using the identification in (6.125) and taking expectation with respect to $\mathbb{P}$, (6.126) yields

$$e^{-\lambda T}\mathbb{E}v(\widetilde{X}(T;x)) - v(x) = -\mathbb{E}\int_0^T e^{-\lambda\zeta}F_0(X(\zeta),\nabla v(X(\zeta))\sigma)\,d\zeta$$
$$+\mathbb{E}\int_0^T e^{-\lambda\zeta}\nabla v(X(\zeta))\zeta R(a(\zeta))\,d\zeta.$$

Recalling that v is bounded, letting $T \to \infty$, we conclude that

$$J^\mu(x;a(\cdot),X(\cdot)) = v(x) - \mathbb{E}\int_0^\infty e^{-\lambda\zeta}F_0(X(\zeta),\nabla v(X(\zeta))\sigma)d\zeta$$
$$-\mathbb{E}\int_0^\infty e^{-\lambda\zeta}\left[\nabla_x v(X(\zeta))\sigma R(a(\zeta)) - l(X(\zeta),a(\zeta))\right]d\zeta.$$

The above equality is known as the *fundamental relation* and immediately implies that $v(x) \le J^\mu(x;a(\cdot))$ and that equality holds if and only if (6.120) holds. $\square$

Theorem 6.55 *Assume Hypothesis 6.52 and that $\lambda > 0$. If $\Gamma(x,z)$ is non-empty for all $x \in H$ and $z \in \Xi^*$ and $\gamma : \Xi^* \to \Lambda$ is a measurable selection of Γ (which exists, see Theorem 8.2.10, in [20]) then there exists a generalized reference probability space $\bar{\mu}$ in which the closed loop equation*

$$\begin{cases} d\overline{X}(\tau) = A\overline{X}(\tau)\,d\tau + \sigma R(\gamma(\nabla v(\overline{X}(\tau))\sigma)\,d\tau + b(\overline{X}(\tau))\,d\tau + \sigma\,dW(\tau), & \tau \ge 0, \\ \overline{X}_0 = x_0 \in H, \end{cases}$$

(6.127)

admits a solution. Moreover, setting $\bar{a}(\tau) = \gamma(\nabla v(\overline{X}(\tau))\sigma)$, the pair $(\bar{a}(\cdot),\overline{X}(\cdot))$ is admissible and optimal for the control problem in the sense that

$$J^{\bar{\mu}}(x;\bar{a}(\cdot),\overline{X}(\cdot)) = v(x).$$

Consequently, we have $v(x) = \overline{V}(x)$.

Proof The point here is to prove the existence of a weak (in the probabilistic sense) solution to Eq. (6.127) in the whole $[0,+\infty)$, see also Sect. 4 in [274]. In order to do this we realize a "canonical"-Ξ-valued Wiener process. We choose a larger Hilbert space $\Xi' \supset \Xi$ in such a way that Ξ is continuously and densely embedded in Ξ' with Hilbert–Schmidt inclusion operator $\mathcal{J}$. By Ω we denote the space $C([0,\infty),\Xi')$ of continuous functions $\omega : [0,\infty) \to \Xi'$ endowed with the standard locally convex topology and by $\mathcal{B}$ its Borel σ-field. Since $\mathcal{J}\mathcal{J}^*$ is nuclear on Ξ' we know (see [180]) that there exists a probability $\mathbb{P}$ on $\mathcal{B}$ such that $W_t'(\omega) := \omega(t)$ is a $\mathcal{J}\mathcal{J}^*$-Wiener process in Ξ' (that is, $t \to \langle W_t',\xi'\rangle_{\Xi'}$ is a real-valued Wiener process for all $\xi' \in \Xi'$ and $\mathbb{E}[\langle W_t',\xi'\rangle_{\Xi'}\langle W_s',\eta'\rangle_{\Xi'}] = \langle \mathcal{J}\mathcal{J}^*\xi',\eta'\rangle_{\Xi'}(t\wedge s)$ for all $\xi',\eta' \in \Xi',t,s \in [0,\infty)$). We denote by $\mathcal{E}$ the $\mathbb{P}$-completion of $\mathcal{B}$ and by $\mathcal{F}_t, t \ge 0$, the $\mathbb{P}$-completion of $\mathcal{B}_t = \sigma(W_s' : s \in [0,t])$.

The Ξ-valued cylindrical Wiener process $\{W_t^\xi : t \ge 0, \xi \in \Xi\}$ can now be defined as follows. For ξ in the image of $\mathcal{J}^*\mathcal{J}$ we take η such that $\xi = \mathcal{J}^*\mathcal{J}\eta$ and define $W_s^\xi = \langle W_s',\mathcal{J}\eta\rangle_{\Xi'}$. Then we observe that $\mathbb{E}|W_t^\xi|^2 = t|\mathcal{J}\eta|_{\Xi'}^2 = t|\xi|_\Xi^2$ and that $\mathcal{J}^*\mathcal{J}\Xi$

is dense in Ξ to deduce that the linear continuous mapping $\xi \to W_s^\xi$ (with values in $L^2(\Omega, \mathscr{F}, \mathbb{P}; \mathbb{R})$) can be extended by continuity to the whole Ξ. An appropriate modification of $\{W_t^\xi : t \geq 0, \xi \in \Xi\}$ gives the required cylindrical Wiener process.

Now let $X \in L_{\mathcal{P},\mathrm{loc}}^p(\Omega, C([0, +\infty), H))$ be the mild solution of

$$\begin{cases} dX(\tau) = AX(\tau)\, d\tau + b(X(\tau))\, d\tau + \sigma\, dW(\tau) \\ X(0) = x \end{cases} \tag{6.128}$$

and let, $\forall T > 0$

$$\rho(T) = \exp\left(-\int_0^T \langle R(\gamma(\nabla v(X(\zeta))\sigma), dW(\zeta)\rangle_\Xi - \frac{1}{2} \int_0^T |R(\gamma(\nabla v(X(\zeta))\sigma)|_\Xi^2 \, d\zeta\right). \tag{6.129}$$

Recall that ∇v is bounded. Thus let $\widehat{\mathbb{P}}_T$ be the probability on $\mathscr{F}_T$ admitting $\rho(T)$ as a density with respect to $\mathbb{P}$. Since Ξ' is a Polish space and $\widehat{\mathbb{P}}_{T+h}$ coincides with $\widehat{\mathbb{P}}_T$ on $\mathcal{B}_T$, $T, h \geq 0$, by known results (see [508], Chap. VIII, Sect. 1, Proposition 1.13) there exists a probability $\widehat{\mathbb{P}}$ on $\mathcal{B}$ such that the restriction on $\mathcal{B}_T$ of $\widehat{\mathbb{P}}_T$ and that of $\widehat{\mathbb{P}}$ coincide, $T \geq 0$. Let $\widehat{\mathcal{E}}$ be the $\widehat{\mathbb{P}}$-completion of $\mathcal{B}$ and $\widehat{\mathscr{F}}_T$ be the $\widehat{\mathbb{P}}$-completion of $\mathcal{B}_T$. Moreover, let

$$\widehat{W}(t) := -\int_0^t R(\gamma(\nabla v(X(\zeta))\sigma)\, d\zeta + W(t).$$

Since, for all $T > 0$, $\{\widehat{W}_t\}_{t \in [0,T]}$ is a Ξ-valued cylindrical Wiener process under $\widehat{\mathbb{P}}_T$ (see again Theorem 6.34) and the restriction of $\widehat{\mathbb{P}}_T$ and of $\widehat{\mathbb{P}}$ coincide on $\mathcal{B}_T$, modifying $\{\widehat{W}_t\}_{t \geq 0}$ in a suitable way on a $\widehat{\mathbb{P}}$-null probability set we can conclude that $\bar{\mu} = (\Omega, \widehat{\mathcal{E}}, \{\widehat{\mathscr{F}}_t\}_{t \geq 0}, \widehat{\mathbb{P}}, \{\widehat{W}_t\}_{t \geq 0})$ is a generalized reference probability space and that if we set $\bar{a}(\tau) = \gamma(\nabla v(X(\tau))\sigma)$ then $(\bar{a}(\cdot), X(\cdot))$ is an admissible pair and (6.127) is satisfied. Indeed, if we rewrite (6.128) in terms of $\{\widehat{W}_t\}_{t \geq 0}$ we get

$$\begin{cases} dX(\tau) = AX(\tau)\, d\tau + b(X(\tau))\, d\tau + G\, [R(\gamma(\nabla v(X(\tau))\sigma)) + d\widehat{W}(\tau)], \\ X_0 = x \end{cases}$$

and this concludes the proof. $\square$

6.11   Application: The Heat Equation with Additive Noise

We show here how the previous results can be applied to a stochastic heat equation with additive white noise in dimension 1. Let, for $t \geq 0$, $\xi \in [0, 1]$:

$$\begin{cases} \dfrac{\partial}{\partial t}x(t,\xi) = \dfrac{\partial^2}{\partial \xi^2}x(t,\xi) + f_0(\xi, x(t,\xi)) + \sigma_0(\xi)r(\xi)\,a(t,\xi) + \sigma_0(\xi)\dfrac{\partial}{\partial t}\mathcal{W}(t,\xi), \\[2mm] x(t,0) = x(t,1) = 0, \\[2mm] x(0,\xi) = x_0(\xi), \end{cases}$$

(6.130)

where $\frac{\partial}{\partial t}\mathcal{W}$ is a space-time white noise on $\mathbb{R}^+ \times [0,1]$. Moreover, we introduce the cost functional:

$$J(x_0, a(\cdot), x(\cdot)) = \mathbb{E}\int_0^\infty \int_0^1 e^{-\lambda t}\left[\ell_0(\xi, x(t,\xi)) + |a(t,\xi)|^2\right]d\xi\, dt \qquad (6.131)$$

which we minimize over all progressive controls $a : [0,\infty) \times [0,1] \to \mathbb{R}$ bounded in $L^2([0,1])$. By this we mean that there exists a suitable constant c_a (depending on the control a) such that:

$$\int_0^1 a^2(t,\xi)d\xi \le c_a, \quad \mathbb{P}\otimes dt\text{-a.s.}$$

To fit the assumptions of our abstract results we will suppose that the functions f_0, σ_0, r, ℓ_0 are all measurable and real-valued and moreover:

(1) f_0 is defined on $[0,1] \times \mathbb{R}$ and $\int_0^1 f_0^2(\xi, 0)d\xi < +\infty$.
   Moreover, for a.a. $\xi \in [0,1]$, we require that $f_0(\xi, \cdot) \in C^1(\mathbb{R})$ and

$$-L_f \le \frac{\partial}{\partial \eta}f(\xi, \eta) \le 0$$

   for a suitable constant $L_f > 0$, almost all $\xi \in [0,1]$, and all $\eta \in \mathbb{R}$.
(2) σ_0 and r are bounded measurable functions from $[0,1]$ to $\mathbb{R}$.
(3) ℓ_0 is defined on $[0,1] \times \mathbb{R}$ and, for a.a. $\xi \in [0,1]$, the map $\ell_0(\xi, \cdot)$ is in $C^1(\mathbb{R}, \mathbb{R})$. Moreover:

$$|\ell_0(\xi, \eta)| \le c_0(\xi), \quad \left|\frac{\partial}{\partial \eta}\ell_0(\xi, \eta)\right| \le c_1(\xi), \quad \text{with } \int_0^1 \left(c_0(\xi) + c_1^2(\xi)\right)d\xi < +\infty.$$

(6.132)

(4) $x_0 \in L^2([0,1])$.

To rewrite the above problem in the abstract way we set (with the notation of Sect. 6.10): $H = \Xi = \Lambda = L^2([0,1])$. By $\{W(t)\}_{t\ge 0}$ we denote a cylindrical Wiener process in $L^2([0,1])$. Moreover, we define the operator A with domain $D(A)$ by:

$$D(A) = W^{2,2}([0,1]) \cap W_0^{1,2}([0,1]), \qquad (Ay)(\xi) = \frac{\partial^2}{\partial \xi^2}y(\xi), \quad \forall y \in D(A),$$

where $W^{2,2}([0,1])$ and $W_0^{1,2}([0,1])$ are the usual Sobolev spaces, and we set

$$b(x)(\xi) = f_0(\xi, x(\xi)), \quad (\sigma z)(\xi) = \sigma_0(\xi)z(\xi), \quad R(a)(\xi) = (Ra)(\xi) = r(\xi)a(\xi),$$
$$l(x, a) = \int_0^1 \left[|a(\xi)|^2 + \ell_0(\xi, x(\xi)) \right] d\xi$$

for all $x, z \in L^2([0, 1])$ $a \in L^\infty([0, 1])$ and a.a. $\xi \in [0, 1]$.

Under the previous assumptions we know, see [177] Sect. 11.2.1, that A, b and σ satisfy Hypothesis 6.40. Moreover, R is a bounded linear operator on $L^2([0, 1])$ and

$$\nabla_x l(x, a)h = \int_0^1 \frac{\partial}{\partial \eta} \ell_0(\xi, x(\xi))h(\xi)d\xi.$$

Hence points 2 and 3 in Hypothesis 6.52 are satisfied.

We also notice that

$$\inf_{a \in H} (|a|_H^2 + z(Ra)) = \inf_{a \in H} (|a|_H^2 + (R^*z)a) = -\frac{1}{4}|R^*z|_{H^*}^2 = -\frac{1}{4} \int_0^1 r^2(\xi)z^2(\xi)d\xi.$$

So $F_0(x, z) = l(x) - \frac{1}{4}|R^*z|^2$ and, taking into account the regularity of ℓ_0, it is immediate to see that point 4 in Hypothesis 6.40 is satisfied. In addition, $\inf_{a \in L^2([0,1])}(|a|_H^2 + z(Ra))$ is a minimum achieved for $a = -\frac{1}{2}rz$.

As a consequence of Theorems 6.54 and 6.55 we have the following result.

Theorem 6.56 *Under the previous assumptions, fixing $\lambda > 0$, there exists at least one generalized reference probability space $\overline{\mu} = (\overline{\Omega}, \overline{\mathscr{F}}, \overline{\mathscr{F}}_s, \overline{\mathbb{P}}, \overline{W})$ and an admissible control pair $(\overline{a}(\cdot), \overline{x}(\cdot))$ for which*

$$\overline{V}(x_0) = J^{\overline{\mu}}(x_0; \overline{a}(\cdot), \overline{x}(\cdot)), \quad x_0 \in L^2([0, 1]).$$

In particular, the triple $(\overline{\mu}, \overline{a}(\cdot), \overline{x}(\cdot))$ is optimal.

The value function $\overline{V}(x_0)$ coincides with the function $v(x_0)$, which is the unique mild solution to the Hamilton–Jacobi–Bellman equation (6.119) in the sense specified by Definition 6.49 (see Theorem 6.51) where (with the standard identifications)

$$F_0(x, \nabla v(x)\sigma) = l(x) - \frac{1}{4}|R^*\nabla v(x)\sigma|_{H^*}^2$$
$$= \int_0^1 \ell_0(\xi, x(\xi))d\xi - \frac{1}{4} \int_0^1 r^2(\xi)\sigma_0^2(\xi)(\nabla v(x)(\xi))^2 d\xi.$$

In the space $\overline{\mu}$ the process $(\overline{x}(s, \cdot))_{s \geq 0}$ is a mild solution to the closed loop equation

$$\begin{cases} \dfrac{\partial}{\partial t}\bar{x}(t, \xi) = \dfrac{\partial^2}{\partial \xi^2}\bar{x}(t, \xi) + f_0(\xi, \bar{x}(t, \xi)) - \dfrac{1}{2}\sigma_0^2(\xi)r^2(\xi)\nabla_x v(t, \bar{x}(t, \cdot))(\xi) + \sigma_0(\xi)\dfrac{\partial}{\partial t}\overline{W}(t, \xi), \\ \bar{x}(t, 0) = \bar{x}(t, 1) = 0, \\ \bar{x}(0, \xi) = x_0(\xi), \end{cases}$$

and the optimal pair $(\overline{a}(t, \cdot), \overline{x}(t, \cdot))$ satisfies the feedback law equality

$$\overline{a}(s,\xi) = -\frac{1}{2}\sigma_0(\xi) r(\xi) \nabla_x v(t, x(t, \cdot))(\xi).$$

6.12 Elliptic HJB Equations with Non-constant Diffusion

In this section we wish to briefly expose the results on the probabilistic representation of the solution to an elliptic HJB equation when the second-order operator $\mathrm{Tr}\left(\sigma(x)\sigma(x)^* D^2\phi(x)\right)$ depends on x. Namely, we will address the resolvability of the following equation:

$$\mathcal{A}u(x) - \lambda\, u(x) = F(x, u(x), \nabla u(x)\, \sigma(x)), \qquad x \in H,$$

where

$$\mathcal{A}\phi(x) = \frac{1}{2}\mathrm{Tr}\left(\sigma(x)\sigma(x)^* D^2\phi(x)\right) + \langle Ax + b(x), D\phi(x)\rangle.$$

The price to pay to allow σ to depend on x is that we will have to assume λ to be large enough.

The detailed proofs of the results reported below can be founded in [285].

Our analysis here will be done on the weighted (in time) spaces that we introduce below.

- $L^p_{\mathcal{P}}(\Omega; L^q_\beta(K))$, defined for $\beta \in \mathbb{R}$ and $p, q \in [1, \infty)$, denotes the space of equivalence classes of processes $\{Y(t)\}_{t \geq 0}$, with values in K, such that the norm

$$|Y|^p_{L^p_{\mathcal{P}}(\Omega; L^q_\beta(K))} = \mathbb{E}\left(\int_0^\infty e^{q\beta s} |Y(s)|^q_K\, ds\right)^{p/q}$$

  is finite, and Y admits a predictable version.
- $\mathcal{K}^p_\beta$ denotes the space $L^p_{\mathcal{P}}(\Omega; L^2_\beta(K)) \times L^p_{\mathcal{P}}(\Omega; L^2_\beta(\mathcal{L}_2(\Xi, K)))$. The norm of an element $(Y, Z) \in \mathcal{K}^p_\beta$ is $|(Y, Z)|_{\mathcal{K}^p_\beta} = |Y|_{L^p_{\mathcal{P}}(\Omega; L^2_\beta(K))} + |Z|_{L^p_{\mathcal{P}}(\Omega; L^2_\beta(\mathcal{L}_2(\Xi, K)))}$.
- $L^q_{\mathcal{P}}(\Omega; C_\eta(K))$, defined for $\eta \in \mathbb{R}$ and $q \in [1, \infty)$, denotes the space of predictable processes $\{Y(t)\}_{t \geq 0}$ with continuous paths in K, such that the norm

$$|Y|^q_{L^q_{\mathcal{P}}(\Omega; C_\eta(K))} = \mathbb{E}\sup_{\tau \geq 0} e^{\eta q \tau}|Y(\tau)|^q_K$$

  is finite. Elements of $L^q_{\mathcal{P}}(\Omega; C_\eta(K))$ are identified up to indistinguishability.
- Finally, for $\eta \in \mathbb{R}$ and $q \in [1, \infty)$, we define $\mathcal{H}^q_\eta$ as the space $L^q_{\mathcal{P}}(\Omega; L^q_\eta(K)) \cap L^q_{\mathcal{P}}(\Omega; C_\eta(K))$, endowed with the norm

$$|Y|_{\mathcal{H}^q_\eta} = |Y|_{L^q_{\mathcal{P}}(\Omega; L^q_\eta(K))} + |Y|_{L^q_{\mathcal{P}}(\Omega; C_\eta(K))}.$$

Clearly, similar definitions and notations also apply to processes with values in other Hilbert spaces, different from K.

As in the previous sections, we denote by $\{W(\tau)\}_{\tau \geq 0}$ a cylindrical Wiener process with values in a Hilbert space Ξ, defined on a complete probability space $(\Omega, \mathscr{F}, \mathbb{P})$. Now we consider the Itô stochastic differential equation for an unknown process $\{X(\tau; x)\}_{\tau \geq 0}$ with values in a Hilbert space H:

$$X(\tau; x) = e^{\tau A} x + \int_0^\tau e^{(\tau-s)A} b(X(s; x)) \, ds + \int_0^\tau e^{(\tau-s)A} \sigma(X(s; x)) \, dW(s), \qquad \tau \geq 0.$$
$$(6.133)$$

Hypothesis 6.57 (i) The operator A is the generator of a strongly continuous semigroup e^{tA}, $t \geq 0$, in the Hilbert space H. We denote by M and a two constants such that $|e^{tA}| \leq M e^{at}$ for $t \geq 0$.

(ii) The mapping $b : H \to H$ satisfies, for some constant $L > 0$,

$$|b(x) - b(y)| \leq L \, |x - y|, \qquad x, y \in H.$$

(iii) σ is a mapping from H to $\mathcal{L}(\Xi, H)$ such that for every $\xi \in \Xi$ the map $\sigma(\cdot)\xi : H \to H$ is measurable, $e^{tA}\sigma(x) \in \mathcal{L}_2(\Xi, H)$ for every $t > 0$ and $x \in H$, and

$$|e^{tA}\sigma(x)|_{\mathcal{L}_2(\Xi, H)} \leq L \, t^{-\gamma} e^{at} (1 + |x|),$$
$$|e^{tA}\sigma(x) - e^{tA}\sigma(y)|_{\mathcal{L}_2(\Xi, H)} \leq L \, t^{-\gamma} e^{at} |x - y|, \qquad t > 0, \ x, y \in H,$$
$$(6.134)$$
$$|\sigma(x)|_{\mathcal{L}(\Xi, H)} \leq L \, (1 + |x|), \qquad x \in H, \qquad (6.135)$$

for some constants $L > 0$ and $\gamma \in [0, 1/2)$.

(iv) For every $t > 0$, we have $b(\cdot) \in \mathcal{G}^1(H, H)$ and $e^{tA}\sigma(\cdot) \in \mathcal{G}^1(H, \mathcal{L}_2(\Xi, H))$.

Proposition 6.58 *Assume that Hypothesis 6.57 holds. Then for all $q \in [1, \infty)$ there exists a constant $\eta(q)$, depending also on γ, L, a, M, with the following properties:*

(i) *For all $x \in H$ the process $X(\cdot; x)$, a solution of (6.133), is in $\mathcal{H}^q_{\eta(q)}$ (here $K = H$).*

(ii) *For a suitable constant $C > 0$ we have*

$$\mathbb{E} \sup_{\tau \geq 0} e^{\eta(q)q\tau} |X(\tau; x)|^q + \mathbb{E} \int_0^\infty e^{\eta(q)qs} |X(s; x)|^q \, ds \leq C(1 + |x|)^q.$$
$$(6.136)$$

(iii) *The map $x \to X(\cdot; x)$ belongs to $\mathcal{G}^1(H, \mathcal{H}^q_{\eta(q)})$ and its derivative is uniformly bounded:*

$$|\nabla X(\cdot; x)h|_{\mathcal{H}^q_{\eta(q)}} \leq C \, |h|, \qquad x, h \in H, \qquad (6.137)$$

for a suitable constant C.

Let us now denote by $\mathscr{F}_\tau$ the natural filtration of $\{W(\tau)\}_{\tau\geq 0}$ augmented in the usual way. We again consider the system of stochastic differential equations: $\mathbb{P}$-a.s., for $0 \leq \tau \leq T < \infty$

$$
\begin{cases}
X(\tau; x) = e^{\tau A}x + \displaystyle\int_0^T e^{(\tau-s)A}b(X(s;x))\,ds + \int_0^T e^{(\tau-s)A}\sigma(X(s;x))\,dW(s), \\
Y(\tau; x) + \displaystyle\int_\tau^T Z(s;x)\,dW(s) + \lambda\int_\tau^T Y(s;x)\,ds \\
\qquad\qquad = \displaystyle\int_\tau^T F(X(s;x), Y(s;x), Z(s;x))\,ds.
\end{cases}
$$

$$(6.138)$$

Y is real-valued and Z takes values in Ξ^*, $F : H \times \mathbb{R} \times \Xi^* \to \mathbb{R}$ is a given measurable function, x is in H and λ is a real number.

For any $q \in [1, \infty)$ we choose $\eta(q)$ as in Proposition 6.58. Then, we know that for every $x \in H$, there exists a unique solution $\{X(\tau; x)\}_{\tau\geq 0}$ in $\mathcal{H}^q_{\eta(q)}$ of the forward equation and the map $x \to X(\cdot; x)$ belongs to $\mathcal{G}^1(H, \mathcal{H}^q_{\eta(q)})$.

Then we fix $p > 2$ and choose q and β satisfying

$$q \geq p(m+1)(m+2), \quad \beta < \eta(q)(m+1)(m+2), \quad \beta < 0. \quad (6.139)$$

On F we shall ask the following

Hypothesis 6.59 (i) There exist $\mu \in \mathbb{R}$ and nonnegative constants L_y, L_z such that

$$
|F(x, y_1, z_1) - F(x, y_2, z_2)| \leq L_y|y_1 - y_2| + L_z|z_1 - z_2|,
$$
$$
\langle F(x, y_1, z) - F(x, y_2, z), y_1 - y_2\rangle_K \leq -\mu|y_1 - y_2|^2,
$$

for every $x \in H$, $y_1, y_2 \in \mathbb{R}$, $z, z_1, z_2 \in \Xi^*$.
(ii) $F \in \mathcal{G}^1(H \times \mathbb{R} \times \Xi^*, K)$.
(iii) There exist $L > 0$ and $m \geq 0$ such that

$$
|\nabla_x F(x, y, z)h| \leq L_x|h|(1+|z|)(1+|x|+|y|)^m,
$$

for every $x, h \in H$, $y \in \mathbb{R}$, $z \in \Xi^*$.

We have the following existence and uniqueness result (in the weighted spaces introduced above).

Proposition 6.60 *Assume that Hypothesis 6.57 holds and that F satisfies the conditions in Hypothesis 6.59. For $p > 2$, β and q satisfying (6.139), and for every $\lambda > \widehat{\lambda} = -(\beta + \mu - L_z^2/2)$, the following holds.*

(i) For every $x \in H$ there exists a unique solution $(X(\cdot; x), Y(\cdot; x), Z(\cdot; x))$ of the forward–backward system (6.138) such that $X(\cdot; x) \in \mathcal{H}^q_{\eta(q)}$ and $(Y(\cdot; x), Z(\cdot; x)) \in \mathcal{K}^p_\beta$ (here $K = \mathbb{R}$ and consequently $\mathcal{L}_2(\Xi, K)$ is Ξ^*). Moreover, $Y(\cdot; x) \in L^p_\mathcal{P}(\Omega; C_\beta(\mathbb{R}))$.

(ii) *The maps* $x \to X(\cdot; x)$, $x \to (Y(\cdot; x), Z(\cdot; x))$, $x \to Y(\cdot; x)$ *belong to the spaces* $\mathcal{G}^1(H, \mathcal{H}^q_{\eta(q)})$, $\mathcal{G}^1(H, \mathcal{K}^p_\beta)$ *and* $\mathcal{G}^1(H, L^p_\mathcal{P}(\Omega; C_\beta(\mathbb{R})))$, *respectively.*

(iii) *Setting* $u(x) = Y(0; x)$, *we have* $u \in \mathcal{G}^1(H, \mathbb{R})$, *and* u *and* ∇u *have polynomial growth. More precisely, there exists a constant* $C > 0$ *such that*

$$|u(x)| \leq C (1 + |x|)^{m+1}, \qquad |\nabla u(x)h| \leq C |h|(1 + |x|)^{[(m+1)^2]}, \qquad x, h \in H.$$

Remark 6.61 Notice that we have shown that the system (6.138) admits a unique solution (in suitable spaces $\mathcal{H}^q_{\eta(q)}, \mathcal{K}^p_\beta$ with parameters satisfying $p > 2$ and condition (6.139)) for all $\lambda > \widehat{\lambda}$ where

$$\widehat{\lambda} = -\mu + L_z^2/2 - \sup\{\eta(q)(m + 1)(m + 2) \wedge 0 : q > 2(m + 1)(m + 2)\}. \tag{6.140}$$

∎

Remark 6.62 If, in addition to Hypothesis 6.59, we suppose that $F(\cdot, 0, 0)$ is bounded and satisfies Hypothesis 6.59 with $m = 0$, then the above results can be improved in the following way. Instead of asking (6.139) it is enough to require: $q > p > 2$ and $\beta < \eta(q) \wedge 0$. Then the conclusions of Proposition 6.60 still hold for $\lambda > -(\beta + \mu - L_z^2/2)$. Thus instead of (6.140) we have

$$\widehat{\lambda} = -\mu + L_z^2/2 - \sup\{\eta(q) \wedge 0 : q > 2\}. \tag{6.141}$$

Moreover, we have $|u(x)| \leq C$ and $|\nabla_x u(x)h| \leq C|h|$ for all $x, h \in H$.   ∎

Assuming that Hypothesis 6.57 holds and denoting by $(X(\tau; x))_{\tau \geq 0}$ the solution of Eq. (6.133), we define in the usual way the transition semigroup $(P_t)_{t \geq 0}$, associated to the process X:

$$P_t[\phi](x) = \mathbb{E} \, \phi(X(t; x)), \qquad x \in H, \tag{6.142}$$

for every bounded measurable function $\phi : H \to \mathbb{R}$. By Proposition 6.57, ϕ can be taken unbounded, with polynomial growth. Formally, the generator $\mathcal{A}$ of (P_t) is the operator

$$\mathcal{A}\phi(x) = \frac{1}{2}\mathrm{Tr}\left(\sigma(x)\sigma(x)^* D^2\phi(x)\right) + \langle Ax + b(x), D\phi(x) \rangle \,.$$

We consider now the solvability of the nonlinear stationary Kolmogorov equation:

$$\mathcal{A}u(x) - \lambda \, u(x) = F(x, u(x), \nabla u(x) \, \sigma(x)), \qquad x \in H, \tag{6.143}$$

where the function $F : H \times \mathbb{R} \times \Xi^* \to \mathbb{R}$ satisfies the conditions in Hypothesis 6.59 (with $K = \mathbb{R}$) and λ is a given number (that will eventually be assumed to be large enough). Note that, for $x \in H$, $\nabla u(x)$ belongs to H^*, so that $\nabla u(x) \, \sigma(x)$ is in Ξ^*.

The definition of a mild solution has to be slightly modified in order to take into account the polynomial growth:

Definition 6.63 We say that a function $u : H \to \mathbb{R}$ is a mild solution of the nonlinear stationary Kolmogorov equation (6.143) if the following conditions hold:

(i) $u \in \mathcal{G}^1(H, \mathbb{R})$;
(ii) for all $x \in H, h \in H$, we have

$$|u(x)| \leq C \, (1 + |x|)^C, \qquad |\nabla_x u(x)h| \leq C \, |h| \, (1 + |x|)^C,$$

for some constant $C > 0$;
(iii) the following equality holds, for every $x \in H$ and $T \geq 0$:

$$u(x) = e^{-\lambda T} \, P_T[u](x) - \int_0^T e^{-\lambda \tau} \, P_\tau \left[F\left(\cdot, u(\cdot), \nabla u(\cdot) \sigma(\cdot) \right) \right](x) \, d\tau.$$
(6.144)

Together with Eq. (6.133) we again consider the backward equation for $0 \leq \tau \leq T < \infty$

$$Y(\tau; x) - Y(T; x) + \int_\tau^T Z(s; x)) \, dW(s) + \lambda \int_\tau^T Y(s; x) \, ds$$
$$= - \int_\tau^T F(X(s; x), Y(s; x), Z(s; x)) \, ds,$$
(6.145)

where $F : H \times \mathbb{R} \times \Xi^* \to \mathbb{R}$ and λ are the same occurring in the nonlinear stationary Kolmogorov equation. Under the stated assumptions, Proposition 6.60 gives a unique solution $\{(X(\tau; x), Y(\tau; x), Z(\tau; x))\}_{\tau \geq 0}$ of the forward–backward system (6.138).

We can now state one of our main results.

Theorem 6.64 *Assume that Hypothesis 6.57 holds and that F satisfies the conditions in Hypothesis 6.59.*

Then there exists a $\widehat{\lambda} \in \mathbb{R}$ such that, for every $\lambda > \widehat{\lambda}$, the nonlinear stationary Kolmogorov equation (6.143) has a unique mild solution. The solution u is given by the formula

$$u(x) = Y(0; x),$$
(6.146)

where $\{(X(\tau; x), Y(\tau; x), Z(\tau; x))\}_{\tau \geq 0}$ is the solution of the backward-forward system 6.138), and it satisfies

$$|u(x)| \leq C \, (1 + |x|)^{m+1}, \qquad |\nabla u(x)h| \leq C \, |h|(1 + |x|)^{[(m+1)^2]},$$

for some constant C and every $x, h \in H$.

Remark 6.65 The constant $\widehat{\lambda}$ in the statement of the theorem can be chosen equal to (6.140). ∎

Remark 6.66 From Remark 6.62 it follows immediately that if, in addition to Hypothesis 6.57 and 6.59, we assume that $F(\cdot, 0, 0)$ is bounded and F satisfies Hypothesis 6.59 with $m = 0$, then $\widehat{\lambda}$ can be chosen equal to (6.141) instead of (6.140). Moreover, in this case, we have $|u(x)| \leq C$, $|\nabla u(x)h| \leq C\,|h|$ for some constant C and every $x, h \in H$.   ∎

Finally, we again apply the above results to a control problem. We mainly wish to show here what frameworks can be covered.

Let again H and Ξ denote real separable Hilbert spaces (the state space and the noise space, respectively) and let Λ be a Polish space (the control space). For $t \in [0, +\infty)$ a *generalized reference probability space* is given by $\mu = (\Omega, \mathscr{F}, \mathscr{F}_s, \mathbb{P}, W)$, where

- $(\Omega, \mathscr{F}, \mathbb{P})$ is a complete probability space;
- $(\mathscr{F}_s)_{s \geq 0}$ is a filtration in it, satisfying the usual conditions;
- $(W(s))_{s \geq 0}$ is a cylindrical $\mathbb{P}$-Wiener process in Ξ, with respect to the filtration $\mathscr{F}_s$.

Given such μ, for every starting point $x \in H$ we will consider the following controlled state equation

$$
\begin{cases}
dX(s; x) = (AX(s; x) + b(X(s; x)) + \sigma(X(s; x))R(X(s; x), a(s)))\ ds \\
\qquad\qquad\qquad\qquad\qquad\qquad + \sigma(X(s; x))\,dW(s), \quad s \in [0, \infty), \\
X(0) = x \in H.
\end{cases}
$$
$$(6.147)$$

In (6.147) and below the equation is understood in the mild sense. $a(\cdot) : \Omega \times [0, +\infty) \to \Lambda$ is the control process, which is always assumed to be progressively measurable with respect to $\{\mathscr{F}_s\}_{s \geq 0}$. On the coefficients A, b, σ, R precise assumptions will be formulated in Hypothesis 6.67 below. As in Sect. 6.5 we will impose on R only measurability and boundedness assumptions. As mentioned, this requires some care in the formulation of the control problem. We again call $(a(\cdot), X(\cdot))$ an *admissible control pair* if $a(\cdot)$ is an $\mathscr{F}_s$-progressively measurable process with values in Λ and $X(\cdot)$ is a mild solution to (6.147) corresponding to $a(\cdot)$. To every admissible control pair we associate the cost:

$$
J^{\mu}(x; a(\cdot), X(\cdot)) = \mathbb{E} \int_0^{\infty} e^{-\lambda s} l(X(s; x), a(s))\, ds,
$$

where l is a suitable real function. As in the parabolic case, see Sect. 6.5, the optimal control problem in the extended strong formulation consists in minimizing the functional $J^{\mu}(x; a(\cdot), X(\cdot))$ over all admissible control pairs (a, X), and characterizing the value function

$$
V^{\mu}(x) = \inf_{(a(\cdot), X(\cdot))} J^{\mu}(x; a(\cdot), X(\cdot; x)).
$$

We will also address the optimal control problem in the extended weak formulation, which consists in further minimizing with respect to all generalized reference probability spaces, i.e., in characterizing the value function

$$\overline{V}(x) = \inf_{\mu} V^{\mu}(x).$$

The corresponding Hamiltonian function is defined for all $x \in H$, $z \in \Xi^*$ setting

$$F_0(x, z) = \inf_{a \in \Lambda} (l(x, a) + z\,R(x, a)). \qquad (6.148)$$

We also define as usual

$$\Gamma(x, z) = \{a \in \Lambda : F_0(x, z) = l(x, a) + z\,R(x, a)\}.$$

We make the following assumption.

Hypothesis 6.67 The following holds:

(1) A, b and σ satisfy Hypothesis 6.57.
(2) $R : H \times \Lambda \to \Xi$ is Borel measurable and $|R(x, a)|_\Xi \leq L_R$ for a suitable constant $L_R > 0$ and all $x \in H$, $a \in \Lambda$.
(3) $l : H \times \Lambda \to \mathbb{R}$ is continuous and satisfies $|l(x, u)| \leq K_l(1 + |x|^{m_l})$ for suitable constants $K_l > 0$, $m_l \geq 0$ and all $x \in H$, $u \in \Lambda$.
(4) F_0 belongs to $\mathcal{G}^1(H \times \Lambda^*, \mathbb{R})$ and satisfies Hypothesis 6.59 (to avoid confusion we denote by m_F the constant m introduced in Hypothesis 6.59) We also notice that by its definition F_0 is Lipschitz with respect to z with Lipschitz constant L_R.
(5) Finally, we fix here $p > 2$, q and β satisfying (6.139) with $m = m_F$, and such that $q > m_F$.

In the following $\eta(q)$ is the constant introduced in Proposition 6.58.

Lemma 6.68 *Assume that $\lambda > 0$ satisfies*

$$\lambda > \frac{L_R m_l}{2(q - m_l)} - \eta(q)\,m_l. \qquad (6.149)$$

Then the cost functional is well defined and $J(x_0; a(\cdot), X(\cdot)) < \infty$ for all $x_0 \in H$ and all generalized reference probability spaces.

By Theorem 6.64, for all $\lambda > \widehat{\lambda}$ (the constant $\widehat{\lambda}$ can be chosen equal to (6.140) with $L_z = L_R$) the stationary Hamilton–Jacobi–Bellman equation relative to the above stated problem, written formally as

$$\mathcal{A}v(x) = \lambda v(x) + F_0(x, \nabla v(x)\sigma(x)), \qquad x \in H, \qquad (6.150)$$

admits a unique mild solution, in the sense of Definition 6.63, which we will denote by v.

We are in a position to solve the control problem:

Theorem 6.69 *Assume Hypothesis 6.67 and suppose that λ satisfies:*

$$\lambda > \left(-\beta + \frac{L_R^2}{2}\right) \vee \left(-\beta + \frac{L_R}{2(p-1)}\right) \vee \left(\frac{L_R m_l}{2(q-m_l)} - \eta(q)\, m_l\right). \quad (6.151)$$

Then the following holds

(1) For all generalized reference probability space μ and all admissible control pairs $(a(\cdot), X(\cdot))$ we have $J^\mu(x; a(\cdot), X(\cdot)) \geq v(x)$.
It follows that $V^\mu(x) \geq v(x)$ for every μ, and so $\overline{V}(x) \geq v(x)$.
(2) For all μ and all admissible control pairs (a, X), the equality $J^\mu(x; a(\cdot), X(\cdot)) = v(x)$ holds if and only if the following feedback law is satisfied:

$$a(s) \in \Gamma(X(s), \nabla_x v(X(s))\, \sigma(X(s))), \quad \mathbb{P}\text{-a.s. for a.a. } s \in [t, T]. \quad (6.152)$$

We again have existence of the optimal control in the extended weak formulation.

Theorem 6.70 *If in addition to the assumptions of the above theorem we suppose that $\Gamma(x, z)$ is non-empty for all $x \in H$ and $z \in \Xi^*$. Let $\gamma : H \times \Xi^* \to \Lambda$ be a measurable selection of Γ (which exists, see Theorem 8.2.10, in [20]). Then there exists at least one generalized reference probability space $\overline{\mu}$ and an admissible control pair $(\overline{a}(\cdot), \overline{X}(\cdot))$ for which (6.152) holds. In particular, it follows that $V_t^{\overline{\mu}}(x) = v(t, x)$ and so $\overline{V}(t, x) = v(t, x)$. In the space $\overline{\mu}$ the process $\overline{X}$ is a mild solution to the closed loop equation:*

$$\begin{cases} d\overline{X}(s) = A\overline{X}(s)\, ds + \sigma(\overline{X}(s))\, R\Big(\overline{X}(s), \gamma(s, \overline{X}(s), \nabla_x v(\overline{X}(s))\, \sigma(\overline{X}(s)))\Big)\, ds \\ \qquad\qquad + b(\overline{X}(s))\, ds + \sigma(\overline{X}(s))\, d\overline{W}(s), \quad s \in [t, T], \\ \overline{X}(0) = x \in H, \end{cases}$$

$$(6.153)$$

the feedback law takes the form

$$\overline{a}(s) = \gamma(\overline{X}(s), \nabla_x v(\overline{X}(s))\, \sigma(\overline{X}(s))), \quad \mathbb{P}\text{-a.s. for a.e. } s \in [0, T],$$

and the pair $(\overline{a}(\cdot), \overline{X}(\cdot))$ is optimal for the control problem in the extended weak formulation.

Remark 6.71 If, in addition to points *1–4* of Hypothesis 6.67, we also assume that l is bounded and Lipschitz in x uniformly in $u \in \mathcal{U}$, then it is easily verified that $F_0(\cdot, 0)$ is bounded and F_0 satisfies Hypothesis 6.59 with $m = 0$. Thus by Remark 6.62 the results of Theorem 6.69 can be improved in the following way.

Instead of Hypothesis 6.67 point 5 it is enough to take $q > p > 2$ and $\beta < \eta(q) \wedge 0$. Moreover, instead of (6.151) it is enough to assume

$$\lambda > -\beta + \left(\frac{L_R^2}{2} \vee \frac{L_R}{2(p-1)} \right).$$

■

6.12.1 The Heat Equation with Multiplicative Noise

Finally, we show how the assumptions on the controlled heat equation in Sect. 6.11 have to be adapted to fit this last framework. We again consider a stochastic heat equation with additive white noise in dimension 1 (for $t \geq 0$, $\xi \in [0, 1]$):

$$\begin{cases} \dfrac{\partial}{\partial t} x(t, \xi) = \dfrac{\partial^2}{\partial \xi^2} x(t, \xi) + f_0(\xi, x(t, \xi)) \\ \qquad\qquad\qquad +\sigma_0(\xi, x(t, \xi)) r(\xi) a(t, \xi) + \sigma_0(\xi) \frac{\partial}{\partial t} W(t, \xi), \\ x(t, 0) = x(t, 1) = 0, \\ x(0, \xi) = x_0(\xi), \end{cases}$$

(6.154)

and the cost functional:

$$J(x_0; a(\cdot), x(\cdot)) = \mathbb{E} \int_0^\infty \int_0^1 e^{-\lambda t} \left[\ell_0(\xi, x(t, \xi)) + |a(t, \xi)|^2 \right] d\xi \, dt. \quad (6.155)$$

The assumptions and notations are the same as in Sect. 6.11 except that:

- σ_0 depends on x as well. We assume that it is bounded, differentiable with respect to x and Lipschitz with respect to x, uniformly in ξ.
- We relax the assumptions on ℓ_0. Namely, we assume that ℓ_0 is defined on $[0, 1] \times \mathbb{R}$. Moreover, for a.a. $\xi \in [0, 1]$, the map $\ell_0(\xi, \cdot)$ is in $C^1(\mathbb{R}, \mathbb{R})$ and

$$|\ell_0(\xi, 0)| \leq c_0(\xi), \quad \left| \frac{\partial}{\partial \eta} \ell_0(\xi, \eta) \right| \leq c_1(\xi), \quad \text{with } \int_0^1 \left(c_0(\xi) + c_1^2(\xi) \right) d\xi < +\infty.$$

(6.156)

- We restrict our analysis to controls taking values in a ball of $L^2([0, 1])$. Namely, we assume:

$$\int_0^1 a^2(t, \xi) d\xi \leq 1, \quad \mathbb{P} \otimes dt\text{-a.s.}$$

The problem can be rewritten in the abstract way exactly as in Sect. 6.11 with the difference that now:

$$\inf_{a \in H : |a| \leq 1} (|a|_H^2 + z(Ra)) = \inf_{a \in H : |a| \leq 1} (|a|_H^2 + (R^*z)a) = \Psi(R^*z),$$

where (with the standard identifications):

$$\Psi(p) = \begin{cases} -(1/4)|p|^2_{L^2((0,1])} & \text{if } |p|_{L^2((0,1])} \leq 2 \\ -|p|+1 & \text{if } |p|_{L^2((0,1])} > 2 \end{cases}, \qquad R^*z = rz.$$

In addition, $\inf_{a \in H}(|a|^2_H + z(Ra))$ is a minimum achieved for $a = \psi(R^*z)$ where

$$\psi(p) = \begin{cases} -(1/2)p & \text{if } |p|_{L^2((0,1])} \leq 2, \\ -p/|p| & \text{if } |p|_{L^2((0,1])} > 2. \end{cases}$$

So $F_0(x, z) = l(x) + \Psi(R^*z)$ belongs to $\mathcal{G}^1(H \times H^*, \mathbb{R})$. As a consequence of Theorems 6.69 and 6.70 we have the following result.

Theorem 6.72 *Under the previous assumption we can find $\hat\lambda$ such that, for all $\lambda > \hat\lambda$, there exists at least one generalized reference probability space $\overline\mu = (\Omega, \mathcal{F}, \mathcal{F}_s, \mathbb{P}, \overline{W})$ and an admissible control pair $(\overline{a}(\cdot), \overline{x}(\cdot))$ for which*

$$\overline{V}(x_0) = J^{\overline\mu}(x_0; \overline{a}(\cdot), \overline{x}(\cdot)), \qquad x_0 \in L^2([0, 1]).$$

In particular, the triple $(\overline\mu, \overline{a}(\cdot), \overline{x}(\cdot))$ is optimal.

The value function $\overline{V}(x_0)$ coincides with the function $v(x_0)$, which is the unique mild solution to the Hamilton–Jacobi–Bellman equation (6.150) in the sense specified by Definition 6.63 (see Theorem 6.64) where (with the standard identifications)

$$F_0(x, \nabla v\sigma) = l(x) + \Psi(R^*\nabla v(x)\sigma(x)) = \ell_0(\cdot, x(\cdot)) + \Psi(r(\cdot)\sigma_0(\cdot, x(\cdot))\nabla v(x)(\cdot)).$$

In the space $\overline\mu$ the process $(\overline{x}(s, \cdot))_{s \geq 0}$ is a mild solution to the closed loop equation

$$\begin{cases} \dfrac{\partial}{\partial t}\overline{x}(t, \xi) = \dfrac{\partial^2}{\partial \xi^2}\overline{x}(t, \xi) + f_0(\xi, \overline{x}(t, \xi)) + \sigma_0(\xi, \overline{x}(\xi))\dfrac{\partial}{\partial t}\mathcal{W}(t, \xi) \\ \qquad\qquad + \sigma_0(\xi, \overline{x}(\xi))r(\xi)\psi\left(r(\cdot)\sigma_0(\cdot, \overline{x}(t, \cdot))\nabla v(\overline{x}(t, \cdot))(\cdot)\right)(\xi)dt, \\ \overline{x}(t, 0) = \overline{x}(t, 1) = 0, \\ \overline{x}(0, \xi) = x_0(\xi), \end{cases}$$

and the optimal pair $(\overline{a}(t, \cdot), \overline{x}(t, \cdot))$ satisfies the feedback law equality

$$\overline{a}(t, \cdot) = \psi\left(r(\cdot)\sigma_0(\cdot, \overline{x}(t, \cdot))\nabla v(\overline{x}(t, \cdot))(\cdot)\right).$$

6.13   Bibliographical Notes

The paper [475] by É. Pardoux and S. Peng is generally recognized as the starting point of the theory of Backward Stochastic Differential Equations (BSDEs): there the authors solved a general nonlinear BSDE under Lipschitz assumptions on the coefficients. Earlier results on the linear case were proved by several authors, in particular by J-.M. Bismut and A. Bensoussan, in connection with the so-called Stochastic Maximum Principle (in the sense of Pontryagin). Since the appearance of [475], the theory began to develop quickly, motivated by applications to stochastic optimal control, partial differential equations and mathematical finance. Some standard references are [211, 420, 477, 575].

Here we limit ourselves to a bibliographical account of the main achievements related to BSDEs driven by a Brownian motion in an infinite-dimensional context, i.e., when at least one of the unknown processes (Y, Z) takes values in an infinite-dimensional space or when the BSDE is coupled with another (forward) stochastic differential equation with infinite-dimensional solution process.

To our knowledge, the first result on BSDEs when the process Y evolves in an infinite-dimensional space is that of Bensoussan [45] concerning the linear case. A highly non-trivial extension of the nonlinear case originally addressed by Pardoux and Peng in the infinite-dimensional context is in [350], followed by [558] and by some results in [284, 285]. The case of dissipative coefficients is considered in [129, 130]. A special class of backward equations, called of Volterra type, are studied in the Hilbert space case in [11, 12].

The Stochastic Maximum Principle, which is not treated in this chapter, remains one of the main sources of interest for studying BSDEs with infinite-dimensional process Y. Although the equation is linear in this case, the occurrence of unbounded coefficients often makes the study technically challenging. After the reference [45] already mentioned, the papers [196, 349] treat the maximum principle for a general controlled evolution equation in a Hilbert space. Applications to concrete controlled stochastic PDEs can be found in [598] for equations linear in the state, and in [280]. The case of a controlled stochastic PDE with additive noise and dissipative drift is treated in [282]. The treatise [414] is entirely devoted to the Stochastic Maximum Principle in infinite dimension.

A special mention is deserved for the study of the stochastic backward Hamilton–Jacobi–Bellman equation, introduced in [481] and further studied in [85]. Representation formulae for equations of similar type are proved in [549].

Many other cases of concrete stochastic PDEs of backward type have been studied, as objects of intrinsic interest and not necessarily related to stochastic optimal control problems, see for instance [197–199, 348, 419, 421, 504, 505, 552], and the subject is developing quickly.

Very often a scalar BSDE (i.e., where the process Y is real-valued) is introduced, coupled with a forward equation representing the dynamics of a controlled process evolving in an infinite-dimensional case, driven by a finite- or infinite-dimensional Brownian motion. This is the situation addressed in this chapter. As seen above, the process Y is then related to the value function of the optimal control problem and, in the Markovian case, it is used to represent or to construct a solution (in an appropriate sense) to the corresponding Hamilton–Jacobi–Bellman (HJB) equation. The first systematic study of this type for controlled stochastic equations in Hilbert space is in [284–286]. More general coefficients (for instance, of dissipative type), or more general growth conditions, were studied in [75–77, 351], see also [593–595].

Often, better results are obtained by a combination of probabilistic arguments on the BSDE and an analytic study of the HJB equation, as in [432, 433]. In [435, 438, 442] very general Hamiltonians are addressed. Smoothing effects of the HJB equation, due to a nondegenerate diffusion coefficient of the controlled equation, were studied in [283, 440].

The case of linear controlled evolution equations and quadratic cost also lead to stochastic backward equations of Riccati type, when the coefficients are perturbed by noise. In the infinite-dimensional framework we cite [333–335, 414, 415].

Applications to models with delay or memory effects can be found in several of the previous references. Memory effects are explicitly studied by BSDE techniques for the heat equation in [131] and for controlled stochastic Volterra equations in [63, 132]. Related results can be found in [600].

A special branch of the literature is devoted to the case when the controlled equation is a stochastic PDE with Brownian noise acting on the boundary conditions, often in combination with a control process on the boundary as well. We mention [181, 332, 437, 591, 592]. We also cite [331] for a version of the Stochastic Maximum Principle in this framework and [62] for the related case of dynamical boundary conditions.

Although in the large majority of the mentioned papers the state space is a Hilbert space, there are a few papers related to extensions to Banach space-valued processes: see [281, 436, 596].

BSDEs can be used to address other stochastic optimization problems, even when the controlled systems evolves in an infinite-dimensional space. In [182, 278] ergodic optimal control problems are studied, whereas applications of BSDEs to the theory of stochastic differential games are given in [274, 275], where games with an infinite number of players are considered.

More specific topics are treated in [273] (connections with conditioned processes in Hilbert spaces) and [330] (strongly coupled infinite-dimensional forward–backward systems, i.e., when the forward equations depends on the unknown pair (Y, Z) solution to the backward equation).

Appendix A
Notation and Function Spaces

In this appendix, we list the main notation and the definitions of the basic function spaces used throughout the book. Definitions of functions spaces that are introduced and used only in specific chapters are not included here.

A.1 Basic Notation

If X is a Banach space we denote its norm by $|\cdot|_X$. If this space is also Hilbert, we denote its inner product by $\langle \cdot, \cdot \rangle_X$. Given $R > 0$, $B_X(\bar{x}, R)$ denotes the closed ball in X centered at $\bar{x}$ of radius R. We will omit the subscript X if the context is clear. The dual space of X, i.e. the space of all continuous linear functionals, will be denoted by X^*. The (operator) norm in X^* will be denoted by $|\cdot|_{X^*}$, and the duality will be denoted by $\langle \cdot, \cdot \rangle_{\langle X^*, X \rangle}$.

If a sequence $(x_n)_{n \in \mathbb{N}} \subset X$ converges to $x \in X$ in the norm topology we write $x_n \to x$. If it converges weakly we write $x_n \rightharpoonup x$.

If X is a Hilbert space and $\{e_k\}_{k \in \mathbb{N}}$ is an orthonormal basis of X we use, for $x \in X$, the notation $x_k := \langle x, e_k \rangle$. Unless stated explicitly, we will always identify its dual X^* with X through the standard Riesz identification.

Given a second Banach space Y with norm $|\cdot|_Y$ (and inner product $\langle \cdot, \cdot \rangle_Y$ if it is also Hilbert) we denote by $\mathcal{L}(X, Y)$ the set of all bounded (continuous) linear operators $T : X \to Y$ with norm $\|T\|_{\mathcal{L}(X,Y)} := \sup_{x \in X, x \neq 0} \frac{|Tx|_Y}{|x|_X}$ (or simply $\|T\|$), using for simplicity the notation $\mathcal{L}(X)$ when $X = Y$. $\mathcal{L}(X)$ is a Banach algebra with identity element I_X (simply I if unambiguous).

Given a linear (possibly unbounded) operator $T : D(T) \subset X \to Y$ such that $D(T)$ is dense in X we will denote its adjoint operator by $T^* : D(T^*) \subset Y^* \to X^*$. If X is a Hilbert space we will denote by $\mathcal{S}(X) \subset \mathcal{L}(X)$ the space of all bounded self-adjoint operators on X.

For $k = 1, 2, \ldots$ we denote by X^k the product space $X \times X \times \cdots \times X$ (k times) endowed with the norm $|(x_1, \ldots, x_k)|_{X^k} := \left(|x_1|^2 + \ldots + |x_k|^2 \right)^{1/2}$ and by $\mathcal{L}^k(X, Y)$

© Springer International Publishing AG 2017
G. Fabbri et al., *Stochastic Optimal Control in Infinite Dimension*,
Probability Theory and Stochastic Modelling 82,
DOI 10.1007/978-3-319-53067-3

the set of all bounded multilinear operators $T : X^k \to Y$ with norm $\|T\|_{\mathcal{L}^k(X,Y)} :=$ $\sup_{x \in X^k, x \neq 0} \frac{|T(x_1,\ldots,x_k)|_Y}{|(x_1,\ldots,x_k)|_{X^k}}$ using for simplicity the notation $\mathcal{L}^k(X)$ when $X = Y$. It is known (see e.g. [266] p. 318, Theorem A.2.6) that $\mathcal{L}^k(X, Y)$ is isometrically isomorphic to the space $\mathcal{L}(X, \mathcal{L}(X, \ldots, \mathcal{L}(X, Y)))$.

Given a complex number $\lambda \in \mathbb{C}$ we denote by $\mathrm{Re}\lambda$ and $\mathrm{Im}\lambda$, respectively, its real and imaginary parts.

For a real number a we write $a^+ = \max(a, 0)$ and $a^- = -\min(a, 0)$ to denote the positive and negative parts of a. The same notation is also used for functions.

A.2   Function Spaces

Let Y and Z be two Banach spaces and let $X \subset Z$ be endowed with the induced topology. We denote by $B(X, Y)$, $B_b(X, Y)$, $C(X, Y)$, $UC(X, Y)$, $C_b(X, Y)$ and $UC_b(X, Y)$ the sets of all functions $\phi : X \to Y$ which are, respectively, Borel measurable, Borel measurable and bounded, continuous, uniformly continuous, continuous and bounded, uniformly continuous and bounded on X. The spaces $B_b(X, Y)$, $C_b(X, Y)$ and $UC_b(X, Y)$ are Banach spaces with the usual norm

$$\|\phi\|_0 = \sup_{x \in X} |\phi(x)|_Y.$$

We denote by $USC(X, Y)$ (respectively, $LSC(X, Y)$) the space of all upper semicontinuous (respectively, lower semicontinuous) functions $f : X \to Y$.

If $Y = \mathbb{R}$, we will simply write $B_b(X)$, $C(X)$, $UC(X)$, $C_b(X)$, $UC_b(X)$, $USC(X)$ and $LSC(X)$ for $B_b(X, \mathbb{R})$, $C(X, \mathbb{R})$, $UC(X, \mathbb{R})$, $C_b(X, \mathbb{R})$, $UC_b(X, \mathbb{R})$, $USC(X, \mathbb{R})$ and $LSC(X, \mathbb{R})$.

For a given $m > 0$ we define $B_m(X, Y)$, (respectively, $C_m(X, Y)$ and $UC_m(X, Y)$) to be the set of all functions $\phi \in B(X, Y)$ such that the function

$$\psi(x) := \frac{\phi(x)}{1 + |x|^m} \tag{A.1}$$

belongs to $B_b(X, Y)$ (respectively, $C_b(X, Y)$ and $UC_b(X, Y)$). These spaces of functions that have at most polynomial growth of order m are Banach spaces when they are endowed with the norm

$$N(\phi) := \sup_{x \in X} \frac{|\phi(x)|}{1 + |x|^m}.$$

We will write $\|\phi\|_{B_m(X,Y)}$, $\|\phi\|_{C_m(X,Y)}$, $\|\phi\|_{UC_m(X,Y)}$ to denote these norms, or simply $\|\phi\|_{B_m}$, $\|\phi\|_{C_m}$, $\|\phi\|_{UC_m}$ when the spaces are clear from the context. The above definition is also meaningful when $m = 0$ and in such case the spaces $B_m(X, Y)$, $C_m(X, Y)$ and $UC_m(X, Y)$ reduce to $B_b(X, Y)$, $C_b(X, Y)$ and $UC_b(X, Y)$. We do not use the notation $B_0(X, Y)$, $C_0(X, Y)$ and $UC_0(X, Y)$ to avoid confusion with respect to the standard notation in the literature. However, we often consider the spaces $B_m(X, Y)$, $C_m(X, Y)$ and $UC_m(X, Y)$ for $m \geq 0$, meaning that the case of the spaces $B_b(X, Y)$, $C_b(X, Y)$ and $UC_b(X, Y)$ is also included.

Let Y be a Banach space and X be an open subset of a Banach space Z. For $k \in \mathbb{N}$, we denote by $C^k(X, Y)$ (respectively, $C_b^k(X, Y)$) the set of all functions $\phi : X \to Y$ which are continuous (respectively, continuous and bounded) on X, together with all their Fréchet derivatives (see Sect. D.2) up to the order k. If $\phi \in C_b^k(X, Y)$ the l-th Fréchet derivative of ϕ is denoted by $D^l\phi$, (or simply $D\phi$ when $l = 1$). We set

$$\|\phi\|_k = \|\phi\|_0 + \sum_{l=1}^{k} \sup_{x \in X} \left\| D^l\phi(x) \right\|_{\mathcal{L}^l(Z,Y)}.$$

Similarly, we define the space $UC^k(X, Y)$ (respectively, $UC_b^k(X, Y)$), to be the set of all functions $\phi : X \to Y$ which are uniformly continuous (respectively, uniformly continuous and bounded) on X together with all their Fréchet derivatives up to the order k.

For $k \in \mathbb{N}$ and $m > 0$ we denote by $C_m^k(X, Y)$) (respectively, $UC_m^k(X, Y)$)) the set of all functions $\phi \in C_m(X, Y) \cap C^k(X, Y)$ (respectively, $\phi \in UC_m(X, Y) \cap C^k(X, Y)$) such that, for all $l = 1, \ldots, k$,

$$\sup_{x \in X} \frac{\|D^l\phi(x)\|_{\mathcal{L}^l(Z,Y)}}{1 + |x|^m} < +\infty, \tag{A.2}$$

requiring also for $\phi \in UC_m^k(X, Y)$ that the maps $x \to \frac{D^l\phi(x)}{1+|x|^m}$ are uniformly continuous for $l = 1, ..., k$. We set

$$\|\phi\|_{C_m^k(X,Y)} = \|\phi\|_{C_m(X,Y)} + \sum_{l=1}^{k} \sup_{x \in X} \frac{\|D^l\phi(x)\|_{\mathcal{L}^l(Z,Y)}}{1 + |x|^m}. \tag{A.3}$$

Equipped with this norm $C_m^k(X, Y)$ and $UC_m^k(X, Y)$ are both Banach spaces. If $\phi \in UC_m^k(X, Y)$ we will denote its norm by $\|\phi\|_{UC_m^k(X,Y)}$.

If $Y = \mathbb{R}$, we write $B_b^k(X)$, $B_m^k(X)$, $C^k(X)$, $C_b^k(X)$, $C_m^k(X)$, $UC^k(X)$, $UC_b^k(X)$, $UC_m^k(X)$ instead of $B_b^k(X, \mathbb{R})$, $B_m^k(X, \mathbb{R})$, $C^k(X, \mathbb{R})$, $C_b^k(X, \mathbb{R})$, $C_m^k(X, \mathbb{R})$, $UC^k(X, \mathbb{R})$, $UC_b^k(X, \mathbb{R})$, $UC_m^k(X, \mathbb{R})$, respectively.

Remark A.1 Our definition of spaces of polynomially growing functions is the one used, for example, in [179] (p. 251), [300].

In some papers (see e.g. [431, 433]) in the definition of spaces of functions with polynomial growth, the weight $(1 + |x|^2)^{m/2}$ is used instead of $1 + |x|^m$. This weight is always Fréchet differentiable when the norm in X is differentiable, whereas the weight $1 + |x|^m$ is not Fréchet differentiable at $x = 0$ when $m \in (0, 1]$. This may create problems when one has to deal with differentials of such weight so it is often more convenient to choose $(1 + |x|^2)^{m/2}$.

If the norm in X is Fréchet differentiable, often the function spaces $C_m^k(X, Y)$ and $UC_m^k(X, Y)$ are defined in a different way. For example, in [102] Sect. 2.1 the space $UC_m^1(X, Y)$ (there for $Y = \mathbb{R}$) is defined as the space of functions $\phi : X \to Y$ such that the functions $\psi(x) := \phi(x)(1 + |x|^m)^{-1}$ belong to $UC_b^1(X, Y)$. A similar definition can be used for $C_m^1(X, Y)$. With this definition the function $\phi(x) = 1 + |x|$ belongs to $UC_1^1(X)$ even if it is not Fréchet differentiable at $x = 0$. On the other hand, if one uses the weight $(1 + |x|^2)^{m/2}$ this problem disappears. Indeed, for each $m > 0$, a function ϕ is Fréchet differentiable in X if and only if the function $\psi(x) = \phi(x)(1 + |x|^2)^{-m/2}$ is Fréchet differentiable in X.

It is easy to see that if the norm in X is Fréchet differentiable then our space $C_m^1(X, Y)$ is equal to the space $\tilde{C}_m^1(X, Y)$ defined as the space of all functions $\phi : X \to Y$ such that the functions $\psi(x) = \phi(x)(1 + |x|^2)^{-m/2}$ belong to $C_b^1(X, Y)$. To simplify the presentation, suppose that $X = H$ is a Hilbert space and $Y = \mathbb{R}$.

Let $\phi \in \tilde{C}_m^1(H, \mathbb{R})$. Then obviously $\phi \in C_m(H, \mathbb{R})$ and moreover

$$D\phi(x) = D\psi(x)(1 + |x|^2)^{m/2} + \psi(x)m(1 + |x|^2)^{m/2-1}x.$$

Since, for all $x \in X$, $(1 + |x|^2)^{-1}m|x| \leq m$, we obtain $D\phi \in C_m(H, H)$. Thus $\phi \in C_m^1(H, \mathbb{R})$.

On the other hand, if $\phi \in C_m^1(H, \mathbb{R})$ then clearly $\psi \in C_b(H, \mathbb{R})$, it is Fréchet differentiable and

$$D\psi(x) = D\phi(x)(1 + |x|^2)^{-m/2} - \phi(x)m(1 + |x|^2)^{-m/2-1}x.$$

Since $(1+|x|^2)^{-1}m|x| \leq m$ we clearly have $D\psi \in C_b(H, H)$, i.e. $\phi \in \tilde{C}_m^1(H, \mathbb{R})$. ∎

For $\alpha \in (0, 1]$ we denote by $C^{0,\alpha}(X, Y)$ the space of all Hölder continuous functions from X to Y endowed with the semi-norm

$$[\varphi]_{0,\alpha} = \sup \left\{ \frac{|\varphi(x) - \varphi(y)|_Y}{|x - y|_Z^\alpha}; \ x, y \in X; \ x \neq y \right\}.$$

The space $C_b^{0,\alpha}(X, Y) := C_b(X, Y) \cap C^{0,\alpha}(X, Y)$ is a Banach space with the norm

$$\|\varphi\|_{0,\alpha} = \|\varphi\|_0 + [\varphi]_{0,\alpha}$$

and is contained in $UC_b(X, Y)$. If $\alpha = 1$ the space $C^{0,1}(X, Y)$ is the space of all Lipschitz continuous functions from X to Y. If X is open, convex and $\varphi \in C^{0,1}(X, Y)$ is Fréchet differentiable in X then the derivative $D\varphi$ is bounded and

$$[\varphi]_{0,1} = \sup_{x \in X} \|D\varphi(x)\|_{\mathcal{L}(Z,Y)}.$$

We set, for $\alpha \in (0, 1]$,

$$C^{1,\alpha}(X, Y) := \{\varphi \in C^1(X, Y) : [D\varphi]_{0,\alpha} < \infty\}.$$

The space $C_b^{1,\alpha}(X, Y) := C_b^1(X, Y) \cap C^{1,\alpha}(X, Y)$ is a Banach space with the norm

$$\|\varphi\|_{1,\alpha} = \|\varphi\|_1 + [D\varphi]_{0,\alpha}.$$

Similarly,

$$C^{2,\alpha}(X, Y) := \{\varphi \in C^2(X, Y) : [D^2\varphi]_{0,\alpha} < \infty\}.$$

The space $C_b^{2,\alpha}(X, Y) := C_b^2(X, Y) \cap C^{2,\alpha}(X, Y)$ is also a Banach space with the norm

$$\|\varphi\|_{2,\alpha} = \|\varphi\|_2 + [D^2\varphi]_{0,\alpha}.$$

If $Y = \mathbb{R}$ the above spaces are denoted by $C^{0,\alpha}(X)$, $C_b^{0,\alpha}(X)$, $C^{1,\alpha}(X)$, $C_b^{1,\alpha}(X)$, $C^{2,\alpha}(X)$, $C_b^{2,\alpha}(X)$.

If $\alpha \in (0, 1]$, we say that $\varphi : X \to Y$ is Hölder (Lipschitz when $\alpha = 1$) continuous on bounded subsets of X, if $\varphi \in C^{0,\alpha}(B(0, R) \cap X, Y)$ for every $R > 0$, i.e. if the semi-norm

$$[\varphi]_{0,\alpha,R} := \sup \left\{ \frac{|\varphi(x) - \varphi(y)|_Y}{|x - y|_Z^\alpha}; \ x, y \in B(0, R) \cap X; \ x \neq y \right\}$$

is finite for every $R > 0$. The space of such functions is denoted by $C_{\mathrm{loc}}^{0,\alpha}(X, Y)$. Similarly we define the spaces $C_{\mathrm{loc}}^{1,\alpha}(X, Y)$ and $C_{\mathrm{loc}}^{2,\alpha}(X, Y)$.

If $Y = \mathbb{R}$ we write $C^{k,\alpha}(X)$ instead of $C^{k,\alpha}(X, \mathbb{R})$ and $C_{\mathrm{loc}}^{k,\alpha}(X)$ instead of $C_{\mathrm{loc}}^{k,\alpha}(X, \mathbb{R})$, for $k = 0, 1, 2$, $\alpha \in (0, 1]$.

Let now X be a real separable Hilbert space and Y be a real Banach space. As usual we set, for an open subset $\mathcal{O}$ of $\mathbb{R}^n$, and $k \in \mathbb{N} \cup \{\infty\}$,

$$C_0^k(\mathcal{O}, Y) := \{f \in C^k(\mathcal{O}, Y) : f \text{ has compact support in } \mathcal{O}\}.$$

Following [120] Sect. 0 and [300] Sect. 2 we define various spaces of cylindrical functions. We set, for $k \in \mathbb{N} \cup \{\infty\}$,

$$\mathcal{F}C_0^k(X, Y) := \{\varphi : X \to Y : \exists n \in \mathbb{N}, \ x_1, \dots, x_n \in X, \ f \in C_0^k(\mathbb{R}^n, Y)$$
$$\text{such that } \varphi(x) = f(\langle x, x_1 \rangle, \dots, \langle x, x_n \rangle), \ \forall x \in X\}.$$

We denote $\mathcal{F}C_0^k(X, \mathbb{R})$ simply by $\mathcal{F}C_0^k(X)$.

Given $k \in \mathbb{N} \cup \{\infty\}$ and a linear closed operator with dense domain $B : D(B) \subset X \to X$, we define

$$\mathcal{F}C_0^{k,B}(X) = \{\varphi : X \to \mathbb{R} : \exists n \in \mathbb{N}, \ x_1, \dots, x_n \in D(B), \ f \in C_0^k(\mathbb{R}^n, \mathbb{R})$$
$$\text{such that } \varphi(x) = f(\langle x, x_1\rangle, \dots, \langle x, x_n\rangle), \ \forall x \in X\}$$
$$(A.4)$$

and

$$\mathcal{F}C_b^{k,B}(X) := \{\varphi : X \to \mathbb{R} : \exists n \in \mathbb{N}, \ x_1, \dots, x_n \in D(B), \ f \in C_b^k(\mathbb{R}^n, \mathbb{R})$$
$$\text{such that } \varphi(x) = f(\langle x, x_1\rangle, \dots, \langle x, x_n\rangle), \ \forall x \in H\}.$$

Let $\mathcal{O}$ be an open subset of $\mathbb{R}^n$ and $p \in [1, +\infty)$. We denote by $L^p(\mathcal{O})$ the set of all real-valued measurable functions[1] $f : \mathcal{O} \to \mathbb{R}$ with $\int_{\mathcal{O}} |f(\xi)|^p d\xi < +\infty$ (classical Lebesgue integral); $L^p(\mathcal{O})$ is a Banach space with the usual norm $|f|_{L^p(\mathcal{O})} := \left[\int_{\mathcal{O}} |f(\xi)|^p d\xi\right]^{1/p}$. We denote by $L^p_{loc}(\mathcal{O})$ the set of all measurable functions $f : \mathcal{O} \to \mathbb{R}$ such that $\int_K |f(\xi)|^p d\xi < +\infty$ for every compact subset K of $\mathcal{O}$. The space $L^\infty(\mathcal{O})$ is the quotient space of $B_b(\mathcal{O})$ with respect to the relation of being equal a.e. and is a Banach space with the usual ess sup norm. (Obviously the above spaces can be defined for more general sets $\mathcal{O}$.)

We denote by $W^{k,p}(\mathcal{O})$ ($k \in \mathbb{N}$, $p \in [1, +\infty]$) the usual Sobolev space of real-valued functions whose distributional derivatives, up to the order k, are p-th power integrable (or essentially bounded if $p = +\infty$). Moreover, $W_0^{k,p}(\mathcal{O})$ is the closure of $C_0^\infty(\mathcal{O})$ in $W^{k,p}(\mathcal{O})$. Following the standard convention, sometimes we will write $H^k(\mathcal{O})$ for $W^{k,2}(\mathcal{O})$ and $H_0^k(\mathcal{O})$ for $W_0^{k,2}(\mathcal{O})$. Similarly, for $\alpha \geq 0$ and $p \in [1, +\infty]$, we denote the fractional Sobolev spaces by $W^{\alpha,p}(\mathcal{O})$, $W_0^{\alpha,p}(\mathcal{O}$ (or $H^\alpha(\mathcal{O})$, $H_0^\alpha(\mathcal{O})$ when $p = 2$) defined in the usual way (see e.g. [1] Chap. VII). By duality then one defines, for every $\alpha > 0$, the negative order spaces $H^{-\alpha}(\mathcal{O})$ setting $(H_0^\alpha(\mathcal{O}))^* = H^{-\alpha}(\mathcal{O})$ (see e.g. [404] Sect. 1.12).

If the boundary $\partial\mathcal{O}$ is a C^∞ manifold of dimension $n - 1$ in $\mathbb{R}^n$ then the spaces $L^p(\partial\mathcal{O})$, $H^\alpha(\partial\mathcal{O})$, $W^{\alpha,p}(\partial\mathcal{O})$ can also be defined. We refer to [404], Sect. 1.7 or [1], Chap. VII, p. 215.

If Y is a real, separable Banach space and $a < b$, we define the space $W^{1,p}(a, b; Y)$ ($p \in [1, \infty]$) to be the set of all functions $f \in L^p(a, b; Y)$ whose weak derivative f' (see [554] Chap. III, Sect. 1) exists and belongs to $L^p(a, b; Y)$. It is a Banach space equipped with the norm $|f|_{W^{1,p}(a,b;Y)} := |f|_{L^p(a,b;Y)} + |f'|_{L^p(a,b;Y)}$.

Let X be a subset of a Banach space Z, Y be a Banach space and I be a subset of $\mathbb{R}^n$ (usually an interval in $\mathbb{R}$). We define the space

$$UC_b^x(I \times X, Y) := \Big\{\varphi \in C_b(I \times X, Y) :$$

$$\varphi(t, x) \text{ is uniformly continuous in } x, \text{ uniformly with respect to } t \in I \Big\}. \quad (A.5)$$

It is equipped with the $\|\cdot\|_0$ norm.

---

[1] That is, the set of all equivalence classes of such functions with respect to the relation of a.e. equality.

If $Y = \mathbb{R}$, we write $UC_b^x(I \times X)$ instead of $UC_b^x(I \times X, \mathbb{R})$.

Remark A.2 We recall some useful properties of $UC_b^x(I \times X, Y)$ (see [108] for more).

(1) If $I \subset \mathbb{R}^n$ is compact and $u \in UC_b^x(I \times X, Y)$, then for every compact set $K \subset X$, the restriction $u|_{I \times K}$ of u to $I \times K$ belongs to $UC_b(I \times K, Y)$. Thus, for every compact set $K \subset X$,

$$u|_{I \times K} \in C_b(I, C_b(K, Y)). \tag{A.6}$$

In particular, $u(\cdot, x)$ is uniformly continuous on I, uniformly with respect to $x \in K$, namely, for every $t, s \in I$, we have

$$\sup_{x \in K} |u(t, x) - u(s, x)|_Y \leq \rho_K(|t - s|), \tag{A.7}$$

where ρ_K is a modulus of continuity (see Appendix D.1) depending on the compact set K.

(2) $UC_b(I \times X, Y) \subset UC_b^x(I \times X, Y) \subset C_b(I \times X, Y)$. In particular, as the uniform continuity is stable with respect to the convergence in the norm $\| \cdot \|_0$, the space $UC_b^x(I \times X, Y)$ is a closed subspace of $C_b(I \times X, Y)$. On the other hand, if X is the whole space, $\varphi \in UC_b(X)$ and $\{e^{tA}, t \geq 0\}$ is a strongly continuous (not uniformly continuous) semigroup on X, then

$$w(t, x) = \varphi(e^{tA}x)$$

is a natural example of a function belonging to $UC_b^x(I \times X, Y)$ but not to $UC_b(I \times X, Y)$.

(3) In view of the above, given $u \in UC_b^x(I \times X, Y)$, the function $I \to UC_b(X, Y)$, $t \to u(t, \cdot)$ may not be continuous and even not be measurable (see on this Lemma 1.21 and the discussion preceding it). ∎

For a given $m > 0$ we define $B_m(I \times X, Y)$ (respectively, $C_m(I \times X, Y)$ and $UC_m(I \times X, Y)$) to be the set of all functions $\phi \in B(I \times X, Y)$ such that the function

$$\psi(t, x) := \frac{\phi(t, x)}{1 + |x|^m} \tag{A.8}$$

belongs to $B_b(I \times X, Y)$ (respectively, $C_b(I \times X, Y)$ and $UC_b(I \times X, Y)$). These spaces of functions that have at most polynomial growth of order m in the variable x are Banach spaces when they are endowed with the norm

$$N(\phi) := \sup_{(t,x) \in I \times X} \frac{|\phi(t, x)|}{1 + |x|^m}.$$

We will write $\|\phi\|_{B_m(I \times X, Y)}$, $\|\phi\|_{C_m(I \times X, Y)}$, $\|\phi\|_{UC_m(I \times X, Y)}$ to denote these norms.

Following [102], we introduce the space

$$UC^x_m(I \times X, Y) := \Big\{ \varphi \in C(I \times X, Y) :$$

$$\psi(t, x) := \frac{\phi(t, x)}{1 + |x|^m} \in UC^x_b(I \times X, Y) \Big\}.$$

$$(A.9)$$

It is equipped with the $\| \cdot \|_{C_m(I \times X, Y)}$ norm. The space has properties similar to those described in Remark A.2 for $UC^x_b(I \times X, Y)$.

Let X be an open subset of a Banach space Z, and Y be a Banach space. Let $I \subset \mathbb{R}$ be open. Given a function $u \in C(I \times X, Y)$ which is l times Fréchet differentiable in t and k times Fréchet differentiable in x, we denote its l-th partial derivative in t by $D^l_t u$, and its k-th partial derivative in x by $D^k_x u$. For the low order derivatives we use the symbols u_t, Du, $D^2 u$, and so on.

For $l, k = 0, 1, \ldots$, we denote by $C^{l,k}(I \times X, Y)$ (respectively, by $C^{l,k}_b(I \times X, Y)$, $UC^{l,k}(I \times X, Y)$, $UC^{l,k}_b(I \times X, Y)$) the space of all functions $\varphi : I \times X \to Y$ that are Fréchet differentiable l times in t and k times in x and which are continuous (respectively, continuous and bounded, uniformly continuous, uniformly continuous and bounded), together with their Fréchet derivatives up to these orders. If I is an interval which is not open then $C^{l,k}(I \times X, Y)$ is the space of functions from $I \times X \to Y$ whose restrictions to $I^o \times X$ (where I^o is the interior of I) belong to $C^{l,k}(I^o \times X, Y)$ and such that the functions and all their derivatives extend continuously to $I \times X$. The spaces $C^{l,k}_b(I \times X, Y)$, $UC^{l,k}(I \times X, Y)$, and $UC^{l,k}_b(I \times X, Y)$, for I not open, are defined similarly.

$C^{l,k}_b(I \times X, Y)$ and $UC^{l,k}_b(I \times X, Y)$ are Banach spaces endowed with the norm

$$\|\varphi\|_{C^{l,k}_b(I \times X, Y)} = \sup_{x \in X} \|\varphi(\cdot, x)\|_l + \sup_{t \in I} \|\varphi(t, \cdot)\|_k.$$

Since the notation $C^{0,1}_b(X, Y)$ is also used to denote the space of bounded Lipschitz continuous functions, to avoid confusion we emphasize that in this book, whenever the first set is equal to the product of a time interval and a subset of a Banach space, $C^{0,1}_b(I \times X, Y)$ will always be the space defined above and not the space of bounded Lipschitz functions on $I \times X$.

For $l, k = 0, 1, \ldots$, and $m > 0$ we denote by $C^{l,k}_m(I \times X, Y)$ (respectively, $UC^{l,k}_m(I \times X, Y)$) the set of all functions $\phi \in C_m(I \times X, Y) \cap C^{l,k}(I \times X, Y)$ (respectively, $\phi \in UC_m(I \times X, Y) \cap UC^{l,k}(I \times X, Y)$) such that, for all $i = 1, \ldots, l$, $j = 1, \ldots k$,

$$\sup_{(t,x) \in I \times X} \frac{\|D^i_t \phi(t, x)\|_{\mathcal{L}^i(\mathbb{R}, Y)}}{1 + |x|^m} < +\infty, \qquad \sup_{(t,x) \in I \times X} \frac{\|D^j_x \phi(t, x)\|_{\mathcal{L}^j(Z, Y)}}{1 + |x|^m} < +\infty.$$

$$(A.10)$$

We set

$$\|\phi\|_{C_m^{l,k}(I\times X,Y)} = \|\phi\|_{C_m(I\times X,Y)}$$

$$+\sum_{i=1}^{l} \sup_{(t,x)\in I\times X} \frac{\|D_t^i\phi(x)\|_{\mathcal{L}^i(\mathbb{R},Y)}}{1+|x|^m} + \sum_{i=1}^{k} \sup_{(t,x)\in I\times X} \frac{\|D_x^j\phi(x)\|_{\mathcal{L}^j(Z,Y)}}{1+|x|^m}. \tag{A.11}$$

Equipped with this norm $C_m^{l,k}(I \times X, Y)$ and $UC_m^{l,k}(I \times X, Y)$ are both Banach spaces. If $\phi \in UC_m^{l,k}(I \times X, Y)$ we will denote its norm by $\|\phi\|_{UC_m^{l,k}(I\times X,Y)}$.

If $Y = \mathbb{R}$ we will use the notation $C^{l,k}(I\times X)$, $C_b^{l,k}(I\times X)$, $C_m^{l,k}(I\times X)$, $UC^{l,k}(I\times X)$, $UC_b^{l,k}(I \times X)$ and $UC_m^{l,k}(I \times X)$ for the above spaces.

Let I be an interval in $\mathbb{R}$, X be a real separable Hilbert space and Y be a real Banach space. Similarly to the time-independent case we set, for $k \in \mathbb{N} \cup \{\infty\}$,

$$C_0^k(I \times \mathbb{R}^n, Y) := \left\{ f \in C^k(I \times \mathbb{R}^n, Y) : f \text{ has compact support in } I \times \mathbb{R}^n \right\}.$$

If $k \in \mathbb{N} \cup \{\infty\}$, we denote by $\mathcal{F}C_0^k(I \times X, Y)$ the space

$$\mathcal{F}C_0^k(I \times X, Y) := \left\{ \varphi : I \times X \to Y : \exists n \in \mathbb{N}, \ x_1, ..x_n \in X, \ f \in C_0^k(I \times \mathbb{R}^n, Y), \right.$$
$$\left. \text{such that } \varphi(t, x) = f(t, \langle x, x_1\rangle, ..., \langle x, x_n\rangle), \ \forall(t, x) \in I \times X \right\}. \tag{A.12}$$

Similarly, for $k \in \mathbb{N} \cup \{\infty\}$ and a linear, densely defined closed operator $B : D(B) \subset X \to X$, we define

$$\mathcal{F}C_0^{k,B}(I \times X) = \left\{ \varphi : I \times X \to \mathbb{R} : \exists n \in \mathbb{N}, \ x_1, ..x_n \in D(B), \ f \in C_0^k(I \times \mathbb{R}^n, \mathbb{R}) \right.$$
$$\left. \text{such that } \varphi(t, x) = f(t, \langle x, x_1\rangle, ..., \langle x, x_n\rangle), \ \forall(t, x) \in I \times X \right\}. \tag{A.13}$$

Appendix B
Linear Operators and C_0-Semigroups

All spaces considered in this book are real. However, the spectral theory has to be done in complex spaces and thus some results presented here require the use of complex spaces. To accommodate real Hilbert and Banach spaces we thus use complexification of spaces and operators, which for Hilbert spaces can be done in a natural way. If H is a real Hilbert space, its complexification H_c is defined by

$$H_c := \{\tilde{x} = x + iy : x, y \in H\}$$

with standard operations

$$(x+iy)+(z+iw) := (x+z)+i(y+w), \quad (a+ib)(x+iy) = (ax-by)+i(bx+ay), \ a, b \in \mathbb{R},$$

and with the inner product

$$\langle (x + iy), (z + iw) \rangle_c := \langle x, z \rangle_H + \langle y, w \rangle_H + i(\langle y, z \rangle_H - \langle x, w \rangle_H).$$

Thus $|x + iy|_{H_c} = |(x, y)|_{H \times H}$. A real Banach space E is complexified in the same way, however, except for special cases, the product norm is no longer a norm because the homogeneity condition fails. To define a norm in E_c we first compute

$$e^{it}(x + iy) = (x \cos t - y \sin t) + i(x \sin t + y \cos t)$$

and then define

$$|x + iy|_{E_c} := \sup_{0 \le t \le 2\pi} |(x \cos t - y \sin t, x \sin t + y \cos t)|_{E \times E}.$$

We refer the reader to [454, 527] for more on complexification.

© Springer International Publishing AG 2017
G. Fabbri et al., *Stochastic Optimal Control in Infinite Dimension*,
Probability Theory and Stochastic Modelling 82,
DOI 10.1007/978-3-319-53067-3

A linear operator $T : D(T) \subset E \to E$ is complexified by setting

$$D(T_c) := \{x + iy : x, y \in D(T)\}, \quad T_c(x + iy) := Tx + iTy.$$

It is easy to see that $T \in \mathcal{L}(E)$ if and only if $T_c \in \mathcal{L}(E_c)$, and moreover $\|T\| = \|T_c\|$. Also T is invertible if and only if T_c is invertible. It is a standard convention, which will not be repeated, that the spectrum and the resolvent set of T are understood to be the spectrum and the resolvent set of T_c. This is how the statements here should be understood in the context of real Hilbert and Banach spaces.

Throughout Appendix B, E will be a Banach space endowed with the norm $|\cdot|_E$ and H will be a Hilbert space endowed with the inner product $\langle \cdot, \cdot \rangle_H$ and the norm $|\cdot|_H$.

B.1   Linear Operators

For an operator T we denote by $D(T)$ its domain, by $R(T)$ its range and by $\ker T$ its kernel (or null space).

Definition B.1 (*Pseudoinverse*) If E is a uniformly convex Banach space, Z is a Banach space, and $T \in \mathcal{L}(E, Z)$, the *pseudoinverse* T^{-1} of T is the linear operator defined on $T(E) \subset Z$ that associates to every element z in $T(E)$ the element in $T^{-1}(z)$ with minimum norm (for the existence of such an element, see [202], II.4.29, p. 74). Notice that if E is a Hilbert space then we have

$$R(T^{-1}) = (\ker T)^{\perp}.$$

The following result is taken from [180], Proposition B.1, p. 429, where the reader can find its proof.

Proposition B.2 *Let E, E_1, E_2 be three Hilbert spaces, let $A_1 : E_1 \to E$, $A_2 : E_2 \to E$ be linear bounded operators, let $A_1^* : E \to E_1$ and $A_2^* : E \to E_2$ be their adjoints and finally let $A_1^{-1} : R(A_1) \subset E \to E_1$, $A_2^{-1} : R(A_2) \subset E \to E_2$ be the respective pseudoinverses. Then we have:*

(i) $R(A_1) \subset R(A_2)$ if and only if there exists a constant $k > 0$ such that

$$|A_1^* x|_{E_1} \leq k |A_2^* x|_{E_2} \quad \forall x \in E.$$

(ii) If

$$|A_1^* x|_{E_1} = |A_2^* x|_{E_2} \quad \forall x \in E,$$

then $R(A_1) = R(A_2)$ and

$$|A_1^{-1} x|_{E_1} = |A_2^{-1} x|_{E_2} \quad \forall x \in R(A_1).$$

Definition B.3 (*Closed operator, Graph norm*) Let E and Z be two Banach spaces. A linear operator $A : D(A) \subset E \to Z$ is said to be *closed* if its graph

$$\{(x, y) \in D(A) \times Z : y = Ax\}$$

is closed in $E \times Z$. Given a closed operator $A : D(A) \subset E \to Z$ the *graph norm* on $D(A)$ is defined as follows:

$$|x|_{D(A)} := \left(|x|_E^2 + |Ax|_Z^2\right)^{\frac{1}{2}} \qquad \text{for all }, x \in D(A).$$

Sometimes an equivalent norm $|x|_{D(A)} := |x|_E + |Ax|_Z$ is also used. If E and Z are Hilbert spaces then $D(A)$ can be endowed with the inner product $\langle x, y \rangle_{D(A)} := \langle x, y \rangle_E + \langle Ax, Ay \rangle_Z$ with respect to which $D(A)$ is a Hilbert space.

Proposition B.4 *Let E and Z be two Banach spaces. If $A : D(A) \subset E \to Z$ is a linear, closed operator, then $D(A)$ with the graph norm is a Banach space. Moreover, if $D(A)$ is endowed with the graph norm, $A : D(A) \to Z$ is continuous.*

Definition B.5 (*Resolvent*) Consider a linear, closed operator $A : D(A) \subset E \to E$ and define, for $\lambda \in \mathbb{C}$, the operator $(\lambda I - A) : D(A) \to E$. The *resolvent set* of A is defined as follows:

$$\varrho(A) := \left\{\lambda \in \mathbb{C} : (\lambda I - A) \text{ is invertible and } (\lambda I - A)^{-1} \in \mathcal{L}(E)\right\}.$$

$$\sigma(A) := \mathbb{C} \setminus \varrho(A)$$

is called the *spectrum* of A. For each $\lambda \in \varrho(A)$ the operator $(\lambda I - A)^{-1}$ is called the resolvent operator of A. The family of resolvent operators satisfies the so-called resolvent identity: for all $\lambda_1, \lambda_2 \in \varrho(A)$ we have

$$(\lambda_1 I - A)^{-1} - (\lambda_2 I - A)^{-1} = (\lambda_2 - \lambda_1)(\lambda_1 I - A)^{-1}(\lambda_2 I - A)^{-1}. \qquad \text{(B.1)}$$

Lemma B.6 *Let $A : D(A) \subset E \to E$ be a linear, closed operator. Then the resolvent set $\varrho(A)$ is open.*

Proof See [576], Theorem 1 p. 201. □

Definition B.7 (*Closable operator, closure of an operator*) Let E and Z be two Banach spaces. A linear operator $A : D(A) \subset E \to Z$ is said to be *closable* if the closure of its graph in $E \times Z$ is the graph of some (closed) operator. Such an operator is called the *closure of A* and it is denoted by $\overline{A}$.

Remark B.8 It is easy to see that $A : D(A) \subset E \to Z$ is closable if and only if, for any sequence x_n in $D(A)$ and any $y \in Z$ such that

$$x_n \xrightarrow[E]{n \to \infty} 0 \quad \text{and} \quad Ax_n \xrightarrow[Z]{n \to \infty} y,$$

we have $y = 0$.   ∎

Definition B.9 (*Core of a closed operator*) Let E and Z be two Banach spaces. Consider a closed linear operator $A : D(A) \subset E \to Z$. A linear subspace Y of $D(A)$ is said to be a *core* for A if it is dense in $D(A)$ (endowed with its graph norm). In other words, Y is a core for A if and only if $Y \subset D(A)$ and, for any $x \in D(A)$, there exists a sequence x_n of elements of Y such that $x_n \to x$ and $Ax_n \to Ax$.

B.2 Dissipative Operators

Definition B.10 (*Duality mapping*) Let E be a Banach space and E^* be its dual. The function $\mathcal{J} : E \to 2^{E^*}$, defined by

$$\mathcal{J}(x) := \left\{ x^* \in E : \langle x^*, x \rangle_{\langle E^*, E \rangle} = |x|_E^2 = |x^*|_{E^*}^2 \right\}, \qquad \text{for } x \in E,$$

is called the *duality mapping*.

The duality mapping is in general multivalued and, for all $x \in E$, $J(x) \neq \emptyset$. If the dual E^* is strictly convex, and in particular if E is a Hilbert space, $\mathcal{J}$ is single-valued (see Sect. 1.1 in [26] and in particular Theorem 1.2).

Lemma B.11 (Kato's Lemma) *Let E be a Banach space and $x, y \in E$. There exists a $w \in \mathcal{J}(x)$ such that $\langle w, y \rangle_{\langle E^*, E \rangle} \geq 0$ if and only if*

$$|x|_E \leq |x + \lambda y|_E$$

for any $\lambda > 0$.

Proof See [26], Lemma 3.1, p. 98.   □

In the remainder of this section A will denote a possibly nonlinear operator.

Definition B.12 (*Dissipative operators*) Let E be a Banach space. An operator $A : D(A) \subset E \to E$ is called *dissipative* if

$$\langle w, A(x) - A(y) \rangle_{E^*, E} \leq 0, \quad \text{for all } x, y \in D(A) \text{ and some } w \in \mathcal{J}(x - y). \quad \text{(B.2)}$$

A dissipative operator A is said to be *m-dissipative* if $R(I - A) = E$. A dissipative operator A is said to be *maximal dissipative* if there does not exist any proper dissipative extension of A.

In the specific case of a linear operator A in a Hilbert space H the expression (B.2) can be rephrased as $\langle Ax, x \rangle_H \leq 0$ for all $x \in D(A)$. If the Hilbert space is complex one considers its real part: Re $\langle Ax, x \rangle_H \leq 0$.

Definition B.13 (*Accretive operators*) Let E be a Banach space. An operator $A :$ $D(A) \subset E \rightarrow E$ is called *accretive* (respectively m-accretive, *maximal accretive*) if $-A$ is dissipative (respectively m-dissipative, maximal dissipative).

Remark B.14 It easily follows from the definition that any m-dissipative operator is maximal dissipative. In case of (possibly nonlinear) operators in Hilbert spaces the two properties are equivalent, see Remark 3.1, p. 101 of [26]. ∎

Proposition B.15 *An operator $A : D(A) \subset E \rightarrow E$ is dissipative if and only if, for any $\lambda > 0$, for any $x, y \in D(A)$,*

$$|x - y|_E \leq |x - y - \lambda(A(x) - A(y))|_E ,$$

or equivalently, if there exists a $\lambda > 0$ with such a property.

Proof See Proposition 3.1, p. 98, of [26]. □

Remark B.16 Proposition B.15 can be rewritten in an obvious way in the following form: $A : D(A) \subset E \rightarrow E$ is dissipative if and only if for any $\lambda > 0$, for any $x, y \in D(A)$,

$$|x - y|_E \leq \frac{1}{\lambda} |(\lambda x - A(x)) - (\lambda y - A(y))|_E ,$$

or equivalently, if there exists a $\lambda > 0$ with such a property. ∎

Proposition B.17 *A dissipative operator $A : D(A) \subset E \rightarrow E$ is m-dissipative if and only if $R(I - \lambda A) = E$ for all (equivalently, for some) $\lambda > 0$.*

Proof See Proposition 3.3, p. 99 of [26]. □

Proposition B.18 *Any linear maximal dissipative operator A in a Hilbert space H is closed if and only if it has a dense domain.*

Proof See [490], Theorem 1.1.1, pp. 200–201 and Lemma 1.1.3, p. 201. □

Proposition B.19 *Let E be a Banach space. Let $\lambda_0 > 0$ and*

$$F: [\lambda_0, +\infty) \rightarrow C^{0,1}(E, E)$$

be a function such that, for any $\lambda \geq \lambda_0$, $F(\lambda)$ is injective and, for any $\lambda, \mu \geq \lambda_0$,

$$F(\lambda) = F(\mu) \circ (I_E + (\mu - \lambda)F(\lambda)) . \tag{B.3}$$

Then there exists a unique operator $A: D(A) \subset E \rightarrow E$ such that, for any $\lambda \geq \lambda_0$,

$$F(\lambda) = (\lambda I - A)^{-1}.$$

If, moreover,

$$|F(\lambda)x - F(\lambda)y| \leq \frac{1}{\lambda}|x - y|, \quad for\ all\ \lambda \geq \lambda_0, x, y \in E, \quad (B.4)$$

then A is m-dissipative. Hence $(\lambda - A)^{-1}$ is well defined for all $\lambda \in (0, +\infty)$ and satisfies the properties (B.3) and (B.4) for any $\lambda, \mu > 0$.

Proof Proposition I.3.3, p. 13 of [146] ensures the existence of the operator A. The second part about the m-dissipativity of A follows from Proposition II.9.6 of [146]. The third part about the extension follows from Propositions I.3.2 and II.9.1 of [146]. □

We refer to Chap. 3 of [26], Appendix D of [180], Chap. 5 of [177], Chap. 3 of [183], [217], [479], [587], [588] for more about dissipative and accretive operators.

B.3 Trace Class and Hilbert–Schmidt Operators

Throughout this section H, U, V will denote real, separable Hilbert spaces, $\langle \cdot, \cdot \rangle_H$, $\langle \cdot, \cdot \rangle_U, \langle \cdot, \cdot \rangle_V, |\cdot|_H, |\cdot|_U, |\cdot|_V$ will be, respectively, the inner products in H, U and V and the related norms.

Definition B.20 A linear operator $T \in \mathcal{L}(U, H)$ is called *nuclear* or *trace class* if T can be represented in the form

$$T(z) = \sum_{k=1}^{+\infty} b_k \langle z, a_k \rangle_U \quad for\ any\ z \in U,$$

where a_k and b_k are two sequences of elements, respectively, in U and H such that $\sum_{k=1}^{+\infty} |a_k|_U |b_k|_H < +\infty$. We denote the set of all nuclear operators from U to H by $\mathcal{L}_1(U, H)$. We write $\mathcal{L}_1(H)$ instead of $\mathcal{L}_1(H, H)$.

Proposition B.21 *$\mathcal{L}_1(U, H)$ is a separable Banach space with respect to the norm*

$$\|T\|_{\mathcal{L}_1(U,H)} := \inf \left\{ \sum_{k=1}^{+\infty} |a_k|_U |b_k|_H : \{a_k\} \subset U, \{b_k\} \subset H, \right.$$

$$\left. and\ T(z) = \sum_{k=1}^{+\infty} b_k \langle z, a_k \rangle_U, \forall z \in U \right\}.$$

$$(B.5)$$

Proof See [523] Proposition 2.8, p. 21 and the subsequent observations. □

Proposition B.22 *Given $T \in \mathcal{L}_1(H)$ and an orthonormal basis $\{e_k\}$ of H, the series*

$$\sum_{k \in \mathbb{N}} \langle Te_k, e_k \rangle_H$$

converges absolutely and its sum does not depend on the choice of the basis $\{e_k\}$.

Proof See [487], pp. 357–358 after the proof of Proposition A.4. □

Definition B.23 Given $T \in \mathcal{L}_1(H)$ and any orthonormal basis $\{e_k\}$ of H,

$$\mathrm{Tr}(T) := \sum_{k \in \mathbb{N}} \langle Te_k, e_k \rangle_H$$

is called the *trace* of T.

We have $|\mathrm{Tr}(T)| \le \|T\|_{\mathcal{L}_1(H)}$ (see e.g. [487], p. 357).

Definition B.24 Let $\{e_k\}_{k \in \mathbb{N}}$ be an orthonormal basis of U. The space of *Hilbert–Schmidt operators* $\mathcal{L}_2(U, H)$ from U to H is defined by

$$\mathcal{L}_2(U, H) := \left\{ T \in \mathcal{L}(U, H) \ : \ \sum_{k \in \mathbb{N}} |Te_k|_H^2 < +\infty \right\}. \tag{B.6}$$

We write $\mathcal{L}_2(H)$ for $\mathcal{L}_2(H, H)$.

Proposition B.25 *The space of Hilbert–Schmidt operators $\mathcal{L}_2(U, H)$ does not depend on the choice of orthonormal basis $\{e_k\}_{k \in \mathbb{N}}$. It is a separable Hilbert space if endowed with the inner product*

$$\langle S, T \rangle_2 := \sum_{k \in \mathbb{N}} \langle Se_k, Te_k \rangle_H , \qquad S, T \in \mathcal{L}_2(U, H).$$

The inner product is independent of the choice of basis.

Proof See Appendix C of [180] after Proposition C.3. □

The following proposition follows easily from the definition of the space of Hilbert–Schmidt operators and elementary calculations.

Proposition B.26 *(i) $T \in \mathcal{L}_2(U, H)$ if and only if $T^* \in \mathcal{L}_2(H, U)$. Moreover, $\|T\|_{\mathcal{L}_2(U,H)} = \|T^*\|_{\mathcal{L}_2(H,U)}$.*
(ii) If $T \in \mathcal{L}_2(U, H)$ and $S \in \mathcal{L}(H, V)$ then $ST \in \mathcal{L}_2(U, V)$ and

$$\|ST\|_{\mathcal{L}_2(U,V)} \le \|S\|_{\mathcal{L}(H,V)} \|T\|_{\mathcal{L}_2(U,H)}.$$

If $T \in \mathcal{L}(U, H)$ and $S \in \mathcal{L}_2(H, V)$ then $ST \in \mathcal{L}_2(U, V)$ and

$$\|ST\|_{\mathcal{L}_2(U,V)} \le \|S\|_{\mathcal{L}_2(H,V)} \|T\|_{\mathcal{L}(U,H)}.$$

Definition B.27 Let $U \subset H$. The embedding $U \subset H$ is said to be *Hilbert–Schmidt* if for some orthonormal basis $\{e_k\}_{k \in \mathbb{N}}$ of U, we have

$$\sum_{k \in \mathbb{N}} |e_k|_H^2 < +\infty.$$

Thanks to Proposition B.25 this definition does not depend on the choice of basis $\{e_k\}$.

Proposition B.28 *The following properties hold:*

(i) *If $T \in \mathcal{L}_1(U, H)$ and $S \in \mathcal{L}(H, V)$ then ST is in $\mathcal{L}_1(U, V)$ and*

$$\|ST\|_{\mathcal{L}_1(U,V)} \leq \|S\|_{\mathcal{L}(H,V)} \|T\|_{\mathcal{L}_1(U,H)}.$$

If $T \in \mathcal{L}(U, H)$ and $S \in \mathcal{L}_1(H, V)$ then ST is in $\mathcal{L}_1(U, V)$ and

$$\|ST\|_{\mathcal{L}_1(U,V)} \leq \|S\|_{\mathcal{L}_1(H,V)} \|T\|_{\mathcal{L}(U,H)}.$$

(ii) *If $T \in \mathcal{L}_1(U, H)$ and $S \in \mathcal{L}(H, U)$ (respectively, $T \in \mathcal{L}(U, H)$ and $S \in \mathcal{L}_1(H, U)$) then ST is in $\mathcal{L}_1(U)$, TS is in $\mathcal{L}_1(H)$ and*

$$\mathrm{Tr}(ST) = \mathrm{Tr}(TS).$$

(iii) *If $T \in \mathcal{L}_2(U, H)$, $S \in \mathcal{L}_2(H, V)$ then $ST \in \mathcal{L}_1(U, V)$ and*

$$\|ST\|_{\mathcal{L}_1(U,V)} \leq \|S\|_{\mathcal{L}_2(H,V)} \|T\|_{\mathcal{L}_2(U,H)}.$$

Moreover, if $U = V$, then $\mathrm{Tr}(ST) = \mathrm{Tr}(TS)$.

(iv) $\mathcal{L}_1(U, H) \subset \mathcal{L}_2(U, H)$.

(v) *If $T \in \mathcal{L}_2(U, H)$ then T is compact.*

Proof Most claims are proved in Appendix A.2 of [487]. More precisely: (i) is proved in Propositions 4, p. 356, (ii) is proved in Proposition A.5-(i), p. 358, while (iv) and (v) are proved in Proposition A.6, p. 359. The proof of (iii) also repeats the proof of Proposition A.5-(ii), p. 358, however we include it here for completeness. Consider an orthonormal basis $\{f_k\}_{k \in \mathbb{N}}$ of H. For every $z \in U$ we have

$$Tz = \sum_{k \in \mathbb{N}} \langle Tz, f_k \rangle_H f_k = \sum_{k \in \mathbb{N}} \langle z, T^* f_k \rangle_U f_k$$

so

$$STz = \sum_{k \in \mathbb{N}} \langle z, T^* f_k \rangle_U Sf_k \tag{B.7}$$

and then, using the definition of $\mathcal{L}_1$-norm given in (B.5) and the Cauchy–Schwarz inequality, we get

$$\|ST\|_{\mathcal{L}_1(U,V)} \le \sum_{k \in \mathbb{N}} |T^* f_k|_U |Sf_k|_V \le \left(\sum_{k \in \mathbb{N}} |T^* f_k|_U^2\right)^{1/2} \left(\sum_{k \in \mathbb{N}} |Sf_k|_V^2\right)^{1/2}$$

$$= \|T^*\|_{\mathcal{L}_2(H,U)} \|S\|_{\mathcal{L}_2(H,V)} = \|T\|_{\mathcal{L}_2(U,H)} \|S\|_{\mathcal{L}_2(H,V)},$$

where in the last step we used Proposition B.26-(i). It also follows from (B.7) and the Parseval identity that, if $U = V$,

$$\text{Tr}(ST) = \sum_{k \in \mathbb{N}} \langle Sf_k, T^* f_k\rangle_U = \sum_{k \in \mathbb{N}} \langle T Sf_k, f_k\rangle_H = \text{Tr}(TS).$$

$\square$

Observe that in general, if $T, S \in \mathcal{L}(H)$ and $TS \in \mathcal{L}_1(H)$, this does not necessarily imply $ST \in \mathcal{L}_1(H)$. Indeed, consider two operators defined on $H \times H$ by

$$T = \begin{pmatrix} I & I \\ -I & -I \end{pmatrix}, \quad S = \begin{pmatrix} -I & I \\ I & -I \end{pmatrix},$$

where I stands for the identity operator. Then

$$TS = 0, \quad ST = 2\begin{pmatrix} -I & -I \\ I & I \end{pmatrix},$$

however ST is not trace class if H is infinite-dimensional. Another example is given in [179], p. 6.

Notation B.29 We set

$$\mathcal{L}^+(H) := \{T \in \mathcal{S}(H) : \langle Tx, x\rangle_H \ge 0 \; \forall x \in H\}$$

and

$$\mathcal{L}_1^+(H) := \mathcal{L}_1(H) \cap \mathcal{L}^+(H).$$

∎

The operators in $\mathcal{L}^+(H)$ are called *positive operators* on H. A positive operator T on H is called *strictly positive* if it satisfies $\langle Tx, x\rangle_H > 0$ for all $x \in H$.

Proposition B.30 *An operator $T \in \mathcal{L}^+(H)$ is nuclear if and only if*

$$\sum_{k \in \mathbb{N}} \langle Te_k, e_k\rangle_H < +\infty$$

for an orthonormal basis $\{e_n\}$ on H. Moreover, in this case $\text{Tr}(T) = \|T\|_{\mathcal{L}_1(H)}$.

Proof See [180], Proposition C.3. $\square$

B.4 C_0-Semigroups and Related Results

B.4.1 Basic Definitions

Definition B.31 (C_0-semigroup) A map $S : [0, +\infty) \to \mathcal{L}(E)$ is called a C_0-semigroup (or a strongly continuous semigroup on E) if the following three conditions are satisfied:

  (i) $S(0) = I$.
  (ii) For all $s, t \in [0, +\infty)$, $S(t)S(s) = S(t + s)$.
  (iii) For all $x \in E$, the map $t \to S(t)x$ is continuous from $[0, +\infty)$ to E.[2]

  For C_0-semigroups we will use the notation $\{S(t), t \geq 0\}$ or simply $S(t)$.

Definition B.32 (*Generator of a C_0-semigroup*) Let $S(t)$ be a C_0-semigroup on E. The linear operator $A : D(A) \subset E \to E$ defined as

$$\begin{cases} D(A) := \left\{ x \in E : \frac{S(t)x-x}{t} \text{ has a limit in } E \text{ when } t \to 0^+ \right\} \\ Ax := \lim_{t \to 0^+} \frac{S(t)x-x}{t} \end{cases}$$

is called the *infinitesimal generator* of $S(t)$.

Proposition B.33 *Let $S(t)$ be a C_0-semigroup on E. Then there exist $M \geq 1$ and $\omega \in \mathbb{R}$ such that*

$$\|S(t)\| \leq M e^{\omega t}, \qquad \text{for } t \geq 0. \tag{B.8}$$

Proof See for instance [479], Theorem 2.2, Chap. 1, p. 4.   □

  The infimum of all ω such that (B.8) is satisfied for some M_ω is called the *type* of the C_0-semigroup $S(t)$ and is denoted by ω_0, see [47], Part II, Sect. 2.2. We have $\omega_0 \in [-\infty, +\infty)$. If $\omega_0 < 0$ we say that the C_0-semigroup $S(t)$ is of *negative type* and if $\omega_0 > 0$ we say that the C_0-semigroup $S(t)$ is of *positive type*.

Definition B.34 (*Contraction semigroup*) A C_0-semigroup $S(t)$ on E is called a C_0-*semigroup of contractions* if (B.8) holds with $M = 1$, $\omega = 0$.

Definition B.35 (*Pseudo-contraction semigroup*) A C_0-semigroup $S(t)$ on E is called a C_0-*semigroup of pseudo-contractions* if (B.8) holds with $M = 1$ for some $\omega \in \mathbb{R}$.

Definition B.36 (*Uniformly bounded semigroup*) A C_0-semigroup $S(t)$ on E is called *uniformly bounded* if (B.8) holds with $\omega = 0$ for some $M \geq 1$.

  We remark that the complexification $S_c(t)$ of a C_0-semigroup on E is a C_0-semigroup on E_c whose generator is the complexification A_c of the generator A of $S(t)$.

---

[2] Equivalently one can ask here that $\lim_{t \searrow 0} S(t)x = x$ for all $x \in E$, see e.g. [479] Corollary 2.3, p. 4.

B.4.2 The Hille–Yosida Theorem and Yosida Approximations

Theorem B.37 (Hille–Yosida) *A linear operator* $A : D(A) \subset E \to E$ *is the infinitesimal generator of a C_0-semigroup $S(t)$ on E satisfying (B.8) if and only if*

(1) A is closed and $D(A)$ is dense in E,
(2) $(\lambda I - A)$ is invertible and $(\lambda I - A)^{-1} \in \mathcal{L}(E)$ for every $\lambda > \omega$, and

$$\|((\lambda I - A)^{-1})^k\| \le M (\lambda - \omega)^{-k} \quad \text{for all } k \in \mathbb{N} \text{ and } \lambda > \omega.$$

Proof See [296], Theorem 2.13, p. 20 or [479], Theorem 5.3. □

In fact, see [479], Remark 5.4, we have the following.

Remark B.38 For complex spaces condition (2) in Theorem B.37 can be replaced by $\{\lambda \in \mathbb{C} : \operatorname{Re}\lambda > \omega\} \subset \varrho(A)$ and, for all $k \in \mathbb{N}$ and $\lambda \in \mathbb{C}$, $\operatorname{Re}\lambda > \omega$,

$$\|((\lambda I - A)^{-1})^k\| \le M (\operatorname{Re}\lambda - \omega)^{-k}.$$

∎

Remark B.39 Considering, if needed, $\tilde{S}(t) := e^{-(\omega+\varepsilon)t} S(t), \varepsilon > 0$, we can always restrict to uniformly bounded C_0-semigroups having invertible generators. In particular, condition (2) of Theorem B.37 can be assumed to hold with $\omega = 0$. ∎

Definition B.40 (*Yosida approximations*) Let $A : D(A) \subset E \to E$ be the infinitesimal generator of a C_0-semigroup $S(t)$ on E satisfying (B.8). For $n \in \mathbb{N}$ greater than ω, define

$$J_n = n(nI - A)^{-1}. \tag{B.9}$$

The *Yosida approximation* of A is defined as follows:

$$A_n := n^2(nI - A)^{-1} - nI = AJ_n \in \mathcal{L}(E). \tag{B.10}$$

Lemma B.41 *Let A and J_n be as in Definition B.40. Then*

$$\|J_n\| \le \frac{Mn}{n - \omega} \quad \text{for all } n > \omega. \tag{B.11}$$

Proof This follows directly from condition (2) ($k = 1$) of Theorem B.37. □

Proposition B.42 *Let $A : D(A) \subset E \to E$ be the infinitesimal generator of a C_0-semigroup on E. Let J_n and A_n be as in Definition B.40. Then*

$$J_n x \xrightarrow[E]{n \to \infty} x \quad \text{for all } x \in E \tag{B.12}$$

and

$$A_n x \xrightarrow[E]{n \to \infty} Ax \quad \text{for all } x \in D(A).$$ (B.13)

Proof See [180], Proposition A.4, p. 409. ☐

Proposition B.43 *Let* $A : D(A) \subset E \to E$ *be the infinitesimal generator of a C_0-semigroup $S(t)$ on E satisfying (B.8). Let A_n be the Yosida approximation of A. Define, for $x \in E$ and $t \geq 0$,*

$$e^{t A_n} x := \sum_{j \in \mathbb{N}} \frac{t^j A_n^j x}{j!}.$$

Then,

$$\| e^{t A_n} \| \leq M e^{t \frac{n\omega}{n-\omega}}$$ (B.14)

and

$$S(t)x = \lim_{n \to \infty} e^{t A_n} x \quad \text{for every } x \in E$$ (B.15)

uniformly on bounded subsets of $[0, +\infty)$.

Proof See [479], Theorem 5.5, p. 21 and [47], Step 2 of the proof of Theorem 2.5, pp. 102–103. ☐

Expression (B.15) shows how to explicitly construct the semigroup generated by a linear operator A.

Notation B.44 The semigroup generated by A will be denoted by $e^{t A}$. ■

Theorem B.45 (Lumer–Phillips) *Let H be a separable Hilbert space. Given a linear operator $A : D(A) \subset H \to H$, the following facts are equivalent:*

(1) A is the generator of a C_0-semigroup of contractions on H.
(2) $\overline{D(A)} = H$ and A is maximal dissipative.
(3) $\overline{D(A)} = H$ and A^* is maximal dissipative.
(4) $\overline{D(A)} = H$, A is dissipative and $R(\lambda_0 I - A) = H$ for some $\lambda_0 > 0$.

Proof See [490], Theorem 1.1.3, p. 203, Theorem 1.4.2, p. 214, and [479], Theorem 4.3, p. 14. □

The following result is a corollary of the Trotter–Kato Theorem.

Proposition B.46 (Trotter–Kato) *Let* $S(t)$, $S_n(t)$, $n \in \mathbb{N}$, *be strongly continuous semigroups on E with generators A and A_n, respectively. Assume that* $D(A) \subset D(A_n)$ *for every $n \in \mathbb{N}$ and that*

$$\|S(t)\|, \|S_n(t)\| \le M e^{\omega t}, \qquad \forall t \ge 0, \ n \in \mathbb{N}$$

for some constants $M \ge 1$ and $\omega \in \mathbb{R}$. If $A_n x \to A x$ for every $x \in D(A)$ then

$$S_n(t)x \to S(t)x, \qquad \forall t \ge 0, \ x \in X,$$

and the limit is uniform in t for t in bounded intervals.

Proof See [479], Theorem 4.5, p. 88, or [217], Theorem 4.8, p. 209. □

Proposition B.47 *Let $S(t)$ be a strongly continuous semigroup on E with the generator A. Let $Y \subset D(A)$ be a subspace of $D(A)$. Assume that*

(i) Y is dense on E,
(ii) $S(t)(Y) \subset Y$ for all $t \ge 0$.

Then Y is a core for A.

Proof See [153], Proposition A.19, p. 204. □

B.4.3 Analytic Semigroups and Fractional Powers of Generators

Throughout this section H is a separable Hilbert space. The material about analytic semigroups requires that H be complex. The statements for real H should be understood with the convention that H, A, $S(t)$ are their complexifications H_c, A_c, $S_c(t)$.

Definition B.48 (*Differentiable semigroup*) A C_0-semigroup $S(t)$ on H is called *differentiable* if for every $x \in H$, $t \to S(t)x$ is differentiable for $t > 0$.

Definition B.49 (*Analytic semigroup*) A C_0-semigroup $S(t)$ on H is called *analytic* if it has an extension $G(z)$ to a sector of the form $\Delta := \{z \in \mathbb{C} \ : \ a < \arg(z) < b\}$ for some $a < 0 < b$ with the following properties:

(i) $z \to G(z)$ is analytic on Δ.
(ii) $G(0) = I$ and $\lim_{\substack{z \in \Delta \\ z \to 0}} G(z)x = x$ for every $x \in H$.
(iii) $G(z_1 + z_2) = G(z_1)G(z_2)$ for $z_1, z_2 \in \Delta$.

Theorem B.50 *Let $S(t)$ be a uniformly bounded C_0-semigroup on H and let A be the generator of $S(t)$. Assume that $0 \in \varrho(A)$. The following are equivalent:*

(1) $S(t)$ can be extended to an analytic semigroup in a sector $\Delta_\delta = \{z \in \mathbb{C} \;:\; |\arg(z)| < \delta\}$, and $\|S(z)\|$ is uniformly bounded in every subsector $\Delta_{\delta'} = \{z \in \mathbb{C} \;:\; |\arg(z)| \le \delta'\}$, $\delta' < \delta$.

(2) There exists a $\delta \in \left(0, \frac{\pi}{2}\right)$ and $B > 0$ such that

$$\Sigma := \left\{\lambda \;:\; |\arg(\lambda)| < \frac{\pi}{2} + \delta\right\} \cup \{0\} \subset \varrho(A)$$

and, for every $\lambda \in \Sigma \setminus \{0\}$,

$$\|(\lambda I - A)^{-1}\| \le \frac{B}{|\lambda|}.$$

(3) $S(t)$ is differentiable and there exists a constant $C > 0$ such that

$$\|A S(t)\| \le \frac{C}{t} \quad \text{for } t > 0.$$

Proof See [479] Theorem 5.2, p. 61.   □

Theorem B.51 *Consider a linear operator $A : D(A) \subset H \to H$ that generates a uniformly bounded analytic C_0-semigroup $S(t)$, and $0 \in \varrho(A)$. Define, for $\alpha < 0$,*

$$(-A)^\alpha := \frac{1}{\Gamma(\alpha)} \int_0^\infty t^{-\alpha-1} S(t)\,dt, \tag{B.16}$$

where $\Gamma(\cdot)$ is the Gamma function, and set $(-A)^0 := I$. Then:

(i) The integral in (B.16) converges in norm and $(-A)^\alpha$ is a well-defined operator in $\mathcal{L}(H)$.

(ii) $(-A)^\alpha(-A)^\beta = (-A)^{\alpha+\beta}$ for $\alpha, \beta \le 0$.

(iii) $(-A)^\alpha$ is injective.

Proof See [479] pp. 70–72.   □

The operator $(-A)^\alpha$ can also be defined by

$$(-A)^\alpha = -\frac{1}{2\pi i} \int_C \lambda^\alpha (\lambda I + A)^{-1}\,d\lambda, \tag{B.17}$$

where the path C is in $\varrho(-A)$ and goes from $\infty e^{-i\theta}$ to $\infty e^{i\theta}$ for $\theta \in \left(\frac{\pi}{2}, \pi\right)$, avoiding the non-positive real axis. Using (B.17) one can define the fractional powers for more general operators, as is done in classical references such as [547] or [479]. We are only interested in the analytic case in this book.

Definition B.52 Let A be as in Theorem B.51. The *fractional powers of* $-A$ are defined as follows:

(i) If $\alpha < 0$, $D((-A)^\alpha) = H$ and $(-A)^\alpha$ is defined by (B.16).
(ii) If $\alpha = 0$, $D((-A)^0) = H$ and $(-A)^0 := I$.
(iii) If $\alpha > 0$, $D((-A)^\alpha) = R((-A)^{-\alpha})$ and $(-A)^\alpha := \left((-A)^{-\alpha}\right)^{-1}$.

Theorem B.53 *Let A be as in Theorem B.51. The fractional powers of $-A$ satisfy the following properties:*

(i) *For all positive α, $D((-A)^\alpha)$ is dense in H and $(-A)^\alpha$ is a closed operator.*
(ii) *If $\alpha \leq \beta$ then $D((-A)^\beta) \subset D((-A)^\alpha)$.*
(iii) *For all real numbers α, β, $(-A)^\alpha(-A)^\beta = (-A)^{\alpha+\beta}$ on $D((-A)^{\beta\vee(\alpha+\beta)})$.*

Proof See [479], Theorem 6.8, p. 72. □

Theorem B.54 *Let A be as in Theorem B.51. The following hold:*

(i) *$e^{tA}(H) \subset D((-A)^\alpha)$ for every $t > 0$ and $\alpha \geq 0$.*
(ii) *$e^{tA}(-A)^\alpha x = (-A)^\alpha e^{tA}x$ for every $x \in D((-A)^\alpha)$ and $\alpha \in \mathbb{R}$.*
(iii) *For every $t > 0$ and $\alpha > 0$ the operator $(-A)^\alpha e^{tA}$ is bounded and there exist $a > 0$ and $M_\alpha > 0$ such that*

$$\|(-A)^\alpha e^{tA}\| \leq M_\alpha t^{-\alpha} e^{-at}. \tag{B.18}$$

(iv) *If $\alpha \in (0, 1]$ then for every $x \in D((-A)^\alpha)$*

$$|e^{tA}x - x|_H \leq C_\alpha t^\alpha |(-A)^\alpha x|_H$$

for some constant C_α independent of x.

Proof See [479], Theorem 6.13, p. 74. □

B.5 π-Convergence, $\mathcal{K}$-Convergence, π- and $\mathcal{K}$-Continuous Semigroups

In this section, where H is always a real separable Hilbert space, we introduce the notions of π-convergence, $\mathcal{K}$-convergence, π-continuous and $\mathcal{K}$-continuous semigroups and we recall their basic properties as well as other related notions. For further results and details we refer the reader to [179, 492, 493, 496], Sect. 6.3, for π-convergence and π-continuous semigroups; to [101, 102, 108] and the appendix of [105], for $\mathcal{K}$-convergence and $\mathcal{K}$-continuous (also called weakly continuous, see e.g. [101]) semigroups. We also recall the paper [170] that deals with semigroups which are not strongly continuous. Finally, we recall the recent paper [243] which develops the theory of equicontinuous semigroups in locally convex spaces which includes, as special cases, both π-continuous and $\mathcal{K}$-continuous semigroups.

Most of the literature on the present subject deals with the spaces $C_b(H)$ and $UC_b(H)$, except for [102, 300] which deal with $UC_m(H)$ and [300] which deals with $C_m(H)$ in the case of $\mathcal{K}$-continuous semigroups, and the final part of [492] (pp. 293–294, see also [493], Sect. 6.5) which shows how to extend the results to the space $B_b(H)$, in the case of π-continuous semigroups. Here we present the results for π-continuous and $\mathcal{K}$-continuous semigroups mainly in the spaces $C_m(H)$ and $UC_m(H)$ because these are the spaces most commonly used in this book. In some cases we will also deal with $B_m(H)$.

B.5.1 π-Convergence and $\mathcal{K}$-Convergence

The definition of π-convergence can be found, for example, in [219], p. 111, where it is called bp-convergence (bounded-pointwise), and in [492]; the former in spaces of continuous and bounded functions, the latter in spaces of uniformly continuous and bounded functions. For $\mathcal{K}$-convergence in $UC_b(H)$ the reader is referred to [101, 108] and, for the $C_m(H)$ (respectively, $UC_m(H)$) framework, to [300] (respectively, $[102]^3$).

We state all the definitions in this section considering $B_m(H)$ $(m \geq 0)$ as the environment space. The same definitions hold if the basic space $B_m(H)$ is replaced by $C_m(H)$ or $UC_m(H)$ $(m \geq 0)$.

Definition B.55 (π-convergence) Let $m \geq 0$. A sequence $(f_n) \subset B_m(H)$ is said to be π-convergent to $f \in B_m(H)$ and we will write

$$f_n \xrightarrow{\pi} f \quad \text{or} \quad f = \pi\text{-}\lim_{n \to +\infty} f_n$$

if the following conditions hold:

(i) $\sup_{n \in \mathbb{N}} \|f_n\|_{B_m} < +\infty$.
(ii) $\lim_{n \to +\infty} f_n(x) = f(x)$ for any $x \in H$.

Moreover, given $I \subset \mathbb{R}$, $t_0 \in I$, a family $(f_t)_{t \in I \setminus \{t_0\}} \subset B_m(H)$ and $f \in B_m(H)$ we write

$$f_t \xrightarrow[t \to t_0]{\pi} f \quad \text{or} \quad f = \pi\text{-}\lim_{t \to t_0} f_n$$

if, for any sequence t_n of elements of $I \setminus \{t_0\}$ converging to t_0, we have $f_{t_n} \xrightarrow{\pi} f$.

Similarly, given $I \subset \mathbb{R}$, a sequence $(f_n) \subset B_m(I \times H)$ is said to be π-convergent to $f \in B_m(I \times H)$ and we will write

$$f_n \xrightarrow[n \to +\infty]{\pi} f \quad \text{or} \quad f = \pi\text{-}\lim_{n \to +\infty} f_n$$

---

[3] In this paper the $\mathcal{K}$-convergence in $UC_m(H)$ is called $\mathcal{K}_m$-convergence.

if the following conditions hold:

(i) $\sup_{n\in\mathbb{N}} \|f_n\|_{B_m} < +\infty$.
(ii) $\lim_{n\to+\infty} f_n(t,x) = f(t,x)$ for any $(t,x) \in I \times H$.

Definition B.56 (*$\mathcal{K}$-convergence*) A sequence (f_n) in $B_m(H)$ is said to be $\mathcal{K}$-convergent to $f \in B_m(H)$ if

$$
\begin{cases}
\sup_{n\in\mathbb{N}} \|f_n\|_{B_m} < +\infty, \\
\lim_{n\to+\infty} \sup_{x\in K} |f_n(x) - f(x)| = 0
\end{cases}
\tag{B.19}
$$

for every compact set $K \subset H$. In this case we will write

$$
f_n \xrightarrow[n\to+\infty]{\mathcal{K}} f \quad\text{or}\quad f = \mathcal{K}\text{-}\lim_{n\to+\infty} f_n.
$$

Moreover, given $I\subset\mathbb{R}$, $t_0 \in I$ and a family $(f_t)_{t\in I\setminus\{t_0\}} \subset B_m(H)$, where $I\subset\mathbb{R}$ and $f \in B_m(H)$, we write

$$
f_t \xrightarrow[t\to t_0]{\mathcal{K}} f \quad\text{or}\quad f = \mathcal{K}\text{-}\lim_{t\to t_0} f_t
$$

if, for any sequence t_n of elements of $I \setminus \{t_0\}$ converging to t_0, we have $f_{t_n} \xrightarrow{\mathcal{K}} f$.

In a similar way, given $I\subset\mathbb{R}$, a sequence (f_n) in $B_m(I \times H)$ is said to be $\mathcal{K}$-convergent to $f \in B_m(I \times H)$ if

$$
\begin{cases}
\sup_{n\in\mathbb{N}} \|f_n\|_{B_m} < +\infty, \\
\lim_{n\to+\infty} \sup_{(t,x)\in I_0\times K} |f_n(t,x) - f(t,x)| = 0
\end{cases}
\tag{B.20}
$$

for all compact sets $I_0\subset I$ and $K \subset H$.[4] In this case we will write, as before,

$$
f_n \xrightarrow[n\to+\infty]{\mathcal{K}} f \quad\text{or}\quad f = \mathcal{K}\text{-}\lim_{n\to+\infty} f_n.
$$

Remark B.57 The notions of π-convergence and $\mathcal{K}$-convergence can be extended in exactly in the same way to sequences in $B_m(H, Y)$ or in $B_m(I \times H, Y)$, where Y is a given Hilbert space. This will be used when we consider convergence of derivatives in Sect. B.7. ∎

---

[4]In the literature (see e.g. [108, 308, 309]) this definition is given taking the supremum over $I \times K$ even when I is not compact. We prefer to use the above definition as it is more coherent with the concept of $\mathcal{K}$-convergence and it does not change anything in the results and in the proofs.

The above convergences induce a series of related concepts like those of closedness and density. In the two following definitions we recall some of them taken from [108, 492, 493].

Definition B.58 Let $m \geq 0$. A subset Y of $B_m(H)$ (respectively, $C_m(H)$, $UC_m(H)$) is said to be π-closed if, for any sequence (f_n) in Y and $f \in B_m(H)$ (respectively, $C_m(H)$, $UC_m(H)$) such that $f_n \xrightarrow{\pi} f$, we have $f \in Y$.

A subset Y of $B_m(H)$ (respectively, $C_m(H)$, $UC_m(H)$) is said to be π-dense in $B_m(H)$ (respectively $C_m(H)$, $UC_m(H)$) if, for any $f \in B_m(H)$ (respectively, $C_m(H)$, $UC_m(H)$), there exists an $(f_n) \subset Y$ such that $f_n \xrightarrow{\pi} f$.

A linear operator $\mathcal{A} : D(\mathcal{A}) \subset B_m(H) \to B_m(H)$ is said to be π-closed if, given a sequence $(f_n) \subset D(\mathcal{A})$, the following condition holds:

$$\left(f_n \xrightarrow{\pi} f \quad \text{and} \quad \mathcal{A}f_n \xrightarrow{\pi} g \right) \Rightarrow \left(f \in D(\mathcal{A}) \quad \text{and} \quad \mathcal{A}f = g \right).$$

Likewise if $\mathcal{A} : D(\mathcal{A}) \subset C_m(H) \to C_m(H)$ or $\mathcal{A} : D(\mathcal{A}) \subset UC_m(H) \to UC_m(H)$.

Definition B.59 Let $m \geq 0$. A subset Y of $B_m(H)$ (respectively, $C_m(H)$, $UC_m(H)$) is said to be $\mathcal{K}$-closed if for any sequence (f_n) in Y and $f \in B_m(H)$ (respectively, $C_m(H)$, $UC_m(H)$) such that

$$\mathcal{K}\text{-}\lim_{n \to +\infty} f_n = f,$$

we have $f \in Y$.

A subset Y of $B_m(H)$ (respectively, $C_m(H)$, $UC_m(H)$) is said to be $\mathcal{K}$-dense if for any $f \in B_m(H)$ (respectively, $C_m(H)$, $UC_m(H)$) there exists a sequence $(f_n) \subset Y$ such that

$$f = \mathcal{K}\text{-}\lim_{n \to +\infty} f_n.$$

A linear operator $\mathcal{A} : D(\mathcal{A}) \subset B_m(H) \to B_m(H)$ is said to be $\mathcal{K}$-closed if, given a sequence (f_n) in $D(\mathcal{A})$ such that

$$\mathcal{K}\text{-}\lim_{n \to +\infty} f_n = f \text{ and } \mathcal{K}\text{-}\lim_{n \to +\infty} \mathcal{A}f_n = g,$$

we have

$$f \in D(\mathcal{A}) \text{ and } \mathcal{A}f = g.$$

Likewise if $\mathcal{A} : D(\mathcal{A}) \subset C_m(H) \to C_m(H)$ or $\mathcal{A} : D(\mathcal{A}) \subset UC_m(H) \to UC_m(H)$.

Let $\mathcal{A} : D(\mathcal{A}) \subset B_m(H) \to B_m(H)$ and $\mathcal{B} : D(\mathcal{B}) \subset B_m(H) \to B_m(H)$ be two linear operators and assume that $\mathcal{A} \subset \mathcal{B}$ and that $\mathcal{B}$ is $\mathcal{K}$-closed. We say that $\mathcal{B}$ is the $\mathcal{K}$-closure of $\mathcal{A}$, and we write $\mathcal{B} = \overline{\mathcal{A}}^{\mathcal{K}}$, if for every $f \in D(\mathcal{B})$ there exists a sequence (f_n) in $D(\mathcal{A})$ such that

$$\begin{cases} \mathcal{K}\text{-}\lim_{n \to +\infty} f_n = f \\[2ex] \mathcal{K}\text{-}\lim_{n \to +\infty} \mathcal{A}f_n = \mathcal{B}f. \end{cases} \qquad (B.21)$$

Likewise if $\mathcal{A} : D(\mathcal{A}) \subset C_m(H) \to C_m(H)$, $\mathcal{B} : D(\mathcal{B}) \subset C_m(H) \to C_m(H)$ or if $\mathcal{A} : D(\mathcal{A}) \subset UC_m(H) \to UC_m(H)$, $\mathcal{B} : D(\mathcal{B}) \subset UC_m(H) \to UC_m(H)$.

Motivated by Theorem 4.5 of [300] we introduce the notions of $\mathcal{K}$-core and π-core.

Definition B.60 ($\mathcal{K}$-*core and* π-*core*) Let $m \geq 0$. Let the operator $\mathcal{A} : D(\mathcal{A}) \subset B_m(H) \to B_m(H)$ be $\mathcal{K}$-closed (respectively, π-closed). A linear subspace Y of $D(\mathcal{A})$ is a $\mathcal{K}$-core (respectively, a π-core) for $\mathcal{A}$ if for any $f \in D(\mathcal{A})$ there exists a sequence f_n of elements of Y such that

$$\mathcal{K}\text{-}\lim_{n \to +\infty} f_n = f \quad and \quad \mathcal{K}\text{-}\lim_{n \to +\infty} \mathcal{A}f_n = \mathcal{A}f$$

(respectively,

$$\pi\text{-}\lim_{n \to +\infty} f_n = f \quad and \quad \pi\text{-}\lim_{n \to +\infty} \mathcal{A}f_n = \mathcal{A}f).$$

The same holds if $\mathcal{A} : D(\mathcal{A}) \subset C_m(H) \to C_m(H)$ or $\mathcal{A} : D(\mathcal{A}) \subset UC_m(H) \to UC_m(H)$.

Remark B.61 Let $(f_n) \subset B_m(H)$ be a sequence which π-converges to a function $f : H \to \mathbb{R}$. Since measurability is preserved over pointwise limits (see e.g. Lemma 1.8) then it must be $f \in B_m(H)$.

Similarly, since converging sequences in H are compact, one can prove that, if $(f_n) \subset C_m(H)$ is a sequence which $\mathcal{K}$-converges to a function $f : H \to \mathbb{R}$, then $f \in C_m(H)$ (i.e. $C_m(H)$ is $\mathcal{K}$-closed in $B_m(H)$).

On the other hand, if $f_n \xrightarrow[n \to \infty]{\mathcal{K}} f$ and $(f_n) \subset UC_m(H)$, it is not true, in general, that $f \in UC_m(H)$. Indeed, using Lemma B.78 below, one easily sees that $UC_m(H)$ is $\mathcal{K}$-dense in $C_m(H)$. Similarly, if $f_n \xrightarrow[n \to \infty]{\pi} f$ and $(f_n) \subset C_m(H)$ it is not true, in general, that $f \in C_m(H)$. ∎

Remark B.62 Concerning π-convergence, in [492], Theorem 2.2 (see also [493], Theorem 6.2.3), the author introduces a "natural" Hausdorff locally convex topology τ_0, not metrizable and not sequentially complete in $UC_b(H)$, whose convergent sequences are exactly π-convergent sequences. As stated in [492] (end of Sect. 5) the same holds in $C_b(H)$. The lack of completeness of τ_0 relies on the fact that continuity (and, a fortiori, uniform continuity) is not preserved under π-convergence, as observed in the previous remark.

Similarly, concerning $\mathcal{K}$-convergence, in [300] it is shown (see, in particular, Proposition 2.3) that in the so-called *mixed topology* $\tau_{\mathcal{M}}$, which is a locally convex and complete one, introduced in [573], convergent sequences are precisely $\mathcal{K}$-convergent sequences. The result is obtained in $C_m(H)$ for $m \geq 0$. Completeness of

$\tau_{\mathcal{M}}$ relies on the fact that continuity is preserved under $\mathcal{K}$-convergence, as observed in the previous remark. However, completeness of such topology is not guaranteed if we consider it on $UC_m(H)$, as observed in the introduction of [300].

In [219] pp. 495–496 a topology in the space $B_b(H)$ is given whose convergent sequences coincide with π-convergent sequences.

It is clear that all the concepts introduced in Definitions B.58, B.59 and B.60 can also be seen as topological concepts in the topologies just described. ∎

Remark B.63 In [423] (Definition 2.1 and Remark 2.6, see also [500] Sect. 5.2) the author also considers π-convergence for multisequences and defines the concepts of π-closedness, π-density (Definition 2.5 there) and π-core (Definition 2.10) with multisequences. It seems that the theory also works using multisequences, since the dominated convergence is still preserved under π-convergent multisequences. Since in this book we do not need such a general setting, we keep the definitions with single sequences which, a priori, are not equivalent to the analogous definitions in [423]. ∎

B.5.2 π- and $\mathcal{K}$-Continuous Semigroups and Their Generators

We use the definitions of the previous section to introduce π-continuous and $\mathcal{K}$-continuous semigroups. The theory of such semigroups has been developed in the literature mainly using $UC_b(H)$ as the environment space, but the definitions and the results can easily be adapted to the $C_b(H)$ (or also to the $C_m(H)$ and $UC_m(H)$) framework. We will present the $UC_b(H)$ setting, making a few remarks on how to generalize the results to $C_b(H)$ (or also to $C_m(H)$ and $UC_m(H)$).

B.5.2.1 The Definitions

Definition B.64 (π-*continuous semigroup*) Let $S(t)$ be a semigroup of bounded linear operators on $UC_b(H)$, namely, for any $f \in UC_b(H)$ and $s, t \in \mathbb{R}^+$, $S(t+s)f = S(t)S(s)f$ and $S(0)f = f$. We say that $S(t)$ is a π-continuous semigroup on $UC_b(H)$ of class $\mathcal{G}_\pi(M, \omega)$ if the following conditions hold:

(i) There exist $M \geq 1$ and $\omega \in \mathbb{R}$ s.t.
$$\|S(t)\|_{\mathcal{L}(UC_b(H))} \leq Me^{\omega t}, \qquad t \geq 0.$$

(ii) For any $(f_n) \subset UC_b(H)$, $f \in UC_b(H)$ s.t. $f_n \xrightarrow{\pi} f$ we have $S(t)f_n \xrightarrow[n\to+\infty]{\pi} S(t)f$ for all $t \geq 0$.

(iii) For any $x \in H$ and $f \in UC_b(H)$, the map $[0, +\infty) \to \mathbb{R}, t \mapsto (S(t)f)(x)$ is continuous.

Definition B.65 (*$\mathcal{K}$-continuous semigroup*) Let $S(t)$ be a semigroup of bounded linear operators on $UC_b(H)$, namely, for any $f \in UC_b(H)$ and $s, t \in \mathbb{R}^+$, $S(t + s)f = S(t)S(s)f$ and $S(0)f = f$. We say that $S(t)$ is a $\mathcal{K}$-continuous semigroup (or a weakly continuous semigroup) of class $\mathcal{G}_{\mathcal{K}}(M, \omega)$ on $UC_b(H)$ if the following conditions hold:

(i) There exist $M \geq 1$ and $\omega \in \mathbb{R}$ s.t.

$$\|S(t)\|_{\mathcal{L}(UC_b(H))} \leq M e^{\omega t}, \qquad t \geq 0.$$

(ii) For any $(f_n) \subset UC_b(H)$, $f \in UC_b(H)$ s.t. $f_n \overset{\mathcal{K}}{\to} f$ we have $S(t)f_n \xrightarrow[n \to +\infty]{\mathcal{K}}$
     $S(t)f$ for all $t \geq 0$. The limit is uniform in $t \in [0, T]$ for any $T > 0$.
(iii) For every $f \in UC_b(H)$ and $t_0 \geq 0$ we have

$$\mathcal{K}\text{-} \lim_{t \to t_0} S(t)f = S(t_0)f.$$

(iv) For any $T > 0$ and $f \in UC_b(H)$, the family of functions

$$\{ S(t)f \, : \, t \in [0, T] \}$$

is equi-uniformly continuous in $UC_b(H)$.

We now give some observations concerning:

- The relationship between the above two definitions and the definition of a strongly continuous semigroup.
- The extension of the above two definitions to more general spaces.
- The relationship of the notions of $\mathcal{K}$ and π-continuous semigroups with strongly continuous semigroups in coarser topologies.
- The main reason why they are introduced: to deal with transition semigroups.

Remark B.66 The above definitions were introduced first in [492, 493] (for π-semigroups) and in [101] (for $\mathcal{K}$-semigroups, there called *weakly continuous*). We observe the following.

(1) In contrast to the case of strongly continuous semigroups (see Proposition B.33), condition (i) has to be included in both definitions above because it is not known if it follows from other assumptions.
(2) Condition (ii) in both definitions gives a kind of continuity with respect to the π or $\mathcal{K}$-convergence. In condition (ii), for a $\mathcal{K}$-continuous semigroup (following [101]) the limit is required to be uniform in $t \in [0, T]$. Such a requirement is avoided in the definition given in [102]: we keep it here since it is verified in all cases we consider. Also this is not the case for π-continuous semigroups. There are examples of π-continuous semigroups where the limit is not uniform (see Remark 6.2.2 in [493]).

(3) Condition (iii) in both definitions are analogues of condition (iii) in Definition
B.31. However, here the equivalence mentioned in the footnote there is not
obvious. Indeed, condition (iii) for a π-continuous semigroup clearly implies
that

$$\pi\text{-} \lim_{t\to 0^+} S(t)f = f, \tag{B.22}$$

while the opposite is not obvious: the above implies right continuity of the map
$t\to (S(t)f)(x)$ but it is not known if left continuity also holds (see Remark 6.2.5
(a) in [493]). In [492] Remark 2.5 (see also Remark 6.2.5 (b) in [493]), the
author proves that the equivalence holds for π-semigroups defined on $C_b(M)$
for a compact metric space M. Similarly, for a $\mathcal{K}$-continuous semigroup is not
obvious if condition (iii) for $t_0 = 0$ implies that the same holds true for all $t_0 \geq 0$.
The reason why in both cases the stronger conditions in the definitions are used
is that the continuity of the map $t\to S(t)f(x)$ is useful to simplify various steps
in the proofs. Moreover, the stronger conditions are satisfied in the cases we
consider in this book. We finally observe that in the original definition of $\mathcal{K}$-
continuous semigroup (see [101], Definition 2.1) the weaker condition (B.22) is
used but it seems that the author still uses continuity of the map $t\to S(t)f(x)$
(proof of Proposition 3.1, [101]).
(4) Condition (iv) for a $\mathcal{K}$-continuous semigroup extends the regularity in time of
the function $(t, x) \to (S(t)f)(x)$ requiring uniformity in time of the modulus
of continuity in x. We point out that in the original definition of $\mathcal{K}$-continuous
semigroups in [101] the uniformity in conditions (ii) and (iv) is required on
$[0, +\infty)$ since in this paper the author deals with semigroups of negative type
which we do not want to assume here.

■

Remark B.67 The definition of a π-continuous semigroup can also be made in $C_b(H)$
or in $B_b(H)$ by simply substituting $UC_b(H)$ with $C_b(H)$ or $B_b(H)$ (see [492] pp.
293–294). Similarly we can define them in $UC_m(H)$, $C_m(H)$, $B_m(H)$.
Concerning $\mathcal{K}$-continuous semigroups, they can be easily defined in $UC_m(H)$
by simply substituting $UC_b(H)$ with $UC_m(H)$, as is done in Sect. 2.2 of [102]. On
the other hand, to define $\mathcal{K}$-continuous semigroups in $C_b(H)$ and in $C_m(H)$ one
should also substantially modify (or erase) condition (iv) of Definition B.65. This
is done in Theorem 4.1 of [300], where (iv) is implicitly substituted with the local
equicontinuity of the family of operators $S(t)$, $t \geq 0$. The extension of the definition
of $\mathcal{K}$-continuous semigroups to spaces $B_b(H)$ or $B_m(H)$ is not considered in the
literature.   ■

Remark B.68 Recalling Remark B.62, consider a π-continuous (respectively, $\mathcal{K}$-
continuous) semigroup $S(t)$ acting on the space $UC_b(H)$ (respectively, on the space
$C_m(H)$) endowed with the topology τ_0 generated by the π-convergence (respectively,
$\tau_\mathcal{M}$ generated by the $\mathcal{K}$-convergence). By Definition B.64-(ii) (respectively, B.65-
(ii)), for every $t \geq 0$, $S(t)$ is sequentially continuous but it is not known if it is also
continuous. In [300] the authors show that the transition semigroups (introduced in

the following section) are strongly continuous with respect to the mixed topology $\tau_\mathcal{M}$ so, in particular, they are continuous in such topology. ∎

Remark B.69 The reason why $\mathcal{K}$-continuous semigroups, and later, π-continuous semigroups, were introduced (in [101] and in [493], respectively) is the need to study[5] Markov transition semigroups associated with finite and infinite-dimensional SDEs since such semigroups are naturally not C_0-semigroups: as shown, for instance, in Example 6.1 of [101], already in spatial dimension 1, the Ornstein–Uhlenbeck semigroup is not strongly continuous. Nevertheless, by a simple application of the dominated convergence theorem, one can see that all Markov transition semigroups defined in (1.99) are π-continuous semigroups (see also Definition 3.5 in [492] and the subsequent comments). Moreover, with a slightly more complicated proof it can also be proved that such semigroups are $\mathcal{K}$-continuous.

On the other hand, it is not true, as one may expect, that all strongly continuous semigroups are also π-continuous or $\mathcal{K}$-continuous. Indeed, it is shown in [493], Proposition 6.2.4 and the subsequent observation, that the class of uniformly continuous semigroups on $UC_b(\mathbb{R})$ is not contained in the class of π-continuous semigroups nor in that of $\mathcal{K}$-continuous semigroups. Moreover, even though the class of π-continuous semigroups has been introduced to study a wider set of problems (see Remark 6.2.2 in [493]), it is not clear if all $\mathcal{K}$-continuous semigroup are also π-continuous semigroups. ∎

B.5.2.2 The Generators

Similarly to what we have done for C_0-semigroups, we can define the generators of π-continuous and $\mathcal{K}$-continuous semigroups. Given a semigroup of bounded operators $S(t)$, we will write

$$\Delta_h := \frac{S(h) - I}{h}.$$

Definition B.70 Let $S(t)$ be a π-continuous semigroup on $UC_b(H)$. We define the infinitesimal generator $\mathcal{A}$ of $S(t)$ as follows:

$$\begin{cases} D(\mathcal{A}) := \{f \in UC_b(H) : \text{there exists a } g \in UC_b(H) \text{ s.t. } \pi\text{- } \lim_{t\to 0^+} \Delta_t f = g\} \\ (\mathcal{A}f)(x) := \lim_{t\to 0^+} \Delta_t f(x), \; for \, x \in H. \end{cases}$$

Definition B.71 Let $S(t)$ be a $\mathcal{K}$-continuous semigroup on $UC_b(H)$. We define the infinitesimal generator $\mathcal{A}$ of $S(t)$ as follows:

---

[5]Other approaches are possible to deal with such semigroups (see, for example, the theory of semigroups on general locally convex spaces [243, 366, 367]) but, as remarked in the introduction of [492], the use of such a theory for the above goal would be much more complicated.

$$\begin{cases} D(\mathcal{A}) := \{f \in UC_b(H) \ : \ \text{there exists a } g \in UC_b(H) \text{ s.t. } \mathcal{K}\text{-} \lim_{t \to 0^+} \Delta_t f = g\} \\ (\mathcal{A}f)(x) := \lim_{t \to 0^+} \Delta_t f(x), \ for \ x \in H. \end{cases}$$

Remark B.72 In [101, 105, 106, 108] the generator $\mathcal{A}$ of a $\mathcal{K}$-continuous semigroup $S(t)$ is defined in a different way, by using the resolvent. Indeed, $\mathcal{A}$ is the unique closed linear operator such that, for all $\lambda > \omega$, $f \in UC_b(H)$, $x \in H$, we have

$$(\lambda I - \mathcal{A})^{-1} f(x) = \int_0^{+\infty} e^{-\lambda t} S(t) f(x) dt.$$

In fact the two definitions are equivalent. The equivalence is implicitly proved for π-continuous semigroups in [493], Proposition 6.2.11 (see also [492], Proposition 2.5), with a proof that easily adapts to $\mathcal{K}$-continuous semigroups too. Moreover, the equivalence of the two definitions is also proved in [300], Remark 4.3, for $\mathcal{K}$-continuous transition semigroups (which is the case of interest in this book).

In [493], Theorem 6.2.13, the author shows that, if a π-continuous semigroup is also $\mathcal{K}$-continuous, the two generators coincide. ∎

We finally observe that the generators can be introduced, in exactly the same way, if we consider, as mentioned in Remark B.67, π- continuous semigroups on $UC_m(H)$, $C_m(H)$, $B_m(H)$ and $\mathcal{K}$-continuous semigroups in $UC_m(H)$, $C_m(H)$, for some $m \geq 0$.

B.5.2.3 Cauchy Problems for π-Continuous and $\mathcal{K}$-Continuous Semigroups

The following propositions establish the relationship between π-continuous/$\mathcal{K}$-continuous semigroups, their generators and homogeneous Cauchy problems.

Proposition B.73 *Let $\mathcal{A}$ be the generator of a π-continuous semigroup $S(t)$ on $UC_b(H)$. Then:*

(i) *$D(\mathcal{A})$ is π-dense in $UC_b(H)$.*
(ii) *$\mathcal{A}$ is a closed and π-closed operator on $UC_b(H)$.*
(iii) *For any $f \in D(\mathcal{A})$:*

   (a) *$S(t)f \in D(\mathcal{A})$ and $\mathcal{A}S(t)f = S(t)\mathcal{A}f$, for any $t \geq 0$.*
   (b) *For any $x \in H$ the mapping*

$$(0, +\infty) \to \mathbb{R}, \quad t \to (S(t)f)(x)$$

   is continuously differentiable and

$$\frac{d}{dt} S(t) f(x) = \mathcal{A}(S(t) f)(x)$$

   for all $t > 0$.

Proof See [493] Proposition 6.2.9 and 6.2.7. Closedness of $\mathcal{A}$ follows from Remark B.72. □

Proposition B.74 *Let $\mathcal{A}$ be the generator of a $\mathcal{K}$-continuous semigroup $S(t)$ on $UC_b(H)$. Then:*

(i) $D(\mathcal{A})$ is $\mathcal{K}$-dense in $UC_b(H)$.
(ii) $\mathcal{A}$ is a closed and $\mathcal{K}$-closed operator on $UC_b(H)$.
(iii) For every $f \in D(\mathcal{A})$:

    (a) $S(t)f \in D(\mathcal{A})$ and $\mathcal{A}S(t)f = S(t)\mathcal{A}f$, for any $t \geq 0$.
    (b) For any $x \in H$ the mapping:

$$(0, +\infty) \to \mathbb{R}, \quad t \to (S(t)f)(x)$$

is continuously differentiable and

$$\frac{d}{dt}S(t)f(x) = \mathcal{A}(S(t)f)(x)$$

for all $t > 0$.

Proof See [108] Proposition 2.9 and Remark 2.10. Closedness of $\mathcal{A}$ follows from Remark B.72. □

Remark B.75 In the case of a $\mathcal{K}$-continuous semigroup with generator $\mathcal{A}$, if we consider the function $u : [0, T] \times H \to \mathbb{R}$ defined as

$$(t, x) \to (S(t)\varphi)(x), \tag{B.23}$$

which should be the natural solution of the homogeneous Cauchy problem corresponding to the operator $\mathcal{A}$, then u belongs to $UC_b^x([0, T] \times H)^6$ and not, in general, to $C([0, T], UC_b(H)) = UC_b([0, T] \times H)$. Indeed, if for every datum $\varphi \in UC_b(H)$, $u \in C([0, T], UC_b(H))$, then $\mathcal{A}$ generates a C_0-semigroup. Similarly, in the case of π-continuous semigroups the function in (B.23) does not belong to $UC_b([0, T] \times H)$ in general, but it belongs in a natural way to a space called $C_\pi([0, T], UC_b(H))$ (requiring global boundedness and separate continuity), introduced in Definition 7.2.1 on p. 161 of [493], which is strictly bigger than $UC_b^x([0, T] \times H)$.

Moreover, in general the $UC_b(H)$-valued function $t \to e^{t\mathcal{A}}\varphi$ is not even measurable. In fact, measurability of this map implies strong continuity of the semigroup (see [576], pp. 233–234 and Proposition B.89 below). ■

The above propositions can be extended, as said in Remark B.67, to the case of π-continuous semigroups in $UC_m(H)$, $C_m(H)$, $B_m(H)$, and $\mathcal{K}$-continuous semigroups in $UC_m(H)$, $C_m(H)$, for some $m \geq 0$.

---

[6]Indeed, as remarked in [108], Remark 2.6, this is equivalent to requiring that points (iii) and (iv) of Definition B.65 be satisfied.

We now consider non-homogeneous Cauchy problems. Similarly to the case of C_0-semigroups (see e.g. [47] Chap. II.1.3), given a π-continuous or a $\mathcal{K}$-continuous semigroup $S(t)$ and the related generator $\mathcal{A}$, one can consider the following non-homogeneous Cauchy problem on $UC_b(H)$[7]:

$$\begin{cases} \frac{d}{dt}v(t) = \mathcal{A}v(t) + f(t), & t \in (0, T], \\ v(0) = \varphi \in UC_b(H), \end{cases} \tag{B.24}$$

where $f: (0, T] \to UC_b(H)$.

In analogy with what is usually done for C_0-semigroups, the mild solution of problem (B.24) is, by definition, the function $u : [0, T] \times H \to \mathbb{R}$ given by

$$u(t, x) = [S(t)\varphi](x) + \int_0^t [S(t - s)f(s)](x)ds. \tag{B.25}$$

Remark B.76 This definition assumes that, for every $t > 0$ and $x \in H$, the map

$$[0, t] \to \mathbb{R}, \qquad s \to [S(t - s)f(s)](x)$$

is measurable. This is true under mild assumptions on f but one has to be careful since, even when f above is constant, as recalled in Remark B.75, for a fixed $t \in [0, T]$, the function

$$[0, t] \to UC_b(H), \qquad s \to S(t - s)f$$

may not be measurable if the semigroup $S(t)$ is not strongly continuous. ∎

Results on uniqueness and regularity of mild solutions and their relationship with other concepts of solutions (in particular, approximations by classical solutions) are contained in [102, 108] for the case of $\mathcal{K}$-continuous semigroups and in Chap. 7 of [493, 496] for π-continuous semigroups. They are applied mainly to obtain suitable approximations of mild solutions of Kolmogorov equations which are used in Chap. 4. Such results are presented in Sect. B.7 for the Kolmogorov equations we are interested in.

Several other results on π-continuous and $\mathcal{K}$-continuous semigroups are obtained in [492, 493, 496] and [101, 102, 108]. We recall in particular the possibility of proving an analogue of the Hille–Yosida theorem in both cases.

---

[7]Clearly, extending the concepts of π-semigroups and $\mathcal{K}$-semigroups it is possible to consider the same problem in the spaces $C_b(H)$, $C_m(H)$, $UC_m(H)$, $B_m(H)$.

B.6 Approximation of Continuous Functions Through $\mathcal{K}$-Convergence

Recall that H is a real separable Hilbert space. When $\dim H = +\infty$, the space $UC_b^2(H)$ is not dense in $UC_b(H)$ (see [457]). We refer to Appendix D.3 for more on approximations in Hilbert spaces. However, in many cases (in particular to prove verification theorems for optimal control problems), we need to be able to approximate functions in $C_m(H)$, $m \geq 0$, (or $UC_m(H)$) by smooth (at least C^2) functions with special properties. We can substitute uniform convergence by π-convergence or $\mathcal{K}$-convergence which are good enough to apply the dominated convergence theorem. We start with the following lemma, which is a slight variation of Lemma 5.2 in [108].

Lemma B.77 *Let A be the generator of the strongly continuous semigroup $S(t) = e^{tA}$ in H. Let J_n be as in (B.9) and let P_n be a sequence of finite-dimensional orthogonal projections on H strongly convergent to the identity operator. Then, for all compact subsets $I_0 \subset \mathbb{R}$ and $K \subset H$, the sets*

$$\mathcal{R}_1(K) =: \{ P_n x : x \in K, \, n \in \mathbb{N} \} \subset H,$$

$$\mathcal{R}_1(I_0 \times K) =: \{ (t, P_n x) : t \in I_0, \, x \in K, \, n \in \mathbb{N} \} \subset \mathbb{R} \times H,$$

$$\mathcal{R}_2(K) =: \{ P_n J_n x : x \in K, \, n \in \mathbb{N} \} \subset H$$

are relatively compact.

Proof We prove the claim for $\mathcal{R}_2(K)$. The argument for $\mathcal{R}_1(K)$ and $\mathcal{R}_1(I_0 \times K)$ is the same and easier. Let $\left(P_{n_j} J_{n_j} x_j \right)_{j \in \mathbb{N}}$ be a sequence in $\mathcal{R}_2(K)$. From compactness of K it follows that there exists an increasing subsequence j_k and $\bar{x} \in K$ such that

$$x_{j_k} \to \bar{x}, \quad \text{as } k \to +\infty.$$

If there exists a $C > 0$ such that $n_{j_k} \leq C$, for all j_k, then we can suppose $n_{j_k} = \bar{n}$, for all j_k, and then

$$\lim_{k \to +\infty} P_{n_{j_k}} J_{n_{j_k}} x_{j_k} = P_{\bar{n}} J_{\bar{n}} \bar{x}.$$

Otherwise, let us suppose $\lim_{k \to +\infty} n_{j_k} = +\infty$. Then, using (B.11) and (B.12),

$$|P_{n_{j_k}} J_{n_{j_k}} x_{j_k} - \bar{x}| \leq |P_{n_{j_k}} J_{n_{j_k}} (x_{j_k} - \bar{x})| + |P_{n_{j_k}} (J_{n_{j_k}} \bar{x} - \bar{x})| + |P_{n_{j_k}} \bar{x} - \bar{x}|$$

$$\leq \frac{M n_{j_k}}{n_{j_k} + \omega} |x_{j_k} - \bar{x}| + |J_{n_{j_k}} \bar{x} - \bar{x}| + |P_{n_{j_k}} \bar{x} - \bar{x}| \to 0 \quad \text{as } k \to +\infty.$$

$\square$

The following lemma generalizes Lemma 2.6, p. 25 in [300].

Lemma B.78 *Let $m \geq 0$ and let $I \subset \mathbb{R}$ be an interval. Then $\mathcal{F}C_0^\infty(H)$ is $\mathcal{K}$-dense in $C_m(H)$ and $\mathcal{F}C_0^\infty(I \times H)$ is $\mathcal{K}$-dense in $C_m(I \times H)$. Moreover, for every closed operator B with dense domain we also have that $\mathcal{F}C_0^{\infty,B}(H)$ is $\mathcal{K}$-dense in $C_m(H)$ and $\mathcal{F}C_0^{\infty,B}(I \times H)$ is $\mathcal{K}$-dense in $C_m(I \times H)$.*

Proof Take an orthonormal basis $\mathcal{E} = \{e_n\}_{n \in \mathbb{N}}$ of H and, for $x \in H$, let $x = \sum_{i=1}^\infty x_i e_i$. For every $n \in \mathbb{N}$ let P_n be the orthogonal projection onto the n-dimensional subspace of H spanned by $\{e_1, \ldots e_n\}$. Define

$$\Pi_n : H \to \mathbb{R}^n, \qquad \Pi x = (x_1, \ldots, x_n),$$

$$Q_n : \mathbb{R}^n \to H, \qquad Q_n(x_1, \ldots, x_n) = x_1 e_1 + \cdots + x_n e_n,$$

and recall that $P_n = Q_n \circ \Pi_n$. Let $\varphi \in C_m(H)$. Given a family of C^∞ mollifiers $\eta_k : \mathbb{R}^n \to \mathbb{R}$ with support in $B(0, 1/k)$, we define the regularizing convolutions (as e.g. [486])

$$\psi_k^n(x) = \int_{\mathbb{R}^n} \varphi(Q_n y) \eta_k(\Pi_n x - y) dy = \int_{\mathbb{R}^n} \varphi(P_n x - Q_n y) \eta_k(y) dy.$$

Observe that, by the definition, we have for $x \in H$,

$$|\psi_n^k(x)| \leq \sup_{|x_1| \leq |P_n x| + 1/k} |\varphi(x_1)| \leq \sup_{|x_1| \leq |x| + 1/k} |\varphi(x_1)|,$$

so

$$\frac{|\psi_k^n(x)|}{1 + |x|^m} \leq \sup_{|x_1| \leq |x| + 1/k} \left\{ \frac{|\varphi(x_1)|}{1 + |x_1|^m} \frac{1 + |x_1|^m}{1 + |x|^m} \right\}$$

$$\leq \|\varphi\|_{C_m(H)} \sup_{|x_1| \leq |x| + 1/k} \frac{1 + |x_1|^m}{1 + |x|^m}.$$

Now it is easy to see that

$$\sup_{|x_1| \leq |x| + 1/k} \frac{1 + |x_1|^m}{1 + |x|^m} = 1 + \rho_m(1/k)$$

for some modulus ρ_m. Hence we get

$$\|\psi_k^n\|_{C_m(H)} \leq \|\varphi\|_{C_m(H)}(1 + \rho_m(1/k)). \tag{B.26}$$

Now, from the properties of finite-dimensional convolutions, we easily observe that, setting $\xi_n(x) := \varphi(P_n x)$, the sequence $(\psi_k^n)_{k \in \mathbb{N}}$ converges to ξ_n uniformly on bounded sets of $\mathbb{R}^n$. For every $n \in \mathbb{N}$, let $k(n) \in \mathbb{N}$ be such that

$$\sup_{|x|\leq n} |\psi^n_{k(n)}(x) - \xi_n(x)| \leq \frac{1}{n}. \tag{B.27}$$

Therefore, if we set $\psi_n = \psi^n_{k(n)}$, we have, for any compact set $K \subset H$ and $n \geq \sup_{x \in K} |x|$,

$$\sup_{x \in K} |\psi_n(x) - \varphi(x)| \leq \sup_{x \in K} |\psi_n(x) - \xi_n(x)| + \sup_{x \in K} |\xi_n(x) - \varphi(x)|$$

$$\leq \frac{1}{n} + \sup_{x \in K} |\xi_n(x) - \varphi(x)|. \tag{B.28}$$

By Lemma B.77 and the continuity of φ it follows that, for all $x \in K$ and $n \in \mathbb{N}$,

$$|\xi_n(x) - \varphi(x)| \leq \rho_{\mathcal{R}_1(K)} |P_n x - x|, \tag{B.29}$$

where $\rho_{\mathcal{R}_1(K)}$ is the modulus of continuity of φ over the compact set $\mathcal{R}_1(K)$. Since $\lim_{n \to +\infty} \sup_{x \in K} |P_n x - x| = 0$, it then follows from (B.26) and (B.28), that

$$\mathcal{K}\text{-}\lim_{n \to +\infty} \psi_n = \varphi. \tag{B.30}$$

To end the proof of the first statement in the time-independent case it is enough to define $\varphi_n(x) := \psi_n(x)\theta(|P_n x|^2/n^2)$, where $\theta \in C^\infty(\mathbb{R})$ is a cut-off function such that $0 \leq \theta \leq 1$, $\theta(r) = 1$ for $|r| \leq 1$ and $\theta(r) = 0$ for $|r| \geq 2$.

Consider now the time-dependent case. Let $f \in C_m(I \times H)$. First let $I = \mathbb{R}$, and define f_n following the same procedure used for φ_n. Given a family of C^∞ mollifiers $\eta_k : \mathbb{R} \times \mathbb{R}^n \to \mathbb{R}$ with support in $B(0, 1/k) \subset \mathbb{R} \times \mathbb{R}^n$, we define the regularizing convolutions

$$g^n_k(t, x) = \int_{\mathbb{R}^{n+1}} f(s, Q_n y)\eta_k(t - s, \Pi_n x - y)ds\,dy$$

$$= \int_{\mathbb{R}^{n+1}} f(t - s, P_n x - Q_n y)\eta_k(s, y)ds\,dy.$$

Similarly to the time-independent case, we prove that

$$\|g^n_k\|_{C_m(I \times H)} \leq \|f\|_{C_m(I \times H)}(1 + \rho_m(1/k)). \tag{B.31}$$

Furthermore, we define $h_n(t, x) := f(t, P_n x)$ and $g_n(t, x) = g^n_{k(n)}(t, x)$ as we did in the time-independent case for ξ_n and ψ_n, respectively. Hence we obtain

$$\mathcal{K}\text{-}\lim_{n \to +\infty} g_n = f. \tag{B.32}$$

Then, defining $f_n(t, x) := g_n(t, x)\theta(|P_n x|^2/n^2)$ with θ as in the time-independent case, we get the claim.

If $I = [a, b]$ for $a < b \in \mathbb{R}$ we first define $\bar{f}(t, x) = f(a, x)$ for $t < a$ and $\bar{f}(t, x) = f(b, x)$ for $t > b$. We then take the approximations $\bar{f}_n$ defined above and finally define f_n as the restrictions of $\bar{f}_n$ to $I \times H$. In other cases, we first take a sequence of suitable approximations. For example, if $I = (a, b)$ is open we first define $\tilde{f}_h$ on $\mathbb{R} \times H$ for $h > 4/(b - a)$ by $\tilde{f}_h(t, x) = \chi_h(t) f(t, x)$, where χ is continuous, $0 \leq \chi_h \leq 1$, $\chi_h(t) = 1$ for $t \in [a + 2/h, b - 2/h]$, $\chi_h(t) = 0$ for $t \notin [a + 1/h, b - 1/h]$. Then for every h we approximate the functions $\tilde{f}_h$ by a sequence $(f_{h,n})_n$ chosen as above. The diagonal sequence $(f_n)_n := (f_{n,n})_n$ $\mathcal{K}$-converges to f. Indeed, given I_0 and K compact subsets of (a, b) and H, respectively, we have

$$\sup_{I_0 \times K} |f_{n,n}(t, x) - f(t, x)| \leq \sup_{I_0 \times K} |f_{n,n}(x) - \tilde{f}_n(x)| + \sup_{I_0 \times K} |\tilde{f}_n(x) - f(x)|.$$

The second term of the right-hand side converges to 0 since, by construction, $\tilde{f}_n$ $\mathcal{K}$-converges to f; for the first term we observe that, replicating (B.28) in this case, it remains to estimate $\sup_{I_0 \times K} |\tilde{f}_n(t, P_n x) - \tilde{f}_n(t, x)|$. By construction, if n is sufficiently large, then $\tilde{f}_n = f$ on $I_0 \times K$. Hence, arguing as in (B.29) and using the compactness of $\mathcal{R}_1(I_0 \times K)$ from Lemma B.77, we get the claim.

Regarding the second statement, it is enough to take an orthonormal basis $\mathcal{E}$ which is contained in $D(B)$, it always exists since $D(B)$ is dense.   □

B.7   Approximation of Solutions of Kolmogorov Equations Through $\mathcal{K}$-Convergence

Let H be a real separable Hilbert space and $T > 0$. We consider the following initial value problem for a linear Kolmogorov equation[8]

$$\begin{cases} u_t = \dfrac{1}{2} \operatorname{Tr}[QD^2 u] + \langle x, A^* Du \rangle + Cu + f(t, x), & (t, x) \in (0, T] \times H, \\[2mm] u(0, x) = \varphi(x), & x \in H, \end{cases}$$

(B.33)

where $C \in \mathbb{R}$, $\varphi \in C(H)$, $f \in C((0, T] \times H)$ and A, Q satisfy the following.

Hypothesis B.79 (i) A is the infinitesimal generator of a strongly continuous semigroup $\{e^{tA}, t \geq 0\}$ on H with $\|e^{tA}\| \leq Me^{\omega t}$ for all $t \geq 0$ for given $M \geq 1$, $\omega \in \mathbb{R}$.

(ii) $Q \in \mathcal{L}^+(H)$ and $e^{sA} Q e^{sA^*} \in \mathcal{L}_1(H)$ for all $s > 0$. Moreover, for all $t \geq 0$,

---

[8]Often (see e.g. [179] Chap. 6) the term $\langle x, A^* Dv \rangle$ is replaced by the term $\langle Ax, Dv \rangle$, which is well defined if $x \in D(A)$. Here we use the former expression because we will look for more regular solutions, having the derivative in $D(A^*)$ (see the notion of *classical solution* used in Sect. 6.2 of [179]).

$$\int_0^t \mathrm{Tr}\left[e^{sA}Qe^{sA^*}\right]ds < +\infty,$$

so the symmetric positive operator

$$Q_t : H \to H, \qquad Q_t := \int_0^t e^{sA}Qe^{sA^*}ds, \qquad (B.34)$$

is of trace class for every $t \geq 0$.

If we let, formally, A be the operator associated to the second and third terms of (B.33) and, still formally, R_t be the corresponding semigroup, then we can rewrite (B.33) in the following mild form

$$u(t, x) = e^{Ct}R_t[\varphi](x) + \int_0^t e^{C(t-s)}R_{t-s}[f(s, \cdot)](x)ds \qquad (B.35)$$

and call the function on the right-hand side the *mild solution* of (B.33). In what follows we provide some useful approximation results for solutions of such equations. Similar results are available in the literature (see e.g. [108, 308, 493, 496]) but only when the data φ and f are bounded and in a slightly different setting. Here we allow the data to have polynomial growth and take a setting which is more suitable for the purpose of this book (see Remark B.97 for more on this).

In the next three subsections we first present (Sect. B.7.1) suitable definitions of classical and strong solutions of (B.33); then (Sect. B.7.2) we define precisely R_t, the mild solutions, and connect the generator A of R_t with (B.33) through the operators A_0 and $\hat{A}_0$; finally, (Sect. B.7.3) we present a useful approximation result.

B.7.1 Classical and Strong Solutions of (B.33)

Let us introduce the operator A_0 in $C(H)$ as follows (compare it with the operators A_1 in Sect. 4.5 and A_0 in Sect. 4.6):

$$\begin{cases} D(A_0) = \left\{\phi \in UC_b^2(H) \; : \; A^*D\phi \in UC_b(H, H), \; D^2\phi \in UC_b(H, \mathcal{L}_1(H))\right\} \\ \\ A_0\phi = \frac{1}{2}\mathrm{Tr}\left[QD^2\phi\right] + \langle x, A^*D\phi\rangle. \end{cases}$$

$$(B.36)$$

We also consider the restriction $\hat{A}_0 \subset A_0$ defined as in [108], Sect. 5 (see also [496], Sect. 5 or [493], Sect. 7.4.3),

$$\begin{cases} D(\hat{\mathcal{A}}_0) = \Big\{ \phi \in UC_b^2(H) \ : \ A^*D\phi \in UC_b(H, H), \\ \qquad\qquad \langle x, A^*D\phi \rangle \in UC_b(H) \text{ and } D^2\phi \in UC_b(H, \mathcal{L}_1(H)) \Big\} \qquad \text{(B.37)} \\ \hat{\mathcal{A}}_0\phi = \tfrac{1}{2}\mathrm{Tr}\,[QD^2\phi] + \langle x, A^*D\phi \rangle. \end{cases}$$

Remark B.80 The operator $\hat{\mathcal{A}}_0$ can be seen as an unbounded operator in $C_m(H)$ for all $m \geq 0$ since, by the definition of $D(\hat{\mathcal{A}}_0)$, one easily sees that $\hat{\mathcal{A}}_0\phi \in C_b(H)$ (and so it also belongs to $C_m(H)$) for all $\phi \in D(\hat{\mathcal{A}}_0)$. On the other hand this is not the case for $\mathcal{A}_0$. Indeed, for a generic $\phi \in D(\mathcal{A}_0)$ we can only say that $\mathcal{A}_0\phi \in C_m(H)$ for $m \geq 1$ since the term $\langle x, A^*D\phi \rangle$ may be unbounded. The same considerations holds if we consider the operators in $UC_m(H)$.

Both operators are used to define suitable approximations of the solution of (B.33). In most of the literature only $\hat{\mathcal{A}}_0$ is used. Here we see that, for approximation purposes, $\mathcal{A}_0$ can also be used with some advantages. ∎

We endow $D(\mathcal{A}_0)$ with the norm

$$\|\phi\|_{D(\mathcal{A}_0)} := \|\phi\|_0 + \|D\phi\|_0 + \|A^*D\phi\|_0 + \sup_{x \in H} \|D^2\phi(x)\|_{\mathcal{L}_1(H)}, \qquad \text{(B.38)}$$

while in $D(\hat{\mathcal{A}}_0)$ we take the norm

$$\|\phi\|_{D(\hat{\mathcal{A}}_0)} := \|\phi\|_0 + \|D\phi\|_0 + \|A^*D\phi\|_0 + \|\langle x, A^*D\phi \rangle \|_0 + \sup_{x \in H} \|D^2\phi(x)\|_{\mathcal{L}_1(H)}.$$
$$\text{(B.39)}$$

Arguing as in Theorem 2.7 of [176], it can be proved that both $D(\mathcal{A}_0)$ and $D(\hat{\mathcal{A}}_0)$ with the above norms are Banach spaces.[9] However, the Banach structure is not essential for our purposes, even if it simplifies the notation.

We now give two notions of classical solutions of (B.33). The first, a more restrictive one, uses the operator $\hat{\mathcal{A}}_0$, and is in line with what is done, for example, in Definition 4.6 of [308] or Definition 4.1 of [309].

Definition B.81 $u \in C_b([0, T] \times H)$ is a classical solution of (B.33) in $D(\hat{\mathcal{A}}_0)$ if

$$\begin{cases} u(\cdot, x) \in C^1([0, T]), \ \forall x \in H \\ u(t, \cdot) \in D(\hat{\mathcal{A}}_0) \text{ for any } t \in [0, T] \text{ and } \sup_{t \in [0,T]} \|u(t, \cdot)\|_{D(\hat{\mathcal{A}}_0)} < +\infty \\ Du, A^*Du \in C_b([0, T] \times H, H), \ D^2u \in C_b([0, T] \times H, \mathcal{L}_1(H)) \\ \hat{\mathcal{A}}_0 u \in C_b([0, T] \times H), \end{cases}$$
$$\text{(B.40)}$$

and u satisfies, for every $(t, x) \in [0, T] \times H$, Eq. (B.33).

The second definition is similar to Definition 4.129 and uses the operator $\mathcal{A}_0$.

---

[9]The definition of the domain of the operator studied in this paper is slightly different but the arguments there can be adapted easily.

Definition B.82 $u \in C_b([0, T] \times H)$ is a classical solution of (B.33) in $D(\mathcal{A}_0)$ if

$$
\begin{cases}
u(\cdot, x) \in C^1([0, T]), \quad \forall x \in H \\
u(t, \cdot) \in D(\mathcal{A}_0) \text{ for any } t \in [0, T] \text{ and } \sup_{t \in [0,T]} \|u(t, \cdot)\|_{D(\mathcal{A}_0)} < +\infty \\
Du, A^*Du \in C_b([0, T] \times H, H), \quad D^2u \in C_b([0, T] \times H, \mathcal{L}_1(H)),
\end{cases}
$$
(B.41)

and u satisfies, for every $(t, x) \in [0, T] \times H$, Eq. (B.33).

Remark B.83 Concerning the above definitions we observe the following.

(1) If we compare these definitions with the one given in Sect. 6.2 of [179], we see that the classical solutions in the sense of both Definitions B.81 and B.82 are more regular. Indeed, the goal here is not (as in Sect. 6.2 of [179]) to prove that mild solutions are classical, but to approximate mild solutions with classical solutions to which we can apply the Itô/Dynkin formula (see Sect. 1.7); hence we want the approximating solutions to be as regular as possible.

(2) The definitions above are the same regardless of whether we work in the spaces $C_m(H)$ (as we mainly do) or $UC_m(H)$, $(m \geq 0)$.

(3) Note that, if u is a classical solution from Definition B.81, then $\hat{\mathcal{A}}_0 u$, and consequently also u_t, belong to $C_b([0, T] \times H)$. On the other hand, if u is a classical solution from Definition B.82, then $\hat{\mathcal{A}}_0 u$, and then also u_t, only belong to $C_1([0, T] \times H)$.

(4) If we compare Definition B.81 with Definition 4.6 of [308] and Definition 4.1 of [309] we see that, in the last two, the functions $u, u_t, Du, D^2u, A^*Du, \hat{\mathcal{A}}_0 u$ are required to belong to $UC_b^x([0, T] \times H)$. The reason is that elements of this space possess, roughly speaking, the maximal joint regularity one can expect even when $f = 0, C = 0$ and $\varphi \in UC_b^\infty(H)$. Indeed, in such case the mild solution is $u(t, x) = R_{T-t}[\varphi](x)$ (where R_t is the Ornstein–Uhlenbeck semigroup corresponding to A and Q) which is in $UC_b^x([0, T] \times H)$ but not in $UC_b([0, T] \times H)$, see Sect. B.7.2. Here we decided to ask for less joint regularity, i.e. $u, Du, D^2u, A^*Du \in C_b([0, T] \times H)$ since this last space is more commonly used in the literature on PDEs in infinite dimension and since all the results we need, in particular the Itô/Dynkin formula, still hold.

(5) The definitions implicitly require regularity of data. Indeed, the second requirement of (B.40) implies that $\varphi \in D(\hat{\mathcal{A}}_0)$ (and similarly for (B.41)) while, from the last requirement we see that necessarily $f \in C_b([0, T] \times H)$.

∎

We now pass to the definition of a strong solution, i.e. approximation of classical solutions. It is substantially a variation of Definition 4.7 of [308] (see also Definition 4.3 of [309]) in the sense that we use, as the underlying space, the space C_m instead of the space UC_b. The definition below is a special case of Definition 4.132 and uses some definitions given in Chap. 4 which, for the reader's convenience, we repeat in the forms needed here.

We recall the definitions of the classes of weights $\mathcal{I}_1$ and $\mathcal{I}_2$ introduced in (4.29)–(4.30):

$$\mathcal{I}_1 := \left\{ \eta : (0, +\infty) \to (0, +\infty) \text{ decreasing and } \eta \in L^1(0, T), \ \forall T > 0 \right\}. \tag{B.42}$$

$$\mathcal{I}_2 := \left\{ \eta \in \mathcal{I}_1 : \exists \lim_{t \searrow 0^+} \frac{1}{\eta(t)} \int_0^t \eta(s) \eta(t-s) ds = 0 \right\}. \tag{B.43}$$

We also recall (see (4.32)) that, given $\eta \in \mathcal{I}_1$, a function $f : (0, T] \times H \to Z$ (where Z is a given real separable Hilbert space) belongs to $C_{m,\eta}((0, T] \times H, Z)$ if

$$f \in C_m([\tau, T] \times H, Z) \ \forall \tau \in (0, T) \text{ and } \eta^{-1} f \in B_m((0, T] \times H, Z). \tag{B.44}$$

When $Z = \mathbb{R}$, we omit it as usual, simply writing $C_{m,\eta}((0, T] \times H)$. Furthermore, we give a variant of the definition of $\mathcal{K}$-convergence (see Definition 4.131).

Definition B.84 Let $m \geq 0$, $\eta \in \mathcal{I}_1$ and let Z be a real Hilbert space. We say that a sequence $(f_n)_{n \in \mathbb{N}} \subset C_{m,\eta}((0, T] \times H, Z)$ $\mathcal{K}$-converges to $f \in C_{m,\eta}((0, T] \times H, Z)$ if

$$\begin{cases} \sup_{n \in \mathbb{N}} \| f_n \|_{C_{m,\eta}((0,T] \times H, Z)} < +\infty, \\ \lim_{n \to +\infty} \sup_{(t,x) \in (0,T] \times K} \eta(t)^{-1} |f_n(t, x) - f(t, x)| = 0, \end{cases} \tag{B.45}$$

for every compact set $K \subset H$. In such case we write $\mathcal{K}\text{-}\lim_{n \to +\infty} f_n = f$ in $C_{m,\eta}((0, T] \times H, Z)$.

Below is the definition of $\mathcal{K}$-strong solution.

Definition B.85 Let $m \geq 0$ and $\eta \in \mathcal{I}_1$. Let $\varphi \in C_m(H)$ and $f \in C_{m,\eta}((0, T] \times H)$. We say that a function $u \in C_m([0, T] \times H)$ is a $\mathcal{K}$-strong solution in $D(\hat{A}_0)$ of (B.33) if $u(t, \cdot)$ is Fréchet differentiable for any $t \in (0, T]$ and there exist three sequences (u_n) in $C_b([0, T] \times H)$, (φ_n) in $D(\hat{A}_0)$, (f_n) in $C_{m,\eta}((0, T] \times H)$ such that:

(i) For every $n \in \mathbb{N}$, u_n is a classical solution in $D(\hat{A}_0)$ (from Definition B.81) of

$$\begin{cases} w_t = \hat{A}_0 w + Cw + f_n \\ w(0) = \varphi_n. \end{cases} \tag{B.46}$$

(ii) The following limits hold

$$\begin{cases} \mathcal{K}\text{-}\lim_{n \to +\infty} \varphi_n = \varphi & in \quad C_m(H) \\ \mathcal{K}\text{-}\lim_{n \to +\infty} u_n = u & in \quad C_m([0, T] \times H) \end{cases}$$

and, for some $\eta_1 \in \mathcal{I}_1$ (possibly different from η),

$$\begin{cases} \mathcal{K}\text{-}\lim_{n\to+\infty} f_n = f & in \quad C_{m,\eta}((0,T]\times H) \\ \mathcal{K}\text{-}\lim_{n\to+\infty} Du_n = Du & in \quad C_{m,\eta_1}((0,T]\times H, H). \end{cases}$$

Finally, we say that a function $u \in C_m([0,T]\times H)$ is a $\mathcal{K}$-strong solution in $D(\mathcal{A}_0)$ of (B.33) if all the above holds substituting $D(\hat{\mathcal{A}}_0)$ with $D(\mathcal{A}_0)$.

Remark B.86 The spirit of this definition is substantially that of the so-called strong solutions in the Friedrichs sense (a terminology dating back to [268]) for abstract Cauchy problems, see e.g. Definition 4.1.1 of [416]. It is useful to connect the concept of a mild solution with that of a classical solution proving that mild solutions are indeed strong solutions. We make two key points here: first, the convergences in (ii) are asked to hold in the $\mathcal{K}$-sense, and second, the convergence of the derivatives is also required. The reason for the former is that uniform convergence in the whole H would be a requirement which is too strong as $UC_b^2(H)$ (and so $D(\mathcal{A}_0)$) is not dense in $UC_b(H)$ when H is infinite-dimensional, see [457] and also [493], Remark 2.2.11. On the other hand the use of π-convergence (as in [496], Sect. 4 or [493], Chap. 7) is possible but in the cases we treat here it would give weaker results, as $\mathcal{K}$-convergence can always be proved when the data are continuous. The latter is motivated by the use of this definition for applications to HJB equations (as in Chap. 4) which are written in the mild form of (B.35) with f depending on u and Du (see e.g. (4.1) and (4.5)). Hence convergence of the derivatives allows us to pass to the limit in such a mild form. ∎

B.7.2 The Ornstein–Uhlenbeck Semigroup Associated to (B.33), the Mild Solutions and Their Properties

Assume that Hypothesis B.79 holds and take a generalized reference probability space $\mu = (\Omega, \mathscr{F}, \{\mathscr{F}_s\}_{s\geq 0}, \mathbb{P}, W)$, where W is a cylindrical Wiener process on H.
Consider the following linear stochastic equation on H

$$\begin{cases} dX(t) = AX(t)\,dt + \sqrt{Q}\,dW(t) \\ X(0) = x. \end{cases} \tag{B.47}$$

Under Hypothesis B.79 the problem (B.47) has an H-valued mild solution from Definition 1.119 (see Proposition 1.147). Such a solution is mean square continuous (see Proposition 1.144 or also [180], Theorem 5.2-(i)) and is denoted by $X(t,x)$.
We denote by $\{R_t, t \geq 0\}$ or simply by R_t, the corresponding (Ornstein–Uhlenbeck) transition semigroup (see Sect. 1.6) on $B_m(H)$:

$$R_t[\varphi](x) = \mathbb{E}\left[\varphi(X(t,x))\right] = \int_H \varphi(y)\,\mathcal{N}(e^{tA}x, Q_t)(dy) = \int_H \varphi(y+e^{tA}x)\,\mathcal{N}(0, Q_t)(dy), \tag{B.48}$$

where $\mathcal{N}(z, Q)$ is the Gaussian measure introduced in Definition 1.58. See Sects. 4.3.1.1 and 4.3.1.2 for more on this.

We immediately see that, when $\varphi \in B_m(H)$ and $f \in B_m([0, T] \times H)$, the function u in (B.35) is well defined. Thus we can give the precise definition of the mild solution of Eq. (B.33).

Definition B.87 Given $\varphi \in B_m(H)$ and $f \in B_m([0, T] \times H)$, the function u given by (B.35) is well defined on $[0, T] \times H$ and is called the mild solution of (B.33).

We now explain the connection between Eq. (B.33) and the notion of the mild solution. We start with the following result.

Lemma B.88 *Let X be a separable Banach space. Let $\{S(t), t \geq 0\}$ be a family of operators in $\mathcal{L}(X)$ satisfying $S(0) = I$ and $S(t + s) = S(t)S(s)$ for all $t, s \in [0, +\infty)$. Assume that $\|S(t)\|_{\mathcal{L}(X)}$ is bounded in $(0, a)$ for some $a > 0$. If for a given $x \in X$, the map $(0, a) \to X$, $s \to S(s)x$, is Borel measurable, then it must be continuous.*

Proof The proof is contained in [201] or in [576] (pp. 233–234). Since both results are slightly different from what we need, we provide a proof.

Let $x \in X$. The function $(0, a) \to X$, $s \to x(s) = S(s)x$ is Borel measurable and bounded since $|x(s)| \leq \|S(s)\|_{\mathcal{L}(X)}|x|$ for all $s \in (0, a)$. Then, for all $0 \leq \alpha < \beta \leq a$ we may define the Bochner integral $\int_\alpha^\beta x(s)ds$ and we have, by (1.4), $\left|\int_\alpha^\beta x(s)ds\right| \leq \int_\alpha^\beta |x(s)|\,ds$. Moreover, for all $t \in [-\alpha, a - \beta]$,

$$\int_\alpha^\beta x(t + s)ds = \int_{\alpha+t}^{\beta+t} x(s)ds.$$

Let $0 \leq \alpha < \eta < \beta < \xi - \varepsilon < \xi$ with $\varepsilon \in (0, \xi - \beta)$. Since $x(\xi) = S(\xi)x = S(\eta)S(\xi - \eta)x = S(\eta)x(\xi - \eta)$, we have

$$(\beta - \alpha)x(\xi) = \int_\alpha^\beta x(\xi)d\eta = \int_\alpha^\beta S(\eta)x(\xi - \eta)d\eta.$$

Hence

$$(\beta - \alpha)\left[x(\xi \pm \varepsilon) - x(\xi)\right] = \int_\alpha^\beta S(\eta)(x(\xi \pm \varepsilon - \eta) - x(\xi - \eta))d\eta,$$

which gives

$$(\beta - \alpha)\,|x(\xi \pm \varepsilon) - x(\xi)| \leq \sup_{s \in (\alpha,\beta)}\ \|S(s)\|_{\mathcal{L}(X)} \int_{\xi-\alpha}^{\xi-\beta} |x(\tau \pm \varepsilon) - x(\tau)|d\tau$$

The right-hand side tends to zero as $\varepsilon \to 0^+$. This may be seen by approximating $x(t)$ by simple functions (for which the right-hand side clearly goes to 0) and using (1.3) and the subsequent remark. $\qquad\square$

Proposition B.89 *Let Hypothesis B.79 hold. Then we have the following.*

(i) *$\{R_t, t \geq 0\}$ is a $\mathcal{K}$-continuous semigroup and a π-continuous semigroup in $C_m(H)$ and $UC_m(H)$ $(m \geq 0)$.*

(ii) *$\{R_t, t \geq 0\}$ is not strongly continuous in these spaces unless $A = 0$. Indeed, as $s \to 0^+$, we have $R_s[\phi] \to \phi$ in $UC_b(H)$ if and only if*

$$\sup_{y \in H} |\phi(e^{sA}y) - \phi(y)| \to 0. \tag{B.49}$$

(iii) *Let $t > 0$. We have $R_{t+s}[\phi] \to R_t[\phi]$ as $s \to 0^+$ in $UC_b(H)$ if and only if*

$$\sup_{x \in H} \left| \int_H \left(\phi(y + e^{(t+s)A}x) - \phi(y + e^{tA}x) \right) \mathcal{N}_{Q_t}(dy) \right| \to 0. \tag{B.50}$$

The above is not satisfied for all $\phi \in UC_b(H)$ (e.g. when $H = \mathbb{R}$, $A = Q = I$, and $\phi(x) = \sin x$). In particular this implies that the map $(0, +\infty) \to UC_b(H)$, $s \to R_s[\phi]$ is in general not measurable.

Proof
Proof of (i). For the $\mathcal{K}$-continuity, see [101], Proposition 6.2, and [493], Sect. 6.3.3, when $m = 0$. Moreover, see [102] for the $UC_m(H)$ case and Theorem 4.1 in [300] for the case of $C_m(H)$ (recalling the change in the definition mentioned in Remark B.67).

For the π-continuity, see [493], Sect. 6.3.3, and [492] (pp. 293–294): there the proof is done in $UC_b(H)$ but it can be easily extended to the cases of $C_m(H)$ and $UC_m(H)$.

Proof of (ii). Regarding the non-strong continuity, see [101], Example 6.1, where an example of a bounded smooth function ϕ is given for which $R_s[\phi] \not\to \phi$ in $UC_b(H)$ as $s \to 0$. The second statement is proved in [179], Proposition 6.3.1.

Proof of (iii). Let $t > 0$, $x \in H$ and let, as in the proof of Proposition 6.3.1 in [179],

$$G_t[\phi](x) := \int_H \phi(y + x) \mathcal{N}_{Q_t}(dy), \qquad \phi \in C_b(H),$$

so $R_t[\phi](x) = G_t[\phi](e^{tA}x)$. As proved in [179] Proposition 6.3.1, the semigroup $\{G_t, t \geq 0\}$ is strongly continuous in $UC_b(H)$.[10] Taking $\phi \in UC_b(H)$ we have, for $t > 0$, $s > 0$, $x \in H$,

---

[10] The semigroup $\{G_t, t \geq 0\}$ is not strongly continuous in $C_b(H)$ as proved, for example, in [179], Theorem 3.1.1.

$$R_{t+s}[\phi](x) - R_t[\phi](x)$$
$$= \big(G_{t+s}[\phi](e^{(t+s)A}x) - G_t[\phi](e^{(t+s)A}x)\big) + \big(G_t[\phi](e^{(t+s)A}x) - G_t[\phi](e^{tA}x)\big).$$

By the strong continuity of G_t, the first term of the right-hand side above converges to 0 as $s \to 0$ uniformly in $x \in H$. Hence $\sup_{x \in H} |R_{t+s}[\phi](x) - R_t[\phi](x)|$ goes to 0 as $s \to 0$ if and only if (B.50) holds.

We now show that (B.50) is not satisfied when $H = \mathbb{R}$, $A = Q = I$, and $\phi(x) = \sin x$. Indeed, in this case we have by standard computations

$$\int_H \Big(\phi(y + e^{(t+s)A}x) - \phi(y + e^{tA}x)\Big) \mathcal{N}_{Q_t}(dy)$$

$$= [\cos(e^{t+s}x) - \cos(e^t x)] \int_{\mathbb{R}} \sin y \, \mathcal{N}_{Q_t}(dy) + [\sin(e^{t+s}x) - \sin(e^t x)] \int_{\mathbb{R}} \cos y \, \mathcal{N}_{Q_t}(dy)$$

$$= [\sin(e^{t+s}x) - \sin(e^t x)] G_t[\cos](0),$$

where in the last equality we used $\int_{\mathbb{R}} \sin y \, \mathcal{N}_{Q_t}(dy) = 0$ and $\int_{\mathbb{R}} \cos y \mathcal{N}_{Q_t}(dy) = G_t[\cos](0)$. We thus obtain

$$\sup_{x \in H} \left| \int_H \big(\phi(y + e^{(t+s)A}x) - \phi(y + e^{tA}x)\big) \mathcal{N}_{Q_t}(dy) \right|$$

$$= |G_t[\cos](0)| \sup_{z \in \mathbb{R}} \left| \sin(e^s z) - \sin z \right|.$$

Since $G_t[\cos](0) \neq 0$ and it is easy to find a sequence $s_n \searrow 0$ such that $\sup_{z \in \mathbb{R}} |\sin(e^{s_n} z) - \sin z| = 2$ (e.g. asking that $e^{s_n}(2n\pi + \pi/2) = 2n\pi + 3\pi/2$) the claim follows.

Concerning the last claim, observe that if the map $(0, +\infty) \to UC_b(H)$, $s \to R_s[\phi]$ is measurable, then by Lemma B.88 it must be continuous, which is not possible for all $\phi \in UC_b(H)$, as we have just proved. $\qquad \square$

From now on $\mathcal{A}^m$ will denote the generator of the transition semigroup R_t as a $\mathcal{K}$-continuous semigroup in $C_m(H)$ (or $UC_m(H)$ when specified), as defined in the previous section. We prove two useful results.

Lemma B.90 *Let Hypothesis B.79 be satisfied and let $\mathcal{A}^m$ be the generator of R_t as a $\mathcal{K}$-continuous semigroup in $C_m(H)$ (or $UC_m(H)$).*

(i) *For all $t \geq 0$, the operator R_t maps $D(\mathcal{A}_0)$ into itself, and there exists a constant $L_1 > 0$ such that (here ω is from Hypothesis B.79)*

$$\|R_t[\phi]\|_{D(\mathcal{A}_0)} \leq L_1 e^{2(\omega \vee 0)t} \|\phi\|_{D(\mathcal{A}_0)}, \quad \forall t \geq 0, \ \phi \in D(\mathcal{A}_0). \tag{B.51}$$

The same holds for $D(\hat{\mathcal{A}}_0)$.

(ii) *For all $m \geq 1$ we have $\mathcal{A}_0 \subset \mathcal{A}^m$ and, for any $\phi \in D(\mathcal{A}_0)$, we have*

$$\frac{d}{dt}(R_t[\phi]) = \mathcal{A}_0 R_t[\phi] = R_t[\mathcal{A}_0 \phi], \quad t \geq 0. \tag{B.52}$$

Similarly, for all $m \geq 0$ $\hat{A}_0 \subset A^m$ and, for any $\phi \in D(\hat{A}_0)$, we have

$$\frac{d}{dt}(R_t[\phi]) = \hat{A}_0 R_t[\phi] = R_t\left[\hat{A}_0\phi\right], \quad t \geq 0. \tag{B.53}$$

(iii) Let $T > 0$. For every $f \in C_b([0, T] \times H)$ such that, for all $t \in [0, T]$, $f(t, \cdot) \in D(A_0)$ and

$$\|f(t, \cdot)\|_{D(A_0)} \leq g_0(t)$$

for a suitable $g_0 \in L^1(0, T; \mathbb{R}^+)$, let us set

$$g(t, x) = \int_0^t R_{t-s}[f(s, \cdot)](x)\,ds, \quad \forall x \in H. \tag{B.54}$$

Then $g \in C_b([0, T] \times H)$ and for every $t \in [0, T]$, $g(t, \cdot) \in D(A_0)$. Moreover,

$$\|g(t, \cdot)\|_{D(A_0)} \leq L_1 e^{2(\omega \vee 0)T} \|g_0\|_{L^1}, \quad \forall t \in [0, T]. \tag{B.55}$$

The same holds if we replace A_0 by $\hat{A}_0$.

(iv) Let $m \geq 0$ and let $\lambda > m(\omega \vee 0)$, where ω is given in Hypothesis B.79. Then $(\lambda I - A^m)^{-1}$ exists and it is a bounded operator. Moreover, it maps $D(A_0)$ into itself. The same is true for $D(\hat{A}_0)$.

Proof We present the proofs of (ii) and (iv) in the case when A^m is the generator of R_t in $C_m(H)$. The case when A^m is the generator of R_t in $UC_m(H)$ is completely analogous and we omit it.

Proof of (i). If $\phi \in D(A_0)$ we have, using the last equality of (B.48) and straightforward computations, for $x, h, k \in H, t \geq 0$,

$$\langle DR_t[\phi](x), h \rangle = \int_H \langle D\phi(y + e^{tA}x), e^{tA}h \rangle \mathcal{N}(0, Q_t)(dy) = R_t[\langle D\phi(\cdot), e^{tA}h \rangle], \tag{B.56}$$

$$\langle A^* DR_t[\phi](x), h \rangle = \int_H \langle A^* D\phi(y + e^{tA}x), e^{tA}h \rangle \mathcal{N}(0, Q_t)(dy) = R_t[\langle A^* D\phi(\cdot), e^{tA}h \rangle], \tag{B.57}$$

$$\langle D^2 R_t[\phi](x)h, k \rangle = \int_H \langle D^2\phi(y + e^{tA}x)e^{tA}k, e^{tA}h \rangle \mathcal{N}(0, Q_t)(dy) = R_t[\langle D^2\phi(\cdot)e^{tA}k, e^{tA}h \rangle], \tag{B.58}$$

and, for a given orthonormal basis $\{e_n\}$,

$$\text{Tr}\left(D^2 R_t[\phi](x)\right) = \sum_{n=0}^{+\infty} \langle D^2 R_t[\phi](x)e_n, e_n \rangle = \int_H \sum_{n=0}^{+\infty} \langle D^2\phi(y+e^{tA}x)e^{tA}e_n, e^{tA}e_n \rangle \mathcal{N}(0, Q_t)(dy). \tag{B.59}$$

Hence, by simple computations, we get that $R_t[\phi] \in D(A_0)$ and

$$\|R_t[\phi]\|_{D(A_0)} \leq \|\phi\|_0 + Me^{\omega t}[\|D\phi\|_0 + \|A^* D\phi\|_0] + M^2 e^{2\omega t} \|D\phi\|_{\mathcal{L}_1(H)},$$

which gives the claim. In the case of $D(\hat{\mathcal{A}}_0)$ we also need to estimate the term $\langle A^* DR_t[\phi](x), x\rangle$. We have, thanks to (B.57),

$$\langle A^* DR_t[\phi](x), x\rangle = \int_H \langle A^* D\phi(y + e^{tA}x), y + e^{tA}x\rangle \mathcal{N}(0, Q_t)(dy)$$

$$- \int_H \langle A^* D\phi(y + e^{tA}x), y\rangle \mathcal{N}(0, Q_t)(dy).$$

Since, by the Hölder inequality and by Proposition 1.59,

$$\left| \int_H \langle A^* D\phi(y + e^{tA}x), y\rangle \mathcal{N}(0, Q_t)(dy) \right| \le \|A^* D\phi\|_0 (\text{Tr}(Q_t))^{1/2},$$

we obtain

$$\sup_{x \in H} | < A^* DR_t[\phi](x), x > | \le \sup_{x \in H} | < A^* D\phi(x), x > | + \|A^* D\phi\|_0 (\text{Tr}(Q_t))^{1/2},$$

which gives the claim thanks to Lemma 4.39.

Proof of (ii). From Dynkin's formula of Proposition 1.168 we easily get that (here $X(s, x)$ is the solution of (B.47))

$$\frac{R_h[\phi](x) - \phi(x)}{t} = \frac{1}{h} \mathbb{E} \left(\phi(X(h, x)) - \phi(x) \right)$$

$$= \frac{1}{h} \mathbb{E} \int_0^h \left(\frac{1}{2} \text{Tr}[QD^2\phi(X(s, x))] + \langle A^* D\phi(X(s, x)), X(s, x)\rangle \right) ds.$$

Now we have $\sup_{s \in [0,T]} \mathbb{E}|X(s; x)|^2 \le C$ for some $C > 0$. Moreover, by the definition of X it is also easy to see that we have $\mathbb{E}|X(s; x) - x|^2 \le \rho(s)$ for some modulus ρ. Thus, arguing similarly as in the proof of Theorem 3.66, we can conclude that when $h \searrow 0$, the first term of the right-hand side of the last formula converges to

$$\frac{1}{2} \text{Tr}[QD^2\phi(x)] + \langle x, A^* D\phi_n(x)\rangle.$$

Hence

$$\lim_{h \to 0} \frac{R_h[\phi](x) - \phi(x)}{t} = \mathcal{A}_0 \phi(x),$$

which, by the definition of a generator (Definition B.71), implies $\mathcal{A}_0 \subset \mathcal{A}^m$ for $m \ge 1$ and $\hat{\mathcal{A}}_0 \subset \mathcal{A}^m$ for $m \ge 0$. Now Eqs. (B.52) and (B.53) follow immediately from Proposition B.74-(iii) and from point (i) of this lemma.

Proof of (iii).[11] Since $f \in C_b([0, T] \times H)$, using the first equality in (B.48) and the mean square continuity of $X(t, x)$, one can prove exactly as in Proposition 4.50-(ii) that also $g \in C_b([0, T] \times H)$. Moreover, since $f(t, \cdot) \in D(\mathcal{A}_0)$, for all $t \in [0, T]$, by part (i) it follows that, for $0 \le s \le t \le T$, $R_{t-s}[f(s, \cdot)] \in D(\mathcal{A}_0)$ and

$$\|R_{t-s}[f(s, \cdot)]\|_{D(\mathcal{A}_0)} \le L_T \|f(s, \cdot)\|_{D(\mathcal{A}_0)} \le L_T g_0(s).$$

Now, using that the derivative operator is closed (see Corollary 4.14), we get that $g(t, \cdot)$ is Fréchet differentiable and, for $t \in [0, T]$, $x, h \in H$,

$$\langle Dg(t, x), h \rangle = \int_0^t \langle DR_{t-s}[f(s, \cdot)](x), h \rangle ds = \int_0^t R_{t-s} \left[\langle Df(s, \cdot), e^{(t-s)A} h \rangle \right] (x) ds,$$
(B.60)

where we exploited (B.56) in the last equality. The integrand is Borel measurable in s since it is the limit of the difference quotients; the continuity of $Dg(t, x)$ is proved again as in Proposition 4.50-(ii). Similarly, using (B.57) we get for $t \in [0, T]$, $x, h \in H$,

$$\langle A^* Dg(t, x), h \rangle = \int_0^t \langle A^* DR_{t-s}[f(s, \cdot)](x), h \rangle ds = \int_0^t R_{t-s} \left[\langle A^* Df(s, \cdot), e^{(t-s)A} h \rangle \right] (x) ds.$$
(B.61)

Moreover, iterating the argument and exploiting (B.56) in the last equality, we prove that $g(t, \cdot)$ is twice Fréchet differentiable and, for $t \in [0, T]$, $x, h, k \in H$,

$$\langle D^2 g(t, x)h, k \rangle = \int_0^t \langle D^2 R_{t-s}[f(s, \cdot)](x)h, k \rangle ds = \int_0^t R_{t-s} \left[\langle D^2 f(s, \cdot) e^{(t-s)A} h, e^{(t-s)A} k \rangle \right] (x) ds.$$
(B.62)

The integrand is Borel measurable in s since it is the limit of the difference quotients and the continuity of $D^2 g(t, x)h$ for each h is proved again as in Proposition 4.50-(ii). This does not even imply the measurability of $Dg(t, x)$ in general, but it is enough to prove the continuity of the trace. Indeed,

$$\mathrm{Tr}(D^2 g(t, x)) = \sum_{n=0}^{+\infty} \langle D^2 g(t, x) e_n, e_n \rangle$$

$$= \int_0^t \int_H \sum_{n=0}^{+\infty} \langle D^2 f(s, y + e^{tA} x) e^{(t-s)A} e_n, e^{(t-s)A} e_n \rangle \mathcal{N}(0, Q_{t-s})(dy) ds.$$
(B.63)

As before the integrand is Borel measurable in (s, y), hence continuity can be proved as in Proposition 4.50-(ii).

---

[11] This proof is partially similar to the proof of Lemma 4.5 in [496].

In the case of $D(\hat{\mathcal{A}}_0)$ we also need to estimate the term $\langle A^* Dg(t, x), x \rangle$. We have, thanks to (B.61),

$$
\begin{aligned}
\langle A^* Dg(t, x), x \rangle &= \int_0^t \langle A^* DR_{t-s}[f(s, \cdot)](x), x \rangle ds = \int_0^t R_{t-s}\left[\langle A^* Df(s, \cdot), e^{(t-s)A}x \rangle \right](x) ds \\
&= \int_0^t \int_H \langle A^* Df(s, y + e^{(t-s)A}x), y + e^{(t-s)A}x \rangle \mathcal{N}(0, Q_{t-s})(dy) ds \\
&\quad - \int_0^t \int_H \langle A^* D\phi(y + e^{(t-s)A}x), y \rangle \mathcal{N}(0, Q_{t-s})(dy) ds.
\end{aligned}
$$

The conclusion now follows as in part (i) using Hölder's inequality and Proposition 1.59.

Proof of (iv). Recall first that, for any $\phi \in C_m(H)$, $x \in H$, the function $t \to e^{-\lambda t} R_t[\phi](x)$ is integrable when $\lambda > m(\omega \vee 0)$, due to the first estimate of Theorem 4.41. Thus by Remark B.72, we have that λ is in the resolvent set of $\mathcal{A}^m$ and

$$
(\lambda I - \mathcal{A}^m)^{-1}\phi = \int_0^{+\infty} e^{-\lambda t} R_t[\phi](x) dt.
$$

Taking $\phi \in D(\mathcal{A}_0)$ or in $D(\hat{\mathcal{A}}_0)$, the required conclusion follows using the same arguments as in the proof of part (iii), which here are even easier since we do not have the time dependency of the integral. □

Proposition B.91 *Let Hypothesis B.79 be satisfied and let $T > 0$.*

(i) *Let $\varphi \in D(\mathcal{A}_0)$ and $f \in C_b([0, T] \times H)$ be such that, for all $t \in [0, T]$, $f(t, \cdot) \in D(\mathcal{A}_0)$ and*

$$
\|f(t, \cdot)\|_{D(\mathcal{A}_0)} \leq g_0(t)
$$

for some $g_0 \in L^1(0, T; \mathbb{R}^+)$. Then the function u defined in (B.35) is a classical solution in $D(\mathcal{A}_0)$ of (B.33).

(ii) *If the assumptions of Part (i) hold with $D(\hat{\mathcal{A}}_0)$ in place of $D(\mathcal{A}_0)$ then the function u defined in (B.35) is a classical solution in $D(\hat{\mathcal{A}}_0)$ of (B.33).*

Proof We take $C = 0$ as the case $C \neq 0$ can be obtained by a straightforward change of variable. We start by proving (i). As a first step we need to prove that u satisfies the regularity required in (B.41). Indeed, by Lemma B.90, in particular (B.51) and (B.55), it immediately follows that the second line of (B.41) is true. The third line of (B.41) is obtained, for the first term in (B.35), from (B.56)–(B.59), and for the convolution term, from (B.60)–(B.63). Hence continuity is immediately deduced. The first line of (B.41) follows from the next computation which also shows that u satisfies (B.33).

We apply Dynkin's formula of Proposition 1.168 to the process $\varphi(X(s, x))$ obtaining, on the interval $[0, t]$,

$$\mathbb{E}\varphi(X(t,x)) = \varphi(x) + \mathbb{E}\int_0^t \langle A^* D\varphi(X(s,x)), X(s,x)\rangle ds$$

$$+ \frac{1}{2}\mathrm{Tr}\left[QD^2\varphi(X(s,x))\right]\Bigg]ds.$$

$$(\mathrm{B.64})$$

We now compute the right derivative of u at $t = 0$. We observe first that, by the definition of u,

$$\frac{u(h,x) - u(0,x)}{h} = \frac{R_h[\varphi](x) - \varphi(x)}{h} + \frac{1}{h}\int_0^h R_{h-s}[f(s,\cdot)](x)ds.$$

Arguing as in the proof of Lemma B.90 we obtain that, when $h \searrow 0$, the first term of the above right-hand side converges to

$$\langle A^* D\varphi(x), x\rangle + \frac{1}{2}\mathrm{Tr}[QD^2\varphi(x)].$$

Again applying the same argument to the integrand in the second term of the right-hand side we get that this term, when $h \searrow 0$, converges to $f(0,x)$. We then have, denoting by D_t^+ the right time derivative,

$$D_t^+ u(0,x) = \lim_{h\searrow 0}\frac{u(h,x) - u(0,x)}{h}$$

$$= \langle A^* D\varphi(x), x\rangle + \frac{1}{2}\mathrm{Tr}[QD^2\varphi(x)] + f(0,x),$$

so the equation is satisfied for $t = 0$. For $t > 0$ we observe that, by the semigroup property (see Theorem 1.157),

$$u(t+h,x) = R_h[u(t,\cdot)](x) + \int_t^{t+h} R_{t+h-s}[f(s,\cdot)](x)ds,$$

hence we have

$$\frac{u(t+h,x) - u(t,x)}{h} = \frac{R_h[u(t,\cdot)](x) - u(t,x)}{h} + \frac{1}{h}\int_t^{t+h} R_{t+h-s}[f(s,\cdot)](x)ds$$

and, arguing as for the case $t = 0$ but replacing φ by $u(t,\cdot)$, we get

$$D_t^+ u(t,x) = \langle x, A^* Du(t,x)\rangle + \frac{1}{2}\mathrm{Tr}\left[QD^2 u(t,x)\right] + f(t,x).$$

Now the right-hand side of the above identity is a continuous function on $[0, T] \times H$ and consequently, by Lemma D.19, $u(\cdot, x)$ is continuously differentiable and satisfies Eq. (B.33).

The proof of part (ii) is almost exactly the same. The only difference is that, thanks to the regularity of the data, $\hat{A}_0 u$ and u_t are also bounded. $\square$

B.7.3   The Approximation Results

We start with the following result about the operators A_0, $\hat{A}_0$ and A^m (recall that ω below is the one given by Hypothesis B.79).

Proposition B.92 *Let Hypothesis B.79 hold. We have the following:*

(i) *For any $m \geq 1$, we have $A_0 \subset A^m$, and $D(A_0)$ is a K-core for A^m in $C_m(H)$. In particular, A_0 is K-closable in $C_m(H)$ and its K-closure $\overline{A_0}^K$ coincides with A^m.*
Moreover, for all $m \geq 0$ and $\lambda > m(\omega \vee 0)$, the set

$$(\lambda I - A^m)^{-1} \left(\mathcal{F}C_0^{\infty, A^*}(H) \right) := \left\{ \phi \in C_m(H) \; : \; (\lambda I - A^m)\phi \in \mathcal{F}C_0^{\infty, A^*}(H) \right\}$$

is contained in $D(A_0)$ and it is always a core for A^m.
All the statements above are also true if $C_m(H)$ is replaced by $UC_m(H)$.
Finally, the set $\mathcal{F}C_0^{\infty, A^*}(H) \subset D(A_0)$ is always a core for A^m in $C_m(H)$ when $m \geq 1$, but not, in general, when $m \in [0, 1)$.

(ii) *For any $m \geq 0$, we have $\hat{A}_0 \subset A^m$, and $D(\hat{A}_0)$ is a K-core for A^m in $C_m(H)$ In particular, $\hat{A}_0$ is K-closable in $C_m(H)$ and its K-closure $\overline{\hat{A}_0}^K$ coincides with A^m.*
Moreover, for all $m \geq 0$ and $\lambda > m(\omega \vee 0)$, the set $(\lambda I - A^m)^{-1} \left(\mathcal{F}C_0^{\infty, A^*}(H) \right)$ is contained in $D(\hat{A}_0)$ and it is always a core for A^m when $m \geq 0$.
Finally, the set $\mathcal{F}C_0^{\infty, A^*}(H)$ is in general not contained in $D(\hat{A}_0)$.
All the statements above are also true if $C_m(H)$ is replaced by $UC_m(H)$.

Proof
Proof of part (i). For any $m \geq 1$, we have $A_0 \subset A^m$ by Lemma B.90-(ii). To prove that $D(A_0)$ is a K-core for A^m in $C_m(H)$ we take any $\phi \in D(A^m)$ and set, for some $\lambda > m(\omega \vee 0)$, $g := (\lambda I - A^m)\phi \in C_m(H)$. Then we consider the approximating sequence (g_n) in $\mathcal{F}C_0^{\infty, A^*}(H) \subset D(A_0)$[12] given by Lemma B.78. By construction we have $g_n \xrightarrow{K} g$. Define $\phi_n := (\lambda I - A^m)^{-1} g_n$. By Lemma B.90-(iv) we have

---

[12]This fact is straightforward: see Step 1 of the proof of Theorem 4.135.

$\phi_n \in D(\mathcal{A}_0)$ and, by the dominated convergence theorem, we get $\phi_n \xrightarrow{\mathcal{K}} \phi$. Moreover,

$$\mathcal{A}_0 \phi_n = \mathcal{A}^m \phi_n = \lambda \phi_n - (\lambda I - \mathcal{A}^m)\phi_n = \lambda \phi_n - g_n,$$

hence also $\mathcal{A}_0 \phi_n \xrightarrow{\mathcal{K}} \mathcal{A}^m \phi$.

As said above $\mathcal{F}C_0^{\infty, A^*}(H) \subset D(\mathcal{A}_0)$ and, since $(\lambda I - \mathcal{A}^m)(D(\mathcal{A}_0)) \subset D(\mathcal{A}_0)$ (Lemma B.90- (iv)),

$$(\lambda I - \mathcal{A}^m)^{-1} \left(\mathcal{F}C_0^{\infty, A^*}(H) \right) \subset D(\mathcal{A}_0).$$

The choice of the sequence above implies that such a set is a core for $\mathcal{A}^m$.

The fact that the set $\mathcal{F}C_0^{\infty, A^*}(H)$ is a core for $\mathcal{A}^m$ when $m \geq 1$ follows using Lemma 4.4 in [300]. This lemma generalizes a well-known result about cores (see e.g. [217], Proposition 1.7, p. 53) implying that a $\mathcal{K}$-dense subspace $\mathcal{D}$ of $C_m(H)$ which is invariant for R_t is always a core. Indeed, let $\phi \in \mathcal{F}C_0^{\infty, A^*}(H)$ and let $f : \mathbb{R}^n \to \mathbb{R}$ and $x_1, \ldots x_n \in D(A^*)$ be such that

$$\phi(x) = f(\langle x, x_1 \rangle, \ldots, \langle x, x_n \rangle).$$

Then

$$R_t[\phi](x) = \int_H \phi(y + e^{tA} x) \mathcal{N}(0, Q_t)(dy)$$

$$= \int_H f\left(\langle y + e^{tA} x, x_1 \rangle, \ldots, \langle y + e^{tA} x, x_n \rangle \right) \mathcal{N}(0, Q_t)(dy)$$

$$= \int_H f\left(\langle y, x_1 \rangle + \langle x, e^{tA^*} x_1 \rangle, \ldots, \langle y, x_n \rangle + \langle x, e^{tA^*} x_n \rangle \right) \mathcal{N}(0, Q_t)(dy)$$

$$= g\left(\langle x, e^{tA^*} x_1 \rangle, \ldots, \langle x, e^{tA^*} x_n \rangle \right)$$

for a suitable function $g \in \mathcal{F}C_0^{\infty, A^*}(H)$. Since $x_1, \ldots x_n \in D(A^*)$ we have $e^{tA^*} x_1, \ldots e^{tA^*} x_n \in D(A^*)$ and the claim is proved.

However, when $m \in [0, 1)$ it is not true in general that, for $\phi \in \mathcal{F}C_0^{\infty, A^*}(H)$, one has $\mathcal{A}_0 \phi \in C_m(H)$, so the core property in general fails, contrary to what is stated in Theorem 4.5 of [300] (see on this [500], Remark 5.11).

Proof of part (ii).

The proof of this part is exactly the same except for the fact that, since in general $\mathcal{F}C_0^{\infty, A^*}(H) \not\subset D(\hat{\mathcal{A}}_0)$, we need to prove directly that

$$(\lambda I - \mathcal{A}^m)^{-1} \left(\mathcal{F}C_0^{\infty, A^*}(H) \right) \subset D(\hat{\mathcal{A}}_0).$$

In fact, we can prove, as in Proposition 4.6 of [496], that

$$(\lambda I - \mathcal{A}^m)^{-1} (D(\mathcal{A}_0)) \subset D(\hat{\mathcal{A}}_0). \tag{B.65}$$

To do this it is enough to show that, given $\phi \in (\lambda I - \mathcal{A}^m)^{-1}(D(\mathcal{A}_0))$, the map $x \to \langle A^* D\phi(x), x \rangle$ is bounded, which is equivalent to proving that $\mathcal{A}_0\phi$ is bounded, since the second-order term is always bounded when $\phi \in D(\mathcal{A}_0)$. Let $f \in D(\mathcal{A}_0)$ and let $\phi := (\lambda I - \mathcal{A}^m)^{-1} f$. Then $f = (\lambda I - \mathcal{A}^m)\phi$. Since $\phi \in D(\mathcal{A}_0)$ we also have $f = (\lambda I - \mathcal{A}_0)\phi$, which gives $\mathcal{A}_0\phi = \lambda\phi - f$. Since the right-hand side is bounded, then $\mathcal{A}_0\phi$ is also bounded.

The proofs of both parts when $C_m(H)$ is replaced by $UC_m(H)$ is exactly the same. Notice that the last statement of part (i) holds only for $C_m(H)$ since its proof is based on Lemma 4.4 of [300] which, up to now, is proved only for $C_m(H)$.   □

We are going to use, in addition to Hypothesis B.79, the following assumption (compare it with Hypotheses 4.29 and 4.32).

Hypothesis B.93 The operators A and Q satisfy the following:

(i) For all $t > 0$ we have $e^{tA}(H) \subset Q_t^{1/2}(H)$.
(ii) Defining $\Gamma(t) := Q_t^{-1/2} e^{tA}$ we have that the map $t \to \|\Gamma(t)\|$ (which is always decreasing) belongs to $\mathcal{I}_1$.

Remark B.94 If Hypothesis B.93 holds, then the semigroup R_t enjoys the smoothing property stated in Theorem 4.41 with the estimate, for all $f \in C_m(H)$,

$$\|DR_t f\|_{C_m(H)} \leq C(m) e^{m(\omega \vee 0)t} \|\Gamma(t)\| \, \|f\|_{C_m(H)}, \tag{B.66}$$

for some constant $C(m) \geq 1$. In particular, Hypothesis B.93 implies that, for $m \geq 0$, $D(\mathcal{A}^m) \subset C_m^1(H)$. Indeed, let $\phi \in D(\mathcal{A}^m)$. Then, taking $\lambda > m(\omega \vee 0)$ we must have, for some $f \in C_m(H)$,

$$\phi(x) = (\lambda I - \mathcal{A}^m)^{-1} f(x) = \int_0^{+\infty} e^{-\lambda t} R_t[f](x) dt, \qquad x \in H.$$

Estimate (B.66) and the closedness of the derivative operator (see Corollary 4.14) then imply

$$D\phi(x) = \int_0^{+\infty} e^{-\lambda t} DR_t[f](x) dt, \qquad x \in H.$$

This fact also implies that the approximating sequence (ϕ_n) of elements in $D(\mathcal{A}_0)$ in the proof of the previous proposition can be chosen such that

$$\phi_n \xrightarrow{\mathcal{K}} \phi, \quad \mathcal{A}_0\phi_n \xrightarrow{\mathcal{K}} \mathcal{A}\phi, \quad D\phi_n \xrightarrow{\mathcal{K}} D\phi.$$

∎

Here is the final result.

Theorem B.95 *Let Hypotheses B.79 and B.93 hold. Let $m \geq 0$ and $\eta(t) = \|\Gamma(t)\|$ for $t > 0$.*

(i) Let $\varphi \in C_m(H)$ and $f \in C_{m,\eta}((0, T] \times H)$. Then the mild solution u (given by (B.35)) of the Cauchy problem (B.33) is also a $\mathcal{K}$-strong solution (in the sense of Definition B.85) of (B.33) both in $D(\mathcal{A}_0)$ and in $D(\hat{\mathcal{A}}_0)$.

(ii) If in addition $\varphi \in C_m^1(H)$, $f \in C_m([0, T] \times H)$, f is differentiable in the x variable and $Df \in C_{m,\eta}((0, T] \times H, H)$, then the approximating sequences u_n, φ_n, f_n defining the $\mathcal{K}$-strong solution u in part (i) (both in $D(\mathcal{A}_0)$ and in $D(\hat{\mathcal{A}}_0)$) can be chosen such that, for some $\eta_1 \in \mathcal{I}_1$,

$$\begin{cases} \mathcal{K}\text{-}\lim_{n \to +\infty} D\varphi_n = D\varphi, & \text{in } C_m(H, H), \\[2mm] \mathcal{K}\text{-}\lim_{n \to +\infty} Df_n = Df, & \text{in } C_{m,\eta}((0, T] \times H, H), \\[2mm] \mathcal{K}\text{-}\lim_{n \to +\infty} Du_n = Du, & \text{in } C_m((0, T] \times H, H), \\[2mm] \mathcal{K}\text{-}\lim_{n \to +\infty} D^2 u_n h = D^2 u h & \text{in } C_{m,\eta_1}((0, T] \times H, H), \quad \forall h \in H. \end{cases}$$

If $\eta \in \mathcal{I}_2$, as defined in (B.43), then we can choose $\eta_1 = \eta$ in (ii) and in the definition of a strong solution in (i).

Proof A similar result is proved in Proposition 4.10 of [308] when $m = 0$. Here we give a complete proof.

Proof of (i). We first approximate φ by a sequence (φ_n) in $\mathcal{F}C_0^{\infty, A^*}(H)$ given by Lemma B.78.

Since f may not belong to $C_m((0, T] \times H)$ due to the singularity at $t = 0$, we have to modify a little the approximation argument of Lemma B.78 (see also the proof of Theorem 4.135). For each sufficiently big $k \in \mathbb{N}$ we define, for $t \in (0, T] \times H$, $\hat{f}_k(t, x) := \chi_k(t) f(t, x)$, where $\chi_k : (0, T] \to [0, 1]$ is a smooth function such that $\chi_k(t) = 1$ for $t \in [2/k, T]$ and $\chi_k(t) = 0$ for $t \in (0, 1/k]$. Then, for every sufficiently big k, we take a sequence $(f_{k,n})_n$ from Lemma B.78 which $\mathcal{K}$-converges to $\hat{f}_k$. Now the diagonal sequence $(f_n)_n := (f_{n,n})_n$ has elements in $\mathcal{F}C_0^{\infty, A^*}([0, T] \times H)$ and can be proved to $\mathcal{K}$-converge to f in $\bar{B}_{m,\gamma_G}([0, T) \times H)$, exactly as in the last part of Lemma B.78. We then set

$$u_n(t, x) := R_t[\varphi_n](x) + \int_0^t R_{t-s}[f_n(s, \cdot)](x) ds. \tag{B.67}$$

By construction we know that φ_n and f_n satisfy the assumptions of part (i) of Proposition B.91 and so u_n is a classical solution of (B.33) in $D(\mathcal{A}_0)$. Moreover, still by construction, we have

$$\begin{cases} \mathcal{K}\text{-}\lim_{n \to +\infty} \varphi_n = \varphi, & \text{in } C_m(H), \\[2mm] \mathcal{K}\text{-}\lim_{n \to +\infty} f_n = f, & \text{in } C_{m,\eta}((0, T] \times H). \end{cases}$$

By Proposition 4.50 (i) and (ii) we get $u \in C_m([0, T] \times H)$ while using dominated convergence we obtain $\mathcal{K}\text{-}\lim_{n\to+\infty} u_n = u$ in $C_m([0, T] \times H)$. Up to now we have not used Hypothesis B.93. We use it now to show that, as required, $u(t, \cdot)$ is Fréchet differentiable for any $t \in (0, T]$ and that Du_n $\mathcal{K}$-converges to Du. Indeed, using Proposition 4.50, Remark 4.52 and the closedness of the derivative operator (see Corollary 4.14) we get the required differentiability and that

$$Du(t, x) = DR_t[\varphi](x) + \int_0^t DR_{t-s}[f(s, \cdot)](x)ds.$$

Note that $Du \in C_{m,\eta_1}([0, T] \times H, H)$, where η_1 is an element of $\mathcal{I}_1$ which is greater than both η (coming from the first term of the right-hand side) and $\int_0^t \eta(s)\eta(t-s)ds$ (coming from the integral term). From Proposition 4.21-(iii) it follows that we can choose

$$\eta_1(t) = 2\eta(t/2)\left[\left(\int_0^T \eta(s)ds\right) \vee 1\right], \qquad t \in (0, T],$$

and set $\eta_1(t) = \eta_1(T)$ for $t \geq T$. Finally, note that if $\eta \in \mathcal{I}_2$, as defined in (B.43), then we have $\lim_{t\searrow 0} \eta(t)^{-1} \int_0^t \eta(s)\eta(t-s)ds = 0$ and hence we can choose $\eta_1 = \eta$. Lastly, the required convergence of Du_n to Du in $C_{m,\eta_1}([0, T] \times H, H)$ follows using the representation formula (4.75) for the derivatives of $R_t[\varphi]$ and $R_{t-s}[f(s, \cdot)]$, and then applying the dominated convergence theorem.

The above approximating sequences u_n, φ_n, f_n are not suitable, in general, for u to be a $\mathcal{K}$-strong solution in $D(\hat{\mathcal{A}}_0)$ since the sequences φ_n and $f_n(t, \cdot)$ may not belong to $D(\hat{\mathcal{A}}_0)$, as we discussed in the proof of Proposition B.92. To get an approximating sequence $\bar{u}_n$ for a $\mathcal{K}$-strong solution in $D(\hat{\mathcal{A}}_0)$ it is enough to define, as in [496], Sect. 4,

$$\bar{u}_n(t, x) := J_n u_n(t, x) = R_t[J_n\varphi_n](x) + \int_0^t R_{t-s}[J_n f_n(s, \cdot)](x)ds,$$

where, as in (B.9), we set $J_n := n(nI - \mathcal{A}^m)^{-1}$. By Lemma B.90-(iv) we easily see that $\bar{u}_n(t, \cdot) \in D(\hat{\mathcal{A}}_0)$ for all $t \in [0, T]$. All the other required properties can be proved exactly as we did for u_n.

Proof of (ii). Assume now that, in addition, $\varphi \in C_m^1(H)$, $f \in C_m([0, T] \times H)$, f is differentiable in the x variable and $Df \in C_{m,\eta}((0, T] \times H, H)$. Take sequences (φ_n) in $\mathcal{F}C_0^{\infty, A^*}(H)$ and (f_n) in $\mathcal{F}C_0^{\infty, A^*}((0, T] \times H)$ given by Lemma B.78 and take u_n given by (B.67). It is not difficult to see, looking at the proof of Lemma B.78, that for such sequences we also have, under the assumptions on φ and f,

$$\begin{cases} \mathcal{K}\text{-}\lim_{n\to+\infty} D\varphi_n = D\varphi, & \text{in } C_m(H, H), \\ \\ \mathcal{K}\text{-}\lim_{n\to+\infty} Df_n = Df, & \text{in } C_{m,\eta}((0, T] \times H, H). \end{cases}$$

To prove the regularity of u and the remaining convergences we observe first that, using (B.56) and (B.60), we have, for $t \in (0, T]$, $x, h \in H$,

$$\langle DR_t[\varphi](x), h \rangle = R_t[\langle D\varphi(\cdot), e^{tA}h \rangle], \tag{B.68}$$

$$\left\langle D \int_0^t R_{t-s}[f(s, \cdot)](x)ds, h \right\rangle = \int_0^t R_{t-s}\left[\langle Df(s, \cdot), e^{(t-s)A}h \rangle \right](x)ds, \tag{B.69}$$

so the required regularity of u follows again using the smoothing property and the estimates of Theorem 4.41. On the other hand, the formulae (B.68) and (B.69) also hold for φ_n and f_n, hence the required convergences simply hold by applying the dominated convergence theorem as in the first part of the proof. Also, the choice of the weight η_1 is the same as in part (i) and again, if $\eta \in \mathcal{I}_2$ (see (B.43)), we can choose $\eta_1 = \eta$.

The proof of (ii) for the strong solution in $D(\hat{\mathcal{A}}_0)$ follows the same arguments, replacing φ_n by $J_n\varphi_n$ and $f_n(s, \cdot)$ by $J_nf_n(s, \cdot)$, $s \in (0, T]$. $\qquad \square$

Remark B.96 In Definition B.85 (of a strong solution), as noted in Remark B.86, we also required convergence of the derivatives. If we did not ask such convergence, then part (i) of Theorem B.95 would be true without using Hypothesis B.93, as noted in the body of the proof. In such a case the first part of Theorem B.95 may be seen as a particular case of a more general result for a class of abstract $\mathcal{K}$-continuous semigroups (see [108] Theorem 4.10 for the case $m = 0$). $\qquad \blacksquare$

Remark B.97 We point out some general observations that can be extracted from the results of this section.

First of all, when one needs to approximate continuous functions in infinite dimension by C^2 functions through $\mathcal{K}$-convergence, a straightforward approach (not the only one, see e.g. [496] Sect. 4 for an alternative) is to use cylindrical functions (as in Lemma B.78). The use of cylindrical functions allows us to find approximating sequences required in the definition of strong solutions, as in Theorem B.95, however it is not adequate to find approximations belonging to $D(\hat{\mathcal{A}}_0)$ since the map

$$x \rightarrow \langle A^*D\phi(x), x \rangle$$

may be unbounded for $\phi \in \mathcal{F}C_0^{\infty, A^*}(H)$. This also implies, as observed in [500] Remark 5.11, that $\mathcal{F}C_0^{\infty, A^*}(H)$ is in general not a $\mathcal{K}$-core for $\mathcal{A}^m$, $m \in [0, 1)$.

Thus we decided to use separately the two operators $\mathcal{A}_0$ and $\hat{\mathcal{A}}_0$. They have their advantages and drawbacks. Indeed, using $\mathcal{A}_0$ allows us to use the cylindrical functions directly to find approximations to the mild solutions of (B.33) but, since the $\mathcal{K}$-closure of $\mathcal{A}_0$ is $\mathcal{A}^m$ only for $m \geq 1$, it is not suitable to look at the generator $\mathcal{A}^m$ for $m \in [0, 1)$ (Proposition B.92). On the other hand, using $\hat{\mathcal{A}}_0$ calls for more complicated approximations but allows us to study the generator $\mathcal{A}^m$ for all $m \geq 0$. For these reasons, $\mathcal{A}_0$ is used when we want to find strong solutions to HJB equations in Sects. 4.5 and 4.7, while $\hat{\mathcal{A}}_0$ is used when we use the generator $\mathcal{A}^m$ in the spaces $C_b(H)$ to study mild solutions to the infinite horizon problem in Sect. 4.6.2. $\qquad \blacksquare$

Appendix C
Parabolic Equations with Non-homogeneous Boundary Conditions

In this section we show how to rewrite some classes of parabolic equations with control and noise on the boundary using the infinite-dimensional formalism. We focus on two particular cases: Dirichlet and Neumann boundary conditions. Throughout the section $\mathcal{O}$ will denote a bounded domain (open and connected) in $\mathbb{R}^d$ with regular (C^∞) boundary $\partial\mathcal{O}$.

We begin introducing and recalling some properties of the Dirichlet and Neumann maps.

C.1 Dirichlet and Neumann Maps

Consider the Laplace equation with Dirichlet boundary condition

$$\begin{cases} \Delta_\xi y(\xi) = 0, \ \xi \in \mathcal{O}, \\ y(\xi) = \gamma(\xi), \ \xi \in \partial\mathcal{O}. \end{cases} \tag{C.1}$$

Theorem C.1 *Given $s \geq 0$ and a boundary condition $\gamma \in H^s(\partial\mathcal{O})$, there exists a unique solution $D\gamma \in H^{s+1/2}(\mathcal{O})$ of the problem (C.1). Moreover, for all $s \geq 0$, the operator*

$$\begin{cases} D : H^s(\partial\mathcal{O}) \to H^{s+1/2}(\mathcal{O}), \\ \gamma \to D\gamma, \end{cases} \tag{C.2}$$

is continuous.

Proof See [404] Theorem 5.4, p. 165, Theorem 6.6, p. 177 and Theorem 7.3, p. 187. $\qquad\square$

Definition C.2 The operator D introduced in (C.2) is called the Dirichlet map.

© Springer International Publishing AG 2017
G. Fabbri et al., *Stochastic Optimal Control in Infinite Dimension,*
Probability Theory and Stochastic Modelling 82,
DOI 10.1007/978-3-319-53067-3

Proposition C.3 *Consider the heat equation with zero Dirichlet boundary condition*

$$\begin{cases} \frac{\partial}{\partial s} y(s, \xi) = \Delta_\xi y(s, \xi), & (s, \xi) \in (0, T) \times \mathcal{O}, \\ y(s, \xi) = 0, & (s, \xi) \in (0, T) \times \partial\mathcal{O}, \\ y(0, \xi) = x(\xi), & \xi \in \mathcal{O}, \end{cases} \tag{C.3}$$

where the initial datum x belongs to $L^2(\mathcal{O})$. For $s \geq 0$ denote by $S_D(s)x$ the solution of (C.3) at time s. Then $S_D(s)$ is a C_0-semigroup on $L^2(\mathcal{O})$. Its generator A_D is given by:

$$\begin{cases} D(A_D) := H_0^1(\mathcal{O}) \cap H^2(\mathcal{O}) \\ A_D x := \Delta x, \quad x \in D(A_D). \end{cases} \tag{C.4}$$

A_D *is maximal dissipative, self-adjoint, invertible, $A_D^{-1} \in \mathcal{L}(L^2(\mathcal{O}))$, and $S_D(s)$ is analytic.*

Proof This is a standard result. We refer for instance to Theorem 12.40 of [507] for the proof that the C_0-semigroup $S_D(s)$ is analytic and that $A_D^{-1} \in \mathcal{L}(L^2(\mathcal{O}))$ (it is included in the definition of an analytic semigroup given there). $\qquad \square$

Proposition C.4 *The Dirichlet map D introduced in (C.2) is continuous as a linear map between the spaces $L^2(\partial\mathcal{O})$ and $D((-A_D)^{1/4-\varepsilon})$ for all $\varepsilon > 0$:*

$$D : L^2(\partial\mathcal{O}) \to D((-A_D)^{1/4-\varepsilon}). \tag{C.5}$$

Proof See [399], Sect. 6.1. $\qquad \square$

Similarly we consider the following problem with Neumann boundary condition:

$$\begin{cases} \Delta_\xi y(\xi) = \lambda y(\xi), & \xi \in \mathcal{O} \\ \frac{\partial}{\partial n} y(\xi) = \gamma(\xi), & \xi \in \partial\mathcal{O}, \end{cases} \tag{C.6}$$

where n is the outward unit normal vector and $\frac{\partial}{\partial n}$ is the normal derivative.

Theorem C.5 *Let $\lambda > 0$. Given any $s \geq 0$ and a boundary condition $\gamma \in H^s(\partial\mathcal{O})$, there exists a unique solution $N_\lambda \gamma \in H^{s+3/2}(\mathcal{O})$ of the problem (C.6). Moreover, for all $s \geq 0$, the operator*

$$\begin{cases} N_\lambda : H^s(\partial\mathcal{O}) \to H^{s+3/2}(\mathcal{O}), \\ \gamma \to N_\lambda \gamma, \end{cases} \tag{C.7}$$

is continuous.

Proof See [404] Theorem 5.4, p. 165, Theorem 6.6, p. 177 and Theorem 7.3, p. 187. $\qquad \square$

Definition C.6 The operator N_λ introduced in (C.7) is called the Neumann map.

Proposition C.7 *Consider the following heat equation with zero Neumann boundary condition*

$$\begin{cases} \frac{\partial}{\partial s} y(s,\xi) = \Delta_\xi y(s,\xi), & (s,\xi) \in (0,T) \times \mathcal{O}, \\ \frac{\partial}{\partial n} y(s,\xi) = 0, & (s,\xi) \in (0,T) \times \partial\mathcal{O}, \\ y(0,\xi) = x(\xi), & \xi \in \mathcal{O}, \end{cases} \quad (C.8)$$

where the initial datum x belongs to $L^2(\mathcal{O})$. For $s \geq 0$ denote by $S_N(s)x$ the solution of (C.8) at time s. Then $S_N(s)$ is a C_0-semigroup on $L^2(\mathcal{O})$. Its generator A_N is given by

$$\begin{cases} D(A_N) := \{x \in H^2(\mathcal{O}) : \frac{\partial x}{\partial n} = 0 \text{ on } \partial\mathcal{O}\} \\ A_N x := \Delta x, \quad x \in D(A_N). \end{cases} \quad (C.9)$$

If $\lambda \geq 0$ then $(A_N - \lambda I)$ is maximal dissipative, self-adjoint, and $e^{s(A_N - \lambda I)}$ is analytic. If $\lambda > 0$ then $(A_N - \lambda I)$ is invertible and $(A_N - \lambda I)^{-1} \in \mathcal{L}(L^2(\mathcal{O}))$.

Proof This is a standard result. For the proof of analyticity, see for instance [2]. □

Proposition C.8 *For any $\varepsilon > 0$, the Neumann map N_λ introduced in (C.7) is continuous as a linear map between the spaces $L^2(\partial\mathcal{O})$ and $D((-A_N + \lambda I)^{3/4-\varepsilon})$:*

$$N_\lambda : L^2(\partial\mathcal{O}) \to D((-A_N + \lambda I)^{3/4-\varepsilon}). \quad (C.10)$$

Proof See [399], Sect. 6.1. □

We remark that, see [399], Sect. 6.1, in fact we have

$$D((-A_D)^\alpha) = H^{2\alpha}(\mathcal{O}), \quad 0 < \alpha < \frac{1}{4},$$
$$D((-A_D)^\alpha) = H_0^{2\alpha}(\mathcal{O}), \quad 0 < \alpha < \frac{3}{4}, \alpha \neq \frac{1}{4}, \quad (C.11)$$
$$D((-A_N + \lambda I)^\alpha) = H^{2\alpha}(\mathcal{O}), \quad 0 < \alpha < \frac{3}{4}, \lambda > 0.$$

Moreover, for $\alpha \geq 0$,

$$|x|_{H^{2\alpha}(\mathcal{O})} \leq C_\alpha |(-A_D)^\alpha x|, \quad x \in D((-A_D)^\alpha), \quad (C.12)$$

$$|x|_{H^{2\alpha}(\mathcal{O})} \leq C_{\alpha,\lambda} |(-A_N + \lambda I)^\alpha x|, \quad x \in D((-A_N + \lambda I)^\alpha), \quad \lambda > 0. \quad (C.13)$$

Remark C.9 The results above also hold in less regular domains, for instance when $\mathcal{O}$ is a rectangular parallelepiped, see e.g. [399] Sect. 6. Examples of problems in such domains are discussed, for instance, in Sects. 2.6.2, 4.3.1.5 and 4.8.3. ∎

We also recall the Sobolev embeddings (see e.g. [1], Theorem 7.57).

Theorem C.10 *Let $\mathcal{O}$ be a domain in $\mathbb{R}^d$ with C^1 boundary, $s > 0, 1 < p < +\infty$. The following embeddings are continuous:*

- *If $d > sp$ then $W^{s,p}(\mathcal{O}) \hookrightarrow L^q(\mathcal{O})$ for $p \leq q \leq dp/(d - sp)$.*
- *If $d = sp$ then $W^{s,p}(\mathcal{O}) \hookrightarrow L^q(\mathcal{O})$ for $p \leq q < +\infty$.*
- *If $d < (s - j)p$ for some nonnegative integer j then $W^{s,p}(\mathcal{O}) \hookrightarrow C_b^j(\mathcal{O})$.*

For more on Sobolev embeddings we refer to [1], Chaps. V and VII.

C.2   Non-zero Boundary Conditions, the Dirichlet Case

In this and in the following subsections, we show how to rewrite some classes of parabolic equations with control and noise on the boundary using the infinite-dimensional formalism.

We will always work on the time interval $[t, T]$ where t and T are such that $0 \leq t < T$. We consider the initial datum at time t instead of time 0 to be consistent with the notation we use in Chap. 2.

We consider the following problem:

$$\begin{cases} \frac{\partial}{\partial s} y(s, \xi) = \Delta_\xi y(s, \xi) + f(s, \xi), & (s, \xi) \in (t, T) \times \mathcal{O}, \\ y(s, \xi) = \gamma(s, \xi), & (s, \xi) \in (t, T) \times \partial \mathcal{O}, \\ y(t, \xi) = x(\xi), & \xi \in \mathcal{O}. \end{cases} \qquad \text{(C.14)}$$

Until the end of the section, H and Λ denote, respectively, the Hilbert spaces $L^2(\mathcal{O})$ and $L^2(\partial \mathcal{O})$, and A_D is the operator defined in (C.4). Recall from Theorem C.1 that the Dirichlet map $D : \Lambda \to H$ is continuous.

Lemma C.11 *Assume that A is the generator of a C_0-semigroup on H. Suppose that $\eta \in W^{1,1}(t, T; H)$, $x \in D(A)$ and*

$$z(s) = e^{(s-t)A}x + \int_t^s e^{(s-r)A}\eta(r)dr, \quad \text{for } s \in [t, T].$$

Then $z \in C^1([t, T], H) \cap C([t, T], D(A))$ and

$$Az(s) = Ae^{(s-t)A}x + \int_t^s e^{(s-r)A}\frac{d}{dr}\eta(r)dr + e^{(s-t)A}\eta(t) - \eta(s), \quad \text{for } s \in [t, T].$$

Proof See [177] Lemma 13.2.2, Chap. 13, p. 242. □

Proposition C.12 *Assume that $y \in C^\infty([t, T] \times \overline{\mathcal{O}})$ is a classical solution of (C.14). Then, with $X(s) = y(s, \cdot)$, $s \in [t, T]$, the solution can be written as*

$$X(s) = e^{(s-t)A_D}x \; - A_D \int_t^s e^{(s-r)A_D} D\gamma(r)dr + \int_t^s e^{(s-r)A_D} f(r)dr, \quad s \in [t, T].$$
$$(C.15)$$

Proof We follow a well-known procedure, see for instance [177] Chap. 13 or [403] Chap. 9, Sect. 1.1.

Since y is smooth, by classical theory we have that $D\gamma(s)$ is smooth and moreover $\frac{d}{ds}D\gamma(s) = D\frac{d}{ds}\gamma(s)$ for $t < s < T$. In particular, the function

$$z(s) := X(s) - D\gamma(s) \qquad \text{for } s \in [t, T] \qquad (C.16)$$

is in $C^1((t, T), H) \cap C([t, T], D(A_D))$ and

$$\frac{d}{ds}z(s) = \frac{d}{ds}X(s) - \frac{d}{ds}D\gamma(s) = A_\xi X(s) - A_\xi D\gamma(s) + A_\xi D\gamma(s)$$
$$- \frac{d}{ds}D\gamma(s) + f(s) = A_D z(s) - D\frac{d}{ds}\gamma(s) + f(s),$$

where in the last equality we used that $A_\xi D\gamma(s) = 0$. Observe that the expression $A_D z(s)$ is well defined because $z(s)$ belongs to $D(A_D)$ while neither $X(s)$ nor $D\gamma(s)$ are contained in $D(A_D)$. In particular, $z(\cdot)$ is a strict solution (see [47], Definition 3.1, p. 129) of the following evolution equation in H:

$$\begin{cases} \frac{d}{ds}z(s) = A_D z(s) - D\frac{d}{ds}\gamma(s) + f(s) \\ z(t) = x - D\gamma(t). \end{cases} \qquad (C.17)$$

The strict solution of (C.17) can be written in the mild form (see e.g. [47] Chap. II.1, Lemma 3.2, p. 135)

$$z(s) = e^{(s-t)A_D}[x - D\gamma(t)] + \int_t^s e^{(s-r)A_D}\left[-D\frac{d}{dr}\gamma(r) + f(r)\right]dr, \quad s \in [t, T].$$

Therefore

$$X(s) = z(s) + D\gamma(s)$$
$$= e^{(s-t)A_D}[x - D\gamma(t)] + \int_t^s e^{(s-r)A_D}\left[-D\frac{d}{dr}\gamma(r) + f(r)\right]dr + D\gamma(s), \quad s \in [t, T],$$

which, upon using Lemma C.11, yields, for $s \in [t, T]$,

$$X(s) = e^{(s-t)A_D}x - A_D \int_t^s e^{(s-r)A_D} D\gamma(r)dr + \int_t^s e^{(s-r)A_D} f(r)dr.$$

$\square$

Observe that Eq. (C.15) can be seen as a mild form of equation

$$\begin{cases} \frac{d}{ds}X(s) = A_D X(s) - A_D D\gamma(s) + f(s) \\ X(t) = x \in H. \end{cases} \tag{C.18}$$

Define for $\varepsilon > 0$

$$G_D := (-A_D)^{1/4-\varepsilon}D. \tag{C.19}$$

By (C.5), $G_D \in \mathcal{L}(\Lambda, H)$. Thus (if $\gamma(\cdot) \in L^1(t, T; \Lambda)$)

$$A_D \int_t^s e^{(s-r)A_D} D\gamma(r)dr = \int_t^s (-A_D)^{3/4+\varepsilon} e^{(s-r)A_D} G_D\gamma(r)dr. \tag{C.20}$$

Hence we can rewrite

$$X(s) = e^{(s-t)A_D}x - \int_t^s (-A_D)^{3/4+\varepsilon} e^{(s-r)A_D} G_D\gamma(r)dr + \int_t^s e^{(s-r)A_D} f(r)dr. \tag{C.21}$$

Notation C.13 Expression (C.21) is called the *mild form* of Eq. (C.18) (and of (C.14)). ∎

Notation C.14 We used the letter X for the unknown in the equation to be consistent with the notation we use in Chap. 2 where the variable y is only used for the equation in the PDE form. ∎

Let $Q \in \mathcal{L}^+(H)$[13] and let $\left(\Omega, \mathscr{F}, \{\mathscr{F}_s^t\}_{s\in[t,T]}, \mathbb{P}, W_Q\right)$ be a generalized reference probability space.

We consider now the following stochastic parabolic equation:

$$\begin{cases} dy(s, \xi) = \left[\Delta_\xi y(s, \xi) + f(s, y(s, \xi))\right]ds + g(s, y(s, \xi))dW_Q(s)(\xi), & \text{on } (t, T) \times \mathcal{O} \\ y(s, \xi) = \gamma(s, \xi), & \text{on } (t, T) \times \partial\mathcal{O} \\ y(t, \xi) = x(\xi), & \text{on } \mathcal{O} \end{cases} \tag{C.22}$$

where $f, g : [t, T] \times \mathbb{R} \times \Omega \to \mathbb{R}$ and $\gamma : [t, T] \times \partial\mathcal{O} \times \Omega \to \mathbb{R}$ are appropriately measurable functions.

Define $b(s, y)(\cdot) := f(s, y(s, \cdot))$ and $[\sigma(s, y)z](\cdot) := g(s, y(s, \cdot))z(\cdot)$ and consider, for $s \in [t, T]$, the following integral equation

$$X(s) = e^{(s-t)A_D}x + \int_t^s e^{(s-r)A_D}b(r, X(r))dr + \int_t^s (-A_D)^{3/4+\varepsilon} e^{(s-r)A_D} G_D\gamma(r)dr$$
$$+ \int_t^s e^{(s-r)A_D}\sigma(r, X(r))dW_Q(r) \quad \mathbb{P}\text{-a.e.} \tag{C.23}$$

---

[13]With the notation of Chap. 1, this means that we assume in this case $\Xi = H$.

Similarly to the deterministic case it can be viewed as the mild form of the equation

$$\begin{cases} dX(s) = [A_D X(s) - A_D D\gamma(s) + b(s, X(s))]\,ds + \sigma(s, X(s))dW_Q(s) \\ X(t) = x \in H. \end{cases}$$

$$(C.24)$$

Notation C.15 Equation (C.23) is called the *mild form* of Eq. (C.22) (and of Eq. (C.24)). Its solution is called *mild solution* of (C.22) (and of (C.24)) and is defined in Definition 1.119, see Remark 1.120. Thanks to (C.19) we can rewrite the term $-A_D D$ in (C.24) as $(-A_D)^{3/4+\varepsilon}G_D$ in (C.23).   ∎

Conditions under which some equations of the form (C.22) (and (C.24)) have unique mild solutions are given in Theorem 1.141. If $\sigma \notin L_2(\Xi_0, H)$ then the stochastic term in (C.23) must be given proper interpretation and the same is also true of (C.32).

C.3   Non-zero Boundary Conditions, the Neumann Case

The Neumann case is similar to the Dirichlet case which was explained in Sect. C.2. We consider the following problem:

$$\begin{cases} \frac{\partial}{\partial s} y(s, \xi) = \Delta_\xi y(s, \xi) + f(s, \xi), & (s, \xi) \in (t, T) \times \mathcal{O} \\ \frac{\partial}{\partial n} y(s, \xi) = \gamma(s, \xi), & (s, \xi) \in (t, T) \times \partial\mathcal{O} \\ y(t, \xi) = x(\xi). \end{cases}$$

$$(C.25)$$

As before, we denote respectively by H and Λ the Hilbert spaces $L^2(\mathcal{O})$ and $L^2(\partial\mathcal{O})$. A_N is the generator of the C_0-semigroup associated to heat equations with zero Neumann boundary conditions defined in (C.9), and $\lambda > 0$. Using the same arguments as those in the proof of Proposition C.12 one can prove the following proposition.

Proposition C.16 *If* $y \in C^\infty([t, T] \times \overline{\mathcal{O}})$ *is a classical solution of (C.25) then* $X(s) := y(s, \cdot)$, $s \in [t, T]$, *can be written as*

$$X(s) = e^{(s-t)A_N} x - (A_N - \lambda I) \int_t^s e^{(s-r)A_N} N_\lambda \gamma(r)dr$$

$$+ \int_t^s e^{(s-r)A_N} f(r)dr.$$

$$(C.26)$$

The previous expression can be seen as the mild form of the equation

$$\begin{cases} \frac{d}{ds} X(s) = A_N X(s) + (\lambda I - A_N) N_\lambda \gamma(s) + f(s) \\ X(t) = x \in H. \end{cases} \tag{C.27}$$

If $\gamma \in L^1(t, T; \Lambda)$, thanks to (C.10) we have

$$- (A_N - \lambda I) \int_t^s e^{(s-r)A_N} N_\lambda \gamma(r) dr = \int_t^s (\lambda I - A_N)^{1/4+\varepsilon} e^{(s-r)A_N} G_N \gamma(r) dr, \tag{C.28}$$

where

$$G_N := (\lambda I - A_N)^{3/4-\varepsilon} N_\lambda \in \mathcal{L}(\Lambda, H). \tag{C.29}$$

Therefore we can rewrite (C.26) as

$$X(s) = e^{(s-t)A_N} x + \int_t^s (\lambda I - A_N)^{1/4+\varepsilon} e^{(s-r)A_N} G_N \gamma(r) dr$$
$$+ \int_t^s e^{(s-r)A_N} f(r) dr, \quad , s \in [t, T]. \tag{C.30}$$

Notation C.17 Equation (C.30) is called the *mild form* of (C.25) and (C.27). ∎

To define a mild form of the stochastic parabolic equation

$$\begin{cases} dy(s, \xi) = \left[\Delta_\xi y(s, \xi) + f(s, y(s, \xi)) \right] ds + g(s, y(s, \xi)) dW_Q(s)(\xi), \text{ on } (t, T) \times \mathcal{O} \\ \frac{\partial}{\partial n} y(s, \xi) = \gamma(s, \xi), \text{ on } (t, T) \times \partial\mathcal{O} \\ y(t, \xi) = x(\xi), \text{ on } \mathcal{O} \end{cases} \tag{C.31}$$

we consider the integral equation on $s \in [t, T]$

$$X(s) = e^{(s-t)A_N} x + \int_t^s e^{(s-r)A_N} b(r, X(r)) dr + \int_t^s (\lambda I - A_N)^{1/4+\varepsilon} e^{(s-r)A_N} G_N \gamma(r) dr$$
$$+ \int_t^s e^{(s-r)A_N} \sigma(r, X(r)) dW_Q(r) \quad \mathbb{P}\text{-a.e.,} \tag{C.32}$$

where $b(s, y)(\cdot) := f(s, y(\cdot))$ and $[\sigma(s, y)z](\cdot) := g(s, y(\cdot))z(\cdot)$. Equation (C.32) is in fact the mild form of the problem

$$\begin{cases} dX(s) = [A_N X(s) + (\lambda I - A_N) N_\lambda \gamma(s) + b(s, X(s))] ds + \sigma(s, X(s)) dW_Q(s) \\ X(t) = x. \end{cases}$$
$$\tag{C.33}$$

Notation C.18 Equation (C.32) is called the *mild form* of Eq. (C.31) (and of Eq. (C.33)) and its solution is called *mild solution*. Thanks to (C.29) we can rewrite the term $(\lambda I - A_N) N_\lambda$ in (C.33) as $(\lambda I - A_N)^{1/4+\varepsilon} G_N$ in (C.32). ∎

Conditions under which some equations of the form (C.31) have unique mild solutions are given in Theorem 1.141.

C.4   Boundary Noise, Neumann Case

Let H, Λ, Q, and $\left(\Omega, \mathscr{F}, \{\mathscr{F}_s^t\}_{s\in[t,T]}, \mathbb{P}, W_Q\right)$ be defined as in Sect. C.2. We consider the following problem:

$$
\begin{cases}
\frac{\partial}{\partial s} y(s, \xi) = \Delta_\xi y(s, \xi) + f(s, y(s, \xi)), & (s, \xi) \in (t, T) \times \mathcal{O} \\[2mm]
\frac{\partial}{\partial n} y(s, \xi) = h(s, y(s, \xi)) \frac{dW_Q(s)}{ds} + g(s, \xi), \ (s, \xi) \in (t, T) \times \partial\mathcal{O} & \text{(C.34)} \\[2mm]
y(t, \xi) = x(\xi)
\end{cases}
$$

where $f, h : [t, T] \times \mathbb{R} \times \Omega \to \mathbb{R}$ and $g : [t, T] \times \partial\mathcal{O} \times \Omega \to \mathbb{R}$ are appropriately measurable functions.

As in Sect. C.3, $G_N := (\lambda I - A_N)^{3/4-\varepsilon} N_\lambda \in \mathcal{L}(\Lambda, H)$, $\lambda > 0$, A_N is defined by (C.9) and N_λ by (C.7).

To rewrite the equation in an infinite-dimensional setting in H, we follow the approach used in Sects. C.2 and C.3. The idea is to consider formally the boundary term as a (particularly irregular) boundary condition corresponding to γ appearing in (C.25). So, defining as before $b(s, y)(\cdot) := f(s, y(\cdot))$ and $[\sigma(s, y)z](\cdot) := h(s, y(\cdot))z(\cdot)$, we define the mild form of (C.34), for $s \in [t, T]$, as

$$
\begin{aligned}
X(s) = e^{(s-t)A_N} x + & \int_t^s e^{(s-r)A_N} b(r, X(r)) dr \\
+ & \int_t^s (\lambda I - A_N)^{1/4+\varepsilon} e^{(s-r)A_N} G_N g(r) dr \\
+ & \int_t^s (\lambda I - A_N)^{1/4+\varepsilon} e^{(s-r)A_N} G_N \sigma(r, X(r)) dW_Q(r) \quad \mathbb{P}\text{-a.e. (C.35)}
\end{aligned}
$$

The above expression can also be referred to as the *mild form* of the infinite-dimensional problem

$$
\begin{cases}
dX(s) = [A_N X(s) + (\lambda I - A_N) N_\lambda g(s) + b(s, X(s))]\, ds \\
\qquad\qquad\qquad\qquad + (\lambda I - A_N) N_\lambda \sigma(s, X(s)) dW_Q(s) \quad \text{(C.36)} \\
X(t) = x \in H,
\end{cases}
$$

where, thanks to (C.29), the term $(\lambda I - A_N) N_\lambda g(s)$ is rewritten as $(\lambda I - A_N)^{1/4+\varepsilon} G_N g(s)$ and the stochastic term is rewritten similarly.

Notation C.19 The solution of (C.35) is called the *mild solution* of (C.34) (and (C.36)). ■

Theorem 1.141 provides conditions under which some equations of the form (C.34) (and (C.36)) have unique continuous mild solutions.

C.5 Boundary Noise, Dirichlet Case

To apply the approach of Sect. C.4 to the Dirichlet-boundary-noise version of the problem (C.34) we would need to give a meaning to the term

$$(-A_D)^{3/4+\varepsilon} \int_t^s e^{(s-r)A_D} G_D \sigma(r, X(r)) dW_Q(r),$$

where A_D and $G_D \in \mathcal{L}(\Lambda, H)$ are introduced respectively in (C.4) and (C.19). However, estimate (B.18) does not allow us to prove the convergence of the integral

$$\int_t^T \left\| (-A_D)^{3/4+\varepsilon} e^{(T-r)A_D} G_D \right\|^2_{\mathcal{L}_2(\Lambda, H)} dr.$$

One way to resolve this problem is to look for a solution of the stochastic PDE in the completion of H in the weaker norm $|x|_* := |(-A_D)^{-\alpha} x|$ for some $\alpha > 0$. It is, roughly speaking, a space of distributions. This approach was used, for example, in [175], see in particular Proposition 3.1, p. 176.

Another way, used, for example, by [9, 65, 225], is to look for a solution in a weighted L^2 space. Consider for example a simple case of the following problem on the positive half-line

$$\begin{cases} \frac{\partial}{\partial s} y(s, \xi) = \frac{\partial^2}{\partial \xi^2} y(s, \xi) + f(s, y(s, \xi)), & (s, \xi) \in (t, T) \times \mathbb{R}^+ \\ y(s, 0) = h(s, y(t, 0)) \frac{dW(s)}{ds} + g(s), & s \in (t, T) \\ y(t, \xi) = x(\xi), \end{cases} \quad (\text{C.37})$$

where W is a one-dimensional Brownian motion.

Define the weight $\eta(\xi) := \min\{1, \xi^{1+\theta}\}$ for some $\theta > 0$ and the weighted L^2 space

$$L^2_\eta := \left\{ p : [0, +\infty) \to \mathbb{R} \text{ measurable} : \int_0^{+\infty} p^2(\xi) \eta(\xi) d\xi < \infty \right\}.$$

It is a real separable Hilbert space. The inner product in L^2_η is given by

$$\langle p, q\rangle_{L^2_\eta} := \int_0^{+\infty} p(\xi)q(\xi)\eta(\xi)d\xi$$

and it induces the usual norm

$$|p|_{L^2_\eta} = \left(\int_0^{+\infty} p(\xi)^2\eta(\xi)d\xi\right)^{1/2}.$$

Proposition C.20 *The heat semigroup with zero Dirichlet-boundary conditions extends to a C_0-semigroup on L^2_η. The semigroup is analytic. Denote its genera-tor by A_η. For $\lambda > 0$, $(\lambda I - A_\eta)$ is invertible and the Dirichlet map*

$$D_\lambda a = \phi \iff \begin{cases} (\lambda I - \partial_x^2)\phi[\xi] = 0 & \text{for all } \xi > 0 \\ \phi(0) = a \end{cases}$$

is linear and continuous from $\mathbb{R}$ to $D((\lambda I - A_\eta)^\alpha)$ for all $\alpha \in [0, 1/2 + \theta/4)$.

Proof See Proposition 2.1 and Lemma 2.2 in [225]. □

Using the above proposition we can proceed in a fashion similar to that of Sect. C.4. We fix $\lambda > 0$ and we choose

$$\alpha_\theta := \frac{1}{2} + \frac{\theta}{8}.$$

We define $G_\eta := (\lambda I - A_\eta)^{\alpha_\theta} D_\lambda \in \mathcal{L}(\mathbb{R}, L^2_\eta)$ and rewrite (C.37) as an integral equation on $s \in [t, T]$

$$X(s) = e^{(s-t)A_\eta}x + \int_t^s e^{(s-r)A_\eta}b(r, X(r))dr$$
$$+ \int_t^s (\lambda I - A_\eta)^{(1-\alpha_\theta)}e^{(s-r)A_\eta}G_\eta g(r)dr$$
$$+ \int_t^s (\lambda I - A_\eta)^{(1-\alpha_\theta)}e^{(s-r)A_\eta}G_\eta\sigma(r, X(r))dW_Q(r) \quad \mathbb{P}\text{-a.e.} \quad (C.38)$$

This integral equation is in fact the mild form of the evolution equation in $L^2(t, T; \mathbb{R})$

$$\begin{cases} dX(s) = [A_\eta X(s) + (\lambda I - A_\eta)^{(1-\alpha_\theta)}G_\eta g(s) + b(s, X(s))]ds \\ \qquad\qquad + (\lambda I - A_\eta)^{(1-\alpha_\theta)}G_\eta\sigma(s, X(s))dW_Q(s) \\ X(t) = x \in L^2(t, T; \mathbb{R}), \end{cases}$$
$$(C.39)$$

and we remark that $(\lambda I - A_\eta)^{(1-\alpha_\theta)}G_\eta = (\lambda I - A_\eta)D_\lambda$.

One can use Theorem 1.141 to obtain the existence and uniqueness of a mild solution $X(\cdot)$ of (C.39) which, similarly to Notation C.19, is also called mild solution of (C.37).

Appendix D
Functions, Derivatives and Approximations

In this appendix we collect some standard definitions related to functions which are used in the text and are recalled here for the reader's convenience. In the second part of this appendix we recall the definition and some properties of the sup-inf convolutions.

D.1 Continuity Properties, Modulus of Continuity

Definition D.1 (*Monotonicity*) Let $I \subset \mathbb{R}$. We say that a function $f : I \to \mathbb{R}$ is increasing (respectively, decreasing) on I if for all $x_1, x_2 \in I$

$$x_1 < x_2 \quad \Longrightarrow \quad f(x_1) \leq f(x_2) \quad (\text{respectively, } f(x_1) \geq f(x_2)).$$

Such functions are also called non-decreasing (respectively, non-increasing).

We say that a function $f : I \to \mathbb{R}$ is strictly increasing (respectively, strictly decreasing) if for all $x_1, x_2 \in I$

$$x_1 < x_2 \quad \Longrightarrow \quad f(x_1) < f(x_2) \quad (\text{respectively, } f(x_1) > f(x_2)).$$

Definition D.2 (*Modulus*) We say that a function $\rho : [0, +\infty) \to [0, +\infty)$ is a *modulus (of continuity)* if ρ is continuous, increasing, subadditive, and $\rho(0) = 0$.

In the literature the subadditivity property in the definition of a modulus is not always required and continuity is sometimes required only at 0. The following theorem shows that one can use Definition D.2 without loss of generality.

Theorem D.3 *Consider* $\eta : \mathbb{R}^+ \to \mathbb{R}^+$, *increasing, such that* $\lim_{s \to 0} \eta(s) = \eta(0) = 0$, *and*

$$\lim_{s \to \infty} \frac{\eta(s)}{s} < +\infty.$$

© Springer International Publishing AG 2017

G. Fabbri et al., *Stochastic Optimal Control in Infinite Dimension*,
Probability Theory and Stochastic Modelling 82,
DOI 10.1007/978-3-319-53067-3

Then there exists a modulus ρ (in the sense of Definition D.2) such that $\rho(s) \geq \eta(s)$ for all $s \in \mathbb{R}^+$.

Proof See Theorem 1, p. 406 in [17].   □

The modulus ρ in Theorem D.3 can also be assumed to be concave. Thus we will always assume that a modulus is concave.

Remark D.4 Thanks to the subadditivity, we have the following property for any modulus $\rho(\cdot)$: given any $\varepsilon > 0$, there exists a $C_\varepsilon > 0$ such that

$$\rho(r) \leq \varepsilon + C_\varepsilon r \qquad \text{for every } r \geq 0. \tag{D.1}$$

                                                                                      ∎

Definition D.5 (*Local Modulus*) A function $\rho : [0, +\infty) \times [0, +\infty) \to [0, +\infty)$ is called a *local modulus* if the following three conditions are satisfied:

  (i) ρ is continuous and increasing in both variables.
  (ii) ρ is subadditive in the first variable.
  (iii) $\rho(0, r) = 0$ for every $r \geq 0$.

Definition D.6 Consider two Banach spaces E_0 and E_1. Given $\varphi \in UC(E_0, E_1)$, we define its *modulus of continuity* $\rho[\varphi](\cdot)$ as follows:

$$\rho[\varphi](\varepsilon) := \sup_{x,y \in E_0} \left\{ |\varphi(x) - \varphi(y)|_{E_1} : |x - y|_{E_0} \leq \varepsilon \right\}, \qquad \text{for } \varepsilon \geq 0. \tag{D.2}$$

Of course $|\varphi(x) - \varphi(y)|_{E_1} \leq \rho[\varphi](|x - y|_{E_0})$ for every $x, y \in E_0$. Any modulus with such property is also called a *modulus of continuity of* φ. In particular, for every $\varphi \in UC(E_0, E_1)$ we can always find two positive constants C_0, C_1 such that

$$|\varphi(x)|_{E_1} \leq C_0 + C_1 |x|_{E_0}, \qquad \text{for every } x \in E_0.$$

Definition D.7 (*Local boundedness from above (below)*) Let $G \subset E_0$. A function $u : G \to \mathbb{R}$ is said to be *locally bounded from above (respectively, below)* if, for every $R > 0$, u is bounded from above (below) on $G \cap (B_{E_0}(0, R))$. u is said to be locally bounded if it is both locally bounded from above and from below.

Definition D.8 (*Local uniform continuity*) Let $G \subset E_0$. A function $u : G \to \mathbb{R}$ is said to be *locally uniformly continuous* if, for every $R > 0$, its restriction to $G \cap (B_{E_0}(0, R))$ is uniformly continuous.

Definition D.9 (*Local uniform convergence*) Given $G \subset E_0$, and $u_n, u \in C(G)$, we say that u_n converge locally uniformly to u if, for every $R > 0$, the restrictions of u_n to $G \cap (B_{E_0}(0, R))$ converge uniformly to the restriction of u to $G \cap (B_{E_0}(0, R))$.

Definition D.10 (*Upper/lower semicontinuous envelope*) Let $G \subset E_0$. Consider a function $u : G \to \overline{\mathbb{R}}$. Its *upper* (*respectively, lower*) *semicontinuous envelope* is defined as follows:

$$
\begin{cases}
u^* : G \to \overline{\mathbb{R}} \\
u^*(x) := \limsup_{\substack{y \in G \\ y \to x}} u(y) \quad (\text{respectively, }, u_*(x) := \liminf_{\substack{y \in G \\ y \to x}} u(y)).
\end{cases}
$$

Definition D.11 (*Strict maximum/minimum*) Let $G \subset E_0$. We say that a function $u : G \to \mathbb{R} \cup \{-\infty\}$ (respectively, $u : G \to \mathbb{R} \cup \{+\infty\}$) has a *strict maximum* (*respectively, strict minimum*) at $x \in G$ if u has a maximum (respectively, minimum) at x and whenever $x_n \in G$ are such that $u(x_n) \to u(x)$ then $x_n \to x$. We define a strict local maximum and a strict local minimum similarly.

Definition D.12 (*Convex set*) Let E be a real Banach space. A subset $G \subset E$ is called *convex* if, for every $x, y \in G$ and $\lambda \in [0, 1]$, $\lambda x + (1 - \lambda)y \in G$.

Definition D.13 (*Convex and strictly convex function*) Let G be a convex subset of a real Banach space E. A function $f : G \to \mathbb{R} \cup \{+\infty\}$ is called *convex* if $f(x) < +\infty$ for at least one $x \in E$ and if, for every $x, y \in G$ and $\lambda \in [0, 1]$,

$$
f(\lambda x + (1 - \lambda)y) \le \lambda f(x) + (1 - \lambda)f(y).
$$

A function $f : G \to \mathbb{R} \cup \{+\infty\}$ such that $f(x) < +\infty$ for at least one $x \in E$ is called *strictly convex* if, for all pairs of distinct points x, y in G with $f(x) < +\infty$ and $f(y) < +\infty$,

$$
f(\lambda x + (1 - \lambda)y) < \lambda f(x) + (1 - \lambda)f(y), \quad \text{for any } \lambda \in (0, 1).
$$

A function f is called concave (respectively, strictly concave) if $-f$ is convex (respectively, strictly convex).

We remark that a function satisfying Definition D.13 is usually called a *proper* convex function in the literature.

D.2 Fréchet and Gâteaux Derivatives

Throughout this section Y, Z are Banach spaces, X is an open subset of Z, and H is a real separable Hilbert space with inner product $\langle \cdot, \cdot \rangle$.

Definition D.14 (*Fréchet derivative*) A function $u : X \to Y$ is said to be *Fréchet differentiable* at a point $\bar{x} \in X$ if there exists a linear functional $Du(\bar{x}) \in \mathcal{L}(Z, Y)$ such that

$$
\lim_{|x - \bar{x}|_Z \to 0} \frac{|u(x) - u(\bar{x}) - Du(\bar{x})(x - \bar{x})|_Y}{|x - \bar{x}|_Z} = 0.
$$

Notation D.15 We consider a continuous function $u : X \to Y$ having Fréchet derivative at all $x \in X$. If the function

$$\begin{cases} X \to \mathcal{L}(Z, Y) \\ \bar{x} \to Du(\bar{x}) \end{cases}$$

is continuous ($\mathcal{L}(Z, Y)$ is endowed with the operator norm) then u is said to be *continuously (Fréchet) differentiable* on X.   ∎

The k-th order Fréchet derivative $D^k u$ of u is defined inductively

$$D^1 u = Du, \qquad D^k u = D(D^{k-1} f), \ k = 1, 2, 3, \ldots$$

(see e.g. [266] p. 186). If $D^k u$ is defined at a point $\bar{x} \in X$ then we say that u is k times Fréchet differentiable at $\bar{x}$ and we call $D^k u(\bar{x})$ the k-th Fréchet derivative of u at $\bar{x}$. Clearly $D^k u(\bar{x}) \in \mathcal{L}(Z, \mathcal{L}(Z, \ldots, \mathcal{L}(Z, Y)) \ldots)$. Since this space is isometrically isomorphic to $\mathcal{L}^k(Z, Y)$, we will always consider $D^k u(\bar{x})$ as an element of this last space endowed with its natural norm $\|T\|_{\mathcal{L}^k(Z,Y)} := \sup_{z \in Z^k, z \neq 0} \frac{|T(z_1, \ldots, z_k)|_Y}{|(z_1, \ldots, z_k)|_{Z^k}}$. If u is k times Fréchet differentiable at $\bar{x}$, the k-linear form $D^k u(\bar{x})$ is symmetric, i.e. $D^k u(\bar{x})(z_1, \ldots, z_k) = D^k u(\bar{x})(z_{\sigma(1)}, \ldots, z_{\sigma(k)})$ for every permutation $\sigma(1), \ldots, \sigma(k)$ of $1, \ldots, k$ (see e.g. [266], statement 3.5.7, p. 192).

If X is an open subset of H and $Y = \mathbb{R}$ then thanks to the Riesz Representation Theorem, we can identify the linear functional $Du(\bar{x})$ with the element $y \in H$ such that $\langle y, x \rangle = Du(\bar{x})(x)$ for all $x \in H$. Abusing the notation slightly, we will denote y by $Du(\bar{x})$. The second-order derivative $D^2 u(\bar{x})$ can be identified with a symmetric bilinear form in $\mathcal{L}^2(H, \mathbb{R})$. So we can identify $D^2 u(\bar{x})$ with the unique $T \in \mathcal{S}(H)$ having the property that

$$\langle Tx, y \rangle = D^2 u(\bar{x})(x, y) \qquad \text{for all } x, y \in H.$$

With an abuse of notation we will again denote T by $D^2 u(\bar{x})$.

The following is a special case of the Generalized Taylor's Theorem.

Theorem D.16 *Let the function* $f : U(x) \subset H \to \mathbb{R}$ *be defined on an open neighborhood* $U(x)$*, of* x*, and let* $f \in C^2(U(x))$*. Then, for all* h *in a neighborhood of the origin in* H*, we have*

$$f(x + h) = f(x) + \langle Df(x), h \rangle + \frac{1}{2} \langle D^2 f(x) h, h \rangle + o(|h|^2).$$

Proof See [586], p. 148.   □

We now introduce the notion of Gâteaux derivative.

Definition D.17 (*Gâteaux derivative*) A function $u : X \to Y$ is said to be *Gâteaux differentiable* at a point $\bar{x} \in X$ if there exists a linear functional $\nabla u(\bar{x}) \in \mathcal{L}(Z, Y)$ such that, for any $y \in Z$ with $|y|_Z = 1$,

$$\lim_{t \to 0^+} \frac{|u(\bar{x} + ty) - u(\bar{x}) - t\nabla u(\bar{x})(y)|_Y}{t} = 0. \tag{D.3}$$

Theorem D.18 *Let the function* $f : U(x) \subset H \to \mathbb{R}$ *be defined on an open, convex neighborhood* $U(x)$ *of* x. *Consider* $h \in Z$ *such that* $x + h \in U(x)$. *Suppose that* f *is Gâteaux differentiable at any point* $x \in U(x)$ *and that the mapping*

$$y \to \nabla f(y)(h)$$

is continuous on $U(x)$. *Then*

$$f(x + h) = f(x) + \int_0^1 \nabla f(x + th)(h)dt.$$

Proof See Theorem 4.A, p. 148 of [586].   □

We finally recall a useful lemma on continuity of derivatives or real functions and a version of the mean value theorem for Banach space-valued functions.

Lemma D.19 *If a continuous real function* f *has a continuous right derivative on an interval* $[0, a)$ *then it is of class* C^1 *on* $[0, a)$.

Proof See Lemma 3.2.4, p. 51 of [179].   □

Theorem D.20 (Mean value theorem) *Let* $-\infty < a < b < +\infty$. *Let* $f : [a, b] \to Y$ *be a continuous map into the Banach space* Y. *Suppose that* $f'(t)$ *exists for all* $t \in [a, b]$; *here* $f'(a)$ *and* $f'(b)$ *are one-sided limits. Then the following are true:*

$$f(b) - f(a) \leq (b - a) \sup_{a < t < b} |f'(t)|_Y,$$

$$f(b) - f(a) - (b - a)f'(t_0) \leq (b - a) \sup_{a < t < b} |f'(t) - f'(t_0)|_Y, \quad \forall t_0 \in [a, b].$$

If f' *is also continuous on* $[a, b]$ *then*

$$f(b) - f(a) = \int_a^b f'(t)dt.$$

Proof See Proposition 3.5, p. 76 of [586].   □

Other types of derivatives are also used in this book. The basic material on directional derivatives can be found in Sect. 6.1.2. The concepts of G-directional derivative, G-Gâteaux derivative and G-Fréchet derivative, used in Chaps. 4 and 5, are introduced in Sect. 4.2.1.

D.3 Inf-Sup Convolutions

We recall here some results from [401] on approximations of bounded uniformly continuous functions in Hilbert spaces. Consider a real and separable Hilbert space H with inner product $\langle \cdot, \cdot \rangle$, and $\dim(H) = +\infty$. Since closed balls are not compact in H, functions in $C_b(H)$ cannot be approximated uniformly on bounded sets by functions in $UC_b(H)$ and consequently by functions in $C_b^1(H)$. It was proved in [393] that $UC_b(H) \cap C^\infty(H)$ is dense in $UC_b(H)$. However, it was observed in [457] that, in contrast to what we have in the finite-dimensional case, $UC_b^2(H)$ is not a dense subset of $UC_b(H)$. Nevertheless, the inclusion $UC_b^{1,1}(H) \subset UC_b(H)$ remains dense. Lasry and P.L. Lions introduced in [401] an explicit way to approximate functions in $UC_b(H)$ by elements of $UC_b^{1,1}(H)$, the so-called *inf-sup-convolutions* and *sup-inf-convolutions*. These explicit approximations have many other interesting properties, for instance they preserve order and commute with translations. More information about them can be found in [179, 401]. We also remark that P.L. Lions proved in [410] that functions in $UC_b^{1,1}(H)$ can be uniformly approximated by functions in $UC_b^{1,1}(H)$ with uniformly continuous second-order partial derivatives, for which Itô's formula can be applied (see the proof of Lemma IV.1 and Lemma III.2 in [410]). The technique of [410] is based on limits of mollifications over increasing finite-dimensional subspaces of H. Also in [493, 494] it is proved that the space $UC_s^2(H)$, the subspace of $UC_b^{1,1}(H)$ admitting a weakly uniformly continuous second Hadamard derivative, is dense in $UC_b(H)$.

Definition D.21 (*Semiconvex and semiconcave functions*) A function $u : H \to \mathbb{R}$ is said to be *semiconcave* (respectively, *semiconvex*) if there exists a constant $M \geq 0$ such that

$$x \to u(x) - M|x|^2 \quad (\text{respectively}, x \to u(x) + M|x|^2)$$

is concave (respectively, convex).

Definition D.22 (*Sup-convolution and Inf-convolution*) Given $u \in C_b(H)$ and $\varepsilon > 0$, we define the *inf-convolution* of u as

$$u_\varepsilon(x) := \inf_{y \in H} \left(u(y) + \frac{|x - y|^2}{2\varepsilon} \right), \quad x \in H$$

and its *sup-convolution* as

$$u^\varepsilon(x) := \sup_{y \in H} \left(u(y) - \frac{|x - y|^2}{2\varepsilon} \right), \quad x \in H.$$

We set $u_0(x) = u^0(x) := u(x)$.

Proposition D.23 *For all* $\varepsilon, \delta \geq 0$ *and* $u \in UC_b(H)$, $u_\varepsilon, u^\varepsilon \in UC_b(H)$, *and* $(u_\varepsilon)_\delta = u_{\varepsilon+\delta}$, $(u^\varepsilon)^\delta = u^{\varepsilon+\delta}$. *Moreover, for all* $u, v \in UC_b(H)$ *and* $\varepsilon > 0$, *we have the following properties:*

(1) $\|u_\varepsilon - v_\varepsilon\|_0 \le \|u - v\|_0$ and $\|u^\varepsilon - v^\varepsilon\|_0 \le \|u - v\|_0$.

(2) $\rho[u_\varepsilon] \le \rho[u]$ and $\rho[u^\varepsilon] \le \rho[u]$.

(3) $\lim_{\varepsilon \to 0} u_\varepsilon = u$ and $\lim_{\varepsilon \to 0} u^\varepsilon = u$ in $UC_b(H)$, more precisely

$$\|u_\varepsilon - u\|_0 \le \rho[u](2\sqrt{\varepsilon \|u\|_0}) \quad \text{and} \quad \|u^\varepsilon - u\|_0 \le \rho[u](2\sqrt{\varepsilon \|u\|_0}).$$

(4) $x \to u_\varepsilon(x) - \frac{|x|^2}{2\varepsilon}$ is concave.

(5) $x \to u^\varepsilon(x) + \frac{|x|^2}{2\varepsilon}$ is convex.

(6) u_ε and u^ε are Lipschitz continuous, and, for all $x, y \in H$,

$$\frac{|u_\varepsilon(x) - u_\varepsilon(y)|}{|x - y|}, \; \frac{|u^\varepsilon(x) - u^\varepsilon(y)|}{|x - y|} \le \frac{2\sqrt{\|u\|_0}}{\sqrt{\varepsilon}}.$$

Proof Most of the claims are proved in [179], Propositions C.3.2, C.3.3, and C.3.4. We sketch the proof of (6) for u^ε.

Let $x, y \in H$. We define the function $v : \mathbb{R} \to \mathbb{R}$ by

$$v(t) := u^\varepsilon(x + t(y - x)).$$

Since u^ε is semiconvex, v is semiconvex and hence locally Lipschitz and differentiable a.e. Let $0 < \bar{t} < 1$ be a point of differentiability of v and let $\varphi \in C^1(\mathbb{R})$ be such that $v - \varphi$ has a strict maximum at $\bar{t}$. By Theorem 3.25, for every n there are $a_n \in \mathbb{R}$, $p_n \in H$, $|a_n| + |p_n| < 1/n$ such that

$$u(z) - \frac{|x + t(y - x) - z|^2}{2\varepsilon} - \varphi(t) + \langle p_n, z \rangle + a_n t \tag{D.4}$$

has a global maximum over $[0, 1] \times H$ at some point (t_n, z_n). Since the maximum of $v - \varphi$ at $\bar{t}$ was strict, it is easy to see that we must have $t_n \to \bar{t}$ and $|z_n| \le C, n \in \mathbb{N}$ for some C. Moreover, setting $z = z_n$ and differentiating the function in (D.4) with respect to t we obtain

$$D\varphi(t_n) = \frac{\langle x + t_n(y - x) - z_n, x - y \rangle}{\varepsilon} + a_n. \tag{D.5}$$

Taking $z = x + t_n(y - x)$ we get, since (t_n, z_n) maximizes (D.4),

$$u(x + t_n(y - x)) + \langle p_n, x + t_n(y - x) \rangle \le u(z_n) - \frac{|x + t_n(y - x) - z_n|^2}{2\varepsilon} + \langle p_n, z_n \rangle,$$

which implies, for some $C_1 > 0$ depending on the norm of x and y,

$$\frac{|x + t_n(y - x) - z_n|^2}{2\varepsilon} \le u(z_n) - u(x + t_n(y - x)) + \langle p_n, z_n - x - t_n(y - x) \rangle \le 2\|u\|_0 + \frac{C_1}{n}.$$

Using this and letting $n \to +\infty$ in (D.5) we thus obtain

$$|D\varphi(\bar{t})| \leq \frac{2\sqrt{\|u\|_0}}{\sqrt{\varepsilon}}|y - x|. \tag{D.6}$$

We can thus conclude that

$$|u^\varepsilon(y) - u^\varepsilon(x)| = |v(1) - v(0)| \leq \int_0^1 |v'(t)|dt \leq \frac{2\sqrt{\|u\|_0}}{\sqrt{\varepsilon}}|y - x|.$$

$\square$

We remark that the above proof shows that one can in fact obtain

$$|u^\varepsilon(y) - u^\varepsilon(x)| \leq \frac{t_\varepsilon}{\sqrt{\varepsilon}}|y - x|,$$

where t_ε is defined in Proposition D.26.

We also remark that it follows from semiconvexity of u^ε that if $v'(t)$ exists, we must have

$$v'(t) = \langle p, y - x\rangle \quad \text{for every } p \in D^-u^\varepsilon(x + t(y - x)),$$

where $D^-u^\varepsilon(x+t(y-x))$ is the subdifferential of u^ε at $(x+t(y-x))$ (see Definition E.1). Moreover, $D^-u^\varepsilon(z)$ is non-empty for every z. In fact, for a semiconvex function $w : H \to \mathbb{R}$,

$$D^-w(z) = \overline{\text{conv}}\{p : Dw(z_n) \rightharpoonup p, z_n \to z\}, \tag{D.7}$$

where the z_n above are points of Fréchet differentiability of w (see [67], p. 522). Lipschitz functions on H are Fréchet differentiable on a dense subset of H by Preiss's theorem.

Finally, we remark that if u^ε is differentiable at some point $\bar{x}$ then $Du^\varepsilon(\bar{x}) = (\bar{y} - \bar{x})/\varepsilon$, where $\bar{y}$ is the unique point such that

$$u^\varepsilon(\bar{x}) = u(\bar{y}) - \frac{|\bar{x} - \bar{y}|^2}{2\varepsilon}. \tag{D.8}$$

To show this let $\varphi \in C^1(H)$ be such that $u^\varepsilon - \varphi$ has a global strict maximum at $\bar{x}$ and $\varphi(x) = |x|^2$ if $|x|$ is large enough. By Theorem 3.25, for every n there are $p_n, q_n \in H, |p_n| + |q_n| < 1/n$ such that

$$u(y) - \frac{|x - y|^2}{2\varepsilon} - \varphi(x) + \langle p_n, x\rangle + \langle q_n, y\rangle \tag{D.9}$$

has a global maximum over $H \times H$ at some point (x_n, y_n). Since the maximum of $u^\varepsilon - \varphi$ at $\bar{x}$ was strict, it is easy to see that we must have $x_n \to \bar{x}$ and

$$u(y_n) - \frac{|x_n - y_n|^2}{2\varepsilon} \to u^\varepsilon(\bar{x})$$

as $n \to +\infty$. Moreover, setting $y = y_n$ and differentiating the function in (D.9) with respect to x we obtain

$$D\varphi(x_n) = -\frac{x_n - y_n}{\varepsilon} + p_n,$$

which implies that $\lim_{n \to +\infty} y_n = \bar{y}$ for some point $\bar{y}$,

$$Du^\varepsilon(\bar{x}) = D\varphi(\bar{x}) = \frac{\bar{y} - \bar{x}}{\varepsilon}, \tag{D.10}$$

and (D.8) is satisfied. Since (D.10) holds for every point $\bar{y}$ satisfying (D.8), it must be unique.

Definition D.24 (*Inf-sup and sup-inf-convolutions*) Given $\varepsilon > 0$ and a function $u \in C_b(H)$, we define the *inf-sup convolution* (respectively, *sup-inf-convolution*) of u as

$$\underline{u}_\varepsilon := (u_\varepsilon)^{\frac{\varepsilon}{2}} \qquad (\text{respectively,} \overline{u}_\varepsilon := (u^\varepsilon)_{\frac{\varepsilon}{2}}).$$

More explicitly we have, for $x \in H$,

$$\underline{u}_\varepsilon(x) = \sup_{z \in H} \inf_{y \in H} \left(u(y) + \frac{1}{2\varepsilon}|z - y|^2 - \frac{1}{\varepsilon}|z - x|^2 \right)$$

(respectively,

$$\overline{u}_\varepsilon(x) = \inf_{z \in H} \sup_{y \in H} \left(u(y) - \frac{1}{2\varepsilon}|z - y|^2 + \frac{1}{\varepsilon}|z - x|^2 \right).)$$

Proposition D.25 *The inf-sup-convolution and the sup-inf-convolution preserve the order. In other words, given $u, v \in UC_b(H)$ such that*

$$u(x) \geq v(x), \qquad \text{for all } x \in H,$$

we have $\underline{u}_\varepsilon(x) \geq \underline{v}_\varepsilon(x)$ and $\overline{u}_\varepsilon(x) \geq \overline{v}_\varepsilon(x)$ for all $x \in H$. Moreover, the inf-sup-convolution and the sup-inf-convolution commute with translations, i.e. for every $y \in H$ and translation $\tau_y : x \to x - y$, we have $\overline{(\tau_y u)}_\varepsilon(x) = (\tau_y \overline{u}_\varepsilon)(x)$ and $\overline{(\tau_y u)}_\varepsilon(x) = (\tau_y \overline{u}_\varepsilon)(x)$ for all $x \in H$.

Proof See [401]. □

Proposition D.26 *Let $u \in UC_b(H)$. Let t_ε be the maximum positive root of the equation*

$$t_\varepsilon^2 = 2\varepsilon \rho[u](t_\varepsilon),$$

which implies that $t_\varepsilon \varepsilon^{-1/2} \to 0$ as $\varepsilon \to 0$. Then $\underline{u}_\varepsilon$ and $\overline{u}_\varepsilon$ belong to $UC_b^{1,1}(H)$ and, for all $x, y \in H$, the following properties hold:

(1) $\inf_{y \in H} u(y) \le \underline{u}_\varepsilon(x) \le u(x) \le \overline{u}_\varepsilon(x) \le \sup_{y \in H} u(y)$.

(2) $|\underline{u}_\varepsilon(x) - \underline{u}_\varepsilon(y)| \le \rho[u](|x - y|)$ and $|\overline{u}_\varepsilon(x) - \overline{u}_\varepsilon(y)| \le \rho[u](|x - y|)$.

(3) $\|\underline{u}_\varepsilon - u\|_0, \|\overline{u}_\varepsilon - u\|_0 \le \rho[u](t_\varepsilon)$.

(4) $\|D\underline{u}_\varepsilon\|_0, \|D\overline{u}_\varepsilon\|_0 \le \dfrac{t_\varepsilon}{\varepsilon}$.

(5) $|D\underline{u}_\varepsilon(x) - D\underline{u}_\varepsilon(y)|, |D\overline{u}_\varepsilon(x) - D\overline{u}_\varepsilon(y)| \le \frac{2}{\varepsilon}|x - y|$.

Proof See [401], pp. 260–261.   ☐

Remark D.27 In fact, $x \to \underline{u}_\varepsilon(x) - \frac{|x|^2}{2\varepsilon}$ is concave and $x \to \underline{u}_\varepsilon(x) + \frac{|x|^2}{\varepsilon}$ is convex. Similarly $x \to \overline{u}_\varepsilon(x) - \frac{|x|^2}{\varepsilon}$ is concave and $x \to \overline{u}_\varepsilon(x) + \frac{|x|^2}{2\varepsilon}$ is convex.   ∎

The inf-sup and sup-inf convolutions can also be used to approximate more general functions.

Proposition D.28 *If $u \in C(H)$ and there exists a $C > 0$ such that $|u(x)| \le C(1 + |x|^2)$ then, for all ε small enough, the inf-sup convolution $\underline{u}_\varepsilon$ and the sup-inf convolution $\overline{u}_\varepsilon$ are well defined, they belong to $C^{1,1}(H)$ and they converge pointwise to u when $\varepsilon \to 0$. Moreover, if u is locally uniformly continuous then $\underline{u}_\varepsilon$ and $\overline{u}_\varepsilon$ converge to u locally uniformly.*

Proof See [401], p. 261 (iii).   ☐

We remark that to obtain pointwise convergence in the above proposition one can replace $u \in C(H)$ by $u \in USC(H)$ for $\overline{u}_\varepsilon$, and by $u \in LSC(H)$ for $\underline{u}_\varepsilon$.

D.4 Two Versions of Gronwall's Lemma

We recall two versions of Gronwall's Lemma. The first is a well-known result while the second is more specialized. (See e.g. [219] or [266] pp. 95–97 for similar versions of Gronwall's Lemma).

Proposition D.29 (Gronwall's Lemma 1) *Let $T \in [0, +\infty) \cup \{+\infty\}$ and $I = [t_0, T)$. Let $a(\cdot)$ be a nonnegative, measurable, increasing function on I. Let $b(\cdot)$ be a non-negative, locally integrable function on I. Suppose that $u(\cdot)$ is a non-negative function such that $b(\cdot)u(\cdot)$ is locally integrable on I. Assume also that*

$$u(s) \le a(s) + \int_{t_0}^s b(r)u(r)dr, \quad \text{for a.e. } s \in I. \tag{D.11}$$

Then

$$u(s) \le a(s) \exp\left(\int_{t_0}^s b(r)dr\right), \quad \text{for a.e. } s \in I.$$

Proof Define

$$v(s) = \exp\left(-\int_{t_0}^{s} b(r)dr\right)\int_{t_0}^{s} b(r)u(r)dr, \quad s \in I.$$

Then, for a.e. $s \in I$, $v'(s)$ exists and

$$v'(s) = \left(u(s) - \int_{t_0}^{s} b(r)u(r)dr\right)b(s)\exp\left(-\int_{t_0}^{s} b(r)dr\right).$$

So, using (D.11) and integrating, we have

$$v(s) \le \int_{t_0}^{s} a(r)b(r)\exp\left(-\int_{t_0}^{r} b(\tau)d\tau\right)dr.$$

Now, since

$$\int_{t_0}^{s} b(r)u(r)dr = v(s)\exp\left(\int_{t_0}^{s} b(r)dr\right), \quad s \in I,$$

by (D.11) we get

$$u(s) \le a(s) + \int_{t_0}^{s} b(r)u(r)dr = a(s) + \exp\left(\int_{t_0}^{s} b(r)dr\right)v(s)$$

$$\le a(s) + \exp\left(\int_{t_0}^{s} b(r)dr\right)\int_{t_0}^{s} a(r)b(r)\exp\left(-\int_{t_0}^{r} b(\tau)d\tau\right)dr$$

$$= a(s) + \int_{t_0}^{s} a(r)b(r)\exp\left(\int_{r}^{s} b(\tau)d\tau\right)dr.$$

Since the function $a(\cdot)$ is increasing the above implies

$$u(s) \le a(s) + \int_{t_0}^{s} a(s)b(r)\exp\left(\int_{r}^{s} b(\tau)d\tau\right)dr$$

$$= a(s) + \left(-a(s)\exp\left(\int_{r}^{s} b(\tau)d\tau\right)\right)\Bigg|_{r=t_0}^{r=s} = a(s)\exp\left(\int_{t_0}^{s} b(r)dr\right),$$

which is the claim.   □

Proposition D.30 (Gronwall's Lemma 2) *Let $T \in [0, +\infty) \cup \{+\infty\}$, $b \ge 0$, $\beta > 0$. Let $a(\cdot)$ be a non-negative, locally integrable function on $[0, T)$. Suppose that $u(\cdot)$ is a non-negative, locally integrable function on $[0, T)$ such that*

$$u(s) \le a(s) + b \int_0^s (s - r)^{\beta-1} u(r) dr, \quad \text{for a.e. } s \in [0, T].$$

Then

$$u(s) \le a(s) + \theta \int_0^s E'_\beta(\theta(s - r)) a(r) dr, \quad \text{for a.e. } s \in [0, T],$$

where, for $s > 0$, $\Gamma(s) := \int_0^{+\infty} r^{s-1} e^{-r} dr$, and θ and $E_\beta(s)$ are defined as

$$\theta := (b\Gamma(\beta))^{1/\beta}$$

and

$$E_\beta(s) := \sum_{n=0}^{+\infty} \frac{s^{n\beta}}{\Gamma(n\beta + 1)}.$$

The function $E'_\beta(s) = \frac{d}{ds} E_\beta(s)$ has the following properties: (i) $E'_\beta(s) = \frac{s^{\beta-1}}{\Gamma(\beta)} + o(s^{\beta-1})$ as $s \to 0^+$, (ii) $E'_\beta(s) = \frac{e^s}{\beta} + o(e^s)$ as $s \to +\infty$.

As a particular case, if $a, b, T \in \mathbb{R}^+$ and $\alpha, \beta \in [0, 1)$, there exists an $M \in \mathbb{R}$ (depending on b, T, α and β) such that any integrable function $u : [0, T] \to \mathbb{R}$ such that

$$0 \le u(s) \le as^{-\alpha} + b \int_0^s (s - r)^{-\beta} u(r) dr, \quad \text{for a.e. } s \in [0, T],$$

satisfies

$$0 \le u(s) \le aMs^{-\alpha}, \quad \text{for a.e. } s \in [0, T].$$

Proof See Lemma 7.1.1, Chap. 7, p. 188 of [341]. The second claim is again in [341], Sect. 1.2.1, p. 6. □

Appendix E
Viscosity Solutions in $\mathbb{R}^N$

We collect some basic definitions and results about viscosity solutions in finite-dimensional spaces. We refer the reader to [139] and the books [40, 41, 263] for more information on the subject.

E.1 Second Order Jets

Let $\mathcal{O}$ be an open subset of $\mathbb{R}^N$.

Definition E.1 (*Sub- and superdifferentials*) Let $u : \mathcal{O} \to \mathbb{R}$ be an upper semicontinuous function. The *superdifferential* of u at a point $\bar{x} \in \mathcal{O}$ is defined as

$$D^+ u(\bar{x}) := \left\{ p \in \mathbb{R}^N : \limsup_{y \to \bar{x}, y \in \mathcal{O}} \frac{u(y) - u(\bar{x}) - \langle p, y - \bar{x} \rangle}{|y - \bar{x}|} \leq 0 \right\}.$$

Similarly, given a lower semicontinuous function $u : \mathcal{O} \to \mathbb{R}$, the *subdifferential* of u at a point $\bar{x} \in \mathcal{O}$ is defined as

$$D^- u(\bar{x}) := \left\{ p \in \mathbb{R}^N : \liminf_{y \to \bar{x}, y \in \mathcal{O}} \frac{u(y) - u(\bar{x}) - \langle p, y - \bar{x} \rangle}{|y - \bar{x}|} \geq 0 \right\}.$$

Lemma E.2 *Let* $u : \mathcal{O} \to \mathbb{R}$ *be an upper semicontinuous function, and* $\bar{x} \in \mathcal{O}$. *Then* $p \in D^+ u(\bar{x})$ *if and only if there exists a function* $\phi \in C^1(\mathbb{R}^N)$ *such that* $u - \phi$ *attains a strict global maximum at* $\bar{x}$ *and*

$$(\phi(\bar{x}), D\phi(\bar{x})) = (u(\bar{x}), p).$$

© Springer International Publishing AG 2017
G. Fabbri et al., *Stochastic Optimal Control in Infinite Dimension*,
Probability Theory and Stochastic Modelling 82,
DOI 10.1007/978-3-319-53067-3

Similarly, if $u : \mathcal{O} \to \mathbb{R}$ is a lower semicontinuous function, and $\bar{x} \in \mathcal{O}$, then $p \in D^- u(\bar{x})$ if and only if there exists a function $\phi \in C^1(\mathbb{R}^N)$ such that $u - \phi$ attains a strict global minimum at $\bar{x}$ and

$$(\phi(\bar{x}), D\phi(\bar{x})) = (u(\bar{x}), p).$$

Proof See [575], Lemma 2.7, p. 173 or [220], p. 544.   □

We remark that Definition E.1 is exactly the same and Lemma E.2 is true if $\mathbb{R}^N$ is replaced by a real Hilbert space.

Definition E.3 (*Second order sub- and superjets*) Let $u : \mathcal{O} \to \mathbb{R}$ be an upper semicontinuous function, and $\bar{x} \in \mathbb{R}^N$. The set

$$J^{2,+} u(\bar{x}) := \left\{ (p, X) \in \mathbb{R}^N \times S(\mathbb{R}^N) : \right.$$

$$\left. \limsup_{y \to \bar{x}, y \in \mathcal{O}} \frac{u(y) - u(\bar{x}) - \langle p, y - \bar{x} \rangle - \frac{1}{2} \langle X(y - \bar{x}), (y - \bar{x}) \rangle}{|y - \bar{x}|^2} \leq 0 \right\}$$

is called the *second-order superjet of u at $\bar{x}$*. Similarly, given a lower semicontinuous function $u : \mathcal{O} \to \mathbb{R}$, and $\bar{x} \in \mathcal{O}$, the set

$$J^{2,-} u(\bar{x}) := \left\{ (p, X) \in \mathbb{R}^N \times S(\mathbb{R}^N) : \right.$$

$$\left. \liminf_{y \to \bar{x}, y \in \mathcal{O}} \frac{u(y) - u(\bar{x}) - \langle p, y - \bar{x} \rangle - \frac{1}{2} \langle X(y - \bar{x}), (y - \bar{x}) \rangle}{|y - \bar{x}|^2} \geq 0 \right\}$$

is called the *second-order subjet of u at $\bar{x}$*.

Lemma E.4 *Let $u : \mathcal{O} \to \mathbb{R}$ be an upper semicontinuous function, and $\bar{x} \in \mathcal{O}$. Then (p, X) belongs to $J^{2,+} u(\bar{x})$ if and only if there exists a function $\phi \in C^2(\mathbb{R}^N)$ such that $u - \phi$ attains a strict global maximum at $\bar{x}$ and*

$$\left(\phi(\bar{x}), D\phi(\bar{x}), D^2\phi(\bar{x}) \right) = (u(\bar{x}), p, X).$$

Similarly, if $u : \mathcal{O} \to \mathbb{R}$ is a lower semicontinuous function, and $\bar{x} \in \mathcal{O}$, then $(p, X) \in J^{2,-} u(\bar{x})$ if and only if there exists a function $\phi \in C^2(\mathbb{R}^N)$ such that $u - \phi$ attains a strict global minimum at $\bar{x}$ and

$$\left(\phi(\bar{x}), D\phi(\bar{x}), D^2\phi(\bar{x}) \right) = (u(\bar{x}), p, X).$$

Proof See [575] Lemma 5.4, p. 193 or [263], Lemma 4.1, p. 211.   □

Definition E.5 (*Closure of second-order sub- and superjets*) Let $u : \mathcal{O} \to \mathbb{R}$ be an upper semicontinuous function, and $\bar{x} \in \mathcal{O}$. We define

$$\overline{J}^{2,+} u(\bar{x}) := \left\{ (p, X) \in \mathbb{R}^N \times S(\mathbb{R}^N) \; : \; \text{there exist } x_n \in \mathcal{O} \right.$$

$$\left. \text{and } (p_n, X_n) \in J^{2,+} u(x_n) \text{ s.t. } (x_n, u(x_n), p_n, X_n) \xrightarrow{n \to \infty} (\bar{x}, u(\bar{x}), p, X) \right\}.$$

Similarly, given a lower semicontinuous function $u : \mathcal{O} \to \mathbb{R}$ and $\bar{x} \in \mathcal{O}$, we define

$$\overline{J}^{2,-} u(\bar{x}) := \left\{ (p, X) \in \mathbb{R}^N \times S(\mathbb{R}^N) \; : \; \text{there exist } x_n \in \mathcal{O} \right.$$

$$\left. \text{and } (p_n, X_n) \in J^{2,-} u(x_n) \text{ s.t. } (x_n, u(x_n), p_n, X_n) \xrightarrow{n \to \infty} (\bar{x}, u(\bar{x}), p, X) \right\}.$$

Remark E.6 Note that the definition is a little different from what one would expect as closures of set-valued mappings. Indeed, we also ask $u(x_n) \to u(\bar{x})$. This form of the closures of the semijets was first introduced in [362]. ∎

Definition E.7 (*Parabolic second-order sub- and superjets*) Let $T > 0$. Let $u : (0, T) \times \mathcal{O} \to \mathbb{R}$ be an upper semicontinuous function, and $(\bar{t}, \bar{x}) \in (0, T) \times \mathcal{O}$. The set

$$\mathcal{P}^{2,+} u(\bar{t}, \bar{x}) := \left\{ (a, p, X) \in \mathbb{R} \times \mathbb{R}^N \times S(\mathbb{R}^N) \; : \right.$$

$$\left. \limsup_{(s,y) \to (\bar{t}, \bar{x})} \frac{u(s, y) - u(\bar{t}, \bar{x}) - a(s - \bar{t}) - \langle p, y - \bar{x} \rangle - \frac{1}{2} \langle X(y - \bar{x}), (y - \bar{x}) \rangle}{|s - \bar{t}| + |y - \bar{x}|^2} \le 0 \right\}$$

is called the *parabolic second-order superjet of u at $(\bar{t}, \bar{x})$*. Similarly, given a lower semicontinuous function $u : (0, T) \times \mathcal{O} \to \mathbb{R}$, and $(\bar{t}, \bar{x}) \in (0, T) \times \mathcal{O}$, the set

$$\mathcal{P}^{2,-} u(\bar{t}, \bar{x}) := \left\{ (a, p, X) \in \mathbb{R} \times \mathbb{R}^N \times S(\mathbb{R}^N) \; : \right.$$

$$\left. \liminf_{(s,y) \to (\bar{t}, \bar{x})} \frac{u(s, y) - u(\bar{t}, \bar{x}) - a(s - \bar{t}) - \langle p, y - \bar{x} \rangle - \frac{1}{2} \langle X(y - \bar{x}), (y - \bar{x}) \rangle}{|s - \bar{t}| + |y - \bar{x}|^2} \ge 0 \right\}$$

is called the *parabolic second-order subjet of u at $(\bar{t}, \bar{x})$*.

The closures of the parabolic second-order sub- and superjets $\overline{\mathcal{P}}^{2,-} u(\bar{t}, \bar{x})$, $\overline{\mathcal{P}}^{2,+} u(\bar{t}, \bar{x})$ are defined in the same way as $\overline{J}^{2,-} u(\bar{x})$, $\overline{J}^{2,+} u(\bar{x})$. A parabolic analogue of Lemma E.4 is also true, see [263], Lemma 4.1, p. 211.

E.2   Definition of Viscosity Solution

Let $\mathcal{O}$ be an open subset of $\mathbb{R}^N$. Consider an equation

$$F(x, u, Du, D^2u) = 0 \quad \text{in } \mathcal{O}, \tag{E.1}$$

where $F : \mathcal{O} \times \mathbb{R} \times \mathbb{R}^N \times S(\mathbb{R}^N) \to \mathbb{R}$ is continuous, increasing in the second variable, and degenerate elliptic, i.e. for every $(x, r, p) \in \mathcal{O} \times \mathbb{R} \times \mathbb{R}^N$, and $X, Y \in S(\mathbb{R}^N)$,

$$F(x, r, p, X) \leq F(x, r, p, Y) \quad \text{if } X \geq Y.$$

Definition E.8 An upper semicontinuous function $u : \mathcal{O} \to \mathbb{R}$ is a viscosity subsolution of (E.1) if

$$F(x, u(x), p, X) \leq 0 \quad \text{if } x \in \mathcal{O} \text{ and } (p, X) \in J^{2,+}u(x).$$

A lower semicontinuous function $u : \mathcal{O} \to \mathbb{R}$ is a viscosity supersolution of (E.1) if

$$F(x, u(x), p, X) \geq 0 \quad \text{if } x \in \mathcal{O} \text{ and } (p, X) \in J^{2,-}u(x).$$

A viscosity solution of (E.1) is a function which is both a viscosity subsolution and a viscosity supersolution of (E.1).

We remark that since F is continuous, we obtain an equivalent definition if $J^{2,+}u(x)$, $J^{2,-}u(x)$ in Definition E.8 are replaced, respectively, by $\overline{J}^{2,+}u(x)$, $\overline{J}^{2,-}u(x)$. Moreover, in light of Lemma E.4, Definition E.8 is equivalent to the following definition using test functions.

Definition E.9 An upper semicontinuous function $u : \mathcal{O} \to \mathbb{R}$ is a viscosity subsolution of (E.1) if whenever $u - \varphi$ has a local maximum at a point $x \in \mathcal{O}$ for a test function $\varphi \in C^2(\mathcal{O})$ then

$$F(x, u(x), D\varphi(x), D^2\varphi(x)) \leq 0.$$

A lower semicontinuous function $u : \mathcal{O} \to \mathbb{R}$ is a viscosity supersolution of (E.1) if whenever $u - \varphi$ has a local minimum at a point $x \in \mathcal{O}$ for a test function $\varphi \in C^2(\mathcal{O})$ then

$$F(x, u(x), D\varphi(x), D^2\varphi(x)) \geq 0.$$

A viscosity solution of (E.1) is a function which is both a viscosity subsolution and a viscosity supersolution of (E.1).

Viscosity sub/supersolutions and solutions of parabolic initial value problems

$$\begin{cases} u_t + F(t, x, u, Du, D^2u) = 0 & \text{in } (0, T) \times \mathcal{O}, \\ u(0, x) = g(x) & \text{on } \mathcal{O} \end{cases} \tag{E.2}$$

are defined in the same way if we replace $J^{2,-}u(\bar{x})$, $J^{2,+}u(\bar{x})$ in Definition E.8 by $\mathcal{P}^{2,-}u(\bar{t}, \bar{x})$, $\mathcal{P}^{2,+}u(\bar{t}, \bar{x})$ and use test functions φ which are once continuously differentiable in t and twice continuously differentiable in x on $(0, T) \times \mathcal{O}$ in Definition E.9.

It is often useful to use the notion of a *discontinuous viscosity solution*. A function u is a discontinuous viscosity subsolution if u^* is a viscosity subsolution, and u is a discontinuous viscosity supersolution if u_* is a viscosity supersolution.

E.3 Finite-Dimensional Maximum Principles

The following form of the finite-dimensional maximum principle was introduced in [138] and is sometimes referred to as the Crandall–Ishii lemma.

Theorem E.10 (Maximum principle) *Let $N \in \mathbb{N}$ and $\mathcal{O}$ be an open subset of $\mathbb{R}^N$. Let $u_i : \mathcal{O} \to \mathbb{R}$, $i = 1, 2$, be two upper semicontinuous functions, and $\phi \in C^2(\mathcal{O} \times \mathcal{O})$. Set, for $x = (x_1, x_2) \in \mathcal{O} \times \mathcal{O}$,*

$$w(x) := u_1(x_1) + u_2(x_2).$$

Suppose that $w - \phi$ has a local maximum at $\bar{x} = (\bar{x}_1, \bar{x}_2) \in \mathcal{O}$. Then, for each $\varepsilon > 0$, there exist $X_i \in S(\mathbb{R}^N)$ such that

$$\left(D_{x_i}\phi(\bar{x}), X_i \right) \in \overline{J}^{2,+}u_i(\bar{x}_i) \quad \text{for } i = 1, 2$$

and

$$-\left(\frac{1}{\varepsilon} + \|D^2\phi(\bar{x})\|\right) I \le \begin{pmatrix} X_1 & 0 \\ 0 & X_2 \end{pmatrix} \le D^2\phi(\bar{x}) + \varepsilon \left(D^2\phi(\bar{x})\right)^2.$$

Proof Theorem E.10 is a particular case of Theorem 3.2 of [139]. Its proof is given in the appendix of [139]. $\qquad\square$

The following is a parabolic version of Theorem E.10 and is taken from [138], see also [139], Theorem 8.2 or [263], Theorem 6.1, p. 216.

Theorem E.11 (Parabolic maximum principle) *Let $T > 0$, $N \in \mathbb{N}$, and $\mathcal{O}$ be an open subset of $\mathbb{R}^N$. Let $u_i : (0, T) \times \mathcal{O} \to \mathbb{R}$, $i = 1, 2$, be two upper semicontinuous functions, and $\phi : (0, T) \times \mathcal{O} \times \mathcal{O} \to \mathbb{R}$ be once continuously differentiable in t and twice continuously differentiable in $x = (x_1, x_2) \in \mathbb{R}^{2N}$. Set, for $(t, x) = (t, x_1, x_2) \in (0, T) \times \mathcal{O} \times \mathcal{O}$,*

$$w(t, x) := u_1(t, x_1) + u_2(t, x_2).$$

Suppose that $w - \phi$ has a local maximum at $(\bar{t}, \bar{x}) = (\bar{t}, \bar{x}_1, \bar{x}_2) \in (0, T) \times \mathbb{R}^{2N}$. Assume, moreover, that there is an $r > 0$ such that for every $M > 0$ there is a $C > 0$ such that for $i = 1, 2$

$$\begin{cases} b_i \leq C \text{ whenever } (b_i, p_i, X_i) \in \mathcal{P}^{2,+} u_i(t, x_i), \\ |x_i - \bar{x}_i| + |t - \bar{t}| \leq r \text{ and } |u_i(t, x_i)| + |p_i| + \|X_i\| \leq M. \end{cases} \quad (E.3)$$

Then, for each $\varepsilon > 0$, there exist $b_i \in \mathbb{R}$, $X_i \in S(\mathbb{R}^N)$ such that

$$\left(b_i, D_{x_i}\phi(\bar{t}, \bar{x}), X_i\right) \in \overline{\mathcal{P}}^{2,+} u_i(\bar{t}, \bar{x}_i) \text{ for } i = 1, 2, \quad b_1 + b_2 = \varphi_t(\bar{t}, \bar{x}),$$

and

$$-\left(\frac{1}{\varepsilon} + \|D^2\phi(\bar{t}, \bar{x})\|\right) I \leq \begin{pmatrix} X_1 & 0 \\ 0 & X_2 \end{pmatrix} \leq D^2\phi(\bar{t}, \bar{x}) + \varepsilon \left(D^2\phi(\bar{t}, \bar{x})\right)^2.$$

We remark that the somewhat strange looking condition E.3 is satisfied if u_1 is a viscosity subsolution and $-u_2$ is a viscosity supersolution of a parabolic equation.

E.4 Perron's Method

Perron's method is an easy and very general procedure to obtain the existence of viscosity solutions. Consider a parabolic initial value problem (E.2), where $\mathcal{O} = \mathbb{R}^N$ and $F : [0, T] \times \mathbb{R}^N \times \mathbb{R} \times \mathbb{R}^N \times S(\mathbb{R}^N) \to \mathbb{R}$ is continuous, increasing in the third variable, and degenerate elliptic. This is the only case that will be used in this book.

Suppose that we have a viscosity supersolution $\bar{u}$ of (E.2) and a viscosity subsolution $\underline{u}$ of (E.2) such that $\underline{u} \leq \bar{u}$ and $\underline{u}(0, x) = \bar{u}(0, x) = g(x)$. Suppose, moreover, that the equation satisfies the following comparison property: If u is a viscosity subsolution of (E.2) and v is a viscosity supersolution of (E.2) such that $\underline{u}_* \leq u, v \leq \bar{u}^*$, then $u \leq v$. We then have the following theorem. Its proof follows standard arguments, see for instance [139], Sect. 4, pp. 22–24.

Theorem E.12 (Perron's method) *If the assumptions of this subsection are satisfied then the function*

$$w(t, x) = \sup\{u(t, x) : \underline{u} \leq u \leq \bar{u}, \ u \text{ is a viscosity subsolution of } (E.2)\}$$

is a viscosity solution of (E.2).

We remark that when applying Perron's method it is often more convenient to use the notion of a discontinuous viscosity solution. The comparison property is then not needed and one always has that the function w, defined as the supremum of discontinuous viscosity subsolutions u such that $\underline{u} \leq u \leq \overline{u}$, is a discontinuous viscosity solution.

References

1. R.A. Adams, *Sobolev Spaces*, Pure and Applied Mathematics, vol. 65 (Academic Press, New York, 1975)
2. S. Agmon, On the eigenfunctions and on the eigenvalues of general elliptic boundary value problems. Comm. Pure Appl. Math. **15**(2), 119–147 (1962)
3. N.U. Ahmed, Optimal control of ∞-dimensional stochastic systems via generalized solutions of HJB equations. Discuss. Math. Differ. Incl. Control Optim. **21**(1), 97–126 (2001)
4. N.U. Ahmed, Generalized solutions of HJB equations applied to stochastic control on Hilbert space. Nonlinear Anal. **54**(3), 495–523 (2003)
5. N.U. Ahmed, K.L. Teo, *Optimal Control of Distributed Parameter Systems* (Elsevier, New York, 1981)
6. R. Aid, S. Federico, H. Pham, B. Villeneuve, Explicit investment rules with time-to-build and uncertainty. J. Econ. Dyn. Control **51**, 240–256 (2015)
7. S. Albeverio, A.-B. Cruzeiro, Global flows with invariant (Gibbs) measures for Euler and Navier-Stokes two dimensional fluids. Commun. Math. Phys. **129**(3), 431–444 (1990)
8. C.D. Aliprantis, K.C. Border, *Infinite Dimensional Analysis: A Hitchhiker's Guide*, 2nd edn. (Springer, Heidelberg, 2006)
9. E. Alòs, S. Bonaccorsi, Stochastic partial differential equations with Dirichlet white-noise boundary conditions. Ann. Inst. H. Poincaré Probab. Statist. **38**(2), 125–154 (2002)
10. A. Ambrosetti, G. Prodi, *A Primer of Nonlinear Analysis*, 2nd edn., Cambridge Studies in Advanced Mathematics, vol. 34 (Cambridge University Press, Cambridge, 1995)
11. V.V. Anh, W. Grecksch, J. Yong, Regularity of backward stochastic Volterra integral equations in Hilbert spaces. Stoch. Anal. Appl. **29**(1), 146–168 (2011)
12. V.V. Anh, J. Yong, Backward stochastic Volterra integral equations in Hilbert spaces, in *Differential and Difference Equations and Applications*, ed. by R.P. Agarwal, K. Perera (Hindawi, New York, 2006), pp. 57–66
13. S. Aniţa, *Analysis and Control of Age-dependent Population Dynamics*, Mathematical Modelling: Theory and Applications, vol. 11 (Kluwer, Dordrecht, 2000)
14. D. Applebaum, On the infinitesimal generators of Ornstein-Uhlenbeck processes with jumps in Hilbert space. Potential Anal. **26**(1), 79–100 (2007)
15. M. Arisawa, H. Ishii, P.L. Lions, A characterization of the existence of solutions for Hamilton-Jacobi equations in ergodic control problems with applications. Appl. Math. Optim. **42**(1), 35–50 (2000)

© Springer International Publishing AG 2017
G. Fabbri et al., *Stochastic Optimal Control in Infinite Dimension*,
Probability Theory and Stochastic Modelling 82,
DOI 10.1007/978-3-319-53067-3

16. L. Arnold, Mathematical models of chemical reactions, in *Stochastic Systems: The Mathematics of Filtering and Identification and Applications*, ed. by M. Hazewinkel, J.C. Willems (Reidel, Dordrecht, 1981), pp. 111–134
17. N. Aronszajn, P. Panitchpakdi, Extension of uniformly continuous transformations and hyperconvex metric spaces. Pacific J. Math. **6**, 405–439 (1956)
18. R.B. Ash, *Probability and Measure Theory*, 2nd edn. (Harcourt/Academic Press, Burlington, 2000)
19. J. Assaad, J.M.A.M. van Neerven, L^2-theory for non-symmetric Ornstein-Uhlenbeck semigroups on domains. J. Evol. Equ. **13**(1), 107–134 (2013)
20. J.-P. Aubin, H. Frankowska, *Set-Valued Analysis*, Modern Birkhäuser Classics (Birkhäuser, Boston, 2009). Reprint of the 1990 edition
21. M. Avellaneda, A. Levy, A. Paras, Pricing and hedging derivative securities in markets with uncertain volatilities. Appl. Math. Finance **2**(2), 73–88 (1995)
22. F. Baghéry, I. Turpin, Y. Ouknine, Some remark on optimal stochastic control with partial information. Stoch. Anal. Appl. **23**(6), 1305–1320 (2005)
23. S. Bahlali, Necessary and sufficient optimality conditions for relaxed and strict control problems. SIAM J. Control Optim. **47**(4), 2078–2095 (2008)
24. M. Bambi, Endogenous growth and time-to-build: the AK case. J. Econ. Dyn. Control **32**(4), 1015–1040 (2008)
25. M. Bambi, G. Fabbri, F. Gozzi, Optimal policy and consumption smoothing effects in the time-to-build AK model. Econ. Theor. **50**(3), 635–669 (2012)
26. V. Barbu, *Nonlinear Differential Equations of Monotone Types in Banach Spaces*, Springer Monographs in Mathematics (Springer, Berlin, 2010)
27. V. Barbu, G. Da Prato, Global existence for the Hamilton-Jacobi equations in Hilbert space. Ann. Scuola Norm. Sup. Pisa Cl. Sci. **8**(2), 257–284 (1981)
28. V. Barbu, G. Da Prato, A direct method for studying the dynamic programming equation for controlled diffusion processes in Hilbert spaces. Numer. Funct. Anal. Optim. **4**(1), 23–43 (1981/82)
29. V. Barbu, G. Da Prato, *Hamilton-Jacobi Equations in Hilbert Spaces*, Pitman Research Notes in Mathematics Series, vol. 86 (Longman, Boston, 1983)
30. V. Barbu, G. Da Prato, Solution of the Bellman equation associated with an infinite-dimensional stochastic control problem and synthesis of optimal control. SIAM J. Control Optim. **21**(4), 531–550 (1983)
31. V. Barbu, G. Da Prato, The stochastic nonlinear damped wave equation. Appl. Math. Optim. **46**(2–3), 125–206 (2002)
32. V. Barbu, G. Da Prato, The two phase stochastic Stefan problem. Probab. Theory Relat. Fields **124**(4), 544–560 (2002)
33. V. Barbu, G. Da Prato, The Kolmogorov equation for a 2D-Navier-Stokes stochastic flow in a channel. Nonlinear Anal. **69**(3), 940–949 (2008)
34. V. Barbu, G. Da Prato, A. Debussche, The Kolmogorov equation associated to the stochastic Navier-Stokes equations in 2D, Infin. Dimens. Anal. Quantum Probab. Relat. Top. **7**(2), 163–182 (2004)
35. V. Barbu, G. Da Prato, C. Popa, Existence and uniqueness of the dynamic programming equation in Hilbert space. Nonlinear Anal. **7**(3), 283–299 (1983)
36. V. Barbu, G. Da Prato, L. Tubaro, Kolmogorov equation associated to the stochastic reflection problem on a smooth convex set of a Hilbert space. Ann. Probab. **37**(4), 1427–1458 (2009)
37. V. Barbu, G. Da Prato, Kolmogorov equation associated to the stochastic reflection problem on a smooth convex set of a Hilbert space II. Ann. Inst. H. Poincaré Probab. Statist. **47**(3), 699–724 (2011)
38. V. Barbu, C. Marinelli, Variational inequalities in Hilbert spaces with measures and optimal stopping problems. Appl. Math. Optim. **57**(2), 237–262 (2008)
39. V. Barbu, T. Precupanu, *Convexity and Optimization in Banach Spaces*, 4th edn., Springer Monographs in Mathematics (Springer, Dordrecht, 2012)

40. M. Bardi, I. Capuzzo-Dolcetta, *Optimal Control and Viscosity Solutions of Hamilton-Jacobi-Bellman Equations*, Systems and Control: Foundations and Applications (Birkhäuser, Boston, 1997)
41. G. Barles, *Solutions de viscosité des équations de Hamilton-Jacobi*, Mathématiques and Applications, vol. 17 (Springer, Paris, 1994)
42. A. Bátkai, S. Piazzera, *Semigroups for Delay Equations*, Research Notes in Mathematics, vol. 10 (Peters, Wellesley, 2005)
43. H.H. Bauschke, P.L. Combettes, *Convex Analysis and Monotone Operator Theory in Hilbert Spaces*, CMS Books in Mathematics (Springer, Berlin, 2011)
44. A. Bensoussan, *Filtrage optimal des systèmes linéaires*, Methodes Mathematiques de l'Informatique, vol. 3 (Dunod, Paris, 1971)
45. A. Bensoussan, Stochastic maximum principle for distributed parameter systems. J. Franklin Inst. **315**(5), 387–406 (1983)
46. A. Bensoussan, *Stochastic Control of Partially Observable Systems* (Cambridge University Press, Cambridge, 1992)
47. A. Bensoussan, G. Da Prato, M.C. Delfour, S.K. Mitter, *Representation and Control of Infinite Dimensional Systems*, 2nd edn., Systems and Control: Foundations and Applications (Birkhäuser, Boston, 2007)
48. A. Bensoussan, J. Frehse, S.C.P. Yam, *Mean Field Games and Mean Field Type Control Theory*, Springer Briefs in Mathematics, vol. 101 (Springer, New York, 2013)
49. A. Bensoussan, J. Frehse, S.C.P. Yam, On the interpretation of the Master Equation, Stochastic Process. Appl. (to appear)
50. A. Bensoussan, The Master equation in mean field theory. J. Math. Pures Appl. **103**(6), 1441–1474 (2015)
51. A. Bensoussan, J.-L. Lions, *Applications of Variational Inequalities in Stochastic Control*, Studies in Mathematics and its Applications, vol. 12 (North-Holland, Amsterdam, 1982)
52. A. Bensoussan, R. Temam, Équations stochastiques du type Navier-Stokes. J. Funct. Anal. **13**(2), 195–222 (1973)
53. D.P. Bertsekas, *Dynamic Programming and Optimal Control*, vol. 1 (Athena Scientific, Belmont, 1995)
54. D.P. Bertsekas, *Dynamic Programming and Optimal Control*, vol. 2 (Athena Scientific, Belmont, 1995)
55. U. Bessi, Existence of solutions of the Master equation in the smooth case. SIAM J. Math. Anal. **48**(1), 204–228 (2016)
56. J. Bierkens, O. van Gaans, S. Lunel, Existence of an invariant measure for stochastic evolutions driven by an eventually compact semigroup. J. Evol. Equ. **9**(4), 771–786 (2009)
57. E. Biffis, F. Gozzi, C. Prosdocimi, *Optimal portfolio choice with path dependent labor income: the infinite horizon case*. In preparation
58. P. Billingsley, *Probability and Measure*, 3rd edn., Wiley Series in Probability and Mathematical Statistics (Wiley, New York, 1995)
59. A. Biryuk, On invariant measures of the 2D Euler equation. J. Stat. Phys. **122**(4), 597–616 (2006)
60. J.-M. Bismut, Martingales, the Malliavin calculus and hypoellipticity under general Hörmander's conditions. Probab. Theory Relat. Fields **56**(4), 469–505 (1981)
61. V.I. Bogachev, *Measure Theory. Vol. I and II* (Springer, Berlin, 2007)
62. S. Bonaccorsi, F. Confortola, E. Mastrogiacomo, Optimal control of stochastic differential equations with dynamical boundary conditions. J. Math. Anal. Appl. **344**(2), 667–681 (2008)
63. S. Bonaccorsi, F. Confortola, E. Mastrogiacomo, Optimal control for stochastic Volterra equations with completely monotone kernels. SIAM J. Control Optim. **50**(2), 748–789 (2012)
64. S. Bonaccorsi, M. Fuhrman, Regularity results for infinite dimensional diffusions. A Malliavin calculus approach. Atti Accad. Naz. Lincei Cl. Sci. Fis. Mat. Natur. Rend. Lincei (9) Mat. Appl. **10**(1), 35–45 (1999)
65. S. Bonaccorsi, G. Guatteri, Stochastic partial differential equations in bounded domains with Dirichlet boundary conditions. Stoch. Stoch. Rep. **74**(1–2), 349–370 (2002)

66. V.S. Borkar, *Optimal Control of Diffusion Processes* (Longman, New York, 1989)
67. J.M. Borwein, D. Preiss, A smooth variational principle with applications to subdifferentiability and to differentiability of convex functions. Trans. Amer. Math. Soc. **303**(2), 517–527 (1987)
68. R. Boucekkine, C. Camacho, G. Fabbri, Spatial dynamics and convergence: the spatial AK model. J. Econ. Theory **148**(6), 2719–2736 (2013)
69. R. Boucekkine, O. Licandro, L.A. Puch, F. del Rio, Vintage capital and the dynamics of the AK model. J. Econ. Theory **120**(1), 39–72 (2005)
70. B. Bouchard, N.-M. Dang, C.-A. Lehalle, Optimal control of trading algorithms: a general impulse control approach. SIAM J. Financ. Math. **2**(1), 404–438 (2011)
71. B. Bouchard, M. Nutz, Weak dynamic programming for generalized state constraints. SIAM J. Control Optim. **50**(6), 3344–3373 (2012)
72. B. Bouchard, N. Touzi, Weak dynamic programming principle for viscosity solutions. SIAM J. Control Optim. **49**(3), 948–962 (2011)
73. N. Bourbaki, *Éléments de mathématique. Intégration. Chapitres 1–4* (Springer, Paris, 2007)
74. A.J.V. Brandão, E. Fernández-Cara, P.M.D. Magalhães, M.A. Rojas-Medar, Theoretical analysis and control results for the Fitz-Hugh-Nagumo equation. Electron. J. Differ. Equ. **164**, 1–20 (2008)
75. P. Briand, F. Confortola, BSDEs with stochastic Lipschitz condition and quadratic PDEs in Hilbert spaces. Stoch. Process. Appl. **118**(5), 818–838 (2008)
76. P. Briand, F. Confortola, Differentiability of backward stochastic differential equations in Hilbert spaces with monotone generators. Appl. Math. Optim. **57**(2), 149–176 (2008)
77. P. Briand, F. Confortola, Quadratic BSDEs with random terminal time and elliptic PDEs in infinite dimension. Electron. J. Probab. **13**(54), 1529–1561 (2008)
78. P. Briand, B. Delyon, Y. Hu, É. Pardoux, L. Stoica, L^p solutions of backward stochastic differential equations. Stoch. Process. Appl. **108**(1), 109–129 (2003)
79. P. Briand, Y. Hu, Stability of BSDEs with random terminal time and homogenization of semilinear elliptic PDEs. J. Funct. Anal. **155**(2), 455–494 (1998)
80. P. Briand, Y. Hu, BSDE with quadratic growth and unbounded terminal value. Probab. Theory Relat. Fields **136**(4), 604–618 (2006)
81. J. Bricmont, A. Kupiainen, R. Lefevere, Exponential mixing of the 2D stochastic Navier-Stokes dynamics. Commun. Math. Phys. **230**(1), 87–132 (2002)
82. Z. Brzeźniak, L. Debbi, B. Goldys, Ergodic properties of fractional stochastic Burgers equation. Glob. Stoch. Anal. **1**(2), 145–174 (2011)
83. Z. Brzeźniak, D. Gatarek, Martingale solutions and invariant measures for stochastic evolution equations in Banach spaces. Stoch. Process. Appl. **84**(2), 187–225 (1999)
84. R. Buckdahn, J. Li, S. Peng, C. Rainer, Mean-field stochastic differential equations and associated PDEs. Ann. Probab. **45**(2), 824–878 (2017)
85. R. Buckdahn, J. Ma, Pathwise stochastic control problems and stochastic HJB equations. SIAM J. Control Optim. **45**(6), 2224–2256 (2007)
86. R. Buckdahn, M. Quincampoix, G. Tessitore, Controlled stochastic differential equations under constraints in infinite dimensional spaces. SIAM J. Control Optim. **47**(1), 218–250 (2008)
87. J.M. Burgers, A mathematical model illustrating the theory of turbulence, in *Advances in Applied Mechanics*, ed. by R. von Mises, T. von Kármán (Academic Press, New York, 1948), pp. 171–199
88. J.M. Burgers, *The Nonlinear Diffusion Equation: Asymptotic Solutions and Statistical Problems* (Springer, Berlin, 1974)
89. P. Cannarsa, G. Da Prato, Second-order Hamilton-Jacobi equations in infinite dimensions. SIAM J. Control Optim. **29**(2), 474–492 (1991)
90. P. Cannarsa, G. Da Prato, Direct solution of a second order Hamilton-Jacobi equation, in *Hilbert Spaces, in Stochastic Partial Differential Equations and Applications*, Pitman Research Notes In Mathematics Series, vol. 268, ed. by G. Da Prato, L. Tubaro (Longman, Harlow, 1992), pp. 72–85

91. P. Cannarsa, G. Di Blasio, A direct approach to infinite-dimensional Hamilton-Jacobi equations and applications to convex control with state constraints. Differ. Integral Equ. **8**(2), 225–246 (1995)
92. P. Cannarsa, H. Frankowska, Value function and optimality conditions for semilinear control problems. Appl. Math. Optim. **26**(2), 139–169 (1992)
93. P. Cannarsa, F. Gozzi, H.M. Soner, A boundary value problem for Hamilton-Jacobi equations in Hilbert spaces. Appl. Math. Optim. **24**(2), 197–220 (1991)
94. P. Cannarsa, F. Gozzi, H.M. Soner, A dynamic programming approach to nonlinear boundary control problems of parabolic type. J. Funct. Anal. **117**(1), 25–61 (1993)
95. P. Cannarsa, C. Sinestrari, *Semiconcave Functions, Hamilton-Jacobi Equations, and Optimal Control*, Progress in Nonlinear Differential Equations and Their Applications (Birkhäuser, Boston, 2004)
96. P. Cannarsa, M.E. Tessitore, Cauchy problem for the dynamic programming equation of boundary control, in *Boundary Control and Variation*, Lecture Notes in Pure and Applied Mathematics, vol. 163, ed. by J.-P. Zolesio (Dekker, New York, 1994), pp. 13–26
97. P. Cannarsa, M.E. Tessitore, Infinite-dimensional Hamilton-Jacobi equations and Dirichlet boundary control problems of parabolic type. SIAM J. Control Optim. **34**(6), 1831–1847 (1996)
98. P. Cardaliaguet, *Notes on mean field games (from P.-L. Lions' lectures at Collège de France)* (2013), https://www.ceremade.dauphine.fr/~cardalia/MFG20130420.pdf
99. P. Cardaliaguet, F. Delarue, J.-M. Lasry, P.-L. Lions, *The master equation and the convergence problem in mean field games*. Preprint (2015), arXiv:1509.02505
100. R. Carmona, F. Delarue, The Master equation for large population equilibriums, in *Stochastic Analysis and Applications*, ed. by D. Crisan, B. Hambly, T. Zariphopoulou (Springer, Berlin, 2014), pp. 77–128
101. S. Cerrai, A Hille-Yosida theorem for weakly continuous semigroups. Semigroup Forum **49**(3), 349–367 (1994)
102. S. Cerrai, Weakly continuous semigroups in the space of functions with polynomial growth. Dynam. Syst. Appl. **4**, 351–372 (1995)
103. S. Cerrai, Differentiability of Markov semigroups for stochastic reaction-diffusion equations and applications to control. Stoch. Proc. Appl. **83**(1), 15–37 (1999)
104. S. Cerrai, Smoothing properties of transition semigroups relative to SDEs with values in Banach spaces. Probab. Theory Relat. Fields **113**(1), 85–114 (1999)
105. S. Cerrai, Optimal control problems for stochastic reaction-diffusion systems with non-Lipschitz coefficients. SIAM J. Control Optim. **39**(6), 1779–1816 (2001)
106. S. Cerrai, *Second Order PDE's in Finite and Infinite Dimension: a Probabilistic Approach*, Lecture Notes in Mathematics, vol. 1762 (Springer, Berlin, 2001)
107. S. Cerrai, Stationary Hamilton-Jacobi equations in Hilbert spaces and applications to a stochastic optimal control problem. SIAM J. Control Optim. **40**(3), 824–852 (2001)
108. S. Cerrai, F. Gozzi, Strong solutions of Cauchy problems associated to weakly continuous semigroups. Differ. Integral Equ. **8**(3), 465–486 (1995)
109. S. Cerrai, M. Röckner, Large deviations for stochastic reaction-diffusion systems with multiplicative noise and non-Lipshitz reaction term. Ann. Probab. **32**(1B), 1100–1139 (2004)
110. S. Cerrai, M. Röckner, Large deviations for invariant measures of stochastic reaction-diffusion systems with multiplicative noise and non-Lipschitz reaction term. Ann. Inst. H. Poincaré Probab. Statist. **41**(1), 69–105 (2005)
111. D.H. Chambers, R.J. Adrian, P. Moin, D.S. Stewart, H.J. Sung, Karhunen-Loéve expansion of Burgers' model of turbulence. Phys. Fluids **31**, 25–73 (1988)
112. C.D. Charalambous, J.L. Hibey, First passage risk-sensitive criterion for stochastic evolutions, in *Proceedings of the IEEE American Control Conference* (IEEE, 1995), pp. 2449–2450
113. C.D. Charalambous, D.S. Naidu, K.L. Moore, Risk-sensitive control, differential games, and limiting problems in infinite dimensions, *Proceedings of the 33rd IEEE Conference on Decision and Control* (IEEE, 1994), pp. 2184–2186

114. J.-F. Chassagneux, D. Crisan, F. Delarue, *A probabilistic approach to classical solutions of the Master equation for large population equilibria*. Preprint (2015), arXiv:1411.3009v2

115. P. Cheridito, H.M. Soner, N. Touzi, The multi-dimensional super-replication problem under gamma constraints. Ann. Inst. H. Poincaré Anal. Non Linéaire **22**(5), 633–666 (2005)

116. M.B. Chiarolla, T. De Angelis, Optimal stopping of a Hilbert space valued diffusion: an infinite dimensional variational inequality. Appl. Math. Optim. **73**(2), 1–42 (2012)

117. H. Choi, R. Temam, P. Moin, J. Kim, Feedback control for unsteady flow and its application to the stochastic Burgers equation. J. Fluid Mech. **253**, 509–543 (1993)

118. A. Chojnowska-Michalik, Representation theorem for general stochastic delay equations. Bull. Acad. Polon. Sci. Sér. Sci. Math. Astronom. Phys. **26**(7), 635–642 (1978)

119. A. Chojnowska-Michalik, Transition semigroups for stochastic semilinear equations on Hilbert spaces. Diss. Math. (Rozprawy Mat.) **396**, 1–59 (2001)

120. A. Chojnowska-Michalik, B. Gołdys, Existence, uniqueness and invariant measures for stochastic semilinear equations on Hilbert spaces. Probab. Theory Relat. Fields **102**(3), 331–356 (1995)

121. A. Chojnowska-Michalik, B. Goldys, On regularity properties of nonsymmetric Ornstein-Uhlenbeck semigroup in L^p spaces. Stoch. Stoch. Rep. **59**(3–4), 183–209 (1996)

122. A. Chojnowska-Michalik, B. Goldys, Symmetric Ornstein-Uhlenbeck semigroups and their generators. Probab. Theory Relat. Fields **124**(4), 459–486 (2002)

123. P.-L. Chow, Infinite-dimensional Kolmogorov equations in Gauss-Sobolev spaces. Stoch. Anal. Appl. **14**(3), 257–282 (1996)

124. P.-L. Chow, *Stochastic partial differential equations*, Chapman & Hall Applied Mathematics and Nonlinear Science Series (Chapman & Hall, Raton, 2007)

125. P.-L. Chow, J.-L. Menaldi, Infinite-dimensional Hamilton-Jacobi-Bellman equations in Gauss-Sobolev spaces. Nonlinear Anal. **29**(4), 415–426 (1997)

126. J. Claisse, D. Talay, X. Tan, A pseudo-Markov property for controlled diffusion processes. SIAM J. Control Optim. **52**, 1017–1029 (2016)

127. F.H. Clarke, *Functional Analysis, Calculus of Variations and Optimal Control*, Graduate Texts in Mathematics, vol. 264 (Springer, London, 2013)

128. F.H. Clarke, Y.S. Ledyaev, R.J. Stern, P.R. Wolenski, *Nonsmooth Analysis and Control Theory*, Graduate Texts in Mathematics, vol. 178 (Springer, New York, 1998)

129. F. Confortola, Dissipative backward stochastic differential equations in infinite dimensions, Infin. Dimens. Anal. Quantum Probab. Relat. Top. **9**(1), 155–168 (2006)

130. F. Confortola, Dissipative backward stochastic differential equations with locally Lipschitz nonlinearity. Stoch. Process. Appl. **117**(5), 613–628 (2007)

131. F. Confortola, E. Mastrogiacomo, Optimal control for stochastic heat equation with memory. Evol. Equ. Control Theory **3**(1), 35–58 (2014)

132. F. Confortola, E. Mastrogiacomo, Feedback optimal control for stochastic Volterra equations with completely monotone kernels. Math. Control Relat. Fields **5**(2), 191–235 (2015)

133. A. Cosso, C. Di Girolami, F. Russo, *Calculus via regularizations in Banach spaces and Kolmogorov-type path-dependent equations*. Contemp. Math. **668**, 43–68 (2016)

134. A. Cosso, S. Federico, F. Gozzi, M. Rosestolato, N. Touzi, *Path-dependent equations and viscosity solutions in infinite dimension*. Preprint (2015), arXiv:1502.05648

135. A. Cosso, F. Russo, *Strong-viscosity solutions: semilinear parabolic PDEs and path-dependent PDEs*, Ann. Prob. (to appear)

136. A. Cosso, F. Russo, Functional and Banach space stochastic calculi: path-dependent Kolmogorov equations associated with the frame of a Brownian motion, in *Stochastics of Environmental and Financial Economics*, Springer Proceedings In Mathematics and Statistics, vol. 138, ed. by F.E. Benth, G. Di Nunno (Springer, Berlin, 2015), pp. 27–80

137. M.G. Crandall, Semidifferentials, quadratic forms and fully nonlinear elliptic equations of second order. Ann. Inst. H. Poincaré Anal. Non Linéaire **6**(6), 419–435 (1989)

138. M.G. Crandall, H. Ishii, The maximum principle for semicontinuous functions. Differ. Integral Equ. **3**(6), 1001–1014 (1990)

139. M.G. Crandall, H. Ishii, P.-L. Lions, User's guide to viscosity solutions of second order partial differential equations. Bull. Amer. Math. Soc. (N.S.) **27**(1), 1–67 (1992)

140. M.G. Crandall, M. Kocan, A. Święch, On partial sup-convolutions, a lemma of P.-L. Lions and viscosity solutions in Hilbert spaces. Adv. Math. Sci. Appl. **3**(Special Issue), 1–15 (1993/94)

141. M.G. Crandall, P.-L. Lions, Viscosity solutions of Hamilton-Jacobi equations in infinite dimensions. IV. Hamiltonians with unbounded linear terms. J. Funct. Anal. **90**(2), 237–283 (1990)

142. M.G. Crandall, P.-L. Lions, Viscosity solutions of Hamilton-Jacobi equations in infinite dimensions. V. Unbounded linear terms and B-continuous solutions. J. Funct. Anal. **97**(2), 417–465 (1991)

143. M.G. Crandall, P.-L. Lions, Hamilton-Jacobi equations, in infinite dimensions. VI. Nonlinear A and Tataru's method refined, in *Evolution Equations, Control Theory, and Biomathematics*, Lecture Notes in Pure and Applied Mathematics, vol. 155, ed. by P. Clément, G. Lumer (Dekker, New York, 1994), pp. 51–89

144. M.G. Crandall, P.-L. Lions, Viscosity solutions of Hamilton-Jacobi equations in infinite dimensions. VII. The HJB equation is not always satisfied. J. Funct. Anal. **125**(1), 111–148 (1994)

145. A. Cretarola, F. Gozzi, H. Pham, P. Tankov, Optimal consumption policies in illiquid markets. Financ. Stoch. **15**(1), 85–115 (2011)

146. G. Da Prato, *Applications croissantes et équations d'évolution dans les espaces de Banach* (Academic Press, New York, 1976)

147. G. Da Prato, Some results on Bellman equation in Hilbert spaces. SIAM J. Control Optim. **23**(1), 61–71 (1985)

148. G. Da Prato, Perturbation of Ornstein–Uhlenbeck semigroups. Rend. Istit. Mat. Univ. Trieste **XXVIII**, 101–126 (1997)

149. G. Da Prato, Regularity results for Kolmogorov equations in $L^2(H, \mu)$ spaces and applications. Ukrainian Math. J. **49**(3), 494–505 (1997)

150. G. Da Prato, The Ornstein-Uhlenbeck generator perturbed by the gradient of a potential. Boll. Un. Mat. Ital. **1**(3), 501–519 (1998)

151. G. Da Prato, Monotone gradient systems in L^2 spaces, in *Seminar on Stochastic Analysis, Random Fields and Applications III*, ed. by R.C. Dalang, M. Dozzi, F. Russo (Basel, Birkhäuser, 2002), pp. 73–88

152. G. Da Prato, *Kolmogorov equations for stochastic PDEs*, Advanced Courses in Mathematics - CRM Barcelona (Birkhäuser, Basel, 2004)

153. G. Da Prato, *An Introduction to Infinite-dimensional Analysis,* Universitext (Springer, Berlin, 2006)

154. G. Da Prato, *Introduction to Stochastic Analysis and Malliavin Calculus* (Edizioni della Normale, Pisa, 2014)

155. G. Da Prato, A. Debussche, Differentiability of the transition semigroup of the stochastic Burgers equation, and application to the corresponding Hamilton–Jacobi equation. Atti Accad. Naz. Lincei Cl. Sci. Fis. Mat. Natur. Rend. Lincei (9) Mat. Appl. **9**(4), 267–277 (1998)

156. G. Da Prato, A. Debussche, Control of the stochastic Burgers model of turbulence. SIAM J. Control Optim. **37**(4), 1123–1149 (1999)

157. G. Da Prato, A. Debussche, Dynamic programming for the stochastic Burgers equation. Ann. Mat. Pura Appl. **178**(1), 143–174 (2000)

158. G. Da Prato, A. Debussche, Dynamic programming for the stochastic Navier-Stokes equations, M2AN Math. Model. Numer. Anal. **34**(2), 459–475 (2000)

159. G. Da Prato, A. Debussche, Maximal dissipativity of the Dirichlet operator corresponding to the Burgers, in *Stochastic Processes, Physics and Geometry: New Interplays*, ed. by F. Gesztesy, H. Holden, J. Jost, S. Paycha, M. Röckner, S. Scarlatti (American Mathematical Society, Providence, 2000), pp. 85–98

160. G. Da Prato, A. Debussche, Two-dimensional Navier-Stokes equations driven by a space-time white noise. J. Funct. Anal. **196**(1), 180–210 (2002)

161. G. Da Prato, A. Debussche, Ergodicity for the 3D stochastic Navier-Stokes equations. J. Math Pure. Appl. **82**(8), 877–947 (2003)

162. G. Da Prato, A. Debussche, m-dissipativity of Kolmogorov operators corresponding to Burgers equations with space-time white noise. Potential Anal. **26**(1), 31–55 (2007)

163. G. Da Prato, A. Debussche, R. Temam, Stochastic Burgers' equation. NoDEA Nonlinear Differ. Equ. Appl. **1**(4), 389–402 (1994)

164. G. Da Prato, D. Gątarek, J. Zabczyk, Invariant measures for semilinear stochastic equations. Stoch. Anal. Appl. **10**(4), 387–408 (1992)

165. G. Da Prato, B. Goldys, J. Zabczyk, Ornstein-Uhlenbeck semigroups in open sets of Hilbert spaces. C. R. Acad. Sci. Paris Sér. I Math. **325**(4), 433–438 (1997)

166. G. Da Prato, A. Lunardi, On the Dirichlet semigroup for Ornstein-Uhlenbeck operators in subsets of Hilbert spaces. J. Funct. Anal. **259**(10), 2642–2672 (2010)

167. G. Da Prato, A. Lunardi, Maximal L^2 regularity for Dirichlet problems in Hilbert spaces. J. Math. Pures Appl. **99**(6), 741–765 (2013)

168. G. Da Prato, A. Lunardi, Maximal Sobolev regularity in Neumann problems for gradient systems in infinite dimensional domains. Ann. Inst. H. Poincaré Probab. Statist. **51**(3), 1102–1123 (2015)

169. G. Da Prato, M. Röckner, B.L. Rozovskii, F.-Y. Wang, Strong solutions of stochastic generalized porous media equations: existence, uniqueness, and ergodicity. Comm. Partial Differ. Equ. **31**(2), 277–291 (2006)

170. G. Da Prato, E. Sinestrari, Differential operators with nondense domain. Ann. Scuola Norm. Sup. Pisa Cl. Sci. **14**(2), 285–344 (1987)

171. G. Da Prato, L. Tubaro, Self-adjointness of some infinite-dimensional elliptic operators and application to stochastic quantization. Probab. Theory Relat. Fields **118**(1), 131–145 (2000)

172. G. Da Prato, L. Tubaro, Some results about dissipativity of Kolmogorov operators. Czech. Math. J. **51**(4), 685–699 (2001)

173. G. Da Prato, J. Zabczyk, Smoothing properties of transition semigroups in Hilbert spaces. Stochastics **35**(2), 63–77 (1991)

174. G. Da Prato, J. Zabczyk, Non-explosion, boundedness, and ergodicity for stochastic semilinear equations. J. Differ. Equ. **98**(1), 181–195 (1992)

175. G. Da Prato, J. Zabczyk, Evolution equations with white-noise boundary conditions. Stoch. Stoch. Rep. **42**(3–4), 167–182 (1993)

176. G. Da Prato, J. Zabczyk, Regular densities of invariant measures in Hilbert spaces. J. Funct. Anal. **130**(2), 427–449 (1995)

177. G. Da Prato, J. Zabczyk, *Ergodicity for Infinite-dimensional Systems*, London Mathematical Society Lecture Note Series, vol. 229 (Cambridge University Press, Cambridge, 1996)

178. G. Da Prato, J. Zabczyk, Differentiability of the Feynman–Kac semigroup and a control application. Atti Accad. Naz. Lincei Cl. Sci. Fis. Mat. Natur. Rend. Lincei (9) Mat. Appl. **8**(3), 183–188 (1997)

179. G. Da Prato, J. Zabczyk, *Second Order Partial Differential Equations in Hilbert Spaces*, London Mathematical Society Lecture Note Series, vol. 293 (Cambridge University Press, Cambridge, 2002)

180. G. Da Prato, J. Zabczyk, *Stochastic Equations in Infinite Dimensions*, Encyclopedia of Mathematics and its Applications, vol. 152 (Cambridge University Press, Cambridge, 2014)

181. A. Debussche, M. Fuhrman, G. Tessitore, Optimal control of a stochastic heat equation with boundary-noise and boundary-control. ESAIM Control Optim. Calc. Var. **13**(1), 178–205 (2007)

182. A. Debussche, Y. Hu, G. Tessitore, Ergodic BSDEs under weak dissipative assumptions. Stoch. Process. Appl. **121**(3), 407–426 (2011)

183. K. Deimling, *Nonlinear Functional Analysis* (Springer, Berlin, 1985)

184. J.-D. Deuschel, D.W. Stroock, *Large Deviations*, revised edn. (Academic Press, Boston, 1989)

185. R. Deville, G. Godefroy, V. Zizler, A smooth variational principle with applications to Hamilton-Jacobi equations in infinite dimensions. J. Funct. Anal. **111**(1), 197–212 (1993)

186. G. Di Blasio, Global solutions for a class of Hamilton-Jacobi equations in Hilbert spaces. Numer. Funct. Anal. Optim. **8**(3–4), 261–300 (1986)

187. G. Di Blasio, Optimal control with infinite horizon for distributed parameter systems with constrained controls. SIAM J. Control Optim. **29**(4), 909–925 (1991)

188. C. Di Girolami, G. Fabbri, F. Russo, The covariation for Banach space valued processes and applications. Metrika **77**(1), 51–104 (2014)

189. C. Di Girolami, F. Gozzi, *Solutions of second order HJB equations in Hilbert spaces via smoothing property.* In preparation

190. J. Diestel, J.J. Uhl, *Vector Measures*, Mathematical Surveys and Monographs, vol. 15 (American Mathematical Society, Providence, 1977)

191. N. Dinculeanu, *Integration on Locally Compact Spaces* (Noordhoff, Leyden, 1974)

192. N. Dinculeanu, *Vector Integration and Stochastic Integration in Banach Spaces*, Pure and Applied Mathematics (Wiley, New York, 2000)

193. S. Dolecki, D.L. Russell, A general theory of observation and control. SIAM J. Control Optim. **15**(2), 185–220 (1977)

194. J.L. Doob, *Measure Theory*, Graduate Texts in Mathematics, vol. 143 (Springer, New York, 1994)

195. R.G. Douglas, On majorization, factorization, and range inclusion of operators on Hilbert space. Proc. Amer. Math. Soc. **17**(2), 413–415 (1966)

196. K. Du, Q. Meng, A maximum principle for optimal control of stochastic evolution equations. SIAM J. Control Optim. **51**(6), 4343–4362 (2013)

197. K. Du, J. Qiu, S. Tang, L^p theory for super-parabolic backward stochastic partial differential equations in the whole space. Appl. Math. Optim. **65**(2), 175–219 (2012)

198. K. Du, S. Tang, Strong solution of backward stochastic partial differential equations in C^2 domains. Probab. Theory Relat. Fields **154**(1–2), 255–285 (2012)

199. K. Du, S. Tang, Q. Zhang, $W^{m,p}$-solution ($p \geq 2$) of linear degenerate backward stochastic partial differential equations in the whole space. J. Differ. Equ. **254**(7), 2877–2904 (2013)

200. E. Duncan, *Probability densities for diffusion processes*, Technical Report Stanford Electronics Labs. 7001-4, University of California, Stanford, May 1967

201. N. Dunford, On one parameter groups of linear transformations. Ann. Math. **39**(3), 569–573 (1938)

202. N. Dunford, J.T. Schwartz, *Linear Operators Part I* (Interscience, New York, 1958)

203. P. Dupuis, W.M. McEneaney, Risk-sensitive and robust escape criteria. SIAM J. Control Optim. **35**(6), 2021–2049 (1997)

204. I. Ekeland, G. Lebourg, Generic Frechet-differentiability and perturbed optimization problems in Banach spaces. Trans. Amer. Math. Soc. **224**(2), 193–216 (1976)

205. I. Ekren, C. Keller, N. Touzi, J. Zhang, On viscosity solutions of path dependent PDEs. Ann. Probab. **42**(1), 204–236 (2014)

206. N. El Karoui, Les aspects probabilistes du contrôle stochastique, in *École d'été de probabilités de Saint-Flour IX-1979*, Lecture Notes in Mathematics, vol. 876, ed. by P.L. Hennequin (Springer, Berlin, 1981), pp. 74–238

207. N. El Karoui, Backward stochastic differential equations: a general introduction, in *Backward Stochastic Differential Equations*, Pitman Research Notes in Mathematics Series, ed. by N. El Karoui, L. Mazliak (Longman, Harlow, 1997), pp. 7–26

208. N. El Karoui, M. Jeanblanc-Picqué, S.E. Shreve, Robustness of the Black and Scholes formula. Math. Financ. **8**(2), 93–126 (1998)

209. N. El Karoui, L. Mazliak (eds.), *Backward Stochastic Differential Equations*, Pitman Research Notes in Mathematics Series, vol. 364 (Longman, Harlow, 1997)

210. N. El Karoui, D. Nguyen, M. Jeanblanc-Picqué, Compactification methods in the control of degenerate diffusions: existence of an optimal control. Stochastics **20**(3), 169–219 (1987)

211. N. El Karoui, S. Peng, M.C. Quenez, Backward stochastic differential equations in finance. Math. Financ. **7**(1), 1–71 (1997)

212. N. El Karoui, X. Tan, *Capacities, measurable selection and dynamic programming part ii: Application in stochastic control problems.* Preprint (2013), arXiv:1310.3364

213. R.J. Elliott, *Stochastic calculus and applications*, Applications of mathematics, vol. 18 (Springer, Berlin, 1982)
214. R.J. Elliott, Filtering and control for point process observations, in *Recent Advances in Stochastic Calculus, Progress in Automation and Information Systems*, eds. by J. Baras, V. Mirelli (Springer, Berlin, 1990), pp. 1–27
215. I. Elsanosi, B. Øksendal, A. Sulem, Some solvable stochastic control problems with delay. Stoch. Stoch. Rep. **71**(1–2), 69–89 (2000)
216. K.D. Elworthy, X.-M. Li, Formulae for the derivatives of heat semigroups. J. Funct. Anal. **125**(1), 252–286 (1994)
217. K.J. Engel, R. Nagel, *One-Parameter Semigroups for Linear Evolution Equations*, Graduate Texts in Mathematics, vol. 194 (Springer, Berlin, 2000)
218. A. Es-Sarhir, M. Scheutzow, J.M. Tölle, O. van Gaans, Invariant measures for monotone SPDEs with multiplicative noise term. Appl. Math. Optim. **68**(2), 275–287 (2013)
219. S.N. Ethier, T.G. Kurtz, *Markov Processes*. Characterization and Convergence, Wiley Series in Probability and Statistics (Wiley, New York, 1986)
220. L.C. Evans, *Partial Differential Equations*, Graduate Studies in Mathematics, vol. 19 (American Mathematical Society, Providence, 1998)
221. G. Fabbri, *First order HJB equations in Hilbert spaces and applications*, Ph.D. thesis, Universià di Roma - La Sapienza (2006)
222. G. Fabbri, A viscosity solution approach to the infinite-dimensional HJB equation related to a boundary control problem in a transport equation. SIAM J. Control Optim. **47**(2), 1022–1052 (2008)
223. G. Fabbri, Geographical structure and convergence: a note on geometry in spatial growth models. J. Econ. Theory **162**, 114–136 (2016)
224. G. Fabbri, International borrowing without commitment and informational lags: choice under uncertainty. J. Math. Econ. **68**, 103–114 (2017)
225. G. Fabbri, B. Goldys, An LQ problem for the heat equation on the halfline with Dirichlet boundary control and noise. SIAM J. Control Optim. **48**(6), 1473–1488 (2009)
226. G. Fabbri, F. Gozzi, Solving optimal growth models with vintage capital: the dynamic programming approach. J. Econ. Theory **143**(1), 331–373 (2008)
227. G. Fabbri, F. Gozzi, A. Święch, Verification theorem and construction of ε-optimal controls for control of abstract evolution equations. J. Convex Anal. **17**(2), 611–642 (2010)
228. G. Fabbri, F. Russo, Infinite dimensional weak Dirichlet processes and convolution type processes. Stoch. Process. Appl. **127**(1), 325–357 (2017)
229. S. Faggian, Boundary-control problems with convex cost and dynamic programming in infinite dimension. I. The maximum principle. Differ. Integral Equ. **17**(9–10), 1149–1174 (2004)
230. S. Faggian, Boundary control problems with convex cost and dynamic programming in infinite dimension. II. Existence for HJB. Discrete Contin. Dyn. Syst. **12**(2), 323–346 (2005)
231. S. Faggian, Regular solutions of first-order Hamilton-Jacobi equations for boundary control problems and applications to economics. Appl. Math. Optim. **51**(2), 123–162 (2005)
232. S. Faggian, Application of dynamic programming to economic problems with vintage capital. Dyn. Contin. Discrete Impuls. Syst., Ser. A, Math. Anal. **15**(4), 527–553 (2008)
233. S. Faggian, Hamilton-Jacobi equations arising from boundary control problems with state constraints. SIAM J. Control Optim. **47**(4), 2157–2178 (2008)
234. S. Faggian, F. Gozzi, Optimal investment models with vintage capital: dynamic programming approach. J. Math. Econ. **46**(4), 416–437 (2010)
235. S. Federico, *Stochastic optimal control problems for pension funds management*, Ph.D. thesis, Scuola Normale Superiore, Pisa (2009)
236. S. Federico, A stochastic control problem with delay arising in a pension fund model. Financ. Stoch. **15**(3), 421–459 (2011)
237. S. Federico, P. Gassiat, F. Gozzi, Impact of time illiquidity in a mixed market without full observation. Math. Financ. **27**(2), 401–437 (2017)
238. S. Federico, B. Goldys, F. Gozzi, HJB equations for the optimal control of differential equations with delays and state constraints, I: regularity of viscosity solutions. SIAM J. Control Optim. **48**(8), 4910–4937 (2010)

239. S. Federico, B. Goldys, F. Gozzi, HJB equations for the optimal control of differential equations with delays and state constraints, II: verification and optimal feedbacks. SIAM J. Control Optim. **49**(6), 2378–2414 (2011)

240. S. Federico, F. Gozzi, *Verification theorems for stochastic optimal control problems in Hilbert spaces by means of a generalized Dynkin formula*. Preprint arXiv:1702.05642, 2017

241. S. Federico, F. Gozzi, *Mild solutions of semilinear elliptic equations in Hilbert spaces*. J. Differ. Equations **262**(5), 3343–3389 (2017)

242. S. Federico, H. Pham, Characterization of the optimal boundaries in reversible investment problems. SIAM J. Control Optim. **52**(4), 2180–2223 (2014)

243. S. Federico, M. Rosestolato, *C_0-sequentially equicontinuous semigroups on locally convex spaces and application to Markov transition semigroups*. Preprint (2015), arXiv:1512.04589

244. S. Federico, E. Tacconi, Dynamic programming for optimal control problems with delays in the control variable. SIAM J. Control Optim. **52**(2), 1203–1236 (2014)

245. S. Federico, P. Tankov, Finite-dimensional representations for controlled diffusions with delay. Appl. Math. Optim. **71**(1), 165–194 (2015)

246. J. Feng, Martingale problems for large deviations of Markov processes. Stoch. Process. Appl. **81**(2), 165–216 (1999)

247. J. Feng, Large deviation for a stochastic Cahn-Hilliard equation. Methods Funct. Anal. Topol. **9**(4), 333–356 (2003)

248. J. Feng, Large deviation for diffusions and Hamilton-Jacobi equation in Hilbert spaces. Ann. Probab. **34**(1), 321–385 (2006)

249. J. Feng, M. Katsoulakis, A comparison principle for Hamilton-Jacobi equations related to controlled gradient flows in infinite dimensions. Arch. Rational Mech. Anal. **192**(2), 275–310 (2009)

250. J. Feng, T.G. Kurtz, *Large Deviations for Stochastic Processes*, Mathematical Surveys and Monographs, vol. 131 (American Mathematical Society, Providence, 2006)

251. J. Feng, T. Nguyen, Hamilton-Jacobi equations in space of measures associated with a system of conservation laws. J. Math. Pures Appl. **97**(4), 318–390 (2012)

252. J. Feng, A. Święch, Optimal control for a mixed flow of Hamiltonian and gradient type in space of probability measures (with Appendix B by Atanas Stefanov). Trans. Am. Math. Soc. **365**(8), 3987–4039 (2013)

253. F. Flandoli, Dissipativity and invariant measures for stochastic Navier-Stokes equations. NoDEA Nonlinear Differ. Equ. Appl. **1**(4), 403–423 (1994)

254. F. Flandoli, An introduction to 3D stochastic fluid dynamics, in *SPDE in Hydrodynamic: Recent Progress and Prospects*, Lecture Notes in Mathematics, vol. 1942, ed. by G. Da Prato, M. Röckner (Springer, Berlin, 2008), pp. 51–150

255. F. Flandoli, F. Gozzi, Kolmogorov equation associated to a stochastic Navier-Stokes equation. J. Funct. Anal. **160**(1), 312–336 (1998)

256. F. Flandoli, B. Maslowski, Ergodicity of the 2-D Navier-Stokes equation under random perturbations. Comm. Math. Phys. **172**(1), 119–141 (1995)

257. F. Flandoli, G. Zanco, An infinite-dimensional approach to path-dependent Kolmogorov's equations. Ann. Probab. **44**(4), 2643–269 (2016)

258. W.H. Fleming, Nonlinear semigroup for controlled partially observed diffusions. SIAM J. Control Optim. **20**(2), 286–301 (1982)

259. W.H. Fleming, M. Nisio, On the existence of optimal stochastic controls. Indiana Univ. Math. J. **15**, 777–794 (1966)

260. W.H. Fleming, M. Nisio, Differential games for stochastic partial differential equations. Nagoya Math. J. **131**, 75–107 (1993)

261. W.H. Fleming, É. Pardoux, Optimal control of partially observed diffusions. SIAM J. Control Optim. **20**(2), 261–285 (1982)

262. W.H. Fleming, R.W. Rishel, *Deterministic and Stochastic Optimal Control*, Applications of Mathematics, vol. 1 (Springer, Berlin, 1975)

263. W.H. Fleming, H.M. Soner, *Controlled Markov Processes and Viscosity Solutions*, 2nd edn., Stochastic Modelling and Applied Probability, vol. 25 (Springer, New York, 2006)

264. W.H. Fleming, P.E. Souganidis, On the existence of value-functions of 2-player, zero-sum stochastic differential-games. Indiana Univ. Math. J. **38**(2), 293–314 (1989)

265. W.H. Fleming, D. Vermes, Convex duality approach to the optimal control of diffusions. SIAM J. Control Optim. **27**(5), 1136–1155 (1989)

266. T.M. Flett, *Differential Analysis: Differentiation, Differential Equations, and Differential Inequalities* (Cambridge University Press, Cambridge, 1980)

267. G.B. Folland, *Real analysis*, 2nd edn. Pure and Applied Mathematics (Wiley, New York, 1999)

268. K.O. Friedrichs, The identity of weak and strong extensions of differential operators. Trans. Am. Math. Soc. **55**(1), 132–151 (1944)

269. K. Frieler, C. Knoche, *Solutions of stochastic differential equations in infinite dimensional Hilbert spaces and their dependence on initial data*, Diploma Thesis, Bielefeld University. BiBoS-Preprint E02-04-083 (2001)

270. M. Fuhrman, Analyticity of transition semigroups and closability of bilinear forms in Hilbert spaces. Studia Math. **115**(1), 53–71 (1995)

271. M. Fuhrman, Smoothing properties of nonlinear stochastic equations in Hilbert spaces. NoDEA Nonlinear Differ. Equ. Appl. **3**(4), 445–464 (1996)

272. M. Fuhrman, On a class of stochastic equations in Hilbert spaces: solvability and smoothing properties. Stoch. Anal. Appl. **17**(1), 43–69 (1999)

273. M. Fuhrman, A class of stochastic optimal control problems in Hilbert spaces: BSDEs and optimal control laws, state constraints, conditioned processes. Stoch. Process. Appl. **108**(2), 263–298 (2003)

274. M. Fuhrman, Y. Hu, Infinite horizon BSDEs in infinite dimensions with continuous driver and applications. J. Evol. Equ. **6**(3), 459–484 (2006)

275. M. Fuhrman, Y. Hu, Backward stochastic differential equations in infinite dimensions with continuous driver and applications. Appl. Math. Optim. **56**(2), 265–302 (2007)

276. M. Fuhrman, Y. Hu, G. Tessitore, On a class of stochastic optimal control problems related to BSDEs with quadratic growth. SIAM J. Control Optim. **45**(4), 1279–1296 (2006)

277. M. Fuhrman, Y. Hu, G. Tessitore, Ergodic BSDES and optimal ergodic control in Banach spaces. SIAM J. Control Optim. **48**(3), 1542–1566 (2009)

278. M. Fuhrman, Y. Hu, G. Tessitore, Ergodic BSDES and optimal ergodic control in Banach spaces. SIAM J. Control Optim. **48**(3), 1542–1566 (2009)

279. M. Fuhrman, Y. Hu, G. Tessitore, Stochastic maximum principle for optimal control of SPDEs. C. R. Acad. Sci. Paris Sér. I Math. **350**(13), 683–688 (2012)

280. M. Fuhrman, Y. Hu, G. Tessitore, Stochastic maximum principle for optimal control of SPDEs. Appl. Math. Optim. **68**(2), 181–217 (2013)

281. M. Fuhrman, F. Masiero, G. Tessitore, Stochastic equations with delay: optimal control via BSDEs and regular solutions of Hamilton-Jacobi-Bellman equations. SIAM J. Control Optim. **48**(7), 4624–4651 (2010)

282. M. Fuhrman, C. Orrieri, Stochastic maximum principle for optimal control of a class of nonlinear SPDEs with dissipative drift. SIAM J. Control Optim. **54**(1), 341–371 (2016)

283. M. Fuhrman, G. Tessitore, The Bismut-Elworthy formula for backward SDEs and applications to nonlinear Kolmogorov equations and control in infinite dimensional spaces. Stoch. Stoch. Rep. **74**(1–2), 429–464 (2002)

284. M. Fuhrman, G. Tessitore, Nonlinear Kolmogorov equations in infinite dimensional spaces: the backward stochastic differential equations approach and applications to optimal control. Ann. Probab. **30**(3), 1397–1465 (2002)

285. M. Fuhrman, G. Tessitore, Infinite horizon backward stochastic differential equations and elliptic equations in Hilbert spaces. Ann. Probab. **32**(1B), 607–660 (2004)

286. M. Fuhrman, G. Tessitore, Generalized directional gradients, backward stochastic differential equations and mild solutions of semilinear parabolic equations. Appl. Math. Optim. **51**(3), 279–332 (2005)

287. T. Funaki, Random motion of strings and related stochastic evolution equations. Nagoya Math. J. **89**, 129–193 (1983)

288. A.V. Fursikov, *Optimal control of distributed systems. Theory and applications*, Translations of Mathematical Monographs, vol. 187, American Mathematical Society, Providence, 2000, Translated from the Russian, originally published by Naucnaya Knyga, Novosibirsk, 1999

289. W. Gangbo, A. Święch, Existence of a solution to an equation arising from the theory of Mean Field Games. J. Differ. Equ. **259**(11), 6573–6643 (2015)

290. P. Gassiat, F. Gozzi, H. Pham, Investment/consumption problem in illiquid markets with regimes switching. SIAM J. Control Optim. **52**(3), 1761–1786 (2014)

291. P. Gassiat, F. Gozzi, H. Pham, Dynamic programming for an investment/consumption problem, in *illiquid markets with regime switching, in Stochastic Analysis, Banach Center Publications*, ed. by A. Chojnowska-Michalik, S. Peszat, L. Stettner, vol. 105 (Institute of Mathematics, Polish Academy of Sciences, Warsaw, 2015), pp. 103–118

292. D. Gątarek, B. Goldys, On invariant measures for diffusions on Banach spaces. Potential Anal. **7**(2), 533–553 (1997)

293. D. Gątarek, A. Święch, Optimal stopping in Hilbert spaces and pricing of American options. Math. Methods Oper. Res. **50**(1), 135–147 (1999)

294. L. Gawarecki, V. Mandrekar, *Stochastic differential equations in infinite dimensions with applications to stochastic partial differential equations,* Probability and its Applications (Springer, Heidelberg, 2011)

295. H. Geman, N. El Karoui, J.-C. Rochet, Changes of numéraire, changes of probability measure and option pricing. J. Appl. Probab. **32**(2), 443–458 (1995)

296. J.A. Goldstein, *Semigroups of Linear Operators and Applications,* Oxford Mathematical Monographs (Oxford University Press, New York, 1985)

297. B. Goldys, On analyticity of Ornstein–Uhlenbeck semigroups. Atti Accad. Naz. Lincei Rend. Cl. Sci. Fis. Mat. Natur. **10**(3), 131–140 (1999)

298. B. Goldys, F. Gozzi, Second order parabolic Hamilton-Jacobi-Bellman equations in Hilbert spaces and stochastic control: L^2_μ approach. Stoch. Process. Appl. **116**(12), 1932–1963 (2006)

299. B. Goldys, F. Gozzi, J.M.A.M. van Neerven, On closability of directional gradients. Potential Anal. **18**(4), 289–310 (2003)

300. B. Goldys, M. Kocan, Diffusion semigroups in spaces of continuous functions with mixed topology. J. Differ. Equ. **173**(1), 17–39 (2001)

301. B. Goldys, B. Maslowski, Ergodic control of semilinear stochastic equations and the Hamilton-Jacobi equation. J. Math. Anal. Appl. **234**(2), 592–631 (1999)

302. B. Goldys, M. Musiela, On partial differential equations related to term structure models Preprint, Univ. New South Wales (1996)

303. B. Goldys, M. Musiela, D. Sondermann, Lognormality of rates and term structure models. Stoch. Anal. Appl. **18**(3), 375–396 (2000)

304. F. Gozzi, Some results for an infinite horizon control problem governed by a semilinear state equation, in *Control and Estimation of Distributed Parameter Systems, International Series of Numerical Mathematics, vol. 91*, ed. by F. Kappel, K. Kunisch (Birkhäuser, Basel, 1989), pp. 145–163

305. F. Gozzi, Some results for an optimal control problem with semilinear state equation. SIAM J. Control Optim. **29**(4), 751–768 (1991)

306. F. Gozzi, Regularity of solutions of a second order Hamilton-Jacobi equation and application to a control problem. Comm. Partial Differ. Equ. **20**(5–6), 775–826 (1995)

307. F. Gozzi, Global regular solutions of second order Hamilton-Jacobi equations in Hilbert spaces with locally Lipschitz nonlinearities. J. Math. Anal. Appl. **198**(2), 399–443 (1996)

308. F. Gozzi, Strong solutions for Kolmogorov equation, in *Hilbert spaces, in Partial Differential Equation Methods in Control and Shape Analysis*, Lecture Notes in Pure and Applied Mathematics, vol. 188, ed. by G. Da Prato, J.-P. Zolesio (Dekker, New York, 1997), pp. 163–187

309. F. Gozzi, *Second order Hamilton–Jacobi equations in Hilbert spaces and stochastic optimal control*, Ph.D. thesis, Scuola Normale Superiore, Pisa (1998)

310. F. Gozzi, Second order Hamilton-Jacobi equations, in *Hilbert spaces and stochastic optimal control, in Stochastic Partial Differential Equations and Applications*, Lecture Notes in Pure and Applied Mathematics, vol. 227, ed. by G. Da Prato, L. Tubaro (Dekker, New York, 2002), pp. 255–285

311. F. Gozzi, Smoothing properties of nonlinear transition semigroups: case of Lipschitz nonlinearities. J. Evol. Equ. **6**(4), 711–743 (2006)
312. F. Gozzi, P. Loreti, Regularity of the minimum time function and minimum energy problems: the linear case. SIAM J. Control Optim. **37**(4), 1195–1221 (1999)
313. F. Gozzi, C. Marinelli, Stochastic optimal control of delay equations arising, in *advertising models, in Stochastic Partial Differential Equations and Applications VII*, Lecture Notes in Pure and Applied Mathematics, vol. 245, ed. by G. Da Prato, L. Tubaro (Chapman & Hall, Raton, 2006), pp. 133–148
314. F. Gozzi, C. Marinelli, S. Savin, On controlled linear diffusions with delay in a model of optimal advertising under uncertainty with memory effects. J. Optim. Theory Appl. **142**(2), 291–321 (2009)
315. F. Gozzi, F. Masiero, *Stochastic boundary control problems: improving the dynamic programming approach*. In preparation
316. F. Gozzi, F. Masiero, *Stochastic optimal control with delay in the control I: solving the HJB equation through partial smoothing*. Preprint (2015), arXiv:1607.06502
317. F. Gozzi, E. Rouy, Regular solutions of second-order stationary Hamilton-Jacobi equations. J. Differ. Equ. **130**(1), 201–234 (1996)
318. F. Gozzi, E. Rouy, A. Święch, Second order Hamilton-Jacobi equations in Hilbert spaces and stochastic boundary control. SIAM J. Control Optim. **38**(2), 400–430 (2000)
319. F. Gozzi, F. Russo, Verification theorems for stochastic optimal control problems via a time dependent Fukushima-Dirichlet decomposition. Stoch. Process. Appl. **116**(11), 1530–1562 (2006)
320. F. Gozzi, F. Russo, Weak Dirichlet processes with a stochastic control perspective. Stoch. Process. Appl. **116**(11), 1563–1583 (2006)
321. F. Gozzi, S.S. Sritharan, A. Święch, Viscosity solutions of dynamic-programming equations for the optimal control of the two-dimensional Navier-Stokes equations. Arch. Rational Mech. Anal. **163**(4), 295–327 (2002)
322. F. Gozzi, S.S. Sritharan, A. Święch, Bellman equations associated to the optimal feedback control of stochastic Navier-Stokes equations. Comm. Pure Appl. Math. **58**(5), 671–700 (2005)
323. F. Gozzi, A. Święch, Hamilton-Jacobi-Bellman equations for the optimal control of the Duncan-Mortensen-Zakai equation. J. Funct. Anal. **172**(2), 466–510 (2000)
324. F. Gozzi, A. Święch, X.Y. Zhou, A corrected proof of the stochastic verification theorem within the framework of viscosity solutions. SIAM J. Control Optim. **43**(6), 2009–2019 (2005)
325. F. Gozzi, A. Święch, X.Y. Zhou, Erratum: "A corrected proof of the stochastic verification theorem within the framework of viscosity solutions". SIAM J. Control Optim. **48**(6), 4177–4179 (2010)
326. F. Gozzi, T. Vargiolu, On the superreplication approach for European interest rates derivatives, in *Stochastic Analysis, Random Fields and Applications III, Progress in Probability, vol. 52*, ed. by C. Dalang, M. Dozzi, F. Russo (Birkhauser, Boston, 2002), pp. 173–188
327. F. Gozzi, T. Vargiolu, Superreplication of European multiasset derivatives with bounded stochastic volatility. Math. Methods Oper. Res. **55**(1), 69–91 (2002)
328. A. Grorud, É. Pardoux, Intégrales Hilbertiennes anticipantes par rapport à un processus de Wiener cylindrique et calcul stochastique associé. Appl. Math. Optim. **25**(1), 31–49 (1992)
329. L. Gross, Potential theory on Hilbert space. J. Funct. Anal. **1**(2), 123–181 (1967)
330. G. Guatteri, On a class of forward-backward stochastic differential systems in infinite dimensions. J. Appl. Math. Stoch. Anal. **42640**, 1–33 (2007)
331. G. Guatteri, Stochastic maximum principle for SPDEs with noise and control on the boundary. Syst. Control Lett. **60**(3), 198–204 (2011)
332. G. Guatteri, F. Masiero, On the existence of optimal controls for SPDEs with boundary noise and boundary control. SIAM J. Control Optim. **51**(3), 1909–1939 (2013)
333. G. Guatteri, G. Tessitore, On the backward stochastic Riccati equation in infinite dimensions. SIAM J. Control Optim. **44**(1), 159–194 (2005)
334. G. Guatteri, G. Tessitore, Backward stochastic Riccati equations and infinite horizon L-Q optimal control with infinite dimensional state space and random coefficients. Appl. Math. Optim. **57**(2), 207–235 (2008)

335. G. Guatteri, G. Tessitore, Well posedness of operator valued backward stochastic Riccati equations in infinite dimensional spaces. SIAM J. Control Optim. **52**(6), 3776–3806 (2014)

336. M. Hairer, J.C. Mattingly, Ergodicity of the 2D Navier-Stokes equations with degenerate stochastic forcing. Ann. of Math. **164**(3), 993–1032 (2006)

337. M. Hairer, J.C. Mattingly, Spectral gaps in Wasserstein distances and the 2D stochastic Navier-Stokes equations. Ann. Probab. **36**(6), 2050–2091 (2008)

338. M. Hairer, J.C. Mattingly, M. Scheutzow, Asymptotic coupling and a general form of Harris' theorem with applications to stochastic delay equations. Probab. Theory Relat. Fields **149**(1–2), 223–259 (2011)

339. U.G. Haussmann, J.-P. Lepeltier, On the existence of optimal controls. SIAM J. Control Optim. **28**(4), 851–902 (1990)

340. T. Havârneanu, Existence for the dynamic programming equation of control diffusion processes in Hilbert space. Nonlinear Anal. **9**(6), 619–629 (1985)

341. D. Henry, *Geometric Theory of Semilinear Parabolic Equations*, Lecture Notes in Mathematics, vol. 840 (Springer, Berlin, 1981)

342. O. Hijab, Partially observed control of Markov processes I. Stochastics **28**(2), 123–144 (1989)

343. O. Hijab, Partially observed control of Markov processes II. Stochastics **28**(3), 247–262 (1989)

344. O. Hijab, Partially observed control of Markov processes III. Ann. Probab. **18**(3), 1099–1125 (1990)

345. O. Hijab, Infinite-dimensional Hamilton-Jacobi equations with large zeroth-order coefficient. J. Funct. Anal. **97**(2), 311–326 (1991)

346. O. Hijab, Partially observed control of Markov processes. IV. J. Funct. Anal. **109**(2), 215–256 (1992)

347. M. Hinze, S. Volkwein, Analysis of instantaneous control for the Burgers equation. Nonlinear Anal. **50**(1), 1–26 (2002)

348. Y. Hu, J. Ma, J. Yong, On semi-linear degenerate backward stochastic partial differential equations. Probab. Theory Relat. Fields **123**(3), 381–411 (2002)

349. Y. Hu, S. Peng, Maximum principle for semilinear stochastic evolution control systems. Stoch. Stoch. Rep. **33**(3–4), 159–180 (1990)

350. Y. Hu, S. Peng, Adapted solution of a backward semilinear stochastic evolution equation. Stoch. Anal. Appl. **9**(4), 445–459 (1991)

351. Y. Hu, G. Tessitore, BSDE on an infinite horizon and elliptic PDEs in infinite dimension. NoDEA Nonlinear Differ. Equ. Appl. **14**(5–6), 825–846 (2007)

352. J. Huang, J. Shi, Maximum principle for optimal control of fully coupled forward-backward stochastic differential delayed equations. ESAIM Control Optim. Calc. Var. **18**(4), 1073–1096 (2012)

353. M. Iannelli, *Mathematical theory of age-structured population dynamics*, Applied Mathematical Monographs - C.N.R., vol. 7, Giardini, Pisa (1995)

354. A. Ichikawa, Absolute stability of a stochastic evolution equationt. Stochastics **11**(1–2), 143–158 (1983)

355. A. Ichikawa, Semilinear stochastic evolution equations: boundedness, stability and invariant measurest. Stochastics **12**(1), 1–39 (1984)

356. N. Ikeda, S. Watanabe, *Stochastic Differential Equations and Diffusion Processes*, 2nd edn., North-Holland Mathematical Library, vol. 24 (North-Holland, Amsterdam, 1989)

357. I. Iscoe, M.B. Marcus, D. McDonald, M. Talagrand, J. Zinn, Continuity of L^2-valued Ornstein-Uhlenbeck processes. Ann. Probab. **18**(1), 68–84 (1990)

358. H. Ishii, Perron's method for Hamilton-Jacobi equations. Duke Math. J. **55**(2), 369–384 (1987)

359. H. Ishii, On uniqueness and existence of viscosity solutions of fully nonlinear second-order elliptic PDEs. Comm. Pure Appl. Math. **42**(1), 15–45 (1989)

360. H. Ishii, Viscosity solutions for a class of Hamilton-Jacobi equations in Hilbert spaces. J. Funct. Anal. **105**(2), 301–341 (1992)

361. H. Ishii, Viscosity solutions of nonlinear second-order partial differential equations in Hilbert spaces. Comm. Partial Differ. Equ. **18**(3–4), 601–650 (1993)

362. H. Ishii, P.-L. Lions, Viscosity solutions of fully nonlinear second-order elliptic partial differential equations. J. Differ. Equ. **83**(1), 26–78 (1990)

363. K. Itô, M. Nisio, On stationary solutions of a stochastic differential equation. J. Math. Kyoto Univ. **4**(1), 1–75 (1964)

364. A.F. Ivanov, Y.I. Kazmerchuk, A.V. Swishchuk, Theory, stochastic stability and applications of stochastic delay differential equations: a survey of results. Differ. Equ. Dynam. Syst. **11**(1–2), 55–115 (2003)

365. M.R. James, J.S. Baras, R.J. Elliott, Output feedback risk-sensitive control and differential games for continuous-time nonlinear systems, in *Proceedings of the 32nd IEEE conference on decision and control* (IEEE, 1993), pp. 3357–3360

366. B. Jefferies, Weakly integrable semigroups on locally convex spaces. J. Funct. Anal. **66**(3), 347–364 (1986)

367. B. Jefferies, The generation of weakly integrable semigroups. J. Funct. Anal. **73**(1), 195–215 (1987)

368. D.-T. Jeng, Forced model equation for turbulence. Phys. Fluids **12**(10), 2006–2010 (1969)

369. R. Jensen, The maximum principle for viscosity solutions of fully nonlinear second order partial differential equations. Arch. Rational Mech. Anal. **101**(1), 1–27 (1988)

370. O. Kallenberg, *Foundations of Modern Probability*, 2nd edn. Probability and its Applications (Springer, New York, 2002)

371. G. Kallianpur, J. Xiong, *Stochastic Differential Equations in Infinite-Dimensional Spaces*, vol. 26 (Hayward, Institute of Mathematical Statistics Lecture Notes (Institute of Mathematical Statistics, 1995)

372. I. Karatzas, S.E. Shreve, *Brownian Motion and Stochastic Calculus*, Graduate Texts in Mathematics, vol. 113 (Springer, New York, 1988)

373. M. Kardar, G. Parisi, Y.C. Zhang, Dynamic scaling of growing interfaces. Phys. Rev. Lett. **56**(9), 889–892 (1986)

374. D. Kelome, *Viscosity solution of second order equations in a separable Hilbert space and applications to stochastic optimal control*, Ph.D. thesis, Georgia Institute of Technology (2002)

375. D. Kelome, A. Święch, Viscosity solutions of an infinite-dimensional Black-Scholes-Barenblatt equation. Appl. Math. Optim. **47**(3), 253–278 (2003)

376. D. Kelome, A. Święch, Perron's method and the method of relaxed limits for "unbounded" PDE in Hilbert spaces. Studia Math. **176**(3), 249–277 (2006)

377. M. Kobylanski, Backward stochastic differential equations and partial differential equations with quadratic growth. Ann. Probab. **28**(2), 558–602 (2000)

378. M. Kocan, *Some aspects of the theory of viscosity solutions of fully nonlinear partial differential equations in infinite dimensions*, Ph.D. thesis, University of California, Santa Barbara (1994)

379. M. Kocan, P. Soravia, A viscosity approach to infinite-dimensional Hamilton-Jacobi equations arising in optimal control with state constraints. SIAM J. Control Optim. **36**(4), 1348–1375 (1998)

380. M. Kocan, A. Święch, Second order unbounded parabolic equations in separated form. Studia Math. **115**(3), 291–310 (1995)

381. M. Kocan, A. Święch, Perturbed optimization on product spaces. Nonlinear Anal. **26**(1), 81–90 (1996)

382. S.M. Kozlov, Equivalence of measures for linear stochastic Ito equations with partial derivatives. Moscow Univ. Sov. Math. Mech **4**, 47–52 (1977)

383. S.M. Kozlov, Some questions of stochastic equations with partial derivatives. Trudy Sem. Petrovsk **4**, 147–172 (1978)

384. N.V. Krylov, *Controlled Diffusion Processes*, Applications of Mathematics, vol. 14 (Springer, New York, 1980). Translated from the Russian, originally published by Nauka, Moscow 1977

385. N.V. Krylov, B.L. Rozovskiĭ, On the Cauchy problem for linear stochastic partial differential equations. Math. USSR, Izv. **11**(6), 1267–1284 (1977)

386. Sovremennye Problemy Matematiki, N.V. Krylov, B.L. Rozovskiĭ, Stochastic evolution equations. J. Sov. Math. 16(4), 1233–1277 (1981). Original paper in Russian. Noveishie Dostizheniya 14, 71–146 (1979)

387. N.V. Krylov, B.L. Rozovskiĭ, Stochastic partial differential equations and diffusion processes. Russ. Math. Surv. 37(6), 81–105 (1982). Original paper in Russian. Uspekhi Mat. Nauk 37–228(6), 75–95 (1982)

388. N.V. Krylov, B.L. Rozovskiĭ, Stochastic evolution equations, in *Stochastic Differential Equations: Theory and Applications*, Interdisciplinary Mathematical Sciences, vol. 2, ed. by P.H. Baxendale, S.V. Lototsky (World Scientific Publishing, Hackensack, 2007), pp. 1–69

389. S. Kuksin, A. Piatniski, A. Shirikyan, A coupling approach to randomly forced nonlinear PDE's. II. Commun. Math. Phys. 230(1), 81–85 (2002)

390. S. Kuksin, A. Shirikyan, *Mathematics of Two-Dimensional Turbulence*, Cambridge Tracts in Mathematics, vol. 194 (Cambridge University Press, Cambridge, 2012)

391. H.H. Kuo, *Gaussian Measures in Hilbert Dpaces*, Lecture Notes in Mathematics, vol. 463 (Springer, Berlin, 1975)

392. T.G. Kurtz, Martingale problems for controlled processes, in *Stochastic Modelling and Filtering*, Lecture Notes in Control and Information Sciences, vol. 91, ed. by A. Germani (Springer, Berlin, 1987), pp. 75–90

393. J. Kurzweil, On approximation in real Banach spaces. Studia Math. 14(2), 214–231 (1954)

394. H.J. Kushner, On the dynamical equations of conditional probability density functions, with applications to optimal stochastic control theory. J. Math. Anal. Appl. 8(2), 332–344 (1964)

395. H.J. Kushner, *Stochastic Stability and Control* (Mathematics in Science and Engineering, vol. 33 (North-Holland, Amsterdam, 1967)

396. O.A. Ladyzhenskaya, On integral inequalities, the convergence of approximate methods, and the solution of linear elliptic operators. Vest. Leningrad State Univ. 7, 60–69 (1958)

397. S. Lang, *Real and Functional Analysis*, Graduate Texts in Mathematics, vol. 142 (Springer, New York, 1993)

398. B. Larssen, N.H. Risebro, When are HJB-equations in stochastic control of delay systems finite dimensional? Stoch. Anal. Appl. 21(3), 643–671 (2003)

399. I. Lasiecka, Unified theory for abstract parabolic boundary problems - a semigroup approach. Appl. Math. Optim. 6(1), 287–333 (1980)

400. I. Lasiecka, R. Triggiani, *Differential and Algebraic Riccati Equations with Application to Boundary/Point Control Problems: Continuous Theory and Approximation Theory*, Lecture Notes in Control and Information Sciences, vol. 164 (Springer, Berlin, 1991)

401. J.-M. Lasry, P.-L. Lions, A remark on regularization in Hilbert spaces. Israel J. Math. 55(3), 257–266 (1986)

402. P. Lescot, M. Röckner, Perturbations of generalized Mehler semigroups and applications to stochastic heat equations with Lévy noise and singular drift. Potential Anal. 20(4), 317–344 (2004)

403. X.J. Li, J.M. Yong, *Optimal Control Theory for Infinite-Dimensional Systems,* Systems and Control: Foundations and Applications (Birkhäuser, Boston, 1995)

404. J.-L. Lions, E. Magenes, *Non-homogeneous boundary value problems and applications I* Die Grundlehren der mathematischen Wissenschaften, vol. 181 (Springer, Berlin, 1972). Translated from the French, originally published by Dunod, Paris, 1968

405. P.-L. Lions, *Mean Field Games*, Course at the Collège de France. Available in video: http://www.college-de-france.fr/site/pierre-louis-lions/_audiovideos.htm

406. P.-L. Lions, Une inégalité pour les opérateurs elliptiques du second ordre. Ann. Mat. Pura Appl. 127(1), 1–11 (1981)

407. P.-L. Lions, *Generalized Solutions of Hamilton-Jacobi Equations*, Pitman Research Notes in Mathematics Series, vol. 69 (Longman, Boston, 1982)

408. P.-L. Lions, Optimal control of diffusion processes and Hamilton-Jacobi-Bellman equations. I. The dynamic programming principle and applications. Comm. Partial. Differ. Equ. 8(10), 1101–1174 (1983)

409. P.-L. Lions, Optimal control of diffusion processes and Hamilton-Jacobi-Bellman equations. II. Viscosity solutions and uniqueness. Comm. Partial. Differ. Equ. **8**(11), 1229–1276 (1983)
410. P.-L. Lions, Viscosity solutions of fully nonlinear second-order equations and optimal stochastic control in infinite dimensions. I. The case of bounded stochastic evolutions. Acta Math. **161**(3–4), 243–278 (1988)
411. P.-L. Lions, Viscosity solutions of fully nonlinear second order equations and optimal stochastic control, in infinite dimensions. II. Optimal control of Zakai's equation, in *Stochastic Partial Differential Equations and Applications II*, ed. by G. Da Prato, L. Tubaro. Lecture Notes in Mathematics, vol. 1390 (Springer, Berlin, 1989), pp. 147–170
412. P.-L. Lions, Viscosity solutions of fully nonlinear second-order equations and optimal stochastic control in infinite dimensions. III. Uniqueness of viscosity solutions for general second-order equations. J. Funct. Anal. **86**(1), 1–18 (1989)
413. W. Liu, M. Röckner, *Stochastic Partial Differential Equations: An Introduction*, Universitext (Springer, Cham, 2015)
414. Q. Lü, X. Zhang, *General Pontryagin-Type Stochastic Maximum Principle and Backward Stochastic Evolution Equations in Infinite Dimensions*, Springer Briefs in Mathematics (Springer, Berlin, 2014)
415. Q. Lü, X. Zhang, Transposition method for backward stochastic evolution equations revisited, and its application. Math. Control Relat. Fields **5**(3), 529–555 (2015)
416. A. Lunardi, *Analytic semigroups and optimal regularity in parabolic problems*, Progress in Nonlinear Differential Equations and their Applications, vol. 16 (Birkhäuser, Basel, 1995)
417. A. Lunardi, On the Ornstein-Uhlenbeck operator in L^2 spaces with respect to invariant measures. Trans. Amer. Math. Soc. **349**(1), 155–169 (1997)
418. T.J. Lyons, Uncertain volatility and the risk-free synthesis of derivatives. Appl. Math. Financ. **2**(2), 117–133 (1995)
419. J. Ma, J. Yong, Adapted solution of a degenerate backward SPDE, with applications. Stoch. Process. Appl. **70**(1), 59–84 (1997)
420. J. Ma, J. Yong, *Forward-Backward Stochastic Differential Equations and Their Applications*, Lecture Notes In Mathematics, vol. 1702 (Springer, Berlin, 1999)
421. J. Ma, J. Yong, On linear, degenerate backward stochastic partial differential equations. Probab. Theory Relat. Fields **113**(2), 135–170 (1999)
422. Z.M. Ma, M. Röckner, *Introduction to the Theory of (Nonsymmetric) Dirichlet Forms*, Universitext (Springer, Berlin, 1992)
423. L. Manca, Kolmogorov equations for measures. J. Evol. Equ. **8**(2), 231–262 (2008)
424. L. Manca, On the dynamic programming approach for the 3D Navier-Stokes equations. Appl. Math. Optim. **57**(3), 329–348 (2008)
425. L. Manca, The Kolmogorov operator associated to a Burgers SPDE in spaces of continuous functions. Potential Anal. **32**(1), 67–99 (2010)
426. R. Marcus, Parabolic Itô equations. Trans. Amer. Math. Soc. **198**, pp. 177–190 (1974)
427. R. Marcus, Parabolic Itô equations with monotone nonlinearities. J. Funct. Anal. **29**(3), 275–286 (1978)
428. C. Marinelli, *On stochastic modelling and optimal control in advertising*, Ph.D. thesis, Graduate School of Business, Columbia University (2004)
429. C. Marinelli, S. Savin, Optimal distributed dynamic advertising. J. Optim. Theory Appl. **137**(3), 569–591 (2008)
430. Carlo Marinelli, Well-posedness and invariant measures for HJM models with deterministic volatility and Lévy noise. Quant. Financ. **10**(1), 39–47 (2010)
431. F. Masiero, *Semilinear Kolmogorov equations and applications to stochastic optimal control*, Ph.D. thesis, Dipartimento di Matematica, Universitá di Milano (2003)
432. F. Masiero, Semilinear Kolmogorov equations and applications to stochastic optimal control. Appl. Math. Optim. **51**(2), 201–250 (2005)
433. F. Masiero, Infinite horizon stochastic optimal control problems with degenerate noise and elliptic equations in Hilbert spaces. Appl. Math. Optim. **55**(3), 285–326 (2007)

434. F. Masiero, Regularizing properties for transition semigroups and semilinear parabolic equations in Banach spaces. Electron. J. Probab. **12**(13), 387–419 (2007)
435. F. Masiero, Stochastic optimal control for the stochastic heat equation with exponentially growing coefficients and with control and noise on a subdomain. Stoch. Anal. Appl. **26**(4), 877–902 (2008)
436. F. Masiero, Stochastic optimal control problems and parabolic equations in Banach spaces. SIAM J. Control Optim. **47**(1), 251–300 (2008)
437. F. Masiero, A stochastic optimal control problem for the heat equation on the halfline with Dirichlet boundary-noise and boundary-control. Appl. Math. Optim. **62**(2), 253–294 (2010)
438. F. Masiero, Hamilton Jacobi Bellman equations in infinite dimensions with quadratic and superquadratic Hamiltonian. Discret. Contin. Dyn. S. **32**(1), 223–263 (2012)
439. F. Masiero, A Bismut Elworthy formula for quadratic BSDEs. Stoch. Process. Appl. **125**(5), 1945–1979 (2015)
440. F. Masiero, HJB equations in infinite dimensions under weak regularizing properties. J. Evol. Equ. **16**(4), 789–824 (2016)
441. F. Masiero, A. Richou, A note on the existence of solutions to Markovian superquadratic BSDEs with an unbounded terminal condition. J. Differ. Equ. **18**(50), 1–15 (2013)
442. F. Masiero, A. Richou, HJB equations in infinite dimension with locally Lipschitz Hamiltonian and unbounded terminal condition. J. Differ. Equ. **257**(6), 1989–2034 (2014)
443. B. Maslowski, J. Seidler, Invariant measures for nonlinear spde's: Uniqueness and stability. Arch. Math. **34**, 153–172 (1998)
444. J.-L. Menaldi, S.S. Sritharan, Stochastic 2D Navier-Stokes equation. Appl. Math. Optim. **46**(1), 31–53 (2002)
445. Q. Meng, P. Shi, Stochastic optimal control for backward stochastic partial differential systems. J. Math. Anal. Appl. **402**(2), 758–771 (2013)
446. G. Metafune, D. Pallara, E. Priola, Spectrum of Ornstein-Uhlenbeck operators in L^p spaces with respect to invariant measures. J. Funct. Anal. **196**(1), 40–60 (2002)
447. M. Métivier, *Semimartingales: A Course on Stochastic Processes*, De Gruyter Studies in Mathematics, vol. 2 (Walter de Gruyter, Berlin, 1982)
448. M. Métivier, J. Pellaumail, *Stochastic Integration, Probability and Mathematical Statistics* (Academic Press, New York, 1980)
449. P.A. Meyer, *Probability and Potentials*, Blaisdell Books in Pure and Applied Mathematics (Blaisdell, New York, 1966)
450. S.E.A. Mohammed, *Stochastic Functional Differential Equations*, Pitman Research Notes in Mathematics Series, vol. 99 (Longman, Boston, 1984)
451. S.E.A. Mohammed, Stochastic differential systems with memory: theory, examples and applications, in *Stochastic Analysis and Related Topics VI*, ed. by L. Decreusefond, J. Gjerde, B. Øksendal, A.S. Üstünel. Progress in Probability, vol. 42 (Birkhäuser, Boston, 1998), pp. 1–77
452. H. Morimoto, *Stochastic Control and Mathematical Modelling, Applications in Economics*, Encyclopedia of Mathematics and its Applications , vol. 131 (Cambridge University Press, Cambridge, 2010)
453. R.E. Mortensen, *Optimal control of continuous-time stochastic systems*, Technical Report Elektronics Research Laboratory Report 66–1 (Univ. of Calif, Berkeley, 1966)
454. G.A. Muñoz, Y. Sarantopoulos, A. Tonge, Complexifications of real Banach spaces, polynomials and multilinear maps. Studia Math. **134**(1), 1–33 (1999)
455. M. Musiela, Stochastic PDEs and term structure models, in *Proceedings of Journées Internationales de Finance, La Baule* ed. By F. Jamshidian. Association Française de Finance, Paris (1993)
456. M. Musiela, M. Rutkowski, *Martingale Methods in Financial Modelling*, 2nd edn., Stochastic Modelling and Applied Probability, vol. 36 (Springer, Berlin, 2005)
457. A.S. Nemirovskiĭ, S.M. Semenov, The polynomial approximation of functions on Hilbert space. Mat. Sb. (N.S.) **92(134)**(2(10)), 257–281 (1973)

458. J. Neveu, *Discrete-Parameter Martingales*, Holland Mathematical Library, vol. 10 (North-Holland, Amsterdam, 1975)

459. M. Nisio, Some remarks on stochastic optimal controls, in *Proceedings of the Third Japan-USSR Symposium on Probability Theory*, ed. by G. Maruyama, J.V. Prokhorov. Lecture Notes in Mathematics, vol. 550 (Springer, Berlin, 1976), pp. 446–460

460. M. Nisio, *Lectures on Stochastic Control Theory*, ISI Lecture Notes, vol. 9 (Macmillan, Delhi, 1981)

461. M. Nisio, Optimal control for stochastic partial differential equations and viscosity solutions of Bellman equations. Nagoya Math. J. **123**, 13–37 (1991)

462. M. Nisio, On sensitive control for stochastic partial differential equations, in *Stochastic Analysis on Infinite-Dimensional Spaces*, ed. by H. Kunita, H.-H. Kuo. Pitman Research Notes in Mathematics Series, vol. 310, (Longman, Harlow, 1994), pp. 231–241

463. M. Nisio, On sensitive control and differential games, in infinite-dimensional spaces, in *Itô's Stochastic Calculus and Probability Theory*, ed. by N. Ikeda, S. Watanabe, H.M. Kunita (Tokyo, Fukushima (Springer, 1996), pp. 281–292

464. M. Nisio, On infinite-dimensional stochastic differential games. Osaka J. Math. **35**(1), 15–33 (1998)

465. M. Nisio, Game approach to risk sensitive control for stochastic evolution systems, in *Stochastic Analysis, Control, Optimization and Applications*, ed. by W.M. McEneaney, G.G. Yin, Q. Zhang. Systems and Control: Foundations and Applications (Birkhäuser, Boston, 1999), pp. 115–134

466. M. Nisio, On value function of stochastic differential games in infinite dimensions and its application to sensitive control. Osaka J. Math. **36**(2), 465–483 (1999)

467. M. Nisio, *Stochastic Control Theory: Dynamic Programming Principle*, Probability Theory and Stochastic Modelling, vol. 72 (Springer, Berlin, 2015)

468. D. Nualart, *The Malliavin Calculus and Related Topics*, Probability and its Applications (Springer, New York, 1995)

469. D. Nualart, É. Pardoux, Stochastic calculus with anticipating integrands. Probab. Theory Related Fields **78**(4), 535–581 (1988)

470. B. Øksendal, A. Sulem, T. Zhang, Optimal control of stochastic delay equations and time-advanced backward stochastic differential equations. Adv. Appl. Probab. **43**(2), 572–596 (2011)

471. M. Ondreját, Uniqueness for stochastic evolution equations in Banach spaces. Dissertationes Math. (Rozprawy Mat.) **426**, 1–63 (2004)

472. É. Pardoux, Stochastic partial differential equations and filtering of diffusion processes. Stochastics **3**(2), 127–167 (1979)

473. É. Pardoux, Equations of non-linear filtering and application to stochastic control with partial observation, in *Nonlinear Filtering and Stochastic Control*, ed. by S.K. Mitter, A. Moro. Lecture Notes in Mathematics, vol. 972 (Springer, Berlin, 1982), pp. 208–248

474. É. Pardoux, Backward stochastic differential equations and viscosity solutions of systems of semilinear parabolic and elliptic PDEs of second order, in *Stochastic Analysis and Related Topics VI*, ed. by L. Decreusefond, J. Gjerde, B. Øksendal, A.S. Üstünel. Progress in Probability, vol. 42 (Birkhäuser, Boston, 1998), pp. 79–127

475. É. Pardoux, S.G. Peng, Adapted solution of a backward stochastic differential equation. Syst. Control Lett. **14**(1), 55–61 (1990)

476. É. Pardoux, A. Răşcanu, Backward stochastic differential equations with subdifferential operator and related variational inequalities. Stoch. Process. Appl. **76**(2), 191–215 (1998)

477. É. Pardoux, A. Răşcanu, *Stochastic Differential Equations, Backward SDEs, Partial Differential Equations*, Stochastic Modelling and Applied Probability, vol. 69 (Springer, Berlin, 2014)

478. K.R. Parthasarathy, *Probability Measures on Metric Spaces*, Probability and Mathematical Statistics, vol. 3 (Academic Press, New York, 1967)

479. A. Pazy, *Semigroups of Linear Operators and Applications to Partial Differential Equations*, Applied Mathematical Sciences, vol. 44 (Springer, New York, 1983)

480. S. Peng, A general stochastic maximum principle for optimal control problems. SIAM J. Control Optim. **28**(4), 966–979 (1990)
481. S. Peng, Stochastic Hamilton-Jacobi-Bellman equations. SIAM J. Control Optim. **30**(2), 284–304 (1992)
482. S. Peng, Backward stochastic differential equations and applications to optimal control. Appl. Math. Optim. **27**(2), 125–144 (1993)
483. S. Peszat, Law equivalence of solutions of some linear stochastic equations in Hilbert spaces. Studia Math. **101**(3), 269–284 (1992)
484. S. Peszat, On a Sobolev space of functions of infinite number of variables. Bull. Pol. Acad. Sci., Math. **41**(1), 55–60 (1993)
485. S. Peszat, Lévy-Ornstein-Uhlenbeck transition semigroup as second quantized operator. J. Funct. Anal. **260**(12), 3457–3473 (2011)
486. S. Peszat, J. Zabczyk, Strong Feller property and irreducibility for diffusions on Hilbert spaces. Ann. Probab. **23**(1), 157–172 (1995)
487. S. Peszat, J. Zabczyk, *Stochastic Partial Differential Equations with Lévy Noise*, Encyclopedia of Mathematics and its Applications, vol. 113 (Cambridge University Press, Cambridge, 2007)
488. B.J. Pettis, On integration in vector spaces. Trans. Amer. Math. Soc. **44**(2), 277–304 (1938)
489. H. Pham, *Continuous-time stochastic control and optimization with financial applications*, Stochastic Modelling and Applied Probability, vol. 61 (Springer, Berlin, 2009)
490. R.S. Phillips, Dissipative operators and hyperbolic systems of partial differential equations. Trans. Amer. Math. Soc. **90**(1), 193–254 (1959)
491. C. Prévôt, M. Röckner, *A Concise Course on Stochastic Partial Differential Equations*, Lecture Notes In Mathematics, vol. 1905 (Springer, Berlin, 2007)
492. E. Priola, On a class of Markov type semigroups in spaces of uniformly continuous and bounded functions. Studia Math. **136**(3), 271–295 (1999)
493. E. Priola, *Partial differential equations with infinitely many variables*, Ph.D. thesis, Universita degli Studi di Milano (1999)
494. E. Priola, Uniform approximation of uniformly continuous and bounded functions on Banach spaces. Dynam. Syst. Appl. **9**(2), 181–197 (2000)
495. E. Priola, A counterexample to Schauder estimates for elliptic operators with unbounded coefficients. Atti Accad. Naz. Lincei Cl. Sci. Fis. Mat. Natur. Rend. Lincei (9) Mat. Appl. **9**, 15–25 (2001)
496. E. Priola, The Cauchy problem for a class of Markov-type semigroups. Commun. Appl. Anal. **5**(1), 49–76 (2001)
497. E. Priola, Schauder estimates for a homogeneous Dirichlet problem in a half-space of a Hilbert space. Nonlinear Anal. **44**(5), 679–702 (2001)
498. E. Priola, Dirichlet problems in a half-space of a Hilbert space. Infin. Dimens. Anal. Quantum Probab. Relat. Top. **5**(2), 257–291 (2002)
499. E. Priola, On a Dirichlet problem involving an Ornstein-Uhlenbeck operator. Potential Anal. **18**(3), 251–287 (2003)
500. E. Priola, S. Tracà, On the Cauchy problem for non-local Ornstein-Uhlenbeck operators. Nonlinear Anal. **131**, 182–205 (2016)
501. E. Priola, J. Zabczyk, Liouville theorems for non-local operators. J. Funct. Anal. **216**(2), 455–490 (2004)
502. E. Priola, J. Zabczyk, Structural properties of semilinear SPDEs driven by cylindrical stable processes. Probab. Theory Relat. Fields **149**(1–2), 97–137 (2011)
503. P.E. Protter, *Stochastic Integration and Differential Equations*, Stochastic Modeling and Applied Probability, vol. 21 (Springer, Berlin, 2004)
504. J. Qiu, S. Tang, Maximum principle for quasi-linear backward stochastic partial differential equations. J. Funct. Anal. **262**(5), 2436–2480 (2012)
505. J. Qiu, S. Tang, Y. You, 2D backward stochastic Navier-Stokes equations with nonlinear forcing. Stoch. Process. Appl. **122**(1), 334–356 (2012)
506. M. Renardy, Polar decomposition of positive operators and a problem of Crandall and Lions. Appl. Anal. **57**(3–4), 383–385 (1995)

507. M. Renardy, R.C. Rogers, *An Introduction to Partial Differential Equations*, vol. 13, 2nd edn., Texts in Applied Mathematics (Springer, New York, 2004)

508. D. Revuz, M. Yor, *Continuous Martingales and Brownian Motion*, 3rd edn., Grundlehren der Mathematischen Wissenschaften, vol. 293 (Springer, Berlin, 1999)

509. M. Röckner, L^p-analysis of finite and infinite-dimensional diffusion operators, in *Stochastic PDE's and Kolmogorov Equations in Infinite Dimensions*, ed. by G. Da Prato. Lecture Notes in Mathematics, vol. 1715 (Springer, Berlin, 1999), pp. 65–116

510. M. Röckner, Z. Sobol, A new approach to Kolmogorov equations in infinite dimensions and applications to stochastic generalized Burgers equations. C. R. Acad. Sci. Paris Sér. I Math. **338**(12), 945–949 (2004)

511. M. Röckner, Z. Sobol, Kolmogorov equations in infinite dimensions: well-posedness and regularity of solutions, with applications to stochastic generalized Burgers equations. Ann. Probab. **34**(2), 663–727 (2006)

512. M. Röckner, Z. Sobol, A new approach to Kolmogorov equations in infinite dimensions and applications to the stochastic 2D Navier-Stokes equation. C. R. Acad. Sci. Paris Sér. I Math. **345**(5), 289–292 (2007)

513. L.C.G. Rogers, D. Williams, *Diffusions, Markov Processes and Martingales. Volume One: Foundations*, Wiley Series in Probability and Mathematical Statistics (Wiley, Chichester, 1994)

514. S. Romagnoli, T. Vargiolu, Robustness of the Black-Scholes approach in the case of options on several assets. Financ. Stoch. **4**(3), 325–341 (2000)

515. M. Romito, Analysis of equilibrium states of Markov solutions to the 3D Navier-Stokes equations driven by additive noise. J. Stat. Phys. **131**(3), 415–444 (2008)

516. M. Rosestolato, *Functional Itô calculus in Hilbert spaces and application to path-dependent Kolmogorov equations*. Preprint (2016), arXiv:1606.06326

517. M. Rosestolato, A. Święch, Partial regularity of viscosity solutions for a class of Kolmogorov equations arising from mathematical finance. J. Differ. Equ. **262**(3), 1897–1930 (2017)

518. M. Royer, BSDEs with a random terminal time driven by a monotone generator and their links with PDEs. Stoch. Stoch. Rep. **76**(4), 281–307 (2004)

519. B.L. Rozovskiĭ, *Stochastic Evolution Systems. Linear Theory and Applications to Non-linear Filtering* (Kluwer, Dordrecht, 1990). Translated from the Russian, originally published by Nauka, Moscow, 1983

520. W. Rudin, *Real and Complex Analysis* (McGraw-Hill, New York, 1970)

521. W. Rudin, *Functional Analysis, McGraw-Hill Series in Higher Mathematics* (McGraw-Hill, New York, 1973)

522. A. Rusinek, *Invariant measures for forward rate HJM model with Lévy noise*. Preprint IM PAN **669**, (2006)

523. R.A. Ryan, *Introduction to Tensor Products of Banach Spaces*, Springer Monographs in Mathematics (Springer, Berlin, 2002)

524. M. Scheutzow, Qualitative behaviour of stochastic delay equations with a bounded memory. Stochastics **12**(1), 41–80 (1984)

525. T.I. Seidman, J. Yong, How violent are fast controls? II. Math. Control Signals Syst. **9**(4), 327–340 (1996)

526. K. Shimano, A class of Hamilton-Jacobi equations with unbounded coefficients in Hilbert spaces. Appl. Math. Optim. **45**(1), 75–98 (2002)

527. I. Singer, *Bases in Banach Spaces I*, Grundlehren der Mathematischen Wissenschaften, vol. 154 (Springer, Berlin, 1970)

528. P.E. Sobolevskiĭ, On equations with operators forming an acute angle, Dokl. Akad. Nauk SSSR (N.S.) **116**, 754–757 (1957)

529. H.M. Soner, N. Touzi, Superreplication under gamma constraints. SIAM J. Control Optim. **39**(1), 73–96 (2000)

530. H.M. Soner, N. Touzi, Dynamic programming for stochastic target problems and geometric flows. J. Eur. Math. Soc. **4**(3), 201–236 (2002)

531. H.M. Soner, N. Touzi, Stochastic target problems, dynamic programming, and viscosity solutions. SIAM J. Control Optim. **41**(2), 404–424 (2002)
532. H.M. Soner, N. Touzi, The problem of super-replication under constraints, *Paris-Princeton Lectures on Mathematical Finance, 2002*, Lecture Notes in Mathematics, vol. 1814 (Springer, Berlin, 2003), pp. 133–172
533. R. Sowers, Large deviations for the invariant measure of a reaction-diffusion equation with non-Gaussian perturbations. Probab. Theory Relat. Fields **92**(3), 393–421 (1992)
534. S.S. Sritharan, An introduction to deterministic and stochastic control of viscous flow, in *Optimal Control of Viscous Flow*, ed. by S.S. Sritharan (Society for Industrial and Applied Mathematics, Philadelphia, 1998), pp. 1–42
535. C. Stegall, Optimization of functions on certain subsets of Banach spaces. Math. Ann. **236**, 171–176 (1978)
536. D.W. Stroock, S.R.S. Varadhan, *Multidimensional Diffusion Processes*, Classics in Mathematics (Springer, Berlin, 2006)
537. A. Święch, *Viscosity solutions of fully nonlinear partial differential equations with "unbounded" terms in infinite dimensions*, Ph.D. thesis, University of California (Santa Barbara) (1993)
538. A. Święch, "Unbounded" second order partial differential equations in infinite-dimensional Hilbert spaces. Comm. Partial Differ. Equ. **19**(11–12), 1999–2036 (1994)
539. A. Święch, The existence of value functions of stochastic differential games for unbounded stochastic evolution, in *Proceedings of the 34th IEEE Conference on Decision and Control* (1995), pp. 2289–2294
540. A. Święch, Risk-sensitive control and differential games in infinite dimensions. Nonlinear Anal. **50**(4), 509–522 (2002)
541. A. Święch, A PDE approach to large deviations in Hilbert spaces. Stoch. Process. Appl. **119**(4), 1081–1123 (2009)
542. A. Święch, E.V. Teixeira, Regularity for obstacle problems in infinite dimensional Hilbert spaces. Adv. Math. **220**(3), 964–983 (2009)
543. A. Święch, J. Zabczyk, Large deviations for stochastic PDE with Lévy noise. J. Funct. Anal. **260**(3), 674–723 (2011)
544. A. Święch, J. Zabczyk, Uniqueness for integro-PDE in Hilbert spaces. Potential Anal. **38**(1), 233–259 (2013)
545. A. Święch, J. Zabczyk, Integro-PDE in Hilbert spaces: existence of viscosity solutions. Potential Anal. **45**(4), 703–736 (2016)
546. A. Talarczyk, Dirichlet problem for parabolic equations on Hilbert spaces. Studia Math. **141**(2), 109–142 (2000)
547. H. Tanabe, *Equations of Evolution*, Monographs and Studies in Mathematics, vol. 6 (Pitman, Boston, 1979). Translated from the Japanese, originally published by Iwanami, Tokyo, 1975
548. H. Tanabe, *Functional Analytic Methods for Partial Differential Equations*, Monographs and Textbooks in Pure and Applied Mathematics, vol. 204 (Dekker, New York, 1997). Translated from the Japanese, originally published by Jikkyo Shuppan, Tokyo, 1981
549. S. Tang, Semi-linear systems of backward stochastic partial differential equations in $\mathbb{R}^n$. Chinese Ann. Math. Ser. B **26**(3), 437–456 (2005)
550. S. Tang, X. Li, Maximum principle for optimal control of distributed parameter stochastic systems with random jumps, in *Differential Equations Dynamical Systems, and Control Science*, ed. by K.D. Elworthy, W.N. Everitt, E.B. Lee. Lecture Notes in Pure and Applied Mathematics, vol. 152 (Dekker, New York, 1993), p. 867
551. S. Tang, X. Li, Necessary conditions for optimal control of stochastic systems with random jumps. SIAM J. Control Optim. **32**(5), 1447–1475 (1994)
552. S. Tang, W. Wei, On the Cauchy problem for backward stochastic partial differential equations in Hölder spaces. Ann. Probab. **44**(1), 360–398 (2016)
553. M. Tehranchi, A note on invariant measures for HJM models. Financ. Stoch. **9**(3), 389–398 (2005)

554. R. Temam, *The Navier-Stokes Equations: Theory and Numerical Analysis* (North-Holland, Amsterdam, 1977)
555. R. Temam, *Navier–Stokes Equations and Nonlinear Functional Analysis*, 2nd edn., CBMS-NSF Regional Conference Series in Applied Mathematics, vol. 66 (Society for Industrial and Applied Mathematics, Philadelphia, 1995)
556. R. Temam, *Navier-Stokes Equations: Theory and Numerical Analysis* (American Mathematical Society, Providence, 2001)
557. R. Temam, *Infinite-Dimensional Dynamical Systems in Mechanics and Physics*, Applied Mathematical Sciences, vol. 68 (Springer, Berlin, 2012)
558. G. Tessitore, Existence, uniqueness and space regularity of the adapted solutions of a backward SPDE. Stoch. Anal. Appl. **14**(4), 461–486 (1996)
559. V.B. Tran, The uniqueness of viscosity solutions of second order nonlinear partial differential equations in a Hilbert space of two-dimensional functions. Acta Math. Vietnam **31**(2), 149–165 (2006)
560. V.B. Tran, D.V. Tran, Viscosity solutions of the Cauchy problem for second-order nonlinear partial differential equations in Hilbert spaces. Electron. J. Differ. Equ. **47**, 1–15 (2006)
561. F. Tröltzsch, *Optimal Control of Partial Differential Equations. Theory, Methods and Applications*, Graduate Studies in Mathematics, vol. 112 (American Mathematical Society, Providence, 2010)
562. J.M.A.M. van Neerven, M. Riedle, A semigroup approach to stochastic delay equations in spaces of continuous functions. Semigroup Forum **74**(2), 227–239 (2007)
563. A.C.M. van Rooij, W.H. Schikhof, *A Second Course on Real Functions* (Cambridge University Press, Cambridge, 1982)
564. T. Vargiolu, *Finite dimensional approximations for the Musiela model in the Gaussian case*. Memoire de DEA Univ, Pierre et Marie Curie - Paris VI, 1997
565. T. Vargiolu, Invariant measures for the Musiela equation with deterministic diffusion term. Financ. Stoch. **3**(4), 483–492 (1999)
566. R.B. Vinter, R.H. Kwong, The infinite time quadratic control problem for linear systems with state and control delays: an evolution equation approach. SIAM J. Control Optim. **19**(1), 139–153 (1981)
567. M.J. Višik, A.V. Fursikov, *Mathematical Problems of Statistical Hydromechanics*, Mathematics and its Applications (Soviet Series), vol. 9 (Springer, Amsterdam, 1988). Translated from the Russian, originally published by Nauka, Moscow, 1980
568. J.B. Walsh, An introduction to stochastic partial differential equations, in *École d'été de probabilités de Saint-Flour XIV-1984*, ed. by P.L. Hennequin. Lecture Notes in Mathematics, vol. 1180 (Springer, Berlin, 1986), pp. 265–439
569. J. Weidmann, *Linear Operators in Hilbert Spaces*, Graduate Texts in Mathematics, vol. 68 (Springer, Berlin, 2012)
570. E. Weinan, K. Khanin, A. Mazel, Y. Sinai, Invariant measures for Burgers equation with stochastic forcing. Ann. of Math. **151**(3), 877–960 (2000)
571. E. Weinan, J.C. Mattingly, Y. Sinai, Gibbsian Dynamics and Ergodicity for the Stochastically Forced Navier-Stokes Equation. Comm. Math. Phys. **224**(1), 83–106 (2001)
572. D. Williams, *Probability with Martingales* (Cambridge University Press, Cambridge, 1991)
573. A. Wiweger, Linear spaces with mixed topology. Studia Math. **20**, 47–68 (1961)
574. D. Yang, Optimal control problems for Lipschitz dissipative systems with boundary-noise and boundary-control. J. Optim. Theory Appl. **165**(1), 14–29 (2015)
575. J. Yong, X.Y. Zhou, *Stochastic Controls, Hamiltonian Systems and HJB Equations*, Applications of Mathematics, vol. 43 (Springer, New York, 1999)
576. K. Yosida, *Functional Analysis*, 6th edn., Grundlehren der Mathematischen Wissenschaften, vol. 123 (Springer, Berlin, 1980)
577. H. Yu, B. Liu, Properties of value function and existence of viscosity solution of HJB equation for stochastic boundary control problems. J. Franklin Inst. **348**(8), 2108–2127 (2011)
578. J. Zabczyk, Infinite dimensional systems in optimal control. B. Int. Stat. Inst. **42**(2), 286–310 (1977). Invited papers (Proceedings of the 41st Session)

579. J. Zabczyk, Linear stochastic systems, in *Hilbert spaces: spectral properties and limit behavior*, in Report of Institute of Mathematics, Polish Academy of Sciences 236, 1981, Also published in Mathematical control theory (C. Olech, B. Jakubczyk, J. Zabczyk, eds.), Banach Center Publications, vol. 14 (Institute of Mathematics, Polish Academy of Sciences, Warsaw, 1985), pp. 591–609

580. J. Zabczyk, *Law equivalence of Ornstein-Uhlenbeck processes and control equivalence of linear systems*. Preprint IM PAN **457**, (1989)

581. J. Zabczyk, Stopping problems on Polish spaces. Ann. Univ. Mariae Curie-Skłodowska Sect. A **51**(1), 181–199 (1997)

582. J. Zabczyk, Parabolic equations on Hilbert spaces, in *Stochastic PDE's and Kolmogorov equations in infinite dimensions*, ed. by G. Da Prato. Lecture Notes in Mathematics, vol. 1715 (Springer, Berlin, 1999), pp. 117–213

583. J. Zabczyk, Bellman's inclusions and excessive measures. Probab. Math. Statist. **21**(1), 101–122 (2001)

584. J. Zabczyk, *Mathematical Control Theory: An Introduction*, Modern Birkhäuser Classics (Birkhäuser, Boston, 2008)

585. M. Zakai, On the optimal filtering of diffusion processes. Z. Wahrscheinlichkeitstheor. Verw. Geb. **11**(3), 230–243 (1969)

586. E. Zeidler, *Nonlinear functional analysis and its applications. Volume I: fixed-point theorems* (Springer, New York, 1986). Translated from German, originally published by Teubner, Leipzig, 1976

587. E. Zeidler, *Nonlinear functional analysis and its applications. Volume II/A: linear monotone operators* (Springer, New York, 1989). Translated from German, originally published by Teubner, Leipzig, 1977

588. E. Zeidler, *Nonlinear functional analysis and its applications. Volume II/B: nonlinear monotone operators* (Springer, New York, 1989) Translated from German, originally published by B. G. Teubner, Leipzig, 1978

589. J. Zhou, *A class of delay optimal control problems and viscosity solutions to associated Hamilton–Jacobi–Bellman equations*, Preprint arXiv:1507.04112

590. J. Zhou, *A class of infinite-horizon delay optimal control problems and a viscosity solution to the associated HJB equation*, Preprint

591. J. Zhou, Optimal control of a stochastic delay heat equation with boundary-noise and boundary-control. Int. J. Control **87**(9), 1808–1821 (2014)

592. J. Zhou, Optimal control of a stochastic delay partial differential equation with boundary-noise and boundary-control. J. Dyn. Control Syst. **20**(4), 503–522 (2014)

593. J. Zhou, The existence and uniqueness of the solution for nonlinear elliptic equations in Hilbert spaces. J. Inequal. Appl. **250**, 1–23 (2015)

594. J. Zhou, B. Liu, Optimal control problem for stochastic evolution equations in Hilbert spaces. Internat. J. Control **83**(9), 1771–1784 (2010)

595. J. Zhou, B. Liu, The existence and uniqueness of the solution for nonlinear Kolmogorov equations. J. Differ. Equ. **253**(11), 2873–2915 (2012)

596. J. Zhou, Z. Zhang, Optimal control problems for stochastic delay evolution equations in Banach spaces. Int. J. Control **84**(8), 1295–1309 (2011)

597. X.Y. Zhou, On the existence of optimal relaxed controls of stochastic partial differential equations. SIAM J. Control Optim. **30**(2), 247–261 (1992)

598. X.Y. Zhou, On the necessary conditions of optimal controls for stochastic partial differential equations. SIAM J. Control Optim. **31**(6), 1462–1478 (1993)

599. X.Y. Zhou, Sufficient conditions of optimality for stochastic systems with controllable diffusions. IEEE Trans. Auto. Control **41**(8), 1176–1179 (1996)

600. X. Zhu, J. Zhou, Infinite horizon optimal control of stochastic delay evolution equations in Hilbert spaces. Abstr. Appl. Anal. **791786**, 1–14 (2013)

601. G. Zitkovic, Dynamic programming for controlled Markov families: abstractly and over martingale measures. SIAM J. Control Optim. **52**(3), 1597–1621 (2014)

Subject Index

© Springer International Publishing AG 2017
G. Fabbri et al., *Stochastic Optimal Control in Infinite Dimension*,
Probability Theory and Stochastic Modelling 82,
DOI 10.1007/978-3-319-53067-3

Notation Index

Functions, processes and operators
A_a, 294
$(-A)^\alpha$, 807
A_D, 844
A_N, 845
A_n, 803
$\overline{A}$, 795
$\mathcal{A}$, 395, 452, 616, 671, 677, 758, 770
$\mathcal{A}_0$, 823
$\mathcal{A}_1$, 477, 640
$\mathcal{A}_1(t)$, 467
$\hat{\mathcal{A}}_0$, 824
$\mathcal{A}^m$, 830
$\mathcal{A}(t)$, 432, 732
$\tilde{\mathcal{A}}$, 671
$a^\mu(\cdot)$, 93
B, 172
B_ρ, 293
C_ρ, 293
C_ρ^*, 293
curl, 149
$D(\cdot)$, 843
$D^+ u(\bar{x})$, 867
$D^- u(\bar{x})$, 867
$D^2 u$, 790
δ, 699
$D^G f$, 374, 375
$D^{G(t,\cdot)} f(t, \cdot)$, 384
$D^2_{H_{-1}} w$, 189
$D_{H_{-1}} w$, 189
$\hat{d} W$, 699
$D^k_x u$, 790
$D^j_t u$, 790
$D^l u$, 785
$D^{\mathcal{M}}_\eta F$, 698
$D^{\mathcal{M}} F$, 698
∇f, 688

$\nabla^G f$, 374, 375
$\nabla^{G(t,\cdot)} f(t, \cdot)$, 384
D_Q, 617, 633
$\overline{D_Q \mathcal{M}_1}$, 638
$\tilde{D}_Q$, 634
Du, 785, 790
e^{tA}, 804
$G_1 f(t)$, 631
$G_2 f(t)$, 631
$\Gamma_G(t)$, 397
$\Gamma_G(t, s, x)$, 400
$\Gamma(t)$, 397
$J^{2,+} u(\bar{x})$, 868
$J^{2,-} u(\bar{x})$, 868
$\overline{J}^{2,+} u(\bar{x})$, 869
$\overline{J}^{2,-} u(\bar{x})$, 869
J^μ, 93
J_n, 803
$\mathcal{L}_t$, 396
$\overline{\mathcal{M}}$, 639
$\mathcal{M}_1$, 637
$\overline{\mathcal{M}}_1$, 638
$\langle M \rangle_t$, 25
$\langle\langle M \rangle\rangle_t$, 25
$\mathscr{M}$, 679
$N_\lambda(\cdot)$, 844
$\overline{\mathcal{P}}^{2,+} u(\bar{t}, \bar{x})$, 869
$\mathcal{P}^{2,+} u(\bar{t}, \bar{x})$, 869
$\overline{\mathcal{P}}^{2,-} u(\bar{t}, \bar{x})$, 869
$\mathcal{P}^{2,-} u(\bar{t}, \bar{x})$, 869
P_s, 78, 610, 671, 677
P_t, 421, 452
P_t, 757
$P_{t,s}$, 77, 424, 432, 731
Q, 27
Q_1, 28
Q_t, 393

© Springer International Publishing AG 2017
G. Fabbri et al., *Stochastic Optimal Control in Infinite Dimension*,
Probability Theory and Stochastic Modelling 82,
DOI 10.1007/978-3-319-53067-3

CPSIA information can be obtained
at www.ICGtesting.com
Printed in the USA
LVHW011434120720
660448LV00018B/1326